MARRIAGES, FAMILIES, and RELATIONSHIPS

Making Choices in a Diverse Society

Fourteenth Edition

Mary Ann Lamanna
University of Nebraska, Omaha

Agnes Riedmann
California State University, Stanislaus

Susan D. Stewart
Iowa State University

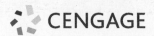

Australia • Brazil • Canada • Mexico • Singapore • United Kingdom • United States

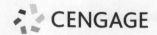

CENGAGE

Marriages, Families, and Relationships:
Making Choices in a Diverse Society,
Fourteenth edition
Mary Ann Lamanna, Agnes Riedmann, and
Susan D. Stewart

Vice President, Higher Education & Skills Product:
Erin Joyner

Product Director: Laura Ross

Product Manager: Kori Alexander

Learning Designer: Emma Guiton

Senior Content Manager: Tim Bailey

Marketing Manager: Tricia Salata

Product Assistant: Shelby Blakey

Digital Delivery Lead: Matt Altieri

Intellectual Property Analyst: Deanna Ettinger

Intellectual Property Project Manager: Carly Belcher

Designer, Creative Studio: Nadine Ballard

Production Service: MPS Limited

Cover Images: Jaromir Chalabala/ShutterStock.com
worradirek/ShutterStock.com
ROSSARINPHOTO/ShutterStock.com
Leon Harris/Cultura Creative (RF)/Alamy

For product information and technology assistance, contact us at
Cengage Customer & Sales Support, 1-800-354-9706
or **support.cengage.com.**
For permission to use material from this text or product,
submit all requests online at **www.cengage.com/permissions.**

Library of Congress Control Number: 2020938334

Student Edition
ISBN: 978-0-357-36874-9

Cengage
200 Pier 4 Boulevard
Boston, MA 02210
USA

Cengage is a leading provider of customized learning solutions with
employees residing in nearly 40 different countries and sales in more
than 125 countries around the world. Find your local representative at
www.cengage.com.

To learn more about Cengage platforms and services, register or access
your online learning solution, or purchase materials for your course, visit
www.cengage.com.

Printed at CLDPC, USA, 06-21

To our families, especially

Larry, Valerie, Sam, Janice, Simon, and Christie

Bill, Beth, Natalie, Alex, and Livia

Gwendolyn, Gene, Lee, Christine, Mom

and Dad

About the Authors

Mary Ann Lamanna is Professor Emerita of Sociology at the University of Nebraska at Omaha. She received her bachelor's degree in political science Phi Beta Kappa from Washington University (St. Louis); master's degree in sociology (minor in psychology) from the University of North Carolina, Chapel Hill; and doctorate in sociology from the University of Notre Dame.

Research and teaching interests include family, reproduction, and gender and law. She is the author of *Emile Durkheim on the Family* (Sage Publications, 2002) and coauthor of a book on Vietnamese refugees. She has articles in journals on law, sociology, and medical humanities. Current research concerns the sociology of literature, specifically "novels of terrorism" and a sociological analysis of Marcel Proust's novel *In Search of Lost Time*. Professor Lamanna has two adult children, Larry and Valerie.

Agnes Riedmann is Professor Emeritus of Sociology at California State University, Stanislaus. She attended Clarke College in Dubuque, Iowa. She received her bachelor's degree from Creighton University and her doctorate from the University of Nebraska. Her professional areas of interest are theory, family, and the sociology of body image. She is the author of *Science That Colonizes: A Critique of Fertility Studies in Africa* (Temple University Press, 1993). Dr. Riedmann spent the academic year 2008–09 as a Fulbright Professor at the Graduate School for Social Research, affiliated with the Polish Academy of Sciences, Warsaw, where she taught courses in family, social policy, and globalization. She has two children, Beth and Bill; two granddaughters, Natalie and Livia; and a grandson, Alex.

Susan D. Stewart is a Professor of Sociology at Iowa State University. She received her bachelor's degree from the State University of New York at Fredonia and her doctorate from Bowling Green State University. Her professional areas of interest are gender, family, and demography and specifically how complex living arrangements affect the physical and emotional health of children, adults, and families. She is the author of *Brave New Stepfamilies* (Sage Publications, 2007), *Co-Sleeping in Families* (Rowan & Littlefield, 2016), and *Multicultural Stepfamilies*, with co-author Gordon Limb (Cognella, 2020). Amazingly, Dr. Stewart discovered she used the third edition of the Lamanna and Riedmann textbook in her undergraduate sociology of families class back in the 1980s! She lives in Ames, Iowa, with her sixteen-year-old daughter and husband and is a member of two generations of stepfamilies.

BRIEF CONTENTS

CONTENTS

1

MAKING FAMILY CHOICES IN A CHANGING SOCIETY 3

2

EXPLORING RELATIONSHIPS AND FAMILIES 29

GENDER IDENTITIES AND FAMILIES 55

OUR SEXUAL SELVES 79

5

LOVE AND CHOOSING A LIFE PARTNER 109

6

NONMARITAL LIFESTYLES: LIVING ALONE, COHABITING, AND OTHER OPTIONS 137

7

MARRIAGE: FROM SOCIAL INSTITUTION TO PRIVATE RELATIONSHIP 163

8

DECIDING ABOUT PARENTHOOD 189

9

RAISING CHILDREN IN A DIVERSE SOCIETY 219

10

WORK AND FAMILY 247

11

COMMUNICATION IN RELATIONSHIPS, MARRIAGES, AND FAMILIES 275

12
POWER AND VIOLENCE
IN FAMILIES 305

13

FAMILY STRESS, CRISIS, AND RESILIENCE 335

14

DIVORCE AND RELATIONSHIP DISSOLUTION 363

15

REMARRIAGES AND STEPFAMILIES 397

16

AGING AND MULTIGENERATIONAL FAMILIES 429

BOXES

PREFACE

As we complete our work on the fourteenth edition of this text, we become aware of how suddenly society and family life can change. If ever there was a dramatic example of how the social environment affects personal and family life, the global pandemic Covid-19 has unfortunately provided it. We had finished revising much of this edition before Covid-19 changed life as we knew it. By the time we were finishing our revision of Chapter 16, however, the virus had quieted cities and overwhelmed hospitals. We recognize the pandemic in Chapter 16 with a new box, "As We Make Choices: Want to Call or Visit an Isolated Senior?"

So many questions and hypotheses come to mind regarding how Covid-19 is likely to impact families. We, your authors, are already beginning to think about how this monumental pandemic will impact the content of future editions. We imagine that Covid-19 will focus greater attention on what family means to us as well as on the critical importance of traditional family functions such as raising children and providing practical and social support to family members. How will Covid-19 impact families' motivation and ability to perform these functions? In what ways might we expect family theory and research respond to Covid-19? Will this pandemic affect our choices about our preferred family forms? Will people be more likely to marry than to cohabit, for instance? Or will their decisions go in other directions? How might parenting concerns, issues, and behaviors change? What about the work–family interface? How do families fare when one or more family members suddenly begin working from home? Or later, when home-based workers return to work?

How will mandatory quarantines affect romantic relationships? It was only a matter of time before researchers would look into the impact of Covid-19 on sex. Only weeks after the virus, psychologist Jessica Zucker explored this in "Health, Sex and Coronavirus: How Does Sexual Intimacy Change During a Pandemic?" In an Instagram poll, whereas 50 percent of respondents said their sex life had improved, 50 percent said their sex life had worsened—the "six feet apart rule" would make sex difficult to achieve for those not already living with a partner. It's important to point out, however, that the research on "baby booms" following natural disasters, such black-outs and hurricanes, is mostly mythology. In general, people avoid bringing children into the world when economic times are uncertain.

Then too, how might family power relations change? We're seeing a divide between how older Americans view this pandemic and how a number of young adults perceive the danger and what it requires of them. We've all seen the images of young people partying on the beach during Spring Break in the midst of social distancing.

Will this divide affect family life? And if so, in what ways? Moreover, unfortunately we're hearing about domestic violence during quarantine. An example is Wendy Patrick's (J.D., Ph.D.) article in the March 19, 2020, online *Psychology Today* blog, "Domestic Abuse During Quarantine: When the Threat is Inside, What Victims Trapped at Home with an Abuser Need to Know" (psychologytoday.com).

How might theory and research on family stress and crisis—which assuredly this pandemic causes!—help us to understand what's going on in ourselves, our families, our communities, and our world? What can research findings tell us about what helps families to pull together during a crisis such as this? How might Covid-19 impact the divorce rate? On the one hand, stress puts added strain on couple relationships, and couples with poor relationship quality who are forced together for months may realize they should not stay together. On the other hand, couples under mandatory quarantine may rediscover what they love about each other and may count their blessings in an uncertain world. This remains to be seen. On another front, in what ways does Covid-19 impact aging families, their younger relatives, and caregivers? We were nearly finished revising Chapter 16 when this pandemic broke out and had time to write a box relating to this unprecedented Covid-19 outbreak: "As We Make Choices: Want to Call or Visit an Isolated Senior?"

Covid-19 aside, we authors look back with pride over thirteen earlier editions. Together, these represent more than forty years spent observing and rethinking American families. Not only have families dramatically changed since we began our first edition but also has social science's interpretation of family life. It is gratifying to be a part of the enterprise dedicated to studying families and sharing this knowledge with students.

Our own perspective on families has developed and changed as well. Indeed, as marriages and families have evolved over the last four decades, so has this text. In the beginning, this text was titled *Marriages and Families*—a title that was the first to purposefully use plurals to recognize the diversity of family forms—a diversity that we noted as early as 1980. Now the text is titled *Marriages, Families, and Relationships*. We added the term *relationships* to recognize the increasing incidence of individuals forming commitments outside of legal marriage. At the same time, we continue to recognize and appreciate the fact that a large majority of Americans—now including same-sex couples—are married or will marry.

Hence, we consciously persist in giving due attention to the values and issues of married couples. Of course, the concept of marriage itself has changed appreciably. No longer necessarily heterosexual, marriage is now an institution to which same-sex couples across the United States and in a growing number of other nations have legal access.

Meanwhile, the book's subtitle, *Making Choices in a Diverse Society*, continues to speak about the significant changes that have taken place since our first edition. To help accomplish our goal of encouraging students to better appreciate the diversity of today's families, we present the latest research and statistical information on varied family forms (including those with lesbian, gay, transgender, and other non-cisgender family members) and families of diverse race and ethnicity, socioeconomic, and immigration status, among other variables.

We continue to take account not only of increasing racial and ethnic diversity but also of the fluidity of the concepts *race* and *ethnicity* themselves. We pay attention to the socially constructed nature of these concepts. We integrate these materials on family diversity throughout the textbook, always with an eye toward avoiding stereotypical and simplistic generalizations and instead explaining data in sociological and sociohistorical contexts. Interested from the beginning in the various ways that gender plays out in families, we have persistently focused on areas in which gender relations have changed and continue to change, as well as on areas in which there has been relatively little change.

In addition to our attention to gender, we have studied demography and history, and we have paid increasing attention to the impact of social structure on family life. We have highlighted the family ecology perspective in keeping with the importance of social context and public policy. We cannot help but be aware of the cultural and political tensions surrounding families today. At the same time, in recent editions and in response to our reviewers, we have given heightened attention to the contributions of biology and psychology and to a social psychological understanding of family interaction and its consequences.

We continue to affirm the power of families as they influence the courses of individual lives. Meanwhile, we give considerable attention to policies needed to provide support for today's families: working parents, families in financial stress, single-parent families, families of varied racial and ethnic backgrounds, stepfamilies, same-sex couples, and other nontraditional families— as well as the classic nuclear family.

We note that, despite changes, marriage and family values continue to be salient in contemporary American life. Our students come to a marriage and family course because family life is important to them. Our aim now, as it has been from the first edition, is to help students question assumptions and reconcile conflicting ideas and values as they make choices throughout their lives. We enjoy and benefit from the contact we've had with faculty and students who have used this book. Their enthusiasm and criticism have stimulated many changes in the book's content. To know that a supportive audience is interested in our approach to the study of families has enabled us to continue our work over a long period.

THE BOOK'S THEMES

We developed the book's themes by looking at the interplay between findings in the social sciences and the experiences of the people around us. Ideas for topics continue to emerge, not only from current research and reliable journalism, but also from the needs and concerns we perceive among our own family members, students, and friends. The attitudes, behaviors, and relationships of real people have a complexity that we have tried to portray. Interwoven with these themes is the concept of the life course—the idea that adults may change by means of reevaluating and restructuring throughout their lives. This emphasis on the life course creates a comprehensive picture of marriages, families, and relationships and encourages us to continue to add topics that are new to family texts. Meanwhile, this book makes these points:

- People's personal problems and their interaction with the social environment change as they and their relationships and families grow older.

- People reexamine their relationships and their expectations for relationships as they and their marriages, relationships, and families mature.

- Because family forms are more flexible today, people may change the type or style of their relationships and families throughout their lives.

These themes appear throughout this text: People are influenced by the society around them as they make choices, social conditions change in ways that may impede or support family life, there is an interplay between individual families and the larger society, and individuals make family-related choices throughout adulthood.

The process of creating and maintaining marriages, families, and relationships requires many personal choices; people continue to make family-related decisions, both big and small, throughout their lives. Making decisions about family life begins in early adulthood and lasts into old age. People choose whether they will adhere to traditional beliefs, values, and attitudes about

gender roles or negotiate more flexible roles and relationships. They may rethink their values about sex and become more informed and comfortable with their sexual choices.

Women and men may choose to remain single, to form heterosexual or same-sex relationships outside of marriage, or to marry. They have the option today of staying single longer before marrying. Single people make choices about their lives ranging from decisions about living arrangements to those about whether to engage in sex only in marriage or committed relationships, to engage in sex for recreation, or to abstain from sex altogether. Many unmarried individuals live as cohabiting couples, often with children. Once individuals form couple relationships, they have to decide how they are going to structure their lives as committed partners. Will they have children? Will other family members live with them—siblings or parents, for example, or adult children later?

Couples will make these decisions not once, but over and over during their lifetimes. Within a committed relationship, partners also choose how they will deal with conflict. Will they try to ignore conflicts? Will they vent their anger in hostile, alienating, or physically violent ways? Or will they practice supportive ways of communicating, disagreeing, and negotiating—ways that emphasize sharing and can deepen intimacy?

How will the partners distribute power in the marriage? Will they work toward relationships in which each family member is more concerned with helping and supporting others than with gaining a power advantage? How will the partners allocate work responsibilities in the home? What value will they place on their sexual lives together? Throughout their experience, family members continually face decisions about how to balance each one's need for individuality with the need for togetherness.

Parents also have choices. In raising their children, they can choose the authoritative parenting style, for example, in which parents take an active role in responsibly guiding and monitoring their children. However, how much guidance is too much? At what point do involved parents become *over* involved parents—that is, "helicopter parents"?

Many partners face decisions about whether to separate or divorce. They weigh the pros and cons, asking themselves which is the better alternative: living together as they are or separating? Even when a couple decides to separate or divorce, there are further decisions to make: Will they cooperate as much as possible or insist on blame and revenge? What living and economic support arrangements will work best for themselves and their children? How will they handle the legal process? The majority of divorced individuals eventually face decisions about forming relationships with new partners. In the absence of firm cultural models, they choose how they will define remarriage and stepfamily relationships.

When families encounter crises—and every family will face *some crises*—members must make additional decisions. Will they view each crisis as a challenge to be met, or will they blame one another? What resources can they use to handle the crisis? Then, too, as more and more Americans live longer, families will "age." As a result, more and more Americans will have not only living grandparents but also great grandparents. And increasingly, we will face issues concerning giving—and receiving—family elder care.

In the past, people tended to emphasize the dutiful performance of social roles in marriages and families for others. Today, people view committed relationships as those in which they expect to find companionship, intimacy, and emotional support for themselves. From its first edition, this book has examined the implications of this shift and placed these implications within social scientific perspective. Individualism, economic pressure, time pressures, social diversity, and an awareness of committed relationships' potential impermanence are features of the social context in which personal decision making takes place. With each edition, we recognize again that, as fewer social guidelines remain fixed, personal decision making becomes both more open and perhaps more challenging.

An emphasis on knowledgeable decision making does not mean that individuals can completely control their lives. People can influence but never directly determine how those around them behave or feel about them. Partners cannot control one another's changes over time, and they cannot avoid all accidents, illnesses, unemployment, separations, or deaths. Society-wide conditions may create unavoidable crises for individual families. However, families can control how they respond to such crises. Their responses will meet their own needs better when they refuse to react automatically and choose instead to act as a consequence of knowledgeable decision making.

Tension frequently exists between individuals and their social environment. Many personal troubles result from societal influences, values, or assumptions; inadequate societal support for family goals; and conflict between family values and individual values. By understanding some of these possible sources of tension and conflict, individuals can perceive their personal troubles more clearly and work constructively toward solutions. They may choose to form or join groups to achieve family goals. They may become involved in the political process to develop state or federal social policy that is supportive of families. The accumulated decisions of individuals and families also shape the social environment.

KEY FEATURES

With its ongoing, thorough updating and inclusion of current research and its emphasis on students' being able to make choices in an increasingly diverse society, this book has become a principal resource for gaining insights into today's marriages, relationships, and families. Over the past twelve editions, we have had four goals in mind for student readers: first, to help them better understand themselves and their family situations; second, to make students more conscious of the personal decisions that they will make throughout their lives and of the societal influences that affect those decisions; third, to help students better appreciate the variety and diversity among families today; and fourth, to encourage them to recognize the need for structural, social policy support for families. To these ends, this text has become recognized for its accessible writing style, up-to-date research, well-written features, and useful chapter learning aids.

Up-to-Date Research and Statistics

As users have come to expect, we have thoroughly updated the text's research base and statistics, emphasizing cutting-edge research that addresses the diversity of marriages and families, as well as all other topics. In accordance with this approach, users will notice several new tables and figures. Revised tables and figures have been updated with the latest available statistics—data from the U.S. Census Bureau and other governmental agencies, as well as survey and other research data.

Box Features

The several themes described earlier are reflected in the special features.

Former users will recognize our box features. The following sections describe our four feature box categories:

As We Make Choices We highlight the theme of making choices with a group of boxes throughout the text—for example, "Rethinking Virginity," "Ten Rules for a Successful Relationship," "Disengaging from Power Struggles," "Selecting a Childcare Facility—Ten Considerations," "Rules for Successful Co-Parenting," "Tips for Step-Grandparents, and" "Want to Call or Visit an Isolated Senior?"

These feature boxes emphasize human agency and are designed to help students through crucial decisions.

A Closer Look at Diversity In addition to integrating information on cultural and ethnic diversity throughout the text proper, we have a series of features that give focused attention to instances of family diversity—for example, "African Americans and 'Jumping the Broom,'" "Diversity and Childcare," "Family Ties and Immigration," "Straight Parents and LGBTQ+ Children," and " Do You Speak Stepfamily?" among others.

Issues for Thought These features are designed to spark students' critical thinking and discussion. As an example, the Issues for Thought box in Chapter 16 explores "Filial Responsibility Laws" and encourages students to consider what might be the benefits and drawbacks of legally mandating filial responsibility. The box "When One Woman's Workplace Is Another's Family" invites students to consider how women's work differs across social class, race and ethnicity, and immigration status.

Facts about Families This feature presents demographic and other factual information on focused topics such as "How Family Researchers Study Religion from Various Theoretical Perspectives," "The Changing Language of Gender," on "Researching at the Kinsey Institute," on "Legal Same-Sex Marriage as a Successful Social Movement," and on "Foster Parenting," among others.

Chapter Learning Aids

A series of chapter learning aids help students comprehend and retain the material.

- Each chapter begins with a list of **learning objectives** specifically formulated for that chapter.

- **Chapter Summaries** are presented in bulleted, point-by-point lists of the key material in the chapter.

- **Key Terms** alert students to the key concepts presented in the chapter. A full glossary is provided at the end of the text.

- **Questions for Review and Reflection** help students review the material. Thought questions encourage students to think critically and to integrate material from other chapters with that presented in the current one. In every chapter, one of these questions is a policy question. This practice is in line with our goal of moving students toward structural analyses regarding marriages, families, and relationships.

KEY CHANGES IN THIS EDITION

In addition to incorporating the latest available research and statistics—and in addition to carefully reviewing every word in the book—we note that this edition includes many key changes, some of which are outlined here. We have worked to make chapter length more uniform throughout the text.

We are streamlining the material presented whenever possible and to ensuring a good flow of ideas. In this edition, we continue to consolidate similar material that had previously been addressed in separate chapters. **Meanwhile, we have substantially revised each and every chapter.** Every chapter is updated with the latest statistics and research throughout. Now that same-sex marriage is legal throughout the United States, we continue to conscientiously revisit all our chapters to make sure we're in line with this major family change. We mention some (but not all!) specific and important changes here.

Chapter 1, Making Family Choices in a Changing Society, continues to present the choices and life course themes of the book, as well as points to the significance for the family of larger social forces. Figure 1.1 is new with data on where Americans find meaning. HINT: their families. Figure 1.3 is new as well. All the boxes have been reworked. We paid special attention to rethinking and reworking the Closer Look at Diversity box in Chapter 1, with updated treatment of immigration due to the immigration crisis at our southern border. As faculty users, students, and casual readers have come to expect, all research and statistics are conscientiously updated. This goes for the entire book.

Chapter 2, Exploring Relationships and Families, continues to portray the integral relationship between family theories and methods for researching families, with new examples designed to better drive home the theoretical perspectives. Examples in the research section of this chapter include more recognition that major surveys are conducted globally, not just in the United States.

Chapter 3, Gender Identities and Families, continues to reflect evolving and expanding understandings of gender and sexual identity as fluid and non-binary, driving by the more progressive attitudes of Millennials and Gen Z. We introduce and define a variety of new terms related to gender and sexuality and discuss, for example, how states are facing political pressure to provide more gender options on birth certificates. We note challenges to toxic masculinity and increased representation of women in politics.

Chapter 4, Our Sexual Selves, continues its exploration into the range of sexual attitudes and behavior that exists in American society with special focus on gender differences, culture, history, politics, and technology. Notable since the last edition is the #MeToo Movement and women increasingly challenging previously taken-for-granted behaviors of men, such as sexual harassment and even sexual assault. In this chapter we broaden our discussion of consent, bystander education, and dispel myths about sexual assault. With increased attention to fluidity in sexual identity and behavior, we discuss the question of what it means to be a virgin. We take a tour of the famous Kinsey Institute and discuss the ethics of conducting sexuality research. Finally, we provide updated statistics on sexual behavior, infidelity, HIV/AIDS, and pornography use.

Chapter 5, Love and Choosing a Life Partner, increases attention to defining love in all its forms and, in particular, the limitations of American's Society's undue focus on romantic love. We continue to examine the changing nature of dating in the United States, not only in terms of new dating patterns, but also dating preferences, such as urban versus rural residence, political ideology, race, and religion. In addition, we draw increased attention to the heteronormative bias in love and dating and include more information on LGBTQ+ couples and gender inequality in relationships. We draw attention to arranged marriages, child marriage, and transnational marriages in the United States. We provide new information on what is known about the link between cohabitation, marital quality, and divorce.

Chapter 6, Nonmarital Lifestyles: Living Alone, Cohabiting, and Other Options, discusses demographic, economic, technological, and cultural reasons for the increasing proportion of unmarrieds, with updated statistics on unmarried men and women in America. New to this edition is a discussion of generational differences in attitudes about the advantages and disadvantages of being single, integrating the attitudes of the youngest generation of Americans, Gen Z, who are just now reaching young adulthood and who have a wide array of lifestyles available to them. We have expanded our discussion of the *transition to adulthood*, which in these tough economic times has continued to lengthen, and is responsible for part of the increase in multi-generational households we are seeing.

Chapter 7, Marriage: From Social Institution to Private Relationship, has been thoroughly updated in accordance with developments after the 2015 legalization of same-sex marriage and also with new statistics and research findings. This chapter explores the changing picture regarding marriage, noting that cohabitation may now be becoming more similar to marriage than it used to be as more couples choose to cohabit. We review the social science debate regarding whether this changing picture represents family change or decline. We thoroughly explore the selection hypothesis versus the experience hypothesis with regard to the benefits of marriage known from research.

Chapter 8, Deciding about Parenthood, continues its focus on the complex process through which couples have children and different infertility patterns by race, ethnicity, class, religion, sexual identity and other variables. We provide data on the rising costs of children with a special focus on childcare. We also have expanded our discussion of the social and emotional costs of children, which has led to an increased number of women

opting to remain childfree. New in this edition is attention to medicalization of childbirth in the United States and our high rate of caesarean sections relative to other industrialized societies. We continue to provide the latest information available on reproductive technologies, adoption, involuntary and nonmarital fertility, adolescent pregnancy and childbearing, multipartnered fertility, contraception, abortion, and the political debate surrounding these issues.

Chapter 9, Raising Children in a Diverse Society, like all the chapters in this edition, has been thoroughly updated with the most current research. As in recent editions, after describing the authoritative parenting style, we note its acceptance by mainstream experts in the parenting field. We then present a critique that questions whether this parenting style is universally appropriate or simply a white, middle-class pattern that may not be so suitable to other social contexts. We also discuss challenges faced by parents who are raising religious- or ethnic-minority children in potentially discriminatory environments.

We continue to emphasize the challenges that all parents face in contemporary America. We have expanded sections on single mothers, single fathers, and nonresident fathers. We have given more attention to relations with young-adult children as more and more of them have "boomeranged" home in this difficult economy.

New to this chapter are recognition and discussions of gender fluidity as related to parenting. For instance, the section "Gender and Parenting" includes discussion of parenting as a lesbian, gay male, or transgender parent. As just one example of something brand new, a fourteenth-edition Issues for Thought box explores the ironic phenomenon of heteronormative bias within the LGBTQ+ community.

Chapter 10, Work and Family. All research and statistics are updated. An example of now incorporating same-sex families into discussions throughout the text involves a study that examined work-home spillover specifically among dual-earner lesbian and gay parents. Concepts introduced for the first time or given considerably more attention due to their growing salience include the idea of the *greedy career*—one that expects 24/7 digital and other availability—coupled with the social development over the past two decades of increasingly intensive (some say relentless) expectations for parenting. These two phenomena, taken together, do much to explain how gender influences the workplace.

Chapter 11, Communication in Relationships, Marriages, and Families, continues its focus on positive communication strategies among couples and families. The mechanisms through which people communicate are rapidly changing and terminology is evolving in response. Since the last edition, research on digital communication and social media has exploded and

we know much more than we did about the positives (online support groups) and negatives (social laziness) of new forms of communication. We extend the implications of digital communication to Gen Z, who have never not known life without social media and truly sees it as an extension of their social identity. Meanwhile, a traditional venue for family communication has always been the evening meal—just make sure to put down your phone or you'll be accused of *phubbing*. As always, John Gottman's research remains a powerful force in understanding interpersonal communication between couples.

Chapter 12, Power and Violence in Families, maintains its ongoing emphasis on power relations within the context of growing family racial and ethnic diversity. This chapter now presents the latest research findings regarding power and decision-making issues among same-sex married couples. Domestic violence among same-sex couples is now explored in this chapter rather than elsewhere in the text. As an example of our keeping not only research findings and statistics up to date but also paying attention to evolving concepts and terminology, we note the development of the concept *coercive control*, formerly termed *intimate terrorism*, itself formerly termed *patriarchal terrorism*. All research and statistics have been thoroughly updated.

Chapter 13, Family Stress, Crisis, and Resilience, continues to emphasize and expand discussion of the growing body of research on resilience in relation to family stress and crises and has been updated with many new examples. As one instance, the chapter expands its exploration of family members' stress related to discrimination against minority race or ethnic groups. Recognizing that family systems are comprised of individuals, this chapter now includes some exploration of individuals' biological stress responses involving complex physiological reactions in the brain and hormonal system. This chapter also addresses what individuals can do to manage personal stress responses—a practice that impacts family responses to crises.

Chapter 14, Divorce and Relationship Dissolution, includes updated statistics on divorce rates, which have continued their decline since the Great Recession, and speculate why this is happening. We continue our discussion of the ever expanding divorce divide and add information on divorce among LGBTQ+ couples, especially those who married after 2015 when marriage became legal across the nation. We have updated all statistics related to divorce as well as information that has changed related to the determinants of divorce, such as cohabitation and women's employment. This chapter continues to highlight the effects of divorce on adults and children and factors that can lessen the negative effects. The implications of different custody arrangements for children and families and child support are

also examined and we include a new section on potential ways of improving divorce outcomes, such divorce mediation.

Chapter 15, Remarriages and Stepfamilies, continues to stress diversity within stepfamilies, reflecting continued growth of nonmarital childbearing, cohabitation, father custody, racial and ethnic diversity, and same-sex couples with stepchildren. We continue to provide the most up-to-date statistics on remarriage, stepfamilies, and living arrangements in the United States. We continue to pay attention to microlevel stepfamily dynamics such as dating with children, the process through which people become stepparents, and the challenges of day-to-day living in stepfamilies. We've enhanced our discussion of the rewards and challenges of relationships between step-grandparents and grandchildren. Finally, in an environment set up for first-married, biological parent families, we provide a comprehensive discussion of financial, legal, and policy issues stepfamilies must grapple with every day, from talking with teachers and doing their taxes to custody decisions and how to divide inheritances.

Chapter 16, Aging and Multigenerational Families, continues to place a thematic emphasis on multigenerational families, ties, and obligations in a cultural content of individualism and includes a discussion of caregiver ambivalence coupled with multigenerational families as safety nets for all generations. Like all the others, this chapter benefits from the most current statistics and research. By the time we reached this chapter in this fourteenth-edition revision process, the global pandemic, Covid-19 and its consequences had become consequential to American families in countless ways, some foreseen at this writing and others yet to be understood. We were able to address Covid-19 to some extent in this chapter and look forward to discussing impacts of this global pandemic thoroughly in our next, the fifteenth, edition.

MindTap for Marriages, Families, and Relationships, Fourteenth Edition

MindTap engages and empowers students to produce their best work—consistently. By seamlessly integrating course material with videos, activities, apps, and much more, MindTap creates a unique learning path that fosters increased comprehension and efficiency.

- MindTap delivers real-world relevance with activities and assignments that help students build critical thinking and analytical skills that will transfer to other courses and their professional lives.
- MindTap helps students stay organized and efficient with a single destination that reflects what's important to the instructor, along with the tools students

need to master the content. MindTap empowers and motivates students with information that shows where they stand at all times—both individually and compared with the highest performers in class.

In addition, MindTap allows instructors to:

- Control what content students see and when they see it with a learning path that can be used as is or matched to their syllabus exactly.
- Create a unique learning path of relevant readings and multimedia and activities that move students up the learning taxonomy from basic knowledge and comprehensions to analysis, application, and critical thinking.
- Integrate their own content into the MindTap Reader using their own documents or pulling from sources like YouTube videos, websites, Google Docs, and more.
- Use powerful analytics and reports that provide a snapshot of class progress, time in course, engagement, and completion.

Instructor Resources

Instructor's Resource Center Available online, the Instructor's Resource Center includes an instructor's manual, a test bank, and PowerPoint slides. The instructor's manual will help instructors organize the course and captivate students' attention. The manual includes key learning objectives, lecture outlines, in-class discussion questions, class activities, extensive lists of reading, video, and online resources, and suggested Internet sites and activities. The test bank includes multiple-choice, true/false, and essay questions, all with answers and text references, for each chapter of the text. The PowerPoints include chapter-specific presentations, including images, figures, and tables, to help instructors build their lectures.

Cengage Testing Powered by Cognero Cognero is a flexible, online system that allows instructors to:

- Import, edit, and manipulate test bank content from the Marriages, Families, and Relationships test bank or elsewhere, including their own favorite test questions.
- Create multiple test versions in an instant.
- Delivery tests from their LMS, classroom, or wherever they want.

ACKNOWLEDGMENTS

This book is a result of a joint effort on our part; we could not have conceptualized or written it alone. We want to thank some of the many people who helped us. Looking back on the long life of this book, we

acknowledge Steve Rutter for his original vision of the project and his faith in us. We also want to thank Sheryl Fullerton and Serina Beauparlant, who saw us through early editions as editors and friends and who had significant importance in shaping the text that you see today.

As has been true of our past editions, the people at Cengage Learning have been professionally competent. Huge thanks go to Elesha Hyde, who provided the constant consultation, encouragement, and feedback to the authors that enabled this edition to come to completion on schedule. We are also grateful to Kori Alexander, Product Manager, who guided this edition, and to Learning Designer Emma Guiton, who guided the development of the MindTap product that accompanies this text. Deanna Ettinger, Intellectual Property Analyst, made sure we were accountable to other authors and publishers when we used their work.

Shelley Ryan, Project Manager for MPS Limited, led a production team whose specialized competence and coordinated efforts have made the book a reality. She was excellent to work with, always available and responsive to our questions, flexible, and ever helpful. She managed a complex production process smoothly and effectively to ensure a timely completion of the project and a book whose look and presentation of content are very pleasing to us—and, we hope, to the reader.

The internal production efforts were managed by Tim Bailey, Content Manager. Copy Editor Richard Camp did an outstanding job of bringing our draft manuscript into conformity with style guidelines. Anjali Kambli, Photo Researcher (Lumina Datamatics), worked with us to find photos that captured the ideas we presented in words.

Nadine Ballard developed the overall design of the book, one we are very pleased with. Once it is completed, our textbook needs to find the faculty and students who will use it. Tricia Salata, Marketing Manager, captured the essence of our book in the various marketing materials that present our book to its prospective audience.

Closer to home, Agnes Riedmann wishes to acknowledge her late mother, Ann Langley Czerwinski, PhD, who helped her significantly with past editions. Agnes would also like to acknowledge family, friends, and professional colleagues who have supported her throughout the thirty-five years that she has worked on this book. Dear friends have helped as well. Agnes would like to specifically recognize Susan Goldstein and Victor Herbert, who often have sent her pertinent articles and engaged her in relevant and stimulating discussions.

Sam Walker has contributed to previous editions of this book through his enthusiasm and encouragement for Mary Ann Lamanna's work on the project. Larry and Valerie Lamanna and other family members have enlarged their mother's perspective on the family by bringing her into personal contact with other family worlds—those beyond the everyday experience of family life among the social scientists!

Mary Ann Lamanna and Agnes Riedmann continue to acknowledge one another as coauthors for forty years. Each of us has brought somewhat different strengths to this process. We are not alike—a fact that has continuously made for a better book, in our opinion. At times, we have lengthy e-mail conversations back and forth over the inclusion of one phrase. Many times, we have disagreed over the course of the past thirty years—over how long to make a section, how much emphasis to give a particular topic, whether a certain citation is the best one to use, occasionally over the tone of an anxious or frustrated e-mail. But we have always agreed on the basic vision and character of this textbook. And we continue to grow in our mutual respect for one another as scholars, writers, and authors. We have now been joined by Susan Stewart as coauthor. She brings a fresh perspective to the book as well as a comprehensive knowledge of research in the field. Her patience and expertise have been especially important to this revision.

Susan Stewart would like to acknowledge Agnes Riedmann and Mary Ann Lamanna for their unwavering support, mentoring, and wisdom as she continues her journey learning the art and science of textbook writing. She would also like to acknowledge her daughter, Gwen, who continues to provide rich experiences that contribute to her understanding about parent–child relationships and adolescent concerns, especially given that she is now a full-fledged member of Gen Z! She acknowledges her parents and sisters, and her ex-spouse and in-laws, as well as her husband, Gene, and stepson, Cameron, and his wife, Anna, who taught her that no amount of reading can replace lived experience. She especially thanks the students in her *Sociology of Intimate Relationships* class who, each and every semester, read this book and act as an important sounding board for the content, both old and new. I thank Dr. David Wahl for his insights into gender and sexuality and his contributions to Chapter 4.

Reviewers gave us many helpful suggestions for revising the book. Although we may not have incorporated all suggestions from reviewers, we have considered them all carefully and used many. The review process makes a substantial, and indeed essential, contribution to each revision of the book.

Fourteenth Edition Reviewers

Amanda Burnam, OCCC; Amy M Smith, Florida State University; Anthony Walker, Indiana State University; Brandon Eddy, UNLV; Carol Campbell, McNese

State University; Cassidy Cooper, University of Mobile; Claudia Hall, Baton Rouge Community College; Darrell Frost, Northern Oklahoma College; David W. Wahl, Iowa State University; Elinor Behana, San Diego State University; Heather Griffiths, Fayetteville State University; Heather Jaffe, San Diego State University; Jacquelyn Benson, University of Missouri; Jennifer George, University of Georgia; Jennifer Valentine, Tidewater Community College; Judy Bohrer, Butler Community College; Julie Taren, California State University, Los Angeles; Kelly Warzinik, University of Missouri; Kevin Dingess, Blue Ridge Community College; Kimberly Michelle Murray, Texas A&M University-Texarkana; Kira Arthurs, Pearl River Community College; Lauren Lewis, Indiana State University; Lee Maria Chancler, University of Central Missouri; Marcia Seddon, Indian Hills Community College; Nicholle Liessmann, San Juan College; Olga Custer, Oregon State University; Rhonda Johnson, Holmes Community College; Yvonne Thai; Leslie Hurt; Linda Behrendt

Thirteenth Edition Reviewers

Cari Beecham-Bautista, Columbia College Chicago; Chris Caldeira, University of California, Davis; Lynda Dickson, University of Colorado at Colorado Springs; Rebecca S. Fahrlander, University of Nebraska at Omaha; Loyd R. Ganey, Jr., College of Southern Nevada; Jamie L. Gusrang, Community College of Philadelphia; Faye Jones, Mississippi Gulf Coast Community College JC Campus; Nancy Reeves, Gloucester County College; Jewrell Rivers, Abraham Baldwin Agricultural College; Chad W. Sexton, State University of New York at Fredonia; and Sharon Wiederstein, Blinn College.

Twelfth Edition Reviewers

Chuck Baker, Delaware County Community College; Adriana Bohm, Delaware County Community College; John Bowman, University of North Carolina at Pembroke; Jennifer Brougham, Arizona State University–Tempe; Shaheen Chowdhury, College of DuPage; Diana Cuchin, Virginia Commonwealth University; James Guinee, University of Central Arkansas; Amy Knudsen, Drake University; Wendy Pank, Bismarck State College; Rita Sakitt, Suffolk County Community College; Tomecia Sobers, Fayetteville Technical Community College; Richard States, Allegany College of Maryland; and Scott Tobias, Kent State University at Stark.

Eleventh Edition Reviewers

Rachel Hagewen, University of Nebraska, Lincoln; Marija Jurcevic, Triton College; Sheila Mehta-Green, Middlesex Community College; Margaret E. Preble, Thomas Nelson Community College; Teresa Rhodes, Walden University.

Of Special Importance

Students and faculty members who tell us of their interest in the book are a special inspiration. To all of the people who gave their time and gave of themselves—interviewees, students, our families and friends—many thanks. We see the fact that this book is going into a fourteenth edition as a result of a truly interactive process between ourselves and students who share their experiences and insights in our classrooms; reviewers who consistently give us good advice; editors and production experts whose input is invaluable; and our family, friends, and colleagues whose support is invaluable.

1

MAKING FAMILY CHOICES IN A CHANGING SOCIETY

Learning Objectives

1. Explain why researchers and policy makers need to define family.

2. Explain the ways that family structure or form is increasingly diverse.

3. Describe the various society-wide structural conditions that impact families.

4. Discuss why the best life course decisions are informed ones made consciously.

5. Explain how families provide individuals with a place to belong.

6. Demonstrate why there is a tension in our culture between familistic values and individualistic values.

7. Identify how global situations and events affect family life in the United States.

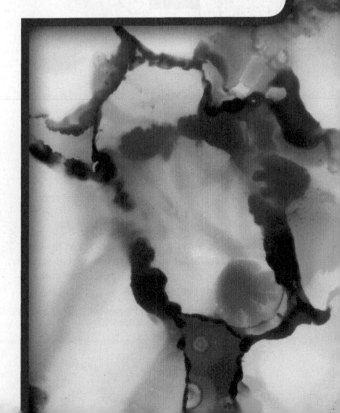

This text is different from others you will read. Although it could help you in a future career, this text has four other goals as well—to help you: (1) appreciate the variety and diversity among families today, (2) become more sensitive to family issues, (3) understand your past and present family situations and anticipate future possibilities, and (4) be more conscious of the personal decisions you make throughout your life and of the societal influences that affect those decisions.

About thirty years ago, stating that "the family constitutes the basic unit of society and therefore warrants special attention," the United Nations designated 1994 as the International Year of the Family. Later, the U.N. proclaimed every May 15th the International Day of Families. Across the world, families are central both to society and to people's everyday lives.

Families worldwide take on the pivotal tasks of raising children and providing family members with support, companionship, affection, and intimacy. As shown in Figure 1.1, national survey results show Americans are most likely to say *family* is what gives them meaning in life (Pew Research Center 2018a). Meanwhile, what many of us think of as family has changed dramatically in recent decades. This chapter explores *family* definitions and notes the varied structures or forms that families

take today. This chapter also describes society-wide conditions that impact families: ever-new biological and communication technologies, economic conditions, historical periods of events, and demographic characteristics such as age, religion, race, and ethnicity.

Later in this chapter, we'll note that when maintaining committed relationships and families, people need to make informed decisions. We end this chapter by discussing four themes that characterize this text. You'll see that these four themes comprise the text's four learning goals, listed in the Preface. We begin with a definition of family—one to keep in mind throughout the course.

DEFINING FAMILY

People make a variety of assumptions about what families are and are not. We've noticed when teaching this course that many students, when asked to list their family members, include their pets. Are dogs, cats, or hamsters family members? On a different note, some individuals who were conceived by artificial insemination with donor sperm are tracking down their "donor siblings"—half brothers and sisters who were conceived using the same man's sperm. They may define their "donor relatives" as family members, although others born under similar circumstances may not. Indeed, *family* has many definitions, not only among laypeople but also among family scientists.

We, your authors, have chosen to define **family** as follows: A family is any sexually expressive, parent—child, or other kin relationship in which people—usually related by ancestry, marriage, or adoption—(1) form an economic or otherwise practical unit and care for any children or other dependents, (2) consider their identity to be significantly attached to the group, and (3) commit to maintaining that group over time.

How did we come to this definition? First, caring for children or other dependents suggests a function that the family is expected to perform. Definitions of many things have both functional and structural components. Functional definitions point to the purpose(s) for which a thing exists—that is, what it does. For example, a functional definition of a smartphone would emphasize that it allows you to make and receive calls, take pictures, connect to the Internet, and access media. Structural definitions emphasize the *form* that a thing takes—what it actually is. To define a smartphone structurally, we might say that it is an electronic device, small enough to be handheld, with a multimedia screen and components that allow sophisticated satellite communication. Concepts of the family comprise both functional and structural aspects. We'll look now at how the family can be recognized by its functions, and then we'll discuss structural definitions of the family.

In an open-ended question, % of Americans who mention__when describing what provides them with a sense of meaning

	Percent
Family	69%
Career	34%
Money	23%
Spirituality and faith	20%
Friends	19%
Activities and hobbies	19%
Health	16%
Home and surroundings	13%
Learning	11%

FIGURE 1.1 Americans are most likely to mention family when asked what provides them with a sense of meaning. What else do these findings suggest to you?

Source: Survey conducted September 14–18, 2017 among U.S adults. Pew Research Center 2018b.

Family Functions

Social scientists usually list three major functions filled by today's families: raising children responsibly, providing members with economic and other practical support, and offering emotional security.

Family Function 1: Raising Children Responsibly

If a society is to persist beyond one generation, adults have to not only bear children but also feed, clothe, and shelter them during their long years of dependency. Furthermore, a society needs new members who are properly trained in the ways of the economy and culture and who will be dependable members of the group. These goals require children to be responsibly raised. Virtually every society assigns this essential task to families.

Traditionally, a related family function has been to control its members' (particularly women's) sexual activity, and this function persists in many parts of the world. Controlling sexuality was historically understood as necessary in order to guarantee responsible childrearing. "Throughout history, marriage has first and foremost been an institution for procreation and raising children. It has provided the cultural tie that seeks to connect the father to his children by binding him to the mother of his children" (Wilcox Marquardt, Popenoe, and Whitehead 2011). However, in the United States and other industrialized societies the child-raising function is more and more often performed by divorced, separated, never-married, or cohabiting parents, and sometimes by grandparents or other relatives. Today researchers talk about "the decoupling of marriage and parenthood" (Hayford, Guzzo, and Smock 2014). Nevertheless, the majority of U.S. births today (about 60 percent) take place within marriage (Martin, Hamilton, Osterman, Driscoll, and Drake 2018, p. 5).

Family Function 2: Providing Economic and Other Practical Support

A second family function involves providing economic support. Historically, the family was primarily a practical economic unit rather than an emotional one (Shorter 1975; Stone 1980). Although the modern family is no longer a self-sufficient economic unit, virtually every family engages in activities aimed

PK Studio/Shutterstock.com

We can define families by their functions—raising children, providing economic support for dependents, and offering emotional support for all family members. This father looks to be doing all that. But functional definitions of family aren't enough. We also need to consider the group's structure. This family consists of a heterosexual couple and their child. They may be married or cohabiting.

at providing for such practical needs as food, clothing, and shelter. Throughout this text, we'll see the varied ways that this function plays out.

Family economic functions now consist of earning a living outside the home, pooling resources, and making consumption decisions together. In assisting one another economically, family members create some sense of material security. For example, family members offer one another a kind of unemployment insurance. If one family member is laid off or can't find work, others may be counted on for help. Family members care for each other in additional practical ways too, such as nursing and transportation during an illness or lending an ear when someone needs to talk.

Family Function 3: Offering Emotional Security

Although historically the family was a pragmatic institution involving material maintenance, in today's world the family has grown increasingly important as a source of emotional security. Thinking of families globally, the United Nations has described the family as a place where "one finds warmth, caring, security, togetherness, tolerance and acceptance" ("International Day of the Family," n.d.). Not just partners or parents but also children, siblings, and extended kin can be important sources of emotional support (Henig 2014; Waite et al. 2011).

This is not to say that families can solve all our longings for affection, companionship, and intimacy. Sometimes, in fact, the family situation itself is a source of stress and pain—as in the case of parental conflict, alcoholism, drug abuse, or domestic violence. But families and committed relationships are expected to provide emotional support. Defining a family by its functions is informative and can be insightful: According to a Chicago Chief Executive Officer, for instance, "To me a family is whoever I can depend on for support, to laugh with, to play with, and to share the challenges and rewards of life with" (Wolf 2018, p. 4).

But defining a family only by its functions would be too vague and misleading. Neighbors or roommates might help with childcare, provide for economic and other practical needs, or offer emotional support, but we might not define them as family. An effective definition of family needs to incorporate structural elements as well.

Structural Family Definitions

Traditionally, both legal and social sciences have specified that the family consists of people related by blood, marriage, or adoption. In their classic work *The Family: From Institution to Companionship*, Ernest Burgess and Harvey Locke (1953 [1945]) specified that family members must "constitute a household," or reside together. Some definitions of the family have gone even further to include economic interdependency and sexual–reproductive relations (Murdock 1949).

The U.S. Census Bureau defines a family as two or more people related by blood, marriage, or adoption and residing together in a household. The Census Bureau defines **household** as any group that resides together. Not all households are families; to be a *family household*, persons sharing a household must also be related by blood, marriage, or adoption. Now that same-sex marriages are legal nationwide, married same-sex couples living together are of course counted as family households. Before the June 26, 2015 U.S. Supreme Court decision legalizing same-sex marriage, lesbian and gay male couples living together were counted as *non*-family households. Cohabiting couples, whether heterosexual or same-sex, continue to be counted as nonfamily households.

Family structure—the form a family takes—varies according to the social environment in which it is embedded. In preindustrial or traditional societies, the family structure involved whole kinship groups. The **extended family** of parents, children, grandparents, and other relatives performed most societal functions, including economic production (e.g., the family farm), protecting family members, providing vocational training, and maintaining social order. In industrial or modern societies, the typical family structure often became the **nuclear family** (husband, wife, children), which was better suited to city life. Until about sixty years ago, social attitudes, religious beliefs, and law converged into a fairly common expectation about what form the American family should take: breadwinner husband, homemaker wife, and children living together in an independent household—the *nuclear-family ideal*.

Nevertheless, the extended family—including adult siblings, a family research topic often neglected—continues to play an important role in many cases, especially among recent immigrants and race and ethnic minorities. To cope with economic hardships more relatives of all races and ethnicities are moving in together to create more multigenerational or otherwise extended-family households. About one-fifth, or 20 percent of Americans live in multigenerational households—about the same percentage as in 1950, but an increase from a low of 12 percent in 1980 (Cohn and Passel 2018). "Accordion" family households that expand or contract with more or fewer family members, depending on family

The extended family—grandparents, aunts, and uncles—can provide occasion for good times as well as an important source of security, its members helping each other, especially during crises.

Ariel Skelley/Getty Images

need, perform important economic and often emotional social functions (Newman 2012).

Meanwhile, today's families are not necessarily bound to one another by legal marriage, blood, or adoption. The term *family* can identify relationships in addition to spouses, parents, children, and extended kin. Individuals fashion and experience intimate relationships and families in many forms. As social scientists take into account this structural variability, it is not uncommon to find them referring to the family as *postmodern*.

Postmodern: There Is No Typical Family

Barely half of U.S. adults are married (U.S. Census Bureau 2019, Table A1). Only about 5 percent of families now resemble the 1950s nuclear family of married couple and children, with a husband-breadwinner and wife-homemaker (Vespa, Lewis, and Kreider 2013, Tables 4, 5). Prompting social scientists to remark on today's "revolution in intimate life relationships," the past several decades have witnessed a proliferation of relationship and family forms: single-parent families, stepfamilies, families with children of more than one father, two-earner couples, stay-at-home fathers, cohabitating heterosexual couples, gay and lesbian marriages and families, three-generation families, and communal households, among others. Individuals construct a myriad of social forms in order to address family functions. Social scientists have typically thought of the nuclear family as the "modern" family form. The more recent term **postmodern family** acknowledges the fact that today's families exhibit multiple of forms as new or altered family forms continue to emerge.

Figure 1.2 displays the types of households in which Americans live today. Only about two-thirds of households contain families. Just 19 percent of households are nuclear families of husband, wife, and children, compared with more than twice that (44 percent) in 1960 (U.S. Census Bureau 2015a, 2018a). The most common household type is married couples without children: Either the children have grown up and left or the couple has not yet had children or doesn't plan to. More households today (28 percent) are maintained by individuals living alone than by married couples with children. "Facts about Families: American Families Today" presents additional information about families. We now see unprecedented diversity in family composition, or form.

Due to this diversity, laws, government agencies, and private corporations such as insurance companies make decisions about what was once taken for granted—that is, what a family is. If rent policies, employee-benefit packages, and insurance policies cover families, decisions need to be made about what relationships or groups of people are to be defined as a family. The

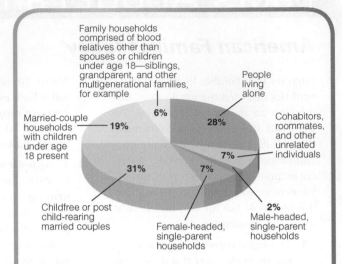

FIGURE 1.2 The many kinds of American households, 2018. A household is one or more persons who occupy a dwelling unit, or domicile. Households can be family or nonfamily. Family households contain persons related by blood, marriage, or adoption. Along with other household types, the Census Bureau classifies opposite- and same-sex unmarried-couple households as nonfamily households. This figure displays nonfamily households in shades of blue and family households in shades of green.

Sources: U.S. Census Bureau 2018b, Tables FG3, FG5, H.

September 11th Victim Compensation Fund of 2001 struggled with this issue in allocating compensation to victims' survivors. As a result, and New York state law was amended to allow awards to unmarried gay and heterosexual partners (Gross 2002). In 2015, the U.S. Supreme Court ruled that same-sex marriage is a nationwide right to be recognized in all fifty states.

Adapting Family Definitions to the Postmodern Family

As family forms have grown increasingly variable, social scientists have proposed—and often struggled with—new, more flexible definitions for the family. Legal definitions of family have become more flexible as well. The 2015 U.S. Supreme Court ruling that legalizes same-sex marriage comes to mind. As another example, a few state legislatures have provided that legal status and rights can be enjoyed by more than two—that is, by three or four—parents in one family. What would be an example of a family like this? Here's one: Two children spend three nights a week with their partnered gay fathers. The other nights they stay with their lesbian mothers, who live nearby (Lovett 2012).

Many employers have redefined family with respect to employee-benefit packages. Just more than half of

Facts About Families

American Families Today

What do U.S. families look like today? Statistics can't tell the whole story, but they are an important beginning. As you read these ten facts, remember that the data presented here are generalizations and do not consider differences among various sectors of society. We explore social diversity throughout this textbook, but for now let's look at some overall statistics.

1. *Marriage is important to Americans— but not to the extent that it was sixty years ago during the "Golden Age of Marriage."* Today about 58 percent of never-married adults say they want to marry someday. Twenty-seven percent are not sure. Another 14 percent don't want to get married (Parker and Stepler 2017). Ten years ago,, 44 percent of 18- to 29-year-olds and 32 percent of Americans age 65 and older saw marriage becoming obsolete (Taylor et al. 2011).

2. *About half of Americans are married.* Just about 50 percent of adults age 18 and older were married in 2018, compared to about three-quarters (72 percent) in 1960.

About 30 percent of Americans today have never married; 10 percent are divorced, and 6 percent widowed (U.S. Census Bureau 2018a, Table A1).

3. *Young people are postponing marriage.* In 2018, the median age at first marriage was 27.8 for women and 29.8—nearly 30—for men, as compared with about 21 for women and 24 for men in 1970. Today's average age at marriage is the highest recorded since the 1890 census (U.S. Census Bureau 2018b, Table MS-2).

4. *With some usually religion-based exceptions, cohabitation has become an acceptable family form (as well as a transitional lifestyle choice).* The number of opposite-sex cohabiting adults increased more than tenfold since 1970—and by 40 percent since 2000. About 40 percent of cohabiting couples live with children under age 18—either their own or those from a previous relationship or marriage. Unmarried couple families are only about 7 percent of American adults at any one time, but more than

50 percent of first marriages are preceded by cohabitation. No longer a minority lifestyle choice, cohabitators are older now, as well as more racially and ethnically diverse, more highly educated, and higher earners (Gurrentz 2019). In fact, for adults ages 18 to 24, living with an unmarried partner is more common than living with a spouse (Gurrentz 2018).

5. *Fertility has declined.* Although there's a slight increase in people who say three or more children would be ideal, fertility is down (Bialik 2018). At 1.77 in 2017, the total fertility rate (TFR)—the average number of births that a woman will have during her lifetime—had dropped by 3 percent from 2016 (Martin et al. 2015). After a high of 3.6 in 1957, the TFR has generally been below replacement level over the past thirty years (Martin, Hamilton, Osterman, Driscoll, and Drake 2018; Matthews, Brady, and Hamilton 2019). A society requires a TFR of at least 2.1 in order for the

---

the Fortune 500 companies, as well as many state and local governments, offer domestic partner benefits to persons in an unmarried couple who have registered their relationship with a civil authority (Appleby 2012). President Barack Obama signed an executive order granting federal employees and their domestic partners some of the rights (but neither health insurance nor retirement benefits) enjoyed by married couples (Miles 2010). If passed in the future, currently proposed federal legislation would extend domestic partner benefits to all federal civilian employees ("Domestic Partnership Benefits and Obligations Act" 2015). Meanwhile, federal practices permit low-income unmarried couples to qualify as families and live in public housing.

We, your authors, began this section with our definition of family. Our definition recognizes the diversity of

postmodern families while paying heed to the essential functions that families are expected to fill. Our definition combines some structural criteria with a more social–psychological sense of family identity. We include the commitment to maintaining a relationship or group over time as a component of our definition because we believe that such a commitment is necessary in fulfilling basic family functions. It also helps to differentiate the family from casual relationships, such as roommates, or groups that easily come and go.

We have worked to balance an appreciation for flexibility and diversity in family structure and relations with the concern that many policy makers and social scientists express about how well today's families perform their functional obligations. Ultimately, there is no one correct answer to the question, "What is a family?"

population numerically to replace itself, so the current TFR is below replacement level.

6. *Particularly among college-educated women, parenthood is often postponed.* The average age for a woman's first birth increased by about 6 years between 1970 and 2017—from age 21 to 27. But the statistics differ according to education with more highly educated women waiting longer to have children (Martin et al. 2018, p. 5). Married women today wait longer after their wedding to conceive than in the past (Hayford, Guzzo, and Smock 2014).

7. *Compared to 4 percent in 1950, the nonmarital birthrate is high* with 40 percent of all U.S. births today being to unmarried mothers. Unlike 1950, however, between one-quarter and one-half of nonmarital births today occur to cohabiting couples (Carter 2009; Martin et al. 2019, p. 6). Seeing marriage as obsolete (as noted in #1 above) may be an overstatement. However, the fact that today "nearly half of U.S. births happen outside marriage" certainly marks a "cultural shift" (Griffin 2018).

8. *Same-sex-couple households increased* by 80 percent between 2000 and 2010 (Homan and Bass 2012). Partly because an unknown number remain "closeted," it is difficult to know how many same-sex-couple households really exist in the United States (Hoffman 2014). According to U.S. Census Bureau estimates, there were approximately 900,000 same-sex households in 2017 (U.S. Census Bureau 2019). Of these, about 500,000 were married, although married-couple same-sex couples comprise less than 1 percent of all U.S. married couples (Cohn 2014; Schwarz 2014). About 17 percent of same-sex households include children (U.S. Census Bureau 2014b, Table 1).

9. *The divorce rate is dropping.* After it doubled between 1965 and 1980, the U.S divorce rate began to drop steadily, falling more than 30 percent from 1980 to 2010 (U.S. Census Bureau 2012a, Table 78). The divorce rate continued to decline through 2016, the year of our most current data. We used to say that about half of marriages end in divorce, but today that figure is closer to one-third (National Center for Health Statistics 2017). This is good news. We need to be aware, though, that fewer and fewer Americans are getting married—and those who do tend to be more highly educated and have high incomes, a category that has traditionally evidenced lower divorce rates.

10. *The remarriage rate has declined in recent decades but remains significant.* About 60 percent of recent marriages are first-time marriages for both spouses. About 40 percent of today's marriages involve a remarriage for at least one spouse. Twenty percent of marriages today and remarriages for both partners. About 4 percent of marrieds wed three or more times. This number rises to 7 percent for those over age fifty (Geiger and Livingston 2019; Lewis and Kreider 2015).

Critical Thinking

What do these statistics tell you about the strengths and weaknesses of the contemporary American family and about family change?

Relaxed Institutional Control over Relationship Choices: "Family Decline" or "Family Change"?

According to public opinion polls, about 30 percent of Americans reject today's trend toward the postmodern family while about the same proportion accept new family forms. Another 37 percent accept some aspects of family change but are concerned about others (Morin 2011). In 2012, 59 percent of Americans found unmarried heterosexual sex to be morally acceptable, but 38 percent saw it as morally wrong. Those numbers had changed from 53 percent and 42 percent in 2001. Sixty-seven percent of Americans today see divorce as morally acceptable, whereas in 2001 that figure was 59 percent. Fifty-four percent of Americans believe having a baby outside marriage is morally acceptable today, compared with 45 percent in 2002 ("Marriage" 2012). Americans are fairly evenly split regarding whether they support same-sex marriage as legally valid, although fewer than 40 percent favored legal same-sex marriage in 2001 (Pew Research Center 2015a). Figure 1.3 shows results of a 2015 national Pew Survey asking respondents what they think about some current trends in family life. As shown in Figure 1.3, about two-thirds of Americans believe that single women having children without a partner is bad for society. Meanwhile, just about half of us think that more unmarried couples raising children is bad for society (Pew Research Center 2018). Americans can be strongly opinionated about family change; we can better understand why if we understand that the family has historically been understood as a **social institution**.

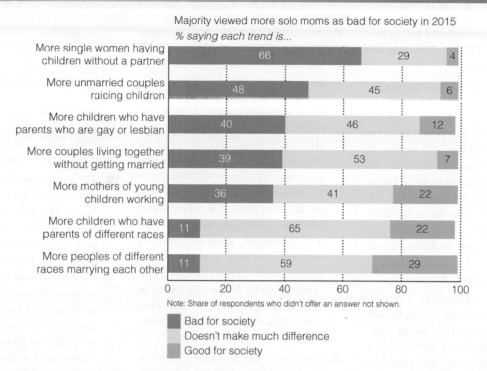

FIGURE 1.3 How Americans view emerging family trends—bad, doesn't matter, or good for society. What do these survey results say about Americans' attitudes about (1) more single women having children without a partner, (2) more people of different races getting married, or (3) more children with gay or lesbian parents—among other indicators of family change?

Source: Survey of U.S. adults conducted February—April, 2015, Pew Research Center 2018a.

Social institutions are patterned and largely predictable ways of thinking and behaving—beliefs, values, attitudes, and norms that are organized around vital aspects of group life and serve essential social functions. Social institutions are meant to meet people's basic needs and enable the society to survive. Earlier in this chapter, we described three basic family functions. Because social institutions prescribe socially accepted beliefs, values, attitudes, and behaviors, they exert considerable social control over individuals.

Beginning in the 1960s, however, family formation became less and less predictable. Demographers noted dramatic social transformations: age at first marriage increased, marital childbearing decreased, nonmarital childbearing increased, divorce rates rose, and cohabitation became common among young adults. Although the most dramatic shifts arose in the 1970s and 1980s, trends established then have continued. Combined with increased longevity and lower fertility rates, these changes have meant that a smaller portion of adulthood is spent in traditionally institutionalized marriages and families (Cherlin 2008).

Critics have described relaxed institutional control over families and relationships as "family decline."

Those with a **family-decline perspective** believe that cultural change toward excessive individualism and self-indulgence has hurt relationships, led to high divorce rates, and undermines responsible parenting (Popenoe and Whitehead 2005):

> According to a marital decline perspective . . . because people no longer wish to be hampered with obligations to others, commitment to traditional institutions that require these obligations, such as marriage, has eroded. As a result, people no longer are willing to remain married through the difficult times, for better or for worse. Instead, marital [or other relationship] commitment lasts only as long as people are happy and feel that their own needs are being met. (Amato 2004, p. 960)

In addition, fewer family households contain children. According to the family-decline perspective, this situation "has reduced the child centeredness of our nation and contributed to the weakening of the institution of marriage" (Popenoe and Whitehead 2005, p. 23; Wilcox, Marquardt, Popenoe, and Whitehead 2011). "Facts about Families: Focus on Children" provides some statistical indicators about the families of contemporary children.

Facts About Families

Focus on Children

In many places throughout this text, we focus particularly on children. Approximately 74 million children under age 18 live in the United States. However, the proportion of today's population that is under age 18—about 23 percent—represents a substantial drop from the 1960s when more than one-third of Americans were children (U.S. Federal Interagency Forum on Child and Family Statistics 2015). Here we look at five statistical indicators regarding U.S. children's living arrangements and well-being.

1. In 1960, 88 percent of U.S. children lived with two married parents. Things have changed considerably. Nonetheless a majority of children today live in two-parent households. In 2018, 69 percent of children under 18 lived with two parents, although not necessarily married. More than one-quarter (27 percent) of children lived with a single parent, the vast majority with their mother. Another 4 percent did not live with either parent (U.S. Federal Interagency Forum on Child and Family Statistics 2019, p. vii).

2. Many children experience a variety of living arrangements while they're young. A child may progress through living in an intact two-parent family, a single-parent household, with a cohabitating parent, and finally in a remarried family. About half of all American children are expected to live in a single-parent household at some point in their lives, most likely in a single-mother household (Kreider and Ellis 2011a).

3. Children are more likely to live with a grandparent today than in the past. In 1970, 3 percent of children lived in a household headed by a grandparent. By 2015 that rate had reached about 8 percent (Child Trends Databank 2015; Wu 2018). In about one-quarter of the cases, grandparents had sole responsibility for raising the child, but many households containing grandparents are extended-family households that may include one or both of the children's parents as well as other relatives (Child Trends Databank 2015; Edwards 2009).

4. Although most parents are employed, children are more likely than the general population to be living in poverty. The 2018 poverty rate for U.S. children under age 18 was 16 percent, compared with 10 percent for Americans age 65 and older. Older Americans typically have had lower poverty rates than children since the 1935 onset of the federal Social Security program—and because older Americans are more likely to vote than young children's parents. Poverty rates in 2018 for nonHispanic white children were 9 percent; for Asian children, 11 percent; for Hispanic children, 24 percent; and for black children, 29 percent (U.S. Census Bureau 2018c, Table 3; DeNavas-Walt and Proctor 2015, Table 3).

5. A growing number of U.S. children have a foreign-born parent. The percentage of children under age 18 living with at least one foreign-born parent rose from 14 percent in 1994 to 26 percent in 2018—over one-quarter of all U.S. children. Twenty-three percent of children were native-born children with at least one foreign-born parent, and 3 percent were foreign-born children with at least one foreign-born parent. In 2016, nearly one quarter (23 percent) of children ages 5 to 17 spoke a language other than English at home, (U.S. Federal Interagency Forum on Child and Family Statistics 2019, p. ix).

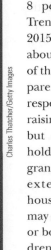

Charles Thatcher/Getty Images

The faces of America's children provide evidence of increasing ethnic diversity. The child population of the United States is more racially and ethnically diverse than the adult population. Making up about one-third of the U.S. population today, racial/ethnic minorities are projected to reach 50 percent of the total population by about 2042. Mostly due to rapid growth in Latino families, the population under age 18 is projected to reach this point by 2023 (Mather 2009).

Critical Thinking

Perhaps the greatest concern Americans have about family change today is its impact on children. What do these family data tell us about the family lives of children?

In a world of demographic, cultural, and political changes, there is no typical family structure. Today's postmodern family includes cohabiting families, single-parent families, lesbian and gay partners and parents, and remarried families. Interracial families are more evident, too, and their increasing social acceptance may result in their experiencing greater community support.

Not everyone concurs that the family is in decline: family change, yes, but not decline (Coontz 2015). Scholars and policy makers with a **family-change perspective** point out that family changes can be for the better. Longer life expectancy can mean more positive years with parents, grandparents, and great-grandparents. Easier access to divorce than was the case fifty years ago offers alternatives to enduring domestic violence. With more than 80 percent of Americans approving black–white marriages (Jones 2011), increasing tolerance for interracial unions means that mixed-race families are likely to experience less hostile communities than in the past. Family flexibility can be functional in times of economic crisis as extended families expand to take in needy relatives.

Family-change scholars argue that we need to view the family from a historical standpoint (Coontz 2015). In the nineteenth and early twentieth centuries, American families were often broken up by illness and death, and children were sent to orphanages, foster homes, or already burdened relatives. Single mothers, as well as wives in lower-class, working-class, and immigrant families, did not stay home with children but went out to labor in factories, workshops, or domestic service. Similar to today's situation, a relatively small proportion of children lived only with their father a century ago (Kreider and Fields 2005).

Family-change scholars posit that today's family forms need to be seen as historically expected adjustments to changing conditions in the wider society, including the decline in well-paid working-class, middle-class, and even upper-middle-class jobs that used to provide solid economic family support (Coontz 2015). Family-change sociologists do not ignore the difficulties that separation, divorce, and nonmarital parenthood present to families, children, and the broader society. However, these social scientists view the family as "an adaptable institution" (Amato et al. 2003, p. 21) and argue that it makes more sense to provide support to families as they exist today rather than to attempt to turn back the clock to an idealized past (Cherlin 2009a; McHale, Waller, and Pearson 2012; Sawhill 2014).

Then too, today's American families struggle with new economic and time pressures that affect their ability to realize their family values. In sharp contrast to the United States, many European countries have paid family-leave policies that enable parents to take time off from work to be with young children and that provide relatively generous economic support for families in general (Human Rights Watch 2011).

Recently the prominent family sociologist Andrew Cherlin (2015) observed an emergent "truce in the war over family":

> [T]he conservative and liberal positions have both shown signs of change. . . . Liberals now seem to acknowledge the downsides of the retreat from marriage. . . . The . . . same-sex marriage [movement] has made it possible for liberals to endorse the importance of marriage without feeling that they have abandoned their commitment to equality. . . . Some conservatives acknowledge that changes in the economy have hurt families, a marked departure from insisting that personal choices are solely to blame. (Cherlin 2015; and see Blankenhorn et al. 2015)

Recognizing that economic and other policy changes have hurt many families involves placing an individual's or family's private troubles within a society-wide context. This way of thinking is the crux of what sociologists call a **sociological imagination**.

A SOCIOLOGICAL IMAGINATION: PERSONAL TROUBLES AND SOME SOCIAL CONDITIONS THAT IMPACT FAMILIES

People's private lives are affected by what is happening in the society around them. In his classic book, *The Sociological Imagination* (1959), sociologist C. Wright Mills developed the principle that private, or personal, troubles are connected to events and patterns in society. Many times what seem to be personal troubles are shared by others, and these troubles often reflect societal influences. For example, when a family breadwinner is laid

Peathegee Inc/Blend Images/Getty Images

off, the cause does not necessarily lie in their poor work performance but in the economy's inability to provide full employment. As another example, the difficulty of juggling work and family is not usually simply a question of an individual's time-management skills but of society-wide influences—work schedules, commuting, and family care in a society that provides limited support for working families. As a final example, the quality of veterans benefits available to soldiers returning from active duty significantly affects their postwar health and readjustment to civilian family life (Finkel 2013).

In this section we'll look at five social factors that affect families:

1. ever-new biological and communication technologies,

2. economic conditions,

3. historical periods or events,

4. demographic characteristics (statistical facts about the makeup of a population) such as age, religion, and race or ethnicity, and

5. family policy.

Ever-New Biological and Communication Technologies

The pace of technological change has never been faster; new technologies will continue to alter not only family relationships but also how we define families. Here we'll look at two broad types of technological change that impact family life—biological and communication technologies.

Biological Technologies A baby is born in Israel and will be raised by gay male parents there. The sperm to conceive the baby came from one of the Israeli fathers. It was frozen and flown to Thailand, where a South African egg donor awaited. After the egg was fertilized, the new embryo went to Nepal, where it was implanted in a surrogate mother, an Asian Indian woman. Nine months later, the two fathers fly from Tel Aviv to Nepal and, thanks to science, claim their infant (Harris 2015).

Since the 1960s invention of the birth-control pill and the 1978 arrival of the first "test-tube baby," modern science has expanded our options regarding both preventing pregnancy and enhancing fertility. Science continues to develop new techniques that offer new options for individuals and couples to have biological children. The more common infertility interventions involve prescription drugs and microscopic surgical procedures to repair a female's fallopian tubes or a male's sperm ducts (Ehrenfeld 2002).

More recent hormone applications and surgical techniques offer transgender options for family members. These developments are further addressed elsewhere in this text, particularly in Chapters 3, 4, 8, and 9. Here we

point out that ever-new biological technologies dramatically impact family members' options and daily lives (Axad 2018; Farrell, VandeVusse, and Ocobock 2012).

Also, *assisted reproductive technology* (ART) offers increasingly innovative reproductive options (Ravitz 2018). Chapter 8 further explores issues regarding ART procedures.. In general, ART involves the manipulation of sperm or egg or both in the absence of sexual intercourse, often in a laboratory. ART procedures include:

- *artificial insemination* (male sperm introduced to a female egg without sexual intercourse),

- *donor insemination* (artificial insemination with sperm from a donor rather than from the man who will be involved in raising the child),

- *in vitro fertilization* (sperm fertilizes egg in a laboratory rather than in the woman's body),

- *gestational surrogacy,* or *surrogacy* (one woman gestates and delivers a baby for another individual who intends to raise the child),

- *egg sale or donation* (by means of a surgical procedure a woman relinquishes some of her eggs for use by others), and

- *embryo transfers* (a laboratory-fertilized embryo is placed into a woman's womb for gestation and delivery).

ART allows singles, infertile heterosexual couples, and LGBTQ+ couples to have biological children. The ability to freeze eggs, sperm, or fertilized embryos enables persons to become pregnant later in life—after careers are launched, after undergoing medical treatments that will leave them infertile, or even after death (Chiu 2019; Rosenblum 2014). Anticipating contact with hazardous materials, catastrophic injury or death, men deployed overseas have banked sperm before they leave. Potential grandparents have financed egg freezing for their adult children (Gootman 2012).

Moreover, now we can determine the DNA blueprint of a fetus months before the baby is born. As a result, thousands of genetic diseases can be detected prenatally, a situation allowing parents to address these conditions while pregnant—by fixing problems, accepting that the child will have a genetic disease, or aborting the fetus (Pollack 2012b).

On a somewhat different note, we can now confirm the paternity of a biological father in the eighth or ninth week of pregnancy. "Besides relieving anxiety, the test results might allow women to terminate a pregnancy if the preferred man is not the father—or to continue it if he is." Many states require fathers to pay child support when DNA testing establishes biological paternity, a situation that can be fraught with conflict. Then too, "men who clearly know they are the father might be more willing to support the woman financially and emotionally during the pregnancy which some studies suggest might lead to healthier babies" (Pollack 2012a).

Biological technologies expand options but also raise thorny relationship and ethical issues (Lewin 2014a). An example of expanded options involves the ability of a spouse to choose gender-reaffirming (formerly called sex- or gender-reassignment) surgery to change their anatomy to better align with their gender identity—and this situation prompts dramatic family readjustment. As another example, new biological technologies raise difficult family issues as at-home genetics tests such as 23andMe and Ancestry.com reveal previously secreted family members who may have been conceived in hidden affairs. ART procedures can result in complex issues. Sperm donors may be sought out by their adolescent or older offspring, a situation that (facilitated by no social script) may be joyful or traumatizing, "and how does the introduction of this new blood relative affect existing relationships with the parents and siblings that a person grew up with?" (Chuck 2018). As one example among many ethical issues, the Catholic Church condemns any form of conception that doesn't involve traditional intercourse (May 2011).

Moreover, due to misunderstandings and sometimes to fraud, ART agreements—particularly those involving surrogacy—have sometimes "delivered heartache" (Lewin 2014a). Frozen embryos can be ruined if a freezer fails (Ravitz 2018). ART parents may discover that the donor is not who they had agreed on with the agency they used. Or—because sperm donation is not well monitored in the United States as opposed to European countries—a donor may have biologically fathered far more offspring than parents imagined—in one case, up to forty-four or more (Cha 2018a; Chuck 2018; Ravitz 2018). What are the chances that these children, often in the same geographical region, might find themselves unwittingly romantically attracted to their half sister or brother?

As yet another example, for some individuals and couples, fertility-enhancing procedures and extensive DNA fetal mapping raise ethical issues surrounding abortion. And in the case of divorce, which spouse gets to own the couple's previously frozen embryos (Cha 2018b)? Policy and ethical issues associated with biological technologies are more fully addressed in Chapter 8.

Communication Technologies Communication technologies have dramatically changed the way family members interact. We video family events on our cell phones and send the images to family members around the world. Texting, e-mail, websites, blogs, Facebook, Skype, and Twitter facilitate communication in ways that we would never have dreamed possible not long ago. Relationships can begin in cyberspace, minimizing the need for geographical proximity at first meeting. With texting and apps such as FindMyFriends or Spoten, parents monitor children wherever they are. Parents monitor their children's driving via technologies in the

These grandparents Skype to keep in touch with their family. Communication technologies have altered family interactions in ways we never imagined a decade or so ago. At the same time, we see a digital divide among America's families. Not all of us have ready access to computers or the Internet.

family car. Some young adults away at college or elsewhere text their parents once or more daily.

Social support for virtually every conceivable challenge—from infertility to living in stepfamilies to caring for someone with a chronic illness, to name just three examples—can be found on the Internet. Social media can enhance family connection, (Padilla-Walker, Coyne, and Fraser 2012). At the same time, the Internet can cause frustration and conflict for partners or parents who experience another family member's emotional absence because of social networking or online game playing. Some families have been faced with Internet pornography or cyber-infidelity. Social networking sites such as Facebook have made breaking up and divorce potentially more hurtful as partners publish details on their pages. Even more sadly, cyberbullying can become painful enough to result in a bullying victim's suicide. Moreover, communication technology results in a "digital divide" between those who have access to computers and the 16 percent of American households that don't and hence cannot access the benefits of computer use, such as filling out online job applications (File and Ryan 2014).

Economic Conditions

As you probably already know from experience—the economy has important consequences for family relationships. The average long-term trend in U.S. household income has been upward (see Figure 1.4). However, that overall upward pattern masks a situation of growing inequality (DeNavas-Walt and Proctor 2015, p. 8.)

Income, Wealth, and Inequality During the post–World War II decades of the 1950s and 1960s, incomes

Rocketclips, Inc./Shutterstock.com

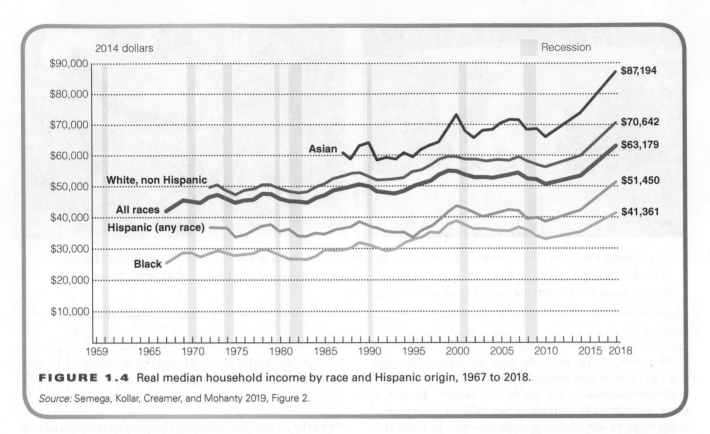

FIGURE 1.4 Real median household income by race and Hispanic origin, 1967 to 2018.

Source: Semega, Kollar, Creamer, and Mohanty 2019, Figure 2.

grew at about the same rate for families at all income levels—almost 3 percent annually. From 1970 to 2000, however, the pattern changed sharply. Incomes of the top 1 percent grew more than threefold (300 percent), while median household income grew less than 15 percent. Today, the poorest 20 percent of the population earn about $25,000 or less annually; the top 20 percent of the population earns $220,000 or more ("Household Income Quintiles" 2019). In 2011, the top one-fifth of U.S. households received more than half (52 percent) of the nation's total income, whereas the poorest one-fifth received just 3.4 percent (DeNavas-Walt, Proctor, and Smith 2012, pp. 8–10 and Table 2). Inequality has continually increased over the past fifty years (Kochkar and Cilluffo 2018).

Robots and other forms of automation, along with job restructuring to employ fewer workers, and outsourcing, or sending jobs to other countries where labor is cheaper, have caused lower wages and diminished job security for middle- and working-class Americans, many of whom struggle to pay their bills even in today's "strong" economy (Long 2019). Multinational corporations "are the new countries" inasmuch as they exist beyond any one nation's borders and detach themselves from any one country's national interest. As one Apple executive interviewed about outsourcing put it, "We don't have an obligation to solve America's problems" (Foroohar 2012).

Furthermore, in percentage terms, the Great Recession that began in 2007 took a far greater toll on the middle and working classes than on the wealthy. The Great Recession may be declared officially over, but it hasn't ended for a significant number of Americans, still looking for work, sometimes homeless, and vulnerable to unmanageable health care insurance premiums and deductibles (Casselman, Cohen, and Burke 2018; Simmons-Duffin 2019; Yarrow 2018).

Then too, wealth gaps between the richest few and the rest have always been greater than income gaps. Household *wealth* differs from income. Income is the annual inflow of wages, interest, profits, or other sources of earning. Wealth is the accumulated sum of assets (houses, cars, savings and checking accounts, stocks and mutual funds, retirement accounts, etc.) minus the sum of debt (mortgages, auto loans, credit card debt, etc.). Wealth gaps have grown to higher and higher levels, resulting in what many economists describe as the shrinking of the middle class (DeNavas-Walt and Proctor 2015, pp. 8–10).

Income varies by gender. Women have gained more than men since about 1980, while men's wages have been largely stagnant. Still, access to a male wage remains an advantage, a situation explored further in Chapters 3 and 10 (Semega, Kollar, Creamer, and Mohanty 2019, Figure 4). Household income varies by family type. Married-couple households have the highest median annual incomes—$93,654 compared to $61,518 for unmarried male-headed family households and $45,128 for unmarried female-headed family households (Semega, Kollar, Creamer, and Mohanty 2019, p. 4).

Dog housing inequality? Yes, indeed. Whether or not an effective definition of *family* can include pets, the lifestyle of the family pooch pretty much matches that of its owner. Economic inequality is rising in the United States. Both lower-income sectors and the middle class are losing ground.

Moreover, "[i]n no state can an individual working full-time at the minimum wage afford . . . a two-bedroom apartment for his or her family." In fact, in many states it takes more than two full-time jobs at minimum wage to afford that apartment—and in California, the District of Columbia, Maryland, New York, and New Jersey, it would take three jobs (Children's Defense Fund 2012, p. 18, Table 9). It would take almost four and one-half full-time jobs at minimum wage to rent that two-bedroom apartment in Hawaii—probably a good start to explaining why Hawaii has the highest proportion of multifamily households (Chandra and Foster 2018; Lofquist et al. 2012, Table 6).

The Great Recession that began in 2007 made things worse for many families. Many Americans lost their jobs and homes. With fewer tax dollars available, state governments cut services, many of them important to poor, working-class, and middle-class families (Long 2019). Although policy makers declared the recession over in 2009, lost jobs and lowered family income have persisted for many (Yarrow 2018). Many Americans have seen a "low-wage recovery" as relatively high-wage jobs that were lost have been "replaced" with those paying much less (National Employment Law Project 2012).

Because many people put off marriage until they can earn enough to support a family, more marriages were delayed or foregone during the recession of about ten years ago, and the birthrate began to decline as well (Mather 2012; Yarrow 2015). Young adults' difficulties in finding jobs mean that more of them are cohabiting rather than marrying (Kreider 2010) or are living in their parents' homes. Job losses and housing evictions have meant not only more homeless families but also more extended-family and intergenerational households as older parents and their adult children move

in together. Today about 21 percent of men and 13 percent of women between ages 25 and 34 live in a parent's home (U.S. Census Bureau 2018a, Table AD1).

Poverty A new and growing category of Americans has emerged—those who live in their cars (Pollard 2018). A significantly larger proportion of Americans, although not necessarily homeless, live in poverty. As a result of President Lyndon Johnson's War on Poverty measures in the 1960s, poverty rates fell significantly during that decade—from about 22 percent in 1959 to about 12 percent ten years later (Semega, Kollar, Creamer, and Mohanty 2019, Figure 7). However, the vast majority of Johnson's War-on-Poverty measures were dismantled by subsequent federal administrations, and until recently the poverty rate has been rising since about 1974. Today the poverty rate of the general population is again about 12 percent. At 10 percent, the poverty rate of Americans age 65 and older is lower than the population average. The *child* poverty rate—that is, the rate for children under age 18—is 16 percent (Semega, Kollar, Creamer, and Mohanty 2019, p. 12). The U.S. child poverty rate is considerably higher than rates in other industrialized nations. Moreover, approximately one-third of all U.S. families (10.2 million) can be classified as "working poor": at least one wage earner is employed fulltime but the family still lives with very low annual income (Roberts, Povich, and Mather 2011–2012).

Income, wealth, and poverty rates diverge by race and ethnicity, along with parents' education. NonHispanic whites had the lowest poverty rate in 2018 (8 percent), followed by Asian Americans (10 percent). Hispanics (18 percent) and African Americans (20.8 percent) have higher rates of poverty. Although the poverty rate of nonHispanic whites is low, they comprise about

40 percent of the total number of persons in poverty because they are a relatively large part of the population (Semega, Kollar, Creamer, and Mohanty 2019, p. 12). **Life chances**—the opportunities one has for education and work, whether one can afford to marry, the schools that children attend, and a family's health care—all depend on family economic resources. Money may not buy happiness, but it expands anyone's options for nutritious food, comfortable residences, better health care, education at quality universities, vacations, household help, and family counseling, among others.

We can think a minute here about student loan debt. Compared to the overall costs of consumer goods, college tuition has risen sharply, outpacing inflation. No wonder more and more Americans see college affordability as a serious national problem. For young-adult-headed households, student loan debt ranks second only to mortgage debt. Thanks to student loan debt, coupled with credit card and car payments, many young adults are financially "maxed out" (Federal Reserve Bank of New York 2013; "Shocking Student Debt Statistics" 2013). Indeed, families are impacted by historical events (dramatically rising college tuition costs, periods of recession, situations of growing inequality, for instance).

Historical Periods and Events

In the early twentieth-century United States, the shift from an agricultural to an industrial economy brought people from farms to cities and thereby helped to change family household composition as well as attitudes and behaviors. Later, family life was experienced differently by people living through the Great Depression of the 1930s, World War II in the 1940s, the optimistic 1950s, the tumultuous 1960s, the economically constricted 1970s and 1980s, the time-crunched 1990s, or war and the threat of terrorism throughout the 2000s (Carlson 2009). For example, during World War II, married women were encouraged to get defense jobs and place their children in day care, and although most were U.S. citizens or long-term residents, Japanese family members, along with some Italians, were sent to internment camps and had their property seized (Taylor 2002b; Tonelli 2004).

After World War II, the 1950s saw an expanding economy and postwar prosperity based on the production of consumer goods. The GI bill enabled returning soldiers to get a college education, and the less educated could get good jobs in automobile and other factories. Most white men earned a "family wage" (enough to support a family), and most white children were cared for by stay-at-home mothers. In those prosperous times, people could afford to get married young and have larger families (Kirmeyer and Hamilton 2011). The expanding economy and government subsidies for housing and education provided a strong foundation for (white, middle-class, heterosexual) married family life. Apparently forgetting minority categories who didn't benefit this way, social scientists dubbed this period—the 1950s that followed World War II—the "golden age" of marriage and the family. Some may have expected this "golden age" to continue, but in hindsight, this era proved to be an historical exception: Such high societal interest in and commitment to living in married and family life has not been replicated either before or since that so-called golden age (Coontz 2000).

In the late 1960s and through the 1970s, marriage rates declined and divorce rates increased dramatically—perhaps in response to a declining job market for working-class men, the increased economic independence of women, and the cultural revolution of the 1960s, which encouraged more individualistic perspectives. The increase in divorce rates slowed the long-term increase of larger families that were more prominent in the 1950s. Additionally, these historical trends, as well as the sexual revolution, also contributed to a dramatic rise in nonmarital births.

Today many soldiers and their families cope with the effects of historic events surrounding the U.S. war against terrorism and deployment in militarized zones abroad. Many returning soldiers, some burdened with unprecedented injuries or posttraumatic stress disorder, struggle to reintegrate into their families (Finkel 2013). Another example of how history impacts individuals involves President Trump's restrictive (some argue inhumane) policies regarding immigrants legally seeking refugee status at our southern border (Cummings 2019; Koerner 2019; Shanker and Mosendz 2019). The 2019 mass shooting that targeted Hispanic El Paso residents was an historical event creating long-lasting fear in many Latino-American families (Nanez, Dianna 2019). Climate change is an historical event affecting us all either directly or indirectly.

Historical periods interact with economics to impact life chances. For instance, most Millennials (born between 1981 and 1996) "came of age and entered the work force facing the height of an economic recession" (Dimock 2019). Career-entry jobs were hard to get, and wages were down.

> As is well documented, many of Millennials life choices, future earnings, and entrance to adulthood have been shaped by this recession in a way that may not be the case for their younger counterparts. The long-term effects of this "slow start" for Millennials will be a factor in American society for decades. (Dimock 2019)

Demographic Characteristics: Age Structure

Increased longevity is a dramatic demographic development. Life expectancy in 1900 was 47 years, but an

American child born in 2017 is expected to live to nearly 79 years—longer than at any time in history (Arias and Xu 2019).

Increased longevity means longer marriages for those who do not divorce, a longer period during which parents and children interact as adults, and a long retirement during which family activities and other interests may be pursued or second careers launched. More of us will have longer relationships with grandparents or grandchildren; some of us will know our great-grandparents or great-grandchildren.

At the same time, the increasing numbers of elderly must be cared for by a smaller group of middle-aged and older adults (Colby and Ortman 2015, p. 7). As the ratio of retired elderly to working-age people grows, so will the problem of funding Social Security and Medicare. At the other end of the age structure, a declining proportion of children is likely to affect social policy support for those families who are raising children. Fewer children may mean less attention and fewer resources devoted to their needs.

Demographic Characteristics: Religion

The historically dominant religion in the United States has been Protestantism, especially "mainstream" denominations such as Presbyterianism and Methodism. Catholics, Latter-day Saints, and Jews have been traditionally present and visible as well. The proportion of religiously unaffiliated Americans has increased over past decades to about 26 percent today. Due to some movement from those identifying as "Christian" to those identifying as "none" regarding religious affiliation, the proportion of Americans who identify as Christian has fallen from nearly 80 percent about fifteen years ago to 65 percent today (Pew Research Center 2019). Immigration from the Middle East and Asia has increased the proportions of Muslims, Hindus, and Buddhists and furthered religious diversity in the United States.

Religious affiliation and practice is a significant influence on family life, ranging from what holidays families celebrate to whether family relations are understood within a moral framework. Religion offers rituals to mark important family milestones such as birth, coming of age, marriage, and death. Religious affiliation provides families with a sense of community, support in times of crisis, and a set of values that give meaning to life. Membership in religious congregations is associated with age and life cycle; young people who have not been actively religious tend to become so as they marry and have children.

Research suggests that "religious couples are less prone to divorce because, on average, they enjoy higher marital satisfaction, face a lower likelihood of domestic violence, and perceive fewer attractive options outside the marriage than their less religious counterparts" (Vaaler, Ellison, and Powers 2009, p. 930). Some studies show that prayer in relationships, especially praying together or for the partner's well-being, is related to greater couple happiness and commitment (Fincham and Beach 2010). Some research suggests that it is not necessarily what religion family members belong to, but the fact that they hold religious beliefs and attend services together (Vaaler, Ellison, and Powers 2009).

Meanwhile, religious beliefs vary and can affect family decisions differently. For instance, Latter-day Saints, evangelical Christians, Jehovah's Witnesses, and Muslims reject homosexuality more strongly than some

AP Images/Susan Walsh

At Arlington National Cemetery, Buddhist monks—their lives dramatically impacted by the historical period in which they live—escort the coffin of an American soldier killed in the Middle East. There has been a Buddhist presence in the United States since at least the nineteenth century, and Buddhist practices have been followed by many Americans of non-Asian backgrounds. But the number of Buddhists more than doubled from 1990 to 2001 as the Asian American population increased through immigration.

other religions. Conservative Protestant Christians, Latter-day Saints, and many Catholics and Muslims are strongly opposed to abortion (Pew Forum on Religion & Public Life 2008, p. 135).

U.S. families of religions out of the mainstream face the challenge of maintaining a religiously proper family life in the context of a culture that does not share their beliefs and may stereotype them (Hirji 2012). Dating, marital choice, child raising, dress, and marital decision making can be religious issues, according to which the morally correct way diverges from mainstream American culture. Muslim (and occasionally other immigrant) families have the added burden of facing suspicion and hostility in the wake of 9/11. For many Americans, finding a balance between participating in the larger society and preserving religious values is challenging.

Demographic Characteristics: Race and Ethnicity

Who could have missed the protest marches and the slogan "Black Lives Matter" in response to the deaths of black men shot by police officers or who died in police custody?

Race is a social construction that reflects how people view varied social groups. "Race is a real cultural, political, and economic concept, but it's not biological," says biology professor Alan Templeton ("Genetically, Race Doesn't Exist" 2003, p. 4). The term *race* implies a biologically distinct group, but scientific thinking rejects the idea that there are separate races clearly distinguished by biological markers. Features such as skin color that Americans used to place someone in a racial group are genetically superficial.

In this text, we use the race and ethnic categories formally adopted by the U.S. government because we draw on statistics collected by the U.S. Census Bureau and other agencies. In the census, racial identity is based on self-reporting, and individuals can indicate belonging to more than one race. The Census Bureau defines *Hispanic* and *Latino* as ethnic, not race, identities. Hispanics may be of any race.

Ethnicity has no biological connotations; instead, it refers to cultural distinctions often based in language, religion, foodways, and history. For census purposes, there are two major categories of ethnicity: Hispanic and nonHispanic. This situation means that data on ethnicities other than Hispanic—Arabs or Portuguese, for example—come from surveys other than those done by the Census Bureau.

Social scientists and policy makers sometimes group African Americans, Hispanics, American Indians, Asians, and other non-whites into a category termed **minority group** or **minority**. This term conveys the idea that persons in non-white race and ethnic categories experience some disadvantage, exclusion, or discrimination in American society when compared to the politically and culturally dominant nonHispanic white group. *Minority* in a sociological context does not have its everyday meaning of less than 50 percent. Regardless of size, if a group is distinguishable and in some way disadvantaged within a society, sociologists consider it a minority group. The term can be controversial, viewed by some as demeaning or ignoring differences among groups and variation in the self-identities of individuals (Gonzalez 2006a). As much as possible, we avoid using it other than when speaking of numerical differences or in reporting Census Bureau data that is so labeled.

Race and Ethnic Diversity Of particular interest is the increasing race and ethnic diversity of U.S. families. About one-fifth of U.S. families (22 percent) speak a language other than or in addition to English at home. Approximately 62 percent of them speak Spanish, with the remaining 48 percent speaking any of forty or more other languages (Duffin 2019; Federal Interagency Forum on Child and Family Statistics 2019).

National population statistics show that in 2018 the nation was 61 percent nonHispanic white, 13 percent black, and 6 percent Asian (U.S. Census Bureau 2018d). In 2012, for the first time, nonHispanic white births accounted for 49.6 percent of all births and hence were no longer the majority (Tavernise 2012). Over the past fifty years, immigration combined with relatively low fertility rates among Asians and nonHispanic whites (compared to higher rates among blacks and Hispanics) have "put the United States on a new demographic path" (Martin et al. 2015; Mather 2009). "A Closer Look at Diversity: Immigration, Public Policy, and Family Ties" discusses immigration further. Hispanics are now 17 percent of the population, surpassing blacks as the largest race and ethnic group after nonHispanic whites. Hispanics and Asians are the fastest-growing segments of the population; the Asian population has grown more because of immigration than high fertility (Colby and Ortman 2015, Table 2).

The 2019 estimate of the child population shows more diversity than our adult population: 51 percent are nonHispanic white, 25 percent Hispanic, 14 percent black, and 5 percent Asian. Five percent of children are American Indian, Alaska Native, Native Hawaiian, or of more than one race (Federal Interagency Forum on Child and Family Statistics 2019, p. ix). Race and ethnic minorities comprise more than one-third of the U.S. population and 48 percent of the child population. By 2060 they are expected to make up 56 percent of the population (Colby and Ortman 2015, Table 2).

No category system can truly capture cultural identity. As race and ethnic categories become more fluid and as the identity choices of individuals with a mixed heritage vary, race and ethnic identities may be understood as

Immigration, Public Policy, and Family Ties

Thanks to immigration, there is now more racial and ethnic diversity among American families than ever before. The U.S. foreign-born population numbered 45 million in 2017 and is expected to reach 65 million by 2040. Expressed in percentages, the foreign-born population was 14 percent in 2017 and is projected to grow to 17 percent by 2040 (Colby and Ortman 2015; U.S. Census Bureau 2017).

Many Americans maintain **transnational families** whose members bridge national borders (Trask 2013). Also, many immigrant families are **binational**, with members having different legal statuses. One partner or spouse may be a legal resident, the other not. Children born in the United States are automatically citizens, even though one or both parents may be undocumented. About 5 million U.S. children come from binational families (Warren 2015). Transnational and binational families are explored in several places throughout this text.

The United States admits approximately 1 million legal immigrants each year. Asia, Latin America, and the Caribbean are the major sending regions, with the highest percentage arriving from Asia (Pew Research Center 2012c). In addition to legal immigrants, approximately 11 million undocumented (not legal) immigrants live in the United States, the majority from Mexico, Central America, and the Caribbean (Krogstad and Passel 2015). Recent Asian immigrants tend to be highly educated professionals (Pew Research Center 2012c). However, many immigrants leave a poorer country for a richer one in hopes of bettering their family's situation (Kapur and McHale 2009).

The earliest voluntary immigrants (African slaves were obviously *not* voluntary immigrants) to the United States were primarily British, German, and Dutch—whites—from northern Europe. In 1875,, the U.S. federal government passed the Page Act, the nation's first immigration law and thereby initiated immigration policy.

What precipitated this first legislation? In about 1850 some native-born, white Protestant Americans argued that the country was being overrun by Catholics, mostly from Ireland—later, from Italy and Poland—who didn't share American values and lacked occupational and language skills. Ironically, the Page Act, followed by the Chinese Exclusion Act, did not exclude Catholics, but it excluded Asians.

Should United States continue as a "melting pot," or should public policy protect it from becoming a "dumping ground"? Such was the nineteenth-century political debate. By 1924, Congress had established a National Origins Formula whereby the number of immigrants from other than non-Northwest European countries was limited "to preserve the ideal" of White "homogeneity" (U.S. Department of State, Office of the Historian n.d.). However, during the civil rights era, the 1965 Immigration and Nationality Act abolished the racially discriminatory National Origins Formula and gave preference to relatives of U.S. citizens, to professionals, and to refuges (Keely 1979).

Since World War II when many displaced persons sought refugee status, the United States (along with other nations) has granted asylum to an annual quota of refugee seekers. Asylum seekers must establish that they fear serious persecution in their home country (Weissbrodt and Danielson 2005). In 2017, the United States granted asylum to 110,000 individuals; in 2019, that number declined to 30,000 (Admissions Reports 2019).

Traditionally after immigrants establish themselves, they send for relatives—in fact, since the 1965 Immigration and Nationality Act, the majority of legal immigrants enter via family sponsorship. Immigrant families pay payroll, Social Security, property, and sales taxes even though some receive limited government benefits (Martin and Midgley 2006). A 2015 poll found 51 percent of Americans agreeing that immigrants "make our country stronger due to their work and talents." Another 41 percent said immigrants "are a burden on our country because they take our jobs, housing, and health care" (Pew Research Center 2015b).

During the Trump administration, policy toward immigrants—legal, undocumented, and refugee or asylum seekers—grew increasingly restrictive. The government made various proposals limiting benefits. Of serious concern are the more than 3 million children who, legal citizens themselves, have seen their undocumented parents deported (American Immigration Council 2018; Golash-Boza 2012; Preston 2007). Due to federal policy, families seeking asylum have faced appalling conditions at our southern border (Associated Press 2019; Cummings 2019; Koerner 2019; Shanker and Mosendz 2019).

Critical Thinking

What are some strengths exhibited by immigrant families? What are some challenges they face? At the society-wide level, how do immigrants benefit the United States? What challenges do immigrants bring?

voluntary—"optional" rather than automatic, especially for young adults (Saulny 2011a). Moreover, considerable diversity exists within major race and ethnic groupings. There are Caribbean and African blacks, for example, as well as those descended from U.S. slave populations. There are Chinese, Japanese, Korean, Indian, and other Asians. There are Salvadoran, Nicaraguan, Costa Rican, Chilean, and other Hispanics. Within-group diversity makes generalizations about race and ethnic groups somewhat questionable. For instance, "Hispanic," "Latina (feminine), "Latino," (masculine) and "Latinx" (pronounced la-tee-neks and indicating either gender) categories are "useful for charting broad demographic changes in the United States . . . [but they] conceal variation in the family characteristics of Latino groups [Cubans and Mexicans, for example] whose differences are often greater than the overall differences between Latinos and non-Latinos" (Baca Zinn and Wells 2007, pp. 422, 424). Then too, there are areas of social life in which race and ethnic differences seem minor—if they exist at all. Little difference in family patterns is apparent between blacks and whites serving in the military, for example (Finkel 2013). Children born to interracial and inter-ethnic unions further add to America's diversity, and the proportion of interracial children is significant (Livingston 2017).

Race and Ethnic Stratification Race and ethnic stratification persists (Pew Research Center 2011a). A history of racial discrimination affects wealth stratification today. As one example, the GI bill mentioned earlier was available to returning black soldiers as well as to whites, but many colleges did not accept African Americans, and one had to be accepted into a college program to qualify for the GI bill's college assistance. Likewise, the GI bill did not officially discriminate against African Americans' desire for home ownership, but the bill was of little use to them because of the many restrictive covenants against black residents and because real estate agents often did not show listed properties to black customers (Reed and Strum 2008). On average, the income and wealth of Asian and of nonHispanic white households are much higher and poverty rates significantly lower than those of African American, Hispanic, and Native American households (DeNavas-Walt and Proctor 2015).

Our experiences are shaped by our **social class** as well as gender, our race and ethnicity. The class position, race and ethnic characteristics of our parents impact our childhood experiences, which will inform the decisions we make and how we experience the world as we mature into adulthood, as well as the advantages or disadvantages that we encounter. Social class can be more important than race or ethnicity in shaping people's families. Yet, race and ethnic heritage—the family's place within our culturally diverse society—affects

attitudes, preferences, options, and decisions, not to mention opportunities. Because they can go about their days without thinking about their race, it may not be surprising that whites are considerably less likely than minority groups such as blacks, Hispanics, and Asians to see their race or ethnicity as central to their identity (Horowitz 2018).

For instance, ethnicity can influence options and decisions about whether or when to marry, where the family will live, employment, wives' work preferences, preferred parenting practices, caring for aging parents, and so on. As the U.S. population changes, policy makers need to recognize the complexity and diversity of the growing minority population. We return to issues of racial and ethnic diversity throughout this textbook.

Individuals' choices depend largely on the alternatives that exist in their social environment and on cultural values and attitudes toward those alternatives. If people are to shape the kinds of families they want, they must not limit their attention just to their own relationships and families. This text assumes that people need to understand themselves and their problems in the context of the larger society. This is a principal reason why we explore social policy issues throughout this course.

Family Policy: A Family Impact Lens

Family policy involves all the procedures, regulations, attitudes, and goals of programs and agencies, workplace, educational institutions, and government that affect families. *Family policy* encompasses policies that directly address the main functions of families—family formation, partner relationships, economic support, childrearing, adoption, childcare, family violence, juvenile crime, and long-term care. Issues regarding same-sex couples' separation, divorce, and child custody, as well as determining the legal status for lesbian parents who used ART, are all social policy matters. Whether the federal government should prohibit farm children under age 16 from driving tractors or working other dangerous agricultural equipment is a matter of family policy—and hotly debated in some states ("Parents Defend. . . ." 2012). Family policy expert Karen Bogenschneider urges that political decisions regarding families be scrutinized through a **family impact lens** (Bogenschneider et al. 2012) by which we ask how the policy in question impacts families—for instance, in what ways might asking whether a Census respondent is a U.S. citizen affect family members (Bierman and Savage 2019; Thomsen 2019)?

Another example: Of concern have been the many young adults whose undocumented parents brought them to the United States when they were children and who are therefore not legal residents but have no connections to their country of origin (Gonzalez 2006b).

In June 2012, President Obama issued an executive order, Deferred Action for Childhood Arrivals, or DACA, affecting some 800,000 youth by allowing them to stay in the United States without fear of deportation (but also without legal citizenship status) and to be able to work (Preston and Cushman Jr. 2012). The subsequent Trump administration worked to revoke the Obama policy (Dwyer 2019).

Looking through the family impact lens reveals that "laws place some families in the margins of society while privileging others" (Henderson 2008, p. 983). Until 2015, federal family policy privileged heterosexual marriages by defining same-sex unions as "not-marriage," a situation that negatively affected many children in LGBTQ+ families who did not have access to a non–adoptive parent's employer-provided health care benefits (Movement Advancement Project 2011). As another example, "racial profiling, mandatory minimum sentences, and especially the disparities in drug laws [which more heavily penalize crimes involving drugs typically used by blacks] have had a dramatic effect on the incarceration rates of young male [family members], especially in urban inner-city neighborhoods" (Clayton and Moore 2003, p. 86; (Wolfers, Leonhardt, and Quealy 2015).

Many social factors condition people's options and choices. One such factor is an individual's place within our culturally diverse society. Here a Seminole elder on a Florida reservation shows children how to see one's destiny on a painted wheel of life. Even within a race and ethnic group, families and individuals may differ in the degree to which they retain their original culture. Some Native Americans live almost solely on a reservation, but many reside in urban settings or go back and forth between a reservation and towns or cities. Less than 1 percent of the population is American Indian or Alaska Native.

Other examples of federal, state, city, or corporate policies that impact families involve:

- gun and ammunition sales (Ducharme 2018),

- Amazon's sale of or refusal to sell books promoting lesbian/gay conversion therapy (Ennis 2019),

- a city's issuing or denying Gay Pride Parade permits (LaBorde 2019),

- the U.S. Department of Justice's position on whether transgender employees should be protected from workplace discrimination (Sopelsa and Moreau 2019), and

- the Trump administration's efforts to limit food assistance (SNAP) and other anti-poverty assistance programs (PBS NewsHour 2019).

Americans disagree on the role government should play vis-à-vis families. Indeed, the diversity of family lifestyles in the United States makes it difficult to develop family policies that would satisfy even most of us. Making well-informed family decisions can mean getting involved in national and local political debates and campaigns. One's role as family member, as much as one's role as citizen, has come to require participation in society-wide decisions to create a desirable context for family life and family choices.

THE FREEDOM AND PRESSURES OF CHOOSING

Social factors influence people's personal choices in three ways. First, it is usually easier to make the common choice. In the 1950s and early 1960s, when people tended to marry earlier than they do now, it felt awkward to remain unmarried past one's mid-twenties. Now parents may pressure their young-adult children *not* to marry until they have finished college, and staying single longer is a more comfortable choice. Similarly, when divorce and nonmarital parenthood were highly stigmatized, it was less common to make these decisions than it is today.

A second way social factors can influence personal choices is by expanding people's options. For example, the availability of effective contraceptives makes limiting family size easier than in the past and enables deferring marriage with less risk of unwanted pregnancy. Meanwhile, social factors can limit people's options. For example, American society has never allowed polygamy (more than one spouse) as a legal option. Those who would like to form plural marriages risk prosecution. As another example, until the 1967 *Loving v. Virginia* U.S. Supreme Court decision, a number of states prohibited racial intermarriage.

WILLARD CULVER/National Geographic Image Collection

The best decisions are informed ones. It helps to know something about all the alternatives; it also helps to know what kinds of social pressures affect our decisions. As we'll see, people are influenced by the beliefs and values of their society. There are **structural constraints**, economic and social forces that limit personal choices. In a very real way, we and our personal decisions and attitudes are products of our environment.

But in just as real a way, people can influence society. Individuals create social change by continually offering new insights to their groups. Sometimes social change occurs because of conversation with others. Sometimes it requires becoming active in organizations that address issues such as abortion, racial equality, immigrant rights, gay rights, or stepfamily supports, for example. Sometimes influencing society involves many people's living their lives according to their values, even when these differ from more generally accepted group or cultural norms.

We can apply this view to the phenomenon of "living together." Fifty years ago, it was widely believed that cohabiting couples were immoral. But in the 1970s, some college students openly challenged university restrictions on cohabitation, and subsequently many more people than before—students and nonstudents, young and old—chose to live together. As cohabitation rates increased, societal attitudes became more favorable. Over time, cohabitation became "mainstream" (Smock and Gupta 2002). Although some religions and individuals continue to object to living together outside

marriage, a majority of Americans today feel that a cohabiting couple who have lived together for five years or more is just as committed as a married couple (Gallup Poll 2012). It is now significantly easier for people to choose this option. We are influenced by the society around us, but we are also free to influence it, which we do every time we make a choice.

Making Informed Decisions

This course about relationships and families can increase your awareness of your alternatives and how a decision you make now may be related to subsequent options and choices. People make choices even when they are not conscious of it. Sometimes we "slide" into a situation rather than make a conscious decision. We can think of these two ways of dealing with choices as **deciding versus sliding** (Stanley 2009) (see Figure 1.5). A good way to make choices is to be well informed— that is, to do so knowledgeably.

An important component of informed decision making involves recognizing as many options as possible. A second component involves recognizing the social pressures that can influence our choices. Some of these pressures are economic; others relate to cultural norms. Sometimes people decide that they agree with socially accepted or prescribed behavior. They concur in the teachings of their religion, for example. Other times, people decide that they strongly disagree with socially prescribed beliefs, values, and standards. Once people

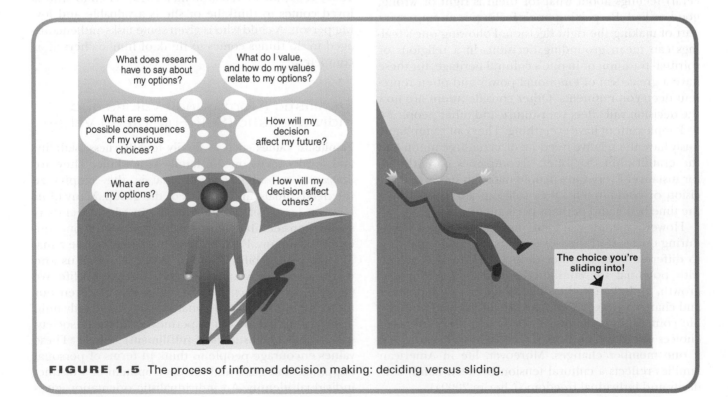

FIGURE 1.5 The process of informed decision making: deciding versus sliding.

recognize the force of social pressures, they can choose whether to act in accordance with them or not.

A third aspect of deciding about, rather than sliding into, a situation involves considering the consequences of each alternative rather than just gravitating toward the one that initially seems easier or most attractive. For example, someone deciding whether to move back into their parents' home may want to list the consequences. In the positive column, moving home might mean being able to help with family finances as well as save money that would have otherwise gone toward separate rent. In the negative column, returning to one's parental home could result in more cramped family space and increased family conflict. Listing positive and negative consequences of alternatives helps one see the larger picture and thus make a more informed decision.

Part of this process might involve finding research studies on your options. It might help to know, for instance, that the well-respected Pew Research Center surveyed young adults who'd moved back home and found that one-quarter said the situation was bad for their relationship with their parents. Another quarter said moving home was good for their relationship, and about half said moving home made no difference (Parker 2012).

If we're going to decide, not slide, we also need to be aware of our values and understand how they relate to each of our options (Meyer 2007). It's important to respect the so-called gut factor—the emotional dimension of decision making. Besides rationally considering alternatives, people have subjective (often almost visceral) feelings about what for them is right or wrong, good or bad. Respecting one's feelings is an important part of making the right decision. Following one's feelings can mean grounding decisions in a religious or spiritual tradition or in one's cultural heritage, for these have a great deal of emotional power and often represent deep commitments. Other considerations are how the decision will affect your future and other people.

People cannot have everything. They can't simultaneously have the relative freedom of a childfree union and the gratification that often accompanies parenthood, for instance. Every time people make an important decision or commitment, they rule out alternatives—for the time being and perhaps permanently.

However, people can focus on some goals and values during one part of their lives, then turn their attention to different ones at other times. Adulthood is a time with potential for continued personal development, growth, and change. In a family setting, development and change involve more than one individual. Multiple life courses must be coordinated, and the values and choices of other members of the family will be affected if one member changes. Moreover, life in American families reflects a cultural tension between family solidarity and individual freedom (Cherlin 2009a).

FAMILIES OF INDIVIDUALS

Americans place a high value on family. It is hardly surprising that a vast majority of Americans report family is extremely important to them (Carroll 2007; "Marriage" 2008). Why?

Families as a Place to Belong

Families create a place to belong, serving as a repository or archive of family memories and traditions (Cieraad 2006). **Family identity**—ideas and feelings about the uniqueness and value of one's family unit—emerge via traditions and rituals: family dinnertime, birthday and holiday celebrations, vacation trips, and perhaps family hobbies such as working together in the garden. Family identities typically include members' cultural heritage. For example, all the children in one family may be given Irish, Hispanic, Asian Indian, or Russian names.

Families provide a setting for the development of an individual's **self-concept**—basic feelings people have about themselves, their abilities, characteristics, and worth. Arising initially in a family setting, self-concept and identity are influenced by significant figures in a young child's life, particularly those in the parent role, together with siblings and other relatives (Wehmeyer 2014).

How family members and others interact with and respond to us continues to impact self-concept, identity, and even our health options throughout life (Cooley 1902, 1909; Mead 1934; Steiner 2019). A child who is loved comes to think he or she is a valuable and loving person. A child who is given some tasks and encouraged to do things comes to think of him- or herself as competent.

Familistic (Communal) Values and Individualistic (Self-Fulfilment) Values

Familistic values such as family togetherness, stability, and loyalty focus on the family as a whole. They are *communal* or *collective* values; that is, they emphasize the needs, goals, and identity of the group. Many of us have an image of the ideal family in which members spend considerable time together enjoying one another's company. Furthermore, the family can be a major source of stability (Connor 2007). Those of us who marry vow publicly to stay with our partners for life. We expect our partners, parents, children, and even our more distant relatives to remain loyal to the family unit.

But just as family values permeate American society, so do **individualistic (self-fulfillment) values**. These values encourage people to think in terms of personal happiness and goals and the development of a distinct individual identity. An individualistic orientation gives

Volunteers at the American Muslim Women's Association work on a craft project to benefit poorer immigrants and refugees. The Arab American population is slightly more than 1.5 million. Contrary to what many think, 65 percent of Arab Americans are Christian, and most are second- or third–generation American citizens. Arabs who have immigrated since the 1950s are likely to be Muslim. Employing a *family-impact policy lens*, media scholar Jack Shaheen examined American movies depicting Arabs or Arab Americans and found that generally they presented negative stereotypes of "barbarism" and "buffoonery" (Beitin, Allen, and Bekheet 2010). Some modern young Muslim women have recently adopted the head scarf to express an intensified identification with Islam in the context of experiences of discrimination or challenges to their religious community.

People as Individuals and Family Members

The changing shape of the family has meant that family lives have become less predictable than they were in the mid-twentieth century. The course of family living results in large part from the decisions two adults make, moving in their own ways and at their own paces through their lives. A consequence of ongoing developmental change in individuals is that the union or family may be put at risk. If one or more individuals change considerably over time, they may grow apart instead of together. A challenge for contemporary relationships is to integrate divergent personal change into the relationship while nurturing any children involved.

How can people make it through their own and each other's changes and stay connected as a family? Two guidelines may be helpful. The first is for family members to take responsibility for their own past choices and decisions rather than blaming previous "mistakes" on others in the family. In addition, it helps to recognize that a changing family situation—for example, a college graduate's returning home to live with parents, a partner's deciding to quit his or her job and attend graduate school, a preteen's getting used to a new stepparent—may mean that family living will be difficult for a while. Family relationships need to be flexible enough to allow for each person's individual changes—to allow family members some degree of freedom. At the same time, it's good to remember the benefits of family living and the commitment necessary to sustain it. Individual happiness and family commitment are not inevitably in conflict; research shows that committed family bonds have significant positive impacts on individual well-being (Waite and Gallagher 2000; Wilcox et al. 2011b).

On the one hand, people value the freedom to leave unhappy unions, correct earlier mistakes, and find greater happiness with new partners. On the other hand, people are concerned about social stability, tradition, and the overall impact of high levels of marital instability on the wellbeing of children. The clash between these two concerns reflects a fundamental contradiction within marriage itself; that is, marriage is designed to promote both institutional and personal

more weight to the expression of individual preferences and the maximization of individual talents and options.

The contradictory pull of both familistic and individualistic values creates tension in society (Amato 2004, Cherlin 2009a)—and tension within ourselves that we must resolve. "It is within the family . . . that the paradox of continuity and change, the problem of balancing individuality and allegiance, is most immediate" (Bengston, Biblarz, and Roberts 2007, p. 323).

American society has never had a remarkably strong tradition of familism, the virtual sacrifice of individual family members' needs and goals for the sake of the larger kin group (Sirjamaki 1948; Lugo Steidel and Contreras 2003). Our national cultural heritage prizes individuality, individual rights, and personal freedom. On the other hand, an overly individualistic orientation puts stress on relationships when there is little emphasis on contributing to other family members' happiness or postponing personal satisfactions in order to attain family goals.

Families are made of individuals, each seeking self-fulfillment and a unique identity, but individuals can find a place to learn and express togetherness, stability, and loyalty within the family. Families also perform the important function associated with providing emotional support—they give us a place to belong. Events, rituals, and histories become intrinsic parts of each individual.

goals. . . . To make marriages with children work effectively, it is necessary for spouses to find the right balance between institutional and individual elements, between obligations to others and obligations to the self. (Amato 2004, p. 962)

Throughout this text we will continue to explore the tension between individualistic and familistic values and discuss creative ways that partners and families can alter committed, ongoing relationships in order to meet their changing needs.

MARRIAGES AND FAMILIES: FOUR THEMES

We have defined the term *family* and discussed diversity and decision making in the context of family living. We can now state explicitly the four themes of this text.

1. Personal decisions must be made throughout the life course. Decision making is a trade-off; once we choose an option, we discard alternatives. No one can have everything. Thus, the best way to make choices is knowledgeably.

2. People are influenced by the society around them. Cultural beliefs and values influence our attitudes and decisions. Societal or structural conditions can limit or expand our options.

3. We live in a society characterized by considerable change, including increased ethnic, economic, and family diversity; by tension between familistic and individualistic values; by decreased marital and family permanence; and by increased political and policy concern about the needs of children and families. This dynamic situation can make personal decision making more challenging than in the past—and more important.

4. Personal decision making feeds into society and changes it. We affect our social environment every time we make a choice. Making family decisions can also mean choosing to become politically involved in order to effect family-related social change. Making family choices consciously, according to our values, gives our family lives greater integrity.

We will revisit these topics throughout this text, and we, your authors, believe that they provide a strong foundation for the subject of marriages and families.

Summary

- Families exist worldwide (of course). Situations and events across the globe increasingly affect family life in the United States—sometimes in our very neighborhoods.

- We, your authors, define family as any sexually expressive, parent-child, or other kin relationship in which people—usually related by ancestry, marriage, or adoption—(1) form an economic or otherwise practical unit and care for any children or other dependents, (2) consider their identity to be significantly attached to the group, and (3) commit to maintaining that group over time.

- Social scientists usually list three major functions served by today's families: raising children responsibly, providing members with economic and other practical support, and offering emotional security.

- With relaxed institutional control, family diversity has progressed to the point that there is no typical family form today.

- Whether we are in an era of "family decline" or "family change" is a matter of debate.

- Families exist in a social context that affects many aspects of family life. Families are affected by ever-new

biological and communication technologies, economic conditions, historical periods, and demographic characteristics such as age, religion, race, and ethnicity.

• Marriages and families are comprised of individuals. Our culture values both families and individuals. Families provide members a place to belong and help ground identity development. Meanwhile, finding personal freedom within families is an ongoing, negotiated process.

• People make choices, either by consciously deciding or by sliding into situations; the best decisions are informed ones consciously made. Our decisions are limited by social structure, and at the same time they are causes for change in that structure.

• Change and development continue throughout adult life. Because adults change, relationships, marriages, and families are far from static.

Questions for Review and Reflection

1. Without looking at ours, write your definition of family and then compare it to ours. How are the two similar? How are they different? Does your definition have some advantages over ours?

2. Why is the family a major social institution? Does your family fulfill each of the family functions identified in the text? If yes, how? If no, why not?

3. What important changes in family patterns do you see today? Do you see positive changes, negative changes, or both? What do they mean for families, in your opinion?

4. What are some examples of a personal or family problem that is at least partly a result of problems in the society? Describe one specific social context of family life as presented in the text. Does what you read match what you see in everyday life?

5. **Policy Question.** What, if any, are some changes in law and social policy that you would like to see put in place to enhance family life?

Key Terms

binational   20
deciding versus sliding   23
ethnicity   19
extended family   6
familistic (communal) values   24
family   4
family-change perspective   12
family-decline perspective   10
family identity   24

family impact lens   21
family policy   21
family structure   6
household   6
individualistic (self-fulfillment) values   24
life chances   17
minority   19
minority group   19

nuclear family   6
postmodern family   7
race   19
self-concept   24
social class   21
social institution   9
sociological imagination   12
structural constraints   23
transnational families   20

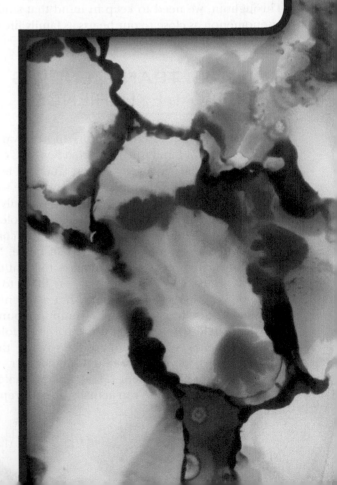

2

EXPLORING RELATIONSHIPS AND FAMILIES

Learning Objectives

1 Demonstrate how scientific knowledge differs from that gained through personal experience.

2 Discuss various theoretical perspectives on families, noting their main contributions and critiques.

3 Describe why rules for research are essential to science.

4 Discuss the most common data-gathering techniques.

5 Describe some ethical principles associated with scientific research.

6 Recognize that social scientists from across the globe research families worldwide.

- "What's happening to the family today?"
- "What's a good family?"
- "How do I make that happen?"

In Chapter 1 we said that the best decisions are informed ones made consciously. Throughout this textbook we, your authors, point out many facts that are supported by evidence about relationships and families. We base what we write on published information that we trust is accurate. We provide citations to the sources of our information, then give the complete reference that goes with each citation in the reference section at the back of this book. If you wish, you can find the article or book that we've cited, then read it for yourself and see whether you agree with our interpretation.

Where does the information in the article or book come from? Mainly, it results from social scientists' use of theoretical perspectives and research methods designed to explore family life. Family research occurs globally, and for the most part researchers worldwide employ similar theories and methods. In this chapter, though, we will focus on the United States. This chapter invites you into the world of social science so that you can understand and share this way of examining family life.

First, we'll discuss how science differs from simply having an opinion or a strongly held belief. Next, we will examine various theoretical perspectives used by social scientists. After that we'll explore some important things to know about scientific research, then discuss various ways that family scientists gather data. Throughout, we need to keep in mind that studying a phenomenon as close to our hearts as family life can be a knotty challenge.

SCIENCE: TRANSCENDING PERSONAL EXPERIENCE

The great variation in family forms and the variety of social settings for family life mean that few of us can rely only on firsthand experience when studying families. Although we "know" about the family because we have lived in one, the beliefs we have about the family based on personal experience may not tell the whole story. We may also be misled by media images and common sense—what "everybody knows." For instance, "everybody knows" that teenagers in families that eat dinner together regularly are happier and less likely to abuse drugs or be otherwise delinquent. But a recent study designed to closely examine this assumption found that this "fact" held true only when the parent–adolescent relationship was strong and positive. When that relationship was weak or fraught with conflict, family dinners were of little benefit (Meier and Musick 2014). What "everybody knows" can actually be a misrepresentation of the facts.

The Blinders of Personal Experience

Although personal experience provides us with information, it may also create blinders. We may assume that our own family is normal or typical. If you grew up in a large family, for example, in which a grandparent or an aunt or uncle shared your home, you probably assumed (for a short time at least) that everyone had a big family. Perceptions like this are usually outgrown at an early age. However, some family styles may be taken for granted or assumed to be universal when they are not.

In looking at family customs around the world, we can easily see the error of assuming that all marriage and family practices are like our own. Common American assumptions about family life not only fail to hold true in other places but also frequently don't even describe our own society well. Lesbian, gay male, or families in which one or more members is transgender; black, Latino, Arab, and Asian families; Jewish, Protestant, Catholic, Latter-day Saints (Mormon), Islamic, Buddhist, and nonreligious families; upper-class, middle-class, and lower-class families; urban and rural families—all represent differences in family lifestyle.

Nevertheless, the tendency to use only our experiential knowledge as a yardstick for measuring things is strong. Therefore, science has developed norms for transcending the blinders of personal experience. The central aim of scientific investigation is to find out what is actually going on as opposed to what we assume is happening. **Science** can be defined as "a logical system

Does this family look like yours? If "yes" or "somewhat yes," in what ways do these folks look like your family? If not, how does your family look? Researchers work to get actual facts about families, not stereotyped images. Some American families do look like this one, but—as discussed in Chapter 1—they are not the numerical majority.

Hogan Imaging/Shutterstock.com

Issues for Thought

Studying Families and Ethnicity

How social scientists go about objectively researching families across races, ethnicities, and immigrant statuses is worth thinking about. Today's family scholars, often from diverse race/ethnic backgrounds, point out how often limited and sometimes biased our theoretical and research perspectives have been (Lopez 2015). For many years, research on African Americans focused almost exclusively on poor, single-parent households in the inner city and ignored middle-class blacks (Hymowitz 2006). Overlooking many other topics, research on Latinos often investigated Mexican immigrants' assumed "patriarchal" culture (Baca Zinn and Wells 2007; Taylor 2007).

Media scholar Professor Jack Shaheen examined more than 1,000 American studio films depicting Arabs or Arab Americans and found a consistent stereotype that presents images of "barbarism" and "buffoonery" (Beitin, Allen, and Bekheet 2010). Following criticism of the earlier, limited portrayal of race/ethnic family differences, researchers began to report on the strengths of families of color, multiracial families, and multi-ethnic families, pointing to strong extended-family support, more egalitarian spousal relationships, and class, regional, and rural–urban diversity (Gottlieb, Pilkauskas, and Garfinkel 2014). For example, a substantial proportion of African American single-mother households contain other adults who take part in raising the children (Laudan 2018; Taylor 2007). Research on extended-family ties illuminates the great amount of instrumental help that Hispanic extended families provide to members (Sarkisian, Gerena, and Gerstel 2006).

Today's research on family and ethnicity tends to be more complex and sophisticated than in the past. Here's a quote from an African American wife telling researchers something about her family's financial and emotional situation:

> It was my birthday and we didn't have any money. We didn't have a vehicle, and I decided I would go to church since we couldn't do anything else. I came home from church and my husband told me that [he] found $2 worth of change in the house. And he walked to Albertson's. . . . [He] bought a Ding Dong [snack cake] because I love Ding Dongs, and a packet of hot chocolate. So when I came home, he had a candle and the Ding Dong with a cup of hot chocolate, told me, "happy birthday." And I've had birthdays since then where he bought me a diamond ring or something, but that was the best birthday I ever had. (Dew, Anderson, Skogrand, and Chaney 2017, p.293)

Concern about family fragility and individual disorganization is (1) typically placed within the family ecology perspective, and (2) balanced by recognition of diversity and of community and family strengths (Henderson et al. 2017; Vesely, Letiecq, and Goodman 2017). Multiple influences on race/ethnic families are acknowledged: (1) mainstream culture, (2) ethnic settings, and (3) the negative impact of disadvantaged neighborhoods or family circumstances that can produce behaviors that are inappropriately viewed as a "minority culture" (S. Hill 2004; Vesely, Letiecq, and Goodman 2017). Structural influences—that is, economic opportunity, deportation risk—are seen as powerful influences on family relations and behavior. At the same time, the role of "agency," or the initiative of families, is recognized: "Families should be seen as settings in which people are agents and actors, coping with, adapting to, and changing social structures to meet their needs" (Baca Zinn and Wells 2007, p. 426).

Critical Thinking

Does your family heritage or your observation of families make you think of family patterns that seem to differ from people's assumptions? How might your insights or observations about families help researchers learn more about them in a variety of sociocultural situations?

that bases knowledge on . . . systematic observation" and on *empirical evidence*—facts we verify with our senses (Macionis 2006, p. 15). The central purpose of the *scientific method* is to overcome researchers' blinders or biases. ("Issues for Thought: Studying Families and Ethnicity," addresses race/ethnic bias in research.) Scientific researchers are ever cognizant of the need to gather data that accurately correspond with reality (Umberson et al. 2015). "We must be dedicated to finding the truth *as it is* rather than as we think it *should be*" (Macionis 2006, p. 18, italics in original).

Scientific Norms

To transcend personal biases, scientists follow certain norms (Babbie 2014; Merton 1973 [1942]). Of course, researchers are expected to be honest and to never fabricate results. Scientists are expected to publish their research. Publishers are required to evaluate submissions only on merit, never taking into account the researcher's social characteristics, such as race/ethnicity, gender, socioeconomic class, religion, or institutional affiliation. To accomplish this, publishers

have reviewers, or "referees," who evaluate submissions "blind" (without knowing the name or anything else about the researcher submitting the article for publication).

Publishing allows research results to be reviewed and critiqued by others. In this way, science becomes *cumulative*. Findings from various research projects build on one another. Over time, a particular conclusion will be seen to have more evidence behind it than others. It is well established, for example, that marriage carries many benefits for the individual, the couple, and their children (Waite and Gallagher 2000; Wilcox et al. 2011; Wilcox Marquardt, Popenoe, and Whitehead 2011). It has also been well established that the arrival of children is associated with at least an initial decline in marital happiness, probably from less leisure time as well as the challenges of child raising and modifications to the couple's relationship (Margolis and Myrskylä 2015).

This last is a conclusion that is not so pleasing to hear, but an important scientific norm involves having *objectivity*: "The ideal of objective inquiry is to let the facts speak for themselves and not be colored by the personal values and biases of the researcher" (Macionis 2006, p. 18). To do this, scientists use rigorous methods that follow a carefully designed research plan. "In reality, of course, total neutrality is impossible for anyone" (Macionis 2006, p. 18). However, following standard research practices and submitting the results to review by other scientists is likely in the long run to correct the biases of individual researchers. At the same time, there are many visions of the family and relationships; what an observer reads into the data depends partly on his or her theoretical perspective (Grzywacz and Middlemiss 2017). We return to the discussion of scientific methods later in this chapter.

THEORETICAL PERSPECTIVES ON THE FAMILY

Theoretical perspectives are ways of viewing reality. As tools of analysis, they are equivalent to lenses through which observers view, organize, and then interpret what they see. A theoretical perspective leads family researchers to identify those aspects of families and relationships that interest them and suggests possible explanations for why patterns and behaviors are the way they are.

There are several different theoretical perspectives on the family. It is useful to think of each as a point of view. As with a physical object such as a building, when we see a family from different angles, we have a better grasp of what it is than if we look at it only from a single fixed position. Often theoretical perspectives

on relationships and families complement one another and may appear together in a single piece of research. We'll point to examples throughout this chapter. In other instances, the perspectives appear contradictory, leading scholars and policy makers into heated debate.

In this section, we describe nine theoretical perspectives related to families:

1. family ecology perspective
2. the family life course development framework
3. the structure–functional perspective
4. the interaction-constructionist perspective
5. exchange theory
6. family systems theory
7. conflict and feminist theory
8. the biosocial perspective
9. attachment theory

Each perspective illuminates our understanding in its own way. Table 2.1 presents a summary of these theoretical perspectives. (Chapter 10's Table 10.1 applies several of these theoretical perspectives to the topic of unpaid household labor.)

The Family Ecology Perspective

The **family ecology perspective** explores how a family is influenced by the surrounding environment. The relationship of work to family life, discussed in Chapter 10, is one example of an ecological focus. Sociologists might look at how nonstandard work schedules, such as working nights or split shifts, or having consistent health insurance benefits affect family relationships, for example (Davis et al. 2008; Lombardi and Coley 2013). We use the family ecology perspective throughout this book when we stress that, although society does not determine family members' behavior, it does present constraints for families as well as opportunities. The concept of *sociological imagination*, introduced in Chapter 1, is in line with the family ecology perspective. Families' lives and choices are affected by economic, educational, religious, and cultural institutions, as well as by historical circumstances such as the development of the Internet, war, recession, and immigration patterns and policies.

We can think of these various outside influences as radiating outward from the family as follows: (1) the neighborhood; (2) the workplace; (3) the community, town, or city; (4) the state, including state laws and policies; (5) the country, including national laws and policies; (6) the world, especially in an era of globalism; and (7) Earth's physical environment. All parts of the model are interrelated and influence

TABLE 2.1	**Theoretical Perspectives on the Family**		
THEORETICAL PERSPECTIVE	**THEME**	**KEY CONCEPTS**	**CURRENT RESEARCH**
Family Ecology	The ecological context of the family affects family life and children's outcomes.	Natural physical-biological environment Social-cultural environment	Effect on families of economic inequality in the United States Race/ethnic and immigration status variations Effect on families of the changing global economy Family policy Neighborhood effects
Family Life Course Development Framework	Families experience predictable changes over time.	Family life course Developmental tasks "On-time" transitions Role sequencing	Emerging adulthood Timing of employment, marriage, and parenthood Pathways to family formation Encore adulthood
Structure–Functional	The family performs essential functions for society.	Social institution Family structure Family functions Functional alternatives	Cross-cultural and historical comparisons Analysis of emerging family structures in regard to their comparative functionality Critique of contemporary family
Interaction–Constructionist	By means of interaction, humans construct sociocultural meanings. The internal dynamics of a group of interacting individuals construct the family.	Interaction Symbol Meaning Role making Social construction of reality Deconstruction Postmodernism	Symbolic meaning assigned to domestic work and other family activities Deconstruction of reified categories
Exchange Theory	The resources that individuals bring to a relationship or family affect the formation, continuation, nature, and power dynamics of a relationship. Social exchanges are compiled to create networks and social capital.	Resources Rewards and costs Family power Social networks Social support	Family power Entry and exit from marriage Family violence Network-derived social support
Systems Theory	The family as a whole is more than the sum of its parts.	System Equilibrium Boundaries Family therapy	Family efficacy and crisis management Family boundaries
Feminist Theory	Gender is central to the analysis of the family. Male dominance in society and in the family is oppressive of women.	Male dominance Power and inequality	Work and family Family power Domestic violence Deconstruction of reified gender categories Deconstruction of definition of marriage as necessarily heterosexual Advocacy of women's issues

(Continued)

TABLE 2.1 (Continued)

THEORETICAL PERSPECTIVE	THEME	KEY CONCEPTS	CURRENT RESEARCH
Biosocial Perspective	Evolution of the human species has put in place certain biological endowments that shape and limit family choices.	Evolutionary heritage Genes, hormones, and brain processes Inclusive fitness	Connections between biological markers and family behavior Evolutionary heritage explanations for gender differences, sexuality, reproduction, and parenting behaviors Development of research methods that can explore the respective influences of "nature" and "nurture"
Attachment Theory	Early childhood experience with caregiver(s) shape psychological attachment styles.	Secure, insecure/anxious, and avoidant attachment styles	Attachment style and mate choice, jealousy, relationship commitment, separation, or divorce

one another (Bronfenbrenner 1979; Trudge et al. 2016; see Figure 2.1).

Earth's physical environment—climate and climate change, soil, plants, animals—provides an essential backdrop against which all family living is played out. Family ecologists stress the interdependence of all the world's families—not only with one another but also with our planet's physical environment (Trask 2013). International social scientists have begun to note the phenomenon of *environmental migrants* or *climate refugees* as people in some parts of the world move to escape natural disasters of previously unknown proportions or rising sea waters that threaten small ocean islands (Goldberg 2019).

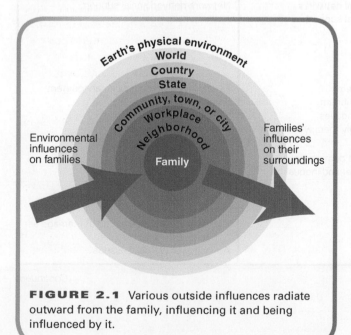

FIGURE 2.1 Various outside influences radiate outward from the family, influencing it and being influenced by it.

The family ecology model also analyzes non-climate-related environments of contemporary families at various levels from the global to the neighborhood (Trask 2013; Yu 2015). On the global level, for instance, changing coffee or sugar prices at your grocery store impact families in coffee- or sugar-growing regions of the world (Alvarez 2010; Valkila, Haaparanta, and Niemi 2010). Closer to home, the September 11, 2001, terrorist attacks on the United States and the subsequent ongoing wars in Afghanistan and Iraq have been part of a global conflict continuously affecting American family life in many ways (Lundquist and Xu 2014; Wadsworth and MacDermid 2010). Then, too, economic globalization—with the increasing outsourcing of jobs to regions with lower labor costs—has affected breadwinning and consumption in many American families (Preston 2015).

On the national level, American policies and culture emphasize a strong military worldwide but seem unable to adequately care for returning veterans, thus affecting many military families (Finkel 2013). As other examples, minimum wage legislation at the federal, state, and city levels affects family life. Families are impacted by federal Head Start, food assistance, and other programs designed to help those in poverty. The national Social Security program, coupled with Medicare, greatly influences elderly family members' retirement and housing choices. Family ecologists stress the importance of workplace, city, state, and national policies on family living (Kalil et al. 2014). Deporting undocumented immigrants affects vulnerable families in many ways, from parenting effectiveness to everyday stress generally (Cardoso, Scott, Faulkner, and Lane 2018; Noah and Landale 2018).

Neighborhoods impact family well-being as well. Homeless children and those raised in poor neighborhoods are at greater risk for negative social, educational,

Hero Images Inc./Superstock

All else being equal, residing in a supportive and helpful neighborhood translates into less family stress than otherwise. Meanwhile, families need to participate in the neighborhood in order to help create a cooperative environment for themselves. This group is planting a community garden. People in this neighborhood join together in activities that benefit families who live there.

economic, and health outcomes, as well as early and indiscriminate death from gun violence (Edin and Kissane 2010; Gültekin 2012). Mothers raising children amid neighborhood poverty express fears about letting them play outdoors (Kimbro and Schachter 2011). Helpful neighborhood resources can reduce the negative effects of violence between intimate partners (Jackson 2016).

Ecologists have also examined the sociocultural settings of relatively privileged families (Swartz 2008). Examining the kinds of economic and social advantages enjoyed by the middle and upper levels of society is uncommon but needs to be encouraged. It may provide insight into the conditions that would enable *all* families to succeed. Moreover, elements in the social–cultural environment of upper-socioeconomic-level families—excessive achievement pressure or the isolation of children from busy, accomplishment-oriented parents—can be problematic. For instance, "the silence (in the community and in academia) surrounding domestic violence in affluent communities jeopardizes the health and safety of [affluent, victimized] mothers and their children . . . " (Haselschwerdt 2012, p. F16).

Contributions and Critiques of the Family Ecology Perspective This perspective first emerged in the late nineteenth century, a period marked by concern about family welfare. The family ecology model resurfaced in the 1960s with the War on Poverty, a federal program directed toward eliminating existing high levels

of poverty. The family ecology perspective makes an important contribution today by challenging the idea that family satisfaction or success depends solely on individual effort. This perspective also turns our attention to family social policy—the various laws and other regulations and policies that impact families—discussed in Chapter 1.

A possible disadvantage of the family ecology perspective is that it is so broad and inclusive that virtually nothing is left out. One research agenda can hardly account for the family's sociocultural environment on all levels from the global to the neighborhood. More and more, however, social scientists are exploring family ecology in concrete settings. For example, Canadian researchers Phyllis Johnson and Kathrin Stoll (2008) investigated how Sudanese refugee men continued to enact the African breadwinner role for their families while resettling alone in western Canada.

The Family Life Course Development Framework

Whereas family ecology analyzes relationships, families, and the broader society as interdependent parts of a whole, the **family life course development framework** focuses on the family itself as the unit of analysis. The concept of the *family life course* is central here, based on the idea that the family changes in fairly predictable ways over time (Martin 2018).

Typical stages in the *family life course* are marked by (1) the addition or subtraction of family members (through birth, death, and leaving home), (2) the various stages that the children go through, and (3) changes in the family's connections with other social institutions (retirement from work, for example, or a child's entry into school). Each stage has requisite *developmental tasks* that must be mastered before family members transition successfully to the next stage. Therefore, this perspective has tended to assume that families perform better when life course stages proceed in orderly fashion.

Traditionally, this perspective assumed that families begin with marriage. The *newly established couple* stage ends when the arrival of the first baby thrusts the couple into the *families of preschoolers* stage, which is followed later by the *families of primary school children stage* and the *families with adolescents stage* (Crosnoe and Cavanagh 2010). *Families in the middle years* help their offspring enter the adult worlds of employment and their own family formation. Later, parents return to a couple focus with the time and money to pursue leisure activities (if they are fortunate). Still later, *aging families* must adjust to retirement and perhaps health crises or debilitating chronic illness. The death of a spouse marks the end of the family life course (Aldous 1996).

Role sequencing, the order in which major life course transitions take place, is important to this perspective.

Nikada/Getty Images

According to the family life course development framework, this father is in the *families of primary school children stage*. Like other family life course stages, this stage has particular tasks that need to be performed—tasks for which previous life course stages, if completed successfully, have helped to prepare him.

The *normative order hypothesis* proposes that the work–marriage–parenthood sequence is thought to be best for mental health and happiness (Jackson 2004; Wilcox, Marquardt et al. 2011a). Then, too, *"on-time" transitions*—those that occur when they are supposed to, rather than "too early" or "late"—are generally considered most likely to result in successful role performance during subsequent life course stages (Booth, Rustenbach, and McHale 2008). The concept of *delayed exits* and the term *boomerang kids* imply that young adult children have not left the parental home "on time" or for good (Sandberg-Thoma, Snyder, and Jang 2015).

Emerging adulthood is a stage in individual development that precedes and affects entry into the family life course. The concept conveys a sense of ongoing development, a period "when the scope of independent exploration of life's possibilities is greater for most people than it will be at any other period of the life course" (Arnett 2000, p. 469). Transition to adulthood is now completed more gradually and later than it has been in the past—usually by age 30 (Furstenberg 2008). As discussed in Chapter 1, it takes longer today to earn enough to support a family (Gibson-Davis 2009). Emerging adulthood is further explored in Chapters 6 and 7.

Researchers using the family life course development framework also extensively study the various transitions, or "pathways," to family formation (Amato et al. 2008). "The scope of research on intimate partnering now includes studies of 'hooking up,' Internet dating, visiting relationships, cohabitation, marriage following childbirth, and serial partnering, as well as more traditional research on transitions into marriage" (Sassler 2010, p. 557). In this vein, researchers note "the continued 'decoupling' of marriage and childbearing" (Smock and Greenland 2010). Noting that experiences of homeless young people hardly resemble the traditional college students who are often the population studied in emerging adult research, researchers have combined the family life course with the family ecology perspective to investigate emerging adulthood among youth living on the street (Williams and Sheehan 2015, p. 129).

More recently, life course theorist Phyllis Moen (2016) introduced the concept *encore adulthood* to recognize unconventional—and unexpected!—ways that many aging individuals fashion their later years, embracing new (to them) forms of social activism or pursuing second or third careers. Marjorie Penn Lasky's *You're Doing What? Older Women's Tales of Achievement and Adventure* (2018) exemplifies Moen's concept. Other life course researchers have tackled the question of how lifelong childlessness affects well-being (Umberson, Pudrovska, and Reczek 2010). Thus, the family life course development framework meets the postmodern family.

Contributions and Critiques of the Family Life Course Development Framework The family life course development framework encourages us to investigate various family behaviors over time. For instance, researchers looked at reasons for calling telephone crisis hotlines across the life course. They found that "issues of loneliness increased with age whereas depression-related calls decreased" (Ingram et al. 2008).

This perspective directs our attention to how particular life course transitions affect family interaction. Researchers have investigated how transitions to parenthood or from cohabitation to marriage affect the time partners spend on housework (Baxter, Hewitt, and Haynes 2008; Yavorsky, Dush, and Schoppe-Sullivan 2015). This perspective also prompts researchers to look at interactions among family members who are in different life course stages, such as a study of ongoing affection between grandparents and young adults (Monserud 2008).

Critics note remnants of the traditional tendency within this perspective toward assumptions of life course standardization, possibly suggesting a white, middle-class bias. Because of economic, ethnic, and cultural differences, two families in the same life cycle stage may be very different. For these reasons, the family development perspective is less popular now than it was half a century or more ago, although the perspective is still

used (Martin 2018). Indeed, the life course perspective is sometimes combined now with the family ecology model: Social scientists Robert Crosnoe and Arya Ansari (2016) researched how family socioeconomic and immigration statuses affect children's transition into school, for instance.

The Structure–Functional Perspective

The **structure–functional perspective** investigates how a given social structure functions to fill basic societal needs. As discussed in Chapter 1, families are principally accountable for three vital family functions: to raise children responsibly, to provide economic support, and to give family members emotional security. *Social structure* refers to the ways that families are patterned or organized—that is, the form that a family may take.

As discussed in Chapter 1, there is no typical American family structure today. Instead, families evidence a variety of forms, including same-sex families, cohabiting families, single-parent families, and transnational families whose members bridge and maintain relationships across national borders. The structure–functional perspective encourages researchers to ask how well a particular family structure performs a basic family function. For example, there is considerable research into how well single mothers, cohabiting couples, or unmarried nonresident fathers perform the function of responsible child raising (Bellamy, Thullen, and Hans 2015). This research is explored in Chapters 6, 7, and 14.

The structure–functional perspective may encourage a family researcher to think in terms of *functional alternatives*—alternate structures that might perform a function traditionally assigned to the nuclear family (Nelson 2013). A study among recent immigrants found that *fictive kin*—relationships "based not on blood or marriage but rather on religious rituals or close friendship ties, that replicates many of the rights and obligations usually associated with family ties"— can serve as a functional alternative to the nuclear family (Ebaugh and Curry 2000). Recent research on black lesbian couples found that in the absence of felt support for their relationship from parents and extended biological kin, the women looked for support among fictive kin in gay and lesbian communities (Glass and Few-Demo 2013).

The term *dysfunction* emerged from the structure–functional perspective (Merton 1968 [1949]) as a focus on social patterns or behaviors that fail to fulfill basic family needs. Obviously, domestic violence is dysfunctional in that it opposes the family function of providing emotional security. Although the term *dysfunctional family* is often used by laypeople and in counseling psychology, sociologists seldom use the term, which is considered too vague and imprecisely defined.

The structure–functional perspective might also encourage one to ask, "Functional for whom?" when examining a particular social structure (Merton 1968 [1949]). For instance, traditional male authority and higher prestige may be functional for fathers—and in some cultures for brothers—but not necessarily for mothers or sisters. Separating may seem to be functional for one or both of the adults involved, but it's not necessarily so for the children (Amato 2000, 2004).

Contributions and Critiques of the Structure–Functional Perspective Virtually all social scientists agree on the one basic premise underlying structure-functionalism: that families are an important social institution performing essential social functions. The structure–functional perspective encourages us to ask how well various family forms do in filling basic family needs. Furthermore, the perspective can be interpreted as encouraging us to examine ways in which functional alternatives to the heterosexual nuclear family may perform basic family functions. However, as it dominated family sociology in the United States during the 1950s, the structure–functional perspective gave us an unrealistic image of smoothly working families characterized only by shared values. Then, too, the structure–functional perspective has generally been understood to define the heterosexual nuclear family as the "normal" or "functional" family structure. As a result, many social scientists, particularly feminists, rebuke this perspective (Anderson and Sabatelli 2007; Stacey 2006). The majority of family sociologists today rarely reference structure-functionalism directly.

The Interaction-Constructionist Perspective

As its name implies, the **interaction-constructionist perspective** focuses on *interaction*, the face-to-face encounters and relationships of individuals who act in awareness of one another. Often this perspective explores the daily conversation, gestures, and other behaviors that go on in families (Glass and Few-Demo 2013). By means of these interchanges, something called "family" appears (Berger and Kellner 1970). Family identity, traditions, and commitment emerge through interaction, with the development of relationships and the generation of *rituals*—recurring practices defined as special and different from the everyday (Byrd 2009; Oswald and Masciadrelli 2008).

Sometimes this perspective explores family *role-making* as partners adapt culturally understood roles—for example, uncle, mother-in-law, grandmother, or

stepfather—to their own situations and preferences. One study looked at how older Chinese and Korean immigrants remade family roles on immigrating to the United States (Wong, Yoo, and Stewart 2006). A Korean grandmother described remaking her mother-in-law role:

> Once I immigrated I realized there are cultural differences between the U.S. and Korea especially when it comes to family dynamics. For example, I can't always say what I would like to say to my daughter-in-law. I follow the American ways and have given up trying to tell her what to do. . . . I would like to tell my daughter-in-law to punish the grandchildren when they misbehave. But in America, us elders do not have the right to say this. I just keep these thoughts to myself. (p. S6)

This point of view also examines how family members interact with the outside world in order to manage family (Glass and Few-Demo 2013). An example is a study of interaction strategies used by couples who had chosen to remain childfree. Feeling potentially stigmatized, some claimed that they were biologically unable to have children. Others aggressively asserted the merits of a childfree lifestyle (Park 2002). The couples worked to construct how others would define their not having children.

Reality as Constructed This approach explores ways that people, by interacting with one another, *construct*, or create, meanings, symbols, and definitions of events or situations. A respondent in the study of Chinese and Korean immigrants saw family photographs as symbols of her changing (reconstructed) family role:

> My children got married and started to have a family of their own. . . . We are now no longer the center, but on the peripheries of their families. Even when we take pictures, we don't stand in the center but on the side. It's totally different in China. Even when we took pictures, parents would be pictured in the middle. (Wong, Yoo, and Stewart 2006, p. S6)

The study of black lesbians mentioned earlier found that biological extended kin sought to "relabel" a lesbian family member as asexual or just going through a phase that she will get over (Glass and Few-Demo 2013).

As people "put out" or *externalize* meanings, these meanings come to be *reified*, or made to seem real. Once a meaning or definition of a situation is reified, people *internalize* it and take it for granted as "real" rather than viewing it as a human creation (Berger and Luckman 1966). For example, many newlyweds take it for granted that a honeymoon should follow their wedding; they don't think about the fact that the idea of a honeymoon is socially constructed (Bulcroft et al. 1997). Sociologists James Holstein and Jaber Gubrium (2008) combine this perspective with the family life course development

framework to investigate how individuals gradually construct their life course.

Exposing the ways that symbols and definitions are constructed is called *deconstruction*, a process typically identified with postmodern theory.

Postmodern Theory **Postmodern theory** can be understood as a special focus within the broader interaction-constructionist perspective (Kools 2008). Having gained recognition in the social sciences since the 1980s, postmodern theory largely analyzes *social discourse* or *narrative* (public or private, written or verbal statements or stories). The analytic purpose is to demonstrate that a phenomenon is socially constructed (Gubrium and Holstein 2009). A principal goal involves debunking *essentialism*—the idea that categories really do exist in nature and are not simply reifications. Examples include analyses of the concepts of *gender* and *race* (see Chapter 1). Formerly taken for granted as essentially "real," these categories are now generally recognized—at least within the social sciences—as social constructions. Chapter 3 further explores the social construction of gender.

When applied to relationships, postmodern theory posits that beliefs about what constitutes a "real" family are nothing more than socially fabricated narratives, having been constructed through public discourse (Barton and Bishop 2014). ("A Closer Look at Diversity: Hetero-Gay Families" illustrates the social construction of a postmodern family form, along with examples of relevant discourse, or narrative.)

Contributions and Critiques of the Interaction-Constructionist Perspective The interaction-constructionist perspective alerts us to the idea that much in our environment is neither "given" nor "natural," but socially constructed by humans—those in the past and those around us now. In this way, the perspective can be liberating. If a social structure, definition, value, or belief is oppressive, it can be challenged: Constructed by human social interaction, phenomena can also be changed by such interaction. As more couples choose to honeymoon at other times than right after their wedding, the taken-for-granted assumption discussed earlier begins to change.

Social movements advocating legalization of same-sex marriage proceed from this beginning point. This perspective leads researchers to focus on family members' interaction patterns, along with emergent definitions, symbols, rituals, and the consequences thereof.

Critics ask, "Where do we go from here?" (Wasserman 2009). Once the taken-for-granted is deconstructed, then what? For one thing, it is virtually impossible to

A Closer Look at Diversity

Hetero-Gay Families

Increasingly we inhabit a postmodern world. Three social scientists tell us about a family form that is known in Israel but may sound new to us: the *hetero-gay family*. In this case, a single heterosexual woman who desires motherhood but not marriage conceives (typically via artificial insemination) one or more children with a gay man. Although they do not reside together, both birth parents actively share financial and parental caregiving responsibilities. Virtually always the children live with their mothers.

Researchers Segal-Engelchin, Erera, and Cwikel (2012) interviewed a small sample of ten Israeli women in hetero-gay families. Seven had one child and three had two children shared with the same father. Although the mothers could have been heterosexually active, none of them happened to be romantically involved at interview time. Five of the study participants' co-parents were living with an intimate male partner; the other five were single gay men. The children ranged in age from 5 months to 9 years.

This path to parenthood is assisted in Israel by the Alternative Parenting Center, a nongovernmental organization established in 1994 for that purpose. The center introduces prospective co-parents and then provides them guidance from their decision to become co-parents through conception, pregnancy, delivery, and the arrival of the newborn into the family unit. The parents negotiate a shared-parenting agreement that determines parental rights and responsibilities, including the child's primary residence, visitation schedules, and child support. Equally shared co-parenting is foreseen to be lifelong.

The study participants' discourse around constructing this family form involved wanting to start a family and the strong belief that having both a mother and a father was best for a child, but not wanting an ongoing sexual—one woman said "sexually charged"—relationship with a heterosexual male. As one explained,

> In a relationship between a straight woman and a gay man, it's an advantage not to have sexual tension. And it's really easy; it makes the whole thing devoid of emotional baggage, devoid of sexual baggage, I mean, we both know ahead of time that we can't fall in love; all that's left is for us to be friends. (p. 397)

The mothers emphasized that children need a "father figure":

> The more I thought about going to the sperm bank . . . and asking and checking with women who did do it through the sperm bank, whose kids are a bit older, the more I decided that I wanted a father for my kid. Kids constantly search for a father figure . . . and when it's the sperm bank and there's nothing, there's just one big emptiness! . . . I wanted a dad who's active, who'll want his kids. I wanted a father that the kids will also know is their dad. (p. 395)

Equally important, the mothers in this study wanted an explicitly understood and absolutely fair division of childcare labor. They were convinced that marriage would likely limit their personal independence and result in an unfair division of household and childcare responsibilities.

Critical Thinking

How do hetero-gay families illustrate (1) postmodern thinking, and (2) postmodern family living? What are some advantages of this family form that are pointed out in these women's narratives? What might be some disadvantages?

---

conduct traditional social science research in the absence of agreed-upon social categories (Cockerham 2007).

Exchange Theory

Exchange theory applies an economic perspective to social relationships. A basic premise is that when individuals are engaged in social exchanges, they prefer to limit their costs and maximize their rewards. We choose among options after calculating potential rewards against costs and weighing our alternatives. Those of us with more resources, such as education or good incomes, have a wider range of options from which to choose. This orientation examines how individuals' personal resources, including physical attractiveness and personality characteristics, affect the formation and continuation of relationships.

According to this perspective, an individual's dependence on and emotional involvement in a relationship affects her or his relative power in the relationship. When alternatives to a relationship seem slim, one wields less power in the relationship. According to the *principle of least interest*, the partner with less commitment to the relationship is the one who has more power (Waller 1951). Those with more resources and options can use them to bargain and secure advantages in

Lawrence Migdale/Science Source

This African American family is celebrating Kwanzaa, created in the 1960s by Ron Karenga and based on African traditions. An estimated 10 million black Americans now celebrate Kwanzaa as a ritual of family, roots, and community. The experience of adopting or creating family rituals fits the interaction-constructionist perspective on the family.

capital, or resources (friendship, people with whom to exchange favors), that results from their social contacts. Social capital is analogous to financial capital, or money, inasmuch as we can "spend" it to acquire rewards, such as a romantic partner, a job, or emotional support (e.g., Qian, Lichter, and Tomin 2018). The social exchange perspective is often—although not always explicitly—combined with the family ecology perspective, described above (Sabatelli, Lee, and Ripoll-Nunez 2018).

Contributions and Critiques of Exchange Theory

The exchange perspective provides a framework from which to draw specific hypotheses about weighing alternatives and making decisions regarding relationships. Furthermore, this perspective leads us to recognize that inequality, or an unfavorable balance of rewards and costs, gradually erodes positive feelings in a relationship. The perspective also encourages us to recognize the social capital brought about by membership in social networks. Exchange theory is subject to the criticism that it assumes a human nature that is unrealistically rational and even cynical at heart about the roles of love and responsibility.

Family Systems Theory

Family systems theory views the family as a whole, or *system*, comprising interrelated parts (the family members) and demarcated by boundaries. Originating in natural science, systems theory was applied to families first by psychotherapists and was then adopted by family scholars.

A *system* is a combination of elements or components that are interrelated and organized into a whole. Like an organic system (the body, for example), the parts of a family compose a working system that behaves fairly predictably. The ways that family members respond to each other can evidence patterns. For example, whenever Jose sulks, Oscar tries to think of something fun for them to do together.

Furthermore, systems seek *equilibrium*, or stable balance and symmetry. Change in one of the parts sets in motion a process to restore equilibrium. For example, in the body system, if one hand becomes disabled, the other must adjust to do the work of both. In family dynamics, this tendency toward equilibrium puts pressure on each member to retain his or her fairly predictable role. A changing family member is subtly encouraged to revert to her or his original behavior within the

relationships. People without resources or alternatives to a relationship typically defer to the preferences of the other and are less likely to leave (Sprecher, Schmeeckle, and Felmlee 2006). From this point of view, responses to domestic violence and decisions to separate or divorce are affected by partners' relative resources.

The relative resources of participants shape power and influence in families and impacts household communication patterns, decision making, and division of labor. Relationships based on exchanges that are equal or equitable (fair, if not actually equal) thrive, whereas those in which the *exchange balance* feels consistently one-sided are more likely to dissolve or be unhappy. Dating relationships, marriage and other committed partnerships, divorce, and even parent–child relationships show signs of being influenced by participants' relative resources (Nakonezny and Denton 2008).

Social Networks Exchange theory also focuses on how everyday social exchanges between and among individuals accumulate to create social networks. Elizabeth drives Juan to the airport, Juan babysits for Maria, Maria proofreads an assignment for Elizabeth, and so forth, until a network of social exchanges emerges. The Internet offers opportunities for building social networks ranging from the local to the international level, such as those on Facebook.

Among other things, *social network theory*, a middle-range subcategory within the exchange perspective, examines how social networks provide individuals with *social*

David Young-Wolff / PhotoEdit

Delayed marriage, high housing costs, student-loan debt and other serious financial pressures make it more common today than twenty years ago for young adults to continue living with their parents or to move back home. Family systems theory tells us that when an adult child moves back home, the family system changes and the entire system of family roles need to readjust in order to maintain balance and restore equilibrium.

2005a)? Sociologist Laura Enriquez (2015) undertook a study that combined family systems theory with the ecological perspective. She examined how undocumented status for some, but not all, family members affected everyone in the mixed-status family system. Undocumented status is an ecological characteristic, a result of national law.

Contributions and Critiques of Family Systems Theory When working with families in therapy, this perspective has proven very useful (Boss 2015). By understanding how their family system operates, individuals can make desired personal or family changes. Also, envisioning the family as a system can be a creative perspective for research. Rather than seeing only the influence of parents on children, for example, system theorists are sensitized to the fact that this is not a one-way relationship and have explored children's influence on family dynamics (Crouter and Booth 2003).

family system. For change to occur, the family system as a whole must change. Indeed, that is the goal of family therapy based on systems theory. The family may see one member as the problem, but if the psychologist draws the whole family into therapy, the family *system* should begin to change.

Social scientists have moved systems theory beyond its therapeutic origins to employ it in a more general analysis of families (e.g., Jahromi 2018). They are especially interested in how family systems process information, deal with challenges, respond to crises, and regulate contact with the outside world. A 2013 study used this perspective to better understand communication among participants in family-operated businesses (Von Schlippe and Frank 2013). Researchers have looked at how parents' feelings of competition at work and home affect others in the family system—that is, their children (Schneider, Wallsworth, and Gutin 2014). Some researchers are interested in the understudied role of sibling relations within a family system (de Bel, Kalmijn, and van Duijn 2019; Senguttuvan, Whiteman, and Jensen 2014).

Moreover, family systems researchers have elaborated and explored concepts such as *family boundaries* (ideas about who is in and who is outside the family system). Stepfamilies have been researched from this point of view: Do children of divorced parents belong to two (or more) families? To what extent are former spouses and their relatives part of the family (Stewart

A criticism of systems theory is that it does not sufficiently consider a family's economic opportunities, race/ethnic and gender stratification, and other features of the larger society that influence internal family relations. When used by therapists, systems theory has been criticized as tending to diffuse responsibility for conflict by attributing dysfunction to the entire system rather than to culpable family members within the system. This situation can lead to "blaming the victim," as well as making it difficult to extend social support to victimized family members while establishing legal accountability for others, as in situations involving incest or domestic violence (Stewart 1984).

Conflict and Feminist Theory

We like to think of families as beneficial for all members. For decades, sociologists ignored the politics of gender and differentials of power and privilege within relationships and families. Beginning in the 1960s, conflict and feminist theorizing and activism began to change that oversight as issues of latent conflict and inequality were brought into the open.

A first way of thinking about the **conflict perspective** is that it is the opposite of structure–functional theory. Not all of a family's practices are good; not all family behaviors contribute to family well–being. Family interaction can include domestic violence as well as holiday rituals—sometimes both on the same day.

Conflict theory calls attention to power—more specifically, unequal power. It explains behavior patterns such as the unequal division of household labor in terms of the distribution of power between husbands and wives. Because power within the family derives from power outside it, conflict theorists are keenly interested in the political and economic organization of the larger society.

The emerging feminist movement applied conflict theory to the sex/gender system—that is, to relationships and power differentials between men and women in the larger society and in the family.

Although there are many variations, the central focus of the **feminist theory** is on gender issues (Sharp and Weaver 2015). A unifying theme is that male dominance in the culture, society, families, and relationships is oppressive to women. *Patriarchy*, the idea that males dominate females in virtually all cultures and societies, is a central concept (Allen 2016).

Unlike the perspectives described earlier, which were developed primarily by scholars, feminist theories emerged from political and social movements over the past fifty years. As such, the mission of feminist theory is to use knowledge to actively confront and end the oppression of women and related patterns of subordination based on social class, race/ethnicity, age, gender identity, or sexual identity/orientation.

The feminist perspective has contributed to political action regarding gender and race discrimination in wages, sexual harassment, divorce laws that disadvantage women, rape and other sexual and physical violence against women and children, and reproductive issues, such as abortion rights and the inclusion of contraception in health insurance. Feminist perspectives promote recognition of women's unpaid work, the greater involvement of men in housework and childcare, efforts to fund quality day care and paid parental leaves, and transformations in family therapy so that counselors recognize the reality of gender inequality in family and sexual life and treat women's concerns with respect (Few-Demo 2014; Weiser 2017). The feminist perspective has combined with the family life course development framework to analyze aging and gender issues (Ross-Sheriff 2008).

In recent years, feminist theory has embraced postmodern analyses, deconstructing formerly taken-for-granted concepts such as *gender as binary* (the idea that there are naturally two very distinct genders) or the idea that marriage must naturally be heterosexual. Having co-opted a pejorative term from the popular culture, some feminists refer to this kind of analysis as *queer theory*.

Contributions and Critiques of Feminist Theoretical Perspectives By calling attention to women's experiences, feminist theory has encouraged us to see things about relationships and family life that had been overlooked before the 1960s. Women's domestic work was largely invisible in social science until the feminist perspective began to treat household labor as work that has economic value. The feminist perspective brought to light issues of child abuse, wife abuse, marital rape, and other forms of domestic violence.

According to some social scientists, feminist theory is too political, value laden, or adversarial to be considered a valid academic approach (Landau 2008; Lloyd, Few, and Allen 2007). The concept of patriarchy has been criticized as being unscientifically vague and ahistorical. Posited to exist in virtually all societies, patriarchy loses meaning as an analytic category when it minimizes differences between America in the twenty-first century and ancient Rome, where husbands allegedly had life-and-death power over women.

The Biosocial Perspective

The **biosocial perspective** is characterized by "concepts linking psychosocial factors to physiology, genetics, and evolution" (Booth, Carver, and Granger 2000, p. 1018). This perspective argues that human physiology, genetics, and hormones predispose individuals to certain behaviors. In other words, biology interacts with the social environment to affect much of human behavior and, more specifically, many family-related behaviors (Middlemiss 2016; Salvatore and Dick 2015).

According to the biosocial (or evolutionary psychology) perspective, much of contemporary human behavior evolved in ways that enable survival and continuation of the human species. Successful behavior patterns are encoded in the genes, and this *evolutionary heritage* is transmitted to succeeding generations. The survival of one's genetic material into future generations is paramount. Hence human behavior has biologically evolved to be oriented to the survival and reproduction of all close kin who carry those genes (D'Onofrio and Lahey 2010).

Evolutionary explanations are offered for many contemporary family patterns. For instance, research suggests that children are more likely to be abused by nonbiologically related parents or caregivers than by biological parents. Nonbiological parent figures are less likely to invest money and time in their children's development and future prospects (Case, Lin, and McLanahan 2000; Wilcox, Marquardt, et al. 2011). The biosocial perspective explains this by arguing that parents "naturally" protect the carriers of their genetic material. From its early days, some proponents of the biosocial perspective have held that certain human behaviors, because they evolved for the purpose of human survival, were both "natural" and difficult to change. It has been asserted, for example, that traditional gender roles evolved from patterns shared with our mammalian

ancestors that were useful in early hunter-gatherer societies. Gender differences—males allegedly more aggressive than females, and mothers more likely than fathers to be primarily responsible for childcare—are seen as anchored in hereditary biology (Rossi 1984; Udry 1994, 2000).

However, biosociologists emphasize that biological predisposition does *not* mean that a person's behavior cannot be influenced or changed by social structure (Salvatore and Dick 2015). "Nature" (genetics, hormones) and "nurture" (culture and social relations) interact to produce human attitudes and behaviors (Horwitz and Neiderhiser 2011).

Contributions and Critiques of the Biosocial Perspective Over the past twenty-five years, the biosocial perspective has emerged as a significant theoretical perspective on the family (Middlemiss 2016; Schlomer et al. 2015). Researchers have employed this point of view to examine such phenomena as gender differences, sexual bonding, mate selection, jealousy, parenting behaviors, marital stability, and male aggression against women (D'Onofrio and Lahey 2010; Dorius et al. 2011; Wright et al. 2012). In the words of two of the perspective's proponents, "Genetically informed studies that have examined family relations have been critical to advancing our understanding of gene-environment interplay" (Horwitz and Neiderhiser 2011, p. 804; Moore and Neiderhiser 2014). However, this perspective was once used to justify gender inequality as biologically based and hence "natural." It is therefore not surprising that many distrust this perspective or that it has been politically and academically controversial. We explore and appraise the biosocial approach, or evolutionary psychology, when discussing gender (Chapter 3), extramarital sex (Chapter 4), childcare (Chapter 10), and children's well-being in stepfamilies (Chapter 15).

These individuals are waiting for dental care in a temporary neighborhood clinic. How might scholars from different theoretical orientations see this photograph? *Family ecologists* might remark on the quality of the facilities—or speculate about the families' homes and neighborhood—and how these factors affect family health and relations. They might compare this temporary clinic, set up for two weeks in an elementary school, with better equipped offices that provide dental care to more privileged Americans. Scholars from the *family life course development framework* would likely note that some of these parents are in the child-raising stage of the family life cycle. *Structure-functionalists* would be quick to note the health-related function(s) that this clinic is performing for society. *Interaction–constructionists* might explore the body language of the people awaiting attention: What are they indicating nonverbally? What might the features of this temporary dental clinic symbolize to them? *Exchange theorists* might speculate about these individuals' personal power and resources relative to others in the United States. *Family system theorists* might point out that most of these persons are part of a family system: Should one person leave or become seriously and chronically ill, for example, the roles and relationships among all members of the family would change and adapt as a result. *Feminist theorists* might point out that typically it is mothers, not fathers, who are primarily responsible for their children's health—and ask why. The answer from a *biosocial perspective* (not universally accepted in social science) might be that women have evolved a stronger nurturing capacity that is partly hormonally based. *Attachment* theorists might ask whether these parents are interacting with children in a way that promotes a secure, insecure/anxious, or avoidant attachment style. How would *you* interpret this photo?

Attachment Theory

Counseling psychologists often analyze individuals' relationship choices in terms of attachment style. **Attachment theory** posits that during infancy and childhood a young person develops a general style of attaching to others (Bowlby 1988; Pittman 2012). Once a youngster's attachment style is established, she or he unconsciously applies that style, or "state of mind," to later adult relationships (Mikulincer and Shaver 2012).

Facts about Families

How Family Researchers Study Religion from Various Theoretical Perspectives

Research topics can be studied from different points of view. Here we see how the topic *religion and family life* has been investigated with different theoretical perspectives and by the use of various research methods.

The Family Ecology Perspective

Researchers perceived religion as one component in families' sociocultural environment in Turkey, a largely Muslim country. Based on secondary analysis of data from Turks in the World Values Survey, the researchers found that highly religious parents in their sample were more likely to foster obedience and good manners in their children and less likely to expect intellectual independence or imagination (Acevedo, Ellison, and Yilmaz 2015).

The Family Life Course Development Framework

Pearce and Davis (2016) conducted secondary analysis on data from a national longitudinal survey and found that women who frequently attended evangelical Protestant services were more likely than others to have first births—both marital and nonmarital—at younger ages. Possible explanations involve the idea that these women's religious experiences introduce beliefs and values related to expected earlier childbearing and (larger) family size. A previous study suggested that early childhood religious exposure influenced later childbearing attitudes. Young adults with Catholic mothers or mothers who frequently attended religious services were especially likely to resist the idea of not having children (Pearce 2002).

The Structure-Functional Perspective

Schottenbauer, Spernak, and Hellstron (2007) found that parents' use and modeling of religiously based coping skills, along with family attendance at religious or spiritual programs, was functional in enhancing children's health, social skills, and overall behavior.

The Interaction-Constructionist Perspective

Hirsch (2008) used naturalistic observation to understand how young, actively Catholic women in Mexico creatively interpret their religion's proscription against birth control while choosing to use it. As one "grassroots theologian" explained, "[E]ven in the bible it says 'help yourself, so I can help you'; even the priests tell you that" (pp. 98–99).

A child's primary caretakers (usually parents and most often the mother) evoke a *style* of attachment in him or her. The three basic attachment styles are *secure*, *insecure/anxious*, and *avoidant*. Children who can trust that a caretaker will be there to attend to their practical and emotional needs develop a secure attachment style. Children who feel uncared for or abandoned develop either an insecure/anxious or an avoidant attachment style (Crespo 2012).

In adulthood, a secure attachment style involves trust that the relationship will provide ongoing emotional and social support. An insecure/anxious attachment style entails concern that the beloved will disappear, a situation often characterized as "fear of abandonment." Someone with an avoidant attachment style dodges emotional closeness either by avoiding relationships altogether or demonstrating ambivalence, seeming preoccupied, or otherwise establishing distance in intimate situations (Knudson-Martin 2012).

Attachment theory has grown in importance and prominence in family studies over the past several decades (Bell 2012; Pittman 2012). Some researchers combine this perspective with the family life course development framework to look at stability or variability of attachment styles throughout an individual's life. Attachment theory

is also used by counseling psychologists. The assumption is that clients who learn to recognize a problematic attachment style can change that style.

Contributions and Critiques of Attachment Theory

This perspective prompts us to look at how personality impacts relationship choices from their initiation to their maintenance. Attachment theory also encourages us to ask what kind of parenting best encourages a secure attachment style. These are important research questions. Critics argue that an attachment style might depend on the situation in which persons find themselves rather than on a consistent personality characteristic (Knudson-Martin 2012). Of course, when therapists employ this point of view, they recognize that even if it is a relatively stable personality characteristic, one's attachment style can be changed over time.

The Relationship between Theory and Research

Theory and research are closely integrated, ideally at least (Fagan 2014). Theory should be used to help direct research questions and to suggest useful concepts. Often when designing their research, scientists employ

Exchange/Network Theory

Longitudinal data show that participation in religious congregations increases the likelihood that adults will marry and that family members will benefit from sharing a network that includes parents, their children, their children's friends and teachers, and their children's friends' parents (McClendon 2016; Smith 2003). A different study, also using longitudinal data, found that young couples who marry are not necessarily more likely to get involved with formal religion, but they are more likely to do so when they have children (Gurrentz 2017).

Family Systems Theory

Lambert and Dollahite (2008) conducted qualitative research with fifty-seven religious couples and concluded that these respondents saw God as a third partner in an otherwise dyadic family system—a third system member whose presence enhanced their marital commitment. Surveying 342 heterosexually married U.S. couples, David and Stafford found that one's having "an individual relationship with God" was related to more couple forgiveness and communication about religion—and consequently to greater marital satisfaction (David and Stafford 2015).

Feminist Theory

Feminist social historians Carr and Van Leuven (1996) edited a cross-cultural anthology whose works examine the implications of religion for female family members of various religious cultures. Overall, the book argues that women's oppression originates not in religion itself but in the exploitation of religion as a subjugation tool by patriarchal religio-cultural systems.

The Biosocial Perspective

Wright (1994) argued that humans have evolved as "the moral animal," a situation that facilitates our species' cooperation toward the goal of survival. Arbel, Rodriguez, and Margolin (2015) researched the role that the hormone cortisol plays during family conflict, a study discussed further in Chapter 11.

Attachment Theory

Reinert (2005) surveyed seventy-five Catholic seminarians, presenting them with "Awareness of God" and "Attachment to Mother" scales. He found that a seminarian's early childhood attachment to his mother is a key influence in the degree of his attachment to a personal God.

one or more theoretical perspectives from which to generate a *hypothesis,* or "educated guess," about the way things are. Scientists then test these hypotheses by gathering data. At other times, to interpret data that have already been gathered, scientists ask themselves what theoretical perspective best explains the facts (Lareau 2012).

Over time, our understanding of family phenomena may change as social scientists undertake new research and modify theoretical perspectives. Even when theory is not directly spelled out in a study, it is likely that the research fits into one or more of the theoretical perspectives previously described. "Facts About Families: How Family Researchers Study Religion from Various Theoretical Perspectives" illustrates ways that researchers have used these theoretical perspectives when studying the broad topic of religion and family. We'll turn our attention now to various methods that researchers use to gather information, or data, on family life.

Students take entire courses on research methods, and obviously we can't cover the details of such methods here. However, we do want to explore some major principles so that you can think critically about published research discussed in this text or in the popular press.

DESIGNING A SCIENTIFIC STUDY: SOME BASIC PRINCIPLES

At the onset of a scientific study, researchers carefully design a detailed research plan. Some research is designed to gather *historical data.* Professor Steven Ruggles (2011) at the University of Minnesota analyzed nineteenth-century U.S. population statistics back to 1850 to examine intergenerational households. He was curious to see whether the majority of intergenerational households formed in order to care for elderly family members or whether, on the other hand, the households evidenced a reciprocal relationship—one in which each generation participated to help take care of the other. He found that nineteenth-century U.S. intergenerational households were mainly reciprocal. Historical studies of marriage and divorce in the United States portray a picture of the past that, contrary to common belief, was not necessarily stable or harmonious (Cott 2000; Hartog 2000). Although family history is an important area of research, this textbook does not allow space for us to fully explore it.

Research designs can also be *cross-cultural*, comparing one or more aspects of family life among different societies. A study described in Chapter 5, that asked students in ten different countries whether it's necessary to be in love with the person you marry, is an example of cross-cultural research (Levine et al. 1995).

Scientists consider many questions when designing their research: Will the study be cross-sectional or longitudinal? Deductive, inductive, or a combination of the two? Will the study be mainly quantitative or qualitative? Will the sample be random and the data generalizable? Because the goal of all research is to transcend our personal blinders or biases, as was discussed at the beginning of this chapter, scientists must meticulously define their terms and take care not to overgeneralize. This section looks briefly at these considerations.

Cross-Sectional versus Longitudinal Data

When designing a study, researchers must decide whether to gather cross-sectional or longitudinal data. *Cross-sectional studies* gather data just once, giving us a snapshot-like, one-time view of behaviors or attitudes. *Longitudinal studies* provide long–term information as researchers continue to gather data over an extended period of time (Cooksey 2016).

For example, to understand how psychotherapy can modify attachment insecurity over the life course, researchers designed longitudinal studies that followed respondents' attachment styles over thirty years from childhood into adulthood (Klohnen and Bera 1998). Other researchers monitored nonresident fathers' involvement with their children for three years and found finances and relations with their children's mothers to be significant causes for changes over time (Ryan, Kalil, and Ziol-Guest 2008). More recently, a longitudinal survey found that teenagers who told surveyors that they expected to marry soon were significantly less likely to have engaged in delinquent behaviors during the following year (Arocho and Dush 2016)

A difficulty encountered in longitudinal studies, besides cost, is the frequent loss of subjects to death, relocation, or loss of interest. Social change occurring over a long period of time may make it difficult to ascertain what precisely has influenced family change (Larzelere and Cox 2013). Yet cross-sectional data (one-time comparison of different groups) cannot show change in the same individuals over time.

Deductive versus Inductive Reasoning

Deductive reasoning in research begins with a hypothesis that has been derived (i.e., deduced) from a theoretical point of view. "Reasoning down" from the abstract to the concrete, a researcher designs a data-gathering strategy to test whether the hypothesis can be supported by observed facts. Researchers who use *inductive reasoning* observe detailed facts and then induce, or "reason up," to arrive at generalizations grounded in the observed data. Inductive studies do not begin with a preconceived hypothesis. Instead, researchers begin their observations with open minds about what they'll see and find. Typically, deductive reasoning is associated with quantitative research and inductive reasoning with qualitative research.

Quantitative versus Qualitative Research

In *quantitative research*, the scientist gathers, analyzes, and reports data that can be quantified or understood in numbers. Quantitative research finds numerical incidences in a population—for example, the average size of a family household or what percentage of Americans are currently cohabiting. Statistical facts and findings, such as those in Chapter 1's "Facts About Families" boxes, are examples of quantitative data. Quantitative research also uses computer-assisted statistical analysis to test for relationships between phenomena. For instance, quantitative analysis has found a statistically demonstrated correlation between being raised by a single parent and teen pregnancy (Albrecht and Teachman 2003).

When performing *qualitative research*, the scientist gathers, analyzes, and reports data primarily in words or stories. For example, sociologist Gina Miranda Samuels (2009) conducted qualitative research with black adults who had been raised by white parents. Samuels's findings are reported in narrative using respondents' own words (p. 89). Here's an example:

> I was in my salon and I didn't even know what a hot comb was. That was my giveaway! And he [the stylist] was like, "Were you raised by white people?" And then, he was like, "OH. I was able to tell that by the way you talked and by the way you carried yourself." (Crystal, 24)

A second example of qualitative data comes in the explanation by a young woman who had not wanted to get pregnant but did:

> [The patch] left a burn mark so I remember taking it off and I was supposed to be starting a different method. . . . I was going to go with the depo shot . . . but I was like, "I'm kind of scared to get that" and he just didn't want to wear a condom. (Reed et al. 2014, p. 255)

Final examples come from a study of families in which a parent is transgender (Diercky, Mortalmans, Motmans, and T'Sjoen 2017). The following is a quote from the nineteen-year-old daughter, Lauren, of a transgender woman; Lauren's parents are still together.

> I would have understood if my mother [would have left my father when he first came out as transgender] 'cause now

I sometimes ask myself if she's still happy. . . .Of course, I was happy then that they stayed together, but now...I just hope she is still happy with that decision. (p. 406)

Another respondent in the same study told researchers:

I walk [with my biological father] in the street and see how people are looking...[but]it doesn't matter that much because I am aware there will always be people who think it is weird and others who will accept it. (p. 406)

The aim of qualitative research is to gain in-depth understandings of people's experiences, as well as the processes they go through when defining, adapting to, and making decisions about their situations. Qualitative research typically employs the interaction-constructionist theoretical perspective described previously.

When designing studies, researchers must also carefully define their terms: What precisely is being studied, and how exactly will it be measured (Bickman and Rog 2009)?

Defining Terms

Researchers scrupulously define the concepts that they intend to investigate, then report those definitions in their published studies so that readers know precisely what was investigated and how. For example, researchers once considered all (heterosexual) cohabitators as fitting one general definition. They found cohabitation before marriage to be statistically related to divorce later (Dush, Cohan, and Amato 2003). However, as definitions of cohabitation were further refined to differentiate serial from one-time cohabitators, results began to show that cohabitating only with your future marriage partner was *not* more likely to end in divorce (Coontz 2015; Lichter and Qian 2008).

In a second example, researchers at Brigham Young University looked at the relationship between maintaining a cluttered home and parenting behaviors. The social scientists felt that, although popular media preaches the negative effects of "home clutter," whether or how "home organization and tidiness" affect family functioning have not been well researched (Thornock et al. 2013, p. 785). The researchers gathered quantitative data from 177 mothers of children between ages 3 and 5. The researchers found that having considerable home clutter was statistically related to a child's fussing and frequent crying. The fussing and crying, in turn, added to the mother's tenseness, a situation that negatively affected her parenting. Whether the clutter caused the crying and tenseness or vice versa cannot be deduced from this study. All we know is that a relationship was found among home clutter, young children's fussiness, and a mother's tenseness.

How did the researchers define home clutter? They used four quantitative survey questions or statements that the parent was asked to answer on a scale from 1 (never) to 5 (very often): (1) "It is often hard to find things when you need them in our household." (2) "We are generally pretty sloppy around the house." (3) "Family members make sure their rooms are neat." (4) "Dishes are usually done immediately after eating." You may or may not agree that these four questions capture the concept of *home clutter*. In any case, the researchers have told you exactly how they measured the concept so that you can decide for yourself what you think of their conclusions.

Samples and Generalization

In the study about transgender parents mentioned previously, the researchers interviewed only 13 children and 15 parents (Dierckx et al. 2017). We cannot expect the situations of these few respondents to correspond with all children or parents in transgender families. For one thing, this study took place in Belgium, a Western European country, not in the United States. We cannot necessarily conclude that children or parents in U.S. transgender families would have the same feelings. The purpose of interviewing these 28 respondents was to learn about the experiences and processes that members of transgender families can go through.

To gather data that can be *generalized* (applied to a population of people other than those directly questioned), a researcher must draw a sample that accurately reflects or represents that population in important characteristics such as age, race, gender, and marital status. Results from a survey in which all respondents are college students, for example, cannot be interpreted as representative of Americans in general.

Gallup polls are examples of research that uses *random samples* that reflect the national adult population. When a Gallup poll reports that most Americans would be unwilling to forgive an unfaithful spouse, we know that the findings from their sample can be generalized to the whole national population with only a small probability of error (Jones 2008). To draw a random sample of a population, everyone in that population must have an equal chance to be selected. The best way to accomplish this is to have a list of every individual in the population and then randomly choose from the list (see Babbie 2014). A national random sample of approximately 1,500 people may validly represent the U.S. population.

Sometimes there are no complete lists of members of a population. For instance, researchers were interested in the ramifications of living with a compulsive hoarder—one who continuously acquires yet fails to discard large numbers of possessions. They located 665 respondents who reported having a family member or friend with hoarding behaviors (Tolin et al. 2008). How did they accomplish this? The researchers had made several national media appearances about hoarding.

As a result, more than 8,000 people had contacted them for guidance or information. Drawing from this group, the researchers e-mailed potential participants, inviting them to take part in the study and asking them to forward the invitation to others in similar situations.

Ultimately, these researchers found that living with the clutter associated with hoarding often causes depression and isolation, partly because one is embarrassed to invite friends home. Although the findings were based on a fairly large sample, they cannot be generalized to all people who live with a compulsive hoarder because the sample was not random: Not everyone who lives with a compulsive hoarder had the same chance of being chosen for the study. It is reasonable to argue that those who contacted the researchers for guidance or information were more distraught than those who didn't. As a result, the findings may show greater difficulty in living with a compulsive hoarder than is generally experienced by all those in this population. Nevertheless, this is valuable research inasmuch as it lends insight into what living with a compulsive hoarder entails, at least for many.

There are many occasions when it is impossible to find a random sample for the topic one wants to investigate. The study of blacks raised by whites previously discussed provides a second case in point. In this instance, the researcher recruited volunteer respondents by using Web-based and print advertisements to African American and multiracial adoption agencies (Samuels 2009). In general, volunteers do not result in a random sample.

In addition to these considerations, designing a study involves decisions about the techniques by which the data will be collected or gathered.

Data-Collection Techniques

We will refer to various **data-collection techniques**—interviews and questionnaires, naturalistic observation, focus groups, experiments and laboratory observation, case studies—throughout this text, so we will briefly describe them here. Each technique has strengths and weaknesses. However, the strengths of one technique can compensate for the weaknesses of another. To get around the drawbacks of a given technique, researchers may combine two or more in one study (Clark et al. 2008).

Interviews, Questionnaires, and Surveys The most common data-gathering technique in family research

Rayes/Getty Images

A research team plans data collection and analysis for a survey of how families spend their time together.

involves personally interviewing respondents or asking them to complete self-report questionnaires about their attitudes and past or present behaviors. When conducting interviews, researchers ask questions in person or by telephone. Gallup polls use telephone surveys. Alternatively, a researcher may distribute paper-and-pencil or Web-based questionnaires that respondents complete by themselves. Increasingly viewed as comparable in reliability to paper-and-pencil questionnaires, Internet surveys use e-mail or Web-based formats. Examples of the latter can be found at Surveymonkey.com, which facilitates the design, distribution, and some analysis of online surveys.

Questions can be *structured* (or *closed-ended*). After a statement such as "I like to go places with my partner," the respondent chooses from a set of fixed answers, such as "always," "usually," "sometimes," or "never." Researchers spend much time and energy on the wording of these types of questions because they want all respondents to interpret them in the same way. Also, word choice can influence responses. For example, respondents tend to be more favorable to the phrase "assistance to the poor" than they are to the term "welfare" (Babbie 2007, p. 251).

Although they can contain some unstructured questions, explained later, a *survey* is mostly a quantitative data-gathering tool that comprises a series of structured, or closed-ended, questions. Once completed, survey responses are tallied and analyzed, usually with computerized coding and statistical programs. Questionnaires are inappropriate for young children, of course. Face-to-face interviews and focus groups offer alternatives in studies involving young children (Freeman and Mathison 2009).

Sociologists often engage in *secondary analysis* of large data sets—the result of fairly comprehensive surveys administered to a national representative sample. Once completed and tallied, the responses are made public, often via the Internet, so that other researchers (who had nothing to do with designing the questions) can analyze the data. A myriad of data sets are available for secondary analysis. The National Survey of Family Growth (NSFG) contains data from a national sample of nearly 11,000 U.S. women between ages 15 and 44. The National Longitudinal Survey of Youth is a nationally representative sample of about 9,000 individuals who were ages 12 to 16 as of December 31, 1996. As the name of this survey suggests, it is longitudinal with surveys being undertaken until as recently as 2011 (Munsch 2015). To investigate the breadwinner role cross-culturally, researchers analyzed data previously collected in the International Social Survey, which is conducted annually in more than twenty countries (Yodanis and Lauer 2007). The European Social Survey interviews about 120,000 people in twenty-three European countries (Ory and Huijts 2015). The longitudinal national survey, Marital Instability Over the Life Course, whose findings are described periodically throughout this text, is also available for secondary analysis. As part of their study design, researchers who use secondary analysis determine how they might analyze answers to already conceived questions in order to yield new information and insights.

A drawback to secondary analysis is that researchers do not design their own questions, so they can only investigate topics and details about which survey questions have been designed by others. Nevertheless, secondary analysis is popular and has made many important contributions to what we know about families (Fagan 2014). Many studies that use secondary analysis on other large data sets are described throughout this textbook.

Surveys are an efficient way to gather data from large numbers of people. Different respondents' standardized answers to structured questions can be readily compared. However, because they allow only predetermined answers to standardized questions, surveys may miss points that respondents would consider important but cannot report. For this reason, some researchers ask unstructured questions.

Unstructured (also called *open-ended*) questions do not offer a limited, preset range of answers. Instead, the purpose is to allow the respondent to talk freely. Interviewers using open-ended questions learn to listen and then probe for more detail. Samuels's study of blacks adopted by whites used unstructured, *open-ended questions*: "I began the interviews by asking participants to share their adoption stories, including what they knew about their birth families. Participants described their childhood communities, how they were raised to think about their racial heritages and adoptions, and if their

insights or identities changed as they became adults" (2009, p. 84).

Questioning respondents—whether quantitatively or qualitatively, whether interviewing or using a self-report questionnaire format—is the most common data-gathering technique used by family researchers. Nevertheless, there are limitations. Valid responses depend on participants' honesty, motivation, and ability to respond. Some individuals—for example, toddlers or people who suffer from the advanced stages of Alzheimer's disease—are not appropriate subjects for questioning.

Naturalistic Observation In **naturalistic observation** (also called "participant observation" or "field research"), the researcher spends extensive time with respondents and carefully records their activities, conversations, gestures, and other aspects of everyday life. This data-gathering technique often accompanies the interaction-constructionist theoretical perspective. The researcher attempts to discern family relationships and communication patterns and to draw implications and conclusions for understanding family behavior in general. A study of eight women living in trailer parks employed naturalistic observation as well as interviews. Over a period of sixteen months, researchers spent twelve to twenty hours in each woman's home, sometimes taking part in family meals (Notter, MacTavish, and Shamah 2008).

The principal advantage of naturalistic observation is that it allows one to view family behavior as it actually happens in its own natural—as opposed to artificial—setting. The most significant disadvantage is that what is recorded and later analyzed depends on what one or a few observers think is significant. Another drawback is that naturalistic observation requires enormous amounts of time to observe only a few families who cannot be assumed to be representative of family living in general. Moreover, not all research topics lend themselves to naturalistic observation. Explaining her decision to interview rather than directly observe blacks raised by whites, Samuels points out that "[t]here are not 'sides of town' or neighborhoods where multiracials or transracially adopted families and individuals reside, in which a researcher can become immersed, gain access, and conduct naturalistic inquiry" (Samuels 2009, p. 84).

Focus Groups A focus group is a form of qualitative research in which, in a group setting, a researcher asks a gathering of ten to twenty people about their attitudes or experiences regarding a situation (Stewart and Shamdasani 2015). As examples, focus groups have explored how parents feel about their overweight children and how hospital staff might better support caregivers of advanced cancer patients, (Jones et al. 2009; Nissim et al. 2017). Participants are free to talk with each other as

well as to the researcher or group leader. Focus group sessions last between one and two hours and are typically electronically recorded and then transcribed or entered into a computer for later analysis. The study of Chinese and Korean immigrants discussed earlier in this chapter is based on data collected in eight San Francisco area focus groups.

Group discussion produces data and insights that would be otherwise less forthcoming. Also, researchers can capture the participants' everyday speech in order to better understand their life situations—and perhaps to include some of their language in subsequent interviews or questionnaires. Focus groups are useful when researchers do not feel they know enough about a topic to design a set of closed-ended questions. Focus groups can also be successful when working with children (Gibson 2012; Stewart and Shamdasani 2015).

The method does have its own disadvantages. Researchers have less control in a group setting than they do in a one-on-one interview, so focus groups can be time consuming. Data can be difficult to analyze because focus group conversation is casual and flows in response to others' comments. Furthermore, the researcher can easily influence responses by inadvertent comments that lead respondents to say things that they may not really mean (Stewart and Shamdasani 2015).

Experiments and Laboratory Observation In an experiment or in laboratory observation, behaviors are carefully monitored or measured under controlled conditions. In an **experiment**, subjects from a pool of similar participants are randomly assigned to groups (*experimental* and *control groups*) that are subjected to different experiences (*treatments*). For example, families with children who are undergoing bone marrow transplants may be asked to participate in an experiment to determine how they may best be helped to cope with the situation. One group of families may be assigned to a support group in which the expression of feelings, even negative ones, is encouraged (the experimental group). A second group may receive no special intervention (the control group). If at the conclusion of the experiment the groups differ according to measures of coping behavior, then the outcome is presumed to be a result of the experimental treatment. Because no other differences are presumed to exist among the randomly assigned groups, the results of the experiment provide evidence of the effects of therapeutic intervention (Babbie 2014).

The experiment just described takes place in a field (real-life) setting, but experiments are often conducted in a laboratory setting because researchers have more control over what will happen. The laboratory setting allows the researcher to plan activities, measure results, determine who is involved, and eliminate outside influences. A true experiment incorporates the features of random assignment and experimental manipulation of the important variable (the treatment).

Laboratory observation, on the other hand, simply means that behavior is observed in a laboratory setting, but it does not involve random assignment or experimental manipulation of a variable. Family members may be asked to discuss a hypothetical problem or play a game while their behavior is observed and recorded. Later those data can be analyzed to assess the family's interaction style and the nature of their relationships. Laboratory methods are useful in measuring physiological changes associated with anger, fear, sexual response, and behavior that is difficult to report verbally. In the 1970s, social psychologist John Gottman began studying newly married couples in a university lab while they talked casually or wrestled with a problem. Video cameras recorded the spouses' gestures, facial expressions, and verbal pitch and tone. Some couples volunteered to let researchers monitor shifts in their heart rates and chemical stress indicators in their blood or urine as a result of their communicating with each other (Gottman Institute 2015). Gottman's findings are discussed in detail in Chapter 11.

Laboratory observation has advantages and disadvantages. An advantage is that social scientists can watch human behavior directly rather than depending on what respondents *tell* them. A disadvantage is that the behaviors being observed often take place in an artificial situation;

fizkes/Shutterstock.com

Conducting focus groups is one way to gather data when doing social science research. Here teenagers are being asked their opinions on a topic of interest to researchers—their attitudes about attending college perhaps.

whether an artificial situation is analogous to real life is debatable. A couple asked to solve a hypothetical problem through group discussion may behave differently in a research laboratory than they would at home.

Clinicians' Case Studies We also obtain information about families from *case studies* compiled by clinicians—psychologists, psychiatrists, marriage counselors, and social workers—who counsel people with marital and family problems. As they see individuals, couples, or whole families over a period of time, these counselors become acquainted with communication patterns and other interactions within families. Clinicians offer us knowledge about family behavior and attitudes by describing cases or reporting conclusions based on a series of cases.

The advantages of case studies are the vivid detail and realistic flavor that enable us to experience vicariously the family life of others. Clinicians' insights can provide hypotheses for further research. However, case studies have notable weaknesses. There is always a subjective or personal element in the way a clinician views the family. It may be hard to believe today, but psychiatrists used to assume that the career interests of women were abnormal and caused sexual problems in marriage (Chesler 2005 [1972]).

Also, people who present themselves for counseling may differ in important ways from those who do not. Most obviously, they may have more problems. For example, throughout the 1950s psychiatrists reported that gays in therapy had many emotional difficulties. Subsequent studies of gay males not in therapy concluded that gays were no more likely to have mental health problems than were heterosexuals (American Psychological Association 2007).

Ideally, a number of scientists examine one topic by using several different methods. The scientific conclusions in this text are the results of many studies from various and complementary research tools. Despite the drawbacks and occasional blinders, the total body of information available from sociological, psychological, and counseling literature provides a reasonably accurate portrayal of marriage and family life today.

Although imperfect, the methods of scientific inquiry bring us better knowledge of the family than does either personal experience or speculation based on media images.

The Ethics of Research on Families

Exploring the lives of families the way social science researchers do carries responsibility. Researchers must do nothing that would negatively impact respondents, a principle summarized as "do no harm." Researchers also must show respect to those being studied and consider the needs of their respondents. Feminist theorists in particular argue that researchers should be attuned to how their findings might help their respondents as well (McGraw, Zvonkovic, and Walker 2000). To help adhere to these standards, most research plans must be reviewed by a board of experts and community representatives called an *institutional review board* (IRB). No federally funded research can proceed without an IRB review, and institutions require one for all research on human subjects. Google "IRB" to see various university IRB websites.

The IRB scrutinizes each research proposal for adherence to professional ethical standards for the protection of human subjects. These standards include *informed consent* (the research participants must be apprised of the nature of the research and then give their consent); lack of coercion; protection from harm; confidentiality of data and identities; the possibility of compensation of participants for their time, risk, and expenses; and the possibility of eventually sharing research results with participants and other appropriate audiences. Other than ensuring that the research is scientifically sound enough to merit the participation of human subjects, IRBs do not focus on evaluating the research topic or methodology.

In this chapter, we've looked at how social scientists explore—that is, think about, research, and study—families. Throughout the remainder of this text you're encouraged to recall perspectives, facts, and issues raised in this chapter as we examine theory and research on a variety of family topics.

Summary

- Scientific investigation—with its ideals of objectivity, cumulative results, and various methodological techniques for gathering empirical data—is designed to provide an effective and accurate way of gathering knowledge about the family.

- Different theoretical perspectives—family ecology, the family life course development framework, structure-functional, interaction-constructionist, exchange, family systems, feminist, biosocial, and attachment theory—illuminate various features of families and provide a foundation for research.

- Research designs can be historical, cross-cultural, cross-sectional or longitudinal, qualitative or quantitative, and inductive or deductive.

- Data-collection techniques include various ways of questioning respondents (interviews, questionnaires, survey instruments), focus groups, laboratory observation and experiments, naturalistic observation, and clinicians' case studies.

- Researchers need to be guided by professional standards and ethical principles of respect for research participants.

Questions for Review and Reflection

1. Choose one of the theoretical perspectives on the family and discuss how you might use it to understand something about life in your family.

2. Choose a magazine photo and analyze its content from one of the perspectives described in this chapter. Analyze the photo from another theoretical perspective. How do your insights differ depending on which theoretical perspective is used?

3. Discuss why science is often considered a better way to gain knowledge than personal experience alone. When might this not be the case?

4. Think of a research topic and then review the data-gathering techniques described in this chapter to decide which of these you might use to investigate your topic.

5. **Policy Question.** What aspect of family life would it be helpful for policy makers to know more about as they make law and design social programs? How might this topic be researched? Is it controversial?

Key Terms

attachment theory 43
biosocial perspective 42
conflict perspective 41
data-collection techniques 48
exchange theory 39
experiment 50
family ecology perspective 32

family life course development
    framework 35
family systems theory 40
feminist theory 42
interaction-constructionist
    perspective 37
naturalistic observation 49

postmodern theory 38
science 30
structure–functional perspective 37
theoretical perspective 32

3

GENDER IDENTITIES AND FAMILIES

Learning Objectives

1 Discuss gender identity as not binary but as fluid and performed.

2 Describe issues surrounding transgender identity and expression.

3 Explain ways in which gender attitudes and norms are socialized and internalized.

4 Describe how gender functions in relation to the religion, government and politics, education, and economics.

5 Discuss how our understanding of gender and women's roles have changed throughout history and in relation to the socio-political climate.

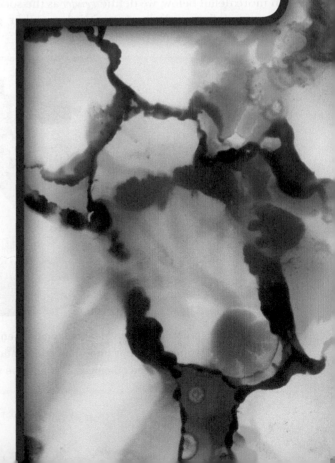

- A 9-year-old girl looking to buy basketball star Stephen Curry's Under Armour athletic shoes, the *Curry 5s*, and finding them only listed in the boys' section, wrote Curry a letter alerting him to the situation. Curry fixed the problem right away and sent her a new pair of sneakers (Lauletta 2018).

- Gillette launches its #TheBestMenCanBe campaign with a commercial during the 2019 Superbowl where men are shown stopping other men from engaging "toxic masculinity." A father intervenes when he sees two young boys fighting and says, "That's not how we treat each other." In another scene, a man stops his friend from catcalling a woman on the street saying, "Not cool, not cool."

- Merriam-Webster adds the non-binary personal pronoun "they" to the dictionary (Schmidt 2019).

- Facebook now provides a wide range of personal profile gender options, including *agender, gender fluid, genderqueer, cisgender woman/man, trans male/female,* among others, as well as the ability to create your own gender identifier.

What's *your* gender identity? While the answer to this question may be straightforward to you, it may be a difficult question for others to answer. The concept of gender identity is complicated and your answer to this question fundamentally shapes your personal relationships, family life, and social, emotional, and physical well-being. This chapter examines various society-wide or *macro* aspects of **gender** as they affect us on the *micro* level as individuals and family members. As discussed in more detail below, we define *gender* as the social construction of masculinity, femininity, or other gender

categories. We will note that the boundary between what's masculine and what's feminine is both culturally prescribed and ambiguous. We'll explore cultural scripts and personality traits typically associated with masculinity and femininity. We will examine how people are socialized into gender attitudes and behaviors. We'll analyze gender in the institutions of religion, politics, education, and economics. We'll look at changes in our understandings of gender throughout history and how we see gender today. To begin, we'll examine what it means to have a gender identity.

OUR GENDER IDENTITIES

Each of us has a gender identity that fundamentally influences our sense of who we are. **Gender identity** refers to the degree to which we see ourselves as feminine, masculine, some other gender, or no gender. Most of us are **cisgender**, which means our gender identity aligns with the sex we were assigned at birth. That is, a person is male and feels masculine, or female and feels feminine (Center for LGBTQIZ+ Student Success, Iowa State University 2019). But gender scholars today see gender identity as fluid, largely socially constructed, and existing along a continuum—an imaginary line along which individuals vary between the opposite poles of femininity and masculinity. This new way of looking at gender is becoming more acceptable to society. As noted above, people have a variety of terms available to them to describe their gender identity. Not surprisingly, younger generations—namely, Millennials (born between 1982 and 1990 and who were age 25 to 38 in 2019) and Gen Z (born after 1996 and who were age 0 to 22 in 2019)—are more accepting of non-gendered language. Thirty five percent of Gen Z and 25 percent of Millennials say they personally know someone who prefers that others refer to them using gender-neutral pronouns, and nearly two-thirds of Gen Z and half of Millennials think that forms and online profiles should include options other than "male" and "female" (Pew 2019; Figure 3.1). Some people find it important to declare to others their gender identity and specify which gender pronouns they prefer: the feminine pronouns "she," "her," and "hers," the masculine pronouns "he," "him," and "his," or the gender-neutral pronouns "they," "them," and "theirs."

In this textbook, we distinguish between the terms *gender* and *sex* for an important reason. The word **sex** is used in reference to male or female anatomy and physiology, which until recently has only been

Mass media presents us with idealized, one-dimensional stereotypes of men and women. This has important connotations for our relationships, how we see ourselves, and how we see each other.

Advertising Archives

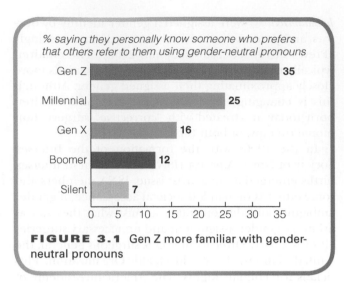

FIGURE 3.1 Gen Z more familiar with gender-neutral pronouns

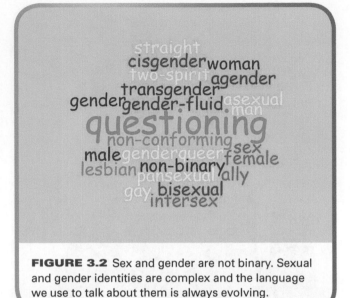

FIGURE 3.2 Sex and gender are not binary. Sexual and gender identities are complex and the language we use to talk about them is always evolving.

thought of in binary terms. Sexual expression, the use of that anatomy to communicate intimately with others, is the subject of Chapter 4. Like gender, sex is a social construction and is assigned at birth by, for the most part, a medical professional's examination of a baby's external genitalia. As discussed in more detail below, this way of determining sex is incomplete and does not take into account genetics or hormones and ignores social factors that shape one's sexual expression.

The terms *sexual orientation* and *sexual identity* refer to the gender of individuals to whom one is sexually attracted. We note here that in the term *LGBTQ+*, refer ring to lesbian, gay, bisexual, transgender, queer/questioning, and others, the first three letters (*LGB*) refer to *sexual* identity (lesbian, gay, bisexual), while the *T* refers to *gender* identity (queer/questioning can refer to either gender identity or sexual identity). **Transgender** refers to people who feel their gender identity does not match the sex assigned to them at birth. Gender identity gets complicated when coupled with sexual identity. A *transgender female*, or *trans female*—a person who transitions from male to female—who is attracted solely to men would typically identify as a heterosexual woman. A *transgender male*, or *trans male*—that is, a person who transitions from female to male—who is attracted solely to men would identify as a gay man. *Transgender people* may be straight (heterosexual), lesbian, gay, or bisexual, or describe their sexuality in some other way. For example, **pansexuality** refers to the potential to be sexually and/or romantically attracted to people of various gender expressions, including those outside the gender-conforming binary. Figure 3.2 provides examples of various sexual and gender identities. Gen Z and Millennials are, however, wary of labels and are reluctant to categorize themselves. Sexual identity and related concepts will be discussed further in Chapter 4.

This chapter uses the term **gender expectations** to describe societal attitudes and behaviors expected of

and associated with an individual's sex assigned at birth. Society expects females to develop a feminine gender identity and males, a masculine one. Meanwhile, a complication occurs because some people do not feel at ease with their sex as ascribed at birth.

Gender Is Fluid

Some cultures offer individuals a third gender option that lies somewhere along the male–female continuum. In South Asian cultures, the term *hijra* applies to a diverse group of more than 6 million people in India, Nepal, Bangladesh, and Pakistan who identify as neither male nor female. Recorded from antiquity, hijra have been thought to bestow good fortune and fertility through dances performed at births and weddings. With England's mid-nineteenth century colonization of South Asia, hijra were stigmatized and were forced to live on society's fringes (Khaleeli 2014). However, India and its neighboring nations now legally recognize hijras as a third gender (Blaustein 2015; Burgos, Kaul, and Newberry 2015). In the United States, the Navajo have a four-gender system: male, female, male with a female essence, and female with a male essence (Nibley 2011). People are increasingly refusing to put themselves in a masculine or feminine box, identifying themselves a *gender non-conforming*, *non-binary*, and/or *gender fluid*. Others consider themselves *agender*, referring to a person who doesn't identify as having a gender identity at all (LGBTQIZ+, Iowa State University 2019). This trend is beginning to make an impact on mainstream society. In 2019, Mattel introduced what they call "the world's first gender-neutral doll," which looks neither male nor female (alternatively, male *and* female) and comes in both dark and light skin tones (Dockterman 2019).

This hijra proudly poses for a camera in Delhi, India. Hijras are legally recognized as a third gender in India. More than 6 million exist in several South Asian cultures, including Nepal, Bangladesh, and Pakistan.

cbimages/Alamy Stock Photo

Intersex Individuals Not all individuals are *cisgender*, a status that most of us take for granted. It has been estimated that about 2 percent of live births are **intersex**—that is, the infants have some anatomical, chromosomal, or hormonal sexual variation from what is considered "normal" (Fausto-Sterling 2000, p. 20). Some scholars consider this definition too broad because it includes people who are not distinguishable from the general population. The common definition of intersex includes only those people whose physical characteristics are not classifiable as either male or female, or whose chromosomal sex and physical sex are inconsistent (Sax 2002).

Any genital ambiguity has traditionally been considered in need of repair. These surgeries were well intentioned and were meant to spare intersex children and their parents the pain of not fitting into a gendered society. In the 1950s, intersex babies (then termed

hermaphrodites) were assigned a gender identity by doctors, and parents were advised to treat them accordingly (Preves 2010; see also Colapinto 2000). The children typically underwent surgery to give them genitals more closely approximating their assigned gender. Although this is changing, most intersex infants and children born today are treated with "corrective" surgery, hormonal therapy, or both (Spack et al. 2012).

In the 1990s with the formation of the Intersex Society of North America (ISNA), treatment of intersex births emerged as an activist issue. ISNA members and some ethicists demanded societal acceptance of gender ambiguity and demonstrated against what they saw as arbitrary gender assignment and unnecessary surgeries (Preves 2010). Three former U.S. surgeons general, the United Nations, the World Health Organization, Physicians for Human Rights, The American Academy of Family Physicians, Human Rights Watch, and Amnesty International have all condemned "gender-normalizing" surgeries on children as have many parents of intersexed children (Compton 2018).

These ideas are taking hold. In 2013, Germany became the first European country to offer a third gender option ("indeterminate") on passports and birth certificates—a measure designed to protect parents of intersex newborns from the pressure of quickly assigning a male or female sex to their infant (Evans 2013). In 2019, New Jersey joined California, Oregon, and Washington in allowing a third gender option on birth certificates. New York City provides a non-binary X option on birth certificates and Washington, D.C., allows gender-neutral driver's licenses (Trimble 2019). The majority of intersex persons continue in the gender identity assigned them at birth. Meanwhile, the terms *transgender* and *transsexual* describe an identity adopted by individuals whose genitalia are clearly male or female but who are uncomfortable with the sex to which they were assigned at birth (Steinmetz 2014).

Transgender Individuals As noted above, *transgender* refers to people who change their gender identity. Between 1 and 1.4 million Americans identify as transgender (Flores, Herman, Gates, and Brown 2016; Meerwijk and Sevelius 2017). Grossman and D'Augelli (2006) assembled a focus group of twenty-four transgender adolescents and young adults ages 15 to 21. The participants were, on average, about 10 years old when they began to feel that their gender identity did not match their biological sex. Their statements reveal an early sense that something was not "right":

- "I used to play baseball and hang out with the boys, but I always felt like a girl."

- "I know that I was biologically a girl, but ever since I was little, I always wanted to be a man so bad. Other people said, 'I want to be a lawyer,' or a doctor, and I said I want to be a man." (pp. 121–122)

Transgender youth may initially think they might be gay or lesbian because, according to their physical anatomy, they are attracted to the same sex. Later, they may realize that they are actually heterosexual but that their anatomy does not coincide with their gender identity (MSNBC 2012).

The concept *transgender* includes transsexuals, transvestites, and cross-dressers. The latter sometimes or regularly adopt the dress and demeanor of the other-sex gender with which they identify for sexual gratification, for fun, or other purposes (Preves 2010). Meanwhile, **transsexuals** may change their physical anatomy and/or physiology through sex-reassignment surgery or less invasive measures such as hormone therapy, electrolysis (hair removal), or breast or calf implants, among others. **Sex-reassignment surgery** (SRS), or **gender-affirming surgery** as it's known in the transgender community, involves surgically altering one's anatomy to resemble that of their gender identity. The first sex-reassignment surgery took place as long ago as 1931 in Berlin, Germany. SRS is major surgery with serious risks, one of which may be loss of sensitivity to the genitals. Because SRS is seldom covered by insurance in the United States, many U.S. transsexuals go abroad for surgery—to Thailand, for example—because it is less expensive there (Gale 2015). The 2013 edition of the American Psychiatric Association's highly influential *Diagnostic and Statistical Manual of Mental Disorders* re-categorized transgender and transsexual people as suffering not from gender identity *disorder*—as it previously asserted—but from *gender dysphoria*, the latter term meaning dissatisfaction with one's body and a difference between one's expressed gender and assigned gender along with significant distress or problems functioning.

Living as a transgender person can be difficult, due in large part to discrimination from the broader society—everything from being called derogatory names to police brutality (Remnick 2015). As of 2017, the rights of transgender Americans were not protected under Title VII, which prevents employment discrimination based on sex, race, color, religion, and national origin, overturning previous interpretations of the law, although some states have their own laws that protect LGBTQ+ employment rights (Bellis 2017; Jarrett, 2017). Transgender people also face barriers to voting in elections with the advent of strict voter ID laws requiring people to prove their identities at the polls (Moreau 2018). For a short time between 2016 and 2019, transgender individuals were allowed to serve openly in the U.S. military, which provided them with access to gender-affirming surgery and psychological care. However, in 2019, the Defense Department reversed this policy and officially barred transgender people from enlisting and mandated that existing personnel diagnosed with gender dysphoria must serve "in their biological sex" and can no longer receive treatment (Liptak 2019). The Bureau of Prison's policy is that the sex on one's birth certificate determines where they will be housed (Benner 2018).

How to treat transgender athletes is new territory. In 2018, The International Association of Athletics Federation ruled that female athletes with hyperandrogenism (increased testosterone production) must take hormone suppressants to compete (Ming 2019). Several states allow transgender athletes to play on teams that correspond with their gender identity rather than the sex listed on school records (Lovett 2013).

A 2011 National Transgender Discrimination Survey of nearly 6,500 respondents showed that nearly 80 percent of young trans people had experienced harassment at school, and 90 percent of workers said they'd experienced it at work. Nearly 20 percent reported having been denied a place to live, and 50 percent said they had been fired, not hired, or denied a promotion because of their gender identity. Transgender youth are over twice as likely to have attempted suicide compared to youth as a whole (Toomy, Syvertsen, and Shramko 2018).

Perhaps it is not surprising that the Internet has been "a revolutionary tool for the trans community, providing answers to questions that previous generations had no one to ask, as well as robust communities of support" (Steinmetz 2014, p. 44). In response to "bathroom bills" introduced by several states that makes it illegal to use a bathroom that does not correspond to the gender listed on one's birth certificate, other states have introduced bills allowing LGBTQ+ to use the bathroom of their choice. A transgender female in California has designed an app called "Refuge Bathrooms" to map gender-neutral bathrooms around the world (Steinmetz 2014). Transgender people are continuing to challenge our social institutions.

Gender and Culture

In all societies that we know of, humans are to some extent differentiated, or thought of as separate and different, according to gender. This **gender differentiation** is apparent in our cultural expectations about how people should behave. You can probably think of "masculine" characteristics associated with being male, along with "feminine" traits associated with being female. Men are assumed to have **agentic** (from the root word *agent*) or **instrumental character traits**—strength, confidence, self-reliance, assertiveness, and ambition—that enable them to accomplish difficult tasks or goals. A "real man" avoids all things feminine or "sissy" (David and Brannon 1979). Accordingly, some coaches and military officers motivate males through insults, calling them "women," "ladies," or "girls." In a national survey of adolescents, 82 percent of boys reporting having heard someone tell a boy he was "acting like a girl," which they took to mean emotional, weak, moody, and other negative traits associated with girls (Plan International 2018).

Heterosexuality is also a cultural requirement for masculinity, as explored in Chapter 4 (Sallee 2011).

A relative absence of agency has traditionally been thought to characterize femininity, and women are expected to embody relationship-oriented or **expressive character traits**: warmth, sensitivity, the ability to express tender feelings, and placing concern about others' welfare above self-interest (Parsons and Bales 1955; Sallee 2011). In one interesting study, communications theorist Sut Jhally (2009) examined how people's hands are portrayed in television, film, and magazines. He noted that female "hands are shown not as assertive or controlling of their environment but as letting the environment control them" (p. 6; also see Goffman 1979).

Masculinities The term **masculinities** refers to the idea that there are varied ways to demonstrate masculinity. Three major culturally defined obligations for men involve (1) group leadership, (2) protecting group territory and weaker or dependent others, and (3) providing resources, typically by means of occupational success (Farrell 1974; Kimmel 2000; Ryle 2015). Many expectations for masculinity are positive: bonding with others and managing conflict through shared activities, humor, and fun; caring for others by providing for and protecting them; developing self- reliance, inner strength, bravery, courage, and heroism; and banding together toward common goals (Kiselica and Englar-Carlson 2010; Stroud 2012). More recently, still another transformation of the ideal male image appeared, partly in response to first responders' widely televised brave behavior after the 9/11 terrorist attacks: the unafraid "can-do" man who tackles traumatic events head-on but also feels free to cry (Adelman 2009, p. 279).

Men are expected to "man up" and to "be a man." In that way, manhood can be seen as something to be *accomplished*, done mainly through protecting and providing but in other ways too. A masculine cultural message glorifies outwitting others in competitions—in combat, contact sports, or barroom brawls (Sullivan and McHugh 2009). These ideas are maintained by rich, powerful, and influential men. Hedge fund billionaire Paul Tudor Jones stated in a speech at the University of Virginia that women don't trade as well as men because becoming a mother is a "killer" to professional focus (Freeland 2013). However, with a decline in male employment alongside women's educational and occupational gains, there are fewer opportunities for men to enact traditional masculine roles, making "manhood" harder to achieve. This loss of status may lead some men to disengage from mainstream society, often turning to social media and even violence (Black 2018).

Primarily in Western industrialized societies, a complementary cultural message emerged: The "new man" was both financially successful and emotionally sensitive, valuing tenderness and equal relationships

The fact that men's crying is no longer necessarily socially stigmatized is evidence that gender expectations are changing, however we are more accepting of tears from a strong, "can-do" man than from a man who has yet to prove himself as masculine.

with women (Messner 1997; Ryle 2015). Nevertheless, whereas society is adopting a broader definition of femininity, masculinity has retained a very narrow set of attributes and expectations. Says one writer, it's this easy to rob a man of his masculinity—"If you want to emasculate a guy friend, when you're at a restaurant, ask him everything that he's going to order, and then when the waitress comes . . . order for him." (Black 2018).

Furthermore, if a masculine-identified individual finds legitimate avenues to occupational success blocked to him due to racial or ethnic discrimination or lack of education, he might compensate, or "make it," through alternative routes such as body building, acts of aggression, intense sports or other highly risky behaviors, subordinating women and nonmasculine men, or striking a "cool pose" (Crook, Thomas, and Cobia 2009; Ezzell 2012). The latter involves dress and postures manifesting fearlessness and detachment and has been adopted by some racial and ethnic minority males for emotional survival in a discriminatory and hostile society. Similarly, white male supremacy movements are widely considered a response to the deteriorating economic and social status of white males (Wade 2019).

Femininities As with masculinities, a variety of cultural messages depict **femininities**—that is, ways of being feminine. Traditionally, the pivotal expectation for a woman has been for her to offer emotional support. The ideal woman has been expected to be physically attractive, not too competitive, a good listener, adaptable, and a man's always supportive helpmate. She was further expected to be a "good mother," putting her family's and children's needs before her own. As an example, a study titled "No

Vacation for Mother" examined gender depictions in 1950s travel literature and focused on cultural expectations surrounding women's domestic labor and the care of husbands and children during family vacations. The author found that the literature "reinforced rigid, traditional gender roles" (Morin 2012, p. 436). Generally, 1950s women were expected "and expected themselves . . . to take care of their men, their families, and their homes. That's what a good wife did, whether or not she earned a living as well" (Risman 2011, p. 20). This cultural message persists today. Overburdened with work and family responsibilities, American women spend significantly less time engaging in leisure activities than do men, a situation referred to as the "leisure gap." Women's free time is also often "contaminated" by household chores and childcare (Mattingly and Bianchi 2003).

As more women entered the workforce and came to value career success, the *professional woman* image emerged: independent, ambitious, self-confident. This cultural expectation combined with the older caregiving and sacrificing one to form the *superwoman* message. In this case, a good wife and mother, "hair flying as she rushed around, attaché case in one arm, a baby in the other," efficiently attained career success *and* supported her children, perhaps by herself. "The Superwoman could have it all, but only if she did it all" (Gottesdiener 2012). Advocates, such as Reshma Saujani, the founder and CEO of *Girls Who Code*, are rejecting the "toxic narrative that many girls face" that pressures them, from a young age, to pursue being the "perfect woman"—successful in the workplace, relationships, and childrearing (Saujani 2018). Yet, whether juggling multiple roles causes women greater stress and lower life satisfaction is unclear, suggesting that women's expanded roles may have positive benefits (e.g., Sumra and Schillaci 2015).

Feminist theorist Dorothy Smith (1987) argues that mainstream culture values masculinity more highly than femininity. For instance, rather than being defined as a form of success in and of itself, family caregiving is culturally defined as secondary. The "ideal" career path has followed a masculine model, according to which occupational dedication is paramount and family caregiving is less central (Slaughter 2012). Smith argues that women live with a **bifurcated consciousness**—a divided perception where she is aware of and troubled by two conflicting messages: first, that caregiving is most important for her; second, that caregiving is not as highly valued across society as is career success. How women navigate this bifurcated consciousness is a question for ongoing research.

The interaction-constructionist perspective (see Chapter 2) prompts us to see expectations about what's gender appropriate as socially constructed. **Dramaturgy**, part of the interactionist-constructionist perspective, sees individuals as enacting culturally fashioned scripts and socially prescribed roles in front of others (everyday-life audiences). However, persons do not blindly follow

a given cultural script. Instead, they ad lib, modifying or renegotiating prescribed roles (Goffman 1959). In other words, we "do" or perform gender (Butler 1988; West and Zimmerman 1987). Sometimes when doing gender, we retain a conventional gender identity but ignore or alter the rules just a bit in a process called *gender bending*. A highly masculine male who paints his finger nails pink is bending his gender performance. An otherwise feminine woman who likes to work on cars is gender bending. The idea of doing—and of bending—gender raises the question: To what extent do people behave in accordance with cultural expectations?

Doing Gender: To What Extent Do Individuals Follow Cultural Expectations?

Research on actual behavioral gender differences suggests that there are fewer than we might think. In fact,

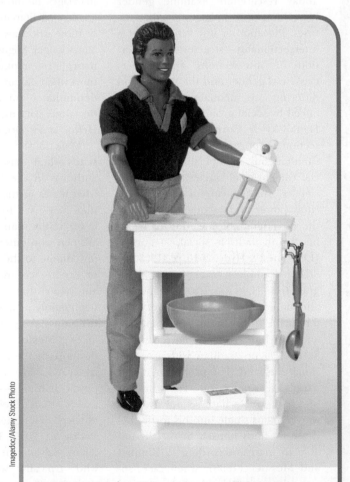

Imagedoc/Alamy Stock Photo

Pink is for girls, and blue is for boys. This image simultaneously shows social change and resistance to change. Ken is cooking: social change. But he's using pink kitchen utensils. Are they Barbie's? Should kitchen utensils be Barbie's and not Ken's? What do you think images like these convey to children?

A Closer Look at Diversity

Racial and Ethnic Differences in Gender Roles

Images of men as instrumental and women as expressive are based primarily on white, middle-class heterosexuals. Especially in past decades, researchers' habit of using middle-class whites as the norm caused ignorance regarding gender and racial and ethnic diversity. Until about 30 years ago, researchers tended to see women as a homogeneous category disadvantaged compared with men, also a homogenous category. Now scholars recognize that men are not equally privileged, nor are women equally disadvantaged. Today researchers examine gender in relation to its linkages to other statuses. Put another way, scholars stress **intersectionality**—structural connections or "intersections" among race, class, and gender and the various and differing ways inequalities are experienced by social groups (Granka 2014; Hankivsky & Cormier, 2011). Intersectional approaches "consider how simultaneous membership across social categories characterize our experiences," paying careful attention to "how power and inequality construct the experience or meaning of membership in multiple social categories" (Else-Quest & Hyde, 2016, p. 319, 332).

For example, a black woman immigrant from Haiti who is a single mother and domestic worker has a very different life from that of her employer, an Asian Indian woman lawyer who is an Ivy League graduate with a professional husband. The professional husband's life is much different from that of a white male high school dropout who remains single into his forties because he finds himself financially unable to marry. According to Kimmel and Messner (1998), when race, social class, sexual identity, physical abilities, and other statuses are taken into account, "male privilege" can be substantially muted. Meanwhile, racial and ethnic differences in gender roles are not as strong as stereotypes suggest; the male provider role is a powerful theme among all racial and ethnic groups (Ashwin and Isupova 2014; Nyman, Reinikainen, and Stocks 2013). In this section, we'll see that a good deal of the racial and ethnic diversity in gender behavior has its roots history.

Immigration

Research finds that, for the most part, the marriages of more recent immigrants are less egalitarian than the marriages of those who have been in the United States longer. Migration tends to change masculine–feminine roles among cultures where family life involves females' dependence on and acceptance of decisions of males (Hengstebeck, Helms, and Rodriguez 2015).

Male household heads typically lose status when masculine privilege and authority in the United States is not what it was in the home country, and they may have to take jobs of much lower status than they had at home. Women who enter the labor force after coming to the United States begin to experience an independence and autonomy that carry over into the negotiation of new gender roles and decision-making patterns. On average among Mexican immigrants, women do more of the housework and childcare and have less decision-making power than men—but they have more power than their counterparts in Mexico (Hengstebeck, Helms, and Rodriguez 2015; Pinto and Coltrane 2009). Salvadoran immigrant women, who typically were employed in their home country, nevertheless remark on their greater autonomy in the United States. They feel freer to come and go without a husband's close monitoring and feel they are more likely to get help dealing with an abusive spouse (Eller 2015), a situation also found among Hispanics, Asians, Arabs, and particularly Muslims (Knapp, Muller, and Quros 2009; Jo 2002; Kibria 2007). As Shireen Zaman of the Muslim Research Institute for Social Policy and Understanding explains, "What we're seeing now in America is . . . a quiet or informal empowerment of women. In many of our home countries, socially or politically it would've been harder for Muslim women to take a leadership role. It's actually quite empowering to be Muslim in America" (Knowlton 2010). Conservative interpretations of Islam religio-cultural traditions are at odds with Western feminist views, but many Muslim women are using the Qur'an to, challenge sharia law, question patriarchy, and demand women's rights (Khurshid 2015; Polgreen 2014).

Hispanics

There are two traditional gender scripts among Hispanic men and women, *machismo* and *marianismo*. Marianismo refers to the expectation that Hispanic girls and women, or Latinas, should be virtuous and chaste, the family and spiritual pillar, and subordinate to others, especially husbands (Piña-Watson, Lorenzo-Blanco, Dornhecker, Martinez, and Nagoshi 2016). Meanwhile, Hispanic boys and men, or Latinos, are expected to follow the *machismo* cultural ideal of hypermasculinity and dominance, although it is important to acknowledge positive attributes of Latinos including *caballerismo*, which includes social responsibility, being chivalrous, nurturing, and emotionally connected to others (Piña-Watson et al. 2016; Ramos-Sánchez and Atkinson 2009).

These ideals are important. Hispanic men and women holding such traditional beliefs (e.g., "wives should respect a man's position" and "Latinas should be morally pure in thought and sexuality") score higher on depressive symptoms, hostility, and anxiety (Nuñez, González, Talavera, Sanchez-Johnsen, Roesch, et al. 2016). In another study, Hispanic men who scored higher on hostile and benevolent sexism also more strongly endorsed traditional gender roles (Bermúdez, Sharp, and Taniguchi 2015). For Latinas, some aspects of marianismo (e.g., acting as a pillar) have a positive effect on their educational aspirations, whereas others (e.g. being subordinate to others) have negative effects. Only caballerismo, and not machismo, was associated with higher educational aspirations of boys (Piña-Watson et al. 2016). Male and female Hispanic immigrants may experience acculturation to the U.S. differently. In a study of Hispanic college students, acculturative stress (e.g., "It bothers me when people don't respect my family's cultural values") had a greater negative impact on the emotional health of men than women. The authors suggest that machismo may discourage men experiencing stress from seeking help which may in turn place them at risk for depression (Castillo et al. 2015). Later marriage and overall lower rates of marriage among Hispanics compared to other racial and ethnic groups will no doubt affect gender roles in the future. For example, a lower percentage of Hispanic high school seniors expect to marry (65 percent) than whites (80 percent) and blacks (72 percent; Alfred 2019). Hispanics of both genders placed a high value on *la familia* or family solidarity, with individual family members' needs and desires subsumed to the collective good. Clearly, culture remains an important to understanding gender roles among Hispanics.

Asian Americans

Asians in the U.S. are a large and diverse population originating from many countries, including China, Japan, Korea, Vietnam, India, and Cambodia, just to name a few, with different religious and cultural traditions for men and women. Male dominance continues to be the characteristic of some recent Asian immigrants, but scholars note increased independence of Asian women in the United States (Eller 2015; Ishii-Kuntz 2000). Asian women are stereotypically viewed as passive, shy, and hyperfeminine. However, the image of Asian women as subordinate to men in patriarchal households was not always the reality. During the Gold Rush of the 1800s, Chinese men were brought to the U.S., often against their will, to be miners and laborers and serve as domestic servants (Wade and Ferree 2015). Asian women subsequently entered the labor force because of the low wages of Asian men. Furthermore, President Franklin D. Roosevelt's authorized internment of Japanese Americans to War Relocation Camps during World War II eroded Japanese husbands' provider role and undercut their authority over women and children. As a result of taking on feminine roles, Asian men tend to be viewed as less masculine than other groups of men. As Asians increase as a proportion of the U.S. population, things may be changing. The 2018 film *Crazy Rich Asians* was both the first to have an all-Asian leading cast and the first to portray Asian Americans as attractive leading men (Kao, Balistreri, and Joyner 2018).

Although there is still evidence of gender division of labor, contemporary Japanese couples evidence greater equality than in the past (Takagi 2002). Similarly, as American Asian Indian women obtain more education and develop their own careers, they demand more help from husbands with domestic chores and childcare (Bhalla 2008;

Kallivayalil 2004; Eller 2015). Today, Chinese and Japanese men's earnings are roughly the same as whites' (Snipp and Cheung 2016).

Native American Indians

Native American Indians have a complex heritage that varies by tribe and may include an organizational structure in which women own the family's house, tools, and land. An important aspect of Native American culture is collectivism and community, and balance and harmony. Historically, Native American women's political power declined with European invasion and the subsequent spread of patriarchally organized social institutions in what is now the United States. Colonial Europeans in influential positions refused to recognize female Native American leaders, and Native Americans' forced movement to reservations further undermined gender equity (Eller 2015; Ryle 2015). Recently, Native American women have begun to regain their power. As is the case for other economically disadvantaged populations, marriage rates among Native Americans are low and out-of-wedlock childbearing and cohabitation is high (Turner, Limb, and Stewart 2020). Sole parenting has provided women with greater power in family life. Although poverty, abuse, and mental health problems are a persistent problem, Native Americans are resilient due to their spirituality, strong extended families, and sense of humor (Goodluck and Willeto 2009). Women are the principal leaders of about one-quarter of the 566 Native American tribes in the United States (National Congress of American Indians 2015). With more women tribal leaders over the last thirty years, it appears that one result has been greater attention to child welfare, other social services, and education (Davey 2006). In 2019, the Cherokee Nation named a woman, Kimberly Teehee, as its first delegate to the U.S. House of Representatives, and she plans to focus on

(*Continued*)

Racial and Ethnic Differences in Gender Roles

health, education, and Indian treaty rights (Brewer 2019).

African Americans

There are numerous stereotypes about African American women and men. For example, black men are often viewed as aggressive, hypersexual, and dangerous; black women are often characterized as unfeminine and physically tough (Wade and Ferree 2015). African American families tend to be communal-oriented and child-centered with permeable family boundaries. Due to a history of slavery in which both genders labored and postslavery discrimination against black men in the labor force, African American women have had higher employment rates than their white counterparts (Corra et al. 2009). Moreover, smaller gender gap in earnings between African Americans than among whites provides African American women with greater power in relationships with men (Snipp and Cheung 2016). Like other economically disadvantaged populations, marriage rates are low among blacks and single motherhood and cohabitation is high. One reason is a so-called lack of "marriageable" men both numerically and socioeconomically, in that many are under or unemployed, incarcerated, deceased, or already married—which has led to greater focus of African American women on their education and careers (Bryant 2020).

Black couples experience and prefer high levels of role flexibility and power sharing, a situation that enhances their relationship (Cowdery et al. 2009; Stanik, McHale, and Crouter 2013). Studies also show that African American boys and girls are fairly equally socialized in employment skills, domestic skills such as cooking, and childcare (Brown et al. 2009; Ryle 2015; Theran 2009). And although marriage may be out of reach for many African Americans, it is still a strong value and greater role flexibility in black families does not lessen the priority of provider-role expectations for men, despite the difficulties encountered by black males in fulfilling this role. Interestingly, this situation may be related to the fact that many African American families see the ability to have male breadwinner and female caregiver gender relationships as evidence of economic success (Cowdery et al. 2009; Eller 2015; Furdyna, Tucker, and James 2008).

"men and women, as well as boys and girls, are more alike than they are different" (Hyde 2005, p. 581). Analysts have found evidence of gender differences in (1) motor performance, especially in boys' greater throwing distance and speed; (2) sexuality, for instance in male's greater incidence of masturbation, use of pornography, and acceptance of casual sex; (3) physical aggressiveness, with males generally more violent than females; (4) use of language; and (5) emotional intelligence, with girls typically scoring higher (Björkqvist 2018; Cabello, Sorrel, Fernández-Pinto, Extremera, and Fernández-Berrocal 2016; Coates 2015; Price and Hyde 2015; Ryle 2015; Thomas and French 1985).

Scientists find girls and women more likely than boys to exhibit sadness and anxiety—"submissive emotions . . . that do not threaten interpersonal interaction in most cases." Boys and men show anger more often than girls do, and in school or business settings, "even laughter at the expense of others." However, there is considerable individual variation in emotional displays. The situational context accounts for much of the difference (Gormley and Lopez 2010; Meier, Hull, and Ortyl 2010). A man may reveal deep sadness at a funeral, for instance, or over a divorce; a woman may get extremely angry should her car be hit by a careless driver.

In *Rage Becomes Her*, author Soraya Chemaly explains that girls and women are taught since birth to repress anger and are more harshly punished than boys if they do. In a survey of adolescents, girls more so than boys reported feeling pressure to always be positive, make sure not to disappoint others, keep everyone happy, and put others' feelings before their own (Plan International 2018). Yet, bottling up negative emotions is associated with mental and physical health problems such as high blood pressure, eating disorders, and cutting (Cox, Stabb, Bruckner 2016; Novaco 2016). Mothers who get angry with their children do not fit the martyrdom associated with motherhood and experience a profound sense of shame and fear at being labeled a "bad mom" (Dubin 2019).

Gender behaviors generally fit a pattern that can be illustrated as two overlapping distribution curves (see Figure 3.3). For instance, although on average men are statistically more likely than women to be physically aggressive, an area of overlap exists in the degree or extent of men's and women's physical aggression, and both genders use verbal aggression equally as much. Differences in sexual behavior and motor skills are relatively small (Björkqvist 2018; Price and Hyde 2015; Thomas and French 1985). It is also true that differences *among* women or *among* men (*within-group*

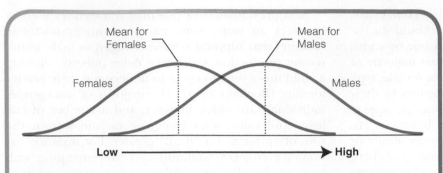

FIGURE 3.3 How females and males differ on one hypothetical trait, aggressiveness, conceptualized as overlapping normal distribution curves. Averages may differ by gender, but trait distributions of men and women share much common ground.

variation) can be greater than the average difference *between* men and women (*between-group variation*). We return to this concept of within-group versus between-group variation again later in this chapter.

GENDER SOCIALIZATION

As people in a given society learn to talk, think, and feel, they *internalize* cultural expectations about how to behave (Mead 1934); that is, they come to accept these expectations as their own. The process by which society influences members to internalize attitudes and expectations is called **socialization**. Interactionists point out that individuals do not automatically absorb; rather, they negotiate cultural attitudes and roles. Furthermore, they do so throughout life. Therefore, gender behavior varies from individual to individual.

Cultural images in language and in the media convey the binary gender expectations. This starts early, even before a child's birth through ultrasounds that can determine a child's gender in the womb as early as four months. "Gender reveal parties" have become popular, where a balloon is popped or a cake sliced to reveal blue or pink confetti (or cake). Soon after birth, most infants receive either a masculine or a feminine name—not to mention a pink or a blue blanket. A new mother may be told that she has either a "lovely girl" or a "strong boy."

Gender Socialization Theories

In the classic **interaction-constructionist perspective** (see Chapter 2), children develop self-concepts based on feedback from those around them. Play is not idle time; it is a significant vehicle through which children develop ideas about appropriate gender roles for themselves both as children and as adults (Mead 1934). In play, preschool children imitate adult roles, such as parent, teacher, or health care provider. Little girls are more likely to play "mommy" with their dolls and kitchen sets, whereas little boys more often play with cars or action figures.

Among school-age children, studies show that boys and girls tend to play separately and differently (Maccoby 2002; McIntyre and Edwards 2009). Girls play in one-to-one relationships or in small groups of two and three; their play is relatively cooperative, emphasizes turn taking, requires little competition, and has relatively few rules. In "feminine" games such as jump rope or hopscotch, the goal is skill rather than winning. Boys more often play in fairly large groups, characterized by more fighting and attempts to affect a hierarchical pecking order. Meanwhile, organized youth sports play a role in socialization. Girls who take part in sports have greater self-esteem and self-confidence (Ryle 2015).

According to **social learning theory**, a theory within the interaction-constructionist perspective, children are taught gender roles by parents, teachers, peers, friends, partners, and the media (Bandura and Walters 1963; Ryle 2015). Children imitate models for behavior and are rewarded by parents and others for whatever is perceived as sex-appropriate behavior. As children grow older, toys, talk of future careers or marriages and parenthood, and admonitions about "sissies" and "tomboys" communicate parents' ideas about appropriate behavior for boys and girls. However subtle, the rewards and punishments that parents and others assign to gender expectations are assumed to be key to behavior patterns. A study of 6,000 children's books published between 1900 and 2000 found that—even moving into the twenty-first century—males were central characters in 57 percent of the stories; females were central in 31 percent. When a child's name was in the book title, the name was most often a boy's (McCabe et al. 2011). In her cleverly titled book, *Cinderella Ate My Daughter*, gender scholar Peggy Orenstein shows how intensive twenty-first-century marketing of a "new girlie-girl culture" encourages girls at "younger and younger ages . . . to define themselves through appearance—from the outside in rather than from the inside out" (Orenstein 2012). In one national survey of adolescents, 53 percent of girls said they look in the mirror and "imagine how others must see them" and one in three do so many times a day. Two-thirds of girls said they are exposed to unrealistic female bodies at least several times a week, and over half say they see female characters on TV or in the movies whose bodies are more important than their brains and abilities; over one in four said they see this every day (Plan International 2018).

Some psychologists think that what comes first is not rules about what boys and girls should do but rather the child's awareness of being a boy or a girl. In this **self-identification theory**, the vast majority of children categorize themselves as male or female, typically by age 3. They then identify behaviors in their families, in the media, or elsewhere that are appropriate to their sex and adopt these behaviors. In effect, children socialize themselves to masculinity or femininity from available cultural materials (Kohlberg 1966; Signorella and Frieze 2008). **Gender schema theory** posits that children develop a framework of knowledge (a gender schema) about what girls and boys typically do (Bem 1981; Ryle 2015). Children then use this schema to organize how they interpret new information and think about gender. Once children have developed a gender schema, the schema influences how they process new information, with gender-consistent information remembered better than gender-inconsistent information. For example, a child with a traditional gender schema might generalize that physicians are men even though the child has sometimes had appointments with female physicians (Signorella and Frieze 2008). In addition to offering theories on the processes by which gender expectations become internalized, social scientists have examined various settings for socialization.

Gender Socialization—The Process

For the most part, gender socialization assumes that gender is binary. Gender fluidity, while becoming increasingly visible, is still considered deviant by many people. Some children as young as toddlers display **gender variance**—that is, they desire to dress and behave more like the "other" gender (Spiegel 2008b). Parents, teachers, and pediatricians wrestle with whether and how early children should be allowed to define themselves (Hoffman 2011; Padawer 2012). Some parents allow their young children to experiment with transgender behaviors—a kindergarten boy sports hot pink socks and sparkly sneakers, for instance, or a 4-year-old girl likes to be called "handsome prince" and has persuaded her mother to get her a Mohawk haircut (Hoffman 2011; Padawer 2012). Swedish preschools are combatting hidden forces that contribute to binary gender roles. Teachers there refer to students as "friends" or their first names rather than "boys and girls" (Barry 2018). American parents of transgender children can find themselves battling schools so that their children may dress how they would like (Frosch 2013). Sometimes parents let the child dress as they like at home but limit what's presented in public. As one father told his kindergarten son, "We'll paint your nails [at home today] and wash them off before you go to school" (Hoffman 2011).

As a pre-adolescent's potential transgender identity becomes an issue, some parents consider hormone blockers that suppress female estrogen or male testosterone production and hence delay puberty, allowing a child more time to come to understand their gender identity (Boghani 2015). The majority of transgender individuals are older, however, and a number of colleges and universities recognize variations from the two, often termed a "third" gender. For instance, "at women's colleges administrators are struggling with how to handle applications from trans women" (Steinmetz 2014, p. 46). Meanwhile, some colleges and universities have adopted student health plans that cover the cost of sex-change treatments (Pérez-Peña 2013; Scelfo 2015). Many have established gender-neutral restrooms, housing, and other accommodations, such as initiating programs to prevent discrimination against trans students. "It's like a constant coming-out process, educating those around you that there is a gender binary, and this is what it means to identify outside of it," said a university senior who identifies as genderqueer and helps to plan university consciousness-raising events (Scelfo 2015).

Transfamilies are families in which one or more family members is or are transgendered. For example, about 40 percent of trans persons are parents (Friedman 2015). Having a loved one who undertakes sexual reassignment can be difficult. As Sabine, a girl whose mother transitioned from female to male when Sabine was 13, said, "It's hard to face the fact that someone who is close to you changes at all—especially a change that big." Trans persons say that finding their true identity can be lifesaving for them, but at the same time, devastating for their children—at least for a period of time. Now age 16, Sabine, whose father still lets her call her "mom," explained, "At first I felt a sense of loss, until a few years later when I saw that my mom was a much happier person. Now I'm cool with this" (James 2012).

Another example of growing acceptance is the father of a New York City firefighter, Brooke, who grew up as George and didn't define herself as a transgender female until she was in college. Her father experienced a difficult, lengthy struggle with Brook's transition. He questioned what he must have done wrong in raising Brooke. He wondered whether things would have been different if he'd spent more time at home. Although Brooke's grandfather is "still having trouble coming to terms with having a granddaughter who used to be his grandson," Brooke's father has concluded, "Brooke is being truly the person she is. This is how she's created" (Nwoye 2015).

Eventually—although assuredly not always—family and friends may become accustomed to and grow to accept the new person. Some scholars argue that as more people choose to define themselves in this way, the larger society may follow with relative levels of acceptance

(Marikar 2009; Winerip 2009). Still, clinical psychiatry professor Richard Friedman (2015, p. 3) asks us to think about this: How many transgender individuals would feel the need to *physically* change gender if they truly felt accepted with whatever gender role they choose?

Everyday Gender Socialization Home, school, daycare, doctor's offices, and anyplace where children interact with others influence their gender attitudes and behavior (Epstein 2011). For example, adolescents in Mexican families that espoused traditional division of labor and attitudes were more likely to espouse traditional gender attitudes themselves (Lam, McHale, and Updergraff 2012). Differential socialization by parents and other family members does continue to exist, although it is typically not conscious (Kane 2012a). For instance, a study of Google searches found that parents are two and a half times more likely to ask "Is my son gifted?" than "Is my daughter gifted?" Conversely, parents are one and a half times more likely to ask whether their daughter is beautiful than whether their son is, and about twice as likely to ask whether she is overweight (Stephens-Davidowitz 2014).

Parents tend to encourage exploratory behavior more in boys than in girls. One way they do this involves the toys they buy. Toys considered appropriate for boys encourage physical activity, independence, and competition, whereas "girl toys" promote caregiving, cooperation, and communication. While family members increasingly encourage girls to develop instrumental skills, boys are still discouraged from or encounter ambivalence about cultivating tenderness (Blakemore and Hill 2008). The toys children play with matter. In a survey of adolescents, boys who played mostly with "boys' toys" as opposed to gender-neutral toys reported thinking more about girls' bodies than their thoughts and personalities, placing less importance on "making the world a better place" (and more on making money, having kids, and getting married), not perceiving sexism as a big problem, and not identifying as a feminist, among other traditional male attitudes (Plan International 2018). Girls who played with mostly "girls' toys" were more likely to report feeling more pressure to dress like an older woman, feeling more pressure to have positive comments on social media, and placing less importance on having a successful career as a life goal, among others. Although playing mostly with male or female toys is likely indicative of a broader patriarchal home environment, the role of gendered toys on promoting stereotypical gender attitudes cannot be ruled out.

Parents more often allow girls to express feelings of anxiety or sadness; boys, on the other hand, are more commonly allowed to express anger (Ryle 2015). Fathers in particular more easily accept a school-age daughter being a tomboy than a son displaying behaviors thought to be feminine. In some ethnic groups, boys are given much more freedom—to explore their neighborhoods or to go out with friends, for example—than are girls (Marks, Lam, and McHale 2009). Half of all boys in a study of adolescents reported hearing their fathers or male family members make sexual jokes or comments about women (Plan International 2018). Boys reporting this also reported feeling pressure to join in when other boys talked about girls in a sexual way, to "hook up" with girls, to feel it is OK to ask a girl for a naked or sexy picture, and to feel pressure to control and dominate others, only somewhat agreeing or disagreeing that women should be equal to men in work, politics, and life.

Still, many parents consciously try to avoid gender stereotyping as they raise their children (Ryle 2015). Yet, even among those who considered gender expectations problematic, their children are confronted with an everyday world teeming with social pressures and judgments from friends, relatives, and even strangers if their kids didn't stick to a narrowly gender path (Kane 2012b). For example, a high school girl in Alabama who chose to wear a tuxedo for her senior portrait found her picture missing from the yearbook (Burke 2019).

Although early family relations shape a child's developing identity, peer groups and teachers become important as children grow to school age. A practice evident among schoolmates involves **borderwork**— the process, or work, of monitoring and maintaining the conceptualized border between appropriately masculine boys or men and acceptably feminine girls or women (Ryle 2015). Educators have observed and documented borderwork students patrolling each other's behavior according to accepted gender expectations and scripts. The most common punishments designed to keep classmates within appropriate gender boundaries

LightField Studios/Shutterstock.com

Toys send messages about gender roles. A consequence of children developing gender-stereotyped toy preferences is an incomplete skill set in adulthood. What does this toy say? What skills does it facilitate and what skills are missing?

were laughter, homophobic name-calling, and social exclusion (Lee and Troop-Gordon 2011; Poteat, O'Dwyer, and Mereish 2012).

Meanwhile, teachers may reinforce the idea that males and females are more different than similar (Bigler, Hayes, and Hamilton 2013). At times, boys and girls interact comfortably in the school band, for example. But teachers often pit girls and boys against each other in spelling bees or math contests, for instance. A study by Pennsylvania State University psychologists found that many (but not all) preschool teachers emphasized dichotomous gender differentiation by the ways that they structure their classrooms—for example, lining up children by gender or having them put their completed work on different bulletin boards. The researchers further found that when teachers differentiate between boys and girls, young children are more likely to negatively stereotype and prefer not to play with the other sex (Hilliard and Liben 2010). Young children engage in fewer stereotypical behaviors when they play in gender-mixed groups (Goble et al. 2012).

Furthermore, studies have shown that teachers pay more attention to males than to females, and males tend to dominate learning environments from nursery school through college. Compared to girls, boys have been more likely to receive a teacher's attention, to call out in class, to demand help or attention from the teacher, to be seen as model students, or to be praised by teachers. Boys are also more likely to be disciplined harshly by teachers (Epstein 2007; Ryle 2015; Zaman 2008). On average, boys have poorer study habits and less concern about doing well in their studies (Klein 2014). More boys fall behind grade level, are suspended, and are placed in special education classes. Boys have a greater incidence of diagnosis of emotional disorders, learning disorders, and attention-deficit disorders, and of teen deaths (Ryle 2015).

Peggy Orenstein, previously mentioned, spent one year observing pupils and teachers in two California middle schools, one mostly white and middle class and the other predominantly African American and Hispanic and of lower socioeconomic status. Orenstein (1994) found that in both schools, girls were subtly encouraged to be quiet and nonassertive whereas boys were rewarded for boisterous and even aggressive behaviors. Interestingly, African American girls were louder and less unassuming than nonHispanic white girls. For example, they called out in class as often as the boys did. But Orenstein noted that teachers' reactions differed. The participation and even antics of white boys in the classroom were considered inevitable and rewarded with extra teacher attention, whereas the assertiveness of African American girls was defined as "menacing, something that, for the sake of order in the classroom, must be squelched" (Orenstein 1994, p. 181; see also Theran 2009). Orenstein further found that Latinas and Asian American girls had special difficulty being heard or even noticed. Probably socialized into a

quiet demeanor at home, these girls' scholastic or leadership abilities largely went unseen. In some cases, their teachers did not even know who the girls were when Orenstein mentioned their names.

Both boys and girls feel the weight of traditional gender roles. In a national survey of adolescents, 72 percent of girls said they felt treated with less respect at least once in a while because they are a girl, 56 percent feel they've been treated unfairly in sports, 36 percent felt they were treated unfairly at school, and 30 percent on social media (Plan International 2018). One in three boys said they felt pressure to "dominate or be in charge of others." Nearly a third said that strength and toughness is what society most values in boys, whereas only 2 percent said honesty and morality and 8 percent said ambition or leadership. The strongest predictor of boys' feeling pressure to be physically strong and ready to fight was exposure to boys making sexual comments and jokes about girls, suggesting that closer monitoring of boys' environments could be an effective means of change.

Although they haven't caught up completely, girls are catching up to boys, even in math and science courses where many teachers stereotyped them as less competent and where they therefore underperformed (Eccles 2011; Riegle-Crumb and Humphries 2012). Analysts concerned about boys note a mismatch between their higher levels of physical activity and school expectations about sitting still and following detailed rules. Accordingly, they propose accepting a certain level of boys' rowdy play as *not* deviant, allowing more physical movement in classrooms, and encouraging activities shared by boys and girls (Orr 2011; Sommers 2013). Others point to the relative lack of male role models in elementary and secondary education. Nine in ten elementary school teachers and two-thirds of secondary teachers are women (Loewus 2017).

Following Traditional Gender Expectations Can Be Costly

Both men and women pay a price for gender as traditionally structured. Biases and gender stereotypes thwart both males' and females' career opportunities, men's confidence in nontraditional family roles, and both genders' ability to communicate supportively with one another (Bobbitt-Zeher 2011; Hancock 2012). Boys act tough to protect themselves, but such a stance interferes with scholastic performance (Kimmel 2001; Ryle 2015). As one indicator of costs to men, we can examine mortality rates. Women live an average of 5 years longer than do men, with a life expectancy of 81 compared to 76 for men (Murphy, Xu, Kochanek, and Arias 2018). Men are far more likely than women to engage in dangerous and risky behaviors such as fast driving, doing drugs, and drinking alcohol. They are more likely to own guns and are averse to seeking mental health care. That said, men have higher rates of deaths from motor vehicle accidents, chronic liver

disease, drug overdoses, homicide, and suicide (Heron 2019; National Institute on Drug Abuse 2019).

Women also suffer costs due to traditional gender structures. Poverty levels are higher for women than for men—12.9 percent for women compared with 10.6 percent for men in 2018. Children tend to stay with their mothers when parental relationships dissolve. The poverty rate for nonmarried single mother households is 25 percent compared to only 4.7 for married couple families and are even higher among African American, Hispanic, and Native American women and children (Semega, Kollar, Creamer, and Mohanty 2019).

People pay a gender price in countless other ways as well. Men's work, childcare, and other family-focused opportunities are limited. Their childcare and housework performances are often trivialized or the focus of humor in the media. Meanwhile, women still do the majority of household labor and childcare even when employed. They assume a disproportionate share of tasks concerning keeping up with extended kin—holiday shopping, remembering birthdays, and getting family photos to relatives, for example (Janning and Scalise 2015). More often than men, women adjust their careers to accommodate family life (Parker 2015). Then, too, gender-differentiated communication patterns can frustrate both sexes, as is discussed in Chapter 11. Family power arrangements can feel problematic, an issue addressed in Chapter 12.

GENDER IDENTITIES IN SOCIAL CONTEXT

A theme of this text is that *socially structured opportunities* have an effect on men's and women's choices and behaviors. Gender differences that we see or think we see often involve men's and women's being assigned to divergent roles. A female secretary, for example, is expected to be compliant and supportive of her male boss's decisions. To observers, *she* seems to have a gentle and submissive personality, while *he* is seen to have leadership qualities (Meier, Hull, and Ortyl 2010; Webster and Rashotte 2009).

Like other personal attributes, gender is embedded in all our institutional structures—family, church, state, the economy, and education—influencing the ways that people enact, or "do," gender. Just as every society has a political structure, such as democracy or monarchy, so too every society has a **gender structure** "from patriarchal to at least hypothetically egalitarian" (Risman 2011). Gender structures also shape the roles that individuals are expected to follow. From the perspective of dramaturgy, gender structures are the writers and directors that create and oversee gender scripts for individuals, relationships, and families.

Social institutions in virtually every society have been characterized by **patriarchy** and **masculine dominance**, in which masculine males exercise authority over females and people they perceive as not masculine enough. On the personal level, masculine dominance involves wielding greater power in a heterosexual relationship. On the societal level, masculine dominance is the assignment to men of greater control and influence over society's institutions and benefits. Only very recently in human history—within the last one hundred years—has patriarchal dominance begun to break down.

Gender expectations are embedded in all our social institutions, and expectations and practices in one institution affect those in the others. Ways that gender is scripted and performed in the institution of education, for instance, affect men's and women's choices about school, attending college, and majors. Those decisions, in turn, affect job opportunities, earnings, and family roles. Throughout this textbook, we examine ways that gender is performed within families. In this section, we address gender structures in four major social institutions other than family.

Religion

Taken as a whole, the institution of religion evidences male dominance (Avishai, Jafar, and Rinaldo 2015). Religion is a strong force in shaping gender expectations and behavior. For example, Christian women with more fundamentalist beliefs (that women are divinely subordinate to men) have a higher likelihood of not working outside the home (Sherkat 2017). Although most U.S. congregations have more female than male participants, men more often hold positions of authority. Nevertheless, there is ample evidence of change. Women have been elected as rabbis, bishops, and denomination leaders in many religions, including Judaism and the African Methodist Episcopal, Anglican, United Methodist, and Presbyterian churches (Chaves, Anderson, and Byassee 2009; Gott 2010). Moreover, there is also a growing feminist movement embedded in virtually all the world's religions. For instance, American Muslim women—both Arabs and blacks—seek to combine their religio-cultural heritage with equal rights for females (Knowlton 2010; Prickett 2015). Similarly, in a movement they term "devoted resistance," Orthodox Jewish women in Israel remain religiously fervent while advocating for greater attention to women's rights (Zion-Waldoks 2015).

The effects of personal religious involvement on daily family lives are complex (Aune 2015). On the one hand, evangelical Protestantism, Catholicism, Islamic fundamentalism, and the Latter-day Saints religion (Mormonism) all teach the traditional family ideal of *male headship*—the husband as provider and decision-making "head" of the family—and a corresponding emphasis on domesticity for women and the ascribing of men's roles to the public sphere and women's roles to the private (Bulanda 2011). Yet, some Mormon mothers

do not comply with traditional gender expectations (has a marriage that is egalitarian) and engage in "ideological compensation" where they accept and defend traditional gender norms (that the man is the head of the household; Leamaster and Bautista 2018). In "I'm Not Your Stereotypical Mormon Girl," researchers Leamaster and Bautista (2018) found many Mormon women who resist traditional gender expectations by not getting married young, working, and by being assertive.

Research shows that households where the husband is more religious than the wife are associated with decreased marital satisfaction, particularly for the wife, and increased risk of divorce (Duba and Watts 2009; Vaaler, Ellison, and Powers 2009). On the other hand, actual practice among conservative Christian couples is more egalitarian than formal doctrine would suggest and there

is a diversity of viewpoints on gender roles (Avishai, Jafar, and Rinaldo 2015; Beaman 2001). For some, "headship has been reorganized along expressive lines, emptying the concept of virtually all of its authoritativeness" (Wilcox 2004, p. 173). The idea of *mutual* submission (rather than simply a wife's submission to her husband) leads to more egalitarian decision making in day-to-day family life. These ideological shifts have far-reaching implications. For example, Americans are less prejudiced toward Muslims if they perceive they are supportive of women's rights (Moss, Blodorn, Van Camp, and O'Brien, 2019).

Government and Politics

With the election of Donald Trump in 2016 came a surge in women's participation in government and

In 2018, a record number of women were elected to Congress, and notably younger women of color. Pictured are Alexandria Ocasio-Cortez (1) representing New York, Rashida Tlaib (2) representing Michigan, Ilhan Omar (3) representing Minnesota, and Ayanna Pressley (4) representing Massachusetts.

politics. A historic number of female candidates won congressional and senate seats in the 2018 mid-term elections. As of October 2019, women held 127 of the 535 seats in U.S. Congress; 25 (25 percent) of the 100 seats in the Senate and 102 (23 percent) of the 435 seats in the House of Representatives. Thirty-seven percent of females in Congress are women of color: 22 African Americans, 13 Latinas, 8 Asians or Pacific Islanders, 2 Native American, 1 Middle Eastern/North African, and 1 multiracial (Center for American Women and Politics 2019). In 2017, the first openly transgender candidate was elected to the Virginia State Legislature, the first openly lesbian mayor was elected in Seattle, and the first openly transgender person of color was elected to public office in Minneapolis (Park 2017).

Three of our nine U.S. Supreme Court justices are women. Women have become more visible in the executive branch of government as well. As of this writing, three women have served as secretary of state, the most recent being Hillary Rodham Clinton, more recently a presidential contender. As of October 2019, 5 of the 19 democratic candidates running for president in 2020 were women. According to political strategists, voters "have grown more accustomed to women in powerful positions" (Toner 2007). Nevertheless, statistics point to the fact that women, although slightly more than 50 percent of the population, are significantly underrepresented in high government positions. This is likely to change. A majority of Americans, especially younger generations, view the increasing number of women running for public office as a "good thing" for society (Pew Research Center 2019).

In a national survey of children and teens age 10 to 19, girls were slightly more likely than boys to say being a leader was a very important life goal, and nearly even percentages of girls and boys reported being a politician one day "had crossed my mind" (26 and 31 percent), indicating a fundamental shift in gender expectations for girls in public life (Plan International USA 2018).

Education

As is the case for girls in elementary, middle, and high school, gender impacts women's experiences in higher education. Women have been the majority of college students since 1979 and were 56% of college students in 2016, despite an increase in male college enrollment (National Center for Education Statistics 2019b). In 2019, women made up just over half of the U.S. college-educated workforce, finally reaching parity with men (Fry 2019). Women earn more than half of all bachelor's degrees, nearly two-thirds of all master's degrees, and half of all doctoral degrees (National Center for Education Statistics 2019a, b). To maintain gender balance, some colleges actively

recruit men, and college readiness programs have been developed to specifically target socioeconomically disadvantaged populations of men (Gordon 2016; Kowarski 2018).

What about the college climate itself? Women have been increasing as a percentage of the professoriate but most of the increase has been in part-time and nontenure track positions. Women make up 49 percent of all college faculty but only 44 percent of *full-time* faculty. Women are less likely to be awarded tenure and move up the ranks. For instance, only 16 percent of full professors (among full-time faculty) are women; a figure largely unchanged since 1993 when it was 15 percent (Flaherty 2016). Moreover, 58 percent of department chairs are men. According to data from the American Association of University Professors Faculty Compensation Survey, female faculty were paid 82 percent less than men during the 2018–2019 academic year (Flaherty 2019).

Although women outnumber men in colleges and universities, most continue to follow traditional gender expectations when choosing majors. Women far outnumber men in social services, liberal arts, and health care majors, while men dominate in business and STEM fields (science, technology, engineering, and mathematics)—fields that typically pay more (Julian 2012; U.S. Census Bureau 2014c). The early socialization of girls discourages women from choosing traditionally male fields of study. In a study of 1,111 college women, those who conformed to more feminine norms, including norms for being relationship-oriented, caring for children, and being domestic, were less likely to choose a major in STEM (Beutel and Borden 2017). Another factor underlying women's underrepresentation in STEM is exposure; a study in which students were asked to a write paper in a particular field were more likely to choose that field as a major (Fricke, Grogger, and Steinmayr 2018).

Faculty in traditionally female or male occupations become role models, nonverbally communicating by their relative numbers what is the appropriate major for students. With fewer female faculty in math, sciences, and engineering, women are less likely to feel encouraged to pursue those topics. With fewer male faculty in early childhood education, nursing, or social work departments, men may feel less comfortable pursuing these options.

Sexual harassment is increasingly being recognized as a problem for women pursuing degrees in male-oriented fields. Law Professor Susan Fortney received funding from the National Science Foundation to develop a self-assessment tool that STEM departments can use to evaluate and improve their organizational structures, policies, and procedures regarding sexual harassment (American Association for the Advancement of Science 2019).

An interesting qualitative study used participant observation in one university aerospace and engineering department to uncover situations that discourage women from entering that field. Sallee (2011) interviewed doctoral students and faculty and concluded that, both as a discipline and as a specific department, aerospace and engineering is characterized by a culture of "invisible masculinity"—that is, by values of hierarchy (winning, being the best) and competition. Tim, a graduate student in engineering, told an interviewer that he'd rather talk with his female than male classmates. "With women, Tim felt that he was free to discuss whatever topics were of interest to either of them. In contrast, with men, 'conversation is more of a contest to win'" (p. 205).

Moreover, women faculty and students suspected as not being truly deserving to be in the department were noticed and remarked on for their appearance. After a new female faculty member's seminar talk, her male colleagues talked of "drooling" over her good looks, an exception in engineering, according to stereotype. As a male student told his interviewer,

> I think my [undergraduate] graduating class [in engineering] had three [girls], but two of them could play linebacker for the Bears. . . . No, engineering is not the place to go if you want to pick up on women. Every once in a while you'll see a cute girl walking down the E-Quad [engineering section of campus]. She's lost. (pp. 206–207)

Uncomfortable in this climate of "invisible masculinity," female faculty and students reported feeling marginalized. Interestingly, departmental efforts to be fairer to women were perceived by the female students as mostly insincere attempts to be politically correct or "careful." This situation pointed to, rather than normalized, the students' gender presence.

We know of no comparable study of subjects in fields numerically dominated by women. But we might speculate that an analogous "invisible femininity" characterizes these disciplines and departments. For instance, social work curricula emphasize collaboration, encouragement, and building on other people's strengths. Perhaps males, more comfortable with a masculine than with a feminine cultural script, are inclined to feel marginalized in departments like these.

Economics

In 2017, the wife/female partner in about a third (31 percent) of married or cohabiting couples earned more than her husband/male partner, compared to only 13 percent among married couples in 1980 (Parker and Stepler 2017). Interestingly, when wives earn more

than their husbands do, both husbands and wives say the husband earns *more* and the wife earns *less*. This tendency was discovered when incomes reported in surveys conducted by the Census Bureau did not match the couple's IRS filings (Heggeness 2018). Despite decades of advances in women's employment, it appears that husbands and wives remain compelled to adhere to traditional gender norms when it comes to earnings.

Although the situation is changing, men on average continue to be dominant economically. As shown in Figure 3.4, women's incomes relative to men's have grown over the decades but appear to have stagnated. Female full-time employees now earn 82 percent of what men do (Bureau of Labor Statistics 2018). Incomes vary drastically by race and ethnicity. Asian women and men earned more than their white, black, and Hispanic counterparts in 2017. Among women, whites ($795) earned 88 percent as much as Asians ($903); blacks ($657) earned 73 percent; and Hispanics ($603) earned 67 percent. Among men, earnings differences were even larger: White men ($971) earned 80 percent as much as Asians ($1,207); blacks ($710) earned 59 percent as much; and Hispanics ($690), 57 percent. (Bureau of Labor Statistics 2018). And although women now make up 49 percent of the paid workforce, the percentage of female chief executive officers (CEOs) at Fortune 500 companies in 2017 was 6.4 percent (up from 4.2 percent in 2013) with only 32 women heading major firms (Horowitz, Igielnik, and Parker 2018).

Some people have argued that men's earnings are higher partly because of employers' assumptions—and perhaps her own assumptions—that a woman will opt out of the labor force to take care of her children or other family members. As a result, an employer may be less likely to select even highly ambitious and fully

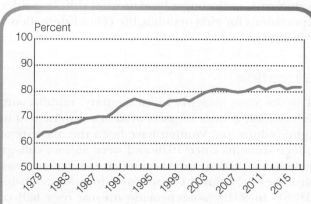

FIGURE 3.4 Women's earnings as a percentage of men's, for fulltime wage and salary workers, 1979–2017 annual averages

committed women employees for further training or positions with advancement potential (Pinto 2009). It's also possible that even ambitious women may pull back from pursuing the high leadership positions, feeling pressured to choose between succeeding at work and being a "good" wife and mother (Sandberg 2013a).

The male–female earnings difference is partly related to women's choice of occupation. However, as evidenced by the differences in pay for the same profession, this explanation does not tell the whole story. The numbers show that in the same occupational categories women continue to earn less on average than do men. In the highest-paying occupation, that of chief executive, women's average salary was $1,920 weekly in 2017 compared to $2,415 for men. Among elementary and middle school teachers, women averaged $987 and men, $1,139 per week (U.S. Bureau of Labor Statistics 2015d). How being married relates to the ratio of women's to men's earnings is worth thinking about. Married women employed fulltime whose spouse is present earn about 79 percent of what all fulltime employed men do, while never-married women earn about 91 percent (U.S. Bureau of Labor Statistics 2018).

Unconscious biases against women in the workplace can limit their occupational success. These biases manifest themselves in different ways, but it is not uncommon for women seeking advancement to hear statements such as, "We've never hired a woman for this before," "I didn't think you'd want that much responsibility," "We need someone who is going to be tough," and "Can you get the coffee?" (American Association of University Women 2019). In a 2016 survey, nearly a quarter of Millennial women believed their gender has prevented them from getting ahead at work. They also reported having to "work harder than male team colleagues to garner recognition and praise" and "feeling constantly under scrutiny for the way they dress." Three-quarters of Millennial women, compared to only 57 percent of men, said more changes are needed to give men and women equality in the workplace (Bentley University 2017).

Not all of the wage gap is explainable by factors such as men's and women's employment histories and skills. There is also evidence that women's lifelong socialization to expected personality characteristics adds to the gender pay gap. Raised to be accommodating and "nice," women are more hesitant than men to ask for a raise (Ludden 2011). "There is always that little voice in your head saying, 'Nice girls don't ask for raises'" (Rosin 2012c). Nevertheless, women leadership advisers increasingly encourage other women to "speak up and let them know you want it," whether it's a raise or a place at the board of directors table or something in between (Barnett and Rivers 2014; Kingsbury 2014). "We can't make change happen by being shrinking violets or apologists" (Butler 2015). Some see these attitudes as blaming women to themselves for their low salaries rather than structural inequalities and implicit biases. For example, data from the Australian Workplace Relations Survey shows that women *do* ask for raises as often as men but they are less likely to *get* them (Artz, Goodall, and Oswald 2018).

GENDER AND SOCIAL CHANGE

Men and women have different experiences with gender inequality, but also have different *perceptions* of gender inequality, with women's perceptions closer to the truth. In a national poll, 86 percent of women compared to 62 percent of men believed that female workers are paid less than men who do similar work (Barone 2019). Men also underestimate women's time on household tasks and childcare. Male respondents reported that women do 20 to 40 percent more unpaid work than men. Female respondents said 60 to 80 percent more (the actual figure is 67 percent more).

These differences in perceptions and recognition that there "could be a problem" with respect to gender equality formed the basis for women's efforts to obtain greater rights and equality.

Feminism and Women's Rights

To be a "feminist" can mean different things to different people, but in general feminists, who can be women *or men*, advocate for social, political, economic, and intellectual equality between people regardless of their gender identity. Feminist movements exist worldwide and are in various stages of development. The "first wave" of feminism in the United States began with an 1848 convention on women's rights at Seneca Falls, New York, and came to an end when a major convention goal, voting rights for women (*women's suffrage*), was achieved in 1920. From about 1920 until the mid-1960s, women made gains in education, and they were encouraged to take employment in factories during the early 1940s when men were fighting in World War II. However, after the war ended in 1945, media glorification of the housewife role for women and the breadwinner role for men helped to make these seem natural and generally desirable. By the 1960s, however, higher levels of education for (white, middle-class) females left college-educated women with a significant gap between their

abilities and the housewife role assigned to them. Betty Friedan's book *The Feminine Mystique* (1963) captured this dissatisfaction and made it a topic of public discussion or discourse, setting off the beginning of "second-wave" feminism, which lasted until the mid-1980s.

During this period, employed women chafed at the unequal pay and sexist conditions in which they worked and began demanding equal opportunity. The civil rights movement of the 1960s provided a model of activism. Among other changes, federal legislation and executive orders declared discrimination against women in federal government contracts illegal and mandated that educational institutions finance sports for females just as they did for males (Title IX). Grassroots feminist groups with a variety of agendas developed across the country. Part of second-wave feminism was the founding of The National Organization for Women (NOW) in 1966. Early on, NOW had multiple goals: opening educational and occupational opportunities to women and establishing support services such as childcare. Today, the organization has expanded their political agenda to include reproductive rights, immigration, equal pay, domestic violence, sexual harassment, sexual assault.

The third wave of feminism began in the early 1990s, marked by the testimony of Anita Hill in front of Congress at the confirmation hearing of Supreme Court Justice Clarence Thomas in 1991. Her claim that Thomas sexually harassed her when she worked for him as a young lawyer fell on deaf ears (he was confirmed), but it encouraged working women in the United States demand more rights. This stage of the feminist movement "tends to be much more pluralistic" and "much less dogmatic" about issues surrounding sexuality (for example, whether pornography is necessarily demeaning to women), personal expression (for example, whether breast-enhancement surgery is demeaning to women), and fashion choices (for example, whether wearing makeup is demeaning to women; Heilmann 2011; Wolf 2012). If *Ms. Magazine*, which celebrated its forty-fifth birthday in 2017, represents second-wave feminism, *Bust Magazine*, established in 1993 "for women with something to get off their chests," characterizes the third wave. The idea that feminism should stress *intersectionality* and inclusiveness is characteristic of third-wave feminism (Few-Demo 2014). For example, First Lady Hillary Clinton, whose agenda included addressing major issues such as universal health care, annoyed stay-at-home mothers around the nation when she said, "I suppose I could have stayed home and baked cookies and had teas," when her career as a lawyer became an issue in the 1992 presidential election.

There is evidence that feminism is currently moving into a "fourth wave" as a result of some very high-profile incidents and the election of Donald Trump as president. One of the highlights being the Women's March in Washington, D.C., on January 21, 2017 (along with similar marches across the country), which involved nearly 5 million people and is thought to be the largest single-day demonstration in American history (Encyclopedia Britannica 2019). Social media—including Facebook, Twitter, and other platforms—is an important feature of fourth-wave feminism. Another example is the #MeToo movement, founded in 2006 by Tarana Burke, a social activist and community organizer to fight sexual abuse among women of color. Starting around 2017, in response to widespread accusations of sexual harassment and sexual assault against women (and men) by prominent actors, movie producers, politicians, newscasters, businessmen, and others, several notable actresses shared their stories online and banded together to spread the hashtag #MeToo to bring attention to the issue. Actress Alissa Milano wrote on Twitter, "If you've been sexually harassed or assaulted write me too as a reply to this tweet" (Pflum 2018).

Movements like #MeToo are having an effect. In a recent national survey of adolescents, 70 percent had heard of the #MeToo movement (Plan International 2018). Over a third (36 percent) of girls and 28 percent of boys said they had a parent who talked to them about how to prevent sexual harassment as a result of #MeToo, and one in four boys and girls reported having a teacher talk to them about #MeToo. #MeToo may be having a positive effect; 55 percent of girls in the survey said the #MeToo movement has made them more likely to tell someone if they had been sexually harassed or assaulted. "Being worried people will not like them," "not believing reporting comments will make a difference," "not be sure if comments are serious enough to report," and "being worried they will not be believed" were listed as reasons why some girls will not report things like boys making sexual comments and jokes. As the researchers point out, these are the same reasons women don't report sexual harassment in the workplace.

Outright displays of **sexism**, prejudice or discrimination on the basis of a person's sex (typically against women), are becoming less frequent in favor of *ambivalent sexism*—a form of sexism marked by "deep ambivalence, rather than a uniform antipathy, toward women" (Glick and Fiske 2018). This type of sexism is comprised of both *benevolent sexism* (positive attitudes toward women in traditional, nurturing roles) and *hostile sexism* (negative attitudes toward women, especially when it comes to moving into traditional male roles; Glick and Fiske 2018). Women remain subjected to **microaggressions** refer to the constant everyday slights, invalidations, insults, and indignities experienced by marginalized groups (e.g., women, people of color, the disabled), often coming from well-intentioned family members, teachers, coworkers, health professionals, and others (Sue 2010). These can include anything from a woman being catcalled as she walks down

the street, to a patient mistaking a female doctor for a nurse, or a coworker referring to a female colleague as "bossy," a term commonly used to describe women in positions of authority but not men. Protests against microaggressions are becoming more common. For example, "manspreading," where men sit with their knees splayed wide apart on public transport and other spaces, is no longer allowed on the New York City subway system (Fitzsimmons 2014).

Some white, working-class women and some women of color believe the women's movement has focused too much on emotional oppression or abuse and on professional women's thwarted career opportunities rather than on "the daily struggle to make ends meet that is faced by working class women" (Aronson 2003, p. 907; Ryle 2015). Many black women have seen the movement as irrelevant to them, because—always having been employed, first under slavery and later because of financial necessity—they were never housewives in large numbers (Baca Zinn, Hondagneu-Sotelo, and Messner 2010; Collins 2000; Lessane 2007). Meanwhile, the media sometimes assert that a younger, "postfeminist" generation does not support a women's movement. The argument is that younger women often have a negative image of feminism, are latently feminist but believe that women's rights' goals have already been achieved, or are just simply too busy with school, work, and family to have time to be active feminists (Anderson 2015).

Gender Today and in the Future

We're all experiencing gender in unprecedented ways. Women couldn't vote anywhere in the world until the late nineteenth century— fewer than 150 years ago, a dot in the long line of human history—and American women became eligible to vote only 100 years ago. Thanks in large part to advances resulting from the women's movement, both women's and men's roles are more flexible than ever before. Support for the breadwinner–homemaker model of families is declining. Data from the General Social Survey indicates that in 2016, only 27 percent of Americans agreed that it's better for men to work and

women to tend to the home compared to about half in 1986. The percentage of Americans who agreed that "a preschool child is likely to suffer if his or her mother works" also declined from about half in 1986 to 28 percent in 2016. However, a gender gap persists, with 32 percent of men and 24 percent of women agreeing with this statement (Allred 2018).

Whereas research comparing wives' with husbands' marital satisfaction used to find husbands generally more satisfied than wives, today's studies find them virtually equally satisfied, although that may a product of couples delaying marriage until they find a truly compatible spouse (Jackson et al. 2014). A decade ago, some argued that we've entered a "postfeminist" era in which gender no longer matters, that we're too busy with work and family to care about it anymore (Read 2011; Showden 2009). Yet, as demonstrated by this chapter, gender inequality remains a persistent feature of American society (Ryle 2015).

On the one hand, examples of females in nontraditional roles abound: women as chief executive officers, astronauts, or military officers and enlisted personnel, one-quarter of them exposed to combat—even before January 2013 when the Pentagon officially allowed women soldiers in combat roles (Pew Research Center 2010b; Rosenthal 2013). We take it for granted that most working-age women, including many mothers, are employed, often in demanding careers.

Bob Peterson/UpperCut Images/Getty Images

Today's postindustrial labor force is more inclusive of women, minimizing gender differentiation and ushering in a new age. For example, more women are entering nontraditional occupations; they currently make up 16 percent of today's military, up from 2 percent in 1973 (Council on Foreign Relations 2018).

Men have moved into traditionally female occupations such as nursing or elementary school teaching, and married men are doing more at home than they used to (Dewan and Gebeloff 2012; Mundy 2012; Ryle 2015). While many fathers are examining how they themselves were fathered and are committing themselves to be more emotionally present, research uncovers heightened levels of caring among male adolescents (Marsiglio 2012; Schalet 2012).

Despite dramatic and unprecedented change over the past fifty years, in many ways we continue to define the **public sphere** (for example, breadwinning and politics) as most important to masculinity, while the **private sphere** (family roles, especially caregiving) remains definitive of femininity. Although most Americans believe making money and cooking family meals should be equally shared by women and men, wives are still far more likely to do the laundry and husbands the yard work (Altintas and Sullivan 2016; Parker et al. 2019). Expected not to jeopardize career success, men often encounter more resistance than women do when they try to exercise "family-friendly" options in the workplace (Miller 2014a). They also may face prejudice when they take jobs traditionally considered women's (Snyder and Green 2008).

In terms of generational change, although they are more socially conscious, Millennials have largely retained traditional gender expectations. In one study, 68 percent of Millennial men expected to be the primary breadwinner, whereas 21 percent of women expected themselves to be. Only 7 percent of Millennial men expected their spouse to be the primary breadwinner compared to 35 percent of women. Millennial women have more realistic expectations that do men: 44 percent saw themselves contributing equally to the household income versus only 25 percent of men (Center for Women and Business at Bentley University, 2017). Gen Z has particularly progressive social views, which bode well for the future of gender equality (Pew 2019; Twenge 2017). In a national survey of adolescents, the vast majority (92 percent) said they believe in gender equality (Plan International 2018). Nevertheless, their attitudes didn't always reflect this supposed belief; 54 percent strongly or somewhat agreed that they are "more comfortable with women having traditional roles in society, such as caring for children and family," and only 51 percent of boys, compared to 64 percent of girls, strongly agreed that they want "equal numbers of men and women to be leaders in work, politics, and life." Only 22 percent of girls said there was gender equality right now compared to 44 percent of boys, and a higher percentage of girls believe sexism is still a problem in society (51 and 19 percent, respectively). Factors associated with traditional gender beliefs among boys and girls included not seeing sexism as a problem, having a Republican parent, not having a parent or teacher who talked about

the #MeToo movement, and more frequent (for boys) and less frequent (for girls) exposure to online pornography. Girls who did not feel judged as a sexual object and who did not feel pressured to "look hot" also had more traditional gender beliefs.

Millennials and Gen Zer workers are diverse in their family structure, race and ethnicity, and immigration status, among other factors, and are demanding inclusive and family-friendly workplaces and safe spaces for women (Bentley University 2017).

Men are beginning to push for more available paternity leave in the U.S. workplace (Lieber 2015; Scheiber 2015). Melinda Gates (wife of Bill Gates) is the founder of *Equality Can't Wait*, an advocacy group for women's rights in the U.S. and around the world. She states, "The good news is that the fight for equality in the U.S. is at a tipping point. Now, more than ever, we have the knowledge, the energy, and the moral insight to crack the patterns of history. Now, more than ever, we know what we need to do to stand up for equality in our homes, in our workplaces, and in our communities" (Equality Can't Wait 2019). Data from the *Gender Gap Report* indicates that, based on current trends in health, education, politics, labor force participation, and family life, American women will not achieve equality with men for another 208 years and out of 149 countries, the U.S. ranks 59th in labor force participation of women and 98th in representation of women in politics (World Economic Forum 2018). However, Melinda Gates says that many steps can made to accelerate the pace of change in all our social institutions, from supporting female candidates to supporting equal pay for the U.S. Women's Soccer Team (Equality Can't Wait 2019).

One way to think about social change regarding gender identities and expectations is to see individuals undoing—or redoing—gender:

> When young wives remain ambitious workers, committed to their own independent economic success, and expect their husbands to be equal partners at home, these women are undoing gender and changing the gender structure. When men use paternity leaves, when they take responsibility for their share of household labor, they, too, are undoing gender, including the male privilege that has defined gender as a stratification system and a structural aspect of our society. (Risman 2011, p. 21)

Gender identities are integral to family life; understanding gender is fundamental to understanding families (e.g., De Henau and Himmelweit 2013; Ledwell and King 2015; Nyman, Reinikainen, and Stocks 2013). Gender expectations and behaviors—both as they have and have not changed—underpin virtually all the topics to be addressed in this course. Gender is important to sexuality, to communication, to parental and other family roles, and to family power, as well as to income and poverty issues. Future chapters explore these topics.

Summary

- The biological, psychological, and social realities of women and men challenge the notion of gender as binary, with clearly demarcated masculine and feminine genders. Today, gender identities are recognized as fluid and self-defined.

- Traditional masculine expectations require that men be confident, self-reliant, and occupationally successful, although it is now acknowledged that men can demonstrate their masculinity in different ways.

- Traditional feminine expectations require that women be physically attractive, emotionally supportive, and that they put others first. It is now being acknowledged that women can engage in typical masculine roles, such as paid employment and sports, and still retain their femininity.

- Although men and women are more alike than different, living in American society is a very different experience for women than it is men, with women's roles less valued.

- Men and women are socialized into their gender identities, expectations, and roles through everyday interaction with others at home, school, and other environments.

- A society's gender structures impact the ways that individuals "do" or "perform" gender. Gender structures include social institutions such as the family, religion, politics and government, education, and the economy.

- Gender must be looked at through an intersectional lens. Gender identities, expectations, and social roles are influenced by race, ethnicity, class, immigration status, and other variables.

- Gender as a sorting variable will continue to evolve as women and men challenge traditional norms.

Questions for Review and Reflection

1. What does it mean to define gender as binary versus fluid? Discuss challenges to that definition, both in the United States and in other cultures.

2. What are some of the characteristics generally associated with boys and men in our society? What traits are associated with girls and women? How do these affect our expectations about the ways that men and women should behave?

3. What are some issues that transgender people and their loved ones face today? Explain how our understanding of transgender people has changed over time.

4. How do gender structures in major social institutions such as politics, religion, education, and the economy affect how people "do" gender? How have these structures influenced your own expectations and choices?

5. **Policy Question.** What gender-related family law and policy changes have occurred in recent years? What policies do you think would be needed to promote greater gender equality and more satisfying lives for men and women?

Key Terms

agentic character traits 59
bifurcated consciousness 61
borderwork 67
cisgender 56
dramaturgy 61
expressive character traits 60
femininities 60
gender 56
gender differentiation 59
gender expectations 57
gender identity 56
gender schema theory 66

gender structure 69
gender variance 66
instrumental character traits 59
interaction–constructionist
    perspective 65
intersectionality 62
intersex 58
masculine dominance 69
masculinities 60
microaggressions 74
pansexuality 57
patriarchy 69

private sphere 76
public sphere 76
self-identification theory 66
sex 56
sexism 74
sex-reassignment surgery/
    gender-affirming surgery 59
social learning theory 65
socialization 65
transfamilies 66
transgender 57
transsexual 59

4

OUR SEXUAL SELVES

Learning Objectives

1 Describe how one's sexual identity develops

2 Explain the different ways in which sexual identity may be expressed.

3 Explain the interpersonal exchange model of sexual satisfaction and the interactionist perspective on human sexuality.

4 Compare sexual values outside and within committed relationships.

5 Discuss current social issues concerning sexuality, such as infidelity, pornography, and sex education.

6 Discuss the role of history and politics in sexuality and sexual expression.

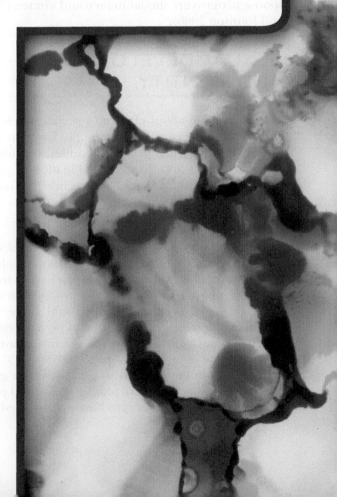

▲ Daly and Newton/Getty Images

From childhood to old age, people are sexual beings. Sexuality has a lot to do with the way we think about ourselves and how we relate to others. It goes without saying that sex plays a vital role in marriages and other intimate relationships. Despite the pleasure it may give, sexuality may be one of the most baffling aspects of our selves, and finding mutually satisfying ways of expressing sexual feelings can be a challenge.

This chapter provides an examination of our "sexual selves." We define sexual identity and discuss the diversity in sexual identities that exists today. We review the historically changing cultural meanings of sexuality. We discuss sex as a pleasure bond that requires open, honest, and supportive communication and look at the role sex plays in intimate relationships over the life course. We will look at some challenges that are associated with sexual expression. What happens when one or both partners has an affair? How have changes in technology and the media altered sexual norms, attitudes, and behavior? Where can people find accurate information about sex?

Before we discuss sexuality in detail, we want to point out our society's tendency to reinforce the differences between women and men and to ignore the common feelings, problems, and joys that make us all human. The truth is that men and women's sexual attitudes and behaviors aren't so different (Peterson and Hyde 2011). Many physiological parts of the male and female genital systems are either alike or analogous, and sexual response patterns are similar in men and women (Masters and Johnson 1966).

SEXUAL DEVELOPMENT AND IDENTITY

Our knowledge about children's sexual development and the emergence of sexual identity has increased in recent years. Just like gender identities, our identities as sexual beings arise from a combination of physical, hormonal, and social factors starting in childhood.

Children's Sexual Development

"Western society does not generally like to think of children as sexual, but sexual development is as much a part of their normal growth and development as is learning language, playing, and getting the proper nutrition in order to grow" (Viglianco-VanPelt and Boyse 2015). Sexual exploration starts at birth and perhaps even earlier. In prenatal ultrasounds, male fetuses have been observed grabbing their penises (Davis 2015). Most children—both boys and girls—play with their genitals fairly regularly by the age of 5 or 6. Almost 100 percent of boys and 25 percent of girls have masturbated to the point of orgasm by age 15. This behavior is generally not associated with sexuality or adult relationships until much later in childhood. This is reassuring to parents, who may be alarmed by their child's behavior (Viglianco-VanPelt and Boyse 2015).

Physically, children, both boys and girls, are maturing about two years earlier than they did a century ago (Brink 2008; Herman-Giddens 2006; Herman-Giddens et al. 2012). According to data from the National Health and Nutrition Examination Study, the average age at **menarche**, a girl's first menstrual cycle, dropped during the twentieth century for all races and ethnicities and is now about age 12 (Papadimitriou 2016). Breast development is also occurring at younger ages (Biro et al. 2018). Reasons for the decline in age at puberty (which has since stabilized) include better nutrition, as well as negative changes such as obesity, decline in physical activity, pollution, and consumption unhealthy foods (Anderson and Must 2005; Carwile 2015; Herman-Giddens 2007; Walvoord 2010).

The effects of earlier sexual maturity are largely negative. It is associated with earlier sexual intercourse, depression and anxiety, behavior problems, smoking, and alcohol use as well as accelerated entrance into cohabitation and marriage (Cavanagh 2011; DeRose et al. 2011; Richards and Oinonen 2011; Shen, Deepthi Varma, Zheng, Boc and Hu 2019; Tondo, Pinna, Serra, De Chiara, and Baldessarini 2017). With puberty beginning earlier in the life cycle and marriage occurring later in the life cycle, there is now a more extended period during which sexual activity and pregnancy may occur among adolescents and unmarried adults.

Sexual Identity

Recall from Chapter 3 that **sexual identity** or *sexual orientation* refers to the gender of whom one is sexually attracted. The first three letters of LGBTQ+ refer to *sexual* identity (lesbian, **gay**, **bisexual**), the *T* to *gender identity* (transgender), and *Q+* to queer/questioning and other identities. **Heterosexual** or *straight* is the traditional way to describe people attracted to opposite-sex partners, whereas **homosexual** or *gay* or *lesbian* is used to describe people attracted to people of their same sex. As with gender identity, sexual identity exists on a spectrum and can change across the life course. For example, **pansexuals** are individuals who have the potential to be sexually attracted to various gender expressions. *Pan* means all, and *pansexual* captures both the fluidity and diversity of one's sexual identity. Pansexual also emphasizes the importance of emotional as opposed to purely physical attraction and the desire to be with a particular person. One student put it best: "It's about hearts, not parts."

Sociologist David Wahl (2020) refers to the plethora of terms used to talk about sexual identity as the "sexual identity alphabet." For instance, **affectional orientation** is a term used in conjunction with sexual identity that encompasses emotional and physical attractions beyond sexual attraction and is preferred by some (LGBTSS, Iowa State University 2015). *Demisexuals* are people who cannot be sexually aroused without a strong emotional bond also being present. *Sapiosexuals* cannot be sexually aroused without a strong intellectual bond. A small number of Americans are **asexual**, meaning that they do not experience sexual attraction to others. This situation differs from *celibacy* or *abstinence*, which as discussed later on, is a *decision* not to have sexual relations, at least for a time, rather than a lack of desire. Asexual individuals may desire intimate relationships with others, just not sexual ones. The Asexual Visibility and Education Network (AVEN) was founded in 2001 as a networking and information resource (www.asexuality.org). This group would like to see asexuality become a recognized sexual identity so that absence of sexual desire is not treated as dysfunctional but as a "normal" alternative (AVEN 2015). As with gender identities, the language surrounding sexual identities reflects the norms and values of the prevailing culture, and our terminology is continually evolving (see Figure 3.2 in Chapter 3).

LGBTQ+ people may have trouble being accepted by others and may be compelled to choose a sexual identity before they are sure. Bisexuals often feel marginalized, that their sexual identity cannot be trusted, or that others think they are not being truthful. As a result, relatively few men and women come out as bisexual (Denizet-Lewis 2014). A common strategy of bisexuals is to let others think they are either straight or gay. Paula, age 22, said, "I would never bring [up my] bisexuality to my father. He might think that there's a chance that I would be straight. I try not to use that label with anyone who is going to desire my heterosexuality" (Scherrer, Kazyak, and Schmitz, 2015, p. 8). In a study of men who have sex with both men and women, most were reluctant to classify themselves ("I guess I'm bisexual," "They say I'm bisexual"), could not or refused to classify themselves, used multiple terms to describe their sexual identity, or, like Paula, used different terms with different people (Baldwin et al. 2014).

Data from the 2015 National Study of Sexual Health and Behavior (NSSHB) shows that the majority of people's sexual attractions and behaviors match their sexual identity, at around 95 percent (Fu et al. 2019). For example, a heterosexual man would be attracted to only women and would engage in sexual activity with only women. However, some people who identify themselves as *straight* sometimes engage in sexual behaviors with people of their own gender (Chandra,

Copen, and Mosher 2013; Kuperberg and Walker 2018). For example, is not that uncommon to see two women kissing and "making out" at bars and parties. What is it that we are witnessing? It may be incorrect to label them lesbians or even bisexual. Using the college lingo, they may be just "bi-curious" or "LUG" (lesbian until graduation). From a *dramaturgical perspective* (see Chapter 2) their behavior might also be viewed as a "performance." Alcohol is often involved, which lowers inhibitions, and often the behavior takes place in public in response to the whoops and cheers of men who reward them with free drinks.

Heterosexual men also sometimes engage in sexual behavior with other men. In her book *Not Gay: Sex Between Straight White Men*, gender and sexuality professor Jane Ward (2015) explores *dudesex* or "straight homosexual sex." She found MSM (men having sex with men) was prevalent in the expected places (e.g., the military, fraternities) but also in unexpected places, including biker gangs and conservative suburban neighborhoods. Rather than threatening masculinity as one might expect, she found that MSM can strengthen it, especially if the men involved are from similar socioeconomic backgrounds. This was the case in Silva's (2017, 2018) study of *bud-sex* among rural, white, straight men, who looked at MSM as "helpin' a buddy out" or "relieving urges." In fact, many of the men found feminine men a turn-off. As Mike said, "If I wanted someone that acts girlish, I got a wife at home" (Singal 2016).

This research helps dispels the pervasive stereotype that gay men are feminine or "flamboyant" and that lesbians are masculine or "butch." Most gay men identify as male and behave in traditionally masculine ways. Most lesbians identify as female and conform to traditional gender norms for women for dress, appearance, and behavior. The reason some individuals develop a gay sexual identity has not been definitively established, but new research indicates there is no one "gay gene" but a combination of many genes that helps determine (along with other factors) sexual orientation (Ganna 2019). The American Psychological Association (APA) takes the position that a variety of factors impact a person's sexuality and that

> sexual orientation is most likely the result of a complex interaction of environmental, cognitive and biological factors . . . is shaped at an early age . . . [and evidence suggests] biology, including genetic or inborn hormonal factors, play a significant role in a person's sexuality.

Sexual identity develops in childhood and progresses through adulthood. In a study of 2,560 California high school students, 11 percent identified as gay or lesbian, 12 percent identified as bisexual, and 5 percent were questioning their sexuality (Russell, Clarke, and Clary 2009). The American Psychological

Lesbian and gay male couples and families have become increasingly visible over the past decade. Meanwhile, discrimination and controversy persist.

Association does not support *sexual orientation change efforts* (SOCE) that try to convert gay men and lesbians to heterosexuals through psychotherapy, support groups, or religious programs, noting negative effects such as loss of sexual feeling, depression, anxiety, and suicidality (American Psychological Association Task Force on Appropriate Therapeutic Responses to Sexual Orientation 2009). So far, 18 states, Puerto Rico, and the District of Columbia (and some cities) forbid licensed professional counselors from engaging in SOCE with children and there has been an effort to ban the practice at the federal level (Sprigg 2015). In 2019, Amazon stopped selling books promoting conversion therapy (Ennis 2019).

Perhaps due to the increased attention in recent years, the American public typically overestimates the percentage of men and women who are gay. In 2019, 54 percent of adults thought that at least 20 percent of the population was gay or lesbian (McCarthy 2019). Estimating the percentage of the population that is gay is difficult. Even deciding who is to be categorized as gay, lesbian, or bisexual for research purposes is not easy: How much experience? How exclusively gay? Poll data from Gallup indicates that the percentage of the population identifying "lesbian, gay, bisexual or transgender" has increased from 3.5 percent in 2012 to 4.5 percent in 2017 (Newport 2018).

Despite increased acceptance and support of the LGBTQ+ community, those with non-heterosexual sexual identities have poorer social, emotional, and physical outcomes than heterosexuals. LGBTQ+ youth and adults have higher rates of anxiety and depression, are in poorer physical health, are more likely to be poor, are less likely to have health insurance, and are more likely to be denied a mortgage (Emlet 2010; Gorman et al. 2015; Russell and Fish 2016; Sun and Gao 2019). Data from the 2017 Youth Risk Behavior Surveillance System (YRBSS) show that gay, lesbian, and bisexual youth have significantly more health risks than straight youth, including alcohol and drug use, smoking cigarettes, dating and sexual violence, suicide attempts, early sexual experiences, and being bullied online (Kann et al. 2017).

How sexual identity unfolds depends on various social factors, including the support of one's family, friends, and community and the accessibility of role models who share one's sexual identity (Rowe 2014). Sexual behavior does not always conform to sexual identity, especially in childhood and adolescence (Kann et al. 2017). There are also a host of social factors that shape sexual norms, attitudes, and behaviors, and many theoretical perspectives on sexuality.

THEORETICAL PERSPECTIVES ON HUMAN SEXUALITY

There are various theoretical perspectives concerning marriage and families, as we saw in Chapter 2. Many of these have been applied to human sexuality. For example, we can look at sexuality using a *structure–functional* perspective. In this case, we see norms as mechanisms designed to regulate sexuality in a way that serves the societal function of reproduction, such as encouraging marriage and children through religious institutions. Those looking at sexuality from a *biosocial* perspective consider that humans—like the species from which they evolved—behave in ways that efficiently transmit their genes to the next generation. This perspective would say that men are naturally promiscuous, seeking multiple partners so they can distribute their genes widely, whereas women, who can generally have only one offspring a year, are inclined to be selective and monogamous (Dawkins 2006). Biology does play a role. An interesting study examining women's hip size found that women with wider

hips had more "one night stands" and more sexual partners than those with narrower hips. The authors theorize that because women's hip size has a direct impact on their risk of injury or death during childbirth, childbearing favors "hippier" women. Moreover, the authors argued that when women have control over their sexual experiences, these experiences reflect this risk, thus connecting larger hip width with riskier sexual behavior (i.e. one-night stands) (Simpson, Brewer, and Hendrie 2014).

Both these perspectives have their limits. Neither tells us anything about the emotions and pleasures of sexual relationships, and the biosocial perspective argues a genetic determinism that is contradicted by historical and cross-cultural variation in sexual behavior and relationships. Two more useful ways of looking at sexual relationships with a sociological perspective are exchange theory and interaction theory, both introduced in Chapter 2.

The Exchange Perspective: Rewards, Costs, and Equality in Sexual Relationships

From an *exchange theory* perspective, women's sexuality and associated fertility are resources that can be exchanged for economic support, protection, and status in society. An exchange theory perspective that brings sex closer to our human experience is the **interpersonal exchange model of sexual satisfaction** (Lawrence and Byers 1995).

Figure 4.1 shows us that in the interpersonal exchange model of sexual satisfaction, satisfaction depends on the *costs* and *rewards* of a sexual relationship, as well as the participant's *comparison level*—what the person expects out of the relationship. Also important

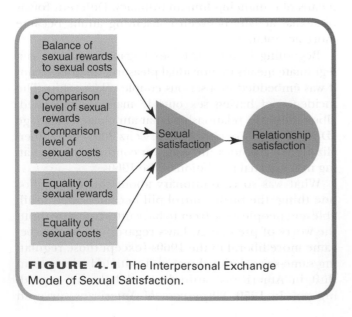

FIGURE 4.1 The Interpersonal Exchange Model of Sexual Satisfaction.

is the *comparison level for alternatives*—what other options are available, and how good are they compared to the current relationship? Increasingly, expectations are likely to include some degree of *equality of rewards* (such as orgasm). Research testing this model found that these elements did indeed predict sexual satisfaction in married, cohabiting, dating, and lesbian couples (Byers and Cohen 2017; Kisler and Christopher 2008; Pascoal et al. 2018), although there was variation by social class (Neff and Harter 2003), race (Stanik and Bryant 2012), and gender (Mark and Murray 2012). Sexual satisfaction is important to relationship satisfaction, and the two are mutually reinforcing (Mark and Jozkowski 2013; Muise et al. 2019).

The Interactionist Perspective: Negotiating Cultural Messages

The *interactionist* perspective emphasizes the interpersonal negotiation of relationships in the context of sexual scripts: "*That* we are sexual is determined by a biological imperative toward reproduction, but *how* we are sexual—where, when, how often, with whom, and why—has to do with cultural learning, with meaning transmitted in a cultural setting" (Fracher and Kimmel 1992). Cultural messages give us legitimate reasons for having sex, as well as expectations regarding who should take the sexual initiative, how long a sexual encounter should last, what positions are acceptable, among other things. Cultural messages about sex are transmitted through parents, schools, peers, movies, television, and social media among other ways.

An **interactionist perspective on human sexuality** holds that women and men are influenced by the *sexual scripts* that they learn from their culture (Simon and Gagnon 1986). **Sexual scripts** are "culturally available messages that define what 'counts' as sex, how to recognize sexual situations, and what to do in a sexual encounter" (Frith and Kitzinger 2001, p. 210). American culture retains largely traditional sexual scripts. For example, college women who watch more television and who find shows more realistic are less assertive in sexual encounters (Seabrook, Ward, Cortina, Giaccardi, and Lippman 2017), and more frequent users of pornography have greater interest in a greater number of sexual behaviors, including violent and/or degrading sexual acts (Bridges Sun, Ezzell, and Johnson 2016). Sex partners assign meaning to their sexual activity—that is, sex is symbolic of something, which might be affection, communication, or recreation. Whether each gives their sexual relationship the same meaning has a lot to do with satisfaction and outcomes. For example, if one is only playing while the other is expressing deep affection, trouble is likely. A relationship goal for couples becoming committed is to establish a joint meaning for their sexual relationship.

CHANGING CULTURAL SCRIPTS

Sexual scripts vary by culture and change over time. For example, in colonial times the purpose of sex in America was reproduction. However, new definitions of sexuality emerged in the nineteenth century and flourished into the present day. It is important to note that sexual scripts vary for different racial, ethnic, and religious groups, and for (and within) the LGBTQ+ community, which is important as the United States becomes more culturally diverse.

Early America: Patriarchal Sex

In a patriarchal society, descent, succession, and inheritance are traced through the male genetic line, and the socioeconomic system is male-dominated. **Patriarchal sexuality** is characterized by beliefs, values, and behaviors that protect the male line of descent. Men control women's sexuality; exclusive sexual possession of a woman by a man in monogamous marriage ensures that her children will be "legitimately" his. Our current view of sexuality is based on this and on the sexual ideals of the Victorian Age and colonial America, which maintained a strict, conservative repression of sexuality (Wahl 2020). For example, nineteenth-century medical science deemed sex to be a dangerous activity, and sexually transmitted infections (STIs) and poor mental health were falsely linked to masturbation. Women caught masturbating or who were considered promiscuous could be diagnosed with a condition called *hysteria* and were sometimes sent off to mental institutions; some were given a clitoridectomy (surgical removal of the clitoris; Garton 2004; Miller and Adams 1996). Even today, the sexual behaviors of men are given far more leeway than those of women (see the *sexual double standard*, later in the chapter). Holding patriarchal attitudes about sex affects sexual behavior. In a study of 434 college freshman, men who held more gender-stereotyped attitudes (e.g., "men should be allowed more sexual freedom than women") had more sexual partners and were less likely to use condoms (Lefkowitz et al. 2014).

Although the patriarchal sexual script has been significantly challenged, it persists to some extent and corresponds to traditional gender expectations. If masculinity is a quality that must be achieved or proven, one arena for doing so is sexual accomplishment or conquest. Data based on 2,000 adults from the 2015 Sexual Exploration in America Study (Herbernick et al. 2017) bears this out. The men and women in the study had roughly similar rates of having had vaginal, oral, and anal sex, and some other activities such as having sex in a public place and role-playing. However, men's rates

of having ever masturbated with someone else were higher (58 percent versus 53 percent), as was using a phone app related to sex (12 percent versus 6 percent), watching pornography (82 percent versus 60 percent), going to a strip club (59 percent versus 30 percent), and having group sex or a threesome (18 percent versus 10 percent). In study of young adults from Mexico and Spain, males were more likely than females to engage in masturbation, oral sex, and anal sex and reported more partners, but men and women were equally likely to report being sexually experienced (Gil-Llario, Giménez, Ballester-Arnal, Cárdenas-López, and Durán-Baca 2018).

There has been substantial movement toward gender equity in sexuality. For example, men and women have similar levels of desire for sexual behaviors not typically associated with male dominance. Men and women are equally likely to find "kissing more often during sex" very or somewhat appealing (86 percent and 87 percent), as well as "saying sweet, romantic things during sex" (80 percent for men and 83 percent for women) and "making the room feel more romantic" (74 percent for men and 80 percent for women).

The Twentieth Century: The Emergence of Expressive Sexuality

In the twentieth century, *expressive sexuality* emerged as the result of societal changes that include the decreasing economic dependence of women, the availability of new methods of birth control, and even the invention of the automobile, which shifted courtship to venues away from the watchful eyes of parents such as drive-in movie theaters (Garcia et al. 2015). **Expressive sexuality** sees sexuality as basic to the humanness of both women and men and is less one-sided. Sex is not primarily or only for reproduction but is an important means of enhancing human intimacy. Different forms of sexual activity between consenting adults became more acceptable.

Beginning in the 1920s, sex began to be seen as a legitimate means to individual pleasure, whether or not it was embedded in a serious couple relationship. The incidence of having sex outside marriage increased, albeit mostly in relationships that anticipated marriage (D'Emilio and Freedman 1998; Zeitz 2003). The liberalization of attitudes and behavior continued, culminating in the sexual revolution of the 1960s.

What was so revolutionary about the sixties? For one thing, the birth-control pill became widely available and people were freer to have intercourse without the worry of pregnancy. Laws regarding sexuality became more liberal in the 1960s (except those regulating same-sex relationships) and initiated a long-term shift in Americans' attitudes and behavior regarding sex. In 1959, 80 percent of Americans surveyed

said they disapproved of sex outside marriage (Smith 1999). By 2019, only 28 percent of Americans felt that it was "morally wrong" (Smith 2019). Not only did attitudes become more liberal, but behaviors changed. Compared to Americans in the 1970s and 1980s, adults in the 2000s reported having more sexual partners and more "casual sex" (Twenge, Sherman, and Wells 2015).

Probably the biggest change concerned the acceptability of premarital sex. In 1973, 43 percent said that premarital sex was *not* morally wrong compared to 71 percent in 2018 (Saad 2019). And in 2019, 42 percent of Americans felt that sex between teenagers was *not* morally wrong (Brenan 2019). Indeed, today most first sexual encounters occur in the teen years and before marriage. A surprising new finding is that young people are having sex later than in the past. The average age of first sexual intercourse for both boys and girls is 18, up from 17 in 2014 (Alan Guttmacher Institute 2014c, 2019c; Bridges and Philbin 2019). Among high school students ages 15 to 19 in 2017, 40 percent reported having had sexual intercourse compared to 47 percent in 2013 (Alan Guttmacher Institute 2019c). Male, black, Hispanic, and older students are more likely to have had sex (Kann et al. 2018). Although the reasons are not known, some possibilities for the so-called *sexual recession* include the use of social media for conducting relationships as opposed to face-to-face interaction, the greater availability of online porn, increased supervision of children by "helicopter parents," and increased pressure and anxiety among Gen Zers to accomplish more educationally and economically. A result of less teen sex, combined with improved sex education and newer long-acting contraceptives, is that rates of teen pregnancy and teen births are at record lows, though they are higher than in other industrialized countries, and socioeconomic disparities in teen pregnancies and births persist (Centers for Disease Control [CDC] 2019).

Choosing to not engage in sexual relations is referred to as **abstinence**. Reasons teens give for abstaining from sex include the absence of love, conservative or religious values, fear of pregnancy and STIs, and parents' disapproval (Ahrold et al. 2011; Diefendorf 2015; Ellison, Wolfinger, and Ramos-Wada 2012; Martinez, Copen, and Abma 2011 Kaye, Sullentrop, and Sloup 2009; Martinez, Copen, and Abma 2011; Rasberry and Goodson 2009). Only 36 percent of students at a conservative Christian college reported having had casual sex versus 64 percent at a state college (Helm, Gondra, and McBride 2015).

Factors associated with a *delay* in the onset of sexual activity are growing up in an intact family, higher family income, religiosity, communicating with parents about sex, having a good relationship with parents, and feeling a greater connection with one's school (Bingenheimer, Asante, and Ahiadeke 2015; Child Trends 2014f). In assessing the decline of teen sexual activity, it is important to note that a lot depends on one's definition of sex. Yes, sexual intercourse is less frequent among adolescents than in the past, but teens have high rates of oral sex: 51 percent of males ages 15 through 19 and 47 percent of females in the same age group have engaged in oral sex (Child Trends 2015a). Many adolescents, however, do not consider oral sex to be "sex." In one study, only 20 percent of college students at one university considered oral sex to be actual sex (Hans, Gillen, and Akande 2010). It does not appear that adolescents are substituting oral sex for vaginal sex though. Only 14 percent of teenagers reported having had oral sex but not intercourse (Child Trends 2015a). Although a "virgin" is still widely considered someone who has not had vaginal intercourse, the definition of virginity is evolving (Child Trends 2015a). In a study of LGBTQ+ women, when asked what it means to have "had sex," 92 percent of respondents who had sex with a woman said receiving oral sex was "sex" versus only 57 of women who had sex with a man (Schick et al. 2016).

Data from the 2015–2017 National Survey of Family Growth provides information on teen sexuality (Alan Guttmacher Institute 2019c). For the majority of female teens who had experienced sex, their first experience was with a romantic partner (75 percent); this was the case for only 48 percent of male teens. Although most sexual experiences are wanted, teens sometimes report mixed feelings or that sex was unwanted, especially among young women; 71 percent of men described their first sexual experience as wanted or that they had mixed feelings (25 percent). A much higher percentage of women than men had mixed feelings (51 percent) as opposed to sex that was wanted (45 percent). Thirteen percent of females and 5 percent of males reported that they had been forced to have vaginal sex (see *Issues for Thought: Sexual Harassment and Sexual Assault in the Age of #MeToo*).

Among young women, alcohol is involved in about half of all first sexual encounters (Orenstein 2016). Early and current sexual activity among teens is associated with a greater risk of engaging in delinquent and externalizing behaviors or "acting out," substance abuse, depression, having multiple partners, having nonconsensual and unwanted sex, having lower educational attainment, and using less contraception (Child Trends 2014e; Kastbom et al. 2015). Sexual activity in adolescence can have consequences in relationships in adulthood. Early sexual activity has been linked to lower marital quality and an increased risk of divorce (Paik 2011; Sassler, Addo, and Lichter 2012).

Sexual Harassment and Sexual Assault in the Age of #MeToo

As discussed in Chapter 3, the worldwide #MeToo movement is having a massive impact on gender relations and our understandings of what is inappropriate and appropriate behavior among men and women in private and public settings. For example, the Supreme Court nomination of Brett Kavanaugh was slowed (but not stopped—he was confirmed in 2018) by allegations that he perpetrated a sexual assault when he was in college, a charge he vehemently denies. Sexual harassment and sexual assault affect women and men from all socioeconomic backgrounds and occurs in the street, across educational institutions, and in every sector of the economy (Raj, Johns, and Jose 2019). In a 2018 national survey of 2,000 individuals, 81 percent of women and 43 percent of men said they had experienced some form of sexual harassment or assault in their lifetime. Verbal sexual harassment was the most common form (77 percent of women and 34 percent of men), followed by unwelcome touching or groping (51 percent of women and 17 percent of men), online sexual harassment (41 percent of women and 22 percent of men), being followed (34 percent of women and 12 percent of men), and genital flashing (30 percent of women and 12 percent of men). The study found that 27 percent of women and 7 percent of men survived a sexual assault (Step Street Harassment 2018).

Students, teachers, employees, and employers are increasingly receiving training and information about what is considered inappropriate behavior (Plan International 2018). Although definitions vary by state, college and university policy, and place of employment, **sexual harassment** can be defined as unwelcome sexual advances, requests for sexual favors, and other verbal or physical harassment of a sexual nature. Sexual harassment can include offensive remarks about a person's sex. For example, it is illegal to harass a woman by making offensive comments about women in general. Victim and the harassers can be of any gender. Laws typically don't prohibit simple teasing, offhand comments, or isolated incidents that are not very serious; however, "harassment is illegal when it is so frequent or severe that it creates a hostile or offensive work environment or when it results in an adverse employment decision (such as the victim being fired or demoted)" (U.S. Equal Employment Opportunity Commission 2019). The harasser can be the victim's supervisor, a supervisor in another area, a co-worker, or someone who is not an employee of the employer, such as a client or customer (U.S. Equal Employment Opportunity Commission 2019).

Sexual harassment and sexual assault are associated with numerous negative outcomes for victims such as depression, anxiety, anger, alcohol problems, stress, chronic diseases, and post-traumatic stress disorder or PTSD (O'Neil, Sojo, Fileborn, Scovelle, and Milner 2018; Peter-Hagene, and Wolff, 2018; Rospenda, and Colaneri 2017). Victims of sexual harassment and assault may have trouble reaching their educational and career goals (Potter, Howard, Murphy, and Moynihan 2018). They have fewer opportunities for job training and advancement and have higher rates of absenteeism or dropping out of school. The most common reaction is to change jobs or schools (Blackstone, McLaughlin, and Uggen 2018; Shaw, Hegewisch, and Hess 2018).

A 2019 survey from the Association of American Universities of 181,752 students at 33 schools, including most of the Ivy League, revealed that 26 percent of female and 7 percent of male undergraduates were survivors of sexual assault or misconduct, an increase of 3 percent and 1 percent, respectively, from 2015 (Association of American Universities 2019). Although definitions vary by state and across institutions, **sexual assault** covers a wide range of unwanted sexual behaviors "that are attempted or completed against a victim's will or when a victim cannot consent because of age, disability, or the influence of alcohol or drugs. Sexual assault may involve actual or threatened physical force, use of weapons, coercion, intimidation, or pressure" and may include penetration (or *rape*), touching, or other behaviors such as undesired exposure to pornography and voyeurism (U.S. Department of Justice 2013).

Data from the National Intimate Partner and Sexual Violence Survey indicates that one in five women and one in fifty men have been raped in their lifetime (Walters, Chen, and Breidig 2013). Data the 2016 National Survey of Porn Use, Relationships, and Sexual Socialization (Herbenick et al. 2019) suggests that "scary sex" is not uncommon. One woman in the study described her experience:

> I gave a guy a ride home and he tried to kiss me good night, he was just a friend, so I tried to stop him, he then got very forceful with me, I truly thought he was going to rape me . . . I used my feet and was able to push him off me, someone came outside right then, so he got out of my car. I found out he did the same thing with another girl I worked with.

Adolescents and young adults between ages 12 and 24 are disproportionately the victims of sexual assault—they are two to three times more likely to be sexually assaulted than adults age 25 and older (Child Trends 2013d). Although state laws vary, sex with someone who cannot legally consent (due to age or disability) is also considered sexual assault.

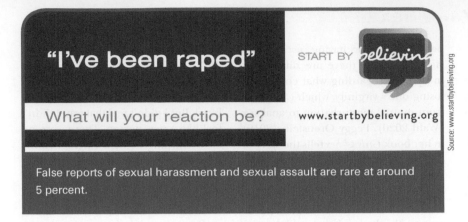

"I've been raped"

START BY *believing*

What will your reaction be?

www.startbybelieving.org

False reports of sexual harassment and sexual assault are rare at around 5 percent.

Ten percent of girls and 6 percent of boys under the age of 16 had their first sexual encounter with someone three or more years older, acts considered to be statutory rape in most places (Child Trends 2013c).

Rape and sexual assault also disproportionately affect women, although men can be victims of sexual assault as well (the perpetrators of sexual assault are almost exclusively men). Among high school students, females are more than twice as likely to report being raped (at 13 percent) as males (5 percent). However, both male and female victims experience similar consequences of rape and sexual assault—from physical injuries and symptoms (fatigue, chronic headaches) to emotional problems, including depression and suicide (Child Trends 2013d).

Contrary to the impression that we are likely to get from news media, most rape victims, especially younger ones, know their rapists (Child Trends 2013d). **Sexual coercion** is common among these perpetrators. Such coercion uses verbal or emotional pressure (tricking, lying, or using guilt), one's position of power (being a boss, teacher, or coach), or other means to manipulate the victim into sexual activity. Sexual coercion is often hard to detect and can include comments such as "We've had sex before, so you can't say no now" and "If you love me, you would have sex with me" (McCoy and

Oelschlager 2013). Alcohol or drugs is often involved in this type of sexual victimization (Foran and O'Leary 2008; Peralta and Cruz 2006). Findings from various research studies over the past decade show that sexually coercive men tend to dismiss women's rejection messages regarding unwanted sex and differ from noncoercive men in their approach to relationships and sexuality: They date more frequently; have higher numbers of sexual partners, especially uncommitted dating relationships; prefer casual encounters; and may "take a predatory approach to their sexual interactions with women."

Many female victims blame themselves, at least partially—a situation that can result in still greater psychological distress (Breitenbecher 2006). One reason victims blame themselves has to do with **rape myths**: beliefs about rape that function to blame the victim and exonerate the rapist (Cowan 2000). Rape myths include the ideas that rapists are violent strangers lurking in the shadows looking for victims; the rape was somehow provoked by the victim (for example, she "led him on" or wore provocative clothes), and that men cannot control their sexual urges or are mentally ill. These beliefs encourage potential victims to feel safe with someone they know, no matter what (Cowan 2000; Littleton et al. 2009). The problem is that women who believe in such myths are less likely to recognize

and leave potentially risky situations and may even interpret a sexual assault as just a "bad hookup" (Franklin et al. 2012; Littleton et al. 2009).

People increasingly understand that the perpetrator rather than the victim is responsible for a sexual assault. This could be the result of rape-prevention information and workshops that reach into male dorms as well as campaigns geared toward men (Cardiff 2013). There is increasing recognition that society bears some responsibility as well. Prevention programs focusing on **bystander education**, such as the Green Dot program, have become more popular and have been shown to be effective in reducing sexual harassment and violence on college campuses (Cameron 2018; Coker et al. 2015; Kettrey, Hensman, and Marx 2019). Bystander education teaches people safe ways to intervene if they think a rape is about to occur. For example, if at a party you might distract the potential perpetrator by spilling a drink on them or by yelling, "The police are here!" Education programs are also emphasizing that partners get clear, unambiguous, verbal consent from one another before having sex, effectively switching the conversation from "no means no" to "yes means yes." This is referred to as **affirmative consent** and is being adopted by colleges across the nation (Perez-Pena and Lovett 2014). It is important for students to familiarize themselves with their college and university policies surrounding sexual harassment, sexual assault, and sexual misconduct. A good resource for help and information on these topics is RAINN, the Rape, Abuse, and Incest Network at www.rainn.org.

Critical Thinking

Do you know your school or employer's policies regarding sexual harassment? What should you do if you or a friend is raped or assaulted? How is affirmative consent different from "no means no?"

As We Make Choices

Rethinking Virginity

By David W. Wahl, Ph.D.

What is virginity? What does it mean to be a virgin? Depending on time, place, culture, and attitudes, sexual behaviors have different meanings for different people. Laura Carpenter (2005) studied the different meanings individuals apply to losing their virginity. First, virginity was seen as a gift—one that could be presented to someone special or, in the case of a new relationship, could be regifted to another as a sort of "born again virginity." Secondly, loss of virginity was viewed as a rite of passage. In the Carpenter study, women were more likely to view loss of virginity as a passage from "girl" to "woman" than men seeing it as a passage of "boy" to "man." Third, virginity was looked upon as an act of worship, in which maintaining one's virginity until marriage is a traditional Christian value that honors God. Finally, loss of virginity carries with it, for some individuals, labels of stigma. A young woman may be stigmatized for losing her virginity prematurely and find herself "slut-shamed." In contrast, young man may

be ridiculed for having *not* lost his virginity. Today there are differences in opinion regarding what constitutes losing one's virginity, which can range from mutual masturbation to anal sex (Wahl 2020). Peggy Orenstein (2016) in her book *Girls & Sex* tells the story of one young woman having to resort to asking Google whether or not she had lost her virginity. At the conclusion of her interviews, Orenstein (2016, p. 78) argues, "In the modern world, 'virginity' as a symbol of sexual initiation is an outdated, meaningless concept" in that most of her research subjects had been sexually active for several years, for example, by engaging in oral sex.

There are different reasons that people decide to maintain their virginity, it may be from a fear of pregnancy, shaming, or STIs. Preservation of virginity may also be a result of one's wish to adhere to religious and family values or to preserve the sanctity of marriage. The choice to lose one's virginity may be an expression of love for one's partner, peer pressure, or the desire to "get it over with" (Orenstein 2016). As one woman in her study said, "I had learned

in school and at church that when you find the 'right person' and you're really in love and you have sex, you will be transformed. Like, this veil would be lifted. But I didn't feel that way. I didn't feel like a new person. There were no birds chirping or bells ringing. And I thought, 'Oh my gosh, maybe it wasn't the right time after all, or maybe we didn't do it right.' I felt like I had been sold a bill of goods" (p. 82).

Currently, adolescents are maintaining their virginity longer than in past generations, with a slight majority of high school students not having sexual intercourse until after graduation (Twenge and Park 2017). Sales of condoms are even down (Beras 2019). Regardless of the cause, the loss of virginity appears to be losing significance as a "rite of passage" and marker of adult status.

Critical Thinking

What does it mean to be a virgin? How important is virginity to you? Why do you think adolescents are choosing to remain virgins for longer period of time than in the past?

The 1980s and 1990s: Challenges to Heterosexism

If the sexual revolution of the 1960s focused on freer attitudes and behaviors among heterosexuals, recent decades have expanded that liberalism to include gay men and lesbians. Until several decades ago, most people thought about sexuality as almost exclusively between men and women. In other words, our thinking was characterized by **heterosexism**—the taken-for-granted system of beliefs, values, and customs that places superior value on heterosexual behavior and denies or stigmatizes non-heterosexual relations. However, fifty years ago the Stonewall riots—sparked by a 1969 police raid on a New York gay bar—galvanized the gay community, and gay males and lesbians not only became increasingly visible but also challenged heterosexism. A number of high-profile people in

government and the media have publicly come out as gay, as have countless actors, actresses, and celebrities. Pete Buttigieg, mayor of South Bend Indiana and a 2020 Democratic contender for president, is married to a man and has openly discussed their desire for children.

The public's attitudes toward gays, lesbians, and sexually fluid people have become more favorable. In the early 1970s, about 70 percent of Americans thought "homosexual relations" were "always wrong" compared to only 35 percent in 2019. As shown in Figure 4.2, 2009 was the year Americans attitudes about the acceptability of gay and lesbian relationships switched from mostly unfavorable to favorable (McCarthy 2019). The vast majority of Americans today (93 percent) agree that gay people should have equal rights in terms of job opportunities, 83 percent say that gay or lesbian

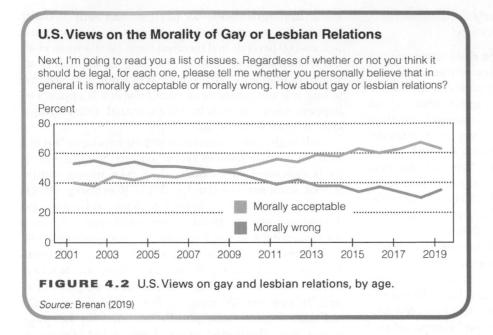

U.S. Views on the Morality of Gay or Lesbian Relations

Next, I'm going to read you a list of issues. Regardless of whether or not you think it should be legal, for each one, please tell me whether you personally believe that in general it is morally acceptable or morally wrong. How about gay or lesbian relations?

FIGURE 4.2 U.S. Views on gay and lesbian relations, by age.

Source: Brenan (2019)

twenty-one states and the District of Columbia have laws protecting the rights of gays and lesbians in the workplace (Bernard 2013; Repa 2014).

People who identify as LGBTQ+ must still deal with prejudice on a daily basis. Forms of prejudice include **homophobia**—viewing of homosexuals with fear, dread, aversion, or hatred—and **microaggressions**—which, recall from Chapter 3, are everyday slights, invalidations, insults, and indignities experienced by marginalized groups (e.g., women, people of color, the disabled), often coming from well-intentioned family members, teachers, co-workers, health professionals, and others (Huber and Solorzano 2013; Sue 2010). Some examples of microaggressions are singling out gays and lesbians to speak on behalf of the whole group, and making statements like, "I've always wanted a gay best friend," or "That's so gay."

relations between consenting adults should be legal, and 75 percent say that gays and lesbians should be allowed to adopt children (McCarthy 2019).

Younger generations—namely, Millennials (born between 1982 and 1996) and Gen Z (born after 1996)—are even more accepting of gay and lesbian relationships. Nearly half of Gen Z and Millennials say that gay and lesbian couples being allowed to marry is a "good thing" for our society compared to a third of Gen Xers (born between 1965 and 1980; Parker et al. 2019). In a 2019 Gallup poll, 63 percent of Americans favored same-sex marriage, with even higher levels of acceptance among younger people and women (McCarthy 2019). Acceptance is growing among young white evangelicals, among whom 47 percent are in favor of gay marriage (Bailey 2017). Since the Supreme Court decision *Obergefell v. Hodges* legalizing gay marriage, nearly two-thirds of same-sex cohabiting couples are married, up from 38 percent before the 2015 ruling (Jones 2017). Americans still view gay couples as less loving than heterosexual couples, an attitude associated with less willingness to grant social recognition to same-sex couples (Doan, Miller, and Loehr 2015).

Americans are divided over whether gays and lesbians choose their sexual orientation, a split that shapes attitudes. In 2019, 49 percent of Americans surveyed thought that being gay or lesbian is something a person is born with, as opposed to it being the result of a person's upbringing or environment (McCarthy 2019). People who see being gay as a choice are less sympathetic to lesbians or gay men regarding jobs and other rights (Loftus 2001). Although employers are "becoming friendlier" to LGBTQ+ employees, and the vast majority of Forbes 500 companies include sexual orientation in their nondiscrimination policies, only

LGBTQ+ people may not be accepted by friends and family. Sara Cunningham became an "unofficial mom" to the LGBTQ+ community by stepping in for parents who refused to attend their children's weddings. She also launched Free Mom Hugs, a national activist group of affirming parents and allies who love the LGBTQ+ community unconditionally (Pelletiere 2019). *Fictive kin* are people perceived as "family" despite a lack of blood relation or legal ties; they are common among racial, sexual, and gender minorities (Eller 2015). **Queer kinship** can help LGBTQ+ people adapt to the loss of family members who are not accepting of their sexual identity. Even when there is no family rejection, queer kinship bonds help provide emotional and social support to marginalized populations (Catalpa and Routon 2018).

Same-sex marriage is legal in North America, most of Western Europe, much of central and South America, as well as South Africa, Australia, and New Zealand (Masci and DeSilver 2019). In 2018, consensual gay sex was legalized in India, the second most populous country in the world with a population of 1.3 billion, overturning 2013 judgement that upheld an 1861 law that categorized gay sex as "unnatural offense," punishable by a ten-year jail term (BBC News 2018). Although there are no laws against same-sex relations in China, the world's largest population at 1.4 billion, a 2019 ruling marriage upheld the restriction on same-sex marriage (Reuters 2019). Overall, societies are becoming more accepting of LGBTQ+ people.

The Twenty-First Century: Risk, Caution—and Intimacy

Although pleasure-seeking characterized sixties sexuality, caution in the face of risk characterizes contemporary times. Starting in the 1980s, a major change was the presence of HIV/AIDS. "Most emerging adults . . . say that fear of AIDS has become the framework for their sexual consciousness, deeply affecting their attitudes toward sex and the way they approach sex with potential partners," (Arnett 2004, p. 91). However, avoiding pregnancy and disease remain the main reason teens say they had not yet had sex (Schalet 2012). Meanwhile, many adolescents and young adults have multiple partners. One in four (23 percent) adult men reported "fifteen or more sexual partners in lifetime" compared to one in ten (9 percent) adult women (Chandra et al. 2011). This is the case among teens as well. According to the 2016–2017 Youth Risk Behavior Surveillance System, 12 percent of male high schoolers reported having had sex with "four or more people" compared to 8 percent of female students. Male students of color were more likely to have more than three partners as were gay, lesbian, and bisexual students (Kann et al. 2018). Nevertheless, most sex occurs alongside other forms of intimacy. Kissing and cuddling are the most common sexual behaviors: 87 percent of respondents in the 2014 NSSHB reported kissing and 70 percent mentioned cuddling at their last sexual encounter (Herbenick et al. 2019). See *Facts about Families: How Do We Know What We Know? The Kinsey Institute and Conducting Sexuality Research* for additional information on sexuality research.

In the digital age, *sexting* is a growing phenomenon and is common among men and women of all ages, genders, sexualities, classes, and belief systems. **Sexting** is sending or posting sexually suggestive text messages and images, including nude or seminude photographs, via phones, computers, and other digital devices (Federal Bureau of Investigation ([FBI] 2012). Statistics on the prevalence of sexting varies widely, but it is thought that the majority of adults and teens have engaged in some form of sexting, with text-only sext more common than sexts with pictures (Drouin, Coupe, and Temple 2016). In a national survey, one in four men and women reported ever sending nude or semi-nude pictures of themselves to someone and 41 percent of men, and 27 percent of women reported receiving them (Kann et al. 2018). In a study of 1,652 undergraduates, 65 percent had sent sexually suggestive texts or photos to a current or potential partner, and 69 percent had received them (Winkelman et al. 2014). In tracing the history of sexting among American youth, Englander (2019) states, "While not all teens sext, sexing it is not a rare or deviant behavior. It appears to happen most frequently within sexual relationships, especially for females" (p. 577).

Although adolescents often regard sexting as "no big deal," the consequences can be serious. Sexting is positively associated with sexual activity, even among middle school students (Hollander 2015; Rosenberg 2014). Sexting can also have important legal consequences. Teenagers under age 18 can be sentenced to juvenile court for possessing and sending nude photos—although law enforcement is abandoning prosecuting adolescents for consensual sexting (Wolak, Finkelhor, and Mitchell 2012). There are social and emotional consequences as well. In one notable case, an 18-year-old girl committed suicide after nude photos intended for her boyfriend were distributed to hundreds of her classmates (FBI 2012).

Newer research indicates that sexing isn't all negative. In a study of 651 teens in Massachusetts and Colorado, teens who sexted cited positive aspects of sexting such as enhanced self-confidence, positive self-image, and strengthening of romantic relationships (Englander, Milosevic, Staksrud 2019). In a study of undergraduate students, half reported positive or neutral consequences from sexting. However, these outcomes vary by gender and type of relationship. Women and those who sexted with a casual as opposed to committed partner identified fewer positive effects (Drouin et al. 2016).

g-stockstudio/Shutterstock.com

Although sexting can be fun, there can be negative consequences such as nude photos being sent to others without one's permission.

Facts about Families

How Do We Know What We Know? The Kinsey Institute and Conducting Sexuality Research

By David W. Wahl, Ph.D.

The pioneer surveys on sex in the United States were conducted by Alfred Kinsey and his associates over seventy years ago (Kinsey, Pomeroy, and Martin 1948, 1953). Kinsey's two-volume study, *Sexual Behavior in the Human Male* and *Sexual Behavior in the Human Female*, provided one of the first looks into the sexual lives of American men and women. It was groundbreaking for its time because it was published during a period of political and cultural conservatism, yet became a best seller within three weeks of its publication. Because Kinsey relied on volunteers, including people from his own social circle, as well as inmates and prostitutes, his research is criticized for not being representative of mainstream sexuality. His work is most useful in terms of representing the broad range of sexual behaviors in American society.

In conducting his dissertation research, Ph.D. candidate David Wahl visited the famous Kinsey Institute at Indiana University. He shares his experience:

Sitting alone in a room in the summer of 2019 atop Morrison Hall, I spent the better part of a week going through the Kinsey archives. Box after dusty box of Dr. Alfred Kinsey's personal papers were put on a desk before me. I was allowed only a pad of paper and a pen; no cell phone to take pictures; no audio-device to record my thoughts. Kinsey has been a primary influence in my interest on human sexuality, and Kinsey's papers proved to be a treasure trove.

As early as fourth grade, I began to seek out answers to questions I had about sex that fell on deaf ears in the conservative environment in which I grew up. Yet, here I was, holding the original, onion skin papers typed by Kinsey himself. Amongst notes to his wife and departmental memos, Kinsey's papers offered a deep insight into the inner workings of his research projects on human sexuality. I read about the research methodology he employed, how he organized his research team, and the philosophy behind the work. I was also able to understand the well-criticized problems in Kinsey's work, both in methodology and ethical concerns, and shape his own work in ways that avoided the same pitfalls. Sitting in Morrison hall, day by day, box by box, paper after paper, I credit the experience as being a significant step forward in my own research abilities. The experience and opportunity was not lost on me. In a letter written my wife, I wrote:

My hands shake slightly as I move from page to page. I am amazed at Dr. Kinsey's care and passion in researching human sexuality. At a time when sex was unspeakable and hidden in the dark, Kinsey's bravery and scientific prowess comes through on every page. We understand what we do now because he got it all started. For that I will forever admire him (Wahl 2020).

As mentioned in Chapter 2, researchers working on projects involving human beings must maintain the principle of "do no harm." This notion is especially necessary in performing sex research with humans. Maintaining confidentiality of subjects' identities and what they say is crucial. Matters of sexuality are considered private, and even casual discussions of sex may make a person feel uncomfortable. Seemingly innocuous questions may trigger traumatic memories for respondents who are survivors of sexual abuse. The dynamics between the researcher and the respondent must always be considered—for example, sometimes the researchers and subjects are matched in terms of gender. Questions about certain sexual behaviors (for example, oral and anal sex) may be asked by way of a questionnaire as opposed to face-to-face. As cited throughout this chapter, there are today many studies of sexual behavior that have used random samples and are representative of sexual behaviors of the U.S. population. Yet, conclusions about sexuality based on survey research must always be qualified by the possibility that respondents have minimized or exaggerated their sexual activity (De Jong and Reis 2015). Kenneth Plummer underscores the sensitive nature of sex research when he says, "Be *passionate* about your research, your teaching, your work: make sure you have a good reason for doing it, find your values, make sure you know where you stand, not just in terms of why you are doing your research and where you want to take it but also in how you treat people" (Barker 2018, p. xvi).

Critical Thinking

Have you ever been curious about the sexual behavior of those around you? To what extent should we trust data on sexuality? If you were given a survey asking personal questions about your sexual behaviors, would you be truthful?

Sexting is not always consensual—many people (mostly females) receive unwanted sexts from others (mostly males). Girls (and some boys) feel pressured or are coerced into sending nude or sexy pictures of themselves. In a national survey of adolescents, one in three girls (33 percent) say all or most of their friends been asked to send a "sexy or naked picture" to someone (Plan International 2018). In a study of college students, approximately one-third of students shared sexts with someone else (Winkelman et al. 2014). Among Belgian

high school students, LGBTQ+ youth were more likely than heterosexual youth to have sent, received, and asked for sexting images. They also felt significantly more pressure to send sexually explicit pictures of themselves (Ouytsel, Walrave, and Ponnet 2019). Negative outcomes, such as worry, regret or trauma, are most strongly linked to being coerced or pressured into sexting. *Connect Safely*, a nonprofit organization that educates technology users about safety, privacy, and security (2018) provides some good advice:

- The safest way to avoid a picture getting into the wrong hands is to never take it or share it. Sadly, there are cases (sometimes called "revenge porn") where someone shares pictures meant only for them — sometimes after a breakup.

- Never take and send an image of yourself under pressure, even from someone you care about.

- If a stranger asks you to take a revealing picture, it could be a scam that could lead to further demands and threats ("sextortion"). Do not respond and consider reporting it to the police, your parents and the CyberTipLine (1-800-843-5678). It could be a criminal who has exploited other people so you're helping others by alerting authorities.

- If a sexting photo arrives on your phone, first, do not send it to anyone else (that's not only a violation of trust, but could be considered distribution of child pornography). Delete the photo(s). If it would help—especially if you're being victimized—talk with a parent or trusted adult. Tell them the full story so they can figure out how to support you. Ask them to keep you involved.

- If the picture is from a friend or someone you know, then someone needs to talk to that friend so he or she is aware of possible harmful consequences. You're actually doing the friend a big favor because of the serious trouble that can happen if the police get involved. Get the friend to delete the photo(s).

- If the photos keep coming, you and a parent might have to speak with your friend's parents, an attorney, or school authorities.

Sexting is part of a broader social phenomenon referred to as *pornification*, a term coined by Pamela Paul (2005) in which sending, receiving, and viewing sexually explicit material is seen as a normal thing for young people to do (see *Pornography* section). Despite the fears of some, however, sexting does not appear to be replacing in-person intimate relationships but is expanding options to express sexuality (Ybarra and Mitchell 2015). Indeed, America men and women engage in a wide range of sexual behaviors,, from whipping (15 percent), spanking (32 percent), and using vibrators (42 percent), to reading erotic stories (58 percent; Herbenick et al. 2017).

A sexual encounter does involve risk. Today's sexually liberated environment is especially disadvantageous for women in terms of experiencing negative emotions, a negative reputation, and unwanted sex (Katz, Tirone, and van der Kloet 2012). At the same time, a more liberal sexual environment offers the potential for expressive sexuality and true sexual intimacy for both women and men. People now have more knowledge of the principles of building good relationships. At the same time, men are becoming more sensitive, less boastful about sexual conquests, and more romantic (Schalet 2012; Seal and Ehrhardt 2003). After the sexual revolution of the 1960s, men became more interested in communicating intimately through sex, whereas women showed more interest than before in physical pleasure (Pietropinto and Simenauer 1977; Seal, O'Sullivan, and Ehrhardt 2007). In the 1970s, pioneering sex researchers Masters and Johnson (1976) argued that more equal gender expectations led to better sex. Now that satisfying sexual relationships seem more attainable, how do men and women negotiate those sexual relationships inside and outside of committed relationships?

SEXUAL VALUES

Because today's young people generally do not marry or form stable, committed partnerships until their late twenties, they have many years, even decades, to experiment sexually. Young adults manage this period of time in different ways. The following sections discuss sexual values or standards of behavior within, and outside of, committed relationships.

Sex Outside Committed Relationships

Sexual intercourse between unmarried men and women is now widely accepted (Gallup 2019). We mainly have sex with people we know and care about, and the majority of men and women have sex within committed relationships (Herbenick et al. 2010b). Most unmarried teens and adults engage in a pattern known as **serial monogamy**, with most partners demonstrating sexual exclusivity (Chandra et al. 2011). However, in the 1990s, a *New York Times* article surprised many by describing a new form of casual sex among adolescents and young adults called **hooking up** (Denizet-Lewis 2004). Hooking up has been defined in different ways, but the defining feature of hooking up is that there is no promise or expectation of further communication, contact, or commitment between participants. Hooking up does not necessarily mean intercourse; oral sex, petting, kissing and hugging are just as common. Hooking up can occur with no prior acquaintance between the parties or among friends.

Friends with benefits is what it sounds like: sex between friends but without expectations for romantic love or commitment (Furman and Shaffer 2011). The basic idea behind hooking up, friends with benefits, and other forms of casual sex is that a sexual encounter means nothing more than just that—sexual activity. An 18-year-old high school senior explained, "Me and her have an agreement that it doesn't go any farther than that. . . . It's so cool like I could actually hook up with her, like she could be my girlfriend. It probably could come down to that because we're so cool about it like we're really good friends . . . but there's no strings" (Lyons et al. 2014, p. 89).

Psychologists Jocelyn J. Wentland and Elke D. Reissing (2011) conducted focus groups with male and female college students in Canada and discovered that hooking up incorporates a range of relationship types. For example, *booty calls* are similar to friends with benefits in that they involve repeated sexual encounters but require less communication, emotional investment, and responsibility. An important aspect of hooking up is the presence of alcohol. Among sexually active adolescents ages 15 to 19, 34 percent reported drinking alcohol or using drugs the last time they had sex (Kann et al. 2014). One 20-year-old single mother said, "I got trashed one night, so it just kinda' happened. I didn't realize it, you know. . . . We were talking . . . but my thinking wasn't clear, and it just happened, so . . . I mean, now that I look back on it, I'm just like . . . whoa. I shouldn't have been drinking that much" (Lyons et al. 2014, p. 88).

Hooking up is most common among college students where hookup culture and "partying" are very much intertwined. This is a lifestyle working-class and poorer students are often shut out of. They work long hours to pay for school, live at home and/or commute to school, and can't afford to skip classes after a late night out (Aronowitz 2016). Hooking up may not be all it's cracked up to be. In a study of 3,000 young adults and college students, 85 percent preferred other options to hooking up, such as spending time with friends or having sex in a serious relationship.

These sexual scripts may have emerged because many of today's young adults—who are delaying marriage, going to college, and developing careers—still want to have sex. They want some intimate connection without risking romantic disappointment and emotional loss. Nearly half of young adults have reported engaging in "sex with someone that you weren't really dating or going out with" (Lyons et al. 2014). Sixteen percent of female teens and 28 percent of male teens had sex for the first time with someone they had just met or someone they considered "just a friend" (Martinez, Copen, and Abma 2011). Dating apps such as Tinder have bolstered this trend by making it easy to get in touch with potential hookups on short notice. On the downside, social media have been implicated in the rise in STIs (Mohney 2015).

Like heterosexuals, gays and lesbians are increasingly using the Internet and social media to meet and interact with sexual partners. In a study of youths ages 13 to 18, gay teens were more likely to interact and meet sexual partners online than heterosexual teens. These differences were especially pronounced for gay men, who were three times more likely to have met their most recent sexual partner that way (Ybarra and Mitchell 2015). However, hooking up is prevalent among people of all ages. For example, Jane, a 59-year-old white heterosexual woman, said, "I think in my twenties I never would have thought about having a friend with benefits. It was either marriage or hit the road kind of thing. . . . I'm involved in a friend with benefits relationship now" (Fahs and Munger 2015).

The Sexual Double Standard

According to the **sexual double standard**, women's sexual behavior must be more conservative than that of men's. Despite the sexual revolution, numerous studies conducted since the 1960s indicate the sexual double standard continues to exist (Crawford and Popp 2003). In its original form, the sexual double standard meant that women should not have sex before or outside marriage, whereas men could. Within the context of marriage and committed relationships, femininity "is typically framed in terms of being sexually desirable rather than sexually desiring whereas masculinity connotes sexual aggression and prowess" (Elliott and Umberson 2008, p. 392). Teen girls weigh social costs of engaging in sexual behavior (fear of parents' reaction, loss of respect of partner, anticipation of guilt) ahead of the physical costs (i.e., pregnancy) and their own pleasure (Guzzo, Lang, and Hayford 2019).

Women are held to a higher standard than men in terms of behavior and appearance. For example, women who drink alcohol and women with tattoos are viewed by men as more sexually available than are other women (Mehta 2013; Riemer et al. 2019). The sexual double standard is reinforced by peers and parents (Heisler 2014; Snapp et al. 2014). At the same time, many young people don't perceive a sexual double standard. In the Harvard study of young adults (Weissbourd et al. 2019), nearly half (48 percent) of respondents either agreed or were neutral about the idea that "society has reached a point that there is no more double standard against women"; 39 percent either agreed or were neutral that it's "rare to see a woman treated in an inappropriately sexualized manner on television." Moreover, 32 percent of male and 22 percent of female respondents thought that men should be dominant in romantic relationships.

The continued existence of the sexual double standard alongside the prevailing pattern of hooking up makes navigating sexual relationships challenging for young women (Bogle 2004; Stepp 2007). The fact that there are currently more women than men on most college campuses, male power is reinforced as heterosexual women are must compete for a smaller group of available men (Berger 2015). Women are more likely than men to feel disrespected after a hookup and have more anxiety about the encounter (Armstrong, England, and Fogarty 2012; Snapp et al. 2014; Townsend and Wasserman 2011). Such feelings are not exclusive to women, however. Many men find these ever-transforming sexual scripts difficult to negotiate. Young men can be as emotionally vulnerable as women and use hookups to find lasting and meaningful relationships (Manning, Giordano, and Longmore 2006). Like sexting, the consequences of hooking up are not wholly negative. In a study of adolescent girls, those with more hookups were not less popular, did not lack friends, and did not have lower self-esteem (Lyons et al. 2011). In a study of college students, men and women had a mixture of positive and negative emotions after hooking up. Whereas some were happy and hopeful, others felt disappointed and used (Strokoff, Owen, and Fincham 2015).

The power imbalance in hooking up can be seen from the perspective of sexual satisfaction. Both female and male college students report that is typical for men to disregard the woman's pleasure in a hookup and that equality is not expected by either party (Armstrong, England, and Fogarty 2012). (Lesbians report greater sexual satisfaction than heterosexual women [Balwin et al. 2019; Holmberg, Blair, and Phillips 2010]). Men are much more attentive to the sexual needs of girlfriends than with casual sex partners and express pride in their ability to bring them to orgasm (Armstrong, England, and Fogarty 2012). Author Peggy Orenstein (2016) conducted more than 70 interviews with young women about their sexual experiences and perceptions of sex. The women described many ways that sex is constructed to serve men. The women described feeling obligated to perform oral sex with no expectation of reciprocation. Total pubic hair removal was standard, and, most importantly, unsatisfying sexual experiences seemed "normal." One young woman, Megan, described her first college hookup as, "pretty terrible. He was the thrusting type, you know, jack-hammering me until I faked an orgasm, and then he went to sleep." Even so, Megan continued to hook up with him "semi-regularly" for the next two months (Orenstein 2016, 126). Similarly, in her *Human Sexuality* class, Debbie Herbernik, a leading sexuality scholar at the Center for Sexual Health Promotion at Indiana University, asked her students to describe "good sex." Whereas men talked about pleasure and orgasm,

"women talk more about the absence of pain. [Studies show that] 30 percent of female college students say they experience pain during their sexual encounters as oppose to five percent of men" (Herbenick cited in Orenstein 2016, p. 70). Women's empowerment regarding sexual satisfaction varies by race, class, religion, and other factors (Cheng et al. 2014). In a study of women from sixteen churches in Atlanta, one African American woman said, "It seems like women compromise so much when it comes to sex... they don't want to lose that person. They aren't really empowered with themselves to know, 'Oh it's OK for me to do this and say this if I don't want to'" (Piper et al. 2019, p. 11–12).

Sex within Committed Relationships

Most sex occurs within marriage and committed relationships. Married couples have sex more often than single individuals, though less often than cohabiting couples (Herbenick et al. 2010a). A significant change resulting from the sexual revolution concerns marital sex. Married couples today have sexual intercourse more often, experience more sexual pleasure, and engage in a greater variety of sexual activities than in the 1950s (Greenberg, Bruess, and Haffner 2002). In the NSSHB study, 83 percent of men said they experienced "quite a bit" to "extreme" pleasure during their most recent sexual event, which was further enhanced when the sex occurred within a committed relationship (Herbenick et al. 2010b). Lately, however, the frequency with which couples have sex is declining—the average adult has sex nine fewer times per year than they did in the 1990s, with the largest declines among married couples, who have sex 16 fewer times a year. Twenge (2017) suggests some reasons for the decline in sex for individuals and couples: the omnipresence of smartphones and social media; the availability of TV entertainment; dating apps that have made seeking relationships less appealing; unemployment among young men, which keeps them from dating; and unreasonable expectations among young people about potential partners.

It used to be that describing sexuality over the course of a marriage would be nearly the same as discussing sex as people grow older. Today, this is not the case. Many couples are remarried, so at age 45 or even 70, a person may be newly married. Nonetheless, we may logically assume that young spouses are in the early years of marriage.

Young Spouses and Partners As one might expect, younger couples have sex more often than older couples (Herbenick et al. 2010a). Although the amount of sexual activity tends to decline over time, married men and women are remaining sexually active longer

than in the past. About 77 percent of married women in their thirties, 70 percent of women in their forties, 54 percent in their fifties, and 42 percent in their sixties had sex a few times a month or more often. Only after age 70 does the figure dip to 21 percent. At least in terms of sexual frequency, it does appear to be the case that "50 is the new 30" (Szalavitz 2011). Young married couples, especially those without children, have fewer distractions and worries. The high frequency of intercourse in this age group may also reflect a self-fulfilling prophecy: These couples may have sex more often partly because society expects them to.

Why, after the first few years, does sexual frequency decline? The sexual intensity of the honeymoon period subsides, and "from then on almost everything—children, jobs, commuting, housework, financial worries—that happens to a couple conspires to reduce the degree of sexual interaction while almost nothing leads to increasing it" (Greenblatt 1983, p. 294). Pregnancy, the presence of small children, and a less than certain birth-control method are factors that reduce sexual activity and satisfaction in young marriages (Maas et al. 2015). The demands of juggling childcare, paid employment, and housework can take a toll on sexual frequency in marriage. Working mothers, in particular, often feel too emotionally and physically exhausted for intercourse. Not that this can't be turned around. Couples in which the husband does more childcare and housework have both more frequent and more *satisfying* sex (Carlson, Hanson, and Fitzroy 2015; Johnson, Galambos, and Anderson 2015).

Spouses and Partners in Middle Age

On average, as people get older, they have sex less often. They also report lower levels of sexual arousal and pleasure, greater trouble with erection and lubrication, and less chance of having an orgasm (Lodge and Umberson 2012; Schick et al. 2010). Physical aging is not the only explanation for the decline of sexual activity and sexual quality over time, although it appears to be the most important one. Relationship satisfaction was the second largest predictor of sexual frequency.

Sexual intimacy is an important component of emotional intimacy, and sexual satisfaction, marital satisfaction, and sexual frequency are intertwined (Birnie-Porter and Lydon 2012). However, "even among couples who rate their marriages as very happy and among those who say they are still 'in love,' frequency of intercourse declines with age," and some of the decline is gendered, with more women reporting lower levels of desire and lower sexual quality than men as they age (T. Smith 2006, p. 13). For example, in a sample of adults age 50 and older from the 2009 NSSHB, the percentage of men who reported

It is important for married and committed couples to make time for each other. Regular "date nights" have been shown to enhance relationship and sexual satisfaction.

"extreme" to "quite a lot" of pleasure with their most recent sexual event was somewhat higher than that of women (78 percent vs. 68 percent). Men also reported higher levels of arousal: 75 percent reported "extreme" to "quite a lot" of arousal compared to 64 percent of women. However, the vast majority of men and women of all ages continue to experience a high level of sexual arousal, pleasure, and orgasm (Schick et al. 2010). Declining frequency appears to be offset by an increase in the quality of their sexual experiences (Lodge and Umberson 2012). It is also important to women's relationship satisfaction (and to a lesser extent, men's) that their partners value both their physical and nonphysical qualities, such as their intelligence, kindness, and supportiveness (Meltzer and McNulty 2014).

There has been a growing societal awareness of the importance of "date night" for sustaining the relationships, especially for couples with children (Weiner 2010). Partners do not have to have intercourse during these times: They should do only what they feel like doing. But scheduling time alone together does mean mutually agreeing to exclude other preoccupations and devote full attention to each other. The point is to have "us time," so this time together should not be spent discussing finances, family, or work. Although it is less common than among adolescents, some married couples engage in sexting to liven up their

relationship. Sending sexy talk messages (as opposed to explicit photos) has been shown to be positively associated with relationship satisfaction (McDaniel and Drouin 2015).

Older Spouses and Partners In our society, images of sex tend to be associated with youth, beauty, and romance; to many young people, sex seems out of place in the lives of older adults. Not too many years ago, public opinion was virtually uniform in seeing sex as unlikely—even inappropriate—for older people. Of course, physical changes associated with aging and declining health do affect sexuality (Christopher and Sprecher 2000, 1002; Schick et al. 2010). However, when Masters and Johnson's work in the 1970s found that many older people were sexually active, public opinion began to swing the other way (Cole 1983). The majority of people in their 50s, 60s, and 70s are sexually active and enjoy sex. However, around age 60, a gender gap becomes evident. Sixty-two percent of men but only 27 percent of women gave sex a high priority (DeLamater and Sill 2005; Lindau et al. 2007; Melby 2010, p. 4).

Sexual satisfaction doesn't have to decline with age. Forbes, Eaton, and Krueger (2017), using data from the Midlife in the United States study, found the negative relationship between age and sexual satisfaction is reduced when thought and effort are put into sex and when relationship quality was taken into account. The authors conclude that older couples acquire what they refer to as *sexual wisdom*: positive age-related changes in sexual skills, beliefs, and attitudes that maintain satisfying sex across the lifespan.

In a nationally representative sample surveyed by the American Association of Retired Persons, a majority (56 percent) of individuals 45 and older agreed that a satisfying sexual relationship is important to one's quality of life. But they rated family and friends, health, being in good spirits, financial security, spiritual well-being, and a good relationship with a partner as more important than a fulfilling sexual connection (Jacoby 2005). Thus, "it stands to reason that individuals and couples . . . who have developed the capacity over the years to experience optimal sexuality have much to teach the rest of us" (Kleinplatz et al. 2009, p. 15). For example, when asked what they would tell younger generations about sexual enjoyment, elderly respondents who had been in committed relationships twenty-five years or longer told researchers that good sex over the long term includes patience and practice (Melby 2010; Kleinplatz et al. 2009). Some older partners shift from intercourse to other sexual activities. According to the NSSHB, among married women ages 60–69, 26 percent engaged in solo masturbation, 9 percent engaged in mutual masturbation, and one in five had given or received oral sex in the previous ninety days (Herbenick et al. 2010a). One 73-year-old businesswoman said it this way: "Sex keeps

An active and satisfying sex life is an important way older couples maintain intimacy and closeness.

you active and alive. . . . I think it's healthy as can be, in fact I know it. That's what kept my husband alive for so long when he was sick. We had excellent sex, and any kind, at any time of day we wanted" (Stein 2012).

Although maintaining a healthy sex drive into the "golden years" is normal, care needs to be taken to ensure that people who choose to accept a decreasing sex drive as they age are not treated as "victims of a pathology" (Marshall 2009, p. 219). Data from the NSSHB showed that a third of men between 57 and 85 had some trouble getting or maintaining and erection during the preceding year (Cornwell and Laumann 2011). However, there is an ever-growing number of remedies available to treat sexual problems, including Viagra for men and estrogen supplements and lubricants for women. In that study, 30 percent of men in their sixties, 23 percent of men in their seventies, and 19 percent of men age 80 and older reported using an erectile medication during their most recent sexual event (Schick et al. 2010). The drug Flibanserin, sold under the trade name as Addyi, has been approved for the treatment of low sexual desire in women, but it is unclear how many women use such medications.

Radius Images/Alamy Stock Photo

Lack of sexual desire *can* be a problem for many men and women. Older age, obesity, smoking, drug use, and diabetes are associated with less sexual desire (Schick et al. 2010). Many antidepressants and blood pressure medications inhibit sexual desire and orgasm (Cohen et al. 2007). Alcohol is associated with less interest in sex and less sexual pleasure (Herbenick et al. 2018). Many women, especially older women, experience pain during intercourse. According to the 2018 NSSHB, only 50 percent of women who experienced pain during sex mentioned it to their partner. Their reasons for not telling them included minimizing and normalizing their pain and prioritizing their partner's pleasure over their own (Carter et al. 2019).

Not receiving treatment for sexual problems affects individuals and couples. Some men are so embarrassed and distressed that they are unable to discuss the issue with their partner or doctor (Lodge and Umberson 2012). In one study, female partners of the men who had erectile dysfunction but did not receive treatment reported less sexual satisfaction and fewer orgasms than the partners of the men who did receive treatment (Fisher et al. 2005). It is also important to consider that men and women may have different sexual desires. In one study, men were more likely than women to value sexual release and pleasing their partner and women were more likely to value intimacy, emotional closeness, and feeling sexually desirable (Mark et al. 2014). Among couples who are "out of sync" sexually, female partners commonly mentioned talking as their main strategy to get "back on track," as well as negotiation, scheduling a time for sex, "having sex anyway," and "trying new things." However, a significant minority of women "gave up" and let sexual problems go unaddressed or unresolved (Herbenick, Mullinax, and Mark 2014). Interest in and treatment of sexual dysfunction should increase as the U.S. population continues to age and as older people remain healthier and desire to be sexually active longer.

CONTEMPORARY ISSUES CONCERNING SEXUALITY

Sexual and reproductive issues have always been controversial. Religious and political conservatives and secular and political liberals regularly confront one another on a range of topics from teen pregnancy, sex education in schools, abortion, and lately, LGBTQ+ rights. Debate about such issues is expected to continue. Take abortion for example. Although 50 percent of Americans feel that abortion was "morally wrong," attitudes vary considerably when they're asked about limiting the number of weeks pregnant women should be allowed to get an abortion, requiring parental consent,

and exceptions for rape or incest, and the majority Americans (78 percent) want abortion to remain legal (Brenan 2019; Saad 2019). Meanwhile, the advent of technology has produced new areas of concern, such as the proliferation of pornography, or have created confusion about previously straightforward sexual standards, such as infidelity.

Sexual infidelity

Marriage and committed relationships typically involve the promise of **monogamy** or sexual exclusivity—that spouses will have sexual relations only with each other. Sexual infidelity is engaging in sexual relations with someone who is not one's own marriage or committed partner. There is a stronger proscription against extramarital sex in the United States than in many other parts of the world due to our strong religious roots. Eighty-nine percent of Americans consider extramarital affairs "morally wrong" (Brenan 2019). Cohabiting and other committed relationships also involve expectations of fidelity, although to a somewhat lesser degree (Norona et al. 2015; Treas and Giesen 2000). Some committed couples have chosen to be *polyamorous* or *consensually nonmonogamous* and have sex with additional partners. Even though these couples are happy in their relationships, they are generally viewed negatively by human services and health care professionals and by society as a whole (Kalata and Bermea 2019). Gay men are more accepting of nonmonogamous relationships and get more sexual and emotional satisfaction from casual encounters compared to lesbians and heterosexuals (Adam 2007; Mark, Garcia, and Fisher 2015). For example, in a national study, 44 percent of gay men who had discussed sex outside the relationship with their partner reported that they had agreed that "under some circumstances it is all right" compared to 5 percent of lesbians, 6 percent of heterosexual males, and 3 percent of heterosexual females (Gotta et al. 2011). As intimate relationships become more diverse and complicated, beliefs and actions do not always quite match, and some people find themselves unable to completely adhere to their own expectations.

Some researchers distinguish between emotional infidelity and sexual infidelity (Blow and Hartnett 2005). Emotional infidelity can be defined as an "intense, primarily emotional, nonsexual relationship" with someone who is not one's own marriage or committed partner (Potter-Efron and Potter-Efron 2008, p. 2). It is thought that whereas "men feel more betrayed by their wives having sex with someone else, women feel more betrayed by their husbands being emotionally involved with someone else" (Glass 1998, p. 35). However, in a study of college students, women and men had similar reactions to scenarios of emotional and sexual infidelity (Wade, Kelley, and Church 2011).

Along with marriage and cohabitation comes an expectation of sexual exclusivity. However, increasing numbers of couples experience an episode of sexual infidelity.

Digital Genetics/Shutterstock.com

Statistics on sexual infidelity are based on what people report. As you might expect, men and women tend to underreport infidelity in surveys. In one study, married women were six times more likely to admit to infidelity in a computer-assisted self-interview than in a face-to-face interview (Whisman and Snyder 2007). Therefore, estimates of prevalence of extramarital sex vary and are not precise. People also tend to believe others are more accepting of infidelity then they themselves are and that there is more cheating going on than there actually is (Boon, Watkins, and Sciban 2014). Indeed, contrary to media hype—which tends to focus on a small number of notable cases of infidelity among famous politicians, actors, and athletes—there is no evidence of an "infidelity epidemic" (Watson 2011). Lifetime prevalence rates of sexual infidelity range somewhere between 20 percent and 25 percent and is higher for men than women (Fincham and May 2017). Infidelity has increased in recent decades, though especially among older men, as a result of medications for erectile dysfunction such as Viagra. Gender differences in infidelity hold across cultures. For example, in a study of Mexican and Spanish young adults, a higher percentage of males (20 percent) than females (13 percent) reported having been unfaithful (Gil-Llario 2017).

The gender gap in infidelity between men and women is closing, especially for younger generations (Fincham and May 2017). Women are working longer hours, are traveling more, and have the same access to cell phones, text messaging, and social media as men, allowing

them to create and nurture intimate connections outside of marriage or a committed relationship (Parker-Pope 2008). In their review of research on infidelity, Fincham and May (2017) found the following individual-level factors to be associated with extramarital affairs for both men and women: neuroticism of the spouse, prior history of infidelity, drug and alcohol use, having a higher number of partners before marriage, psychological distress, insecure attachment, permissive attitudes toward sex, having a parent who was unfaithful. Protective factors against infidelity include high relationship satisfaction and investment, perceptions of few alternatives to the relationship, value placed on shared possessions and shared experiences, homogamy (partners coming from similar backgrounds), religious service attendance, and belief in the Bible. Contextual variables include having just one spouse in the paid workforce, number of days spent traveling for work, having a job with contact with a large number of potential sex partners, and time spent on Internet sites having to do with sex or facilitating infidelity, such as AshleyMadison.com and SeekingArrangement.com.

Despite popular belief, most affairs are *not* spontaneous (the result of too much alcohol, for example) nor are they the consequence of overwhelming romantic passion (Treas and Giesen 2000). Rather, entering an extramarital affair is a rational decision. In a recent study, loneliness was found to be an important factor in one's decision to be unfaithful. Military veterans also have higher levels of infidelity, possibly a result of long-term separations from their spouse. In one study, 32 percent of veterans reported having had an extramarital sexual relationship compared to 17 percent of nonveterans (London, Allen, and Wilmoth 2013). Sexual infidelity is also more common among couples where (1) the wife is pregnant, (2) partners spend little time together, (3) the husband is economically dependent on his wife, and (4) there is violence (DeMaris 2009; Munsch 2015; Treas and Giesen 2000; Whisman, Gordon, and Chatav 2007).

In her 2017 book, *State of Affairs*, marriage therapist Esther Perel explores the reasons why people cheat,

especially why increasing numbers of women are unfaithful. Interviews with husbands and wives revealed that wives' expectations for marital equality often do not mesh with reality, resulting in resentments over childcare and housework and constant selflessness associated with being a wife and mother, which erodes sexual desire. Modern women, taught since birth that they could "have it all," find themselves bored and unappreciated in their marriage and go outside for attention and excitement. Perel says that younger generations, having been taught since birth to pursue personal happiness and who experience FOMO (fear of missing out) puts them on a "hedonistic treadmill"—endlessly searching for something, *or someone*, better.

Although people in unhappy marriages may seek out other partners (i.e., cheating is just a "symptom" of bigger relationship issues), happily married people also sometimes cheat. Many do so in the midst of an identity crisis where they are trying to recapture the person they once were or explore "the road not taken" (Perel 2017). Perel quotes Priya, who is happily married yet in the midst of an intense affair with "not someone I would ever date—ever, ever." Priya says, "I've always been good. Good daughter, good wife, good mother. Dutiful. Straight As." She says the affair makes her "feel like a teenager with a boyfriend" (p. 156–158.)

What constitutes "cheating" is increasingly unclear. Advances in computer technology have given rise to a new brand of marital infidelity—"digital infidelity." E-mail, smartphones, dating apps, and other forms of social media have created ample opportunities for individuals to develop secret relationships (Hertlein 2012; Smith 2011). Facebook, Instagram, and Snapchat are common mechanisms for cheating. Researchers at the University of Indiana found that Facebook is sometimes used to maintain "backburner" relationships in case the current relationship should fall apart (Dewey 2014). The emotional connection may lead to a meeting—and then perhaps to a sexual relationship. Even if the couple does not meet, most men and women think of online relationships as a form of betrayal (Hertlein and Piercy 2006; Schneider, Weiss, and Samenow 2012; Whitty 2005). Addiction to cybersex, (sexually based conversations and social exchanges occurring online and/or sexually

based material available online), including among those in committed relationships, is also increasing (Ross, Månsson, and Daneback 2012).

Infidelity usually has long-term negative impacts on committed relationships, including emotional distress, depression, and post-traumatic stress disorder (Gordon, Baucom, and Snyder 2004; Leeker and Carlozzi 2012; Meier, Hull, and Ortyl 2009; Previti and Amato 2003). About half of men and half of women who have had extramarital sex eventually separate or divorce (Allen and Atkins 2012). An affair *can* have positive effects, such as encouraging closer relationships, paying greater attention to couple communication, and placing a higher value on the family (Olson et al. 2002), but only a very small percentage of couples experience an improvement (Blow and Hartnett 2005). Not only has trust been eroded and feelings hurt, but the uninvolved spouse may have been exposed to various STIs, which heightens anger and turmoil over affairs (Crooks and Baur 2005). The uninvolved partner may feel financially exploited as well, because of the money spent on the affair such as hotels and dinners out (T. Smith 2006).

Given that affairs do occur, people will have much to think about if they discover that their spouse has had or is having one. The uninvolved mate will need to consider how important the affair is relative to the marital relationship as a whole. Can she or he regain trust? Whether trust can be reestablished depends on several factors. One is how much trust there is in the first place. For this reason, new relationships may be especially

Photographee.eu/Shutterstock.com

Although unfaithfulness presents a serious challenge to marriages and committed partnerships, some couples can overcome the experience with the help of an experienced counselor.

vulnerable to breaking up after an affair. Couples who have been together longer may have an easier time. As one wife said, "Our relationship involved a big investment in time, memories, and property, and we didn't want to forfeit any of that. We worked out that we both were very motivated as we had three kids, a house, and lots of good memories" (Abrahamson et al. 2011, p. 1503). Many couples benefit from relationship counseling (Baucom et al. 2006). According to individual and couples therapist Shirley Glass, "The affair creates a loss of innocence and some scar tissue. I tell couples things will never be the same. But the relationship may be stronger" (Glass 1998, p. 44).

Pornography

Pornography—sexually explicit images and descriptions in books, magazines, film, and cyberspace—has seeped into American life to the extent that many of us are no longer disturbed by it, a phenomenon referred to as **pornification** (Paul 2005). The main consumers of pornography are not who you would expect. The average Internet porn user is a middle-aged, married man of median income (Frontline 2012). Among adolescents and young adults, men are more likely to view sexually explicit material than women (Morgan 2011; Shaughnessy, Beyers, and Walsh 2011). In a study of college-age men, 90 percent said that movies, television, magazines, pornography, or the Internet is where they saw their first sexual image, and many reported seeing their first sexual images before age 10. One-third thought, "This is great," but one-quarter said they felt "unprepared" (Allen and Lavender-Stott 2015). Women also now read and view porn, as sales of the 2011 bestselling erotic novel *Fifty Shades of Grey* attest (James 2011).

Acceptance of pornography is increasing. Forty-three percent of U.S. adults think pornography is morally acceptable, up from 36 percent just one year prior (Dugan 2018). Still, the majority of Americans feel that pornography is "morally wrong," with greater acceptability among men than women (Brenan 2019). However, some women who view pornography report several benefits: It helps them explore their sexuality, gives them ideas to use in real-life sex, and helps them learn how to look and act sexy (Paul 2005). Furthermore, pornography, along with other sexual games such as role-playing, sometimes enhances and "livens up" sexual encounters and may increase sexual satisfaction and overall intimacy (Hertlein 2012; Pratt, Brody, and Gu 2011). There is a small but growing literature on sexual subcultures such as BDSM (bondage and discipline, sadism and masochism), but our knowledge remains limited (Bauer 2014). Finally, several studies indicate that users of pornography have more progressive, as opposed to traditional, attitudes about sex (Kohut, Baer, and Watts 2015). Pornhub, a popular porn site, provides seemingly limitless genres of pornography, and analysis indicates that search terms "trend" over time (such as "anal," which grew by 120 percent between 2009 and 2015), and vary by age, gender, sexual identity, and state, with upticks of certain search terms on holidays (The Cut 2017).

On the other hand, pornography may contribute to a heterosexual couple's gender inequality and the objectification of the female partner. Most mainstream pornography can be considered **misogynistic**—that is, exhibiting hatred, dislike, mistrust, mistreatment, or general disregard for women. The sex portrayed is generally that of a dominant male in charge of a sexual encounter with a submissive female doing as she is told (Klaassen and Peter 2014). Furthermore, pornographic sex typically reflects the manner in which males are more likely to come to orgasm as opposed to the way that most females experience orgasm. Viewing pornography can alter one's sexual preferences: more-frequent viewers are more likely to want to engage in sex acts portrayed online (Morgan 2011).

Viewing pornography can also alter people's understandings of what normal bodies should look like because many performers' bodies have been surgically altered. In one study, women who viewed pictures of surgically modified vaginas (a procedure referred to as *labiaplasty*) were later more likely to rate them as "closer to society's ideal" than women who viewed images of ones that hadn't undergone surgery (Hilmantel 2013). Some women report that having pornography around the house makes them uncomfortable with their own bodies and body image, but they feel unable to complain.

Moreover, some men admit that their regular, perhaps addictive, use of porn has hurt their ability to enjoy sex in real life (Paul 2005). Among college students, more frequent viewing of pornography was associated with lower age at first intercourse, more frequent casual sex, a greater number of partners, and lower sexual and relationship satisfaction (Braithwaite et al. 2015; Kastbom et al. 2015; Morgan 2011). Viewing pornography can be especially risky for individuals who are sad or lonely. Among unhappy individuals, pornography exposure was found to increase the odds of engaging in casual sex by more than seven times (Wright 2015). There is even evidence that the use of pornography may deter men from pursuing committed relationships (Ferdman 2014). One study found that men who use pornography are less committed to their partners and are more likely to be unfaithful (Streep 2014).

Men and women can become victims of "revenge porn." Revenge porn is the distribution of sexually

explicit images without consent and is a form of sexual and emotional abuse. Ruvalcaba and Eaton (2019) investigated the topic in an online survey of 3,044 U.S. adults. One in 12 respondents reported having been a victim of revenge porn (9 percent of women; 7 percent of men) and one in 20 reported having been a perpetrator (3 percent of women; 7 percent of men). Gay men had higher rates of victimization. Victims of revenge porn experience psychological distress and physical symptoms such as headaches, with worse effects for women than men, but the vast majority (73 percent) did not seek help.

Sex Education

Deciding how much and what kind of information children, adolescents, and young adults should be provided with about sex, *and by whom*, has always been controversial. Overall, the United States does not do a very good job educating youth about sexuality and sexual relationships. Where do youth learn about sex? At home, both boys and girls receive most of their sexual knowledge from their mothers and female family members (Grange, Brubaker, Cand Corneille 2011; Hutchinson and Cedarbaum 2011). However, they are often provided with only the most basic information and advice about sex ("Use a condom") and much of this was communicated through innuendo ("Be careful," "Make sure you take care of that girl") and intimidation ("Don't do anything you're going to regret") rather than through direct messages. Parents were often unsure about the appropriate time to have a discussion or waited until they thought their child was having sex (Hyde et al. 2013). In a study of mothers of young female adolescents, a sizable proportion—especially those with less sexual knowledge, comfort, and self-efficacy—had no intention of discussing sexual health with their children in the next six months (Byers and Sears 2012). Among college students, gynecologic knowledge (female anatomy, pregnancy risk, etc.) is low; only about half of students knew what a Pap smear was. Knowledge was even lower among men and among those whose parents who did not discuss anatomy with them (Volck et al. 2013).

In a study of 3,000 young adults and college students conducted by Harvard University, 70 percent wished they had received more information from their parents about the emotional side of relationships such as "how to have a more mature relationship" (38 percent) and "how to deal with breakups" (36 percent; Weissbourd, Anderson, Sashin, and McIntyre 2019). Children and young adults whose parents talked to them about sexual health, emotional support, direct communication about sexual risks, and encouragement of goals for employment or education were more likely to use a condom

use during sex (Kogan et al. 2013; Wright, Herbenick, and Paul 2019). Children can benefit from discussions with their fathers, even if those discussions are incomplete (Hutchinson and Cedarbaum 2011). One woman explains:

> He gave me a philosophy about sex that has served me very well. His philosophy was that sex is not dirty but rather a serious issue with lots of emotional and physical ramifications. He didn't tell me not to have sex but rather to know what I was getting myself into and to consider heavily the consequences of my decision. (Hutchinson and Cedarbaum 2011, p. 559)

However, pornography remains a main source of information about sex, especially for boys (Tanton et al. 2015). Girls receive more sexual education than boys, both in school and from parents, especially concerning menstruation. Allen, Kaestle, and Goldberg (2011) collected twenty-three written narratives from male college students on the topic of menstruation. The following comes from Kyle, who first heard about menstruation when he was about 10 years old:

> The elementary school I was attending decided to separate the guys and girls into two separate classes for the afternoon. The boys' class was supervised while we played board games and we were told that the girls were doing the same thing. This was, of course, the school's weak attempt at covering up sex education for the girls, where they were being taught about menstruation. After the class, school was out and while on the bus many of my girlfriends that were in the class were talking about it. Looking back it was pretty funny because they were trying to tell me that girls had periods. But they were too shy to actually describe it; they just kept saying, "you know; a period, girls have periods." All I could think about was the punctuation mark because up till then that's all I thought about when I heard the word period. (Allen, Kaestle, and Goldberg 2011, p. 141)

They found that men who had one-to-one conversations with sisters, girlfriends, and parents were more empathetic and were less likely to have disdainful attitudes, crack jokes about menstruation, and state that periods were "gross."

Not surprisingly, then, both parents and children say they want schools to provide sex education and school is increasingly the main source of information (McKay et al. 2014; Rose et al. 2014; Tanton et al. 2015). The vast majority of U.S. adults and the American Academy of Pediatrics and the American Medical Association support age-appropriate and medically accurate **comprehensive sex education (CSE)**, which includes information about contraception, abstinence, healthy relationships, and sexual orientation (NARAL 2015). However, the content of sexual education programs varies greatly from state to state and remains politically controversial (see Table 4.1 for national standards developed by the Future of Sex

	BY THE END OF EACH GRADE, STUDENTS SHOULD BE ABLE TO	
GRADE LEVEL	**CORE CONCEPTS AND INFORMATION**	**ANALYZING INFLUENCES, COMMUNICATION, DECISIONS, GOALS, SELF-MANAGEMENT, AND ADVOCACY**
Grade K–2	Use proper names for male and female body parts; explain how living things reproduce.	Provide examples of how friends, family, media, and society influence the ways that we think and act; identify characteristics of a good friend, "good touch, bad touch," and saying "no"; identify adults they can trust and talk to.
Grade 3–5	Describe medically accurate information about male and female reproductive systems; understand puberty and personal hygiene.	Understand societal influences on body image; how to manage physical and emotional changes associated with puberty; demonstrate how to treat others with respect, ways of working together, and positive communication; help others take action against bullying and abuse.
Grade 6–8	Identify accurate sources of information about sexual health, contraception, and pregnancy; differentiate between gender identity, gender expression, and sexual orientation; explain range of gender roles; define sexual abstinence; examine how alcohol, friends, family, the media, and society affect sexual choices.	Analyze how external forces impact attitudes about gender identity, sexual orientation, and gender roles; develop a plan to promote dignity and respect in school community; learn how to support others' choices; develop strategies to stay safe when using social media.
Grade 9–12	Describe the human sexual response cycle; understand advantages and disadvantages of various forms of contraception; understand how to access emergency contraception; understand "consent"; access accurate information about pregnancy options; access skills and resources needed to become a parent.	Apply a decision-making model to situations relating to sexual health; demonstrate effective ways of communicating about sexual behavior; advocate for STI and HIV testing; know how to access information about sexual assault, harassment, and dating violence; advocate for school policies that promote dignity and respect.

Source: Adapted from Future of Sex Education Initiative (2012).

Education Initiative). Although less common and controversial, in an effort to reduce teen pregnancy, some schools are providing sexual and reproductive health care, including contraceptives for low-income and uninsured adolescents in the form of school-based health centers (Boonstra 2015).

The federal government's support of comprehensive sex education varies depending on the political orientation of the executive branch. In 1996, as part of welfare reform, the federal government took the official position that abstention from sexual relations unless in a monogamous marriage is the only protection against STIs and pregnancy—and that abstinence is the only morally and rationally appropriate principle of sexual conduct. These **abstinence-only-until-marriage** (AOUM) **programs**, which emphasize no sex until marriage, received nearly $1.5 billion from the federal government between 2001 and 2009 (NARAL 2015). Abstaining from sex can have positive effects. For example, delaying first intercourse increases the

likelihood that adolescent girls will graduate high school (Sabia and Rees 2009). However, there is no evidence that abstinence-only *programs* are effective in delaying sex or preventing pregnancy. The CDC examined 66 comprehensive "risk reduction" programs and 23 AOUM programs. There is no evidence that AOUM programs have any effect on abstinence, sexual behavior, or other sexual health outcomes (Chin et al. 2012). In fact, AOUM programs have been found to increase adolescent pregnancy rates in some states (e.g., Fox, Himmelstein, Khalid, and Howell 2019).

After decades of federal support for AOUM programs, President Obama in 2010 eliminated all funding for community-based abstinence-only education. In its place, he allocated $200 million to a teen pregnancy prevention initiative, the Teenage Pregnancy Prevention Program, or TPPP (Kappeler and Farb 2014). However, state policies regarding the provision and content of sex education vary, and many states have ignored the recommendations of the CDC and

do not cover their "19 Critical Sex Education Topics" (CDC 2016). These include the expected information on types of contraceptives but also topics such as goal-setting and decision-making, gender roles, communication, and the media.

In 2017, the Society for Adolescent Health and Medicine took a strong negative position on abstinence-only programs, stating that such programs "undermine public health goals and the safe transitioning of young people into sexually healthy adults" (Santelli et al. 2017, p. 402). Upon his election in 2016, Donald Trump and his administration have set out to "refocus" Obama's TPPP to better align with AOUM ideology, cut funding for the TPPP, and proposed five million dollars be allocated to AOUM programs. At the time of this writing, several states and TPPP grantees have filed lawsuits against the Trump administration for illegal termination of TPPP grant periods (SIECUS 2019). As of 2019, 39 states and the District of Columbia mandate sex education; 27 states and D.C. mandate both sex education and HIV education, 2 states mandate only sex education, and 10 states mandate only HIV education. Forty states and D.C. require parents to be involved in sex education—for example, by requiring parental notification and consent for their child's participation. States vary widely as to the content of sex education. For example, some states require programs be inclusive of different sexual orientations and cultures, that they teach about dating and sexual violence, and are required to be research-based, whereas other states emphasize heterosexuality, abstinence, and sex only within marriage (Alan Guttmacher Institute 2019). Sex education in schools is also constructed in terms of the risks and dangers of sex and boys' sexual response. Ann, shares her experience: "In seventh grade health class we went over the anatomy of the human body. I remember learning about the vagina and the many parts of it, but *never was the clitoris mentioned*" (emphasis added; Waskul, Vannini, and Wiesin 2007, p. 157).

Sexual Health

Understanding how to prevent pregnancy and avoid STIs are important in the time leading up to committing to a sexual partner and deciding whether or not to have children. The United States has a higher rate of unintended pregnancy (45 percent of all pregnancies) than many other industrialized countries. Although declining, teen pregnancy rates are also higher in the United States than in Europe, which can be attributed to less access to birth control, less consistent use of birth control, and the use of less effective methods of birth control (Child Trends 2014b). There is widespread support for birth control in the United States; 92 percent of Americans personally believe that

birth control is "morally acceptable" (Brenan 2019). In 2017, 86 percent of teens reported using a method of contraception the first time they had sex (Kann et al. 2018). By far the most common method was the condom (54 percent), followed by the pill (21 percent) and other methods such as the IUD, the patch, injectables, Nuva-Ring, or Implanon. Access to contraception and abortion is worsening with declining state and federal funding and the closing of many women's health clinics across the nation, especially in rural and low-income areas. An increase in unintended pregnancy is a likely outcome of these budget cuts.

STIs are viruses (e.g., HPV, herpes, HIV/AIDS), bacteria (e.g., *Chlamydia*, Gonorrhea), or parasites (e.g., trichomoniasis, public lice or "crabs") that can be transmitted through sexual contact. Coinciding with a drop in federal funding for women's health mentioned above, the prevalence STIs increased sharply between 2013 and 2017. "We've been sliding backwards," says Dr. Gail Bolan of the Centers for Disease Control; the U.S. has the highest rate of STIs in the industrialized world and some are calling on President Trump to declare STIs in America a public health crisis (Harris 2018).

The HIV that produces AIDS has been with us for over four decades. If untreated, an HIV infection eventually progresses to full-blown AIDS, a viral disease that destroys the immune system. With a lowered resistance to disease, a person with AIDS becomes vulnerable to infections and other diseases that other people easily fight off. However, with proper treatment, including antiretroviral drugs, today the life expectancy of people with HIV is nearly the same as everyone else's (Gueler et al. 2017).

Nevertheless, despite a rapid decline in deaths from HIV/AIDS since the early 1990s, it remains an important public health concern, especially among gay and bisexual men, young adults, the poor, and racial and ethnic minorities. About 1.1 million people in the United States are living with HIV and men comprise 80 percent of cases (CDC 2018). Most cases of HIV are attributable to male-to-male sexual contact, followed by heterosexual contact and injected drug use. You can see from Figure 4.3 that gay and bisexual men have been the most severely affected populations, accounting for 66 percent of all new HIV infections. African Americans are the most severely affected regardless of mode of transmission. Although they represent only 13 percent of the U.S. population, they accounted for 43 percent of all new HIV infections in 2017, a rate that is eight times that of whites. Like African Americans, Hispanics have a disproportionately high rate of HIV, accounted for 26 percent of HIV diagnoses.

An amazing new advance against HIV and AIDS is the introduction of highly effective pre-exposure prophylaxis (PrEP), or medications that reduce the risk of

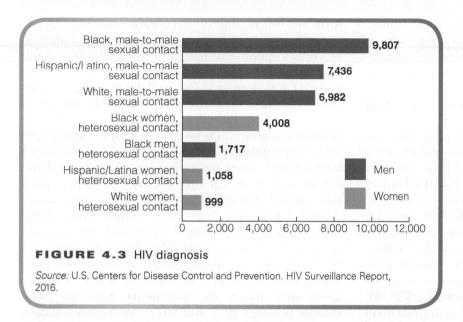

FIGURE 4.3 HIV diagnosis

Source: U.S. Centers for Disease Control and Prevention. HIV Surveillance Report, 2016.

getting HIV from an infected partner or intravenous drug use. These medications, taken daily, are designed for high-risk populations such as gay men, drug users, and sex workers (CDC 2019). Such medications are so new it's unclear how widely available they are, especially to low-income populations, who are at the highest risk. In 2015, the White House released an updated National HIV/AIDS Strategy for the United States. This document is the federal government's five-year plan for reducing new infections, increasing access to health care, improving outcomes, reducing HIV-related health disparities by gender, race, sexual identity, and income, and achieving a more coordinated national response. New strategies include using social media such Facebook, YouTube, and Twitter to spread messages about testing and safe sex (HIV.gov. 2019). States are enacting policies as well. For example, Michigan requires HIV-infected people to disclose their status, and public health officials use a confidential database of HIV-infected persons to identify their previous sexual partners (Hoppe 2013).

During the 2019 State of the Union address, President Trump announced the administration's new "Ending the HIV Epidemic: A Plan for America" and budgeted 291 million dollars toward this effort. This will be a ten-year initiative beginning in 2020 with the goal of reducing new HIV infections by 75 percent by 2025 and 90 percent by 2030. This level would essentially mean that HIV transmissions would be rare and meet the definition of ending the epidemic (HIV.gov 2019).

The U.S. Centers for Disease Control and Prevention (www.cdc.gov/std/default.htm), the Alan Guttmacher Institute (https://www.guttmacher.org/), and Planned Parenthood https://www.plannedparenthood.org/learn /stds-hiv-safer-sex,https://www.plannedparenthood.org/learn/stds-hiv-safer-sex) are excellent sources of information on all matters related to sexual health, including contraception, pregnancy, abortion, STIs, and relationships. Pregnancy, contraception, and abortion are discussed further in Chapter 8.

SEXUAL RESPONSIBILITY

Sex is largely a positive and enjoyable activity and has numerous physical and emotional health benefits in addition to creating children. For example, one recent study of 159 married couples in various industries found that having sex boosted their work performance the next day, as well as improved their overall mood (Leavitt, Barnes, Watkins, and Wagner 2019). However, people make decisions about sex in a climate characterized by social and political conflict. Premarital and other nonmarital sex, same-sex relationships, abortion, sex education, and contraception represent political issues as well as personal choices. Public and private communication must rise to new levels. HIV and AIDS have brought the importance of sexual responsibility to our attention in a dramatic way.

Making knowledgeable choices is a must. Because there are various standards concerning sex, all individuals must determine what sexual standard they value, which is not always easy. Making these choices and feeling comfortable with them requires recognizing and respecting one's own values instead of just being influenced by others in a sexual situation. Anxiety may accompany the choice to develop a sexual relationship, and there is considerable potential for misunderstanding between partners. Relationships between men and women are also becoming increasingly diverse and can be strictly platonic or include varying degrees of sexual interest and involvement. Yet old notions of male–female relationships are still adhered to. For example, it is not uncommon for platonic friends to be "challenged" by others to admit they are romantically involved (Schoonover and McEwan 2014).

In sharing sexual pleasure, partners realize that sex is something partners do with each other, not to or for each other. Each partner participates actively as an equal in the sexual union. This includes communicating with partners or potential sexual partners. As we've seen in this chapter, sex may mean many different things to different people. A sexual encounter may mean love and intimacy to one

partner and be a source of achievement or relaxation to the other. Honesty lessens the potential for misunderstanding and hurt feelings between partners.

Further, each partner is responsible for his or her own sexual response. When this happens, the stage is set for conscious, mutual cooperation. As mentioned previously, among heterosexual couples, both men and women place more value on the male partner's orgasm. Results from five national sex surveys in Finland indicated that less than half (46 percent) of women "always or nearly always" had an orgasm during sex and only 6 percent of women "always" had an orgasm during sex (Kontula and Miettinen 2016). Women are beginning to speak out about the "orgasm gap" between men and women (Compton 2019; Halton 2019; Mintz 2015). As rapper Nicki Minaj said, "I demand that I climax. I think women should demand that. I'm a pleaser, but it's 50-50" (Valenti 2015). It is possible to flip this sexual script in favor of women——for instance, by making sure the woman has an orgasm before the man (Muehlenhard and Shippee 2010).

Men could benefit from more education on women's pleasure. For example, many men don't know that women generally cannot have an orgasm through intercourse alone (Herbenick, Fu, Arter, Sanders, and Dodge 2018). Women who rated their partner more

sexually "skilled" had significantly more orgasms, as did women whose partners engaged in a variety of techniques and who had sex longer (Kontula and Miettinen 2016). Rather than "faking it," it is a woman's responsibility to communicate to her partner what "turns her on." Studies have found mental and relationship factors to be the most important determinants of women's orgasms (and receiving oral sex). These include communication between partners, women's sexual assertiveness, women who rated their relationships "happier" and considered sex important to the relationship, and having higher "sexual self-esteem" (i.e., they thought of themselves as "good in bed"; Backstrom, Armstrong, and Puentes 2012; Bay-Cheng and Fava 2011; Kontula and Miettinen 2016).

In expressing sexuality today, each of us must make decisions according to our own values. A person may choose to follow values held on the basis of religious commitment or as put forth by ethicists, psychologists, or counselors. People's values change over the course of their lives, and what's right at one time may not appear to be later. Despite the confusion caused both by internal changes as our personalities develop and by the social changes going on around us, it is important for individuals to make thoughtful decisions about sexual relationships.

Summary

- Social attitudes and values play an important role in the forms of sexual expression that people find appropriate and enjoyable.

- Sexual identity develops over time and is shaped by both genetic and social factors. Recent decades have witnessed increased acceptance of LGBTQ+ individuals, though some disapproval, discrimination, and hostility remain.

- Whatever one's sexual identity, sexual expression is negotiated amid cultural messages about what is sexually permissible or desirable. In the United States, these cultural messages have moved from patriarchal sex to a message that encourages sexual expressiveness in myriad ways.

- Values outside of committed relationships include abstinence, recreational sex, and the double standard—the latter diminished since the 1960s but still alive.

- Sex between committed partners changes throughout the life course. Young spouses have sex more often than do older mates. Although the frequency of sexual intercourse declines over time and through the length of a marriage, many older adults have active and satisfying sex lives.

- Making sex a pleasure bond, whether a couple is married or not, involves cooperation in a nurturing, caring relationship. Partners need to develop high self-esteem, break free from restrictive gendered stereotypes, communicate openly, and make time for each other.

- Infidelity, pornography, sex education, and sexual health are important issues that affect sexual relationships.

- Whatever the philosophical or religious grounding of one's perspective on sexuality, there are certain guidelines for personal sexual responsibility that we should all heed.

Questions for Review and Reflection

1. How would you categorize your own sexual identity? Have you always been certain of it or has your sexual identity evolved over time?

2. Describe your own "sexual script" in comparison to those of previous generations.

3. What behavior constitutes infidelity in your mind? For example, what would you do if you found out your partner was having an online affair?

4. Discuss what you've learned about the nature of sexual relationships. What did you find useful and relevant to everyday life? What seems remote from real-world experience?

5. **Policy Question**. What role, if any, should government policy play in sex education and research and disseminating that knowledge to the public?

Key Terms

abstinence-only-until-marriage (AOUM) programs 102
abstinence 85
affectional orientation 81
affirmative consent 87
asexual 81
bisexual 80
bystander education 87
comprehensive sex education (CSE) 101
expressive sexuality 84
friends with benefits 93
gay 80
heterosexism 88
heterosexual 80

HIV/AIDS 90
homophobia 89
homosexual 80
hooking up 92
interactionist perspective on human sexuality 83
interpersonal exchange model of sexual satisfaction 83
lesbian 80
LGBTQ+ 80
menarche 80
microaggressions 89
misogynistic 100
monogamy 97
pansexual 80

patriarchal sexuality 84
pornification 100
pornography 100
programs 102
queer kinship 89
rape myths 87
serial monogamy 92
sexting 90
sexual assault 86
sexual coercion 87
sexual double standard 93
sexual identity 80
sexual scripts 83

5

LOVE AND CHOOSING A LIFE PARTNER

Learning Objectives

1 Discuss various ways of defining love and the different models of love.

2 Compare arranged and free-choice marriages.

3 Define assortative mating and the process of finding a committed partner.

4 Describe dating from early adolescence to adulthood.

5 Describe how technology has changed dating relationships.

6 Discuss ways of nurturing committed relationships and things to watch out for.

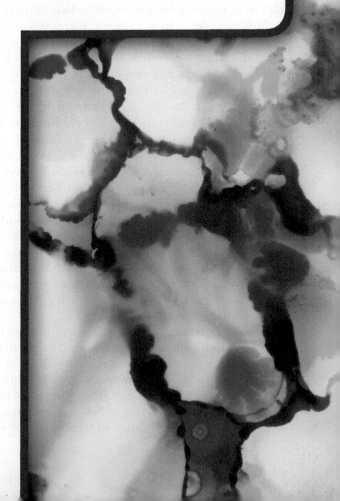

Almost three-quarters of young adults believe in "one true love" (Robison 2003; Whitehead and Popenoe 2001), and most people have said "I love you" to just one or two partners in their lifetime (Chalabi 2015). We all want to be loved, and most of us expect to be in a committed relationship—if not now, then in the future. Being in a committed relationship involves selecting someone with whom to become emotionally and sexually intimate and, often, with whom to raise children. Accordingly, the choice of a lifelong partner is a major life course decision. Figure 1.5 in Chapter 1 illustrates that making intentional, knowledgeable decisions requires an awareness of one's personal beliefs and values, as well as conscious consideration of alternatives and serious thought about the probable consequences. **Mate selection** is the process of choosing a partner with the intention of forming a marriage or committed relationship. Although more people are choosing to remain single, most everyone at some point wonders, "Who will I spend my life with?" Research suggests that the best way to choose a life partner is to look for someone who is socially responsible, respectful, and emotionally supportive. It is also important that the person is committed both to the relationship and to the value of staying together. It helps if that person also demonstrates good communication and problem-solving skills (Bradbury and Karney 2004; Hetherington 2003). Equally important is looking for a mate with values that resemble one's own because similar values and attitudes are strong predictors of ongoing happiness and relationship stability.

Most people, at least in Western countries, get married because they are in love. They also get married to meet cultural expectations, such as the social norm that children be born and raised within a marriage. Institutional factors, such as laws and religion, play a role. Research bears this out. In a national survey, nine in ten Americans said "love" was a very important reason to get married, followed by making a lifelong commitment, companionship, having children, financial stability, having a relationship *recognized in a religious ceremony*, and for *legal rights and benefits* (Pew Research Center 2010). Expectations are also high that married men (and men in committed relationships) be a good financial provider (Parker and Stepler 2017). Nevertheless, love and mutual attraction is consistently ranked as the most important factors in selecting a future mate, followed by likability, dependability, trustworthiness, intelligence, and pleasing personality (Boxer, Noonan, and Whelan 2015).

People are taking longer to make the decisions about marriage and lifelong partnerships. The median age at marriage has increased to an all-time high of 30 for men and 28 for women, compared to 23 for men and 20 for women in 1950 (U.S. Census Bureau 2018). But Americans still value marriage. Nearly three-quarters of women and almost two-thirds of men marry by age 30;

by age 40, more than 80 percent of Americans have married (Kreider and Ellis 2011b). Moreover, the majority of men and women who divorce marry for a second time and nearly a quarter of married people had been married before (Livingston, Parker, and Rohal 2014). Therefore, the topic of choosing a marriage partner is critically important. We need to note here that research and published counseling advice on choosing a committed life partner have focused primarily on mate selection into *marriage*. However, increasingly men and women are establishing committed, lifelong relationships outside of marriage, mainly within cohabitation but sometimes not. Although the majority of people marry, the number of U.S. adults in cohabiting relationships has climbed, increasing 29 percent since 2007 (Stepler 2017).

Research and books on mate selection have also focused almost solely on *heterosexual* mate selection. This leaves out committed partnerships among gay men and lesbians both outside of and within marriage. Like the general public, people who identify as LGBTQ+ list love as a very important reason to marry, followed by companionship, and making a lifelong commitment. However, there are differences too. Nearly half of LGBTQ+ people listed "for legal rights and benefits" as a very important reason to marry compared to one-quarter of the general public. A lower proportion of LGBTQ+ people listed "having children" and "having a relationship recognized in a religious ceremony" as very important reasons, although these data were collected before the legalization of gay marriage in 2015 (Pew Research Center 2013). Researchers and counselors are therefore turning their focus to committed gay and lesbian couples and committed, but unmarried, heterosexual couples.

In this chapter, we'll look at some factors that influence the choice of a life partner. We will examine theories and research on how the process of selecting a potential mate unfolds. Within potential partnerships, we will examine how a relationship develops and proceeds from first meeting to commitment and how technology is influencing these processes. We will also discuss interreligious, interracial, and inter-ethnic unions. Finally, we'll discuss features of unhealthy and healthy relationships, the process of breaking up, and intimate partner violence. To begin, we explore some things that we know about love.

LOVE AND COMMITMENT

When most people think of love, they think of "romantic love" as opposed to love between parents and children, family members, and friends. This is the kind of love people most associate with "falling in love" (Hemesath 2020). Choosing a partner for a lifelong commitment

usually starts with love. Yet, "[w]ith respect to love, the gap between everyday people and family scholars is surprisingly wide. . . . [M]ost researchers have avoided the topic. . . . Yet attitude surveys reveal that the great majority of Americans view love as the primary reason for getting and staying married" (Amato 2007, p. 306). Why the lack of research on love?

Defining Love

Love is exceedingly difficult to define, although many have tried. Researchers tend to feel more comfortable measuring the concepts of attachment, intimacy, compassion, and infatuation, which are *related to* and often treated as *measures* of love, as opposed to love itself (Langeslag, Muris, and Franken 2012; Neff and Beretvas 2013). There is an emerging area of "love" research from the physical sciences, such as studies of levels of oxytocin and serotonin in the blood of romantic couples (Langeslag, van der Veen, and Fekkes 2012; Schneiderman et al. 2012).

The definition of love many of us are familiar with is commonly heard at Christian wedding ceremonies and comes from the Bible:

> Love is patient, love is kind. It does not envy, it does not boast, it is not proud. It does not dishonor others, it is not self-seeking, it is not easily angered, it keeps no record of wrongs. Love does not delight in evil but rejoices with the truth. It always protects, always trusts, always hopes, always perseveres (Corinthians 13:4, New International Edition).

One family scholar's definition of love is quite similar. Paul Amato (2007) defines love as "a strong emotional bond with another person that involves sexual desire, a longing to be with the person, a preference to put the other person's interests ahead of one's own, and a willingness to forgive the other person's transgressions" (p. 206). Sexual desire is also important. For her book *Falling Out of Romantic Love* (2020), Crystal Hemesath conducted in-depth interviews with men and women on the topic of romantic love. She found that sexual desire and physical attraction combined with emotional attachment typifies most romantic relationships. Adam put it this way, "I think you can love somebody on an emotional level, um, I think you can lust somebody on a physical level . . . but when you have both of those, that's where the magic happens" (p. 122). The importance of sexual attraction is exactly why romantic relationships can be so unstable. Hemesath uses the metaphor of a tornado to describe romantic love. Although humans crave stability, calm, and predictability, "as with the weather, instability in love makes things interesting" despite the risk (p. 67). The problem is people often ignore warning signs when it comes to romantic love. One young woman put it this way, "I fell in love with someone else who is not available . . . we are perfect in every way . . . it's physical, it's chemical, it's intellectual . . . on every level it's a perfect fit . . . I just ache to see him . . . any tiniest little bit of contact . . . which is making me go crazy because I love him . . . (Barri and Morgan 2011, p. 16).

Western culture's definition of love reflects a **heteronormative bias**, which means love is thought about as something that can occur only between *opposite sex* partners. This "love bias" starts as early as preschool! Sociologist Heidi Gansen (2017) spent ten months observing preschool children in Michigan. She found that in some classrooms, teachers activity encouraged "crushes" and kissing between boys and girls. She saw boys kissing girls without their consent brushed off as a "sweet" gesture. Whereas friendship and affection expressed between boys and girls was assumed to be "romantic" (a boy and girl holding hands was referred to as a *budding romance* by one teacher), same-sex friendships were assumed to be just that—friendships. More research on the development of heteronormativity with respect to love, and research that explores loving relationships among same sex couples, is greatly needed.

Commitment

With the obvious exceptions of physical and emotional abuse, loving involves the acceptance of partners for themselves and "not for their ability to change themselves or to meet another's requirements to play a role" (Dahms 1976, p. 100). People are free to be themselves in a loving relationship and to expose their feelings, frailties, and strengths (Armstrong 2003). Related to this acceptance is having *empathy* toward one's partner, which includes understanding them from *their* frame of reference rather than one's own. This also may the concern a person has for the partner's growth and the willingness to "affirm [the partner's] potentialities" (Jaksch 2002; May 1975, p. 116).

Maintaining a loving relationship also requires commitment of both partners (Dixon 2007). **Commitment** is a willingness to work through problems and conflicts as opposed to calling it quits when problems arise. In this view, commitment involves consciously investing in the relationship (Etcheverry and Le 2005). Committed lovers have fun together, but they also share more tedious times. Committed partners view their relationship as worth keeping, and they work to maintain it despite difficulties or disagreements (Amato 2007; Love 2001). Then, too, committed partners "regularly, routinely, and predictably attend to each other and their relationship no matter how they feel" (Peck 1978, p. 118). Is this how young people view commitment? In a study of African American teens from a Boys' and Girls' Club in a public housing community by Barton and colleagues (2017), adolescents were asked, "What does it mean to commit to someone else?" To provide support and care,

to be honest, and to remain sexually faithful were their most common responses. However, their definition of commitment was different in different relational contexts. The ultimate level of commitment was marriage. Says Jayden, "[I want to get married in future] 'cause I always have a girl right there by my side, she [will] always be there for me . . . a girl can break up with you, but your wife, she made a commitment and she will always be there for you" (p. 24). Fidelity also loomed large, especially within marriage. As Robbie explained, "The way she probably take it is you're not married, so you might go out there and you might do something else [but] if you in a marriage, she be like well, you know, he's committed, he made his vows so you supposed to be there. Til death you know" (p. 126).

Gender Differences in Love

Love—and the need for love—is generally thought to be the domain of women. However, an emerging field of research examining love from the perspective of men is calling this assumption into question. Amy Schalet, the author of *Not Under My Roof: Parents, Teens, and the Culture of Sex* (2011), examined sex and relationships among teens from the United States and the Netherlands. Her findings suggest that boys, at least a subset of them, are becoming less focused on casual sex and more focused on romance. As one boy from her study says, "[My girlfriend] is the only one I ever want to have from now on. . . . We're just so happy together and I couldn't imagine being with anyone else. My first priority is being in love with my girlfriend, and giving her everything I can" (Schalet 2011, p. 158). Another study of 1,316 adolescent boys and girls from Toledo, Ohio, by Giordano, Longmore, and Manning (2006) similarly challenges the notion that girls want romance and boys want sex. The boys and girls in their study scored equally on a scale of "passionate love" developed by psychologist Elaine Hatfield and sociologist Susan Sprecher more than twenty-five years ago (Hatfield 2013; Hatfield and Sprecher 2010). Giordano and colleagues also conducted in-depth interviews with some of the boys in the study. A 17-year-old boy responded to the question "How important is your relationship to [your girlfriend] in your life?" by saying:

> As important as you get. You know, well, you think of it as this way, you give up your whole life, you know, know, to save Jenny's life, right? That's how I feel. I'd give up my whole life, to save any of my friends' life too. But it's a different way. Like, if I could save Jon's life, and give up my own, I would, because that is something you should, have in a friend, but I wouldn't want to live without Jenny, does that make some sense? (Giordano et al., p. 277)

Men also fall in love more quickly than women. Whereas women take an average of 134 days to say "I love you," men take about half that time. Thirty-nine percent of men say "I love you" within the first month of seeing someone compared to 23 percent of women (DeLacey 2013). And when it comes to love and breakups, women are more resilient (McClintock 2014b; Simon and Barrett 2010). As one researcher said, "It appears that young men benefit more than women from support, and that they are more harmed than women by strain in ongoing romantic relationships" (Paul 2010). In another study, men were more likely than women to look at letters from former lovers, suggesting they are more sentimental about past relationships (Janning 2015). These studies indicate that, despite gendered expectations of a tough exterior, love is important and meaningful in the lives of men.

There are also societal expectations that men show their love. Sociologists David Schweingruber, Sine Ahahita, and Nancy Berns (2004) found that in the typical

david f/Getty Images

Contrary to the popular notion that women are consumed with love, men are actually more "romantic." Men expend a great deal of energy pursuing romantic partners, and their well-being is highly dependent on maintaining romantic relationships.

engagement proposal, men are responsible for and successfully perform various "romantic rituals" (asking the father's permission, getting down on one knee), as well as executing elaborate demonstrations of love, often in a public setting. A study of online dating indicates that men who focus on their partner rather than themselves received more "clicks" from potential dates (McFarland, Jurafsky, and Rawlings 2013). On the downside, in a recent study of 1,387 college students, 26 percent agreed that love "brainwashes women" (Wade 2015a).

Sternberg's Triangle Theory of Love

In research on relationships varying in length from one month to thirty-six years, psychologist Robert Sternberg (1988a, 1988b; Sternberg and Sternberg 2008) found three components necessary to authentic love: intimacy, passion, and commitment. According to **Sternberg's triangle theory of love**, *intimacy* "refers to close, connected, and bonded feelings in a loving relationship. It includes feelings that create the experience of warmth in a loving relationship . . . [such as] experiencing happiness with the loved one; . . . sharing one's self and one's possessions with the loved one; receiving . . . and giving emotional support to the loved one; [and] having intimate communication with the loved one" (Sternberg 1988a, pp. 120–121).

Passion "refers to the drives that lead to romance, physical attraction, sexual consummation, and the like in a loving relationship" (Sternberg 1988a, pp. 120–121). *Commitment*—the "decision/commitment component of love"—consists of not only deciding to love someone but also deciding to maintain that love. **Consummate love** (see Figure 5.1), composed of all three components,

is "complete love, . . . a kind of love toward which many of us strive, especially in romantic relationships" (Sternberg 1988a, pp. 120–121). As the theory goes, relationships in which these three components are present and occur in equal measure are more successful than those with only one or two components or those in which the components are out of balance.

The three components of consummate love develop at different times as love grows and changes. "Passion is the quickest to develop, and the quickest to fade. . . . Intimacy develops more slowly, and commitment more gradually still" (Sternberg, quoted in Goleman 1985). Passion, or "chemistry," peaks early in the relationship but generally continues at a stable, although fluctuating, level and remains important both to our good health (Kluger 2004) and to the long-term maintenance of the relationship (Love 2001). More so than intimacy and passion, commitment is associated with reproductive success (Sorokowsk et al. 2017). Intimacy, which includes conveying and understanding each other's needs, listening to and supporting each other, and sharing common values, becomes increasingly important as time goes on. In fact, psychologist and marriage counselor Gary Smalley (2000) argues that a couple is typically together for about six years before the two feel safe enough to share their deepest relational needs with one another. Commitment is essential; however, commitment without intimacy and some level of passion is hollow. Because these components not only develop at different rates but also exist in various combinations of intensity, a relationship is always changing, if only subtly.

Attachment Theory and Loving Relationships

Recall from Chapter 2 that attachment theory posits that during infancy and childhood a young person develops a general style of attaching to others (Ainsworth et al. 1978; Bowlby 1988; Pittman et al. 2011). Applying attachment theory to loving relationships, we can presume that people with more secure attachment styles would have less ambivalence about emotional closeness and commitment. "Secure attachment, in part, depends on regarding the relationship partner as being available in times of need and as trustworthy" (Kurdek 2006, p. 510).

We might therefore conclude that those with a *secure* attachment style have stronger interpersonal skills and are better prospects for a committed relationship (Jenkins-Guarnien, Wright, and Hudiburgh 2012; Rauer and Volling 2007). An *insecure/anxious* attachment style entails "fear of abandonment" with consequent possible negative behaviors such as unwarranted jealousy or attempts to control one's partner. An *avoidant* attachment style leads one to pass up or shun closeness

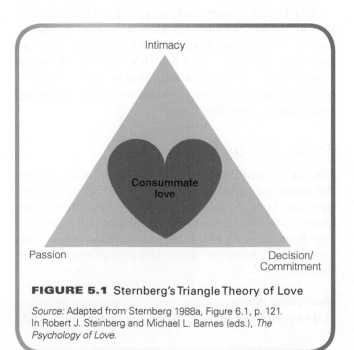

FIGURE 5.1 Sternberg's Triangle Theory of Love

Source: Adapted from Sternberg 1988a, Figure 6.1, p. 121. In Robert J. Steinberg and Michael L. Barnes (eds.), *The Psychology of Love.*

Fact about Families

Six Love Styles

Relationships evidence different characteristics or personalities. John Alan Lee (1973) classified six love styles, initially based on interviews with 120 white, heterosexual respondents of both genders. Lee subsequently applied his typology to same-sex relationships (Lee 1981). Researchers then developed a Love Attitudes Scale (LAS): eighteen to twenty-four questions that measure Lee's typology (Hendrick, Hendrick, and Dicke 1998). Although not all subsequent research has found all six dimensions, this typology of love styles has withstood the test of time, has proven to be more than hypothetical, and may even have cross-cultural relevance (Lacey et al. 2004; Le 2005; Masanori, Daibo, and Kanemasa 2004).

Love styles are sets of distinctive characteristics that loving or lovelike relationships take. The word *lovelike* is included in this definition because not all love styles amount to genuine loving as defined in this chapter. People may incorporate different aspects of several styles into their relationships. What are Lee's six love styles?

1. *Eros* (AIR-ohs) is a Greek word meaning "love"; it forms the root of our word *erotic*. This love style is characterized by intense emotional attachment and powerful sexual feelings or desires. Sustained relationships established by erotic couples are characterized by continued and emotionally intense sexual interest. A sample question on the LAS designed to measure eros asks respondents to agree or disagree with the following statement: "My partner and I have the right chemistry between us" (Hendrick, Hendrick, and Dicke 1998).

2. *Storge* (STOR gay) is an affectionate, companionate style of loving. This love style focuses on deepening mutual commitment, respect, friendship over time, and common goals. Storgic lovers' basic attitudes to their partners are one of familiarity: "I've known you a long time, seen you in many moods" (Lee 1973, p. 87). Storgic lovers are likely to agree that "I always expect to be friends with the one I love" (Hendrick, Hendrick, and Dicke 1998).

3. *Pragma* (PRAG-mah) is the root word for *pragmatic*. Pragmatic love emphasizes the practical element in human relationships and rational assessment of a potential partner's assets and liabilities. Arranged marriages are often examples of pragma, as is a person who decides rationally to get married to a suitable partner. The following is one LAS statement that measures pragma: "A main consideration in choosing a partner is/was how he/she would reflect on my family" (Hendrick, Hendrick, and Dicke 1998).

and intimacy by evading relationships, demonstrating ambivalence, seeming preoccupied, or, among men, rejecting romance and expressing hostile attitudes toward women (Fletcher 2002; Hart et al. 2012). Attachment style has been found to be associated with college students' motives for "hooking up" (discussed in Chapter 4). Compared to students who had secure attachment styles, students who had avoidant and anxious styles were more likely to use hooking up as a means to cope and deal with disappointment, and, among men, to boost self-confidence and impress peers (Snapp et al. 2014). Attachment style is also associated with young adults' use of communication technology in romantic relationships. Those with the avoidant style were more likely to use e-mail as opposed to phone calls and texting (Morey et al. 2013). Those with an insecure/anxious attachment style texted more often and were more likely to look at their partners' Facebook pages (Marshall et al. 2013; Nowinski 2014)—perhaps to check up on their partners?

Attachment theory and Sternberg's triangular theory of love are not the only ways of looking at love, of course. "Facts about Families: Six Love Styles" analyzes love in yet a third way. A fourth way to better understand love is to think about what love is *not*. We turn now to an examination of three things love isn't: martyring, manipulating, and limerence.

Three Things Love Is Not

Love is not inordinate self-sacrifice. And loving is not the continual attempt to get others to feel or do what we want them to—although each of these ideas is frequently mistaken for love. Nor is love all the crazy feelings you get when you can't get someone out of your mind. We'll examine these misconceptions in some detail.

Martyring Martyring involves maintaining relationships by consistently minimizing one's own needs while

4. *Agape* (ah-GAH-pay) is a Greek word meaning "love feast." Agape emphasizes unselfish concern for a beloved's needs even when that requires personal sacrifice. Sometimes called *altruistic love* or *compassionate love*, agape emphasizes nurturing others with little conscious desire for a return other than the intrinsic satisfaction of having loved and cared for someone else (Fehr, Harasymchuk, and Sprecher 2014). Agapic lovers would likely agree that "I try to always help my partner through difficult times" (Hendrick, Hendrick, and Dicke 1998).

5. *Ludus* (LEWD-us) focuses on love as play or fun. Ludus emphasizes the recreational aspects of sexuality and enjoying many sexual partners rather than searching for one serious relationship. Of course, ludic flirtation and playful sexuality may be part of a more committed relationship based on one of the other love styles. LAS statements designed to measure ludus include "I enjoy playing the game of love with a number of different partners" (Hendrick, Hendrick, and Dicke 1998).

6. *Mania*, a Greek word, designates a wild or violent mental disorder, an obsession, or a craze. Mania involves strong sexual attraction and emotional intensity, as does eros. However, mania differs from eros in that manic partners are extremely jealous and moody, and their need for attention and affection is insatiable. Manic lovers alternate between euphoria and depression. The slightest lack of response from a love partner causes anxiety and resentment. Manic lovers would be likely to say, "When my partner doesn't pay attention to me, I feel sick all over" or "I cannot relax if I feel my partner is with someone else" (Hendrick, Hendrick, and Dicke 1998). Because one of its principal characteristics is extreme jealousy, we may learn of manic love in the news when a relationship ends violently. Of Lee's six love styles, mania least fits our definition of love as described earlier.

How do these love styles influence relationship satisfaction and continuity? Psychologists Marilyn Montgomery and Gwendolyn Sorell (1997) administered the LAS to 250 single college students and married adults of all ages. They found that eros can last throughout marriage and is related to high satisfaction. Agape is also positively associated with relationship satisfaction (Neimark 2003). Interestingly, Montgomery and Sorell found storge to be important only in marriages with children. Ludus did not necessarily diminish relationship satisfaction among those who are mutually uncommitted. However, ludic attitudes have been empirically associated with diminished long-term relationship and marital satisfaction (Le 2005; Montgomery and Sorell 1997).

Critical Thinking

Thinking about relationships (if you are in one), what is your and your partner's "love style"? Is this a different love style than you have had in other relationships?

trying to satisfy those of one's partner. Periods of self-sacrifice are necessary through difficult times. However, excessive self-sacrifice or martyring is unworkable. Martyrs may have good intentions, believing that love involves doing unselfishly for others without voicing their own needs in return. However, because martyrs give their power away, they may feel helpless and grow angry and resentful (Garcy 2013). A martyr's reluctance to express his or her needs is damaging to a relationship because it prevents openness and intimacy and encourages dependency on the relationship.

Manipulating **Manipulating** means working to control the feelings, attitudes, and behavior of your partner or partners in underhanded ways rather than by directly stating your case. A manipulator may act one way in private and another in front of different people, make excessive demands, and doesn't take "no" for an answer (Ni 2014). When not getting their way, manipulators are likely to find fault with a partner, sometimes with verbal abuse. "You don't really love me," they may accuse. Manipulating, like martyring, can destroy a relationship.

Limerence Have you ever been so taken with someone that you couldn't get him or her out of your mind? Although the object of your attention may be unaware of your feelings, you review every detail of the last time you saw him or her and fantasize about how you might actually develop a relationship. Psychologist Dorothy Tennov ([1979] 1999) named this situation **limerence** (LIM-er-ence). She makes the following points: First, limerence is not just "lust" or sexual attraction. People in limerence fantasize about being with the limerent object in all kinds of situations—not just sexual ones. Second, many of us have experienced limerence—it has biochemical origins related to the hypothalamus and pituitary gland and the chemicals they produce including dopamine, the feel-good hormone. Third, limerence can possibly turn into genuine love, but more often than not, it doesn't. What to do if you are experiencing

these three issues in a romantic relationship? Some advice, in addition to openly communicating how you are feeling, is to assert yourself (if you are a martyr), back-off (if you see yourself engaging in manipulative behaviors), and do a "reality check" (if experiencing limerence). Overall, ask yourself what your relationship is really about.

People discover love; they don't simply find it. The term *discovering* implies a process—developing and maintaining a loving relationship require seeing the relationship as valuable; committing to mutual needs, satisfaction, and self-disclosure; engaging in supportive communication; and spending time together. We now turn to factors that affect how that love plays into the selection of a life partner.

MATE SELECTION: THE PROCESS OF SELECTING A COMMITTED PARTNER

The popular notion of finding one's lifelong partner is that it resembles being struck by Cupid's arrow—it happens suddenly, unexpectedly, and by chance. Realistically, settling on a mate is a much more complex process. Imagine a large marketplace in which people come with goods to exchange for other items. In traditional societies, a person may go to market with a few chickens to trade for some vegetables. In more industrialized societies, people attend hockey-equipment swaps, for example, trading outgrown skates for larger ones. People choose partners in much the same way: They enter the **marriage market** armed with resources—personal and social characteristics—and then they bargain for the best "buy" they can get.

Arranged versus Free-Choice Marriages

In much of the world, particularly in parts of Asia and Africa that are less Westernized, parents have traditionally arranged their children's marriages. In **arranged marriage**, future spouses can be brought together in various ways. In India, parents typically check prospective partners' astrological charts to ensure future compatibility. Traditionally, the parents of both prospective partners (often with other relatives' or a paid matchmaker's help) worked out the details and then announced the upcoming marriage to their children. Although arranged marriage has declined in India, newly married couples still show hallmarks of arranged marriage, such as spouses not meeting until their wedding day (Allendorf and Pandian 2016). Today it is more common for the children to marry only when they themselves accept their parents' choice. Unions

like these, sometimes called *assisted* or *semi-arranged marriages*, can be found among some Muslim groups and other recent immigrants to the United States (Ingoldsby 2006b; MacFarquhar 2006). And in Nepal, although most marriages remain arranged, couples often talk or meet beforehand, facilitated by technology (Diamond Smith, Minakshi Dahal, Puri, and Weiser 2019).

The majority of young couples in cultures that have traditionally practiced arranged marriage continue to heed extended family members' opinions about a prospective mate (Zhang and Kline 2009). The fact that marriages are arranged doesn't mean that love is ignored by parents. Indeed, marital love may be highly valued. However, couples in arranged marriages are expected to develop a loving relationship *after* the marriage, not before (Tepperman and Wilson 1993). How having an arranged marriage versus a freely chosen one affects marital satisfaction is unclear. Whereas some studies show higher marital satisfaction among couples in arranged marriages than "love marriages" (Nadia and Fatima 2015), others show that couples arranged marriages are less happy or that type of marriage made no difference (Madathil and Benshoff 2008; Myers, Madathil, and Tingle 2005). With increasing globalization, arranged marriages have become less common, especially among those with higher education (D. Jones 2006; Zang 2008).

There has been increasing concern both in the United States and around the world that children, usually girls, are being forced into arranged marriages, mostly with older men. One in three girls in the developing world marries before age 18. A new United Nations initiative aims to end child marriage by 2030 (UN News 2016). Thousands of girls in the United States get married every year. In 2014, nearly 60,000 15- to 17-year-olds were married (McClendon and Sandstrom 2016). Some religious and cultural groups exploit the fact that many states have low age requirements for marriage, and loopholes that allow parents to get around age requirements such as getting judicial "consent." Lawmakers in many states are working on laws banning the marriage of children. For example, Connecticut banned marriage before age 16, New York raised its minimum marriage are from 14 to 17, and Texas set new requirements limiting marriage to only those minors who have been emancipated from their parents (Tsui 2017).

The United States is an example of what cross-cultural researchers call a **free-choice culture**: People choose their own mates, although often they seek parents' and other family members' support for their decision. Immigrants who come to the United States from more collectivist cultures, in which arranged marriages have been the tradition, may face the situation of living with a divergent set of expectations for selecting a mate. Some immigrant parents from India, Pakistan, and

other countries arrange for spouses from their home country to marry their offspring. Either the future spouse comes to the United States to marry the young person, or the young person travels to the home country for a wedding ceremony, after which the newlyweds usually live in the United States (Dugger 1998). In these marriages, it is typical for that one partner to be more westernized, so the spouse needs to adjust not only to marriage but also to a new culture. The incidence of **transnational marriage**, the marriage of people from two different countries, may decline due to the recent tightening of immigration laws making it harder for the immigrating spouse to obtain a visa. According to Bélanger and Flynn (2018), migration for the purpose of marriage is today viewed with suspicion and is associated with gender stereotypes such as women looking for "a free meal ticket" and men looking for "mail order brides," yet "gender is entrenched into the marriage migration phenomenon" and must be studied further (p. 197).

Regarding arranged marriages, parents go through a bargaining process not unlike what takes place at a traditional village market. They make rationally calculated choices after determining the social status or position, health, temperament, and, sometimes, physical attractiveness of their prospective son- or daughter-in-law. Sometimes, as in the Hmong culture, the exchange involves a *bride-price*, money or property that the future groom pays the future bride's family so that he can marry her. More often, the exchange is accompanied by a *dowry*, a sum of money or property the female brings to the marriage in the form of cash, jewelry, furniture, electronics, and other household items. Typically, the higher the dowry amount the higher the woman's status in the household (Makino 2019). Dowries are associated with child marriage, domestic abuse, and daughters' families being sentenced to lifelong debt and payments to the groom's family. Dowries are illegal in many countries, including India and Bangladesh (Rahman et al. 2018). Nevertheless, even in countries where dowries are illegal, the practice remains widespread (Zakaria 2018).

With arranged marriage, the bargaining is obvious. The difference between arranged marriages and marriages in free-choice cultures may seem so great that we are inclined to overlook an important similarity: *Both* involve bargaining. What has changed in free-choice societies is that individuals, not family members, do the bargaining.

Social Exchange

The idea of bargaining resources in relationships comes from exchange theory, which is discussed in Chapter 2. A basic idea of exchange theory is that whether relationships form or continue depends on the rewards and costs they provide to the partners. Individuals, it is presumed, want to maximize their rewards and avoid costs, so when they have choices they will pick the relationship that is most rewarding or least costly. This analogy is to economics, but in relationships, individuals are thought to have other sorts of resources to bargain with besides money: physical attractiveness, intelligence, educational attainment, family status, and so on. Individuals may also have costly attributes, such as belonging to a different social class, religion, or racial/ethnic group; being irritable or demanding; and being geographically inaccessible (less of a consideration with technological advances but still a major consideration in modern society). Already having children or having been previously married, for example, can be a costly attribute (Goldscheider, Kaufman, and Sassler 2009; Qian and Lichter 2018). Men and women are more selective in choosing a mate the more they perceive they could find a desirable partner (Sprecher, Econie, and Treger 2019).

An unbalanced sex ratio—an imbalance in the number of single males and females—can also affect the marriage market. For example, the preference for sons in China and India has skewed the sex ratio, making it difficult for men to find wives (*The Economist* 2016). The Communist Youth League sponsored a dating event to address "the marriage problem" in China—that women have become "pickier" and men have had to work harder to find a wife (Rauhula 2018). In the United States, there are currently twice as many women as men on college campuses, one of the main venues for finding a mate (Marano 2015). A lack of men both numerically and in terms of "marriagability" (having adequate income potential) is a particular problem among African Americans. Higher rates of homicide, incarceration, and unemployment among black men and higher rates of college completion among black women diminish the dating pool available to black women, driving marriage rates down (Sawhill and Venator 2015). Yet, even in the context of unbalanced sex ratios that would favor marriage outside one's racial, cultural, or ethnic group, men and women still tend to marry within their own grouping (Kalmijn and Van Turbergen 2010).

The Traditional Exchange Historically, women have traded their ability to bear and raise children, coupled with domestic duties, sexual accessibility, and physical attractiveness, for a man's protection, status, and economic support; this is referred to as the **exchange model** of mate selection. Evidence from dating websites shows that the traditional exchange of money for beauty is still important. England and McClintock (2009) note "the double of standard of aging," in which age more negatively affects marriage prospects for women than for men (Boxer et al. 2011; McFarland, Jurafsky, and Rawlings 2013). Exchanges vary by gender in the expected direction, with more women than men being

Although the arranged marriage of this couple in northern India (right) may seem to be a world apart from the more freely chosen marriage of this couple in the United States (left), bargaining has occurred in both unions. In arranged marriages, families and community do the bargaining based on assets such as status, possessions, and dowry. In freely chosen marriages, the individuals perform a more subtle form of bargaining, weighing the costs and benefits of personal characteristics, economic status, and education.

more willing to marry someone older by five or more years, who already had children, who would earn much more than themselves, who had been married before, who had more education, and who was not "good looking." Men were more willing to marry someone younger by five or more years, not likely to hold a steady job, and who would earn much less (Bech-Sørensen, and Pollet 2016). However, the magnitude of the differences is less than in the past, as is one's willingness to marry someone of a different race or education level. Lower status, lower earning men may go outside the country in search of a traditional exchange. In one study of transnational marriages between U.S. men and Eastern European women, wives were an average of nine years younger than their husbands, four times greater than the age difference among U.S. couples (Levchenko and Solheim 2013). An analysis of data from the American Community Survey by Balistreri, Joyner, and Koa (2017) provides evidence of "trading youth for citizenship." They found that immigrants to the United States after marrying a U.S. citizen tend to have much older spouses, especially women. Even though over two-thirds of women contribute equally to the household income and one-third earn more than their partner, the majority of men *and women* (about 70 percent) still feel that it is very important for a man to be able to support a family to be a good husband or partner. Only 25 percent of men and 39 percent of women think the same is true for women (Parker and Stepler 2017).

As indicated above, mate selection based on status exchanges may be declining. Today, many wives have more education than their spouses (Lamidi, Brown, and Manning 2015b) and men are placing more emphasis

on women's financial prospects. On the other hand, women are *not* placing more value on men's ability to be good at childcare and housework, indicating that women's work continues to be devalued in the marriage market (Boxer, et al. 2015). There is evidence that the beauty–financial exchange may have been exaggerated because previous studies did not control for **matching**, which is the tendency of individuals to select partners with characteristics similar to their own. Men and women tend to agree that a spouse or partner be well educated to make a good spouse or partner (Parker and Stepler 2017). Another study showed no evidence of beauty–status exchanges among young adults once matching is taken into account (McClintock 2014a). Older daters are more likely to follow traditional dating exchanges (McWilliams and Barrett 2014).

ASSORTATIVE MATING: A FILTERING-OUT PROCESS

Individuals gradually filter or sort out those they think would not make the best life partner or spouse. Research has consistently shown that people are willing to date and live with a wider range of individuals than they would marry (Jepsen and Jepsen 2002; Manning and Smock 2002). Social psychologists call this process **assortative mating**. The assortative mating theory posits that mate selection involves narrowing down the possibilities until a suitable partner is found. The process operates not unlike a funnel, as illustrated in Figure 5.2. Some people, of course, do not make it through these stages and therefore start the process again until

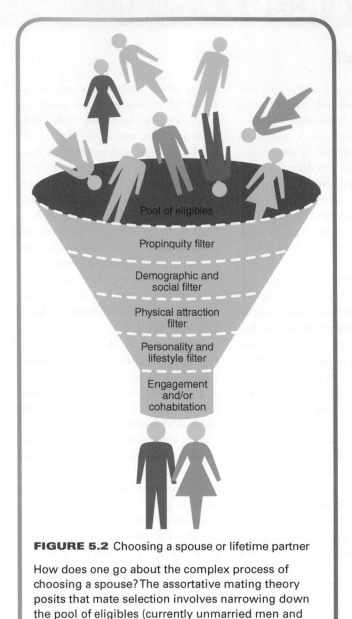

FIGURE 5.2 Choosing a spouse or lifetime partner

How does one go about the complex process of choosing a spouse? The assortative mating theory posits that mate selection involves narrowing down the pool of eligibles (currently unmarried men and women) using a series of filters until a suitable partner is found. Finally, the couple announces their intentions by getting engaged, moving in together, or both.

the desirable mate is found. Assortative mating raises another factor that shapes partner choice—the tendency of people to form committed, and especially marital, relationships with others with whom they share certain social characteristics. Social scientists term this phenomenon **homogamy**.

Individuals tend to make relationship choices in socially patterned ways, viewing only certain others as potentially suitable. The market analogy would be to choose only certain stores or websites at which to shop. Each shopper has a socially defined **pool of eligibles**: a group of individuals who are unmarried or unpartnered. The pool of prospective mates starts out large but is made smaller by geographical availability, demographic and social characteristics, physical attraction, and personal and lifestyle factors.

Geographic Availability Geographic availability, or *propinquity,* has historically been an important factor in finding a mate—that is, marrying "the girl next door" (Harmanci 2006; Travis 2006). Propinquity remains important even today. In a study of marriage license applications in Philadelphia, one-third of couples lived within five blocks of each other (Margalit 2014). This is why singles in rural areas are disadvantaged. As Rachel Monroe (2016) wrote in "What it's like to play Tinder in Rural America," dating apps give the impression that there are many more potential mates than there are. She interviewed Gabriel, age 24, from a small town in Texas, who gave up looking for a match. She explained, "Gabriel signed in and instructed the app to look for men ages 18 to 30 within a ten-mile radius. Nothing. Then a thirty-mile radius. Tinder's red circle blinked forlornly, like a radar seeking out a target that didn't exist." Elizabeth, age 35, who lives in Missouri near the Arkansas border stated, "It's slim pickings." Chances are that they, a friend, or family member already dated them. There are positives to small town dating. For example, you know their dating history and if they cheat, you'll know that too (Caldwell 2016)!

Geographic segregation, which can result from either discrimination or strong community ties, contributes to homogamous marriages (Iceland and Nelson 2008; Lichter et al. 2007). Intermarriage patterns within the American Jewish community are an example. Only about 6 percent of Jews married non-Jews in the late 1950s. Now that the barriers that once excluded Jews from certain residential areas and colleges are gone, about half marry gentiles (Sussman 2006). As the sizes of various immigrant communities in the United States grow, the geographic availability of eligibles of the same ethnicity increases, resulting in greater ethnic homogamy (Gowan 2009; Qian and Lichter 2007).

Geographic availability helps account for some educational and social class homogamy. Middle-class people tend to socialize together and send their children to the same schools; upper-and lower-class people do the same. Princeton alumna Susan Patton sparked controversy by recommending that women find their husband in college, which was widely misinterpreted as recommending that women marry young (Lara 2013). In reality, she was simply describing the current marriage market for women. "For most of you, the cornerstone of your future and happiness will be inextricably linked to the man you marry," Patton warns. "You will never again have this concentration of men who are worthy of you. Here's what nobody is telling you: Find a husband on campus before you graduate" (Lucas 2013).

As We Make Choices

Looking for Love on the Internet

Before the Internet, e-mail, and social media, the process of finding a future mate proceeded in a relatively straightforward fashion: A couple would meet through friends, family, classes, bars, or "by chance" and then might chat about their backgrounds and common interests, at which time they might set up a date to get to know each other better. These days, options for finding romantic partners have been greatly expanded by technology, through dating websites, social media such as Facebook, various apps that arrange meetings between individuals and groups of friends, online video games, and the chat rooms of various interest and support groups that allow real-time interaction between strangers in different communities, states, and countries.

Such dating mechanisms have many advantages. Technology improves dating efficiency (Slater 2013). Dating sites allow participants to select mates with desirable characteristics before meeting so that participants don't "waste time" on unsuitable partners. There are special websites that cater to men and women wanting a partner who shares their religion, age, ethnicity, or educational level. Dating sites encourage homogamy by first matching partners on a range of characteristics, including education, income, desire for children, hobbies and interests, and even preferences for a particular height and body type. Only then are partners allowed to get to know each other through e-mails, instant messaging, or face-to-face meetings.

These advantages may be particularly attractive to certain groups. For example, young people who have relocated to a new community for employment are often cut off from friends and relatives who would help facilitate meetings with suitable partners. Single men and women who are parents usually lack both time and the child care necessary to get out and meet people in traditional ways; instead, they can meet online. Indeed, populations in the "thin" marriage market (e.g., divorced people, physically disabled) are especially likely to find partners online (Barrow 2010; Rosenfeld and Thomas 2012). Typical settings for finding mates remain geared toward young adults and revolve around alcohol, which not everyone, especially older singles, is comfortable with. Likewise, some people are ill at ease in public settings, which hampers their success on the dating scene. Women may feel more comfortable reaching out to men online as opposed to in person (Kreager et al. 2014). Online dating may provide a mechanism for gay men and lesbians to meet a wide range of potential partners with various relationship intentions from short-term relationships to long-term, monogamous ones (Potârcă, Mills, and Neberich 2015).

Dating online also has disadvantages. Online profiles are incomplete—participants play up positive qualities and downplay or omit negative ones (Hertlein 2012). "Chemistry" and physical attraction can be difficult to assess. Many elements of attraction are biological and unconscious in their origins and may have developed as human beings evolved. Online dating requires good written communication skills, which individuals possess in varying degrees (Bucior 2012; Weidman et al. 2015). Prescreening potential

Today, first encounters may occur online, and people meet others as far away as other continents. Some websites facilitate matches between men and women of different races and ethnicities. However, the Internet may actually encourage pairings within religious or racial/ethnic groups, who can advertise online for homogamous dating partners (see also Desmond-Harris 2010; Wimmer and Lewis 2010). Illustrating these points, Russel K. Robinson, a black gay man writing in the *Fordham Law Review,* describes his experience as follows:

Although I lived on the wealthy, predominantly white west side of [Los Angeles], the Internet created opportunities for me to interact with men in [less wealthy areas of the city]—men I almost certainly would not meet randomly while going through my daily routine. . . . Even as the Internet increases romantic opportunity, it also channels interactions. . . . Like many dating websites, Match.com prompts the user to indicate which races he will and will not date. . . . [I]f a white user is interested only in white romantic partners, he can easily structure his screen so that he never even has to view nonwhite profiles. (Robinson 2008, 2791–2792)

Recently, websites and apps have been developed to facilitate pairings between political conservatives and Republicans. One is "Donald Daters"—an app for Donald Trump supporters with the tagline *Make America Date Again* (Del Valle 2018).

Demographic and Social Filter People tend to form committed relationships with people of similar race, age, education, religious background, and social class

mates on demographic, physical (e.g., blonde hair, blue eyes), and social factors may cause participants to miss out on partners with whom they might mesh in terms of personality, life goals, and dreams for the future. In a study of 1,855 people from a subscription-based dating site, the list of "deal breakers" is long: age, height, weight, education, marital status, number of children, and smoking and drinking habits. They found that women were less likely to browse the profile of a shorter man, and men were less likely to browse the profiles of heavier women (Bruch, Feinberg, and Lee 2016). On other hand, because profiles are public, members can be inundated with texts and e-mails from people they are not interested in. Apps such as Tinder allow men and women to scroll through the pictures of potential dates, swiping right to "like" people they are interested in and swiping left to continue their search (Knox 2014). Although physical attraction is important, dating apps based on superficial qualities such as appearance can lead to hurt feelings, and rejection received through social media can be harsh. Moreover, women who use these apps are more often the victims of online

harassment. Thirty-five percent of teen girls have blocked or unfriended someone whose "flirting" made them uncomfortable (Lenhart, Anderson, and Smith 2015). It pays off to look beyond physical attributes. In a study of keywords from the dating site OKCupid, women were much more responsive to men who noted their interests and complimented their online profile than to those who mentioned their appearance (Wade 2015b). In a speed-dating experiment with undergraduates, women were more attracted to dates that were high on "mindfulness," meaning their dates showed them their undivided attention; not so for men (Janz, Pepping, and Halford 2015).

Dating websites also encourage continuous dating rather than marriage. Meeting online as opposed to offline is also associated with higher breakup rates (Paul 2014). As opposed to being known by people in one's social network, online dates are "virtual strangers" who may introduce risks of physical, emotional, or financial harm (Rosenfeld and Thomas 2012). Therefore, it is important to meet potential dates in a public place, get their full name, and let a family member or friend know who they are

Men and women of all ages are increasingly "looking for love" online.

going out with and when and where. State sex-offender registries and online public court records should be used to check out dates before a first meeting. Most singles don't rely exclusively on technology. Online dating may be best thought of as part of a single person's "tool kit" in the quest for love.

Critical Thinking

Have you, or someone you know, ever tried online dating? Would you recommend it to a single friend or family member? Do you think dating websites take the romance out of dating?

(Braithwaite et al. 2015; Lin and Lundquist 2013). As an example, the Protestant, Catholic, and Jewish religions, as well as the Muslim and Hindu religions, have all traditionally encouraged **endogamy**: marrying within one's own social group.

Why is it that men and women with similar characteristics tend to partner with each other? For one thing, people often find it easier to communicate and feel more at home with others like themselves (Lewin 2005). They are likely to have attitudes, mannerisms, and vocabulary similar to one another. They feel comfortable in one another's surroundings. There is also social pressure from friends and family members (Dubbs, Buunk, and Li 2012). Friendship networks tend to be highly homogeneous, encouraging similar people to meet and interact (Wimmer and

Lewis 2010). Inter-ethnic relationships are more likely to develop when young adults are independent of parental influence or when one's parents have an ethnically diverse network of friends (Rosenfeld and Byung-Soo 2005). Despite increased immigration, most native- and foreign-born people marry within their own nativity group (Lichter, Qian, and Tumin 2015). Populations with smaller dating pools are more open to differences, perhaps out of necessity (Conway et al. 2015). Older people are less selective when choosing a mate than are younger people, and LGBTQ+ couples are less homogenous than different-sex couples (Rosenfeld and Kim 2005; Schwartz and Graf 2009). People who have been previously married also "cast a wider net" resulting in greater heterogamy among remarried couples (Qian and Lichter 2018).

Physical Attraction Filter Physical attraction is a step in assortative mating that is difficult to measure because it involves both social and biological factors. Physical attractiveness is a powerful resource in the marriage market. Physically attractive men, for instance, use this as a mating strategy—they are found to have a greater number of sex partners than less attractive men (McClintock 2011). On the other hand, more physically attractive women are more successful than less attractive women in delaying sexual intercourse and securing committed relationships.

A look at wedding photos, as well as a large body of empirical research, indicates that most individuals marry a partner of similar physical attractiveness as their own. Couples even tend to look alike, as a result of the preference for people to look like themselves and even their parents (Ducharme 2019). Based on the National Longitudinal Study of Adolescent Health's Romantic Pair data, Carmalt, Cawley, Joyner, and Sobal (2008) found physical attraction to be the most important mate-selection criterion. This extended not only to facial appearance but also to body weight. Being obese was found to be more detrimental to women than to men, and it was more of a barrier to white women than to black women in obtaining an attractive partner. However, this can be offset by greater education, a good personality, and attention to grooming.

Although physical attractiveness is important, it is not the only criterion people use when picking a mate. A study of the preferences of adults ages 18 to 40 suggests that physical attractiveness is less important as people grow older (Sprecher et al. 2019). In a study of men and women in their twenties, care, likability, and extroversion ranked higher than attractiveness (Boxer et al. 2015). Today, it is not unusual for a romantic relationship to start out as a friendship. These couples tend to place less emphasis on attractiveness than couples who meet in more conventional ways (Hunt, Eastwick, and Finkel 2015).

Personality and Lifestyle Filter Personality traits can set the tone of the "emotional climate" of marriages. Partners who experience negative patterns of interaction during courtship have less-satisfying marriages (Markman et al. 2010; Wilson and Huston 2013). For relationships to flourish in the long term, couples need to be not only physically and emotionally compatible but also intellectually, ideologically, and spiritually compatible (De La Lama, De La Lama, and Wittgenstein 2012).

The Final Filter: Cohabitation and Engagement Engagement is the way most couples indicate to others that they are "serious" in their intentions toward one another (Schweingruber et al. 2004). As such, engagement activities often take place in public settings, from picking out rings to the proposal itself. The couples' engagement provides family and friends a final opportunity to approve the relationship before going forward with the marriage. Cohabitation has become an increasingly important "filter factor" in the decision to marry and some cohabitors have definite plans to marry their partner. The stigma attached to cohabitation has declined considerably. The majority of Americans today feel that a cohabiting couple who have lived together for five years or more is just as committed as a married couple, and 69 percent of American teens ages 13 to 17 approve of living together before marriage. Older, more conservative, and more religious individuals remain less supportive of cohabitation (Gallup Poll 2012; Lyons 2004).

The average length of a cohabiting relationship is longer than in the past. Demographers Esther Lamidi, Wendy Manning, and Susan Brown's (2019) analysis of the National Survey of Family Growth showed that cohabitations formed between 2006 and 2013 lasted an average of eighteen months, compared to an average of twelve months for cohabitations formed in the mid-1980s. They attribute this to fewer cohabiting couples transitioning to marriage as opposed to breaking up. Marriage rates are falling but the percent of U.S. adults in cohabiting relationships has either stayed the same (for those under 30 and 40 and older) or has risen (for those in their thirties; Saad 2015).

"Today, cohabitation is increasingly thought of as part of the marriage process" (Guzzo 2009a, p. 198). Since the 1970s, the proportion of marriages preceded by cohabitation has grown steadily. Seven in ten women who married between 2010 and 2014 cohabited with their spouse prior to their wedding (Hermez and Manning 2017). Although some cohabitors have no intention of marrying their partner, many cohabitors mention the importance of "testing out" the relationship before making the final step toward marriage. One study found that cohabitors who had talked about future marriage had "generally been living with their partners for about two years, indicating that the issue of greater permanence in their relationships surfaces over time" (Sassler 2004, p. 501). Young people are increasingly seeing cohabitation as just another step in the courtship process. In 1976, only 40 percent of high school seniors viewed cohabitation as a "testing ground" for marriage compared to 71 percent in 2016 (Anderson 2016).

As cohabitation grew, researchers have carefully studied whether cohabitation affects subsequent marital success. Many young people today follow the intuitive belief that "cohabitation is a worthwhile experiment for evaluating the compatibility of a potential spouse, [and therefore] one would expect those who cohabit first to have even more stable marriages than those who marry without cohabiting" (Seltzer 2000, p. 1,252). Among high school seniors, about two-thirds agreed that "[i]t is

usually a good idea for a couple to live together before getting married in order to find out whether they really get along" (National Marriage Project 2009).

In contrast to the notion that cohabitation makes marriages more stable, many studies found living together before marriage increased the likelihood of divorce (Jose, O'Leary, and Moyer 2010). The proposed reason is that when cohabitation was less common, cohabitors "selected" themselves into cohabitation, referred to as the **selection hypothesis**. This assumes that people who cohabit are different from those who do not; these differences translate into higher divorce rates. For example, serial cohabitors tend to have relatively low levels of education and income, less effective problem-solving and communication skills, and more negative views of marriage—factors related to divorce (Amato et al. 2008; Lichter and Qian 2008; Rhoades, Stanley, and Markman 2009; Thornton, Axinn, and Xie 2007). Since cohabitation is now so common, it was thought that the positive relationship between cohabitation and divorce would diminish over time, referred to as the **normalization hypothesis** (Rosenfeld and Roesler 2018). For example, in Norway, where the characteristics of cohabiting and married couples are largely similar and where they receive the same government benefits, premarital cohabitation had no negative effect on marital stability (Liefbroer and Dourleijn 2006).

Another theory has been used to explain higher divorce rates among cohabitors. The **experience hypothesis** posits that cohabiting experiences themselves affect individuals so that, once married, they are more likely to divorce (Amato 2010). For instance, cohabitors may discover they like the lower level of commitment required by cohabitation (Thomson and Colella 1992). There is also evidence that "young adults become more tolerant of divorce as a result of cohabiting, whatever their initial views were," possibly because "cohabiting exposes people to a wider range of attitudes about family arrangements than those who marry without first living together" (Seltzer 2000, p. 1,253). The experience of cohabitation can also be positive—for instance, couples learn to adapt to one another's tastes, likes, and dislikes. Additionally, housework is negotiated, and couples decide how to spend time together, how much sex to have, and so forth (*positive adaptation*). Rosenfeld and Roesler (2019) decided to put these hypotheses to the test using a six different cohorts of U.S. women ages 18 to 44 in first marriages between 1970 and 2015. They did not find support for the normalization hypothesis or positive adaptation hypothesis. They found that the positive relationship between premarital cohabitation and divorce remained. Why all the conflicting findings? They argue that early researchers did not account for marital duration and that "[t]he benefits of cohabitation experience in the first year of marriage has misled scholars into thinking that the most recent marriage cohorts will not experience heightened marital dissolution due to premarital cohabitation" (p. 1).

There are numerous factors that affect the relationship between cohabitation and marital dissolution. Education, income, race and ethnicity, and number of previous cohabitations affect how cohabitation affects subsequent relationship stability (Miller, Sassler, and Kusi-Appouh 2011; Phillips and Sweeney 2005; Reinhold 2010; Xu, Hudspeth, and Bartkowski 2006). Cohabitors who marry after having a child have lower marital relationship quality than do cohabitors who did not (Tach and Halpern-Meekin 2009); such couples may marry because they feel that they should rather than make more deliberative decisions (Stanley 2009). Level of commitment matters as well. In one study, cohabitors who are engaged have a lower risk of divorce (Manning and Cohen 2012). In general, cohabiting couples with similar understandings of the nature and goals of the relationship have a lower likelihood of divorce (Wilson and Huston 2013). Cohabitation is discussed in more detail in Chapter 6.

HETEROGAMY IN RELATIONSHIPS

The opposite of endogamy is **exogamy**, marrying outside one's group, or **heterogamy**—that is, choosing someone dissimilar in race, age, education, religion, or social class. With "the loosening of relationship conventions," more older women, for example, are dating or marrying men at least five years younger—the media-hyped "cougar" phenomenon (Kershaw 2009b). In a national survey, 13 percent of women reported having had sex with a man five years younger in the preceding year (Alarie and Carmichael 2015). Even so, in only 15 percent of married couples the wife was older than the husband (Lamidi, Brown, and Manning 2015a). With regard to socioeconomic class and education, although people today are marrying across small class distinctions, they still are not doing so across large ones. For instance, individuals of established wealth or high education levels seldom marry those who are poor or who have low educational achievement (Fu and Heaton 2008). This "loosening of relationship conventions" evidences itself in other types of heterogamy as well—namely, relationships that cross racial, ethnic, and religious lines.

Interracial and Inter-Ethnic Unions

As young adults experience increased independence from family influence, we can expect a rise in interracial and inter-ethnic unions (Rosenfeld 2008). **Interracial unions** include unions (married and

cohabiting) between partners of the white, black, Asian, or Native American races with a spouse outside their own race. Interracial marriage has existed in the United States throughout our history (Maillard 2008). However, not until 1967 (*Loving v. Virginia*) did the U.S. Supreme Court declare that interracial marriages must be considered legally valid in all states. It is important to note here that the U.S. Census Bureau defines Hispanics not as a race but, rather, an ethnic group. Unions between Hispanics and others, as well as between different Asian/Pacific Islander, Hispanic, or black ethnic groups (such as Thai–Chinese, Puerto Rican–Cuban, or African American–black Caribbean), are considered **inter-ethnic**.

Interracial and inter-ethnic marriage, now referred to simply as **intermarriage**, has grown substantially in recent years. In 1967, only 3 percent of newlyweds married somebody of a different race or ethnicity, compared to 17 percent in 2015 (Geiger and Livingston 2019). Note that cultural diversity can exist within racial and ethnic categories. For example, Asians include individuals from a variety of nations and cultures. Today, immigration has an important part to play in the marriage market. Most immigrants form racially homogenous unions, but there is also a high rate of marriage, and an even higher rate of cohabitation, between recent arrivals to the United States and native-born Americans (Qian, Glick, and Batson 2012).

Whereas intermarriage may have benefits, there can be disadvantages such as disapproval from parents (Farr 2011). Among Native Americans, children resulting from interracial unions are vulnerable to losing their federal tribal benefits because they must be at least one-quarter Indian (Ahtone 2011). There are substantial racial and ethnic differences in intermarriage. As shown in Figure 5.3, American Indians have the highest rate of intermarriage, followed by Asians, Hispanics, blacks, and whites. The most common racial or ethnic pairing is between Hispanics and whites, representing 42 percent of all newly intermarried couples. There are also gender differences in interracial marriages. Most black–white marriages involve black men married to white women and most Asian–white marriages involve white men and Asian women (Livingston and Brown 2017). The percentage of racially or ethnically heterogeneous couples is similar among married and cohabiting couples (Livingston and Brown 2017).

Reasons for Interracial and Inter-Ethnic Relationships

Much attention has been devoted to why people intermarry or form interracial or inter-ethnic relationships when society's norms promote homogamous unions. One reason among racial/ethnic groups that are relatively small in number is simply that they have a smaller pool of eligible partners in their own race or ethnicity (Choi and Tienda 2017; Strully 2014). Intermarriages

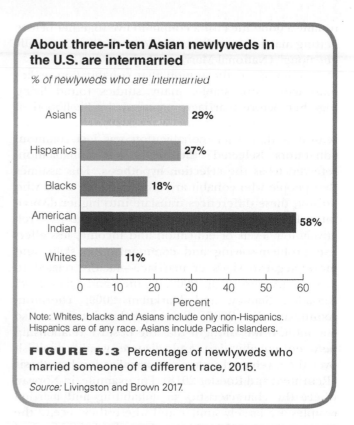

About three-in-ten Asian newlyweds in the U.S. are intermarried

% of newlyweds who are intermarried

Asians	29%
Hispanics	27%
Blacks	18%
American Indian	58%
Whites	11%

Note: Whites, blacks and Asians include only non-Hispanics. Hispanics are of any race. Asians include Pacific Islanders.

FIGURE 5.3 Percentage of newlyweds who married someone of a different race, 2015.

Source: Livingston and Brown 2017.

are also more common in metropolitan areas and in Western states where there is more racial and ethnic diversity (Livingston and Brown 2017).

Another explanation is the **status exchange hypothesis**—the argument that an individual might trade his or her socially defined superior racial/ethnic status for the economically or educationally superior status of a partner in a less-privileged racial/ethnic group (Kalmijn 2010). In this regard, racial stereotypes may play a part:

> [A] society dominated by Euro-Americans will unsurprisingly privilege a standard of beauty and cultural styles that is a mirror image of itself, even if that image is a media distortion. . . . [I]nterviews with Asian men and women [found] that a sizable minority of respondents preferred whites as potential or current mates because of their preference for "European" traits including tallness, round eyes, buffness for men and more ample breasts for women. (Gallagher 2006, p. 150)

Applying the status exchange hypothesis to black–white intermarriage would suggest that whites marry "up" by trading their socially defined superior racial status for the economically superior status of a middle-or upper-middle-class black partner. Studies of intermarriage among native Hawaiians, Japanese, Filipinos, and Caucasians in Hawaii and among white, brown, and black individuals in Brazil support this hypothesis (Fu and Heaton 2000; Gullickson and Torche 2014).

Some African Americans have expressed concern about black men—especially educated black men—choosing

spouses from other races (Crowder and Tolnay 2000; Staples 1994, 1999). Robert Davis, a past president of the Association of Black Sociologists, believes that black men are inclined to see white women as "the prize" (Davis, quoted in "Why Interracial Marriages" 1996). Some African American women view black males' interracial relationships as "selling out"—sacrificing allegiance to one's racial heritage to date someone of higher racial status (Paset and Taylor 1991). This issue can be a problem on college campuses, where black men, especially athletes, are perceived as only pursuing relationships with white women (Wilkins 2012). According to one black student on a football scholarship: "I'm not gonna lie, bein' young, I thought that havin' sex with white women was also a sense of accomplishment . . . because my whole life I've been told that is off-limits" (Wilkins 2012, p. 175).

Because of the "shortage" of black men discussed earlier, the Council on Contemporary Families (2011) suggests that intermarriage may be a good thing for black women as well—and possibly a better route to marriage.

It is often assumed that interracial couples marry *in spite of* their partner being of a different race or ethnicity. In fact, some men and women purposefully marry someone of a different race, religion, or culture. One woman explains:

> Part of the fact that he speaks a different language and part of the fact that, like, a whole culture and that difference is what I like about him so much . . . the whole element of a different culture and different languages is a huge part of what attracts me to him. (Yodanis, Lauer, and Ota 2012, pp. 1029–1030)

In a study of online daters, biracial men and women were sometimes preferred over those who were of just one race (Curington, Lin, and Lundquist 2015). Women (and men) may seek a partner of a different race to escape what they perceive as patriarchal gender roles in their own culture. One Asian woman said, "I try and put in my effort, but he didn't see that cause he's always on the computer and then he go to work, and come home and expects, for me to be a homemaker and cook and clean and do all the stuff. At the time I was going to school, I was going to work." Another said, "I can't even picture myself with an Asian guy you know. I tried to stick to Asian, I really tried, but I found that the best respect I got was from my White boys, you know?" (Morgan 2015, p. 13). On the other hand, Asian men are often rejected by women of other cultures for being too feminine looking (Kao et al. 2018).

Acceptance of interracial and interracial marriage is growing among the public, especially among younger generations. According to data from the American Community Survey, half of Gen Zers (born after 1996) and Millennials (born between 1981 and 1996) say people of different races marrying each other is "a good thing" for society, compared to 41 percent of Gen Xers (born

Some ethnic groups, particularly those consisting of a large proportion of recent immigrants, strongly value homogamy. Nevertheless, an increasing number of Americans enter into ethnically heterogamous unions. Although "feeling at home"—a factor that encourages homogamy—may be difficult at first, some individuals thrive on the cultural variety characteristic of interracial or interethnic relationships.

between 1965 and 1980), 30 percent of Baby Boomers (born between 1946 and 1964), and 20 percent of people in the Silent Generation (born between 1928 and 1945). Acceptance is also higher among people of color, those with college degrees, and those who identify as Democrat as opposed to Republican (Geiger and Livingston 2019). However, intermarried couples still struggle with getting strange looks, slights, and other microaggressions. A 31-year-old consultant in Chicago recalled being at a party months after her engagement to a man from the Middle East. During a conversation with an acquaintance, the man, who had been drinking, said, "So you're getting married? Wow! When did you realize that he wasn't a terrorist?" Taken aback, she said dryly, "I think what you meant to say was congratulations on your recent engagement" (Foster 2016).

Interfaith Relationships

In 2010, Riley (2013) commissioned a poll of a nationally representative sample of 2,450 Americans that included an oversample of **interfaith marriages**. She found that approximately 45 percent of married couples married outside their religion (Riley 2013). Although religious involvement in Western countries has declined, whether or not to marry outside one's faith remains an important consideration for adults contemplating marriage. In a study of Australian university students by psychologists Silam Yahya and Simon Boag (2014), 20 percent perceived that their parents would disapprove if they married outside their religion or culture. One Muslim

man said, "I dated girls that weren't from my culture or faith but I always knew it wouldn't be serious. I can't marry them. . . . My family wouldn't accept it" (p. 764). Preserving religious traditions was also an important factor. As one student said, "People have always tried to destroy us Jews. . . . In a way marrying somebody from a different faith would be fulfilling Hitler's wish in my opinion" (p. 764). Other students were worried about the complications of raising children when the parents are of different religions or cultures.

Being highly educated seems to lessen individuals' commitment to religious homogamy (Petersen 1994). Religions that see themselves as the one true faith and people who adhere to a religion as an integral component of their ethnic/cultural identity (e.g., some Catholics, Jews, and Muslims) are more likely to encourage homogamy, sometimes by pressing a prospective spouse to convert (Bukhari 2004). Often religious bodies are concerned that children born into the marriage will not be raised in their religion (Sussman 2006). Because Americans are becoming less religious, perhaps religious homogamy can be expected to decline further.

Heterogamy and Relationship Quality and Stability

How does marrying someone from a different religion, social class, or race/ethnicity affect a person's chances for a happy union? In general, research suggests that marriages that are homogamous in age, education, religion, and race are the happiest and most stable (Bratter and King 2008; Jones 2010; Larson and Hickman 2004). This is also true for cohabiting couples (Hohmann-Marriott and Amato 2008).

Marriages that are homogamous are more likely to be stable because partners are more likely to share the same values and attitudes when they come from similar backgrounds (Durodoye and Coker 2008; Lincoln, Taylor, and Jackson 2008). Heterogamous marriages may create conflict between the partners and other groups such as parents, relatives, and friends. In these highly socially polarized times, the pressure to form homogenous marriages now includes political party. In 1958, one-third of Democrats and one-quarter of Republicans wanted their son or daughter to marry within their own party, compared to two-thirds of each group in 2016 (Enoch 2017). Continual discriminatory pressure from families and the broader society may create undue psychological and marital distress that can increase the risk of divorce (Bratter and Eschbach 2006; Childs 2008; Yancey 2007; Zhang and Van Hook 2009). Inter-ethnic and interracial marriages have a higher rate of divorce than homogamous ones (Wang 2012). Higher rates of divorce may also reflect the fact that these partners are likely to

be less conventional in their values and behavior and may divorce more readily than others (see Hohmann-Marriott and Amato 2008).

However, several studies do not support worse outcomes for interracial and inter-ethnic relationships. One study of unmarried interracial couples in college found *higher* relationship satisfaction compared to same race couples (Troy, Lewis-Smith, and Laurenceau 2006). A comparison of Mexican American–nonHispanic white marriages with those of homogamous white and homogamous Mexican American couples found little difference in marital satisfaction among the three groups (Negy and Snyder 2000). Moreover, a study of 23,139 married couples "failed to provide evidence that interracial marriage per se is associated with an elevated risk of marital dissolution." Instead, the risk of divorce or separation among the interracial couples sampled was similar to that of the race of the spouse from the more divorce-prone race. Accordingly, "[m] ixed marriages involving blacks were the least stable followed by Hispanics, whereas mixed marriages involving Asians were even more stable than endogamous white marriages" (Zhang and Van Hook 2009, p. 104). Because growth in intermarriage has occurred relatively recently, it's unclear how life unfolds as these couples age. In a Canadian study, compared to same-race unions, intermarried couples age 45 and older received less assistance from others with transportation, suggesting they may continue to face challenges. On the other hand, they received similar amounts of help with housework and greater emotional support (Penning and Wu 2013).

Interreligious marriages also tend to be more stressful and less stable than homogamous ones (Mahoney 2005; Riley 2013). One probable reason that religious homogamy improves chances for marital success involves value consensus. Religion and class-based values may come into play when negotiating leisure activities, child-raising methods, money, and spousal roles (Lambert and Dollahite 2006; Streib 2015). Some research shows a declining effect of religious differences on marital satisfaction over the past several decades due to the greater effect of couples' gender, work, and co-parenting roles (Myers 2006; Williams and Lawler 2003).

Minimizing Mate Selection Risk

Other chapters in this text focus on various aspects of partner interactions and social support. Here we focus on choosing a partner who is best predisposed psychologically to maintain a stable and committed relationship.

Psychologists and counselors advise choosing a partner who is integrated into society by means of school,

employment, and a network of friends and who fairly consistently demonstrates supportive communication and problem-solving skills (Cotton, Burton, and Rushing 2003). We're reminded that "[h]eavy or risky drinking is associated with a host of marital difficulties including infidelity, divorce, violence and conflict" (Roberts 2005, p. F13). The same can be said for other forms of substance abuse (Kaye 2005, F15). Another step in minimizing mate selection risk is to let go of misconceptions we might have about love and choosing a partner. According to psychologist Beverly Smallwood (2013), some of the most common love myths are:

- *The right person will meet all my needs.* In reality, it is too much to expect one person to meet your every need. One needs "God, friends, a strong sense of purpose, healthy self-esteem, and a willingness to take responsibility for your own happiness."

- *I can change my partner.* There is only one person you can change: yourself.

- *Love will conquer all.* Face differences in values, behavioral choices, backgrounds, and personal habits *before* making a lifelong commitment.

- *Love is a feeling.* According to Smallwood, "love is a verb. It's about doing—even in those temporary times when you inconveniently don't have wonderful feelings to stimulate the positive action."

- *We'll live happily ever after.* This is the idea that "real love" is something you won't have to work at. Smallwood says, "A marriage certificate is really a work permit." She says that "even the best relationships have potholes, tragedies, and disappointments" and that it is important to keep your eyes wide open.

Some couples go to counseling to assess their future compatibility and commitment (Carlson et al. 2012). Others may access marital compatibility tests on the Internet. Although we, your authors, are unable to attest to the efficacy of these, they do stimulate couple discussions about important topics.

Again, we see that private troubles—or choices—are intertwined with public issues. Some ethnic and religious groups strongly value homogamy. Meanwhile, it is also true that if people are able to cross racial, class, or religious boundaries while simultaneously sharing important values, they may open doors to a varied and exciting relationship. Chapter 9 explores raising children in interracial families. In our society, choice of life partners—whether homogamous or heterogamous—typically involves developing an intimate relationship and establishing mutual commitment. The next section examines these processes.

MEANDERING TOWARD MARRIAGE: DEVELOPING THE RELATIONSHIP AND MOVING TOWARD COMMITMENT

Social scientists have been interested in the process through which a couple develops their relationship and mutual commitment. What first brings people together? What keeps them together?

Young people today "meander toward marriage," feeling that they'll be ready to marry when they reach their late twenties or so (Arnett 2004, p. 197). Assuming that adolescents begin exploring and pursuing romantic relationships in their teen years, the period in which premarital relationships can occur may last five, ten, or even twenty years and sometimes longer (Wilson 2009; see also Meier and Allen 2009; Wallace 2007). Experiencing unprecedented freedom, today's young adults often express the need to explore as many options as possible before "settling down." As one young woman explained:

> I think everyone should experience everything they want to experience before they get tied down, because if you wanted to date a black person, a white person, an Asian person, a tall person, short, fat, whatever, as long as you know you've accomplished all that, and you are happy with who you are with, then I think everything would be OK. I want to experience life and know that when that right person comes, I won't have any regrets. (Quoted in Arnett 2004, p. 113)

In Chapter 4, you read about *sexual scripts* or norms governing sexual behavior. There are also **dating scripts** that govern behavior in the getting-to-know-you stage of dating relationships. These are highly gendered, with men and women having far different expectations about what happens during and after a date (Littleton et al. 2009). Whereas men are more likely to desire and pursue sexual activity, women are more likely to look at dating in terms of the possibility of a committed relationship. Moreover, dating scripts can vary by race and ethnicity and other factors. For example, deaf university students' dating scripts were mostly similar to hearing university students, except deaf students mentioned more group activities and had lower sexual expectations (Gilbert, Clark, and Anderson 2012). Compared to whites, African Americans were more likely to include "meeting the family" and exhibited larger gender differences in dating expectations, especially regarding sexual behavior (Jackson et al. 2011). African American males' internalization of the rigid "street code"—intense concern about toughness, respect, reputation, revenge (Anderson 1999)—is associated with less commitment and satisfaction, and greater hostility and

conflict in romantic relationships (Barr, Simons, and Stewart 2013). Although they are lessening, the values of *machismo* and *marianismo*, which emphasizes traditional dating scripts (he must be a "gentleman" and she a "good girl"), remain a part of Hispanic dating relationships, as discussed in Chapter 3 (Bermúdez, Sharp, and Taniguchi 2015).

Contemporary Dating

Now that the period of dating has been greatly extended, research on dating has flourished. But what is "dating"? These days there is considerable variation in premarital romantic relationships, making dating especially difficult to define. Therefore, dating tends to be defined broadly by researchers. For example, after extensive pretesting and taking into account gender, class, and race differences in adolescents' understandings of premarital romantic relationships, Giordano, Longmore, and Manning (2006) defined it this way: "Now we are interested in your own experiences with dating and the opposite sex. When we ask about 'dating' we mean when you like a guy, and he likes you back. This does not have to mean going on a formal date" (p. 268). These researchers' interviews with young adults in Toledo, Ohio, revealed something interesting. This definition not only captured "dating" relationships but also captured what they refer to as "nondating" relationships. This form of dating includes some elements of traditional dating (namely, sexual behavior) but lacks the commitment associated with being "boyfriend" and "girlfriend." These situations are discussed in the following section.

Dating versus "Nondating"

The traditional dating script that most of us think of as "dating" emerged in the middle of the twentieth century. Dating was facilitated by widespread access to automobiles, which removed interaction between young men and women from the front parlor of the young woman's home, under the watchful eye of parents, to places free from parental control such as movies or the local malt shop. Therefore, dating replaced old-fashioned "courtship" as the process of finding a mate. These days, the key difference between courtship and dating is that, for the latter, finding a marriage partner may or may not be the end goal.

With all the attention to the current "hooking up culture" and casual sex discussed in Chapter 4, it is easy to forget that young couples still engage in traditional dating. Dating is still a progression from mixed-sex group outings in middle school or high school, to pairing off within a group, to going on dates with one another (Child Trends 2014c). Perhaps due to technology, dating has declined for all age groups since the early 1990s.

Child Trend's *Monitoring the Future Survey* (2019) asked high school students "On average, how often do you go out on a date?" About 70 percent of eighth graders, 55 percent of tenth graders, and half of twelfth graders say they never date. These days, what does it mean to "date?" Bartoli and Clark (2006) asked college students to describe a typical date. Both men and women in the study said that a typical date involved three phases: (1) initiation—meeting in a public place (class, a party, or bar), casual talking, finding common interests, and calling for a date. (2) The date itself—movies, dinner, or engaging in a shared interest. And (3) an outcome—going back to the house, kissing goodnight, going home, and developing a relationship. Adolescent boys and girls find the traditional script equally desirable (Giordano et al. 2006). There are differences between men and women, though, that reflect traditional gender roles. For example, when it comes to eating out, women were significantly more likely than men to eat nonfattening and nonmessy foods like salad (Amiraian and Sobal 2009). Men have higher sexual expectations for the first date, especially if they pay for the date (Emmers-Sommer et al. 2010).

It is also easy to forget that parents are involved in overseeing their children's behavior, including dating relationships. Schalet (2011), in her study of U.S. and Dutch adolescents, found that American parents tended to view their children as unable to control their impulses and therefore requiring a great deal of protection and control, which often led to the children keeping their dating lives secret. One girl said this:

> They don't want to know that I'm doing it. It's kind of like, "Oh, my god, my little girl is having sex kind of thing." . . . [It's] just really overwhelming to them to know that their little girl is in their house having sex with a guy. That's just scary to them. . . . [They] won't even let me have a guy in my room without the door open. (Schalet 2011, p. 113)

Dutch parents, on the other hand, had higher expectations for their children's ability to control themselves and make good decisions on their own. One Dutch teenager explained:

> [My parents] do not say, "You are not allowed to [smoke, drink, do drugs]." They know: if kids want to do it, they will do it anyway. I am free to do it. They give me a lot of information about what drugs, alcohol, and smoking do to you. Not to disapprove, but . . . to reach your own conclusions that it is bad for you. (Schalet 2011, pp. 148–149)

Another girl explained her mother's attitude about birth control. She said, "If you want to go on the pill, I will allow you to. Because I'd rather you go on the pill than come home pregnant really young" (Schalet 2011, p. 141).

Many adolescents and young adults engage in what's been referred to as *nondating*. Nondating is generally sexual in nature and takes various forms such as

"hooking up" or "friends with benefits" (discussed in Chapter 4). Among adolescents from the Toledo study, there were key differences between daters and nondaters (Manning, Giordano, and Longmore 2006). Daters referred to themselves as boyfriend and girlfriend, were close in age, reported that sex brought them closer, had relationships that tended to last a few months, and told their friends about the relationship. Whereas daters followed the traditional pattern, nondaters did not. Nondaters reported not wanting a boyfriend or girlfriend; their partners included friends and exes; their relationships lasted days or years; there was a greater gap in ages; sex did not bring the couples closer; the couples didn't always tell their friends; half the sexual unions were for one time only; and the relationships were less exclusive. Whereas traditional dating has been found to have positive effects such as higher self-esteem and better grades, hooking up is associated with risky behaviors such as alcohol abuse and engaging in sexual intercourse without using a condom (Child Trends 2014c; Downing-Matibag and Geisinger 2009; Norris et al. 2013).

Because of high rates of divorce and union dissolution, dating occurs across the life course. In a study of older singles (ages 57 to 85), 14 percent reported they were currently dating (Brown and Shinohara 2013). After about age 30, dating declines with age and declines the longer a person remains single, especially for women (Rapp 2018). It is also important to note that dating patterns vary between men and women of different racial, class, and religious groups (Jackson et al. 2011).

Technology and Dating

The rapid increase in technology has had a profound effect on dating relationships (Ellin 2009). The first impact has to do with how couples meet. Dating websites that match couples on demographic and social traits take credit in their advertising for creating thousands if not millions of relationships and marriages. Researchers are currently studying whether meeting through a dating website affects marital quality and stability. A survey of 4,002 couples found no difference in relationship quality between couples who met online as opposed to in person (Rosenfeld and Thomas 2012). Despite the growing use of technology in dating and romantic relationships, the majority of couples meet for the first time in face-to-face encounters, such as at school, work, a game, or a party (Lenhart, Anderson, and Smith 2015).

Once a couple starts to date, technology continues to play an important role in their relationship. For some time, texting and instant messaging have been common forms of communication in romantic relationships both in the United States and around the world (Huang and Leung 2009). Now, 75 percent of all U.S. teens text. Between 2009 and 2011, the average number of texts sent on a typical day rose from fifty to sixty (Lenhart 2012). When initiating a relationship, teens will text to share something funny or interesting or to flirt with a potential partner. Among dating couples, texting and social media are viewed as places to make an emotional connection and publicize their romance (Emery et al. 2015). Although romantic couples text to "express affection," they may also use texting to avoid confrontation or to hurt one another (Coyne et al. 2011). In a study of 354 undergraduates, half reported that their romantic partner used an electronic method (text, e-mail) to deliver good news, such as saying "I love you" or telling them about a job promotion. On the other hand, the same percentage of students reported receiving bad news through electronic communication, such as "I think we should break up" or "I cheated." Males more so than females used electronic communication to deliver this kind of bad news (Knox 2014). Teens are largely disapproving of breaking up through electronic means, although 62 percent admit they have broken up with someone this way and 47 percent said they have been broken up with this way (Lenhart, Anderson, and Smith 2015). Discontinuing contact with no explanation, referred to as *ghosting*, is a particularly difficult way to learn a relationship is over (see "The Possibility of Breaking Up").

Because it is so new, the question of whether technology is good or bad for romantic relationships is unclear. It likely depends on several factors, including gender, age, and race or ethnicity. In a study of young adults ages 18 to 25 in committed relationships, texting was associated with higher perceived relationship stability among women but negatively associated with perceived relationship stability among men. As one man put it, "I'm afraid to date a woman three times since she will start this drama of, 'Why didn't you text me when you first woke up this morning.'" On the other hand, higher frequency of texts from a male partner is associated with lower relationship satisfaction among women (Schade et al. 2013). The authors reasoned that more frequent texts from male partners may indicate jealousy. Or frequent texting may be the result of less face-to-face contact and disengagement from the relationship. Other research indicates that online communication enhances commitment and trust because there is greater self-disclosure than can take place in person (Hertlein 2012).

Researchers are also investigating the extent to which digital communication, specifically texting, might be replacing in-person communication for couples. The *stimulation hypothesis* suggests that the two go hand-in-hand and that more texting would be positively associated with other forms of communication. Thus far, studies do not support this idea. Texting appears to be

replacing other forms of communication, supporting the *substitution hypothesis* (Luo 2014). Even so, in a study of 1,000 young adults ages 18 to 25, most believed that "talking," "hanging out," and "sharing intimate details" are more important than communication technologies when establishing a relationship (Rappleyea, Taylor, and Fang 2014). Research on the role of technology in dating and relationships is in its early stages and many questions remain.

From Dating to Commitment

However individuals meet, what is it that draws and keeps them together? One way to explore this process involves the idea of a developing relationship as moving around a wheel.

From an interaction-constructionist perspective (see Chapter 2), qualitative research with serious dating couples shows that they pass through a series of fairly predictable stages by which they further define their relationship. Hinting, testing, negotiating, joking, and scrutinizing the partner's words and behavior characterize this process. As one example, a 32-year-old emergency room worker described reading her partner's joking about marriage—and his use of the word *yet*—as a possible sign that he had considered marrying her: "It was right here in our kitchen I put it (the food) down on his plate and I'm like 'prison food.' And he said 'Um gee and we're not even married yet.' And that was like the first joke. It stuck in my mind" (Sniezek 2007).

As the relationship progresses toward an eventual wedding, the "marriage conversation" is introduced. In the research being described here, women were more likely to initiate marriage talk—cautiously and indirectly: "Yeah it's like so what are you thinking? Where is this relationship heading?" In other cases, the marriage conversation began more directly. One woman raised the question of marriage when her partner suggested that they live together: "When he asked me to move in with him, I told him I felt uncomfortable living with someone and not being married—a moral issue for me. So we talked about [the probability of getting married] at that time" (Sniezek 2007).

Once marriage talk is initiated, the couple faces negotiating a joint definition of the relationship as premarital. If one partner rejects the idea that the relationship should lead to marriage, then several responses can occur. In some cases, one partner's marriage hopes may be relinquished, although the relationship continues. In other cases, a partner may deliver an ultimatum. Sometimes an ultimatum results in marriage; in other cases, it causes an irreparable rift in the relationship.

Finally, most couples do not define themselves as "really" engaged until one or more ritualized practices take place—buying rings, setting a wedding date, making public announcements to family and friends,

holding engagement parties, and so on. These practices make the redefinition of the relationship increasingly public and "hardened" (Sniezek 2007). Marriage results from both conscious decision making and the natural progression of the relationship over time. Couples who develop a "shared reality" of their courtship experience by spending time together and talking about their relationship have fewer relationship problems and lower risk of divorce (Vennum et al. 2015).

The Wheel of Love According to this theory, the development of love has four stages in a circular process—a **wheel of love**—that can continue indefinitely. The four stages—rapport, self-revelation, mutual dependency, and personality need fulfillment—are shown in Figure 5.4 and describe the span from attraction to love.

Feelings of **rapport** rest on mutual trust and respect. A principal factor that makes people more likely to establish rapport is similarity of values, interests, and background (McFarland, Jurafsky, and Rawlings 2013). The outside circle in Figure 5.4 conveys this point. However, rapport can also be established between people of different backgrounds, who may perceive one another as an interesting contrast to themselves or see qualities in one another that they admire.

Self-revelation, or *self-disclosure*, involves gradually sharing intimate information about oneself. In one study, research participants were put in a lab and asked increasingly personal questions of one another.

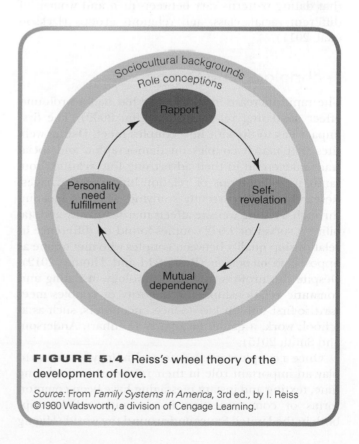

FIGURE 5.4 Reiss's wheel theory of the development of love.

Source: From *Family Systems in America*, 3rd ed., by I. Reiss ©1980 Wadsworth, a division of Cengage Learning.

Participants who shared information were more likely to fall in love than those who did not (Catron 2015; Sprecher and Treger 2015). People have internalized different views about how much self-revelation is proper. The middle circle of Figure 5.4, role conceptions, signifies that ideas about class-, ethnic-, or gender-appropriate behaviors influence how partners self-disclose and respond to each other's self-revelations and other activities.

For most of us, love's early stages produce anxiety. We may fear that our love won't be returned. Maybe we worry about being exploited or are afraid of becoming too dependent. Ironically, one way of dealing with these anxieties is to let others see us as we really are and to share our motives, beliefs, and feelings (Peck and Peck 2006). As reciprocal self-revelation continues, an intimate relationship may develop while a couple progresses to the third stage in the wheel of love: developing interdependence or mutual dependency.

In the **mutual dependency** stage of a relationship, the two people desire to spend more time together and thereby develop interdependence or, in Reiss's terminology, *mutual dependency*. Partners develop habits that require the presence of both partners. Consequently, they begin to depend on or need each other. For example, watching a good movie may now seem lonely without the other person because enjoyment has come to depend not only on the movie but also on sharing it with the other. Interdependency leads to the fourth stage: a degree of mutual personality need fulfillment.

In the **need fulfillment** stage of relationship development, two people find that they satisfy a majority of each other's emotional needs. As a result, rapport increases, leading to deeper self-revelation, more mutually dependent habits, and still greater need satisfaction. The relationship is one of ongoing emotional exchange and mutual support. Along this line, social scientist Robert Winch (1958) once proposed a theory of complementary needs: We are attracted to partners whose needs complement our own (Malakh-Pines 2005). Sometimes people take this idea to mean that "opposites attract." This may make intuitive sense to some of us, but needs theorists more often argue that we are attracted to others whose strengths are *harmonious* with our own (Klohnen and Mendelsohn 1998; Schwartz 2006). As partners develop mutual need satisfaction and interdependence, they gradually define their relationship. Part of defining a relationship should involve, according to counselors, talking about serious questions (Pendley 2006). What are some things that deserve discussing?

Some Things to Talk About Talking about the relationship that you and your partner want may bring up

differences, many of which can be worked out. If value differences are uncovered and cannot be worked out—for example, about whether or not to have children—it might be better to end the relationship before committing to a union that cannot satisfy either partner. Openly and honestly discussing matters like the following is important to successful mate selection (Pendley 2006):

- When should a relationship be dissolved and under what circumstances? How long and in what ways would you work on an unsatisfactory relationship before dissolving it?

- What are your expectations, attitudes, and preferences regarding sex?

- Do you want children? If so, how many? Whose responsibility is birth control?

- If you have children, how will you allocate child-rearing responsibilities? Do the two of you agree on child-raising practices, such as whether to spank a child?

- What is the financial situation of each partner as he or she comes into the union? How will the couple manage any previous debt, credit problems, and existing financial situations in general?

- Will partners equally share breadwinning and homemaking responsibilities or not? How will money be allocated? Who will be the owner of family property, such as family businesses, farms, or other partnerships?

- Do you expect your partner to share your religion? Will you attend religious services together? If you are of a religion different from that of your mate, where will you worship? What about the children's religion?

- What are your educational goals? How about your prospective partner's?

- How will each of you relate to your own and to your partner's relatives?

- What is your attitude toward friendships with people of the opposite sex? How about cyberfriends? Would you ever consider having sex with someone other than your mate? How would you react if your partner were to have sex with another person?

- How much time alone do you need? How much are you willing to allow your partner?

- Will you purposely set aside time for each other? If communication becomes difficult, will you go to a marriage counselor?

- What are your own and your partner's personal definitions of *intimacy*, *commitment*, and *responsibility*?

Discussing topics such as these is an important part of defining a couple's relationships. We turn now to an even more serious issue—dating violence.

DATING VIOLENCE: A SERIOUS SIGN OF TROUBLE

Romantic love is a wonderful feeling, but idealizing love and one's partner can lead people into unsatisfying, if not destructive, relationships. Barri and Morgan (2011) studied young singles in New Zealand. As one woman from their study explains:

> It took me six months to get the 'rose-colored glasses' off and to stop living the dream of being a couple. A lot of people I knew, who also knew him, said, Don't get involved, but I did . . . he'd been married in the past and had been so hurt . . . everybody looking at this person saw that he was just flitting . . . I got so caught up in the dream of being a couple I thought . . . it will be different for me . . . I thought that being consistent and loyal with my love would make the difference . . . and he would change . . . he might have flitted with other people but not with me. . . . it caused the greatest pain, but it caused the greatest learning as well . . . (p. 15)

Researchers have examined the relationship between "soulmate theory" and intimate partner violence. It is a myth that soulmates "ride off into the sunset" so to speak. Believing in soulmates, for example, reduces intimate partner violence among partners who are not particularly well matched, *but only early in the relationship* (Franiuk, Shain, Bieritz, and Murray 2012). That said, sometimes we need to make decisions about continuing or ending a relationship that is characterized by physical violence, verbal abuse, or controlling and threatening behaviors. In a study of 5,647 youth from ten different schools, 23 percent of females and 35 percent of males reported being victims of physical violence (Zweig, et al. 2014). (Interestingly, males reported more violence.) LBGTQ+ people have a higher incidence of experiencing physical and sexual violence and emotional abuse in dating relationships than do heterosexuals and cisgender people (Decker, Littleton, and Edwards 2018; Reuter and Whitton 2018).

Most interpersonal violence (IPV) is *reciprocal*, with both genders engaging in physical aggression toward one another (Cui et al. 2013; Renner and Whitney 2012; Zweig, et al. 2014). Among heterosexuals, by far the more serious injuries result from male violence against females (Johnson and Ferraro 2000). Furthermore, women are more inclined to "hit back" once a partner has precipitated the violence rather than to physically strike out first (Luthra and Gidycz 2006). Some risk factors associated with IPV for both men and women include childhood sexual abuse or neglect, low self-esteem, and perpetration of youth violence. IPV is discussed further in Chapter 12. Sexual harassment and abuse are discussed in Chapter 4.

Dating violence typically begins with and is accompanied by verbal or psychological abuse (Lento 2006) and tends to occur over jealousy, with a refusal of sex, after illegal drug use or excessive alcohol consumption, or when arguing about drinking behavior (Cogan and Ballinger 2006; Ryan, Weikel, and Sprechini 2008). Teens who perpetrate violence in dating relationships are more likely to have experienced violence themselves—for example, being spanked or hit by parents (Gershoff 2018). They are also more likely to have adopted a social script (from parents, communities, etc.) in which conflict is handled through aggressive means (Reyes et al. 2015). Many women and men are victims of controlling behavior that may or may not include physical violence. These days, abusive partners often use technology to monitor their partner's activities and whereabouts, such as by installing spyware on their partner's computer or a tracking system on their phone. Among teens ages 13 to 17, 37 percent had a current or former partner who checked up on them multiple times a day, asking them where they were, who they were with, or what they were doing. Twenty-one percent reported their partner read their text messages without permission, 16 percent had partners who required them to remove former girlfriends and boyfriends from their friends on social media, and 13 percent had partners who demanded to know the passwords for their e-mail and social media accounts (Lenhart, Anderson, and Smith 2015).

About half of abusive dating relationships continue rather than being broken off, and IPV tends to continue into cohabitation and marriage (Cui et al. 2013). Given that the economic and social constraints of marriage are not usually applicable to dating, researchers have wondered why violent dating relationships persist. Evidence suggests that having experienced violence in one's family of origin—even chronic verbal abuse in the absence of physical violence—is significantly related to both being abusive and accepting abuse as normal (Cyr, McDuff, and Wright 2006; Tshann et al. 2009). A recent qualitative study of twenty-eight female undergraduates in abusive dating relationships found that some of these women felt "stuck" with their partner (Few and Rosen 2005). A majority had assumed a "caretaker identity" similar to martyring. As one explained: "I always was a rescuer in my family. I felt that I was rescuing him [boyfriend] and taking care of him. He never knew what it was like to have a good, positive home environment, so I was working hard to create that for him" (p. 272).

Others felt stuck because they wanted to be married and their dating partners appeared to be their only prospects: "I think near the end, one of the reasons I was scared to let go was: 'Oh, my God, I'm twenty-seven.'

I was worried that I was going to be like some lonely old maid" (p. 274).

What are some early indicators that a dating partner is likely to become violent? A date who is likely to become physically violent often exhibits one or more of the following characteristics:

- handles ordinary disagreements or disappointments with inappropriate anger or rage;
- has to struggle to maintain self-control when some little thing triggers anger;
- goes into tirades;
- is quick to criticize or be verbally mean;
- appears unduly jealous, restricting, and controlling; and
- has been violent in previous relationships.

Dating violence is never acceptable. Making conscious decisions about whether to marry a certain person raises the possibility of not marrying him or her. Letting go of a relationship can be painful, however. This means looking at the possibility of breaking up.

THE POSSIBILITY OF BREAKING UP

Returning to Reiss's wheel theory of love, we note that once people fall in love, they may not necessarily stay in love. A recent survey of 986 U.S. adults indicated that although the average person has only one or two long-term relationships in his or her life, nearly a third have three or more (Chalabi 2015). Therefore, the possibility of breaking up is always a concern. Sometimes breaking up is a good thing: "Perhaps the hardest part of a relationship is knowing when to salvage things and when not to" (Sternberg 1988a, p. 242). In terms of causing the least amount of distress, the best breakups are straightforward and involve "open confrontation" rather than simply breaking off communication ("ghosting"), starting fights, or using friends to communicate the desire to break up (Collins and Gillath 2012). Not having a clear understanding of why the relationship ended or not having "closure" can prevent people from moving on. "Simply disappearing may look like the easy way out, but it corrodes self-respect" says Lisa Phillips (2019) in her article "The Endless Breakup." She suggests the two parties have an "exit talk" to discuss what happened in an honest, calm, and respectful way. Finally, she says, "unfriend and block each other on social media and take each other's contact information out of your phones. It's not mean. It's enforcing a healthy boundary to permit healing" (p. 73).

Some couples find themselves having to continue to reside together post-breakup, for example, if the couple cannot afford to live separately or cannot get out of their lease. While most couples would find this living situation awkward and confusing, others have said living together eased the transition to single status. Or, it can be a growing experience. Luke, age 28, explains:

> Following the split, we ended up living together for another six months, predominantly sharing the same bed (but I'd sometimes sleep on the couch). While some may think this was an odd decision, it was actually fine. Our situation worked out because we were both focusing on our careers, still got on really well with our housemates, and weren't looking to date anyone else. We were quite lucky—we didn't experience many awkward moments and ultimately I learned that sometimes things don't work out but if you respect each other and handle things the right way, you can avoid the drama of an awkward break up and still remain good friends. (O'Malley 2019)

Committed love will require some sacrifices over the course of time. However, as one therapist put it, "love should not hurt" (Doble 2006). Breaking up is difficult, and men and women may remain emotionally invested in their partners for some time (Graham, Keneski, and Loving 2014; Spielmann, MacDonald, and Tackett 2012). Partners must regain their sense of self after a breakup; failure to do so contributes to prolonged emotional distress (Mason et al. 2012). While it can be useful to reflect on what went wrong, too much "wallowing" after a breakup can delay healing. It helps to focus on the good aspects of the relationship and how the relationship encouraged personal growth (Singh 2015). Here are ten warning signs that it may be time to move on from a romantic relationship (Hemesath 2020, p. 172):

1. I have lost emotional connection with and/or physical desire for my partner.
2. The costs of my relationship seem to outweigh the rewards.
3. My partner and/or I have changed in significant ways.
4. I experience my partner's behavior as increasingly frustrating, annoying, or poor.
5. Emerging realizations have led me to have negative feelings about myself, my partner, and/or our relationship.
6. I no longer feel my relationship with my partner is special.
7. Single life has become more appealing to me and/or I have interest in other romantic partners.
8. I have struggled with feelings of sadness, hopelessness, and/or indifference about the relationship.

Love is a process of discovery that involves continual exploration and sharing. Choosing a supportive partner is an important factor in developing a satisfying long-term relationship. Creating and maintaining that union involves recognizing that challenges will arise and committing to face and overcome them.

9. I have no desire or plans to work on the relationship and/or I am considering ending the relationship.

10. There is nothing more my partner or I can do to change our situation.

NURTURING LOVING AND COMMITTED RELATIONSHIPS

Let us return to the idea that each of us only has "one true love," sometimes referred to as a *soulmate*. It turns out that framing romantic relationships in an idealistic way rather than viewing love as a "journey" is associated with less ability to handle conflict and lower relationship satisfaction (Shpancer 2014). Therefore, it is important for couples to have realistic expectations for relationships. Maintaining a satisfying long-term relationship is challenging, if only because two people, two imaginations, and two sets of needs are involved. Differences *will* arise because no two individuals have exactly the same points of view. From a developmental perspective, it's normal for adolescent and young adult relationships to be unstable as partners explore their own identities, personal goals, and what they want out of a relationship (Klimstra et al. 2013). For example, it is not uncommon for young couples to be "on-again, off-again," a pattern relationship experts refer to as relationship *churning* (Dailey, Brody and Knapp 2015; Halpern-Meekin et al. 2013).

However, couples who establish a positive relationship while dating have more successful marriages later on. Can people learn how to love? School-based "Healthy Relationship" programs have been shown to help boys and girls develop a "tool kit" for handling disagreements, building trust and respect, and communicating effectively, skills that promote high-quality, lasting relationships in adulthood (Kerpelman 2014). Young people are clamoring for more guidance on sex, love, and relationships. In a Harvard study, 70 percent of 18- to 25-year-olds wished they had received more information from their parents about emotional aspects of romantic relationships, and 65 percent wanted this information to be provided at school (Weissbourd et al. 2017). Choosing a supportive partner is an important factor in developing this kind of long-term love and relationship satisfaction, and the vast majority of people list their romantic partner as the greatest source of happiness in their lives (Smith 2013).

Digital Vision/Getty Images

Summary

- Most people equate love with romantic love. Romantic love is difficult to define, but generally includes passion, intimacy, and commitment to the relationship.

- Our understanding of love is biased toward love among heterosexual and married couples.

- Love should not be confused with martyring, manipulating, or limerence, and emotional and physical abuse should never be tolerated.

- People discover love; they don't simply find it. The term *discovering* implies a process—developing and maintaining a loving relationship requires seeing

the relationship as valuable, committing to mutual needs satisfaction and self-disclosure, engaging in supportive communication, and spending time together.

- Historically, marriages were arranged in the marriage market as business deals. In less Westernized areas of the world, some marriages are still arranged. Some groups practice "assisted" marriage.

- Whether marriage partners are arranged, "assisted," or more freely chosen, social scientists typically view people as choosing marriage partners in a marriage market; armed with resources (personal and social characteristics), they bargain for the best deal they can get.

- Although gender roles and expectations are certainly changing, some aspects of the traditional marriage exchange remain, such as a man's providing financial support in exchange for the woman's bearing and raising children, her domestic services, and her sexual availability. Nevertheless, couples today

are increasingly likely to value both partners' potential for financial contribution to the union.

- An important factor shaping partner choice is homogamy, the tendency of people to select others with whom they share certain social characteristics. Despite the trend toward declining homogamy, it is still a strong force, encouraged by geographical availability, social pressure, and feeling at home with people like ourselves.

- Dating has changed over time and incorporates new technologies for meeting and establishing relationships with romantic partners.

- Committed relationships develop through building rapport and gradually negotiating the relationship as premarital, leading to marriage.

- Couples today "meander toward marriage" and usually date a number of potential partners and cohabit before settling on a permanent mate.

- Committed relationships require steady nurturing and expressions of love toward one's partner.

Questions for Review and Reflection

1. Sternberg offers the triangle theory of love (Figure 5.1). What are its components? Are they useful concepts in analyzing any love experience(s) you have had?

2. Discuss the reasons why marriages are likely to be homogamous. Why do you think homogamous unions are more stable than heterogamous ones? How might the stability of interracial or inter-ethnic relationships change as society becomes more tolerant of these?

3. If possible, talk to a few married couples you know who lived together before marrying and ask them how their cohabiting experience influenced their

transition to marriage. How do their answers compare with the research presented in this chapter?

4. Think about your own romantic relationships. Were you engaged in dating or "nondating"? What ways do you use technology in your dating relationships?

5. This chapter lists topics that are important to discuss before and throughout one's marriage. Which do you think are the most important? Which do you think are the least important? Why?

6. **Policy Question.** What policies might be enacted to discourage dating violence? Do some research as to where to go for help in your own community.

Key Terms

arranged marriage 116
assortative mating 118
commitment 111
consummate love 113
dating scripts 127
endogamy 121
exchange model 117
exogamy 123
experience hypothesis 123
free-choice culture 116
heterogamy 123
heteronormative bias 111

homogamy 119
interfaith marriages 125
inter-ethnic 124
interracial unions 123
intermarriage 124
limerence 115
manipulating 115
marriage market 116
martyring 114
matching 118
mate selection 110
mutual dependency 131

need fulfillment 131
normalization hypothesis 123
pool of eligibles 119
rapport 130
selection hypothesis 123
self-revelation 130
status exchange hypothesis 124
Sternberg's triangle theory
    of love 113
transnational marriage 117
wheel of love 130

6

NONMARITAL LIFESTYLES: LIVING ALONE, COHABITING, AND OTHER OPTIONS

Learning Objectives

1 Describe current statistical trends regarding the proportion of singles in the United States.

2 Discuss reasons for the increasing proportion of singles in the United States today.

3 Describe various living arrangements for singles.

4 Describe advantages and disadvantages of being single.

5 Compare cohabitation practices in different cultures.

6 Describe the effect of cohabitation on children.

7 Discuss ways singles can promote and maintain supportive social networks.

Danaher, Luke, Rick, and Shyaporn are four heterosexual, thirty-something New York City guys who have lived together for eighteen years and plan to go on doing so. When they decide as a group that they need to move, the four look for housing together. Explains social scientist Bella DePaulo, author of the book *Singled Out* (2006), "there are so many variations on how to live" these days, and many adults now follow this "friendship model," according to which committed roommate arrangements become family or family-like (Howard 2012). In the words of family sociologist Judith Stacey, these men are "unhitched," because, for one thing, living among friends means that "the vagaries of sexual attraction don't disrupt your security and stability" (Stacey, in Howard 2012; Stacey 2011).

As Eric Klinenberg, author of *Going Solo: The Extraordinary Rise and Surprising Appeal of Living Alone* states, "Living alone is one of the least discussed and, consequently, most poorly understood issues of our time" (2012, p. 21). He maintains that American society "problematizes" singlehood and considers the rising rate of singlehood a sign of "narcissism, fragmentation, and diminished public life" (p. 22). In reality, men and women are finding that there are many advantages to living alone, such as being able to freely pursue one's personal and career interests. Moreover, changes in technology now allow social relationships and access to entertainment without having to leave one's home. Many other countries have embraced singlehood as a valid lifestyle choice. Rates of living alone are higher in most of Western Europe and Japan than in the United States.

When asked, "How important is family in your life?" nearly 100 percent of Americans say it's "the most important" (76 percent) or "one of the most important" (22 percent) elements (Pew Research Center 2010a, p. 41). Yet, we saw in Chapter 1 that today's *postmodern* family is characterized by a diversity of family forms. For example, the Millennial Generation (born between 1981 and 1996) were more likely to rate "being a good parent" above "having a successful marriage" as one of the most important things in their lives (52 percent vs. 30 percent), a difference of 22 percent. When those of the previous generation, Generation X (born between 1965 and 1980), were of same age, they rated children and marriage of similar importance: 42 percent felt being a good parent and 35 percent thought having a successful marriage were among the most important things in life, a difference of only 7 percent (Wang and Taylor 2011). So, while family life remains important, the place of *marriage* in family life has become less central. In this chapter, we will examine various living arrangements of unmarried people: living alone or with one's parents, living communally or in groups, and living with (or living apart from) an opposite-sex or same-sex partner.

WHAT DOES IT MEAN TO BE SINGLE?

Many college students take "being single" as not being in a romantic relationship. By this way of thinking, a person in a serious dating relationship or who is cohabiting would *not* be single. There is also the popular assumption that being single means to have *never been married*. However, singles can include people who have been divorced and widowed, groups that are increasing as a percentage of all singles. To the U.S. Census Bureau, "single" means *unmarried*. Single people may live alone, they may live with their children, and/or they may live with roommates or relatives. In official government statistics, to be "single" also means to be *unpartnered*. In 1990, the census began providing direct counts of unmarried men and women who are cohabitating by asking householders to identify "unmarried partners" in the home. In this chapter, we will examine what social scientists know about the large and growing number of singles and people in nonmarital, cohabiting relationships. To begin exploring these ideas, we'll examine some reasons for the increasing proportion of unmarrieds in our society.

REASONS FOR MORE UNMARRIEDS

Figure 6.1 shows the proportions of married, never-married, divorced, and widowed women and men since 1950. The percent of Americans who are currently unmarried is considerably higher than in past decades. For example, you can see the percentage of women who are married fell from 66 percent in 1950 to 51 percent in 2019. The percentage of married men followed a similar pattern although slightly more men are married because they tend to marry younger women and therefore have a larger dating pool. More recently, in 1970, 33 percent of men and 38 percent of women were unmarried (never-married, divorced, or widowed) compared to 46 percent of men and 49 percent of women in 2019. Never-marrieds, divorced, and widowed now make up nearly half the U.S. population, a figure that is unprecedented (U.S. Census Bureau 2019b). There are substantial race differences in the proportion married. Asian Americans are most likely to be married and least likely to be divorced. African Americans are most likely have never married, followed by Hispanics and American Indians (Raley, Sweeney, and Wondra 2015). These figures don't account for cohabitation; 17 percent of unmarried women and 16 percent of unmarried men reside with a cohabiting partner (Nugent and Daugherty 2018). Cohabitation is discussed later in the chapter.

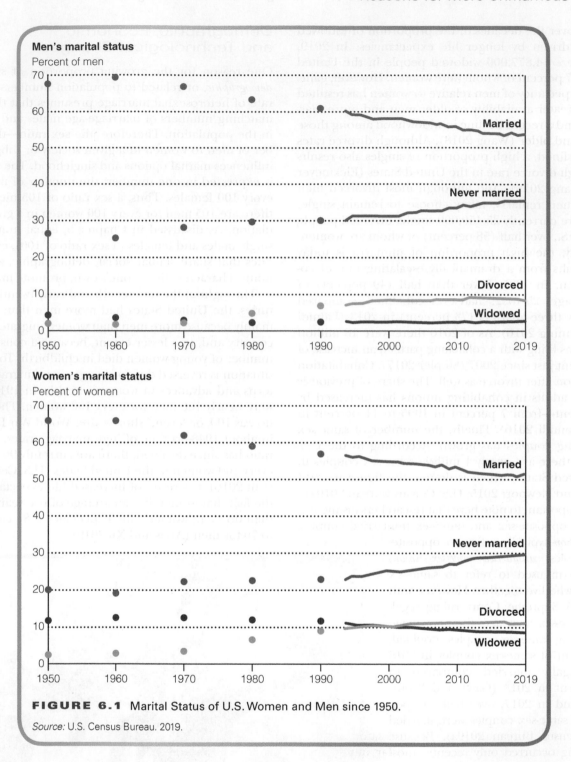

FIGURE 6.1 Marital Status of U.S. Women and Men since 1950.

Source: U.S. Census Bureau. 2019.

There are several reasons for this change. First, as discussed in Chapter 5, most men and women wait until their late twenties and early thirties to get married. The median age at marriage has increased to 30 for men and 28 for women (U.S. Census Bureau 2018). As a consequence of postponing marriage, the proportion of singles in their twenties and early thirties has risen dramatically. In 1990, 44 percent of those 18 to 34 were married, compared to only 26 percent in 2016

(Wang 2018). As we explain below, marriage at a young age has both become less appealing than it once was and, among those who want to be married, financially inaccessible.

Increasingly, adults are choosing to not marry at all. The percentage of all adults who has never married has never been higher: 23 percent of men and 17 percent of women age 25 and older have never been married (Wang and Parker 2014). Another factor has been the

growth over the decades in the proportion of widowed elderly, driven by longer life expectancies. In 2019, there were 14,877,000 widowed people in the United States, 77 percent of whom were women. However, gains in life expectancy of men relative to women has resulted in the greater availability of older men and more marriages, and a recent decline in widowhood among those age 65 and older (Wang 2018). Although divorce rates have declined, a high proportion of singles also results from high divorce rate in the United States (Eickmeyer 2016; Wang 2018). And although most divorced men and women remarry, many choose to remain single. There are currently 25,459,000 single, divorced people in the U.S., over half (58 percent) of whom are women.

Finally, the high proportion of unmarrieds today also results from a dramatically escalating rate of cohabitation. In 1995, fewer than half (49 percent) of women ages 25 to 29 had ever cohabited, compared to nearly three-quarters (73 percent) in 2013 (Lamidi and Manning 2016). As of 2016 there were 18 million Americans living with a cohabiting partner, an increase of 29 percent just since 2007 (Stepler 2017). Cohabitation is common after divorce as well. The share of previously married adults in cohabiting unions has increased by 57 percent—from 7 percent in 1995 to 11 percent in 2014 (Lamidi 2016). Finally, the number of same-sex cohabiting couples has grown. According to the U.S. Census, there are nearly 1 million same-sex couples in the United States, which is double the figure in 2013 (Gates and Newport 2015; U.S. Census Bureau 2019a).

It is important to note here that researchers distinguish between opposite-sex and same-sex unmarried couples. *Cohabitation* typically refers to opposite-sex couples; *unmarried same-sex unions* is the term used to refer to same-sex couples who live together. However, with the 2015 Supreme Court ruling legalizing same-sex marriage, the percentage of *married* same-sex couples doubled; 45 percent of same-sex couples in 2015 were legally married compared to 21 percent in 2013 (Gates and Brown 2015), and in 2017, over half (55 percent) of same-sex couples were married (U.S. Census Bureau 2019a). Because the ruling occurred only recently, most studies of same-sex unions are based on unmarried couples. This research is covered here. What is known about married same-sex couples is covered in Chapter 7.

Several social factors—demographic, economic, technological, social, and cultural—encourage Americans to postpone marriage, not marry at all, or divorce rather than stay married.

Demographic, Economic, and Technological Changes

One reason for the growing proportion of singles is *demographic*, or related to population numbers. A high rate of heterosexual marriage presumes that there are matching numbers of marriage-age males and females in the population. Therefore, the **sex ratio**—the number of men to women in a given society or subgroup—influences marital options and singlehood. The sex ratio is expressed in one number: the number of males for every 100 females. Thus, a sex ratio of 105 means that there are 105 men for every 100 women in a given population. As discussed in Chapter 5, equal numbers of single males and females (a sex ratio of 100), especially ones that share similar social, demographic, and economic characteristics to one's own, promote marriage.

Throughout the nineteenth and early twentieth centuries, the United States had more men than women, mainly because more men than women migrated to this country and, to a lesser extent, because a considerable number of young women died in childbirth. Today, this situation is reversed due to changes in immigration patterns and advances in women's health. In 1910, there were nearly 106 men for every 100 women. The sex ratio was 100, or "even," shortly after World War II ended, in about 1948—a time of high marriage rates. The sex ratio has since declined; there are currently 96 men for every 100 women in the United States (U.S. Census Bureau 2016) as a result of increased life expectancy and the fact that women live an average of five years longer than do men; women's life expectancy is 81 compared to 76 for men (Arias and Xu 2019).

More adults are choosing to "go solo." Singlehood has many benefits, including greater freedom to pursue career opportunities and more time to spend with friends.

William Perugini/Shutterstock.com

Specialized sex ratios may be calculated—for example, the sex ratio for specific racial and ethnic categories at various ages. Beginning with middle age, there are increasingly fewer men than women in every racial and ethnic category. Sex ratios differ somewhat for various races and ethnicities, however. For instance, high rates of homicide and incarceration combined with lower income and education of black men relative to black women reduce marriage rates within that group (Sawhill and Venator 2015; Schneider, Harknett, and Stimpson 2018). As the status of women in a group or category rises, the ratio of "marriageable" or desirable males to females declines, as does marriage (Bolick 2011b; Stanley 2011).

In addition to demographics, *economic* factors have increased the proportion of nonmarrieds. For one thing, expanded educational and career options for college-educated women over the past several decades have encouraged many of them to postpone marriage (Blossfeld and Kiernan 2017). Then, too, many middle-aged and older career or divorced women tend to look on marriage skeptically, viewing it as a bad bargain once they have gained financial and sexual independence (Levaro 2009).

In addition, people increasingly view marriage as a status that needs to be financially affordable (Graf 2019). Although it might appear at first glance that getting married should not be more expensive than living together, couples tend to have a number of specific concerns. They feel they should be able to afford the trappings of a middle-class lifestyle (e.g., the ability to purchase a home) and a reasonable wedding (rather than go downtown to the courthouse), achieve financial stability, be debt free, and demonstrate fiscal responsibility (Gibson-Davis, Edin, and McLanahan 2005; Smock and Greenland 2010). Difficulty in finding jobs and burdensome student loans mean that more young adults who would otherwise have married are currently cohabiting or are remaining single (Fry 2014; Yarrow 2015). Although young adults' debt burdens have declined since the onset of the recession in 2007, individual Americans between ages 18 and 35 owe a median average of $15,473; 15 percent of the total debt in that age group is the result of student loan debt (Fry 2013). In 2019, Americans owed a collective total of 1.5 trillion dollars in student loans (Friedman 2019), and roughly one in ten borrowers is in default (U.S. Department of Education 2019). Finally, there are gender expectations for marriage that are these days difficult to meet. Seventy-eight percent of never-married women feel that it is "very important" that their spouse or partner has a steady job (Wang and Parker 2014).

In addition to economics, *technological* changes over the past sixty-five years have affected the proportion of singles. Beginning with the introduction of the birth-control pill in the 1960s, improved contraception has contributed to the decision to delay or forgo marriage. With effective contraception, sexual relationships outside marriage and without great risk of unwanted pregnancy became possible (Gaughan 2002; Coontz 2005b). Moreover, as described in Chapter 1, reproductive technologies such as artificial insemination offer the possibility for planned pregnancy to unpartnered women as well as to same-sex couples. In addition to these structural reasons for the increasing proportion of nonmarrieds, cultural changes have played a part.

Social and Cultural Changes

Social scientists have recognized a fairly new life cycle stage called **emerging adulthood**. *Emerging adults* are typically defined as young adults ages 18 to 29 or 34. Compared to young adults in the past, today's men and women in their twenties spend more time in higher education or exploring options regarding work, career, and family making than in the past (Arnett 2000; Furstenberg 2008). Although all age groups contribute to the greater proportion of singles, emerging adulthood accounts for many unmarrieds today. This period of life is profoundly different from that of previous generations. In the 1950s, one's early twenties was typically consumed by marriage and childrearing. Today's young people are pushing marriage further and further out in the future. In 1976, only 26 percent of high school seniors said their ideal time to marry was "over 5 years from now" versus 52 percent in 2014 (Anderson 2016). Today, young adults spend their twenties exploring what may seem like an endless array of choices available to them with respect to love, education, work, and even where to live. In his book *Emerging Adulthood: The Winding Road from the Late Teens through the Twenties* (2014), Jeffrey Arnett writes, "Such freedom to explore different options is exciting, and this is a time of high hopes and big dreams. However, it is also a time of anxiety, because the lives of young people are so unsettled, and many of them have no idea where their explorations will lead" (p. 5).

Several other cultural changes over the past few decades also account for the growing proportion of singles. First, attitudes toward nonmarital sex have changed dramatically over past decades. With 71 percent of adults of all ages approving, sexual intercourse outside marriage has become fairly widely accepted (Brenan 2019). "Hooking up"—discussed in Chapter 4 and called *recreational sex* in the 1960s—among singles is widespread (Lyons et al. 2014).

With proscriptions against nonmarital sex relaxed, and as American culture gives greater weight to personal autonomy, many find that—at least "for now"—singlehood is more desirable than marriage

(Furstenberg 2008; Meier and Allen 2009). As one young man explained:

> One day I was at work and my friend called me up from Florida and said, "What are you doing?" I'm like, "Just working," and he said, "Can you come down?" . . . So I took a week off all of a sudden and went down to Florida. And I know I'd never be able to do that if I was married. (Arnett 2004, p. 101)

A young woman evidences a similar attitude:

> I hope to be married by the time I'm 30. I mean, I don't see it being any time before that. . . . Just go out and enjoy life and then settle down, and you'll know you've done everything possible that you wanted to do, and you won't regret getting married. (Arnett 2004, p. 103)

And from a 25-year-old single mother:

> I think it's just that I'm not out there really looking for someone right now. I'm self-centered . . . right now. I want to get through school, and I want to get employment. . . . And I want to get things going for myself before I do anything. . . . It's like just for me and my son. I got a son to raise. (Manning et al. 2010, p. 94)

Although these days it might be hard to believe, being single used to be considered deviant, not the acceptable option that it is today. During the 1950s, people, including social scientists, tended to characterize the never-married as selfish, neurotic, or unattractive. The divorced were also stigmatized. It probably goes without saying that these views have changed among the majority of Americans of all ages. In 2019, only 20 percent of Americans considered divorce morally wrong (Brenan 2019).

Then, too, getting married is no longer just about the only way to gain adult status. Before about 1940, the most legitimate reason for leaving home, at least for women, was to get married. Today, young men and women leave home for other reasons, such as to attend college or "to gain independence." That begs the question, What does it mean to be an adult? In a Clark University poll of emerging adults (ages 18–29), 36 percent said that "accepting responsibility for yourself" was the key to becoming an adult, followed by "becoming financially independent" (30 percent) and "finishing education" (16 percent). Only 4 percent said "getting married" was the key to adult status (Clark University 2013a). Moreover, cohabitation has become socially accepted. Forty-two percent of U.S. adults believe that an unmarried couple that has lived together for one year is just as committed as a couple that has been married for one year (Gallup Poll 2012; Newport 2015c). In another survey, sixty-nine percent of Americans said cohabitation is acceptable even if the couple doesn't have plans to marry (Graf 2019), although recent immigrants and members of more conservative religions are less accepting of cohabitation. Young adults are also influenced

by their parents' attitudes toward marriage—and young adults experience considerably greater independence and less parental pressure to marry than in the past (Arnett 2004; McLanahan 2010; Willoughby and Carroll 2012).

Finally, the diminished permanence of marriage may render it less desirable now (Cherlin 2009a). Marriage has become less strongly defined as permanent, and some singles fear a possible future divorce of their own (Miller, Sassler, and Kusi-Appouh 2011). In one recent qualitative study, some women said they were reluctant to marry, not because of any stigma associated with divorce, but because they feared the divorce process itself: "If I get married [and then divorce], I don't want to have to go through all that stuff" (Manning et al. 2010, p. 94). Fear such as this may reduce the likelihood of marriage (Waller and Peters 2008). If marriage is losing its permanent status, then "[i]ndividuals, as a result, have less faith that a successful marriage is possible, and they transfer support for marriage into support for other coupling arrangements, such as cohabitation—arrangements that are easier to dissolve if (and when) problems arise" (Willetts 2006, p. 125). Parental divorce is a "push factor" of children leaving home and forming cohabiting unions as opposed to married ones (Feldhaus and Heintz-Martin 2015).

We can apply the exchange theoretical perspective (see Chapter 2) to this issue of fewer compelling reasons to marry. Overall, as people weigh the costs against the benefits of being married, marriage now seems to offer fewer benefits relative to being single—although as Chapter 7 points out, there *are* significant benefits to marriage. If fewer Americans are married today, what are singles' various living arrangements?

SINGLES: THEIR VARIOUS LIVING ARRANGEMENTS

Singles make a variety of choices about how to live. Some live alone, others with parents, and still others in groups or communally. Some unmarrieds cohabit with partners of the same or the opposite sex. It is important to note that living alone can be less of a choice than a result of circumstance, such as not being able to afford a place of your own or the death of a spouse. This section explores these living arrangements. Single parents are addressed at length in Chapter 9 as well as elsewhere throughout this text.

Living Alone

The number of one-person households has increased dramatically in recent decades. Eric Klinenberg (2016), mentioned previously, stated, "[T]he extraordinary rise of living alone is among the most significant

social changes of the modern world." In 2018, there were 36 million single person households, making up over one-quarter (28 percent) of all U.S. households—up from just 8 percent in 1940 (U.S. Census 2018f). The likelihood of living alone increases with age for all racial and ethnic groups and is markedly higher for older women than for older men (Vespa, Lewis, and Kreider 2013).

Asians and Hispanics of all ages are less likely to live alone than are blacks or nonHispanic whites. More collectivist cultures, including Asians and Hispanics, discourage being unmarried and living alone. As Kelly, age 56, said:

> As someone who is both of Nigerian descent and a Christian, people tend to be totally shocked when I tell them I don't want to get married. Both these cultures are extremely patriarchal and tend to judge a woman's worth in terms of her relationship to men—with the role of wife and mother being seen as the ultimate crown of womanhood . . . I grew up witnessing multiple female role models either forfeit their dreams, stay in abusive relationships, or operate from a place of low self-worth due to these religious, cultural, and social constructs. Because of this, marriage . . . came to symbolize entrapment, restriction, and the loss of identity very early on in life. (Lusinski 2018)

Significantly less likely to be married than other racial and ethnic groups, blacks of all social classes are more likely than others to be living by themselves, particularly in older age groups.

Living alone has both advantages and disadvantages. Sociologist Elyakim Kislev (2019) interviewed 142 single people in the United States and Europe. Sarah, one of the women in his study, said this:

> There are two times and situations when I don't like living alone. The first one is when I can't open something. I had serious thoughts the other day of throwing a brand-new jar of salsa onto the tile floor with some force and then eating around the glass. But, I am a woman, and I tapped and twisted and got out towels and hit the edges and emerged victorious, albeit with very sore hands, five minutes after I commenced opening said jar. The second time is when I'm sick. There is, quite frankly, nothing worse than being sick and alone. . . . (p. 106–107)

For some, living alone is associated with social isolation and loneliness, especially for the elderly, who are frail and cannot get out of the house and whose social network has diminished through the death of family and friends. Their adult children are also increasingly likely to live far away. Social isolation depends on where you live, with less isolation and loneliness in neighborhoods with denser housing, busy sidewalks, commercial activity, well-maintained public spaces, and community organizations (Klinenberg 2016).

On the other hand, living alone can spell freedom from what some consider the drudgery of marriage and childrearing. Sixty-two-year-old Jane says:

> I was a housewife. I spent ten years driving the car, taking my children from one school to another school. I spent all day doing something, and at the end of the day, I was waiting for everybody to go to sleep so I could sit by myself, to write in my journal—have some time. But I had to wake up in the morning and start again. You are doing something, but then you sit by yourself and you think, "Why?" (Kislev 2019, p. 147)

In *Going Solo,* Klinenberg (2016) describes how many of his subjects told him that "nothing had made them feel lonelier than being in a bad marriage" (p. 786). There are alternatives, however, for singles who desire both an intimate relationship and freedom at home. One is *living apart together.*

Living Apart Together

An emerging lifestyle choice is **living apart together (LAT)**. Here a couple is committed to a long-term relationship, but each partner also maintains a separate dwelling (DePaulo 2012). The number of these relationships is difficult to ascertain because they are hard to define and because the U.S. Census Bureau does not measure them (Cherlin 2010). However, according to David Popenoe, codirector of the National Marriage Project at Rutgers University, LAT is clearly an emerging trend in the United States. LAT is at least partly motivated by a desire to retain autonomy. As one LAT woman said, "I like my own life, my own identity and want to keep it. I like having the things I love around me." As one man put it, "I am as devoted as any husband to her, . . . but I like my alone time and being around my stuff, not [hers]" (Brooke 2006).

LAT relationships can be difficult for people others to understand because, as Sharon Hyman, a filmmaker making a documentary on LAT, put it, "The assumption is that if you really love each other, you will live together. My question is, 'says who?" (Ansberry 2019). For an older adult, LAT "allows for unencumbered contact with adult children from previous relationships while protecting their inheritance and offering freedom from caregiving as a prescribed duty. . . . Separate homes also allow a tangible line of demarcation in terms of gender equity and the distribution of household labor" (Levaro 2009, p. F10). Sociologist Susan Brown calls LAT a "new frontier in partnered relationships" (Brown, Manning, Payne, and Wu 2016). The reasons people LAT varies by gender—men want to protect their leisure time and women want to protect their autonomy (Ansberry 2019). Some people in LAT relationships are already burdened with enough actual or potential family caregiving—for children still at home, parents, or a disabled sibling, for instance—that they hesitate to take on more caregiving demands that could compromise the ones they have already (Duncan and Phillips 2011). Men and women

in LAT relationships fulfill many functions of the family, such as taking care of one another during times of illness, sharing meals, and spending time together. As one 83-year-old widow stated, "We are very, very committed. He cares for me deeply," while her 87-year-old partner, a widower, stated, "Without Luci, I would be a very lonesome person" (Ansberry 2019). It's not just older and divorced adults who engage in LAT. Many young adults are in LAT relationships. One difference is that they might be residing with their parents as opposed to on their own (S. Smith 2006).

Living with Parents

Since 2000, the proportion of adult children living with their parents has risen. In 2018, 33 percent of young adults ages 18 to 34 lived in a parent's home, up from 29 percent in 2007. As one would expect, living with parents is more common among younger adults than older. Over half (56 percent) of adults ages 18 to 24 lived at home, compared to 23 percent of adults ages 25 to 29, and 12 percent of adults ages 30 to 34 (Payne 2019). In 2014, 43 percent of men and 36 percent of women ages 18 through 34 lived with their parents (or relatives). Among younger men and women ages 18 through 24, more than half (59 percent of men and 55 percent of women) lived with a parent (U.S. Federal Interagency Forum on Child and Family Statistics 2014). Some adults who live with their parents have never moved out, but others—called **boomerangers**—have left home and then returned. This is not uncommon after a divorce. In 2018, 46 percent of 18- to 24-year-olds, 26 percent of 25- to 29-year-olds, and 19 percent of 30- to 34-year-olds who were previously married lived at home (Payne 2019). In one study, emerging adults who boomeranged back to their parents' home (as opposed to having never left) were found to have more depressive symptoms (Copp, Giordano, and Longmore 2015). There were higher levels of depression among young adults living at home as a result of employment problems as opposed to enjoying living with their parents.

As Figure 6.2 indicates, living at home is more common in young adulthood among more recent generations. Fifteen percent of Millennials lived at home, compared to 10 percent of Gen Xers when they were that same age, 10 percent of Baby Boomers (born between 1946 and 1964), and 8 percent of the Silent Generation (born between 1928 and 1945). On the other hand, prior to the Silent Generation, adult children living at home was nothing unusual. In 1940, the proportion of adults under age 30 living with

their parents was quite high. Sociologists Paul Glick and Sung-Ling Lin suggest why:

> The economic depression of the 1930s had made it difficult for young men and women to obtain employment on a regular basis, and this must have discouraged many of them from establishing new homes. Also, the birthrate had been low for several years; this means that fewer homes were crowded with numerous young children, and that left more space for young adult sons and daughters to occupy. (Glick and Lin 1986, p. 108)

These same reasons apply to many young people today.

As discussed in Chapter 1, the economic recession that began in 2007 has led to an increase in the number of adults living with their parents (Jacobsen and Mather 2011; Pew Research Center 2012b). Moreover, even before that, housing in urban areas was too expensive for many singles to maintain their own apartments. Despite the end of the Great Recession, many adults do not have the financial means to afford their own place. Living with parents is twice as high among young adults without a college degree (20 percent) than those with one (10 percent; Fry 2017). It's also the case that for some adult children, living with parents is their preferred place to live (Payne 2019). As discussed in Chapter 8, more recent generations of adult children were raised by parents heavily involved in their lives: emotionally, financially, and in practical terms (e.g., making major decisions). It makes sense that this pattern would continue into adulthood.

Many believe that in multigenerational households, the adult child is there to care for their elderly parents.

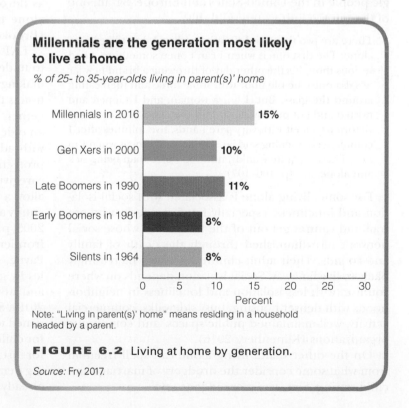

Millennials are the generation most likely to live at home

% of 25- to 35-year-olds living in parent(s)' home

Millennials in 2016	15%
Gen Xers in 2000	10%
Late Boomers in 1990	11%
Early Boomers in 1981	8%
Silents in 1964	8%

Note: "Living in parent(s)' home" means residing in a household headed by a parent.

FIGURE 6.2 Living at home by generation.

Source: Fry 2017.

Some do, of course. However, the majority of the time it is the adult child who is receiving the assistance. In one study, more than half (53 percent) of adults in their twenties and thirties have received financial assistance from parents since turning 21, including help paying bills and cell phone expenses, rent, health insurance, gas and groceries, childcare, and help with student loans. The average Millennial receives $11,011 per year in financial support and unpaid labor each year (Seligson 2019). In contrast to the stereotype of emerging adults as "freeloaders," in another study, the majority of emerging adults surveyed (69 percent) said they receive little or no financial support from their parents or only occasional support when needed (Clark University 2012b).

Social commentators lament that young people seem increasingly unprepared for adulthood. According to Burnette and Hardesty (2018) in their book, *Adulting 101: #Wisdom4Life*, **adulting** is "being skillful in the art of living." Yes, this includes getting to work on time and paying bills, but also practical matters such as knowing how to write out a personal check, make scrambled eggs, return an item to the store, and buy plane tickets. Many high schools today offer "life skills" classes where students learn the basics of banking, how to write a complaint letter, and filling out tax forms. At one Kentucky high school, seminars were offered on "dorm room cooking," "physical fitness after high school," "writing a resume, cover letter, and filling out an application," and "when to seek medical treatment and level of care (ER versus doctor visit)," among others (Holohan 2019).

How does "returning to the nest" affect parent-child relationships? The answer may depend on whether you are the child or the parent. In the Pew survey, among young adults who moved back home, one-quarter said the situation was bad for their relationship with their parents. Another quarter said moving home was good for their relationship, and about half said moving home made no difference (Parker 2012). In the Clark University poll, 61 percent of parents of emerging adults said that having their adult child living at home was "mostly positive," 67 percent felt that their relationship became closer emotionally, and 66 percent felt more companionship with their child as a result of them living at home (Clark University 2013b).

Several recent books argue that today's young adults have been coddled and spoiled since they were toddlers and have not developed the necessary skills to transition to responsible adulthood. Typical of these books is Sally Koslow's *Slouching Toward Adulthood* (2012). Koslow is the mother of two boomerang college-graduate sons, whom—along with the rest of their generation—she dubs "adultescents." On the other hand, young people choosing a slower path to adulthood, engaging in so-called *slow life strategies*, can be a smart choice because doing so means more time to prepare for career and family responsibilities. The many challenges facing emerging adults is not unique to the United States. Young adults in many European countries also struggle with unemployment, lack of affordable housing, gender discrimination, and cultural practices discouraging independent living. Policy makers suggest adopting a life course perspective that sees a successful transition to adulthood "a consequence of a series of intertwined life events" with the goal of young adults reaching residential, economic, and psychological independence (Berrington, Billari, Thevenon, and Vono de Vilhena 2017).

Attitudes about when adulthood begins have changed. Clark University psychology professor Jeffrey Arnett, mentioned previously, has been studying emerging adulthood for more than a decade. In 2012, he and his research team created the Clark University poll to understand this group. Researchers asked them (and their parents) questions on an array of topics: their living situations, amount of college debt, relationships with parents, and aspirations for the future. Only 29 percent of parents of 18- to 21-year-olds agreed that their children had reached adulthood (though this rises to 75 percent for the parents of 26- to 29-year-olds). The path to adulthood is getting longer, but is this a good thing? Some parents don't seem to think so. Forty-three percent said that the lengthened period of transitioning to adulthood is "mostly negative" or is "both positive and negative." Only 13 percent of parents considered it "mostly positive" (Clark University 2013b). Adult children who live with their parents who are employed or who are in school, are in a relationship (dating, cohabiting, or married), and those who are closer to their parents have less depression (Copp et al. 2017). Living with parents can put a cramp in dating—as Maria Kefalas, professor at St. Joseph's University stated, "Courtship becomes a challenge when you're living in your parents' basement" (Halpert 2018).

Adult children may choose to live with parents in a variety of circumstances. Overall, our society highly values individuality and independence, but some ethnic groups such as the Hmong, *expect* single women to reside with their parents until marriage. Women who have babies, especially those who became mothers in their teens, also may be living with parents. Formerly married young men and women sometimes return to their parental home after divorce. In general, adults return home when they experience "turning points" in their lives such as completing their education, breaking up with a spouse or partner, or becoming unemployed (Stone, Berrington, and Falkingham 2014).

Although it is mostly the adult child on the receiving end of financial assistance, many adult children who live with their parents contribute what they can, financially and otherwise, to the family household (McLanahan 2010). Parents often appreciate the help and companionship of their live-in adult children (Parker 2012; Straus 2009). Laura Napolitano (2015) interviewed

Many adult children "boomerang" to their childhood home after major life transitions such as graduating from college or a divorce.

young adults about the kinds of assistance they provide to their parents. Although the majority of the men and women in her sample were struggling financially, they often provided financial support. Pete, 27 and working full-time as a janitor, said this:

> I pay the electric. The water. We don't have a gas bill, we've moved over to the electric to kind of focus it all into one bill. And the electric can be high sometimes. They [freaking] get me with the cable bill. . . . It's kind of freelance with the way I pay bills, too, because they'll tell me what bills they need me to pay, and I'll pay them. . . . So what will happen is, I'll think I'm good, because I get my money Friday, I pay the electric or whatever, and then I won't get asked for money. So I'll be like, all right, good, I paid my bill, I'm good. Then out of nowhere, "Oh, I need you to pay the [cable] bill." And I can't tell them no, but they always got the best timing. It's always just when I got a couple dollars to make through something and they want me to pay a bill. (p. 341)

This can put a strain on family relationships. Still, the obligation to help family can be strong. David is 23 years old and wants to move out. But, he said:

> I don't want to leave [my house] and leave my mom like stranded. That's one of my things. But if she—as soon as she gets a job or whatever, because she's been looking, I'm pretty much gone. . . . If I left now, she'd probably have to put this house up for sale or something like that, and I'd feel awful. So I'm not going to leave her high and dry. (p. 345)

Young adults may provide their families with what Napolitano refers to as "instrumental support," such as helping with repairs or watching younger siblings. Liz, 21, dropped out of high school during her junior year and has been working as an administrative assistant. During this time, she took responsibility for her

newborn nephew when her sister, a drug addict, could not. She explained, "I raised my nephew for like seven, eight months when he was born. I took care of him and he lived with me" (p. 342).

Parents and emerging adults today have more in common in attitudes and values than did Baby Boomers and their parents (Jayson 2006). In the words of two other experts on this issue, "Our research shows that the closer bonds between young adults and their parents should be celebrated, and do not necessarily compromise the independence of the next generation" (Fingerman and Furstenberg 2012). Today's parents may expect to serve as "collaborators in their children's transition to adulthood, acting as scaffolding systems to help young people reach their goals and as safety nets to catch them before they fall too far" (Swartz et al. 2011, p. 427). On the other hand, conflict with parents can precipitate the decision to move out and perhaps take up residence with a romantic partner (Sassler 2004). Just as economic considerations, desire for emotional support, or the need for help with child raising may lead young singles to choose to live with parents, similar pressures may encourage singles to fashion group or communal living arrangements.

Group or Communal Living

Groups of single adults and perhaps children may live together. Often these are simple roommate arrangements. But some group houses purposefully share aspects of life in common. **Communes**—that is, situations or places characterized by group living—have existed in American society throughout its history and have widely varied in their structure and family arrangements.

In some communes, such as the nineteenth-century American Shakers, the Oneida colony, and the traditional twentieth-century Israeli kibbutzim, economic resources have been shared (Kephart 1971; Kern 1981; Spiro 1956). Work is organized by the commune, and commune members are fed, housed, and clothed by the community. Other communes may have some private property. Sexual arrangements also vary among communes, ranging from celibacy to monogamous couples (as in the kibbutzim and some communes in the United States) to the open sexual sharing found in both the Oneida colony and some modern American groups. Children may be under the control and supervision of a parent, or they may be raised more communally, with

biological relationships deemphasized and responsibility for discipline and care vested in the entire community.

Living communally has declined in the United States since its highly visible status in the 1960s, when many communes were established as ideological retreats from what their founders saw as the misguided American life characteristic of the 1950s. However, some communes established then still exist. New forms of communal living have emerged. The recession that began in 2007 prompted more single adults in their twenties and thirties to acquire a roommate or take in boarders (Wang and Morin 2009). According to the U.S. Census (2018d), there were nearly 9 million households with two or more unrelated persons in 2018. More families are doubling up—that is, relatives are moving in together to create more multigenerational or otherwise extended-family households. Roughly one-in-five Americans lives in a multigenerational household compared to only 12 percent in 1980. Increases have been seen among all racial and ethnic groups (Cohn and Passel 2018). "Accordion" family households that expand or contract around more or fewer family members, depending on family needs, perform important economic and often emotional and social functions (Newman 2012).

Communal living, either in single houses or in **cohousing** complexes that combine private areas with communal kitchens and "family rooms," may be one way to cope with some of the problems of aging, unattached singlehood, or single parenthood ("Cohousing in Today's Real Estate Market" 2006; Scott 2018). In a small but growing number of cohousing complexes, people of diverse races, ethnicities, and ages choose to reside together, sharing some meals and recreational activities. Communal living is designed to provide enhanced opportunities for social support and companionship. Cohousing communities are on the rise. Karin Hoskin, executive director of the Cohousing Association of the United States, says there are currently 165 cohousing communities with 140 more in the planning stages. She says, "Cohousing provides the privacy we've all become accustomed to with the community we seek" (Scott 2018, p. 8). Cohousing communities have been developed for families with special-needs children, LGBTQ+ individuals, seniors, and returning veterans. Forty-two-year old Malik Scott, a single dad, spent fifteen years serving in the Middle East and grapples with depression and Post Traumatic Stress Disorder (PTSD) and lives in a planned community that provides counseling, meditation, legal services, babysitting, art therapy, and more. He says, "We all pitch in and support one another. It's like the military but not. It's a little village here, like the old days" (p. 9).

Urban planners have been busy developing housing alternatives that meets the needs of the growing population of singles. Microhousing is one such option. Microhousing consists of small, usually 200 to 400 square feet, individual housing units with a super-efficient design.

Singles have become disillusioned with traditional housing, which is expensive and requires a lease or mortgage, and staying in one place. "Tiny houses" are low-cost, environmentally friendly, and mobile—perfect for the lifestyles of today's singles.

Microhousing communities provide social support, safety, and are an affordable housing option for singles priced out of the housing market, especially on the West coast (Kislev 2019). Tiny houses, very small stand-alone, mobile houses that are typically under 400 square feet, provide similar benefits, along with the pride of home-ownership and the ability to design your own home.

We turn now to a discussion of "living together," or cohabitation.

COHABITATION AND FAMILY LIFE

Cohabitation—unmarried couples living together, being financially obligated to one another, sharing time together, and having a sexual relationship—is one of the most significant changes in American families in the last fifty years. Not only in the United States but also in other industrialized nations, "living together" has dramatically increased. In this country, the cohabitation trend spread widely in the 1960s, took off sharply in the 1970s, and has risen steadily ever since. In 1987, only 33 percent of women ages 19 to 44 had ever cohabited. By 2013, this figure had nearly doubled to 65 percent. Moreover, cohabitation increased across all age groups, with the greatest increases seen among older adults (Hermez and Manning 2017; Manning and Stykes 2016; Stepler 2017). There are currently 8.5 million unmarried couple households in the United States (U.S. Census Bureau 2018e). This number may be an undercount because cohabiters do not necessarily live alone. Instead, they may live together in a parental home or reside with roommates and therefore would

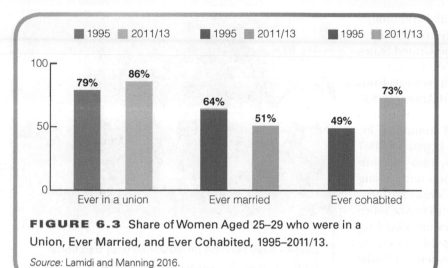

FIGURE 6.3 Share of Women Aged 25–29 who were in a Union, Ever Married, and Ever Cohabited, 1995–2011/13.

Source: Lamidi and Manning 2016.

not be included in a census count (Manning and Smock 2005; Pilkauskas 2012).

Today, it is now more common to have cohabited than to have married (Graf 2019; Gurrentz 2018; Lamidi and Manning 2017). As shown in Figure 6.3, whereas the percentage of women ages 25 to 29 who had ever married declined between 1995 and 2013, from 64 percent to 51 percent, the percentage of women of that age who ever cohabited increased from 49 percent to 73 percent. The authors of the study point out that young adults are not "retreating from union formation but are choosing cohabitation rather than marriage." For example, the percentage of women ages 25 to 29 in a union (married or cohabiting) increased during the period from 79 percent to 86 percent (Lamidi and Manning 2017). The question is, why are more people cohabiting?

In the early twentieth century, living together outside marriage was illegal in every state and was uncommon. Due to laws enacted at the turn of the twentieth century, unmarried cohabitation has remained illegal in a handful of states, although the laws are seldom enforced. The American Civil Liberties Union has sued to overturn anticohabitation laws in states where they still exist (Alternatives to Marriage Project 2012; Jones 2006). In Florida, where cohabitation was illegal until just recently, a 2011 push to abandon the law was met with opposition from social conservatives, who declared that they were "not ready to give up on monogamy and a cultural statement that marriage still matters" (M. Hartman 2011; see also Haughney 2011). However, in 2016, Governor Rick Scott signed a bill repealing the 1868 law. Now just two states, Mississippi and Michigan, have laws against unmarried cohabitation (Hanson 2016). As "A Closer Look at Diversity: The Different Meanings of Cohabitation for Various Racial and Ethnic Groups" suggests, cohabiting means different things to different people. Generally, however, cohabitation "is very much a family status, but one in which the levels of certainty about commitment are less than in marriage" (Bumpass, Sweet, and Cherlin 1991, p. 913). Put more strongly, cohabitation "does not institutionalize commitment in a way that is easily understood and honored by romantic partners and their friends and family" (Wilcox et al. 2011).

Nevertheless, the number of cohabiting adults has increased more than ten times since 1970—and by 40 percent just since 2000 (Lofquist et al. 2012; U.S. Census Bureau 2012c). By 2013, an estimated

Gender roles are less traditional in cohabitation than in marriage. Cohabiting couples must sort out who does what, potentially leading to conflict, but also to more freedom to do what one chooses.

The Different Meanings of Cohabitation for Various Race/Ethnic Groups

You probably have an idea of what cohabitation means to you, and you may assume that living together signifies the same thing to all of us. But researchers who have analyzed national survey data to study cohabitation among various racial and ethnic groups have uncovered interesting differences (Manning 2004; Ojeda 2011). Cohabiting means different things in different cultures and ethnic groups (Fomby and Estacion 2011; Landale, Schoen, and Daniels 2009; Liu and Reczek 2012).

For instance, Puerto Rican women have a long history of **consensual marriages** (heterosexual conjugal unions that have not gone through a legal marriage ceremony). The tradition of consensual marriages probably began among Puerto Ricans because of a lack of economic resources necessary for marriage licenses and weddings: "Although nonmarital unions were never considered the cultural ideal, they were recognized as a form of marriage and they typically produced children" (Manning and Landale 1996, p. 65). Therefore, for Puerto Ricans in the continental United States, cohabitation symbolizes a committed union much like marriage, and they don't necessarily feel the need to marry legally should the woman become pregnant because they have already defined themselves as (consensually) married. Compared to Mexican Americans,

Puerto Ricans are more likely to agree that "[i]t's all right for an unmarried couple to live together if they have no plans to marry" (Oropesa 1996).

Meanwhile, compared to Puerto Ricans (and nonHispanic whites), Mexican Americans were more likely to agree that "[i]t is better to get married than go through life being single." Mexican Americans weigh marriage very positively, and a couple's plans to marry significantly increase the likelihood that Mexican Americans will approve of cohabitation for that couple. These findings are especially strong for foreign-born Mexican Americans. However, economic barriers to marriage apparently induce many low-income Mexican Americans to continue to cohabit rather than marry and to raise several children in cohabiting families (Lloyd 2006; Wildsmith and Raley 2006). Furthermore, as a result of their exposure to general cultural values and attitudes in the United States, we can expect second- and third-generation Mexican Americans to embrace cohabitation in increasing proportions (Fomby and Estacion 2011; Oropesa and Landale 2004).

Similarly, African Americans consistently tell pollsters that they value marriage highly. However, during slavery, African Americans were not legally able to marry and could be easily sold away from their homes, leading to a

greater incidence of "marriage-like" unions (Cherlin 2009a). The difference in meaning between legal marriage and cohabitation may not be as great for blacks as it is for other racial and ethnic groups (Liu and Reczek 2012).

For Puerto Ricans, "living together" is likely to symbolize a committed union virtually equal to marriage. Among Mexican Americans, cohabitation is less valued than marriage but allowable if the couple plans to marry—although economic barriers to marriage can thwart those plans (Fomby and Estacion 2011). Our diverse American society encompasses many ethnic groups with family norms that sometimes differ from one another. Together with structural and economic factors, cultural meaning systems play a part in how people define cohabitation.

Critical Thinking

The U.S. Census Bureau groups Puerto Ricans and Mexican Americans, along with other Hispanics, into one ethnic category. What does the previously presented information tell you about diversity *within* the ethnic category of Hispanic? Would you suppose that the various groups of Asians, such as the Hmong or Asian Indians, who are also categorized together, differ as well? What about African Americans or whites?

69 percent of 40- to 44-year-old women had lived with an opposite-sex partner at some time in their lives—up from 33 percent in 1987 (Manning and Stykes 2016). Fifty-four percent of women and 40 percent of men have cohabited by age 25 (Payne 2016). Associated with the recession that began in 2007,

cohabitation rates increased more than usual as people postponed marriage until they could better afford it (Cherlin et al. 2013). In particular, credit card debt and student loans have caused couples to delay marriage in favor of cohabitation (Addo 2014).

The incidence of cohabitation is expected to escalate further as new generations grow up in cohabiting families or in families with permissive attitudes about cohabitation; hence, these new generations may be socialized to take cohabiting for granted (Willoughby and Carroll 2012). Over two-thirds (69 percent) of Americans say it's acceptable for unmarried couples to live together even if they don't plan to get married; only 14 percent say it's never acceptable and 16 percent say it's acceptable only if they plan to get married (Graf 2019). As respondent Alan, age 27, told an interviewer, "My great grandma said you got to test drive the car before you buy it. So cohabitation is a good way to really get to know someone" (Manning, Cohen, and Smock 2011, p. 138). Alan's great grandma aside, younger people are far more accepting of cohabitation than are older people. The majority of Americans (65 percent) say they would like cohabiting couples the same legal rights as marrieds when it comes to health insurance, inheritances, and tax benefits, as is the case in many European countries such as Sweden (Graf 2019).

Characteristics of Cohabitors

Comparing marrieds to cohabiters, analyses of data from several sources show that, on average, cohabiters are younger, are less educated, earn less income, are less likely to own their own homes, are more likely to have experienced multiple living arrangements as children, and are likely to have relatively permissive attitudes toward sex, and are more likely to have had an unintended birth (Fry and Cohn 2011; Gurrentz 2018; Nugent and Daugherty 2018; Ryan et al. 2009, Schoen et al. 2009). Research shows that some of these trends begin in childhood: Cohabiting women were found to have had lower academic achievement and fewer parental resources in adolescence (Amato et al. 2008; Willoughby and Carroll 2012). Especially in Southern states, relatively conservative religious affiliation is negatively related to cohabitation, and the highest cohabitation rate is among those without any religious affiliation (Eggebeen and Dew 2009; Gault-Sherman 2012). Nevertheless, people from all social classes, educational categories, and religious persuasions have cohabited. Although in previous years cohabitation was more common among African Americans than among Hispanics and nonHispanic whites, today nonHispanic whites have slightly higher rates of cohabitation (Hermez and Manning 2017; Manning and Stykes 2016). Although approximately three-quarters of cohabiters (71 percent) are younger than 45 (U.S. Census Bureau 2015a), the proportion of middle-aged cohabiters has increased over the past two decades. For example, the number of cohabiting adults age 50 and older increased by 75 percent between 2007 and 2016 and now totals 18 million (Brown, Bulanda, and Lee 2012). The majority (55 percent) of cohabitors

of this age are divorced, followed by 27 percent who were never married, 13 percent who were widowed, and 6 percent who were separated (Stepler 2017). Similar to LAT couples discussed previously, cohabitation offers many advantages, such as maintaining control over one's own finances. The many reasons why cohabitors cohabit are discussed below.

Why Do People Cohabit?

When cohabitation first emerged, researchers wanted to know why some couples "chose" cohabitation and others "chose" marriage. There are a variety of reasons why cohabitors say they are cohabiting, from practical reasons such as convenience and that is makes sense financially, to emotional reasons such as wanting love and companionship. Many cohabitors say they moved in together to "test the relationship" (Graf 2019). Indeed, a large percentage of cohabitors go on to marry, suggesting that cohabitation serves as a trial run to determine compatibility. More recently, another perspective emerged: that cohabitation is not necessarily a conscious decision. In other words, the idea of marriage never enters into the equation. It turns out that cohabitation serves different purposes for different people. Living together "may be a precursor to marriage, a trial marriage, a substitute for marriage, or simply a serious boyfriend–girlfriend relationship" (Bianchi and Casper 2000, p. 17). Although most adults ages 18 to 44 who have cohabited (62 percent) have only ever lived with one partner, over a third (38 percent) have lived with two or more partners (Graf 2019). This implies cohabitation is a useful living arrangement across the life course.

Cohabitation as a Prelude to Marriage The strongest piece of evidence that couples use cohabitation as a *prelude to marriage* is the number of married couples who lived together first. More than two-thirds (69 percent) of marriages are preceded by cohabitation (Manning and Stykes 2016). In a 2016 poll of high school seniors, 71 percent viewed cohabitation as a "testing ground" for marriage (Anderson 2016).

Data from the National Survey of Family Growth indicate that nearly half of all first cohabitations begin with the intention to marry, and many individuals view cohabitation as part of the marriage process (Guzzo, cited in Rose-Greenland and Smock 2013). And according to a new Pew research study of cohabiting couples (Graf 2019), two-thirds of adults who lived with their spouse before they were married (and who were not yet engaged when they moved in together) saw cohabitation as "a step toward marriage." In their Marriage and Cohabitation in America study, Wendy Manning and Pamela Smock conducted focus groups with cohabiting adults in their twenties and thirties about their motivations for cohabiting. One young man said, "From my

personal experience, the reason why I moved in with my girlfriend [is] because [I was] looking at this woman as the person who I would consider marrying and being with for the rest of my life." One young woman responded that cohabitation "allow[s] the partners to work through issues or habits before marriage" (Smock et al. 2016 p. 26).

Does living with one's future spouse before marriage ensure a happy marriage? Over half (48 percent) of U.S. adults believe that couples who live together before marriage have a better chance of having a successful marriage, 13 percent say they have a worse chance, and 38 percent say it doesn't make much difference (Graf 2019). As discussed in Chapter 5, couples who live together before marriage are slightly more likely to divorce, regardless of level of education and income and other factors that "select" couples into cohabitation, suggesting that cohabitation may have a destabilizing effect on the relationship (Rosenfeld and Roesler 2019). However, some cohabiting couples do not transition into marriage as readily as others. That is, cohabitation may be a legitimate lifestyle of its own.

Cohabitation as an Alternative to Marriage

In addition to functioning as a trial marriage, there is evidence that cohabitation is becoming a socially acceptable *alternative to marriage*. Couples are less likely to feel it's necessary to marry upon pregnancy or childbirth, and people routinely take an "unmarried partner," their "significant other," or "plus one" to work or family get-togethers. Some of these couples construct their own definition of commitment: one that doesn't include marriage (Arnett 2004; Byrd 2009). As one cohabiter explained: "We've been together 10 years. We met at college. . . . We graduated and started living together. We never say never, but we certainly don't have any plans to marry. We're very happy being unmarried to each other" (Sachs, Solot, and Miller 2003).

A 32-year-old woman who has been in a monogamous relationship for ten years put it this way:

I have friends who have been married and divorced already in the time that we have been together. . . . And I think I like the luxury of the fact that every day that we are together I know we are together because we both choose to be and not because we feel some artificial obligation to be together. (Byrd 2009)

People's reasons for living together as an alternative to marrying may include the belief that marriage is too confining, signifies loss of identity, or stifles partners' equality and communication (Gold 2012; Moore, McCabe, and Brink 2001; Willetts 2003). Others are morally or politically opposed to the institution of marriage or feel that marriage is a religious institution they want no part

of. Others find the "ownership" aspect of marriage distasteful (Hatch 2015).

More commonly, cohabiters who do not plan to marry tend to see marriage as beyond them financially. The most common reason cohabitors give for not getting married is that they and/or their partner is not ready financially (Graf 2019). Indeed, as alluded to previously, many cohabiting couples feel they need to "save" for marriage. Pamela Smock says that marriage is increasingly being viewed as a *capstone* (Smock, Manning, and Porter 2005). That is, young couples see marriage as the endpoint of a series of life decisions regarding education, employment, and where to live. Only after these benchmarks are reached do couples feel financially ready to marry. So strong is this belief that marriage is only for those who are financially secure that many couples even buy a home and have children first. One young woman explained:

He gave me a ring, he asked me to marry him a long time ago. We talked about getting married, it's just that we don't have anyone to pay for the wedding and I don't want to go downtown. So, it's on us . . . and when we came down to the decision of do we want kids or do we want the marriage and for now we wanted the kids. . . . (Smock, Manning, and Porter 2005, p. 688)

The decision to marry often hinges on the financial resources of the male partner and whether he feels ready to support a family. For example, Henry, a 33-year-old information systems manager, reflected: "Had I been in a financial position where I was able to take care of myself and a family, then it might have moved things along quicker." Jamal, 27, said, "What would make me ready? Knowing that I could provide." Similarly, Victor stated that "the male's financially responsible for like, you know, the household, paying the bills" (Smock, Manning, and Porter 2005, p. 691).

Middle-aged and older cohabiters are generally in relationships of longer duration and—as you might expect—are more likely to have been divorced (Brown, Bulanda, and Lee 2012). Older singles express less desire to marry (or remarry) than do younger singles (Mahay and Lewin 2007; Swarns 2012a). Older couples have found that living together absent legal marriage may be economically advantageous because they can retain some financial benefits that are contingent on not being married (Ebeling 2007). "When it comes to marriage and money, widows often say straight out that they are unwilling to risk the financial security of a departed husband's pension." Moreover, many cohabiting older couples "are aware of their children's fear for their inheritance and for that reason may . . . refrain from marriage" (Levaro 2009, p. F10). "Even if your will leaves everything to your kids, a second spouse can claim what's known as an 'elective share' of your estate—typically, a third. After you're

married, it takes more lawyering to disinherit a spouse" (Ebeling 2007).

Another piece of evidence to suggest that cohabitation is an alternative to marriage is the presence of children in the household. Until relatively recently, marriage was considered the only appropriate context in which to raise children. However, we are currently seeing the decoupling of marriage and parenthood (Hayford, Guzzo, and Smock 2014). Roughly one-third of opposite-sex unmarried couples have children in the household. Although many of these children are from previous relationships, many are the biological children of the couple (U.S. Census Bureau 2018e). There are two reasons for this. First, because the stigma against nonmarital childbearing has weakened, a decreasing number of cohabiting couples choose to "legitimize" an unplanned pregnancy by getting married (McLanahan and Sawhill 2015; Su, Dunifon, and Sassler 2015). Second, an increasing number of births to cohabiting couples are planned. Among cohabiting couples with children, 29 percent had intended to have a child (Curtin, Ventura, and Martinez 2014).

Still, most women and men anticipate getting married—and most do. High school seniors are only slightly more likely than thirty years ago to report that they are questioning marriage as a way of life; 30 percent questioned marriage as a way of life in 1976 compared to 35 percent in 2008 (National Center for Marriage and Family Research 2010). Moreover, legal and social differences remain between marriage and "just" living together. To be a true marriage alternative, cohabitation and marriage would have to become virtually indistinguishable. In many European countries, cohabitors are extended the same legal rights as marrieds, and children are more likely to be born into cohabitation than into marriage. The Netherlands, Sweden, and Norway are examples of places where cohabitation is common and virtually as institutionalized as marriage (Hansen, Moum, and Shapiro 2007). Still, even in those places, a significant majority of cohabiting women marry their child's father after their first birth, "suggesting that marriage remains the predominant institution for raising children" (Perelli-Harris et al. 2012).

Cohabitation as an Alternative to Being Single

Recall from Chapter 1 that many life course decisions are made through a process called *sliding versus deciding*. In other words, for better or worse, family patterns are not necessarily the result of conscious decisions and a careful weighing of options. Besides being considered a prelude or an alternative to marriage, cohabitation can also be viewed as an *alternative to being single*.

In a qualitative study of 120 heterosexual cohabiters, nearly two-thirds ranked the reason "I wanted to spend more time with my partner" as the most important reason for moving in together (Rhoades, Stanley, and Markman 2009). Some couples begin to live together shortly after their first date; others wait for months or longer (Sassler 2004). Accounts of how cohabitation begins suggest that cohabiting does not always result from a well-considered decision. As one 23-year-old woman who had been living with her parents explained:

I was looking for my own apartment at this time. . . . He was like, "Why don't you just move in with me?" I was like, "Let's give it some time," or whatever. So I dated him for like a month and then finally all my stuff ended up in his house. (Sassler 2004, p. 496)

Some cohabitants view living together as an alternative to dating or unattached singlehood. As one respondent told her interviewer:

Um, he had come over, and we had talked and . . . he had spent the night and then from then on he had stayed the night, so basically . . . he just honestly never went home. I guess he had just got out of a relationship, the person he was living with before, he was staying with an uncle and then once we met, it was like love at first sight or whatever and um, he never went home, he stayed with me. (Manning and Smock 2005, p. 995)

Psychologist Jeffrey Arnett (2004) has dubbed those who live together as an alternative to being single *uncommitted cohabiters*. Manning and Smock conducted 115 in-depth interviews with men and women about their most recent cohabitation experience. Tim, a construction worker, described it this way: "I wasn't thinking of marrying Denise. I wasn't thinking about even like staying there. I was just thinking she's a cool girl. I liked being around her, and I ended up just staying around there all the time and you know, taking showers over there, I would call it living there" (Manning and Smock 2005, p. 999).

For these kinds of couples, cohabitation is especially difficult to measure. In interviews and on surveys, cohabitors often have trouble pinpointing when they started living together, and there can be disagreement between partners as to whether or not they even are cohabitating (Waller and McLanahan 2005). Another young man from Manning and Smock's study, when asked when he and his partner starting living together, said this:

It began by an attrition of this thing at her parents' house. In other words, she stayed at my house more and more from spending the night once to not going home to her parents' house for a week at a time and then you know further, um, so there was no official starting date. I did take note when the frilly fufu soaps showed up in my bathroom that she [had] probably moved in at that point. (Manning and Smock 2005, p. 995)

Clearly, our understanding of cohabitation is still evolving. Nevertheless, researchers have discovered

interesting things about cohabiting relationships, especially compared to married ones.

The Cohabiting Relationship

As a category, heterosexual cohabiting couples differ from married couples in several ways. First, cohabiters are less homogamous or alike in social characteristics than marrieds. At 18 percent, cohabiting couples are about twice as likely as marrieds to be in a union with a partner or spouse of a different race or ethnicity (Livingston 2017). Compared with married women, cohabiting women are more likely to earn more and be several years older than their partners (Fields 2004). Cohabiting men spend fewer hours in the labor force than do married men (Kuperberg 2012). Cohabiters have been more likely than marrieds to be nontraditional in many ways, including attitudes about gender roles, and to have parents with nontraditional attitudes (Davis, Greenstein, and Gertelsen Marks 2007; Kuperberg 2012). Perhaps this is because children raised in nontraditional families (e.g., single-parent families, stepfamilies) are more likely to cohabit as adults (Valle and Tillman 2014).

Cohabitors have lower levels of trust and relationship satisfaction than marrieds. Married adults are more likely than cohabitors to say that things are going "very well" in their relationship (58 percent vs. 41 percent). They also expressed higher levels of satisfaction with daily matters such as how to divide household chores, balancing time between work and home, communication, and their approach to parenting (for those with children). Marrieds and cohabitors, however, were equally satisfied with their sex lives (Graf 2019).

Uncertainty about commitment, together with less well-defined norms for the relationship, may be reasons that cohabitants pool their finances to a lesser extent than marrieds (Hamplova and LeBourdais 2009). Couples who live together as opposed to getting married accumulate far less wealth, even among cohabitors who eventually married their partner (Britt-Lutter, Dorius, and Lawson 2018). Sociologist Cassandra Dorius, one of the authors of the study, says this: "Cohabiting relationships tend to be more short-term and unstable, and you keep starting over every time. That is difficult for wealth generation." Cohabitors are less likely than marrieds to say that they are happy with their relationships, find them less fair with regard to sharing finances (Dew 2011; Rhoades, Stanley, and Markman 2012), place greater importance on sexual frequency (Yabiku and Gager 2009), and are less likely to be sexually faithful (Treas and Giesen 2000).

However, the relationship quality of *long-term* cohabiting couples (together for at least four years) differed little from that of marrieds in conflict levels, amount of interaction together, or relationship satisfaction. One thing did differ, however: For both marrieds and long-term cohabiters, relationship satisfaction declined with the addition of children to the household, but this decline was more pronounced for cohabiters (Willetts 2006). Other research has found that older cohabiting couples generally report higher relationship quality than younger cohabiters. A lack of plans for marriage is associated with lower relationship satisfaction (Brown, Manning, and Payne 2015). Research shows that engaged cohabiting couples who have agreed-on marriage plans have less conflict than do couples experiencing ambiguity about future marriage (Willoughby, Carroll, and Busby 2011). Interestingly, cohabiting men with intentions to marry their partner do more housework than other cohabiting males (Ciabattari 2004; Kuperberg 2012).

Cohabitation and Interpersonal Violence As is the case for dating relationships, discussed in Chapter 5, there is considerable domestic violence, or *interpersonal violence* (IPV), in cohabiting relationships (Cui et al. 2013; Johnson et al. 2015). **Selection effects**, the situation in which individuals "select" themselves into a category being investigated—in this case, cohabitation—largely account for these findings. As we have seen, individuals who live together without marrying tend to be less well educated and poorer than marrieds, and—although domestic violence occurs at all economic and education levels—low income and education are statistically associated with higher levels of couple conflict and domestic violence (Anderson 2010; Dush 2011). Furthermore, differences in selection *out* of cohabitation and into marriage may help account for the statistical difference because the better cohabiting relationships can be expected to transition into marriage, a situation that leaves a disproportionate share of violent couples among cohabiters (Kenney and McLanahan 2006).

Although not as often as one might expect, IPV can result in the dissolution of the relationship. In one study, psychological aggression was a stronger predictor of the couple breaking up than physical aggression (Curtis, Epstein, and Wheeler 2015). Research on the economic consequences of cohabiter breakups finds similarities to getting divorced. On average, men experience moderate financial decline, whereas women's economic decline is more pronounced (Tach and Eads 2015). Counselors stress the importance of being fairly independent before deciding to cohabit, understanding one's motives, having clear goals and expectations, and being honest with and sensitive to the needs of both oneself and one's partner. This is especially necessary when children are involved. "As We Make Choices: Some Things to Know about the Legal Side of Living Together"

Some Things to Know about the Legal Side of Living Together

When unmarried partners move in together, they may encounter regulations, customs, and laws that cause them problems; being prepared for these situations can help (Clark 2010). Consulting a lawyer is strongly advised, as is becoming familiar with the laws in your state. The following are potential trouble spots.

Domestic Partners

- In many areas and employment sites, opposite- and same-sex unmarried couples may register their partnerships and then enjoy certain rights, benefits, and entitlements that have traditionally been reserved for marrieds, such as access to joint health insurance.

- Registering as domestic partners usually requires joint residence and finances, plus a statement of loyalty and commitment.

Residence

When two unmarried people are renting, landlords may ask each to sign the lease—a legally binding contract—so that each is held responsible for all rent and associated costs.

Bank Accounts

Any couple can open a joint bank account, but one partner can then withdraw money without the other's approval.

Power of Attorney for Finances

In the event one partner should become incapacitated, a document establishing power of attorney for finances will allow the authorized partner to pay bills, run the partner's business, file taxes, and so on.

Credit Cards

If an unmarried couple shares a credit card, both partners are legally responsible for all charges made by either of them, even after the relationship ends. Creditors generally will not remove one person's name from an account until it is paid in full.

Property

A couple can have a written agreement about what happens to property that was purchased together should the relationship end.

Insurance

- The routine extension of auto and home insurance policies to "residents of the household" cannot be presumed to include nonrelatives.

- Anyone may name anyone else as the beneficiary on a life insurance policy. However, insurance companies sometimes require an "insurable interest," which is generally interpreted to mean a conventional family tie.

Wills and Living Trusts

- If you have no will or living trust when you die, your property will

discusses the legal implications of cohabiting in the United States today.

Cohabiting Parents and Outcomes for Children

Only four in ten U.S. adults today think that it is "morally wrong" to have a baby outside of marriage (Newport 2015). Indeed, 40 percent of all U.S. births are born to unmarried mothers compared to only 4 percent in 1950 (Carter 2009; Martin, Hamilton, Osterman, Driscoll, and Drake 2018). It is important to consider that more than half (58 percent) of these nonmarital births occur to cohabiting mothers (Child Trends 2015c). As shown in Figure 6.4, the number of cohabiting couples, and cohabiting couples with children, has grown, especially in the last decade. In 2018, over one-third (36 percent) of cohabiting heterosexual households contained children under age 18—a proportion only 3 percentage points lower than that of heterosexual married couples (U.S. Census Bureau 2018c,d). Seven percent of children under age 18 (about 5.5 million) live with a cohabiting parent or parents; of children who live with cohabiting couples, over half (56 percent) live with both of their unmarried biological or adoptive parents (U.S. Census Bureau 2018b). Children are therefore often an important part of the life of cohabiting couples.

One study showed that among urban couples who had a child while cohabiting, 28 percent married within five years, another 28 percent separated, and 45 percent remained in the cohabiting relationship (McClain 2011). Although having a child while cohabiting does not necessarily increase a couple's odds of staying together, conceiving a child during cohabitation and then marrying before the baby is born apparently does increase union stability (Rackin and Gibson-Davis 2012). Why would this be? "Although birth in cohabitation indicates a decision to remain together during pregnancy, it also represents a decision not to commit to

pass to individuals designated by state law—usually, legal spouses and blood relatives. A surviving partner could inherit nothing, not even property that he or she paid for.

- In some states, a cohabiting couple who have filed a joint tax return or used the same last name may be considered married in "common law." In this case, a cohabiting partner could claim a share—perhaps as much as one-third—of your estate, despite a will saying you wanted it to go to your children. It's best to sign a joint statement saying that you do not intend a common law marriage (Ebeling 2007).

Health Care Decision Making

Anyone too ill to be legally competent should have an agent to act on his or her behalf in medical decision making. Many cohabitants want their partners to play this role. To be sure that medical personnel honor this desire, designate your partner as decision maker through a durable power of attorney for health care document.

Children

With the 1972 case of *Stanley v. Illinois* (405 U.S. 645), an unmarried mother is no longer entitled to sole disposition of the child in many states. Although courts have placement discretion, unmarried couples should stipulate in writing that custody is to go to the partner if the other parent dies. Note also that financial obligations for child support do not depend on marital status.

Some—but not all—courts grant visitation rights to a nonbiological same-sex co-parent should the relationship end. Unmarried parents to a partner's child should consider three documents:

- a co-parenting agreement that spells out the rights and responsibilities of each partner;
- a nomination of guardianship that adds language to a will or living trust; and

- a consent to medical treatment that allows the co-parent to authorize the child's medical procedures.

Breaking Up

Ending a cohabiting relationship is not to be taken lightly:

- Couples who do not stipulate in writing—and preferably with an attorney's assistance—paternity, property, and other agreements can expect legal hassles. See an attorney about the laws in your state.
- Ending a registered domestic partner agreement in California and some other states requires a formal property settlement agreement and a dissolution proceeding in court (California Secretary of State's Business Programs Division 2011).

Critical Thinking

Does having to worry about the legal aspects of cohabitation lessen what appear to be some of the advantages of living together? Why or why not?

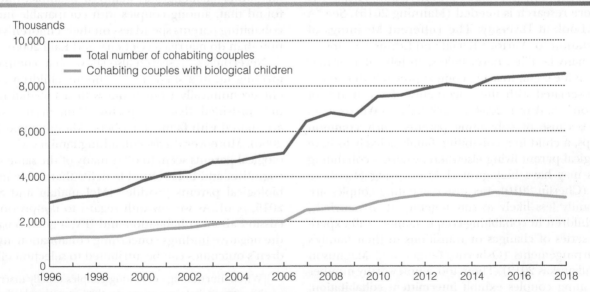

FIGURE 6.4 Number of opposite-sex cohabiting couples (in thousands), with and without children under 18, 1996–2015.

Source: Child Trends 2015d; U.S. Census Bureau 2018d.
Note: Children under 18 include biological children of either partner, who have never been married.

marriage" (Manning 2004, p. 677). Cohabiting parents who together see a father's involvement in parenting as critically important are also more likely to stay together (Hohmann-Marriott 2009; McClain 2011). *Unintended fertility*, having children who were unplanned, has been found to be generally disruptive for couples, whether cohabiting or married (Guzzo and Hayford 2012). Perhaps ironically, the fear of divorce among unmarried parents may reduce their likelihood of marrying (Waller and Peters 2008).

Children's Outcomes As pointed out in Chapter 2, with regard to studying families, the *structure–functional theoretical perspective* sometimes asks us to think in terms of *functional alternatives*—alternate structures that might (or might not) well perform a function traditionally assigned to the nuclear family. With regard to living together without marriage, many researchers have asked, how well do cohabiting adults fill the parenting role when compared with married couples? In this regard, a recent study among Latina mothers whose ethnic origin is in Mexico, Puerto Rico, or the Dominican Republic concluded that cohabitating among many Latinos may be similar enough to marriage that children's outcomes do not significantly differ according to whether their parents are married or cohabiting. The researchers suggest that in ethnic groups where cohabitation is more likely to be considered similar to marriage, cohabiting couples enjoy the benefits of less ambiguity regarding their relationship and more support from extended kin (Fomby and Estacion 2011). However, few studies have examined whether cohabitation affects children in similar or different ways for blacks, whites, and Hispanics, and more research is needed (Manning 2015). See "A Closer Look at Diversity: The Different Meanings of Cohabitation for Various Racial and Ethnic Groups."

For many families, nevertheless, "[c]ohabitation may not be an ideal childrearing context precisely due to the stress associated with the uncertainty of the future of the union" (S. Brown 2004, p. 353; see also Wilcox et al. 2011). Because of cohabiting parents' fairly common breakups, a child in a cohabiting family is likely to have a biological parent living elsewhere, whereas cohabiting parents may have biological children living elsewhere as well (Cherlin 2010). Because cohabiting couples are significantly less likely to stay together than marrieds, many children in cohabiting-couple families will experience a series of changes or transitions in their family's living arrangements (Osborne, Berger, and Magnuson 2012; Pilkauskas 2012). Then, too, a noteworthy number of parenting couples exhibit **intermittent cohabitation**. Over time, they move in together, then out, then back in (Cross-Barnet, Cherlin, and Burton 2011). Sociologist Wendy Manning puts it this way: "[C]ohabitation is often a marker for marital instability, and family instability is

strongly associated with poorer outcomes for children. In this way, being born into a cohabiting family sets the state for later instability" (Manning 2015, p. 2).

The cumulative instability of cohabitating families is related to a wide range of problematic outcomes—from lower scholastic performance to greater incidence of marijuana use (Cavanagh 2008; Mandara, Rogers, and Zinbarg 2011; Sun and Li 2011). "Residential and other household changes associated with the formation of new partnerships may disrupt well-established patterns of [parental] supervision" (Thomson et al. 2001, p. 378; see also Magnuson and Berger 2009). Other research has found a relationship between a mother's overall stress (which negatively affects parenting) and her forming a co-residential relationship with a nonbiological father to her child (Cooper, McLanahan et al. 2009; Osborne, Berger, and Magnuson 2012).

When compared to those living with two married biological parents, adolescents who have lived with a cohabiting parent are more likely to experience earlier premarital intercourse, higher rates of school suspension, and antisocial and delinquent behaviors coupled with lower academic achievement and expectations for college (Albrecht and Teachman 2003; S. Brown 2004; Carlson 2006). Having a cohabiting male in the household who is not the biological father appears not to improve adolescents' outcomes when compared with living in a single-mother household (Manning and Lamb 2003). In fact, children living with a single parent who has a cohabiting partner in the household have significantly higher rates of abuse and neglect than do children growing up in other family forms (Sedlak et al. 2010). Children growing up in cohabiting households face other challenges as well. For example, one study found that, among couples with comparable incomes, cohabiting parents spend less on their children's education than do marrieds (DeLeire and Kalil 2005).

Nevertheless, research also shows that, compared to growing up in a single-parent home, children do benefit economically from living with a cohabiting partner, provided that the partner's financial resources are shared with family members (Manning and Brown 2006). Moreover, "*stable* cohabiting families with two biological parents seem to offer many of the same health, cognitive, and behavioural benefits that stable married biological parents provide" (McLanahan and Sawhill 2015, p. 6). As we saw with regard to the previous discussion about cohabitation and IPV, a good portion of the negative findings concerning cohabitation and children's outcomes can be attributed to selection effects.

[W]ell-functioning cohabiting couples usually marry when the woman gets pregnant or they consciously decide to start a family. . . . So most couples who have the social and personal characteristics that foster both stable relationships and good parenting move on to marriage. This leaves a larger pool of cohabiting couples with economic,

psychological, and relationship issues that simultaneously inhibit them from marrying and make them less effective parents—issues like financial instability, low education, infidelity, conflict, violence and lack of trust. (Coontz 2012)

In the end, the effect of cohabitation on children is complex, and its effects on children and adults depend on a range of social, economic, and emotional variables, including race and ethnicity, income, education, and relationship satisfaction, as well as the characteristics of the children, including their age and gender (Manning 2015). The high and rising incidence of cohabitation reminds us that today's postmodern family encompasses many forms. We turn now to another relationship type that illustrates this point: same-sex couples.

Cohabiting Same-Sex Couples

Google LGBTQ+ organizations, and you'll find websites for men and women, African Americans, Latinos and Latinas, Jews, and Muslims, parents, and other groups. Contrary to what we're likely to see on television, LGBTQ+ singles make up a diverse category of all ages and racial and ethnic groups (Family Equality Council 2017). Moreover, same-sex couples living together in long-term, committed relationships are not a recent development. According to social historian Samuel Kader (1999), same-sex committed couples date back to ancient times, and commitment ceremonies between same-sex partners were not unknown in early Christianity. More recently, scholars

> have uncovered a long and complicated history of gay relationships in nineteenth-century America. Sometimes women passed as men to form straight-seeming relationships; sometimes men or women lived together as housemates but were really lovers; sometimes individuals would marry [heterosexually] but still carry on romantic, sometimes lifelong same-sex intimate relationships. (Seidman 2003, p. 124)

As mentioned at the beginning of the chapter, today there are approximately 1 million same-sex couple households in the United States, about evenly divided between gay male and lesbian couples, and nearly one-fifth have children. Over twice as many female same-sex couples (24 percent) as male same-sex couples (9 percent) have children (U.S. Census Bureau 2018a). It is unclear if the legalization of marriage for same-sex couples will affect their choices to have children. Same-sex cohabiting couples are a diverse lot.

One family studies professor describes the diversity apparent in her own lesbian family as follows:

> My partner and I live with our two sons. Our older son was conceived in my former heterosexual marriage. At first, our blended family consisted of a lesbian couple and a child from one partner's previous marriage. After several years, our circumstances changed. My brother's life

Vicky Kasala/Getty Images

Increased singlehood is not limited to younger generations. An increasing percentage of older people are unmarried. In addition to living alone, many older couples opt to "live apart together," maintaining separate residences to avoid intermingling finances.

partner became the donor and father to our second son, who is my partner's biological child. My partner and I draw a boundary around our lesbian-headed family in which we share a household consisting of two moms and two sons, but our extended family consists of additional kin groups. For example, my former husband and his wife have an infant son, who is my biological son's second brother. All four sets of grandparents and extended kin related to our sons' biological parents are involved in all our lives to varying degrees. These kin comprise a diversity of heterosexual and gay identities as well as long-term married, ever-single, and divorced individuals. (Allen 1997, p. 213)

As a second example of same-sex family diversity, Carol conceived Griffin through in vitro fertilization (IVF) with sperm from her friend George. Monday, Tuesday, Thursday, and Sunday nights, George stays in the spare room at Carol's apartment. The other three

nights he spends with his domestic partner, David. "It's not like Heather has two mommies," George quips. "It's George has two families" (Kleinfield 2011). Interestingly, as mentioned in Chapter 1, a handful of states have legislated that parental legal status and rights can be enjoyed by more than two parents in one family (Lovett 2012).

Before same-sex marriage became legal in 2015, many same-sex couples took part in a **civil union** or registered as domestic partners. The terms **domestic partnership** and **civil union** both refer to officially recognized unions in which unmarried cohabiting couples enjoy some (although not all) rights and benefits ordinarily reserved for marrieds. However, the term *civil union* is ordinarily used to refer to same-sex couples with a legal status somewhat like marriage. The term *domestic partnership* is typically used to refer to a formal status according to which same- or opposite-sex unmarried partners enjoy benefits (such as health insurance) offered by some employers, cities, counties, and states. Generally, domestic partnerships grant couples lesser status and fewer benefits than do civil unions.

Given the legalization of gay marriage, the future of domestic partnerships and civil unions is currently unknown. It is unclear how many couples would choose legal marriage over these options. Although discrimination assuredly persists, attitudes toward the LGBTQ+ community have become more accepting over the past fifty years. According to the Gallup organization (McCarthy 2019), in 2019, 93 percent of Americans polled said that gay people should have equal rights in terms of job opportunities, 83 percent said that gay or lesbian relationships should be legal, 75 percent said that gays and lesbians should be allowed to adopt children, and 49 percent said that being gay or lesbian is something a person is born with, compared to 56 percent, 43 percent, 14 percent, and 13 percent in 1977. Support for gay marriage is at about 63 percent, up from 27 percent in 1996, when the question was first asked. Younger Americans and women are more accepting of gay and lesbian relationships than older Americans and men. LGBTQ+ issues are discussed in detail in Chapter 3. Same-sex couples who are married are discussed in more detail in Chapter 7, and those who are parenting are discussed in Chapter 9.

MAINTAINING SUPPORTIVE SOCIAL NETWORKS AND LIFE SATISFACTION

With the rapid growth of singlehood in recent decades, one may wonder about the future of marriage. Have young adults given up on the idea? As discussed in Chapter 1, marriage remains important to Americans,

although not to the extent that it was in the past. The majority of high school students, even among Hispanics and blacks for whom marriage rates are low, expect to marry (Anderson 2016). In the Clark University poll of emerging adults, 86 percent of surveyed 18- to 29-year-olds expected their marriages would last a lifetime (Clark University 2012a). Thus, it appears that young adults still value marriage.

Research and polls consistently find that marrieds as a group are happier than singles (Brown and Jones 2012; Lee and Ono 2012; Wienke and Hill 2009). Research also shows that, regardless of whether they are legally married, people in secure interpersonal heterosexual or same-sex relationships—and those who socialize often with friends and family—are happier than those who spend considerable time alone (Adamczyk and Segrin 2015; Lee and Ono 2012). *Satisfaction* with one's relationship status is more important to personal happiness than whether one is single, married, or cohabiting (Lehmann et al. 2015).

Perhaps it is not surprising that life satisfaction is associated with income as well as marital status and that many singles and single parents, particularly women, just do not make enough money (Cheung and Lucas 2015). Many work more than one low-paying job and then take care of their homes and children (Huston and Melz 2004). For them, "career advancement" means hoping for a small raise or just hanging on to a job in the face of growing economic insecurity. Pursuing higher educational opportunities means rushing to one or more classes and working full-time. Moreover, research shows that poor women have less-effective private safety nets than do others because their families and friends are also most likely to be poor and overburdened financially and emotionally (Harknett 2006).

If we think of the various living arrangements of unmarrieds as forming a *continuum of social attachment*, we realize that not all singles are socially unattached, disconnected, or isolated (Ross 1995). Research consistently shows that people in close, supportive relationships—whether heterosexually married, same-sex married, heterosexually cohabiting, same-sex cohabiting, or living alone—tend to enjoy better health and be happier and less depressed than those with no intimate partner at all (Liu and Reczek 2012; Musick and Bumpass 2012; Oswald and Lazarevic 2011).

Living alone does not necessarily imply a lack of social integration or meaningful connections with others (Trimberger 2005). For some unattached singles, living alone can be lonesome, and unattached singles have tended to report feeling lonely more often than marrieds (Harter and Arora 2008; Pelham 2008). Poor and older singles are especially likely to be lonely, perhaps because the low incomes and ill health that tend to accompany old age make socializing difficult (Kim and McKenry 2002; Klinenberg 2016). Sociologist

E. Kay Trimberger (2005) argues that the "heaviest thing" for unattached, middle-aged women is the "idea of the couple, and that's so internalized." Trimberger identifies the following "pillars of support" for unattached single women: a nurturing home, satisfying work, satisfaction with their sexuality, connections to the next generation, a network of friends and possibly family members, and a feeling of community. However, marriage also involves a set of obligations and the responsibility of coping with both the burdens of other family members and the disappointments that come with family life. Valuing personal autonomy, Americans may find these obligations emotionally stressful (Gove, Style, and Hughes 1990). There are some areas in which nonmarrieds may feel better off than marrieds. Less irritation and a greater sense of control over one's life can be among the advantages of being single (Hughes and Gove 1989). Singles are freer to explore different career opportunities, to travel, volunteer, and invest in close friendships.

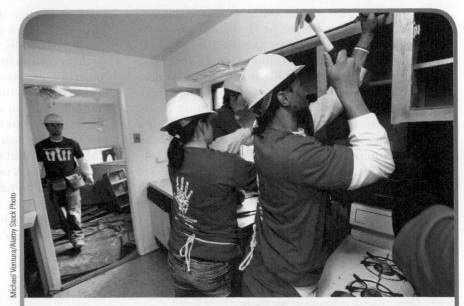

Michael Ventura/Alamy Stock Photo

When we think of singlehood as a continuum, we realize that not all singles—even those who live alone—are socially unattached, disconnected, or isolated. It's important to develop and maintain supportive social networks of friends and family. Single people place high value on friendships, and they are also major contributors to community services and volunteer work.

Actress Emma Watson, of *Harry Potter* fame and currently a UN Goodwill ambassador, on the eve of her 30th birthday declared herself "self-partnered." Kate Bolik, author of *Spinster: Making a Life of One's Own* (2015), said, "I was delighted with the news. . . . [w]hen I was Watson's age, there wasn't a public conversation about the very real benefits of being alone. It heartens me to think that young people today are getting this message, and that perhaps society is on the road toward not being obsessed with coupledom" (Pinsheta 2019).

All of us need support from people we are close to and who care about us. Isolation increases feelings of unhappiness, depression, and anxiety (Umberson et al. 1996), whereas being socially connected "seems to keep stress responses . . . from running amok," according to UCLA psychologist Shelley Taylor (quoted in "Save the Date" 2004). Among unmarried mothers, a social network that can function as "a parent safety net" is important to positive parenting practices (Ryan, Kalil, and Leininger 2009). Moreover, it is not uncommon for young unmarrieds to become overwhelmed by the massive array of choices available to them. Millennials may experience what is sometimes referred to as a *quarter-life crisis*—a great deal of difficulty figuring out "who they want to be."

Maintaining close relationships with parents, siblings, and friends is associated with positive adjustment and life satisfaction (Kurdek 2006; Soons and Liefbroer 2009; Spitze and Trent 2006). A study of family and community support experienced by same-sex couples concludes with the following policy advice:

> Community initiatives to strengthen families could emphasize the importance of staying in touch and getting along despite disagreements among family members; these campaigns could identify sexual orientation as a topic where adult family members can learn to disagree with each other while agreeing to provide a loving environment. . . . (Oswald and Lazarevic 2011, p. 383)

A crucial part of one's support network involves valued friendships. If a friend has ever "broken up with you," you understand that pain, which some argue can be as bad, if not worse, than the pain of a breakup with an intimate partner. Whereas romantic relationships are assumed to be tumultuous and fickle, friends are thought of able to "withstand all," according to psychotherapist Meg Josephson. She goes on to say, "we are bombarded with messages, 'Friends 'till the end, true blue friends, through thick and thin . . . we grow up with the idea that if a friend is a good friend, they will unconditionally accept us and be there" (Glanz 2019). We are left unprepared from the sudden loss of a friend, and it can be hard for adults to develop new ones.

Despite changing gender roles, men remain less likely than women to cultivate psychologically intimate relationships with siblings or same-sex friends (Levy 2005; Weaver, Coleman, and Ganong 2003). Indeed, a man may be more open and disclosing with a woman friend (Wagner-Raphael, Seal, and Ehrhardt 2001). One study of men in the construction industry found that rather than building truly supportive friendships, many of them engaged in horseplay and talked about alcohol consumption, risk taking, and physical prowess and generally engaged in one-upmanship (Iacuone 2005). Men (as well as women) who do not establish friendships based on emotional honesty run the risk of feeling socially isolated. In addition to friendships, other sources of support for singles include co-living situations, religious fellowships, and volunteer work (Mustillo, Wilson, and Lynch 2004).

A survey conducted by the Pew Research Center during the recession found that young-adult single men and women are mostly optimistic about their futures despite hard economic times (Pew Research Center 2012b). Singles' optimism can no doubt be attributed partly to their reaching out to families and friends. "There are so many ways to live and love. The sentimentalized image of Mom, Dad, and the kids gathered around the hearth has had its day. A new American experiment has begun. We're not all going nuclear anymore" (DePaulo 2012). However a person chooses to live the single life, establishing a sense of belonging by maintaining supportive social networks is crucial.

Summary

- Since the 1960s, the number of unmarrieds has risen dramatically. Much of this increase has resulted from young adults postponing marriage coupled with a marked rise in cohabitation.

- One reason people are postponing marriage today is that the increasing alternatives may make marriage less attractive. The recession that began in 2007 has led many to postpone marriage.

- The low sex ratio—fewer men for women of marriageable age—has also caused some women to postpone marriage or put it off entirely.

- Attitudes toward marriage and singlehood have changed so that being unmarried is more often viewed as preferable, at least "for now."

- More and more unmarrieds are living in their parents' homes, usually at least partly as a result of economic constraints.

- Emerging adulthood has recently been recognized by researchers as an important life stage.

- Some unmarrieds have chosen to live in communal homes, termed *cohousing*.

- Many more men and women are cohabiting than in the past and cohabit for a variety of reasons.

- Increasing numbers of cohabiting households include children either born to the union or from a previous relationship.

- The relative instability of heterosexual cohabiting unions has led to concern for the outcomes of children living in cohabiting families.

- Same-sex couples are a diverse population and may be legally married, unmarried, parents, nonparents, and people of all races, ethnicities, and educational and income levels.

- However one chooses to live the single life, it is important to maintain supportive social networks.

Questions for Review and Reflection

1. Individual choices take place within a broader social spectrum—that is, within society. How do social factors influence an unmarried individual's decision regarding his or her living arrangements?

2. What are some advantages and disadvantages of being unmarried compared to married?

3. What does current research tell us about the outcomes of children raised in homes with two married biological parents compared to those raised in cohabiting families?

4. **Policy Question.** Should cohabiting couples be afforded the same legal rights as married couples? Provide some facts to support both sides of the argument.

Key Terms

7

MARRIAGE: FROM SOCIAL INSTITUTION TO PRIVATE RELATIONSHIP

Learning Objectives

1 Analyze the value that Americans place on marriage as attitudes about it change.

2 Describe the marriage premise and its components and expectations for permanence and for monogamy.

3 Explain why marriage has historically been a public rather than private ceremony.

4 Discuss same-sex marriage as legal in the United States and elsewhere.

5 Define marriage as deinstitutionalized, using individualized marriage and long-term cohabitation as an example.

6 Contrast the selection hypothesis with the experience hypothesis regarding research-established benefits of marriage.

7 Discuss positive effects correlated with marriage.

8 Describe the debate between policy makers and scholars who see marriage as "in decline" versus those who see marriage as "changing" but not necessarily in decline.

Why a party for a wedding? A wedding marks a public announcement that, as opposed to other options, the couple has chosen marital commitment to define themselves and their relationship. A wedding ceremony publicly declares that marriage is important to society. Marriage is valuable to individual Americans as well as to society as a whole—although not to the extent that it was a few decades ago. About half of Americans are married now, compared with nearly three-quarters in 1960 (U.S. Census Bureau 2018a, Table A1). Between 1/2 and 60 percent of never-married adults say they want to marry someday. However, between one-quarter and one-third aren't sure, and about 14 percent tell pollsters they don't want to get married (Parker and Stepler 2017; Wang and Parker 2014).

Meanwhile, for decades we've watched same-sex couples campaign for the right to marry legally. U.S. couples won that right in June 2015, with the Supreme Court decision in the case *Obergefell v. Hodges.* "Facts about Families: Legal Same-Sex Marriage as a Successful Social Movement" describes this struggle. Both same-sex and heterosexual Americans say the top three reasons to marry are for love, to make a lifelong commitment, and for companionship (Geiger and Livingston 2019).

Academic research and opinion polls have traditionally shown marrieds to be happier, healthier, and wealthier than unmarrieds and more satisfied with family life than singles, including cohabitors (Institute for American Values 2015). Yet increasingly, Americans, both younger and older, are choosing, not to marry, but instead to cohabit—possibly in a long-term relationship.

In his majority opinion granting legal marriage to same-sex couples, U.S. Supreme Court Associate Justice Anthony Kennedy wrote that marriage is a "keystone of our social order" (Liptak 2015). Although the situation is much less definitive now than it was several decades ago (and more apparent today among the highly educated and wealthier), marriage remains the most acceptable gateway to committed family life in the United States (Gibson-Davis and Rackin 2014; Institute for American Values 2015).

Nevertheless, some feel they don't really need to be married. Others believe they can't afford to marry (Horowitz 2019). This chapter explores marriage as an evolving institution. We'll examine what distinguishes marriage from other couple relationships, then look at its changing nature. We'll discuss research on the benefits associated with marriage and review various ways social scientists explain this correlation. We will see that getting married—as opposed to cohabiting, for instance—is both a private relationship and a publicly proclaimed commitment. This chapter ends with an exploration of some tasks involved in establishing long-term committed relationships. We begin by looking at marital status today.

MARITAL STATUS: THE CHANGING PICTURE

Marriage has changed and continues to evolve. Historically, lower percentages of Americans are married today, and the value of marriage to many Americans has apparently declined. Meanwhile, in fighting to legalize same-sex marriage, LGBQ activists brought renewed public attention to the value of marriage. Moreover, having won the battle for legal marriage, same-sex couples have successfully challenged the **heteronormative bias**—the idea that heterosexuality is the only normal, acceptable, or "real" marriage option.

Fewer Heterosexual Married Couples

Due to divorce, postponing marriage until older ages, and increased cohabiting, the proportion of Americans age 18 and over who are married declined significantly over the past fifty years—from almost three-quarters (72 percent) in 1960 to just under half (49 percent) in 2018. Some of this decline can be attributed to our aging population and the fact that older Americans are more likely to be widowed. However, the proportion who are married has declined for Americans in all age groups (U.S. Census Bureau 2018a, Table A1).

Throughout the first half of the twentieth century, the trend was for more people to marry and at increasingly younger ages. Median ages for a first marriage fell from 1890 (the first time these data were collected) until 1960, when they began to rise again. Around 1950, family sociologists noted a standard pattern of marriage at about age 20 for women and 22 for men (Aldous 1978). As pointed out in Chapters 1 and 6, however, today's average age at marriage is the highest recorded since 1890: 27.8 for women and 29.8—nearly 30—for men in 2018 (U.S. Census Bureau 2018b, Table MS-2).

An important reason for change in rates and age at marriage is economic: Increasingly, Americans want to have finished their education, perhaps launched a career, and assured themselves that they can finance being married before they "tie the knot." Poorer individuals are forgoing marriage until they feel that they can afford it (Lundquist and Xu 2014; Smock and Greenland 2010). According to sociologist Andrew Cherlin, "Marriage used to be the first step into adulthood. Now it is often the last. For many couples, marriage is something you do when you have the whole rest of your personal life in order. Then you bring family and friends together to celebrate" (Rabin 2018).

The disparity in marriage rates between the poor and the not-poor has become significant enough that social scientists have coined a term for this situation—the **marriage gap**. Some statistics indicating a marriage gap look like this: Among U.S. heterosexual

males between 30 and 34 years old, 53 percent are married. Among those with earnings of $5,000 annually or less, one-quarter or fewer are married. Among those earning $100,000 and above, about three-quarters are married (U.S. Census Bureau 2018b, Table A1). At least among heterosexuals, a significantly smaller proportion of Americans are married today than in the past, and this trend is more pronounced among those with less education and income (Parker and Stepler 2017). What about same-sex couples?

Legal Same-Sex Marriage

According to U.S. Census Bureau estimates, there are close to a million same-sex households. Of these, between 1/2 and 60 percent were married couples, who make up less than 1 percent of all U.S. married couples (U.S. Census Bureau 2019). About 16 percent of same-sex households—10 percent of male and 23.7 percent of female—include children (Masci, Brown, and Kiley 2019).

In 2000, the Netherlands became the first country to allow same-sex partners to legally marry. Today, at least twenty other countries—including Argentina, Canada, Ireland, Norway, Portugal, South Africa, and Spain—allow legal same-sex marriage (Partners Task Force for Gay and Lesbian Couples 2015). In 2004, Massachusetts became the first U.S. state to legalize same-sex marriage. Eleven years later, in 2015, the United States became the twenty-first nation to legalize same-sex marriage.

Before same-sex couples could marry legally, many identified as spouses anyway (Gates 2009). Sometimes they publicly declared their commitment in ceremonies among friends or in some congregations and churches, such as the Unitarian Universalist Association or the Metropolitan Community Church, the latter expressly dedicated to serving the LGBTQ+ community. As one observer put it, with legal marriage, same-sex couples have "moved from 'outlaw' status (wherein homosexuality, or 'sodomy,' was criminalized) to 'in-law' status" (Shulman, Gotta, and Green 2012, p. 160). On a micro level, same-sex partners can now benefit from legal status if they so choose (Ocobock 2018).

"Facts about Families: Legal Same-Sex Marriage as a Successful Social Movement" describes the struggle for legalized same-sex marriage. Among LGBTQ+ couples, legal marital status is associated with lower perceived prejudice, reduced discrimination, and consequently

iStock.com/SolStock

Having first emerged as a remote possibility in the 1970s, legal marriage for same-sex couples became reality on June 26, 2015, with the U.S. Supreme Court decision in the case *Obergefell v. Hodges.*

better mental health (Cao, Zhou, Fine, Liang, Li, and Mills-Koonce 2017; LeBlanc, Frost, and Bowen 2018; Liu and Wilkinson 2017). Meanwhile, some lesbian and gay individuals find it unfortunate that, due to legal same-sex marriage and otherwise greater social inclusion by the larger society now, what they had experienced as a collective gay identity and community, largely prompted by their second-class-citizen status, has declined in salience for LGBTQ+ persons (Ocobock 2018). Moreover, not all same-sex couples want to marry.

Not All Same-Sex Couples Want to Marry Interestingly, although virtually all lesbians and gays support the claim for same-sex marriage in principle, some same-sex couples say "marriage is made 'not by some sort of legal sanctions' but by the commitment of the people in the relationship to each other" (Porche and Purvin 2008, p. 155). Actually, some lesbians and gays have argued that legalizing same-sex unions further stigmatizes sex outside marriage, with unmarried lesbians and gay men facing heightened discrimination ("Monogamy: Is It for Us?" 1998; Seidman 2003). We can't say how representative the following statement is, but it does illustrate the viewpoint of at least one lesbian when contemplating the possibility of legalized same-sex marriage: "I don't want to get married, so this marriage thing is going to make it harder for me to find a person to be in a relationship with. I know that because I don't want to get married, women will think I'm not a good potential partner, and move on" (Lannutti 2007, p. 145). Individuals who do want to marry value becoming part of a time-honored marriage premise.

Legal Same-Sex Marriage as a Successful Social Movement

A social movement is a concerted group action for social change. The struggle to win legal same-sex marriage has been a successful social movement in the United States and elsewhere. Before the 2015 U.S. Supreme Court ruling that legalized same-sex marriage nationally, at least thirty-five states had laws or state constitutional provisions that defined marriage as between one man and one woman. But beginning in the late 1990s, same-sex couples in several states filed lawsuits claiming that barring lesbians and gays from legal marriage was unconstitutional because it discriminated against them. Some but not all courts agreed and ordered their state legislatures to pass nondiscriminatory laws. As a result, Massachusetts in 2004 became the first state to allow gay and lesbian couples to marry legally. Many states followed. Activist organizations such as the American Civil Liberties Union and Partners Task Force for Gay and Lesbian Couples continued to agitate for change.

Having first emerged as a remote possibility in the 1970s, legal marriage for gay and lesbian couples "became a frontline issue" after 1991 when gay activists formed the Equal Rights Marriage Fund (Seidman 2003; Taylor et al. 2009). Among other arguments, activists asserted that denying lesbians and gay men the right to marry legally violated the U.S. Constitution because it discriminated against a category of citizens (Schwartz 2010, 2011b). As one lesbian spouse, having been legally married in Massachusetts, put it: "[Without being able to marry legally,] I always felt oppressed and not a part of America, not really. But this [being legally married in Massachusetts] seems like finally there is a light in the dark, like finally . . . the government is saying that my relationship counts and I count, too" (Lannutti 2007, p. 141).

However, for Christian fundamentalists, Islamic and Orthodox Jewish congregations, Mormons (LDS), some Catholic dioceses, and other conservative groups, the move to legalize same-sex marriage has been an "attempt to deconstruct traditional morality" (Egelko 2008; Goodstein 2015; Wilson 2001). Proponents of this view argued that heterosexual marriage alone has deep roots in the Judeo-Christian tradition. They further claimed that only heterosexual, married parents can provide the optimal family environment for raising children and that legalizing same-sex marriage would weaken an institution already threatened by single-parent families, cohabitation, and divorce (Blankenhorn 2007; McKinley and Schwartz 2010).

Opponents of same-sex marriage proposed a federal amendment to the U.S. Constitution that would have defined marriage as between one man and one woman and banned same-sex marriage in the United States, but passing a U.S. constitutional amendment requires a two-thirds majority in both the U.S. House of Representatives and the Senate and then

THE TIME-HONORED MARRIAGE PREMISE: PERMANENCE AND SEXUAL EXCLUSIVITY

Why does a marriage require witnesses and a license from the state? About 400 years ago in Western Europe, the government, representing the community, officially became involved in marriage (Halsall 2001; Thornton 2009). For about a century before that, Roman Catholic Canon Law included rules, or canons, that regulated European marriage—although the canons, difficult to enforce in widely separated rural villages, were often ignored (House 2002; Therborn 2004).

The Netherlands first enacted a civil marriage law in 1590 (Gomes 2004). England passed its first marriage act in 1653 but did not require a legal marriage license until 1754 (House 2002). Shortly after Europeans established colonies in the United States, they enacted rules for marriage similar to those they had known in Europe (Cott 2000). In the 400 years since then, our federal and state governments have generated a massive number of marriage-related laws and court decisions. For instance, polygamy was declared illegal in the United States in 1878. Also, before issuing a marriage license, some states require blood tests for various communicable diseases. Many states have waiting periods ranging from seventy-two hours to six days between the license application date and the wedding ("Chart: State Marriage License and Blood Test Requirements" 2015).

Even in the absence of law, communities throughout the world, represented by kinship groups or extended families, had claimed a stake in two important marriage

approval by three-fourths of the states. With such stiff requirements, the proposed amendment did not pass. Then June 2015 brought the U.S. Supreme Court decision that legalized same-sex marriage across the nation.

In 1975, Boulder County (Colorado) clerk Clela Rorex (left) began issuing marriage licenses to lesbian and gay male couples—on her own initiative, without state sanction—because she felt that denying them the right to marry discriminated against them. Her actions gained national media attention and bolstered the emerging social movement for same-sex marriage.

In 2015, Rowan County (Kentucky) clerk Kim Davis (right) gained international media attention when, citing her conscience, she refused to obey a court order requiring that she issue marriage licenses for same-sex couples. This followed the June 2015 U.S. Supreme Court ruling that made same-sex marriage legal across the United States.

and family functions: (1) guaranteeing property rights and otherwise providing economically for family members, and (2) ensuring the responsible upbringing of children (Ingoldsby 2006a).

Partly because of these two social functions, also discussed in Chapter 1, social scientists have defined the family as a **social institution**—a fundamental component of social organization in which individuals, occupying defined statuses, are "regulated by social norms, public opinion, law and religion" (Amato 2004, p. 961). In the vast majority of cultures around the world, a wedding marks a couple's passage into institutionalized family roles, often well monitored by in-laws and extended kin. The concept *social institution* is also addressed in Chapter 1.

Marriage marks the joining, not just of two individuals but two kinship groups (Sherif-Trask 2003;

Thornton 2009; Zerubavel 2013). From the couple's perspective, marriage had much to do with "getting good in-laws and increasing one's family labor force" (Coontz 2005b, p. 6; see also Coontz 2015). Family as a social institution has historically rested on the time-honored **marriage premise** of permanence, coupled in our society with expectations for monogamous sexual exclusivity.

The Expectation of Permanence

With few cross-cultural or historical exceptions, marriages have been expected to be lifelong undertakings—"until death do us part." **Expectations of permanence** derive from the fact that marriage was historically a practical institution (Coontz 2005b). Economic agreements between partners' extended families, as well as

Marking a couple's commitment, weddings are public events because the community has a stake in marriage as a social institution. Publicly proclaiming commitment to the marriage premise can help to enforce a couple's mutual trust in the permanence of their union. Additionally, weddings are an opportunity for a couple to symbolically represent their commitment to one another in a way that is unique to them. More and more, however, a wedding is also a status symbol, seemingly reserved for the middle and upper classes.

an amazing array of permissible sexual arrangements. **Polygamy** (having more than one spouse) is culturally accepted in many parts of the world. Polygamy can be divided into two types. *Polygyny*, a form of polygamy whereby a man can have multiple wives, "is a marriage form found in more places and at more times than any other" (Coontz 2005b, p. 10). Polygyny is not frequent because many men cannot afford multiple wives. *Polyandry*—the second type of polygamy, in which a woman has multiple husbands—is even less frequent (Ingoldsby 2006a; Stephens 1963).

Marriage in the United States legally disallows both forms of polygamy and requires monogamy, along with **expectations of sexual exclusivity**, in which spouses promise to have sexual relations only with each other. (There are exceptions to our cultural expectation for monogamy, however, and several of these exceptions are touched on in "Issues for Thought: Three Subcultures with Norms Contrary to Sexual Exclusivity.")

In Europe, requirements for women's sexual exclusivity emerged to maintain the patriarchal line of descent; the bride's wedding ring symbolized this expectation. The Judeo-Christian tradition eventually extended expectations of sexual exclusivity to include not only wives but also husbands. Over the last century, as "the self-disclosure involved in sexuality [came to] symbolize the love relationship," couples began to see sexual exclusivity as a mark of commitment (Reiss 1986, p. 56). Polls find that as many as 92 percent of Americans believe adultery to be morally wrong (Dugan 2015b; Gallup Poll 2012). Furthermore, expectations of sexual exclusivity have broadened from the purely physical to include expectations of emotional centrality or putting one's partner first. Marriage counselors now talk about "emotional affairs" (Herring 2005; Meier, Hull, and Ortyl 2009) and cyber affairs.

We'll note that, although the vast majority of Americans say they disapprove of extramarital sex for themselves and their partners, marital infidelity rates show that the situation is somewhat different in practice

society's need for responsible child raising, required marriages to be "so long as we both shall live."

In the United States today, marriage seldom involves merging two families' properties. In other ways, too, marriage is less critically important for economic security. Furthermore, we've seen that marriage today is less decisively associated with raising children than in the past. Meanwhile, another function of marriage—providing love and ongoing emotional support—has become key for most people (Cherlin 2004; Coontz 2005b). With about one-third of marriages ending in divorce, marriage is considerably less permanent now than historically—although it is generally more often permanent than cohabiting relationships (Cherlin 2009a; Miller, Sassler, and Kusi-Appouh 2011). Despite changes, more than any other non-blood relationship, marriage holds the potential for life-long togetherness. We turn to the second component of the time-honored marriage premise: sexual exclusivity.

The Expectation of Sexual Exclusivity

Every society and culture that we know of has exercised control over sexual behavior. Put another way, sexual activity has virtually never been allowed simply on impulse or at random. Meanwhile, anthropologists have found

Three Very Different Subcultures with Norms Contrary to Sexual Exclusivity

Although a substantial majority of Americans value monogamy as a cultural standard, there are subcultural exceptions. One Canadian study found marrieds more tolerant today of marital sexual infidelity than in the past, although very seldom in their own unions (Green, Valleriani, and Adam 2016). Here we look at three of subcultural exceptions to marital sexual exclusivity: polygamy, polyamory, and swinging (Society for Human Sexuality 2015).

Polygamy

Polygamy has been illegal in the United States since 1878, when the U.S. Supreme Court ruled that freedom to practice the Mormon religion did not extend to having multiple wives (*Reynolds v. United States* 1878). Today, some activists are pursuing legalization of polygamy in the United States (Stacey and Meadow 2009; Whitehurst 2012).

The Church of Jesus Christ of Latter-day Saints (LDS) officially outlawed polygamy in 1890. Nevertheless, dissident Mormons remain who are not recognized as LDS by the mainstream church but who follow the traditional teachings and take multiple wives (Whitehurst 2012; Woodward 2001). Some multiple wives have argued that polygyny is a feminist arrangement because the sharing of domestic responsibilities benefits working women (Johnson 1991; Joseph 1991; and see Egan 2014; Mack 2011).

Federal law prohibits prospective immigrants who practice polygamy from entering the United States. Civil libertarians argue that the Supreme Court should rescind its *Reynolds* decision on the grounds that the right to privacy permits this choice of domestic lifestyle as much as any other (Baltzly 2012; Schwartz 2011a).

Polyamory

Polyamory means "many loves" and refers to marriages in which one or both spouses retain the option to sexually love others in addition to their spouse (Polyamory Society n.d.).

Deriving their philosophy from the sexually open marriage movement, which received considerable publicity in the late 1960s and 1970s, polyamorous spouses agree that each may have openly acknowledged sexual relationships with others while keeping their marriage relationship primary (Society for Human Sexuality 2015). Within this philosophy, outside relationships can be emotional, as well as sexual. Couples usually establish limits on the degree of sexual or emotional involvement of the outside relationship, along with ground rules concerning honesty and what details to tell each other (Veaux and Rickert 2015). "Polyamorists are more committed to emotional fulfillment and family building than recreational swingers" (Rubin 2001, p. 721; see also Sheff 2014).

The Polyamory Society's Children Educational Branch offers advice for polyamorous parents and maintains a PolyFamily scholarship fund, as well as the Internet-based "PolyKids Zine" and "PolyTeens Zine," both designed to present "uplifting PolyFamily stories and lessons about PolyFamily ethical living" (Polyamory Society n.d.). Some polyamorists want to establish legally sanctioned group marriages and have begun to organize in that direction (Veaux and Rickert 2015). Conservative groups such as the Institute for American Values see such moves as threatening American values and harmful to children (Kurtz 2006; Marquardt n.d., p. 30).

Swinging

Swinging is a marriage arrangement in which couples exchange partners to engage in purely recreational sex. Swinging gained media and research attention as one of several "alternative lifestyles" in the late 1960s and early 1970s (Rubin 2001). At that time, it was estimated that about 2 percent of adults in the United States had participated in swinging at least once (Gilmartin 1977).

Although they receive less research attention now than in their heyday of the 1970s, swingers continue to exist as a minority subculture. Michael, a 28-year-old construction worker, and Sara, 24, who works in a doctor's office, have been in a committed relationship for more than a year but they do "full swaps," complete with intercourse, although they refuse to kiss strangers. "Sex is more of a primal, more of an urge-based," Michael said. "The kissing is more intimate so we like to keep that for us" (Chang and Lieberman 2012).

Swingers often face the challenge of managing jealousy, but they emphasize what they see as the positive effects—variety, for example (Chang and Lieberman 2012; DeVisser and McDonald 2007). Condoms are typically available at private parties and swing clubs (Rubin 2001; Society for Human Sexuality 2015).

Critical Thinking

What do you think about these exceptions to monogamy? Do you see them as threatening to American values? If so, why? If not, why not? Does one or more of these exceptions seem reasonable to you while others do not? If so, why? If not, why not?

(Fincham and Beach 2010). Meanwhile, a 2016 qualitative Canadian study found respondents fairly tolerant of nonmonogamous marital behavior in the abstract, but they did not apply such tolerance to their own relationship. As one respondent put it, "I'll say that it's different for everyone, and you have to find what works for you. You're committed to each other and you're married, but then you guys decide every Friday night we're going to swinger parties. . . . [If] that excitement is what brings you together, then awesome. But is it going to be for me? No" (Green, Valleriani, and Adam 2016, p. 423).

In the researchers' conclusion,

> [T]he increasing separation of sexuality from institutions of social control, with the consequences that intimate life—including marriage—is increasingly reflexive, less bound by religious-based norms of fidelity, and more individualized with respect to the significance given to the particular needs and wishes of the partners. (Green, Valleriani, and Adam 2016, p. 426)

Sexual infidelity is also addressed in Chapter 4.

To summarize this section, expectations for sexual exclusivity have gradually been extended to include emotional centrality. Simultaneously, there is early evidence that expectations for sexual exclusivity have become more likely to be interpreted loosely, at least in the abstract. Meanwhile, marital permanence—still a strong expectation—has become a weakened norm over past decades. Although its importance has diminished throughout the past sixty years or so, the marriage premise continues to bolster long-term, monogamous couple commitment. Next, we look in detail at the marriage premise as changing.

FROM "YOKE MATES" TO "SOUL MATES": A CHANGING MARRIAGE PREMISE

Chapter 1 points to an individualist orientation in our society. In eighteenth-century Europe, **individualism** emerged as a way to think about ourselves. No longer were we necessarily governed by rules of community. Societies changed from **communal**, or **collectivist**, to **individualistic**. In individualistic societies, one's own self-actualization and interests are a valid concern. In collectivist societies, people identify with and conform to the expectations of their extended kin. Western societies are characterized as individualistic, and individualism is positively associated with valuing romantic love (Dion and Dion 1991; Goode 2007 [1982]). (By *Western*, we mean the culture that developed in Western Europe and now characterizes that region and Canada, the United States, Australia, New Zealand, and some other societies.)

The Industrial Revolution and its opportunities for paid work outside the home, particularly in the growing cities and independently of one's kinship group, gave people opportunities for jobs and lives separate from the family. In Europe and the North American colonies, people increasingly entertained thoughts of equality, independence, and even the radically new idea that individuals had a birthright to "the pursuit of happiness" (Coontz 2005b). These ideas were manifested in dramatically unprecedented political events of the late 1700s, especially the U.S. Declaration of Independence and the French Revolution.

The emergent individualistic orientation meant diminished obedience to group authority because people increasingly saw themselves as separate individuals rather than as intrinsic members of a group or collective. Individuals began to expect self-fulfillment and satisfaction, personal achievement, and happiness. With regard to marriage, an emergent individualist orientation resulted in three interrelated developments:

1. The authority of kin and extended family weakened.
2. Individuals began to find their own marriage partners.
3. Romantic love came to be associated with marriage.

Weakened Kinship Authority

Kin, or extended family, includes parents and other relatives such as in-laws, grandparents, aunts and uncles, and cousins. Some groups, such as Italian Americans, African Americans, Hispanics, and gay male and lesbian families, also have "fictive" or "virtual" kin—friends who are so close that they are hardly distinguished from actual relatives (Eller 2015; Furstenberg 2005; Sarkisian, Gerena, and Gerstel 2006). In collectivist, or communal, cultures, kin have exercised considerable authority over a married couple. For instance, in traditional African societies, a mother-in-law may have more to say about how many children her daughter-in-law should bear than the daughter-in-law does herself (Caldwell 1982).

In Westernized societies, however, kinship authority is weaker. By the 1940s in the United States, at least among white, middle-class Americans, the husband–wife dyad was expected to take precedence over other family relationships. Sociologist Talcott Parsons (1943) noted that the American kinship system was not based on extended-family ties. Instead, he saw U.S. kinship as comprised of "interlocking conjugal families" in which married people are members of both their **family of orientation** (the family they grew up in) and their **family of procreation** (the one formed by marrying and having children). Parsons viewed the husband–wife bond and the resulting family of procreation as the most meaningful "inner circle" of Americans' kin relations

The married couples embedded in this family of Eastern European immigrants who arrived in New York City in 1902 may be in love, but they were not *expected* to find love in marriage. Instead, their union was held together by strong expectations of permanence as bolstered by the social control of the kinship group.

surrounded by decreasingly important outer circles. However, Parsons pointed out that his model mainly characterized the American middle class. Recent immigrants and lower socioeconomic classes, as well as upper-class families, still relied on meaningful ties to their extended kin.

Although the situation is changing, the extended family (as opposed to the couple or nuclear family) has been the basic family unit in the majority of non-European countries (Ingoldsby and Smith 2005). In the United States, extended families continue to be important for various European ethnic families, such as Italians, and for Native Americans, blacks, Hispanics, and Asian Americans, as well as other immigrant families (Eller 2015; Richardson 2009). In research on a largely ignored topic, researchers find that adult siblings can be important sources of emotional and practical support (Henig 2014).

Norms about extended-family ties derive both from cultural influences and from economic or other practical circumstances (Hamon and Ingoldsby 2003; Wong, Yoo, and Stewart 2006). Immigrants from many less-developed nations work in the United States and send money to extended kin in their home countries (Ha 2006). More and more Hispanics today value the conjugal bond as equal to or even more important than the extended family (Eller 2015; Hengstebeck, Helms, and Rodriguez 2015). Still, among Hispanics, *la familia* ("the family") means the extended as well as the nuclear

family (Sarkisian, Gerena, and Gerstel 2006).

We note that extended kin can be helpful—and they can also add stress. A focus group participant of Mexican American origin explained:

[T]he husband comes [to the United States] first and later brings his wife. But while he was waiting to earn money to bring his wife over, he brought his cousins and nephews. So this woman is living with . . . five of her relatives. So she is playing the role of wife, cousin, friend, and servant in the house, making food for all these people. And then, she has to go to work so that the husband can pay off the money for her trip here. . . . [I]n Mexico, everyone lives in difficult conditions [too]. But at least they live in their little shacks, . . . and just . . . the husband, [wife,] and the kids [live there]. And here they have to get used to living with fifteen people. (Helms, Supple, and Proulx 2011, p. 82)

All this is not to say that extended-family members are irrelevant to non-Hispanic white families in the United States. Nuclear families maintain significant emotional and practical ties with extended kin and parents-in-law. A qualitative study with a sample that was 95 percent white showed that uncles often mentor nephews or nieces (Milardo 2005). Extensive data from the Longitudinal Study of Generations show that young adults highly value extended family members (Bengston, Biblarz, and Roberts 2007). Recent research shows that adult siblings can be important sources of emotional support and happiness, especially as people get older (Henig 2014). However, as individuals and couples increasingly become more urban and geographically mobile, the power of kin to exercise social control over family members declines. If an individualist orientation has weakened kinship authority, it has also led to the desire to find a spouse on one's own.

Finding A Spouse on One's Own

Finding a romantic partner is fully addressed in Chapter 5. Here we want to state that arranged marriage has characterized collectivist societies (Ingoldsby 2006b; MacFarquhar 2006; Sherif-Trask 2003). Because a marriage joined extended families, selecting a suitable mate was a "huge responsibility" not to be left to the young people themselves (Tepperman and Wilson 1993, p. 73).

Analyzing arranged marriage in twentieth-century Bangladesh, sociologist Ashraf Uddin Ahmed notes that an individual finding his or her own spouse has been

"thought to be disruptive to family ties and is viewed as a child's transference of the loyalty from a family orientation to a single person, ignoring obligations to the family and kin group for personal goals" (Ahmed, quoted in Tepperman and Wilson 1993, p. 76). Moreover, there is concern that an infatuated young person might choose a partner who would make a poor spouse.

Ahmed argues that the arranged marriage system functioned not only to consolidate family property but also to keep the family's traditions and values intact. But as urban economies developed in eighteenth-century Europe and more young people worked away from home, arranged marriages began to give way to those in which individuals selected their own mates. Love rather than property became the basis for unions (Coontz 2005b).

Love and Marriage

Throughout the first 5,000 years of human history in all the world's cultures that we know of, people probably fell in love, but they weren't expected to do so with their spouses. Marriage was thought to be "too vital an economic and political institution to be entered into solely on the basis of something as irrational as love" (Coontz 2005b, p. 7). Love—an intense, often unpredictable, and possibly transitory emotion—was viewed as threatening to the practical institution of marriage. Valuing romance could lead individuals to ignore or challenge their social responsibilities.

With time, however, the ideology of romantic love came to be expected of middle-class marriages (Meier, Hull, and Ortyl 2009). In family historian Stephanie Coontz's words, basing marriage on love and companionship

> represented a break with thousands of years of tradition. . . . Critics of the love match argued . . . that the values of free choice and egalitarianism could easily spin out of control. If the choice of a marriage partner was a personal decision, . . . what would prevent young people . . . from choosing unwisely? If people were encouraged to expect marriage to be the best and happiest experience of their lives, what would hold a marriage together if things were "for worse" rather than "for better"? (2005b, pp. 149–50)

To use Coontz's metaphor, couples were no longer yoked together (like field oxen). A successful marriage came to be measured by how well the union met its members' emotional needs.

To summarize, emergent individualism in eighteenth-century Europe meant that people, increasingly valuing personal satisfaction, began to associate romantic love with marriage and hence to want to find a marriage partner on their own, a practice that both resulted from and further caused weakened kinship authority. Couples were no longer bound by the yoke of kin control. As you might guess, the nature of marriage changed.

DEINSTITUTIONALIZED MARRIAGE

Coontz (2005b) wrote that love and expectations for intimacy "conquered marriage." What did she mean? Coontz was talking about what family sociologist Andrew Cherlin (2004) calls the **deinstitutionalization of marriage**—a situation in which time-honored family definitions and social norms associated with the marriage premise "count for far less" than in the past (p. 853). For instance, childbearing outside marriage, once severely stigmatized, now "carries little stigma" (Cherlin 2009a, p. 919; Sawhill 2014a, 2014b). The following sections explore Cherlin's analysis of the shift from *institutional* to *companionate* to *individualized* marriage.

Institutional Marriage

We have witnessed a gradual historical change in Western and Westernized societies away from **institutional marriage**—that is, marriage as a social institution based on dutiful adherence to the marriage premise, described above (Cherlin 2004, 2009a; Coontz 2005b).

Institutional marriage offered practical and economic security, along with the rewards that we often associate with custom and tradition (knowing what to expect in almost any situation, for example). With few exceptions over the past 5,000 years, institutional marriage was organized according to patriarchal authority, requiring a wife's obedience to her husband and the kinship group. It is also true that, legally, institutional marriage could involve what today we define as wife and child abuse or neglect. Child and wife abuse were not recognized as social problems in this country until the 1960s and 1970s, respectively.

Companionate Marriage

By the 1920s, family sociologists in the United States had begun to note a shift away from institutional marriage, and in 1945 the first sociology textbook on the American family (by Ernest Burgess and Harvey Locke) was titled *The Family: From Institution to Companionship*. By **companionate marriage**,

> Burgess was referring to the single-earner, breadwinner-homemaker marriage that flourished in the 1950s. Although husbands and wives in the companionate marriage usually adhered to a sharp division of labor, they were supposed to be each other's companions—friends, lovers—to an extent not imagined by the spouses in the institutional marriages of the previous era. (Cherlin 2004, p. 851)

With companionate marriage, middle-class Americans often dreamed of attaining "the white picket fence"—that is, they saw marriage as an opportunity for idealized domestic happiness within their own single-family home.

(This is why we have drawn a picket fence to symbolize the companionate marriage bond in Figure 7.1A to C.)

> Companionate marriages of the 1950s were exceptional in many ways. Until that decade, relying on a single breadwinner had been rare. For thousands of years, most women and children had shared the tasks of breadwinning with men. . . . Also new in the 1950s was the cultural consensus that everyone should marry, and that people should do so at a young age. The baby boom of the 1950s was likewise a departure from the past, because birth rates in Western Europe and North America had fallen steadily during the previous 100 years. (Coontz 2005c)

Meanwhile, women's increasing educational and work options, coupled with their expectations for more intimate marital love, sowed the seeds for the demise of companionate marriage (Cherlin 2004, 2012; Coontz 2005c, 2015). An individualistic orientation views each person in a married couple as having talents that deserve to be actualized. In this climate, women in companionate marriages began to embrace self-actualization and to expect a husband's support for their doing so (Jackson 2007). Furthermore, women challenged centuries of previously ignored domestic violence. Given the tension between gender inequality and expectations for emotionally supported self-actualization, the companionate marriage "lost ground" (Cherlin 2004, p. 852). As one result, critics began to warn that American culture was becoming "narcissistic": Individuals appeared less focused on commitment or concern for future generations (Bellah et al. 1985; Lasch 1980).

Feminists defined this situation somewhat differently: Attention to domestic abuse, unequal couple decision making, and unfair division of household labor—as well as a wife's ability to more easily leave an intolerable situation through divorce—could be good things (Hackstaff 2007; Stacey 1996). However one saw it, by the late 1980s, companionate marriage—which had lasted for but a minute in the long hours of human history—had largely given way to its successor: individualized marriage.

Individualized Marriage

Relating this discussion back to Chapter 1, we can say that individualized marriage is associated with the postmodern family. Four characteristics distinguish **individualized marriage**:

1. It is optional.
2. Spouses' roles are flexible—negotiable and renegotiable.
3. Its expected rewards involve love, communication, and emotional intimacy.
4. It exists in conjunction with a vast diversity of family forms.

FIGURE 7.1A The institutional marriage bond. Couples are "yoked" together by high expectations for permanence and bolstered by the strong social control of extended kin and community.

© Designed by Agnes Riedmann. Cengage Learning 2015

FIGURE 7.1B The companionate marriage bond. Couples are bound together by companionship, coupled with a gendered division of labor, pride in performing spousal and parenting roles, and hopes for "the American dream"—a home of their own and a comfortable domestic life together.

© Designed by Agnes Riedmann. Cengage Learning 2015

FIGURE 7.1C The individualized marriage bond. Spouses in individualized marriages remain together because they find self-actualization, intimacy, and expressively communicated emotional support in their unions.

© Designed by Agnes Riedmann. Cengage Learning 2015

As an ideal type, the *companionate marriage* that characterized most of the twentieth century emphasized love and compatibility, as well as separate gender roles. However, in reality, couples represent this ideal type to varying degrees. Although this couple, who own and together operate a bakery, illustrate companionate marriage in *some* ways, they do not fit the definition of companionate marriage in at least one important way: They share the family provider role.

Partly because marriage is optional today, brides, grooms, and long-married couples have come to expect different rewards from marriage than people did in the past. They continue to value being good partners and, optionally, parents. However, today's spouses are less likely to find their only or definitive rewards in performing these roles well (Byrd 2009). More than in companionate marriages, partners now expect love and emotional intimacy, open communication, role flexibility, gender equality, and personal growth (Cherlin 2004, 2009a; Meier, Hull, and Ortyl 2009). Perhaps they expect more personal autonomy as well: One recent study found that today's couples are less likely than in the past to pool finances, although a majority of marrieds still do and, in fact, show many other examples of interdependent and integrated couple behavior (Lauer and Yodanis 2011; Yodanis and Lauer 2014). Over the course of about three centuries, nonetheless, couples have moved "from yoke mates to soul mates" (Coontz 2005b, p. 124). "How good is your relationship?" has become a question equal in importance to "Are you married?" (Giddens 2007).

Social theorist Anthony Giddens (2007) argues that expectations for a relationship based on intimate communication to the extent that "the rewards derived from such communication are the main basis for the relationship to continue" (p. 30) can lead to disappointment.

Unrealistically high expectations may be associated with deciding not to marry because the perfect match doesn't come along. In a national poll, 60 percent of never-married Americans said they weren't married because they hadn't met the right person (Parker and Stepler 2017). Unrealistic expectations may also precipitate divorce (although assuredly there are other reasons for divorce, as described in Chapter 14). As marriage becomes individualized, it begins to resemble cohabitation.

Marriage and Cohabitation Begin to Look Alike

As more persons cohabit, living together without being married becomes an increasingly legitimate option. Also, cohabitors more closely resemble the general population than they once did (Gurrentz 2019). As a result, although significant differences remain, cohabitation begins to look increasingly like marriage. For instance, in Norway where cohabiting is even more common and acceptable than in the United States, a study compared cohabiting with married persons' relationship with their partner's parents. The study found that having children increased the strength of a person's connection with their cohabiting partner's parents, while getting married did not make any additional difference (Wiik and Bernhardt 2017).

In a study in Australia, where cohabitation is also common and increasingly legitimate, researchers found that, upon moving in together, cohabiting couples who envisioned staying together long-term were more likely to undertake joint financial planning. Subsequently deciding to marry failed to create further changes in this regard. These authors see couple commitment as developing *during* long-term cohabitation, not necessarily because of a decision to marry (Fulda and Lersch 2018). A U.S. study showed that cohabiting enhanced parental school involvement for mothers of primary school children. However, getting married did not additionally increase participation (Ressler, Smith, Cavanagh, and Crosnoe 2017).

And finally, research by sociologist Jennifer Kane (2016) at the University of California, Irvine analyzed national survey data to examine whether being married during pregnancy encouraged healthier prenatal behaviors among expectant mothers. Kane used birthweight as a measure; low birthweight would indicate

less healthy behaviors on the mother's part. Kane found that, on average, infants born to non-partnered women had lower birth weight than those born to marrieds. However, there was no significant difference between babies born to married and cohabiting women. Kane concludes: "The lack of a salient cohabiting-married disparity in birth weight suggests exposure to a cohabiting union does not disadvantage infants. . . . This finding could indicate a growing convergence in child well-being across family types as family diversity grows, but more research is needed to evaluate this possibility" (Kane 2016, p. 224).

The marriage rate is dropping, cohabitation rates are increasing, and the two relationship statuses have begun to resemble one another. Meanwhile, 40 percent of Americans believe that living together outside marriage is "bad for society," while just 7 percent see cohabitation as a good thing society-wide. More than half say that more couples living together without getting married "doesn't make much difference" (Pew Research Center 2018a). The next section examines evidence regarding this latter belief.

DEINSTITUTIONALIZED MARRIAGE: EXAMINING THE CONSEQUENCES

In her classic 1995 presidential address to the Population Association of America, family demographer Linda Waite (1995) asked rhetorically, "Does marriage matter?" Like 40 percent of Americans, she concluded that indeed it does, for both adults and children. After reviewing research comparing the well-being of family members in married households with those in unmarried, including cohabiting ones, Waite reported that spouses generally had greater income and wealth, better sex and lower rates of substance abuse, and were more likely to be healthy and to exhibit healthy behaviors. Waite further found that children in married households were significantly less likely to live in poverty or drop out of school and had better relationships with their parents.

Health, Happiness, Finances, and Child Outcomes: Does Marriage Matter?

Since her address, social scientists have further researched Waite's findings. Research continues to show that spouses in enduring, supportive marriages generally have better physical and mental health than unmarrieds (Choi and Marks 2013; Fincham and Beach 2010; Miller et al. 2013). Furthermore, at least among heterosexuals, opinion polls find marrieds more likely to say they are "very happy" and to score higher on well-being indicators such as physical and emotional health (Brown and Jones 2012; Jackson, Miller, Oka, and Henry 2014). Additionally, national income data supports Waite's argument that marrieds are financially better off. For instance, the median income for married-couple families in 2018 was $93,329 compared with $54,336 and $40,233 for unmarried male- and unmarried female-headed households, respectively (U.S. Census Bureau 2018b, Table F-7). Clearly, higher income is positively correlated with marriage.

Furthermore, research supports Waite's conclusion that growing up with married parents correlates with better outcomes for children (Magnuson and Berger 2009; Sun and Li 2011; Wilcox, Marquardt et al. 2011; Wu, Schimmele, and Hou 2015). At least among heterosexual families, teens living in single-parent and cohabiting families, when compared with those living with married biological parents, are considerably more likely to experience earlier premarital intercourse, lower academic achievement, and lower expectations for college, together with higher rates of school suspension, delinquency, and marijuana use (Carlson 2006; Mandara, Rogers, and Zinbarg 2011; Manning and Lamb 2003; VanDorn, Bowen, and Blau 2006). An analysis of national data found that (among heterosexuals) married mothers exhibited the healthiest prenatal behaviors when compared to those in other family forms (Kimbro 2008). Part of the advantage that children in married households enjoy has to do with better financial circumstances (Fagan 2011; Lowrey 2014).

Comparing outcomes for children in same-sex married households with those of same-sex cohabiting couples' children is a topic for future research. We turn now to an examination of marital status and poverty.

Marital Status and Poverty

Children growing up in poverty want for enough nutritious food, are more likely to live in environmentally unhealthy neighborhoods, have more health and behavioral problems, do less well academically, and are more likely to drop out of school (Children's Defense Fund 2014). Twelve million (12 percent of) U.S. children under age 18 live at or below the poverty line (U.S. Census Bureau 2018b, Table POV01). An additional 16 million live in "near poor" conditions—above the official poverty line but at less than 125 percent of the poverty level. Twelve percent of nonHispanic white, 14 percent of Asian, 35 percent of Hispanic, and 43 percent of black children age 5 and under live below 125 percent of the poverty level.

Six percent of *married-couple* families live below the poverty line compared to 17 percent of single-male householder families and 34 percent of single-female

A Closer Look at Diversity

African Americans and "Jumping the Broom"

Nationally representative surveys show that among blacks, husbands and wives, like other Americans, are more likely than unmarrieds to report being "very happy" and satisfied with their finances and family life (Blackman et al. 2006). A significant proportion of African American couples have strong, enduring marriages (Marks et al. 2008). Meanwhile, with 34 percent of black men and 27 percent of black women currently married, African Americans are significantly less likely to be wed than are other U.S. race/ethnic groups (U.S. Census Bureau 2015b, Table A1). A large body of literature, written by both blacks and whites, is accumulating on the structural and cultural reasons for this situation (Chambers and Kravitz 2011; Coates 2015; McAdoo 2007). Contrary to some people's opinions, research also indicates that the availability of welfare is not a significant factor in a black woman's decision not to marry (Berlin 2007).

Answering a Gallup poll, 69 percent of African Americans said that it is "very important" for a couple to marry when they plan to spend the rest of their lives together. Asked, "When an unmarried man and woman have a child together, how important is it to you that they legally marry?" college-educated African Americans were *more* inclined than either Hispanics or non-Hispanic whites to say that marrying in this situation is "very important." The figures were 55 percent of blacks, 46 percent of Hispanics, and 37 percent of nonHispanic whites (Saad 2006). Attitude surveys consistently show that African Americans value marriage, perhaps more than nonHispanic whites do (Banks 2011).

The news media have focused so frequently on poverty-level African Americans and on the relatively low proportion of married blacks of all social classes that we may forget about the 9.4 million (30 percent of) African Americans who *are* married. In fact, among blacks between ages 45 and 49, 53 percent of males and 48 percent of females are married. Among black males earning $75,000 and above, 68 percent are married (U.S. Census Bureau 2015b, Table A1).

For African American brides and grooms, the significance of the broom originated among the Asante in what is now the West African country of Ghana. Used to sweep courtyards, the handmade Asante broom was also symbolic of sweeping away past wrongs or warding off evil. Brooms played a part in Asante weddings as well. To culminate their wedding ceremony, a couple might jump over a broom lying on the ground or leaning across a doorway. Jumping the broom symbolized the wife's commitment to her new household, and it was sometimes said that whoever jumped higher over the broom would be the family decision maker (DiStefano 2001; Prahlad 2006).

Among slaves brought to the Americas, jumping the broom continued. Not allowed to marry legally, slaves sometimes jumped a broom as an alternative ceremony to mark their marital commitment. The association of jumping the broom with slavery has stigmatized the tradition for some African Americans. However, the ritual is coming back as more middle-class blacks plan

householder families. We can safely conclude that being married is associated with lower poverty rates, but this association between marriage and poverty is not the whole story. For one thing, the majority of unmarried families are *not* in poverty.

Table 7.1 shows that, despite being married, 4 percent of nonHispanic white, 7 percent of black, 10 percent of Hispanic, 6 percent of Asian, and 16 percent of American Indians or Alaska Natives live in poverty (U.S. Census Bureau 2018b, Table 4). We conclude that marriage alone is not sufficient to alleviate poverty. In addition to marital status, the number of children in the household, along with low education and wages, and with poor mental or physical health care, among other situations, contribute to poverty (Crowley, Lichter, and Qian 2006). We can conclude that being married is associated with a family's economic well-being, but a child does not necessarily need to live with married parents in

order to grow up above the poverty line. Moreover, factors other than marriage influence positive outcomes.

Additional Influences on Outcomes

Clearly, there is a correlation between marriage and positive outcomes for both adults and children. However, this correlation is not the whole story. For one thing, variables in addition to marital status have been shown to influence health, happiness, finances, and children's behavior outcomes. Besides being married, higher education, a comfortable income, and not being faced with prejudice or discrimination improve mental health (Bierman, Fazio, and Milkie 2006; Mandara et al. 2008). A study on Hispanic children showed that the child's neighborhood and peers, family conflict, parental nurturance and involvement in the child's school activities, parents' participation in religious services, and

culturally relevant wedding celebrations (African Wedding Guide n.d.).

Today African Americans are increasingly divided between a middle class that has benefited from the opportunities opened by the civil rights movement and a substantial sector that remains disadvantaged. Childbearing and child rearing are increasingly separate from marriage. True of all race/ethnic groups, this trend is especially pronounced among blacks, with 71.5 percent of births in 2014 to unmarried mothers (Child Trends 2015), and black couples are far more likely than the national average to have never married. As a consequence, only 35 percent of African American children are living with their biological married parents, compared with 75 percent of nonHispanic white and 64 percent of Hispanic children (U.S. Federal Interagency Forum 2015). As recently as the 1960s, more than 70 percent of black families were headed by married couples, whereas 45 percent were in 2014 (Billingsley 1968; U.S. Census Bureau 2015b, Table F7). Given values that are similar to other Americans regarding marriage, research consistently suggests that causes for African Americans' differing marriage patterns are both cultural

(resulting from enforced family behaviors during slavery) and economic (poverty, high unemployment, high black male incarceration rates, for instance) (Banks 2011; Burton et al. 2009; Coates 2015).

Meanwhile, scholars have noted the strengths of black families (Hill 2003 [1972]; Hill 2004; Taylor 2007), especially strong kinship bonds. Single-parent families or unmarried individuals are embedded in extended families and experience family-oriented daily lives.

Although somewhat controversial because it can be a reminder of slavery, jumping the broom at African American weddings is becoming more common again as black couples plan wedding celebration rituals designed to incorporate their cultural heritage. Attitude surveys show that African Americans value marriage highly. However, popular media as well as scholarship have focused on the low proportion of married blacks, and we tend to ignore the approximately 30 percent of African American adults who *are* married.

Critical Thinking

How does jumping the broom symbolize the marriage premise? Why would it be important to incorporate traditions that are relevant to one's own culture into a wedding ceremony? Why do you think we hear relatively little about African Americans' weddings or marriages?

TABLE 7.1 Percentage of U.S. Families Below Poverty Level, 2018

FAMILY TYPE	MARRIED COUPLE	MALE HOUSEHOLDER, NO SPOUSE PRESENT	FEMALE HOUSEHOLDER, NO SPOUSE PRESENT
All races	6	17	34
NonHispanic white	4	10	23
Black	7	21	30
Hispanic	10	13	31
Asian	6	9	19
American Indian/ Alaska Native	16	*	27

*No data available

Source: U.S. Census Bureau 2018b, Table 4.

parents' available social support affected children's outcomes (Broman, Li, and Reckase 2008; Ryan, Kalil, and Leininger 2009; Wen 2008; Wu and Hou 2008).

Furthermore, **transitions** to and from various family structures have been found to result in poorer outcomes for children (Amato 2012; Hadfield, Amos, Ungar, Gosselin, Canong 2018). "The next step in creating a . . . research agenda on family instability and child well-being is to answer the 'how?' question by identifying and empirically testing a variety of explanations and mechanisms at play" (Smith, Crosnoe, and Cavanagh 2017, p. 608.).

Frequent family instability is consistently predictive of higher levels of behavior problems for children born to unmarried mothers, largely caused by stress (Fomby and Osborne 2017). "Theoretical development in the area of family structure change should reflect the multiple relationships in children's lives that are shaped by the dynamic and complex nature of contemporary family formation" (Fomby and Osborne 2017, p. 89). Children with a history of family instability also experience greater financial hardship, harsher discipline, and greater residential mobility at age 9 compared to children who experienced no union status change (Fomby and Osborne 2017, p. 90).

It is interesting that there is evidence of alternatives to marital support among some categories of Americans. Research based on a national representative sample of more than 10,000 U.S. teens found that the negative effects of time lived with a single mother were less serious for black and Hispanic adolescents than for whites (Liu and Reczek 2012; Mandara, Rogers, and Zimbarg 2011). Why might this be? "The common history among blacks [and among Hispanics] allows for the emergence and primacy of social supports, such as women-centered kinship networks, co-residence with extended family, and strong ties to the church, which can buffer the negative effects of stress caused by family instability" (Heard 2007, p. 336). For more on blacks and marriage, see "A Closer Look at Diversity: African Americans and 'Jumping the Broom.'" But how do we explain the correlation that does exist between positive outcomes and marriage?

Selection versus Experience Effects

You may recall that Chapter 5 examines selection versus experience effects regarding the relationship between divorce and cohabiting before marriage. Here we examine the selection and experience hypotheses regarding the apparent benefits of marriage.

A criticism of Waite's claims is that much—although not all—of the association between marriage and positive outcomes is due to selection effects: People may "select" themselves into a category—in this case, marriage—and this self-selection can yield the results for which a

researcher was testing. Increasingly, individuals with higher education, incomes, and physical and mental health are more likely to marry (Goodwin, McGill, and Chandra 2009; Lowrey 2014; Teitler and Reichman 2008). Put another way, the **selection** posits that benefits associated with marriage—greater wealth, better health actually result, not from being married, but from the personal characteristics of those who choose to marry (Cherlin 2003; Tumin and Zheng 2018). For instance, married women are more likely than those who are cohabiting or heading single-family households to inherit wealth (Ozawa and Lee 2006). Being positioned to inherit wealth from one's family of origin is a personal characteristic that *precedes* getting married.

But not *all* the benefits associated with marriage are accounted for by selection effects. In contrast to the selection hypothesis, the **experience hypothesis** holds that something about the married experience itself causes these benefits. Figure 7.2 illustrates the selection and the experience hypotheses. The experience hypothesis posits that one reason children of married parents, as a category, evidence better outcomes than those of non-married parents may be the *experience* of growing up in a married-couple household.

Individualized marriage may be beginning to resemble long-term cohabitation, and vice-versa, but there is considerable evidence that people, both heterosexual and same-sex partners, do see getting married as enhancing commitment (Schecter 2008). With its presumption of permanent commitment to the family as a whole, marriage "allows caregivers to make relationship-specific investments in the couple's children—investments of time and effort that, unlike strengthening one's job skills, would not be easily portable to another relationship" (Cherlin 2004, p. 855; and see Blankenhorn et al., 2015; Wilcox, Marquardt et al. 2011). One study found that among heterosexual couples with comparable incomes, married parents spent more on their children's education (and less on alcohol and tobacco) than did cohabiting parents (DeLeire and Kalil 2005).

According to several studies, the experience of being married has been shown to heighten feelings of commitment for same-sex couples (Ocobock 2018). One aspect of the experience of being married by same-sex couples is status recognition and diminished institutional discrimination. Being in a legal marriage, research shows, is associated with lower perceived unequal social recognition and hence better mental health, whereas prior to same sex marriage being legalized being in a domestic partnership was associated with the opposite—unequal recognition consistently associated with worse mental health (LeBlanc, Frost, and Bowen 2018).

Recent research has begun to try to distinguish outcomes correlated with marriage from those correlated with households where parents live in long-term cohabiting relationships. We saw earlier in this chapter that

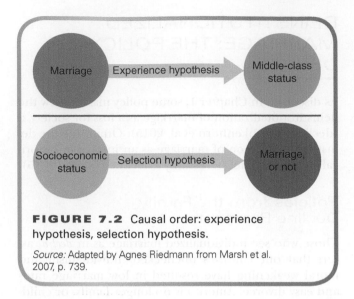

FIGURE 7.2 Causal order: experience hypothesis, selection hypothesis.

Source: Adapted by Agnes Riedmann from Marsh et al. 2007, p. 739.

Kane's (2016) study found little difference in birth weight of infants born to married compared with cohabiting expectant mothers. However, Kane also noted that the experiences of pregnant women in married and long-term cohabiting relationships differed from those of single, unattached expectant mothers. For instance, Kane pregnant women in either married or long-term cohabiting relationships were significantly more likely to feel supported in their pregnancy, more likely to feel that their partners encouraged healthy behavior, including quitting smoking and more likely to say their infant was wanted.

Researchers hypothesize that selection effects explain much—although not all—of marriage's advantage for children. Individuals who marry are better educated and have higher incomes. They live in neighborhoods more conducive to successful child raising. Less likely to be stressed because of financial problems, they are more likely to practice effective parenting behaviors (Manning and Brown 2006; Teachman 2008b). In her address to the Population Association of America, Waite (1995) acknowledged the contribution of selection effects. She added, however, that "we have been too quick to assign *all* the responsibility to selectivity here, and not quick enough to consider the possibility that marriage *causes* some of the better outcomes we see" (p. 497, italics in original).

Again we face the question of whether these findings result from selection effects or experience—that is, are happier people with a greater sense of well-being more likely to get married? (Yes.) Or is there something about the experience of being married that enhances happiness and well-being? (Yes.) Research shows that both are true.

For one thing, there are pragmatic reasons that spouses (and cohabitors, but perhaps to a lesser degree) benefit from an *economy of scale.* Think of the saying, "Two can live as cheaply as one." Although this principle is not entirely true, some expenses, such as rent, do not necessarily increase when a second adult joins the household (Goode 2007 [1982]; Thomas and Sawhill 2005; Waite 1995). Then, too, the promise of permanence associated with the marriage premise accords spouses the security to develop some skills and neglect others because they can count on working in complementary ways with their partners (Goode 2007 [1982]; Nock 2005). Furthermore, "[s]pouses act as a sort of small insurance pool against life's uncertainties, reducing their need to protect themselves *by themselves* from unexpected events" (Waite 1995, p. 498).

Furthermore, marriage offers *continuity,* the experience of building a relationship over time and resulting in a uniquely shared history. Marriage is the most likely family form to encourage the willingness to sacrifice for togetherness communion in intimate partnerships (Johnson, Horne, and Neyer 2018). And, finally, marriage provides individuals with a sense of obligation, not only to their families but also to the broader community (Goode 2007 [1982]; Wolfinger and Wolfinger 2008). This, in and of itself, gives life meaning (Waite 1995, p. 498).

In addition, marriage offers enhanced social support (Blankenhorn et al. 2015; Lee and Ono 2012). Marriage can connect people to in-laws and members of a widened extended family who may be able to help when needed—for instance, with childcare, transportation, a down payment on a house, or just an emotionally supportive phone call (Rittenour and Soliz 2009; Wilcox et al. 2011; Wilcox, Marquard et al. 2001). The enhanced social support that often accompanies marriage works to encourage the union's permanence (Giddens 2007; Lee and Ono 2012). For example, family and friends send anniversary cards, celebrations of the years the couple has spent together, and reminders of the couple's vow of commitment. Beginning with a public ceremony, marriage makes for what sociologist Andrew Cherlin (2004) calls *enforced trust:*

> Marriage still requires a public commitment to a long-term, possibly lifelong relationship. This commitment is usually expressed in front of relatives, friends, and religious congregants. . . . Therefore, marriage . . . lowers the risk that one's partner will renege on agreements that have been made. . . . It allows individuals to invest in the partnership with less fear of abandonment. (p. 854)

To summarize, a large body of research shows that marriage is correlated with benefits for adults and children. This relationship is complex, however, and much—but not all—of it is due to selection effects. Furthermore, with individualized marriage beginning to resemble long-term cohabitation, more research is needed to determine whether and how experience

Rhoberazzi/Getty Images

Hispanics and African Americans have higher percentages, or *rates*, of mother-headed, single-parent families. Nevertheless, the majority of mother-headed, single-parent families are nonHispanic white. Also, Hispanics and African American families have higher poverty *rates*, but the majority of families in poverty are nonHispanic white. Although about one-third of mother-headed, single-parent families live below the official poverty level, nearly two-thirds do not.

effects are unique to marriage or can be associated with both marriage and living together without marriage. For instance, the situation *economy of scale* may to some extent apply to living together, whether married or not. Meanwhile, the situation of *enforced trust,* because it is largely based on a public proclamation of coupe commitment, is unique to marriage.

As we saw early in this chapter, people increasingly—especially those with lower incomes and education—see marriage as unattainable, not really necessary, or both, while the highly educated and wealthy get married in relatively high percentages. Just as researchers have responded in various ways to Waite's address, discussed above, policy makers have had conflicting reactions.

DEINSTITUTIONALIZED MARRIAGE: THE POLICY DEBATE

As discussed in Chapter 1, some policy makers view the deinstitutionalization of marriage as a loss for society, a "decline" (Blankenhorn et al. 2015). Others see the deinstitutionalization of marriage as an inevitable historical change and not necessarily a decline (Coontz 2015).

Policies from the Family Decline Perspective

Those who see individualized marriage as in *decline* assert that our culture's unchecked individualism and moral weakening have resulted in low marriage rates and easy divorce. America is no longer family- or child-centered (Whitehead and Popenoe 2008). From this point of view, the American family is breaking down. The causes are cultural: self-indulgent attitudes and values focused on short-term gratification to the detriment of marital commitment and responsible child rearing (Giele 2007; Popenoe 2008).

Decline policy makers tend to be political conservatives. Concerns about "family breakdown" involve the high number of federal dollars spent on "welfare" for poverty-level single mothers, coupled with the irresponsible socialization of children (Giele 2007; Parkinson 2011; Rector 2012). A goal is to return to a society more in line with the values and norms of companionate marriage.

As a means to that end, advocates have developed government, religion-based, and private programs to motivate lower-income people to marry and to teach anger management and other communication skills that could result in greater marital permanence (Nadir 2009). The federally funded Healthy Marriage Initiative is an example (Hawkins, Amato, and Kinghorn 2013).

Critics of programs like these argue that low-income Americans value marriage and would like to marry, but marriage is difficult to achieve for many (Huston and Melz 2004; Trail and Karney 2012; see also Tumin and Zheng 2018). Low-income single mothers want trustworthy, steadily employed husbands (Burton et al. 2009; Joshi, Quane, and Cherlin 2009). Program evaluation research shows that teaching low-income couples communication skills may help to stabilize their relationship somewhat, but economics remains the major obstacle to marriage (Amato 2014; Johnson 2014; Hawkins 2014). Particularly in poor neighborhoods, many men cannot promise a steady income (Burton and Tucker 2009; Coates 2015; Harris and Parisi 2008). Poor women are not necessarily good marriage prospects either. They may have less-than-desirable economic histories, as well as mental, physical, or substance abuse issues. Also, they

The War on Poverty

As shown in Figure 7.3, the poverty rate for children ages 18 and younger is 18 percent. It was about 27 percent in 1959, but beginning with President Lyndon Johnson's **War on Poverty** in the 1960s, it dropped consistently during the 1970s to a low of about 14 percent. You may have heard of War on Poverty programs such as the Job Corps or the Neighborhood Youth Corps, Head Start, or Adult Basic Education. Although the majority of War on Poverty measures have ended, Head Start and the Job Corps continue to exist.

The War on Poverty offered strategies and programs to decrease poverty—community meal programs and health centers, legal services, summer youth programs, senior centers, neighborhood development, adult education, job training, and family planning (Garson n.d.). Commitment to the War on Poverty diminished after the 1970s, with national rhetoric shifting to debates focused on individual responsibility. Today, however, scholars and some policy makers are again urging U.S. taxpayers and the government to pay attention to society-wide supports for children and families regardless of—or in addition to—concerns about changing family structure (Lowrey 2014).

Critical Thinking

In your opinion, should the United States enact new anti-poverty strategies similar to the War on Poverty programs of the 1960s? Why or why not?

---

may have children by more than one biological father—a situation that renders them less desirable in the eyes of a future male partner (Manning et al. 2010).

The future of the Healthy Marriage Initiative is in question, although similar programs exist despite critics (Dupree, Whiting, and Harris 2016; McCormick, Hsueh, Merrilees, Chou, and Cummings 2017; Randles and Woodward 2018).

Policy leaders associated with the *decline* and "family breakdown" perspective once hoped to effect a "family turnaround" (Whitehead and Popenoe 2003). Meanwhile, policy makers who see marriage simply as *changing* recognize that many families are struggling but criticize the solutions offered by conservatives and propose their own.

Policies from the Family Change Perspective

Some policy makers see the deinstitutionalization of marriage as resulting from inevitable social *change*. These thinkers argue that people who look back with nostalgia to the good old days may be imagining incorrectly the situation that characterized marriage throughout most of the nineteenth and twentieth centuries.

Progressives more likely focus on people's individual rights. Over the past forty years, virtually uncompromising debate has characterized this divide as conservatives blamed a deteriorating culture (poor values and attitudes) for the dropping marriage rate, while progressives saw structural forces (deteriorating neighborhoods, the economy, racism, high incarceration rates) as more problematic.

Policy advocates from a marital *change* perspective more often focus, not on Americans' retreat from institutionalized marriage, but on economic and related challenges as causing difficult adult and child-raising environments with resulting negative outcomes for some poor children. Chapter 2 presents the family ecology perspective. From this viewpoint, family struggle results from society-wide conditions such as high incarceration rates for low-income males of color (Coates 2015), unemployment, low wages, difficult work schedules, lack of necessary mass transit, and expensive day care. Accordingly, these spokespersons argue for ecological, including federal, state, workplace, and neighborhood-level, solutions (Amato 2014; Trail and Karney 2012).

Many policy makers point out that Americans are struggling with economic and time pressures that get in the way of their ability to realize family values (Copeland and Snyder 2011; Kalil et al. 2014). Suggested remedies include job training, support for education, good jobs, drug rehabilitation, neighborhood improvements, small business development, and parenting skills education (Amato 2014; Baker 2015; Cherlin 2009a; Lombardi and Coley 2013). "Facts about Families: The War on Poverty" illustrates an ecological approach.

Current Policy Measures: Take-Aways from Both the Decline and the Change Perspectives

A study of U.S. military personnel concluded that the unusually high marriage rate in the armed services results from situations that indirectly encourage marriage. Gainfully employed, members of the military

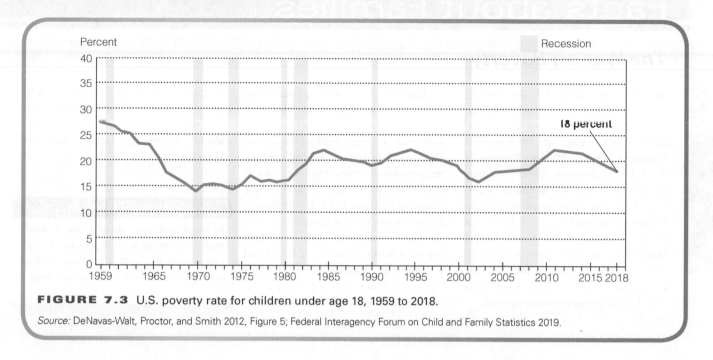

Percent · Recession

FIGURE 7.3 U.S. poverty rate for children under age 18, 1959 to 2018.

Source: DeNavas-Walt, Proctor, and Smith 2012, Figure 5; Federal Interagency Forum on Child and Family Statistics 2019.

marry to share housing (and get out of the barracks), to be deployed overseas together, and to have survivors' benefits, among other reasons. In the authors' words, "The biggest policy implication of our research relates to all families, not just military families. . . . [O]n the basis of the military example, marriage is widespread in part due to stable, decent-paying jobs. . . . A serious jobs creation program [would likely bolster marriage rates among non-military Americans]" (Lundquist and Xu 2014, p. 1078–79).

Our class-based marriage gap may be an issue for both political camps. As Blankenhorn and colleagues (2015) argue, "Progressives are coming to realize that they need to be concerned about family structure if they care about social justice." At the same time conservatives are coming to see that "aspirations to family formation are being stymied by wage stagnation and disappointing job prospects among working-class and less-educated men. . . . Instead of being asked to choose between the decline and the change agenda today America has little choice but to pursue both agendas at once."

One example of simultaneously pursuing both agendas involves President Obama's announcement of a $200 million program,

The disparity in marriage rates between the poor and those who are not poor has become significant enough that social scientists have coined a term for this situation— the *marriage gap*. Some experts see marriage as a partial antidote to poverty. However, low-income Americans generally value marriage and would like to marry, but marriage appears difficult to achieve for many of them, although certainly not for all.

funded by private-sector foundations, designed to bolster education, work skills, and job opportunities for young, low-income black men (Shear 2014). As he announced this structural initiative, President Obama was also working on changing values and attitudes when he said publicly, "There are a lot of men out there who need to stop acting like boys, who need to realize that responsibility does not end at conception, who need to know that what makes you a man is not the ability to have a child but the courage to raise one" (Porter 2014).

According to conservative social scientists and policy maker from the family decline perspective Isabel Sawhill (2014a, 2014b), "Younger people must begin to take greater responsibility for their choices. The old social norm was, 'Don't have a child outside of marriage.' The new norm needs to be 'Don't have a child until you and our partner are ready to be parents.'" At the same time, Sawhill (2014a, 2014b) contends that, although "government alone can't solve this problem," we do "need more (and better quality) childcare and a higher minimum wage, as well as serious education and training for those who are struggling to care for their families."

Andrew Cherlin (2009a, 2012) has argued, not for *marital* stability necessarily, but for *family* stability—supporting children and their parents in whatever family form they inhabit. "[I]t is time to get away from trying to reshape families into some historic ideal and turn our attention to helping the kind of complicated families that raise children in today's flawed world" (Porter 2014). Policy measures in this vein include community educational interventions designed to promote healthy and supportive co-parenting relationships regardless of marital status (Brown, Manning, and Stykes 2015; McHale, Waller, and Pearson 2012).

COUPLE SATISFACTION AND CHOICES THROUGHOUT LIFE

Our theme of making choices throughout life applies both to couples deciding about whether to marry and also to decisions made during the early years of living together, whether married or not. We'll examine these topics now.

Preparation for Marriage

Given today's high divorce rate, clergy, teachers, parents, policy makers, and others are concerned that individuals be prepared for marriage. High school and college family life education courses are designed to prepare individuals for marriage (Coalition for Marriage, Family and Couples Education 2009; Fincham and Beach 2010). Premarital counseling typically takes place at churches or with private counselors. The Unitarian Universalist Church offered commitment counseling for same-sex couples. Illustrating the connection between private lives and public interest, some states have established incentives to encourage couples to seek premarital counseling. For example, Minnesota discounts the price of a marriage license when a couple agrees to premarital counseling (Meyer 2011).

Premarital counseling goals involve helping the couple evaluate whether their relationship should lead to marriage, develop a realistic yet hopeful and positive vision of their future marriage, recognize potential problems, and learn positive problem-solving and other communication skills. Research shows that premarital education programs improve a couple's communication skills and relationship quality, at least in the short term (Blanchard et al. 2009; Fincham and Beach 2010). However, we have little data on the relationship between premarital counseling and long-term relationship *stability*. Nevertheless, psychologist Scott Stanley identifies four benefits of premarital education:

> (a) [I]t can slow couples down to foster deliberation, (b) it sends a message that marriage matters, (c) it can help couples learn of options if they need help later, and (d) there is evidence that providing some couples with some types of premarital training . . . can lower their risks for subsequent marital distress or termination. (Markman, Stanley, and Blumberg 2001, p. 272)

Counseling for partners going into cohabitation with hopes for long-term commitment is also available.

The First Years Together

At least for marrieds, research shows that although couple relationships are important to happiness and well-being *over* the life course, the first years together tend to be the happiest, with gradual declines afterward (Jackson, Miller, Oka, and Henry 2014). Why this is true is not clear. One explanation points to life cycle stresses as children arrive and economic pressures intensify. At least for those without children, partners' roles are relatively similar. Couples tend to share household tasks and, because of similar experiences, may be better able to empathize with each other. Other explanations argue that falling in love and new relationships are periods of emotional intensity from which there is an inevitable decline (Glenn 1998; Whyte 1990). Couples who are newly together also engage in role-making.

Role-Making From the interaction-constructionist theoretical perspective (see Chapter 2), **role-making** refers to personalizing a role by modifying or adjusting the expectations and obligations traditionally associated with it. New couples negotiate expectations for sex and intimacy (Blumenstock and Papp 2017), establish

communication and decision-making patterns, balance relationship with job or school responsibilities, and come to agreement about becoming parents or not, as well as how they will handle money. When children are present, role-making involves negotiation about how to parent.

During this period, couples request changes and negotiate resolutions. The couple constructs relationships and interprets events in ways that reinforce their sense of themselves as a couple. In many respects, same-sex relationships are similar to heterosexual ones. Like heterosexuals, same-sex partners highly value love, faithfulness, and commitment (Meier, Hull, and Ortyl 2009). Research indicates that the need to resolve issues of sexual exclusivity, power, and decision making is not much different in same-sex pairings than among heterosexual partners. However, same-sex partners of both genders may be more likely to evidence more equality and role-sharing than heterosexual couples (Gotta et al. 2011; Kurdek 2006, 2007; Parker-Pope 2008). Developing supportive relationships but not interfering in relations with in-laws can be important (Rittenour and Soliz 2009). Role making among cohabiting couples needs to be researched.

A couple's task in early marriage is to begin to identify as a committed couple and to fashion couple connection.

Creating Couple Connection Role-making involves creating an identity as a couple. Listen to the following respondent:

A big part of my identity as a gay man in my 20s and 30s was my ability to be attractive to other men. Even though we were in a monogamous relationship it was still important for me to feel desirable. But in that first year of being married we steered away from that into a different sort of socializing pattern, and that was really hard to me to give up. . . .

Being married made me feel like "you're at a certain age"—I was 34—"you have a house, a husband, that is what you should be turning your attention to now." I matured a lot in the sense that I willing and joyfully gave that up because I got all this other great stuff. (Ocobock 2018, p. 373)

Creating couple connection may involve giving up things associated with a former lifestyle in order to get "other great stuff." It also involves communicating mutual affection and respect, learning more and more about a partner's inner emotions, and finding new ways to know each other and to feel known and interested in the other (Gottman and Gottman 2017). Role-making continues throughout a couple's relationship, during which partners learn more about one another and what their identity is as a unified couple, instead of two separate individuals.

To have enduring unions, partners need to make their relationship a high priority. Spending time together, communicating supportively, and pursuing leisure activities together are related to couple satisfaction (Gager and Sanchez 2003; Johnson et al. 2005; Kurdek 2005). Couples who make time for shared new experiences are likely to be happier (Burpee and Langer 2005; Dew 2008). An important psychologist and expert on couple communication, John Gottman (1994), offers this advice:

Happy, solid couples nourish their marriages with plenty of positive moments together. . . . Too often, families lead complex—even grueling—lives. . . . But if you want to keep your marriage alive, it's essential to rediscover—or perhaps simply make time for—those experiences that make you feel good about your spouse and your marriage. (p. 223)

Then, too, research has focused on how religious convictions can facilitate lasting couple connection,

Fuse (RF)/Jupiter Images

Spouses are freer to plan a future together because, more than others, married couples can count on continuity bolstered by mutual commitment and enforced trust.

as well as positive communication patterns (David and Stafford 2015; Fincham and Beach 2010). For one thing, a religious belief system called **marital sanctification** encourages partners to see their marriages as ordained by God and hence of divine significance. Sanctification promotes couple bonding, fosters positive emotions and diminishes negative ones, and facilitates positive attitudes and overall resilience when encountering stress (Ellison et al. 2011).

Noting that "[l]ove is not an express lane concept," observers suggest creating daily "connecting moments" when couples can be alone together and pay attention to their relationship (Brennan 2003; Brotherson 2003). Keeping one's relationship meaningful requires that partners develop identities as married individuals, as well as consciously and continuously building couple commitment—being willing to invest in their union long-term and to persist together through trying times

(Byrd 2009). Satisfaction with the committed relationship has a great deal to do with values, beliefs, and the choices that partners make accordingly (Willoughby, Hall, and Luczak 2015). One important set of decisions involves commitment to developing a mutually supportive relationship, as well as practicing positive communication skills, addressed in Chapter 11.

One theme of this text is that society influences people's options and thereby impacts their decisions. To the extent that they are able, people today organize their personal, romantic, and family lives as they see fit (Byrd 2009). In this climate of individualized marriage, increased cohabitation, and the postmodern family, a wide variety of family forms emerge. Meanwhile, in the words of David Blankenhorn and colleagues (2015), "Warts and all, [marriage may be] today's best bet if you are seeking faithfulness and lasting love."

Summary

- The marriage situation in the United States and other industrialized nations has witnessed two developments over recent decades: First, fewer adults are marrying; second, LGBTQ+ activist couples won the right to legal same-sex marriage on June 26, 2015.

- The marriage premise involves permanence and—for some societies, particularly Western cultures—monogamous sexual exclusivity.

- New norms for love-based marriage gradually became prevalent in Europe throughout the 1700s and 1800s. Expectations for personal happiness and love in marriage have somewhat weakened the marriage premise over the past 100 years, although it remains important.

- Marriages and families have become deinstitutionalized. Marriage has changed from institutionalized to companionate and now to individualistic.

- Researchers consistently find a significant correlation between marriage and many positive outcomes for both adults and children, but the relationship is more complicated than it first appears because some of the benefits associated with marriage are the result of selection effects.

- Scholars and policy makers who view individualized marriage and the postmodern family as indications that marriage is in "decline" see America's

individualistic culture as responsible and have proposed ways to try turning things around. One of these has been the federally funded Healthy Marriage Initiative.

- Scholars and policy makers who view individualized marriage and the postmodern family as results of inevitable historical change, resulting in struggle for some families more than for others, have proposed ecological solutions to poverty and family struggle—such as neighborhood development, an adequate minimum wage, and improved employment opportunities.

- Today some policy makers recognize that raising the marriage rate necessitates both ecological and cultural initiatives.

- Optional and less permanent than in the past, marriage still continues to offer benefits; married adults are more likely than others to say that they are happy with their lives. Being married continues to help bolster the marriage premise, largely because family and community social support results in enforced trust.

- Partners in early committed relationships engage in role-making, a process that includes—among other things—negotiating issues surrounding money, sexual frequency, and time together.

Questions for Review and Reflection

1. Discuss the marriage premise with its expectations of permanence and sexual exclusivity.

2. Describe how love has changed the marriage premise over the past three centuries. What does family demographer Stephanie Coontz mean when she says that "love conquered marriage"?

3. Recognizing the possibility of selection effects, what are some ways that the experience of being married can enhance happiness and life satisfaction?

4. Describe the social movement that resulted in legalized same-sex marriage in the United States in June 2015.

5. **Policy Question.** If you were in a position to devise solutions to child poverty, what would they be? Would they include encouraging more Americans to marry? Why or why not?

Key Terms

collectivist society   170
communal society   170
companionate marriage   172
deinstitutionalization
   of marriage   172
expectations of permanence   167
expectations of sexual exclusivity   168
experience hypothesis   178
family of orientation   170

family of procreation   170
heteronormative bias   164
individualism   170
individualistic society   170
individualized marriage   173
institutional marriage   172
kin   170
marital sanctification   185
marriage gap   164

marriage premise   167
polyamory   169
polygamy   168
role-making   183
selection hypothesis   178
social institution   167
swinging   169
transitions   178
War on Poverty   181

8

DECIDING ABOUT PARENTHOOD

Learning Objectives

1 Discuss U.S. trends in fertility rates.

2 Describe how educational attainment and income affect fertility decisions.

3 Define the concepts of value of children and opportunity costs of parenthood.

4 Discuss the options and circumstances regarding decisions about parenthood.

5 Discuss the debate between pro-life and pro-choice activists.

6 Discuss the causes and possible solutions of involuntary infertility.

7 Explain adoption options available in the United States today.

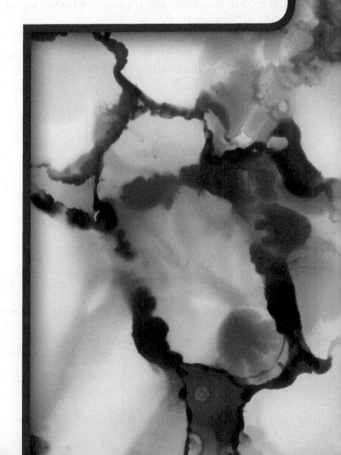

▲ LPETTET/Getty Images

There are many things to consider when thinking about becoming a parent. For example, you may know someone who has been adopted, perhaps by parents of another race. You may know someone who is considering getting pregnant with a first child, who is undergoing infertility treatment, who is deciding whether or not to have an abortion, or who has chosen to not have children. Each decision is highly personal but is also influenced by society. Decisions about parenthood are affected by people's economic circumstances, cultural beliefs, and societal norms. Decisions about parenthood also proceed from individual needs, values, and attitudes. Throughout this chapter, we'll be looking at many aspects of individuals' and couples' decisions about having children.

We address these issues as options related to informed decision making, but we should note that, when asked why they decided to have their first child, nearly half (47 percent) of a Pew Research national representative sample answered, "There wasn't a reason, it just happened" (Livingston and Cohn 2010). In fact, nearly half of pregnancies are unintended—but "it may make more sense to conceptualize pregnancies not as two mutually exclusive categories denoted planned or unplanned, but as points on a continuum of intendedness" (Shreffler et al. 2011, p. F8). Put another way, many women are ambivalent about having children: "The strong assumption that all women plan and therefore that all unintended pregnancies (and births) are unwanted obscures variation among women in degrees of planning pregnancy (or not) as well as variation from pregnancy to pregnancy among the same women" (Shreffler et al. 2011, p. F8). For example, when women who were part of the National Survey of Fertility Barriers were asked if they wanted to get pregnant, nearly one-fourth (23 percent) said they were "okay either way" (McQuillan, Greil, and Shreffler 2011). Nevertheless, much of this chapter assumes informed decision making, and as you study it, you may want to refer back to Chapter 1's discussion on how we best make informed decisions.

In this chapter, we'll look at the rewards and costs of having children, along with how children affect a couple's happiness. We'll examine several options regarding having children—choosing to be childfree,

having only one child, and so on. We will also address various circumstances under which Americans have children today. For instance, we'll discuss nonmarital births, consciously choosing to be a single parent, whether to have children at a younger or older age, and having children with more than one partner. We'll also look at issues concerning pregnancy, childbirth, infertility, and adoption. These issues are relevant to gay and lesbian individuals and same sex couples. Unfortunately, this research is sparse compared to what is known about heterosexual men, women, and couples. However, we highlight what we know thus far. There is also not nearly the research on men's fertility and childbearing decisions as there is on women's. We note similarities and differences when we can. To begin, we'll examine fertility trends in the United States. Then we'll address decision making about whether or not to become a parent.

FERTILITY TRENDS IN THE UNITED STATES

Beginning early in the nineteenth century, America began to witness significant changes in childbearing or fertility patterns. Demographers use the term **fertility** to refer to live births. The **total fertility rate (TFR)** is the number of live births a typical woman will have during her lifetime. As shown in Figure 8.1, the TFR for the United States dropped sharply from a high of 3.5 in the late 1950s to the lowest level ever recorded—1.7 in 1976, a year marked by an economic recession with dramatically rising fuel and related costs that accompanied

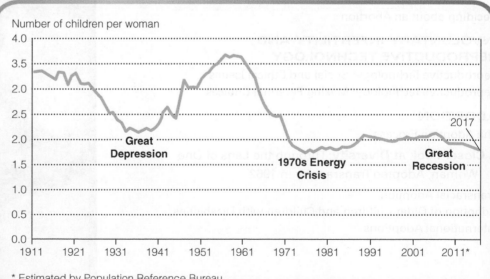

Number of children per woman

* Estimated by Population Reference Bureau.

FIGURE 8.1 United States total fertility rate (TFR), 1911 to 2011. U.S. fertility has declined during periods of economic slowdown.

Source: Mather 2012; Population Reference Bureau 2015; Mathews and Hamilton 2019.

the energy crisis of the 1970s. In recent years, the TFR has fluctuated around 2.0. People forgo or put off having children in difficult and uncertain economic times, such as the Great Recession of 2007. Despite declarations that the recession has come to an end, fertility has not recovered to pre-recession levels. In 2017, the TFR for the United States was 1.8 children per woman (Martin, Hamilton, Osterman, Driscoll, and Drake 2018).

The TFR is currently lower among all racial and ethnic groups than it was during the baby boom era (1946 to 1964). The overall U.S. fertility rate today is below **replacement level fertility**, the level of fertility necessary for a society to replace its population. A society needs two children in order to replace every two adults, and demographers peg the replacement-level TFR at 2.1 to take into account that not all infants survive and not every woman will reproduce. Many European countries and Japan are experiencing declining populations for that reason. The fertility rate in the United States has been below replacement level for a decade (Mathews and Hamilton 2019; World Bank 2019). In fact, the U.S. population would be in decline were it not for immigration. We talk below about factors underlying this unprecedented decline in fertility, which some commentators consider alarming.

Historical Patterns of Fertility and Family Size

U.S. fertility decline shows a continuous pattern dating back to the early 1800s. As discussed in Chapter 1, historical periods or events affect individuals' options and decisions. In a pre-industrial economy, women could combine productive work on the farm or in a home artisan shop with motherhood. But when work moved from home to factory, the roles of worker and mother were not so compatible. As women's employment increased, fertility declined. Moreover, as a result of improved living conditions, infant mortality declined. Gradually, it was no longer necessary to bear many children in order to ensure the survival of a few.

In fact, today large families are looked down on. Some mothers of large families have reported feeling stigmatized. They feel that others see them as uneducated, insufficiently attentive to their children, messy housekeepers, and ignorant about birth control (Hagewen and Morgan 2005). One mother of four reported feeling like "a walking freak show." Daily comments from others, often strangers, range from the harmless ("You must you're your hands full!") to the judgmental ("Were they all planned?," "When are you going to get fixed?," and "You don't want anymore, right?") (Chesser 2017). Many larger families are religious and find acceptance in their religious communities. A pro–large family movement known as *Quiverfull* is situated in some fundamentalist Christian churches (McGowin 2018). Thinking of their offspring as "an army they're building for God,"

Quiverfull families aim for six or more children (Joyce 2006, p. St-1). The Quiverfull movement has been criticized for being antifeminist because it preaches "biblical patriarchy" and because the movement has been linked to child sexual abuse (Blumberg 2015).

In the face of the long-term decline over the past two centuries, the upswing in fertility in the late 1940s and 1950s requires explanation. As adults, those who had grown up during the Great Depression (see Figure 8.1), when family size was limited by economic factors, found themselves in an affluent post–World War II economy. Perhaps compensating for economic deprivations they suffered as children, they later fulfilled dreams of a relatively abundant family life (Easterlin 1987). Marriage and motherhood became dominant cultural goals for American women; men also concentrated their attention on family life. Couples in this generation averaged more than three children (Kirmeyer and Hamilton 2011).

Today, two children constitute Americans' ideal family size, and this two-child preference has been evident for at least fifty years. According to a 2018 Gallup poll, nearly half of Americans (47 percent) said two is their ideal number of children, followed by three (26 percent), four or more (15 percent), and none or one child at 4 percent (Saad 2018). American couples typically want one boy and one girl, although there is a preference for boys among men, with 43 percent of men preferring a boy compared and 28 percent preferring a girl (if they could only have one). Women are equally likely to desire to have boys and girls, at 30 percent and 31 percent (Newport 2018). Interestingly, the desire for large families (three children or more) increased 33 percent in the last decade. However, women's extended educational and employment options, the rising costs of raising a child, and increased work and family demands are keeping overall fertility rates low. Furthermore, women are waiting longer to have first babies, a situation that ultimately reduces family size because it shortens the years during which a female is most naturally fertile. The mismatch between desired number of children and actual number of children suggest that Americans are having fewer children than they'd like (Stone 2018).

Regarding age and fertility, teen births have declined to its lowest rate in seventy years of record keeping, a development discussed later in this chapter. Birthrates peak among women in their late twenties, and births to women in their twenties constitute about half of all births. However, by 2013, the birthrate of women in their twenties reached a record low that has continued. Meanwhile, birthrates for women in their thirties and forties have risen (Livingston 2018a; Martin et al. 2017). As a result, the mean age of first birth for women increased steadily from 21 in 1972, to 25 in 2000, to a high of 27 in 2017. Similarly, the birthrates for men in their twenties have declined, while the birthrates for men in their late thirties and older have increased (Martin et al. 2017). Recognizing overall

For many decades now, the ideal family size in the United States is two children. These days, large families of four or more children are a distinct minority and are subject to negative stereotypes.

Jaren Jai Wicklund/Shutterstock.com

childbearing trends, we also note that fertility rates differ among segments of the population. These differences are described in the following sections.

Differential Fertility Rates by Education, Income, Race, and Ethnicity

Interestingly, women and couples with higher education and income tend to want, and have, fewer children (Saad 2018). Given that these couples are the best able to afford them, why would this be? One reason is that children are not only financially costly. There are also high **opportunity costs** associated with children. Opportunity costs are the things parents must give up in order to have children such as applying for a promotion that requires longer hours or travel, taking a vacation, or simply pursuing a hobby or interest. People with greater education and income have more alternatives to parenthood than do those with less education and income (Duffin 2019; Livingston 2018a)—for example, the opportunity to pursue a demanding career, enjoy world travel, or go to the gym. When people with these options for self-actualization weigh them against having children, the latter may lose out.

Birthrates also reflect beliefs and values about having children, which vary among racial

and ethnic groups. As Figure 8.2 indicates, fertility among U.S. women all four racial and ethnic categories has declined since 1990 for all groups shown. However, these rates slightly differ according to race and ethnicity, mostly driven by lower educational attainment and incomes among women of color (aside from Asians, who have lower fertility rates). In 2017, Asians had the lowest TFR of the four groups at 1.6, followed by nonHispanic whites at 1.7, and nonHispanic blacks at 1.8. Hispanics and other Pacific Islanders had the highest TFRs at 2.0 and 2.1. Among Hispanics, the TFR for Mexican American women is the highest, followed by women of Central American and South American background, and then women of Puerto Rican and Cuban backgrounds (Martin et al. 2018). Many Hispanics have migrated from cultures characterized by high birthrates and by rural and Catholic traditions that encourage large families. As immigrants assimilate, their birthrates and other family behaviors (e.g., divorce) tend to converge with those of the general population (Hoffman, Breck, and Beasley 2020).

The fertility rates of Native Americans/Alaska Natives have declined by more than 20 percent since 1990 and is similar to nonHispanic whites and Asians

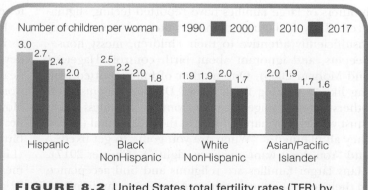

FIGURE 8.2 United States total fertility rates (TFR) by race/ethnicity, 1990, 2000, 2010, and 2017.

Source: Martin et al., 2012; Martin et al., 2018.

Native Americans/Alaska Natives currently have the lowest fertility rates in the United States, proving the exception to the rule that lower education and income are associated with higher fertility rates.

at 1.7 (Martin et al. 2017; Mather 2012). With Native Americans/Alaska Natives evidencing lower average educational attainment and income levels, their fertility pattern is an exception to the rule that lower education and income levels are associated with *high* fertility although Native American women who live on reservations have higher fertility than those who do not.

Variations in birthrates reflect decisions shaped by cultural values and individual attitudes about having children. We turn now to a discussion of things to consider when deciding about parenthood.

THINGS TO CONSIDER WHEN DECIDING ABOUT PARENTHOOD

It is not easy to choose how many children to have (if any), when to have them, or even *how* to have them. "Issues for Thought: The Medicalization of Childbirth and Caesarean Sections: Should a Delivery Be Planned for Convenience?" explores, for example, considerations regarding delivery by caesarean section and other medical interventions during childbirth, as well as media portrayals of childbirth.

As we saw in the introduction to this chapter, people do not always consciously choose to have or not have a child. According to the Guttmacher

Institute (2019e), nearly half (45 percent) of all pregnancies are unintended, and of these, 60 percent were "mistimed" (they wanted to have children in the future, but the pregnancy happened before they wanted it to) and 40 percent were unwanted (they did not want to become pregnant then or anytime in the future). Fortunately, unintended pregnancy has been declining: in 2008, a little more than half (51 percent) of all pregnancies to American women were unintended. This can be at least partially attributable to increased sex education, contraceptive use, and availability of highly effective contraceptives. Nevertheless, there are large disparities in unintended pregnancies in the United States, with higher rates among younger, lower income, less educated women as well as African American and cohabiting women (Guttmacher Institute 2019e). Researchers estimate that roughly one-third of the decline in fertility is due to a decline in unintended pregnancies, primarily among younger women (Sharkey 2019). Because some women with unintended pregnancies have abortions, the percentage of *births* that are unintended is slightly lower. One in three births are to women who did not intend to get pregnant, which is problematic because these children have lower birth weights, are less likely to be breastfed, and are less likely to receive prenatal care (Kost and Lindberg 2015).

Moreover, social pressure to have children—a cultural phenomenon called **pronatalist bias** (Hagewen and Morgan 2005)—might influence fertility decisions. This is higher in the United States than many other Western countries. For example, the share of American women who, by the end of her childbearing years, has ever had a baby is 86 percent, compared to

About 90 percent of babies born in the United States are delivered in hospitals attended by doctors, unlike most of Europe, where the majority of babies are delivered at birthing centers attended by midwives. Yet, the U.S. has much higher rates of maternal and infant mortality.

The Medicalization of Childbirth and Caesarean Sections: Should a Delivery Be Planned for Convenience?

If the baby is too large or if the mother's or baby's physical condition makes the stress of childbirth dangerous, a physician may decide to deliver the child by **caesarean section** (or C-section), so called after Julius Caesar, who was supposedly born in this way. A *caesarean section* is a surgical operation in which a physician makes an incision in the mother's abdomen and uterine wall and removes the infant.

Greater education and income among women, greater ability of women to control their fertility (i.e., ability to delay having children and have fewer children), prenatal care, hospital delivery, and a doctor's assistance in childbirth contributed to the decrease in infant and maternal deaths during childbirth throughout the twentieth century. The overwhelming majority of births in the United States occur in hospitals and are attended by physicians, unlike other developed countries such as the United Kingdom, where most births occur in "birthing centers" attended by midwives. Yet, childbirth is safer there and in European countries than in the United States (Womersley 2017). In the United States, hospital births attended by a certified nurse midwife (about 8 percent) is increasing. Only a small percentage of births (about 1 percent) occur at home under the care of a certified or traditional midwife. Home births are most common among non-Hispanic white, educated women with a low risk of complications who desire a "natural birth" with no drugs or medical interventions (Declercq 2015; Martin, Hamilton, and Osterman 2014).

Meanwhile, the rising incidence of caesarean sections has raised concerns among health experts. In 2017, about one-third (32 percent) of all U.S. births were delivered this way. Although the percentage of births through caesarean deliveries rose nearly 60 percent from 1996 to 2009, this rate has since leveled off (Martin et al. 2018). Even so, the United States' rate of C-sections ranks 15 among 137 countries in a study conducted by the World Health Organization (WHO 2015). Because they are riskier than a vaginal delivery for both mothers and infants, experts express concern at today's high rate of caesarean births in the United States and around the world. The WHO recommends that no country exceed a rate of 10 percent after finding that rates higher than that level are not associated with reductions in maternal and newborn mortality rates. This would make over half of C-sections performed in the United States medically unnecessary.

The reasons for today's high rates of C-sections include women having children at older ages and increased obesity, both of which increase the risks to mothers and babies. Another factor that is not health related is the routine use of fetal monitoring, the results of which may trigger unnecessary intervention. Difficult births may require surgical intervention, and there is no question that caesareans are often life-saving procedures for both mothers and infants, but a C-section is major surgery that introduces substantial risks to mothers such as complications with anesthesia and infection. Women

take longer to recover, are often in pain and cannot take pain-reducing drugs if they want to breastfeed, and need help caring for their newborn. There are few resources for women who have undergone a traumatic birth, and there is a sense of shame among women who seek help for physical pain and emotional problems when their attention is expected to be directed toward the baby (McLaren 2017).

The increased rate of C-sections in the United States is part of the "medicalization" of childbirth that occurred in the twentieth century and continues today. Prior to that time, women gave birth at home under the care of a female relative or midwife. This shifted with the advent of medical science and application of this "science" to childbearing. Midwives were replaced by physicians, who at that time were nearly 100 percent male. Medical interventions were introduced and women were taught to seek the advice of "experts" on anything to do with pregnancy and childbirth (Luce et al. 2016). For example, it was not uncommon for a woman to be unconscious or put in "twilight sleep" during the birth of her child, to receive an enema, have their pubic hair removed, and be given an episiotomy (a surgical enlargement of the vaginal opening) as a matter of routine. Today, most women give birth lying down with their feet in stirrups and hooked up to monitors, rendering them powerless in their own birth experience (for example, they are not able to get up and walk around, which helps childbirth move along). Women who are sedated are not able to push as

only 75 percent of German women (Livingston 2018b). Interestingly, as same-sex unions become more visible and accepted, some lesbian and gay male couples join straights in feeling pressured to be parents. As one gay man told an interviewer, "Everyone's asking: What's your timetable? What's your plan?" In this case, "everyone" included his parents, his husband's parents, their friends, and work colleagues (Swarns 2012a).

effectively, requiring doctors to use forceps or suction to get the baby out of the womb, which can lead to cranial injuries to the child (Ehrenreich 2018).

Other reasons for the increase in C-sections might be pragmatic. For example, a woman may live a substantial distance from a hospital, and her tendency to deliver quickly might make it difficult to reach the hospital when labor begins. Increasingly, parents, many of them are highly educated and in demanding careers, seek to control the birth process as much as possible in order to get through it with minimal discomfort and inconvenience (Declercq, Menacker, and MacDorman 2004). Or a physician may believe that she or he can do the best job when time of delivery is chosen rather than occurring in the middle of the night after a long day of medical practice.

Caesarean deliveries are more convenient to physicians. In Brazil, roughly half of all babies are delivered through caesareans. One woman was pressured to have a caesarean by her doctor who said, "I was at a birthday party, and want this done fast because I want to go back and finish my whisky" (Khazan 2014). Private insurance, especially fee-for-service reimbursement schemes, makes caesareans in that country more lucrative because physicians can schedule as many as eight procedures a day. Culture plays a role as well. In China, with a similarly high rate of C-sections, couples may schedule one so their child has a "lucky" birthday (Khazan 2014). Yet, an elected C-section raises women's chances of death by 60 percent (Doucleff 2018).

Women are also influenced by the media, who portray childbirth as dangerous and extremely painful and risky, and carry the message that women can't have a baby safely without a doctor's oversight (Bick 2010; Liechty, Coyne, Collier, and Sharp 2018; Luce et al. 2016). Vicki Elson examines media representations of childbirth in her documentary, *Laboring Under an Illusion: Mass Media Childbirth vs. the Real Thing* (Newman 2009). Fathers are portrayed as "panicky," and women "screaming for drugs" or are otherwise are portrayed as "wilted and powerless." Yet, the vast majority of women choose to receive an epidural to make childbirth more comfortable and most births are straightforward. Rates of maternal mortality (dying as a result of complications during childbirth) in the United States are low at 19 per 100,000 live births (but higher than in other industrialized countries). For reasons not fully known, maternal mortality increased 27 percent between 2000 and 2014 (MacDorman, Declercq, Cabral, and Morton 2016). Women are three times more likely to die in childbirth than are women in Canada, and six times more likely than women in Scandinavia (Martin and Montagne 2017).

Maternal mortality (and infant mortality) vary dramatically by race, ethnicity, education, and poverty status, with black women dying at a rate over three times higher than nonHispanic white women, mainly due to poor management of pre-existing conditions and lack of pre- and postnatal care (Somer, Holdt, Sinkey, and Bryant 2017). Some point to greater economic and family stress among African American women, greater incidence of postpartum depression and racism and discrimination during their pregnancy and birth and less attention paid to African American women when they sense "there is something wrong" (Martin and Montagne 2017). Today, the Internet and social media has a powerful influence on decision making among pregnant women. In one study, 83 percent of women wanted to have more control over decisions affecting their pregnancy and two-thirds sought information and support from Internet sources about what a "normal" pregnancy should look like (Luce et al. 2016).

Meanwhile, premature birth (less than thirty-seven weeks' gestation) and low birth weight (less than about five and a half pounds) are leading causes of infant disabilities and deaths. The rate of preterm births rose slightly between 2016 and 2017 from and stands at about 10 percent (Martin et al. 2018). However, these rates are significantly higher than in the early 1980s (Bakalar 2010; Hamilton, Martin, and Ventura 2009). Reasons for these trends are not clear but may be related to increased use of induced labor and caesarean deliveries (Martin, Osterman, and Sutton 2010). The medical community is increasingly concerned about the "alarming" rate of C-Sections in the United States. In Chicago, doctors have begun tracking the conditions under which women are given C-sections and are reconsidering how staffs manage labor (Bowen 2017). One idea is for hospitals to rely more heavily on midwives, who are far less expensive than physicians, who use fewer interventions, and produce similar if not better outcomes for mother and child (Stewart 1998).

Critical Thinking

What do you think? Given the benefits of pain control and physicians' ability to control timing, is there anything wrong with planning to have a delivery by caesarian section? What would be some benefits of doing so? What are some potential costs?

Meanwhile, some observers argue that U.S. society is characterized by **structural antinatalism**—that is, not doing what it could to support parents and their children (Huber 1980). Critics of American family policy point out that nutrition, housing, health insurance, daycare, and social, financial, and education programs directly affecting the welfare of children are not adequate compared to those of other nations at our economic

level (The Children's Defense Fund 2017). Nor do we provide paid parental leave or other support for parents of young children as virtually all our counterparts do. Children in the United States are more likely to be poor than children in other developed nations, and the United States has a higher rate of poverty than even some less advanced economies including Russia and Mexico (OECD 2019). Given strong antinatalist forces that dilute societal support for parents, some scholars have asked, "Why do people choose to have any children at all?" (Overall 2013). In answer, we look at rewards and costs associated with parenthood.

Rewards and Costs of Parenthood

"From the day children are born they become a source of joy and a source of burdens for their parents" (Nomaguchi and Milkie 2003, p. 372). According to the **value of children perspective**, children historically were economic assets. Even when very young, they provided more working hands in the fields and kitchens. Gradually, the shift from an agricultural to an industrial society and the development of compulsory education transformed children from economic assets to economic liabilities. At school rather than at work in the factory, needing school clothes and fees, children no longer added financially to the family. Instead, they cost their families. But as their economic value declined, children's emotional significance to parents increased, partly because declining infant mortality rates made it safer to become attached to them (Hoffman and Manis 1979; Lamanna 1977). Parents' desire was for "a child to love" and the "joy that comes from watching a child grow" (Morgan and King 2001, p. 11).

Children, of course, bring many benefits to parents. Becoming a parent can certify one's attainment of adulthood, although there are other avenues to adult status as well (see Chapter 6). For men as well as women, gay as well as straight, parenthood is an extremely meaningful aspect of personal identity (Parker, Horowitz, and Rohal 2015). In a national survey of marrieds, 57 percent of wives and 45 percent of husbands with children at home strongly agreed that "life has an important purpose." These figures compare to just 40 percent of wives and 35 percent of husbands with no children in the home (Wilcox, Marquard et al. 2011, p. 16). Young people remain supportive of having children. In 2008, 70 percent of high school seniors agreed that fatherhood is one of the most fulfilling experiences for a man, and 60 percent agreed that motherhood is one of the most fulfilling experiences for a woman (NCMFR 2010). Parenthood can also provide a sense of commitment and meaning in an uncertain social world (South and Crowder 2010).

Although the rewards of children are often immeasurable, one thing we do know is that raising children is costly. In fact, one recent study concluded that parents psychologically exaggerate to themselves the joys of parenthood in order emotionally to counter the costs (Wilkinson 2011). The most obvious costs associated with children are financial. Parents' finances take a hit, and the financial costs of children can offset the rewards of parenthood. For example, the average monthly cost of licensed infant childcare in the United States is $1,230, which accounts for a third of the take-home pay of a worker earning $50,000 per year (Mekouar 2019; Workman and Jessen-Howard 2018). Factor in a family's average monthly mortgage payment ($1,500; monthly rental amounts vary considerably by state), utility bills, home repairs, automobile and transportation costs, food, clothing, medical insurance, health-related costs (prescription drugs, co-pays), and student loan payments, there is not much, if anything, left over.

And the situation is worsening. Adjusted for inflation, the average weekly childcare expenses for families with working mothers rose more than 70 percent between 1985 and 2011, far faster than incomes. In every region of the country, a year of daycare is now about double the amount of a year of public college tuition (Keshner 2019). There is also a problem with simply finding care. Many communities have long waiting lists for a small number of day care providers, including in-home or "family" care, most of which is not licensed by the state. And the vast majority of U.S. companies do not offer childcare on-site or financial assistance with childcare. Two-thirds of parents say it is either very or somewhat hard to find good, affordable childcare (Parker, Horowitz, and Rohal 2015).

Every year, the United States Department of Agriculture estimates how much it costs the average American family to have a child. In 2015, for a middle-income family, the average cost of raising a child to age 18 was $233,610, not including college (Lino 2017). The largest expenditures are housing and transportation, followed by childcare and education, food, clothing, and health care. These figures vary considerably by region, income, and number of children (additional children cost less). Parents sacrifice many things for their children. For a single mother in poverty, feeding her children well may mean risking her own health as she skips meals and chooses for herself less-nutritious foods, both of which foster obesity (Martin and Lippert 2012).

In a 2018 survey conducted by *The New York Times* (Miller 2018), 1,858 men and women ages 20 to 45 were asked why they had (or expected to have) fewer children than their ideal. Four out of the five most common responses were financial: "child care is too expensive" (64 percent) followed by "want more time to focus on the children I have" (54 percent) and "worried about the economy" (49 percent), "can't afford more children" (44 percent), and "waited because of financial instability" (43 percent). As David, age 29 says, "Wages

are not growing in proportion to the cost of living, and with student loans on top of that, it's just really hard to get your financial footing—even if you've gone to college, work in a corporate job and have dual incomes." He and his wife have $100,000 in student loan debt between them.

Added to the direct costs of parenting are opportunity costs: opportunities for wage earning and investments as well as opportunities to pursue personal interests that don't include children. Whereas the financial costs of raising children are more important to men's life satisfaction, because they are the primary caretakers of children, a women's life satisfaction is more affected by the reduction in time to do other things (Pollman-Schult 2014). Mothers who work full-time as opposed to part-time or not at all feel the greatest lack of time for friends and hobbies (Parker, Horowitz, and Rohal 2015). In Miller's (2018) article, Jessica, age 26, stated she "has a long list of things she'd rather spend time doing than raising children: being with her family and her fiancé; traveling; focusing on her job as a nurse; getting a master's degree; and playing with her cats." She says, "My parents got married right out of high school and had me and they were miserable. But now we have a choice." A common theme, too, is the fear of divorce, especially for the millions of adult children who grew up with divorced parents (Miller 2018).

A woman's career advancement may suffer as a consequence of becoming a mother, especially in a society that does not provide adequate day care or a flexible workplace (Slaughter 2012). Some career women report hiding their pregnancies for as long as possible to avoid workplace discrimination, such as failure to be considered for promotion (Quart 2012). A 2012 study in Sweden found that men and women were more likely to plan a pregnancy when their workplace policies were more supportive of parent–employee needs (Kaufman and Bernhardt 2012).

A couple in which one partner quits work to stay home with a child or children typically faces a substantial loss in family income. The spouse who quits work (more often the woman) also faces individual losses in retirement income, a pension, and Social Security benefits later. All in all, in our society there is "a heavy financial penalty on anyone who chooses to spend any serious amount of time with children" (Crittenden 2001, p. 6). The rising cost of childcare may even help explain the recent increase in mothers not working outside the home (Cohn, Livingston, and Wang 2014). Studies associate wanting to get pregnant with greater appreciation for parental rewards and decreased emphasis on the costs (East, Chien, and Barber 2012), and 68 percent of male and 70 percent of female parents strongly agree that the "rewards of being a parent are worth it despite the cost and work it takes" (Martinez et al. 2006).

szefei/Shutterstock.com

Children can bring vitality and a sense of purpose into a household. Having a child also broadens a parent's role in the world: Mothers and fathers become nurturers, advocates, authority figures, counselors, caregivers, and playmates.

How Children Affect Couple Happiness

Research on how children affect a couple's relationship satisfaction and happiness has for the most part focused on married, heterosexual couples. Research is building as to the extent to which the findings reported in this section apply to same-sex couples, cohabitors, or other unmarried couples. Among marrieds, evidence shows that children, especially young ones, stabilize a union; that is, parents are less likely to divorce (Stykes 2015). But a stable marriage is not necessarily a happy one: "[C]hildren have the paradoxical effect of increasing the stability of the marriage while decreasing its quality" (Bradbury, Fincham, and Beach 2000, p. 969). Couple conflict and strain are common costs of having children and relationship functioning declines; this has been found to be true for same sex and cohabiting couples as well as opposite sex married couples. The research on adoptive parents is mixed with studies showing less of a decline or no decline and even relationship improvements (Doss and Rhoades 2017).

Along with a major review of the research in this area, survey data show that not only do parents report lower marital satisfaction than nonparents, but also the more children they have, the lower their marital happiness (Doss et al. 2009; Wilcox, Marquard et al. 2011). Research involving 1,000 families concluded that even with highly anticipated births, most couples (upon becoming parents) become much more traditional in their approach to housework and childcare. No matter how much they think the tasks will be shared, most

women wind up doing more housework than they did before the birth and more of the childcare than they expected. The discrepancy between what the couples hoped for and the reality of wives having to take on a "second shift" at home leads to feelings of tension, depression, and sometimes anger in both partners (Cowan and Cowan 2009). In a study of dual earner couples using time-diaries, having a child was associated with an additional 2 hours of work for women compared to only 40 minutes for men (Yovorsky, Kamp Dush, and Schoppe-Sullivan 2015).

Marital satisfaction tends to decline over time regardless of whether couples have children. But conflicts over each partner's employment and home responsibilities may erupt when a child arrives, especially (and perhaps ironically) in "soul-mate" relationships that have to be "nurtured and coddled in order to thrive" (Whitehead and Popenoe 2008, p. 8). In particular, fathers may feel a reduced sense of confidence in their place in the family after the birth of a child, whereas mothers experience higher levels of couple conflict (Doss et al. 2009). The negative effect of children on parents' relationship may be temporary. Over time, parents' and non-parents' relationship begin to converge—that is, relationship satisfaction rebounds as children get older (Doss and Rhoades 2017).

Numerous studies have also shown lower levels of psychological well-being among parents than nonparents (e.g., Cohen and Janicki-Deverts 2012; Hildingsson and Thomas, 2014). In the late 1960s, social scientist Alice Rossi (1968) wrote that the *transition to parenthood* is difficult for several reasons. With little experience, first-time parents must abruptly assume twenty-four-hour duty caring for a fragile, dependent baby. Babies interrupt parents' sleep, work, and leisure time. As one new mother said, "[Y]ou have to run to bathe yourself because the child is going to wake up" (Ornelas et al. 2009, p. 1464).

Younger parents may have more trouble coping with day-to-day stressors and have more worries, as do parents in nontraditional families, single parents, and cohabitors (Amato 2010; Bures, Koropeckyj-Cox, and Loree 2009; Parker et al. 2015). As men's involvement in parenting increases, understanding their perspectives becomes crucial, especially because men experience stress differently than women and can be a major source of support for new mothers (Miller 2019; Widarsson et al. 2013). Factors that exacerbate the negative effect of children on the couple relationship is lower income, difficulty with communication, parental anxiety and depression, and characteristics of the child such as them being a poor sleeper. Girls are associated with greater declines in marital quality than are boys (Doss and Rhoades 2017). Interventions such as co-parenting programs can help mitigate these negative effects (Doss, Cicila, Hsueh, Morrison, and Carhart 2014), as can

having a sense of humor (Theisen, Ogolsky, Simpson, and Rholes 2019).

One review of the research through 2000 concluded that the negative effects of children on marital satisfaction seem to be stronger for today's young adults (Twenge, Campbell, and Foster 2003). Perhaps when children arrive, couples today experience a greater "before–after" contrast than in the past. Couples today often marry and become parents later in their lives than couples did in earlier decades, after several years in which they experienced a great deal of personal freedom and a career focus that children interrupt. Moreover, the increased individualism of our culture may make day-to-day responsibility for the care of young children seem less natural than in the 1950s, when social obligations were culturally dominant (Turner 1976). The drop in life satisfaction associated with having a child has been shown to reduce the likelihood of having another (Margolis and Myrskylä 2015). Given these circumstances, some adults decide against parenthood.

Choosing to Be Childfree

As a result of the same forces associated with lower fertility, more women are choosing not to have children at all. This section examines **voluntary childlessness,** or being *voluntarily childfree*. It's important to note that government statistics on fertility are generally limited to biological children; women who have not given birth may be mothers to adopted children for example and "childfree" women may struggle with infertility (both discussed later). And most research on childlessness focuses on women. Nevertheless, here's what we know: Nearly one-half of women between the ages of 40 and 44 who do not have children because they don't want to (Livingston, Parker, and Rohal 2015). It is expected that 14 percent of American women will reach the end of their childbearing years without having had a biological child. This rate varies greatly by marital status. Among never married women, 45 percent will not have had a child compared to 90 percent of women who have been married. Remaining childless is more common among whites (17 percent) than blacks (15 percent), Asians (14 percent), and Hispanics (10 percent). These figures are roughly two times greater than in 1976 (Livingston 2018a).

Different countries and cultures have different norms regarding voluntary childlessness (Merz and Liefbroer 2012). Although not so openly as in the 1970s, negative stereotypes of the voluntarily childfree persist (Abma and Martinez 2006; Kelly 2009). At least one contemporary commentator (in Canada) has called the "trend of couples not having children just plain selfish" (O'Connor 2012). This view is shared by Pope Francis of the Catholic Church (Neuman 2015). However, in 1990, about 70 percent of Americans told

pollsters that children were very important to a successful marriage; by 2007, that percentage had dropped to 41 percent (Pew Research Center 2007). The United States appears to have strong, although weakening, fertility norms that discourage childlessness and only-child families. Men and women who desire sterilization (vasectomy or tubal ligation) are sometimes met with resistance by medical providers (Hintz and Brown 2019). Attitudes are changing. The American College of Obstetricians and Gynecologists (2017) makes the following recommendation: "A request for sterilization in a young woman without children should not automatically trigger a mental health consultation. Although physicians understandably wish to avoid precipitating sterilization regret in women, they should avoid paternalism as well."

Partly as a result of these norms, along with social pressure to have children, many women are profoundly ambivalent about their decision. Women without children can feel "grief" and "relief" simultaneously (Baldwin 2019). For many, it is a choice that develops gradually over time. For others, remaining childless is a consequence of putting off parenting until it feels "too late now" or until one is less likely to become pregnant because of age (Whitehead and Popenoe 2008). On the other hand, for many younger women, not having children represents early commitment to continuing nonreproductivity. "Firm choices to have no children may signal an increasing proportion of women who see the costs of childbearing as too high" (Hagewen and Morgan 2005, p. 522). As one 40-something foreign correspondent wrote, "I had zero desire to replicate my mother's life—divorce, two children and a shortage of cash" (Baldwin 2019).

In the 1970s, feminism challenged the inevitability of the mother role. More than 70 percent of women surveyed in 2001 said "no" to the question of whether "a woman need[s] the experience of motherhood to have a complete life" (Center for the Advancement of Women 2003). As we saw earlier in this chapter, research finds childfree couples to be more satisfied with their relationships than parenting couples are. It is not just the absence of children; the voluntarily childless have more education and are more likely to have professional employment and higher incomes. They are more urban, less traditional in gender roles, less likely to have a religious affiliation, and less conventional than their counterparts (Kingston 2009; Lundquist et al. 2009). Childfree women tend to be more invested in a satisfying career (Jayson 2011; Livingston 2015). Couples value their relative freedom to change jobs, pursuing any endeavor they find interesting (Kelly 2009; Park 2005). A greater ability to control fertility, greater participation of women in paid employment, concern about overpopulation and the environment, or an ideological rejection of the traditional family

provide the social context for some people's decisions to remain childless (Kingston 2009). With the growth of *attachment* or *intensive parenting*, a child-centered and immersive style of parenting, raising children has become much more demanding and time intensive, especially for middle-class parents pressured to "cultivate" their child's talents and abilities through sports, music lessons, and travel (Parker, Horowitz, and Rohal 2015; Sandler 2013). Indeed, 37 percent of adults under age 50 say they are not likely to have kids in the future (Livingston and Horowitz 2018).

How happy with being childfree are individuals once they reach older ages? Eric Klinenberg, author of *Going Solo* (2012), says childlessness is toughest for people in their late thirties and forties, when same-age peers are consumed with child rearing. Some women report feeling shame—"for being selfish, unfeminine, or unable to nurture" (Safer cited in Gilbert 2015). There online groups for the voluntarily childless, such as Facebook's *Childfree...is Not a Dirty Word*, which has 37,000 followers. Provided that being childless is voluntary, the childfree elderly are as satisfied with their lives and less stressed than parents; in fact, some studies show they have lower levels of depression than their counterparts who did have children (Bures, Koropeckyj-Cox, and Loree 2009), and have developed social support networks other than those provided by children (Dykstra and Hagestad 2007; Park 2005). However, perhaps it is not a surprise that studies show that childless women who had wanted children are concerned about their identities as women and find some things difficult, such as holidays and family gatherings because, not having children, they feel left out or sad (McQuillan et al. 2012). Childless women in the United Kingdom who sought infertility treatment (and who were unsuccessful) had elevated levels of depression and shorter life expectancies (Agerbo, Mortensen, and Munk-Olsen 2012). Meanwhile, individuals who do have children face an array of options and circumstances.

HAVING CHILDREN: OPTIONS AND CIRCUMSTANCES

Discussions about having children once evoked images of a newly married couple. More and more, however, decisions about becoming parents are being made in a much wider variety of circumstances. In this section, we address childbearing with reference to postponing parenthood, child spacing, having one child only, nonmarital childbearing, and multipartnered fertility (MPF). Deciding about parenthood in stepfamilies is addressed in Chapter 15. Much of the material in this chapter may apply not only to heterosexual but also to LGBTQ+ parents.

Timing Parenthood: Earlier versus Later

Until a few decades ago, the vast majority of U.S. mothers had their first child in their early twenties. Applying the *family life course development framework*, this situation was considered "on-time" parenthood. Today, because Americans are postponing having children, we consider childbearing in one's early twenties (or younger) as "early." Marrying at later ages and men and women's desire to complete their education and become established in a career are two of the main reasons for delaying having children until their thirties and even later. This is made possible with the availability of reliable contraception and the promise of assisted reproduction technology. We'll look at benefits and drawbacks of having children earlier versus later.

Earlier Parenthood Choosing early parenthood means greater certainty of being physically able to have children. Furthermore, parenting early means greater freedom to pursue other activities after the children are raised (Christopherson 2006). In addition, earlier parenthood means a greater likelihood of having more time with grandchildren and even great-grandchildren.

Annika Erickson/Getty Images

Many couples are postponing parenthood into their thirties, sometimes later, but the timing of parenthood is a trade-off: Having children when you're younger gives you more potential time with your children, but waiting until you're older allows you time to become more psychologically mature and hence a more emotionally steady parent.

In one qualitative study, early mothers said they felt more spontaneity as youthful parents (Walter 1986). "We wanted to be young parents [said one mother]. . . . We didn't want to be sixty when they got out of high school" (Poniewozik 2002, pp. 56–57). In a study based on interviews with 114 Canadian expectant mothers, the younger pregnant women spoke of their physical health as an asset, as well as the health of their parents—they expected to rely on their own parents for help with their children (Dion 1995).

However, relatively young parents may have to forgo some education and may get a slower start up the career ladder. Early parenthood can create strains on a marriage if the breadwinner's need to support the family means little time to spend at home or if young parents lack the maturity needed to cope with family responsibilities (Gillmore et al. 2008). Moreover, couples who have children early usually begin saving for college or retirement later in their adult lives. Often having relatively low incomes, they have to work harder and longer to meet family needs (Jayson 2010; Joshi, Quane, and Cherlin 2009).

Later Parenthood Older mothers tell researchers that they benefited by waiting to have children because they needed a period of time for personal development—not just career development—for themselves and their partners. In 2013, 42 percent of childless women ages 40 to 44 said they desired to have more children and 54 percent said they would be a little, some, or a great deal bothered if they didn't (Guzzo 2018). Although overall women are waiting longer to have children, childlessness has declined slightly among older women (Livingstone 2018a). These women have typically reached a professional level that provides them with job flexibility, allowing them to leave work for school plays and doctor's appointments, They are likely to be married to highly educated men, who have more progressive gender role attitudes and are more likely to take on extra housework and childcare responsibilities, which are associated with greater intentions to have children (Okun and Raz-Yurovich 2019).

Delaying parenthood can also mean greater maturity and preparation for parenthood (Tyre 2004). This goes for both mothers and fathers. Older fathers see themselves as more patient (Vinciguerra 2007). Some evidence shows that older men and women find more joy in parenthood than do their younger counterparts (Paul 2011). Older mothers felt that they had more confidence in their ability to manage their changed lives because of the organizational skills they had developed in their work. They also had more money with which to arrange support services, such as babysitters and house cleaners, and they felt confident about their ability as parents (Jayson 2010).

However, for both men and women, fertility does decline with age, although less dramatically for men. Older age in both women and men also increases the likelihood of reproductive risks (Brandt, Ithier, Rosen, and Ashkinadze 2019). Older mothers have higher rates of premature or low-weight babies and multiple births—all risks for learning disabilities and health problems. Older mothers also have higher rates of miscarriage, as well as health problems such as diabetes and hypertension (Crawford and Steiner 2015). Some physicians nevertheless advise that, "while a lot of complications of labor and pregnancy are increased . . . the vast majority of [older mothers] do perfectly fine" (Dr. William Gilbert, quoted in "Older Moms' Birth Risks" 1999).

Nevertheless, economist Sylvia Ann Hewlett's book *Creating a Life* (2002), based on her survey of 1,168 older women in the top 10 percent of earners, reports a high rate of childlessness among successful managerial and professional career women, most of whom had not intended to be childless. Hewlett faults women for focusing on careers based on the assumption that it will be easy to have children later in life. An article in the *Journal of Family and Reproductive Health* noted, "As women delay childbearing, there is now an unrealistic expectation that medical science can undo the effects of aging" (Karimzadeh and Ghandi 2008, p. 62). Medical reasons (41 percent) followed by age (25 percent) top the list of why parents who want more children say they are unlikely to have children in the future (Livingston and Horowitz 2018). Family policy scholars (e.g., Pollitt 2002; Mishel, Bernstein, and Shierholz 2009) argue that the real problem is the failure of society-wide or structural support for working families so that younger women could better manage both motherhood and career building.

As they age, older parents often experience a sense of limited time with their children that increases pleasure in parenting: "Everything is more precious." However, it also creates anxiety about the future: That he could die before his daughter reaches adulthood "is a reality that I live with," said one 59-year-old father when his daughter was born (Vinciguerra 2007, p. ST-1). That he may not live to see grandchildren is another reality.

Being born to older parents affects children's lives as well. They usually benefit from the financial and emotional stability that older parents can provide and the attention given by parents who have waited a long time to have children. But children of older parents often experience anxiety about their parents' health and mortality (Vinciguerra 2007). Older parents may become frail before their children have established themselves in their adult lives and before they themselves can serve as active grandparents.

A Word about Child Spacing When parents begin having children later, their options for child spacing may be limited because their desired number of children needs to be born before the proverbial biological clock runs out. Parents who have children while younger may want them close together because they believe that having children close in age will encourage them to play together and make parenting easier as the children take part in age-similar activities. However, experts report that for the physical and intellectual health of both mother and child, the optimal spacing of children is between 18 and 24 months, but less than 5 years (Mayo Clinic Staff 2017).

Moreover, there is evidence that waiting longer to have the next child benefits the older one academically. "This positive effect of a larger gap between kids may be the result of increased time or resources spent on the older child or other factors, but the research, based on a panel of more than 12,000 children, looks pretty solid" (Dell'Antonia 2011). For prospective parents interested in the timing of their parenthood, it's important to have an awareness of the trade-offs. Meanwhile, some prospective parents not only consider the challenges of parenthood to be monumental but also reject the idea of childlessness. For them, the solution is the one-child family.

Having Only One Child

The number of one-child families continues a steady increase, the number of women who end their childbearing years has doubled between 1976 and 2014, from 11 percent to 22 percent (Parker et al. 2015). The proportion of one-child families in the United States is growing for the same reasons that women are having fewer children: (1) women's increasing career opportunities and aspirations in a context of inadequate domestic support (childcare, etc.), (2) individuals' desire to parent in a context of inadequate workplace support for the parenting role, (3) the high cost of raising a child through college, and (4) greater peer support—that is, the choice to have just one child becomes easier to make as more couples do so. Divorced, separated, or widowed people who do not form a new reproductive partnership may end up with only one child because the relationship ended before more children were born.

Negative stereotypes present only children as "socially unskilled, dependent, anxious, and generally maladjusted" (Hagewen and Morgan 2005, p. 514). To find out whether there was any basis for this image, psychologists in the 1970s produced a staggering number of studies that generally found no negative effects of being an only child (Pines 1981). More recent studies have found only children to be indistinguishable from other children and where there are differences, only children outperform their peers, especially in the area of academic achievement, motivation, and intelligence (Mancillas 2006).

Moreover, research reports only children have better leadership ability, and better health and life satisfaction both as children and as adults (Hagewen and Morgan 2005; Newman 2011; Watson 2012). One study found that parents of only children had higher educational expectations for their child, were more likely to know their child's friends and the friends' parents, and had more money saved for their child's college education. They reported that they could enjoy parenthood without feeling overwhelmed and tied down, had more free time, and were better off financially than they would have been with more children (Deveny 2008). They were more likely than parents in larger families to share domestic chores and could afford to do more things together (Newman 2011).

There are disadvantages and challenges to having just one child. For parents, disadvantages include the fear that the only child might be seriously hurt or die and the feeling, in some cases, that they have only one chance to prove themselves good parents. For the children, disadvantages include lack of opportunity to experience sibling relationships, not only in childhood but also as adults. For example, the 2010 census indicated that roughly 3 million adult siblings live together (U.S. Census 2012e). Siblings provide social support to each other and can serve as confidantes who have shared the same struggles in life (Roberts and Blanton 2001) as well as exchanges of material assistance and someone to rely on in emergencies (Lee 2006; Riedmann and White 1996). Only children may face extra pressure

from parents to succeed, and they are sometimes under an uncomfortable amount of parental scrutiny. As adults, they have no help in caring for their aging parents (Watson 2012).

Nonmarital Births

As discussed in Chapter 6, we are currently seeing the "decoupling" of marriage and parenthood, meaning that more women (and men) are choosing parenthood but not marriage (Hayford, Guzzo, and Smock 2014). About 40 percent of all births are to unmarried mothers (Martin et al. 2019). Although 42 percent are to unpartnered women, 58 percent of all nonmarital births occur to cohabiting mothers (Child Trends 2015c). About half of these mothers are raising their child alone with no or minimal financial, social, or emotional support from their child's father (Huang 2009; U.S. Census Bureau 2012a). Low education is a leading factor associated with nonmarital births, especially among minorities (Carlson, VanOrman, and Pilkauskas 2013). Black women are also less likely to use contraception than white women (Barber, Yarger, and Gatny 2015).

From 1940 to the early 1960s, only 4 to 5 percent of births were to unmarried women. By 1980, 18 percent were. Today that rate is about 40 percent. Nonmarital births vary dramatically by race and ethnicity. As Figure 8.3 shows, in 2017, 69 percent of African American births, 69 percent of American Indian/Alaska Native births, 52 percent of Hispanic births, 28 percent of non-Hispanic white births, and 12 percent of Asian births occurred outside marriage (Martin et al. 2018).

The growing proportion of nonmarital births accompanies changing societal attitudes. In 2002, 45 percent of Americans told pollsters that they felt it was morally acceptable to have a baby outside of marriage. Today, more than half—62 percent—find doing so morally acceptable (Newport 2015c). Individuals are now much less likely to marry with the discovery of a nonmarital pregnancy (Su, Dunifon, and Sassler 2015; McLanahan and Sawhill 2015). Meanwhile, a substantial proportion of women with unplanned pregnancy choose to begin or continue to cohabit

Sisters who get along well can provide companionship and support for each other as they go through life, especially when it comes to becoming a parent.

Hiroko Masuike/The New York Times/Redux

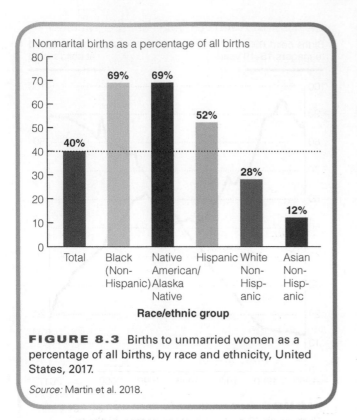

Nonmarital births as a percentage of all births

FIGURE 8.3 Births to unmarried women as a percentage of all births, by race and ethnicity, United States, 2017.

Source: Martin et al. 2018.

(Rackin and Gibson-Davis 2012). In fact, births to women who are living with a cohabiting partner underlies nearly all the growth in births outside of marriage (Parker et al. 2015). Birthrates for never-married cohabitants are virtually the same as those for married women (Child Trends 2015c). Among younger generations (Millennials and Gen Xers), 75 percent were supportive of bearing and rearing children in cohabiting unions (Stykes 2015a).

We note here that unmarried parents, including those not living together, may have a more regular relationship with each other than was previously thought. Some are intermittent cohabitators: The children's father moves into and out of the mother's domicile as the couple experiences separations and reunifications (Cross-Barnet, Cherlin, and Burton 2011). The Fragile Families and Child Well-Being Study is a national longitudinal study of unmarried and married parents conducted in twenty large U.S. cities. At the time the baby was born, the majority of unmarried parents described themselves as "romantically involved on a steady basis." Of these, 50 percent were cohabiting and 33 percent "visiting." Many fathers from that study engage in "on and off parenting" (Turney and Halpern-Meekin 2017).

Fathers visited the hospital during delivery or otherwise helped mothers during pregnancy and childbirth. Nearly all fathers said they wanted to be involved in their children's lives, and 93 percent of mothers agreed that they should be. "The myth that unwed fathers are

not around at the time of the birth could not be further from the truth" (McLanahan et al. 2001, p. 217). Nevertheless, father involvement is lower among fathers who are cohabiting or are intermittently in contact with the mother, and involvement of these fathers declined more so than it did for dads who were married to their children's mother (Mitchell, Booth, and King 2009).

As opportunities grow for women to support themselves and the permanence of marriage becomes less certain, there is less motivation for a woman to avoid giving birth outside of marriage. Furthermore, stigma and discrimination against unwed mothers have lessened. Still, because the burden of responsibility for support and care of the child remains on the mother, overall "the economic situation of older, single mothers is closer to that of teen mothers than that of married childbearers the same age" (Foster, Jones, and Hoffman 1998, p. 163). Single-parent households are 4.5 times more likely than married couple households to be living in poverty (Annie E. Casey Foundation 2019). In 2017, 41 percent of children in single-mother households lived in poverty compared to only 8 percent of children in married-couple households. These rates are even higher for black and Hispanic children at 57 and 53 percent (Child Trends 2019).

Single Mothers by Choice

Although unwed birthrates are highest among young women in their twenties, they have increased significantly for older women (Martin et al. 2018). Many of these women are **single mothers by choice**. There are many books, blogs, and Facebook groups that provide information and support to women who want to parent on their own and support for women choosing to raise a child alone is building (Scott, Wilder, and Bennett 2019). The image is that of an older woman with an education, an established job, and economic resources who has made a choice to become a single mother. Not having found a stable life partner, yet wanting to parent, a woman makes this choice as she sees time running out on her "biological clock" (Lehmann-Haupt 2009). Sociologist Rosanna Hertz (2006) interviewed sixty-five single mothers who had their first child at age 20 or older and who, more significantly, were self-sufficient economically. Not having the "chance" to be in stable, child-rearing marriages, they became mothers through various routes: biological pregnancy, artificial insemination by known or unknown donor, or adoption.

"For the women in this study, single motherhood was never a snap decision" (Hertz 2006, p. 26):

> I always had in the back of my mind that if I was 30 and not married, then I'd have children on my own. Then it was when I was 32. Then it was when I went back to school at Princeton to get my master's degree. Then it was 36 and I had just broken up with another man. (p. 26)

Women were often surprised to find themselves taking what they saw as an unconventional step:

> Daring to consider getting pregnant on my own just seemed like such an outrageous thing to do. And from that point of thinking about it, to doing it, was the longest stretch because I was kind of shocked that I would think that way, and I wasn't sure of what I really wanted to do. (p. 27)

Once these single women became mothers, their parenting practices by choice and sense of family were very traditional. In fact, they saw themselves as exemplifying family values by having chosen parenthood.

Indeed, single mothers by choice see themselves as responsible, emotionally mature, and financially capable of raising a child and conforming to normal family goals, although some described themselves as "moral pioneers" when it comes to defining family. The decisions to become a single mother is often well accepted by the family, friends, employers, clergy, and physicians (Bock 2000; Graham 2018; Mannis 1999). There are trade-offs. Single mothers by choice must often cut back on work hours and downgrade their lifestyle. However, they often lean on friends and family for instrumental support like childcare (Hertz, Rivas, and Jociles 2016). In studies, children raised by single mothers by choice were no different from other children in social and emotional adjustment (Golombok, Zadeh, Imrie, Smith, and Freeman 2016).

These were white, middle-class, educated women who insisted on the great difference between themselves and "welfare" or teen mothers. What, in fact, are the realities of teen parenthood today?

Births to Adolescents Public concerns about outcomes for the children of unmarried parents intensify when the mother is a teenager. The term *teenage pregnancy* has been associated with the word *problem* for as long as most of us can remember. Adolescent birthrates rose in the late 1960s as sexual behavior liberalized. However, by the time a "teen pregnancy epidemic" was identified, adolescent pregnancies and births had already begun to decline as shown in Figure 8.4. Teenage childbearing declined 77 percent between 1957 and 2015 (Dorius 2018). The teen pregnancy and birth rate are today are at record lows, even among sexually experienced teens (Centers for Disease Control and Prevention [CDC] 2019b; Child Trends 2018b). It is thought that the decline in teen births might be the result of better and more effective use of contraception as well as the return to comprehensive sex education in schools (as opposed to education focused on abstinence). Teenagers today are also less sexually active than previous generations of teens and are waiting longer when they do have sex (Bridges and Philbin 2019; Guttmacher Institute 2019d). These issues are discussed in more detail in Chapter 4.

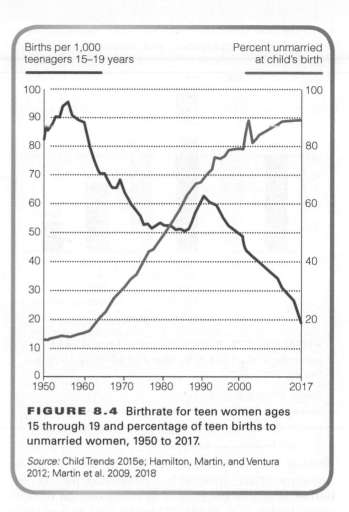

FIGURE 8.4 Birthrate for teen women ages 15 through 19 and percentage of teen births to unmarried women, 1950 to 2017.

Source: Child Trends 2015e; Hamilton, Martin, and Ventura 2012; Martin et al. 2009, 2018

Declines in the adolescent birthrates are happening among all racial and ethnic groups, although rates for black and Hispanic teens are still more than twice those of whites (Annie E. Casey Foundation 2019). Some agencies are piloting technologies to women of color, such as the mobile app *Pulse*, designed for black and Latinx women ages 18 to 20. According to its developers, Pulse is

> a web-based mobile health application [that] provides comprehensive, medically accurate sexual and reproductive health information to young women, in English and Spanish, through highly engaging interactive and multimedia features. These features include dynamic text and graphics, self-assessments, comics that pose various scenarios, and videos of racially diverse peers that model real-life scenarios, including short films promoting birth control use and clinic use. (Center for Evidence & Innovation 2019)

In a randomized controlled trial of the app with 1,124 participants, users were less likely to report having unprotected sex, had more accurate knowledge about contraceptives, and were more confident about their ability to use contraceptives effectively (Manlove et al. 2019). Public support for such programs is important.

It has been estimated that the teen pregnancy rate would be 75 percent higher if it were not for publicly funded family planning programs and services (Guttmacher Institute 2019d).

Along with adolescent pregnancies and births, the adolescent abortion rate has declined dramatically from a high of 43 per 1,000 females ages 15 to 19 in the 1980s to 11 per 1,000 by 2013. Currently, about one-quarter of all teen pregnancies end in abortion. Teen abortion rates also vary by age and race and ethnicity, with higher abortion rates among older teens and blacks and Hispanics (Child Trends 2018a). Although still high when compared to those of other wealthy, industrialized nations, a downward trend over time is encouraging.

As Figure 8.4 indicates, the vast majority of teen women giving birth are not married. In fact, the majority of teen parents are cohabiting (Manning and Cohen 2015). Either way, unmarried teen mothers are likely to lack consistent economic and social support of a coparent. Additionally, teenage parents, especially those with more than one child, face a bleak educational future, limited job prospects, and a strong chance of living in poverty compared to peers who do not become parents as teenagers. (Diaz and Fiel 2016; Kane et al. 2013; Perper, Peterson, and Manlove 2010). They have a lower likelihood of marriage and, if they do marry, a greater likelihood of divorce (Child Trends 2015e). Being the child of a teen mother has short- and long-term negative effects, including pre-term births and low birth weight. Children born to teens tend to have poorer academic achievement, worse behavioral outcomes, and are more likely to engage in sexual activity and become teen parents themselves (Annie E. Casey Foundation 2019).

Yet, the outcomes of teen parenthood vary and are not by any means uniformly negative. One longitudinal study of black teen mothers from low-income families in Baltimore concluded that although early childbearing increases the risk of ill effects for mother and child, it does not justify the popular image of the adolescent mother as an unemployed woman living on welfare with a number of poorly cared-for children. To be sure, teenage mothers do not manage as well as women who delay childbearing, but most studies have shown that there is great variation in the effects of teenage childbearing (Furstenberg, Brooks-Gunn, and Morgan 1987). According to a study of workers who work with at-risk youth (Boustani, Frazier, Harley, Meinzer, and Hedemann 2015), teens perceive many benefits to early parenthood, including cementing bonds with a romantic partner, proving oneself as an independent adult, and creating an emotional connection with a baby. One worker said, teens think having a baby will help them keep their partner "to actually stick around in their life this time cuz most of them like their parents are in and out, out of jail or just in out of their life period. So if they have a child by somebody else, that would give them a reason to be around." A baby provides youth with the ability to control some aspect of their life, "perhaps to compensate for the long-standing instability that characterized their family of origin. Workers also reported that youth wanted a baby as something that 'would be their own,' 'something that actually belongs to them,' reflecting a desire for something stable—'this is that one thing they can control.'" Many of these youth had a fear of abandonment, and a baby would compensate for family attachments they were denied growing up in a home characterized by dysfunction and want: "They've never had love, someone to love" (p. 86). However, the workers also felt that the youth minimized the potential costs and magnified the benefits.

Multipartnered Fertility Multipartnered fertility (MPF)—having children with more than one biological partner—is a new area of research for family scientists. MPF is not new, but has grown. In 1955, in 6 percent of married couples, one or both spouses had had a child with a different partner. Thirty years later, this figure had doubled (Guzzo 2014; Zobl and Smock 2015). The census estimates that one out of every ten adults have had children with more than one partner; 20 percent of parents with two or more children has MPF (Monte 2017). The forces underlying MPF differed between these two periods. In the 1950s, most MPF was a result of widowed or divorced women and men having additional children in subsequent marriages. Today, most MPF is a result of nonmarital childbearing. For example, 31 percent of families with children with no partner present have MPF, as does 44 percent of families with children with a cohabiting partner, compared to only 23 percent of families with married parents (Monte 2019). MPF within the context of remarriage and stepfamilies is covered in Chapter 15.

Many of the urban parents from the Fragile Families Study, particularly those unmarried at the birth, went on to have children with new partners (Guzzo 2014). African American men and women are more likely than those of other groups to have children by more than one partner, as are people who started having children at younger ages (Monte 2019). MPF seems likely to lead to especially complex family systems but weaker ties with extended families. Indeed, MPF is associated with less financial, housing, and childcare support from kin networks and a greater likelihood of living in poverty (Carlson and Furstenberg 2006; Harknett and Knab 2007; Monte 2019). "[H]aving children by different fathers can present daunting challenges for young mothers. Having to negotiate paternal support and involvement with different men is stressful and may result in different levels of involvement for children who live in the same household but do not share the same father" (p. 37).

In a study using data from the National Longitudinal Study of Adolescent Health, Guzzo and Furstenberg (2007) paid particular attention to the policy implications of MPF. The lower the levels of education in men, the more likely they are to have fathered multiple children outside of marriage or committed relationships (Bronte-Tinkew, Horowitz, and Scott 2009). Much of the foregoing discussion of fertility issues, and especially reports of declining birthrates in some sectors, leads us to the topic of preventing unwanted pregnancies.

PREVENTING PREGNANCY

The dramatic decline in fertility over the last fifty years was achieved through "modern" methods of birth control, also called *contraceptives*. Today, there are many more birth control options than even ten or twenty years ago. All come with advantages and disadvantages and women should work with a health care provider to determine which is best for them (for an side-by-side evaluation of alternatives, see https://www.plannedparenthood.org /learn/birth-control). Birth-control methods involve (1) physical barriers (for example, condoms) that prevent male sperm from reaching and thereby fertilizing a female egg, (2) short-acting hormonal substances (for example, "the pill," the "patch") that prevent ovulation, (3) longer-acting forms of birth control (intrauterine devices [IUDs], injectables, implants), and (4) "natural" methods, including "fertility awareness" (not having sex during "fertile" weeks) and withdrawal ("pulling out"). There are also pharmaceutical products ("Plan B," or "the morning after pill") that prevents fertilization and/or prevent a zygote from implanting itself in the uterine wall. Finally, female sterilization or tubal ligation (a woman having her "tubes tied") prevents eggs from being released for fertilization. Men can have a vasectomy, an operation to keep a man's sperm from going into his penis, which unlike female sterilization (which is major surgery) takes just a few minutes in a doctor's office. Both procedures should be considered permanent (although in some cases vasectomy can be surgically reversed). The most effective methods for preventing pregnancy is sterilization, followed by long-acting methods such as the implant and IUD, followed by short-acting methods including the pill, injectables, the Patch, and the Ring. The least effective methods are barrier (condoms) and natural methods (CDC 2019b). The choice of birth control varies by the woman's social class, with lower-income and less educated women using less effective methods because long- and short-acting methods are more expensive and require a visit to the doctor. Women with lower levels of education are also more likely to use birth control "sporadically" as a result of both the financial costs and forgetting to take or administer it as prescribed (Sassler and Miller 2014).

As early as 1832, a book describing birth-control techniques and devices was published in the United States. The diaphragm and the condom were introduced in the late 1800s, but it was not until the contraceptive pill became available in the 1960s that women could be more certain of controlling fertility and did not need male cooperation to do so. Today, the pill is the most common birth-control method, followed by surgical sterilization, particularly for women in their thirties and older who feel they have finished with childbearing (Guttmacher Institute 2015b).

The long-awaited male pill remains . . . long-awaited. However, researchers are running clinical trials of hormonal male contraception. In one, men rub a gel containing synthetic hormones into their upper arms and shoulders once a day. Thus far, the gel has been shown to suppress sperm levels for up to 72 hours (MacMillan 2017). A male contraceptive pill is also being tested by researchers at the Clinical and Translational Science Institute at Los Angeles Biomed Research Institute. "Safe, reversible hormonal male contraception should be available in about 10 years," predicts Dr. Christina Wang, co-investigator of the study (The Endocrine Society 2019).

Only physical barriers help to prevent HIV/AIDS or other sexually transmitted infections (STIs); these are covered in Chapter 4. For the most effective protection against both pregnancy and STIs, experts advise combining a barrier with a hormonal contraceptive method. For personally useful information on contraception and STIs, see one or more of the following websites: (1) Centers for Disease Control and Prevention, (2) Planned Parenthood, and (3) the Guttmacher Institute.

Because the physical and opportunity costs of childbearing tend to be higher for women than for men, family-planning services have always been oriented toward women. However, with the possible exception of surgical sterilization, contraception takes place in a sexual encounter or relational context that affects not only choice of methods but also whether contraception is used at all. The fact that not only birthrates but also abortions have fallen over the past several decades speaks to the increasingly effective use of contraception by all age and racial and categories. One reason for this development may be that family-planning organizations are increasingly reaching out to adolescent young men with contraceptive and health information (Ball and Moore 2008).

ABORTION

Effective use of contraception prevents unwanted pregnancy. When contraception is not used or fails, a woman who does not want to bear a child may choose to have an induced **abortion**—the surgically or pharmaceutically caused expulsion of a fertilized embryo or fetus from

the uterus. Abortion remains an important method of fertility control, as half of patients who obtained an abortion report using a contraceptive method in the month they became pregnant (Guttmacher Institute 2019a). An estimated one in four American women will have an abortion by the age of 45. Mainly due to increased use of long-acting methods of birth control combined with comprehensive sex education in schools, the rate of unplanned pregnancy is down. Correspondingly, the rate of abortions has been falling since 1980 and is at a historic low and is continuing to decline; abortions are down 7 percent since 2014 and declined another 2 percent between 2015 and 2016 (Guttmacher Institute 2019a; Jatlaoui et al. 2019).

There are many misconceptions about abortion, including who is getting them and when, and their safety. According to the abortion surveillance system of the CDC (Jatlaoui et al. 2019), in 2016, two-thirds (66 percent) of abortions occurred at less than 8 weeks gestation, and nearly all were performed within the first trimester (91 percent). Medication or "chemical" abortions (as opposed to surgical) comprise an increasing percentage of all abortions and accounted for 39 percent of all abortions in 2017 compared to 29 percent in 2014 (Guttmacher Institute 2019a).

The vast majority (85 percent) of abortions are obtained by unmarried women, which includes 29 percent of abortions obtained by cohabiting women. Most abortions are to older women. More than half (59 percent) are obtained by women in their twenties, with only 9 percent by teens. NonHispanic white women, as opposed to women of color, account for the greatest number of abortions and 36 percent of all abortions (Guttmacher Institute 2019a; Jatlaoui et al. 2019). On television and in the media in general, women's risk of death as a result of an abortion is greatly exaggerated (Crockett 2015). Abortion, whether medical or surgical is very safe. In 2015, two women were identified as having died as a result of complications from legal induced abortion, and for one additional death, whether the woman died from an induced or spontaneous abortion was unclear (Jatlaoui et al. 2019).

Overall, researchers conclude that "women who have abortions are diverse, and unintended pregnancy leading to abortion is common in all population subgroups" (Jones, Darroch, and Henshaw 2002, p. 232). Still, women in poverty account for a disproportionate share of abortions. About 43 percent of women who have had an abortion have had a previous abortion, occurring mostly to women in poverty who have inconsistent access to effective birth control (Jatlaoui et al. 2019). Reasons for abortion reported in surveys and interviews at abortion sites include the following:

- concern for or responsibility to another individual (55 percent);

- having a child would interfere with the woman's education, work, or ability to care for dependents (74 percent);

- not being able to afford a baby at this time (73 percent);

- not wishing to be a single mother or having relationship problems (48 percent);

- the woman or couple had completed childbearing (38 percent); and

- the woman or couple were not ready to have a child (33 percent) (Boonstra et al. 2006; Finer et al. 2005; Guttmacher Institute 2014a).

"Although women who have abortions and women who have children are often perceived as two distinct groups, in reality they are the same women at different points in their lives" (Guttmacher Institute 2006, p. 9). In 2016, over half (59 percent) of women who had an abortion were mothers (Jatlaoui et al. 2019). Rhetorically asking why this "mother majority" of those getting an abortion is seldom discussed, *Slate* blogger Lauren Sandler posits that in a culture seeking to discredit abortion and those getting abortions, "antiabortionists have successfully depicted women who choose to terminate a pregnancy as sexually indiscriminate. It's much easier than to demonize the mother who is struggling to support the kid she already has" (Sandler 2011).

The Politics of Family Planning, Contraception, and Abortion

As discussed in Chapter 4, there is debate over the content of sex education in schools, despite research that "abstinence only" education (as opposed to comprehensive sex education) results in hundreds of thousands of excess unplanned, mistimed, and/or unwanted pregnancies, abortions, and births in the United States. Ordinarily, however, the most heated conflict concerns abortion.

Access to abortion, both nationally and internationally, increasingly depends on which political party, Democrat or Republican, holds the majority in Congress, holds the most seats on the Supreme Court, and holds the office of the President. This is the case at the state level too. The terms and conditions under which women can get an abortion depends on who holds the most power in the legislative, judicial, and executive branches of government. There is a large gap in attitudes toward abortion between Democrats (who are more favorable) and Republicans (who are less favorable) (Pew Research Center 2019).

This debate is not new. Abortion has existed throughout history and was not legally prohibited in the United States until the mid-nineteenth century. Laws prohibiting abortion stood relatively unchallenged until the

1960s, when an abortion reform movement resulted in the 1973 U.S. Supreme Court's *Roe v. Wade* decision. *Roe v. Wade* legalized induced abortion without question in the first three months, or first trimester, of pregnancy. But abortion remains subject to regulation in the second trimester and may be outlawed by states after fetal viability (the point at which the fetus is able to live outside the womb), which occurs in the third trimester. Only 1 percent of abortions take place after 5 months (Jatlaoui et al. 2019). Pro-choice and pro-life activists—those who favor or oppose legal abortion, respectively—have made abortion a major political issue. In 2016, 46 percent of Americans said that abortion was "one of many important factors" in choosing a political leader (Gallup Poll 2016). Legislation and other public policy responses to abortion have been shaped by this struggle, as has been the availability of abortion services (Pickert 2013). Currently, the federal government does not "fund" or pay for abortions—the Hyde Amendment, in place since 1977, has prevented the use of federal funds, like Medicaid, for abortions, except when a woman might die or if her pregnancy is the result of rape or incest. One in five women of reproductive age is on Medicaid, which provides health insurance to low income Americans. Federal health insurance plans, such as for military personnel and their families, inmates, and undocumented women in custody, place heavy restrictions on abortion (Planned Parenthood 2019).

Although more than half of Americans have favorable attitudes toward Planned Parenthood, describe themselves as pro-choice, and say abortion should be legal in all or most cases, many states have placed serious restrictions on access to abortion (Gallup Poll 2016; Pew Research Center 2019). Through a variety of mechanisms, the number of clinics providing abortion declined from 839 in 2011 to 808 in 2017. Access to abortion largely depends on where you live. Fifty clinics in the South were closed during that time period, with twenty-five closures in Texas alone. Clinic closures are accomplished though TRAP (targeted regulation of abortion providers), such as requiring that a clinic be within close distance to a hospital (Guttmacher Institute 2019c). Many clinics have been shut down in recent years by a plethora of bureaucratic state rules mandating, for instance, the size of recovery, procedure, and janitorial space—requirements that facilities were unable to meet (Levy 2011).

As of 2017, North Dakota had only one abortion clinic in the entire state (Guttmacher Institute 2019b), and 89 percent of U.S. counties did not have abortion providers (Jones, Witwer, and Jerman 2019). Other state restrictions on abortion include twenty-four-hour waiting periods, state-mandated counseling that provides misleading information about risks (that abortion increases the risk of suicide and breast cancer), parental involvement for minors seeking an abortion, and restricting the conditions under which private insurance

will pay for an abortion, among many others (Stotland 2018). State legislators have introduced some extreme bill regarding abortion. In Ohio, a bill was introduced that, in addition to outlawing abortion outright, requires doctors to "reimplant" an ectopic pregnancy (when the fertilized egg implants in the fallopian tube rather than the uterus, which is dangerous to the mother) or face "abortion murder" charges. "I don't believe I'm typing this again but, that's impossible," wrote Ohio obstetrician and gynecologist Dr. David Hackney on Twitter. "We'll all be going to jail," he said (Glenza 2019). Several states, including Ohio, currently have "heartbeat bills," which ban abortion after 6 weeks, before many women know they are pregnant. Other state-level bills include ones that require the burial of aborted fetuses (a measure subsequently blocked in the courts) and others such as one that allows dads to "veto" abortions (Fernandez 2018; Weill 2017). Some states have ruled such restrictions unconstitutional (Deprez 2015). More than half of Americans say they are either "somewhat dissatisfied" or "very dissatisfied" with the nation's policies on abortion (Gallup Poll 2016). Although the U.S. Supreme Court has upheld many state restrictions on abortion, it has not outlawed the procedure. According national polls, the majority of Americans (70 percent) believe that *Roe v. Wade* should remain the law of the land. Public support for abortion is heavily qualified, however, as you can see in Table 8.1.

The Right to Life movement has claimed that abortion is a threat to women's physical and emotional health, including the ability to have children in the future and an increased risk of breast cancer (National Right to Life 2019). Research has found that abortion has no impact on a woman's ability to become pregnant later (Guttmacher Institute 2014a). A recent meta-analysis

TABLE 8.1 Percentage of U.S. Adults in 2019 Saying

ABORTION SHOULD BE...	PERCENTAGE
Legal under any circumstances*	25
Legal only under certain circumstances**	53
Illegal in all circumstances	19
No opinion	2

*Circumstances might involve the woman's physical or mental health or life being endangered, when the pregnancy was caused by rape or incest, or when there is evidence that the baby may be physically impaired.

**The poll questions did not ask about specific circumstances.

Source: Gallup Poll 2019.

of twenty-five studies found no association between induced abortion and breast cancer (Deng, Xu, and Zeng 2018). Restricting women's access to abortion does not promote greater use of contraceptives and unplanned pregnancies (Felkey and Lybecker 2017).

Deciding about an Abortion

For most women and for many of their male partners, abortion is an emotionally charged, often upsetting, experience. Some women report feeling guilty or frightened. Emotional stress is more pronounced for second-trimester than earlier abortions and for women who are uncertain about their decision. Women from religious denominations or ethnic cultures that strongly oppose abortion may have more negative and mixed feelings after an abortion, but outcomes also depend on the woman's emotional well-being before the abortion (Major et al. 2009).

Some women have reported that the decision to abort enhanced their sense of personal empowerment (Kliff 2010). A study of 757 women recruited from abortion facilities in the United States by Upadhyay, Biggs, and Foster (2015) found that women who were denied an abortion (i.e., they were over the term limit) were

less likely to have aspirational one-year plans than those who obtained an abortion. Those who were denied an abortion were more likely to have neutral or negative expectations about their future. The one-year plans of women sought and/or received an abortion most often mentioned educational and employment goals, changes in living situation, residence, relationships, and changes related to emotions and existing children. These included "I hope that I will be back in school," "I just want to be happy," "to have a less stressful [life]," "to have a better job," "to be more financially stable," "I am hoping to be able to support me and my daughter on my own," "I won't live with my parents anymore," "I'll be married," and "As long as I stay away from the person I was with, I'll be 100 percent better." Research has found positive educational, economic, and social outcomes for young women who resolve pregnancies by abortion rather than giving birth (Holmes 1990; Fergusson, Boden, and Horwood 2007). The decision to abort is often extremely difficult to make and act on (Major et al. 2009; Steinberg and Russo 2008), but the emotional distress involved in making the decision and having the abortion does not typically lead to severe or long-lasting emotional problems. A study of 956

KAREN BLEIER/AFP/Getty Images

Alex Wong/Getty Images

In our society, sexuality and reproduction have become increasingly politicized. Nowhere is this more apparent than in the intensely heated pro-life versus pro-choice debate over abortion, one of the most polarizing issues in America today.

U.S. women who sought an abortion conducted by researchers at the University of California, San Francisco, showed similar or lower levels of depression, anxiety, self-esteem, and life satisfaction compared to women denied an abortion (Foster, Steinberg, Roberts, Neuhaus, and Biggs 2015).

The question of abortion can also arise for couples who, through prenatal diagnostic techniques, find out that a fetus has a serious defect. Recent years have seen extraordinary scientific advances in monitoring fetal development, including ultrasound, amniocentesis, and blood-screening techniques used to assess a fetus's risk of abnormality. These tests becoming more widely used as women postpone childbearing to older ages. Common concerns are Down syndrome, cystic fibrosis, and spina bifida, the latter possibly leading to severe mental and physical disability. In some cases, fetal surgery can correct problems discovered by prenatal testing. Because detection of an abnormal fetus gives prospective parents the option to choose abortion, antiabortion groups have objected strenuously to prenatal screening and genetic counseling. Some advocates for Down syndrome children "see expanded testing as a step toward a society where children like theirs would be unwelcome" (Harmon 2007a). Making the decision to end a pregnancy under these conditions can be exceedingly difficult, especially for women with strong religious and moral beliefs that abortion is wrong (Luscombe 2019).

But some parents who would reject abortion under any circumstances have undertaken prenatal screening with the thought that, should testing reveal an abnormality, they would have time to prepare to care for their infant. Prenatal testing and the accompanying abortion decision remain ethically and personally difficult choices. Individuals dealing with pregnancy decisions should seek expert medical and ethnical or spiritual advice regarding methods and their risks. For some, concern about fertility means avoiding unwanted births. Other couples and individuals face a different problem: They want to have a child, but either they cannot conceive or they cannot sustain a full-term pregnancy. We turn now to the issue of involuntary infertility.

INVOLUNTARY INFERTILITY AND REPRODUCTIVE TECHNOLOGY

Involuntary infertility is the condition of wanting to conceive and bear a child but being physically unable to do so. Infertility, defined by doctors as not being about to conceive after twelve months of regular and unprotected intercourse, affects an estimated 13 percent of women and 10 percent of men of childbearing age (Datta et al. 2016).

Although infertility is commonly thought of as a "female problem," male infertility accounts for 50 percent of infertility cases (Vander Borght and Wyns 2018). For example, sperm counts among men are going down, related to factors such as stress, obesity, and exposure to alcohol and drugs (Rodprasert 2019). Infertility has become more as more couples postpone childbearing until their thirties or forties and who are financially able to seek treatment.

Infertility treatments can involve drug therapy to induce ovulation, donor insemination, in vitro fertilization, and related techniques. Roughly 13 percent of women ages 15 to 49 have ever received these services or have sought advice or testing for infertility (National Center for Health Statistics 2019). Louise Brown, the first "test-tube" baby, gave birth to *her* first child in 2006 ("World's 1st" 2007). This look back reminds us of how astonishing the first developments in reproductive technology were. Today, **assisted reproductive technology (ART)** has become an accepted reproductive option. ART, which refers to any procedure in which the egg and/or sperm is manipulated with the intention of pregnancy, is further discussed in the section "Ever-New Biological and Communication Techniques" in Chapter 1. When faced with involuntary infertility, an individual or a couple experiences a loss of control over life plans and may feel helpless, defective, angry, and often guilty. A number of studies have found that infertile couples have higher incidences of anxiety and depression than do fertile couples (Fallahzadeh et al. 2019). These emotions also commonly follow miscarriage (which affects 10 to 15 percent of all pregnancies) the spontaneous *nonvoluntary* loss of a fetus before its ability to survive outside the womb (Mutiso, Murage, and Mukaindo 2018). Although research studies concur that infertility is generally more stressful for women than for men, both are highly affected emotionally by the challenge to taken-for-granted life plans and sense of manhood or womanhood (Himanek 2011). Taylor (2018) interviewed African American couples about their infertility experiences. As one husband shared, "As a guy, the first thing is everyone is like, 'Oh you got low sperm count? You can't get your soldiers to march?' Just the cracks, 'You ain't feeling like a man today?' It's always an attack on your manhood, when you need to talk about fertility (p. 365)." Another wife said, "I would say it was definitely moments where I'm like I'm less than a woman. Like when can we have a kid? How am I going to not have a kid and my husband comes from a very big family?" (p. 364). When successful, couples face decisions about how much to tell colleagues and friends—and, later, their child—about how they became pregnant (Rauscher and Fine 2012). When infertility treatment is not successful, couples are faced with the decision whether or when to quit trying.

Reproductive Technology: Social and Ethical Issues

Reproductive technologies enhance choices and can reward infertile couples with much-desired parenthood. They enable same-sex couples or uncoupled individuals to become biological parents. They allow heterosexual men and women to delay childbearing. But reproductive technologies raise thorny questions as well. For one thing, ART, the fertility-enhancing procedures, raise abortion issues.

ART and Abortion In ART procedures, several eggs typically are fertilized, but only one or sometimes two or three zygotes are allowed to develop. Remaining fertilized eggs may be discarded or used in research. Those who view human life as beginning at conception define this practice as abortion. Therefore, some ART parents carry all zygotes to term—hence the increase in multiple births, even multiple births in large numbers of five or more infants. As one mother said, "I, personally, could not bear to think of 'flushing' them [the fertilized eggs]" (Frith et al. 2011, p. 3333). The average number of embryos transferred varies from clinic to clinic. American Society for Reproductive Medicine (2014) found that the average number of embryos that nearly all clinics transferred to women younger than age 35 ranged from one to three. The American Society for Reproductive Medicine and the Society for Assisted Reproductive Technology discourage the transfer of a large number of embryos because of "the increased likelihood of multiple-fetus pregnancies, which increase the probability of premature births and related health problems" (p. 2).

Participating in a (Christian-initiated) "embryo adoption program," other couples work to place ART embryos with future parents who will use them in their own "family-building" endeavors: "We felt responsible that we had created these lives and were responsible for finding them good homes if we could not be that home" (Frith et al. 2011, p. 3334). Some of the relinquishing parents hoped to keep communication options open for future contact with a child produced from their relinquished embryo. A relinquishing couple might imagine "for our son to have 'brothers or sisters,' with whom I hoped he could possibly have a special relationship later in life" (Frith et al. 2011, p. 3335). In the words of another relinquishing couple, "We don't want to insert ourselves in [the embryo adopters'] lives but after 18 years old, we want them to have our information so they can get answers to questions. It will also let [our twins] have their questions answered" (Frith et al. 2011, p. 3335).

Inequality Issues Going through infertility treatment is costly and not often covered by health insurance and therefore infertility treatments are generally reserved for the upper classes. Said one woman, "There need to be some options for people like us who don't have money sitting in the bank" (Becker 2000, p. 20). There are many options for couples who can afford it. Higher income professionals who want to wait to have children until they've become well-established in their careers can freeze eggs, sperm, and embryos to be used later on. Tech companies such as Facebook and Apple now offer egg freezing as a perk of employment. In 2019, more than 9,000 women froze their eggs. A single egg-freezing procedure costs between $6,500 and $10,200, not including drugs to induce ovulation and storage. Despite the cost, only 10 to 15 percent of women had returned to the clinic to thaw their egg. The technology is so new it's unclear how long frozen eggs remain viable. However according to one survey, one-quarter of the women felt "more relaxed, focused, less desperate and with more time to find the *right* partner" (Richards 2019). It is recommended that women and men undergoing cancer treatments freeze their eggs or sperm beforehand.

Another aspect of inequality issues raised by ART involves the individuals who can afford to buy a woman's eggs, for instance, or hire a surrogate mother to gestate a child for them. Women who sell their eggs or serve as a surrogate are generally less well-off compared to those who buy their services. A surrogate mother may also feel herself to be under the strict surveillance of the woman hiring her—told not to pump her own gas, for instance, because the fumes might damage the fetus (Ali and Kelley 2008). On the other hand, surrogates are also able to exercise power in the relationship. In one case, the would-be parents demanded that their surrogate abort the fetus she was carrying after ultrasounds indicated that the child would be born with serious birth defects. The surrogate refused and delivered the baby (who was since adopted) (Neel 2013).

In the United States, a surrogate birth can cost couples between $90,000 and $150,000, with some amount typically paid in cash to the surrogate. Expenses include medical, legal, and agency costs and fees, as well as maternity clothes for the surrogate or other incidentals (Hinders 2012). To keep costs down and avoid legal repercussions if the surrogacy arrangement does not go as planned, a growing number of American women are choosing to work with Asian Indian surrogates overseas. A surrogate in India costs substantially less, between $12,000 and $25,000 (Sloan 2017). However, this approach is controversial. Many people worry that poor women are being exploited by the surrogacy arrangement. There are also concerns that standards of medical care may not be as high as they are in the United States.

There are no national laws regulating surrogacy, and the Supreme Court recently rejected hearing the case of an Iowa woman who signed an agreement to carry a child for another couple (with a donor egg) and who wanted to keep the child after the birth. The lower court enforced the contract and ruled in favor of the couple

(the husband was the biological parent to the child). Surrogacy is regulated by states, which vary greatly in terms of their breadth and restrictions. All but four states (Michigan, New York, Indiana, and New Jersey) allow surrogacy. Several states (Iowa, California, New Hampshire, Virginia, and Washington) have laws preventing surrogates from going back on their contracts, if they are not biologically related to the baby (Pitt 2018; Sloan 2017). Given the multitude of ways to create a baby, society's definition of "parenthood" is continuing to evolve.

Who Is a Parent? Reproductive technology creates "family" relationships that depart considerably from what is possible through unassisted biology (Hammel 2012; Zernike 2012). Surrogacy, along with embryo transfer, creates the possibility that a child could have three mothers (the genetic mother, the gestational mother, and the "social mother," the child's caretaker), as well as two fathers (genetic and social). There is also the question of who gets the embryos should the couple's union dissolve. Judges most often rule in favor of the party who does not want the embryo developed. But this may be changing. An Arizona law went into effect in 2019 that gives the embryos to the party that wants to "develop the birth" (Cha 2018d). In situations like these, how do courts define the "real" parents?

Other concerns involve sperm and egg donation. For one thing, a paid male donor who sells his sperm to one or more fertility clinics or sperm banks in a geographical region may biologically father tens or even hundreds of offspring, all of them genetic half siblings. Women, too, can produce a number of half siblings (Cha 2018a). What if they should unknowingly meet in adolescence or adulthood and then together have a biologically or genetically compromised child through an incestuous relationship (Mroz 2011)? A second issue involves a donor's financial, social, or emotional responsibility for his offspring. In one reported instance, a male donor, Mr. Russell, provided sperm to a female friend. The resulting child, Griffin, knows his biological father as "Uncle George." But at some point the mother

> intends to tell her son the truth. Mr. Russell worries about that moment. He never wanted to be a parent; he saw the sperm donation as a favor to a friend. He did not attend the birth or Griffin's first birthday party. His four sisters were trying to figure out whether they were aunts. (Kleinfield 2011)

An interesting recent development is children seeking the identities of their biological fathers (and finding them) (CBS News 2016). Today there are many organizations, such as the Donor Sibling Registry, that provide this service. DNA test kits such as Ancestry.com provide information that allows the identification of close genetic relations (Chuck 2018). Many states have laws by which sperm donors, with the exception of the husband

or partner of the mother, have no parental rights, but this barrier between sperm donors and their biological children is gradually being broken. In fact, the American Society for Reproductive Medicine recommends against secrecy: "It's no longer possible to think of sperm donation without thinking of what the child it produces may someday want" (Talbot 2001, p. 88).

What Kind of Child? As technology advances, the potential to create a child with certain traits expands. Embryo screening—a technology for examining fertilized eggs before implantation to choose or eliminate certain ones—is a boon for prospective parents whose family heritage includes disabling genetic conditions. But embryo screening also raises the possibility of sex selection, especially in places such as China and India, but also in the United States, where boys are preferred (Cha 2018b; Newport 2018). Parents can select other traits such as hair and eye color. Those who use sperm donors are already scanning the records to find evidence of traits they would like in their children. Philosophers ponder the implications for parents and children when children can be "made to order."

Commercialization of Reproduction A general concern is that the new techniques commercialize reproduction when performed for profit. Prospective parents and their bodies are treated as products and thereby dehumanized (Rothman 1999). Examples include the selling of eggs or sperm to for-profit fertility clinics and the marketing of sperm or eggs with certain donor characteristics such as intelligence, education, physical attractiveness, and athletic ability.

Unlike many other countries, fertility clinics and ART in the United States are largely unregulated. Reports of fraud, overstatement of positive outcomes, failure to warn about the risk of multiple births, and other professional violations make it important to understand that an individual seeking treatment is, in fact, a consumer and should interview the doctor and investigate the facility (Leigh 2004). Although success rates vary from clinic to clinic, overall less than 20 percent of ART cycles result in a full-term, normal birth weight child (CDC 2019a). We should note that these concerns about reproductive technology are primarily articulated by medical and public health professionals, academics, policy analysts, and ethicists. For the most part, prospective parents are mostly focused on their desire for a child.

Reproductive Technology: Making Personal Choices

Choosing to use reproductive technology depends on one's values and circumstances. Religious beliefs and cultural values may influence decisions. The Catholic Church and Judaism, for instance, take a negative view of

nontraditional means of getting pregnant (Cha 2018c). Furthermore, fertility treatment can be financially, physically, and emotionally draining. The need for frequent physician's visits can interfere with job obligations, and infertility treatment can lead to tensions in a marriage.

On the positive side, the vast majority of children born by means of ART are thoroughly normal, although they have slightly elevated incidences of preterm birth and low birth weight (Goisis, Remes, Martikainen, Klemetti, and Myrsklä 2019). Whether or not a child was conceived naturally or through ART does not appear to negatively affect the mother's psychological health, such as post-partum depression (Amirchaghmaghi, Malekzadeh, Chehrazi, Ezabadi, and Sabeti 2020).

Meanwhile, infertility treatments do not work for all who hope to become parents. Coming to terms with infertility has been likened to the grief process, in which initial denial is followed by anger, depression, and usually ultimate acceptance. Of course, not everyone has the same experience: "When I finally found out that I absolutely could not have children . . . it was a tremendous relief. I could get on with my life" (Bouton 1987, p. 92). Some people gradually choose to define themselves as permanently and comfortably childfree (Koropeckyj-Cox and Pendell 2007). Another way to get on with life yet retain the hope of parenthood is through adoption. Some couples explore their adoption options even as they continue infertility treatments.

ADOPTION

There are about 1.5 million children with at least one adoptive parent in the United States, representing about 7 percent of all children. The total number of adoptions declined slightly between 2007 and 2014, mainly as a result of a decline in intercountry adoption. The number of children adopted from the foster care system has increased, but so has the number of children in foster care (Jones and Placek 2017). More girls than boys are adopted. Women, especially single women, prefer to adopt girls, and girls are more likely to be available for adoption. Due to the One Child Policy (since dissolved) and preference for sons, 95 percent of Chinese babies who are adopted are girls (Kreider 2003; U.S. Census Bureau 2007a).

Most adoptive families are married couples (73 percent), followed by female-headed households (21 percent), male-headed households (7 percent), and heterosexual cohabiting couples (5 percent) (Kreider and Lofquist 2014). Adoption is expensive, and parents who adopt are older, more educated, and more financially secure than other parents (Zill 2017). They also tend to have no other children and have used infertility services (Kreider and Lofquist 2014). More gay couples are adopting children. A recent qualitative study

looked at gay-male couples' motivations for becoming parents. One respondent said, "We see so many kids that . . . haven't gotten a break. . . . If we could make a difference in just one kid's life, you know, wouldn't it be sad if we didn't?" (Goldberg, Downing, and Moyer 2012, p. 165). With the legalization of same-sex marriage, the Supreme Court in 2017 ruled that same-sex couples should be treated no differently than opposite-sex couples, making adoption by same-sex couples legal in all fifty states.

It has been estimated that nearly half of adoptions are "kinship" adoptions, the adoption of a child by a biological relative or a stepparent (Jones and Placek 2017; Stewart 2010). Some children are adopted informally—that is, the children are taken into a parent's home but the adoption is not legally formalized. **Informal adoption** is most common among Alaska Natives, African Americans, and Hispanics (Stewart and Limb 2020).

The Adoption Process

The experience of legal adoption varies widely across the country, partly because it is subject to differing state laws. Adoptions may be public or private. *Public adoptions* take place through licensed agencies. These make up 92 percent of all adoptions (Jones and Placek 2017). *Private adoptions* (also called *independent adoptions*) are arranged between the adoptive parent(s) and the birth mother, usually through an attorney. Legal fees and the birth mother's medical costs are usually paid by the adopting couple.

More and more, adoptions are open—that is, the birth and adoptive parents meet or have some knowledge of each other's identities. Even when an adoption is closed, as adoptions used to be, some states now have laws permitting adoptees access to records at a certain age or under specified conditions. With regard to attitudes about open adoption, an interesting recent study compared heterosexual with same-sex adopting couples. The study found that both couple types favored open adoption but for different reasons. Heterosexuals saw open adoption as virtually their only option because closed adoptions are less and less available. Same-sex couples embraced open adoption because they appreciated "the philosophy of openness whereby they were not encouraged to lie about their sexual orientation in order to adopt" (Goldberg et al. 2011, p. 502).

A concern that has arisen because of some high-profile cases is whether birth parents can claim rights to a biological child after the child has been adopted (Markon 2010). However, of all domestic adoptions in the recent past, less than 1 percent has been contested by biological parents (Stewart 2007). Another concern of prospective adoptive parents has to do with the adjustment of adopted children—are they likely to have more problems than other children? Considerable research suggests that adopted children are at higher

A Closer Look at Diversity

Through the Lens of One Woman, Adopted Transracially in 1962

By Cathy L. Wong, Ph.D.

In history today we are studying about China. I do not know anyone there, but they might know of me. The teacher seemed to think that I should know about "my people." I tell her I am adopted and she mumbles cautiously that it was "probably for the best." She continues, "at least you will fit in if you were to visit the homeland someday." I'm confused, but wonder if she had stated that because I don't "fit" into America? Or was she implying I did not fit in to this community? I'm confused and saddened by what I think I understand. I look around and see no one who looks like me. I find myself staring intensely at the pictures of all those people climbing the Great Wall and wonder if anyone misses me. I come to realize that I cannot speak Chinese—I feel like a foreigner within both worlds. (Cathy L. Wong, personal journal, 1972, age 11)

I was born in Hong Kong in 1962 and, as an infant, adopted by Americans. Despite the love, attention, and caring of two well-meaning adoptive parents, as long as I can recall, I was left with many unanswered questions as a transracial adoptee. In the midst of all of my unanswered questions and the voids in my own narrative, I have constructed meaning, finding a place for myself within the larger context of the society in which I live.

I am Chinese, my mother is of Swiss and Italian descent, and my father is of Greek origin. In my case, like those of many other transracial adoptees, history played an important role. War and economic strife led to the abandonment of

thousands of children in Hong Kong—mostly girls like me—at the end of World War II. This is how I ended up here, in the United States.

In 1972, debates arose in the United States regarding adoptions of children across racial and ethnic lines. Questions were raised such as, "Can white people properly raise children of color?" and "Is this in the best interest of adopted children?" The debate continues today.

At the heart of the debate are white, predominately middle- or upper-middle- class households in first-world countries who are adopting children from second- and third-world countries.

My transracial adoption story is my own and not necessarily reflective of other transracial adoptee experiences. I think that today things are different for many transracial adoptees, because our culture may now be more sensitive to respecting differences.

Critical Thinking

What are some ways that a society's sensitivity to cultural diversity might influence transracial adoptees' comfort or discomfort in their adoptive family? What would be your answer to the question, "Can white people properly raise children of color?"

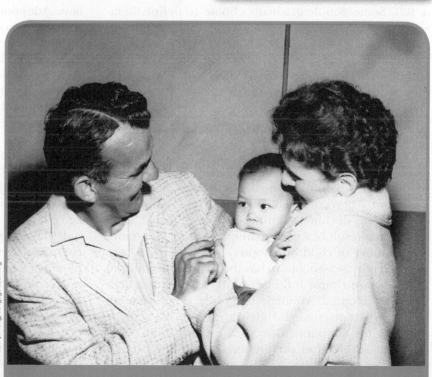

Dr. Kathy Wong/Cengage Learning

1962. Cathy Wong, as an infant, arrives in the United States from Hong Kong after being adopted by her nonHispanic white parents.

risk of problems in school achievement and behavior, psychological well-being, and substance use, especially among boys, children adopted at older ages, and those from foster care (Brown, Waters, and Shelton 2017; Simmel, Barth, and Brooks 2007). Reviews of studies on outcomes of adopted children have concluded that the overall body of research supports "the view that most adoptive families are resilient" and that the majority of adopted children are well-adjusted (Hornfeck et al. 2019; O'Brien and Zamostny 2003, p. 679).

Transracial Adoption

Although adoptive parents are disproportionally white (77 percent), more parents are adopting children of a different race than in the past (Zill 2017). The percentage of adopted kindergarteners being raised by a mother of a different race or ethnic group rose 50 percent between 1999 and 2011. NonHispanic white children made up 39 percent of all adopted children in kindergarten in 2011, followed by Hispanics, Asians, multi-racial, black, and Native American (at 23 percent, 17 percent, 11 percent, 9 percent, and 1 percent, respectively) (Jones and Placek 2017). Family diversity created by transracial adoption seems in tune with the increasing diversity of American society. Yet it has been controversial, as "A Closer Look at Diversity: Through the Lens of One Woman, Adopted Transracially in 1962" suggests.

In 1971, agencies placed more than one-third of their black infants with white parents (Nazario 1990). At that time, the number of black adoptive homes was much smaller than the number of available children, whereas the reverse was true for whites. But interracial adoptions were temporarily curtailed in the 1970s when the National Association of Black Social Workers strongly objected. Suggesting that transracial adoption amounted to cultural genocide, minority advocates expressed concern about identity problems and the loss of children from the black community ("Preserving Families of African Ancestry" 2003). Native American activists have often—but not always—successfully asserted tribal rights and collective interest in Indian children (Liptak 2012; Rayman 2013). The Indian Child Welfare Act of 1978 requires that "adoptive placement be made with (1) members of the child's extended family, (2) other members of the same tribe, or (3) other Indian families so as "to protect the rights of the Indian child as an Indian and the rights of the Indian community and tribe in retaining its children in its society." In practice, outcomes of contested adoption cases have depended on the parents' attachment to the reservation and other circumstances (Liptak 2012; Rayman 2013).

In one situation, a Native American father, on finding out that his (non–Native American) girlfriend was pregnant, renounced his parental rights in a text message to her. The girlfriend placed her baby for adoption with non–Native American parents. Three years later, the father wanted his child back under the 1978 Indian Child Welfare Act. The case went to the U.S. Supreme Court, which ruled that—under law—an Indian court could revisit the case, possibly removing the child from its adoptive home. However, the tribal court ruled that the child could remain with her adoptive parents, saying, "It would have been cruel to take [her] from the only mother [she] knew" (Liptak 2012; and see Rayman 2013 for a recent similar case).

Long-term studies suggest that transracial adoption has proven successful for most parents and children, including with regard to racial issues. Sociologist Rita Simon and social work professor Howard Altstein followed interracial adoptees from their infancy in 1972 to adulthood. They concluded that, as adolescents and later, transracially adopted children "clearly were aware of and comfortable with their racial identity" (p. 222).

Another longitudinal study of transracial (white parents and African American, Asian, and Latino children) and in-race adoptions (white parents, white children) followed the children from the mid-1970s to 1993, when they were in their early twenties. There were no differences in general adjustment or problem behavior between the two groups. What adjustment difficulties that did exist among the transracially adopted children tended to be connected to racial issues—discrimination and "differentness" of appearance. It is not surprising then that researchers found that neighborhood made a difference within the transracial adoptee group; those who were reared in mixed-race neighborhoods were more confident in their racial identity (Feigelman 2000).

Some researchers have suggested that, rather than causing serious problems, transracial adoptions may produce individuals with heightened skills at bridging cultures. "The message of our findings is that transracial adoption should not be excluded as a permanent placement when no appropriate permanent in-racial placement is available" (Simon 1990).

Adoption of Older Children and Children with Disabilities

Children with disabilities are 15 percent of all adopted children, which is a higher percentage than that of biological children (Kreider and Lofquist 2014). Special needs adoptions are pursued by couples who are infertile but who have altruistic motives. Gay men have adopted infants with HIV/AIDS, for example (Morrow 1992).

Older children have a lower chance of being adopted. The majority of adoptions of older children and children with disabilities work out well. Disruption and dissolution rates rise with the child's age at adoption. Among adoptions generally, only about 2 percent of agency adoptions end up being *disrupted adoptions* (the child is returned to the agency before the adoption is legally final) or *dissolved adoptions* (the child is returned after the adoption is final) (Festinger 2005).

What causes these disrupted and dissolved adoptions? For one thing, some children available for adoption may have social and emotional problems or are developmentally impaired due to drug- or alcohol-addicted biological parents, physical abuse from biological or foster parents, or previous broken attachments as they have been moved from one foster home to another. Adoption professionals point out that parents are willing

to adopt children with problems as long as they know what they are getting into (Vandivere, Malm, and Radel 2009). Agencies have increasingly tried to gain information about the circumstances of the pregnancy and the child's early life and to match children's backgrounds with couples who know how to help them (Ward 1997). An unfortunate side effect of the lack of this information is the "rehoming" of adopted children, or "giving away" these children to another family. There is an effort underway to curb this trend through the Protecting Adopted Children Act, sponsored by Representative James Langevin of Rhode Island. He said, "All children deserve a safe, loving home, and yet innocent children are being traded on the Internet with no oversight and little concern for their best interests" (Wetzstein 2015). According to Congressman Langevn's website, in July of 2017, the Protecting Adopted Children Act was presented in conjunction with the Supporting Adopted Children and Families Act in an effort to establish preventative measures that would end the "rehoming" of adopted children. These measures include addressing the causes of "rehoming," such as: "[a] lack of preparation and support services to ensure families never reach a point where they feel they can no longer care for - and thus must give up custody - of their adopted child" (Morente 2017).

International Adoptions

International adoptions grew dramatically through 2003 but have since slowed. The number of children outside the United States available for adoption has declined. China's One Child Policy has ended and Russia no longer allows U.S. parents to adopt (Zill 2017). Roughly one in five adopted children are foreign-born; about half of all children adopted from overseas by American parents were from Asia as well as India, Guatemala, and Russia (Kreider and Lofquist 2014). Parents who have adopted internationally have encountered challenges: the expense of travel to a foreign country—and getting time off from work to go to the child's country for an extended stay; difficulty with negotiations and paperwork in a foreign language and the need to rely on translators and brokers; the uncertainty about being able to choose a child as opposed to having one thrust on the parent; the occasional unexpected expansion of adoption fees or expected charitable contributions; the ambivalence and reluctance of a nation to place its children abroad; and the complete failure to bring home a child.

The biological mother's consent is an issue in overseas adoptions because it is more difficult to be sure that the mother has willingly placed her child for adoption rather than being coerced or misled by a baby broker. Romania placed a moratorium on adoptions, fearing corruption of the nation's entire system. Guatemala is revising its process to comply with the Hague Convention on Intercountry Adoption. China has moved to impose new rules on foreign adoptions, including not only establishing an age requirement for adoptive parents (less than fifty) and stable marriage specifications but also ruling out obese prospective parents (Clemetson 2007; "The Hague Convention" 2010; Gross 2007; Yin 2007).

Prospective adoptive U.S. parents who want to adopt internationally face the frustration and emotional pain of getting into the adoption process and even meeting and beginning to bond with a child from overseas, only to have the process aborted by international politics. Situations like this developed with Russian adoptions (Herszenhorn and Kramer 2012). Russia has since barred adoption by U.S. parents (U.S. Citizenship and Immigration Services 2020).

Meanwhile, international adoptions can pose some of the same problems as the adoption of older children. Conditions in homes and institutions overseas may not be ideal beginnings, and children may have mental and physical health problems (Elias 2005; Gross 2006a; Vandivere, Malm, and Radel 2009). But the vast majority of international adoptions are successful (Tanner 2005). A meta-analysis of around one hundred studies found that adopted children are referred to mental health services more often than nonadoptees are, perhaps a function of adjustment concerns and high-income parents more than troubling behavior. "Most international adoptees are well adjusted" (Juffer and van Uzendoorn 2005, p. 2,501). They "are underrepresented in juvenile court and adult mental health placements," according to Dr. Laurie C. Miller, editor of the *Handbook of International Adoptive Medicine* (Miller 2005b, p. 2,533; see also Miller 2005a).

Those who adopt internationally say they made this choice for several reasons. They are more apt to be able to adopt a healthy infant, face a shorter wait, and encounter fewer limits in terms of age or marital status. The adoption is perceived to be less risky in that there is little likelihood of a birth mother seeking to reclaim the child (Clemetson 2006; Zuang 2004). To what degree racial preferences enter into the choice of international adoption is difficult to determine—that is, whether wanting a white Eastern European rather than an African American child, for instance, motivates overseas adoptions.

This chapter has reviewed topics and issues concerning decisions about parenthood. We'll note here that studies are paying more attention to men's attitudes and feelings about becoming parents, couples delaying childbearing and becoming parents at older ages, infertility and use of ARTs and surrogacy, the use of medical interventions in childbirth, LGBTQ+ couples and single men and women pursuing parenthood, and couples' decisions to remain "childfree." Chapter 9 explores the rewards and challenges of life once children arrive.

Summary

- Birthrates have declined for married women, and many women are waiting longer to have their first child. Having a child outside of marriage has become increasingly acceptable. Although nonmarital birthrates have risen in recent decades, teen birthrates have declined.

- Today, individuals have more choices than ever about whether and when to have children and how many to have.

- Although parenthood has become a choice, the majority of Americans continue to value parenthood. Only a small percentage expects to be childless by choice.

- Nevertheless, it is likely that changing values concerning parenthood, a wider range of alternatives for women, the desire to postpone childbearing, and the availability of modern contraceptives and legal abortion will result in a higher proportion of

Americans remaining childless or having only one child in the future.

- Some observers believe that societal support for children is so lacking in the United States that it amounts to *structural antinatalism*. They point to the absence of workplace inflexibility, lack of affordable quality day care, and the absence of paid maternal or paternal leave as are provided in Europe and elsewhere.

- Children can add a fulfilling and highly rewarding experience to people's lives, but they also impose complications and stresses, both financial and emotional.

- Deciding about parenthood today can include consideration of postponing parenthood, having a one-child family, engaging in nonmarital births, having new biological children in stepfamilies, adopting, and taking advantage of infertility treatment.

Questions for Review and Reflection

1. What are some reasons that there aren't as many large families now as there used to be?

2. Discuss the advantages and disadvantages of having children. Which do you think are the strongest reasons for having children? Which do you think are the strongest reasons for *not* having children?

3. How would you react to becoming the parent of twins? Triplets? More? If your choice is to take fertility treatments that pose a risk of multiple births

or to not have children at all, what would you do—and why?

4. Which reproductive technology would you be willing to use? In what circumstances?

5. **Policy Question**. How is a pronatalist bias shown in our society? Are there antinatalist pressures? What policies might be developed to support parents? Are there any special policy needs of nonparents?

Key Terms

abortion   206
assisted reproductive technology (ART)   210
caesarean section   194
fertility   190
informal adoption   213
involuntary infertility   210
multipartnered fertility (MPF)   205
opportunity costs   192
pronatalist bias   193
replacement level fertility   191
single mothers by choice   203
structural antinatalism   195
total fertility rate (TFR)   190
value of children perspective   196
voluntary childlessness   198

9
RAISING CHILDREN IN A DIVERSE SOCIETY

Learning Objectives

1 Describe how the traditional roles of mother and father are influenced by gendered expectations.

2 Describe the ways that single motherhood differs from married motherhood.

3 Discuss the diversity among fathers.

4 Distinguish between *authoritative* parenting and *authoritarian* parenting.

5 Describe two differing positions regarding spanking.

6 Identify parenting issues unique to race/ethnic, religious, and sexual-identity minorities in the United States.

7 Contrast the concerted cultivation parenting model with the accomplishment of natural growth parenting model.

8 Discuss nonbiological parent families.

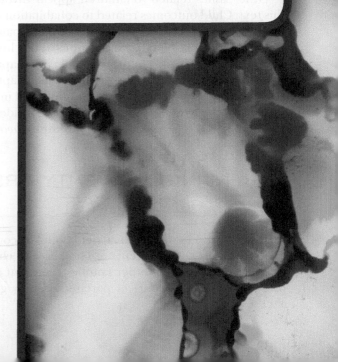

▲ Ariel Skelley/Getty Images

"Becoming a parent is one of the most sudden and dramatic changes in adult life" (Senior 2014, p. 1). Raising children can be joyful and fulfilling, but parenting takes place in a social context that makes it difficult. For most of human history, adults raised children simply by living with them and modeling adult roles. Children shared the everyday world of adults, working beside and dressing like them. At least in Europe, the concept of childhood as different from adulthood did not emerge until about the seventeenth century (Ariès 1962). Today, we regard children as needing special training, guidance, and care.

In this chapter, we will discuss the parenting process in the United States. As you study, we encourage you to think about how parenting is influenced by ways that gender, racial and ethnicity, social class, and sexual identity intersect within family structure. Within the single-parent family, for example, parenting is a different experience for a low-income father of color than for a middle-class, nonHispanic white mother who may have deliberately chosen single parenthood. Among married couples, the parental experience of a middle-class heterosexual couple differs from that of gay men and lesbians.

We'll begin by looking at some general characteristics of the parenting process. Next, we'll examine how gender affects parenting. We will then describe parenting styles, noting that the authoritative parenting style is advised by child development experts. We'll address ways that parenting differs according to racial and ethnicity, religious diversity, and sexual identity. We'll also describe grandparents who serve as parents.

Other issues related to children appear throughout this text. Child outcomes related to cohabitation are addressed in Chapter 6. Combining work and parenting is explored in Chapter 10. Suggestions about how best to communicate with children appear in Chapter 11. Violence against children is discussed in Chapter 12. Divorce and children's outcomes are addressed in Chapter 14. Issues unique to stepparents are considered in Chapter 15. Here we address the parenting *process* in a variety of social circumstances.

PARENTING IN TWENTY-FIRST CENTURY AMERICA

As shown in Figure 9.1, married couples comprise about two-thirds (65 percent) of families with children under age 18. Single-mother families represent almost one quarter (24 percent) of parenting family groups.

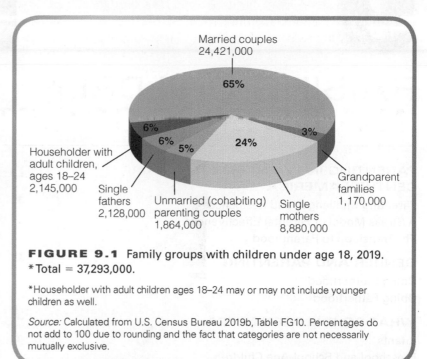

FIGURE 9.1 Family groups with children under age 18, 2019.
*Total = 37,293,000.

*Householder with adult children ages 18–24 may or may not include younger children as well.

Source: Calculated from U.S. Census Bureau 2019b, Table FG10. Percentages do not add to 100 due to rounding and the fact that categories are not necessarily mutually exclusive.

The remaining parental family groups include single fathers, unmarried (cohabiting) couples, and grandparent families. Six percent of parenting family households include young adult children (U.S. Census Bureau 2019b, Table FG10). It is important to note that according to Census Bureau definitions, a single parent can be either cohabiting or not. In fact, Americans may be more creative in fashioning their parenting arrangements than is the U.S. Census. For instance, the census has no count for co-parenting siblings who live together (Gonzalez 2016).

Many parents display a marked fluidity in living arrangements and family structures. Parents move into and out of marriage and cohabitation. A single mother may move from her own apartment to live with her mother and then move again to reside with other relatives or a romantic partner. Then, too, **multipartnered fertility** (a person's having children with more than one partner; see Chapter 8) can mean that a father resides, perhaps temporarily, with one or more of his children but not with others (Lindberg, Kost, and Maddow-Zimet 2017). Hence parenting situations are often fluid or changeable, a situation that social scientists characterize as *family instability*, discussed in Chapter 7.

Regardless of their living arrangements, parents face questions characteristic of our times: Breast or bottle? Should baby boys be circumcised? How much fast food is too much? Is my child bullying others? Being bullied? Should I believe the teacher who says my child needs medication? Does my youngster spend too much time on their iPhone? How can I keep my child safe while online? Should I let my child walk to school? Does my teen

text while driving? What should I tell my child about terrorism? Is my teen depressed? Do I need to worry about vaping? Should I have a parenting app? Do I, as a parent, spend too much time on *my* cell phone?

Parenting Challenges and Resilience

We would not want to point out the difficulties facing parents without noting some positives. In general, parents today have more education and are likely to have had some exposure to formal knowledge about child development and child-raising techniques. Many fathers are more emotionally involved than several decades ago (Fagan et al. 2014; Palm 2014). Despite media attention to street crime, fewer children are exposed to violent crimes today than twenty years ago (Goode 2012). Cell phones and social media make being in touch with other family members far more likely than a generation

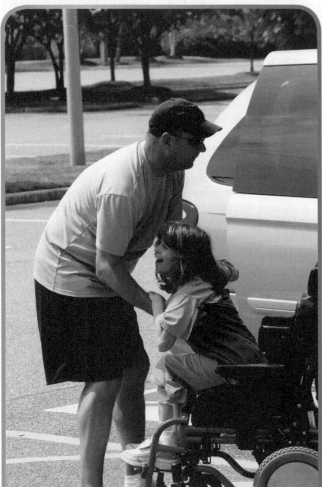

Raising a child with disabilities reminds us that parents and children can evidence resilience, which is enhanced by feeling of empowerment, significant father involvement, and strong familial bonds (Fox et al. 2015).

iStock.com/ktaylorg

ago (Khrais 2012). The Internet offers information for parents dealing with just about any situation.

Nevertheless, parents can make mistakes—even doing their best! Sometimes they blame themselves when outcomes for their children are not what they would have wanted (Moses 2010). It helps, though, to know that children can be remarkably **resilient**: Children and adults can demonstrate the capacity to recover from adverse situations and events (Criss et al. 2015; Masten and Monn 2015). There is evidence that one caring, conscientious adult can generate a resilient child (Johnson 2000; Soukhanov 1996). We return to the topic of resiliency in Chapter 13. Meanwhile, the family ecology perspective (see Chapter 2) leads us to look at ways that society challenges parents today. Here we list seven societal features that make parenting difficult:

1. In our society, parenting often conflicts with employment, and employers typically place work demands first (Sullivan 2015). A majority of parents worry about juggling work and family demands and wish they had more time with their children (Pew Research Center 2015d).

2. Today's parents raise their children in a society characterized by conflicting values and points of view. Think of expecting parents who sponsor elaborate gender-reveal parties. Now think of parents raising "theybies"—children purposefully reared without gender designation (Compton 2018). Moreover, schools, peers, television, movies, music, and the Internet influence youngsters. You may know parents who have responded to this situation by homeschooling their children. Concerns about outside influences may be greater for immigrant parents whose values differ from some that they encounter in the United States (Killoren et al. 2011).

3. Although lots of parenting advice is useful, the emphasis on how much parents influence their children's health, weight, eating habits, intellectual abilities, behaviors, and self-esteem can be overwhelming (Valencia 2015). Furthermore, experts' advice can be contradictory. Some experts advise against children's sharing a parent's bed, for instance, while other experts find bed sharing a positive experience (Stewart 2017).

4. Today's parents are often sandwiched between simultaneously caring for children and elderly parents. Although caregiving can increase life satisfaction, lack of sleep adds to building stress as parents juggle employment, housework, and childcare, as well as caring for elderly parents (Livingston 2018).

5. Over the past fifty years, parenting has become one lifestyle choice among many, and society-wide support has diminished for the child-raising role, once taken for granted as central (Livingston and

Horowitz 2018). As a declining proportion of Americans are raising children, some communities show less willingness to support publicly funded child-centered facilities and programs for children (Wilcox, Marquardt et al. 2011).

6. Today's parents are given full responsibility for successfully raising "good" children, but their authority is often questioned. The state may intervene in parental decisions about schooling, discipline and punishment, vaccinations and other medical care, children's safety as automobile passengers, and even whether a child can be trusted to walk home from school, play in a nearby park, or walk the family dog alone (Hanson 2018; Wergin 2015). Immigrant parents from cultures that espouse parenting practices that are illegal or discouraged in the United States face difficulties.

7. Experts have characterized twenty-first-century parenting as "relentless." For one thing, there seems to be constant need to monitor and "navigate [children's] screen time device distractions" (Jiang 2018). And even a child's weight can be seen as reflecting a parent's morality and sense of responsibility (Elliott and Bowen 2018). Today's parents, especially mothers, are expected to do intensive parenting, putting in more time and effort than twenty years ago, and being judged harshly personally and in social media (Geiger, Livingston, and Bialik 2019). Due to these factors, among others that you may add, it's no wonder that parents are more stressed than nonparents (Sullivan 2015).

A Stress Model of Parental Effectiveness

As explored in Chapter 7, considerable research shows that growing up with married parents is statistically related to better child outcomes than having unmarried parents, the latter statistically associated with low income and family instability. According to the **stress model of parental effectiveness** (see Figure 9.2), parental stress—from job demands, problems with childcare, financial worries, concerns about neighborhood safety, feeling stigmatized due to living in a negatively stereotyped family form, or race, or ethnic discrimination—causes parental frustration, anger, and depression, increasing the likelihood of household conflict. Parental depression, aggression toward each other, and general household conflict, in turn, lead to poorer parenting practices—inconsistent discipline, limited parental warmth or involvement, and lower levels of parent-child trust and communication (Nomaguchi, Johnson, Minter, and Aldrich 2017). As a result of these poorer parenting practices, research has found poorer child outcomes. (Simons, Wickrama, Lee, Landers-Potts, Cutrona, and Conger 2016).

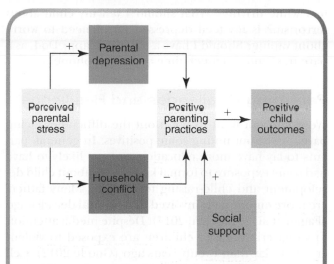

FIGURE 9.2 Stress model of effective parenting. In this figure, plus signs (+) depict positive relationships between variables, and minus signs (−) depict negative ones. Greater use of positive parenting practices results in more positive outcomes for children. However, higher stress levels result in more (+) parental depression and more (+) household conflict. Parents might feel stressed due to job or educational demands, financial difficulties, or concerns about neighborhood safety, or they may feel stigmatized as a result of racial/ethnic discrimination or negative stereotyping associated with nonmarital family forms. Increased parental depression and household conflict result in diminished (−) use of positive parenting practices. Meanwhile, higher levels of perceived social support—due to high levels of family cohesion, private safety nets, or policies and programs that support parents—are positively related (+) to effective parenting practices, hence to positive child outcomes.

Source: This figure was designed by Agnes Riedmann and derived from research findings from the following: Baxter 1989; Benner and Kim 2010; Broman, Li, and Reckase 2008; Bronte-Tinkew, Horowitz, and Scott 2009; Brush 2008; Burrell and Roosa 2009; Cordova, Ciofu, and Cervantes 2014; Gassman-Pines 2011; Goodman et al. 2011; Goosby 2007; Jackson, Choi, and Bentler 2009; Joshi and Bogen 2007; Landers-Potts et al. 2015; Lee et al. 2009; Mackler et al. 2015; Riley, Scaramella, and McGoron 2014.

Here are some questions from a parental stress scale:

- The major source of stress in my life is my child(ren).
- I feel overwhelmed by the responsibility of being a parent.
- Having children leaves little time and flexibility in my life.
- Having children has been a financial burden.

Respondents answer on a scale of 1 through 5 with 1 being strongly disagree and 5 being strongly agree (Zelman and Ferro 2018).

Having social support from relatives, friends, Internet sites, parenting classes and groups, or social activities that include both parents and children diminishes this adverse relationship (Freeman and Dodson 2014).

The Transition to Parenthood

Forty-five years ago in a classic analysis, social scientist Alice Rossi (1968) wrote that the **transition to parenthood** is difficult for several reasons. With little experience, many first-time parents, heterosexual and same-sex, abruptly assume twenty-four-hour duty, caring for a fragile, dependent baby who interrupts parents' sleep, work, and leisure time (Wright and Deaver 2017). Relationship satisfaction may decline with the transition to parenthood. This situation largely results from having less time together alone as a couple (Goldberg and Sayer 2006).

Then, too, for same-sex couples, raising children in a *heteronormative society* that defines heterosexuality as *the* correct way to be can cause new-parent stress. In "a society fixated on labels," lesbian couples may be asked, "Who is the real mom?" (Chabot and Ames 2004, p. 354). Adoptive gay fathers out in public report strangers asking where mom is the moment their baby cries (Vinjamuri 2015).

Whether same-sex or not, new parents may be geographically distant from their own parents and other relatives who might give advice and help. For example, a Mexican immigrant mother told an interviewer:

> In Mexico, when you have a baby, well your mother is always there, or your family is there. . . . They help you . . . like . . . to pick him up, to hold him, to change him. All of that. So, you come here, and you find yourself alone and with a little baby that you don't even know how to pick up. (in Ornelas et al. 2009, p. 1,568)

More disconnected from others than before the baby's arrival, nonimmigrant new mothers also report feeling isolated (Paris and Dubus 2005). Employed mothers of infants, especially those who have jobs with difficult hours and little opportunity for advancement, are more likely to feel stressed (Marshall and Tracy 2009). Fathers are likely to feel stress, too; more than their own fathers, new married fathers spend considerable time—but not as much as mothers—caring for the new baby (Kotila, Schoppe-Sullivan, and Dush 2013). It helps to know that babies differ even at birth; the fact that a baby cries a lot does not necessarily mean that she or he is receiving the wrong kind of care (Rankin 2005). Some are "easy," responding positively to new foods, people, and situations, and transmitting consistent cues (such as tired cry or hungry cry). Other infants are more "difficult." They have irregular sleeping or eating habits, adapt slowly to new situations, and may cry for extended periods for no apparent reason (Thomas, Chess, and Birch 1968; Komsi et al. 2006).

Meanwhile, women who are more pleased about their pregnancy are less likely later to view parenting as burdensome (East, Chien, and Barber 2012; Trillingsgaard, Baucom, and Heyman 2014). In fact, although many couples find the transition to parenthood to be difficult or upsetting to their relationship, not all committed couples do (Don and Mickelson 2014; Holmes, Sasaki, and Hazen 2013). Nevertheless, becoming a parent typically involves what one researcher has called the *paradox of parenting.* New parents feel overwhelmed, but the motivation to overcome their stress and do their best proceeds from the stressor itself—the child as a source of love, joy, and satisfaction (Coles 2009).

For the majority of couples, the transition to parenthood means less time spent relaxing together and declines in their emotional and sexual relationship (Clayton and Perry-Jenkins 2008). Couples who have established fairly egalitarian relationships may find their roles becoming more traditional, particularly if one partner quits or reduces employment (Yavorsky, Dush, and Schoppe-Sullivan 2015). Moreover, employed parents face the challenge of finding affordable, quality childcare, a topic explored more fully in Chapter 10. Couples who had been more focused on the romantic quality of their relationship may find the transition more difficult (Doss et al. 2009).

Transition to parenthood can be difficult for many reasons, including upset schedules and lack of sleep. It's a paradox that (1) new parents feel overwhelmed while (2) inspiration to overcome their stress and do their best is provided by the stressor itself—the child as a source of profound delight.

Among parents who rated their relationship high in quality before becoming parents, the transition was easier, even with an unusually fussy baby (Schoppe-Sullivan et al. 2007). Coworkers' support helps employed mothers (Perry-Jenkins et al. 2011). Partly because it is a source of community support, religious attendance reduces declines in marital satisfaction after a baby's birth (Wilcox et al. 2011b; Wright and Deaver 1017). When a new mother's expectations are met concerning how much support she will receive or how much her partner will be involved with the baby, the transition to parenthood is easier (Cohen, Pentel, Boeding, and Baucom 2019; DeMaris and Mahoney 2017; Gillis, Gabriel, Galdiolo, and Roskam 2019).

GENDER AND PARENTING

Historically, fathers have been expected to be breadwinners and not necessarily competent in day-to-day childcare. Today, however, our culture prescribes that "good" fathers not only assume considerable (often primary) financial responsibility but also actively participate in the child's care. Indeed, research indicates that mothers and many fathers are becoming more similar in the amount of time they spend and in the kinds of things they do with their children (Knop and Brewster 2016; Livingston and Parker 2019). Meanwhile, according to tradition, mothers still assume primary responsibility for raising children. Just 17 percent of stay-at-home parents are fathers. Moreover, compared to one fourth of stay-at-home fathers, three fourths of mothers do so to care for their family. Of fathers, 40 percent are at home because they are ill or disabled (Livingston 2018a,b). Employed or not, a mother is expected to be the child's primary **psychological parent**, assuming—with self-sacrifice whenever necessary—major emotional responsibility for her children's upbringing (Lockman 2019; Miller 2018).

An obvious exception involves gay male parents, who "must cope with the fact that they will be challenging societal notions regarding the absence of a woman as the primary caregiver" (Berkowitz and Marsiglio 2007, p. 367; Carroll 2018). "Gay and bisexual fathers and transgender and gender-fluid parents challenge assumptions about gender, sexualities, and families and highlight the need for theories that move beyond gender categories while retraining insights provided by feminist theory" (Doucet and Lee 2014, p. 363). How do cultural expectations regarding mother- and fatherhood correspond with the daily experiences of mothers and fathers? Despite change, studies continue to show that motherhood differs from fatherhood (Palkovitz, Trask, and Adamsons 2014). How does a parent's gender identity influence how they "do" parenthood?

Doing Motherhood

Although heterosexual parents' gender expectations and behaviors are becoming more alike, mothers still do more hands-on parenting than fathers, and they take primary responsibility for their children's upbringing (Pew Research Center 2015d; Miller 2018). A study of employed, coupled heterosexual parents found that mothers often define "quality time" differently from fathers. Fathers are more likely to see quality time with their children as being home and available if needed. Mothers more often see quality time as having heart-to-heart talks with their children or engaging in child-centered activities (Snyder 2007).

Some women quit successful careers to accommodate their mothering role. For women of color who have made this choice, an organization called Mocha Moms (www.mochamoms.org) provides social support (Crowley and Curenton 2011). As more and more moms have entered the labor force, we have redefined men's roles to include a larger part in the day-to-day care of their families (Pew Research Center 2013b, 2015d). When mothers see fathers as competent parents—and when fathers believe that their child's mother has confidence in them—fathers are more likely to be highly involved (Fagan and Barnett 2003). Though stressful for virtually all mothers, mothering as a single parent is generally even more so.

Single Mothers About 40 percent of all births occur to unmarried women, about half of these to cohabiting mothers (Martin, Hamilton, Osterman, and Driscoll. 2019)). We can therefore conclude that between 15 and 20 percent of all births occur to uncoupled, single mothers. As you can see from Figure 9.1, the intersection of gender with family form is evident in the fact that single women, dramatically more than single men, assume responsibility for child raising (U.S. Census Bureau 2019b, Table FG10). Especially when there is only one adult in the household, interaction in single-mother families can be highly intense emotionally as parent and child(ren) work to negotiate their respective roles—parent or friend? Child or listener/problem solver? (Nixon, Greene, and Hogan 2015).

The category *single mother* is diverse by racial and ethnicity, immigration experience, education, and socioeconomic class. Some women have purposefully decided to raise a child as a single parent. However, the vast majority of single mothers never intended to raise their children without a partner. Some women, becoming mothers while married or cohabiting, believed that their relationship would last but have since divorced or separated. Others realized that their relationship was not permanent and the pregnancy may have been unplanned, but they chose to bear the child anyway: "My heart felt ready for the baby. I knew it was something

I could do with or without a spouse" (in Holland 2009, p. 173; and see Carter 2009).

As a category, single mothers' median family incomes are considerably lower than those for either married mothers or single fathers, and single mothers are more likely to live in poverty, as shown in Table 9.1. Some single mothers may have wanted to marry but did not find a man whom they considered acceptable (Holland 2009). Some single mothers keep the fathers of their children at a distance due to poor relationships with them, safety concerns for their children, or apprehension about the father's illegal activities, or because they see him as generally unreliable (England and Edin 2007).

In a society that strongly advocates a two-parent model, single mothers report feeling stigmatized (LaRossa 2009; Mollborn 2009). But single mothers evidence creativity and resilience as they construct support networks to help with finances, housing, childcare, and other needs (King, Mitchell, and Hawkins 2010). For instance, a service called CoAbode (www.coabode.com) facilitates house and friendship sharing for single mothers. In addition, single mothers maintain Web-based support groups that offer practical assistance, such as providing business clothes for those in need. Single mothers may rely on brothers, brothers-in-law, grandfathers, uncles, or male cousins to serve as father figures for their children (Richardson 2009).

A **private safety net**, or social support from family and friends, is associated with children's better adjustment (Ryan, Kalil, and Leininger 2009). Ironically, those most in need of financial or other practical assistance are often in neighborhood, family, or friendship networks that are least able to help because their own circumstances preclude doing so (Gordon and Cui 2014; Harknett and Hartnett 2011; Pilkauskas, Campbell, and Wimer 2017). Moreover, social support from extended family is not always without cost. For example, a single mother told of her father's offer to pay for her family's medical insurance but only on the (unspoken) condition that she listen while he regularly criticized her (Brush 2008, p. 128).

To improve life for themselves and their children, many single mothers choose further education. This decision is not without added stress, however. Given work and parenting obligations, finding time to attend assigned off-campus activities or meetings to plan group projects poses problems (in Duquaine-Watson 2007, p. 234).

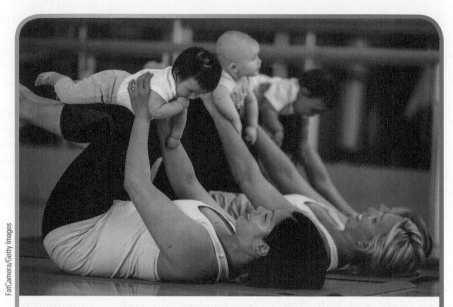

FatCamera/Getty Images

Whether a mother is partnered or not, finding time for herself along with work and household and childcare responsibilities can be daunting. Exercise classes that include baby are one community response to mothers' time pressures.

In accordance with the stress model of parental effectiveness, the average result of single mothers' time constraints, generally poorer economic resources, and resultant higher depression levels—not family structure per se—is less effective parenting behaviors (Riley, Scaramella, and McGoron 2014). Instead of between-group comparisons of children raised by single mothers with those raised in two-parent homes, researchers sometimes make within-group comparisons of children raised in various single-mother families. In these latter cases, variations in income, education, stress, and depression levels largely explain child-outcome differences (Crawford and Novak 2008). It is not surprising that stress is less pronounced among single mothers with relatively higher education, fewer children, better jobs, and more personal resources (Riley, Scaramella, and McGoron 2014).

Doing Fatherhood

In general, research shows that—provided the relationship is positive—a father's involvement in his child's upbringing is related to positive cognitive, emotional, and behavioral outcomes from infancy into adolescence (e.g., Lee and Schoppe-Sullivan 2017; Mallette et al. 2015; Yoder, Brisson, and Lopez 2016). Especially for boys, father absence has generally been associated with adverse effects on children's cognitive, moral, and social development (Leonhardt 2014; Mitchell, Booth, and King 2009). We note that "father absence" can be

other than simply residential absence and can include psychological absence, or indifference with minimal positive father-child interaction. Of course, residential fathers have more opportunities to develop psychological presence than do nonresident fathers (Cabrera et al. 2014; Krampe 2009).

However, incidences of a father's substance abuse and of father-perpetrated partner violence and child abuse remind us that encouraging father contact is not *always* best for children (Osborne and Berger 2009; Salisbury, Henning, and Holdford 2009). Furthermore, **social fathers** (nonbiological fathers in the role of father, such as stepfathers) do not seem to improve adolescents' outcomes when compared with living in a single-mother household (Bzostek 2008). Indeed, a mother's male relatives—for example, the child's uncle or grandfather—may be more effective parent figures than a mother's romantic partner (Jayakody and Kalil 2002). Nevertheless, research does show that, compared to growing up in a single-mother home, younger children benefit economically from living with a social father, provided he shares his financial resources with the family (Manning and Brown 2006).

Married fathers are increasingly involved and invested in their children's daily lives from birth as they engage in breadwinning, planning, sharing activities, and teaching their children (Kotila, Schoppe-Sullivan, and Dush 2013; Livingston and Parker 2019). Nevertheless, more than half—about 60 percent—of fathers say they don't spend enough time with their children, mostly due to employment demands (Livingston 2018a). Furthermore, children's interactions with fathers often differ from those with their mothers; fathers more typically play with or engage in leisure activities with their children than do mothers (Cancian and Oliker 2000). Although there's a tendency to stereotype low-income fathers of color as unmarried and absent, interviews with young African American and Hispanic fathers in New York City uncovered married fathers who were actively involved with their children. Better educated fathers with more satisfying jobs and men who wanted the pregnancy show higher levels of parental engagement. Experiencing high unemployment, money problems, or workplace pressure adds to fathers' stress, resulting in less effective parenting (Wilkinson et al. 2009).

Married-Couple Families with a Stay-at-Home Father The number of stay-at-home fathers has grown to about 2.5 million over the past decade or so (Livingston 2014). Fathers point out the relative lack of status associated with full-time parenting, at least in some circles, as well as in the media, where fathers caring for babies are often the butt of jokes (Beaubien 2018; Poniewozik 2012). One stay-at-home father, laid off from a Fortune 500 company, told a *New York Times* reporter: "To go from the 24-7, high-end, deal-making prestige of working for places that are written about in newspapers to this, it took a long time to get comfortable. . . . It's humbling" (Kershaw 2009a).

Single Fathers Compared to single mothers, the proportion of single fathers is dramatically small. Among families with children under age 19, about 2.4 percent are single-father families—4 percent for blacks, 3 percent for Hispanics, 2 percent for nonHispanic whites, and 1 percent for Asians (U.S. Census Bureau 2019b Table FG10).

Single fathers often assumed their role when they "stepped up" in difficult and unforeseen circumstances (Coles 2009). Single fathers often know that extended family support is available but may not rely on it: "I call my sister occasionally for advice, but I have a strong autonomous streak in me. I'd rather do it myself" (in Coles 2009, p. 1,329).

Whether single or married, poor or financially better off, fathers as primary parents report fighting stereotypes that regard them as odd, unmasculine, or weak (Troilo and Coleman 2008). Watched and evaluated as parents, they feel that they have to prove themselves capable (Coles 2009). In response, single fathers have

Linda Coan O'Kresik/The New York Times/Redux

A parent's deployment overseas means considerable changes in the family system. Responsibility for household chores may change; the family may move to be with extended kin; the nondeployed parent may experience higher stress and anxiety; or children may develop new behavior problems (Schlomer et al. 2012). Available online, "Flat Daddies" and "Flat Mommies," cardboard cutouts of a parent deployed overseas, serve as symbolic placeholders in the families left behind. Flat Daddies and Flat Mommies can go to school, sports events, and holiday dinner tables. They remind family members of their ongoing connection during deployment (www.flatdaddies.com).

organized various support groups (e.g., the National At-Home Dad Network).

Nonresident Fathers Nonresident fathers are biological or, less often, adoptive fathers who do not live with one or more of their children. Less true for divorced fathers, never-married nonresidential fathers move in and out of their children's lives, a situation generally viewed as problematic by researchers who find that family instability is related to problems for children (Fomby and Osborne 2017). Due to multipartnered fatherhood, a father may be living with one or more of his biological children but be "nonresident" with regard to others. Then, too, a nonresident father may be serving as a social father to one or more children whom he did not conceive, usually because he lives with a woman who had at least one child from a previous relationship (Rackin and Gibson-Davis 2012).

Research finds that "cooperative coparenting does not occur in most nonresident father families" (McGene and King 2012). In some cases, cooperative parenting does exist with nonresident fathers, but the father's active involvement is challenging (Carlson, VanOrman, and Turner 2017). Meanwhile, although often stereotyped as "absent" and disinterested, many nonresident fathers express love and genuine concern for their children (Karre and Mounts 2012). The majority of nonresident fathers maintain some presence in their children's lives and provide them with various kinds of practical support, such as diapers, formula, or childcare (Amato, Meyers, and Every 2009; England and Edin 2007). Researchers have found that whether a nonresident father is involved depends on his employment status, age, education, religious participation, and substance abuse history as well as family background (Goldscheider et al. 2009). Some research shows that a nonresident father is more involved when his child is male (Bronte-Tinkew and Horowitz 2010). A nonresident father's involvement also depends on a conflict-free relationship with his child's mother and, to a lesser extent, her extended family (Fagan and Lee 2011; Gonzalez and Barnett 2014).

Nevertheless, given limited resources, involvement of nonresident fathers typically declines over their children's lifetimes, especially for daughters (Bronte-Tinkew and Horowitz 2010; Mitchell, Booth, and King 2009). Many researchers encourage policy support—programs to bolster nonresident father employment and self-esteem as well as encourage their parenting involvement—for these single-parent families, especially in communities where father absence is commonplace (Cowan et al. 2014; Porter 2014). The next section explores what children need, regardless of the family form in which they reside.

WHAT DO CHILDREN NEED?

Children of all ages need parental acceptance, encouragement, adequate nutrition and shelter, parental interest in their schooling, and consistency in rules and expectations. Parental guidance should be congruent with the child's age or development level (Li and Meier 2017; Mental Health America 2009). Children's needs differ according to age.

Infants

Infants need to bond with a consistent and dependable caregiver. To develop emotionally and intellectually, they need affectionate, intimate relationships as well as conversation and variety in their environment. Discipline is never appropriate for babies because they cannot understand its purpose and are unable to change their behavior in response (Brazelton and Greenspan 2000; Wang et al. 2015).

Preschool and School-Age Children

Preschool children need opportunities to practice motor development as well as wide exposure to language, especially when people talk directly to them. They also need consistent, clear definitions of what behavior is unacceptable (U.S. Centers for Disease Control and Prevention 2015b). School-age children need to practice accomplishing goals appropriate to their abilities and to learn how to get along with others. To better accept criticism as they get older, they need realistic feedback regarding task performance—neither exaggerated praise nor aggressive criticism. They also need to feel that they are contributing family members by being assigned tasks and taught how to do them (Hall 2008; Jayson 2005).

Teenagers

Unfortunately social scientists, mental health professionals, teachers, and parents are increasingly concerned about adolescents' rising rates of anxiety, depression, suicides, suicide attempts, and overall unhappiness over the past ten years or more (Child Trends 2018). Generally, experts trace causes to the demise of local communities and of unstructured time or activities (Richtel 2015; Schrobsdorff 2016). We return to this point later in this chapter, when we discuss the "hurried child."

Although the majority of teenagers do not cause familial "storm and stress" (Kantrowitz and Springen 2005), the teen years do have special potential for reducing marital quality and sparking parent–child conflict (Cui and Donnellan 2009). As they begin to define

The children in this family have different needs that correspond with their varied ages. Meanwhile, all children need encouragement along with consistent parental expectations and rules. Authoritative parents are emotionally involved with their children, setting limits while encouraging them to develop and practice their talents. The children in this extended family will no doubt benefit from the love and attention of aunts and uncles.

Karan Kapoor/Riser/Getty Images

who they are and will be as adults, adolescents need to be listened to; they need firm guidance coupled with parental accessibility and emotional support (Guilamo-Ramos et al. 2006; McWey et al. 2015). Teens also need to learn effective methods for resolving conflict (Tucker, McHale, and Crouter 2003).

Despite stereotypical ideas to the contrary, influences from teens' peers are not necessarily negative and can be positive (Hall 2008). Furthermore, parents can and do influence their teenagers' behavior (Longmore et al. 2009). It's important for parents to remember "the obvious fact that most adolescents make it to adulthood relatively unscathed and prepared to accept and assume adult roles" (Furstenberg 2000, p. 903).

EXPERTS ADVISE AUTHORITATIVE PARENTING

Parents gradually establish a *parenting style*—a general manner of relating to and disciplining their children. Parenting styles combine two dimensions—(1) parental warmth and (2) parental expectations—coupled with monitoring of their children. As shown in Table 9.1, we can distinguish among permissive, *authoritarian,*

and *authoritative* parenting styles using these two dimensions (Baumrind 1978; Maccoby and Martin 1983).

The **authoritarian parenting style** is low on emotional warmth and nurturing but high on parental direction and control. The authoritarian parent's attitude is, "I am in charge and set/enforce the rules, no matter what" (Gaertner et al. 2007). Some authoritarian parents evidence a hostile parenting style, criticizing a child's ideas, insulting or swearing at them, shouting or yelling in anger, for example (Wu et al. 2014). Parents who employ this style are more likely to spank their children or use otherwise harsh punishment (Grogan-Kaylor and Otis 2007). Parental negativity together with unnecessarily high parental direction or control has been associated with adolescent delinquency and with children's decreased sense of personal effectiveness or mastery over a situation, even among children as young as 4 (Moorman and Pomerantz 2008; Wu et al. 2014; Zemp, Merrilees, and Bodenmann 2014).

The **permissive parenting style** gives children little parental guidance. Although low on parental direction or control, permissive parenting may be high on emotional nurturing—a situation characterized as *indulgent* that leads to the classic "spoiled child." A second variant of the permissive style is low on *both* parental direction

TABLE 9.1 Parenting Styles

		PARENTAL WARMTH	
		Low	High
PARENTAL MONITORING	High	Authoritarian	Authoritative/ positive
	Low	Permissive-emotional neglect	Permissive-indulgent

© Table designed by Agnes Riedmann

Straight Parents and LGBTQ+ Children

University of California Professor of Gender and Sexuality Studies Brandon Robinson (2018) labels some families *conditional families*. These are families where parents accept their children on the *condition* of the child's *heteronormativity*, or being cisgender and straight. A tragic result for LGBTQ+ youth may be family expulsion and homelessness. Things don't have to be this way, however.

Parenting a lesbian, gay, bisexual, transsexual, queer, or generally genderfluid (LGBTQ+) child can be both broadening and challenging. When an LGBTQ+ child comes out to family members, some of them may feel confused, ambivalent, alone, embarrassed, and/or angry. Even parents who see themselves as progressive may be surprised at their disappointment and grief (Chrisler 2017). If homosexuality is against the parent's religion, a child coming out can be all the more disconcerting (Schwartz 2012). It helps to recognize the following:

- All cultures and historical periods include individuals who have identified themselves as lesbian, gay, bisexual, or transgender.

- The American Psychological Association and the American Pediatric Association do not consider being LGBTQ+ as a psychological disorder.

- LGBTQ+ adolescents and young adults may feel guilty about their sexual orientation, worried about responses from their families and friends, and fearful of discrimination in clubs, sports, college, or the workplace.

- Hiding one's sexual orientation can be extremely stressful and isolating. Adolescents and adult children come out because they want to live their lives openly and honestly without deception (Huckaby 2009).

- The physical and mental health of LGBTQ+ youth is better when they feel social support (Hughes, Heiden-Rootes, Weingard, and Bono 2016).

- Parents' acceptance of their child's sexual identity allows for open discussions about the child's dating relationships and related issues, such as ways to deal with prejudice and

discrimination or how to reduce risks associated with HIV/AIDS.

LGBTQ+ children need assurance that their parents love them as they are. Those who feel parental acceptance evidence better mental health (Olson, Durwood, DeMeules, and McLaughlin 2018; Russell, Pollitt, Li, and Grossan 2018). As parents work through their feelings, they may want to talk with a counselor. Also, Parents, Families, and Friends of Lesbians and Gays (PFLAG) is a national organization comprising local educational and support groups and providing Internet resources.

Critical Thinking

How might the stress model of effective parenting be applied to the situation of an LGBTQ+ child's coming out to his or her parents? How might an authoritarian parent's reaction to a child coming out differ from the response of an authoritative parent?

Source: Parents, Families, and Friends of Lesbians and Gays (PFLAG), 2015 (http://www.pflag.org).

---

and emotional support—a situation of *emotional neglect.* Authoritarian and permissive parenting styles are associated with children's and adolescents' depression and otherwise poor mental health, low school performance, behavior problems, high rates of teen sexuality and pregnancy, and juvenile delinquency (Hall 2008; Waldfogel 2006).

When both warmth and monitoring are high, parents are said to exhibit an **authoritative parenting style** that is sometimes called *positive parenting*. At least for white, middle-class children, research consistently shows that an authoritative parenting style is the most effective of the four possible styles (Neppl et al. 2015; (Schlomer, Fosco, et al. 2015). Moreover, a recent study of Mexican-origin families found that adolescents did better in school when their parents combined messages about hard work and school success with maternal

warmth (Suizzo et al. 2012). This style, characterized as warm, firm, and fair, combines emotional support with parental direction (although not excessive control). Authoritative parents would agree with the statement, "I consider my child's wishes and opinions along with my own when making decisions" (Manisses Communications Group 2000, p. S1).

Authoritative parenting involves encouraging the child's individuality, talents, and emerging independence while also consciously setting limits and clearly communicating and enforcing rules (Brooks and Goldstein 2001; Ginott, Ginott, and Goddard 2003). Limits are best set as house rules and stated objectively in third-person terms: "The time to be home is 10 o'clock." Regardless of family structure, authoritative parents are more likely than others to have children who do better in school and are socially competent, with relatively

high self-esteem and cooperative yet independent personalities (Crawford and Novak 2008; Fivush et al. 2009; Tramonte, Gauthier, and Wilms 2015). Positive effects of authoritative parenting last into adulthood (Schwartz et al. 2009). When two parents are involved, their collaboration or working together renders them more effective, especially when both parents use the authoritative parenting style (Kjobli and Hagen 2009; Simons and Conger 2009). "A Closer Look at Diversity: Straight Parents and LGBTQ+ Children" asks you to consider how an authoritative parenting style would apply to this situation.

Is Spanking Ever Appropriate?

Spanking refers to hitting a child with an open hand without causing physical injury. A little more than half of American parents say they never spank their children— 58 percent of Hispanics, 55 percent of nonHispanic Whites, and 31 percent of Blacks (Pew Research Center 2015). Whether spanking children is ever a good idea is controversial.

According to the late domestic violence researcher Murray Straus (2007, 2010), parents should never hit children of any age under any circumstances. Spanking or vigorously shaking an infant can lead to permanent damage and even death. At least among nonHispanic white American youngsters, being frequently spanked in childhood is linked to later behavior problems, as well as depression, suicide, alcohol or drug abuse, physical aggression against one's parents in adolescence, and later abuse of one's own children and intimate partner violence (Gromoske and Maguire-Jack 2012). Physical punishment can be "harsh" (involving pushing, grabbing, shoving, slapping, or hitting), and disciplining in these ways has been linked to mood and anxiety disorders in adulthood (Afifi et al. 2012).

However, some researchers and pediatricians contend that Straus, although highly regarded, overstated the case (Saadeh, Rizzo, and Roberts 2002). For one thing, we have become an increasingly diverse society of immigrants from cultures that believe in spanking. One study found that some West African immigrant parents hope to send their children back to relatives in West Africa where they can be effectively disciplined (spanked) because they feel their inadequately disciplined children have become unruly "mad cows" in the United States (Bledsoe and Sow 2011). Furthermore, spanking Psychologist Marjorie Gunnoe (cited in Gilbert 1997) theorizes that spanking is most likely to have negative results for children only when they perceive being spanked as an aggressive act (Gunroe, cited in Gilbert 1997). For instance, children of conservative Protestant parents who espouse spanking may not see being spanked as demeaning because their religious environment defines spanking as normal and necessary

(Ellison, Musick, and Holden 2011). Somewhat similarly, spanking is apparently more common and taken-for-granted among Black children, although at least one study found spanking to have negative outcomes for black as well as for white children (Gershoff and Grogan-Kaylor 2016).

As shown in Table 9.2, the American Academy of Pediatrics recently strengthened its position against spanking, saying that spanking is never appropriate and advising parents, "Don't do it, ever." Meanwhile, a separate organization, the American College of Pediatricians, distinguishes between disciplinary spanking and physical abuse (see Table 9.2) and offers guidelines for the former.

SOCIAL CLASS AND PARENTING

We live in an increasingly unequal society—a country where rich families build "gyms, soccer fields and baseball batting cages at home to develop their student-athlete skills" while other families live in their cars. Not to be ignored, such inequality is harmful to children, their parents, and to society in general (McLaughlin 2018).

Nevertheless, there are effective and ineffective parents in all social classes (Jackson, Choi, and Bentler 2009). Meanwhile, this section examines some ways that social class impacts parental alternatives and choices. You'll recall a theme of this text: Decisions are influenced by social conditions that expand or limit one's options. Virtually all opportunities and experiences, or *life chances*, are influenced by **socioeconomic status (SES)**—one's position in society measured by educational achievement, occupation, or income. Parenting is no exception (Lareau 2006).

Research shows that family education and income have more influence on parenting behaviors and children's outcomes than do racial and ethnicity or family structure in and of itself. We have seen that parents who are less stressed and relatively content practice more positive child-raising behaviors. Higher education, along with reduced stress and emotional well-being, in turn, are statistically correlated with higher socioeconomic status.

Middle- and Upper-Middle-Class Parents

Compared with lower-SES parents, middle-class, and especially upper-middle-class, parents can better afford to provide for their children's needs and wants. Higher-SES parents have the resources to hire household help, purchase devices like baby monitors, or hire academic or athletic coaches for their children (Gregory 2017;

TABLE 9.2 The American Academy of Pediatrics Position against Spanking versus the American College of Pediatricians' Distinction between Disciplinary Spanking and Corporal Punishment

| | AMERICAN ACADEMY OF PEDIATRICS | AMERICAN COLLEGE OF PEDIATRICIANS | |
	SPANKING	DISCIPLINARY SPANKING	PHYSICAL ABUSE
The Act	Hitting a child with an open hand without causing physical injury	Spanking: one or two swats to the buttocks of a child. Spanking should never cause physical injury and "leave only transient redness of the skin." *	Physical assault, including to beat, kick, punch, choke, etc.
The Intent	Modify behavior	Training: to modify behavior	Violence: physical force intended to injure, intimidate, get revenge, or abuse
The Attitude	Often, parental frustration	Love and concern	Anger and malice
The Effects	Later emotional and behavior problems: depression, suicide, substance abuse, physical aggression against parents in adolescence, intimate partner violence in adulthood.	Mild to moderate discomfort; behavioral correction	Physical and emotional injury
Is spanking ever appropriate?	No, never. Don't do it, ever.	Yes, if: used for clear, deliberate misbehavior, milder forms of discipline have resulted in noncompliance, not administered on impulse or when out of control—always planned, the child is forewarned and the reasons for the spanking explained.	

Source: Table created by Agnes Riedmann with data from the following: Jenkins 2018 reporting on American Academy of Pediatrics; American College of Pediatricians 2018.

McLaughlin 2018). Furthermore, they reside in neighborhoods conducive to successfully raising and educating their children.

More highly educated parents are older than other parents, on average, have fewer children, show less anxiety regarding their parenting skills, and are likely to emphasize **concerted cultivation** of their child's talents and overall development (Bui and Miller 2018; Kee 2018; Lareau 2003b, 2011). According to this parenting model, they more often praise their children; play with or talk to them "just for fun"; read to them; create and enforce rules about watching television; engage their children in extracurricular lessons, clubs, and sports; take them on outings; enroll them in private or charter schools; and say that there are people in the neighborhood whom they can count on. And—back to our gender-and-parenting discussion—"mothers [have been]

the ones expected to be doing the constant cultivation" (Miller 2018).

More so than in low-income neighborhoods, middle- and upper-class children are likely to have neighborhood and school friends who share their parents' values and can therefore serve as parallel socialization agents (Hall 2008). Also, volunteering at their children's schools and monitoring other students' and even teachers' behavior, highly educated parents secure educational advantages for their children. Should problems arise at school, higher-SES parents are likely to have network contacts with community professionals who can help and may even challenge school officials' decisions (Hassrick and Schneider 2009).

Higher-SES parents are likely to get parenting information from professional sources such as books or the Internet (Radey and Randolph 2009). Often using

the authoritative parenting style, they negotiate with their children in ways meant to foster language and critical thinking skills, self-direction, initiative, and self-advocacy (Lareau 2006; Shinn and O'Brien 2008). This parenting model well prepares children for high academic achievement and success in the broader society because schools and professions value the self-direction, critical thinking, and self-advocacy that these children learn at home (Harding, Morris, and Hughes 2015). But can parents take concerted cultivation too far?

Hyperparenting: Intensive Parenting the "Hurried Child," "Helicopter Parents," and "Snowplow" or "Lawnmower" Parents According to some observers, many parents, particularly higher-SES parents, engage in **hyperparenting**, also called **intensive parenting**. Dubbed "helicopter parents," they hover over and meddle excessively in their children's lives (Warner 2006). "Snowplow" (or "lawnmower") parents plow (or mow) obstacles away for their children—a behavior that sometimes continues into college and perhaps even beyond Quealy and Miller 2019). Unfortunately, these kids don't have opportunity to develop tools to handle life's challenges. A related issue involves parents and schools "hurrying" children with sports training for toddlers and with the increased use of standardized testing and decreased emphasis on music, art, or time for spontaneity and play (Richtel 2015; Tugend 2011).

Parents' and children's life experiences are significantly related to their socioeconomic status. Middle- and upper-middle-class parents tend to emphasize concerted cultivation of their children's development and talents. Ironically, although possibly enhancing parental freedom, baby-monitoring devices can also increase anxiety levels because they encourage defining the infant as extremely fragile (Nelson 2008). Can you think of other parental aids that might have similar effects?

Critics have warned that often higher-income parents engage their children in too many scheduled activities—private lessons, extracurricular activities associated with school or church, and organized recreational programs. Expecting achievement in these endeavors, parents may place too many demands on their children (Zernike 2011).

The overscheduled or "hurried child" (Elkind 2007a,b) is denied free time while encouraged to assume too many challenges and responsibilities too soon (Stout 2011; Tugend 2011). Overscheduled, yet spending thirty hours or more a week on their iPhone, the hurried child is overprotected—not allowed to play unsupervised, for instance, because "a preoccupation with safety has stripped childhood of independence, risk taking, and discovery" (Rosin 2014b). In one author's opinion, children turn to iPhone and social media because it's one area left to them where they can explore the world relatively unsupervised (Buttigieg 2018). In another expert's language, "We have ruined childhood.... For youngsters these days, an hour of free play is like a drop of water in the desert. Of course they're miserable" (Brooks 2019). "The key to happiness might be as simple as [unstructured time in] a library or a park" (Buttigieg 2018). You may have heard opinion makers calling for a return to a less hovering parenting style similar to the 1950s—today dubbed "free-range parenting" (Wergin 2015).

Compared with life in higher-SES families, "childhood looks different" in families of lower socioeconomic status where children have grown up with less supervised play, fewer scheduled activities, and more unplanned interaction with extended family members and friends (Lareau 2003a). Sociologists Elliot Weininger and Annette Lareau (2009) have noted an inconsistency. Higher-SES parents determine to promote self-direction in their children but exercise subtle forms of control that can undermine their intent. Conversely, lower-SES parents tend to espouse obedience but often—at least in nonviolent neighborhoods—grant children considerable autonomy (as they play outside unsupervised, for example), thereby limiting emphasis on conformity.

Working-Class Parents

Working-class parents work at construction, manufacturing, repair, installation, and service jobs such as health care assistants that require at least a high school education and pay higher than minimum wage. More so than with higher-SES parents, working-class parents have suffered the negative effects of declining factory work and union power, decreased wages and benefits, insecurities associated with temporary work, and escalating housing, utilities, and transportation costs. Children in this social class are more likely to watch a lot of television, a

sedentary activity, and to eat fast food—both situations related to obesity and generally relatively poor health (Augustine, Prickett, and Kimbro 2017).

Although economic strains affect daily life, working-class families can be less harried by tight scheduling, and children are able to live at a slower pace. They can relax and have free time to create their own play. They can interact equally with cousins, for instance, and spend time with other relatives instead of being on a constant forced march to do well academically and rack up trophies or ribbons in extracurricular activities. They can be children.

Working-class parents do not necessarily view the concerted development parenting model as good parenting. In fact, they may view this model as negative, creating demanding-children (Guzzo and Lee 2008). Instead, they tend to follow the **accomplishment of natural growth parenting model**, according to which children's abilities are allowed to develop naturally (Lareau 2006). Among other things, this means that working-class children spend more time watching television and playing video games than do children of highly educated parents (Richtel 2012).

Some working-class parents employ the authoritative parenting style. Nevertheless, much parent–child communication tends to be authoritarian, emphasizing obedience and conformity and less often eliciting children's feelings or opinions (Lareau 2003a, 2006). Working-class parents are more likely to tell their children what to do rather than trying to persuade them with reasoning. When dealing with professionals (e.g., doctors, religious clergy, teachers, public officials), working-class parents are likely to encourage their children to keep their thoughts and questions to themselves (Shinn and O'Brien 2008).

Many working-class parents are involved in their children's schools and do promote academic success in their children (Cooper, Crosnoe, et al. 2009). However, the natural growth parenting model, coupled with the authoritarian parenting style, does not correspond well with the middle-class culture and expectations of schools and professions. Although higher-SES children appear to gain a sense of entitlement, working-class children are likely to grow up with feelings of discomfort, constraint, and distrust regarding their school and work experiences (Lucas 2007). For children from working-class families who embark on professional careers, this sense of not fitting in can persist (Lubrano 2003).

Low-Income and Poverty-Level Parents

Few of us think about the parents and their children, stooped in U.S. fields. For low pay they labor to bring affordable fruits and vegetables to our tables (Coates and Fernandez 2019). More generally, at least

12 million (about 17.5 percent of) U.S. children age 18 and younger live below the poverty line (Federal Interagency Forum on Child and Family Statistics 2019). Twelve percent of nonHispanic white, 14 percent of Asian, 35 percent of Hispanic, and 43 percent of black children age 5 and under are poor—living below 125 percent of the poverty level (U.S. Census Bureau 2018b, Table POV01).

The majority of low-income and poverty-level parents work at minimum- or less-than-minimum-wage jobs with irregular and unpredictable hours and no employer-subsidized medical insurance or other benefits. Because more and more low-income jobs are part-time and because working even full-time at minimum wage does not pay enough to live above the poverty level, many low-income parents have two or three jobs.

Irregular work schedules in low-wage jobs with little autonomy and high supervisor surveillance, coupled with housing or neighborhood troubles as well as financial worries, cause stress. Living in rented homes, apartments, or motel rooms, these parents struggle with rent burdens, utility payments, and housing instability (Vanderbilt-Adriance et al. 2015). One single mother who had recently moved in hopes of better living conditions told her interviewer:

> I had to deal with junkies and drugs in the hallways, and there was a shooting outside right before we left. And I was like, I don't care if I have to work two other part-time jobs, if I have to make this work, I will never let my son live like that again. (Freeman 2017, p. 680)

Many poor families move from city to city to live with relatives or to search for jobs. In sharp contrast to higher-SES parents, low-income parents struggle to provide their children with not only clean clothes or lunch money, but with a few "extras," such as a "respectable" birthday party, school field trips, or a high school class ring (Dawson 2018; Lee, Katras, and Bauer 2009; Mistry et al. 2008; Vera 2019). Analysis of data from the National Longitudinal Survey of Youth shows that single mothers who work part-time in low-wage jobs with nonstandard hours generally raise their children in more cluttered homes with fewer books (Lleras 2008).

With fewer resources, lower-income parents are less likely to have high-speed Internet or to live in neighborhoods that value education or encourage high achievement (Anderson and Perrin 2018). In fact, advantages that middle-class Americans take for granted, such as money and facilities for consistently clean clothes or relatively safe, gang-free neighborhoods, are often unavailable (Frech and Kimbro 2011; Zimmerman and Pogarsky 2011), and parental control is more difficult to achieve in neighborhoods characterized by antisocial behavior (Gayles et al. 2009; Moore et al. 2009). Furthermore, poverty-level families are more likely than others to live with air pollution and to have poorer

nutrition, more illnesses such as asthma, schools that are less safe, and limited access to quality medical care (Children's Defense Fund 2017). Children living in poverty—more often disabled or chronically ill than other children—have expensive health care needs that welfare or other social services do not always or completely cover (Cohen and Bloom 2005; Levine 2009). It doesn't help that schools can be insensitive to poor parents' and their children's situations. A Pennsylvania school district—yes, it was in debt—sent a letter home telling parents to pay their back lunch debt or their children would be sent to foster care (Vera 2019).

About 8 percent of children who are raised in poverty (compared with about 5 percent of children raised in families that are not poor) have emotional or behavioral difficulties (Simons et al. 2006, p. 3; Teachman 2008b). Economic hardship in childhood is related to lowered emotional well-being in early adulthood even in a two-parent married family, particularly when it lasts for a long time or occurs in adolescence (Sobolewski and Amato 2005; Vandewater and Lansford 2005). Not having enough money causes stress, which often leads to mothers' depression, parental conflict, and general household turbulence. Household turbulence and mothers' depression, in turn, are associated with lower parent-child (especially teen) relationship quality, a situation that results in a child's lower psychological well-being (Jackson, Choi, and Bentler 2009).

Homeless Families Over the past three decades, extreme poverty, a shortage of affordable housing, job erosion, home foreclosures, declining public assistance, lack of affordable health care, domestic violence, substance addiction, and mental illness have helped to create a significant number of homeless families—a phenomenon that would have been unthinkable forty years ago (Children's Defense Fund 2017).

Families with children are among the fastest-growing segments of the homeless population, a situation that has become more pronounced since the beginning of the recession that began in 2007. Approximately 35 percent of the homeless are in family groups; children under age 18 making up about one-quarter of the homeless. Fathers, some of whom are single parents, are also found among the homeless (National Coalition for the Homeless 2019).

Homeless parents move often and have little in the way of a helpful social network. Getting children to school and supervising their homework are difficult for homeless parents. Although families benefit from entering shelters, life in a homeless shelter is itself stressful. Some shelters require that families leave during the day, regardless of the weather: "How can this mother go out and look for a job or even look for a place to live when she's got three kids, and it's raining, or it's cold?" Problematic rules involve bedtimes, mealtimes, keeping children quiet, and the requirement that children be with their parents at all times. Other stressors occur as well. For example, one mother told of a single male resident "getting fresh with my older girl" (Lindsey 1998, p. 248).

Although patterns of inequality are found along racial and ethnic lines, social class may be more important than race and ethnicity in terms of parental values and interactions with children (Lareau 2003b, 2006; Rich 2014). Middle-class parents of all race/ethnic groups are more alike than different, and so are poverty-level parents. Upper-middle-class black parents perform their role differently than do working-class black parents or those living below poverty level (Peters 2007). At the same time, social scientists do look at how various ethnic groups evidence culturally specific parenting behaviors (Cohen, Tran, and Rhee 2007). The major focus of the following section is on parenting behaviors and challenges that are specific to various racial, ethnic, and religious minorities.

PARENTING AND DIVERSITY: SEXUAL IDENTITY, RACIAL AND ETHNICITY AND RELIGION

As a beginning, we need to note two factors. First, there is considerable overlap among class and race/ethnic categories. For instance, although the upper middle class now includes substantial numbers of people of color, particularly Asians, it is still largely nonHispanic white. Many African American and Hispanic families are now solidly middle class, but these race/ethnic groups remain overrepresented in low-income and poverty categories. A second factor to note is that, as pointed out in Chapter 1, there is considerable ethnic diversity *within* the following groups. For instance, Asian Americans include a broad range of ethnicities, including Chinese, Japanese, Korean, Vietnamese, and Asian Indians, among others. Similarly, the category "black" or "African American" includes not only descendants of African slaves historically brought to the United States but also recent immigrants from Africa and the Caribbean.

Same-Sex Parents

The research on same-sex parents is relatively small compared to that on straight parents, but it's growing (Umberson et al. 2015). Many same-sex parents emphasize their similarity to heterosexual parents: "We go to story time at the library and worry about all the same food groups" (in Bell 2003). Most report that their children have regular (i.e., at least monthly) contact with one or more grandparents, as well as with other adult friends and relatives of both genders,

As with other race/ethnic minorities, African Americans' parental attitudes and behaviors are similar to other parents in their socioeconomic status (SES). Nevertheless, the intersection of gender and race with SES means that this father, culturally expected to be an effective breadwinner, also risks race discrimination as he navigates a job search in a minimum-wage economy. He is pictured here at a New York state employment services office in Brooklyn.

Chris Hondros/Getty Images

school. As one lesbian mother, unhappy with the local public schools and wanting to send her daughter to a private school, said:

> We were very upset about the lack of good choices that we had. [Local religious private schools] teach in the Bible that homosexuality is a sin, so, we went to Montessori because it was most middle of the road and they accept [my partner] and I as partners. (Goldberg, Allen, Black, Frost, and Manley 2018, p. 692)

Children of same-sex parents have formed a support group called COLAGE (Children of Lesbians and Gays Everywhere) and maintain a website, www .COLAGE.org.

"What we really need in this field is for strong skeptics to study gay, stable parents and compare them directly to a similar group of heterosexual, stable parents," concluded New York University sociologist Judith Stacey (in Carey 2012). Now that same-sex couples can marry legally, we might expect more of this research.

We know that being stigmatized for whatever reason is associated with depression. Family cohesion, meanwhile, and close peer and friendships groups lessen this association (Van Gelderen, Gartrell, Box, and Hermanns 2013). Moreover, children of same-sex parents generally view the legalization of same-sex marriage as having given them added security and comfort. Interestingly, a recent qualitative study on adoptive same-sex parents' relationships with their children's heterosexual birth parents found that several birth parents specifically chose a same-sex couple to raise their child. Usually they did so because they knew a lesbian or gay friend or relative or couple and saw them as worthy and potentially admirable parents (Farr, Ravvina, and Grotevant 2018). "Issues for Thought: Heteronormative Bias within the LGBTQ+ Community.

African American Parents

Evidence suggests that African American parents' attitudes, behaviors, and hopes for their children are similar to those of other parents in their social class. Nevertheless, the impact of race remains important. For instance, blacks are far more likely than other race or ethnic categories to see race or ethnicity as central to

sometimes even ex-husbands (Patterson 2000). A lesbian mother who, like her partner, brought a daughter into the relationship from a heterosexual previous marriage, explains: [B]oth of [the girls'] fathers live very close.... So, the fathers were always there visiting and taking care of [the girls], especially my ex-husband when . . . I was back in school (Hequembourg 2007, p. 169). Research generally finds children of lesbian and gay male parents to be well adjusted without significant differences from children of heterosexual parents in school performance, behavior, emotional development, gender identity, or sexual orientation (Amato 2012; Biblarz and Savci 2010; Goldberg 2010). Research undertaken before same-sex couples could legally marry found that same-sex parents, especially those who identify as spouses, were much like married heterosexuals in their parenting practices (Goldberg 2010). Now that same-sex marriage is legal throughout the United States, research comparing parenting practices of married same-sex and straight parents can move forward. The American Academy of Pediatrics (2013) supports same-sex parenthood.

Meanwhile, like children of other minority groups, those in same-sex families may experience stereotyping, prejudice, and discrimination from friends, classmates, or teachers. Same-sex parents are particularly aware of potential stigma when choosing their child's

Issues for Thought

Heteronormative Bias within the LGBTQ+ Community

Two studies suggest that heteronormative, race, and class biases can affect us all, including LGBTQ+ individuals themselves. Let's look at these two studies:

1. "The Experiences of Sexual Minority Mothers with Trans Children," (Kuvalanka, Allen, Munroe, Goldberg, and Weiner 2018) is based on findings from unstructured interviews with eight lesbian mothers of trans children. The researchers wondered (a) how society and how the mothers' own sexual identities influenced their responses to their trans children. In many cases, the mothers said they tried to curb their child's trans messages and behaviors:

   When he was two, about the time he started really talking, he would sometimes identify as a girl, and at that point it was kind of cute and funny. [We said], "no, you're not a girl.... Over the next year he was in preschool . . . . I would start getting reports from either teachers or other children saying, "He's saying he's a girl," , , , We were like, "knock it off, because you're throwing people off balance here," you know? (p. 77).

   Respondents reported trying to control their children's clothing choices When her assigned-at-birth son said they wanted to wear a Halloween princess costume, mom thought, "I just didn't think I could do it. I remember [saying]: "Let's be an animal or let's be something else" (p. 78). Gradually

these mothers accepted their child as trans. As one mother explained, "For me it was just really this up and down, kind of like, 'Well, we can't let her transition, this is a bad idea,' to 'Maybe we have to, maybe this is a good idea'" (p. 78). Another mother described her self-reflection:

   Being a lesbian, I'm already in this community. [But] it's still challenging, and it's embarrassing to me because . . . I never really paid attention to the *T* part of [LGBTQ+]. And, the transgender people that I've known, I feel, have been kind of weird . . . . And so, to me, that's just been: "Oh, that's just because they are transgender." . . . That's been a really hard thing [to process] my internal prejudice towards transgender people. I mean, I can't have a transgender child and hold on to that. (p. 80)

2. Carroll, Megan. 2018. "Gay Fathers on the Margins" (Carroll 2018) is based on participant observation among gay dads in California, Texas, and Utah. Carroll interviewed 56 gay fathers. Among other results, Carroll found that single gay fathers, those of color, and those of lower socioeconomic status can feel marginalized within the gay fatherhood community. Media stories and images of gay fatherhood "reproduce [nuclear] family normativity." As one single dad put it, "All of the publicity that you see is coupled dads. And there's a great deal of us that are single dads" (p. 110).

Interestingly, single gay dads in public with their children were more likely than coupled dads to confront heterosexual assumptions, such as being asked by strangers whether it was mom's day off.

A coupled black father explained:

   We're in the gay community and the gay community itself is segregated. So we're in the Black gays, you know . . . . We're in the Black section with children in the gay community. We don't see our image around anywhere." (p. 111)

Gay fathers also felt marginalized due to social class:

   [My support group includes] a lot of different ethnicities . . . [but] I feel like . . . they all have a home. They all are living the "American Dream" while I'm living in a room and I'm in deb . . . and I feel like I'll never get out. I feel like I'm stuck . . . within the "teen dad" [category], I'm stuck within the "Mexican" [category]. I can't rise above it. (p. 113)

LGBTQ+ individuals share gender minority status. Meanwhile these two studies suggest that lesbian and gay parents can also experience internalized heteronormative bias, along with race and class biases as well.

Critical Thinking

Can you think of examples of your own heteronormative, race, class, or other biases? What are some examples that you find in society?

their identity (Horowitz 2019). It's little surprise that black parents take race and the possibilities of racial discrimination into account when deciding about their child's school options (Williams, Banerjee, Lozada-Smith, Lambouths III, and Rowley 2017).

We saw earlier in this chapter that black parents are more likely than European American or Hispanic American parents to spank their children. This is true

even when social class is taken into account. Among blacks, physical punishment is more acceptable and hence more likely to be viewed by both parent and child as an appropriate display of positive parenting (Jackson-Newsom, Buchanan, and McDonald 2008). Miami University Professor Anthony James (2016) argues that physical punishment by black parents goes back to slavery when children could possibly lose their lives due to

behavior defined as disobedient or disrespectful toward or in the presence of whites. Hence harsh punishment on the part of caring parents.

Even higher-SES African Americans remain vulnerable to discrimination (Lacy 2007; Welborn 2006). A mother in the previously mentioned Mocha Moms support group described a situation in which her young son tried to get on a tire swing at his school and another boy told him that the swing was only for people with light skin (Crowley and Curenton 2011, p. 8). In addition to incidences of direct racism, other things are difficult. So simple a matter as buying toys becomes problematic. Black dolls only? Should the child choose? What if the choice is a white Barbie doll? For middle-class African American parents, forging a unique *black middle-class* identity and then instilling this identity into their children is a major undertaking (Lacy 2007).

Native American Parents

Native American parents have been described as exercising a permissive parenting style that some critics have viewed as bordering on neglectful. However, describing Native American parenting in this way smacks of Eurocentrism (Seideman et al. 1994). Traditionally, Native American culture has emphasized personal autonomy and individual choice for children as well as for adults. Before the arrival of Europeans and for some time thereafter, Native Americans successfully raised their children by using nonverbal teaching examples and "light discipline," possibly coupled with "persuasion, ridicule, or shaming in opposition to corporal punishment or coercion." Native Americans continue to respond with warmth to their children's needs and also to "respect children enough to allow them to work things out in their own manner" (John 1998, p. 400).

Valuing their cultural heritage, many Native Americans have been reluctant to assimilate into the broader society, and this reluctance may mean a rejection of the authoritative parenting style advised by European American psychologists. The use of tribal elders "to help mitigate the loss of parental involvement and early nurturant figures in the lives of Native American adolescents" is important to many tribes (John 1998, p. 404).

Meanwhile, researchers have noted that Native American parents and children demonstrate resilience. For example, a longitudinal study of twenty-nine Navajo Native American mothers who as teenagers bore infants found that, twelve to fifteen years later, many had completed or gone beyond high school (Dalla et al. 2009). Although single-parent households do occur in Native American communities, the extended family serves as an instrument of group solidarity by reinforcing cultural standards and expectations and lending practical assistance. With symbolic and actual leadership status in family communities, Native American grandparents often monitor grandchildren and may fully adopt the parenting role when necessary (Letiecq, Bailey, and Kurtz 2008).

Hispanic Parents

Research on Mexican-origin parents finds that, especially among recent immigrants, parents exhibit "complementary" roles: Fathers are authority figures, and mothers serve as principal caretakers (Updegraff et al. 2012). Hispanic parents have been described as more authoritarian than their nonHispanic counterparts. However, as with other race/ethnic minorities, it may be that this description is Eurocentric and therefore inaccurate. The concept of **hierarchical parenting**, which combines warm emotional support for children with a demand for significant respect for parents, older extended-family members, and other authority figures, may more aptly apply to Hispanic parents. Hierarchical parenting is designed to instill in children a more collective value system rather than the relatively high individualism that European Americans favor (McLoyd et al. 2000).

This collectivism has been found to be functional. For instance, a study that compared Mexican American with European American parents in Southern California found that family cohesion (*familismo*) lessened the relationship between economic stress and negative parenting (Behnke et al. 2008; see also Martyn et al. 2009). Research shows similar positive effects of family cohesion for Vietnamese Americans as well as for families of other ethnicities, including European Americans (Lam 2005; Vandeleur et al. 2009).

Hispanic parents teach their children the traditions and values of their cultures of origin while often coping with a generation gap that includes differential fluency and different attitudes toward speaking Spanish (Knight et al. 2011; Pew Research Center 2009). As in other bicultural families, intergenerational conflicts may extend into many matters of everyday life. For instance, bicultural children are often expected to translate for their monolingual parents, but this role can be stressful as the following quote from a Latino teen focus group participant explains:

> My dad use to take me to, like, cause he didn't speak English, and . . . I was supposed to go with to the lawyer, like, and they would use some big words in English that I don't know like what's that. And my dad is like, "I brought you so you could translate!" And I'm like, yeah but those are some big words. I used to get nervous 'cause I would be like oh my God he's gonna get mad, he's gonna get mad! (Cordova, Ciofu, and Cervantes 2014, p. 698)

Intergenerational conflicts arise as the younger generation becomes more assimilated into U.S. culture. For example, "My mother would give me these silly dresses to wear to school, not jeans," complained a 15-year-old Mexican American female (Suro 1992, p. A-11). And another example from a Latina adolescent in the focus group just mentioned:

Well, I had a friend that she was a person that they gave her a lot of freedom and my mom thought that I should not have that friendship because they gave her a lot of freedom, her mom, and because she is this or the other. And that hurts you a lot because you love that friend. . . . Then that hurts a lot because . . . parents want you to have friends. [But] when you have one, they don't let you, and that is not fair. (Cordova, Ciofu, and Cervantes 2014, p. 700)

As Hispanic immigrant parents adjust to U.S. culture, they are likely to place less emphasis on *familismo* and become increasingly permissive, the consequences of which can be detrimental to adolescents' behavior (Baer and Schmitz 2007; Driscoll, Russell, and Crockett 2008).

Asian American Parents

Despite the fact that, as a category, Asian Americans have the highest average income of any race/ethnic group measured by the U.S. Census, Asian Americans are also found in lower socioeconomic strata and may experience economic hardship (Ishii-Kuntz et al. 2009). Nevertheless, compared to the average of 35 percent for all Americans older than 24, 56 percent of Asian Americans are college graduates or have advanced degrees (U.S. Census Bureau 2018b, Table 3). At 88 percent, nonHispanic Asian-American mothers are more likely to be married than the U.S. average of 60 percent (Driscoll 2019, p. 4).

The Asian American parenting style is often characterized as authoritarian, emphasizing obedience and possibly using physical punishment, but coupled with more praise and hugs than in mainstream American society (Lindner Gunnoe, Hetherington, and Reiss 2006). Research findings are mixed regarding the success of the expert-preferred authoritative parenting style among Asian Americans, again suggesting that the model may be Eurocentric (Pong, Johnston, and Chen 2010).

Social scientists have offered an alternative parenting concept, the **Confucian training doctrine**. This parenting model is named after the sixth-century Chinese social philosopher Confucius, who stressed (among other things) honesty, sacrifice, familial loyalty, and respect for parents and all elders. The Confucian training doctrine blends parental love, concern, involvement, and physical closeness with strict and firm control, or "training" (Chao 1994; McBride-Chang and Chang 1998).

"Training" may involve parents' use of guilt, shame, and moral obligation (defined negatively as "shaming" in mainstream culture) to control their children's behavior (Farver et al. 2007):

By their own lights, Asian Americans sometimes go overboard in stressing hard work. Nearly four in ten (39 percent) say that Asian-American parents from their country of origin subgroup put too much pressure on their children to do well in school. Just 9 percent say the same about all American parents. On the flip-side of the same coin, about six-in-ten Asian Americans say American parents put too little pressure on their children to succeed in school while just 9 percent say the same about Asian American parents. (Pew Research Center 2012c, p. 4)

However, the extremely strong ties between Asian American parents and their children have been found to lessen parent-child conflict, along with potential negative effects of shaming (Benner and Kim 2009; Park, Vo, and Tsong 2009).

Like other ethnic minorities, Asian Americans have suffered from discrimination (Lau, Takeuchi, and Alegria 2006; Tong 2004). Moreover, Asian immigrant parents may face conflicts with their children when expecting traditional behavior characteristic of the homeland while their children assimilate into the American culture and no longer adhere to traditional expectations regarding dating, for example, or marital monogamy (Ahn, Kim, and Yong 2008; Farver et al. 2007).

Parents of Multiracial Children

One in seven (14 percent) of U.S. babies born in 2015 is multiracial. Of mixed-race Americans, about three-quarters are white in combination with at least one other race, mainly Hispanic; about 16 percent of multiracial infants born in 2015 are black in combination with another race (Livingston 2017). As more and more multiracial individuals reach childbearing age and as racial heterogamy loses its taboo, the number of multiracial births is expected to continue to climb (Jones and Bullock 2012).

Raising biracial or multiracial children has unique rewards and challenges (Kennedy and Romo 2013; Rockquemore and Laszloffy 2005). One challenge involves insensitive remarks from strangers: "How come she's so white and you're so dark?" (Saulny 2011b). Another challenge may be tension between parents and children—and between the parents and extended family members—over cultural values and attitudes (Lorenzo-Blanco, Bares, and Delva 2013).

A psychologist surveyed multiracial adults and asked whether they thought their parents had been prepared to raise children of mixed race. The majority did not believe so (Dunnewind 2003; Lorenzo-Blanco, Bares, and Delva 2013). Today, however, many schools are more sensitive to the needs of multiracial children, and more

resources are available for parents raising multiracial children. According to one fairly recent study, multiracial and multi-ethnic families that consciously foster a family identity as multicultural, multiracial, or multi-ethnic have happier, better-adjusted children (Soliz, Thorson, and Rittenour 2009).

Parents in Transnational Families

As pointed out in Chapter 1, more and more American families are *transnational*. Due to the emigration of one or more individuals, family members in different countries maintain relationships across national borders. Typically, family members work to keep their home culture alive; often they nourish dreams of taking their children to visit the home country so they might better understand their native culture (Dominguez and Lubitow 2008).

According to a study published in the *Harvard Educational Review*, "fear and vigilance" characterize the home lives of undocumented parents, with the result that they are less likely to engage with their children's teachers or be active in their communities (Preston 2011). U.S. immigration policy, particularly in states such as Arizona that have passed legislation hostile to undocumented immigrants, has resulted in the deportation of undocumented parents and hence their possible separation from their children, American citizens born in the United States.

Research speaks to "deep and irreversible harm" done to the children whose parent(s) is/are exported: "Having a parent ripped away permanently, without warning, is one of the most devastating and traumatic experiences in human development" (Yoshikawa and Kholoptseva 2013). U.C. Irvine sociologist Laura Enriquez (2015) calls this practice *multigenerational punishment,* "a distinct form of legal violence wherein the sanctions intended for a specific population [that is, unauthorized immigrant parents] spill over to negatively affect individuals who are not targeted by laws [that is, children born in this country and thus legal citizens]" (p. 939). How these children fare, who cares for them, and what will be their outcomes as they grow and mature are questions for future research.

Meanwhile, journalists have featured children who were born and raised in the United States and then accompanied their deported parents to Mexico, where they miss their American lifestyle and friends, face adjusting to a different culture, and often feel like outsiders. "These kinds of changes are really traumatic for kids," according to a Marta Tienda, a Princeton sociologist born to migrant Mexican workers. "It's going to stick with them" (Cave 2012). Among other research questions that transnational families prompt, studies are beginning to emerge on parent–child relations between emigrating parents who have left their home country and their "left-behind children" who did not accompany them (Graham and Jordan 2011).

Religious Minority Parents

Recent studies suggest that regardless of the particular religion, children in families who adhere to a religious belief system tend to be better adjusted. We might think of the religious benefits as *spiritual capital*—coping "resources of faith and values derived from commitment to a religious tradition" (Grace 2002, p. 236). The thinking is that it is less about religious beliefs and more about being a member of a community that is important for family life (Good and Willoughby 2006; Lees and Horwath 2009).

Ethnicity is often associated with religious belief. For example, Chinese Americans are likely to be Buddhists; Asian Indian Americans are likely to be Hindus or Sikhs. In a Christian dominant culture, diverse ethnoreligious affiliations affect parenting for many Americans. For instance, Muslims have their own holy days, such as Ramadan, which may not be taken into account in public schools' scheduling. Wearing flowing robes and, more often, head scarves or veils (called *hijab*), Muslims report that they fear ridicule and face discrimination from employers and others (American Moslem Society 2010).

Those religions such as Islam or Judaism that depart from a Christian tradition have the added burden of raising children in a society that does not support their faith; the same could be said of evangelical Protestant groups as well. American parents of religious minorities generally hope that their children will remain true to their religious heritage amid a majority culture that seldom understands and is sometimes threatening ("Muslim Parents Seek Cooperation" 2005). One solution has been the emergence of religion-based summer camps for children of Baha'i, Buddhist, Catholic, Hindu, Jewish, Mormon, Mennonite, Muslim, and Sikh parents, among others.

Raising Children of Minority Race, Ethnic, Religious, or Gender Identity in a Prejudicial and Discriminatory Society

The felt stigma for LGBTQ+ adolescents and young adults is associated with a suicide rate five times higher than for the heterosexual population (Hughes, Heiden-Rootes, Weingard, and Bono 2016). Furthermore,

> [i]ndividuals who fall into multiple minority statuses (e.g., gay, Latino male) are subject of the societal discrimination and prejudice associated with each minority status and likely in a more complex way given the rejection (or acceptance) that could come for both inside and outside their racial community. . . . Several protective factors seem

At this festival marking Eid, the end of Ramadan, this Muslim community in central Texas gathers for afternoon prayers. Muslim parents hope that their children will remain true to their religious tradition. Meanwhile, like parents of other minority religions in the United States, they must help their children face fear of ridicule and actual discrimination.

to reduce major health risks. Parents and significant caregivers play a crucial role in adolescents' well-being and health. (Cao, Ahou, Fine, Liang, Li, and Mills-Koonce. 2017, pg. 1262)

A parent's or child's experiencing fear or harm due to negative stereotypes and discrimination adds pain and stress to an already demanding parenting process (Alexander 2014; Benner and Kim 2009). An illustration:

My son wants an answer. He is 10 years old, and he wants me to tell him that he doesn't need to worry. He is a black boy. . . . His eyes are wide and holding my gaze, silently begging me to say: No, sweetheart, you have no need to worry. Most officers . . . would not shoot you. . . . I am stammering. (Alexander 2014)

Families of color and or religious minorities attempt to serve as an insulating environment, shielding children from or confronting injustices (Brody et al. 2008; Cooper, Smalls-Glover, Metzger, and Griffin 2015). Muslim parents emphasize to their children that "terrorists are not representative of us" (Ingber 2015). Meanwhile, parents of black offspring worry about their safety on the streets (for example, see Levin 2015). Most race, ethnic, and religious minority parents engage in **race socialization**—developing children's pride in their cultural heritage while preparing them to encounter discrimination (Caughy 2011; Cooper and McLoyd 2011; Hansen 2012). A review of 21 studies on race socialization among black families concluded that messages emphasizing racial pride and culture are more consistently associated with their children's positive mental health outcomes than are messages that focus on preparing for racial bias (Reynolds and Gonzales-Backen 2017).

Valuing one's cultural heritage while simultaneously being required to deny or "rise above" it in order to advance in mainstream society can pose problems between parents and their children. For instance, Native Americans must often choose between living on the reservation or living an urban life that is often alienating but presents greater economic opportunities. (Seccombe 2007). Hispanics may see a threat to deeply cherished values of family and community in the competitive individualism of the mainstream American achievement path (McLoyd et al. 2000). Asian Americans may follow the "model minority" route to success but experience emotional estrangement from their culturally traditional parents (Kibria 2000).

As one response, minority parents may encourage their children to participate successfully in the larger society with regard to occupation and education while maintaining their original cultural values with regard to religion and family norms (Portes and Zhou 1993). The desire of ethnic minority parents to preserve their culture tends to be sharpened when they feel their culture is devalued or that mainstream culture encourages negative behaviors in their children (Kalmijn and van Tubergen 2010). We note that at least one research study documents white progressive fathers' commitment to socializing their own children to "*antir*acist" attitudes and behaviors (Hagerman 2017). We turn now to a discussion of another variation in the parenting experience—grandparents as parents.

GRANDPARENTS AS PARENTS

As shown in Figure 9.1, grandparent families comprise about 3 percent of all family groups with children age 18 or younger. About 3 million or 11 percent of U.S. grandparents are raising grandchildren (Mendoza and Fruhauf 2015). More than 3.6 million children under age 18 are living in a grandparent's household, a few with only their grandfather, many with two grandparents, and many more with only their grandmother (King, Mitchell, and Hawkins 2010). Having risen dramatically over the past ten years, the number of grandparents residing with grandchildren (about 7 million) and the number both living with and responsible for

Facts about Families

Foster Parenting

States monitor parents' behavior toward their children. When officials determine that someone under age 18 is being abused or neglected, they can take temporary or permanent custody of the child and remove him or her from the parental home for placement in **foster care**. As wards of the court, foster children are financially supported by the state.

About 443,000 children are in foster care in the United States (Kids Count 2019). There would be more, but there are not enough foster parents or other facilities to fill current needs. Seventy-four percent of foster care takes place in a licensed foster parent's home—46 percent with nonrelatives and 28 percent with relatives. The remainder of children in foster care live in various arrangements, including group homes (6 percent) or institutional settings (8 percent) such as Nebraska's Boystown (which also accepts girls) (Child Welfare Information Gateway 2019; U.S. Department of Health and Human Services 2012a; "Numbers and Trends" 2015).

The mean age of children in foster care is about eight years. Children stay in foster care for an average of about two years, but 20 percent stay for only one to five months, and nearly 10 percent remain until age 18, when they "age out." Although the rate of children in foster care is higher for African Americans than for other race/ethnic groups, the highest percentage of children in foster care are white (44 percent), followed by 23 percent for blacks, 21 percent for Hispanics, 9 percent for mixed-race children, 2 percent for Native Americans, and 1 percent for Asian children (Child Welfare Information Gateway 2019).

Very often without family support, those who are neither reunited with family members nor adopted "age out" of the system. For the most part, these youth were older when they became foster children and were the developmentally neediest; they face serious challenges as they work toward assuming adult roles. They are more likely than other young adults to become imprisoned, homeless, unemployed, or pregnant outside marriage (Koch 2009; Nunn 2012).

Motivations for becoming a foster parent include fulfilling religious principles, wanting to help fill the community's need for foster homes, enjoying children and hoping to help them, providing a companion for one's only child or for oneself, and earning money. Technically not salaried, foster parents are "reimbursed" in regular monthly stipends by the government.

Foster caregiving can be stressful (Richardson and Futris 2019). Nevertheless, some foster parents see fostering as a step toward adopting (Baum, Crase, and Crase 2001). Although the ultimate goal in half the cases is reunification of foster children with their parents or principal caretakers, about one-quarter of foster children are available for adoption (U.S. Department of Health and Human Services 2012a). The National Foster Parent Association and the Foster Care and Adoptive Community provide online education and support.

Critical Thinking

From the structure–functional perspective discussed in Chapter 2, foster parents are functional alternatives to biological or adoptive parents. What are some ways that the foster parent system is functional? What are some instances in which it could be dysfunctional?

---

raising children (about 3 million) is expected to continue to rise (Mendoza and Fruhauf 2015).

Taken together, unmarried parenthood, divorce or separation, poverty, substance abuse, HIV/AIDS, domestic violence, abandonment, and incarceration account for a large majority of families headed by grandparents—that is, grandparent families, or **grandfamilies** (Reddock, Caldwell, and Antonucci 2015). "The transition to parenting a second time around requires many grandparents to reconsider their work-family roles and make significant, unexpected changes" (Bailey, Letiecq, and Vannatta 2011, p. F18). One study found that grandparents' coping strategies involved relying on their religious faith as well as imagining the situation would somehow "just go away" (Lumpkin 2008).

Becoming a primary parent requires considerable adjustment for grandparents. Living with children in the house is a significant change after years of not doing so (Mendoza and Fruhauf 2015). A grandparent's circle of friends and work life may change. They may retire early, reduce work hours, or try to negotiate more flexible ones. On the other hand, a grandparent may return to work to finance raising the child(ren) while finances suffer while paying for items such as additional beds, food, and clothing (Bailey, Letiecq, and Vannatta 2011).

To help, under the **formal kinship care** system, some states offer financial compensation to grandparents (or other relatives, such as aunts) who raise their grandchildren as state-licensed foster parents. However, some

grandparents report having trouble navigating their state's kinship care system due to their fear and distrust of the child welfare system and daunting bureaucratic regulations, among other reasons (Letiecq, Bailey, and Porterfield 2008). "Facts about Families: Foster Parenting" further discusses foster parenting.

Grandparents raising grandchildren is characterized by ambivalence (Bailey, Letiecq, and Vannatta 2011). Unsure whether or when their grandchildren will return to the parental home, grandparents may "learn a . . . stance of detachment to cope with the shifts they are sure to experience and probably even applaud" (Nelson 2006, p. 822). Furthermore, there are often questions about the possible legal termination of the parent's parental rights (McWey, Henderson, and Alexander 2008). When parental rights are not terminated, grandparents who are responsible for the children in their care lack legal rights over them (LaPierre 2011).

Some grandchildren being raised by a grandparent see one or both parents either regularly or sporadically; but generally these relationships are complicated and often marked with difficulties. In a qualitative study with white, black, and mixed-race children being raised by grandparents, some children hoped for reunification with their parents, while the majority had accepted their situations likely as permanent (Dolbin-MacNab and Keiley 2009; Mendoza and Fruhauf 2015).

A grandparent's living in the home of a poor single mother is advantageous because it can add income—from Social Security benefits, for example. In addition, grandparents tend to provide stability, family cohesiveness, and solidarity while possibly enhancing young children's cognitive development (Spradling 2009). However, not all grandparents raising grandchildren employ effective parenting practices (Barnett 2008). Grandmothers have been found to be most sensitive and beneficial to infants (Dunifon and Kowaleski-Jones 2007).

Research on the feelings of adults who were raised by their grandparents found that some were grateful and felt a strong bond with their grandmothers while others evidenced distance and distrust (Dolbin-MacNab and Keiley 2009). Social service agencies have initiated educational and coping programs for grandfamilies. The National Center on Grandfamilies promotes awareness of grandfamilies and gives advice on how to help grandparents meet their various needs.

PARENTING YOUNG ADULT CHILDREN

We tend to expect that family members love and communicate with one another unconditionally, but research that examines intergenerational family relationships have found that "relationships between parents and children in

adulthood are complex and diverse" (Blake 2017, p. 521). In fact, some young adult children and their parents are estranged, a situation in which parents typically may feel misunderstood and children, generally unsupported. Noting these situations, we'll say that they are the minority.

Throughout their twenties and beyond, individuals benefit from their parents' practical guidance, emotional support, and positive attitude toward whatever challenges they're facing (Neppl et al. 2015). As young adults transition to adult roles, parent-child relations may grow less conflicted (Straus 2009). At the same time, both parents and young adults may be angry or depressed over student loans due to rising college costs, lingering childhood issues, or difficulties with assuming adult roles (Arnett 2004; Sullivan 2019).

Meanwhile, concerns over the young adult's delayed transition to adulthood can cause parental ambivalence and parent–child conflict. Parents who see their grown children as needing too much support report poorer life satisfaction when compared with other parents (Fingerman et al. 2012; Gilligan et al. 2015). Young-adult children who receive aid may themselves be ambivalent about doing so. One study found that, while receiving parental assistance is useful and helpful, recipients evidence higher levels of depression and lower self-esteem, compared to those not receiving help from parents (Johnson 2013).

Parents in virtually all income brackets help out their young-adult children, offering services such as babysitting, housing in their home, or financial help (Fingerman et al. 2015). Families with more children likely give both more services and a higher proportion of their income to young-adult children, although each child probably gets less than children in smaller and higher-income families (Goodsell et al. 2015; Neppl et al. 2015). A significant majority of higher-SES parents lend or give their children money to repay student loans, buy a car, help with rent or credit card debt, or put a down payment on a house (Ray 2012; Wightman, Schoeni, and Robinson 2012). One interesting study found that parents tend to provide money not only to their neediest but also to their most successful children, the latter in anticipation of help from the child as the parent grows older (Fingerman et al. 2009).

Chapter 6 more thoroughly explores young adult children's living with their parents. Here we point out that underemployment, along with a decline in affordable housing, can make launching oneself into independent adulthood especially difficult (Davidson 2014). "It's the financial riddle of the 30-something years. How does anyone, even those with a stable, upwardly mobile job, let alone a family, afford to live in places like New York City, Los Angeles, Boston, Chicago, San Francisco or Washington, D.C.?" (Seligson 2019). As one result, adult children increasingly do not leave the family home or "boomerang," returning to it after college, divorce, or on finding their first jobs unsatisfactory (Parker 2012).

"Young adulthood in America [today]: children are grown, but parenting doesn't stop" (Quealy and Miller 2019). Some counselors advise parents to *expect* children to move back home—but not to micromanage their offspring's career or to sacrifice too much. "I see too many parents, especially mothers, helping out grown children when they should be squirreling away more money for their own retirement," said the president of the nonprofit Women's Institute for a Secure Retirement (Kobliner 2010). Allowing grown children to live with parents while they get on their occupational and separate-housing feet can be good for them (Sandberg-Thoma, Snyder, and Jang 2015). It's important, though, that parent–child expectations be set with both parties explicitly understanding the context of the child's return home: a quick stay? a prolonged stay? a stay with underlying medical issues like addiction or depression (Sullivan 2019)?

Families might negotiate a parent-adult child residence-sharing agreement. The following are likely issues to negotiate:

1. How much money will the adult child contribute to the household?
2. What are the standards for neatness?
3. Who is responsible for cleaning what and when?
4. Who will cook what and when?
5. How will laundry tasks be divided?
6. What about noise levels?
7. When are guests welcome and in what rooms of the house?
8. What are the expectations about informing other family members of one's whereabouts?
9. If the grown child has returned home with one or more children, who is responsible for their care?

Several residence-sharing agreements are available on the Internet; google "residence sharing young adult child." Experts advise professional counseling for thorny or more serious challenges (Sullivan 2019).

Finally, we turn to a discussion of how society might better support good parenting.

TOWARD BETTER PARENTING

The family ecology model, discussed in Chapter 2, suggests that one way to improve parenting would be to build stronger family–school partnerships (Sheridan and Wheeler 2017). In an article titled, "Teachers Are Serving as First Responders to the Opioid Crisis," a sixth-grade West Virginia instructor told a reporter, "My job as a teacher is to be a first responder to poverty" (Klein 2018). We can't expect teachers alone to address poverty.

However, recent community crises have occasioned new and positive examples of school–family partnerships. For example, teachers, and staff have gone on standby to help children in the aftermath of ICE raids (Fowler 2019). As another example, West Side High School in Newark, New Jersey, added a free laundromat and detergent so that poor, perhaps homeless, students can wash their clothes at school, hence avoiding the humiliation and shame associated with being bullied over having dirty clothes; the principal argued that shame often keeps impoverished children from attending school (Dawson 2019).

Furthermore, Studies show that optimal parenting involves the following factors:

- supportive family communication (Leidy et al. 2009; Lindsey et al. 2009);
- involvement in a child's life and school (Cooper, Crosnoe, et al. 2009);
- private safety nets—that is, support from family or friends (Lee et al. 2009; Ryan, Kalil, and Leininger 2009);
- adequate economic resources (Guzzo and Lee 2008);
- workplace policies that facilitate a healthy work–family balance and support parenting in other ways as well (Aber 2007; Bass et al. 2009);
- safe and healthy neighborhoods that encourage positive parenting, school achievement, and reciprocal social support (Byrnes and Miller 2012; Gordon and Cui 2014); and
- society-wide policies that bolster all parents (Marshall and Tracy 2009).

Chapter 11 explores the first factor listed: supportive family communication. Here we note some existing programs designed to improve child raising, and then we discuss further ways to promote better parenting.

Over the past several decades, many national organizations have emerged to help parents (Lam and Kwong 2012). Some programs address parental needs in specific life cycle stages, such as the transition to parenthood (Jones, Feinberg, Hostetler, Roettger, Paul, and Ehrenthal 2018). Other programs serve parents in general. One of these is Thomas Gordon's Parent Effectiveness Training (PET) (Gordon 2000). Another is Systematic Training for Effective Parenting (STEP) (Center for the Improvement of Child Caring n.d.). Both STEP and PET combine instruction on effective communication techniques with emotional support for parents. Some other parent education classes incorporate anger management training (Fetsch, Yang, and Pettit 2008). Some curricula are designed for particular race/ethnic groups (Center for the Improvement of Child Caring n.d.; Kumpfer and Tait 2000; Mandara et al. 2012). Many programs are available in several languages (Cox 2017).

Good parenting involves having adequate economic resources, being involved with the child, using supportive communication, and having support from family or friends, along with workplace and broader social policies that bolster all families.

home, especially during the first year. Research by the National Institute of Child Health and Human Development (NICHD) suggests that intervention programs that might be effective in enhancing parenting are important for development (Vandell et al. 2010, p. 738).

Other childcare scholars agree with the NICHD researchers that subsidies should be available to permit parents to cut back work hours. The stress model of effective parenting suggests that reducing parents' stress would improve parenting. Accordingly, "improving the socioeconomic conditions of parents, particularly among the most vulnerable, might improve parenting outcomes across all relationship types" (Guzzo and Lee 2008, p. 58; Neppl et al. 2015).

Moreover, because free time to engage in leisurely social interaction and activities is crucial to psychological well-being, work and society-wide policies aimed at freeing up time for mothers and fathers would improve parenting. More concerted attention on the part of employers to parental child-raising needs and responsibilities would help, along with their greater recognition that good parenting is essential to a civil society (Mackler et al. 2015).

We've seen that informal social support has been found to mitigate stress and hence to be related to more positive parenting (Lee et al. 2009; Ryan, Kalil, and Leininger 2009). But policy makers also urge greater civic and community activism on the part of parents. Some parent-education programs include instruction on how to become more civically involved or engage in community activism (Doherty, Jacob, and Cutting 2009). "Pediatrics is politics," the late pediatrician Benjamin Spock once said (quoted in Maier 1998). He meant that good parenting involves working for better neighborhoods, communities, and family-centered social policies—and these, in turn, result in better parenting.

Then, too, "higher levels of father involvement and a positive coparenting relationship may keep couples together, which allows children to spend their early years with both biological parents in the household" (McClain 2011, p. 889). Hence, there is research and there are programs intended to study or increase fathers' positive involvement, some programs fairly successful (Bellamy, Thullen, and Hans 2015; Cowan et al. 2009). We need to note, however, that benefits of a father's involvement depend absolutely on the quality of the father's parenting and on his relationship with his child(ren) (Yoder, Brisson, and Lopez 2016). Some programs that target adolescent fathers aim to teach skills that improve coparent relations with the child's mother, whether he continues to be romantically involved with her or not (McHale, Waller, and Pearson 2012).

Childcare researchers consider the policy implications of the research. Belsky (2002) argues for tax or other policies to support full-time parental care in the

Summary

- Parenting can be difficult today for several reasons, one of which is that work and parent roles often conflict.

- The family ecology theoretical perspective reminds us that society-wide conditions influence the

- parent–child relationship, and these factors can place emotional and financial strains on parents.

- The stress model of effective parenting posits that stressors of various sorts lead to parental depression

and household conflict, which in turn result in less-positive parenting practices and ultimately in poorer child outcomes.

- Although more fathers are involved in childcare today, mothers are the primary parent in the vast majority of cases and continue to do the majority of day-to-day childcare.

- Child psychologists prefer the authoritative parenting style, although some scholars describe the authoritarian, permissive, and authoritative parenting style typology as ethnocentric or Eurocentric.

- Family form, gender, socioeconomic class, and racial and ethnicity intersect to result in parenting experiences for individuals.

- Higher-SES parents tend to follow the concerted cultivation parenting model, whereas working-class parents are more likely to adhere to the accomplishment of natural growth model.

- A trend over the past several decades has been for an increasing number of grandparents to serve as primary parents, often as a result of some crisis in the child's immediate family.

- More than 400,500 children are in foster care today, many of them in formal kinship care.

- To have better relationships with their children, parents are encouraged to accept help from others (friends and the community at large as well as professional caregivers), to build and maintain supportive family relationships (the subject matter for Chapter 11), and to engage in community or civic activism.

Questions for Review and Reflection

1. Describe reasons why parenting can be difficult today. Can you think of others besides those presented in this chapter?

2. Compare these three parenting styles: authoritarian, authoritative, and permissive. What are some empirical outcomes of each? Which one is recommended by most experts? Why?

3. How does parenting differ according to social class? Use the family ecology theoretical perspective to explain some of these differences.

4. What unique challenges do African American, Native American, Hispanic, and Asian American parents face today, regardless of their social class? How would *you* prepare an immigrant child or a child of color to face possible discrimination?

5. **Policy Question.** Describe some social policies that could benefit all low-income parents, regardless of their gender, racial and ethnicity, or family structure.

Key Terms

accomplishment of natural growth parenting model 233
authoritarian parenting style 228
authoritative parenting style (positive parenting) 229
concerted cultivation 231
Confucian training doctrine 238
formal kinship care 241
foster care 241
grandfamilies 241
hierarchical parenting 237
hyperparenting 232
intensive parenting 232
multipartnered fertility 220
permissive parenting style 228
private safety net 225
psychological parent 224
race socialization 240
resilient 221
social fathers 226
socioeconomic status (SES) 230
stress model of parental effectiveness 222
transition to parenthood 223

10

WORK AND FAMILY

Learning Objectives

1 Describe how work–family expectations for men and for women in co-resident relationships have changed over the past several decades.

2 Discuss how "greedy work" can affect family life.

3 Discuss alternatives available to two-earner couples.

4 Discuss division-of-labor changes regarding unpaid family work.

5 Explain why women in heterosexual relationships do more unpaid family work than men do.

6 Describe the causal feedback loop regarding work and family living.

7 Discuss family policy regarding the work–family interface.

8 Explain how feelings of fairness regarding work–family roles impact relationship satisfaction.

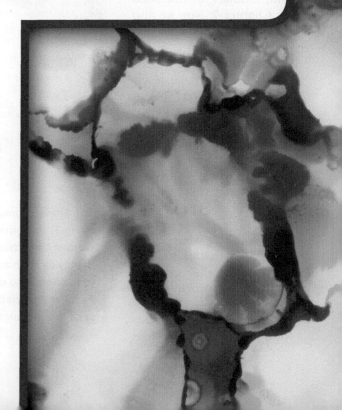

▲ Ariel Skelley/Getty Images

The three-generation Lee family occupies a four-story building in New York City. With his two grown children, the widowed Chinese-immigrant grandfather Gung Gung, 86, bought the building fifteen years ago. Now with Gung Gung, two married couples, and seven children in the household, the family splits the mortgage, food, and repair bills. Gung Gung's son, Warren, makes dough and fillings for the family bakery. Warren also cooks for the family, lights his father's Buddhist altar each morning, and gets him to medical appointments. Warren's wife, Jen, is a gym teacher. Gung Gung's daughter, May, is assistant principal at a nearby school. May's husband, Ben, drives for FedEx. The family rents out the building's lowest floor to a Mexican restaurant. The Lees employ two Cantonese-speaking babysitters (S. Kramer 2011).

Every Monday, financial executive Karen Cangas leaves home for the airport. She'll be back Thursday night. Until then she'll work with clients in various cities. Her husband drops off their two children at day care on his way to work. Evenings, the family visits with Karen on Skype (Cullen 2007).

In different ways, these two American families (like all families) inhabit the intersection, or *interface*, between work and family. In both scenarios, family members work to provide resources; they organize their family lives according to available economic options and opportunities. Until fairly recently in human history, cooperative labor for survival was the dominant purpose of marriages and families, as discussed in Chapter 7. Now we also expect love and emotional support from our families.

In addition to expecting love in marriage, a second development began with the eighteenth-and nineteenth-century Industrial Revolution in Europe and the United States when economic production moved to factories and offices. Wage earners and a **labor force**—people who are employed or who are looking for paid work—emerged. Given these two historical developments—(1) wanting emotional satisfaction from family life and (2) leaving home for employment—combining work and family has become characterized by multiple options and decisions to make.

This chapter examines partners' traditional and changing roles in the workplace and family. We'll look at how couples juggle—and struggle—to balance employment with unpaid family labor. We'll see that, as argued in Chapter 3, aspects of traditional gender roles persist despite social change. The persistent gendered divergence in earnings, as well as the (slow) trend toward earnings convergence that we have seen over the past several decades, have affected the division of household labor.

The family ecology perspective (which suggests that structural forces outside families influence what goes on inside them) operates within a context of gender expectations and structures that influence options and decisions (Garey and Hansen 2011; Perry-Jenkins and Wadsworth 2017). This chapter investigates the relationship, often in conflict, between work and family, then considers what is needed to resolve work–family conflict. To begin, we will explore the often-gendered interface of work and family life—that space where the social institutions of family and the economy meet.

iStock.com/Mie Ahmt

The *family ecology perspective* tells us that workplace requirements impact family living. This mom's military career certainly affects her family life. Four in ten military women have children. Even new mothers that are breast-feeding are required to deploy as soon as four months after a birth. Some mothers return after deployment to toddlers who don't remember them. Spokespersons with the National Military Families Association say that military moms should not be deployed overseas until at least twelve months after a birth (Browder 2010; Mann 2008).

THE INTERFACE OF WORK AND FAMILY LIFE

The concepts *sociological imagination* and the *family ecology perspective*, discussed in Chapters 1 and 2, hold that family life is influenced by cultural expectations and social structures external to it. One such influence is today's economy, recovering from the Great Recession that began in 2007. Nevertheless, today's jobs often pay less than before the Great Recession and are more likely to be part-time and/or temporary without benefits equivalent to before the Great Recession (Casselman, Cohen, and Burke 2018; Kosanovick 2018. Some Americans—about 5 percent—even teachers, work two or three jobs (Foster 2018; Reilly 2018). For example, in today's gig economy, characterized by a growing number of temporary, flexible jobs, independent workers spend evenings and weekends delivering restaurant take-out or driving Uber. With the federal minimum wage well under $10.00/hour, one 40-hour/week job will not cover an individual's housing and other living expenses.

Blue-collar, white-collar, and professional jobs such as those in accounting and publishing have moved overseas (Preston 2015). Student loan debt is historically higher than ever before (Fry 2014). All these situations make for work–family stress. Furthermore, the workplace itself influences everyday family life. Workplace policies, such as the availability of maternity or paternity leave, impact fertility decisions, worker stress and health and parents' labor force participation (Grzywacz and Smith 2016; Sullivan 2015).

Meanwhile, family members can influence workplace policies and conditions. In 2010, the WalMart Corporation paid $86 million in damages for gender discrimination to more than 200,000 California female employees and family members. WalMart had paid them less than it paid its male employees, and the female employees sued for damages (Stempel 2010). More recently, female McDonald's employees walked out to protest sexual harassment on the job (Rushe 2018). These are *macro* or society-wide examples. We can also think in terms of *micro*, or smaller group examples.

On the micro level, family researchers note that a substantial majority of Americans like to be working and mostly like their jobs. Researchers also look at the **spillover** from work situations into family life—how pleasures or stresses associated with work affect interaction within the family (e.g., Hill et al. 2013; Miller and Chang 2015). Research shows significant spillover from employed mothers' perceived workload and from the quality of mothers' interactions with supervisors to interactions with their children. Feeling overloaded or enduring supervisor criticism are positively correlated with harsh and withdrawn mother-child interactions; not feeling overloaded or enjoying supervisor praise were positively associated with warm mother-child interactions (Gassman-Pines 2013). (This research is also indicative of the *stress model of parental effectiveness*, discussed in Chapter 9.)

As another example, a study of heterosexual marrieds found a positive correlation between mothers' working fulltime at a primary job plus a nonstandard shift at a second job and their children's being overweight or obese (Miller and Chang 2015). Can you hypothesize about why this might be?

Not just mothers are stressed. Increasingly, fathers experience work–family conflict too (Coleman 2017). Meanwhile, a study that examined spillover among dual-earner lesbian and gay parents found that having a job in an LGBTQ+ friendly workplace reduced parental anxiety at home, while an unfriendly work atmosphere had the opposite effect (Goldberg and Smith 2013).

Spillover can be studied in the other direction, too. A Singapore study found that workers with distressed marriages were depressed at work and had lower employee motivation and output (Sandberg et al. 2012). Similarly, a study of economically disadvantaged mothers found that an adolescent child's delinquency negatively impacted the mother's work stability and performance (Coley, Ribar, and Votruba-Drzal 2011).

Gender and the Work–Family Interface

A 2012 Census Bureau publication called "Who's Minding the Kids" defines a mother as the "designated parent" but a father as a "childcare arrangement" (Women's Law Project 2012; see also Laughlin and Smith 2015). A 2017 poll found 77 percent of women saying they felt a lot of pressure to be an involved parent, compared with 49 percent of men; meanwhile, 44 percent of women felt pressured to be successful in their job or career, compared with 68 percent of men (Geiger, Livingston, and Bialik 2019).

Men have traditionally been expected to show *instrumental* character traits, such as self-reliance and ambition, while women have been supposed to be *expressive*, relationship-oriented, and supportive helpmates (Parsons and Bales 1955; Sallee 2011). At least in the American middle-class, husbands have been more likely to be in the labor force, an instrumental role. Wives have been more likely to be mother-homemakers, an expressive role. As long as middle-class nonHispanic white women remained out of the labor force, this division of labor was taken for granted by most researchers and policy makers.

However, over time, societal trends have been changing. A 2012 Pew Research survey found for the first time that women ages 18 to 34 topped young men in desiring high-paying careers. Nearly two-thirds (66 percent) of young women said that "being successful in a

high-paying career or profession" was "very important" or "one of the most important things" in their lives. This figure compares to 56 percent in 1997 (Pew Research Center 2012d).

A 1977 survey found fewer than half of respondents (48 percent) saying marriages work better when both spouses have jobs. By 2010, close to two-thirds (62 percent) thought so (Pew Research Center 2010a,).

Despite social change, however, evidence of traditional expectations and behaviors persists. In 2017, when asked what makes a good husband, 71 percent of Americans said a good husband needs to be able to support a family, while only 32 percent said that about a good wife (Stepler 2019). The Gallup poll has been asking about U.S. men's and women's work–family preferences for many years. About 75 percent of men have consistently over the past several decades said they prefer a job outside the home. About 22 percent of men would "prefer to stay at home to take care of the house and family" (Saad 2012a).

Greedy Work, Paid, and Unpaid Americans spend considerably more time at paid work than they did two decades ago (Pew Research Center 2016). Their commutes are longer, too, as housing costs increase, especially in central cities where the better jobs tend to be concentrated (Siddigui 2018). Moreover, high-level careers—professional and managerial jobs—with long, inflexible hours now pay noticeably more to workers with round-the-clock availability, expectedly answering weekend e-mails, for instance (Hardy 2018). "The mere expectation of being in contact 24/7 is enough to increase strain for employees and their families" (Hardy 2018). But such positions create an "overwork premium" for those willing to work fifty or more hours per week (Miller 2019). Could this be partly why 22 percent of millennials, whether married or not, say they have "no friends" with even more (27 percent) saying they have "no close friends" (Resnick 2019)?

In addition to paid work, modern parenting expectations are increasingly relentless, as discussed in Chapter 9 (Miller 2018a). Compared to 10 hours per week in 1965, mothers in 2016 spent 14 hours weekly on child care. Fathers' child care hours increased too—from 2.5 hours per week to 8 (Geiger, Livingston, and Bialik 2019). And it appears that at least some middle-class children, their time filled with lessons and homework, put in less time help parents out with household chores than a generation ago (Anderson 2018).

Role Conflict When social institutions are not well integrated (that is, when they do not work together), individuals who play roles in both institutions experience **role conflict**: meeting the demands of one institution conflicts with meeting the simultaneous but different demands of another. For example, an employed parent needs to be (1) at home to monitor a teenager's after-school behavior *and* (2) at work. Often the career world tends to view someone who takes time from work for family as less than professional, yet family norms encourage parents—most often mothers, but increasingly fathers—to do exactly that. Work–family conflict on the macro, or society-wide, level often results in personal experiences of role conflict. A study of medical students in the United States and Great Britain found that female students anticipated role conflict. They expected family demands to hamper their careers, while males seemed less influenced by family concerns (Riska 2011). In a 2015 Pew survey, twice as many mothers as fathers (41 compared with 20 percent) said that being a working parent had made career advancement harder (Pew Research Center 2015d, Question 34). Not to mention getting enough sleep (Harrison and Oh 2019).

MEN'S WORK AND FAMILY ROLES

Whether single or married, parent or not, men on average are employed more hours than women. Among heterosexual marrieds, husbands are more likely (88 percent) to work fulltime than wives (76 percent) (U.S. Bureau of Labor Statistics 2019 Table 21). Historically and today, the male **provider role**, in which family men are expected to supply resources (shelter and food, for instance), is evident in all social classes. In this country the male provider role emerged during the 1830s. Before then, a man was expected to be "a good steady worker," but "the idea that he was *the* provider would hardly ring true," because in a farm economy husband and wife together produce family income (Bernard 1986, p. 126).

The male provider role (and its counterpart, the female homemaker, or housewife) predominated into the 1970s. Since then the proportion of heterosexual married-couple families in which only the husband is employed has gradually declined—from 42 percent in 1960 to 19 percent in 2018. As Figure 10.1 shows, both spouses are employed in the majority (58 percent) of married couples (U.S. Bureau of Labor Statistics 2019, Table 24). If we consider all couples, not just marrieds, with children under age 18, about two-thirds are dual income (Livingston and Parker 2019).

Meanwhile, heterosexual men continue to be *primary* breadwinners—that is, they are employed fulltime while their partners are employed part-time or not at all—in 43 percent of relationships. As shown in Table 10.1, fathers outearn mothers in most two-parent households, even when the mother is employed fulltime (Pew Research Center 2015d). Furthermore, whether a family

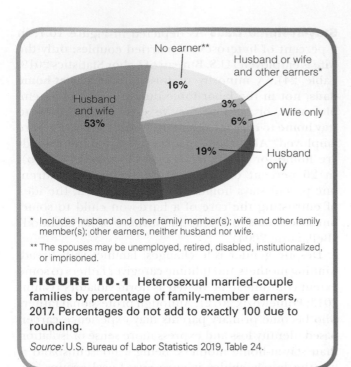

* Includes husband and other family member(s); wife and other family member(s); other earners, neither husband nor wife.

** The spouses may be unemployed, retired, disabled, institutionalized, or imprisoned.

FIGURE 10.1 Heterosexual married-couple families by percentage of family-member earners, 2017. Percentages do not add to exactly 100 due to rounding.

Source: U.S. Bureau of Labor Statistics 2019, Table 24.

relocates is often a career decision, and in a heterosexual marriage "typically the man's is prioritized, with the secondary career subsequently constrained" (Lersch 2016, p. 230). Anthropologist Nicholas Townsend interviewed thirty-nine nonHispanic white, Hispanic, and Asian American men who graduated high school in the 1970s. Regardless of ethnicity, the men described their life goals as "the package deal": marriage, children, home ownership, and a steady job. Work was viewed as essential to being a good father: Although these men (like many men today) desired to spend more time with their children, in reality their time was devoted to paid work (Livingston 2018a). Many had two jobs or put in extensive overtime.

"Good Providers" and "Involved Fathers"

Since about the late 1970s there have been two models for the husband-father role. More traditional fathers—researchers have named them **good providers**—emphasize the provider role as defining their merit, and they work *more* hours than men with no children (Glauber and Gozjolko 2011). On the other hand, in order to spend more time with their children, **involved fathers** work fewer hours than do childless men so they can participate more at home (Kaufman and Uhlenberg 2000). Indeed, more fathers today than in past decades take off work following the birth of a child, and they are more visible in parenting classes, in pediatricians' offices, and dropping off and picking up children at day care centers (Livingston and Parker 2011).

Although two-thirds of fathers say parenting is extremely important to their identity and half say it's rewarding all the time, two-thirds of them feel they spend too little time with their children (Livingston and Parker 2019). Largely due to work requirements—many fathers "aren't the dads they thought they'd be" (Miller 2015). Being an involved father is not easy.

TABLE 10.1   Fathers Outearn Mothers in Most Two-Parent Households

RESPONSE TO THE QUESTION "WHO EARNS MORE?"	RESPONSES OF WORKING PARENTS IN TWO-PARENT HOUSEHOLDS (PERCENTAGE)	
	BOTH WORK FULLTIME	FATHER WORKS FULLTIME, MOTHER WORKS PART-TIME
Father	50	83
Mother	22	3
Earn about the same	26	14
Don't know/Ref.	2	*

*No data.

Note: Based on respondents who are employed part-or fulltime and are married to or living with a partner who is employed fulltime or part-time and is the parent of at least one of the respondent's children (n = 811).

Source: Modified from Pew Research Center 2015d.

Workplace Obstacles to Being an Involved Father

Fathers' providing childcare "is highly dependent on the father's employment situation and the available time they have for childcare" (Laughlin and Smith 2015, p. 1). High-powered careers often expect "professional dedication," which requires an employee to work long hours and even to be available electronically in off hours (Cooke and Fuller 2018). Employers may not believe that employees, especially males, should allow family responsibilities to interfere with work (Haines et al. 2013). Many men who would prefer to work less than they do "may feel that they have no real option if their jobs demand longer than their desired hours" (Shafer 2011, p. 261). A man who gives priority to family may deal with coworkers' resentment or challenges to his masculinity.

As one Texas wife, pregnant with the couple's first baby, explained to an interviewer, "Where he works, the men don't take off to take care of sick children. I hope to split time missed from work with my husband, but that battle is yet to be fought. . . . He's flexible; it's his boss that may not be" (Stanley-Stevens and Kaiser 2011, p. 121). As a result, some men are reluctant to take advantage of technically available family benefits (Fuhrmans 2018). Admittedly, however, the resistance men encounter may be partly a self-imposed perception that they will be viewed as less committed employees if they access options such as paternity leave (Lieber 2015).

Some husbands report having lied to bosses or taken other evasive steps at work to hide conflicts between job and family. One man tells his boss that he has "another meeting" so that he can leave the office each day at 6 p.m.: "I never say it's a meeting with my family" (Jacobs and Gerson 2004, p. 6). This situation illustrates the *family ecology perspective.* The greediness of work, especially for men, "has real implications for gender inequality, in that wives may exit the labor force because the husbands' careers make their attempts to combine work and family too difficult" (Coontz 2013; Lieber 2015; Shafer 2011, p. 261).

Nevertheless, in the view of San Francisco Worklife Law Professor Joan C. Williams, "The huge thing that's changed only in about the past five years is suddenly [some] men feel entitled to take time off for family. They're willing to put their careers on the line to live up to that idea. It's revolutionary" (Scheiber 2015b). We should note, too, that, provided employers allow it, communication technology facilitates working from home and negotiating flexible hours for at least some fathers.

Stay-at-Home Dads

As depicted in Figure 10.1, in 6 percent of heterosexual married couples, only the wife is employed (U.S. Bureau of Labor Statistics 2019, Table 24). A minority of husbands is **stay-at-home dads**, not in the labor force due to disability, unemployment, or because they've specifically chosen to stay home to care for their family while their wives are employed. About one-quarter of stay-at-home dads are home for this latter reason (Livingston 2018b). In 26 percent of gay male couples with children, one parent stays home: "To some gay men, the idea of entrusting the care of a hard-won child to someone else seems to defeat the purpose of parenthood" (Bellafante 2004).

Despite gender role changes, families with breadwinning mothers and fulltime caregiver fathers to some extent continue to be viewed as norm violators (Gaunt 2013; Poniewozik 2012). Although professional fathers who become primary parents may experience career-based identity loss and express more sense of isolation than stay-at-home mothers, being a househusband is not the lonely choice it once was. Local groups, national organizations, and Internet chat rooms bring househusbands together, and stay-at-home mothers are more welcoming of their male counterparts than they used to be. As with many aspects of family life, choice is the key to a man's satisfaction with the househusband role, as is the couple's mutual understanding about the specifics of their division of labor (Gerson 2010; Lewin 2009). Stay-at-home fathers are also discussed in Chapter 9.

David Sacks/Litesize/Jupiter Images

Many men today expect—and are expected by their partners—to be involved fathers, working at home doing childcare or domestic labor, as well as holding a job.

WOMEN'S WORK AND FAMILY ROLES

As the Industrial Revolution got under way, middle-class white women generally remained at home and engaged in domestic labor and "homemaking." Women in lower social classes, immigrant women, and women of color often supported themselves and their families by taking in laundry, marketing baked goods, working as domestic labor in other people's homes, and housing boarders. Some women worked in factories, but it was largely men who held formal "jobs" and were visible in economic production.

Women in the Labor Force

Gradually, beginning around 1890, more women began to enter the labor force. Industrialization gave rise to bureaucratic corporations, which depended heavily on paperwork. Clerical workers were needed, and not enough men were available. Textile industries sought workers with a dexterity thought to be possessed by women. The expanding economy needed more workers, and women were drawn into the labor force in significant numbers. As Figure 10.2 shows, women's participation in the labor force has increased greatly since the beginning of the nineteenth century.

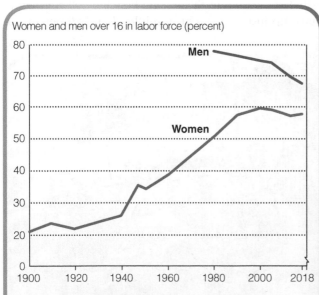

FIGURE 10.2 The participation of women and men over age 16 in the labor force, 1900–2018. Women's and men's labor force participation declined between 2005 and 2010 due to the Great Recession that began in 2007.

Source: U.S. Bureau of Labor Statistics 2014b, Table 1; U.S. Bureau of Labor Statistics 2019, Table 2.

This trend accelerated during World War I, the Great Depression that followed, and World War II. The trend slowed for a time following the end of World War II in 1945. As soldiers came home, the government encouraged women to return to their kitchens. Despite these cultural pressures, the number of wage-earning women would rise once more. As material expectations increased for housing and consumer goods and as more families began to imagine financing a child's college education, wives' wages grew more important.

Beginning in about 1970, the number of employed nonHispanic white women began to increase rapidly. By 1976, stagnant and declining real earnings led more families to rely on a second earner. The growth in the divorce rate left women uncertain about the wisdom of remaining out of the labor force and dependent on a husband's earnings. The women's movement emerged and was a strong force for anti–sex-discrimination laws that opened occupations formerly closed to females. The feminist movement also altered societal attitudes regarding women's careers with the result that they began to seem normative. By 1979, a majority of married women were employed outside the home.

Historically, black women have been more likely than nonHispanic white women to work for wages. Today, white women—with labor force participation of 54.5 percent—are catching up to black women at 56.6 percent. Fifty-five percent of Asian and 54 percent of Hispanic women are employed (U.S. Bureau of Labor Statistics 2019, Table 3).

Mothers in the Labor Force Mothers of young children were the last women to move into the labor force. Although many mothers remained at home while their children were small, by 1970 half of wives with children between ages 6 and 17 earned wages, and that figure increased to 76.5 percent in 2018 (U.S. Bureau of Labor Statistics 2019, Table 5). Today, 60 percent of mothers are employed by the time their children are 3 years old, the majority working fulltime (U.S. Bureau of Labor Statistics 2019, Table 6). In fact, 58 percent of married women with children under age 1 have joined the labor force. Even larger proportions of single mothers are employed: 79 percent of those with children ages 6 to 17, and 68 percent of those with children under age 6 (U.S. Census Bureau 2012a, Tables 599, 600).

Women's Occupations

Jobs that women more often hold differ from those of men. The tendency for men and women to be employed in different job types is termed **occupational segregation**. Occupational segregation has declined somewhat since 1960. Women are physicians, attorneys, astronauts, engineers, and military officers. Men have become nurses, dental hygienists, and elementary

school teachers. However, occupational segregation remains the case generally. Women are 31 percent of dentists, but 98 percent of the dental hygienists; 33 percent of lawyers, but 86 percent of paralegals and legal assistants; 36 percent of physicians, but 90 percent of registered nurses (U.S. Bureau of Labor Statistics 2019, Tables 10, 11). The jobs women are more likely to hold often pay less. Although it doesn't account for all the difference, occupational segregation contributes to the discrepancy between women's and men's average earnings (Gauchat, Kelly, and Wallace 2012).

The Female–Male Wage Gap

Research shows that daughters do more household chores than sons, but get paid less for similar tasks and have smaller allowances (Miller 2018e). The pattern begins early. In 1970, just 4 percent of wives outearned their husbands; about one-third did in 2012 (U.S. Bureau of Labor Statistics 2014b, Table 26). This figure is hard to ascertain, though, because when a wife outearns her husband, she is likely to underestimate her earnings to researchers while he is likely to overestimate his—more evidence that traditional gender roles persist (Bernard 2018; Miller 2018b).

Meanwhile, in 2018 women on average earned 81 percent of what men earned (U.S. Bureau of Labor Statistics 2019). This ratio is higher than in past decades, having risen steadily since the 1970s. Chapter 3 points out that some (assuredly not all) of the wage convergence between women and men is explained by falling wages for men—as well as by rising wages for women with higher education.

The **wage gap** (the difference in earnings between men and women) varies considerably, depending on occupation, and tends to be greater in the more elite, higher-paying positions. For instance, in 2018 women corporate chief executive officers (CEOs) averaged $1,736 weekly compared with $2,488 for men. Among accountants and auditors, women averaged $1,108 weekly compared with $1,404 for men. Female physicians earned $1,677 weekly compared with $2,513 for men (U.S. Bureau of Labor Statistics 2019, Table 18). "American women of working age are the most educated ever. Yet it's the most educated women who face the biggest gender gaps in seniority and pay" (Miller 2019). Compared to an overall average of 81 cents on the dollar, among the highly educated, women earn 74 cents for every dollar men make (Day 2019).

Greedy Work and the Female–Male Wage Gap What does high education and greedy work have to do with the female–male wage gap? Here's one way to think about it:

> Daniela Jampel and Matthew Schneid met in college at Cornell, and both later earned law degrees. They both got jobs at big law firms, the kind that reward people who make partner with seven-figure pay packages.

One marriage and 10 years later, she works 21 hours a week as a lawyer for New York City, a job that enables her to spend two days a week at home with their children, ages 5 and 1, and to shuffle her hours if something urgent comes up. He's a partner at a midsize law firm and works 60 hour weeks—up to 80 if he's closing a big deal—and is on call nights and weekends. He earns four to six times what she does, depending on the year.

It isn't the way they'd imagined splitting the breadwinning and the caregiving. But he's been able to be so financially successful in part because of her flexibility, they said. "I'm here if he needs to work late or go out with clients," Ms. Janpel said. "Snow days are not an issue. I do all the doctor appointments on my days off. Really, the benefit is he doesn't have to think about it. If he has to work late or on weekends, he's not like, 'Oh my gosh, who's going to watch the children?' The thought never crosses his mind." (Miller 2019)

Despite the fact that "the number of women running Fortune 500 companies is at a record high," men dominate corporate America. Women comprise 45 percent of the labor force in the richest 500 U.S. corporations. Nevertheless, women in those same Fortune 500 companies make up just 37 percent of midlevel and 26 percent of

Steadily more women have entered the labor force since the 1960s. This woman may be mothering children. Women in blue-collar jobs are still a minority, although more women are entering these jobs, which tend to pay better than traditional women's jobs in service or clerical work.

Vadim Ratnikov/Shutterstock.com

senior-level managers. Women hold 21 percent of board-of-director positions and are slightly less than 6 percent of CEOs (Connley 2019; "Women in S&P 500 Companies" 2020). Racism blocks the path to management for nonwhite and Hispanic men; both racism and sexism block the path for nonwhite and Hispanic women, who hold relatively few executive and board-of-director positions in major corporations. In 2010, the first African American woman, Ursula Burns of Xerox, became CEO in the Fortune 500 list of top companies (Angelo 2010).

Some argue that employed women on average earn less than men partly because women—socialized to be cooperative rather than competitive—don't negotiate for high salaries as well as men do.

> When women advocate for themselves, they have to navigate more than a higher salary: They're managing their reputation, too. Women worry that pushing for more money will damage their image. Research shows they're right to be concerned: Both male and female managers are less likely to want to work with women who negotiate during a job interview. (Milne-Tyte 2014)

A few companies hope to lessen the wage gap by eliminating salary negations (Noguchi 2015).

We've seen that coupled heterosexual women are more likely than men to limit their labor force participation in order to care for family members. But even if she has no plans to do so, a woman contends with employers' *assumptions* that she will opt out of the labor force to take care of her children or other family members. As a result, an employer may be less likely to select even highly ambitious and fully committed women employees for further training or for positions with advancement potential (Pinto 2009).

The concept **motherhood penalty** describes the fact that motherhood has a significant negative lifetime impact on female income—a situation that creates a long-term gender-earnings gap. The motherhood penalty persists despite women's increasing education and dedicated labor force involvement (Coontz 2013; National Women's Law Center 2010). Afraid that impending motherhood will jeopardize workplace status or promotion, some professional women hide their pregnancies for as long as possible or even hire surrogates to carry their children (Quart 2012; Richards 2014).

Stay-at-Home Moms

About 27 percent of heterosexual, coupled mothers with children under age 15 are not in the labor force, a figure compared with 46 percent in 1970 (Livingston 2018b). While one-quarter of stay-at-home

fathers say they're home specifically to care for the family, more than three-quarters (78 percent) of mothers stay at home for this reason (Livingston 2018b).

For upper-middle-class nonHispanic white women, we might think of stay-at-home mothers as comprising **neotraditional families**—that is, families reminiscent of 1950s norms and values, but with the new (neo) aspect that the model is consciously selected from several options—options that the vast majority of wives in the 1950s did not have.

> This [neotraditional] order is appealing to men and women who are discontented with . . . family modernization, the lack of clarity in gender roles . . . , and the pressures associated with combining two fulltime careers. It is also appealing to women who continue to identify with the domestic sphere, who wish to see homemaking and nurturing accorded high value, and who wish to have husbands who share their commitment to family life. (Wilcox 2004, p. 209)

Today's mothers who can afford it are more likely than their own mothers to plan career pauses or to limit their working hours. Many of these women had planned to work outside their homes "but are increasingly caught off guard by the time and effort it takes to raise children." [Then too,] "they saw their parents struggle while working fulltime or leave the labor force altogether, and wanted a different option (Miller 2018c)." Unfortunately, dropping out of the labor force, even for a relatively short time, is costly because reentering the professional labor force at the level equivalent to which one left it is typically difficult (Miller 2015a, 2018c).

Bokan/Shutterstock.com

Some women opt out of the labor force to raise their children at home. This former executive may hope to return to the labor force when her children are older. Research shows, however, that professional mothers who leave the labor force, temporarily pausing their careers, pay a price inasmuch as they may not be able to return to work at the same level at which they left.

Despite media interest in highly educated stay-at-home moms, however, the majority of stay-at-home mothers have no more than a high school diploma and have relatively low household incomes (Saad 2012b). Many stay-at-home mothers are immigrants who are following models that are traditional in their home countries' cultures (Frank and Hou 2015). For instance, Arab American women are employed at lower-than-average rates. Scholars suggest the reasons for this are varied, but traditional gender roles are emphasized in many Arab American families, and women are encouraged to stay out of the labor force because they are considered "bases of security and stability" for other family members (Gold and Bozorgmehr 2007, p. 52).

TWO-EARNER PARTNERSHIPS AND WORK–FAMILY OPTIONS

In 1968, there were equal proportions of two-earner and provider–housewife couples: 45 percent of each (Hayghe 1982). Today, **two-earner partnerships** in which both partners work are the majority. Partners display considerable flexibility in how they design their two-earner unions. Arrangements are ever-changing and flexible, varying with the arrival and ages of children and with partners' job options and preferences. In this section, we examine four ways that two-earner couples navigate the work–family interface: two-career partnerships, part-time employment, shift work, and working at home.

Two-Career Partnerships

Careers differ from *jobs* in that careers hold the promise of advancement, are considered important in themselves—not just a source of money—and demand a high degree of commitment. Career men and women work in occupations that typically require education beyond a bachelor's degree—for example, medicine, law, academia, financial services, and corporate management. One-third of professional or managerial men work fifty or more hours per week, and one in six women do (Jacobs and Gerson 2004; Usdansky 2011).

The vast majority of two-earner partnerships would not be classified as *two career* because one or both partners' employment does not have the features of a *career*. Nevertheless, the two-career couple is a powerful image. Most of today's college students view the **two-career relationship** as an available and workable option. For two-career couples with children, however, family life can be hectic as partners juggle schedules, chores, and childcare (Sullivan 2015). Despite intermittent chaos and frenzy, the majority of working parents believe two-career employment, however organized, is the best option for them (Horowitz 2019b).

Working Part-Time

About 24 percent of women and 12 percent of men work part-time (U.S. Bureau of Labor Statistics 2019, Table 21). Of part-time workers who had once worked fulltime, 70 percent say they changed to part-time because of childcare problems (U.S. Census Bureau 2012a, Table 613). Evidencing traditional expectations, 70 percent of Americans in a 2012 Pew Research Center survey said the ideal situation for a father with young children is working fulltime; 20 percent said the ideal situation for fathers would be working part-time. Regarding mothers with young children, 16 percent said working fulltime is ideal, and 42 percent felt that working part-time is preferable: "Women more than men adjust their careers for family life" (Parker 2015; Young and Schieman 2018). This situation is particularly true for nonHispanic white women (Florian 2018). Less work–family conflict and more family and personal time are clear benefits of part-time employment. Part-time mothers are significantly less likely to say they always feel rushed, and they evidence better health and more sensitive and involved parenting than do other mothers, both employed and stay-at-home moms (Buehler and O'Brien 2011).

Fifty-eight percent of moms employed fulltime believe they spend the right amount of time with their children compared with 77 percent of mothers employed part-time (Pew Research Center 2015d, Questions 2, 12). However, part-time employment has its costs. As it exists now, part-time work seldom offers job security. And part-time pay is rarely proportionate to that of fulltime jobs. For example, a part-time teacher or secretary usually earns well below the equivalent hourly wage paid to fulltime staff. In higher-level professional and managerial jobs, a different problem appears. To work part-time as an attorney, accountant, or aspiring manager is to forgo the salary, status, security, and promotions of a fulltime position.

Shift Work and Variations

Sometimes one or both partners engage in **shift work**, defined by the Bureau of Labor Statistics as any work schedule in which more than half an employee's hours are before 8 a.m. or after 4 p.m. It has been estimated that in one-quarter of all two-earner couples, at least one partner does shift work; one in three if they have young children. Some couples use shift work to ease childcare arrangements, addressed later in this chapter.

Variations on shift work involve "just-in-time" scheduling and "clopenings." *Just-in-time scheduling* entails large corporations, such as Starbucks, timing shifts exactly when employees are needed, sometimes in very small time segments, even at the last minute. An employee may be called in to check out customers during a two-hour busy period, then sent home when the store is less busy, only to be called back later that day.

Just-in-time scheduling "can wreak havoc on the lives of workers who can't plan [childcare or anything else] around work obligations that might pop up at any time" (DePillis 2015). "Clopening" scheduling involves requiring a worker to close a store late at night and then open it early the next morning. There is evidence that these kinds of work scheduling are harmful to children's well-being (Scheiber 2015c). After receiving public criticism for its just-in-time and clopening scheduling practices, Starbucks pledged to stop (Scheiber 2015d). Some states and cities have enacted fair scheduling laws that prohibit clopening or just-in-time scheduling (National Women's Law Center 2015).

Shift work reduces the overlap of family members' leisure time, which can negatively affect the relationship: "To the extent that social interaction among family members provides the 'glue' that binds them together, we would expect that the more time partners have with one another, the more likely they are to develop a strong commitment to their marriage and feel happy with it" (Presser 2000, p. 94; Craig and Brown 2917). Unsurprisingly, then, shift work is associated with a decrease in marital stability. Shift work is also associated with negative spillover effects on mother-child interactions and—probably due to fatigue—with less positive parental mood, poorer maternal sensitivity, and consequent lower children's academic achievement and more negative child behaviors (Gassman-Pines 2011; Grzywacz et al. 2011; Han and Fox 2011). An alternative for some parents involves working at home.

Doing Paid Work at Home

Home-based work—working from home, either completely or in part, either for oneself or for an employer—increased significantly over the past few decades (Tozzi 2010). Many parents find themselves logging into work-related meetings from home. "The mute button is amazing on a conference call, so they can't hear that you're actually in the bathroom with a 2-year-old over the potty who's saying, 'I'm pooping!'" (Wirecutter Staff 2019).

Historically, home-based work involved *piecework*—sewing or making artificial flowers, for example. This mode of home production is declining due to competition from low-wage workers overseas. It still exists, particularly in the assembly of medical kits, circuit boards, jewelry, and some textile work, but many home-based workers are educated and are engaged in professional services such as law, accounting, computer programming, consulting, marketing, finance, and so on (Peck

2015). Other home-based businesses include selling cosmetics, kitchenware, or other products.

Home-based work also includes working from home for an employer by connecting to the office, customers, clients, or others via the Internet—that is, telecommuting. About half of telecommuters are women (Peck 2015). Many of them, especially those with children under age 6, say that they work from home to "coordinate work with personal/family needs" (Wight and Raley 2009, Table 1; Guynn 2013).

Telecommuting helps parents coordinate work and family obligations, but it may not be a panacea. As the author of a study of women in a home-based direct-selling business noted, "many women soon discovered . . . that they had exchanged one set of challenges for another" (Gudmunson et al. 2009). For one thing, parents employed at home may simultaneously be expected to do unpaid family work.

UNPAID FAMILY WORK

Social scientists interested in the division of household labor have used many of the theoretical perspectives discussed in Chapter 2 to study unpaid family work. To illustrate this, Table 10.2 applies several of these to the topic and gives examples from the Lee family, whose story opened this chapter. **Unpaid family work** involves

Working from home is one way to manage the work–family interface. Can you create a hypothesis about some positive consequences of this choice? Some less positive consequences?

TABLE 10.2 Theoretical Perspectives Applied to Unpaid Household Labor

THEORETICAL PERSPECTIVE	APPLICATION TO UNPAID HOUSEHOLD LABOR	EXAMPLES FROM THE LEE FAMILY
Family ecology	Distribution of household labor is influenced by the environment that surrounds the family, particularly the workplace and work–family policy.	The Lee family's urban work options include employment opportunities in local schools, running a family business, and renting out family-owned property, among others.
Family life course development framework	In our society, adults are assigned more unpaid household labor than are children or the elderly.	The Lee family's teenagers are expected to be good students but not expected to work outside the household. Gung Gung, the Lee family's 86-year-old patriarch, does less work now than he once did.
Structure-functionalism	What might be some *functional alternatives* to the traditional male breadwinner and female homemaker roles?	Functional alternatives to the traditional division of labor include: May works outside the home as a school principal. Warren's wife, Jen, works outside the home as a gym teacher. Warren does considerable unpaid family work—for example, cooking and getting his father to doctor appointments.
Interaction-constructionist	Individuals, couples, groups, and societies invest housework with meanings (Sallee 2011). According to the *gender-performance* hypothesis, women "do gender" by doing housework, whereas men "do gender" by avoiding housework (Schneider 2011).	The family has defined Warren's primary family role as caregiver to his father.
Exchange or bargaining	Individuals who provide more family income can exchange resources brought into the household for the right to be excused from more undesirable household tasks.	May and Jen work outside the home, one as a school principal and the other as a gym teacher. They exchange the earnings they bring to the household for not having to prepare meals.
Family systems theory	Family members design and perform their respective roles in response to how others in the family system play their roles.	May's husband, Ben, who drives for FedEx, is out of the family building from 7 a.m. until after 8 p.m. His wife adjusts to this situation by assuming the majority of the discipline for the couple's children.

caring for dependent family members, such as the elderly or children, as well as maintaining the family domicile. Chapters 13 and 16 address meeting the needs of ill, disabled, or elderly family members. Chapter 9 explores expectations and behaviors associated with mother and father roles. Here we focus primarily on household labor associated with maintaining the family domicile.

Household Labor

It may be hard to believe now, but utopians and social engineers once shared a hope that advancing technology and changed social arrangements would make the need for families to cook and clean obsolete (Hayden 1981). But collective arrangements proposed by utopians and early feminists never caught on. As opportunities opened up in the labor force, servants, who had done much of the work for earlier middle-class housewives, entered factory work or took other jobs. Middle-class women were left to do their own housework (Cowan 1983). Then, especially after World War II and during the 1950s, technology raised standards even as it made some housework less time-consuming. With automatic washers, we began to change clothes daily, for example, a practice that creates lots more laundry to do (Newport 2008).

Less Housework Is Being Done Now Than in the Past Today, sometimes assisted by fast food, prepared take-out meals, laundry, and other paid services—and perhaps "living with a little more dust"—twenty-first-century

couples often adjust to women's labor force participation by scaling down what was considered necessary housework a few decades ago, even though they may not want to (Kornrich and Roberts 2018). Women put about thirty-two hours weekly into housework in 1965, and men put in about four. In 2016, women put eighteen hours weekly into housework while men put ten (Geiger, Livingston, and Bialik 2019).

Men's Share of Housework Is Greater Than in the Past While the trend has been for women to do less household work, over the past several decades men on average have done more. And among heterosexual couples, men's time spent on childcare has tripled since the mid-1960s. However, although fathers are doing more household labor than they did in the past, mothers continue to do about twice as much as fathers do (Geiger, Livingston, and Bialik 2019d).

Women Still Average More Household-Labor Hours Than Men Do Although the gap has lessened, nearly two-thirds of heterosexual mothers say they're more likely than their partners to manage their children's schedules or activities and to take care of the kids when they are sick. Half of mothers believe they do more of the household chores, while one-third of fathers say they themselves do more chores (Pew Research Center 2015d, Question 35a, b, e).

A good deal of women's unpaid family labor goes unnoted and unmeasured. For instance, *health behavior work*—promoting family member's healthy behaviors, such as making family members' dental appointments or monitoring their eating habits—is a virtually invisible aspect of caregiving that is usually assigned to women (Reczek and Umberson 2012). More often than do men, women assume the work of coordinating paid services when a family hires them (Coontz 2007). Women disproportionately organize family activities and care for or "keep" kin—**kin keeping**—maintaining contact, remembering anniversaries and birthdays, sending cards, and shopping for gifts (Coontz 2007; Hagestad 1986). Of course, it's possible that "the work men do to maintain relations as fathers, brothers, and uncles may not be captured by questionnaires or analyses that focus on emotional support or caregiving tasks" (McCann 2012, p. 254). Nevertheless, it's safe to conclude that, overall, women do significantly more unpaid household labor than do men. Why might this be?

Why Do Women Do More Household Labor?

Whether or not they are employed outside the home, wives and cohabiting women in heterosexual relationships do more unpaid family labor than men do (Parker and Wang 2013). However, it's important to note that

the imbalance may disappear when *hours* employed are counted rather than simply employment *status*. Husbands spend fewer hours in housework than wives do, but husbands spend more hours in paid employment than do wives (Konigsberg 2011; Parker and Wang 2013). Wives with husbands who earn more and work more hours do more housework. Also, wives do more housework when they have husbands who bring work home or are preoccupied with work-related problems when at home (Sullivan and Coltrane 2008). Social scientists have proposed alternative hypotheses for why women do more unpaid household labor (Cooke and Hook 2018).

Hypothesis 1: Partners' Relative Earnings Influence Their Division of Household Labor Men's participation in household labor had long been thought to be related to their relative earnings and the proportionate share of household income produced by each partner (Hartmann, English, and Hayes 2010; Miller 2019). On average, employed wives contribute significantly to financial resources—more than a third (37 percent) of a family's income today and up from 27 percent in 1970 (U.S. Bureau of Labor Statistics 2019, Table 25). Nevertheless, wives contribute less than half of family incomes, as we have seen, and this situation may be why they put in more hours of unpaid household labor (Carlson and Lynch 2017; De Henau and Himmelweit 2013). "[A]s married women increase their earnings share toward equality with their husbands, they reduce the amount of time they spend each day on routine housework tasks" (Schneider 2011, p. 857). Heterosexual women who outearn their partners have mates who contribute more in unpaid family labor (Crary 2008; Sullivan and Coltrane 2008).

Hypothesis 2: Gender Roles Influence Partners' Division of Household Labor Even with increased equality in partners' division of household labor, they tend to split household tasks according to traditional gender expectations. For example, men say their household tasks include car upkeep, yard work, and investment decisions, whereas women tend to do more meal preparation, dishwashing, grocery shopping, house cleaning, laundry, and childcare (Newport 2008). The persistence of "male" and "female" household tasks suggests that household labor can have meaning beyond simple maintenance. "Housework is not just the performance of basic household tasks but . . . also a symbolic expression of gender relations, particularly between wives and husbands" (Artis and Pavalko 2003, p. 748; Gilbert 2008).

Virtually all of us judge our own and others' behavior according to whether it agrees with gender norms (normative behavior) or does not (gender deviance) (Gaunt 2013). Doing tasks that are considered traditionally feminine or masculine can reinforce—or threaten—a partner's feminine or masculine gender identity.

A man's cleaning the garage may reinforce his masculine gender identity (DeHenau and Himmelweit 2013). Doing the dishes, however, may threaten his gender identity, so he may avoid this activity.

Furthermore, if a husband feels threatened by his wife's earnings, he may do *less* housework in an effort to psychologically neutralize his feelings of gender deviance. It has also been proposed that wives who earn more than their husbands may neutralize their gender role deviance by doing more of the housework (Kluwer 2011; Sullivan 2011a, 2011b). This hypothesis, although interesting, lacks empirical support (Schneider 2011). We should note, too, that the meaning of unpaid family labor may have changed somewhat, as have gender expectations. That is, over the past several decades, it has become more socially acceptable for men to be involved in childcare, cooking, and cleaning (McClintock 2017).

Hypothesis 3: The Partner with More Power in the Relationship Can Escape Undesirable Household Labor

Social scientists studying this topic have long associated power in relationships with household work, particularly with undesirable tasks like cleaning the toilet. It looks like men, on average, may be more willing to endorse gender equality at work than at home. "At home, men are more resistant to change because it really means surrendering privilege," according to sociology professor David Cotter (Miller 2018d; and see Lockman 2019). Research shows that generally the partner with more control over the other does less around the house (Lam, McHale, and Crouter 2012; Sullivan and Coltrane 2008).

Hypothesis 4: The Relative Time Available to Each Partner Influences the Division of Household Labor

The more hours men work, the less housework they do. Men who work fewer hours do more housework. Unemployed men, presumably with more available time, spend almost double what employed men do on household labor (Parker and Wang 2013). This evidence provides support for the idea that, rather than avoiding housework to mitigate any gender deviance created through unemployment, these men are responding to reduced work time by increasing housework, behavior that appears more consonant with the predictions of time availability theory (Schneider 2011, p. 857).

At least one writer concludes that the "widespread belief that working mothers have it the worst . . . is simply not the open-and-shut case it once was" (Konigsberg 2011). Then, too, sharing unpaid family labor varies by race and ethnicity and by sexual identity.

Diversity and Household Labor

In some ethnic groups, such as the Vietnamese and Laotian, housework is significantly shared, if not by husbands, then by household members other than the child's mother (Johnson 1998). Among African Americans, adult children living at home, extended kin, and nonresident fathers are likely to share housework and childcare and to help with repairs (Gerson 2010).

Research on racial and ethnic differences finds that the pattern of men's spending less time than women in housework occurs in white, black, Asian Indian, and Hispanic families. However, black men spend more time in unpaid family work than do nonHispanic white men. One explanation is that black men have more egalitarian attitudes regarding unpaid labor and that black wives are more likely than other racial and ethnic categories to have high earnings relative to their husbands' earnings (Gerson 2010; Glauber and Gozjolko 2011). The differences among nonHispanic white, black, and Hispanic men's household labor time also may reflect other (as yet unknown) differences among them (Sullivan and Coltrane 2008).

Lesbians, Gay Men, and Household Labor Given that gay couples involve people of the same gender, how does their household division of labor work out and what impact does that have on the relationship? A small qualitative study of lesbian and gay male couples explored these questions. Each partner was employed fulltime, and there were no children in the household.

fizkes/Shutterstock.com

Do you think this man cleans the toilets at home? Changes diapers? Folds laundry? Social scientists have proposed four hypotheses to explain why, on average, husbands do less household labor than wives. First, husbands who earn more than their wives may exchange their financial input for doing less housework. Second, some tasks may seem like "women's work," and, protecting their gender identity, men are reluctant to do them. Third, the partner with more power in the relationship does less around the house. Finally, the one who spends more hours at paid employment has less time available and therefore does less household labor. Which of these hypotheses do you think is or are more likely to be supported by empirical evidence?

The study looked at who performed some traditionally female tasks. Generally, lesbian couples' division of labor was more egalitarian than that of gay male couples. Perceived equality was closely tied to relationship satisfaction and that, in turn, to relationship stability (Kurdek 2007). As with heterosexual couples, research shows that the number of hours individual members of a couple work in paid labor has the greatest influence on the division of household labor (Sutphin 2006). The next section investigates how partners juggle employment and unpaid family labor.

JUGGLING EMPLOYMENT AND FAMILY WORK

The term *juggling* implies hectic and stressful situations. Typically, two-earner and single-parent families are hectic indeed. Ways that families manage vary. Many two-earner couples, especially in upper-income families, hire household help and may purchase the services of immigrant, racial and ethnic minority, and working-class people for cleaning, childcare, and, to a lesser extent, cooking (Killewald 2011). In contrast to fifty years ago, children are less often required to do household chores and more likely to spend time on their mobile devices (Pena, Menendez, and Torio 2010; Saint Louis 2015).

Virtually every researcher studying the work–family interface hears expressions of time pressures and feeling rushed and stressed (Miller 2015b; Pew Research Center 2015d; Sullivan 2015). Stress is most pronounced for parents during children's early years, especially for single mothers and those with demanding careers (Jacobs and Gerson 2004). A Gallup poll asked individuals whether they have enough time to do what they want, and between 55 and 58 percent of adults under age 54 said no (Carroll 2008).

Employed parents put in a **second shift** of unpaid family work that amounts to an extra month of work each year (Hochschild 1989). Nearly twenty-five years ago when sociologist Arlie Hochschild's book by that title was published, she applied the term *second shift* to wives, not husbands. The women Hochschild interviewed talked about being overtired, sick, and "emotionally drained," as well as how much sleep they could "get by on." (Hochschild 1989, p. 9). Today, husbands feel pressured, too (Konigsberg 2011). The second shift means forfeiting leisure time and sleep to accomplish unpaid family work. For a majority of Americans, rest and relaxation time and time for friends, hobbies, and sleep is not what they would like it to be (Craig and Brown 2017; Saad 2004).

Two-Earner Families and Children's Well-Being

Before mothers entered the work force in large numbers, their employment outside the home was considered problematic by child development experts and the public. However, by 2001, survey data found more than 90 percent of women agreed that one can be a good mother and have a successful career (Center for the Advancement of Women 2003). Research supports women's opinions in this regard, concluding that maternal employment does not cause behavior problems in children and is actually good for them. For one thing, the economic benefit to children of working mothers cannot be overlooked (Agee, Atkinson, and Crocker 2008; Miller 2015c). "Issues for Thought: When One Woman's Workplace Is Another's Family" examines social class and workplace issues surrounding professional working parents and their hired nannies.

Furthermore, before the era of working mothers, those who stayed home did not spend all their time interacting with their children. They devoted more time than today's mothers to housework (including home decorating), leisure activities, or volunteer work. A variety of studies indicates that parents today spend as much

Hero Images/Getty Images

Although spouses increasingly share the provider role today, they actually spend more time with their children than parents did several decades ago when mothers were less likely to be in the labor force. Today's parents, perhaps more family-centered than their grandparents were, spend less leisure time on their own and share more family activities with their children.

Issues for Thought

When One Woman's Workplace Is Another's Family

"Every weekday evening in affluent homes across America, two groups of women trade places. Mothers who follow careers come home and the women who are paid to care for their children prepare to depart or step aside" (Rimer 1988). Corporate and professional mothers (and some fathers) change places with less financially fortunate women, many of them immigrants, some undocumented. A Manhattan public relations executive says she is "completely dependent on" her nanny from Trinidad: "We couldn't earn a living without her" (Rimer 1988).

A minority of paid caregivers immigrate to the United States from poorer countries, work for several years, then return to their home countries with money sufficient to join the middle class there (Barrionuevo 2011). From the employers' perspective, paid caregivers have "tremendous power"—"the tyranny of the nannies," as one executive put it (Rimer 1988). For the employer, the nanny is all powerful: A caregiver's not arriving on time (or at all) is out of her employer's control but will assuredly impact the employer's work day.

Tyranny of the Nannies

Many employers say they worry constantly that their nannies might quit. Or arrive late. Or take time off to care for their own sick children. As one executive mother said, "When child care breaks down, everything else breaks down" (Rimer 1988). Sometimes nanny arrangements do fall apart. Some arrangements last just a few days. Even in ongoing arrangements, things can come up. A middle-aged Iranian immigrant, having been with an Oakland, California, family for ten years, quit in order to return to Iran and care for her aging mother.

Employers say they anxiously strive to keep their nannies happy: "You think, if the nanny is happy the baby is happy. If the baby's happy, you're happy" (Rimer 1988). But when upper middle-class women hire poorer women, many of them of different racial and ethnic or national background, conflict is possible over different mothering ideals, as discussed in Chapter 9 (Macdonald 2011). What about the professional nutritionist mom whose nanny takes the 3-year-old to McDonald's weekly? "You have to constantly make compromises," says one parent-employer. "This is one of the most important relationships I'll have in my entire life. I work at it all the time" (Rimer 1988). But there's another side to the story as well.

---

or more time with children as in the past. Figure 10.3 presents the time parents spend caring for their children doing routine care and child-enrichment activities. Time spent was measured in time diary studies conducted by various universities using samples ranging from 1,200 to more than 5,000 individuals. In this data-gathering method, respondents keep daily diaries of time spent in certain activities. Consistent with the finding that younger generations are more family-oriented, mothers and fathers spend more time taking care of their children today than did previous parents going back to 1965 (Bianchi, Robinson, and Milkie 2006).

How did parents, especially mothers, accomplish an increase in time caring for their children and also increase their employment hours? They cut back on housework and spent less time doing leisurely activities by themselves or only with partners or friends. Parents combined time in children's activities with their own leisure time and with time spent with each other (Craig 2015; Craig and Brown 2017). The "thing about the reallocation of mothers' time to employment is that it appears to have been accomplished with little effect on

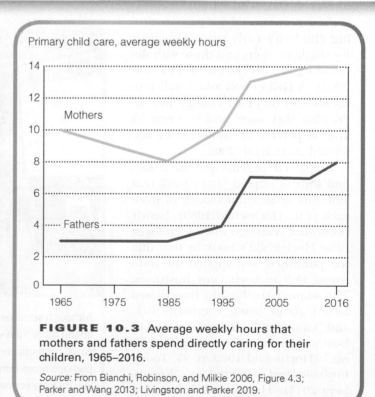

FIGURE 10.3 Average weekly hours that mothers and fathers spend directly caring for their children, 1965–2016.

Source: From Bianchi, Robinson, and Milkie 2006, Figure 4.3; Parker and Wang 2013; Livingston and Parker 2019.

Tyranny of Caregivers' Employers

In-home workers are at the low end of the pay scale, many making minimum wage. Typically, they work long hours and may be exploited regarding their days off. "When you travel with them, it's twenty-four hours," explained a Jamaican nanny in Aspen, Colorado. "Sometimes it's two or three weeks before you get time off" (Rimer 1988). Undocumented immigrant domestic workers are reluctant to complain because they fear being reported to U.S. immigration (ICE) and deported.

Tyranny of the Neighborhood

On such mothers' Internet forums as UrbanBaby, stay-at-home mothers complain that they dislike having to socialize with employed parents' nannies. Public spaces, such as playgrounds, have to be shared. But does a nanny belong at a child's birthday party in a neighborhood home? Many nannies say that, paid to be there, they try to be civil while feeling unwelcome (Bindley 2011).

Sometimes the nannies are the casualties of this conflict. If they can't get along in the neighborhood, they may be let go for another paid caregiver who (hopefully) can (Bindley 2011).

Intersectionality

As pointed out in Chapter 3, contemporary feminism stresses *intersectionality*—structural connections among race, class, and gender (Baca Zinn, Hondagneu-Sotelo, and Messner 2007). "Housework is ascribed on the basis of gender, and it is further divided along class lines and, in most cases, by race and ethnicity. Domestic service accentuates the contradiction of race and class in feminism, with privileged women of one class using the labor of another woman to escape aspects of sexism" (Romero 2002, p. 45).

Solutions

In 2011 New York state passed a Domestic Workers' Bill of Rights that allows disability benefits for fulltime home workers and legislates the possibility of compensation for workplace sexual harassment or discrimination. The legislation requires employers to pay time and a half for overtime work, and to provide three vacation days annually as well as twenty-four hours off for every seven days worked. A legal workday is defined as eight hours and a workweek as forty hours—or forty-four for live-in workers (Bufkin 2012).

> ## Critical Thinking
>
> How does the employer-nanny relationship illustrate *intersectionality*? What might be some aspects of work–family conflict in this relationship that are not discussed here? In your opinion, is regulating the employer-nanny relationship as a tradesperson-client agreement a good solution to the problems raised here?

children's well-being," noted sociologist Suzanne Bianchi (2000) in her presidential address to the Population Association of America. These results present an optimistic view of how employed mothers' children are faring. Concerns remain, however. For instance, the schedules of parents and their children have become increasingly frantic with decreasing unstructured or "down" time, an issue explored in Chapter 9 (Craig 2015; Elkind 2010; Kreider 2012). Decreased time spent by couples just for themselves is concerning: "Of significance is that individuals experience greater happiness and meaning and less stress during time spent with a spouse opposed to time spent apart" (Flood and Genadek 2016, p. 142).

SOCIAL POLICY, WORK, AND FAMILY

According to the structure–functionalist model, explored in Chapters 1 and 2, families are accountable for three vital functions: providing economic support, raising children responsibly, and giving family members emotional support and security. Regarding work and family issues, social policy asks whether providing economic support (participating in the labor force) enhances or threatens the other two family functions. Furthermore, this perspective encourages us to ask about functional alternatives to ways that work–family life is structured. Policy issues center on three questions:

1. What are the issues?
2. What is needed to address the issues?
3. Who will provide what's needed to address the issues?

We'll look at these three questions now.

What Are the Issues?

We address three work–family issues in this section: (1) undesirable financial conditions for many Americans, (2) parental stress due to work–family conflict, and (3) persistent although diminishing gender inequality in today's work–family division of labor.

Inadequate Economic Resources Even with two or more jobs, many Americans do not make enough money to support a family. Low-income and poverty-level parents often work at minimum-or less-than-minimum-wage jobs with unpredictable hours and no benefits. They struggle with housing and transportation instability due to unaffordable rents and utility, gasoline, and car repair costs (Cook et al. 2009).

Not just the poor but also the middle class have suffered as a result of a globalizing economy, the Great Recession that began in 2007, and a current "economic recovery stuck in low gear" (Schwartz 2012; and see Frank 2011; Pew Research Center 2012e). Laid off and too long unemployed, many took jobs at lower pay. Some resorted to temporary ("temp") work. With few if any benefits, temp work is often fraught with the additional stress of landing another placement when the current one ends (Weeks 2010).

Parental Stress Due to Work–Family Conflict We have seen throughout this chapter that many parents struggle with high stress levels due to work–family conflict and spillover—stress that can damage parents' health and lead to less effective parenting. Having multiple roles (such as parent, partner, employee, student, DoorDash driver) does not necessarily increase stress and in fact may enhance personal happiness—provided there's enough time to accomplish everything. However, coupled with financial worries, parents' irregular work schedules in low-wage jobs with little autonomy and high supervisor surveillance increase anxiety and stress (Jacobs and Gerson 2004). And let's not forget that high-paying careers are increasingly demanding, with some positions expecting 24/7 availability.

Persistent Gender Inequality Workplace characteristics impact within-family interactions and decision making—and within-family interactions and decision making impact the workplace. We can think of these reciprocal situations as making a **causal feedback loop** (see Figure 10.4). Regarding the work–family interface, economic and workplace expectations, demands, and practices influence decisions made within families and vice versa. We've seen that the traditionally expected American career path has required giving priority to one's profession while making family caregiving secondary. Opposed to her male partner, a heterosexual woman faced with work–family conflict is likely to minimize her work rather than her family role.

When women temporarily leave the labor force to care for children, elderly parents, or other relatives, they forego opportunities for career development and reduce their lifetime earnings (Liu and Hynes 2012; Usdansky 2011). Similarly, when women choose occupations that are more "family friendly"—shorter or more flexible hours—they choose employment that

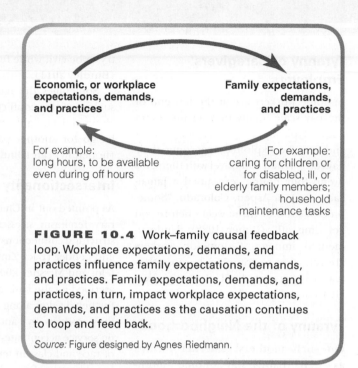

FIGURE 10.4 Work–family causal feedback loop. Workplace expectations, demands, and practices influence family expectations, demands, and practices. Family expectations, demands, and practices, in turn, impact workplace expectations, demands, and practices as the causation continues to loop and feed back.

Source: Figure designed by Agnes Riedmann.

tends to pay less. As a result, wives risk lower pensions and inadequate support should the marriage break up, and they may not have the careers they might have had otherwise.

Largely because they are paid less on average than their male partners, employed heterosexual women are the more likely to adjust their work schedules to accommodate caregiving (Coltrane 2000; Parker and Wang 2013). Then, because wives more often than husbands adjust their employment time to accommodate family needs—taking off to care for a sick child, for example—employers feel justified in offering women lower wages and fewer promotions. As a result,

> Women are still paid less than men, especially at high education levels. Women are less likely than men to hold jobs that offer flexibility or family-friendly benefits. When they become mothers, they face more scrutiny and prejudice on the job than fathers do.

So, especially when women are married to men who work long hours, it often seems to both parties that they have no choice. Female professionals are twice as likely to quit work as other married mothers when their husbands work 50 hours or more a week and more than three times more likely to quit when their husbands work 60 hours or more. (Coontz 2013)

What's Needed to Address the Issues?

On the one hand, the question "What's needed to address these issues?" can be answered from the micro, more psychological, or individualist perspective. On the

other hand, the question can be addressed from a structural perspective, using a sociological imagination. The sociological focus of this textbook leads us to concentrate on structural solutions. However, first we'll examine some proposed individual remedies.

Individual Remedies Individual solutions to issues focus on what individuals themselves can do to address the issue in question. Inadequate economic resources might be addressed, for instance, by better budgeting or refusing to live beyond or above one's means by cutting up credit cards.

Individual solutions proposed to address stress due to work–family conflict involve scheduling fewer family activities (Becker 2014; Kreider 2012). Psychotherapist Stephanie Sarkis advises that people shouldn't wear being overextended and overscheduled "like a badge of honor." "Put your phone down for the day. Take a walk. See where the day takes you. Take a 'mental health day.' Learn the word 'no', and use it. . . . Is what you're so stressed about (and overscheduled with) really worth it?" (Sarkis 2011).

Individual solutions to persistent gender inequality include urging women to "be more proactive and confident: We can't make change happen by being shrinking violets or apologists" (Butler 2015). Facebook executive Sheryl Sandberg advises women to "lean in" to their work roles in order to get noticed and promoted (Luscombe 2013; Sandberg 2013c). That means "go the extra mile," volunteer to do extra at work, work late, and show up early.

Is Sandberg's advice "hard on [individual] women and soft on [structural] sexism" (Corrigan 2013)? Policy makers from a more sociological viewpoint are in general agreement that families would benefit from necessary structural, or societal, changes: (1) adequate wages, (2) quality and affordable elder and childcare, (3) family leave, and (4) flexible employment scheduling (Ruppanner 2013). We'll examine each of these in turn.

1. Adequate Wages If the federal minimum wage had kept pace with the rise in executive salaries since 1990, America's poorest paid workers would be making more than $23 an hour (National Employment Law Project 2013). In his 2013 State of the Union address, President Obama proposed raising the hourly federal minimum wage from $7.25 to $9.00. In his words,

> Today, a fulltime worker making the minimum wage earns $14,500 a year. . . . [A] family that earns the minimum wage still lives below the poverty line. That's wrong. That's why, since the last time this congress raised the minimum wage, nineteen states have chosen to bump theirs even higher. Tonight, let's declare that in the wealthiest nation on Earth, no one who works fulltime should have to live in poverty. (Franke-Ruta 2013)

There is evidence from cities and states that have raised the minimum wage above federal mandates that doing so not only improves workers' lives somewhat but also does not cost jobs (Wasikowska 2011). Nevertheless, at the time of this writing, an increase in the federal minimum wage isn't highly anticipated. A second condition that would benefit many struggling families is quality, affordable childcare.

2. Quality, Affordable Elder and Childcare Many employees have elderly parents in need of their attention and care. Some workers have retired early or just quit to care for parents, whereas others have turned down promotions, switched to part-time work, taken leaves of absence, or simply taken time off from work (Livingston 2018c; Porter 2019). The need for companies to offer employees help with elderly dependents has lately become more recognized, and about 25 percent of U.S. companies offer elder care benefits as a recruiting tool (McNamara 2009; Woldt 2010). Caring for elderly parents, as well as issues pertaining to the "sandwich generation"—sandwiched between elder and childcare responsibilities—are explored in Chapter 16.

Policy researchers define **childcare** as the fulltime care and education of children under age 6, care before and after school and during school vacations for older children, and overnight care when employed parents must travel. Childcare may be unpaid or paid and provided by parents, relatives, or paid childcare workers. Many heterosexual couples, even two-earner couples, prefer that the mother care for the children. Sometimes influenced by strong conservative religious beliefs, the couple maintains a traditional division of labor as much as possible, with an employed mother working while the children are in school or working nights while the children are sleeping (Hertz 1997).

Table 10.3 shows childcare arrangements for young children of employed mothers. As you can see in the table, 36 percent of children under age 5 whose mothers are employed are cared for by the other parent—either mom or dad. Often this arrangement involves a "tag team" approach.

Tag-Team Handoffs "Tag team" is a somewhat whimsical term used when two-earner couples exchange childcare and work roles daily or even more often than that. The term is also sometimes applied to single mothers who juggle multiple friend and family childcare resources in a given day (Gornick, Presser, and Batzdorf 2009). The term suggests the clockwork timing necessary to accomplish exchanges.

Tag-team handoffs are tightly scheduled. For example, one parent drives to a predetermined location with the children he's taken care of during the morning hours; he gets out and takes his wife's car to work, while she gets into his car and takes the children

TABLE 10.3 Percent of Employed Mothers' Children Under Age 5 in Various Childcare Arrangements, 2019

CHILDCARE ARRANGEMENT	PERCENT IN ARRANGEMENT
Other parent	36
Grandparent	56
Sibling	3
Other relative	23
Licensed day care center, nursery school, or Head Start	47
Paid nonrelative in the child's home	5
Family day care—paid daycare in another household	11
Self-care	1
Multiple arrangements	29

Source: Malik 2019; U.S. Census Bureau 2011d, Table 1B. Percentages add to more than 100 because some children (29 percent) are in multiple arrangements.

home where she will stay with them for the afternoon (Shellenbarger 2009). Other couples, determined to share childcare, structure their work to this end. If they can afford it, partners may reduce their paid work hours. In blue-collar and in lower-income families, shift work or periodic male unemployment enables a shared-parenting approach.

Childcare by Relatives, Especially Grandparents

In another approach, employed parents arrange for relatives, often grandmothers, to care for their children. As Table 10.3 shows, about 30 percent of children under age 5 whose mothers are employed are cared for by grandparents, with another 13 percent cared for by an older sibling or another relative. Never-married, Hispanic, black, and low-income mothers are more likely to ask grandparents and other relatives to help with childcare. Despite the prevalence of African American grandparent caregivers, black mothers—who still rely heavily on kin networks—saw that option decline in the 1990s, as grandmothers themselves were often in the

labor force (Brewster and Padavic 2002). As a result, black children are now more likely than Hispanics to be in center care (Capizzano, Adams, and Ost 2006). Chapter 9 addresses grandparents raising their grandchildren; Chapter 16 discusses grandparents helping with childcare.

Paid, Nonrelative Childcare Arrangements By the time they enter school, an estimated 44 percent of children have been in a paid, nonrelative childcare arrangement. There are essentially three types of paid, nonrelative childcare: (1) a paid *in-home caregiver* who lives in or comes to the house daily; (2) *family childcare,* in which care is provided in the caregiver's home, often by a woman who has chosen to remain out of the labor force to care for her own children; and (3) *center care,* which provides group care in childcare centers.

Parents who prefer family care seem to be seeking a family-like atmosphere, with a smaller-scale, less routinized setting. They may also desire social similarity of caregiver and parent to better ensure that their children are socialized according to their own values. Center care serves more children per setting, tends to be less expensive, is typically more structured, and can feel more bureaucratized. Family day care tends to be relatively expensive and hence used by more highly educated and wealthier parents. You may want to see "Facts About Families: Selecting a Childcare Facility—Ten Considerations."

Paid Childcare Costs and Concerns Paid care is more common for children in higher-income families. Indeed, paid childcare is unaffordable for many American families (Fraga 2019; Malik 2019). Annual childcare

Optimally for families and for the society as a whole, quality daycare would be available to all parents. However, for many American families finding quality daycare is challenging and expensive.

Goran Bogicevic/Shutterstock.com

Facts about Families

Selecting a Childcare Facility—Ten Considerations How to go about Choosing a Day Care Facility?

State laws establish minimal standards, and professional organizations like the American Academy of Pediatrics have guidelines for quality childcare. The National Association for the Education of Young Children (NAEYC) is an accrediting agency for childcare centers. Check the NAEYC website for accredited centers near you. Although not all good childcare centers have taken this step, accreditation by the NAEYC is a good sign. The facility should:

1. be state licensed;

2. have a low child-to-staff ratio;

3. have a well-trained staff with warm personalities and little turnover;

4. demonstrate cultural sensitivity and knowledge about the diverse racial and ethnic, religious, and social class cultures, as well as the diversity of family forms in this society and potentially at the day care facility;

5. give age-appropriate attention to all the children, including the babies;

6. provide age-appropriate and stimulating activities and play spaces without depending on television watching;

7. use time-outs for discipline rather than physical punishment;

8. welcome parental involvement, even in the form of unannounced visits;

9. accurately respond to questions about practical and financial considerations, such as what happens when the child does not attend as usual because of illness or travel, for example; and

10. provide names of other parents for references.

Critical Thinking

Rank order these suggestions according to what you think are most important and explain why.

Source: Child Care Aware 2019; National Association of Child Care Resource and Referral Agencies (NACCRRA) 2019.

---

costs for a 4-year old range from about $4,000 in rural South Carolina to more than $17,700 in Washington, D.C.—still considerably less than paid infant care yet more than in-state tuition at a public four-year college in about half our states (Hill 2015). Among all working, two-parent families that pay for childcare for children under age 5, average child-care spending is about 10 percent of family income—40 percent higher than the U.S. Health and Human Services definition of affordability (Malik 2019). Less affluent parents who pay for child care spend larger proportions of their income on it. Child care can cost single-parent families an average of 36 percent of their household income (Fraga 2019, p. 4).

On average, married mothers who use paid childcare services for children under age 5 spend 22 percent—more than one-fifth—of their income on childcare. For single mothers with less than a high school education, the comparable figure is up to 36 percent—more than one-third of a woman's earnings. For those living below poverty level, childcare costs can run as much as 61 percent—between one-half and two-thirds—of a parent's income (U.S. Census Bureau 2011d, Table 6). Indeed, average monthly childcare costs exceed what a low-income family might spend on food.

Other parental concerns involve childcare facilities' scheduling that does not meet parents' needs given today's workforce demands. Most family day care and many childcare centers are open weekdays only and close by 7 p.m. Some parents, such as single parents in shift work or those who travel, need access to twenty-four-hour care centers, and some (although not yet enough) are opening (Tavernise 2012). Childcare is also difficult to find for mildly ill youngsters too sick to go to their regular day care facility, although there are now centers beginning to fill this need (National Association for Sick Child Day Care 2010).

As they struggle to find quality, affordable childcare, many parents make more than one arrangement for each child. A system of patched-together childcare arrangements becomes increasingly unpredictable and stressful ("Parents and the High Price" 2009; Roberts 2013). Meanwhile, adequate and affordable childcare accommodations are important not only to parents but also to children's long-term welfare. (See "Facts About Families: Selecting a Childcare Facility—Ten Considerations.")

Self-Care In 2010, of children between ages 5 and 14 whose mothers were employed, nearly one-third (31.6 percent) were cared for by a parent. Another

As We Make Choices

Self-Care (Home Alone) Kids

With parents' difficulties finding available and affordable childcare, many children care for themselves after school or at other times. If the child and the parent each have a smart phone, this option becomes less risky because they can be in touch by texting and/or by the parent's using an app such as Family Tracker to know where the child is.

According to the American Academy of Pediatrics, children should not be self in self-care until they are in fourth or fifth grade. Some states have laws regulating the minimum age that a child can be left home alone, and parents considering this option should become familiar with their state's legislation. Here are some things for parents to consider before choosing this option:

1. Are there neighborhood resources offering support for children in self-care?

2. How does the child feel about being home alone?

3. Is the child able to follow directions and solve problems on their own?

4. How available will a parent or other trusted adult be if needed?

5. How safe is the neighborhood?

Here are some things to consider doing after choosing this option:

1. Set specific rules for the child to follow.

2. Give the child specific instructions about how to reach a parent or other trusted adult at all times.

3. Show the child what to do in certain potential situations, such as if someone comes to the door or the electricity goes out.

4. Make sure the child knows their full name, address, and phone numbers—as well as yours and how to help a third party reach you, if need be.

5. Make sure the child knows how to get help in an emergency, including how to call 911.

6. Show the child appropriate ways to carry a house key so that it is hidden and secure.

7. Provide a daily or weekly schedule of homework and other activities for your child to follow.

8. Have a first-aid kit available and teach the child how to use it.

9. Teach the child key safety tips, such as:

10. **a.** Don't enter the house if there's an open door or window.

b. Lock the door after entering the house.

c. Call or text a parent or parent figure immediately upon getting home to let a responsible adult know.

d. Don't open the door for anyone you do not know.

11. Make time to discuss the day's events with your child when you get home.

For more information, visit the National Center for Missing and Exploited Children website.

Critical Thinking

Under what circumstances do you think it is reasonable to leave a child home alone? Under what circumstances would it not be reasonable? How might family finances affect the decision to let a child stay home alone? How might the availability of extended kin affect the decision to let a child stay home alone?

Source: Child Welfare Information Gateway 2013; National Center for Missing and Exploited Children and U.S. Department of Justice 2013.

---

17 percent were being watched by a grandparent, 7 percent were in after-school programs, and another 5 percent were in day care centers. In addition, about 10 percent of children ages 9 to 11 and 36 percent of those ages 12 to 14 whose mothers were employed were in **self-care**—that is, without adult supervision—for an average of seven hours per week (U.S. Census Bureau 2011d, Tables 3B, 4).

Self-care is more common in nonHispanic white upper-middle-and middle-class families than in black, Latino, or low-income settings, perhaps due to different cultural attitudes concerning children's capabilities and also due to differences in neighborhood safety (U.S. Census Bureau 2011d, Table 3B). Also, nonHispanic white parents are less inclined to use relative childcare arrangements. Self-care is also evident in poverty-level families where parents cannot afford paid childcare arrangements and may not have access to this type of support from family members. Probably due to the high cost of childcare, the proportion of school-age children home alone after school has risen since prior to the recession that began in 2007 (McClure 2010). "As We Make Choices: Self-Care (Home Alone) Kids" gives suggestions for parents considering self-care.

3. Family Leave Family leave involves employees' being able to take extended periods of time from work, either paid or unpaid, for the purpose of caring for their own health needs or for newborns, newly adopted or seriously ill children, or elderly parents—with the guarantee of a job on returning. The concept of family leave incorporates maternity, paternity, ill-child, and elder care leaves.

The 1993 U.S. Family and Medical Leave Act mandates up to twelve weeks of unpaid family leave for workers in companies with at least fifty employees. But unpaid leave will not solve the problem for a vast majority of employees because most working parents need the income. The federal government offers up to twelve weeks of paid family leave for federal employees, and a majority of the top one hundred American companies offer one or more weeks of paid maternity, paternity, or family leave, although this is seldom available to working-class or minimum-wage workers (Cabiness 2019; Rush 2016). It is not surprising that the workers with the highest rates of available paid family leave are in management, professional, and related fields.

4. Fair and Flexible Scheduling Fair scheduling is addressed early in this chapter. In a 2009 poll, nearly half of American men said that companies should provide more flexible work schedules to both men and women (Halpin and Teixeira 2009). **Flexible scheduling** involves such options as **job sharing** (two people share one position), working at home or telecommuting, compressed workweeks, flextime, and personal days (days off for the purpose of attending to a personal matter such as a doctor's appointment or a child's school program). Compressed workweeks allow an employee to concentrate the workweek into three or four or sometimes slightly longer days. **Flextime** involves flexible starting and ending times, with required core hours.

Flexible scheduling, although not a panacea, can help parents share childcare or be at home before and after an older child's school hours (Jang, Zippay, and Park 2012). Some types of work do not lend themselves to flexible scheduling, but the practice has been adopted by some companies and the federal government because it offers employee-recruiting advantages, reduces turnover, and frees up office space when some employees work at home.

Although telecommuting had been considered the unquestioned way of the future—actually the way of the *present*—and although workers who have flexible hours report enhanced job satisfaction and loyalty to their employer, telecommuting came under corporate scrutiny and renewed national debate when Yahoo changed its policy to prohibit the practice. Several

corporate executives outside Yahoo argued that the move was "a backward step in an age when remote working is easier and more effective than ever" (Guynn 2013; Pofeldt 2013). The fact that Marissa Mayer, Yahoo's CEO, is a married mother of a young child was not lost in the debate. "When a working mother is standing behind this [denying employees telecommuting privileges], you know we are a long way from a culture that will honor the thankless sacrifices that women too often make," read one e-mail sent to a technology blog (Guynn 2013).

Ironically, while many workers lobby for more opportunity to telecommute, others feel strapped by career expectations that they be digitally available round the clock. A few European countries—France and Germany, for example—have enacted "right to disconnect" laws assuring that employees cannot suffer negative sanctions if digitally unavailable evenings or weekends (Hardy 2018).

Who Will Provide What's Needed to Meet the Challenges?

Policy experts, lawmakers, employers, parents, and citizens disagree over who has the responsibility to provide what is needed regarding work–family solutions. A principal conflict concerns whether such solutions as childcare or family leave should be government policy or constitute privileges for which a worker negotiates with an employer.

Most European countries remain committed to *paid* family leave policies (Human Rights Watch 2011). These nations have more pronatalist and social-welfare orientations than the United States and view family benefits as a right belonging to all citizens (Lewis 2008). Accustomed to a lack of family policy at the federal level, American parents sometimes turn their attention to local schools as a source of help for the care of older children in after-school programs and younger children in preschool programs and all-day kindergarten (Swanbrow 2000). Evidence that children do better when supervised after school than when left alone has led policy makers to call for expansion of after-school programs (Aizer 2004; McClure 2010).

In the private sector, some large corporations demonstrate interest in effecting **family-friendly workplace policies** that are supportive of employee efforts to combine family and work commitments. Such policies include on-site childcare centers, sick-child care, subsidies for childcare services or childcare locator services, flexible schedules, parental or family leaves, workplace seminars and counseling programs, and support groups for employed parents.

But family-friendly policies are hardly available to all American workers (Rush 2016). As previously mentioned, professionals and managers are much more likely than technical and clerical workers to have access to leave policies, telecommuting, or flexible scheduling ("National Compensation Survey" 2007, Table 19). "At the high end, the big corporations are stepping up to provide benefits to help families, and at the lower end, as women leave welfare, there's now much more support for the idea that they deserve help with childcare. But the blue-collar families, the K-Mart cashier, get nothing" (work–family policy expert Kathleen Sylvester, quoted in Lewin 2001).

In another wrinkle, workers without children argue for a better "work–life balance." Some complain about what they see as the privileging of parents when they themselves may have family caregiving needs: for elderly parents, siblings, or friends with whom they maintain caregiving relationships. Furthermore, they may find it onerous to cover for parent coworkers who are on leave or out of the office. They may feel that fairness should permit schedule flexibility for other than specifically caregiving needs. Some companies have begun to accommodate these workers by redefining policies previously characterized as "family-friendly" and instituting sabbaticals, "flexible culture," and "employee-friendly" policies.

We would like to think that family- or employee-friendly companies represent the future of work. After all, "children . . . are 'public goods'; society profits greatly from future generations as stable, well-adjusted adults, as well as future employees and tax payers" (Avellar and Smock 2003, p. 605). Nevertheless, these voluntary programs and benefits do depend on cost constraints and corporate self-interest and are likely to be less available during economic downturns or restructuring. Although family-friendly programs need to be more widely available to all workers, the family life course theoretical framework, described in Chapter 2, directs us to keep in mind that most workers need extensive family support only while parenting young children. From that perspective, the challenge looks less daunting.

In general, multiple roles can be negatively conflicting or positively synergistic, complementing each other. Work and family roles can conflict—when work demands are unreasonable, when childcare is difficult to organize or afford, or when family demands are too high as when one is sandwiched between caring for children and also for aging parents. Work and family roles can also be (positively) synergistic—when money earned enhances family life or when what happens at work is uplifting or enhances self-esteem and good feelings spill over into family interactions.

Family-Friendly Workplace Policies and Unintended Consequences

Sometimes social policies that are designed to help working families have unintended consequences. For instance, requiring that companies offer maternity leave may dissuade them from hiring women; requiring that businesses provide childcare can result in their offering lower salaries. Researchers say that there is no easy way to prevent family-friendly policies from backfiring like this. One precaution is not to make employers foot the bill for family-friendly policies but rather to finance them by means of employee payroll taxes, for example—as is the case in a few states including California and New Jersey (Miller 2014b, 2015d).

Entering the political arena to work toward the kinds of changes families want is one aspect of creating satisfying marriages and families. We need to focus on "the ways that macrolevel discourses convey the priorities of the powerful from top down and become objects of struggle from the bottom up" (Ferree 2010, p. 433). We have historical evidence that social activism, coupled with government intervention in the service of families, can be effective. European countries evidence ways to successfully alleviate work–family conflict. In the United States, the War on Poverty of the 1960s, which created Head Start programs, among other things, successfully reduced poverty levels (DeNavas-Walt, Proctor, and Smith 2012, Figure 5). But employed couples also want to know what *they* can do themselves to maintain happy marriages. We now turn to that topic.

THE TWO-EARNER COUPLE'S RELATIONSHIP

Policy analyst Margaret Usdansky has noted a paradoxical incongruity regarding social class and the work–family interface. Many spouses at the upper end of the social-class continuum desire to share paid and unpaid labor equally, whereas less educated couples with less money may prefer traditionally specialized breadwinner-homemaker roles. But we've seen that, hampered by structural impediments, many couples' behavior does not fully align with their values or desires. Feminists see this development as a *stalled revolution*: Attitudes and behaviors became increasingly egalitarian since the 1970s, then the trend toward more egalitarian behavior slowed during the 1990s (Coontz 2013; Fairchild 2014). Why?

Finding it difficult to manage two careers, house care, and childcare, middle-class couples may decide that the less well-paid partner (usually the wife) will

leave the labor force or reduce her hours to part-time: "So I decided to quit, and this was a really, really big deal . . . because I never envisioned myself not working" (a female marketing executive in Usdansky 2011, p. 163; Miller 2019). On the other hand, many working-class wives, who may not have wanted to, have joined the labor force, because either the family needs more income than what the husband alone can provide or the husband has been laid off or cannot find work (Usdansky 2011).

Relatedly, human development scholar Jing Zhang and colleagues investigated what they term "gender role disruption"—conflict between ideas about how gender should be performed and how it is actually enacted. This research was conducted among forty university student couples, about half of them with children. They had come to the United States from China for the husband's further education. In China, wives expect to be employed outside the home and to earn wages equal to their husbands'. After giving up their employment to follow their student husband to the United States, they found themselves unable to speak English, unemployed, at home all day, and feeling "useless." Their situations and self-esteem were hardly improved by their husbands' culturally conditioned negative attitudes about stay-at-home wives (Zhang et al. 2011).

Frances Roberts/Alamy Stock Photo

This couple is returning home from a long workday, plus time getting to and from work on mass transit. One effect of the trend toward two-earner families is increasing inequality between families with high-status, high-paying *careers* and those with two poorly paid jobs. The tendency of college-educated professionals to marry other college-educated professions enforces this difference (Usdansky 2011). However, for two-earner couples of all social classes, longer workdays can mean lack of sleep.

Fairness and Couple Happiness

According to a national Pew Research poll of marrieds, sharing household chores is a "very important" aspect of marriage—56 percent of respondents said so. The only two aspects that got higher percentages were having shared interests (64 percent) and a satisfying sexual relationship (61 percent) (Geiger 2018). Two kinds of changes are involved in today's developing work–family interface: In heterosexual unions, women are increasingly sharing the provider role, and men take more responsibility for unpaid household work.

In considering the provider role, we turn to the notion of *meaning*: Is women's sharing of the provider role a *threat*, so that men fear losing masculine identity, women's domestic services, and power? Or is a woman's sharing the provider role a *benefit* because men benefit materially from wives' employment and earnings and from a partner's enthusiasm for the wider world? Some men may view a partner's earning more than they do as a threat. Actually, research shows that when wives earn more than husbands do, both can be reluctant to admit it publicly (Bernard 2018; Miller 2018b). Other men may see this situation as a nonissue (Coughlin and Wade 2012).

Generally, whether partners think their work–family arrangement is fair is strongly associated with the couple's happiness (Frisco and Williams 2003; Wilcox, Marquard et al. 2011). For instance, in households where the husband is more religious than the wife, particularly in Islam and evangelical Protestantism, strict, traditionally gendered divisions of labor can lead to family tension and decreased marital satisfaction for the wife (Duba and Watts 2009; Vaaler, Ellison, and Powers 2009). For the most part, however, social scientists see a "gender convergence" in current attitudes regarding work and family roles. Both men and women want a balance of work and family in their lives (Coleman 2017). Then, too, one's "fair share" of household labor may depend on gendered expectations or on hours spent in the labor force (Konigsberg 2011). Men are doing more than they did in the past, and women are unloading some of their former responsibilities. Today we see relatively comparable men's and women's total hours of paid and unpaid household labor (Coleman 2017; Livingston and Parker 2019). Partners may conceptually combine housework and employment to conclude that the total burden of family responsibility is fair. Both husbands and wives are significantly more likely to say that they are "very happy" when both feel that the housework is shared equitably (Wilcox, Marquard et al. 2011, p. 38).

Husbands may now carry a greater share of the family work than in the past, and when the transition proceeds from a desire to achieve a more equitable relationship, the result can be greater intimacy.

Summary

- Traditionally in heterosexual unions, the man's role was to be the provider and the woman's to be the homemaker. Expectations concerning this division of labor have changed and continue to change as more and more women have entered the workforce.

- Nevertheless, women remain segregated occupationally, continue to average lower incomes than men, and do more unpaid household labor than men do.

- Paid work is not often structured to allow time for household responsibilities, and women more than men continue to adjust their time to accomplish both paid and unpaid work.

- In recent years, men have been increasing their share of the housework, and men and women now have a balance in total work hours—that is, time spent in employment plus time spent on unpaid family labor.

- Household work and childcare are pressure points as heterosexual women enter the labor force and two-earner partnerships become the norm. To make it work, either the structure of work must be changed, social policy must support working families, or women and men must change their household role patterns—very probably all three.

- Regarding the work–family interface, a causal feedback loop operates, according to which workplace policies affect within-family decision making. For example, lower average pay for women may lead to deciding that the (lower-paid) wife will reduce employment hours when faced with work–family conflict.

- Cultural expectations and public policy affect people's options. Provided that individuals increasingly come to realize this, we can hope for pressure on public officials and corporations to meet the needs of working families by providing supportive policies: parental leave, childcare, and flextime.

- To be more successful, two-earner partnerships will require social policy support and workplace flexibility.

Questions for Review and Reflection

1. Discuss to what extent distinctions between men's and women's work and family caregiving responsibilities are disappearing.

2. Write two hypotheses regarding the division of labor in lesbian or gay male families.

3. Discuss research-based reasons why in general women do more household labor than men do.

4. What work–family conflicts do you see around you? Interview some married or single-parent friends of yours for concrete examples and for some suggestions for resolving such conflicts.

5. **Policy Question.** What family-friendly workplace policies would you like to see more of? What are some ways you could activate for these?

Key Terms

causal feedback loop   264
childcare   265
family-friendly workplace
   policies   269
Family leave   269
Flexible scheduling   269
Flextime   269
good providers   251
involved fathers   251

job sharing   269
kin keeping   259
labor force   248
motherhood penalty   255
neotraditional families   255
occupational segregation   253
provider role   250
role conflict   250
second shift   261

self-care   268
shift work   256
spillover   249
stay-at-home dads   252
two-career relationship   256
two-earner partnerships   256
unpaid family work   257
wage gap   254

11

COMMUNICATION IN RELATIONSHIPS, MARRIAGES, AND FAMILIES

Learning Objectives

1   List six characteristics of *cohesive families*.

2   Describe the Four Horsemen of the Apocalypse and why they should be avoided.

3   Discuss gender differences in couple and family communication.

4   List ten guidelines for effective interpersonal communication.

5   Describe how technology and social media are influencing the way we communicate.

6   Explain how communication with "outside others" is associated with good social and emotional health.

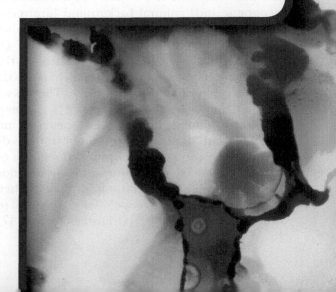

We all want good relationships, ones that we can trust to continue in supportive ways (Gottman 2011). A firm attachment to parents and caregivers helps children develop the capacity for intimate relationships later on. Adolescent and young adults experiment with different types of relationships, some of which may evolve into permanent ones. Midlife parents and family members hope to maintain positive relations while faced with difficult challenges such as putting food on the table or getting their children through school. Older adults focus on maintaining already established bonds with spouses, friends, and other family members.

Families are powerful environments. Virtually nowhere else in our society is there such capacity to support, hurt, comfort, denigrate, reassure, ridicule, hate, and love. Research from a variety of samples and pertaining to a variety of family situations overwhelmingly supports what may be intuitively obvious: Conveying affection for one's partner and other family members, and receiving affection, is a vitally important determinant of relationship and family happiness, and one's social and emotional well-being (Fagan 2009; Soliz, Thorson, and Rittenour 2009). For example, in a study of 281 couples, negative couple interaction was associated with lower work satisfaction and poorer health for men and elevated depression for both women and men (Sandberg et al. 2013). And although conflict is a natural part of every relationship, developing positive communication skills can help family members to resolve conflicts in constructive ways.

Some scholars have criticized family communication research. They have argued that most family communication studies are designed as if only the couple or family members are involved. This indeed used to be the case. Communication research has rapidly expanded to include studies of cohabiting and dating couples, parents and children, siblings, and whole families, including nontraditional family forms such as LGBTQ+, multiracial, and stepfamilies. Incorporating the family ecology model (see Chapter 2), some communication researchers have begun to take a *dyadic approach* in which "elements of the macroenvironment, such as [economic and] cultural background [are recognized as interacting] directly and indirectly with spouses' individual characteristics and marital behavior" (Helms, Supple, and Proulx 2011; Trail and Karney 2012). Importantly, in modern societies the mechanisms through which we communicate evolve quickly. There is still a lot we don't know about how technology (e.g., computers, the Internet, mobile devices, social media) is influencing family and couple communication, and it is difficult for researchers to keep up. Even the language of communication has changed. Anymore, *talking* does not necessarily mean a conversation was had. Rather, it's a synonym for "seeing someone" as in, "We're not serious, Mom. We're just talking." (Orenstein 2016, p. 47).

This chapter emphasizes the importance of communicating affection as well as addressing conflict in positive ways. We'll examine connections between communication and relationship satisfaction. We'll discuss gender differences in couple and family communication. We will review ten guidelines suggested for addressing family and couple conflicts. And we'll consider how technology, including social media, has changed communication patterns. To begin, we'll look at the characteristics of cohesive families.

FAMILY COHESION AND CONFLICT

Many Americans believe their marriages and family relationships are closer today than in the past. Fifty-one percent of marrieds feel that their spousal relationships are closer than their parents' were, and 40 percent say that the families they have today feel closer than their childhood families did (Pew Research Center 2010a). **Family Cohesion** represents the sense of unity, togetherness, and closeness within a family, and is related to the extent to which family members are affectionate, respectful, and warm with each other and can be (Demby, Riggs, and Kaminski 2017). A couple or family can have too much cohesion (an **enmeshed couple or family**) or too little (a **disengaged couple or family**). In enmeshed families, all things are discussed openly—nothing is off limits. However, parents who talk about money or in-law problems at the dinner table can worry and upset their children. Some parents hamper their children's autonomy by "checking in" several times a day via text or apps and by tracking their whereabouts with their phone. Families who are disengaged may share too little. In these families, children may feel pressure to keep "family secrets" such as physical or emotional abuse. The negative effects of bullying, for instance, can be lessened by communicating with a parent about what's going on, and children in disengaged families may have to look elsewhere for support or not at all (Ledwell and King 2015).

Experts advise a *balanced level of cohesion*—one that combines a reasonable and mutually satisfying degree of emotional bonding with individual family members' need for autonomy. In this chapter, we will use the term *family cohesion* to refer to a balanced degree of cohesion—neither enmeshed nor disengaged. Before going further, we should recognize that for different families—and for families of different races and ethnicities—the definition of *balance* with regard to family cohesion varies. "If a couple's/family's expectations or subcultural group norms support more extreme [cohesion levels], families can function well as long as all family members desire the family to function [at that level]" (Olson and Gorall 2003, p. 522). For instance,

in Mexican American families, a relatively high level of family cohesion has been related to positive outcomes for adolescents (Behnke et al. 2008; Martyn et al. 2009). New technologies can challenge and complicate family cohesiveness. For example, whereas some couples may be okay with sharing passwords and email accounts (as 67 percent and 27 percent, respectively, do), others may not be (Pew Research Center 2014).

Characteristics of Cohesive Families

To find out what makes families cohesive, social scientist Nick Stinnett researched 130 "strong families" in rural and urban areas throughout Oklahoma (Stinnett 1985, 2008). Obviously, this limited sample, selected with help from home economics extension agents, has no claim to representativeness. Furthermore, the concept *strong family* is subjective. Various individuals or groups have their own ideas about just what a strong family is. But Stinnett's research helped to advance ideas about what makes for couple and family cohesion. In general, Stinnett's families constructed their lives in ways that enhanced family relationships. Instead of drifting into relationship habits by default, they made knowledgeable choices, each member playing an active part in carrying out family commitments. When Stinnett made his observations, the following six qualities stood out. More recent research has corroborated Stinnett's findings and is included in the following discussion. Stinnett's observations:

1. Both verbally and nonverbally, family members often openly expressed their *appreciation for one another.* Intent on finding the best in one another, they "built each other up psychologically" (Stinnett 2008; see also Niehuis et al. 2011; Samek and Reuter 2011).

2. Members of cohesive families had a *high degree of commitment* to the family group as a whole (Stinnett 2008). Families like this create a shared family identity and reality (Offer 2013; Rueter and Koerner 2008). From a symbolic interactionist perspective, families create a shared reality through frequent, spontaneous, and unconstrained conversations that allow family members to participate together in defining family beliefs, values, situations, events, and rituals. Social media and video Internet communication technologies such as Skype allow for developing or ongoing family connectedness across distances and even across continents (Furukawa and Driessnack 2013; O'Neal and Mancini 2017). Even family members who live relatively close to one another take advantage of technology. As one father explained,

I keep a family messaging chat open with my wife and children. My kids—both in their 20s—live in Brooklyn, which is close to where we live, but over an hour away. However, we all participate in the chat, often several times in a day. We share pictures, links, stories, plans. It is simply much lower friction than how I managed to remain in contact—or didn't, really—with my parents when I was in my 20s. (Anderson and Rainie 2018)

There are many questions regarding how communication technology is being used by families and how it affects, and is affected by, family dynamics. For example, 25 percent of married or partnered adults have texted their spouse or partner when they were both home together (Pew Research Center 2014). Could this be a form of avoidance, or is it simply more convenient than walking down the hall?

3. Dinnertime together is important to family cohesiveness and to positive child and adolescent behaviors (Goldfarb et al. 2015). Moreover, rituals surrounding holiday celebrations, birthdays, and rites of passage (e.g., graduations) are important venues for encouraging communication in families (Galvin and Braithwaite 2014). Simply spending leisure time together (for instance, having regular movie or game nights) is also important for family members (Clayton and Perry-Jenkins 2008; Smith, Freeman, and Zabriskie 2009). In his study, Stinnett found that, on a regular basis, members of "strong families" *arranged their personal schedules* so that they could do things together. Today, 11 percent of adults report sharing an online family calendar (Pew Research Center 2014). In Stinnett's study, when life got so hectic that members didn't have enough time for their families, they listed the activities they were involved in, found those that weren't worth their time, and scratched them off their lists (Stinnett 2008). In a college-student sample, those who reported more shared time with their grandparents had more satisfying relationships with them (Mansson, Myers, and Turner 2010).

Unfortunately, with both parents often working and children's busy schedules, it is challenging for family members to spend time together. As Figure 11.1 indicates, the days of families regularly sitting down together for the evening meal are over. Family meals are especially difficult to accomplish as children get older and become more involved in extracurricular activities. Whereas 63 percent of children age 5 and younger ate dinner with their families six or seven days a week, only 53 percent of children ages 6 to 11 and 37 percent of children ages 12 to 17 did so (Murphey 2012). Family dinners have less positive effects for children in nontraditional family forms, such as single-parent families and stepfamilies (Stewart 2010). Fortunately, if shared dinners aren't possible, *breakfasts* together can produce similar positive effects on families, adults, and children.

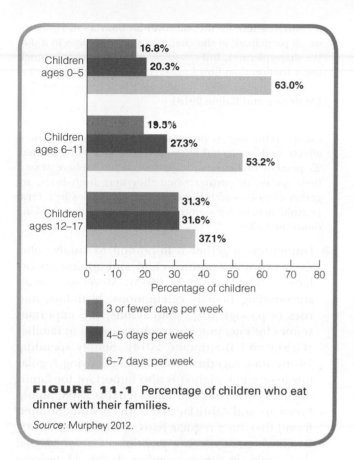

FIGURE 11.1 Percentage of children who eat dinner with their families.

Source: Murphey 2012.

4. Stinnett found that strong families were able to *deal positively with crises*. Family members were able to see something good in bad situations, even if it was just gratitude that they had each other and could face the crisis together (Stinnett 2008). Chapter 13 addresses dealing creatively with stress and crises.

5. Many of the families that Stinnett studied had a *spiritual orientation*. Although they were not necessarily members of any organized religion, some of these families had a sense that their union and family had some "sanctified" purpose and power greater than themselves (Ellison et al. 2011). Typically, these families had a "hopeful attitude toward life" (DeFrain 2002). Having a "personal relationship with God" is associated with higher marital quality, to the extent that it encourages religious conversations between partners (David and Stafford 2015; Olson et al. 2015).

6. These families had positive communication patterns. Members of families like these talk with and listen to one another, conveying respect and interest (Schrodt 2009). They confirm, validate, and accept one another (Dailey 2009). Technology can help and is especially used by younger generations; 41 percent of 18- to 29-year-olds in serious relationships say they felt closer to their partner because of online or text message conversations (Pew Research Center, 2014).

In addition to the previously mentioned characteristics, cohesive families and supportive couple relationships involve fairness, prudence, humility, tolerance, gratitude, justice, charity, and forgiveness (Day and Acock 2013; Fincham, Stanley, and Beach 2007).

As noted above, maintaining family cohesion can be challenging for families who are separated by long distances or who otherwise live apart. Divorce, incarceration, immigration, and work-related travel can reduce opportunities for the activities listed above, even with advances in technology. In "Families at Arm's Length," Catherine O'Neal and Jay Mancini (2017) argue that there are many ways to "do family" at a distance. For example, there are 3.5 million commuter marriages—marriages in which spouses are apart three to four times a week—in the United States. This type of marriage is not indicative of a troubled marriage or precursor to divorce as people sometimes assume. Although couples in commuter marriages may find their marriages less satisfying, they report more satisfaction with their work lives and less stress overall (Bergen 2017). Communications professor Erin Parcell and colleagues (Parcell, Baker, and Johnson 2017) provide advice for military families facing deployment, ranging from talking about the upcoming separation beforehand (what does each person expect, desire, or fear?), signing up for a family support group, adapting expectations when unforeseen circumstances arise, and preparing for possible relationship turbulence upon the soldier's return. Synchronous (e.g., phone, FaceTime) and asynchronous (e.g., texting, email) technology can help divorced parents stay in touch with their children. Divorced parents can use technology to reduce conflict and reduces the possibility of children having to be the "go between" (Markham 2017).

Children, Family Cohesion, and Unresolved Conflict

An individual's family life growing up can influence how he or she communicates with others in adulthood. In a study in which researchers observed couples' communication in a lab, male partners who came from a "risky family background" displayed greater negativity toward their female partners than men who did not (Maleck and Papp 2015). Another study indicated that adolescents who enjoyed greater parental support and supervision as children experienced less relationship conflict with intimate partners (Valle and Tillman 2014). Regardless of family structure or background, however, a family characterized by warmth, cohesion, and generally supportive communication is better for children (Demby et al. 2017; Froyen et al. 2013; Offer 2013). Siblings matter, too. If a child has one or more siblings, feeling close to them is positively associated with healthy social and emotional development

Among other things, cohesive families have high levels of commitment and positive communication patterns. Making time to be together, they build one another up psychologically.

Blend Images/Alamy Stock Photo

Buehler et al. (1998) sampled 337 sixth through eighth graders, mostly from white, married, middle-class household. The students were asked to fill out questionnaires that assessed their behavior and any conflict between their parents. *Externalizing behavior problems* (associated with "acting out," aggression toward others, or rule breaking) were measured by students' agreeing or disagreeing with statements such as "I cheat a lot" or "I tease others a lot." *Internalizing behavior problems* (those associated with emotional or psychological problems) were measured by students' agreeing or disagreeing with statements such as "I am unhappy a lot" or "I worry a lot." The children were also asked about their parents' conflict. Negative *overt parental conflict styles* involved such parental behaviors as calling each other names, telling each other to shut up, or threatening each other in front of the child. Negative *covert parental conflict styles* included behaviors such as trying to get the child to side with one parent and asking the child to relay a message from one parent to the other because the parents refused to speak to each other.

Although conflict between parents was not the only cause of children's behavior problems, there was a strong positive correlation between interparental conflict and behavior problem for both girls and boys. When parents used an overtly negative style, the youth were more likely to report externalizing behavior problems. When parents used a covert, negative style, the youth were more likely to report internalizing behavior problems.

Recent research has investigated the effects of parental and general family conflict on sibling relationships. Relationships with siblings are generally a person's longest and most enduring relationships.

(Samek and Reuter 2011). The quality of sibling relationships can also affect romantic relationships. In a study of adolescents by Doughty, McHale, and Feinberg (2015), conflict with a sibling was associated with less perceived power in romantic relationships, whereas closeness with a sibling was associated with greater perceived power.

Conversely, a home characterized by significant, unresolved, and ongoing conflict negatively impacts children (Schoppe-Sullivan, Schermerhorn, and Cummings 2007). However, contrary to the idea that all conflicts in the home are necessarily detrimental to children, conflicts can end in constructive ways from the children's perspective. "[C]hildren [can] feel an increased sense of well-being resulting from the confidence of knowing that although their parents disagree, their relationship is safe and will endure" (Goeke-Morey, Cummings, and Papp 2007, p. 751).

However, a climate of *unresolved* marital conflict, especially when accompanied by parental depression, which it often is, correlates with children's emotional insecurity (Kouros, Merrilees, and Cummings 2008). A Hong Kong study of children's responses to ongoing parental conflict found that the children felt anxious over the future of their parents' relationship as well as feeling that they had to mediate the conflict (Lee et al. 2010). Other studies showed a link between unresolved parental conflict and children's behavior problems (Feinberg, Kan, and Hetherington 2007; Teachman 2009). In one study,

Unresolved Family Conflict and Sibling Relationships Close sibling relationships can provide helpful emotional and practical support over the life course. Cohesive families foster sibling closeness (Samek and Reuter 2011). But is there another aspect of this story?

Communicating with Children—How to Talk So Kids Will Listen and Listen So Kids Will Talk

There are more as well as less effective ways to communicate with children, and a knowledgeable choice would involve more effective ways. What are some of these methods?

Helping Children Deal with Their Feelings

Children—including adult children—need to have their feelings accepted and respected.

1. *You can listen quietly and attentively.*

2. *You can acknowledge their feelings with a word.* "Oh . . . mmm . . . I see. . . ."

3. *You can give the feeling a name.* "That sounds frustrating!"

4. *You can note that all feelings are accepted, but certain actions must be limited.* "I can see how angry you are at your brother. Tell him what you want with words, not fists."

Engaging a Child's Cooperation

1. *Describe what you see, or describe the problem.* "There's a wet towel on the bed."

2. *Give information.* "The towel is getting my blanket wet."

3. *Describe what you feel.* "I don't like sleeping in a wet bed!"

4. *Write a note (above towel rack):* Please put me back so I can dry. Thanks! Your Towel

Instead of Punishment

1. *Express your feelings strongly—without attacking character.* "I'm furious that my saw was left outside to rust in the rain!"

2. *State your expectations.* "I expect my tools to be returned after they've been borrowed."

3. *Show the child how to make amends.* "What this saw needs now is a little steel wool and a lot of elbow grease."

4. *Give the child a choice.* "You can borrow my tools and return them, or you can give up the privilege of using them. You decide."

Encouraging Autonomy

1. *Let children make choices.* "Are you in the mood for your gray pants today or your red pants?"

2. *Show respect for a child's struggle.* "A jar can be hard to open."

3. *Don't ask too many questions.* "Glad to see you. Welcome home."

4. *Don't rush to answer questions.* "That's an interesting question. What do you think?"

5. *Encourage children to use sources outside the home.* "Maybe the pet shop owner would have a suggestion."

6. *Don't take away hope.* "So you're thinking of trying out for the play! That should be an experience."

Praise and Self-Esteem

Instead of evaluating, describe.

1. *Describe what you see.* "I see your car parked exactly where we agreed it would be."

2. *Describe what you feel.* "It's a pleasure to walk into this room!"

3. *Sum up the child's praiseworthy behavior with a word.* "You sorted out your pencils, crayons, and pens, and put them in separate boxes. That's what I call *organization!*"

Freeing Children by Playing Roles

1. *Look for opportunities to show the child a new picture of himself or herself.* "You've had that toy since you were three, and it looks almost like new!"

2. *Put children in situations in which they can see themselves differently.* "Sara, would you take the screwdriver and tighten the pulls on these drawers?"

3. *Let children overhear you say something positive about them.* "He held his arm steady even though the shot hurt."

4. *Model the behavior you'd like to see.* "It's hard to lose, but I'll try to be a sport about it. Congratulations!"

5. *Be a storehouse for your child's special moments.* "I remember the time you. . . ."

Critical Thinking

What bit of advice given here might you choose to practice when communicating with the child(ren) in your life? Why is it important to encourage children to talk? Why is it important to listen to children? Why does how we talk to children matter?

Source: Faber and Mazlish (2012)

A study of mothers, fathers, and adolescents from 200 middle- and working-class, mostly white families found parental conflict to be causally associated with parents' differential treatment of their children. A possible reason is because ongoing, unresolved parental conflict encourages a parent to form a parent–child alliance with one sibling—a situation that leaves other siblings out. Perceived differential treatment among siblings leads to sibling conflict and underlying resentments that can linger into adulthood (Kan, McHale, and Crouter 2008).

Another study looked at a racially and ethnically diverse sample of elderly mothers with at least two adult children and found that a mother's perceived favoritism while the children were growing up reduced siblings' closeness in adulthood. "Further, mothers' favoritism appeared to reduce closeness regardless of which child was favored, suggesting that siblings' relationships are shaped . . . by principles of equity" (Suitor et al. 2009, p. 1032). According to linguist Deborah Tannen (2006), many grown daughters continue to feel rejected because their mothers persist in showing preference to their brothers.

However, a different study found that, in childhood and adolescence, the sibling who felt slighted was likely to be depressed (Shanahan et al. 2008). Meanwhile a study of 246 two-parent Mexican American families found that adolescents in families with a solution-oriented conflict-management style, rather than ongoing unresolved conflict, had better sibling relationships (Killoren, Thayer, and Updegraff 2008).

Connecting vs. "Spiraling" Of course, children's behavior also depends on how the parents communicate with the children themselves, even when family conflict exists (Schrodt et al. 2009). This can be difficult and what works can even be counterintuitive. For example, parents often try to relate to their children using statements such as "When I was your age . . ." However, saying, "When I was your age, I always did my homework right after school" is less helpful than simply showing empathy: "When I was your age, I was stressed out about homework too. What can we do to work on that?" (Fraga 2017). "As We Make Choices: Communicating with Children—How to Talk So Kids Will Listen and Listen So Kids Will Talk" describes some effective ways to communicate with children.

By now you may have surmised that in families headed by couples, family communication tends to be influenced by the degree of supportiveness or negativity in the couple relationship itself (Doohan et al. 2009). Distressed couples tend toward negative exchanges that put family relationships on a downward spiral (Driver and Gottman 2004). Partners who communicate mutual affection create a positive "spiraling effect" so that the atmosphere becomes one of emotional support (White

1999). Meanwhile, as we'll see in the following section, not all supportive couples and families are alike.

COMMUNICATION AND RELATIONSHIP SATISFACTION

Couples demonstrate different **relationship ideologies**—expectations for closeness or distance as well as ideas about how partners should play their roles. A pivotal task for all couples is to balance each partner's need for autonomy with the simultaneous need for intimacy, togetherness, and support (Brock and Lawrence 2009; Lavy et al. 2009).

Couples also differ in their attitudes toward conflict. Some expect to engage in conflict only over big issues. Others argue more often. Still others expect a relationship that largely avoids not only conflict but also demonstrations of affection (Fitzpatrick 1995). All of these couple types can be happy with their relationship. What matters is whether the partners' actual interaction matches their ideology.

Meanwhile, unhappy relationships have some common features: less positive and more negative verbal and nonverbal communication (for instance, refusing to smile or glaring at the partner during conflict), together with more reciprocity of negative—but not of positive—messages (Gottman and Levenson 2000; Noller and Fitzpatrick 1991; and see Patterson et al. 2012).

Affection and Antagonism

Having gathered data on married couples, researchers Ted Huston and Heidi Melz (2004), in their study of married, heterosexual couples, classified relationships into four types: *warm* or friendly; *tempestuous* or stormy; *bland* or empty shell; and *hostile* or distressed (p. 951). The extent to which their findings and conclusions apply to otherwise committed couples, such as cohabiting or LGBTQ+ couples, is unknown and needs investigation (Lincoln, Taylor, and Jackson 2008).

Warm relationships are high on visible signs of love and affection while low on antagonism. Tempestuous unions are high on both affection and antagonism. Bland marriages are low on affection as well as on antagonism. Hostile marriages are low on love and affection but high on antagonism.

We can assume that warm and friendly relationships best fill the family function of providing emotional security. We can also conclude that hostile ones are undesirable. Huston and Melz called both bland and tempestuous unions "mixed blessing" relationships because these two types evidenced only one of two desirable attributes. Although bland relationships have little antagonism, they lack displays of affection. And although tempestuous couples intermittently show

affection, they deal with conflicts in aggressive, or antagonistic, ways.

Regarding married couples, we note that some maintain flirtatious and playful communication patterns throughout their marriage (Frisby and Booth-Butterfield 2012). According to researchers Huston and Melz (2004), a more common finding is that after the honeymoon stage, there is a "coming down to earth" stage in a marriage. Interestingly, however, in the years after "the honeymoon's over," couples did not necessarily argue more. Instead, their marriages showed a decline in signs of love and affection. "One year into marriage, the average spouse says 'I love you,' hugs and kisses their partner, makes their partner laugh, and has sexual intercourse about half as often as when they were newly wed." Although marriages do not necessarily "become more antagonistic as time passes, the unpleasant exchanges that *do* occur are embedded in a less affectionate context, and thus, the spouses are likely to come to feel that their marriage is less of 'a haven in a heartless world'" (p. 951).

> If our goal is to identify the early signs of a marital rupture, our research suggests that we look to the loss of love and affection early in marriage as symptomatic. . . . This loss of good feelings, rather than the emergence of conflict early in marriage, seems to be what sends relationships into a downward spiral, no doubt eventually leading to increased bickering and fighting and, ultimately, to the collapse of the union. (Huston and Melz 2004, pp. 951–52)

This situation helps to explain the general finding, discussed in Chapter 7, that for many couples the early years of marriage are the happiest. Of course, partners can change this by making informed decisions (see Chapter 1) about communicating intimacy.

Traditional gender role expectations—although rigid and limiting—can be a way of expressing love and caring. When a wife cooks her husband's favorite meal or a husband can pay for family travel, each feels cared about. Many people have noted the potential of shared work and of shared provider and caregiving roles for enriching a marriage (Galinsky, Aumann, and Bond 2009; Van der Lippe 2010). As partners relinquish some traditional behaviors, they need to create new ways of letting each other know they care.

Communicate Positive Feelings

Other research, conducted by widely recognized communication psychologist John Gottman and his colleagues, found that "[t]he absence of positive affect and not the presence of negative affect . . . was most predictive of later divorcing" (Gottman and Levenson 2000, p. 743). **Positive affect** involves verbal and nonverbal expression of affection. Gottman further argued that, at least for the middle-class couples in his sample, he could predict a married couple's later divorce by examining how well the spouses showed that they were interested in each other:

> In a careful viewing of the videotapes, we noticed that there were critical moments during the events-of-the-day conversation that could be called either "requited" [returned, acknowledged, or reciprocated] or "unrequited" interest and excitement. For example, in one couple, the wife reported excitedly about something their young son had done that day, but she was met with her husband's disinterest. After a time of talking about errands that needed doing, he talked excitedly about something important that happened to him that day at work, but she responded with disinterest and irritation. No doubt this kind of interaction pattern carried over into the rest of their interaction, forming a pattern for "turning away" from one another. (Gottman and Levenson 2000, p. 744)

Affection can be expressed verbally, nonverbally (a smile, a touch), or symbolically (Demby et al. 2017). A husband's sharing in housework carries a symbolic meaning for a wife, indicating that her work is recognized and appreciated (Wilcox et al. 2011). Technology has

An important characteristic of happy couples involves disclosure of feelings and showing affection for one another. Meanwhile, even the happiest couples experience conflicts.

Tetra Images/Getty Images

Facts about Families

Ten Rules for Successful Relationships

Psychologists Nathaniel Branden and Robert Sternberg have developed some rules for nourishing relationships. Here are ten. The first seven can be applied to all family relationships. The final three pertain to romantic couple relationships.

In All Family Relationships

1. *Express your love verbally.* Say "I love you."

2. *Be physically affectionate.* Offer (and accept) a touch or hug that says, "I care," "I'm sorry," or "I understand."

3. *Express your appreciation.* Tell your loved ones what you like, enjoy, and admire about one another. Listen with interest.

4. *Help the relationship or family to become an emotional support system.* Be there for each other in times of illness, difficulty, and crisis; be generally helpful and nurturing—devoted to each other's well-being.

5. *Express your affection in material ways.* Send cards or give presents on more than just expected occasions. Lighten a family member's burden once in a while by doing more than your agreed-upon share of the chores.

6. *Accept your family members' shortcomings.* We are not talking about putting up with physical or verbal abuse here. But harmless shortcomings are part of every relationship. Love your family members, not an unattainable idealization of them.

7. *Do unto each other as you would have the other do unto you.* Unconsciously, we sometimes want to give less than we get, or to be treated in positive ways that we fail to offer our family members. Try to see things from another family member's viewpoint.

In Romantic Couple Relationships

1. *Share more about yourself with your partner than you do with any other person.* In other words, keep each other primary.

2. *Make time to be alone together.* This time should be exclusively devoted to the two of you as a couple. Understand that love requires attention and leisure.

3. *Do not take your relationship for granted.* Make your relationship your first priority and actively seek to meet each other's needs.

Source: Branden 1988, pp. 225–228; Sternberg 1988b, pp. 272–277

Critical Thinking

Often, we read a list like the previous one and think about whether our partner or other family members are doing them, not whether we ourselves are. How many of the items on this list do you yourself do? Which two or three items might you begin to incorporate into a relationship?

---

created new ways to show affection. Anthropologist Marcel Danesi at the University of Toronto says *emojis*, ideograms (graphic symbols that represent an idea or concept), and "smileys" can convey emotions that written language cannot always do. In 2015, the "Tears of Joy" emoji (☺) was chosen by *The Oxford Dictionary* as the Word of the Year. On the other hand, the same emoji can have different meanings to different people and the "short-hand" of emojis in the absence of words can lead to misunderstandings (Danesi 2016).

Relatedly, University of California Los Angeles (UCLA) psychologist Shelly Gable studied how one partner responds when something positive happens to the other one, such as a promotion at work (Gable et al. 2004). A partner might respond enthusiastically. ("That's wonderful, and it's because you've had so many good ideas in the past few months.") But they could instead respond in a less-than-enthusiastic manner ("Hmmm, that's nice."), seem uninterested ("Did you see the score of the Yankees game?"), or point out the downsides. ("I suppose it's good news, but it wasn't much of a raise.") According to Gable's research, the only "correct" reaction—the response that's correlated with intimacy, satisfaction, trust, and continued commitment—is the first response: the enthusiastic, active one (Lawson 2004a). "Facts About Families: Ten Rules for Successful Relationships" presents ideas on how to show positive affect.

Letting someone know you care about them is risky in that you must "put yourself out there" and may end up rejected or hurt. In two studies roughly 50 heterosexual dating couples (Luerssen, Jhita, and Ayduk 2017), researchers examined how self-esteem (a feeling of worth or value) influences how partners give and receive affection. They found that compared to partners with high self-esteem, partners with low self-esteem were more uncomfortable giving and receiving affection (in the form of a compliment) and projected their discomfort onto their partners (i.e., they tended to assume their

partner was similarly uncomfortable). The partners of those with lower self-esteem also received less benefit than those with partners with higher self-esteem. The authors concluded, "Given that expressing affection involves putting oneself on the line, we hypothesized that people with lower self-esteem are less likely to take this risk and are half hearted when attempting to do so" (p. 592). Fortunately, most people *are* successful in their pursuit of a satisfying, long-term relationship—the majority of people get married, for example, and most of those stay married and rate their marriage as "happy."

STRESS, COPING, AND CONFLICT IN RELATIONSHIPS

Daily stresses make for communication challenges; they call for coping on the part of family members as individuals and also as parts in the family system (Bernard 2012; Duba et al. 2012; Falconier and Epstein 2011; Ledermann et al. 2010). Stresses might involve how to balance work and family commitments, conflict between parents over child-raising practices, illness or health concerns about a family member, or neighborhood issues. Bad credit, college loans, and credit card debt on one partner's part can strain relationships. One study shows that the greater the mismatch between partners' two credit scores, the greater the chance their relationship will dissolve (Quinn 2018). As a further example, communication challenges can surface in same-sex unions as partners may have to negotiate whether and how to "come out" to relatives when one partner wishes to but the other does not (Lannutti 2013). Sometimes stress involves how to talk with a family member about necessary lifestyle changes after a heart attack or how to discuss with an older family member the need to give up driving privileges (Goldsmith, Bute, and Lindholm 2012).

Couples have to negotiate differences related to their children, money, household chores, in-laws and other relatives, how to allocate their time, and irritation with one another's habits. Another challenging topic involves family communication itself, often with one family member feeling that another is not paying attention or understanding (Papp, Cummings, and Goeke-Morey 2009). Meanwhile, daily stresses interact with communication patterns to make the stressful situation feel either better or worse (Ledermann et al. 2010). Partners can help or hinder one another in dealing with stresses and strain. Experts advise **relationship-focused coping**—that is, coping in which "each partner attempts to cope with his or her own strain in ways that do not harm the relationship and also [attend] to the other's emotional needs" (Falconier and Epstein 2011, p. 307). These strategies can vary by such factors as culture and age (Guilherme et al. 2019)

and social context. For example, players of team sports are more likely to use collaborative methods of coping with stress (Leprince, Fabienne D'Arripe-Longueville, and Doron 2018).

Conflict among Happy Couples

Through it all, anger and conflict should be viewed as challenges to be met rather than avoided (Schechtman and Schechtman 2003). Sociologist Judith Wallerstein conducted lengthy interviews with fifty predominantly nonHispanic white, middle-class married couples in northern California (Wallerstein and Blakeslee 1995). The shortest marriage was ten years and the longest forty years. To participate, both husband and wife had to define their marriage as happy. When discussing what she found, Wallerstein wrote this:

> [E]very married person knows that "conflict-free marriage" is an oxymoron. In reality it is neither possible nor desirable. . . . [I]n a contemporary marriage it is expected that husbands and wives will have different opinions. More important, they can't avoid having serious collisions on big issues that defy compromise. (p. 143)

The couples in Wallerstein's research quarreled:

> In one marriage the husband and wife sat in the car to argue, to avoid upsetting the children. She told him that passive smoke was a proven carcinogen, and while the children were young he could not smoke in their home. He could do what he wanted outside. The man admitted that the request was reasonable, but he was furious. He punished her by not talking to her except when absolutely necessary for three months. Then he accepted the injunction on his smoking and they resumed their customary relationship. (p. 148)

Wallerstein concluded,

> The happily married couples I spoke with were frank in acknowledging their serious differences over the years. . . . What emerged from these interviews was not only that conflict is ubiquitous but that these couples considered learning to disagree and to stand one's ground one of the gifts of a good marriage. (p. 144)

The way couples handle conflict has implications for couples' sexual relationships. Is a benefit of an argument really good "make-up sex"? Israeli researchers Birnbaum, Mikulincer, and Austerlitz (2013) videotaped sixty-one heterosexual couples in steady monogamous relationships who they instructed to discuss "a major problem in their relationship" and their "daily routine." Whereas discussing a problem was negatively associated women's motivation to have sex, it was positively associated with men's motivation. In addition, relational conflict (i.e., discussing a problem) increased men's perception of their partner as sexually attractive, but decreased women's perception of their partner as sexually attractive. It is therefore important to keep in mind that although some

people may desire sex to restore emotional intimacy after a fight, others may need a cooling-off period.

Counselors generally advise that an important aspect of learning to disagree involves expressing anger directly, a skill explored later in this chapter. Here we turn to an examination of *indirect* expressions of anger.

Indirect Expressions of Anger

Learning to express anger and deal with conflict early in a relationship is a challenge to be met rather than avoided. Acknowledging and resolving conflict is painful, but it often strengthens the couple's union in the long run. Meanwhile, feeling uncertain about how invested one's relationship partner is can cause psychological turmoil and pain (Knobloch , Solomon, and Theiss 2013). A key to keeping a relationship healthy is often to share positive events and feelings so that angry arguments occur within an overall context of couple satisfaction and mutual trust.

Many of us may feel uncomfortable about expressing our anger directly. As a result, we can find ourselves engaging in **passive-aggression**—that is, expressing anger indirectly. Chronic criticism, nagging, nitpicking, and sarcasm are all forms of passive-aggression. Procrastination, especially when you have promised a partner that you will do something, may be a form of passive-aggression (Ferrari and Emmons 1994). These behaviors create unnecessary distance and pain in relationships. Most people who use sarcasm do so unthinkingly, unaware of its hurtful consequences. But being the target of sarcastic or otherwise hurtful remarks can result in partners feeling alienated from each other (Murphy and Oberlin 2006). Then, too, sex and other expressions of intimacy become arenas for ongoing conflict when partners passive-aggressively withhold them. For example, a partner makes a disparaging comment in front of company. The hurt spouse says nothing at the time but rejects the other's sexual advances later.

Other forms of indirect anger include sabotage and displacement. **Sabotage**, a means of getting revenge or "payback" (Boon, Deveau, and Alibhal 2009), involves one partner's attempts to spoil or undermine some activity that the other has planned. For example, the partner who is angry because the other invited friends over when he or she wanted to relax may sabotage the evening by acting bored. In **displacement**, a person directs anger at people or things that the other cherishes. An individual who is angry with a partner for spending too much time on a career may hate the partner's expensive car.

Typically, individuals express anger indirectly because they are afraid of conflict, either generally or with reference to a specific person or persons (Murphy and Oberlin 2006; Oyamot, Fuglestad, and Snyder 2010). Advising partners to express anger directly rests on assumptions of equitable power and feelings of security in a relationship (Knobloch and Knobloch-Fedders 2010; Knudson-Martin and Mahoney 2009). Nevertheless, partners and family members who do not express their anger directly risk emotional or sexual detachment (Gottman and Levenson 2000). Of course, it is important to recognize that direct expressions of anger can go too far, resulting in violence (Rehman et al. 2009), discussed in Chapter 12. We turn now to what an important research team has to say about conflict management.

JOHN GOTTMAN'S RESEARCH ON COUPLE COMMUNICATION AND CONFLICT MANAGEMENT

Social psychologist John Gottman (1979, 1994, 1996; Gottman and Notarius 2000, 2003) has earned his reputation in the field of marital communication. He has completed twelve longitudinal studies of more than 3,000 couples, the longest of which lasted twenty years. As a result of this work, Gottman claims that he can predict which (heterosexual) couples will divorce with 90 percent accuracy (Gottman and Levenson 2002). The extent to which Gottman and colleagues' findings apply to same-sex couples and the dissolution of cohabiting relationships is uncertain and more research is needed (Kim, Capaldi, and Crosby 2007).

In the 1970s, applying an interactionist perspective to partner communication, he began studying newly married couples in a university lab while they talked casually, discussed issues that they disagreed about, or tried to solve problems. Video cameras recorded the spouses' gestures, facial expressions, and verbal pitch and tone. After he began this research, Gottman kept in contact with more than 650 of the couples, some for as long as fourteen years. Typically, the couples were videotaped intermittently. Some couples volunteered for laboratory observation that monitored shifts in their heart rate and chemical stress indicators in their blood and urine as a result of their communicating with each other (Gottman 1996).

Studying marital communication in this detail, Gottman and his colleagues were able to chart the effects of small gestures. For example, early in his career he reported that when a spouse—particularly the wife—rolled her eyes while the other was talking, divorce was likely to follow sometime in the future, even if the couple was not thinking about divorce at the time (Gottman and Krotkoff 1989). Other research supports Gottman's methods, and similar patterns were seen among gay and lesbian couples (Gottman et al. 2003; see also Houts and Horne 2008).

In a study of college students, objective raters observed couples completing a drawing task. Even though the observations lasted a very short time—between three and five minutes—raters were able to

Eye rolling on the part of the female partner is highly predictive of union dissolution.

accurately identify pairs in which one or the other had cheated on their partner with someone else (Lambert, Mulder, and Fincham 2014). Recognizing the need for continued research in this area, we present Gottman's highly influential findings here.

The Four Horsemen of the Apocalypse

Gottman's research (1994) showed that conflict and anger themselves did not predict divorce, but processes that he called the **Four Horsemen of the Apocalypse** did. The word *apocalypse* refers to the biblical idea that the world is soon to end, being destroyed by fire. The Four Horsemen are allegorical figures representing war, famine, and death, with the fourth uncertain (*Concise Columbia Encyclopedia* 1994, p. 309). Gottman used the phrase to indicate attitudes and behaviors that foreshadow impending divorce.

The Four Horsemen of the Apocalypse are contempt, criticism, defensiveness, and stonewalling. Rolling one's eyes indicates **contempt**, a feeling that one's spouse is inferior or undesirable. **Criticism** involves making disapproving judgments or evaluations of one's partner. **Defensiveness** means preparing to defend oneself against what one presumes is an upcoming attack. **Stonewalling** involves resistance—refusing to take a partner's complaints seriously. Stonewallers react to their partner's attempts to raise tension-producing issues by refusing to entertain them. Avoiding or evading an argument is an example of stonewalling. Argument evaders use several tactics to avoid fighting, such as vacating the scene when an argument threatens; turning sullen and refusing

to talk; declaring, "I can't take it when you yell at me"; using the hit-and-run tactic of filing a complaint, then leaving no time for an answer or resolution; and saying "OK, you win" without meaning it. Chronic stonewallers may fear rejection or retaliation and therefore hesitate to acknowledge their own or their partner's angry emotions. In several of Gottman's studies, these behaviors identified those who would divorce. Later, after more research, Gottman added **belligerence**, a behavior—as in the above example—that challenges the other's power or authority (for example, "What can you do if I do go drinking with Dave? What are you going to do about it?") (Gottman et al. 1998, p. 6). To illustrate how Gottman's horsemen make their ways into communication, consider the following exchange:

Partner A: I can't find my phone.

Partner B: It's never on when I call you anyway.

Partner A: What's that got to do with anything?

Partner B: So look for it.

Partner A: So help me.

Partner B: You're just like your dad—always expecting somebody to do things for you.

Partner A: "#!@#*!"

In this scenario, Partner A mentions having misplaced a cell phone, and an argument develops, illustrating contempt, defensiveness, and belligerence. When Partner A announces the need for the phone, Partner B raises a complaint: "It's never on when I try to call you anyway." In a less-distressed couple, A might respond to B's complaint. However, A fails to deescalate the interchange and does not acknowledge B's complaint. Less-distressed couples might stop this negative spiral with shared humor or some sign of affection. However, Partner A subsequently requests that B help look for the phone. Partner B's reply is contemptuous and critical: "You're just like your dad—always expecting somebody to do things for you." Again, the couple fails to deescalate the negative affect. This time A calls B a name, maybe with an expletive. Name-calling is contemptuous.

It appears the couple has forgotten what the fight is about. In fact, one wonders whether they ever knew what the fight was about. Counselors point out that distressed couples, like the couple depicted here, may

Nicoleta Ionescu/Shutterstock.com

unconsciously allow trivial issues to become decoys so that they evade the real area of conflict and leave it unresolved. In sum, contempt, criticism, defensiveness, stonewalling, and belligerence characterize unhappy marriages and may signal impending divorce (Gottman and Levenson 2002).

Positive vs. Negative Affect

Gottman and his colleagues videotaped 130 newlywed couples as they spent fifteen minutes discussing a problem that was causing ongoing disagreement in their marriages (Gottman et al. 1998). Each couple's communication was coded in one-second sequences and then synchronized with each spouse's heart rate data, which was being collected at the same time. The heart rate data would indicate each partner's physiological stress.

The researchers examined all of the interaction sequences in which one partner first expressed **negative affect**: anger, sadness, whining, disgust, tension and fear, belligerence, contempt, or defensiveness. Belligerence, contempt, and defensiveness (three of Gottman's indicators of impending divorce) were coded as *high-intensity, negative affect*. The other emotions listed previously (anger, sadness, whining, and so on) were coded as *low-intensity negative affect*.

Next, the researchers watched what happened immediately after a partner had expressed negative affect or raised a complaint. Sometimes the partner reciprocated with negative affect in kind, either low or high intensity. As examples, Partner A whines, and Partner B whines back; A expresses anger, and B responds with tension and fear; or A is contemptuous, and B immediately becomes defensive.

At other times, one partner's first negative expression was reciprocated with an escalation of the negativity. As examples, Partner A whines, and Partner B grows belligerent; or A expresses anger, and B becomes defensive. Gottman and his colleagues called this kind of interchange *refusing-to-accept influence*, because the spouse on the receiving end of the other's complaint refuses to consider it and instead escalates the fight. Here's an example:

> A thirty-eight-year-old woman from a Chicago suburb had a husband who was addicted to porn. She said, "He would come home from work, slide food around on his plate during dinner, play for maybe half an hour with the kids, and then go into his home office, shut the door, and surf Internet porn for hours. . . . I would continually confront him on this. There were times I would be so angry I would cry and cry and tell him how much it hurt. . . . It got to the point where he stopped even making excuses. It was more or less 'I know you know and I don't really care. What are you going to do about it?'" (Paul 2005, p. 3)

Negative escalation is also evidenced in the cell-phone conflict described previously.

In Gottman's study, many couples were likely to communicate with positive affect, responding to each other warmly with interest, affection, or shared (not mean or contemptuous) humor. Positive affect typically deescalated conflict (Gottman and Levenson 2000, 2002). Gottman and his colleagues found that "[t]he only variable that predicted both marital stability and marital happiness among stable couples was the amount of positive affect in the conflict" (1998, p. 17). In stable, happy couples, shared humor and expressions of warmth, interest, and affection were apparent even in conflict situations and therefore deescalated the argument.

The researchers "found no evidence . . . to support the [idea that] anger is the destructive emotion in marriages" (1998, p. 16). Instead, they found that contempt, belligerence, and defensiveness were the destructive attitudes and behaviors. Furthermore, Gottman and his colleagues concluded that the interaction pattern best predicting (heterosexual) divorce was a wife's raising a complaint, followed by her husband's refusing-to-accept influence, followed, in turn, by the wife's reciprocating her husband's escalated negativity, and the absence of any deescalation by means of positive affect. Despite changing gender roles, researchers continue to observe substantial gender differences in communication patterns.

GENDER DIFFERENCES IN COMMUNICATION

Before the nineteenth century, men's and women's domestic activities involved economic production, not personal intimacy. With the development of separate gender spheres in industrializing societies during the late nineteenth century, expressions of emotion became the domain of middle-class women, whereas work was defined as more appropriate to masculinity. As a result of this historical legacy, we see men as less well equipped than women for emotional relatedness (Real 2002). Empirical evidence shows that, on average, females in heterosexual relationships do more social and emotional labor than males to keep the relationship in existence and satisfying (Malinen, Tolvanen, and Ronka 2012). More often than a man, a woman acts as a "relationship barometer"—continually taking the pressure of the relationship and making adjustments accordingly (Faulkner, Davey, and Davey 2005; Fleming and Cordova 2012). Additionally, mothers more so than fathers set the tone for all the relationships in the family system. Based on a sample of eighty-six largely white, middle-class families with children between the ages of 8 and 11 years old, psychologist Shelley Riggs

and colleagues (Demby et al. 2017) found that secure attachment to mothers, combined with positive affect, was associated with overall positive family functioning and better adjustment and adaptation in the child (i.e., ability to handle stress, express emotions appropriately, and behave in prosocial ways with parents and peers). Attachment to fathers was not predictive of children's outcomes. The authors theorize that mothers have a stronger impact because they are the "hands-on" caretakers of children and whereas fathers take on less influential "playmate" role.

Nearly thirty years ago, sociologist Francesca Cancian (1987) expanded on these points to argue that men are equally loving but that women, not men, are made to feel primarily responsible for love's endurance or success. Furthermore, expressions of love are defined and perceived mostly on feminine terms—that is, verbally—and women are the more verbal sex. Expressions of love that men may make, such as doing favors or reducing their partners' burdens, are not credited as love (Cancian 1985). Other research appears to support Cancian's analysis. For instance, a study with 453 heterosexual couples drawn from a national representative survey looked at changes that women would like in their partners compared with changes that men would like. The women were more likely than the men to want increases in a partner's demonstrations of positive emotion (Heyman et al. 2009).

In Cancian's analysis, "The consequences of love would be more positive if love were the responsibility of men as well as women and if love were defined more broadly to include instrumental help as well as emotional expression" (1985, p. 262). Cancian has also argued that a more balanced view of how love is expressed—one that includes masculine as well as feminine elements—would find men equally loving and emotionally profound. It is important to acknowledge, too, that men are capable of expressing love in what we think of as "feminine" ways. For example, contrary to popular belief, men are not more likely than women to fall asleep first after sex. Rather, men and women are equally likely to desire post-coital cuddling, as opposed to sleeping (Kruger and Hughes 2011).

It turns out that "pillow talk" can provide insight into the quality of the couples' relationship. In one study, couples who engaged in longer post-sex affection like snuggling reported higher sexual and relationship satisfaction (Muise, Giang, and Impett 2014). Hormones released during sex also play a role. Women who orgasm, for example, disclose deeper and more intimate thoughts to their partner than women who don't (Denes and Afifi 2014) and men and women who orgasmed assessed greater benefits and fewer risks to self-disclosure, which was associated with higher relationship satisfaction (Denes 2018). Women with higher levels of testosterone (a suppressor of oxytocin, the "feel good" hormone) were less likely to perceive the benefits of post-sex communication and engaged in less positive disclosures and perceived more "risk" in talking (Denes, Afifi, and Granger 2017).

Overall though, women are more likely to view intimacy in terms of breaking down boundaries with self-disclosure, talking, and sharing feelings than are men. This is true for women in lesbian relationships as well. Umberson, Thomeer, and Lodge (2015) conducted interviews with fifteen lesbian, fifteen gay, and twenty heterosexual couples. Danielle described her relationship with Gretchen as "ideal." She explains, "There is just nothing that we can't talk about . . . this one issue that I shouldn't go into or whatever. We don't have that" (p. 546). The gay male couples in the study defined intimacy somewhat differently. Aidan said of his partner, "Max is truly a comfort. Because he is so self-sufficient . . . sometimes I don't feel as though I am giving him the support that I think he needs. . . . But he assures me that he is getting everything he needs. But he doesn't ask for it" (p. 549).

More recently, Deborah Tannen's book *You Just Don't Understand* (2013) suggested that men typically engage in **report talk**, conversation aimed mainly at conveying information. Women, on the other hand, are likely to engage in **rapport talk**, speaking to gain or reinforce intimacy or connection with others. Men are likely to bring up problems, for instance, only when hoping to trigger suggestions for solution. Women, on the other hand, are likely to talk about problems simply to share or foster rapport. These gendered differences

> lead to an imbalance in many families. If the mother is telling about troubles she confronted during her day but the father is not, the result is that mothers come across as more problem-ridden and insecure than fathers. And many men, because they don't tend to talk in this way, understandably assume that a woman who recounts a problem must be seeking help solving it; why else would she talk about it? That's why they generously provide solutions. So the woman's conversational gambit ends up being refracted through the man's point of view. This misunderstanding of women's rapport-talk often results in mothers appearing to their families as less confident, or even less competent, than their husbands. (Tannen 2006, pp. 83–84)

Moreover, some researchers speculate that women, being more expressive and attuned to the emotional quality of a relationship, are more likely than men to suggest marriage counseling or a marriage enrichment program (Fleming and Cordova 2012). Women are also more likely to bring conflict into the open, sometimes in an attention-getting negative tone (Cui et al. 2005). Men try to minimize the impending conflict by either conciliatory gestures or stonewalling. The male's minimization of conflict may appear to the female as failure to recognize her emotional needs (Canary and Dindia 1998;

Noller and Fitzpatrick 1991). In the following, a husband describes this situation:

> The more I try to be cool and calm her the worse it gets. I swear, I can't figure her out, I'll keep trying to tell her not to get so excited, but there's nothing I can do. Anything I say just makes it worse. So then I try to keep quiet, but . . . wow the explosion is like crazy, just nuts. (in Rubin 2007, p. 323)

We might compare this to a wife, who told her interviewers that "I can't stand that he's so damned unemotional and expects me to be the same. He lives in his head all of [the] time, and he acts like anything that's emotional isn't worth dealing with" (in Rubin 2007, p. 322).

Reviews of research on couple communication (Gottman and Notarius 2000; Bradbury, Fincham, and Beach 2000) concluded that men and women differ in their responses to negative affect in close relationships. When faced with a complaint from a partner, men tend to withdraw emotionally, whereas women do not (Green and Addis 2012). Indeed, Gottman's research (1994) showed that in heterosexual relationships, 80 percent of the time it was the woman who brought up issues in the relationship. Conversely, 85 percent of stonewallers were men (Gottman 1994). Researchers have found this pattern to be common enough that some call it the **female-demand/male-withdraw interaction pattern** (Gottman and Levenson 2000). In distressed unions, this pattern becomes a repeated cycle of negative verbal expression by one partner and withdrawal by the other that can spiral out of control (Bradbury, Fincham, and Beach 2000).

Many researchers and therapists agree that generally there is a female-demand/male-withdraw pattern—at least among middle-class white heterosexually married couples like the ones Gottman studied (Miller and Roloff 2005; Weger 2005). However, an alternative view argues that "it is not gender per se but the nature of the marital discussion—for example, whether it is the wife or the husband who desires a change—that may determine who is demanding and who is withdrawing" (Roberts 2000, p. 702; see also Kim, Capaldi, and Crosby 2007). Research on same-sex couples has found the same pattern—that is, one partner demands while the other withdraws (Parker-Pope 2008). And research in stepfamilies suggests that neither partner is likely to demand as much as in many first marriages, while both partners are more likely to withdraw from a conflict (Halford, Nicholson, and Sanders 2007).

Obviously, this pattern leads to both partners feeling misunderstood, thereby decreasing marital satisfaction (Weger 2005). Gottman and his colleagues (1998) concluded that wives and husbands have different goals when they disagree: "The wife wants to resolve the disagreement so that she feels closer to the husband and respected by him. The husband, though, just wants to avoid a blowup. The husband doesn't see the disagreement as an opportunity for closeness, but for trouble" (p. 17).

In one husband's words, "I just got mad and I'd take off—go out with the guys and have a few beers or something. When I'd get back, things would be even worse." From his wife's perspective, "The more I screamed, the more he'd withdraw, until finally I'd go kind of crazy. Then he'd leave and not come back until two or three in the morning sometimes" (Rubin 1976, pp. 77, 79).

Gottman and his colleagues sought to better understand this pattern. You'll recall that the researchers monitored spouses' heart rates as indicators of physiological stress during conflict. They hypothesized that, "it is likely that the biological, stress-related response of men is more rapid and recovery is slower than that of women, and that this response is related to the greater emotional withdrawal of men than women in distressed families" (Gottman et al. 1998, p. 19). That is, when confronted with conflict from an intimate, men may experience more intense and uncomfortable physical symptoms of stress than women do. Men may therefore be more likely than women to withdraw emotionally or physically (contrary to Birnbaum et al., 2013 findings on "make up sex").

An alternative—or complementary—view is that men have been socialized to withdraw (Green and Addis 2012). The cultural options for masculinity include "no 'sissy' stuff," according to which men are expected to distance themselves from anything considered feminine. We guess this could include taking a wife's complaints seriously. In *I Don't Want to Talk About It*, a book on men and communication, therapist Terrence Real (2002) attributes males' withdrawal to a "secret legacy of depression" brought on by men's traditional socialization, particularly society's refusal to let them grieve over losses (e.g., "Don't cry over nothing").

Peggy Orenstein, author of *Girls & Sex* (2016), for her book *Boys & Sex* (2020) interviewed over one hundred college and college-bound boys and men about sex, including hook-up culture, pornography, relationships in the age of #MeToo, and what it means to be a man. She found substantial differences in how girls and boys communicate, and how they are *taught* (or *not taught*) to communicate. Whereas, talking about feelings is encouraged for girls and women, boys and men are taught to not communicate their feelings. When asked to define an "ideal guy," Tristan, age 18, said an ideal guy is "reserved": "You can't flaunt your emotions. You have to be strong. Emotionally and physically. If I have issues, if I have something wrong, that's my problem. I have to deal with it." (p. 9). Ryan, also 18, said, "The biggest determining factor is assertiveness. If I am dominating other people, I am being masculine." (p. 10). It important to note that there are times when girls and women

don't communicate their feelings: in situations that might result in conflict or "making waves," especially when it comes to boys and men. A high school English teacher interviewed by Orenstein explained:

> I talk to a hundred girls a month who are superassertive, feminist, who can correct their teachers about the symbolism of a novel in class. Then they're at a party and some dude's hand is on their leg—or between their legs—and they feel like duct tape is over their mouths. The literally can't say, "Can you move your hand?"

Issues of consent, sexual assault, and sexual harassment are discussed in detail in Chapter 4.

Race and ethnicity also have a powerful effect on boys' communication (and girls) styles. Orenstein (2020) says that black boys in predominantly white schools are taught from an early age how to talk and behave in a manner that is nonthreatening and won't "scare" people. "You've got to have that innocence broken in at a young age," said Xavier, who attends a private high school in an East Coast city, "Especially when you're going to be around people who are so powerful and rich" (p. 138). Overall, what she found in through her interviews surprised her. Despite the boys' descriptions of the "ideal man" as stoic, smooth, "being able to handle yourself," and "an asshole" to get girls, the boys wanted to talk. About their feelings, about their insecurities, pressures, and pain. Often, they paused and said, "I've never told anybody this," or "F*** it, I'm just going to tell you this." (p. 3). Boys' shutting down their feelings can have far reaching effects on the boys themselves, their relationships, and society. Cole's girlfriend broke up with him saying, "You're a really nice person, but you do a lot of things for yourself." (p. 35).

Gay men receive the same messages about manhood, burying emotions, and keeping feelings inside. When he was in college, Zane's mother tried various ways to get him to come out, such as, "You're different than you used to be." Even though he knew they'd be supportive, he was reluctant until finally one day he finally blurted out, "I'm not straight!" (Orenstein 2020, p. 110). Not that those conversations come easily to lesbian girls. One young woman in Orenstein's study (2016) met her mother in a coffee shop. Unable to utter the words "I'm gay," she gave her mom a letter she wrote earlier that started with, "I love you and don't want to disappoint you . . ." (p. 156).

In the end, regardless of gender, Gary Chapman (2014), the author of *The Five Love Languages*, maintains that individuals each have their own "love language," or expressions of love that make them feel loved and secure in their relationships. Different "love languages" can include (1) words of affirmation, (2) quality time, (3) receiving gifts, (4) acts of service, and (5) physical touch. These can vary, say, with women more likely to desire words of affirmation and men desiring physical touch. He says that it less important that couples have similar love styles than that they *understand* each other's particular love language and therefore strive to meet their partner's needs as opposed to their own. Moreover, as couples get older, they are more likely to avoid conflict and argue less (Holley, Haase, and Levenson 2013).

Finally, Gottman and his colleagues (1998) suggest that, as we have already seen, it is important for couples to think about communicating with positive affect more often in their daily living and not just during times of conflict (Gottman and DeClaire 2001). "Facts about Families: Ten Rules for Successful Relationships" suggests ways to do this. Then too, as we have seen, cohesive families have arguments. Most arguments end; most conflicts are resolved. We turn to some guidelines for helping make this happen.

WORKING THROUGH CONFLICTS IN POSITIVE WAYS—TEN GUIDELINES

In addition to the theoretical and empirical contributions of family communication scholars, researchers have focused on developing and teaching communication skills to individuals, couples, and families (Galvin and Braithwaite 2014). First, it is important to consider that different cultural groups vary in their endorsement of openly expressing emotion and directly expressing conflicts (Matsunaga and Imahori 2009; May, Kamble, and Fincham 2015) and that the guidelines suggested in this chapter, which accent direct communication styles, may be ethnocentric or Eurocentric (Tannen 2013). Nevertheless, counselors do advise that there are better (and not-so-good) ways that virtually all couples and family members can resolve differences.

Second, we want to point out that not all negative facts and feelings need to be communicated. Before voicing a complaint, we might ask ourselves, "How important is it?" (Sanford 2006). Counselors suggest that if, after giving it some time and thought, we believe that raising a particular grievance is important, then we should. Similarly, when offering negative information, it is important to ask ourselves why we want to and whether the other person really needs to know. We turn now to ten specific guidelines for constructive conflict management.

Guideline 1: Express Anger Directly and with Kindness

Family members may have the false belief that their intimates automatically know—or should know—what they think and how they feel. This incorrect idea is

detrimental to relationships (Hamamci 2005). When complaints are not addressed directly, conflict goes unresolved, with lingering grievances sparked again and again by "subtle triggers." Consider the following family situation:

> An ongoing point of contention in this family is the mother's belief that her teenage daughter, Joyce, spends too much money on clothes and makeup, which she buys in upscale stores rather than more economical stores, like Wal-Mart. So when the father, who is scanning a newspaper, remarks, "I see Wal-Mart set a record for sales yesterday," the seed is planted for an argument to sprout. (Tannen 2006, p. 123)

The underlying conflict is voiced as follows:

Mom: So? We don't shop at Walmart, so what's the point?

Dad: Okay.

Joyce: What does that have to do with anything?

Mom: Okay, I'm just saying—

Joyce: Saying what?

Mom: Yeah, so what's the point?

Joyce: What point, Mom? You don't shop there either, Mom.

Mom: Yes, I do. You could shop there for toiletries.

Joyce: For clothes you shop there, Mom?

Mom: No.

Joyce: See, so why should we go shopping there for toiletries? . . . I don't go shopping for toiletries anywhere because you buy them for me.

Mom: No, but you buy makeup.

Dad: Well, this year we can do all our Christmas shopping at Wal-Mart. (Tannen 2006, pp. 123–125)

Tension and conflict go unresolved.

Counselors advise expressing anger directly because doing so makes way for resolution (Bernstein and Magee 2004). For example, the mother might say, "I feel that you've been spending more than we can afford on makeup." Counselors further advise that a grievance will be less threatening to the receiver when positive feelings are conveyed at the same time that the grievance is voiced. "If you're angry and resentful, requests for

It may sound impossible to fight more fairly when you're angry, but "practice makes better." Using "I" statements, avoiding mixed messages, focusing your anger on specific issues, and being willing to change are some guidelines worth trying.

Noel Hendrickson/Getty Images

change will be met with resistance and countercharge efforts: 'It's not my problem; it's your problem.' But if you learn to approach each other with acceptance and empathy, you can create a collaborative context, and often people will make spontaneous changes" ("Loving Your Partner" 2000).

So, even better, the mother might say, "You always look nice, and I like the way that you choose to wear your makeup, but I feel that you're spending more than we can afford on it." Being direct is not the same as being unnecessarily critical. In fact, it's possible to be simultaneously direct, sensitive, and kind.

Guideline 2: Check Out Your Interpretation of Others' Behaviors

Because family members and partners in distressed relationships seldom understand each other as well as they think they do, a good habit is to ask for feedback by a process of *checking it out*: asking the other person whether your perception of her or his feelings or of the present situation is accurate. Checking it out often helps to avoid unnecessary hurt feelings or imagining trouble that may not exist, as the following example illustrates:

Family Member A: I think you're mad about something. (*checking it out*) Is it because it's my class night and I haven't made dinner?

Family Member B: No, I'm irritated because I was tied up in traffic an extra half hour on my way home.

Digital Communication and the Rise of Social Media

Technological advances in computing and mobile technology have fundamentally altered the way Americans and people around the globe communicate. Whereas e-mail has been around since the 1980s, *social media* is a more recent development, emerging only in the last decade or so. **Social media** can be generally defined as a group of Internet applications that allow the creation and exchange of *user-generated* content (Kaplan and Michael 2010). Probably the most notable form of social media is Facebook, invented in 2002 by Harvard University students Mark Zuckerberg and Eduardo Saverin. Originally limited to students at Harvard, by 2006 Facebook was available to everyone 13 years of age and older with a valid e-mail address (Abram 2006; Facebook 2015). Today, there are many forms of social media, including blogs, business networks, Internet forums, photo sharing, reviews of products and services, social gaming, video-sharing services, and virtual worlds (Aichner and Jacob 2015). Because new forms of social media are constantly being created, the number of types of social media is impossible to estimate.

According to eBizMBA (2020), the top fifteen social networking sites and apps are Facebook, YouTube, Instagram, Twitter, WhatsApp, Pintrest, Reddit, Ask.fm, Tumblr, Flickr, SnapChat, VK, LinkedIn, Tagged, and Meetup. Through these mechanisms, people can share photos and videos, share links to websites, instant message one another, and make business contacts, among other things. The number of ways to digitally "connect" with others is growing every day, and the possibilities seem endless. Users of Fitbit, an excise-monitoring system worn on the wrist, can encourage friends or family members who are also Fitbit users to "get in their 10,000 steps" (the number of daily steps recommended by the U.S. Surgeon General to maintain cardiac health) by sending "challenges" to them. Various forms of social media are also connected to one another. For example, people who join Twitter or Instagram can check which of their Facebook friends are also users and add them to their contact list.

The effects of the proliferation of social media are unclear although research on the topic has exploded. Is digital communication replacing in-person communication? Are people going to lose their ability to articulate their thoughts orally? One thing is certain: Digital communication lacks nonverbal cues, which can lead to misunderstandings (Golbeck 2014). "Technoference," the unwanted intrusion of technology into relationships, is also a growing problem (McDaniel and Coyne 2016). "Phubbing" refers to snubbing someone you're talking to look at a smartphone (Chotpitayasunondh and Douglas 2016). In a study of 143 married or cohabiting couples, McDaniel and Coyne (2016) found that 62 percent of women said their partners picked up their phones during what was supposed to be "couple leisure time" at least once a day, and one-third said their partners took out their phones during a meal or in the middle of a conversation. Phubbing is negatively associated with perceived communication quality and relationship satisfaction and belongingness (Chotpitayasunondh and Douglas 2018). In the NBC News *2015 State of Kindness Poll* (Raymond 2015), 70 percent of Americans surveyed agreed that technology is weakening relationships, and 68 percent thought that relationships suffer from social media. Psychologist Maggie Mulqueen (2019) argues that technology, namely texting, has led to "social laziness," arguing it has become too easy to be late for appointments, cancel plans, and toss off a quick "sorry" to apologize for hurt feelings. Does a smiley face emoji or champagne-and-cake birthday meme require the same level of thoughtfulness and care as a card or phone call? As one woman says, "Even limiting my friends on Facebook to people I know or knew well personally, I realize that over time we talk and see each other less now that we can merely 'like' or comment on each other's Facebook pages to give the impression we're close" (Anderson and Rainie 2018).

Psychologist Jean M. Twenge, in her book *IGen* (2017) blames the Internet and social media with its "relentless positivity" for unprecedented levels of anxiety, depression, and self-harm among today's teens and young adults, or *GenZ*, who she refers to as "mentally fragile," although research is far from definitive that social media is responsible. As a result of being "connected" 24/7, she worries about a whole generation may end up with a diminished capacity to form in-person connections with flesh and blood human beings. As one young woman wrote,

> Ever since my younger sister got her own Instagram and Twitter accounts, she has spent our car rides silently scrolling, head down, her face lit up with the blue-white of the 5.44 by 2.64-inch cell phone screen. I try to engage in conversation with her, and she responds with absent-minded, one-word answers. I don't blame her for this, because I'm guilty of the same thing. Rather, I'm saddened by the fact that our online lives have become more important than our real ones. (Twenge 2017, p. 291)

Constant exposure to up-to-the-minute posts of pictures and videos of friends having fun, she says, can lead to FOMO (Fear of Missing Out). Indeed, today's teens are more likely to feel lonely and left out than previous generations of teens. Frequent viewing of carefully curated images can leave teens feeling inadequate. What can we do? The Internet and social media aren't going anywhere and simply "unplugging" isn't an option. Most high school and college classes require students to interact on social media, access materials online, or use online mechanisms to communicate with teachers. Nevertheless, Twenge advises kids to "put down the phone" (at least some of the time) and that parents hold off giving their children cell phones as long as possible, regulate their use,

and encourage kids to have unstructured "down-time" with friends.

The same goes for *parents*. In a survey by Common Sense Media, a nonprofit children's advocacy and media ratings organization, four in ten teenagers are concerned about their parents' phone use. The number of parents who report feeling "addicted" to their phone nearly doubled between 2016 and 2019, from 27 percent to 45 percent; 52 percent of parents thought they spent too much time on their mobile devices (Chokshi 2019).

Although some studies have found that social media is eroding young adults' writing (and spelling!) ability, others indicate that use of social media may actually boost literacy levels (Harris and Dilts 2015). As discussed in Chapter 4, digital communication has given rise to Internet predators—people who pretend to be someone they're not to victimize others, especially children. Parents and schools must be constantly vigilant to protect their kids from harm. Growing technology and Internet use worries many parents (Sorbring 2014). Because children tend to be more technologically savvy, parents often have difficulty staying "ahead of the game." Parents who try to regulate their children's use of technology generally face considerable pushback (Fletcher and Blair 2014). On the positive side, school systems are increasingly using social media to bring together students for projects, share documents, and allow parents to see what their kids are working on in the classroom.

The Internet and social media have many benefits. We are exposed to a broader range of people, information, and ideas. There are online support groups for parents of children with chronic illnesses, people with substance abuse problems, and new mothers, which can alleviate feelings of isolation. Adolescents and young adults can use social media to reach out for support. Anna, age 26, had recently broken up with a boyfriend, was homesick and had issues with low self-esteem. It was too late to call her therapist or stop by and see a friend. She stated, "I used to be very shy about posting personal stuff on Facebook because I didn't want people judging me. But that night, I was in such a bad place; I was desperate, and I thought anything would help." Almost immediately, she says, three people who she met while playing Quidditch (a sport based on the Harry Potter books) responded with offers to talk. She talked to two of them until she fell asleep (Jacewicz 2017).

Social media can be especially useful for "relationship maintenance," or keeping up with friends and family who may otherwise fade away (Golbeck 2014). For example, digital communication is often used by college students to stay in touch with siblings, which in turn is associated with greater closeness, especially among siblings who text or instant message in real time (Lindell, Campione-Barr, and Killoren 2015). Moreover, the asynchronous nature of some types of digital communication can help people manage conflict by providing a cooling-off period after disagreements (Golbeck 2014). Shy people can use social media to connect; singles can find partners through dating websites.

On the negative side, because people are more likely to post positive experiences (a vacation, a perfect report card, or a birthday celebration) than negative ones (an illness, an argument, or a poor grade on a test), frequent use of social media can lead to depression and low self-esteem (Tromholt 2016; Vogel et al. 2014). Online bullying and harassment through social media has become a serious problem. Nearly half of the respondents from the NBC poll said they have had to block or unfriend someone because they were being unkind (Raymond 2015). Others may disagree with what their friends post, such as their political or religious views, leading them to cut off relations (Fahrlander 2015). Even Silicon Valley, the birthplace of social media, is developing a consciousness about its potential negative effects and recommends limiting children's time on mobile devices and apps (Bowles 2018).

Are these college students having dinner "together"? Commentators worry that digital communication is replacing in-person conversations and that it is eroding intimacy.

Guideline 3: To Avoid Attacks, Use "I" Statements

Attacks, sometimes interpreted as blame, involve insults or assaults on another's character or self-esteem, which should be considered a "shared relationship resource"—that is, both partners are happier when each one's self-esteem is high, rather than low (Robinson and Cameron 2011). Needless to say, attacks do not help either to enhance self-esteem or to bond a couple (Sinclair and Monk 2004). A rule in avoiding attack is to use the word *I* rather than *you* or *why*. For example, instead of declaring "You're late," or asking "Why are you late?"—both of which can smack of blame—making a statement such as "I was worried because you hadn't arrived" may allow for more positive dialogue. "And while comments like 'Are you trying to put us in the poorhouse?' may be emotionally satisfying in the moment, they're ineffective in the long run" (Mangla 2013). The receiver is more likely to perceive "I" statements as an attempt to recognize and communicate feelings; "you" and "why" statements are more likely to be perceived as attacks, even when not intended as such.

Of course, making "I" statements may be too much to ask in the heat of an argument. One social psychologist has admitted what many of us may have experienced: "It is impossible to make an 'I-statement' when you are in the . . . 'wanting-revenge, feeling-stung-and-needing-to-sting-back' state of mind" (Gottman et al. 1998, p. 18). Of course, this is partly the point. Keeping in mind the possibility of expressing a complaint—at least *beginning* a confrontation—with an "I" statement can discourage family members from getting to that wanting-revenge state of mind in the first place.

Guideline 4: Avoid Mixed, or Double, Messages

Mixed, or double, messages contradict each other. Contradictory messages may be verbal, or one may be verbal and one nonverbal. For example, a family member offers to take the family to a movie yet sighs and says that they are exhausted after a really hard day at work. Or a partner insists, "Of course I love you" while picking an invisible speck from his or her sleeve in a gesture of indifference.

Senders of mixed messages may not be aware of what they are doing, and mixed messages can be extremely subtle. They sometimes result from simultaneously wanting to recognize and deny conflict or tension. A classic example is the *silent treatment*. One partner becomes aware that she or he has said or done something upsetting and asks what's wrong. "Oh, nothing," the other replies without much feeling, but everything about the partner's face, body, attitude, and posture suggests that something is indeed wrong (Lerner 2001).

Communication scholars and counselors point out that there are two major aspects of any communication: *what* is said (the verbal message) and *how* it is said or interpreted (the nonverbal "meta-message"). The meta-message depends on tone of voice, inflection, and body language, as well as on the receiver (Nierenberg and Calero 1973). In a mixed message, the verbal message does not correspond with the meta-message.

Moreover, communication involves both a sender and a receiver. Just as the sender gives both an overt message and an underlying meta-message, so also does a receiver give cues about how seriously she or he is taking the message. For example, listening while continuing to do chores sends the nonverbal message that what is being heard is not particularly important.

Guideline 5: When You Can, Choose the Time and Place Carefully

Arguments are less likely to be constructive if the complainant raises grievances at the wrong time. One partner may be ready to argue about an issue when the other is almost asleep or working on an important assignment, for instance. At such times, the person who picked the fight may get more—or less—than they had expected.

Family members might negotiate a time and place for addressing issues. Arguing "by appointment" may sound silly and be difficult to arrange, but doing so has advantages. For one thing, complainants can organize their thoughts and feelings more calmly and deliberately, increasing the likelihood that they will be heard. Also, recipients of complaints have time before the argument to prepare themselves to hear some criticism.

A qualitative British study of teenagers and their parents found that some parents and teens use mobile phones to raise sensitive issues that they intend to later pursue face-to-face. One teen said, "Well, I think mobiles can be really good if you've got something you don't wanna tell straight away," and another young respondent said, "I'd maybe text if it's something that I can't, I dunno, something I can't get across [face-to-face] and stuff" (Devitt and Roker 2009, p. 192; see also Lasen and Casado 2012). This goes for adult relationships. In a national study, one in ten married or cohabiting partners said they have resolved an argument on-line. Younger couples are more likely to resolve conflicts this way, at 23 percent. On the other hand, 8 percent of couples in committed relationships have argued about the amount of time one of them was spending on-line and 4 percent said they got upset at something they found their partner doing on-line (Pew Research Center 2014). Some worry that reliance on phones and other social media may result in people's failure to learn how to communicate in person about touchy issues (Moore 2010) and can undercut the benefits derived from in-person

interactions (Kushlev, Dwyer, and Dunn 2019). For more on this, see "Issues for Thought: Digital Communication and the Rise of Social Media."

Guideline 6: Address a Specific Issue, Ask for a Specific Change, and Be Open to Compromise

Constructive relationships aim to resolve current, specific problems. Recipients of complaints need to feel that they can do something specific to help resolve the problem raised. This will be difficult if they feel overwhelmed by old gripes. Furthermore, complainants should be ready to propose one or more solutions. Recipients might come up with possible solutions themselves. When family members can entertain potential solutions to a definite problem at hand, they are better able to negotiate alternatives. A specific issue in a household might be a lost TV remote. Rather than family members spending time blaming each other for it being lost, which does nothing to solve the problem, they should decide on a place for the remote to be kept, such as on the coffee table and come up with a list of consequences for people not putting it in its place.

Guideline 7: Be Willing to Change Yourself

The principle that couples or family members should accept each other as they are sometimes merges with the idea that individuals should be exactly what they choose to be. The result is an erroneous assumption that if someone loves you, he or she will accept you just as you are and not ask for even minor changes. In truth, partners need to be willing to be influenced by their loved ones and to change themselves (Lerner 2001).

Therapists note that, in some relationships, each person expects the other one to do the changing: "You have to understand, she's [or he's] impossible to live with" (Ball and Kivisto 2006, p. 155). One counselor team (Christensen and Jacobson 1999) has suggested "acceptance therapy," helping individuals accept their partners and other family members as they are instead of demanding change—although these counselors also suggest that, paradoxically, showing acceptance can lead to a partner's changing behavior. We need to balance acceptance of another against not being a doormat, but being willing to change ourselves is key.

Guideline 8: Don't Try to Win

Counselors encourage us to recognize that there are probably several ways to solve a particular problem, and backing others into a corner with ultimatums and counter-ultimatums is not negotiation but attack. Moreover, wanting to win a dispute with a loved one typically encourages us to use unnecessarily hurtful language, which nonproductively increases the recipient's stress (Priem, McLaren, and Solomon 2010). We're reminded that recipients of painful messages typically see them as more hurtful than do the senders (Zhang 2009). How we say things impacts how others perceive them (Young 2010). Even hurtful information can be conveyed with sensitivity.

Modern, capitalist societies such as ours that emphasize competition encourage people to see almost everything they do in terms of winning or losing (Fromm 1956). In an editorial entitled "Americans Are Now Utterly Intolerant of Ever Being Told They're Wrong About Almost Anything," Tom Nichols argues that Americans have become increasingly opinionated, are increasingly unwilling to engage in conversations with people who have different beliefs, and operate on a binary, winner-take-all basis, even in the face of evidence to the contrary (Nichols 2017).

Yet research clearly indicates that tactics associated with winning in a particular conflict are also those associated with lower relationship satisfaction (Clunis and Green 2005; Heene, Buysse, and Van Oost 2007; Houts and Horne 2008). Losing lessens a person's self-esteem, increases resentment, and adds strain to the relationship. On the other hand, everyone wins when family members mutually agree on solutions to their differences (Carroll, Badger, and Yang 2006).

Guideline 9: Practice Forgiveness

A growing number of therapists suggest that being willing to forgive is critical to ongoing happy relationships (Fincham, Hall, and Beach 2006). Forgiveness "is the idea of a change whereby one becomes less motivated to think, feel, and behave negatively (e.g., retaliate, withdraw) in regard to the offender." Forgiveness is not something to which the offender is necessarily entitled but is granted nevertheless.

Contrary to what many individuals believe, however, forgiveness does not require that the offended partner minimize or condone the offense. Rather, "an individual forgives despite the wrongful nature of the offense and the fact that the offender is not entitled to forgiveness." Further, "forgiveness is distinct from denial (an unwillingness to perceive the injury) . . . or forgetting (removes awareness of offence from consciousness)" (Fincham, Hall, and Beach 2006, p. 416).

Forgiveness is often a process that takes time, rather than one specific decision or act of the will. Being willing to forgive has been associated in research with marital satisfaction, lessened ambivalence toward a partner, conflict resolution, enhanced commitment, and greater empathy (Fincham, Hall, and Beach 2006; David and Stafford 2015).

Guideline 10: End the Argument

Ending the argument is important. Sometimes when individuals are too hurt to continue, they need to stop arguing before they reach a resolution. A family member may signal that he or she feels too distressed to go on by calling for a time-out. Or it could help to bargain about whether the fight should continue at all.

The happily married couples interviewed by Gottman and his colleagues, as well as by Wallerstein, knew how and when to stop fighting. Arguments can end with compromise, apology, submission, or agreement to disagree (Goeke-Morey, Cummings, and Papp 2007). Ideally, a fight ends when there has been a mutually satisfactory airing of each partner's views. Should couples avoid "going to bed angry" as the saying goes? One study suggests that people who keep their anger in, as opposed to letting it out or staying calm under pressure, have worse sleep (Young 2017). So the answer would be yes—make up before bed if you can or put a hold on the argument.

Couples *can* change their fighting habits. The key to staying happily together is to make knowledgeable choices—about not avoiding conflict but dealing with it openly, or directly, and in supportive ways. The goal isn't necessarily agreement, but acknowledgment, insight, and understanding.

Relationship and Family Counseling

Relationship and family counseling is a professional service having two goals: (1) helping individuals, couples, and families gain insight into the actually or potentially troublesome dynamics of their relationship(s), and (2) teaching clients more effective and supportive communication techniques. According to the American Association for Marriage and Family Therapy (AAMFT), this type of counseling is meant to be "solution-focused; specific, with attainable therapeutic goals; [and] designed with the 'end in mind'" (AAMFT 2020). It might be helpful for some people to look at a therapist as a "family coach" (Hudziak and Ivanova 2016).

Experts advise couples or families to visit a counselor when communication is typically hostile or conflict goes unresolved, when they cannot figure out how to resolve a family problem themselves, or when a partner is thinking of leaving a committed relationship. However, counseling is also appropriate—and perhaps more effective—as a preventive technique, undertaken at the onset of family stress or when a couple or family sees a potentially troublesome transition ahead.

Families are made up of individuals, each one seeking not only a unique identity but also a cohesive place to belong. These sisters are sharing memories recorded in digital photos. They're good friends, but the main reason that the majority of adult siblings in one small study gave for continuing their relationship was because "we're family" (Myers 2011).

People go to counselors for help in working through premarital and engagement issues, as well as issues related to cultural clashes, sexual identity, cohabitation, infidelity, divorce, substance abuse, finances, unemployment, co-parenting conflict, infertility, sexual difficulties, and changing roles such as with retirement, remarriage, and stepfamilies, among other concerns. Certain types of interventions and programs may work better for certain families, so it might take some time to find the right "fit" (Wong 2017).

Qualifications of Counselors The qualifications of counselors vary. A counselor who is a member of the AAMFT has a graduate degree and at least three years of clinical training under a senior counselor's supervision. The safest way to choose a qualified counselor is to select one who belongs to the AAMFT. To do so, check the organization's website. Personal recommendations from family members, friends, or both may also be helpful.

It is important to have a counselor you like and trust and who empathizes with you. It is also important that the counselor respect your religious and personal values. Even well-trained counselors can be capable of unintentional bias that may get in the way of productive therapy (Hoch 2019; Wing, Neville, and Smith 2019). For example, couples who come to therapy are often assumed to be heterosexual and cisgender (when a person's gender identity matches their sex assignment at birth; see Chapter 4) (Spenger, DeVore, Spengler, and Lee 2019). If after three or four sessions you do not feel comfortable with the counselor or don't believe the counselor is effective, then it might be a good idea to try someone else. Experts advise interviewing a prospective counselor before beginning therapy. The Mayo Clinic Staff (2020) advises asking lots of questions, including the following:

- Are you a clinical member of the AAMFT, licensed by the state, or both?
- What is your educational and training background?
- What is your experience with my type of issue?
- How much do you charge?
- Are your services covered by my health insurance?
- Where is your office and what are your hours?
- How long is each session?
- How often are sessions scheduled?
- How many sessions should I expect to have?
- What is your policy on canceled sessions?
- How can I contact you if I have an emergency?

Will Counseling Save a Relationship? Despite its substantiated benefits, the extent to which counseling "saves" a relationship is difficult to measure. One issue is that it

takes the average couple six years to begin seeking help for marital problems, which can be too late to resolve long-standing problems (Gottman 1994). Counseling is also based on the presumption that partners are willing to cooperate, and it is possible that one's partner may not be willing. No counselor can or will attempt to change a person to a partner's liking without active cooperation from all involved (Rasheed, Rasheed, and Marley 2010). Nevertheless, a review of studies indicates that couples' therapy positively impacts 70 percent of couples receiving treatment, which is a similar level of effectiveness as individual therapy (Lebow et al. 2012). Positively impacts how? Marital satisfaction? There are a few things to keep in mind regarding the effectiveness of couples' therapy. First, couples' therapy will only work if both members of the couple are motivated by love and the desire to do better (Shafer, Jensen, and Larson 2014). The therapist must understand that each member of the couple contributes to the problems in the relationship. The primary task of the couple is to take a good, hard look at themselves (Psychology Today 2012).

TALK TO A STRANGER: THE IMPORTANCE OF "OUTSIDE OTHERS"

Adult, child, and family well-being is greatly enhanced when they have positive communication habits among family members. However, interaction and communication with friends, community members, and even complete strangers is also important to our social, emotional, and physical health. Thinking of the family ecology model, family cohesion can be further enhanced when families are embedded in supportive social structures and institutions, as family relationships unfold in the context of work, school, church, and recreational and extracurricular activities. Even a brief interaction, with the desk attendant at the gym, for example, or a compliment from a stranger can make your day!

Research on the importance of "outside others" to one's well-being is growing. A happy life is not just based on the *quality* of interactions but also the *quantity*, even if those interactions are fleeting. Psychologists Gillian Sandstrom and Elizabeth Dunn (2013) wanted to find out how people's day-to-day interactions with others affect their happiness and sense of belonging. To find out, the researchers designed a field experiment in which they recruited 60 participants from in front of a Starbucks in a busy, urban shopping district. They randomly assigned the participants to have an interaction with the barista that was either "social" (they were instructed to make eye contact, smile, and have a brief conversation) or "efficient" (they were instructed

to make the transaction as efficient as possible, have their money ready, and avoid conversation). Participants were interviewed immediately following their interaction. What they found was striking. Participants in the "social" group scored significantly higher on positive emotions and belonging than the "efficient" group. They concluded that even seemingly trivial interactions can bolster one's well-being. Your "Have a nice day!" may actually help make a person's day nice, especially for people experiencing stress or difficulties at home.

Technology, and specifically the smartphone, is having a worrying effect on social interaction between strangers according to research (Kushlev et al. 2019). Kushlev and colleagues (Kushlev, Prouix, and Dunn 2016) conducted a study in which college students were asked to find their way to an unfamiliar building on campus. One group was given a smartphone for the task and one group had to use other means, which could include stopping people to ask for directions, using campus signs or maps, or simply wandering around. Although the smartphone users located the building more easily, they felt less socially connected than those who did not use a smartphone; 80 percent of those who did not use a phone talked to another person compared to only 10 percent of smartphone users. In a second study, they found using a smartphone had both positive and negative effects on mood. Participants were less tense because they found the building easier, but were in a less positive mood as a result of having minimal social interaction. They ask, is the benefit of technology worth the loss in social connection? A subsequent study by these authors (Kushlev, Hunter, Proulx, Pressman, and Dunn 2018) conducted in a lab experiment in which a pair of two strangers was instructed to wait in a room together for 10 minutes after being told the experiment was "running behind." Videotapes of their faces showed that pairs who were allowed to keep their smartphones exhibited significantly fewer "Duchenne smiles" (true, authentic smiles, as opposed to "fake" or polite smiles) than pairs who did not have their phones, and significantly fewer smiles of any kind.

This loss of social connection is worrying given what some public health officials refer to as "an epidemic of loneliness" in our society, especially among older adults (Hafner 2016). There is shame and stigma associated with admitting one is lonely, so the problem remains hidden. According to Dr. Carla Perissinotto, "The profound effects of loneliness on health and independence are a critical public health problem." Loneliness is associated with cognitive decline, physical illness, increased secretion of cortisol (a stress hormone), depression, and early death (Hafner 2016). However, communities are increasingly recognizing loneliness, especially among older people. The Friendship Line is a 24-hour

call-in line run by the Institute on Aging in San Francisco. "Men's Sheds," is an international program with the goal of bringing men together and combat isolation. As discussed previously, socialization has hampered men's ability to talk about their feelings directly. Gathering men and "putting them to work," for example, in a woodworking shop making toys, fixing bicycles, or cooking together, provides a safe environment for men to talk indirectly as they work together on community projects (Fallik 2018). Living alone has become increasingly common among all age groups; According to the U.S. Census, one in ten American adults live alone and over one-quarter of all U.S. households contain just one person (Stebbins 2018), suggesting an increasing the need for such programs.

Similarly, community initiatives to strengthen family cohesion. With regard to same-sex families, for instance,

> [c]ommunity initiatives to strengthen families could emphasize the importance of staying in touch and getting along despite disagreements among family members; these campaigns could identify sexual orientation as a topic where adult family members can learn to disagree with each other while agreeing to provide a loving environment for children that is free from conflicts between adults. (Oswald and Lazarevic 2011)

Doing so requires working on ourselves as well as on our relationships. A first step involves consciously recognizing how important the relationship is to us. A second step is to set realistic expectations about the relationship (Cloud and Townsend 2005). As one married woman put it,

> You just have this idealized version of getting married, you know, everybody plays it up as so romantic and so wonderful and sweet. Now that I am married and now that I have gotten older and hit the real world I'm kind of like. . . . It's a lot more hands-on, you know, getting stuff done . . . than it is that idealized romantic notion that you get as a girl. (Fairchild 2006, p. 13)

A third step involves improving our own (1) **emotional intelligence (EI)**, the awareness of what we're feeling so that we can express our feelings more authentically; (2) ability and willingness to repair our moods, not unnecessarily nursing our hurt feelings; (3) healthy balance between controlling rash impulses and being candid and spontaneous; and (4) sensitivity to the feelings and needs of others (Keaten and Kelly 2008). "People who have high EI tend to have stronger relationships and they can manage difficulties calmly and effectively" according to Mind Tools (2020), a career-building skills organization. People are capable of developing greater flexibility of thought, learning to think of several alternative workable solutions to problems and to have several ways of responding to a situation, not just one that habitually comes up by default (Koesten, Schrodt, and

Ford 2009). There are many websites where you can assess your own EI.

Counselors encourage making time for play and incorporating new activities into relationships (Lawson 2004b; Smith, Freeman, and Zabriskie 2009). Social psychologist John Crosby points out that people may misinterpret the idea of "working at" committed relationships: Instead of working *at* relationships, "we may, with all good intentions, end up making work *of*" them (Crosby 1991, p. 287).

IMPROVING COMMUNICATION, SETTING BOUNDARIES, AND TAKING CARE OF YOURSELF

As we saw previously in this chapter, many Americans see their current couple and family relationships as closer than what they experienced in childhood. Due to communication research over the past several decades, people know more today about how to nurture supportive relationships than they did in the past, and they may more actively value doing so. Put another way, they realize that keeping a loving relationship or creating a cohesive family is not automatic. Meanwhile, the quality of a relationship is pivotal to a union's survival (Cherlin 2009a; Ledermann et al. 2010).

Meanwhile, some of us have grown up with poor role modeling on the part of our parents (Ledbetter 2009; Rovers 2006; Schrodt et al. 2009). Regardless of how our parents behaved, we can choose to change how we communicate (Braithwaite and Baxter 2006; Turner and West 2006; Wright 2006). Training programs in couples and family communication, often conducted by counseling psychologists and sometimes designed for specific religions (e.g., Catholic), life course stages (e.g., new parents), or family forms (e.g., stepfamilies), have proven effective in helping to change negative communication patterns (Blanchard et al. 2009; Lucier-Greer and Adler-Baeder 2012; Trillingsgaard et al. 2012). Some family life education and communication education programs are available online (Hughes et al. 2012).

One program for married and cohabiting couples is PREPARE/ENRICH, originally developed by social psychologist David Olson at the University of Minnesota (PREPARE/ENRICH 2020). This program, often administered by counselors or clergy, involves spouses or unmarried partners taking an online assessment that identifies existing strengths and potential weaknesses in their relationship with the goal of improving communication skills. A similar program is PREP (the Prevention and Relationship Enhancement Program), developed by marital communication psychologists Scott Stanley and Howard Markman with the overall aim of strengthening marriages and preventing divorces (Smart Marriages 2020). Marriage Encounter and similar organizations offer weekend workshops designed for mostly satisfied marrieds who want to improve their relationships (Marriage Encounter 2020). Advertising "psychological care for the whole family," the Family Success Consortium offers programs for all couples, whether or not they are married (Family Success Consortium 2020).

Men's groups aimed at encouraging their expressions of emotional intimacy have been shown to enhance couple and family relationships (Garfield 2010). Some conflict-management programs have been developed for child or adolescent siblings (Kennedy and Kramer 2008; Thomas and Roberts 2009) or for families of particular races and ethnicities (e.g., Soll, McHale, and Feinberg 2009). Some programs have been designed for same-sex couples (Heffner 2003; Unitarian Universalist Association n.d.). As mentioned in Chapter 9, some parenting-enhancement programs incorporate anger management components (Dixon et al. 2012; Fleming and Cordova 2012).

Couples or family members who want to work for change on their own might practice the previously mentioned guidelines for conflict resolution. As partners and family members grow accustomed to voicing grievances regularly and in more respectful or caring ways, their disagreements less often become full-fledged fights: Family members gradually learn to incorporate many irritations and requests into their normal conversations, arguing in normal tones of voice and even with humor. Although these suggestions may help, learning to fight fair is not easy. Sometimes couples and families feel that they need outside help, and they may decide to engage a counselor, discussed later in this chapter. There are many books and online resources available that could help. As examples, there are books on overcoming passive-aggressive behavior (Murphy and Oberlin 2006), recognizing how we sabotage our relationships (Matta 2006), and changing habits that can thwart a satisfying life in general (Kagan and Einbund 2008). Some books, such as *Person to Person: Positive Relationships Don't Just Happen* (Hanna, Suggett, and Radtke 2008), focus on both individual self-improvement and couple communication.

Susan Halpern's *Finding the Words: Candid Conversations with Loved Ones* (2009) covers topics such as cultivating conscious conversations as a couple, communicating in ways that might lessen the disruptive effects of divorce, and improving communication between parents and their adult children. John Gottman's research, already described in this chapter, comes to life in his readable *The Seven Principles for Making Your Marriage Work* (Gottman and Silver 2015). There are books on communication designed specifically for same-sex couples (Clunis and Green 2005). And, of course, there

are university courses and textbooks on interpersonal communication and relationships.

In addition, there are a vast number of good (and perhaps not-so-good, so be selective) online resources on communication of all sorts. Among others, topics range from managing unresolved family conflict, to parent–child communication, opening the door to renewed communication between estranged siblings, and—of course—communication between partners. Family relationships are dynamic and can change for the better. For example, an adult woman told this story about her improving relationship with her sister:

[We] spent some time together. . . . We hadn't done that in three or four years. . . . It was . . . getting to the point where . . . we could just continue to stick our head in the sand or we could . . . try this again. Because this is the only family. . . . So [now, after beginning to repair the relationship], it's sort of inching along like that. A little better, a little better. (Connidis 2007, p. 489)

Finally, there is the option of relationship or family counseling.

"Generally, marriages that have built up positive emotional bank accounts through respect, mutual support, and affirmation of each person's worth are more likely to survive" (Hetherington 2003, p. 322). Relationships require attention (Cole 2011). It is the *rewards* of a long-term relationship—love, respect, friendship, and good communication—that are most effective in keeping marriages and other relationships together. It follows that people need to keep their relationship rewarding for one another.

Many observers strongly criticize the way that American culture tends to equate love with infatuation, or chemistry, but infatuation "merely brings the players together. . . . Relationships live on time" (Lewis, Amini, and Lannon 2000, pp. 205–207).

Ben Glass/Warner Bros. Pictures/Everett Collection

When Is It Okay to Let Go?

Despite what is known about various techniques for improving relationship communication, some relationships remain broken. Although difficult to quantify, **estrangement** between family members is not as uncommon as one would think. In a study of 1,300 Danish adults, one in three had experienced a conflict with a close family member that led to relations being broken off (Copenhagen Post 2014). Why do family members sever contact with one another? Megan Gilligan and colleagues (Gilligan, Suitor, and Pillemer 2015) studied estrangement between 2,013 mothers and their adult children. Estrangement was measured in terms of not having contact with a child in any way for at least one year. They found that mothers and their children often become estranged due to their differing values. One example is Ruth, a 75-year-old devout Catholic mother of two sons and one daughter. She explained, "Up until his difficultly in his [first] marriage and his divorce, we were very similar. After that it became a matter of both religion and social differences. It's a difficult situation. Now he has remarried and made a new family. So it's painful for me to be judgmental, but I have religion in the way and my own morals and social ideas" (p. 8). Other parents in the study became estranged from their children as a result of their children's substance abuse or criminal activity. Clearly, in some circumstances, such as in the case of physical or emotional abuse, dissolving the relationship is the best outcome. However, in many cases, estrangement can be averted.

With the advent of the texting and social media, ending relationships can be as easy and quick as the click of a button. However, is this the best way to handle it? As discussed in Chapter 5, "ghosting" is when a person ends a relationship by cutting off all contact with no explanation (Freedman, Powell, Le, and Williams 2018). Ghosting has become a popular, and unfortunate, "exit strategy" for intimate relationships. Most teens and young adults these days have ghosting experience, having either ghosted someone or been themselves ghosted. Dating apps such as Tinder encourage ghosting behavior because the sense of investment and obligation to any one person is minimal and they often have no mutual friends or previous ties (LeFebvre 2017; LeFebvre et al. 2019). The unilateral ceasing of communication may be easier on the initiator than talking directly, although they might experience guilt, but leaves the receiver confused and hurt. It is never pleasant or comfortable to break up with someone or end a friendship, but an in-person or phone conversation is best (unless one fears for their own personal safety) in that it provides an opportunity for personal growth and reflection that will benefit both parties.

Meanwhile, even the fairest fighters hit below the belt once in a while, and just about all fighting involves some degree of frustration and hurt feelings. Moreover, some individuals have a partner who chooses not to learn to face conflict positively. Sometimes attending a relationship-enhancement program can end in one or both partners' disappointment when one partner doesn't seem to be motivated or trying during the program itself or when what seemed to be progress during the program itself is not carried out or followed through. Disappointed partners may feel hurt or anguished, try to make their partner feel guilty, or ignore the problem—these are coping mechanisms (Dixon et al. 2012).

Even when both partners develop constructive habits, all their problems will not necessarily be resolved. Although a complainant may feel that he or she is being fair in bringing up a grievance and discussing it openly and calmly, the recipient may view the complaint as critical and punitive and may not want to bargain about the issue. Family cohesiveness and supportive couple relationships, has much to do with commitment, gentleness, and humor and on letting our loved ones know how much we care about and appreciate them. Importantly, it is about *listening* as well as talking. Sandstrom and colleagues (2016) installed the mobile app "My Social Ties" on the phones of thirty-six undergraduate students and recorded their conversations. They found that conversations in which a person spent a smaller percentage of their time speaking enjoyed their conversations more.

The introduction to this chapter pointed out that family communication takes place within and is influenced by the family's external social environment (Trail and Karney 2012). Here's something to think about from this point of view:

To improve or further enhance marital functioning, people should try to reduce and cope—individually or dyadically—with both high levels of external stress that tends to spill over into the relationship and high levels of relationship stress. To reduce the level of external stress, employers are required to provide safe working conditions and fair wages. In addition, governmental and other social service programs should pay special attention to the needs of low-income couples and help them to overcome external strains, as they often experience more stress and face greater problems in building and maintaining a healthy intimate relationship than better off couples.

Meanwhile, finding effective ways to deal with stress occurring inside the relationship is important to stave off deterioration of marital functioning on both the individual and dyadic levels. Couple programs that teach coping skills have demonstrated promising results in improving aspects of marital functioning. (Ledermann et al. 2010, p. 204)

Meanwhile, to close this chapter, we note that not every conflict can be resolved, even between the fairest and most mature individuals. If an unresolved conflict is not crucial, then the two may simply have to accept their inability to resolve that particular issue and let it go. In relationships where one wants to change and the other doesn't, sometimes much can be gained if just one partner begins to communicate more positively. Other times, however, positive changes in one individual do not spur growth in the other. Situations like this may end in alienation, separation, or divorce. Sometimes, it is best for partners to go their separate ways. Divorce and union dissolution are discussed in Chapter 14. We need to note that communication patterns can convey messages of dominance and disaffiliation (McLaren, Solomon, and Priem 2012). The next chapter examines issues of dominance and power in relationships.

Summary

- Members of cohesive families express their appreciation for each other, have a high level of commitment to the family group as a whole, do things together, know how to deal positively with stress or crises, and evidence positive communication patterns.

- From a family ecology perspective, cohesive families depend on supportive external social environments (the economy and work environments, for instance), together with positive communication patterns.

- Technology has fundamentally altered the way that families communicate and according to researchers can have both positive and negative effects.

- Research on couple communication indicates the importance to relationships of both positive communication and the avoidance of a spiral of negativity.

- Although some family communication patterns are decidedly negative, family conflict itself is an inevitable part of normal family life.

- Although arguing is a normal part of the most loving relationships, there are better and worse ways to manage conflict.

- Alienating practices such as belligerence and employing the Four Horsemen of the Apocalypse—contempt, criticism, defensiveness, and stonewalling—should be avoided.

- Constructive arguing habits may not only resolve issues but also bring participants closer together.
- Constructive arguments are characterized by efforts to be gentle and by deescalating negativity. No one loses.

- It is important to forge and maintain relationships with people outside the family unit and even a pleasant "hello" to a stranger can boost your well-being, and theirs.

Questions for Review and Reflection

1. We often hear that communication is important to maintaining family relationships. Can you discuss specific reasons why this is true?

2. Describe the Four Horsemen of the Apocalypse. If someone you care for treated you this way in a disagreement, how would you feel? Do you ever treat others with one or more of the Four Horsemen?

3. Discuss your reactions to each of the ten guidelines proposed in this chapter for constructive arguing. What would you add—or subtract?

4. Discuss how technological advances have changed communication. What in your opinion is good about technology for relationships? What is not so good?

5. **Policy Question.** Besides the suggestions in "Facts about Families: Ten Rules for Successful Relationships," what *society-wide* ideas might you offer for enhancing loving relationships?

Key Terms

belligerence 286
contempt 286
criticism 286
defensiveness 286
disengaged couple
  or family 276
displacement 285
emotional intelligence 298
enmeshed couple or family 276

estrangement 300
family cohesion 276
female-demand/male-withdraw
  interaction pattern 289
Four Horsemen of the
  Apocalypse 286
mixed, or double,
  messages 294
negative affect 287

passive-aggression 285
positive affect 282
rapport talk 288
relationship-focused coping 284
relationship ideologies 281
report talk 288
sabotage 285
social media 292
stonewalling 286

12
POWER AND VIOLENCE IN FAMILIES

Learning Objectives

1 Describe the social psychological bases of relationship power.

2 Discuss how cultural context impacts the resource hypothesis.

3 Summarize the incidence of family violence, including trends and reasons for the trends.

4 Distinguish between coercive control and situational couple violence.

5 Distinguish among child abuse, child neglect, and willful child neglect.

6 Describe the following three approaches to stopping family violence: separating victim from perpetrator, engaging the criminal justice system, and psychological therapy.

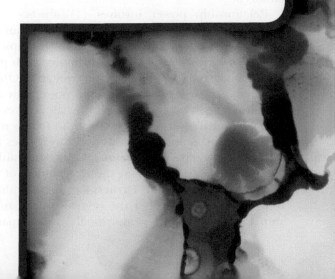

- Sarah applies for a promotion, and accepting it will mean moving to another city; Sarah's partner does not want to relocate.

- Antonio wants a new stereo for his truck; his partner would prefer to spend the money on ski equipment.

- Marietta would like to talk to her partner about what each of them does (and doesn't do) around the house, but her partner is too busy to discuss it.

- Greg feels that he gives more and is more committed to developing a satisfying, cooperative marriage than his wife is.

- Nicole and Paul go to counseling and parenting education classes in efforts to curtail violence in their family.

Each scenario illustrates power in a relationship. This chapter examines family power, a central concept in relationships even if ironically not often explicitly researched by social scientists (Farrell Orina, and Rothman 2015; Monk, Ogolsky, and Whittaker 2019). We will look at a classic study on marital decision-making, one indicator of **conjugal power** (marital power), then explore contemporary research on conjugal power. We'll discuss why playing power politics is harmful to intimacy and look at alternatives. We will then explore family violence as an unfortunate, too often tragic, consequence of abuse of power. We will explore intimate partner violence (IPV), then address child maltreatment—a form of family violence that is often present along with IPV. We close this chapter with a discussion of ways that family violence can be addressed and lessened. We begin by defining power.

WHAT IS POWER?

Power is the ability to exercise one's will. Power exercised over oneself is *personal power*, or autonomy. Having a comfortable degree of personal power is important to self-development. *Social power* is the ability to exercise one's will over others. Social power may be exerted in different realms, including within relationships. Parental power, for instance, operates between parents and children.

Relationship power involves (1) *objective measures of power*—who makes more, or more important, decisions, or who does more housework, for example—and (2) *subjective measures of fairness*—whether each partner feels their arrangement is fair or equitable. Objective and subjective measures of couple power often yield similar results. This is not always the case, however. For example, a career-invested spouse who makes most of the important family decisions and rarely does housework may perceive the relationship as fair—although from an objective viewpoint the distribution of power is not equal. Feelings of fairness can be thought of in terms of **equity**—whether the rewards of the relationship feel subjectively proportional to each partner's

contributions, which may not necessarily be equal. Partners' subjective perceptions of fairness influence relationship satisfaction and commitment. When partners perceive themselves as reciprocally respected, listened to, and supported, they are more apt to define themselves as equal partners. They are also less depressed, generally happier, and more satisfied with their relationship (Corra et al. 2009; Greenstein 2009; Sullivan and Coltrane 2008). Another way to think about relationship or family power is to consider the concept of *power bases*.

Power Bases

Psychologists John French and Bertram Raven (1959) developed a typology of six bases, or sources, of social power: reward, coercive, expert, informational, referent, and legitimate power. We'll use these bases of social power in this chapter to analyze family (mostly couple) power relationships (see Table 12.1).

Reward power is based on an individual's ability and willingness to give material or nonmaterial gifts and favors, ranging from emotional support—for example, listening, eye contact, a smile, a gentle hand on a shoulder—to financial support or gifts. **Coercive power** is based on the dominant person's ability and willingness to punish the partner with psychological–emotional abuse or physical violence or, more subtly, by withholding favors or affection (Davies, Ford-Gilboe, and Hammerton 2009, p. 28). Slapping a mate and spanking a child are examples of *coercive power*; so is refusing to talk to the other person—the silent treatment.

Expert power stems from the perception that the more powerful person has superior ability, knowledge, or judgment. Although this is certainly changing, our society traditionally attributed expertise in such important matters as finances to men, while women were attributed expertise in the domestic sphere. **Informational power** is based on the persuasive content of what the more powerful person tells the other. A partner may be persuaded to charge less on a credit card when the other shares information on the card's high interest rate. Or a partner may be persuaded to choose a particular Netflix movie because the other has persuasive information that it has gotten impressive reviews.

Referent power is based on a person's emotional identification with the partner. A partner who attends a social function when he or she would rather not "because my loved one wanted to go and so I wanted to go too" has been swayed by *referent power*. Referent power can enhance couple commitment (Zhang and Tsang 2013). Research shows that in happy relationships, *referent power* increases as partners grow older together (Pyke and Adams 2010).

Finally, **legitimate power** stems from the more powerful person's ability to claim authority, or the right to expect compliance. *Legitimate power* in traditional, heterosexual partnerships would involve some degree of acceptance

TABLE 12.1 Bases of Social Power as Applied to the Family

TYPE OF POWER	SOURCE OF POWER	EXAMPLE
Coercive power	Ability and willingness to punish the partner	Emotional abuse —Partner sulks, refuses to talk, withholds sex, puts you down; physical abuse/violence
Reward power	Ability and willingness to give partner material or nonmaterial gifts and favors	Partner gives affection, attention, praise, respect, and assistance in realizing goals (e.g., takes over unpleasant tasks so partner can study)
Expert power	Knowledge, ability, judgment	Savings and investment decisions shaped by partner with more education or experience in financial matters
Informational power	Knows more about child rearing, travel destination, housing, health issues, or other items and can thus persuade partner	Persuades other parent about most effective mode of child discipline, citing experts; persuades partner to see *Hamilton* after studying reviews
Referent power	Emotional identification with partner	Partner agrees to home purchase or travel plans preferred by the other because they want to make partner happy
Legitimate power	Society and culture authorize the power of one or the other partner or both	In traditional marriage, husband exercises authority as household head. The current ideal is more equal partners.

Source: Typology of power concepts from French and Raven (1959). Specific wording of definitions and the examples are by Agnes Riedmann.

by both partners of the male's role as central. The concept, *legitimate power* urges us to recognize cultural influences on relationships: that is, society and culture define power as legitimate in some circumstances, often related to gender.

Although not the case for all U.S. families, today's ideal in mainstream culture is an egalitarian couple partnership (Bennett 2015; Coontz 2013). Heterosexual relationships are becoming more egalitarian, partly resulting from partners' sharing similar education levels more often than in the past (Miller 2019; Schwartz 2014).

Throughout this chapter, we will see the various power bases at work. For Example, the consistent research finding that economic dependence of one partner on the other results in the dependent partner being less powerful may be explained by understanding the interplay of both *reward power* and *coercive power.* If I can reward you with financial support—or threaten to take it away—then I am more able to exert power over you. We also note that the exercise of power is situational: Partner A may apply informational power about seeing a certain movie, for instance,

while Partner B employs expert power about whether to deduct a new bicycle when doing the family taxes.

We turn now to look at research on marital power and the theoretical perspectives used to explain couple power relationships.

Although an older generation may hold more tightly to traditionally legitimated patriarchal power, the next generation may strive to renegotiate and consciously change those roles.

Monkey Business Images/Shutterstock.com

THE RESOURCE HYPOTHESIS: A CLASSICAL PERSPECTIVE ON MARITAL POWER

Research on marital power began in the 1950s when social scientists Robert Blood and Donald Wolfe (1960) were curious about how married couples made decisions. Blood and Wolfe's research popularized the resource hypothesis, derived from the exchange theoretical perspective described in Chapter 2. The **resource hypothesis** posits that the partner with more resources has greater power in the relationship. Resources primarily include earnings and education, the latter resulting in informational and expert power. In Blood and Wolfe's research, the resource hypothesis was supported by the finding that the spouse with higher earnings and educational attainment made more decisions. Relatedly, Blood and Wolfe found that a wife's power was greater when she had no young children and when she worked outside the home—situations that made her less financially dependent on her husband. Today's research continues to show that wives' relative income increases their say in important decisions.

Resources and Gender

The Blood and Wolfe study made a major contribution by encouraging researchers to see couple power as based on each partner's relative resources. However, the study was strongly criticized, primarily because it ignored sources of power other than individual resources—namely, the power of gender expectations, norms, and socialization (Safilios-Rothschild 1970; Tichenor 2010). Accordingly, Blood and Wolfe came under heavy fire for their naïve conclusion that a patriarchal power structure had been replaced by egalitarian marriages.

Feminist Dair Gillespie (1971) pointed to a situation that persists, although to a diminishing extent: Power-granting resources remain importantly socially structured by gender and hence unevenly distributed. Despite social change, the majority of heterosexual men earn more than their partners and hence have greater access to economic *resources*. As explored later in this chapter, men's greater physical strength can be an important resource, granting actual or potential coercive power. Although couples are moving toward **egalitarian relationships** (i.e., relationships that are equal), research continues to show that American heterosexual unions are not fully equal (Few-Demo and Allen 2020).

Resources in Cultural Context

In some cases, patriarchal norms remain strong enough to override personal resources and preferences. The concept of **resources in cultural context** stresses that society-wide gender structures (see Chapter 3) influence conjugal power. Individual resources fully influence conjugal power only when there is no cultural norm for conjugal power—either an **egalitarian norm** or a **patriarchal norm**. When traditional norms of male superiority are strong, husbands will likely exercise more power. Also, if an egalitarian norm were to be thoroughly accepted society-wide, partners would share equal power regardless of their relative economic, educational, or other resources. Only in societies or situations where neither patriarchal nor egalitarian norms are firmly entrenched is power freely negotiated by couples according to their relative preferences and resources (Wight, Bianchi, and Hunt 2012).

CURRENT RESEARCH ON COUPLE POWER

Current research measures couple power in the following four ways:

1. *Decision-making:* Who gets to make decisions about everything from where the couple will live to how they will spend their leisure time?

2. *Division of labor:* Who provides income? Who does the household labor? Who takes primary responsibility for childcare if there are children?

3. *Allocation of money earned by either or both partners:* Who controls household spending? Who has personal spending money?

4. *Ability to influence the other partner* and feeling comfortable in raising complaints about the relationship.

We will address each of these in this section. Division of household labor is the focus of Chapter 10.

Decision-making

A national survey in 2000 compared conjugal decision-making in 1980 with that in 2000 and found that in 2000, "respondents were significantly more likely to report equal-decision-making" (Amato et al. 2003, p. 9). Even wives in evangelical families often have more decision-making power than their formal submission to the male family head would indicate (Perry 2015). In fact, some research shows that "co-parenting and joint decision-making are more common in evangelical homes than in secular and mainline religious households" (Bartkowski and Read 2003, p. 88; Vaaler, Ellison, and Powers 2009).

On the other hand, a study of wives who earn more than their husbands suggests that "the gender structure exerts an influence that is independent of breadwinning

or relative financial contributions" (Tichenor 2005, p. 117). A residual sense of the propriety of traditional male privilege—that is, *legitimate power*—ascribed more authority to men even in situations where they had fewer relative resources. "Just as women's income does not buy them either relief from domestic labor or greater financial power . . . , it does not give them dominion in decision-making" (p. 117; see also Treas and Tai 2012; Few-Demo and Allen 2020).

In an interesting twist, women can sometimes gain power from their greater knowledge of the household. They can use this *informational power* to shape decisions about purchases and household arrangements, as we noted previously.

Division of Household Labor

Social scientists use housework as one criterion of power on the assumption that (1) doing more of the family housework results in a partner's earning less money, and (2) no one really wants to do it (Few-Demo and Allen 2020). As discussed in Chapter 10, women's satisfaction with the fairness of how household labor is allocated is strongly associated with relationship happiness and commitment. Conversely, when a wife has more egalitarian expectations than her spouse fulfills, depression and marital conflict likely follow (Gerson 2010; Greenstein 2009).

Recent decades have seen a significant increase in men's share of housework (Livingston and Parker 2019). Nevertheless, the fact that women continue to do more housework than men do is seen as an objective indicator of their relatively less conjugal power (Geist and Ruppanner 2018; Hess, McPhil, and Hayes 2020; Miller 2020). As shown in Figure 12.1, when the three categories are calculated together, mothers and fathers now spend about the same number of hours in paid work, housework, and childcare. In fact, fathers average four more total hours than mothers—sixty-one versus fifty-seven.

Partners may be inclined to see this division of total paid and unpaid labor as fair. Note, however, that such a division of labor does not readily facilitate a woman to maximize her career potential or advance into top management positions. Being saddled with nearly twice the housework and childcare responsibilities takes time from women's career participation and development. Accepting this situation renders wives and mothers— even potential wives and mothers—less effective in the labor force (Few-Demo and Allen 2020, p. 331).

> [D]espite widespread attitudinal support for "fairness" in the gender division of labor, and the apparently approximately equal overall gender burden of paid plus unpaid work, ...the equal-but-different composition of overall gendered work time implies a situation of unfairness in terms of economic life chances....This in turn

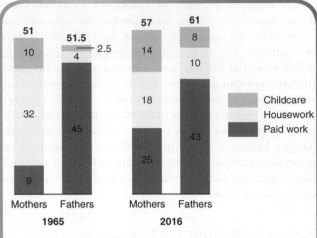

FIGURE 12.1 How mothers and fathers spend their time, 1965 and 2016: Average number of hours per week spent on . . .

Source: Livingston and Parker 2019.

affects the gender wage gap, the disadvantage women experience in respect of the opportunities for career advancement, earnings, and ultimately pensions. If men spend more time in paid work than their female partners do, they also accumulate more human capital, that is, more earning power in the long term. And if women stay at home to care for the children while their male partners work longer hours at their jobs, the earnings-capability gap widens. (Sullivan, Gershuny, and Robinson 2012, p. 276)

Money Allocation

Scholarly interest on couples' **money allocation systems**—whether they pool their money and who controls pooled or separate money—is fairly recent. In the industrial era—from about the 1800s through about the 1950s—a family's allocation system was typically one of complete control of his earnings by the male breadwinner, who doled out a housekeeping allowance to his homemaker wife. The allowance was often rather skimpy, while the husband was privileged to take money "off the top" for personal spending and recreation (Pahl 1989).

By the 1970s, feminists had begun to criticize the husband-controlled joint pool system (Allen and Jaramillo-Sierra 2015). With more women earning income, separate financial accounts and control began to be seen as a favored alternative, with each partner making proportionately equitable contributions to running the household (Vogler, Lyonette, and Wiggins 2008). Cohabitants and those who have been previously divorced are especially likely to maintain separate money (Bennett 2013; Kenney 2006).

Currently, a variety of allocation systems operate in American unions and involve two dimensions: (1) whether to pool their money, and (2) which partner controls. Gender continues to play a strong role in heterosexual unions, and men seem to retain more ultimate control over the family's spending—at least on major purchases (Carlson and Lynch 2017). Even otherwise egalitarian-oriented men may assume they have "veto power" over major financial decisions (Vogler, Lyonette, and Wiggins 2008). This would seem to be an example of the ongoing influence of traditional *legitimate power* ascribed to a male overriding the wife's *resource power.*

Ability to Influence the Other

In supportive and satisfying relationships, each partner feels confident that he or she can be heard, even when raising uncomfortable issues. Partners can air concerns without fear of being dismissed or otherwise treated badly. Each has power to influence the other. Where this is not the case, the couple's power balance is unequal. Rather than risking confrontation, "[t]he spouse with less power typically spends more time aligning emotions with [the spouse's] expectations" (Coltrane 1998, p. 201).

Either partner may wield considerable power by means of referent power—one's love for the other. Then, too, contemporary men have been influenced by the cultural expectation that they respond favorably to their wives' raising concerns (Connell 2005). As discussed in Chapters 3 and 11, women have traditionally been socialized to be more attuned to relationship dynamics. However, men today *are* encouraged to engage in "emotion work," "to express emotion to their wives, to be attentive to the dynamics of their relationship and the needs of their wives, [and] . . . to set aside time for activities focused especially on the relationship" (Wilcox and Nock 2006, p. 1322).

Diversity and Marital Power

Because the United States is a pluralistic society, we expect to find varied visions and enactments of marital power. One example of diversity and conjugal power is described in "A Closer Look at Diversity: Mobile Phones, Migrant Mothers, and Conjugal Power." We'll look at three general types of conjugal power now: egalitarian, neotraditional, and gender-modified egalitarian unions. If we were to place these on a continuum from most to least equal, egalitarian unions would be most equal and neotraditional unions least so. Gender-modified egalitarian unions would fall somewhere in between.

Egalitarian Unions In egalitarian unions partners share equally in the four components of couple power: decision-making; division of household labor, particularly housework; money allocation; and ability to raise relationship issues. Whether or not egalitarianism is fully realized, social scientists have generally assumed that egalitarian unions are the most sought after by the majority of U.S. couples.

A possibility is that society-wide equality *norms* could become so strong that partners would share power equally regardless of their respective individual resources. This situation would be a new form of legitimate power—one that justifies and supports equality regardless of resources. Currently, lesbian couples seem to have come closest to attaining this ideal. The resources that each brings to the relationship affect each person's power minimally, if at all (Jeong and Horne 2009). As gender norms move from traditional to egalitarian, *all* family members' interests and preferences gain legitimacy. This situation

This family is having breakfast in a household where roles may be somewhat differentiated by gender. Or, perhaps they are not. What do you see in this photo? Is mom standing so she can more easily serve whoever might need something? Is dad tending to the children?

Pixland/Jupiter Images

A Closer Look at Diversity

Mobile Phones, Migrant Mothers, and Conjugal Power

A study in the journal *New Media & Society* looks at conjugal power in transnational Filipino families, power between "left-behind" fathers and migrant mothers who leave the Philippines to get better-paying jobs (often in health care fields) in wealthier nations.

Researchers visited with the fathers and their children in ten Filipino homes. The researchers concluded that, "while the mobile phone can lead to increasing cooperation between left-behind fathers and migrant mothers, it has mostly resulted in exacerbating the already tremendous chasms that divide them" (Cabanes and Acedera 2012, p. 916). How did the authors come to this conclusion? For one thing, the mobile phone can facilitate ongoing conflict. Said one husband,

> [My wife and I] always fight because of texting. She keeps on saying that every time I text her, there is nothing else I talk about except financial problems.... But sometimes, the only thing I want is

for her to show some sympathy. It can be very overwhelming when you're alone, you know. (Cabanes and Acedera 2012, p. 923)

The authors explain that mobile phones and related technologies, such as Skype, FaceTime, or Viber, allow family members to be virtually in two places at once. Family members who are nations apart can readily phone or text each other, a development that allows them to communicate daily, hourly, or even more often. Meanwhile, a migrating mother reverses traditional Filipino conjugal power because she becomes the breadwinner, sending money back to her family. Her left-behind husband becomes responsible for homemaking and childcare.

> The mothers' crossing over to the role of provider tends to be very bruising to the fathers' egos. And the mobile phone appears to be central to this experience. Many fathers share that their wives "abuse" mobile phone calls .... Said one father, "When that thing rings, it does not

matter what time it is or where I am. I have to answer. And I have to answer well." (Cabanes and Acedera 2012, p. 923)

The mothers call regularly to help ensure that the money they're sending home isn't going waste.

Despite the authors' conclusion that the mobile phone mostly increased power struggles between these spouses, the mothers also often called to ask their children to help their fathers with the household chores. And fathers texted their wives for help with children's homework or discipline (Cabanes and Acedera 2012).

Critical Thinking

Can you find evidence of traditional gender expectations in the situation described here? How about changing gender expectations? Has the iPhone changed the power dynamics in any of your relationships? If yes, how? If no, why not? If you do see changes, how would you describe them?

---

means that more and more family decisions must now be consciously negotiated. Although a possible outcome of such conscious negotiating can be increased conflict, another is greater intimacy.

Gender-Modified Egalitarian Unions In the **gender-modified egalitarian model**, absolute equality is diminished by the symbolic importance of maintaining fairly traditional and familiar gender roles. Partners compromise with the fully egalitarian ideal.

In their study of more than 5,000 couples drawn from the 1992–94 National Survey of Families and Households, Wilcox and Nock found support for the hypothesis that "the gendered character of marriage seems to remain sufficiently powerful as a tacit ideal" (2006, pp. 1339–40).

Neotraditional Unions A second model favors a traditional division of labor and male family leadership. What makes this model different from the traditional

models of the nineteenth and early twentieth centuries is the melding of traditional ideals with a new (neo) egalitarian spirit. Evangelical Christians and other conservative religious groups tend to embrace this model, characterized by a gendered division of labor, formal male dominance in decision-making, and an "egalitarian spirit."

Although a husband's dominant power is legitimated in this milieu, marital power in practice is often negotiated (Perry 2015). First "articulated by evangelical feminists," the "mutual submission" (of husband and wife to each other) has become increasingly popular because it justifies the shared decision-making that characterizes many evangelical marriages (Dolan 2008, p. 32; Vaaler, Ellison, and Powers 2009). Another way in which a norm of equality is represented in these ostensibly husband-dominant marriages is in an "economy of gratitude" (Pugh 2009, p. 6; Wilcox 2004, p. 154), as husbands display appreciation for their wives' "gift" of household work. When compared to traditional

conjugal power, the sharp edges of male dominance are softened in the **neotraditional family**, as the title of sociologist Bradford Wilcox's book—*Soft Patriarchs, New Men* (2004)—suggests.

Immigration and Conjugal Power Recent immigrants from traditional cultures such as those in Eastern Europe, Asia, and Central and South America typically arrive in the United States having fairly traditional gender roles and conjugal power relations.

> Migration can also strain couple relationships for couples who live together in the destination country. Gender roles within couples may initially reflect those in immigrants' sending countries and may be modeled from their parents' relationships . . . , but these patterns fade with time in the destination country . . . . These changes may not come easily . . . . For example, in a study of Iranian immigrants in Canada, men faced challenges adjusting to new gender roles and shifts in division of labor at home in response to women's labor force participation . . . . (Van Hook and Glick 2020, p. 328)

As they assimilate, subsequent immigrant generations tend to adopt less traditional gender-role patterns according to which "husband–wife relationships are

It seems from these partners' body language that they share various aspects of couple power on a fairly equal basis. What conditions might influence which one does more housework? Makes more of the important couple decisions? Has more say in how their money is spent or where to go on their next date night?

more flexible and negotiated . . . [and] socioeconomic achievements become the basis for negotiation within the family" (Cooney et al. 1982, p. 622; Harper and Martin 2014).

Even among native-born Americans, we must recognize the continuing salience of tradition and the assumption that to some degree for a majority of heterosexual couples it is legitimate for men to wield some aspect of family authority.

POWER POLITICS VERSUS FREELY COOPERATIVE RELATIONSHIPS

We've seen that relationships perceived as fair and equitable are more apt to be stable and satisfying. Power asymmetry often characterizes dissatisfied couples (Gottman et al. 1998). Not seeking to "win" an argument and highly respecting each other facilitate satisfying relationships. Put another way, supportive partners avoid **power politics**.

In a union characterized by power politics, partners lock into a relationship-damaging cycle of behaviors. Sometimes (or often) hinting at leaving the marriage, partners alternate in acting sulky, critical, or distant. The sulking, critical, or distant partner carries on this behavior until she or he fears that the mate will "stop dancing" if it goes on much longer. Then it's the other's turn. This seesawing creates alienation and loneliness for both partners.

There are alternatives to this kind of power struggle. As discussed in Chapter 11, communication expert John Gottman and his colleagues advise partners to share power if they want to be happy together (Coan and Gottman 2007). "As We Make Choices: Domination and Submission in Couple Communication Patterns" illustrates communication patterns of dominance and submission.

Partners who see themselves as respected, equally committed, and listened to when they raise concerns are more likely to see their relationship as egalitarian and are more satisfied overall with their relationship. Changing power patterns can be difficult, because usually power patterns have been established from the earliest days of the relationship. Certain behaviors not only become expected but also come to have symbolic meaning: "She always grocery shops and cooks for me, and therefore I know she loves me," for example. Sociologist William Goode offers the following insight:

> Men have . . . occupied the center of the stage, and women's attention was focused on them . . . . [But] the center of attention shifts to women more now than in the past. I believe

As We Make Choices

Domination and Submission in Couple Communication Patterns

Our communication patterns reflect the power relationships between ourselves and whomever we are communicating with. Communication patterns illustrate (1) dominance, (2) submission, or (3) mutual cooperation. Below are a few questions that researchers use to measure dominance or submission.

Measuring Dominance— Agree or Disagree

1. When we disagree, my goal is to convince my partner that I am right.
2. When we argue or fight, I try to win.
3. I try to take control when we argue.
4. I rarely let my partner win an argument.

5. When we argue, I let my partner know I am in charge.

Measuring Submission— Agree or Disagree

1. I give in to my partner's wishes to settle arguments on my partners' terms.
2. When we have conflict, I usually give in to my partner.
3. I surrender to my partner when we disagree on an issue. Sometimes I agree with my partner just so the conflict will end.
4. When we argue, I usually try to satisfy my partner's needs rather than my own (Zacchilli, Hendrick, and Hendrick 2010, p. 1081).

Elements of equality and mutual cooperation

1. We try to stay attuned to one another.
2. We both take responsibility for maintaining our relationship.
3. We usually respond to each other's needs, including bids for attention
4. We directly express our positive regard for each other.
5. Chores feel fair (Marano 2014).

Critical Thinking

Do you recognize yourself in any of these questions? What are some ways that partners might change their dominant or submissive communication patterns, do you think? How might partners increase their mutual cooperation?

---

that this shift troubles men far more, and creates more of their resistance, than the women's demand for equal opportunity and pay in employment. (Goode 1982, p. 140)

The above is an old quote; still it may ring true for many. More recent research suggests that, typically, one person in a relationship doesn't *always* in every situation exercise more power than the other. Rather, a partner is likely to wield power in some, of several, *power domains*, such as decisions about:

- How much time to spend together, and how to spend it
- Career/moving decisions
- Childrearing
- Demonstrations of affection/sexual relations
- Division of labor/household tasks
- Family and friends
- Finances
- Philosophy of life/religion/values (Farrell, Simpson, and Rothman 2015)

From this point of view, partners can find themselves in a power struggle in one domain (or several), but probably not in all of these. It's also likely that one partner appears to exercise more power in one or several domains, but not in all of them (Alman 2018; Didonato 2015).

How to address uncomfortable power struggles? One thing you can do is google "relationship power" and find several Internet sources by credentialed psychologists that may seem helpful (e.g., Alman 2018; Gage 2017; Holmes 2018; Howard 2018). Among useful ideas:

- Something bothering you? Try not to complain, especially to a third party, like your good friend. Complaining reinforces the idea that you can't do anything to change things. Instead of complaining, ask your partner for a specific change (pleasantly is best). Put another way, tell your partner the truth about you're feeling regarding the issue.

- Pay attention to your feelings; don't dismiss or discredit them, telling yourself they don't really matter or aren't that important. Your feelings are good indicators of what's disturbing you, and they deserve to be noticed.

- Take responsibility for your part in the power struggle, and figure out what you might do better (while honoring your feelings and telling the truth).

313

- Know your own worth—and/or work on knowing it better. Recognize that your felt needs are as important as your partner's.

- Remember that you have options and make choices. Giving in is a choice. Speaking up is a choice. Doing it gently is a choice. Being mean is a choice.

A good way to work through power changes is to use communication techniques described in Chapter 11. At the same time, it's good to remember that, although it promises a more rewarding relationship in the long run, changing an uncomfortable power balance is a challenge to any relationship and can be painful for both partners. One option for handing power change is to pursue family or relationship counseling. Whether on their own or with help, partners can choose cooperation over playing power politics. Unfortunately, when freely cooperatively relationships do not develop—when one partner experiences low relationship power and then finds themselves in a situation they'd like to control but can't seem to—one result is aggression, or family violence (Overall, Hammond, McNulty, and Finkel 2016). We turn to that topic now.

FAMILY VIOLENCE

Emotional and physical violence have been used to manifest power in families throughout history, but only in the last fifty years has family violence been labeled a social problem (Hardesty et al. 2015). First, child abuse was recognized as a social problem in the 1960s (Kempe et al. 1962). Then—in conjunction with the second wave of the women's movement in the 1970s (see Chapter 3)—"wife abuse," at that time typically called "wife beating" or "wife battering," came into focus. Gradually researchers and policy makers expanded the concept *wife abuse* to **intimate partner violence (IPV)**—the physical or emotional abuse of partners of any gender or sexual orientation: current or ex- cohabiting or noncohabiting heterosexual or same-sex relationship partners. The federal government includes all these forms of couple violence in its reports on IPV.

With the 1980s came concern about elder abuse and, later, husband abuse. More recently, researchers have begun to address sibling violence and child-to-parent violence. Dating violence is discussed in Chapter 5; elder abuse and neglect in Chapter 16. We discuss many of the remaining forms of family violence here. We'll look first at sources of data on intimate partner violence.

IPV Data Sources

Data on physical intimate partner violence come from four types of sources: (1) violent crime reports filed

Gay and lesbian couples—especially lesbian couples—are more likely to share power and domestic duties than heterosexual couples. At the same time, some same-sex couples experience family violence, a situation that went ignored for many decades but has now begun to be researched and addressed.

in police departments and then reported to the FBI; (2) national surveys of the general population; (3) smaller research studies, often qualitative, that relate persons' experience with family violence; and (4) reports from social workers, counselors, or volunteers at hospital emergency rooms or in other settings. "Facts About Families: Major Sources of Family Violence Data" gives detail on major family-violence data sources. Different IPV data types give somewhat divergent pictures of IPV, a point we will return to later in this chapter. Now we turn our attention to pioneering survey research that helped bring family violence to public attention and remains influential.

The National Family Violence Surveys The 1970s women's movement raised attention to domestic violence against women, particularly to serious injuries observed in hospital emergency rooms. Meanwhile, the early and continuing research of Murray Straus, Richard Gelles, and their colleagues in their **National Family Violence Surveys** shaped scientific research on family violence among the general population. The research group undertook a national household survey in 1975, followed by another one ten years later (Straus, Gelles, and Steinmetz 1980; Straus and Gelles 1986, 1988, 1995; Gelles and Straus 1988).

Interested in *physical* domestic violence, the research team defined family violence as "an act carried out with the intention of causing physical pain or injury to another person." The researchers developed a measure of family violence called the **Conflict Tactics Scale**.

Facts about Families

Major Sources of Family-Violence Data

In addition to many smaller data sets, there are several important national sources of family-violence data. The work of Murray Straus, Richard Gelles, and colleagues in their National Family Violence Surveys pioneered and shaped the scientific study of family violence. Having developed the Conflict Tactics Scale, the research group undertook a household survey in 1975, followed by another ten years later. Together, the surveys produced data from more than 8,000 husbands, wives, and cohabitors (Gelles and Straus 1988; Straus and Gelles 1986, 1988, 1995; Straus, Gelles, and Steinmetz 1980). Gelles and Straus laid the groundwork, and you'll still find them cited in the domestic violence literature. The following are four more recent or ongoing large, national data sources:

1. The Uniform Crime Reports, begun in 1929 and issued by the Federal Bureau of Investigation (FBI), provide data on criminal incidents reported to the police by many (although not all) state, city, college campus, and other law enforcement agencies. A weakness of these data is that many crimes are not reported to the police, including an estimated half of IPV crimes. However, because most homicides are reported, the homicide data are considered more valid (Loftin, McDowall, and Fetzer 2008).

2. The National Intimate Partner and Sexual Violence Survey (NISVS), commissioned by the Centers for Disease Control and Prevention (CDC) is an ongoing, nationally representative telephone survey of women and men victims of physical violence, rape, and stalking, first conducted in 2010. Although this survey used random digit dialing and therefore ignored people without either a ground or cell phone, NISVS is an important resource on IPV—physical violence, emotional aggression, sexual violence (rape, attempted rape, sexual coercion, any unwanted sexual contact), control of reproductive or sexual health by an intimate partner), and stalking (including text messages, emails, and monitoring devices known or unknown to the victim).

3. The National Crime Victimization Survey (NCVS), conducted every two years by the federal Bureau of Justice Statistics (BJS), is a national survey that interviews asks a representative sample of about 250,000 respondents about any criminal victimization they've experienced, from physical violence to being burglarized to identity theft. The first interview is typically face-to-face; subsequent interviews occur by phone. This survey reports on intimate partner violence (IPV). Spouses, ex-spouses, and current or former boyfriends or girlfriends, including same-sex partners, are considered *intimate partners*.

4. The National Child Abuse and Neglect Data System, directed by the U.S. Department of Health and Human Services, collects data from government child-protection agencies and other community agencies voluntarily submitted by states. The collected data result in an annual report called *Child Maltreatment*.

Critical Thinking

Why might a survey that asks a nationally representative sample of U.S. couples about whether they ever get violent with each other—including throwing something or shoving—yield significantly different results regarding IPV than a survey asking whether the respondent had even been an IPV crime victim?

Respondents were asked whether they had done the following and how often:

- threw something at the other person;
- pushed, grabbed, or shoved;
- slapped or spanked (a child);
- kicked, bit, or hit with a fist;
- hit or tried to hit with something;
- beat up the other person;
- burned or scalded (children) or choked (spouses);
- threatened with a knife or gun; and
- used a knife or gun (Straus and Gelles 1988, p. 152).

Severe violence was defined as acts that have a relatively high probability of causing injury: kicking, biting, punching, hitting with an object, choking, beating, threatening with a knife or gun, using a knife or gun—and, for violence by parents against children, burning or scalding the child (Straus and Gelles 1988, p. 16). To this day, many family-violence researchers use the Conflict Tactics Scale, which has since been modified to include sexual assault (Straus et al. 1996). It's important to note that the Conflict Tactics Scale is different from and broader than the crime categories of assault and homicide that form the basis of criminal justice system statistics. For instance, behaviors that tend to be less severe—throwing things, pushing, grabbing, or shoving—are counted in this research as

violent acts. We return to this point later in this chapter. At this point, we turn to the statistics on violence between intimate partners.

The Incidence of Intimate Partner Violence (IPV)

First, some good news: Family violence—intimate partner violence (IPV) and child maltreatment—has declined considerably since first being addressed in the 1970s (Catalano 2013). These welcome declines show that efforts to combat family violence pay off. The overall rate of nonfatal IPV declined by 70 percent between 1994 and 2018—from 10 victimizations per 1,000 persons age 12 or older to 3 per 1,000 (Catalano 2013; Morgan and Oudekerk 2019, Table 1). Figure 12.2 shows declining victimization rates, primarily for women and slightly for men.

Nevertheless, despite declining significantly, IPV comprises about one third of police-recorded violence (Catalano 2013). More than half of these cases were serious violent crimes: rapes, sexual and other assaults, and crimes involving serious injuries or weapons (Truman and Langton 2015, Table 1). Furthermore, for reasons unknown at this writing, IPV increased recently—from 635,000 cases reported by the U.S. Bureau of Justice Statistics in 2014 to 850,000 cases in 2017 (Morgan and Oudekerk 2019, Table 1).

Family violence crosses all generational, social, cultural, class, and religious groups. Nevertheless, IPV rates vary by race/ethnicity. Victimization rates are relatively high (11 per 1,000) for Native American women. Black women's rates are somewhat high (5 per 1000). NonHispanic White and Hispanic females have moderate rates (4 per 1000), whereas Asians have very low IPV rates. White, black, and Hispanic male victimization rates are low, whereas those of Native American men are relatively high (Catalano 2013).

Correlates of Family Violence

IPV perpetrators and victims tend to be young adults (Shortt et al. 2013). The rate of violence between cohabiting partners is significantly higher than that of marrieds (Anderson 2010; Dush 2011; Martin et al. 2013). Overall, cohabitors are younger, less integrated into family and community, and more likely to have psychobehavioral problems such as depression and alcohol abuse—all factors associated with family violence. Another possibility is that there is less institutional control over cohabiting than married couples—regular contact with or monitoring by in-laws and other relatives, for instance (Ellis 2006). Still another thesis is that nonviolent cohabitors are more likely to marry whereas marrieds experiencing IPV are more likely to separate or divorce. This situation sharpens the difference between cohabiting and married couples (Kenney and McLanahan 2006).

For both victims and perpetrators, having experienced or witnessed IPV while growing up—either physical or psychological abuse—correlates with family violence as an adult, both against a partner and the family children (Hendy et al. 2012). Children who grow up with family violence may have trouble emotionally attaching to others and come to regard violence as normal (Lee et al. 2012; Shortt, Tiberio, Capaldi, and Low 2019). Unfortunately, children reside in an estimated 35 percent of households where IPV takes place (Catalano 2013). All this does not mean that abused children are predestined to be abusive parents or partners. Gelles and Cavanaugh (2005) report an intergenerational transmission rate of 30 percent. That is much higher than the general average of 2 percent to 4 percent. Nevertheless, "[t]he most typical outcome for individuals exposed to violence in their families of origin is to be nonviolent in their adult families. This is the case for both men and women" (Heyman and Slep 2002, p. 870).

Stress is another correlate of family violence (Zielinski 2009). Overload—that is, having too much to deal with or to worry about—creates stress that can lead to family violence, including child abuse. Other causes of family stress are changing lifestyles and standards of living, children's misbehavior, and a parent's feeling unrealistic pressure to do an excellent job on all occasions.

Poverty and economic stressors, as well as neighborhoods with higher concentrations of drug offense arrests and bars and liquor stores and higher rates of violence in general are correlated with IPV (Anderson 2010). The tendency to be concentrated in disadvantaged neighborhoods and

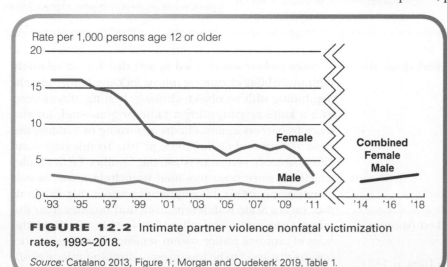

FIGURE 12.2 Intimate partner violence nonfatal victimization rates, 1993–2018.

Source: Catalano 2013, Figure 1; Morgan and Oudekerk 2019, Table 1.

lack of affordable housing options are also issues confronting more vulnerable segments of our population. These factors help to explain race/ethnic disparities in children's exposure to violence as well (Zimmerman and Messner 2013). We note here that, on average, Native Americans, African Americans, and Latinx are disproportionately more likely to encounter greater stressors than nonHispanic whites or Asian Americans, including work-related problems and financial strain, as well as interpersonal and institutional racism (e.g., at the hands of police, schools, public officials, lenders and landlords, and others). The subject matter of Chapter 13, dealing with family stressors is also addressed in Chapter 9.

Local advocacy groups draw attention to efforts to prevent domestic violence and to the need for more resources. In many localities, there are still not enough shelters to meet the needs of battered women and their children. Shelters and other services for male IPV victims are rare.

In addition to stress, substance abuse, including heavy alcohol use and binge drinking—often resulting in impulsivity, together with the tendency to feel put down by another's fairly innocent remarks—is a factor in IPV (Kilmer et al. 2013; Wiersma et al. 2010). Depression and other mental disorders are also associated with being either a victim or perpetrator of family violence (Trevillion et al. 2012). Some studies have identified a pattern of assortative partnering, whereby individuals at higher risk for family violence are mutually attracted and mirror each other's IPV risk factors for IPV—with high-risk individuals assortatively selecting other high-risk individuals while low-risk individuals likely choose low-risk partners (Anderson 2010).

GENDER AND INTIMATE PARTNER VIOLENCE (IPV)

The National Intimate Partner and Sexual Violence Survey, described in "Facts About Families: Major Sources of Family Violence Data" reports that intimate partner violence against women is dramatically higher than against men in every racial and ethnic category—about one in four women are victimized by an intimate partner, compared for about one in ten men (Smith, Zhang, Basille, Merrick, Wang, Kresnow, and Chen

2018, p. 7). Separated or divorced women are more likely to be IPV victims (Catalano 2012, Figures 3 and 6). One explanation for this gender disparity is that males who experience generally low relationship power are likely to become aggressive in situations where they feel especially powerless. In our culture, males are more often expected to exercise greater authority or power in heterosexual relationships; hence men are more likely to feel frustrated, sometimes violently so, when they feel they don't (Overall, Hammond, McNulty, and Finkel 2018).

However, studies using the Conflict Tactics Scale find approximately equal amounts of IPV perpetrated by men and women (Fountain et al. 2009; Hardesty and Ogolsky 2020). These conflicting findings have set off an unresolved dispute about whether IPV is *asymmetrical* (with women primarily the victims of male aggression) or whether couple violence is *symmetrical* (with men and women perpetrating IPV at about the same rate) (Anderson 2013; Fincham et al. 2013). The assertion that IPV is asymmetrical and that males are far more often perpetrators than victims is a strongly held position that goes back to the 1970s when feminists first raised the issue of wife abuse, or "battering" (Allen and Jaramillo-Sierra 2015). Whether women's violence toward men is mostly in self-defense, as feminist violence researchers argue, is part of the debate about IPV gender differences.

Self-Defense? Feminist scholars are likely to argue that IPV by women is largely in self-defense (or at least retaliation for) rather than initiation of a violent attack, especially in the case of homicides committed by heterosexual women against their partners (Belknap 2012). Sociologist Murray Straus, on the other hand, refuted this claim, arguing that wives often strike out first (Straus 2008; Straus and Ramirez 2007). Straus and his colleagues acknowledged, however, that violence perpetrated by women produces far fewer serious injuries and deaths than does that perpetrated by men. Nevertheless, injuries to men should not be dismissed and will be addressed later in this chapter (Straus 2008).

A principal reason for contradictory findings is that questions on the Conflict Tactics Scale include a broader range of acts that may be considered violent according to the survey questions but are not serious enough to be reported to crime authorities (Hardesty and Ogolsky 2020). Today researchers propose two distinct forms of IPV—(1) *coercive control* (formerly termed *intimate terrorism*, itself formerly termed *patriarchal terrorism*), and (2) *situational couple violence* (Hardesty et al. 2015; Johnson 2008). We'll look at situational couple violence first.

Situational Couple Violence

Situational couple violence refers to symmetrical (mutual, perpetrated by women as well as by men) violence between partners that occurs in conjunction with a specific argument, tends to be less severe in terms of injuries, and is unlikely to escalate as the relationship progresses (Hardesty and Ogolsky 2020). Situational couple violence appears to be relatively more common than coercive control—especially among young, cohabiting couples. Situational couple violence typically erupts during a fight and is often accompanied by heavy drinking (Anderson 2010). The following admission by one young woman, speaking to a researcher, may shed some light on this situation:

> I was like quick-tempered. It was little things that set me off. I used to get in a lot of fights with my brother. And my mom, we used to argue a lot . . . . My temper is still bad and I take things out on him [her partner]. (Smith et al. 2011)

Situational couple violence can result in serious injury. Nevertheless, coercive control is the more dangerous form of IPV—even leading in some cases to homicide. Table 12.2 compares situational couple violence with coercive control.

TABLE 12.2 Two Forms of IPV: Coercive control vs. Situational Couple Violence

COERCIVE CONTROL	SITUATIONAL COUPLE VIOLENCE
Explained by feminist theory (see Chapter 2); also called *intimate terrorism*	Explained by family systems theory (see Chapter 2)
Less common	More common
More often (but not exclusively) found among marrieds	More often (but not exclusively) found among cohabitors
Almost always perpetrated by men (asymmetrical violence)	Perpetrated by both women and men fairly equally (symmetrical violence)
Once it begins, may occur frequently in the course of the relationship and is progressive—that is, occurs more and more often	May occur less frequently over the course of the relationship
Motivated by the need to dominate or possess the partner; aimed at controlling the partner through intimidation and physical, emotional, or sexual abuse	Sparked by frustration and anger and aimed at winning a particular fight through physical or emotional abuse
Follows a "cycle of violence"	Does not necessarily follow a predictable pattern or cycle
Severe injuries probable, even homicide	Less severe injuries on average

Source: Hardesty and Ogolsky 2020; Johnson 2008.

Coercive Controlling Violence

Coercive control refers to abuse that is decisively oriented to controlling one's partner through fear and intimidation—for example, "he was carrying on [threatening] before I was going out and forced a love bite on my neck" (in Hearn 2013, p. 156). Feminists maintain and research supports that coercive control is almost entirely perpetrated by males (Anderson 2010). Not focused on a particular matter of dispute between the partners, coercive control is intended to establish an overall pattern of dominance and, at least among heterosexuals, appears to occur more often in marriage than among cohabitors. Furthermore, coercive control is likely to escalate and, especially in more advanced stages, is more likely than situational couple violence to produce serious injury.

Contrary to what many victims think, coercive control is not about the perpetrator's loss of control. Nor is coercive control necessarily precipitated by an outburst of anger. Rather, coercive control is a method of establishing and maintaining power. As depicted in Figure 12.3, coercive control often (although not always) follows a cycle often identified as the **cycle of violence**, consisting of three consecutive phases:

phase one—the violent episode itself, including physical, emotional, or sexual abuse, and typically growing more violent over the course of the relationship;

phase two—a calm period, sometimes termed the *honeymoon* phase, during which the abuser may ignore or deny the violence; blame the episode on the victim; or act genuinely sorry, sending cards and flowers—hence the expression *honeymoon* phase;

FIGURE 12.3 Coercive control and the cycle of violence.

Source: Adapted from cycle-of-violence literature by Agnes Riedmann and publisher's artists.

and phase three—tension buildup during which the victim feels increasingly disappointed and intimidated while the abuser's behavior is unpredictable and threatening.

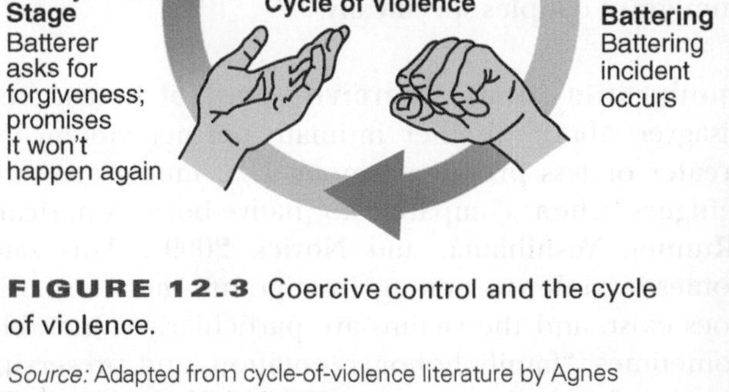

In the absence of therapy—and perhaps even despite therapy—the cycle of violence repeats itself, usually with phase two shortening while phases one and three lengthen. An Internet search for "cycle of violence" yields several graphic depictions of this cycle. Here we note that physical abuse is just one of the tools an intimate terrorist uses; emotional abuse is frequent as well. "Facts about Families: Signs of Coercive control" gives specific examples of intimate partner terrorism.

Emotional Abuse **Emotional abuse** includes frequent comments that damage a partner's self-esteem (Meyer 2013). Emotional abuse includes verbal abuse such as name-calling and demeaning verbal attacks; threats (to take away the victim's children, for instance, or to harm a victim's family or friends); and attacks on pets or the victim's property. Emotional abuse also includes constant criticism, isolating the victim, intimidation, causing a partner's sleep deprivation, and withholding money or other basic necessities. Trashing someone on Facebook or other social media is an example of emotional abuse. Cyberstalking a partner by means of location apps is emotional abuse. Often part of a pattern of control and domination, emotional abuse typically accompanies physical violence or threats of violence.

Marital Rape and Reproductive Coercion Estimates are that between 10 and 14 percent of women experience marital rape (Bergen 2006; Ferro, Cermele, and Saltzman 2008, p. 765). These sexual assaults often involve other violence as well. The feminist movement defined and publicized **marital rape** in the 1970s. Under traditional common law, a husband's sexual assault or forceful coercion of his wife was not considered rape because marriage meant the husband was entitled to unlimited sexual access to his wife (Jackson 2015). However, as a result of feminist political activity, all states have outlawed marital rape since 1993 (McMahon-Howard, Clay-Warner, and Renzulli 2009; National Clearinghouse on Marital and Date Rape n.d.).

Several small studies have reported that pregnancy increased the likelihood of male-to-female IPV (Brownridge et al. 2011; Cox 2008; Martin et al. 2004). On the other hand, research using national samples has found that when age was controlled, there was no increased risk of being abused with pregnancy. Still, the many studies that have found an association between pregnancy and violence have kept this hypothesis alive, along with a possible explanation—the male partner's jealousy, specifically that the new baby would interfere with the wife's attention to and care of the man (Sarkar 2008).

Signs of Coercive control

The most telling sign you're a victim of coercive control is feeling anxious or fearful regarding your partner. "If you feel like you have to walk on eggshells around your partner—constantly watching what you say and do in order to avoid a blow-up—chances are your relationship is unhealthy and abusive" (Smith and Segal 2012a). Tactics used by coercive controllers include:

1. dominance—making unilateral decisions for other family members, telling them what to do and expecting unquestioned obedience;

2. humiliation—making remarks or gestures likely to encourage others to feel bad about themselves or defective in some way that includes insults, name-calling, shaming, and put-downs, sometimes public;

3. invading your privacy—checking your email, social media platforms, text messages, or recent iPhone calls; reading your journal or snail mail; listening to your phone conversations;

4. isolation—cutting you off from work, acquaintances, friends, or relatives increasing your dependence on the abuser;

5. threats—threatening other family members or pets; threatening to commit suicide if challenged; and

6. intimidation—making frightening looks or gestures, smashing things, displaying weapons with the message that disobedience could result in violence toward you.

A coercive controller is not changing if they:

1. minimize the abuse or deny the seriousness of its consequences;

2. blame others;

3. insist the victim *owes* them another chance;

4. resist staying in treatment or threaten to quit;

5. say they can't make it if their partner leaves;

6. expect something from the victim in exchange for promises to get help (Smith and Segal 2012b).

- If you believe you're in an abusive relationship, information at these websites can help: mayoclinic.org; healthline. com; webmd.com; loveisrespect.org. If you believe someone you know is being abused, a first step is to say something like this: "I want to be there for you. How can I help?" Expressing to the victim your anger with the abuser will likely not be productive (Welch 2011).

- If you believe you are behaving abusively in one or more of your relationships, you can take steps to change things. See the websites listed above and/or find a counselor/therapist.

Critical Thinking

If you believe you or someone you know is being physical, emotionally, or financially abused, what first or next step will you take to address the situation?

The American College of Obstetricians and Gynecologists recognizes the phenomenon of **reproductive coercion**—behavior related to reproductive health that is used to maintain power and control in a relationship. For instance,

[a] partner may sabotage efforts at contraception, refuse to practice safe sex, intentionally expose a partner to a sexually transmitted infection (STI) or human immunodeficiency virus (HIV), control the outcome of a pregnancy (by forcing the woman to continue the pregnancy or to have an abortion or to injure her in a way to cause a miscarriage), forbid sterilization, or control access to other reproductive health services. (American College of Obstetricians and Gynecologists 2012)

About 20 percent of women who seek care in family-planning clinics and also report a history of IPV have experienced reproductive coercion (with the majority reporting birth-control sabotage) (American College of Obstetricians and Gynecologists 2012; and see Hess and Del Rosario 2018).

Figure 12.4, the "Control Wheel," further illustrates how perpetrators of intimate terrorism have a need for control than can result in emotional and physical violence. The inner circle describes coercive control generally. The outer circle applies to coercive control with immigrant couples specifically.

Immigrants and Coercive control Researchers disagree about whether intimate partner violence is greater or less prevalent among U.S. immigrants and refugees when compared to native-born Americans (Runner, Yoshihama, and Novick 2009). This said, domestic violence among immigrants most certainly does exist, and the victims are particularly vulnerable. Sometimes, "family honor, reputation, and preserving harmony" are primary values that impede seeking help.

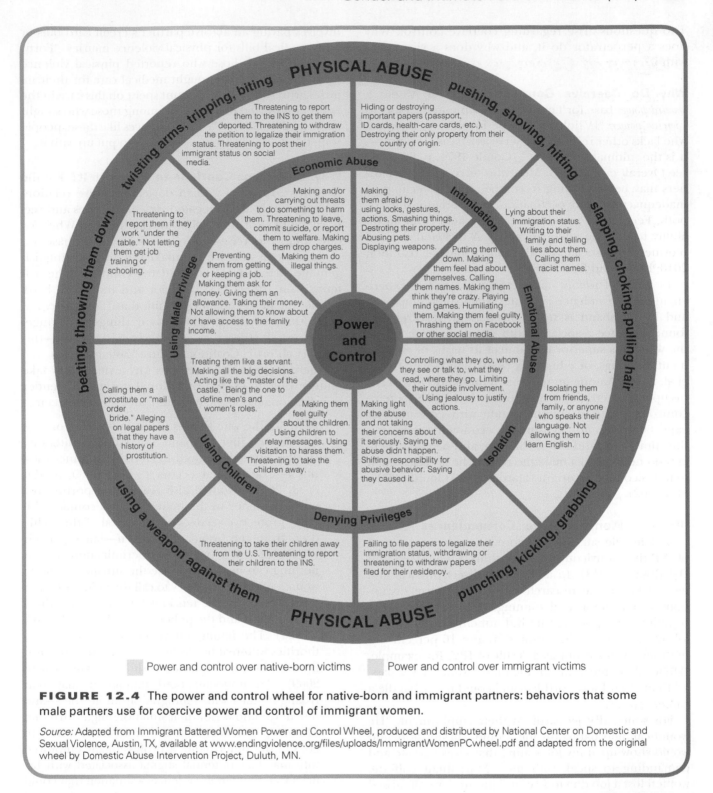

FIGURE 12.4 The power and control wheel for native-born and immigrant partners: behaviors that some male partners use for coercive power and control of immigrant women.

Source: Adapted from Immigrant Battered Women Power and Control Wheel, produced and distributed by National Center on Domestic and Sexual Violence, Austin, TX, available at www.endingviolence.org/files/uploads/ImmigrantWomenPCwheel.pdf and adapted from the original wheel by Domestic Abuse Intervention Project, Duluth, MN.

Immigrant women may have limited English skills and be socially isolated. They may be living with in-laws who support the abusive husband; a woman's own family may urge her to remain in the marriage despite the abuse.

The federal 1996 **Violence Against Women Act,** reauthorized in 2019, allowed immigrant victims to file independently for legal status if victimized by domestic violence (U.S. Citizenship and Immigration Services 2019). However, a victim who is not yet a citizen may be unaware of this legislation and fear that seeking help could result in deportation. Programs have emerged to assist immigrant women victimized by family violence.

Two questions arise regarding coercive control: Why does a perpetrator do it, and why does a victim live with it?

Why Do Coercive Controllers Do It?

Absent a *reward power* base for family power, some men resort to *coercive power:* "[V]iolence will be invoked by a person who lacks other resources to serve as a basis for power—it is the 'ultimate resource'" (Goode 1971, p. 628; and see Overall et al. 2016). Men who terrorize their partners may be attempting to compensate for feelings of inadequacy in their occupation, their relationship, or both. Feelings of powerlessness may stem from an inability to earn an adequate salary. A husband's unemployment is associated with family violence (Condon 2010; Lauby and Else 2008).

In terms of *relative* status, a woman's risk of experiencing severe violence is greater when she is employed and her husband is not. Considerable research has found violence associated with status reversal, where the woman is superior to the man in terms of employment, earnings, or education (Kaukinen 2004; Overall et al. 2016). Among immigrants, a husband's loss of status upon immigration could be associated with coercive control when jobs commensurate with education or expectations do not measure up, economic hardship is the family's lot, and wives, children, and people in general do not accord a male the respect he is accustomed to in his country of origin (Hardesty and Ogolsky 2020; Min 2002).

IPV and Women—Some Consequences

A study of 164 female survivors explored some consequences of IVP that nonvictims may not think about (Hess and Del Rosario 2018). In addition to direct physical and mental harm, the research showed long-term disruptions in education and training, with nearly half the respondents reporting they did not enroll in school or job training when they wanted to and 16 percent saying they dropped out as a result of IPV: for example, "He would demean me and tell me I would never pass, and state that I was worthless and could never handle a career" (p. 21).

For some, IPV jeopardized their employment: "He would call and harass my job, would call nonstop, and would show up at my job stating that I was his wife and demanding to speak with me." More than half the women lost a job due to IPV, and just under half missed work. About one fifth felt they missed out on a promotion due to IPV.

Hess and Del Rosario also found incidences of economic abuse, including stolen money, stolen or purposely damaged property, damaged credit ("My ex-partner put large amount of debt onto my credit cards"), and "coerced debt." Forty-four percent said they often had money taken from them. "Coerced debt" involves paying an abusive partner's credit card bills or high medical bills for physical violence injuries. "Forty-four percent of those who reported physical violence from a partner…have sought medical care for their injuries, with the average amount spent on this care in the last year reported to be $1,252 among those who sought care" (p. 29). Given IPV consequences like these, people sometimes ask why victims continue to put up with it.

Why Do Victims Continue to Live with It?

For the most part, battered women do leave abusive relationships, but only after repeated violent episodes and reconciliations (Roberts, Wolfer, and Mele 2008). Why? For one thing, a coercive controller's behavior in phase two of the cycle of violence can make it hard to distinguish genuine change from manipulative conduct. "The batterer is on [their] best behavior and the victim is reminded of all the qualities in him that [they] love. . . . More than anything, [they] want things to change. [They] want [them] to mean what [they] say—this time" (California Coalition Against Violence, n.d.).

There are additional reasons why a victim may take quite a while to leave: fear, cultural norms and gender socialization, economic hardship, and low self-esteem.

- **Fear.** "[T]he wife figures if she calls police or files for divorce, her husband will kill her—literally" (Gelles, quoted in Booth 1977, p. 7). In some cases, this is exactly what occurs (Seager 2009; Snider et al. 2009). Women also fear that reporting domestic violence to the police will risk contact with child protective services and removal of their children from the home. All women—but especially women of color—may fear discrimination or that nothing can be done to ease the situation. Among women of color, hesitancy to call the police also may derive from historic tensions between race/ethnic communities and the police force (Wolf et al. 2003, p. 124). The immigrant victim may fear that authorities will treat her or her partner with insensitivity, even hostility (Runner, Yoshihama, and Novick 2009). An undocumented immigrant may fear risking contact with immigration authorities—and one never knows how that situation might turn out. Then, too, a victim may fear that she has nowhere to go. On the other hand, a professionally employed wife may fear potential stigma associated with being a victim of family violence, a situation that could result in negative consequences to her professional reputation and career (Hess and Del Rosario 2018, Figure 8, p. 34; Kaukinen, Meyer, and Akers 2013).

- **Cultural norms and gender socialization.** Historically, women were encouraged to put up with abuse. English common law, the basis of the American legal structure, asserted that a husband had the right to physically "chastise" an errant wife. Although

the legal right to physically abuse women has long since disappeared, our cultural heritage continues to have an influence on whether victims seek help (Ellison et al. 2007). For one thing, cultural norms have traditionally suggested that it is primarily a woman's responsibility to keep a marriage intact. Believing this, wives are often convinced that their emotional support may lead husbands to reform (Roberts, Wolfer, and Mele 2008). On the other hand, the U.S. culture of violence, including violent pornography, denigrates and may frighten women.

- **Economic hardship.** Economic uncertainty and concerns about future finances can discourage victims from leaving (Hess and Del Rosario 2018). Leaving— or just pressing charges—can mean loss of a partner's income or damage to their professional reputation and future success. Hess and Del Rosario's respondents reported not having another place to live (83 percent), not having a job (63 percent), inability to afford childcare (50 percent), not having access to affordable transportation (49 percent), inability to pay for legal help regarding the separation (26 percent), and the fact that, if they left the relationship, they would lose access to their partner's health insurance coverage (15 percent) (Hess and Del Rosario 2018, Figure 8, p. 34). The prospect of financial problems is heightened when an abused partner looks ahead to single parenthood (Kaukinen, Meyer, and Akers 2013).

- **Low self-esteem.** Low self-esteem interacts with fear, depression, confusion, anxiety, feelings of self-blame, and loss of a sense of personal control (Umberson et al. 1998) to create **battered woman syndrome,** in which a wife cannot see a way out of her situation ("Battered Woman Syndrome" 2017; Walker 2009).

Social scientists have applied *exchange theory* to an abused woman's decision to stay or leave (McDonough 2010). As Figure 12.5 illustrates, an abused partner weighs such things as investment in the relationship, their (dis)satisfaction with the relationship, the possibility and quality of alternatives, and beliefs about whether it is appropriate to leave ("subjective norm") against such questions as whether they will be better off after leaving (might the perpetrator retaliate, for example?) and whether leaving is actually possible. The victim's personal resources along with available community, or structural, resources, such as shelters or other forms of assistance, further affect the decision. Personal barriers might involve not having a job with adequate pay or an extended family that can help. Structural barriers might include the lack of professional community services for practical help.

Much of this discussion has assumed that in a heterosexual relationship, the male is the perpetrator of family violence. However, this is not always the case; men can be domestic-violence victims as well (Fiebert 2012; Mayo Clinic Staff 2020).

Male Victims of Heterosexual Terrorism

The butt of jokes, male IPV victims are often considered unmanly. Some social scientists question the extent to which males, socialized to be the strong ones, would acknowledge to researchers that they had been battered by females (Brown 2008; Dutton and Nicholls 2005). If males can be suspected of minimizing IPV victimization, so also can some family-violence researchers (Gaille 2017). Often, for instance,

> female-perpetrated abuse is minimized and understood as either defensive or situational in nature, an isolated

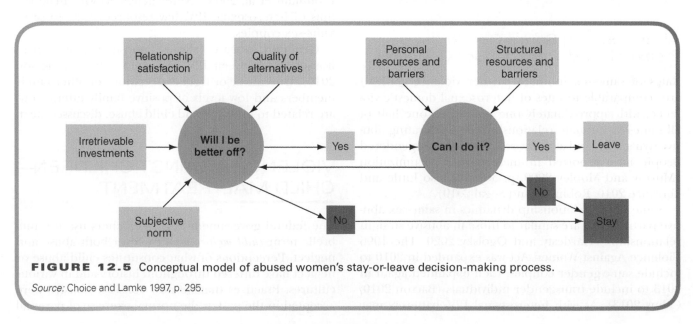

FIGURE 12.5 Conceptual model of abused women's stay-or-leave decision-making process.

Source: Choice and Lamke 1997, p. 295.

expression of frustration in communicating with an unsympathetic partner, in contrast to the presumably intentional, pervasive, and generally controlling behaviors exhibited by men. (Hamel and Nicholls 2006; see also Johnson and Leone 2005)

If male victims do call police, they may themselves be arrested as suspected perpetrators. The police appear ill-equipped to deal with female abuse of males, going so far as to downplay violence against men even when they're called to the scene (Brown 2008).

Funding to combat IPV has overwhelmingly been spent on programs designed to support only women victims (Gaille 2017; Watson 2010). Meanwhile A small number of shelters—the Valley Oasis shelter in Antelope Valley, California, is one of those few—provide a full range of shelter services to men.

> There is no doubt that domestic violence against men can be reduced; the domestic violence initiatives of the past [50] years have brought a hidden crime to light and provided protection for millions of women. The next step is to admit that domestic violence is not a male or female problem, but rather a human problem, and that a lasting solution must address the cruelty—and suffering—of both sexes. (Watson 2010)

Furthermore, "Admitting the problem and seeking help doesn't mean you have failed as a man or as a husband. You are not to blame, and you are not weak." Sharing details of your abuse "can offer a sense of relief and provide some much-needed support" (Robinson and Segal 2019).

Reasons abused men don't leave involve feeling ashamed, lacking resources, fearing that reporting the violence or leaving will mean losing access to their children, wanting to protect their children, and being in denial (Robinson and Segal 2019). Men can find help online at Help for Abused Men (www.HelpGuide.org and/or at valleyoasis.org).

Abuse among Same-Gender, Bisexual, and Transgender Couples

Rates of same-sex intimate partner violence (SSIPV) are "comparable to rates of heterosexual domestic violence, with approximately one quarter to one half of all same-sex intimate relationships demonstrating abusive dynamics," and as many as half of all transgendered people have reported intimate partner victimization (Murray and Mobley 2009, p. 361; see also Little and Terrance 2010; Robinson and Segal 2019).

Some of the relationship dynamics in same-sex abusive partnerships are similar to those in abusive straight relationships (Hardesty and Ogolsky 2020. The 1996 Violence Against Women Act was expanded in 2010 to include same-gender couples, then expanded again in 2013 to include transgender individuals (Barron 2010; Calms 2013). As with heterosexual IPV, batterers may abuse drugs or alcohol or have a history of childhood violence. Also, like heterosexuals involved in IPV, the gay or lesbian couple is likely to deny or minimize the violence, along with believing that the violence is the victim's fault.

Although sexual minorities experience all of the same threats as heterosexual victims, they have an additional concern—the abusive partner can threaten to "out" them to employers, family members, and friends. Hate crimes, discrimination, internalized homophobia, and fear of being "outed" are all stressors that can impact LGBTQ+ relationships and lead to SSIPV (Brown 2008; Fountain et al. 2009; Hardesty and Ogolsky 2020). According to a Mayo Clinic publication, if you're LGBTQ+ you can experience IPV if your partner:

- threatens to tell friends, family, or others your sexual orientation or gender identity;

- tells you authorities won't help a gay, bisexual, or transgender person;

- tells you that leaving the relationship means you're admitting that gay, bisexual, or transgender relationships are deviant;

- justifies abuse by telling you that you're not "really" gay, bisexual, or transgender; or

- says men are *naturally* violent, or women *can't be* violent (Mayo Clinic Staff 2017).

LGBTQ+ individuals may be afraid to go to the police—or to use any domestic violence intervention services—for fear of having their gay identity revealed or receiving a hostile response (Simpson and Helfrich 2005). Domestic violence services oriented to LGBTQ+ individuals or couples are now somewhat available in large cities with substantial gay/lesbian communities (Fountain et al. 2009). Nevertheless, as with male victims of heterosexual IPV, few resources exist to serve same-sex couples.

Generally, research shows a substantial overlap in households between IPV and child abuse (Anderson 2010). Unrealistic or rigid expectations of other family members and low levels of positive family interactions are related to both IPV and child abuse, discussed next.

VIOLENCE AGAINST CHILDREN— CHILD MALTREATMENT

The federal government and researchers use the umbrella term *child maltreatment* to cover both abuse and neglect. Perceptions of what constitutes child abuse or neglect have differed throughout history and in various cultures. Practices that we now consider abusive were accepted in the past as the normal exercise of parental

rights or as appropriate discipline. Today, standards of acceptable childcare vary according to culture and social class. What some groups consider mild abuse others consider right and proper discipline. In 1974, however, Congress provided a legal definition of *child maltreatment* in the Child Abuse Prevention and Treatment Act.

The act defines **child maltreatment** as the "physical or mental injury, sexual abuse, or negligent treatment of a child under the age of 18 by a person who is responsible for the child's welfare under circumstances that indicate that the child's health or welfare is harmed or threatened" (U.S. Department of Health, Education, and Welfare 1975, p. 3). According to this definition, nearly 3 million children—about one in twenty-five—experience maltreatment annually (Sedlak et al. 2010).

Estimates from the 2018 U.S. Department of Health and Human Services report *Child Maltreatment* are based on state reports of child abuse. Of the reported cases of child maltreatment in 2018, approximately 79 percent were of neglect, 18 percent of physical abuse, 9 percent of sexual abuse, 8 percent of psychological mistreatment, and 2 percent of other forms of neglect such as threatened abuse or a parent's drug abuse (U.S. Department of Health and Human Services 2020).

Neglect and Abuse

Child neglect is by far the most common form of child maltreatment and involves failing to provide adequate physical or emotional care. Often neglect is grounded in parents' or guardians' economic problems, mental health issues, lack of parenting skills, or personal histories of childhood neglect or abuse. Less common, *willful* neglect involves purposeful failure to provide care even when resources are available. Physically neglected children often show signs of malnutrition, lack immunization against childhood disease, lack proper clothing, attend school irregularly, and need medical attention for such conditions as poor eyesight or bad teeth.

Emotional neglect involves a parent's or guardian's often being overly harsh and critical, belittling a child, failing to provide guidance, or being uninterested in a child's needs. Some activists have begun to label more specific forms of emotional neglect such as *educational neglect* (failure to see that a school-age child gets to school regularly) and *medical neglect* (failure to obtain necessary medical care for a child).

The term **child abuse** refers to overt acts of aggression—excessive verbal derogation (emotional child abuse) or physical abuse such as beating, whipping, punching, kicking, hitting with a heavy object, burning or scalding, or threatening with or using a knife or gun. Data collected by the U.S. Department of Health and Human Services between 1993 and 2018 show a 29 percent decline in the rate of physical child abuse. The incidence of children victimized by sexual

Diego Cervo/Shutterstock.com

Where does discipline end and child abuse begin? For one thing, physical discipline should not be undertaken when a parent or parent figure is angry, because it is too easy to go beyond reasonable limits.

abuse decreased by 47 percent during that time. The incidence of emotionally abused children decreased by 48 percent.

However, the estimated number of *emotionally neglected* children more than doubled during that interval, rising from about 500,000 in 1993 to more than 1 million today (U.S. Department of Health and Human Services 2020). An estimated 1,770 children died from abuse or neglect in 2018, with 71 percent younger than age 3 and 80 percent of the deaths caused by a parent (U.S. Department of Health and Human Services 2020, pp. 46–48).

Child Sexual Abuse Another form of child abuse is **sexual abuse** in which a child is forced, tricked, or coerced by an older person into sexual behavior—exposure, unwanted kissing, fondling of sexual organs, intercourse, rape, incest, prostitution, and pornography—for purposes of sexual gratification or financial gain. Of abused children, between 10 and 20 percent are sexually abused (U.S. Department of Health and

Human Services 2020). Internet-facilitated sexual abuse is assuredly a real and dangerous possibility (Internet Watch Foundation 2018). As with other forms of family violence, however, data collected between 1992 and today show that child sexual abuse overall has actually declined by about 50 percent (U.S. Department of Health and Human Services 2020).

Incest involves sexual relations between related individuals. The most common forms are sibling incest followed by father–daughter incest. The definition of child sexual abuse *excludes* mutually desired sex play between or among siblings close in age, but coerced sex by stronger or older siblings is sexual abuse and is more widespread than parent–child incest (Kiselica and Morrill-Richards 2007; Thompson 2009).

Incest is the most emotionally charged form of sexual abuse and also the most difficult to detect. Childhood incest appears in the background of a variety of sexual, emotional, and physical problems among adults (Banyard et al. 2009; Carlson, Maciol, and Schneider 2006; Sachs-Ericsson et al. 2010).

We see occasional media stories about female sex abusers, but research indicates that the vast majority of sexual abusers are male (Peter 2009). About 47 percent of sexual assaults of children are by relatives; 49 percent by others such as clergy, teachers, coaches, or neighbors; and only 4 percent by strangers (Hernandez 2001).

How Extensive Is Child Maltreatment?

Abused children live in families of all socioeconomic levels, races, nationalities, and religious groups. Nevertheless, child maltreatment is reported much more frequently among poor and nonwhite families than among middle-and upper-class white families (Department of Health and Human Services 2020).

Differences in neglect rates are largely attributed to insufficient income. Meanwhile, differences in all maltreatment rates may be partly due to reporting differences, with poor children more likely to be seen by social welfare authorities (Gelles and Cavanaugh 2005). Another reason may be unconscious race-or class-based discrimination on the part of medical personnel and others who report abuse and neglect (Lane et al. 2002). There are also real differences, however, and the fact that low-income or race/ethnic minority parents can

At least 500,000 children are used annually in pornography or prostitution. Child victims of organized sexual exploitation in the United States are often runaways. Many have left home to escape family violence (*Domestic Sex Trafficking of Minors* n.d.; Harris 2009).

be highly stressed is often offered as one explanation (Gelles and Cavanaugh 2005).

The percentages of male (48.3 percent) and female (51.3 percent) victims are not significantly different. The youngest children (through age 3) are more vulnerable than older children. A child also faces the greatest risk of becoming a victim of homicide during the first year of life (U.S. Department of Health and Human Services 20108).

About 54 percent of perpetrators are women. Approximately 80 percent of abused or neglected children are mistreated by at least one parent: by mother only in 38.3 percent of cases, by father only in 18 percent of cases, and by both mother and father in nearly 18 percent of cases. In 10 percent of cases, children were mistreated by other caregivers: foster parents or legal guardians, day care workers, or unmarried partners of a parent (U.S. Department of Health and Human Services 2020).

Risk Factors for Child Abuse When exaggerated, the following factors can encourage even well-intentioned parents to mistreat their children:

- Having learned to view children as requiring physical punishment in order to develop properly (Milner et al. 2010, p. 335).

- Having unrealistic expectations about what a child is capable of and being unknowledgeable regarding children's physical and emotional abilities and needs (Letarte, Normandeau, and Allard 2010). For example, slapping a bawling toddler to stop her or his crying is completely unrealistic, as is too-early toilet training.

- Feeling highly stressed or helpless in the parent or provider role (Milner et al. 2010).

- Being a young adult and inexperienced with child-care or managing stress (Milner et al. 2010).

- Experiencing marital discord or divorce, especially when coupled with children perceived as unusually demanding or otherwise difficult (Milner et al. 2010).

- Abusing alcohol or other substances (Milner et al. 2010).

- A mother's cohabiting with a male partner, who could potentially abuse the child—and is significantly more likely to do so than the child's biological father (Crary 2007a).

- Having a stepfather—because stepfathers are more likely than biological fathers to abuse children (Crary 2007a; Sedlak et al. 2010). Still, it is important to remember that 80 percent of child maltreatment perpetrators are biological parents, whereas cohabitants and stepparents make up just 4 percent each (U.S. Department of Health and Human Services 2015).

Abuse versus "Normal" Child Raising Issues surrounding spanking children are addressed at length in Chapter 9. Here we note that it is too easy for parents to go beyond reasonable limits when angry or distraught and to include "discipline" that most observers would define as abuse (Baumrind, Larzelere, and Cowan 2002; Feigelman et al. 2009). Hence, child abuse must be seen as a potential behavior in many families (Feigelman et al. 2009). Moreover, immigrant families may come from cultures where rather severe physical punishment is considered necessary for good child rearing. Those parents may not be aware that what these forms of parental discipline are illegal in this country. They may instead view themselves as responsible parents (Renteln 2004, pp. 54–57).

Then, too, immigrant parents may be mistakenly identified as having abused children because of certain cultural practices not initially understood in this country. There are healing practices in certain cultures that can produce what looks like evidence of injuries to an American physician or social service worker. Southeast Asians employ a practice known as *coining* whereby they rub the edge of a coin along the skin. This leaves marks that can appear to be those of a whip (Child Abuse

Prevention Council of Sacramento, n.d.). Similarly, Asian children may have *Mongolian spots* on their skin, a natural phenomenon, but one that appears as bruising to an uninformed health practitioner (Families with Children from China 1999). The issue whether spanking children should be considered abusive is addressed in Chapter 9. We turn now to two more recently recognized forms of family violence: sibling violence and child-to-parent violence.

SIBLING VIOLENCE

Sibling violence—violent acts perpetrated by one sibling against another—is often overlooked and rarely studied (Butler 2006; Finkelhor et al. 2005), even though the early National Family Violence Survey found it to be the most pervasive form of family violence (Hardesty and Ogolsky 2020); Straus, Gelles, and Steinmetz 1980). Furthermore, sibling violence is not simply "harmless" teasing ("UF Study" 2004). A fairly recent national study found that 35 percent of children had been hit or attacked by siblings in the previous year. Fourteen percent were repeatedly attacked, 5 percent of them attacked hard enough to sustain injuries such as bruises, cuts, chipped teeth, and sometimes broken bones. Two percent were hit with rocks, toys, broom handles, shovels, or knives (Butler 2006).

Child psychologist John Caffaro (in Butler 2006) sees sibling abuse as situational, not personality driven. When parents are frequently physically or emotionally absent from the home, or when they have their own problems, sibling violence is more apt to occur. Failure to intervene effectively also plays a part, as does parental favoritism of one child over another. Trauma, anxiety, and depression are likely to result from experiencing sibling violence, as well as an increased likelihood to perpetrate violence as an adult and to have relationship problems (Butler 2006; Hoffman and Edwards 2004; Noland et al. 2004).

Perpetrators of sibling violence are more likely than others to become perpetrators of dating violence, according to a study of more than 500 men and women at a Florida community college. "Siblings learn violence as a form of sibling manipulation and control as they compete with each other for family resources. . . . They carry these bullying behaviors into dating, the next peer relationship in which they have an emotional investment" (researcher Virginia Noland in "UF Study" 2004). Furthermore, adolescent dating violence tends to continue at least into young adulthood (Cui et al. 2013). Yet sibling violence has received comparatively little research attention, and even less attention has been given to preventive or therapeutic responses. Noland et al. (2004) recommends that sibling violence be taken more seriously and that anger management

programs be implemented while potentially violent individuals are still children ("UF Study" 2004).

CHILD-TO-PARENT VIOLENCE

Relatively little research has been done on **child-to-parent violence** (Cottrell and Monk 2004; Holt 2013). Yet, like other forms of family violence, child-to-parent abuse has been there all along (Calvette et al. 2014).

This section relies heavily on a review article by Cottrell and Monk (2004) and on the book *Adolescent-to-Parent Abuse* (2013) by criminal justice professor Amanda Holt. Data suggest that 9 to 14 percent of parents have been abused by adolescent children, with injuries that include bruises, cuts, and broken bones. Types of assaults have included kicking, punching, biting, and the use of weapons. Mothers, especially single mothers, and elderly parents of youth are the most frequent victims (Holt 2013).

Adolescent boys are the most frequent perpetrators, with their growth in size and strength associated with increases in violence. Although there are no clear findings of differences in race/ethnicity or social class, poverty and other family stressors are related to child-to-parent violence. Abusive children may exhibit diminished emotional attachments to parents. The child may have been abused by the parent or witnessed intimate partner abuse in the household. Overly permissive parents and those who abandon their authority in response to the violence tend to see more of it. Parents whose child-raising styles contradict each other are also at risk. Drug use by the adolescent may play a role (Holt 2013). Parents who are victims of assaults by their adolescent children often engage in denial. Unfortunately, at the moment, few if any support services exist, and the criminal justice system has not responded systematically (Cottrell and Monk 2004).

STOPPING FAMILY VIOLENCE

Stopping family violence involves policy action on both the micro (relationship or personal) and the macro (structural) levels (Larrivée, Brabant, and Lessard 2012). First, we will explore micro or relationship approaches. On the relationship level, three major approaches to combating family violence involve (1) separating victim from perpetrator, (2) the criminal justice approach, and (3) the therapeutic approach (Hardesty and Ogolsky 2020).

Separating Victim from Perpetrator

With regard to both IPV and child maltreatment, separation of the victim from the perpetrator is one approach to stopping family violence. Separating victims from perpetrators involves escaping to shelters for IPV victims and removing an abused child from the home in the case of child abuse.

First established in the 1970s by feminists for female abuse victims, a network of **shelters** now provides a woman (and often her children) with temporary housing, food, and clothing to alleviate the problems of economic dependency and physical safety (MacFarquhar 2019). Shelter staff also provide counseling to encourage a stronger self-concept so that the woman can view herself as worthy of better treatment and being capable of making her way alone if need be. Finally, shelters may provide guidance in obtaining employment, legal assistance, or family counseling. In two studies on correlates of domestic homicide in large U.S. cities, researchers found that the availability of shelter and hotline services for domestic violence and more aggressive arrest and prosecution lowered homicide rates between intimate partners (Anderson 2010).

Protecting abused or neglected children may involve removing them from their family homes and placing them in foster care, which is discussed at greater length in Chapter 9. This practice is controversial because foster parents have been abusive in some cases, and there are not enough foster parents to go around in many regions of the country. Moreover, removal from the home can be traumatic to children, who are often attached to their parents despite the abuse (Kaufman 2006). They may blame themselves for the breaking up of the family (Gelles and Cavanaugh 2005).

An alternative to child removal is **family preservation,** in which a child protective services worker is able to "leave the child with [the offending] family and provide support in the form of housekeeping help or drug treatment, and then visit frequently to monitor progress" (Kaufman 2006, p. A12). The family-preservation approach is not appropriate if harm to the child appears imminent. Family preservation is a controversial strategy (Gelles 2005; Wexler 2005), but both removing the child from the home and the family-preservation approach carry risks. Sometimes separating a victim from a perpetrator is accompanied by a criminal justice response.

THE CRIMINAL JUSTICE RESPONSE

Until the 1970s there was little legal protection for battered women. Today, experts believe that arrest can deter future intimate partner violence, at least by perpetrators who are employed and married—spouses with a "stake in conformity." Those who are unemployed or not married to the person they abused may react

to arrest with increased violence (Dugan, Nagin, and Rosenfeld 2004).

Another problem with the arrest strategy has meant that literal readings of mandatory arrest laws have resulted in the arrest of victims, along with perpetrators, when victims resisted with violent force. In some cases, women claimed that the exchange between the perpetrator and the police officer was characterized by "male bonding," in which the perpetrator's story overrode the woman's complaint of violence. Today, more victims report positive and protective experiences to researchers (Wolf et al. 2003).

Several interventions to address partner violence have been implemented through the criminal justice system, including specialized domestic violence courts, coordinated community responses, mandatory arrest and prosecution policies, and court-ordered batterer treatment. Civil protective orders have also been used, more as a preventive measure, the goal being to prevent future violence rather than punish for past acts of violence (Logan, Walker, and Hoyt 2012).

Meanwhile, some advocates favor the **criminal justice response**—that is, the punitive approach—for perpetrators of child maltreatment. These advocates believe that one or both parents should be held legally responsible for child abuse. Although all states have criminalized child abuse, the approach to child protection has gradually shifted from punitive to therapeutic. Not all who work with abused children are happy with this shift. They reject the family system approach to therapy because it implies distribution of responsibility for change to all family members. Nevertheless, social workers and clinicians—rather than the police and the court system—increasingly investigate and treat abusive or neglectful parents.

A complicated issue related to this approach involves holding battered women criminally responsible for "failing to act" to prevent such abuse at the hands of their male partners. Fathers are typically *not* held accountable for child abuse committed by female partners (Liptak 2002). Meanwhile, legal advocates question whether the law should hold a battered woman responsible for failing to prevent harm to her children when, as a battered woman, she cannot even defend herself: "When the law punishes a battered woman for failing to protect her child against a batterer, it may be punishing her for failing to do something she was incapable of doing . . . . She is then being punished for the crime of the person who has victimized her" (Erickson 1991, pp. 208–209). An alternative—or in some cases a complement—to the criminal justice approach is the therapeutic approach.

The Therapeutic Approach

The therapeutic approach involves establishing counseling and educational programs designed to define

and treat offenders as needing educational and psychological guidance rather than as criminals (Hardesty and Ogolsky 2020; Whiting and Merchant 2014). As an example, preventive programs that address relationship skills among teen mothers have reduced IPV levels (Langhinrichsen-Rohling and Capaldi 2012). Programs designed to prevent child maltreatment might include participation of perpetrators' mothers and siblings "to enhance development of more peaceful conflict resolution patterns within and outside the family" (Hendy et al. 2012).

Although therapy was earlier thought to be ineffective, several therapy programs for perpetrators of family violence have now emerged. Many IPV perpetrators have difficulty controlling anger and frustration, dealing with stress, and relinquishing control over others. Some therapists' data now suggest that supportive group counseling can sometimes help create a setting in which abusers learn constructive ways to cope with anger and recognize their partner's right to autonomy and respect.

Another tactic involves teaching empathy—the ability to experience the condition of another person vicariously and to care about how the other person feels in that situation (Barlinska, Szuster, and Winiewski 2013). This solution has been used to counter bullying. Then too, because research shows that spirituality can be important to many people who have experienced IPV, religious communities have strengthened their family violence counseling (Austin and Falconier 2012; Ellison et al. 2007). In some cases, couples therapy or individual counseling programs may help women

> assert themselves in an appropriate but determined way early in relationships, rather than being passive in the face of mounting assaults on their integrity and autonomy. The hope is that if women can create interpersonal limits, for both their own and their partner's speech and actions, they will be less likely to overlook incremental escalations of offending behavior. (Stein 2013, p. 190)

We note, however, that couples therapy programs designed to stop IPV remain controversial because they proceed from the premise that a couple's staying together without violence after an abusive past is possible. Feminist scholars have argued that this assumption is dangerous and that the danger women face in violent relationships is highly underestimated (Kulkin et al. 2008; Simpson and Helfrich 2005).

The therapeutic approach with regard to child maltreatment involves increasing parents' self-esteem and their knowledge of children's development. Programs attempt to reach stressed parents before they hurt their children, and many operate twenty-four-hour hotlines for parents under stress. High school classes on family life, child development, and parenting are now virtually universal and are thought to have helped reduce child

abuse by teaching parents what to expect from children at different ages.

Parent education directed toward new immigrant parents might decrease the number of situations that end in removal of children from their home. For example, if some immigrant families do not realize that their traditional disciplinary practices constitute criminal child abuse in the United States, parent education offered through refugee service centers could anticipate that problem (Gonzalez and O'Connor 2002; Renteln 2004, pp. 54–58).

Macro or Structural Approaches

The family ecology model, introduced and described at length in Chapter 2, argues that family dynamics are influenced by all that goes on outside families—from neighborhood to workplace, to state and federal legislation, and even to global developments such as migration patterns. When looking at stopping family violence, a family ecology approach urges us to consider how more macro, or social structural, programs and efforts impact more micro, relationship solutions (Asay et al. 2014).

A macro approach notes the social, cultural, and economic context of family violence and then provides programs and services to help reduce or otherwise address it. For instance, housing assistance and subsidized childcare might dissuade a low-income or socially isolated parent from neglectfully leaving children alone while working (U.S. Department of Health and Human Services 2013). After-school programs have been known to positively impact adolescents who live in violent homes (Gardner, Browning, and Brooks-Gunn 2012).

Note again that family violence, involving both intimate partners and children, has declined since the 1990s. This decline may well be a consequence of the support and treatment programs developed since the 1970s. Shelter options, women's increased employment and ability to support themselves, cultural change that takes domestic violence seriously and endorses women's taking self-protective actions, and increased interest in and understanding of domestic violence on the part of law enforcement agencies are all developments that may account for the decrease in fatal and nonfatal violence against women by their intimate partners. Although a dramatic rise in women's employment and earnings may prove threatening to low-earning husbands in the short run, mutual awareness of a woman's potential economic independence may deter wife abuse by changing the family power dynamic in the long run (Yakushko and Espin 2010).

The subsequently broadened Violence Against Women Act resulted in greater perpetrator accountability for rape and stalking; in increasing rates of prosecution, conviction, and sentencing of offenders; in more easily granting a victim's protection order; and

in ensuring that police respond to crisis calls. The act also established the National Domestic Violence Hotline, which has answered more than 3 million calls since it was established. ("Factsheet: The Violence Against Women Act" n.d.).

Information, advice, and referral help for IPV and other forms of family violence is available on the Internet—for example from the National Coalition Against Domestic Violence, Helpguide.org, and the National Domestic Violence Hotline. A 2012 national Internet survey of rape crisis centers, the Rape Crisis Center Survey, found that as many as 80 percent of rape victims suffer from post-traumatic stress disorder, and a significant percent miss time at work or school. Studies show that when victims receive advocate-assisted services, they do better in short-and long-term recovery. Unfortunately, however, statewide and regional budget cuts, along with increasing competition for donor funding, have resulted in program closings. Today, programs have waiting lists for counseling services and support groups. The consequences for abuse victims may be increasingly serious as money for shelters and related services dries up. A qualitative study of thirteen women who had entered shelters for abuse victims found that, in a follow-up two years later, "long after women leave the shelter and their abusive relationships they will need the continued support of vocational psychologists to provide career assessment and counseling with focus on long-term career and educational opportunities" (Brown, Trangsrud, and Linnemeyer 2009).

Structural approaches to combat child maltreatment would involve energetic national measures to raise living standards, keep children out of poverty, and include children of immigrants (Svevo-Cianci and Velazquez 2010). Poverty makes things worse for children growing up in violent homes, yet these children can and do show resilience, especially when aided by community resources and support (Yoo and Huang 2012). Encouraging greater neighborhood cohesiveness would also help. In neighborhoods with support systems and tight social networks of community-related friends—where other adults are somewhat involved in the activities of the family—child abuse and neglect are much more likely to be noticed and stopped (Falconer et al. 2008; U.S. Department of Health and Human Services 2020).

Social policy and programs to combat child maltreatment, first established in the 1960s and 1970s, have contributed to the decline in the incidence of child abuse and willful neglect. For instance, there are community-support options such as volunteers available to babysit with potentially abused children to give stressed parents a break. Another community resource is the *crisis nursery,* where parents may take their children when they need to get away for a few hours. Ideally, crisis nurseries are open twenty-four hours a day and accept children at any hour without prearrangement.

People find meaning and belonging in family. Family life is supportive when partners strive for no-power relationships, characterized by mutual respect, feeling of fairness, attention to others' physical and emotional needs, and the willingness to be vulnerable, to be influenced by and accommodate each another, and to repair conflicts by listening sincerely to the other's point of view.

We need programmatic support for male victims of spouse abuse: "Compassion for victims of violence is not a zero sum game. . . . Reasonable people would rationally want to extend compassion, support, and intervention to all victims of violence" (Kimmel 2002, p. 1,354). Indeed, the male victim of violence has few resources and often little sympathy (Cook 2009).

Power disparities discourage intimacy, which is based on honesty, sharing, and mutual respect. For most, therefore, attainment of the American ideal of equality in marriage would seem to support the development of intimacy in marital relationships. We close this chapter with a reminder of everyone's basic right to be respected—and not to be physically, emotionally, or sexually abused—in any relationship. And we end on an optimistic note, as most forms of family violence show evidence of declining rather than increasing.

Summary

- Power, the ability to exercise one's will, may rest on cultural authority, economic and personal resources that are gender-based, love and emotional dependence, interpersonal dynamics, or emotional or physical violence.

- Marital, or conjugal, power, or power in other intimate partner relationships, includes decision-making, control over money, the division of household labor, and a sense of empowerment in the relationship. American marriages experience tension between egalitarianism on the one hand and gender identities that effectively preserve male authority on the other.

- The relative power of a husband and wife within a marriage or other intimate partnership varies by education, social class, religion, race/ethnicity, age, immigration status, and other factors. It varies by whether or not the woman works and the presence and age of children. Studies of married couples, cohabiting couples, and gay and lesbian couples illustrate the significance of economically based power and norms about who should have power.

- Couples can consciously work toward more egalitarian marriages or intimate partner relationships and relinquish "power politics." Changing gender roles, as they affect marital and intimate relationship power, necessitate negotiation and communication.

- Researchers do not agree on whether intimate partner violence is primarily perpetrated by males or whether males and females are equally likely to abuse their partners.

- The effects of intimate partner violence indicate that victimization of women is the more crucial social problem, and it has received the most programmatic attention.

- Some programs have now been developed for male abusers. Studies suggesting that arrest is sometimes a deterrent to further wife abuse illustrate the importance of public policies in this area.

- Economic hardships and other stress factors (among parents of all social classes and races) can lead to physical or emotional child abuse, as can lack of understanding of children's developmental needs and

abilities. One difficulty in eliminating child abuse is drawing a clear distinction between "normal" child rearing and abuse.

- Physical, verbal, and emotional abuse, as well as sexual abuse and child neglect, are forms of violence against children. Sibling violence is an often-overlooked form of child abuse.

- Child-to-parent abuse is a recently "discovered" form of family violence. It may grow out of previous abuse of a child.

- Criminal justice, therapeutic, and social welfare approaches are ways of addressing family violence from IPV to child maltreatment.

Questions for Review and Reflection

1. How is gender related to power in marriage? How do you think ongoing social change will affect power in marriage?

2. Do you think that power in a marriage or other couple relationship depends on who earns how much money? Or does it depend on emotions? Is it possible for a couple to develop a no-power relationship?

3. Looking at domestic violence, why might women remain with the men who batter them? Do you think that shelters provide an adequate way out for these women? What about arresting abusers? Should intimate partner violence against men receive more attention in the form of social programs? Why or why not?

4. What factors might play a role when well-intentioned parents abuse their children?

5. **Policy Question.** What can we as a society do to combat child neglect that is really due to family poverty?

Key Terms

battered woman syndrome 323
child abuse 325
child maltreatment 325
child neglect 325
child-to-parent violence 328
coercive power 306
Conflict Tactics Scale 314
conjugal power 306
criminal justice response 329
cycle of violence 319
egalitarian norm 308
egalitarian relationships 308
emotional abuse 319
emotional neglect 325

equity 306
expert power 306
family preservation 328
gender-modified egalitarian
    model 311
incest 326
informational power 306
intimate partner violence (IPV) 314
coercive control 319
legitimate power 306
marital rape 319
money allocation systems 309
National Family Violence
    Surveys 314

neotraditional family 312
patriarchal norm 308
power 306
power politics 312
referent power 306
reproductive coercion 320
resource hypothesis 308
resources in cultural context 308
reward power 306
sexual abuse 325
shelters 328
sibling violence 327
situational couple violence 318
Violence Against Women Act 321

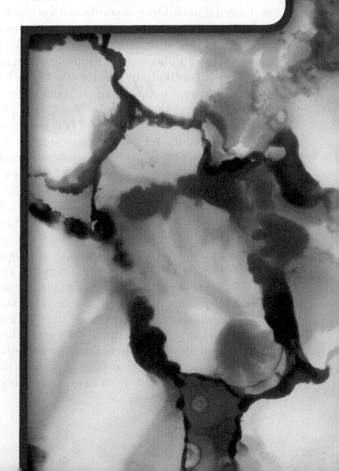

13

FAMILY STRESS, CRISIS, AND RESILIENCE

Learning Objectives

1 Distinguish between family stress and family crisis.

2 Describe the different types of stressors that exist.

3 Explain stressor overload and stressor pileup.

4 Describe the course of a family crisis.

5 Discuss how family members' appraisals of a stressor situation affect the outcome or course of the crisis.

6 List family characteristics associated with resilience.

7 List family resources associated with meeting crises creatively.

Americans are stressed—about money and their own and their children's financial prospects, about gun violence, about terrorism, about global contagious diseases, and about how to protect our environment, among many other things. A recent Gallup poll concluded that Americans are among the most stressed peoples in the world (Ray 2018; Saad 2017). Black males in particular, as well as Muslim Americans, are stressed due to prejudice and sometimes deadly discrimination. With the 2020 outbreak of COVID-19, the novel corona virus that first infected humans in China, Asian Americans, particularly Chinese Americans, became increasingly vulnerable to hostile discrimination. Same-sex families may experience stigma and discrimination (Prendergast and MacPhee 2018).

Polls by the American Psychological Association (2015) show that more than half of U.S. adults lie awake at night with worry at least once a month, a situation dangerous to health and more common to parents and younger adults, the latter mostly worried about future work or career opportunities as well as paying student loans.

Teens are stressed—perhaps more than adults—with nearly half telling one national poll they're stressed "all the time" (After School 2018; Aleccia 2014). A 2019 Pew Research poll found 70 percent of teenage respondents saying anxiety and depression is a major problem among their peers. Anxiety/depression was the most often mentioned problem—after bullying, drug addiction, and others (Horowitz and Graf 2019).

Pre-teen children are worried as well. A poll by the national organization KidsHealth asked children ages 9 to 13 whether they worry and, if so, what about. Results showed that children as young as 9 worry about the health of family members and getting into automobile accidents, among other concerns ("KidsPoll" 2008). An unfortunate finding was that some kindergarteners worry about their body image ("Body Image" 2008). Children with deployed military parents worry about whether their parents will come home and when or whether their parents will be injured (Lucier-Greer et al. 2015; National Child Traumatic Stress Network 2014).

We can think of families as continuously balancing the demands put on them against their capacity to meet those demands. This chapter addresses family stress, crisis, and resilience. We will review various theoretical perspectives on the family and discuss how these can be applied to family stress and crises. We'll discuss what precipitates family stress or crisis and then look at how families define or interpret stressful situations and how their definitions affect the course of a family crisis.

In several places throughout this text, we say families are more likely to be happy when they work toward mutually supportive relationships—and when they have the resources to do so. Nowhere does this become more apparent than in a discussion of how families manage stress and crises. To begin, we'll define the concepts of family *stress*, *crisis*, and *resilience*.

DEFINING FAMILY STRESS, CRISIS, AND RESILIENCE

Stress is a state of tension that results from the need to respond to change. This chapter is about *family* stress. However, because family systems are comprised of individuals (as discussed in Chapter 1) we will begin with a brief exploration of individual stress.

Individual Stress

Individual stress is physically and emotionally experienced in reaction to a change that requires a response or adjustment. An individual's biological stress response involves a complex physiological reaction in the brain and hormonal system as a response to perceived threat or danger (Ha and Granger 2016; Repetti and Robles 2016). The threatened brain automatically responds by prompting the body to secrete cortisol, an adrenal hormone that effects physical reactions designed to help the individual (1) fight back against the perceived threat, or (2) get away (the "fight or flight" response).

Stresses can come from any situation or concern—even positive changes—that makes one feel anxious, frustrated, nervous, or excited. Having your best friend as a house guest for a period of time can be stressful. Finally landing the job you've been working toward can be stressful as well.

The Holmes–Rahe Life Stress Inventory Scale provides a way to measure an individual adult's or child's stress level. Many of the scale items—including death of a spouse, death of a parent, divorce, divorce of parents, marital separation, death of a close family member, marriage, marital reconciliation, change in health of a family member, pregnancy, fathering a pregnancy out of marriage, gaining a new family member, a child leaving home, and trouble with in-laws—are actually family stressors and result in not only individual stress but also family stress or crisis.

The authors rank various life events according to how difficult they are to cope with. For instance, death of a spouse, the most stressful life event on the adult scale, is equivalent to one hundred *life change units*. Divorce is equivalent to seventy-three life change units, while trouble with in-laws is equivalent to twenty-nine. For children, divorce of parents is equivalent to seventy-seven life change units, while hospitalization of a parent is equivalent to fifty-five units and loss of a job by a parent is equivalent to forty-six. The Holmes–Rahe Stress Inventory can be found at the American Institute of Stress website (www.stress.org).

Reasonable levels of individual stress are normal—part of being human. Ongoing high stress, though, can be physically and emotionally harmful. Signs of

individual high stress include breaking out with acne, headaches, frequent colds or other sicknesses, tiredness and decreased energy, insomnia, changes in libido or sex drive, digestive issues, appetite changes, depression, sweating, and rapid heartbeat. If minor problems or disappoints upset you excessively, you may be under an intolerable level of personal stress. The same is true for feeling like you don't enjoy life's daily pleasures, the inability to stop thinking about your problems, feeling inadequate, or experiencing sudden flashes of anger over situations that didn't used to bother you (Cardinal 2019).

Research on children and stress shows that growing up in poverty, or experiencing neighborhood or school dangers, ongoing family tension or conflict are stressors, possibly having lasting effects on physical and mental health into adulthood (Perry and Conners-Burrow 2016; Repetti and Robles 2016). "Facts About Families: Stress and Children" further explores stress and children.

Just a few decades ago, we assumed that the human brain was pretty unchangeable once individuals reached adulthood. Not anymore. Now we think of the brain as changing throughout life. New physical pathways form in the brain through life. Put another way, we experience—and can benefit from *neuroplasticity*—the fact that the neurons in our brains can change. We can, not just emotionally but physically change our brains, retrain or reinforce our brains to think more positively, slow down, relax, stop pestering us with worry about things we can't do anything about (Cardinal 2019; Greenberg 2017; Helmstetter 2013). This takes work, practice. Just knowing about it isn't enough. Books, articles, and online resources explain neuroplasticity and how you can change your thinking and brain for the better. One online resource is Sherry Cardinal's "Stress Proofing Your Life" (www.criticalincidentstress.com /stress_proofing_your_life). Among many others, a good book on this topic is Shad Helmstetter's *The Power of Neuroplasticity* (2013).

We've given attention to how individuals experience stress because families are systems comprised of interacting individuals, and each family member's response influences how the family system reacts. A family where most members see a stressor as challenging but manageable is likely to be resilient, or able to respond well. A family comprised of individuals who tend to blame each other or envision only negative outcomes is less likely to be resilient (Lee and Roberts 2018).

Family Stress

Family stress is a state of tension that arises when demands test or tax a family's capabilities. As with individual stress, situations that we think of as good as well bad are capable of creating stress in our families. Moving to a different neighborhood, taking a new job, getting a promotion, or bringing a baby home might be examples of "good" situations that create family stress. As sociologist Pauline Boss (1997) reminds us,

> Perhaps the first thing to realize about stress is that it's not always a bad thing to have in families. In fact it can make family life exciting—being busy, working, playing hard, competing in contests, being involved in community activities, and even arguing when you don't agree with other family members. Stress means change. It is the force exerted on a family by demands. (p. 1)

Family stress might also be caused by potentially harmful, ambiguous, or difficult situations such as trying to find adequate housing on a limited budget, financing children's education on a middle-class income, being laid off, or wondering how to pay a huge medical bill. A family member's injury or death stresses the family system. Responding to the needs of aging parents causes family stress (see Chapter 16). Undergoing infertility treatments or losing a pregnancy are stressors (see Chapter 8) (Shreffler, Greil, and McQuillan 2011). Living as a cancer survivor is a stressor (Marshall 2010). Living with discrimination due to race, class, gender, religion, gender identity, or sexual preference is a stressor (e.g. Cao et al. 2017; Liu and Wilkinson 2017). Many topics addressed in this text are family stressors: weddings, moving in together as cohabitors, forming multigenerational households, immigrating, transitioning to parenthood, parenting, separation or divorce, juggling family needs against workplace obligations, family violence, poverty—can you think of others?

Family stress calls for family adjustment. As an example of adjustment, often in response to financial pressures, more and more family members of various ages have moved in together, forming multigenerational households (Cohn and Passel. 2018). Family households that expand to include more family members—often young adults returning home, as discussed in Chapter 6—and then possibly contract when additional family members leave are sometimes called *accordion families* and demonstrate often remarkable adjustments (Newman 2012; Newman and Knapp 2013).

When adjustments are not easy to come by, family stress can lead to a **family crisis**: "a situation in which the usual behavior patterns are ineffective and new ones are called for immediately" (National Ag Safety Database n.d., p. 1; Patterson 2002a). We can think of a family crisis as a sharper jolt to a family than more ordinary family stress. The definition of *crisis* encompasses three interrelated ideas:

1. Crises necessarily involve change.

2. A crisis is a turning point with the potential for positive effects, negative effects, or both.

3. A crisis is a time of relative instability.

Stress and Children

We can never completely shield our children from natural disasters, news of terrorism, possible school shootings, alarming contagious illnesses, or other dangers. And it's appropriate for children to feel anxious about parents and other relatives serving in the military, being hospitalized, imprisoned, addicted to drugs or alcohol, engaged in chronic family discord or violence, or going through other anxiety-provoking situations.

We can, however, recognize signs of stress in children and address them. Some signs of children's stress:

- Change in appetite
- Reluctance to go to school
- Headache
- Stomach ache
- Nail biting
- Bedwetting
- Tantrums or meltdowns
- Nervous tics, such as blinking, leg shaking, nail biting
- Sleep disturbances.
- Repeatedly using words such as *tired*, *mad*, *scared*, *sad*, or *worried*
- Saying things such as "nobody likes me" or "nothing is fun."

Some ways to address children's stress:

- As much as possible, keep the child's routine consistent and predictable.
- Inform your child ahead of time—as soon as possible—about upcoming changes in the child's routine.
- Schedule time for relaxed and calm activities.
- Don't overschedule your child.
- Encourage the child and remember to show affection.
- Manage your own stress; if you are stressed, watch your temper.
- Listen to your child closely and with care.

Various websites offer information addressed to children and stress, including the following:

- The American Academy of Pediatrics has developed a guide for pediatricians.
- The Department of Defense Education Activity offers information on school safety, the children of military personnel, and other programs.
- The Federal Emergency Management Agency (FEMA) has material on preparing children for natural disasters and national security emergencies.
- The KidsHealth organization has material designed for parents and children regarding things that children worry about and how parents can help.
- The National Child Traumatic Stress Network is another good resource.

"Knowing that their child is going through a tough time can be overwhelming for any parent. But remember, kids draw their strength from their parents, a guardian, a favorite teacher, or a relative. You can help your child deal with a stressful situation, whether at home or at school, if you are alert and compassionate" ("9 Signs of Stress in Children…." 2018).

Critical Thinking

When or how might you make use of these or similar resources in your family life? How might family dynamics in the face of a stressor event such as an earthquake or explosion mitigate, or lessen, a child's stress? How might family dynamics exacerbate, or worsen, a child's stress?

Source: Curejoy Editorial 2018; Gavin 2018; "Post-traumatic Stress Disorder in Children," Centers for Disease Control and Prevention 2019; "Signs and Symptoms of Stress in Children." Nd, Family Wizard.

Family crises are turning points that require some change in the way family members think and act to meet a new situation (Hansen and Hill 1964; McCubbin and McCubbin 1991; Patterson 2002a). In the words of social worker and crisis researcher Ronald Pitzer:

Crisis occurs when you or your family face an important problem or task that you cannot easily solve. A crisis consists of the problem and your reaction to it. It's a turning point for better or worse. Things will never be quite the same again. They may not necessarily be worse; perhaps they will be better, but they will definitely be different. (Pitzer 1997a, p. 1)

In part, whether things get better depends on a family's level of resilience—the ability to recover from challenging situations (MacPhee, Lunkenheimer, and Riggs 2015; McCubbin et al. 2001). We return to the topic of resilience later in this chapter. Meanwhile, we note several ways that social science theory gives insight into family stress, crisis, and resilience.

THEORETICAL PERSPECTIVES ON FAMILY STRESS AND CRISES

We saw in Chapter 2 that there are various theoretical perspectives concerning marriages and families. Throughout this chapter we will apply several of these theoretical perspectives to family stress and crises. Here we give a brief review of several theoretical perspectives that are typically used when examining family stress and crises.

You may recall that the *structure–functional* perspective views the family as a social institution that performs essential functions for society—raising children responsibly and providing economic and emotional security to family members. From this point of view, a family crisis threatens to disrupt the family's ability to perform these critical functions (Patterson 2002a).

The *family development,* or *family life course,* perspective sees a family as changing in predictable ways over time. This perspective typically analyzes **family transitions**—expected or *predictable* changes in the course of family life—as family stressors that can

precipitate a family crisis (Carter and McGoldrick 1988). For example, having a baby or sending the youngest child off to college taxes a family's resources and brings about significant changes in family relationships and expectations. Over the course of family living, people may form cohabiting relationships, marry, become parents, break up or divorce, remarry, and make transitions to retirement and widow- or widowerhood. All these transitions are stressors (Cooper, McLanahan, et al. 2009).

In addition, the family development perspective focuses on the fact that predictable family transitions, such as an adult child's becoming financially independent, are expected to occur within an appropriate time period, although the window of acceptable time has lengthened over that past several decades (Arnett 2004; Furstenberg et al. 2004). As discussed in Chapter 2, transitions that are "outside of expected time" create greater stress than those that are "on time" (Hagestad 1996; Rogers and Hogan 2003). Partly for this reason, teenage pregnancy is often a family stressor. Another example, explored in Chapter 9, involves a grandparent's filling the parent role (Davis-Sowers 2012). "A Closer Look at Diversity: Young Caregivers" provides a third example of assuming a role outside of expected time.

The *family ecology* perspective explores how a family is influenced by the environments that surround it. From this point of view, many causes of family stress originate outside the family—in the neighborhood, workplace, and national or international environment (Boss 2002; Lee and Roberts 2018). Living in a violent neighborhood causes family stress and has potential for sparking family crises. Conflict between work and family roles, largely created by workplace demands, is another example of an environmental factor that can cause family stress. Natural disasters such as tornadoes, hurricanes, and earthquakes also create family stress and crises (Taft et al. 2009). Moreover, as we'll see later in this chapter, our family's external environment offers or denies us resources for dealing with stressors.

The *family system* theoretical framework looks at the family as a system—like a computer system or an organic system such as a living plant or the human body. In a system, each component or part influences all the other

For decades some family members, many of them U.S. citizens, have been divided by the U.S.–Mexico border. Since the 1990s, years before Donald Trump's presidency began, the United States has been erecting a wall and fencing along the border. Hoping to catch glimpses of one another through cracks in a fortified fence, these family members gather on both sides of the barrier. Being divided in this way constitutes a family crisis. The family ecology perspective focuses on how factors external to the family, such as economic conditions and immigration policies in both countries, can result in family crisis. From a family systems perspective, all the members in this family have to adapt to being separated.

John Moore/Getty Images

A Closer Look at Diversity

Young Caregivers

We tend to think of caregivers as middle aged or older, and a large majority of them are. But caregivers are diverse in age; some of them are children. For young caregivers, the role involves additional stress because it does not take place "on time." Here's what one thirty-something caregiver wrote in her blog:

Being thrust into a caregiver role at a younger age, when my mom at the age of 57 had a debilitating stroke, I was faced with all the "common" caregiver challenges but at a time in my life when it was least expected and with absolutely no warning. I immediately left my career, my home, my friends to move back home (2,000 miles away) to do everything that was humanly and sometimes inhumanly possible to help my mom. . . . Being a caregiver, especially at a young age, is a huge sacrifice. I don't regret it, but sometimes I can't help but feel that I am missing out on some of the best years of my life.

During my [twenties], I mostly focused on my career. I was always a very driven person and while I had one or two serious relationships during that time, I was not ready to "settle down." In my mind, I felt like that's what my [thirties] would be for. Had I been able to predict the future, I would've married my college boyfriend and started having babies immediately. OK, maybe not, but the idea of it sure sounds good now (laugh). So, here I am, one year into caregiving and I just started working again (my career had to be redefined too). . . .

Meanwhile, my friends and acquaintances are getting married, having babies, buying houses, etc. Sometimes I feel like everyone is moving forward, and I am frozen in time. . . . I wonder if and when will I have the opportunity to fulfill my own hopes and dreams. As a young caregiver, and in my particular situation, this is my biggest challenge and fear. . . . So to all the young caregivers out there—whether you are caring for your spouse/significant other, a sibling, or a parent—You are not alone. (Caregiver Support Blog 2008)

In addition to caregivers in their twenties and thirties, an estimated 1.5 million U.S. children under age 18 serve as caregivers to family members (Shifren 2009; "Young Caregivers" 2009). Experts expect the numbers to grow as chronically ill patients leave hospitals sooner and live longer, as the recession compels patients to forgo paid help, and as more returning veterans need home care (Belluck 2009).

Child and teen caregivers are often responsible for keeping the care recipient company. In addition, they shop, do household chores, and help with meal preparation. Some assist the care recipient with eating, getting in and out of bed, getting dressed, taking a bath, or going to the bathroom. Some administer medications; help the care recipient communicate with doctors, nurses, or other medical professionals; make appointments; or arrange for others to help the care recipient (Hunt, Levine, and Naiditch 2005; see also Champion et al. 2009).

Caretaking can give purpose to a young person's life (Shapiro 2006a). Although some childcare givers do well, others grow depressed or angry as they sacrifice social and extracurricular activities. Some miss—or even quit—school (Belluck 2009). Policy makers urge further research on questions, such as how to improve support groups for young caregivers, how teachers and schools can assist them, and how educational, social, and career opportunities can be fostered within the context of caregiving (Hunt, Levine, and Naiditch 2005).

Support organizations for young caregivers include the American Association of Caregiving Youth (AACY), the National Alliance for Caregiving (NAC), the National Family Caregivers Association (NFCA), the Family Caregiver Alliance (FCA), the Children of Aging Parents (CAPS), and the Caregiving Youth Project, sponsored by AACY (Belluck 2009).

Critical Thinking

From a policy point of view, what might be done to assist young caregivers? What might your local community do to help?

---

parts. When one family member changes a role, all the family members must adapt and change as well. As an example, when a family member becomes addicted to steroids, or alcohol or other mind-altering drugs, the entire family system is affected (El-Sheikh and Flanagan 2001; Foster and Brooks-Gunn 2009). As another example, when one sibling has a disability, other siblings—as well as parents and other relatives—have to adjust (Neely-Barnes and Graff 2011; Saxena and Adamsons 2013).

Finally, exploring the discussions, gestures, and actions that go on in families, the *interactionist perspective* views families as shaping family traditions and family members' self-concepts and identities. By interacting with one another, family members struggle to create shared family meanings that define stressful or potentially stressful situations—for example, as good or bad, disaster or challenge, someone's fault or no one's fault. As we will explore later in this chapter, "a family's

shared meanings about the demands they are experiencing can render them more or less vulnerable in how they respond" (Patterson 2002a, p. 355).

WHAT PRECIPITATES A FAMILY CRISIS?

Demands put on a family cause stress and sometimes precipitate a family crisis. Social scientists call such demands **stressors**—a precipitating event or events that create stress. Stressors vary in kind and degree, and their nature is one factor that affects how a family responds. In general, stressors are less difficult to cope with when they (1) are expected, (2) are brief, (3) are seen as not so serious, and (4) gradually improve over time.

Types of Stressors

There are several types of stressors (Figure 13.1). We will briefly examine nine of the most common here.

Addition of a Family Member Adding a member to the family—for example, through birth, adoption (Bird, Peterson, and Miller 2002), marriage, remarriage, or the onset of cohabitation—is a stressor. You may recall Chapter 9's discussion on why the transition to parenthood is stressful. As another example, the addition of adult family members may bring people who are very different from one another in values and life experience into intimate social contact. Furthermore, not only are in-laws (and increasingly stepparents, step-grandparents, and step-siblings) added through marriage or cohabitation but also a whole array of their kin come into the family. Then, too, having an adult child return home or a member of one's extended family move into the household because of financial problems is a stressor. Transition to a multigeneration household is a stressor but can have positive outcomes for some children of unmarried mothers who move into parental homes (Augustine and Raley 2012).

Like any system, a family has boundaries. Family members need to know "who is in and who is outside the family" (Boss 1997, p. 4). Adding a family member is stressful because doing so involves family boundary changes; that is, family boundaries have to shift to include or "make room for" new people or to adapt to the loss of a family member (Boss 1980). This situation applies to the addition of a cohabiting partner to the family, as well as his or her departure from the household (Cherlin 2009a).

Loss of a Family Member The death of a family member is, of course, a stressor. Family systems theory reminds us that children as well as adults grieve the absence of a family member, and their grief needs to be recognized and addressed (Boss 1980; Monroe and Kraus 2010; Saint Louis 2012). The same is true for other family members who are sometimes "forgotten grievers," such as grandparents on occasion, or aunts and uncles, among others (Gilrane-McGarry and O'Grady 2012).

It is interesting to note that the likelihood of death in a society can influence how people define a death in the family. For instance, under the mortality conditions that existed in this country in 1900, half of all families with three children could expect to have one die before reaching age 15. Social historians have argued that parents defined the loss of a child as almost natural or predictable and, consequently, may have suffered less emotionally than parents do today (Wells 1985). In contrast, family members who lose a child of any age today do so "outside of expected time," a situation that exacerbates, or adds to, their grief. The long-term effects of grieving such a loss may negatively affect a couple's intimacy (Gottlieb, Lang, and Amsel 1996).

Loss of potential children through miscarriage or stillbirth has the possible added strain of family disorientation (Shreffler, Greil, and McQuillan 2011). Attachment to the fetus may vary substantially so that the loss may be grieved greatly or little. Add to that the generally minimal display of bereavement customary in the

| Addition of a family member | Loss of a family member | Ambiguous loss | Sudden, unexpected change | Ongoing family conflict | Caring for a dependent, ill, or disabled family member | Demoralizing event | Daily family hassles | Anxieties about children in a culture of fear |

FIGURE 13.1 Types of stressors.

United States and the omission of funerals or support rituals for perinatal (birth process) loss, and

> all these ambiguities mean that a family may have to cope with sharply different feelings among family members . . . [and] the family as a whole may have to cope with the fact that they as a family have a very different reaction to loss than do the people around them. (Rosenblatt and Burns 1986, p. 238)

In addition to permanent loss, the temporary loss of a family member, such as through an older sibling's going away to college or a parent's leaving for long periods due to work demands, is a stressor. Temporary losses that not only create change in family structure but also introduce fear of the unknown are forms of ambiguous loss (Huebner et al. 2007; Whealin and Pivar 2006).

Ambiguous Loss There are many kinds of ambiguous loss (Boss 2016). The loss of a family member is ambiguous when it is uncertain whether the family member is "really" gone:

> Ambiguous loss is a loss that remains unclear. . . . [U]ncertainty or a lack of information about the whereabouts or status of a loved one as absent or present, as dead or alive, is traumatizing for most individuals, couples, and families. The ambiguity freezes the grief process and prevents cognition, thus blocking coping and decision-making processes. Closure is impossible. (Boss 2007, p. 105)

Grieving over a secret loss or one that society fails to publicly recognize, such as having survived sexual assault, is an ambiguous loss (Bordere 2017). Having a military family member deployed overseas or missing in action are situations of ambiguous loss (Boss 2007; Robins 2015). Having a family member in the reserves during a historical period when they are likely to be called into active duty is a significant stressor (Lane et al. 2012; Oshri et al. 2015).

Moreover, a family member may be physically present but psychologically absent, as in the case of family members with alcoholism or mental illness, those suffering from Alzheimer's disease or brain injury, or children with cognitive impairment or severe disabilities (Altman and Blackwell 2014; Farrell, Bowen, and Swick 2014). The ambiguity of postseparation or postdivorce family boundaries can be stressful. A nonresident father whose relationship with the child's mother—and hence with his future child—is uncertain may experience ambiguous loss (Leite 2007).

From the family systems perspective, ambiguous loss is uniquely difficult to deal with because it creates family **boundary ambiguity** (see Figure 13.2)—"confused perceptions about who is in or out of a particular family" (Boss 2004, p. 553; Carroll, Olson, and Buckmiller 2007):

> With a clear-cut loss, there is more clarity—a death certificate, mourning rituals, and the opportunity to honor and dispose [of the] remains. With ambiguous loss, none of these markers exists. The clarity needed for boundary maintenance (in the sociological sense) or closure (in the psychological sense) is unattainable. . . . [P]arenting roles are ignored, decisions are put on hold, daily tasks are undone, family members are ignored or cut off, and rituals and celebrations are canceled even though they are the glue of family life. (Boss 2004, p. 553)

Boundary ambiguity makes reintegration of a formerly absent military member difficult. On the one hand, the reintegrating person may want to feel that nothing changed in their absence; on the other hand, the remaining family members may fully realize that much has changed; reintegration will take conscious work (Hartman 2015; Hollingsworth, Dolbin-MacNab, and Marek 2016; Sherman, Hawkey, and Borden 2015). A 2019 Pew Research Survey of military veterans found that 52 percent said the military prepared them very well for life in the military, but only 16 percent felt very well prepared for transition back to civilian life (Parker 2019).

Unanticipated Change An unexpected or unanticipated change in the family's income, social status, or definition of itself is a stressor. A family member's receiving a cancer diagnosis or having a heart attack or a child's running away are examples of unexpected change). Job loss is another example. Having a family member—parent, child, sibling—come out as LGBTQ+ can be an unexpected change for the family system (Pollitt, Muraco, Grossman, and Russell 2017). Natural disasters, mentioned previously, cause unanticipated change. Family members being unexpectedly quarantined on a cruise ship or elsewhere in 2020 because of the spreading covid-19 (corona) virus was a stressor, both for the actually quarantined and for others family members worrying about them. Most people think of stressors as negative, and some sudden changes are. Some changes may be defined as negative by some family members, but not necessarily so by others. Meanwhile, positive changes, such as winning the lottery (don't you wish?) or getting a significant promotion, can cause stress too. (Why might this be, do you think?)

Ongoing Family Conflict Ongoing, unresolved conflict among family members is a stressor (Hammen, Brennan, and Shih 2004). Deciding how children should be disciplined can bring to the surface divisive differences over parenting roles, for example. The role of an adult child living with parents is often unclear and can be a source of unresolved conflict. Watching an adult grandchild go through family conflict can be a stressor for a grandparent. If children of teenagers or of divorced adult children are involved, the

Situations
of physical
absence &
psychological
presence

Situations
of physical
presence &
psychological
absence

Situations of physical absence and psychological presence
There is a preoccupation with thinking of the absent member. The process of grieving and restructuring cannot begin because the facts surrounding the loss of the person are not clear.

Catastophic and unexpected situations:
• war (missing soldiers)
• natural disaster (missing persons)
• kidnapping, hostage taking, terrorism
• incarceration
• desertion, mysterious disappearance
• missing body (murder, plane crash, etc.)

More common situations:
• divorce
• military deployment
• young adults leaving home
• elderly mate moving to a nursing home

Situations of physical presence and psychological absence
Families where a member is physically present but not emotionally available to the system. The family is intact, but a member is psychologically preoccupied with something outside the system.

Catastophic and unexpected situations:
• Alzheimer's disease and other dementias
• chronic mental illness
• addictions (alcohol, drugs, gambling, etc.)
• traumatic head injury, brain injury
• coma, unconsciousness

More common situations:
• preoccupation with work
• obsession with computer games, Internet, TV

FIGURE 13.2 Two common forms of boundary ambiguity: (1) a family member's physical absence coupled with psychological presence, and (2) a family member's physical presence coupled with psychological absence. "Sometimes a family experiences an event or situation that makes it difficult—or even impossible—for them to determine precisely who is in their family system" (Boss 1997, pp. 2–3).

situation becomes even more challenging (Hall and Cummings 1997).

Caring for a Dependent, Ill, or Disabled Family Member Caring for a dependent or disabled family member is seen as genuinely meaningful by a large majority of those who do it—more meaningful than whatever we do in our leisure time—but it's also a stressor (Lindo, Kliemann, Combes, and Frank 2016; Livinston 2018d). This is especially true when work demands conflict with those of family care (Stewart 2013). Examples involve being responsible for an adult child or a sibling with mental illness or physical or developmental disabilities (Fields 2010; Lowe and Cohen 2010;

see also Sutorius 2015). Due mainly to advancing medical technology, the number of dependent people and the severity of their disabilities have steadily increased over recent decades (Altman and Blackwell 2014). For instance, more babies today survive low birth weight and birth defects. Caring for a disabled child can be stressful enough for parents that some decide not to have another (Woodman 2014). Analysis of national data found that mothers of firstborn children with a disability were statistically less likely to have a second child (MacInnes 2008).

Also, more people now survive serious accidents, and many seriously injured soldiers deployed abroad have survived to require extensive ongoing care and medical

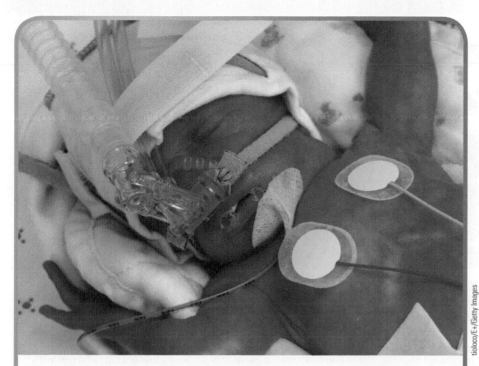

Sometimes a situation may be classified as more than one type of stressor. Due to advanced medical technology, for instance, more newborns today survive low birth weight or birth defects but may need ongoing remedial attention. Therefore, adding a baby to the family may also mean caring for a medically fragile child.

attention. "Issues for Thought: Caring for Patients at Home—A Family Stressor" discusses how recent technological advances, coupled with the goal of containing medical care costs, have created new stressors for families who are increasingly expected to care for very ill patients at home.

In addition, parents may be raising children with chronic physical conditions such as asthma, diabetes, epilepsy, or autism (Rao and Beidel 2009; Wong et al. 2014). Families may need to see their children through bone marrow, kidney, or liver transplants, sometimes requiring several months' residence at a medical center away from home (LoBiondo-Wood, Williams, and McGhee 2004). Adults with advanced AIDS may return home to be taken care of by family members. A family may have to care for a disabled or terminally ill member, which can be a stressor for young children, who may exhibit behavior problems as a response, as well as for the adults in the household (Toot et al. 2013). Chapter 16 addresses the issue of the sandwich generation of caring for one's own children as well as for aging parents.

Demoralizing Events Stressors may be demoralizing events—those that signal some loss of family morale (Early, Gregoire, and McDonald 2002; Wildeman 2012). Demoralization can accompany many stressors

already described. But, among other things, this category also includes financial troubles, poverty, homelessness, having one's child placed in foster care, juvenile delinquency or criminal prosecution, scandal, family violence, mental illness, incarceration, or suicide (Bricker et al. 2012; Wharff, Ginnis, and Ross 2012). Being the brunt of racist or other prejudicial treatment can certainly be demoralizing for both adults and children (Murry et al. 2001; Trent, Dooley, and Douge 2019). Grandparents' raising grandchildren is a situation that is often—although not always—associated with demoralizing events (Henderson et al. 2009; Smith, Cichy, and Montoro-Rodriguez 2015). Recently the phenomenon of preschools expelling rambunctious toddlers for jumping around too much, hitting other children, or ignoring safety precautions, among other things, has come to public attention—a demoralizing family event with possible long-term negative social, emotional, and academic consequences for the "tiny troublemaker" (Elnhorn 2019). A similar, demoralizing issue involves children—many younger than 10—being arrested, some handcuffed, for misbehaviors at school.

Physical, mental, or emotional illnesses or disorders, such as an eating disorder in a family can feel demoralizing (Lee and Roberts 2018). Alzheimer's disease or brain injury, in which a beloved family member seems to have become a different person, can be heartbreaking (Godwin, Lukow, and Lichiello 2015). Some illnesses can be especially demoralizing when they are associated with the possibility of being socially stigmatized. Autism, HIV/AIDS, attention deficit/hyperactivity disorder (ADHD), anorexia nervosa, and bulimia are examples (Hall and Graff 2012). In military personnel who have served during wartime, post-traumatic stress disorder (PTSD) can be demoralizing, causing "family members [to] feel hurt, alienated, or discouraged, and then become angry or distant toward the partner" ("PTSD and Relationships" 2006; Sherman, Hawkey, and Borden 2015).

Everyday Family Hassles Everyday family hassles are stressors. Balancing work against family demands, working odd hours, being regularly stuck in traffic on long commutes to work, and arranging

Issues for Thought

Caring for Patients at Home—A Family Stressor

Between 20 million and 50 million family members in the United States today are providing care that medical professionals once performed in hospitals. Family caregivers—mostly wives and daughters, but also partners, siblings, husbands, sons, grandchildren, and grandparents—provide about 80 percent of all care for ill or disabled relatives, which represents an estimated $13 billion in unpaid caregiving services annually (Brody 2008; Guberman et al. 2005). We can expect need for family caregiving to increase as the population ages, the incidence of chronic disease such as diabetes rises, the number of day surgeries grows, and modern medicine is able to save more and more lives, resulting in more special-needs infants and returning veterans who need care, among others (Brody 2008; Guberman et al. 2005).

It is somewhat disconcerting to imagine an activity taking place in the hospital and then displacing this same activity to the home. In the hospital . . . the patient is in a supposedly sterile environment. Diet and medications are completely controlled by hospital staff. Indeed, a patient who asks to keep and self-administer his or her medications is refused. An interdisciplinary professional team is present, and when any one member is confronted with a problem (leaking IV tube, patient discomfort, apparatus malfunction), the members of the team are backed up by specialists (IV technicians, specialized doctors, technicians, and so on) and by a team of people responsible for the organization of the instrumental activities of daily living (meals, toileting, and so on). . . .

Now transfer this to the home setting. . . . The IV pole is squeezed in between the bed and the night table, and there is almost no room to move because of the addition of a small table that is used to lay out equipment. The patient frequently gets caught in the line and loosens the catheter in his or her vein. When it starts bleeding at the site, the home care nurse has already come and gone. What to do? The caregiver makes an adjustment. He or she is abused for hurting the patient, but the IV starts to flow again and the bleeding stops. There is another dispute between patient and caregiver concerning hygiene around the IV. What does keeping a sterile area mean? Can the dog sit on the bed? Does the caregiver have to wear gloves? Are these questions important enough to disturb medical personnel for answers? Who should be called—the hospital, the home care nurse, or the [twenty-four-hour] medical-information line? . . .

[P]erhaps the most unsettling aspect of the transfer of care responsibilities to patients and their families is the anxiety and insecurity of assuming this care without sufficient supervision and emergency backup. In the hospital, you have an emergency call button if something goes wrong. But what replaces this button when you are being cared for at home? Indeed, the home is psychologically, and sometimes physically, very far from immediate help in the case of an emergency or an unforeseen development. [In this study, the] majority of patients and caregivers assuming complex care felt alone and abandoned, causing high levels of stress and anguish and conflicts within couples and within families. . . .

Based on our study, we raise serious questions about the legitimacy of the transfer of high-tech care to the family.

Critical Thinking

Can you apply the family ecology theoretical perspective to this situation? What are some creative ways that a family might deal with high-tech caregiving at home? In what ways might community activism play a part in addressing this situation?

Source: Largely excerpted from Guberman et al. 2005, pp. 247–72; also Brody 2008.

---

transportation or childcare—not to mention childcare breakdowns!—all these are everyday family stressors (Adamo 2013; Pilarz and Hill 2017). Raising children in a child-centered cultural environment that today calls for seemingly relentless "intensive parenting" that "demands great parental time, financial, and emotional investments in childrearing" is a stressor (Nomaguchi and Milkie 2020). Protecting children from danger, especially in neighborhoods characterized by violence is an everyday stressor. A recent study found that, compared to married mothers, single mothers were more likely to feel greater stress and feelings of inadequacy, evidenced by migraines or chronic back pain, when experiencing daily hassles associated with their children's allergies or frequent colds (Ontai et al. 2008). "Facts about Families: ADHD, Autism, Stigma, and Stress" explores this point further. Parenting a chronically ill child is a family stressor (Zelman and Ferro 2018). Then too, parents can have physical disabilities or chronic illness themselves and therefore face the everyday stresses commonly associated with parenting simultaneously with managing their own health (Shandra and Penner 2017; Turney and Hardie 2018).

FADHD, Autism, Stigma, and Stress

The Centers for Disease Control and Prevention estimate that about 1 in 59 children under age 8 in the United States has some degree of autism (Baio, Wiggins, and Christensen et al. 2018). According to estimates by the American Psychiatric Association, attention deficit/hyperactivity disorder has been diagnosed in 5 to 10 percent of U.S. schoolchildren, more often in boys than in girls. ADHD can also be an adult diagnosis (Retz and Klein 2010). Autism and ADHD are similar inasmuch as children with either disease risk being stigmatized.

Families with children diagnosed with autism or ADHD face ongoing stressors that can be understood as daily hassles, some severe and demoralizing (Vogan et al. 2014). Parents report that, due to the child's behavior, they often feel interrupted; miss social events because they are hesitant to leave their child with a babysitter; are anxious about taking the child out in public; must deal with complaints from other parents, neighbors, teachers, or school bus drivers; spend excessive amounts of time with the child's homework; worry that the child will get into trouble or be injured; have difficulty finding adequate after-school placement for a child and are unable to find or afford other professional or school services for the child; lose patience due to especially trying morning routines; must address siblings' resentment of parents' extra time and attention spent with the child diagnosed with ADHD; miss work; face lack of sleep due to disrupted bedtimes; and do not have enough time for themselves (Firmin and Phillips 2009; Hall and Graff 2012; Ramisch 2012; Reader, Stewart, and Johnson 2009). Regarding airplane travel, parents face potential troubles with boarding or in-flight regulations when taking an autistic child who may not readily follow the rules (Swarns 2012b).

Raising a child diagnosed with autism or ADHD can involve feeling embarrassed as a result of specific instances of misbehavior and also as a consequence of being stigmatized. Stigmatizing others involves prejudice or discrimination based on others' perceived negative characteristics, status, or behaviors (Goffman 1963). **Courtesy stigma** refers to a situation in which not only the initially stigmatized individual but also her or his intimates are stigmatized by association (Goffman 1963; Koro-Ljungberg and Bussing 2009).

Focus groups (see Chapter 2) with thirty parents of children diagnosed with ADHD revealed that the parents often received unsolicited advice and felt negatively judged by both extended family members and strangers. Indeed, public debate over whether the diagnosis itself is legitimate or simply a convenient label for bad behavior increases the possibility of stigma (Koro-Ljungberg and Bussing 2009).

Some parents have informed others of the diagnosis to ward off potential criticism of their child's behavior. Parents who were able to resist or shrug off negativity from neighbors, extended kin, or community were better able to cope. Some parents mentioned spirituality as a resource. As one mother said, "I just leave it in God's hand because the only thing I can do is just pray for him. Just pray and ask God to shield and protect him" (in Koro-Ljungberg and Bussing 2009, p. 1192).

For many parents, however, management of courtesy stigma may involve withdrawal on the one hand coupled with activism on the other hand. The majority of the parents in the focus group research managed stigma mainly by avoiding potentially stressful situations. They kept the diagnosis to themselves, interacting primarily or only with families of children who demonstrated behaviors similar to their child's. Many admitted doing homework and school projects for their children to reduce the possibility of being stigmatized by their child's teacher or by other parents (Koro-Ljungberg and Bussing 2009).

Parents also engaged in activism. Some volunteered at school to advocate for their children. They pressed for special school services for children diagnosed with ADHD, and they politicized ADHD by demanding more public education and increased awareness that could help reduce the stigma associated with ADHD (Koro-Ljungberg and Bussing 2009).

A variation of the Strategies to Enhance Positive Parenting program (STEPP) has been developed specifically to address parenting children diagnosed with ADHD (Chacko et al. 2008). Although sponsored by the pharmaceutical company Shire, the website ADHDactionguide.com includes nonpharmaceutically based tips for managing adult ADHD (see also Manning, Wainwright, and Bennett 2011; Retz and Klein 2010). Not surprisingly, formal and informal social support has been found to positively impact the health of parents and other caregivers to children with ADHD or autism (Gouin, Da Estrela, Desmarais, and Barker 2016).

Critical Thinking

Do you know an adult, child, or parents of a child who has been diagnosed with autism or ADHD? Could you have added to their feelings of being stigmatized? If you are parenting a child diagnosed with autism or ADHD, have you experienced courtesy stigma? If so, how have you handled it? How might you handle it in the future?

Scholars have investigated everyday stressors that are unique to certain professions. Especially in recent years, military families "are subjected to unique stressors, such as repeated relocations that often include international sites, frequent separations of service members from families, and subsequent reorganizations of family life during reunions" (Drummet, Coleman, and Cable 2003, p. 279; Hawkins et al. 2012; Lowe et al. 2012; Schlomer et al. 2012; Wadsworth and McDermid 2012). Families of Protestant clergy experience not only the stressors of ministry demands but also family criticism and situations in which members of the congregation "intrusively assume that the minister will fulfill their expectations without due consideration of the minister's priorities" (Lee and Iverson-Gilbert 2003, p. 251). Research interestingly shows that attention to social media is stressful in several ways: feeling left out when discovering that friends got together without you, feeling jealous of other's accomplishments, or worrying about media friends' posted problems (Hampton, Rainie, Lu, Shin, and Purcell 2015).

Anxieties about Children, Parents, or Other Relatives or Friends in a "Culture of Fear" A final stressor involves living with chronic anxiety regarding family safety. Increased media portrayal of various dangers seems to have led to a general "culture of fear" (Glassner 1999), which makes "anxiety about children ... a central matter in twentieth-century American culture" (Fass 2003; Stearns 2003). High-profile kidnappings and school shootings certainly inspire worry, but parental fear may exceed the reality of the risk. Misrepresented by the media as high or rising, in reality many perceived threats to children are statistically low. And although one wouldn't know it by watching televised news, the odds that a stranger will kidnap a child are extremely low—a fraction of 1 percent of children who have gone missing were taken by strangers.

We don't mean to imply that kidnappings, crimes at school, or other feared threats to children will never occur. A realistic analysis of dangers, collecting information on strategies appropriate to living safely in our neighborhoods, a plan for talking with children about protection from actual risks, and a "check it out" attitude toward frightening media stories are good parental approaches.

As you read this section, you may have noted that sometimes a single event can be classified as more than one type of stressor or can combine stressor types. Adopting a child with special needs often involves both adding a family member and caring for a disabled child (Schweiger and O'Brien 2005). As another example, raising a child with emotional or behavior problems adds to everyday family hassles, can be demoralizing, and may precipitate family conflict (Chacko et al. 2008; Talan 2009). Having immigrated to the United States

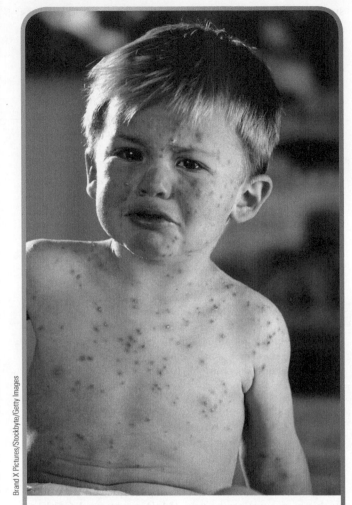

Brand X Pictures/Stockbyte/Getty Images

Daily family hassles, such as a child's getting chicken pox, put demands on a family. Sometimes everyday hassles pile up to result in *stressor overload*. This is especially true when a new stressor is added to already difficult daily family life.

means having lost the physical companionship of family members in one's home country as well as struggling with daily hassles and challenges, perhaps regarding a new language, documentation, or financial issues (Raffaeli et al. 2012). From another point of view, having parents who emigrate, leaving children or elderly parents in the home country, causes stress for those left at home (Bodrug-Lungu and Kostina-Ritchey 2013; Cabanes and Acedera 2012).

The September 11, 2001, attacks on New York City and Washington, D.C., were events that can be classified as sudden changes in the family environment—changes that, among other things, sparked parents' need to consider how to talk with their children about terrorism (Boss 2004). For many, the attacks also constituted demoralizing events. For others, the events were not only

sudden and demoralizing but also sadly marked the loss of a family member, as did subsequent gun violence and terrorist attacks.

Stressor Overload

One type of stressor overload is the addition of psychological depression to an earlier stressor such as chronic poverty or an adolescent family member's living with epilepsy (Seaton and Taylor 2003).

A family may be stressed not just by one serious, chronic problem but also by a series of large or small, related or unrelated stressors that build on one another too rapidly for the family members to cope effectively (McCubbin, Thompson, and McCubbin 1996). This situation is called **stressor overload**, or *pileup*:

> Even small events, not enough by themselves to cause any real stress, can take a toll when they come one after another. First an unplanned pregnancy, then a move, then a financial problem that results in having to borrow several thousand dollars, then the big row with the new neighbors over keeping the dog tied up, and finally little Jimmy breaking his arm in a bicycle accident, all in three months, finally becomes too much. (Broderick 1979, p. 352)

In some cases, stressor overload characterizes the primary stressor event itself. For instance, experiencing a natural disaster such as a wildfire, hurricane, or tornado may require evacuation, finding shelter, getting children to school from a new location, getting appraisals of home damage, worrying about whether belongings left at home are secure, and going without familiar clothing, cosmetics, or toys (McDermott and Cobham 2012).

Then, too, stressor overload can creep up on people without their realizing it. Even though it may be difficult to point to any single precipitating factor, an unrelenting series of relatively small stressors can add up to a crisis. In today's economy, characterized by longer working hours, a family's need to rely on more than one paycheck, fewer high-paying jobs, fewer benefits, and little job security, stressor overload may be more common than in the past.

A third example might involve the ambiguous loss of a family member deployed overseas, followed by the stressors associated with the family member's return home, possibly compounded by the soldier's serious physical injuries or post-traumatic stress disorder (England 2009; Hoge 2010; Johnson 2010). As a final example, a parent's promotion and consequent relocation to another part of the country involves packing, moving, finding new doctors and pharmacies, locating new schools for the children, and the need to make friends and forge an informal support system in the new locale. We'll return to the idea of stressor pileup shortly. Now, however, with an understanding of the various kinds of events that cause family stress and can precipitate a family crisis, we turn to a discussion of the course of a family crisis.

THE COURSE OF A FAMILY CRISIS

Although family stress simply involves "pressure put on the family," a family *crisis* results from an "imbalance between pressure and supports" (Boss 1997, p. 1). A family crisis ordinarily follows a fairly predictable course, similar to the truncated roller coaster shown in Figure 13.3. Three distinct phases can be identified: the event that causes the crisis, the period of disorganization that follows, and the reorganizing or recovery phase after the family reaches a low point. Families have a certain level of organization before a crisis; that is, they function at a certain level of effectiveness—higher for some families, lower for others.

Families that are having difficulties or functioning less than effectively before the onset of additional stressors or demands are said to be **vulnerable families**; families capable of "doing well in the face of adversity" are called **resilient families** (Patterson 2002a, p. 350). It's important to note here that the definition of a family's "doing well" (that is, family well-being) may differ according to whether the family views its own well-being from a more individualistic or more collectivist set of values (see Chapter 1). From a more individualistic perspective, family well-being may emphasize economic, educational, and occupational resilience. From a more collectivist perspective, however, family well-being may emphasize the ability of family members to get together and a return to practicing cultural traditions (McCubbin et al. 2013).

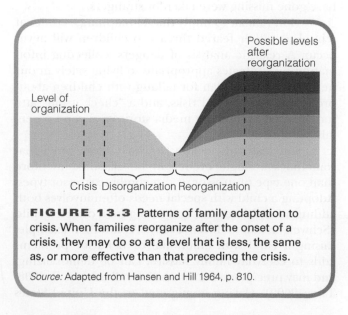

FIGURE 13.3 Patterns of family adaptation to crisis. When families reorganize after the onset of a crisis, they may do so at a level that is less, the same as, or more effective than that preceding the crisis.

Source: Adapted from Hansen and Hill 1964, p. 810.

Spencer Grant/age fotostock

Resilient families do well in the face of adversity. Greater financial resources are advantageous in coping with family stress and crises, but low-income families are often creatively resilient in locating resources.

In the period of disorganization following the crisis, family functioning declines from its initial level. Families reorganize, and after the reorganization is complete, (1) they may function at about the same level as before; (2) they may have been so weakened by the crisis that they function only at a reduced level (more often the case with vulnerable families); or (3) they may have been stimulated by the crisis to reorganize in a way that makes them more effective (a characteristic of resilient families).

At the onset of a crisis, it may seem that no adjustment is required at all. A family may be confused by a member's alcoholism or numbed by the new or sudden stress and, in a process of denial, go about their business as if the event had not occurred. Gradually, however, the family begins to assimilate the reality of the crisis and to appraise the situation. Then the **period of family disorganization** sets in.

THE PERIOD OF DISORGANIZATION

At this time, family organization slumps, habitual roles and routines become nebulous and confused, and members carry out their responsibilities with less enthusiasm. Although not always the case, this period of disorganization may be "so severe that the family structure collapses and is immobilized for a time. The family can no longer function. For a time, no one goes to

work; no one cooks or even wants to eat; and no one performs the usual family tasks" (Boss 1997, p. 1). Typically, and legitimately, family members may begin to feel angry and resentful.

Expressive relationships within the family change, some growing stronger and more supportive perhaps, and others more distant. In the words of social psychologist Benjamin Karney:

> [S]tressful events occurring outside of a relationship interfere with couples' ability to maintain an intimate bond within the relationship. First, stress outside the relationship changes what couples need to talk about and the time available to talk about it. . . . Time that couples spend deciding how they are going to cut back to get their bills paid, or negotiating who is going to take off work to care for a sick relative, is time that is not spent on other activities, like having sex or participating in shared interests that are more likely to promote closeness. (2011, p. F2)

Other research shows that sexual activity, one of the most sensitive aspects of a relationship, often changes sharply and may temporarily cease. Parent-child relations may also change. As one example, a child described life with his mother during the period of disorganization after his father was deployed: "I could tell my mom was getting like really depressed and since she wouldn't talk, I wouldn't talk. And so around the house everyone was just kind of depressed for a little while and you could tell because they didn't speak a lot" (in Huebner et al. 2007, p. 117).

Relations between family members and their outside friends, as well as the extended kin network, may also change during this phase. Some families withdraw from all outside activities until the crisis is over; as a result, they may become more private or isolated than before the crisis began. As we shall see, withdrawing from friends and kin often weakens rather than strengthens a family's ability to meet a crisis.

At the **nadir**, or low point, of family disorganization, conflicts may develop over how the situation should be handled. For example, in families with a seriously ill member, the healthy members are likely either to overestimate or underestimate the sick person's incapacitation and, accordingly, act with either more sympathy or with less tolerance than the ill member wants (Conner 2000; Pyke and Bengston 1996). Reaching the optimal balance between nurturance and encouragement of the ill person's self-sufficiency may take time, sensitivity, and judgment.

During the period of disorganization, family members face the decision of whether to express or smother

any angry feelings they may have. Expressing anger as blame typically sharpens hostilities; laying blame on a family member for the difficulties being faced will not help to solve the problem and will only make things worse (Stratton 2003). At the same time, when family members opt to repress their anger, they risk allowing it to smolder, thus creating tension and increasingly strained relations. How members cope with conflict at this point—for instance, whether they use communication techniques that foster bonding or cause hurtful conflict or distancing—greatly influences the family's overall level of recovery (McDermott and Cobham 2012; Masarik et al. 2016).

Recovery

Once the crisis hits bottom, things often begin to improve. By trial and error or by thoughtful planning, family members usually arrive at new routines and reciprocal expectations. They are able to look past the time of crisis to envision a return to some state of normalcy and to reach some agreements about the future. Some families do not recover intact, as today's high divorce and separation rates illustrate. Divorce or the separation of a cohabiting relationship can be seen as an adjustment to a previous or ongoing family crisis such as a partner's infidelity or substance abuse and can be seen as a family crisis in itself (see Figure 13.4).

Other families stay together, although at lower levels of organization or mutual support than before the crisis. As Figure 13.4 shows, some families remain at a low level of recovery, with members continuing to interact much as they did at the low point of disorganization. This interaction often involves a series of circles in which one member is viewed as deliberately causing the trouble and the others blame that individual and nag him or her to stop. This is true of many families in which one member is alcoholic or is otherwise

chemically dependent or a chronic gambler, for example. Rather than directly expressing anger about being blamed and nagged, the offending member persists in the unwanted behavior.

In some instances, social structural, or environmental conditions limit a family's odds of recovery. For instance, former prisoners are often denied work opportunities from potential employers who ask about prior convictions on employment applications (Tierney 2013). Although this remains the situation for many who have been incarcerated, in 2012 the Equal Employment Opportunity Commission "approved an updated policy making it more difficult for employers to use background checks to systematically rule out hiring anyone with a criminal conviction" (Greenhouse 2012).

Some families match the level of organization they had maintained before the onset of the crisis, whereas others rise to levels above what they experienced before the crisis (McCubbin 1995). For example, a family member's attempted suicide might motivate all family members to reexamine their relationships.

Reorganization at higher levels of mutual support may also result from less dramatic crises. For instance, partners in midlife might view boredom with their relationship as a challenge and revise their lifestyle to add some zest—by traveling more or planning to spend more time together rather than in activities with the whole family, for example. Research on family reorganization after a young child has been diagnosed with asthma found that families that were more cohesive and operated on a higher level of supportive functioning prior to the diagnoses recovered more positively than did families that were less supportive of one another before the diagnosis (Spagnola and Fiese 2010). Family support reduces depression after a child has come out at bisexual (Pollitt, Muraco, Grossman, and Russell 2017).

We have suggested that separation or divorce is a family crisis, and one study suggests that former partners' recovery trajectories evidence a similar range of possibilities. A longitudinal study of 216 men's and 238 women's recovery "pathways" over the course of ten years after divorce found that about one-fifth of the sample "grew more competent, well adjusted, and fulfilled" than they had been before the divorce. Meanwhile, the largest group had fashioned lives about as satisfying as what they'd had before divorce. Others, about one-tenth of the sample, evidenced low social responsibility and self-esteem and were more depressed and defeated than they had been before they divorced (Hetherington 2003, pp. 324–325). Now that we have examined the course of family crises, we will turn our attention to a theoretical model specifically designed to explain family stress, crisis, adjustment, and adaptation.

FIGURE 13.4 Divorce or separation as an adjustment to a previous or ongoing family crisis—for example, partner's infidelity, substance abuse, or violence—and as a crisis in itself.

FAMILY STRESS, CRISIS, ADJUSTMENT, AND ADAPTATION: A THEORETICAL MODEL

Some decades ago, sociologist Reuben Hill proposed the ABC-X family crisis model, and much of what we've already noted about stressors is based on the research of Hill, his colleagues, and his successors (Hill 1958; Hansen and Hill 1964). The **ABC-X model** states that A (the stressor event) interacting with B (the family's ability to cope with a crisis) interacting with C (the family's appraisal of the stressor event) produces X (the crisis) (see Sussman, Steinmetz, and Peterson 1999). In Figure 13.5, A would be the demands put on a family, B would be the family's capabilities (resources and coping behaviors), and C would be the meanings that the family creates to explain the demands.

As Figure 13.5 illustrates, families continuously balance the demands put on them against their capabilities to meet those demands. When demands become heavy, families engage their resources to meet them while also appraising their situation—that is, they create meanings to explain and address their demands. When demands outweigh resources, family adjustment is in jeopardy, and a family crisis may develop. Through the course of a family crisis, some level of adaptation occurs (Patterson 2002a).

Stressor Pileup

Building on the ABC-X model, Hamilton McCubbin and Joan Patterson (1983) advanced the *double* ABC-X model to better describe family adjustment to crises. In Hill's original model, the A factor was the stressor event; in the double ABC-X model, A becomes *Aa*, or "family pileup." *Pileup* includes not just the stressor but also previously existing family strains and future hardships induced by the stressor event.

When a family experiences a new stressor, prior strains that may have gone unnoticed—or been barely managed—come to the fore. Prior strains might be any residual family tensions that linger from unresolved stressors or are inherent in ongoing family roles, such as being a single parent or a partner in a two-career family. For example, ongoing but ignored family conflict may intensify when parents or stepparents must deal with a child who is underachieving in school, has joined a criminal gang, or is abusing drugs. As another example, financial and time constraints typical of single-parent families may assume crisis-inducing importance with the addition of a stressor such as caring for an injured child.

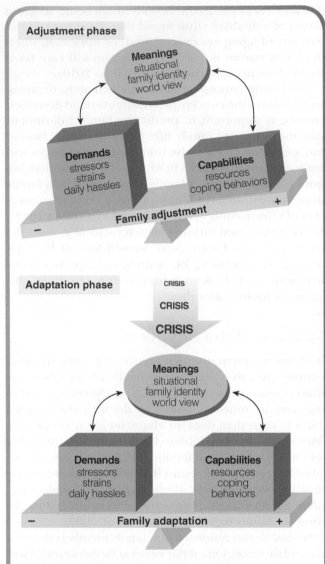

FIGURE 13.5 Family stress, crisis, adjustment, and adaptation. Families continuously balance the demands put on them with their capabilities to meet those demands. When demands become heavy, families engage their resources to meet them while also appraising their situation—that is, they create meanings to explain and address their demands. When demands outweigh resources, family adjustment is in jeopardy, and a family crisis may develop. Through the course of a family crisis, some level of adaptation occurs (Patterson 2002a).

Source: Patterson 2002a, p. 351.

An example of future demands precipitated by the stressor event would be the unpaid bills and possibly even loss of the family domicile that follow a parent's loss of a job (a stressor). A study of adolescents found that a teen's prior risk factors, such as depression or low academic performance, resulted in heightened family stress when a parent was deployed overseas

(Lucier-Greer et al. 2015). A study about being an aging parent of a disabled adult found that the stress and uncertainty of aging are exacerbated, or worsened, when an elderly parent is worried about who will care for a disabled offspring in the future (Fujiura 2010).

The pileup concept of family-life demands, or stressors (similar to the concept of stressor overload described earlier), is important in predicting family adjustment over the course of family life. Social scientists believe that, generally, an excessive number of life changes and strains occurring within a brief time, perhaps a year, are more likely to disrupt a family. Pileup renders a family more vulnerable to emerging from a crisis at a lower level of effectiveness (McCubbin and McCubbin 1989). We have examined various characteristics of stressor demands put on a family. Next, we will look at how the family makes meaning of, defines, or appraises those demands. We'll look at crisis-meeting resources and coping behaviors after that.

Appraising the Situation

From an interactionist perspective, the meaning that a family gives to a situation—how family members appraise, define, or interpret a crisis-precipitating event—can have as much or more to do with the family's ability to cope than does the character of the event itself (McCubbin and McCubbin 1991; Patterson 2002a,b). For example, a study of families faced with caring for an aging family member found that some families felt ambivalent, even negative, about having to provide care; other families saw caregiving as one more chance to bring the family together (Roscoe et al. 2009).

Several factors influence how family members define a stressful situation. One is the *nature of the stressor itself.* How serious is it? How long can it be expected to last? Is the stressor event likely to improve or get worse? A study of families coping with a preschooler who had asthma found that the severity of the child's asthma attacks affected how smooth the family's adjustment period was the family adjusted (Spagnola and Fiese 2010). Sometimes in the case of ambiguous loss, families do not know whether a missing family member will ever return or whether a chronically ill or a chemically dependent family member will ever recover. A stressor event that puts a family in limbo this way is especially difficult to manage.

In addition to the nature of the stressor itself, a second factor that influences a family's appraisal of the situation is *the degree of hardship or the kind of problems the stressor creates.* Temporary unemployment at age 16 is less a hardship than not finding a position after college graduation or being laid off at age 45. Being victimized by a crime is always a stressful event, but coming home to find one's house burglarized may be less traumatic than being robbed at gunpoint. Caregiving for a short time after a family member's surgery differs from the long-term caregiving of indefinite length associated with serious chronic illness.

A third factor is *the family's previous successful experience with crises,* particularly those of a similar nature. If family members have had experience in nursing a sick member back to health, they will feel less bewildered and more capable of handling a new, similar situation. Believing from the start that demands are surmountable, and that the family has the ability to cope collectively, may make adjustment somewhat easier (Pitzer 1997b; Wells, Widmer, and McCoy 2004). Family members' interpretations of a crisis event shape their responses in subsequent stages of the crisis (Zehner Ourada and Walker 2014). Meanwhile, the family's crisis-meeting resources affect its appraisal of the situation.

A fourth, related factor that influences a family's appraisal of a stressor involves the *adult family members' legacies from their childhoods* (Carter and McGoldrick 1988; Fosco and Grych 2012). For example, growing up in a family that tended to define anything that went wrong as a catastrophe or a "punishment" from God might lead the family to define the current stressor more negatively. On the other hand, growing up in a family that tended to define demands simply as problems to be solved or as challenges to be overcome might mean defining the current stressor more positively.

Defining the Situation as Catastrophic or Not A family tendency to define events as catastrophic or not, as only negative or not, as occasion for greater warmth and demonstrations of caring or not is often learned in childhood (Fosco and Grych 2012; see also Parade, Leerkes, and Helms 2013). In their model of family stress and crisis, social workers Betty Carter and Monica McGoldrick (1988, p. 9) see "family patterns, myths, secrets [and] legacies" as *vertical stressors*—because they come down from the previous generations. These authors call the type of stressors that we have been discussing in this chapter *horizontal stressors*—that is, stressors that occur in real, present time and do not involve family legacies from the past.

Whether or not taught in childhood, family members can focus on and learn resilience—ways to think about stressors that render them less difficult to handle. Research on parents with a school-aged child with autism, for instance, found that **reframing**—redefining stressful events to make them more manageable—was associated with more positive family functioning (Manning, Wainwright, and Bennett 2011). In this case, reframing involved accepting the child's condition, striving to lead as normal a life as possible under the circumstances, including the autistic child in the family's daily life, focusing on valued aspects of the child, and viewing the child as an occasion for other family members to learn and develop—although assuredly having a child with autism does increase parental and other family members'

stress (Manning, Wainwright, and Bennett 2011; Vogan et al. 2014).

Not All Family Members Necessarily Agree in Their Appraisal of the Situation Although we are discussing the family's appraisal, or definition, of the situation, it is important to remember the possibility that each family member experiences a stressful event in a unique way:

> These unique meanings may enable family members to work together toward crisis resolution or they may prevent resolution from being achieved. That is, an individual's response to a stressor may enhance or impede the family's progress toward common goals, may embellish or reduce family cohesion, may encourage or interfere with collective efficacy. (Walker 1985, pp. 832–833)

A dramatic example involves a family member's need for a sibling to donate an organ, such as a kidney or stem cells to fight a fatal illness (Begley and Piggott 2012). Some family members will see the donation, which requires surgical risk and time off work, as a taken-for-granted requirement to save a brother's or sister's life. However, the sibling who is expected to donate may see things differently and feel ambivalent. The same may be true for other family members. For instance, "Sarah has three children, and her husband is unhappy about her becoming a donor. He is going for promotion, the children have important school years ahead of them and he does not want a disruption to family life. Sarah feels trapped" (Begley and Piggott 2012, p. 184). Feeling trapped is a stressor. Resolving situations like these involve honest family communication, facilitated when possible by counseling.

Crisis-Meeting Resources

A family's crisis-meeting capabilities—resources and coping behaviors—constitute its ability to prevent a stressor from creating severe disharmony or disruption (Masarik, et al. 2016). We categorize a family's crisis-meeting resources into three types: personal or individual, family, and community.

The personal resources of each family member (for example, intelligence, problem-solving skills, and physical and emotional health, ability to emphasize) are important (Lee and Roberts 2018). At the same time, the family *as family* or family system has a level of resources, including bonds of trust, appreciation,

and support (family harmony); sound finances and financial management and health practices; positive communication patterns; healthy leisure activities; and overall satisfaction with the family and quality of life (Boss 2002; Lee and Roberts 2018; Patterson 2002a).

Family rituals (see Chapter 1) are resources (Boss 2004; Oswald and Masciadrelli 2008). A study of families with alcoholism found that adult children of alcoholics who came from families that had maintained family dinner and other rituals (or who married into families that did) were less likely to become alcoholics themselves (Bennett, Wolin, and Reiss 1988; Goleman 1992).

And, of course, *money* is a family resource. For instance, a fairly recent study found that whether a family can afford diabetes medication impacts their efficacy in fighting the disease (Novak, Anderson Johnson, Walker, Wilcox, Lewis, and Robbins (2017). We assume the same would be true for all diseases.

As another example, a breadwinner's losing his or her job is less difficult to deal with when the family has substantial savings. In a qualitative study among U.S. working-poor rural families, one respondent

Andrew Burton/Getty Images News/Getty Images

A positive outlook, spiritual values, supportive communication, adaptability, public services, and informal social support—along with extended family and community resources—are all factors in family resilience or meeting a crisis creatively. These scientific findings, while recommended and possible, can be more easily said than done, however. This is a picture taken in 2015 of Newtok, Alaska. Newtok is one of about thirty Alaskan villages that are disappearing, threatened by rising ocean levels due to melting polar ice. You can see children running on recently placed boardwalks designed to keep Newtok residents from sinking into the melting permafrost. Today, *climate refugees,* residents around the globe of villages like Newtok, are vacating familiar and beloved birthplaces to relocate inland—assuredly an individual, family, and community crisis (Taylor 2015).

explained,"I had absolutely nothing after I paid my bills to feed my kids. I scrounged just so that they could eat something, and I had to shortchange my landlord so that I could feed them, too, which put me behind in rent." Another said, "I felt overwhelmed and stressed because every time I get paid, I just don't have money for everything . . . because I have two . . . children [with medical problems]" (Dolan, Braun, and Murphy 2003, p. F14). At the other end of the financial spectrum are families who can afford to send troubled adolescents to costly "wilderness camps," for example, or other residential treatment facilities for illegal drug use or otherwise negative behaviors. Parents have told evaluation researchers that facilities such as these help to abate or relieve a family crisis and also stabilize the family (Harper 2009).

The family ecology perspective alerts us to the fact that *community resources* are consequential as well (Socha and Stamp 2009). Increasingly aware of this, medical and family practice professionals have in recent years designed a wide variety of community-based programs to help families adapt to medically related family demands, such as a partner's cancer or a child's diabetes, congenital heart disease, and other illnesses (Marshall 2010; Tak and McCubbin 2002). In fact, in many instances, family members have become community activists, working to create community-based resources to aid them in dealing with a particular family stressor or crisis. Parents have been a driving force in shaping services and laws related to individuals with mental challenges (Swenson and Lakin 2014). As a second example, parent groups and adults with disabilities worked together to help pass the Americans with Disabilities Act (Bryan 2010).

As a third example, school districts vary in the amount and quality of community resources they offer. A news story produced by Columbia University School of Journalism followed two NYC middle school boys with similar learning disabilities. Each boy had parent advocates, but only one was poor, nonwhite, and in public school. The other boy, attending a private school, got tax-financed help, while the boy attending public school did not (Elsen-Rooney 2020).

Vulnerable vs. Resilient Families Ultimately, the family either successfully adapts or becomes exhausted and vulnerable to continuing crises. Family systems may be high or low in vulnerability, a situation that affects how positively the family faces demands; this enables us to predict or explain the family's poor or good adjustment to stressor events (Patterson 2002a).

More prone to poor adjustment from crisis-provoking events, *vulnerable families* evidence a lower sense of common purpose and feel less in control of what happens to them. They may cope with problems by showing diminished respect or understanding for one another.

Vulnerable families are also less experienced in shifting responsibilities among family members and are more resistant to compromise. There is little emphasis on family routines or predictable time together (McCubbin and McCubbin 1991).

From a social psychological point of view, *resilient families* tend to emphasize mutual acceptance, respect, and shared values (Masten 2016). Family members rely on one another for support. Generally accepting difficulties, they work together to solve problems with members, feeling that they have input into major decisions (McCubbin et al. 2001). It may be apparent that these behaviors are less difficult to foster when a family has sufficient economic resources. The next section discusses factors that help families meet crises creatively.

MEETING CRISES CREATIVELY

Meeting crises creatively means that after reaching the nadir in the course of the crisis, the family rises to a level of reorganization and emotional support equal to or higher than that which preceded the crisis. For some families—for example, those experiencing the crisis of domestic violence—breaking up may be the most beneficial (and perhaps the only workable) way to reorganize. Other families stay together and find ways to meet crises effectively (Masarik, Martin, Ferrer, Lorenz, Conger, and Conger 2016). What factors differentiate resilient families that reorganize creatively from those that do not?

A Positive Outlook

In times of crisis, family members make many choices, one of the most significant of which is whether to blame one member for the hardship. Casting blame, even when it seems deserved, is less productive than viewing the crisis primarily as a challenge (Stratton 2003).

The more that family members strive to maintain a positive outlook, the more a person or family is likely to meet a crisis constructively (Burns 2010; Lee and Roberts 2018). Developing more open, supportive family communication—especially in times of conflict—also helps individuals and families meet crises constructively (Stinnett, Hilliard, and Stinnett 2000). Families that meet a crisis with an accepting attitude, focusing on the positive aspects of their lives, do better than those that feel they have been singled out for misfortune (Bluth et al. 2013; Burns 2010). For example, many chronic illnesses have downward trajectories, so both partners may realistically expect that the ill mate's health will only grow worse. Some couples are remarkably able to adjust to this, "either because of immense closeness to each other or because they are

grateful for what little life and relationship remains" (Strauss and Glaser 1975, p. 64).

Spiritual Values and Support Groups

"Spirituality, however the family defines it, can be a strong comfort during crisis" (Thomason 2005, p. F11). Many authors have argued that strong religious faith is related to high family cohesiveness and helps people manage demands or crises, partly because it provides a positive way of looking at suffering (Ellison et al. 2011). A spiritual outlook may be fostered through Buddhist, Jewish, Christian, Muslim, and other religious or philosophical traditions. Nevertheless, a sense of spirituality—that is, a conviction that there is some power or entity greater than oneself—need not be associated with membership in any organized religion (Bluth et al. 2013). Self-help groups such as Alcoholics Anonymous and Al-Anon for families of alcoholics incorporate a "higher power" and can help people take a positive, spiritual approach to family crises. Informal support groups that are not necessarily spiritually based are also helpful (Freeman and Dodson 2014).

Open, Supportive Communication

Families whose members interact openly and supportively meet crises more creatively (Lee and Roberts 2018; Olson and Gorall 2009). Free-flowing communication opens the way to understanding. Expressions of support from parents help children to cope with daily stress (Centers for Disease Control and Prevention 2019; Valiente et al. 2004). As another example, better-adjusted husbands with multiple sclerosis believed that even though they were embarrassed when they fell in public or were incontinent, they could freely discuss these situations with their families and feel confident that their families understood (Power 1979). As a final example, talking openly and supportively with an elderly parent who is dying about what that parent wants—in terms of medical treatment, hospice, and burial—can help (Fein 1997).

Knowing how to indicate the specific kind of support that one needs is important at stressful times. Differentiating between—and knowing how to request— simply listening as opposed to problem-solving discussion can help reduce misunderstandings (Stinnett, Hilliard, and Stinnett 2000; Tannen 1990). Families whose communication is characterized by a sense of humor, as well as a sense of family history, togetherness, and common values, evidence greater resilience (Henry, Morris, and Harrist 2015; Thomason 2005). Interesting research with women experiencing breast cancer showed that support from others was more likely to be offered and more often from the healthy spouse or other family members when the woman suffering was able to feel

positive enough to express appreciative feelings toward her caregivers (Sheridan et al. 2010).

Adaptability

Adaptable families are better able to respond to stress and crises effectively (Boss 2002; Uruk, Sayger, and Cogdal 2007). Families are more adaptable when conjugal power is fairly egalitarian (see Chapter 12). When one family member wields authoritarian power, the whole family suffers if the authoritarian leader does not make effective decisions during a crisis—and allows no one else to move into a position of leadership (McCubbin and McCubbin 1994). A partner who feels comfortable only as the family leader may resent their loss of power, and this resentment may continue to cause problems when the crisis is over.

Family adaptability in aspects other than leadership is also important (Burr, Klein, and McCubbin 1995). A recent study shows that a father's willingness and ability to adapt his identity to having a child with disabilities is a factor in the family's smoother functioning (Fox et al. 2015). Families that can adapt their schedules and use of space, family activities and rituals, and connections with the outside world to the limitations and possibilities posed by the crisis will cope more effectively than families committed to preserving sameness. A study of married parents caring for a disabled adult child found that when their division of labor was adapted to feel fair, both parents experienced greater marital satisfaction and less stress (Essex and Hong 2005).

Informal Social Support

It's easier to cope with crises when a person doesn't feel alone (Lee and Roberts 2018; Tak and McCubbin 2002). Polls show that time spent with others is necessary to individuals' emotional well-being (Harter and Arora 2008). Families may find helpful support in times of crisis from kin, good friends, neighbors, and even acquaintances such as work colleagues (Johnson 2010). The caregiver needs support, too. Spouse caregivers of the elderly, disabled, or chronically ill themselves need social support; working part-time can help when possible (Fujiura 2010; Glauber and Day 2018). Family members in hospitals need support from nurses and other hospital staff (Cypress and Frederickson 2017; Lind et al. 2012). Caregivers of burn patients showed symptoms of anxiety and need of counseling and support up to a year after the burn patient was injured (Backstrom, Armstrong, and Puentes 2013). We need support services of all types, including for parents of disabled or emotionally or behaviorally disturbed children (Vaughan et al. 2012).

A qualitative study of family members with a close relative hospitalized in intensive care found that

feeling supported by staff helped: "The nurse's tone was so peaceful, which reassured me. I was so touched by her kindness" (Cypress and Frederickson 2017, p. 209). Analysis of data from the National Survey of Black Americans found that many of them in times of crisis received support from fellow church members (Taylor, Lincoln, and Chatters 2005). These various relationships provide a wide array of help—from lending money in financial emergencies to helping with childcare to just being there for emotional support. Research on families in poverty shows that, although the informal social support they receive rarely helps to lift them out of poverty, it does help them to cope with their economic circumstances (Henly, Danziger, and Offer 2005). Even continued contact with more casual acquaintances may be helpful, as they often offer useful information, along with enhancing one's sense of community (Orthner, Jones-Sanpei, and Williamson 2004).

And, of course, the Internet offers information, social media options, and support for many stressors (Gilkey, Carey, and Wade 2009). A qualitative study that recruited participants by means of Web pages asked the seventy-seven respondents who answered an Internet-based survey about the advantages and disadvantages of Internet support compared with face-to-face social networks (Colvin et al. 2004). Respondents mentioned two main Internet advantages—anonymity and the ease of connecting with others in the same situation despite geographical distance. Disadvantages related to lack of physical contact: "No one can hold your hand or give you a Kleenex when the tears are flowing" (p. 53).

An Extended Family

Grandparents, siblings, and other kin networks can be a valuable source of support in times of crisis (Gonzalez 2016; Ryan, Kalil, and Leininger 2009). Although extended families as residential groupings represent a small proportion of family households, kin ties remain salient (Furstenberg 2005; Reyes 2018). One aspect of all this involves reciprocal friendship and support among adult siblings (DeBel, Kalmijn, and Duijn. 2019; Gonzalez 2016; White and Riedmann 1992). In times of family stress or crisis, new immigrants, as well as African Americans, may rely on **fictive kin**—relationships based not on blood or marriage but on "close friendship ties that replicate many of the rights and obligations usually

Many—although not all—turn to their extended family for social support in times of stress. Kin may be able to provide emotional and monetary support as well as practical help. This is not always the case, however. Sometimes extended family do not support an individual family member's wants or needs. Can you think of instances when this might be the case?

Blend Images/Superstock

associated with family ties" (Ebaugh and Curry 2000; Nelson 2013).

We need to be cautious, though, not to romanticize the extended family as a resource. A study that compared mothers who had children with more than one father found that the women received less support from their kin networks than did single mothers who did not have multipartnered births. The researchers concluded that "smaller and denser kin networks seem to be superior to broader but weaker kin ties in terms of perceived instrumental support" (Harknett and Knab 2007). Among the poor, extended kin may not have the resources to offer much practical help—and when they do, they are more likely to offer it to female than to male family members (Mazelis and Mykyta 2011; Reyes 2018; White and Riedmann 1992).

Moreover, along with some previous research, a small study of low-income families living in two trailer parks along the mid-Atlantic coast concluded that "low-income families do not share housing and other resources within a flexible and fluctuating network of extended and fictive kin as regularly as previously assumed." Extended family members may not get along, or individuals may be too embarrassed to ask their kin for help. One woman explained that neither her parents nor any one of her five siblings could help her because "they all have problems of their own." A Hispanic mother told the interviewer, "I know you've probably heard that Hispanic families are close-knit,

well, hmmph! [It's not necessarily true.]" (Edwards 2004, p. 523). Then, too, among some recent immigrant groups such as Asians and Hispanics expectations of the extended family may clash with the more individualistic values of a more Americanized family member who needs help.

Community Resources

In 2008, Nebraska became the last state in the United States to adopt a safe-haven law whereby parents can abandon their children at hospitals without fear of prosecution. Intended as a way to save unwanted newborns from being murdered or left in dumpsters or motel rooms, Nebraska's safe-haven law failed to limit the ages of children who could be legally abandoned. During the month after the legislation passed, more than thirty youngsters were left at Nebraska hospitals. Most of them were older than eleven. Some had extremely severe mental and behavioral problems (Hansen and Spenser 2009). Some had been transported to Nebraska by overwhelmed parents from outside the state. A month later, Nebraska amended its law to require that legally abandoned children had to be younger than thirty days old (Italie 2008; Jenkins 2008). The story became fuel for jokes on late-night television and afternoon talk shows.

> But . . . something is wrong when so many parents are so eager to abandon so many children. . . . Because what happened in Nebraska constitutes a message from overstressed parents, one we ignore at our own peril. It is not a complicated message. On the contrary it is as simple and succinct as a word: Help. (Pitts 2008)

After a special session of the Nebraska legislature addressed this issue, more than one legislator indicated that the state would have to examine the accessibility of social services for older children and their families (Eckholm 2008). Subsequent reviews showed that in some cases the state had failed to help desperate parents who did not receive necessary services for their children until after the children had been dropped off. In other cases, the parents themselves did not seem to know where to turn—although appropriate services were available—until they heard about the safe-haven law (Hansen and Spenser 2009).

The success with which families meet the demands placed on them depends at least partly on the availability of community resources, coupled with families' knowledge of and ability to access the **community-based resources** available to help (Karney 2011).

> Community-based resources are defined as all of those characteristics, competencies and means of persons, groups and institutions outside the family which the family may call upon, access, and use to meet their demands. This includes a whole range of services, such as medical and health care services. The services of other institutions in the family's . . . environment, such as schools, churches, employers, etc.[,] are also resources to the family. At the more macro level, government policies that enhance and support families can be viewed as community resources. (McCubbin and McCubbin 1991, p. 19)

Among others, community-based resources include schools and school personnel; social workers and family welfare agencies; foster childcare; church programs that provide food, clothing, or shelter to poor or homeless families; twelve-step and other support programs for substance abusers and their families; programs for crime or abuse victims and their families; support groups for people with serious diseases such as cancer or AIDS, for parents and other relatives of disabled or terminally ill children, or for caregivers of disabled family members or those with cancer or Alzheimer's disease; and community pregnancy prevention or parent education programs. An Oregon study of nonHispanic white and Hispanic teen mothers found that a government-funded home-visitation program increased family functioning, especially for the Hispanics in this sample (Middlemiss and McGuigan 2005). Other research shows that community programs—for family members with serious mental illness, for example—result in more positive family interactions and in relatives' feeling less burdened (Pernice-Duca et al. 2015).

Strength-Based Programs As an example of community-based resources, family empowerment, or *strength-based*, programs recognize and build on a family's positive attributes (strengths) to foster resilience (Cleek et al. 2012). A unique example of family empowerment and parent education programs, mandated by the U.S. government in 1995, involves federal prison inmates. Parent inmates learn general skills such as how to talk to their children. They also learn ways to create positive parent–child interaction from prison—such as games they can play with a child through the mail—as well as suggestions on what to do when returning home after release (Coffman and Markstrom-Adams 1995; Kohl 2012; Loper, Clarke, and Dallaire 2019). "Issues for Thought: When a Parent Is in Prison" further describes some of these programs.

Family Counseling Another community resource, family counseling (see Chapter 11) can help families after a crisis occurs, such as a family member's suffering from post-traumatic stress disorder. Counseling can also help when families foresee a family change or future new demand. For instance, a couple might visit a counselor when expecting or adopting a baby, when deciding about work commitments and family needs, when the youngest child is about to leave home, or when a partner is about to retire. Family counseling is not just

When a Parent Is in Prison

Nearly 3 million children have a parent who is in jail or prison (Cochran, Siennick, and Mears 2018)—a demoralizing family stressor event, coupled with boundary ambiguity (Wildeman 2012). Incarceration rates for black and Hispanic males have risen sharply over the past several decades (Coates 2015). A child's risk of parental incarceration increased more than 60 percent between 1978 and 2000 (Roberts 2012). Although the risk remains low for nonHispanic white children, calculations show that 14 percent to 15 percent of black children born in 1978 and 25 percent to 28 percent "of black children born in 1990 had a parent imprisoned by the time the child was 14" (Wildeman 2009, p. 271).

Prior to their imprisonment, 79 percent of mothers and 53 percent of fathers were living with their children (National Resource Center on Children and Families of the Incarcerated 2009a). While mothers are in prison, about one-quarter of their children live with their fathers. Grandparents care for about half of all children with incarcerated mothers. NonHispanic white children are more likely to be in nonfamily foster care (see Chapter 10) than are African American or Hispanic children (Lee, Genty, and Laver 2005). This may be because black and Hispanic communities have had more of a tradition of shared care of children (e.g., Stack 1974), a situation facilitating making arrangements that place children with adult relatives, often grandparents (Enos 2001; Poehlmann 2005). The children's caregivers often

> feel compelled to lie about their loved one's whereabouts. If the children are young, their mother may explain the father's absence by saying that "Daddy's away on a long trip" or "He's working on a job in another state." One caregiver . . . explained to her nephews that their father was away at "super-hero school." Older children who know the truth may feel that they need to be careful not to discuss it at school or with friends. (Arditti 2003, p. F15)

Children's visiting an incarcerated parent can be expensive and otherwise difficult to arrange because prisons are often far from their homes (McManus 2006; National Resource Center on Children and Families of the Incarcerated 2009b). One study found that half of children of women prisoners did not visit at all during their mother's incarceration. However, including phone calls and letters, 78 percent of mothers and 62 percent of fathers had at least monthly contact with children (Mumola 2000).

More and more, policy makers have realized that disrupted family ties have a severe and negative impact on the next generation (Casey, Shlafer, and Masten 2015; Cochran, Siennick and Mears 2018; Turney and Lanuza 2017). Consequently, a number of correctional systems, including the Federal Bureau of Prisons, have developed visitation programs to facilitate parent–child contact (Geller 2013). Many correctional facilities have returned to an earlier practice of permitting babies born in prison to remain with their mothers for a time (Rutgers University School of Criminal Justice and the New Jersey Institute for Social Justice 2006; Comfort 2008). Although visitation programs were initially oriented solely to mothers, prisons have more recently developed programs for fathers as well (Enos 2001; McManus

for relationships that are in trouble but also is a resource that can enhance family dynamics. Increasingly, counselors and social workers emphasize empowering families—that is, emphasizing and building on a family's strengths (Burns 2010; Mullins et al. 2015). In addition to counseling, resources include books and many online resources on various subjects related to family stress and crises, as mentioned above.

The other side of the community-based resource story, however, is that there just aren't enough of them. For instance, families struggling with caring for a disabled family member often face workplace insensitivity, career challenges, and financial difficulties (Stewart 2013). Moreover, federal and state budget cuts have resulted in reduction of services and programs for families struggling with difficult stressors, such as caring for a disabled child, among many other stressors (Streitfeld 2010). In the words of social psychology professor Benjamin Karney,

> [F]amilies will benefit from policies that make their lives easier. Higher wages, more job security, and access to health care—to the extent that these policies would reduce the stress of modern life—would also promote the stability and quality of relationships directly. [Furthermore], even in the absence of serious changes to their lives, couples might be encouraged to recognize the ways that stress affects their relationships and assisted in developing communication patterns and concrete resources to help them cope with stress effectively when it arises. (Karney 2011)

As discussed in Chapter 1, political decisions (such as cutting family-centered federal, state, county, or city

2006). Additional programs for families of inmates exist as well (Eddy and Poehlmann-Tynan 2019). Several such programs can be found online (e.g., "7 Helpful Programs for Children of Incarcerated Parents" 2016).

We focus here on children's needs, but imprisonment demoralizes other family members as well and usually has a negative economic impact on the family system, not only when the prisoner has been an essential breadwinner but also due to costs associated with visiting the prisoner and making long-distance family telephone calls, among others (Arditti, Lambert-Shute, and Joest 2003; Geller and Franklin 2014). Moreover, because of stigma associated with incarceration, prisoners' families receive little community support (Arditti 2003; Turney 2015).

Strong family bonds appear to reduce children's negative behaviors (Poehlmann 2005), although "incarceration can undermine social bonds, [and] strain marital and other family relationships" (Western and McLanahan 2000, p. 323). Policy analysts argue that "[a]n over reliance on incarceration as punishment, particularly for nonviolent offenders, is not good family policy" (Arditti 2003, p. F17; Wildeman 2009;

see also Dyer, Pleck, and McBride 2012). They propose alternatives to incarceration, such as home confinement with work release (Comfort 2008; see also sentencingproject.org).

Critical Thinking

Can you apply the ABC-X model to this situation of having an incarcerated family member?

Gerhard Joren/LightRocket/Getty Images

Having a family member in prison or jail is a crisis that a growing number of families face today. Family stress and adjustment experts tell us that virtually all family crises have some potential for positive as well as negative effects. Can you think of any possible positive effects in this case? What community supports might help this family? What might be some alternatives to incarcerating parents who have been actively involved in raising their children?

programs) need to be evaluated through a *family impact lens* (Bogenschneider et al. 2012). From this point of view, most family policy scholars are apt to argue that program cutting—even such apparently small things as closing city swimming pools—unfortunately adds to family stress.

CRISIS: DISASTER OR OPPORTUNITY?

A family crisis is a turning point that requires members to change (McCubbin and McCubbin 1991, 1994). We tend to think of *crisis* as synonymous with *disaster*, but the word comes from the Greek for *decision*. Although

we cannot control the occurrence of many crises, we can decide how to cope with them. Even the most unfortunate crises may have potential for positive as well as negative effects (Patterson 2002b). In fact, for couples and families that have the "resources to cope effectively...

> Stress provides an opportunity to exercise those abilities and draw upon those resources. Doing so may well have lasting positive benefits for couples [and families] in terms of increased self-efficacy, trust, and confidence that the relationship can weather new challenges in the future. (Karney 2015)

Whether a family emerges from a crisis with a greater capacity for supportive family interaction depends at least partly on how family members choose to define

It's important to remember that not all stressors are unhappy ones. Happy events, such as moving into a new house, can also be family stressors.

iStock.com/SDI Productions

the crisis (e.g., Manning, Wainwright, and Bennett 2011). A major theme of this text is that, given the opportunities and limitations posed by society, people create their families and relationships based on the choices they make. However, although they have options and choices, family members do not have absolute control over their lives. Many family troubles are really the results of public issues. The serious family disorganization that results from incarceration or poverty is as much a social as a private problem (Trask et al. 2005). Moreover, most American families have some handicaps in

meeting crises creatively, partly because the majority of American families are under a high level of stress at all times. Furthermore, many family crises are more difficult to bear when communities lack adequate resources to help families meet them (Byrnes and Miller 2012; Pitts 2008).

One response to this situation is for individual family members to learn more about the human brain's neuroplasticity, then practice physically and psychologically calming behaviors. There are many resources available in bookstores and on the net in this regard.

A second response might be for couples and family members to touch base regularly during the course of a crisis to check in with one another about what went well, what specifically is appreciated in each family member, what each might do better, and who's to be in charge of each next step (Bagley 2019).

A third response is to engage in community activism (Bryan 2010). For example, one couple, frustrated by the lack of organized community support available to them and their autistic child, founded Autism Speaks. A project of Autism Speaks is developing a central database of children with autism that will provide standardized medical records that researchers need to conduct effective research (Roth and Barson 2010; Wright 2005). When families act collectively toward the goal of obtaining needed resources for effectively meeting the demands placed on them, family adjustment can be expected to improve overall.

Summary

- Throughout the course of family living, all families are faced with demands, transitions, and stress.
- Family stress is a state of tension that arises when demands test, or tax, a family's resources.
- Family systems are comprised of individuals, and so how constructively individuals address their personal stresses is important in meeting family stress creatively.
- A sharper jolt to a family than more ordinary family stress, a family crisis encompasses three interrelated factors: (1) family change, (2) a turning point with the potential for positive or negative effects or both, and (3) a time of relative instability.

- Demands, or stressors, are of various types and have varied characteristics. Generally, stressors that are expected, brief, and improving are less difficult to cope with.
- The predictable changes of individuals and families—parenthood, midlife transitions, postparenthood, retirement, and loss of a spouse—are all family transitions that may be viewed as stressors.
- A common pattern can be traced in families that are experiencing family crisis. Three distinct phases can be identified: (1) the stressor event that causes the

crisis, (2) the period of disorganization that follows, and (3) the reorganizing or recovery phase after the family reaches a low point.

- The eventual level of reorganization a family reaches depends on a number of factors, including the type of stressor, the degree of stress it imposes, whether it is accompanied by other stressors, the family's appraisal or definition of the crisis situation, and the family's available resources.

- Meeting crises creatively means resuming daily functioning at or above the level that existed before the crisis.

- Several factors can help families meet family stress or crises more creatively: a positive outlook, spiritual values, the presence of support groups, high self-esteem, open and supportive communication within the family, adaptability, counseling, and the presence of a kin network.

Questions for Review and Reflection

1. Compare the concepts *family stress* and *family crisis*, giving examples and explaining how a family crisis differs from family stress.

2. Differentiate among the types of stressors. How are these single events different from stressor overload? How might economic recession cause stressor overload?

3. Discuss issues addressed in other chapters of this text (e.g., work–family issues, parenting, separation, divorce, and remarriage) in terms of the ABC-X model of family crisis.

4. What factors help some families recover from crisis while others remain in the disorganization phase?

5. **Policy Question.** In your opinion, what more, if anything, could/should government do to help families experiencing stress? Families experiencing crisis?

Key Terms

ABC-X model   351
boundary ambiguity   342
community-based resources   357
courtesy stigma   347
family crisis   337
family stress   337

family transitions   339
fictive kin   356
nadir (of family disorganization)   349
period of family disorganization   349

reframing   352
resilient families   348
stress   336
stressors   341
stressor overload (pileup)   348
vulnerable families   348

14

DIVORCE AND RELATIONSHIP DISSOLUTION

Learning Objectives

1   Describe historical trends in divorce and relationship dissolution.

2   Describe current trends in divorce and relationship dissolution.

3   Identify economic, sociocultural, legal, and demographic factors associated with divorce and relationship dissolution.

4   Contrast the economic, social, and emotional consequences of divorce for women, men, and children.

5   Discuss the process by which couples decide to divorce and decide on custody arrangements for their children.

6   Describe the rules for successful co-parenting.

7   Explain ways to reduce the negative effects of divorce on adults, children, and families.

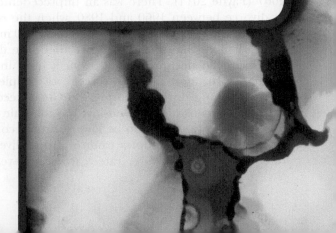

The statistic is well known: Roughly half of all first marriages in the United States will end in divorce (Kennedy and Ruggles 2014). Although divorce is a common experience in this country, a married couple's probability of divorce depends on a host of factors, including their education, income, race and ethnicity, and even region of the country where they live. In this chapter, we'll examine the characteristics of people who divorce, factors that affect people's decisions to divorce, and the experience of divorce itself. Divorce and union instability are associated with worse social, emotional, and physical health for parents, children, and can take a toll on relationships with extended family members and friends. We identify conditions that can mitigate these negative effects and discuss various policies and programs designed to improve divorce-related outcomes.

Although the research on relationship dissolution has been conducted primarily on the dissolution of marriages, many of the dynamics apply to breakups of committed nonmarital relationships. The dissolution of nonmarital romantic relationships is discussed in Chapters 5 and 6. The rate of relationship dissolution for dating and cohabiting couples is considerably higher than that for marrieds. Divorce research, which has for the most part focused on heterosexual couples, may also apply to some extent to LGBTQ+ couples, although much less is known about their process of breaking up, especially given that same-sex couples have only recently been allowed to legally marry (and therefore divorce). But we'll talk about what is known so far. We'll begin by looking at divorce rates in the United States, which are among the highest in the world.

TODAY'S DIVORCE RATE

Even though divorce was much less common in past centuries than today, divorce is not new. The divorce rate started its upward swing in the late nineteenth century (Preston and McDonald 1979). The frequency of divorce increased throughout most of the twentieth century, as Figure 14.1 shows, with dips and upswings surrounding historical events such as the Great Depression, major wars, and the Great Recession beginning in 2007 (Payne 2014). There was an unprecedented rise in divorce between 1960 and 1980, when the **refined divorce rate** (the number of divorces per 1,000 married women) more than doubled. Since then, the divorce rate has been declining, aside from a brief increase around 2010, most likely as a result of couples putting off getting divorced during the Great Recession. Divorce is expensive and is sensitive to economic conditions; divorce rates tend to be lower during economic downturns because couples can't afford to live separately. In 2018, the divorce rate reached a 40-year low

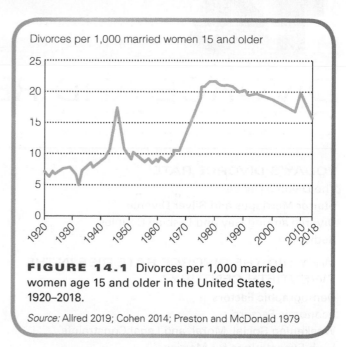

FIGURE 14.1 Divorces per 1,000 married women age 15 and older in the United States, 1920–2018.

Source: Allred 2019; Cohen 2014; Preston and McDonald 1979

(Allred 2019b). Still, for every two marriages that takes place, there is one divorce (Schweizer 2019).

Another way to look at divorce trends is to observe the **crude divorce rate**, the number of divorces per 1,000 population (see Figure 14.2). This rate includes portions of the population—children and the unmarried—who are not at risk for divorce. Despite its limitations, the crude divorce rate is sometimes used for comparisons over time because these data are often the only long-term annual data available. The crude divorce rate has declined almost 44 percent since its peak in 1980, and, like the refined divorce rate, is at its lowest point since the divorce rate began its ascent in the 1960s (National Center for Health Statistics 2018).

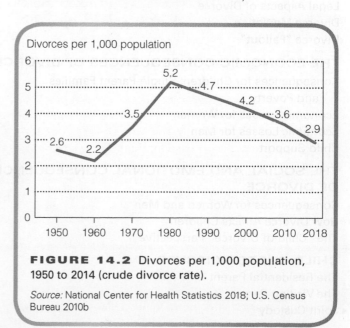

FIGURE 14.2 Divorces per 1,000 population, 1950 to 2014 (crude divorce rate).

Source: National Center for Health Statistics 2018; U.S. Census Bureau 2010b

The question is, will the divorce rate drop even further? To know that, it's important to understand the reasons for the recent decline. Some say it's due to Millennials, a huge generation of over 75 million Americans born between 1981 and 1996, "tying the knot" at older ages, after they have finished school and established their careers—all factors associated with greater marital stability. Along with that, now that cohabitation is an option, people who do marry are primarily those who are best equipped for the stresses and strains of marriage, both financially and emotionally. Family demographer Susan Brown says this, "The change among young people is particularly striking. The characteristics of young married couples today signal a sustained decline in divorce rates in the coming years" (Steverman 2018).

The Divorce Divide

Despite the 50 percent divorce rate, the risk of divorce varies greatly by social class, especially as measured by education. For example, in the last 25 years the crude divorce rate declined among college graduates but remained stable for men and women with less education (Amato 2010; Elliot and Simmons 2011). This holds true for African Americans and whites (Kim 2012). In particular, completing a college degree can be seen as an "insurance policy" against divorce. Whereas only 27 percent of college graduates have ever experienced a divorce, that proportion is almost double for those who did not complete high school. Men and women who did not have a college degree had a similar risk of divorce regardless of whether they had only a high school education or had some college (43 percent and 48 percent, respectively) (Aughinbaugh, Robles, and Sun 2013). This has produced what sociologist Steven Martin calls the **divorce divide**, the large disparity in divorce rates between those with and without a college degree (Ono 2009; Schweizer 2019). This divide is similar to the *marriage gap* discussed in Chapter 7, which is disparity in marriage rates between higher income and lower income people and, along with that, those with and without college degrees (Schneider, Harknett and Stimpson 2018). Indeed, divorce rates also vary by income. Compared with people who earn under $25,000 annually, people who earn more than $50,000 have a 30 percent lower risk of divorce (National Marriage Project and the Institute for American Values 2012).

Starter Marriages and Silver Divorces

Most divorces occur relatively early in marriage—a large percentage end within five years, do not involve children, and are followed shortly by remarriage— situations that led journalist Pamela Paul to coin the term **starter marriage** (Paul 2002). A *starter marriage* is a first marriage that ends in divorce within the first few years, typically before the couple has children. In her interviews, young divorced men and women listed the desire to move out of their parents' house, the desire to have a fairytale wedding, and the inevitability of marriage after a long period of dating as reasons why they got married (and why their marriage did not last).

That said, the median length of a first marriage that ends in divorce is about five years (Hoy and London 2017). Marriages lasting weeks or months (as opposed to years) are rare. For example, in Iowa there were only seventy-four marriages from 2000 to 2009 that lasted seventy-two or fewer days; the average length of a marriage (including remarriages) that ended in divorce during that timeframe was eleven years (Kilen 2011).

Meanwhile, the proportion of divorces among older couples and those in long-term marriages has increased. This came as a surprise to many demographers, who refer to this phenomenon as a **silver**, or **gray, divorce** (Brown and Lin 2012). This trend is particularly noticeable for the Baby Boom generation, those born between 1946 and 1964. Data from the American Community Survey of the U.S. Census indicates that between 1990 and 2017, whereas the divorce rate declined for people under age 45, it increased between those 45 and older (Allred 2019a; Cohen 2019). Moreover, the divorce rate among adults aged 55 to 64 doubled between 1990 and 2015 and the divorce rate tripled during that period for those 65 and older (Allred 2019c; Wu 2017). A college degree does not appear to have the protective effect observed among younger couples against later-life divorce (Brown, Lin, and Payne 2014a). Maybe all the baby boomers who wanted to divorce already have—it appears that the surge in gray divorce may be abating (Cohen 2019).

What explains the jump in divorce among baby boomers? First, the baby boom generation is one of the largest in history, with the largest number of marriages and divorces. Second, the high rate of remarriage among this age group also contributes to the high rate of divorce. The divorce rate of those aged 50 and older is more than two times higher in remarriages than first marriages (Brown and Lin 2012). The authors of the study stated that the "complex marital biographies" of baby boomers play a role in increasing the probability of divorce later in life. They identify several life transitions that may cause older adults to "take stock," such as an "empty nest" (adult children leaving home) and impending retirement (Brown 2013). Although married couples have been making such life transitions for decades, divorce was heavily stigmatized in the past and therefore was not considered an option, which is less the case today. Longer life expectancies among more recent generations of older couples mean many more years enduring an unhappy marriage unless it is dissolved. Another difference between previous and

Divorce has declined in recent years for all couples except for those over age 50, who've seen an increase in divorce rates.

current generations of older couples is society's increased focus on individual self-fulfillment (Thomas 2012). Like all divorce, silver divorce can have negative financial consequences, especially for women, who, along with never-married women, often live in poverty in their later years (Lin, Brown, and Hammersmith 2015). The risk of poverty is highest among what Brown and Lin refer to as "divorce careerists," men and women who have experienced a divorce before age 50 and after age 50 (10 percent of all those who have ever divorced), because they have accrued fewer assets (Spangler et al. 2016).

Divorce among Gay and Lesbian Couples

Like heterosexual couples, same-sex couples sometimes break up. With the legalization of same-sex marriage across the entire United States in 2015, it is likely that more gay and lesbian couples will choose to marry. With more same-sex couples marrying, we can expect more same-sex couples to divorce. In fact, divorce is one of the benefits of same-sex marriage in that it provides for a formal, clearly recognized way to address couple breakup issues regarding property, child custody, and so on (Allen 2007). In the past, there were many legal barriers to divorce for gay and lesbian married couples if, for example, the couple moved to a state that didn't recognize their marriage (Van Eeden-Moorefield et al. 2011). Although this barrier has been removed in the United States, same-sex couples who marry in the United States may be unable to divorce if one or both spouses does not meet the residency requirements (should one or both not be a U.S. citizen) or if the couple moves to

a country in which same-sex marriages are not recognized (Butler and Kirkby 2013).

Because the legalization of same-sex marriage is so recent, research on divorce among gay and lesbian couples is just getting started, and it is too early to tell how divorce among same-sex couples will affect overall divorce rates. Moreover, trends in divorce among the "B," "T," and "Q+" part of the LGBTQ+ community are largely unknown. Many LGBTQ+ people marry, and always have, it's just that they often remained hidden in marriages where they could not express their true gender and sexual identities Although research is just beginning on these groups, one study indicates that marriages in which one spouse identifies as bisexual are shorter in duration than both opposite-sex and same-sex marriages (Hoy and London 2017). Another found that among transmen who were married or cohabiting before their transition, half of their relationships dissolved either during or after their transition (Meier et al. 2013).

The number or percentage of same-sex marriages that will end in divorce is difficult to estimate, with research hampered by lack of data and small sample sizes with only a few studies based on population-level data (Manning and Joyner 2019). There may be a temporary rise in same-sex divorce among unhappy couples who were legally unable to divorce previously because of the legal hurdles cited earlier (Bellware 2015). The results of research on same-sex divorce is inconsistent (Reczek 2020), but one study of same-sex partners (and those who exhibit sexual behavior with same-sex partners) shows that they have marriages that are shorter in duration than opposite-sex marriages by about a year-and-a-half (Hoy and London 2017). Research varies regarding the risk of divorce for female same-sex couples versus male same-sex couples (Raley and Sweeney 2020; Reczek 2020). Clearly, more research is needed.

Redivorce

In summing up the statistics, we must note that a high divorce rate does not mean that Americans have given up on marriage. Culturally, Americans revere marriage and have higher marriage (and remarriage) rates most other Western countries, especially those in Northern Europe, where cohabitation is more common. Although marriage has declined among lower income and less educated people (who opt for cohabitation), marriage as an institution has maintained its privileged status in American Society (Cherlin 2020). Even those

who have been married (and divorced) twice often continue their quest for a happy marriage. Among ever-married men and women, roughly 13 percent have married two or more times (Lewis and Kreider 2015). But a consequence of remarriages—which have higher divorce rates than first marriages—is *redivorce* or experiencing multiple divorces. Data from the 2009 American Community Survey indicated that one-third of men and women who divorced in the preceding twelve months had been previously divorced; 26 percent had one prior divorce and 9 percent had two or more prior divorces (Elliot and Simmons 2011). Although not unheard of, divorced couples remarrying each other a second time is extremely rare (Kale 2019). As discussed in Kale (2019), according to Nancy Kalish of California State University, many of these couples had divorced as a result of "situational reasons" and stressors such as long work hours, issues that have since disappeared. The article goes on to provide an excerpt from an interview with a man named Damian, who remarried his ex-wife Amanda:

> "The five years we'd spent apart, I'd learned to become a better person. With maturity comes patience and tolerance. We probably understand and appreciate each other's needs a lot more now." Chris is also self-critical: "I wasn't really a nice person, the first time around. And back then, Dee was very quiet and passive. The second time around, I'd grown up and got a bit softer, and Dee had got more assertive, and confident with dealing with me. We just blended straight away. (Kale 2019)

In a context that combines a high though declining divorce rate with a positive view of marriage, why get divorced? The following section identifies economic, sociocultural, legal, interpersonal, and demographic factors associated with divorce.

Demographic Factors

Certain demographic factors have been shown to be important risk factors for divorce. Many of these same factors are associated with the breakup of cohabiting unions. They include the following:

- Remarriage: Remarried couples are more likely than first-married couples to divorce, especially if there are stepchildren in the home (Aughinbaugh, Robles, and Sun 2013; DeLongis and Zeicker 2017).

- Young age at first marriage, especially marrying as a teenager (Copen et al. 2012; Kuperberg 2014): Divorce also generally declines with age, although the divorce rate has risen recently among adults 45 and older (Allred 2019a).

- Slightly higher divorce rates among women compared to men: Whereas 42 percent of women have ever divorced by age 46, only 36 percent of men of that same age have (Aughinbaugh, Robles, and Sun 2013).

- Heterogamous marriage: Marrying someone of a different race, ethnicity, or religion is associated with a higher risk of divorce, and, interestingly, when the wife is older than the husband (Fogle 2015; Fu and Wolfinger 2011; Wright, Rosato, and O'Reilly 2017).

- Cohabitation: As shown in Figure 14.3, about two-thirds of Gen-Xers (born between 1965 and 1980) and Millennials agreed that living together before marriage may help prevent divorce (Eickmeyer 2015). Not surprisingly, cohabitors were more likely than non-cohabitors to feel this way. But are their feelings correct? It makes sense that living together before marriage should help couples decide

WHY DID THE DIVORCE RATE RISE IN THE TWENTIETH CENTURY?

Various factors can bind married couples together: economic interdependence between spouses; spouses' having similar social and demographic characteristics; legal, social, and moral constraints; and the spouses' relationship itself. Yet, the binding strength of these factors has begun to unravel.

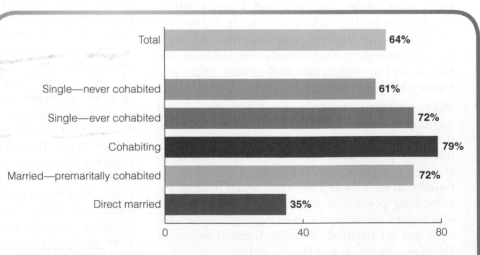

FIGURE 14.3 Percent of GenX-ers and Millennials age 15 to 44 who believe living together may help prevent divorce, by relationship status.

Source: Eickmeyer 2015 (2011–2013 National Survey of Family Growth).

whether they're compatible for marriage leading to higher marital stability. But this is not the case. In fact, couples who have cohabited have higher divorce rates (Rosenfeld and Roesler 2019). As discussed in Chapter 5, one reason has to do with selectivity of less divorce-adverse people into cohabitation. There is also the possibility that one's marital search is "truncated" as a result of living with cohabiting partner with no consideration of long-term compatibility, and then marry out of inertia or a premarital pregnancy (Stanley, Rhoades, and Markman 2006). Additionally, couples who cohabited before marriage tend to do so at an early age, which is associated with higher rates of dissolution (Kuperberg 2014).

- Premarital sex, premarital pregnancy, and premarital childbearing: These factors increase the risk of divorce in a subsequent marriage (Bellido, Molina, Solaz, and Stancanelli 2016; Child Trends 2015d).

- Having no or older children, compared to young children, increases divorce risk (Bellido et al. 2016; Stykes 2015b). Whereas one study found that having daughters versus sons does not increase a couple's risk of divorce (Hamoudi and Nobles 2014), other research indicates higher divorce rates among couples with two daughters versus two sons (Fogle 2015).

- Having parents and grandparents who divorced increases one's own likelihood of divorce (Amato 2010).

- Race and ethnicity: Blacks, Hispanics, and Native Americans have higher divorce rates compared to whites. Among Hispanics, immigration status is also important. Hispanics born outside the United States have lower divorce rates than those who are native-born (Raley, Sweeney, and Wondra 2015). The reasons for the racial disparities in divorce are complex, stemming from a mix of individual, family, and community-level forces, cultural norms and beliefs, and low education and income (Barr and Simons 2018; Raley et al. 2015).

- Military service: The stresses and demands placed on military couples may lead to higher divorce rates, though one study found that military marriages are not more prone to divorce than civilian marriages (Karney, Loughran, and Pollard 2012). Yet, another found that military deployments, which typically involve long periods of separation, were associated with higher divorce rates, even among couples who were not yet married when deployment occurred (Negrusa, Negrusa, and Hosek, 2016).

- Pre-wedding jitters: When the wife, but not the husband, has "cold feet" before the wedding (Fogle 2015).

Economic Factors

Traditionally, as we've seen, the family was a self-sufficient productive unit, such as in the case of the family farm. Survival was far more difficult outside of families, so members remained economically bound to one another. But today, because family members no longer need one another for basic necessities, they are freer to divorce than they once were. This is especially true for women, who today have access to an education, are legally allowed to work, and who can support themselves financially outside of marriage (and their parents' home). Nevertheless, families are still interdependent economically. Even though marriage "has become less economically necessary . . . it remains economically advantageous in most cases" (Wilcox and Marquardt 2009, p. 42). As long as marriage continues to offer practical benefits, economic interdependence will help hold marriages together. The economic practicality of marriages varies according to several conditions.

Overall, the higher a couple's social class as defined by education, employment, income, and wealth (including home ownership), the less likely they are to divorce (Raley and Sweeney 2020). In particular, a husband's full-time employment lowers the risk of divorce (Killewald 2016). On the other hand, a loss of income has been found to increase the likelihood of divorce, especially when it is the male's income (Wilcox and Marquardt 2009). Although the Great Recession of 2007 resulted in fewer divorces, it also resulted in many home foreclosures, which were found to be associated with an increased risk of divorce (Cohen 2014).

What about women's earnings? The answer is complicated. The upward trend of divorce and the upward trend of women in the labor force have accompanied each other historically. A woman's earnings might give an unhappily married woman the economic power, increased independence, and self-confidence that help her decide on divorce. For instance, in *unhappy* marriages women's employment increases divorce risk (Sayer et al. 2011; Schoen et al. 2002), and Amato et al. (2007) found wives' work hours increased their spouses' perceptions of marital problems. Another study found that wives who were dependent on their husband for health insurance had lower divorce rates (Sohn 2015). However, this **independence effect** may hold for white women only (Teachman 2010) and for women whose marriages began before 1975, when married women's employment was not yet widespread (Killewald 2016). Similarly, wives with higher relative earnings and who earned more than their husbands were more likely to divorce for couples married in the 1960s and 1970s but not in the 1990s (Schwartz and Gonalons-Pons 2016). It is thought that any impact of wives' employment and earnings has weakened considerably.

Another perspective is that a wife's education and earnings may actually help to hold the marriage together by counteracting the negative effects of economic insecurity and related economic stressors (Aughinbaugh et al. 2013; Schoen, Rogers, and Amato 2006). Whether low earnings by itself increases divorce risk is unclear, as research shows that the couple's financial well-being is less predictive of divorce than financial disagreements between spouses (Dew, Britt, and Huston 2012). Rather than reducing stress and increasing marital quality, Boertien and Härkönen (2018) argue that education, income, and earnings present a barrier to divorce, in the sense that leaving the marriage would threaten the couple's lifestyle (see "'What's Stopping Me?' Barriers to Divorce").

Weakening Social, Moral, and Legal Constraints

There are far fewer social, moral, and legal constraints on divorce than in the past. Today, 77 percent of Americans see divorce as morally acceptable compared to only 62 in 2009 (Gallup 2019). Unmarried people and those who have cohabited are more accepting of divorce than marrieds. Attitudes about divorce are similar among men and women. Younger generations are more accepting of divorce. Gen-Xers, Millennials, and Gen Z tend to have similar views on social issues such as divorce (Eickmeyer 2015; Parker et al. 2019).

These changes in views have become widespread. Nationally, conservative Protestant and Catholic churches, traditionally strongly opposed to divorce, have become more tolerant of divorced people and often explicitly welcome them into the congregation (Cherlin 2009a). Rural and nonmetropolitan areas, known for their more conservative family values and religious beliefs, have traditionally had lower divorce rates. Yet now, divorce rates in those areas have grown to such an extent that the rural–urban distinction in divorce has completely disappeared (Tavernise and Gebeloff 2011). In Southern states, sometimes referred to as the "Bible Belt" and known for their traditional cultural values, many young adults feel pressure to marry young (Hetter 2011). It may therefore come as a surprise that divorce rates are higher there than in the more liberal Northeast (Allred 2019b; Glass and Levchak 2014).

To say that societal constraints against divorce no longer exist would be an overstatement. Religious views and differences in culture still affect people's attitudes toward divorce. For example, Hispanics have more negative attitudes toward divorce, despite their higher divorce rates (Ellison, Wolfinger, and Ramos-Wada 2012). Most states have waiting periods of various lengths before a divorce can be granted, and it may take many more months for divorcing couples to come to an agreement about how to divide assets (and debts), whether financial support will be paid, and who will have child custody. Despite greater societal acceptance of divorce, social pressure against divorce among affluent groups has recently been growing, leading sociologist Andrew Cherlin to remark, "The condemnation of divorce is also coming from the group that is most confident it can make its marriages succeed, and that allows them to be dismissive of divorce" (cited in Paul 2011).

With greater societal acceptance of divorce came lessening legal restrictions on divorce, such as no-fault divorce laws, which were widely adopted throughout the United States in the 1970s and 1980s. This revision of divorce law was intended to reduce the hostility of the partners and to permit an individual to end a failed marriage readily. Before the 1970s, the *fault system* predominated. A fault divorce required a legal determination that one party was guilty and the other innocent. Parties seeking divorce had to prove that they had "grounds" for divorce, such as the spouse's adultery, mental cruelty, or desertion. Obtaining a divorce might require falsifying these facts. The one judged guilty rarely received custody of the children, and the judgment largely influenced property settlement and alimony awards, as well as the opinions of friends and family (Coontz 2010a; Stevenson and Wolfers 2007). Such a protracted legal battle of adversaries increased hostility and diminished chances for a civil post-divorce relationship and successful co-parenting. With no-fault divorce, a marriage became legally dissolvable when one or both partners declared it to be "irretrievably broken" or characterized by "irreconcilable differences." No-fault divorce is sometimes termed **unilateral divorce** because one partner can secure the divorce even if the other wants to continue the marriage. Researchers have studied whether no-fault divorce laws lead to higher divorce rates. This might have been the case in the early days of no-fault divorce, but today divorce rates have fallen despite the relative ease of getting a divorce (Kneip, Bauer, and Reinhold 2014; Wolfers 2006). In 2010, New York became the last state to allow no-fault divorce, making us "no-fault nation" (Coontz 2014).

High Expectations for Marriage

As discussed in Chapter 7, the focus of institutional marriages is on meeting obligations to one's spouse, family, and community—love, especially romantic love, was not required or even necessarily expected. In companionate marriage, couples are expected to be "in love." But love, especially romantic love, tends to wane over time, making it a rather risky basis for marriage. The subsequent move from companionate to individualistic marriage in the mid-twentieth century put added pressure on marriage to meet each spouse's need for love but also personal happiness and life satisfaction (Cherlin 2009b; Demo and Fine 2010).

If pressure from extended families, communities, churches, and the legal system can no longer be counted on to preserve marital stability, the quality of the relationship becomes central to the survival of a marriage (Karney and Bradbury 2020). Research has found that couples whose expectations are more practical are more satisfied with their marriages than those who expect completely loving and expressive relationships (Becker 2012). Although many couples part for serious and specific reasons, others may do so because of unrealized expectations and general discontent. Sociologist Andrew Cherlin (2009a) suggests that the United States' high divorce rates may be in part a reflection of a broad sense of "restlessness" unique to American culture.

Interpersonal Dynamics

Along with larger social forces, researchers have examined aspects of marital relationships that increase divorce risk. In general, relationship dissatisfaction, especially on the part of the female, is positively associated with the risk of divorce (Røsand et al. 2014). Marital complaints associated with divorce include a partner's poor communication, handling of money, unwillingness to work on the relationship, infidelity, alcohol or drug abuse, jealousy, moodiness, violence, low levels of trust, and, much less often, homosexuality, as well as perceived incompatibility and growing apart (Crouch and Dickes 2016; Fincham and Beach 2010a; Leonard, Smith, and Homish 2014; Williamson et al. 2015).

Counselors suggest that some common complaints—about money, sex, and in-laws, for example—are really arenas for acting out deeper conflicts, such as who will be the more powerful partner, how much autonomy each partner should have, and how emotions are expressed (Amato 2010; Amato and Hohmann-Marriot 2007). Amato cautions that "not all couples display a pattern of relationship dysfunction prior to divorce" (Amato 2010, p. 653), and Karney and Bradbury (2020) point out that poor communication and conflict are not consistent predictors of marital distress and divorce. Moreover, even couples who are moderately happy and have low levels of conflict can be at risk of divorce in the presence of other risk factors. Yet, many social trends in recent years are helping to decrease the divorce rate. Some of these are discussed in the following section.

WHY IS THE DIVORCE RATE DROPPING?

The divorce rate has stabilized since its height in the 1980s, but a decline is currently being observed (Allred 2019b). Social scientists propose several reasons for this development. First, fewer people are marrying at younger ages—the median age at first marriage is 28

for women and 30 for men (U.S. Census Bureau 2018). Those who wait are likely to make better choices and to have the maturity and commitment to work through problems. In addition, married people are more likely to be college educated than in the past and may be better able to communicate and work out any issues. Egalitarian marriages and stabilization of gender role expectations has reduced conflict over housework and childcare (Cherlin 2017). The standard of living has improved over the past few decades for two-earner families with good jobs, a situation that leads to less tension at home and lower probability of divorce (Wilcox and Marquardt 2009). At the same time, the bad economy and real estate market and the high costs of obtaining a legal divorce also keep people married (Bilefsky 2012). Couples today are more inclined to seek marriage or family counseling, which may prevent them from splitting up. Children of divorced parents may possess an increased determination to make their own marriages work (Crary 2007a; Teachman, Tedrow, and Hall 2006).

THINKING ABOUT DIVORCE: WEIGHING THE ALTERNATIVES

Not everyone who thinks about divorce actually gets one. Because divorce is now an available option, spouses may compare the benefits of their current union to the projected consequences of not being married. One model of deciding about divorce derived from exchange theory (see Chapter 2) is **Levinger's model of divorce decisions**, proposed by social psychologist George Levinger. He posits that spouses assess their marriage in terms of the *barriers* to divorce, *alternatives* to marriage, and *rewards* of marriage (Levinger 1976).

"What's Stopping Me?" Barriers to Divorce

Children, religion, and lack of financial resources are common barriers to divorce (Knoester and Booth 2000; Previti and Amato 2003). Both mothers and fathers anticipated that "divorce would worsen their economic situation and their abilities to fulfill the responsibilities of being a parent" (Poortman and Seltzer 2007, p. 265). Most couples are dependent upon two incomes and a divorce would put a definite crimp in a couple's (and their children's) lifestyle, if not plunge them into poverty (Boertien and Härkönen 2018). Indeed, affection for their children and concern about the children's welfare after divorce discourages some parents from dissolving their marriage. This concern sometimes leads to delaying an intended divorce (Furstenberg and Kiernan 2001; Heaton 2002; Poortman and Seltzer 2007). Longer marriages are less likely to end in divorce

Deciding to divorce is difficult. Couples struggle with feelings about their past hopes, current unhappiness, and an uncertain future.

Image Source/Corbis

(Kulu 2014). One reason for this, in addition to the marital bond itself, is that common economic interests and friendship networks increase over time and help stabilize the marriage during times of tension (Brown, Orbuch, and Maharaj 2010).

"Would I Be Happier?" Alternatives to the Marriage

Alternatives, another of Levinger's concepts, were found to be the least important in decisions to divorce (Previti and Amato 2003). Yet, some married people may ask themselves whether they would be happier if they were to divorce. Some people may prefer to stay single after divorce, but many partners weigh their chances for a remarriage or a new relationship.

Leaving a bad marriage may have a positive outcome regardless of whether the individual remarries. A British study found people to be less happy one year after separation, but by one year after the divorce, both men and women were happier than they had been while married (Gardner and Oswald 2006). Nevertheless, marriage, if happy, can and often does provide emotional support, sexual gratification, companionship, and economic and practical benefits, including better health (Dahl, Hansen, and Vignes 2015; Karney and Bradbury 2020). The poor health of either partner may reduce the benefits of marriage. For instance, a serious illness (cancer, heart disease, lung disease, or stroke) on the part of the wife was found to raise a couple's risk of divorce (Karraker and Latham 2015).

"Can This Marriage Be Saved?" Rewards of the Current Marriage

Some partners respond to marital distress by trying to improve their relationship and by focusing on the rewards of their current marriage (the third component of Levinger's theory). Many studies have shown that marital quality declines over time, with a "typical honeymoon then years of blandness" (Aron, Norman, Aron, and Lewandowski 2002; p. 182). But is this really the case? A ten-year longitudinal study of newly married couples indicates that marital satisfaction stabilizes and may even increase in later years (Sullivan et al. 2010). Improvements in those marriages came about through the passage of time (children got older, jobs or other problems improved); because partners' efforts to work on problems, make changes, and communicate better were effective; or because individual partners made personal changes (travel, work, hobbies, or emotional disengagement) that enabled them to live relatively happily despite an unsatisfying marriage. And a recent review of longitudinal studies of marital satisfaction by psychologists Benjamin Karney and Thomas Bradbury (2020) suggests that "only a minority of couples actually experience high initial marital satisfaction that declines steadily and significantly during the course of their marriage. Instead, most couples who start their marriage happy stay happy for long periods of time" (p. 102).

One must decide whether divorce represents a healthy step away from an unhappy relationship that cannot be improved. A marriage counselor may help partners become aware of the consequences of divorce so that they can make informed decisions. If a couple decides to divorce, they may engage in what has been referred to as "conscious uncoupling," which "takes place when a couple believes that they have both tried to work through problems in their relationship or marriage in a way that causes the least possible damage to themselves, their integrity, and their children" (Moriarty 2014). Katherine Woodward Thomas, the originator of the concept, explains in her book *Conscious Uncoupling: The Five Steps to Living Happily* Even *After* (2015) that

When parents consider divorce, they often think about the potential impact on their children—and that is a barrier to divorce.

process involves "harnessing negative emotions" and partners each taking responsibility for their part in the separation. She acknowledges the process isn't easy. According to Thomas: "[M]ost of the people I work with are dealing with deep betrayal, horrific losses and damage that feels like they're in danger of dimming down for the rest of their lives because of how shattering the pain is. The goal is to learn from the experience and to go on and have healthier and happier relationships" (Saner 2018).

Other Solutions to Marital Distress

Not all unhappy couples divorce. The following sections present alternative solutions to marital distress.

Marital Separation Nearly 2 percent of adults in the United States reported their marital status as "separated" (U.S. Census Bureau 2019c). According to the 2009 American Community Survey, the average length of separation among men and women who eventually divorce is about ten months (Kreider and Ellis 2011b). The duration between separation and divorce for first marriages varies by race and ethnicity, with longer separations among Hispanics and blacks than among whites (Copen et al. 2012). These groups are also more likely than whites to dissolve their marriage through permanent marital separation rather than divorce (Bramlett and Mosher 2002). Husbands (but not wives) who become unemployed during the period of separation are less likely to divorce and are more likely to prolong the divorce process, leaving some unhappily married

couples "in limbo" (Tumin and Qian 2015).

Little research covers the relationship of the partners during the period between marital separation and divorce. Some marital partners who have separated do make efforts to reconcile. In one of the few studies of marital separation, Howard Wineberg (1996) used a sample of white women from the National Survey of Families and Households and found that 44 percent of the separated women attempted reconciliation. Half of the resumptions of marriage that followed took place within a month, suggesting that those separations may have been impulsive and soon regretted. Virtually no marriages were resumed after eight months of separation. Only one-third of the reconciliations "took"—that is, resulted in a continued marriage. The author cautions that "not all separated couples should be encouraged to reconcile because a reconciliation does not ensure a happy marriage or that the couple will be married for very long" (p. 308).

Stable Unhappy Marriages From time to time, researchers have taken up the question of what happens to couples who are distanced, unhappy, or in conflict if they don't divorce (Waite, Luo, and Lewin 2009). It is, however, surprising that "long-term low-quality marriage . . . has received relatively little attention" (Hawkins and Booth 2005, p. 451). In a unique study, Hawkins and Booth (2005) followed unhappy marriages for twelve years and compared people in unhappy marriages to divorced single and remarried individuals. "Divorced individuals who remarry have greater overall happiness, and those who divorce and remain unmarried have greater levels of life satisfaction, self-esteem, and overall health than unhappily married people. . . . We suggest that unhappily married people who dissolve low-quality marriages likely have greater odds of improving their well-being than those remaining in such unions" (p. 468). Yet, other research suggests that people who remain in an unsatisfying marriage are less depressed than those who leave their marriages (Waite, Luo, and Lewin 2009) and a study of 374 continuously married couples found that happy and unhappy couples had similar levels of health and emotional well-being (Margelisch, Schneewind, Violette, and Perrig-Chiello 2017). Some couples may prefer to stay with "the devil they know" even if their marriage is less than ideal. Then, too, unhappy couples may have religious or moral objections to divorce and don't consider divorce an option.

iStock.com/Netta Collection

GETTING THE DIVORCE

One of the scariest things about divorce is uncertainty and lack of knowledge about what the future may hold in terms of one's finances, living situation, and children. A much less mentioned stressor is the divorce process itself.

The "Black Box" of Divorce

Whereas we know a great deal about how to get married—buy a ring, propose, have an engagement party, plan the wedding, and so on—how to get divorced is a "black box." We know quite a bit about relationship dynamics before divorce (inputs) and after the divorce (outputs), but not about the inner workings of the divorce process itself. Hearing "I want a divorce" from one's spouse involves entering a world where no one wants to go and learning a whole new language of petitioners and respondents, of financial disclosures and waiting periods, of parenting plans and spousal support. Use caution before taking the advice of friends and relatives. What should one do when one's spouse wants a divorce? Consult an attorney who has expertise in divorce law.

Initiating a Divorce

Not all divorced people wanted to or were ready to end their marriage. It may have been their spouse's choice. Initiating and non-initiating partners tend to talk about their reasons in different terms. The initiator of the divorce typically invokes a "vocabulary of individual needs" while the non-initiating partner speaks in terms of "familial commitment" (Hopper 1993). Indeed, partners contemplating divorce must grapple with the contradictory pull of commitment to one's family and one's personal happiness and individual goals (Amato 2004; Cherlin 2009a).

Women more often initiate a divorce, as they are often the ones under-benefitting from the marriage (Coontz 2010b; Thomas 2012). It is not surprising that research shows that the degree of trauma a divorcing person suffers usually depends on whether that person or the spouse wanted the dissolution. The one "left" experiences a greater loss of control and has much mourning yet to do—the divorce-seeking spouse may have already worked through his or her sadness and distress (Amato 2000; Braver, Shapiro, and Goodman 2006). In a study of forty divorced men and women, wives were determined to be more attuned to emerging relationship problems than were husbands and were more likely to bring them to the husbands' attention (Williamson et al. 2015). As one wife stated, "It's something that had to happen, and it wasn't something that either one of us really controlled. It was just an awful situation that we had to get out of, and I recognized it and he didn't" (Hopper 2001; p. 438). The negative effects of a spouse initiating a separation (as opposed to oneself or jointly) include both physical and mental health challenges and affect both men and women (Hewitt and Turrell 2011). Even for those who actively choose to divorce, however, divorce and its aftermath may be unexpectedly painful.

Legal Aspects of Divorce

A legal divorce is the dissolution of the marriage by the state through a court order terminating the marriage. The principal purpose of the legal divorce is to dissolve the marriage contract so that emotionally divorced spouses can conduct economically separate lives and be free to remarry. Divorce is regulated by the states, and states have different law regarding waiting periods, counseling, and parenting classes. The Internet is full of materials on "do it yourself" divorce, which is less expensive but requires a high level of agreement of both parties. Couples contemplating divorce should be clear on the laws in their particular state.

Two aspects of the legal divorce make marital breakup painful. First, divorce, like death, creates the need to grieve. But the usual divorce in court is a rational, unceremonial exchange that takes only a few minutes. Divorcing individuals may feel frustrated by their lack of control over a process in which the lawyers are the principals.

A second aspect of the legal divorce that aggravates conflict and misery is the *adversary system*. Under our judicial system, lawyers advocate for their clients' interests only and are eager to get the most for their clients and protect their rights. Opposing attorneys are not trained to and ethically are not even supposed to balance the interests of the parties and strive for the outcome that promises the most mutual benefit.

A final note on the legal divorce is that, by definition, it applies only to marriage. There is no legal forum in which cohabitants, whether heterosexual or gay or lesbian, may obtain a divorce. Some couples may be cohabiting precisely to avoid the prospect of going to court should their relationship sour, especially couples who have been divorced once already. However, they are likely to find that the absence of a venue in which to resolve separation-related disputes in a standardized way is also a problem. Many cohabiting couples must go to small claims court to sort out belongings, assets, and debts (the legal side of living together is discussed in Chapter 6).

Divorce Mediation

Divorce mediation is an alternative, non-adversarial means of dispute resolution by which a couple, with the assistance of a mediator or mediators (frequently a lawyer/therapist team) who negotiate the settlement of

their custody, support, and property. In the process, they hope to learn a pattern of dealing with each other that will enable them to resolve future disputes. Mediation is either recommended or mandatory in all states for child-custody and visitation disputes before litigation can be commenced (Winestone 2015). Mediation has been increasing since the early 2000s (Olmer and Brown 2016).

Studies indicate that couples who utilize divorce mediation have less hostility and tension, feel more satisfied with the process and the results, and report better relationships with ex-spouses and children (Bailey and McCarty 2009; Holtzworth-Munroe, Applegate, and D'Onofrio 2009). Mediation can vary in quality, however, with agreements that are "clear, fair, comprehensive, and 'tailor-fit'" resulting in higher post-divorce well-being among adults (Baitar et al. 2012, p. 65). Having a mediation professional who displayed "problem-solving behaviors"—including (1) structuring the mediation process; (2) noticing relevant arrangement details; and (3) being sensitive to the emotional reactions of conflicting parties—was also associated with more positive outcomes (Baitar et al. 2012). Mediation allows parties to develop options, consider alternatives, and reach an agreement in an effort to minimize harm to the child. Yet, only a minority of disputes are successfully mediated by a neutral third party (Mnookin & Weisberg, 2014).

There are arguments for and against divorce mediation. Women's advocacy groups have claimed that mediation may be biased against females in that they may be less assertive in negotiations (Marlow and Sauber 2013). Others have argued that mediators take insufficient account of prior domestic violence (Comerford 2006; Freeman 2008). A woman is more likely to concede wealth over the risk of losing her child, even if the risk is very low. Mediation does not necessarily work better for same-sex couples, who are assumed to be "egalitarian," when vulnerabilities may exist (Ben-Asher 2017). On the other hand, a mediator trained to recognize power imbalances can help ensure women's voices are heard in negotiations (Mirzaie 2016). Whereas mediation can be useful for parents, the effect of mediation on reducing stress in children remains unclear (Kwiatkowski et al. 2014; Wong, Ma, and Xia 2019). Moreover, it does seem that "[m]ediation produces higher levels of compliance [with court decisions] and lower relitigation rates than litigation or attorney-negotiated settlement." It is also less costly and generally less time-consuming than litigation (Comerford 2006; Crary 2007b).

Divorce "Fallout"

Marriage is a public announcement to the community that two individuals have joined their lives. Marriage usually also joins extended families and friendship networks and simultaneously removes individuals from the world of dating and mate seeking. **Divorce fallout** refers to ruptures of relationships and changes in social networks that come about as a result of divorce. At the same time, divorce provides the opportunity for forming new ties. Divorce fallout also involves financial fallout or the economic consequences of divorce.

Kin No More? Researchers are increasingly recognizing the importance of kin relationships beyond the nuclear family. They provide emotional and instrumental support, such as childcare (Furstenberg 2020). When a couple divorces, relationships with family members can be "reinterpreted" in the following ways: (1) kin promotion, redefinition of a distant relative to a close relative; (2) kin exchange, reclassifying a family relationship (such as a sibling relationship becoming more like a parent–child relationship); (3) nonkin conversion, turning friends and colleagues into family members; and (4) kin retention, in which ex-in-laws are kept as family members (Allen, Blieszner, and Roberto 2011). What to do about "ex" family members is a vexing problem, especially given that they remain the family members of one's children. Another problem is how to incorporate new family members.

The Divorce-Extended Family A surprising phenomenon encountered by those who research divorced families is the expansion of the kinship system that is produced by links between ex-spouses and their new spouses and significant others and beyond to their extended kin, including grandparents, aunts, uncles, cousins, and friends. This kinship system is referred to as the *divorce-extended family*. Ahrons (1994) believes it is important for parents to "accept that your child's family will expand to include nonbiological kin" (1994, p. 252) and that the "relationships formed when a parent remarries also tend to be more rewarding for the children as their kinship system expands rather than contracts" (Ahrons 2007, p. 64).

Some post-divorce extended families find they can enjoy and benefit from connections to one another even when they have a history of conflict and old tensions sometimes resurface (Kleinfield 2003). "Letting go of old resentments, whether between ex-spouses, parent and child, or stepparent and child, is the most challenging part of" creating divorce-extended families, but doing so presents additional resources, friendship, and love to individuals and families (Bernstein 2007, p. 74).

Grandparenting after Divorce Given the frequency of divorce, most grandparents find themselves touched by it. Grandparents fear losing touch with grandchildren, and this does happen. In response, all fifty states have passed grandparent visitation laws. However, a Supreme Court decision struck down Washington's law (*Troxel v. Granville* 2000) because it was considered to interfere with parents' rights to determine how their children are to be raised. The status of other states' laws is uncertain;

These eight grandparents, all connected to the young basketball player by marriage, divorce, and remarriage, come together to cheer him on and enjoy the game.

some courts have allowed grandparents visitation rights in certain circumstances (Stewart and Timothy 2020).

In favorable circumstances, grandparents and grandchildren become closer as adult children turn to grandparents for help and emotional support (Henderson et al. 2009; Ruiz and Silverstein 2007; Timonen, Doyle, and O'Dwyer 2011). The strength of the grandparent–grandchild bond is associated with better subjective well-being in grandchildren. This is especially true for children with divorce parents (Jappens and Van Bavel 2019). Researchers and therapists have concluded that

> these relationships work best when family members do not take sides in the divorce and make their primary commitment to the children. Grandparents can play a particular role, especially if their marriages are intact: symbolic generational continuity and living proof to children that relationships can be lasting, reliable, and dependable. Grandparents also convey a sense of tradition and a special commitment to the young. . . . Their encouragement, friendship, and affection has special meaning for children of divorce; it specifically counteracts the children's sense that all relationships are unhappy and transient. (Wallerstein and Blakeslee 1989, p. 111; see also Henderson et al. 2009)

Indeed, children who were close to their grandparents had fewer problems adjusting to their parents' divorce

(Connidis 2009; Ruiz and Silverstein 2007). Of course, more and more grandparents' own marriages are not intact today. Nevertheless, one can assume that even a loving, divorced grandparent could add to the support system of a grandchild of divorce.

As the "kin-keepers" of the family, women are more likely than men to retain in-law relationships after divorce, particularly if they had been in close contact before the divorce and if the in-law approves of the divorce (Connidis 2009). Relationships between former in-laws are more likely to continue when children are involved, but conflict between parents can erode these relationships and lead to a discontinuation of grandparent-grandchild contact (Westphal, Poortman, and Van der Lippe 2015). After divorce, grandchildren are more likely to remain close to their maternal than to their paternal grandparents because children tend to reside with their mothers and mothers typically grow closer to and rely more on their own parents after divorce (Connidis 2009; Henderson et al. 2009; Jappens and Van Bavel 2015). A good relationship between a mother and her ex-mother-in-law is associated with closer grandparent–grandchild relationships (Attar-Schwartz and Fuller-Thomson 2017).

Divorce affects the extended family as well as the nuclear one. In some families, grandparents may lose touch with grandchildren, whereas in others, they may become more central figures of support and stability.

Friends No More? A change in marital status is likely to mean changes in one's community of friends. Divorced people may feel uncomfortable with their friends who are still married because activities are done in pairs; the newly single person may also feel awkward. Couple friends may fear becoming involved in a conflict over allegiances, and they may experience their own sense of loss. Moreover, if married friends have some ambivalence about their own marriages, a divorce in their social circle may cause them to feel anxious and uncomfortable. A common outcome is a mutual withdrawal.

Like many newly married people, those who are newly divorced must find new communities to replace old friendships that are no longer mutually satisfying. The initiative for change may in fact come not only from rejection or awkwardness in old friendships but also from the divorced person's finding friends who share with him or her the new concerns and emotions of the divorce experience. Priority may also go to new relationships with people of the opposite sex; for the majority of divorced and widowed people, building a new community involves dating again.

Deciding knowledgeably whether to divorce means weighing what we know about the consequences of divorce. The next section examines the economic consequences of divorce.

THE ECONOMIC CONSEQUENCES OF DIVORCE

Upon divorce, a couple must become two distinct economic units, each with its own property, income, control of expenditures, and responsibility for taxes, debts, and so on. The economic fallout of divorce can be severe. No one "wins" financially in a divorce. Everyone's standard of living suffers, especially children's. The economic consequences of divorce vary for husbands and wives, however, with wives (and children) experiencing greater, and more enduring, losses (Sayer 2006; Tach and Eads 2015). Although many parents help out financially when their adult child gets divorced, these are generally "one time" gifts and not enough to prevent subsequent economic decline (Leopold and Schneider 2011; Mazelis and Mykyta 2011; Timonen, Doyle, and O'Dwyer 2011). Older couples who divorce also face substantial losses.

Among those 50 and older, household income drops 23 percent for men and 41 percent for women (Fried 2016). Among people who divorce and do not repartner, women are over twice as likely to live in poverty (27 percent) as men (12 percent) (Lin, Brown, & Hammersmith 2017). As one financial planner says,

Patrick Pihl/Alamy Stock Photo

Both mothers and fathers experience a substantial decline in their standard of living after a divorce. They may need to move to less expensive—and less desirable—housing and away from their former neighborhood, school, and friends.

"When you divorce in your [fifties and sixties], you lose the luxury of time to recover from any financial shortfalls or mistakes" (Yip 2012).

Consequences for Children: Single-Parent Families and Poverty

The continuation of a high incidence of divorce and the dissolution of cohabiting unions contributes to the increased prevalence of single-parent families. Children's living arrangements vary greatly by race and ethnicity, as Figure 14.4 indicates. Based on the most recent available data, nonHispanic white children (75 percent) and Asian children (85 percent) are most apt to be living in two-parent families (with biological parents or a parent and stepparent). Only 68 percent of Hispanic children

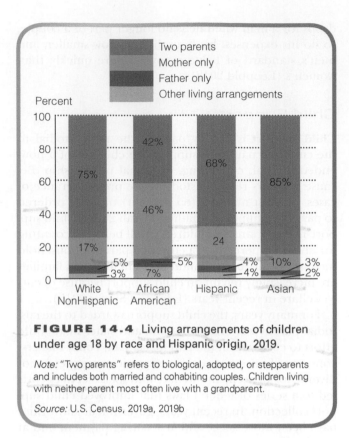

FIGURE 14.4 Living arrangements of children under age 18 by race and Hispanic origin, 2019.

Note: "Two parents" refers to biological, adopted, or stepparents and includes both married and cohabiting couples. Children living with neither parent most often live with a grandparent.

Source: U.S. Census, 2019a, 2019b

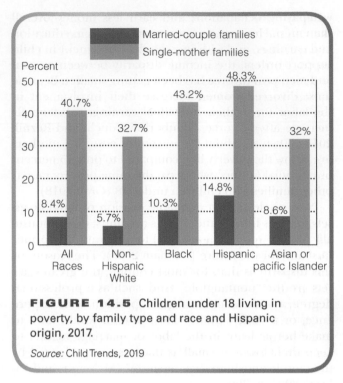

FIGURE 14.5 Children under 18 living in poverty, by family type and race and Hispanic origin, 2017.

Source: Child Trends, 2019

and 42 percent of black children live with two parents (U.S. Census Bureau, 2019a,b). There are many reasons why so many African American and Hispanic children do not live in couple households. One is the lower level of education of these mothers relative to other groups. More than a third of children with mothers without a high school diploma or who have a GED live in single-mother or cohabiting households; this figure is 11 percent for the children of mothers with a college degree (Stykes and Williams 2013).

How do children living with single fathers fare relative to children living with single mothers? Single fathers are more likely than single mothers to have a bachelor's degree (23 percent versus 18 percent) and children in single father household are half as likely to live in poverty as children living with single mothers (U.S. Census Bureau 2016). Children in single father families are also more likely to have private medical insurance than are children in single-mother families, at 54 percent compared to 35 percent (Eickmeyer 2017).

Family structure and poverty are interrelated. Figure 14.5 shows the proportions of children living in poverty in 2019, comparing poverty rates by race and ethnicity and family type. As you can see, 41 percent of all children who reside with a single mother (with no husband or partner present) live in poverty, whereas only 8 percent of those living in married-couple families do. The relationship between family type and

poverty is consistent across racial and ethnic categories, with minorities experiencing the highest levels of poverty.

Economic Losses for Women

Women's labor force participation and earnings have increased in recent years, so divorce is not as financially devastating to women as it once was (Raley and Sweeney 2020). Even so, women sustain much greater economic losses than men, and single mothers and their children have the highest level of poverty of any other type of family (Cooper and Pugh 2020; Leopold and Kalmijn 2016). Decades ago, Sociologist Lenore Weitzman researched the financial plight of divorced women and their children in her landmark book, *The Divorce Revolution* (1985). She found that wives' (and children's) income declined considerably after divorce while husbands' disposable incomes increased. Although the size of the effect was later questioned, studies continue to show that women's incomes decline after divorce (Sayer 2006). Another statistic used to examine divorced women's economic decline is the **income-to-needs ratio**—that is, how well income meets financial needs. Women and their children experience a decline of 20 percent to 36 percent in their income-to-needs ratio (Meadows, McLanahan, and Knab 2009; Sayer 2006).

A fundamental reason for the income disparity between ex-husbands and their former wives is men's and women's unequal wages and different work patterns. As discussed in Chapter 10, women on average work in less

well-paying occupations and earn less money overall than men. However, as women make gains in education and earnings, and with increased enforcement in child support orders, the income disparity between women and men post-divorce has been getting smaller. Because most divorced women increase their involvement in the labor force, their individual earnings substantially increase after divorce (Tamborini, Couch, and Reznik 2015). Even so, about 30 percent of custodial mothers live below the poverty line compared to only 15 percent of custodial fathers, who have the same poverty rate as other families with children under 18 (Grall 2018).

Nevertheless, even equitable division of tangible assets, such as houses and savings accounts, does not usually place a woman on an equal footing with her former husband for the future (Oldham 2008). The reason for this disparity is that, for most couples, the greatest assets are the "nontangible" kind, such as a professional degree, a business or managerial position, work experience, or a skilled trade. "Most women would have to make heroic leaps in the labor or marriage market to keep their losses as small as the losses experienced by the men from whom they separate" (McManus and DiPrete 2001, p. 266).

Returning to the labor force or increasing work hours does not allow most women—especially those with children—to fully recoup the economic losses from divorce. Remarriage or cohabitation with a new partner remains a more certain strategy to recovering the family's standard of living (Raley and Sweeney 2020), even though the pool of eligible partners available to divorced women tend to have lower incomes. Women who are custodial parents must also depend on child support from the other parent to meet their new single-parent family's expenses. Child support amounts are set relatively low, and much child support remains unpaid (as discussed later).

Economic Losses for Men

Although men's economic losses after divorce are not as great as those of women, men's incomes also decline. An analysis of data from the Panel Study of Income Dynamics indicated that six years after their marriage ended, divorced men's incomes were 23 percent lower than they would have been had they remained married (De Vaus, Gray, Qu, and Stanton 2017). The chief reason for their declining standard of living is the loss of the partner's income, but this is compounded by a higher tax rate for men who become the noncustodial parent and thus have no or fewer dependents to claim (Braver, Shapiro, and Goodman 2006). It's also possible that divorced men may be discriminated against in the workplace. Killewald (2013) found that men outside of traditional family structures are paid significantly less than married men. However, though family income

drops for a man when he is no longer part of a couple, so do his expenses; his household is now smaller, and men's standard of living recovers more quickly than women's (Leopold 2018).

Child Support

Child support is money paid by the noncustodial to the custodial parent to support the children of a now-ended marital, cohabiting, or sexual relationship. Because mothers retain custody in the preponderance of cases, the vast majority (85 percent) of those ordered to pay child support are fathers (Grall 2018). Child support is important to children's well-being, accounting for about 70 percent of a custodial parent's household income (Grall 2016). Children in single parent families are increasingly reliant on child support because of cuts to welfare in recent years (Edin and Shafer 2015).

For many years, the child support awarded to the custodial parent was often not paid, and states made little effort to collect it on the parent's behalf. Policy makers' concerns about poverty, the economic consequences of divorce for women, and welfare and social services costs led to a series of federal laws that improved child support collection. In recent years, government authorities have been more successful at securing payment and at standardized amounts that are often higher than previously provided. Nevertheless, the poverty rate for custodial parents has remained unchanged in the last two decades (Grall 2018).

Despite the fact that child support amounts are increasing overall, child support awards continue to be small, given what it costs to raise a child (see Chapter 8). The Current Population Survey of the U.S. Census provides information on custodial parents and child support receipt (Grall 2018). Even when fully paid, child support amounts, which average $287 per month, are not impressive. The result is that half of custodial mothers and one-third of custodial fathers receive government assistance. Although the majority of custodial parents are women, fathers have increased as a percentage of all custodial parents over the last two decades, from 16 percent in 1994 to 20 percent in 2016. Custodial fathers are less likely to receive full child support payments than are custodial mothers, likely the result of non-custodial mothers' lower incomes and less stable employment, which is a main reason why they don't have custody.

Some noncustodial parents do make additional "in kind" contributions in the form of gifts, clothes, food, medical expenses, and childcare. About 61 percent of custodial parents reported receiving some form of non-cash support for their children (Grall 2018). However, it is a myth that noncustodial parents substitute in-kind support for cash support. Parents who receive child support are more likely to receive noncash support as well

(Garasky et al. 2010). Nonetheless, lower-income men with fewer financial resources may substitute in-kind support for formal child support, which can constitute a large proportion of their income (Kane, Nelson, and Edin 2015).

A complicating factor is that only about half of custodial mothers have legal child support agreements. Common reasons for not seeking child support include the custodial parent feeling that "the other parent provides what he or she can," that they "did not feel the need to make it legal," and that the "other parent could not afford to pay." Child support agreements were also less likely among couples who were not married (Grall 2016b).

Why Do Some Parents Not Pay Their Child Support Obligations? In 2015, the most recent data available, roughly one-third of custodial parents who were due child support did not receive any payments, and less than half (44 percent) of those due child support received the full amount (Grall 2018). The principal reason for a noncustodial parent's failure to pay remains unemployment or underemployment (Sorensen 2010). Among families in which the nonresident parent has been employed during the entire previous year, payment rates are 80 percent or more. Not so when unemployment is involved. "The key to reducing poverty [among single-parent families] thus appears to be the old and unglamorous one, of solving un- and underemployment, both for the fathers and the mothers" (Braver, Fitzpatrick, and Bay 1991, pp. 184–185). Compliance with court-ordered child support decrees is higher among noncustodial parents with a college degree, those who are older, and those who were married to their children's other parent (Grall 2018). Compliance is also related to the noncustodial parent's involvement in the child's life. Compliance rates are higher when they have either joint custody or visitation arrangements (Grall 2018). Sometimes fathers withhold child support from the mother as a "power play" to see their children more often, and mothers may prevent visitation in order to ensure the child support due to them is paid (Moore 2012). Another problem is that mothers and fathers often go on to have more children with a new spouse or partner, a situation associated with lower child support payments to "first families" (Manning and Smock 2000; Meyer and Cancian 2012). Nonresident fathers sometimes have children with different mothers and owe child support to each, and a father's child support burden is typically higher when the children live in multiple households, and older children are typically given priority. How to balance the needs of multiple sets of children is an ongoing discussion (Berger and Carlson 2020).

Two suggested solutions to the problem of nonpayment of child support are government-guaranteed child support and a children's allowance. Both are based on the principle of society-wide responsibility for all children. With guaranteed child support, a policy adopted in France and Sweden, the government sends to the custodial parent the full amount of support awarded to the child. It then becomes the government's task to collect the money from the parent who owes it. A second alternative, a children's allowance, provides a government grant to all families—married or single-parent, regardless of income—based on the number of children they have. All industrialized countries except the United States have some version of a children's allowance. In the present political and economic climate in the United States, however, such measures seem unlikely to be adopted.

Low-income fathers are typically expected to pay a higher portion of their income in child support than middle-class fathers, resulting in a spiral of expanded debt and often withdrawal from their children (Huang, Mincy, and Garfinkle 2005). Recognizing this situation, some programs provide employment services, family support, and mediation services to low-income fathers. As another approach to securing payment of child support, some states have begun experimenting with "responsible fatherhood" programs, often supported by government grants (Dion, Zaveri, and Holcomb 2015; Fagan and Kaufman 2015). Although many fathers are highly motivated to take part in such programs, the effectiveness of these is unclear. A recent review of such programs showed a small increase in father–child contact but no effect on employment or child support payments (Holmes, Hawkins, Egginton, Robbins, and Shafer 2018). Similarly, programs that integrate case management, review and adjustment of child support obligations, and employment and parenting services were effective in increased contact and a feeling of responsibility toward children but did not affect compliance with child support orders (Cancian, Meyer, and Wood 2019). Noncustodial fathers continue to face numerous challenges to maintaining social and financial engagement with their children (Gearing 2015).

We have been speaking of child support in the context of marriage and divorce and heterosexual couples. Like opposite-sex couples, courts have awarded child support when same-sex couples who have been raising children together break up (Graham 2008). Given the recency of the Supreme Court's decision to allow same-sex marriage, it is unclear how same-sex parents will navigate laws surrounding marital dissolution and child support (Debele 2015). One issue is the primacy given to genetic ties. The child is the biological offspring of just one parent, and in custody disputes discrimination against the non-genetic parent is a concern (Feinberg 2016). We turn now from the economics of post-divorce family support to a broader examination of the aftermath of divorce.

THE SOCIAL AND EMOTIONAL CONSEQUENCES OF DIVORCE

People who divorce—and their children—can expect several emotionally significant transitions in family structure and lifestyle. For obvious reasons, most research has focused on how divorce affects children. However, divorce has important effects on adults as well. We will examine those effects first.

Consequences for Women and Men

Numerous studies over several decades indicate that, compared to married individuals, people who have divorced have more health problems, more symptoms of anxiety and depression, more substance use, and greater overall mortality (Raley and Sweeney 2020; Tamborini, Reznik, and Couch 2016). Researchers have been interested in the extent to which the negative outcomes are caused by the divorce itself or are the result of selection. The selection perspective posits that individuals have preexisting traits called *selection factors* that sort them into groups with higher divorce risks as well as negative outcomes after divorce (Idstad et al. 2015).

Although several studies provide evidence in favor of selection (for example, divorce is more common among those with poor mental health), there is strong evidence that divorce has a negative effect on the health and well-being of adults regardless of such predispositions (Raley and Sweeney 2020). Researchers have identified a number of contributors, including the loss of social support and companionship, a decline in standard of living, the need to change residences, feelings of anger and sadness, and discontinuance of measures associated with a healthy lifestyle such as yearly doctor's visits (Amato 2010). These factors could be the reason that divorce is associated with increased disabilities, even as far out as 20 years after the divorce (Couch, Tamborini, and Reznik 2015). The negative effects of divorce on life satisfaction is greater for men and women with children, especially if the children are young (Leopold and Kalmijn 2016). Overall, divorce is thought to be tougher psychologically on men than on women. Divorced men report greater happiness in their marriages prior to dissolution than divorced women and, without their wives to encourage healthy behaviors, may fall into unhealthy habits such as smoking and alcohol consumption (Reczek et al. 2016; Schoppe-Sullivan and Fagan 2020). Men also have fewer friendships and sources of social and emotional support than do women (McClintock 2014b).

The negative effects of divorce vary for different racial and ethnic groups. Compared to whites, divorce was associated with higher levels of psychological distress among Hispanics, although this effect was not observed for men. Divorce goes against cultural expectations and created conflict with family members, which was related to greater distress (Darghouth, Brody, and Alegría 2015). Older Americans who divorce may have to prolong their working life and reenter the labor force and marriage market. Nevertheless, according to a study of divorced men and women ages 40 to 79 conducted by the American Association of Retired Persons, 80 percent considered themselves happy, scoring more than five on a ten-point happiness scale, with 56 percent scoring an eight, nine, or ten (Thomas 2012). Several studies indicate that after an initial crisis period, divorced men and women adapt to their new circumstances (Raley and Sweeney 2020).

Divorce and Stress-Related Growth Scholars and clinicians have begun to talk about **stress-related growth**. There is now more emphasis on the diversity of outcomes related to divorce. Stress-related growth can take different paths. A crisis-related pathway results when coping with a traumatic event makes the person stronger (see Chapter 13). A stress-relief pathway results when, for example, the end of a marriage and its problems brings relief to one or both of the partners. Kinds of growth include personal growth in the self, growth in interpersonal relationships (becoming closer to family and friends), and growth or change in philosophy of life.

> Research on stress-related growth indicates that most individuals who have experienced traumatic events report positive life changes. . . . One thing that is clear from the existing research . . . is that it is at least as common to experience positive outcomes following divorce as negative one[s], and that positive outcomes can coexist with even substantial pain and stress. (Tashiro, Frazier, and Berman 2006, pp. 362, 364)

Divorce provides an escape from marital behaviors that may be more harmful than divorce itself, such as a partner's alcoholism or drug abuse (Coontz 1997). Several studies show divorce leads to short-term improvements in physical health (e.g., Leopold 2018). An interesting study by economists Betsey Stevenson and Justin Wolfers (2004, 2007) found that no-fault divorce was associated with declines in suicide rates for women, domestic violence against both men and women, and intimate partner homicides of women. The existence of an escape route seems to change the balance of power and reduce violence.

Another way to look at stress-related growth—as well as less happy outcomes—comes from E. Mavis Hetherington's Virginia Longitudinal Study of Divorce and Remarriage. She and her colleagues followed 144 couples for twenty years; half were divorced initially and half not. Additional families were added as time

Divorce is difficult but can also be liberating. The psychological impact of divorce depends on many factors, including level income, extent of conflict during the marriage, and whether or not the couple has children.

went by. Families were interviewed at various points, but our interest here is in the ten-year point. A previously developed typology of post-divorce adaptive patterns was used to assess the adjustment of divorced adults. Hetherington found 20 percent of those studied to have "enhanced" lives, while 40 percent had "good enough adjustment" (Hetherington 2003; Hetherington and Kelly 2002).

Perhaps the best overall assessment of the outcomes of divorce is from Paul Amato:

> On one side are those who see divorce as an important contributor to many social problems. On the other side are those who see divorce as a largely benign force that provides adults with a second chance for happiness and rescues children from dysfunctional and aversive home environments. It is reasonable to conclude that . . . [d]ivorce benefits some individuals, leads others to experience temporary decrements in well-being that improve over time, and forces others on a downward cycle from which they might never fully recover. (Amato 2000, p. 1282)

Whether a divorce can have a positive effect on adult adjustment depends on a range of moderating factors such as income, the level of distress in the marriage, and the presence of children. Social and emotional support from family and friends as well positive religious beliefs have been shown to reduce the negative effects of divorce (Krumrei, Mahoney, and Pargament 2011; Webb et al. 2010). Among divorced parents with infants, maintaining a positive relationship with one's ex-spouse or partner is associated with less depression (Paulson, Dauber, and Leiferman 2011). A divorce might evoke feelings of elation, at least at first. "Divorce parties" have become popular, as have "divorce rings," which some newly divorced women buy for themselves to replace their missing wedding bands (Fishman 2013; White 2012). For most, however, divorce is an overwhelming and draining experience, often involving enormous changes in family life for parents and children.

How Divorce Affects Children

A large number of American children will experience their parents' divorce, either as children or as adults. More than half of all divorces involve children under 18, and about 40 percent of children born to married parents will experience marital disruption (Amato 2000). How do separation and divorce affect children? Outcomes for children depend a great deal on the circumstances before and after the divorce. Although the divorce experience is psychologically stressful and, in most cases, financially disadvantageous for children, children in high-conflict marriages may benefit from a divorce (Kalmijn 2015a). Living in an intact family characterized by unresolved tension and alienating conflict can cause as great or greater emotional stress and a lower sense of self-worth in children than living in a supportive single-parent family (Barber and Demo 2006). When the conflict level in the home has been low, however, children have poorer post-divorce outcomes. They are likely surprised by a divorce and seem to suffer more emotionally. Among other things, it is difficult for them to see the divorce as necessary (Stevenson and Wolfers 2007).

The "Child of Divorce" Perspective

In the late twentieth century, psychologist Judith Wallerstein and her colleagues had been influential in defining the situation of children of divorce for both professionals and the public. For example, in their longitudinal study of children's post-divorce adjustment, Judith Wallerstein and Joan Kelly interviewed all of the members of some sixty families with one or more children who had entered counseling at the time of the parents' separation in 1971. They then re-interviewed children at one year, two years, five years, ten years, and, in some cases, fifteen years later and finally again at the twenty-five-year point (Wallerstein and Lewis 2007, 2008).

In the initial aftermath of the divorce, children appeared worst in terms of their psychological adjustment at one year after separation. By two years post-divorce,

households had generally stabilized. At five years, many of the 131 children seemed to have come through the experience fairly well: 34 percent "coped well"; 29 percent were in a middle range of adequate, though uneven, functioning; and 37 percent were not coping well, with anger playing a significant part in the emotional life of many of them (Wallerstein and Kelly 1980).

Their research and that of other researchers indicates that children of divorced parents have less money available for their needs, which is especially significant because some of the negative impact of divorce can be attributed to economic deprivation (Raley and Sweeney 2020). Compared with continuously married parents, divorced parents provide less economic assistance to their adult children (Shapiro and Remle 2010). In the Wallerstein study, financing college was found to be especially problematic, even among those who were middle-class. Sixty percent of those children were likely to receive less education than their fathers; 45 percent were likely to receive less than their mothers. Even divorced fathers who had retained close ties, who had the money or could save it, and who ascribed importance to education seemed to feel less obligated to support their children through college (Wallerstein and Blakeslee 1989; Wallerstein and Lewis 2007, 2008).

Wallerstein's research is well known for its methodological problems: It was a small, unrepresentative sample recruited by offering free counseling to the family, it lacked a control group, there was difficulty separating family troubles and mental health concerns that predated the separation and divorce from those that might be effects of divorce, and the study took place when divorce was still relatively new. Still, research supports Wallerstein's conclusion that overall, divorce has negative and long-term effects on children in the areas of academic success, conduct, psychological adjustment, social competence, and self-concept, and they have more troubled marriages and weaker ties to parents, especially fathers, that can extend into adulthood (Raley and Sweeney 2020).

But that is not the whole story. Researchers have moved away from studying divorce (and remarriage) as "isolated events in children's lives and toward considering cumulative histories of instability and diversity in children's family and living arrangements" (Raley and Sweeney 2020, p. 88). That is, researchers are finding that it is the *accumulation* of family transitions and changes *associated with* divorce (parents' subsequent cohabitation or remarriage, the introduction of new step- and half-siblings, moving to a new home) that are responsible for most of this effect as oppose to the divorce itself. Numerous studies make it clear that children's well-being declines as the number of family transitions they face increases (Raley and Sweeney 2020).

Racial and Ethnic Differences in Effects of Divorce on Children Research suggests that the negative effects of divorce are less serious for black and Hispanic (and socioeconomically disadvantaged) children than they are for whites and those who are more economically well-off (Bernardi and Boertien 2016; Cross 2019; Raley and Sweeney 2020). Why might this be? One possibility is that union instability is more common in those populations and therefore less unexpected. The stigma of having divorced parents may be minimal within the black community, where single-parent families are more normative (Heard 2007; Liu and Reczek 2012). Accordingly, "[t]he common history among blacks allows for the emergence and primacy of social supports, such as women-centered kinship networks, coresidence with extended family, and strong ties to the church, which can buffer the negative effects of stress caused by family instability" (Heard 2007, p. 336; Chatters et al. 2015). Analyzing data on 867 African American children from the Family and Community Health Study, Simons and her colleagues found that "child behavior problems were no greater in either mother–grandmother or mother–relative families than in those in intact nuclear families." At least among blacks, these researchers found mother–grandmother families to be "functionally equivalent" (Simons et al. 2006, p. 818). Then too, other extended kin in black families—uncles, for example—may be involved in childcare, which may offset some of the negative effects of father absence (Richardson 2009) and support from church communities among African Americans has also been found to be beneficial (Taylor et al. 2013).

Reasons for Negative Effects of Divorce on Children Researchers and theorists offer a variety of explanations for why and how divorce could adversely affect children. There are various theoretical perspectives concerning the reasons for negative outcomes (Amato 1993; Ribar 2015).

1. The **life stress perspective** assumes that, just as divorce is known to be a stressful life event for adults, it must also be so for children. Furthermore, divorce is not one single event but a process of associated events that may include moving—often to a poorer neighborhood—changing schools, giving up pets, and losing contact with grandparents and other relatives (Osborne, Berger, and Magnuson 2012; Westphal, Poortman, and Van der Lippe 2015). Custodial parents who weren't working or who worked part-time generally return to the labor force, resulting in greater work–family conflict (Nomaguchi and Milkie 2020). This perspective holds that an accumulation of negative stressors results in problems for children of divorce.

2. The **parental loss perspective** assumes that a family with both parents living in the same household is the optimal environment for children's development. Both parents are important resources, providing children love, emotional support, practical assistance, guidance, and supervision, as well as modeling social skills such as cooperation, negotiation, and compromise. Accordingly, the absence of a parent from the household is problematic for children's socialization. Single parents also have less time to spend with their children than married parents. For example, teenagers, especially boys, in single-parent households spend more time in unsupervised activities than do teens in two-parent households (Ribar 2015). Children in single-parent households also have less access to parents' social networks, relatives, and friends (Westphal, Poortman, and Van der Lippe 2015).

3. The **parental adjustment perspective** notes the importance of the custodial parent's psychological adjustment and the quality of parenting. Supportive and appropriately disciplining parents facilitate their children's well-being. However, the stress of divorce and related problems and adjustments may impair a parent's child-raising skills (Ribar 2015). Compared to married parents, divorced parents are "less supportive, have fewer rules, dispense harsher discipline, provide less supervision, and engage in more conflict with their children" (Amato 2000, p. 1,279). A study using the National Longitudinal Study of Adolescent Health found that the daughters of single mothers had sex earlier than the daughters of married mothers, which was attributable to less communication and fewer controls (e.g., curfews) among single mothers as well as their more permissive attitudes toward sex and teenage pregnancy (Zito and De Coster 2016). Single parenting is associated with greater parenting strain, stress, and fatigue (Nomaguchi and Milkie 2020). Yet, Raley and Sweeney (2020) argue that differences in parenting between divorced and married parents probably play a small role in child outcomes relative to other social factors.

4. The **economic hardship perspective** assumes that economic hardship brought about by marital dissolution is primarily responsible for the problems faced by children whose parents divorce (Brown, Manning, and Stykes 2015; Fomby, Goode, and Mollborn 2016; Thomson and McLanahan 2012). Divorced households have lower incomes, greater food insecurity, lower rates of homeownership, and less savings, and are less likely to have health insurance (Ribar 2015). Nevertheless, controlling for income, children in remarried, cohabiting, or single-parent families still lag behind children from two-parent families on various outcomes (Amato 2010; Fomby, Goode, and Mollborn 2016). Poorer parents may have to lower their expectations for their children, which is related to lower academic motivation and educational attainment among these kids (Wu, Schimmele, and Hou 2015). In contrast, Devor, Stewart, and Dorius (2018) found no evidence that divorced parents have lower expectations that their children go to college than married parents.

5. The **interparental conflict perspective** holds that conflict between parents is responsible for the lowered well-being of children of divorce. Many studies, including that of Wallerstein, indicate that some negative results for children may not result simply from divorce per se, but from exposure to parental conflict prior to, during, and after the divorce (Raley and Sweeney 2020).

6. The **selection perspective**, discussed previously in relation to adults, says that at least some of the child's problems after the divorce were present before the marriage. Some of these might be the result of the forthcoming marital breakdown, but some may be due to earlier dysfunctional family patterns, behaviors, or personality traits on the part of the child's parents (Idstad et al. 2015; Uecker and Ellison 2012), although the negative effects of divorce on children persist even after accounting for predivorce factors (Kim 2011).

7. The **family instability perspective** stresses that the number of transitions in and out of various family settings is the key to children's adjustment (Manning 2015; Mitchell et al. 2015). In their book, *Beyond the Average Divorce* (2010), David Demo and Mark Fine refer to this as **family fluidity**, meaning "the frequency and rate of changes in family related experiences and outcomes" (p. 7). The logic of the family fluidity or instability hypothesis is this:

Transitions may include parents' separation; a cohabiting romantic partner's move into, or out of, the home of a single parent; the remarriage of a single (noncohabiting) parent; or the disruption of a remarriage. The underlying assumption is that children and their parents, whether single or partnered, form a functioning family system and that repeated disruption of this system may be more distressing than its long-term continuation. . . . Stable single-parent households or stepfamilies, in contrast, do not require that children readjust repeatedly to the loss of coresident parents and parent-figures or the introduction of cohabiting parents and stepparents. (Fomby and Cherlin 2007, p. 182)

Divorce and Sibling Relationships With so much focus on how divorce affects relationships between parents and children, we sometimes forget how divorce

affects relationships between other family members such as siblings. Siblings may respond to their parents' divorce in two ways. One hypothesis is that sibling rivalry and conflict increase as a result of scarcity of the mother's time and attention. A second hypothesis is that siblings become closer and more supportive in the face of unstable and unreliable relationships with adults. Most available research unfortunately supports the first hypothesis. Biological sibling relationships have been found to be more conflictual, negative, and ambivalent in divorced and remarried homes than in original two-parent families (McHale, Updegraff, and Whiteman 2012). Worse sibling relationships among children with divorced parents are less the result of the divorce itself than being raised by parents in lower quality marriages who may be hostile to one another (Milevsky 2019).

A More Optimistic Look at Outcomes for Children of Divorce

Having considered the reasons for the negative effects of divorce on children, we now try to assess just how important divorce is in the lives of affected children. We've given considerable attention to the research of Wallerstein and her colleagues because it has been highly influential. "Judith Wallerstein's research on the long-term effects of divorce on children has had a profound effect on scholarly work, clinical practice, social policy, and the general public's views of divorce" (Amato 2003, p. 332).

Nevertheless, concern that children of divorce are disadvantaged does not rest solely on Wallerstein's research and what many see as her exaggerated presentation of the dangers of divorce (Cherlin 1999). A "persuasive body of evidence supports a moderate version of [her] thesis" (Amato 2003, pp. 338–339; Hetherington 2005). The remaining question, then, is this: How *much* does divorce negatively affect children?

E. Mavis Hetherington has been studying divorcing families for about the same length of time as Judith Wallerstein, and she has a much more optimistic view of the outcomes for children—and adults. Starting in 1974 in Virginia with parents of 4-year-olds, forty-eight divorced and forty-eight married-couple families, her research ultimately included 1,400 stable and dissolved marriages and the children of those marriages, some followed for almost thirty years. Hetherington found that 25 percent of these children of divorced parents had long-term social, emotional, or psychological problems, compared to 10 percent of those whose parents had not divorced. However, in assessing the impact of divorce, she would emphasize the 75 to 80 percent of children are coping reasonably well (Hetherington and Kelly 2002).

Constance Ahrons (2004) also followed children of divorce into adulthood. Twenty years after their parents' divorce, 79 percent thought their parents' decision to divorce was a good one, and 78 percent felt that they are either better off than they would have been or else not that affected. Twenty percent, however, did not do so well, having "emotional scars that didn't heal" (p. 44). Less attention is paid to the effects of divorce on adult children whose parents divorced after decades of marriage. Children whose parents appeared to be perfectly happy can be especially devastated. One young woman said, "I get flashbacks to idyllic family holidays and think, was that all just a front?" (Carroll 2015, p. 15).

Clinical Problems vs. Psychological Pain Paul Amato (2010), based on the work of Laumann-Billings and Emery (2000), draws a distinction between clinical problems and psychological pain. That is, although children of divorced parents have an increased risk of social and emotional problems, they are not more likely than children with continuously married parents to exhibit serious issues such as being in the clinical range for depression or anxiety.

Other scholars agree: "On average, parental divorce and remarriage have only a small negative impact on the well-being of children" (Barber and Demo 2006, p. 291; see also Demo, Aquilino, and Fine 2005). Experiencing adversity in childhood, such as parental divorce, can lead to a stronger sense of control and more positive expectations for the future in adulthood (Kim and Woo 2011; Schafer, Ferraro, and Mustillo 2011).

This does not mean, however, that children of divorce do not experience substantial stress emanating from the divorce, such as sadness over not seeing one of their parents, nervousness about being left home alone, or worry about where to spend the holidays. Children of divorce also are more likely to have been "parentified," meaning they were forced to take on adult responsibilities before they were developmentally mature enough to handle them. **Parentification** has been linked to worse child outcomes (Shaffer and Egeland 2011). These types of horizontal relationships happen even in households where parents work hard to keep the distinction between "parent" and "child" (Nixon, Greene, and Hogan 2012). A 13-year-old girl living alone with a single mother explains: "I pretend that I'm not sad sometimes because it just worries her so much. . . . I don't like to worry her. I think if I'm sad that's mine to deal with; she doesn't have to worry about it all the time." A 12-year-old boy from the same study says, "If she's stressed at work, if she has to meet a deadline or something, she can come back and I can help her out with stuff around the house and everything. . . . I'd bring her a cup of tea or whatever and put my brother to bed" (Nixon, Greene, and Hogan 2012, p. 149).

All in all, divorce researchers seem to be moving to a middle ground in which they acknowledge that children of divorce are disadvantaged compared to those of married parents—and that those whose parents were

not engaged in serious marital conflict have especially lost the advantage of an intact parental home. But many have moved away from simplistic or overly negative views of the outcomes of divorce. A theme that runs through virtually all studies on the impact of marriage and divorce on the well-being of children rests on the behavior of the parents, toward the children and each other (Ahrons 2004; Barber and Demo 2006; Freeman 2008). If a divorced couple continues to have a relationship that is fraught with conflict, the outcomes for their children will be negative. All children, regardless of family structure, thrive if they feel nurtured, loved, and supported by parents and reside in families who engage in conflict resolution and work hard at getting along with one another.

We now turn to issues of custody, the setting in which children will live after the divorce.

CHILD-CUSTODY ISSUES

A basic issue in a divorce of parents is the determination of which parent will take custody—that is, assume primary responsibility for caring for the children and making decisions about their upbringing and general welfare. **Legal custody** refers to who has the right to make decisions with respect to a child's upbringing (e.g., health, religion, education), and **physical custody** refers to where a child will live (Stewart 2007). In **joint custody**, both divorced parents continue to take equal responsibility for important decisions regarding the child's general upbringing. Joint custody may be legal, physical, or both. In joint *physical* custody, children spend time living with each parent. It doesn't have to be a 50-50 split—most states define custody as joint if the children live with their other parent at least 35 percent of the time (Nielsen 2018b). Arrangements can vary from the child spending one week with one parent and the next with the other, to cycles of several days with one parent followed by several days with the other, or other arrangements such as weekdays with one parent and weekends with the other. The second variation is joint *legal* custody—in which both parents have an equal right to participate in important decisions and retain a symbolically important legal authority.

Custody patterns and preferences in law have changed over time. For most of U.S. history and until the beginning of the twentieth century, women and children were considered the property of fathers and husbands and had no legal rights. The twentieth century brought the advent of the *tender years doctrine*, which advanced the idea that mothers alone are best suited to care for children and mothers were nearly always awarded custody. In the 1970s, states reformed their divorce laws and incorporated new ideas about men, women, and parenthood; custody criteria were made gender neutral. The tender years doctrine was gradually replaced with the *best interests of the child*, with determinations based on who in the family could best meet the child's essential needs to grow and develop and achieve their full capabilities as adults (Kruk 2015). This doctrine has become standard in Western countries and provides a potential mechanism for fathers to gain custody and visitation of minor children. In 1990, The United Nations Convention on the Rights of the Child established that the welfare and best interests of the child should be paramount in laws and policies concerning the lives of children (Atkin 2008). Recall from Chapter 1 that many people consider pets to be members of the family. In 2016, Alaska became the first state to allow judges to provide for the well-being of pets in divorce proceedings, as opposed to considering them marital "property" (Animal Legal & Historical Center 2019). This, too, is seen by many as a positive advance.

Many states have adopted the *presumption of joint custody*. What this means is that joint physical and legal custody is awarded automatically by a judge unless it can be proven that one of the parents is unfit in some way. In general, the child in question is given an age-appropriate opportunity to weigh in with respect to where they live, usually at about age 14.

Although more fathers are receiving joint or sole custody of their children, mothers still receive sole physical custody of their children the vast majority of the time. Eighty percent of custodial parents are mothers; 20 percent are fathers. Not all custodial mothers in government statistics have been divorced; 43 percent were never married to their child's father (Grall 2018). It is important to note that laws regarding relationship dissolution and custody may be different for divorced versus unmarried couples (Cyr, Stefano, and Desjardins 2013).

However, there has been a dramatic increase in fathers who have joint physical and legal custody in recent years, even among unmarried parents (Chen 2015) and children's living arrangements after divorce have become much more diverse (Havermans, Vanassche, Matthijs 2017). Sociologists Maria Cancian, Daniel Meyer, and Eunhee Han (2012) have analyzed court records from Wisconsin (Cancian et al. 2014; Cancian, Meyer, and Han 2012). The percentage of couples awarded joint physical custody increased from 11 percent in 1989 to 50 percent today (Meyer, Cancian, and Cook 2017). Father custody (shared or sole) is associated with higher incomes, having older children, and having only boys. However, should shared custody become more common, such factors will likely decline in importance (Cancian et al. 2014). The majority of lesbian couples opt for joint custody as well (Gartrell et al. 2011). Joint physical custody among unmarried couples has also increased. A tiny minority of joint custody couples set up "bird's nests" for their children in which the parents,

rather than the children, move between households (Luscombe 2011). Split custody, in which each parent has physical custody of at least one child, remains uncommon and only occurs in about 2 percent to 4 percent of divorce cases (Cancian et al. 2014). Another arrangement seen recently among low-income families is for couples who are no longer romantically involved to reside in the same households to share the day-to-day care of the children (Cross-Barnet, Cherlin, and Burton 2011).

Growth in father custody (whether joint or sole) is likely the result of men's changing social roles and their increasing involvement with children in terms of direct care and emotional closeness. It also may be the case that mothers have become less inclined to insist on sole custody. Fatherhood scholar James Levine thinks that "[w]e're seeing some weakening of the constraints on women to feel they can only be successful if they are successful mothers," so they are more willing to concede custody to willing fathers (quoted in Fritsch 2001, p. 4). Other research paints a different picture of father custody, with some men utilizing gender-neutral laws about custody to continue to exert power and control over their ex-spouses and avoid paying child support (Elizabeth, Gavey, and Tolmie 2012). Several studies show that residence patterns of children in father custody are more unstable than those of children in mother custody (Stewart 2007). Children in father custody are more likely than children in mother custody to "drift" to their other parent's home. Because returning to court is costly, most mothers whose children have returned to them do not seek an adjustment in child support awards or other changes in divorce decrees, leaving them at a financial disadvantage. The effect of joint custody on children's outcomes is discussed in the following sections.

The Residential Parent

As previously discussed, most custodial parents are women. Single parenthood is a role associated with stressor overload, or stress pileup (see Chapter 13). Custodial parents face the challenge of being solely responsible for their children's care. One-third of children with a nonresident parent had no contact with him or her in the previous year (Grall 2018). Custodial parents often feel overwhelmed and exhausted, and they have little time for meeting their own needs for rest, socializing, exercise, and hobbies, which negatively affects their mental and physical health. In a Danish study, mothers who shared physical custody of their children with their

Divorced mothers, most of whom are the residential parent, often face financial worries and emotional overload as they try to be the complete parent for the children.

RubberBall Productions/Getty Images

children's father reported less time pressure than those with sole physical custody (van der Heijden, Poortman, and Van der Lippe 2016). Given that child support awards are typically low, they must grapple with "making ends meet" financially. As discussed in Chapter 15, they must also balance the needs of new spouses and partners with their children and often must play "mediator" between the two.

The Visiting Parent

Most nonresident parents are fathers. Though estimates vary—depending on the data source and the way that nonresident fatherhood was assessed and because of underreporting of nonresident children—it is safe to say that about one in ten men have a child under the age of 18 living elsewhere (Stykes, Manning, and Brown 2013). They often find it difficult to construct a satisfying parent-child relationship. During the marriage, a father's authority in the family gave weight to his parental role, but this vanishes in a nonresidential situation. Custodial mothers may act effectively as gatekeepers, facilitating or refusing to facilitate the noncustodial father's relationship with his children (Adamsons and Pasley 2006; Moore 2012; Sano, Richards, and Zvonkovic 2008). As one noncustodial father stated, "Nothing ever happens to her. She basically goes along and does whatever she wants. . . . She denies access for a month and I actually got to the point where I went to the police and I said I want to lay a charge and they're like, we don't deal with this; it's a civil matter" (Kruk 2015, p. 85).

Maintaining a relationship with children who live in another household is a challenge. Nonresident fathers'

involvement with their children is variable, with some fathers becoming less involved over time, some becoming more involved, and some fathers "persistently in or out of contact with their children" (Schoppe-Sullivan and Fagan 2020, p. 185). Other factors associated with low visitation between nonresident fathers and their children include not having been married to the child's mother, low level of education, geographical distance, conflict with the mother, and nonpayment of child support (Carlson, VanOrman, & Turner, 2017; Goldberg 2015; Kalmijn 2015b; Schoppe-Sullivan and Fagan 2020). Swiss and Le Bourdais (2009) discuss the material conditions that impact the noncustodial father's relationship negatively. For example, they point out that working-class fathers, especially those who earn lower incomes, "are more likely to be working in low-paying, part-time, or shift-oriented work where they may not be available to their children when the latter are free, that is, in the evenings and weekends" (p. 644).

A new marriage or cohabiting relationship may be a factor decreasing visitation as well as the quality of the co-parental relationship (Goldberg and Carlson 2015). The presence of children in the father's new family is also associated with a decline in visitation (Manning and Smock 2000). In general, fathers seem to find it difficult to parent children across two families (Schoppe-Sullivan and Fagan 2020).

Nevertheless, noncustodial fathers are more involved with their children than in the past. Data from the National Survey of America's Families (NSAF) indicate that a third of children had at least weekly contact with their fathers, although overnight and extended visits were still rare (Stewart 2010). Many nonresident fathers embrace their new parenting role, if only on a part-time basis. Some fathers felt they had become even more involved with their children than they had been, for example, by taking their children to doctor's appointments. One father with a 5-year-old daughter explains, "I make breakfast for her. She likes that, she sits at the counter while I'm cooking and we talk" (Troilo and Coleman 2012, p. 607). Other fathers expressed frustration with their new "part-time fulltime role":

> I pick her up around 5:30 [on Wednesday evening], then before you know it, it's almost 8, and I'm telling her, "Ok, let's start getting toward going to bed. Do you know what you're going to wear tomorrow?" If not, I wash clothes if needed, make sure her teeth are brushed, and see if she needs a bath, read stories, and you know 8:30, 9 is gone before you know it. You're talking 2.5 hours on a weeknight. The mornings

are pretty much getting her up and out the door. There's no quality time there. (Troilo and Coleman 2012, p. 610)

E-mail, texting, and social media can be used to increase communication between nonresident parents and children, but they may come with difficulties (such as children "defriending" a parent). Nevertheless, one father who communicates with his daughter daily says, "with email and text, it's just like you're there is some ways" (Troilo and Coleman 2012, p. 606). Technology may be helpful to children in other ways. For example, ex-spouses can use a shared online calendar to manage their children's activities rather than asking the child to relay information back and forth between parents.

How Nonresident Father Involvement Affects Children Nonresident fathers affect their children in important ways. In two different studies of disrupted families in the Netherlands by Matthijs Kalmijn (2015a, 2015b), divorced, nonresident fathers had as much influence on their children's educational attainment as did married, resident fathers. The more involved these fathers were with their children during the marriage, the better the relationship with their children later. Authoritative parenting (see Chapter 9) on the part of the nonresident father is especially important to children's well-being, especially among adolescent boys (Karre and Mounts 2012). Indeed, the quality of the father–child relationship is more important to children's well-being than how often the child sees his or her father or whether the father lives in the household (Booth, Scott, and King 2010; Stewart 2007). Relationship quality,

Monkey Business Images/Shutterstock.com

Divorced fathers, most of whom are the nonresidential parent, face the loss of time with children, as well as a more general loneliness. Being the "visiting parent" is often difficult, but maintaining the father-child bond is significant in a child's adjustment to divorce.

closeness, and responsive parenting (that fathers consider the child's point of view and explain decisions) were found to reduce adolescents' "internalizing" (depression, self-esteem) and "externalizing" (aggression, antisocial behavior, drug use) problems and was associated with higher grades (Booth, Scott, and King 2010; Stewart 2003). Father involvement positively affects the well-being of even very young children. In a study of infants born outside of marriage, visitation from a nonresident father was associated with a greater likelihood of the child being rated as in "excellent health" (Tracey and Polachek 2015).

Yet, a nonresident father is not always the best role model if, for example, he abuses drugs or alcohol or is violent (see Chapter 9). In those cases, it may be best if the father is not involved (Osborne and Berger 2009; Salisbury, Henning, and Holdford 2009). In such situations, the courts may step in to limit or prevent fathers from having contact with the mother and children.

Nonresident Mothers In only about 10 percent of divorce cases does the father gain sole physical custody of the children so that the mother becomes the nonresident parent (Cancian et al. 2014). The popular stereotype is that noncustodial mothers are deviant in some way. Whereas some of those mothers have lost custody of children due to child abuse, child neglect, or problems with substance abuse or their mental health, most noncustodial mothers voluntarily surrender custody, mainly because they don't feel they can financially support their children (Stewart 2007). It is important to recognize that about half of children with nonresident mothers are not living with their fathers but living with grandparents or relatives or in foster care (Stewart 2010). Many of these placements are temporary, and children resume living with their mothers when their mothers are financially stable.

Most research and discussion on visiting parents has concerned fathers, but one study did compare nonresident mothers and fathers. Using two national datasets, the National Survey of Families and Households (NSFH) and NSAF, Stewart (1999a, 1999b) compared visitation and child support payments between children with noncustodial mothers versus fathers. Whereas noncustodial mothers exhibited more day-to-day visitation, phone calls, and extended visits, noncustodial fathers were more likely to pay child support and paid higher amounts.

Like being a nonresident father, being a nonresident mother is a stressful parenting role. Kruk (2015) interviewed comparable samples of divorced noncustodial mothers and fathers in British Columbia. He found their experiences to be similar in many ways. Common difficulties include loss of attachment, feelings of grief and loss, denial of access to the child by the resident parent, and financial problems. Issues specific to noncustodial mothers included feelings of shame, stigma, and humiliation at not being the primary caretaker of their children, a lack of support services and adequate legal representation, and physical and emotional abuse. One mother stated,

> I felt helpless because I also needed to develop a new life, and I went back to school and I had to look for a job and I wish I had some access to some resources that would have allowed me . . . to have my children in my life. I wish I was offered some sort of financial support so I could actually do something rather than sit helplessly, rather than not know what I would or should do. Social and legal support, I really didn't know anyone who could advise me, or could help me with my grief of loss. (Kruk 2015, p. 87)

In another study, noncustodial mothers were unable to be traditional mothers but found a "mother-as-friend" role insufficient and uncomfortable (Eicher-Catt 2004). The author advises that noncustodial mothers focus on building a relationship rather than thinking of the traditional maternal role. Regarding child outcomes, studies have also found few or no difference in well-being between children living with only their fathers (apart from their mothers) and living only with their mothers (apart from their fathers) (Stewart 2007).

Joint Custody

Recall that joint custody means that both divorced parents continue to take equal responsibility for important decisions regarding the child's general upbringing. When parents live close to each other and when both are committed, joint custody can bring the experiences of the two parents closer together, providing advantages to each. Both parents may feel they have the opportunity to pass their own beliefs and values on to their children. In addition, joint custody gives each parent some downtime from parenting (Lee 2002). Table 14.1 lists advantages and disadvantages of joint custody from a father's perspective. Shared custody gives children the chance for a more realistic and normal relationship with each parent (Arditti and Keith 1993). It results in more father involvement and in closer relationships with both parents (Kelly 2007).

The high rate of geographic mobility in the United States can make joint physical custody difficult. Joint custody can also be expensive. Each parent must maintain housing, equipment, toys, and often a separate set of clothes for the children and must sometimes pay for travel between homes if they are geographically distant. Mothers, more than fathers, find it difficult to maintain a family household without child support, which is often not awarded when custody is shared. There may be situations—an abusive parent, other domestic violence, or extremely high levels of parental conflict, for example—where sole custody is preferable (Hardesty and Chung 2006).

TABLE 14.1	Joint Custody from a Father's Perspective

John is an engineer who shares physical custody of his 13-year-old son, Dustin. Dustin lives with his dad during the week and with his mom and stepdad on weekends, except for Wednesday nights, when he stays overnight with his mom. John and Dustin's mom, who owns a hair salon with her husband, live in the same town, and Dustin transitions to and from each parent's house in the afternoons by way of the school bus. Here John talks about the advantages and disadvantages of joint custody.

ADVANTAGES	DISADVANTAGES
Shared custody schedule gives me much more time with my son than a traditional, limited-visitation schedule.	Dustin sometimes uses the rotating schedule to avoid following rules he does not like, such as consistently doing his homework.
I get to be a "real dad" instead of a weekend guest in my child's life.	I feel sometimes "out of the loop" regarding events going on in Dustin's life, such as knowing who his friends are and who he is taking to the eighth-grade dance.
I think the time we spend physically together could not be replaced with phone conversations and texting.	Every time Dustin comes back from his mom's, he has to spend a bit of time relearning my house rules.
Our time together happens at a "normal" pace instead of being crammed into a short visitation.	Bedtime and chores are often a struggle—I probably let Dustin stay up later than he should because I often work late and want to spend time with him.
Me and his mom each get to have "child-free" time. I have more time with my girlfriend and with friends. I can take weekend trips on my motorcycle.	I worry that Dustin never gets fully settled in either house.
Dustin gets to have two regular households instead of a regular house and a "foreign" house.	With joint custody, I am tied to a small geographic area, which affects my career. I would go up for a promotion at my company except most opportunities for advancement would take me out of state.
Dustin has a room, clothes, and his things at each house and doesn't have to "pack up" to move between households.	His mom, her husband, and I have to work harder to get along.
Dustin has become accustomed to following two sets of rules, one for each house. For example, he goes to church and youth group when at his mom's.	
Both households have equal weight, so there's no power play between me and his mom.	
Direct expenses for Dustin (clothes, lunch money, etc.) are more evenly shared.	

How do children in joint physical custody fare? Some commentators feel that a "cookie cutter approach" to child custody decisions does not work and that decisions should be made on a case-by-case basis with individualized plans for children's care. However, others point to research showing children's outcomes are better for children in joint physical custody, suggesting that overarching policies can be developed. Indeed, the majority of studies indicate that most children benefit from the arrangement (Nielson 2018a,b; Stienbach 2019). Fathers' greater involvement with children, including shared custody, has been shown to "lighten the load" on mothers (Nomaguchi, Brown, and Lehman 2017; Van Heijden, Poortman, and van der Lippe 2016).

Melinda Stafford Markham and Marilyn Coleman's (2012) interviews with mothers who shared custody reveal that not all joint-custody parents get along—they disagree about child support and parenting style and often have trouble communicating. An early study on joint custody by Maccoby and Mnookin (1992) found that children in joint custody and mother custody did similarly well: "[T]he welfare of kids following a divorce did not depend on who got custody, but on how

the household was managed and how the parents cooperated" (psychologist Eleanor Maccoby in Kimmel 2000, p. 141). A review of ten studies (Birnbaum and Saini 2015) found that although most children in joint custody were satisfied with the arrangement and their parents' interactions, some had difficulty making the transition between homes and wanted more input. These studies revealed the complex feelings of children in joint custody. As one child stated, "it is exhausting . . . no doubt about that . . . but at the same time . . . I think it's bad spending very little time with one of your parents . . . you lose contact with them . . . I think it's worth the stress" (p. 125). The true effect of joint custody on children's well-being is difficult to assess, however, because parents who opt for joint custody tend to be those who get along (Schoppe-Sullivan and Fagan 2020).

What about the effect of joint physical custody on parents? On the one hand, researchers have found better father–child relationships, less parenting stress, lower parental conflict, and better overall adjustment among parents with joint custody than with sole custody. On the other hand, parents were less satisfied with custody arrangements (Bauserman 2012; Cyr, Di Stefano, and

Desjardins 2013). Nevertheless, most of the noncustodial mothers and fathers interviewed by Kruk (2015) felt that shared custody would work best for them.

STYLES OF PARENTAL RELATIONSHIPS AFTER DIVORCE

Sociologist Constance Ahrons (1994) led one of the first explorations of relationships between parents after divorce, the Binuclear Family Study. In a **binuclear family**, the child is the "nucleus" in two households within one family. The study is based on interviews with ninety-eight divorced couples beginning approximately one year after their divorce. Ninety percent of them were followed to the five-year point, and each couple was interviewed three times. These were primarily white, middle-class couples from one Wisconsin county in which the mother had primary custody of the children.

At the one-year point, 50 percent of the ex-spouses had amicable relations; the other 50 percent did not. In half of those cases, the divorce was a bad one and harmful to family members; in the other half, the divorcing spouses had "preserved family ties and provided children with two parents and healthy families" (p. 16).

The ninety-eight couples represented a broad range of post-divorce relationships. Half of the couples exhibited what she considered "cooperative" parenting. Ahrons identified two styles of cooperative parenting, which she termed *perfect pals* and *cooperative colleagues.*

Perfect pals (12 percent) were friends who called each other often and brought their common children and new families together on holidays or for outings or other activities. This pattern occurred in a minority of the "good divorces." More often (38 percent), the couples were cooperative colleagues who worked well together but did not attempt to share holidays or be in constant touch—occasionally, they might share children's important events such as birthdays. Ex-spouses might talk about extended family, friends, or work. They still had areas of conflict but were able to compartmentalize them and keep them out of the collaboration that they wanted to maintain for their children (Ahrons 1994, 2007). Cooperative parenting is associated with higher social and emotional well-being for children, although the effect may be small and may only apply to certain outcomes (Amato, Kane, and James 2011).

Other research (Amato, Kane, and James 2011; Maccoby and Mnookin 1992) has identified a post-divorce parental relationship they refer to as **parallel parents**, parents who parented alongside each other but with minimal contact or communication or conflict. Relative to both cooperative parenting and parenting without any involvement from the ex-spouse, parallel parenting is associated with worse child outcomes (Amato, Kane, and James 2011; Pryor 2011). Returning to the Ahrons study, other divorcing couples were the angry associates (25 percent) or fiery foes (25 percent) that we may think of in conjunction with divorce.

○ **Co-parenting**, like cooperative parenting, is a team approach to raising children after divorce. In this model of parenting, parents strive to work together as "colleagues," with the goal of meeting their children's emotional and financial needs. Co-parenting should not be confused with joint custody, which refers to couples' legal arrangements for where the children should live and how decisions about the children are made. That is, parents with sole or joint custody can be co-parents. Although divorced parents for the most part are allowed to spend time and interact with their children as they want, co-parenting is considered the best model for enhancing children's well-being and future success.

Most states offer or require parent education classes that teach co-parenting skills to divorcing parents. In Canada, several provinces require parents to take part in "parenting coordination" programs after divorce (Cyr, Di Stefano, and Desjardins 2013). In some communities, the children also meet in groups with

When divorcing parents continue to engage in conflict and especially when children are drawn into it, a child's adjustment is poorer. Interparental conflict does tend to diminish with the passage of time.

Wavebreakmedia/Getty Images

a teacher or mental health professional (Pollet and Lombreglia 2008). The idea is that parents will continue to raise their children as co-parents, and they are likely to need help in meeting this new challenge. Evaluation forms completed after the sessions have shown predominantly positive responses, but more research is needed on the effect of co-parenting on children's well-being (Amato, Kane, and James 2011).

The Our Family Wizard website can help facilitate co-parenting relationships by minimizing "face time" between parents. Randy Kessler, chair of the American Bar Association's Family Law Section explains, "People don't want to talk to their exes because the sounds [sic] of their voice is irritating. But they can e-mail. They can share an online calendar. They can use any number of resources on the Internet. There are even divorce apps" (Paul 2012).

Communication technology does not make things easier if the relationship between ex-spouses is a contentious one (Ganong et al. 2012). Overall, however, the effect has been positive. Andrew Cherlin explains, "Before this electronic media, noncustodial parents had very formalized, appointment-driven communication with their kids. Electronic media may help noncustodial parents by informalizing the process of communicating with their children. They become less dependent on schedules and therefore more consistent with an easy, informal flow of information, which may be what teenagers like" (Meyers 2011).

Unfortunately, most divorced parents are not engaging in co-parenting. Most have low (70 percent) or medium (17 percent) levels of co-parenting (defined as at least monthly discussions about the children and "a great deal" or "some" influence from the father in child-rearing decisions) as opposed to high (12 percent)

levels and the degree of co-parenting declines over time (McGene and King 2012).

The factors that seem to affect co-parenting success are rather straightforward: a previous good co-parenting relationship during the marriage, a mediated rather than hostile divorce process, a reasonably good post-divorce relationship between ex-spouses, and some length of time since the divorce (Adamsons and Pasley 2006). Parents who share physical or legal custody or both and who are thought to get along better than other divorced parents are not always consistently nice and easy with one another. They, too, display various patterns of co-parenting, including continuously contentious, always amicable, and "bad to better" (Markham and Coleman 2012). One bad-to-better mother explains:

> I think always keeping [my son] at the focus of what I was trying to do as a person, as a parent you have to let some of that go. And since it's been working so far, I have no reason to rock the boat. Anything that I would do to be vengeful against my ex wouldn't do anything but damage my son. . . . I've let a lot of anger go. (Markham and Coleman 2012, p. 593)

"As We Make Choices: Rules for Successful Co-Parenting" provides some general guidelines for divorcing parents who want to cooperate in parenting their children.

IMPROVING DIVORCE OUTCOMES

In the United States, most divorces are adversarial and unpleasant. But does it have to be that way? As mentioned previously, mediation allows parties to develop options, consider alternatives, and reach an agreement in an effort to minimize harm to the child. Many professionals advocate **collaborative divorce**, which is a model based on mutual respect with the goal of preserving the emotional and financial resources of the entire family and where both parties agree not to litigate. An added advantage is that it is less expensive and time consuming than traditional divorce (Hemesath 2020).

What about child custody and visitation? Children should be given a bigger role in decisions about their care. In a unique study, The Children and Young People in Separated Families Project (Carson et al. 2018), Australian researchers tracked children and families' actual experiences in court proceedings. Three-quarters of the children said they wanted greater communication, information, and their parents to listen more

These divorced parents have both come to meet with their child's teacher. When parents work together to co-parent their children, they continue to have a sense of "family."

Michael Newman/Photo Edit

As We Make Choices

Rules for Successful Co-Parenting

Between Two Homes is a national organization that is "committed to supporting adults in raising children between two homes and to offering tools to the professionals assisting these families . . . dealing with transitions such as separation, divorce, or other family matters involving two homes" (Between Two Homes 2020). The organization provides co-parenting classes and information for divorcing couples with children. Such classes are mandatory in many states—a couple's divorce will not be granted without them. Here are their guidelines for parents:

1. Do not talk negatively, or allow others to talk negatively, about the other parent, their family and friends, or their home in hearing range of the child. This would include belittling remarks, ridicule, or bringing up valid or invalid allegations about adult issues.

2. Do not question the children about the other parent or the activities of the other parent (their personal lives). In specific terms, do not use the child to spy on the other parent.

3. Do not argue or have heated conversations when the children are present or during exchanges.

4. Do not make promises to the children to try and win them over at the expense of the other parent.

5. Do communicate with the other parent and make similar rules in reference to discipline, bedtime routines, sleeping arrangements, and schedules. Appropriate discipline should be exercised by mutually agreed-upon adults.

6. At all times, the decisions made by the parents will be for the child's psychological, spiritual, and physical well-being and safety.

7. Schedule changes will be made and confirmed beforehand between the parents without involving the child in order to avoid any false hopes or causing any disappointments or resentments toward the other parent.

8. Do notify each other in a timely manner of the need to deviate from the court order, including canceling time with the child, rescheduling, and being prompt.

9. Do not schedule activities for the child during your child's time with the other parent without the other parent's consent. Do work together to allow the child to be involved in extracurricular activities.

10. Do keep the other parent informed of any scholastic, medical, psychiatric, or extracurricular activities or appointments of the child.

11. Do keep the other parent informed at all times of your address and telephone number. If you are out of town with the child, do provide the other parent the address and phone number where the children may be reached in case of an emergency.

12. Do refer to the other parent as the child's mother or father in conversation rather than using the parent's first or last name.

13. Do not involve the child in adult issues and adult conversations about custody, the court, or the other parent.

14. Do not ask the child where he or she wants to live. But do encourage the child to understand he or she has two homes.

15. Do not attempt to alienate the other parent from the child's life.

16. Do not allow stepparents or others to negatively alter or modify your relationship with the other parent.

17. Do not use phrases that draw the children into your issues or make them feel guilty about the time spent with the other parent. For example, rather than saying "I miss you!" say "I love you!"

Critical Thinking

Do you know any divorced parents who follow, or do not follow, the rules of successful co-parenting? What do you think it would be like to be a "child of divorce"? How well do you think your parents would get along?

Source: Modified from *Between Two Homes* 2020.

regarding their living arrangements. Parent–child communication was found to help children accept their new living situation and build and re-build post-separation relationships. Only about half felt their views were acknowledged by legal services, family consultants, and other non-legal services such as counselors, and some found dealing with such people unhelpful, frustrating, and stressful. This research suggests that a child-inclusive approach is effective when parents' separate and that "giving children a bigger voice more of the time" should continue to be emphasized moving forward (Carson et al. 2018, p. ix).

Decisions regarding children should continue to be based on the best interest of the child. Lawyers,

mediators, and judges could research the history of the relationship, ask the children their preferences, do psychological assessments of family members, bring in expert witnesses, and utilize the research of family scientists (Mason et al. 2002). These measures would help reduce judges' discretion in court cases and would produce more predictable results (Malia 2008). States' increasing adoption of the presumption of shared custody is a positive advance. However, laws and policies have not kept pace with this trend. For example, children are not allowed to be "shared" across households on tax returns and some government programs, such as food stamps, require that children reside in the household fulltime (Berger and Carlson 2020).

There are advantages and disadvantages to these suggestions however. Advantages are improved family cohesion and commitment, increased opportunity for children to have two involved parents, increased contact with and economic support from nonresident parents, less stress, and reduced family conflict. Some disadvantages include increased red tape and bureaucracy, and possible increased conflict because parents must interact more frequently, and the stress of children having to navigate complex and fluid living situations. Custody arrangements should also be revisited along the way as the child gets older. Oftentimes in adolescence, children opt to stay in one place and the parent who is not "chosen" should be understanding of children's developmental stage and need for stability. Parents should also not make threats to send children to their other parent to live. Children need to be secure in the knowledge that their home lives will not be disrupted should they misbehave or if the family should go through a rough patch.

Although there are many divorce-related resources for parents, there are few books and other materials designed for children. It was not until 2012 that Sesame Street introduced a character, Abby Cadabby, whose parents are divorced, as part of their project *Little Children, Big Challenges* (West 2012). Shows like this can reduce children's fears about divorce and educate other children about the challenges their friends may face. Claire Masurel (2002) wrote the book *Two Homes* after talking to a child who was sad about her parents' divorce. She explains, "To comfort her, I talked about her two homes, and all the many things she could do in them. It was a positive way of helping her accept the changes in her life, focusing not on what was missed, but on the abundance of good times — and love—that she would continue to share with her mom and dad" (Google Books 2020).

The title of Constance Ahrons's book—*We're Still Family* (2004)—characterizes the positive outcomes of divorce that she has found among many of the families she studied. She argues that we must "recognize families of divorce as legitimate." To encourage more "good divorces," it is important to dispel the "myth that only in a nuclear family can we raise healthy children" (Ahrons 1994, p. 4). People often find what they expect, and social models of a functional post-divorce family have been lacking. In Chapter 15, we turn to a consideration of that common step for many divorced people: remarrying or forming a stepfamily or both.

Summary

- Divorce rates rose sharply in the twentieth century, and divorce rates in the United States are now among the highest in the world. Since around 1980, however, divorce rates have declined substantially.

- Among the reasons divorce rates have increased to the present level are changes in society. Economic interdependence and legal, moral, and social constraints are lessening. Expectations for intimacy have risen, while expectations of permanence are declining.

- People's personal decisions to divorce involve weighing the advantages of the marriage against marital complaints in a context of weakening barriers to divorce and an assessment of the possible consequences of divorce.

- Two consequences that receive a great deal of consideration are how a divorce will affect any children and whether it will cause serious financial difficulties.

- Economically, divorce is more damaging for women than for men, and this is especially so for custodial mothers many of whom are poor and receive no child support from their children's father.

- More couples are choosing to share physical custody of the children. Although going back and forth between two homes can be a challenge, maintaining close relationships with both parents is beneficial to children.

- Researchers have proposed several theories to explain negative effects of divorce on children. These include the life stress perspective, the parental loss

perspective, the parental adjustment perspective, the economic hardship perspective, the interparental conflict perspective, the family instability perspective, and the selection perspective.

● Debate continues among family scholars and policy makers concerning how important a threat of divorce is to children today. Some see divorce as part of a set of broad social changes, the implications of which must be addressed in ways other than turning

back the clock. Now there is also a moderate view of the impact of divorce on children: Yes, there is some disadvantage; no, divorce is not the most powerful influence on children's lives.

● New norms and new forms of the post-divorce family seem to be developing. Some post-divorce families can share family occasions and attachments and work together civilly and realistically to foster a "good divorce" and a binuclear family.

Questions for Review and Reflection

1. What factors bind marriages and families together? How have these factors changed, and how has the divorce rate been affected?

2. How is the experience of divorce different for men and women? How are these differences related to society's gender expectations?

3. In what situation(s), in your opinion, would divorce be the best option for a family and its children?

4. Do you think couples are too quick to divorce? What are your reasons for thinking so?

5. **Policy Question.** Should divorced parents with children be required to remain in the same community? Permitted to move only by court authorization? Be free to choose whether to be geographically mobile?

Key Terms

binuclear family   390
child support   378
collaborative divorce   391
co-parenting   390
crude divorce rate   364
divorce divide   365
divorce fallout   374
divorce mediation   373
economic hardship
   perspective   383
family fluidity   383

family instability perspective   383
income-to-needs ratio   377
independence effect   368
interparental conflict
   perspective   383
joint custody   385
legal custody   385
Levinger's model of divorce
   decisions   370
life stress perspective   382
parallel parents   390

parental adjustment
   perspective   383
parental loss perspective   383
parentification   384
physical custody   385
refined divorce rate   364
selection perspective   383
silver or gray divorce   365
starter marriage   365
stress-related growth   380
unilateral divorce   369

15

REMARRIAGES AND STEPFAMILIES

Learning Objectives

1 Describe the various stepfamily types.

2 Describe the process through which adults and children become members of stepfamilies.

3 Compare happiness, satisfaction, and stability in remarriages and stepfamilies relative to first marriages and original, two-parent families.

4 Identify the challenges and positive aspects of stepfamily roles and relationships.

5 Compare adult and child well-being in stepfamilies relative to that in original two-parent families.

6 Explain the characteristics of a supportive and healthy stepfamily environment.

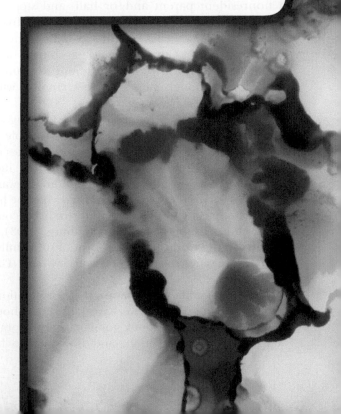

Today, roughly one in every four marriages is a remarriage for one or both partners (Payne 2018a). Nearly two-thirds (63 percent) of remarriages include children from a previous relationship or **stepchildren** (Lewis and Kreider 2015). Families that include stepchildren are called **stepfamilies**. Not all stepfamilies are the result of remarriages, and stepfamilies come in a variety of forms. Since the early 1990s, there has been a 60% increase in births to unmarried mothers (National Center for Family and Marriage Research [NCMFR], 2017). In 14 percent of *first* marriages, at least one member of the couple has had children with someone other than their current spouse (Stykes and Guzzo 2015).

Marriage and remarriage have declined as pathways into stepfamily life in favor of cohabitation (Allred 2018; Guzzo 2017). Between 2008 and 2016, the marriage rate declined 10% and the remarriage rate declined 15% (Payne, 2018b). Meanwhile, the percentage of women who reported ever having cohabited nearly doubled between 1987 and 2013, and two-thirds of married stepfamilies now begin with cohabitation (Kreider & Lofquist, 2014; NCFMR, 2017). Today, roughly half of children in stepfamilies are living with a parent's cohabiting partner as opposed to spouse (NCFMR, 2017). Overall, about three-quarters (78%) of U.S. children whose parents' union dissolved will acquire a stepparent and 40% of all children will spend at least some of their youth living in a stepfamily household (Andersson, 2002; Andersson and Philipov 2002; Andersson, Thomson, and Duntava 2017).

It is important to note that figures generally do not include children in single parent and stepfamily households who have stepfamily members residing in a different household—the spouse or partner of a nonresident parent and/or half- and stepsiblings. Many same-sex couples are raising children from previous heterosexual relationships as well (Ganong and Coleman 2018). Government data sources and national surveys typically do not attempt to count these groupings although this may be changing with the advent of legal same-sex marriage.

Overall, 42% of all Americans report having at least one step-relative. This figure is 50% for people under age 30 (Parker, 2011). The effect of stepfamilies on the United States cannot be overstated and the increase in stepfamilies is a global phenomenon (Stewart and Limb 2020). Although the United States has the highest level of re-partnering among industrialized countries (Heuveline, Timberlake, & Furstenburg, 2003), every country in the world has some form of stepfamily, with divorce and remarriage increasing everywhere (Ganong & Coleman, 2018).

Most stepfamilies are happy in their relationships and lives. Much of what we have said throughout this book—for example, about good parenting practices, supportive communication, and how best to handle family stress—applies to remarriages and stepfamilies. At the same time, stepfamily members experience unique challenges. Foremost, forming a cohesive and satisfying stepfamily requires "leaving behind traditional views of [family] belonging" (Sky 2009).

In this chapter, we will discuss how new family patterns have created various types of stepfamilies and complicated stepfamily dynamics. We'll look at negative stereotypes of stepfamilies, their ongoing stigmatization, and problems resulting from their lack of legal recognition. We will talk about the delicate process of stepfamily formation; ambiguity in norms, roles, and family boundaries in stepfamilies; and what is known about how stepfamily relationships change over time. We'll discuss happiness and stability in remarriages and among couples with stepchildren and how living in a stepfamily affects the well-being of its members. We'll explore some challenges typically associated with stepfamily life and ideas for creating supportive remarriages and stepfamilies. Throughout this chapter, we will point out how stepfamily processes vary for different types of stepfamilies including those of different races, religions, and ethnicities. We'll begin with a discussion of how stepfamilies are currently defined and measured.

DEFINING AND MEASURING STEPFAMILIES

This section describes changes in how stepfamilies are defined and sources of diversity in stepfamilies, providing statistics on the prevalence of remarriages and stepfamilies in the United States.

What Makes a Stepfamily?

The seemingly straightforward question "What is a stepfamily?" is not as easy to answer as you might think (see "Issues for Thought: What Makes a Stepfamily?"). Traditionally, a stepfamily has been defined as a household containing a remarried couple and one or both spouses' children from a previous marriage. However, as mentioned previously, important social and demographic shifts in recent decades have altered family and stepfamily life alike. These shifts include growth in the number of women having children outside of marriage, an increase in unmarried couples with children living together, increasing involvement of nonresident parents in their children's lives, and increasing support of parenting by same-sex couples, among other factors (Stewart 2007). The result is that today there are many different types of families and stepfamilies (Coontz 2015). Although life within these stepfamilies can be quite different, they have one thing in common—the absence of a biological relationship between one's children and one's romantic

Issues for Thought

What Makes a Stepfamily?

What comes to mind when you hear the word *stepfamily*? Perhaps *The Brady Bunch*, the iconic 1960s TV family in which a widower with three sons and a widow with three daughters marry and merge households. The following are stories of five real-life families from this century.

Rhonda and her boyfriend, Al, have a child, Emily. The couple's relationship doesn't work out and Al moves out of state. He has no contact with his daughter and pays no child support. Rhonda meets Peter and they marry. Peter adopts Emily.

Carol and Randy are divorced. Carol has custody of their two children, Emma and Sophia. After a few months of dating, Carol's boyfriend, Roger, moves in with Carol and the girls.

Bobby lives with his mother, Elaine. Bobby's father, Doug, has remarried Leslie. The couple lives with Leslie's children from her first marriage, Teddy, Austin, and Abbey. Bobby sees his father every other weekend, usually for a movie and a bite to eat. Sometimes Leslie and her children go along (depending on what's showing).

Janet is married to Ron, and the couple has two sons, Billy and Justin. Janet falls in love with Ann, a coworker, and divorces Ron. Ann moves in with Janet, Billy, and Justin. The boys refer to her as "Aunt Ann."

Rosemary is divorced and has three college-aged children, Sarah, Ben, and Lily. Rosemary meets James, a widower, at a retirement party and after a two-week courtship, they fly off to Las Vegas and get married. James has a grown son, Todd. The kids get together and plan a reception for the couple.

Critical Thinking

Do you know any families that are similar to these?

Would you consider any of these a stepfamily? Why or why not?

---

partner (stepchildren who have been adopted by a stepparent are discussed here as well).

Pathways to Stepfamily Living Sociologist Kathryn Tillman (2007) has described the various pathways that lead to stepfamily living (see Figure 15.1). For example, a stepfamily can originate with a birth to a married or cohabiting couple (House A). Stepfamilies can also originate from a birth to a single mother who is neither married nor cohabiting (House B). Children born to married or cohabiting parents may later experience the death of a parent (D) or their parents' divorce or union dissolution (C), after which a parent may marry or remarry, forming a married stepfamily (F). Alternatively, rather than marrying, a parent may go on to form a cohabiting union with a new partner (G). A cohabiting stepfamily may be a permanent arrangement or it may transition to a married or remarried stepfamily, depending upon whether either partner has been married previously. Children born outside of a union may later experience their mother's marriage, remarriage, or cohabitation, resulting in either a married or a cohabiting stepfamily. Multipartnered parenthood (having children with more than one partner) and subsequent union dissolutions and formations add even greater complexity to these pathways (Monte 2007).

The various pathways stepfamilies follow can result in vastly different experiences for stepfamily members. For instance, stepfamilies formed by previously married spouses are likely to include relationships with ex-spouses and relatives of ex-spouses. Although stepfamilies that form after the death of one parent do not have relationships with ex-partners, they may have ongoing ties with relatives of the deceased parent. The complexity that results from various stepfamily-formation pathways is just one factor that influences the diversity of the stepfamily experience.

Various Types of Stepfamilies

The family formation patterns described previously have created many different types of stepfamilies, and each type varies in its frequency. For example, whereas stepfamilies formed through divorce and remarriage remain common, stepfamilies with gay and lesbian parents, though they have become more prevalent in recent years, represent only a fraction of all stepfamilies. In the following section, we describe different kinds of stepfamilies and provide the most recent statistics on each.

Stepfamilies Created by Widowhood or Divorce Followed by Remarriage Historically, well into the twentieth century, almost all stepfamilies were formed after the death of a spouse (Strow and Strow 2006). The term *stepparent* originally meant a person who replaces a dead parent, not an additional parent figure (Bray 1999). Remarriage after the death of a spouse was common throughout American history. Analyzing 1689 census data from the Plymouth Colony town of Bristol, John Demos (1966) estimated that, among people at least 50 years old, 40 percent of men and 26 percent of women remarried at least once. These remarriages

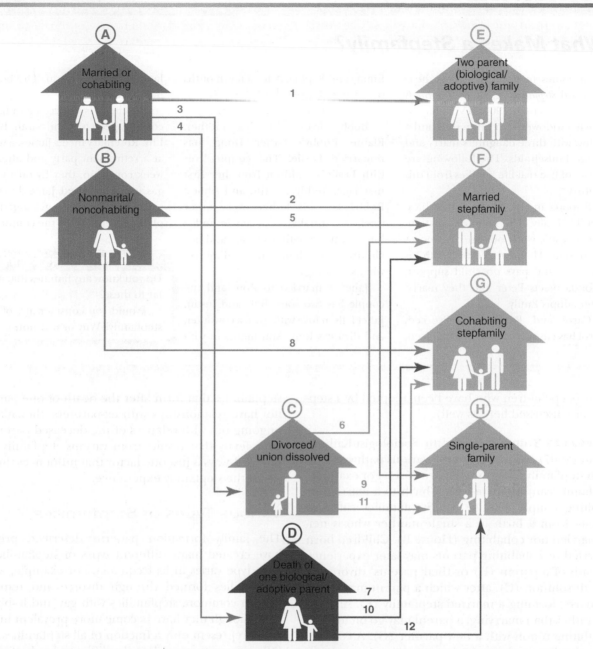

FIGURE 15.1 Pathways to stepfamily living and other family types. The various possible pathways to stepfamily living begin with a baby being born to (or being adopted by) a married or cohabiting couple (House A) or to a single (unmarried, not cohabiting) parent (House B). The couple in House A may take Path 1 and stay together (House E). The parent residing in House B may take Path 2 and choose to not marry or repartner (House H). The couple from House A may take Path 3 and divorce or break up (House C), or they may take Path 4 as a result of one of the spouses or partners dying (House D). The single parents in House B, House C, House D may choose to take Path 5, Path 6, or Path 7 and marry (or remarry), resulting in a married stepfamily (House F). Alternatively, these parents may take Path 8, Path 9, or Path 10 and choose to cohabit, resulting in a cohabiting stepfamily (House G). The previously coupled parents in House C and House D may also take Path 11 or Path 12 and not marry or repartner, resulting in a single parent family (House H). There is further complexity, based on the gender of the biological or adoptive parent, the continued involvement of ex-spouses and partners, and subsequent dissolutions and unions.

Source: Adapted from Tillman 2007, Figure 1, 383–424.

took place quickly, usually within a year of a spouse's death. Data from the English village of Clayworth (Nottinghamshire) indicate that in 1688 roughly one in six households was a stepfamily (Phillips 1997).

Today, many websites are dedicated to remarital wedding planning and services (e.g., I Do! Take Two), an indication that remarriages remain common. However, the vast majority of remarried couples today have been divorced rather than widowed. Although the divorce rate has declined in recent years, roughly half of all marriages will still end in divorce (Kennedy and Ruggles 2014).

As noted above and in Chapter 7, the rate of marriage has fallen. This is also the case for the remarriage rate, which is half of what it was in 1950, with an especially steep decline since 1980 (Payne 2018b; Schweizer 2019). Nevertheless, 60 percent of divorced people remarry, although remarriage rates for men are twice those of women (Livingston, Parker, and Robal 2014; Schweizer 2019). Widowed men also remarry at higher rates than widowed women (Sweeney 2010). Marrying three or more times, sometimes referred to as *serial remarriage*, remains uncommon—only 8 percent of newly married adults have been married more than twice (Livingston, Parker, and Robal 2014). Like marriage, the likelihood of remarriage is substantially higher for men and women with more education (Payne 2018a). Related to their more limited dating pool available to divorced people, there is a larger age gap between remarried husbands and wives than between first-marrieds. In one-third of remarriages, the husband is six or more years older than the wife, as compared to 14 percent of first marriages (Livingston, Parker, and Robal 2014). Remarried couples are also more likely to come from different religious backgrounds and racial backgrounds (Schramm et al. 2012; Stewart and Limb 2020). Remarriage rates also vary by race and ethnicity. Remarriage is more prevalent among Asians and Hispanics and lowest among whites and blacks. Remarriage is also higher among those who are foreign-born as opposed to native-born (Payne 2018a).

About 10 percent of all children (over 5 million) live in married stepfamilies with one biological parent and one stepparent (Kreider and Lofquist 2014). Remarried mothers are more likely to have their children living with them than remarried fathers. Although joint custody is increasing, mothers receive sole physical custody of their children in about 80 percent of divorce cases (Grall 2018). However, the stepfamily is no longer merely the product of divorce or the death of a spouse. Some new kinds of stepfamilies have emerged.

Stepfamilies Created by Nonmarital Childbearing

Stepfamilies can originate from a nonmarital birth, followed by a first marriage, remarriage, or cohabitation. About 40 percent of all births in the United States are born to unmarried mothers (Martin et al. 2019) and more than a third of women who had a child outside of marriage formed a married or cohabiting union with a new partner within five years (Bzostek, McLanahan, and Carlson 2012). It is important to note that childbearing outside of marriage can occur to previously married women, too.

Stepfamilies Created by Cohabitation As discussed in Chapter 6, growth in cohabitation has transformed American families, including stepfamilies. The majority of young men and women today will cohabit at some point in their lives (Manning and Stykes 2015). Cohabitation is particularly common among the previously married and couples in which one or both partners was previously married represent more than half (53 percent) of all cohabiting couples (Lamidi 2015). Most of these couples have children from previous relationships. Between 1988 and 2013, the percentage of stepfamilies comprised of cohabiting couples doubled, increasing from 16 percent to 38 percent (Guzzo 2015).

Living Arrangements of Couples with Stepchildren In one out of five married or cohabiting couples, one or both partners have children from a previous relationship (Monte 2017). These children do not necessarily have to live in the couple's household and may live with their other parent some or all of the time. Figure 15.2 shows the family composition of remarried

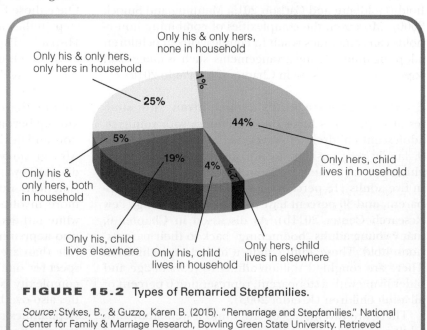

FIGURE 15.2 Types of Remarried Stepfamilies.

Source: Stykes, B., & Guzzo, Karen B. (2015). "Remarriage and Stepfamilies." National Center for Family & Marriage Research, Bowling Green State University. Retrieved February 21, 2020 (https://www.bgsu.edu/ncfmr/resources/data/family-profiles/stykes -guzzo-remarriage-stepfamilies-fp-15-10.html).

couples with stepchildren. By far, the most prevalent stepfamily type is *only hers*, that where the wife's biological children are living in the household (44 percent). The second most prevalent type is *his and hers* with hers living in the household and his living elsewhere (25 percent), followed by *only his* with his children living elsewhere (19 percent). The rest of remarried couples have various other constellations of children (Stykes and Guzzo 2015). Keep in mind that these statistics describe a situation at one point in time and do not reflect changes in stepchildren's living arrangements, which can be quite unstable. Note that this graph does not include information on complex families in which the couple has gone on to have shared biological children (i.e., the half-siblings of stepchildren).

Multiple-Household Stepfamilies Determining where people in stepfamilies live is not always easy. We tend think of families and households as synonymous, but in stepfamilies it is not unusual for members of the same family to live in different households or for family members to move back and forth between households, such as in the case of joint-custody (see Chapter 14). One of the most common scenarios is for children living with their mothers to visit their fathers on holidays, weekends, and during the summer months; only about one-third of children with a nonresident parent see him or her at least weekly (Stewart 2010). Because most nonresident parents eventually cohabit or marry, these visits often include stepparents. Visitation with nonresident parents can fluctuate over time but tends to decline when the nonresident parent forms a new relationship and has biological children or stepchildren in their new household (Goldberg and Carlson 2015; Manning and Smock 2000). Moreover, the complexities of combining households can sometimes result in couples with stepchildren adopting unique living arrangements, such as *living apart together (LAT)* discussed in Chapter 6 (DePaulo 2012).

Stepfamilies with Adult Stepchildren Most studies of stepfamilies focus on stepfamilies with young or adolescent children. However, as discussed in Chapter 9, family relationships do not simply disappear when children reach their eighteenth birthday. Nearly one in five adults (18 percent) have at least one living stepparent, and 30 percent have a step- or half-sibling (Pew Research Center 2011b). As discussed in Chapter 6, many young adults "boomerang" back to their parents' household. These adult children include stepchildren. There are roughly 1 million adults 18 years of age and older living with a stepparent, representing 6 percent of all adult children (Kreider 2003).

There are two modes of entry into a stepfamily with adult stepchildren: (1) the "aging" of stepfamilies that were formed when the children were young and (2) the parents of adult children forming unions with new spouses and partners. The number of such families is increasing for two reasons. First, the American population is aging, meaning that a growing percentage of Americans are reaching retirement age and becoming elderly. A substantial number of older Americans, especially Baby Boomers (born between 1946 and 1964), have experienced divorce, remarriage, and stepfamily life. The stepfamilies formed by this generation are aging along with the population, resulting in an unprecedented number of older Americans who are stepparents (Stewart 2007). These trends will continue as the children and stepchildren of the so-called divorce generation become adults themselves and are in the process of forming (and sometimes dissolving) their own families. Second, the divorce rate has risen among older adults (Brown and Lin 2012), as has cohabitation (Vespa 2012). These trends are producing an increasing number of stepfamilies formed later in life.

Cultural and Religious Diversity in Stepfamilies Just as the U.S. population is becoming more racially and ethnically diverse, so, too, are stepfamilies. Unfortunately, few studies have examined stepfamilies of races, ethnicities, religions, and cultures. One main reason for our lack of knowledge is that some groups, such as Asians and Muslims, are underrepresented in stepfamilies because they have low rates of divorce, nonmarital childbearing, and cohabitation. Another is that most studies of stepfamilies focus on those formed through divorce and remarriage and some racial and ethnic groups, namely blacks and Hispanics, are more likely to enter stepfamilies though nonmarital childbearing, first marriage, and cohabitation (Stykes 2020). Once those families are accounted for, it turns out that stepfamilies are actually disproportionately Black and Hispanic (Papernow 2013; Stewart 2007).

Research on stepfamilies of different races and ethnicities reveals similarities as well as differences. For example, stepfamilies are more easily incorporated into cultures that emphasize intergenerational relationships, kinship beyond the nuclear family, a collectivist orientation, and flexible family boundaries, as is the case among African Americans, Asians, Hispanics, and American Indians (Stewart and Limb 2020). For example, African American nonresident fathers and grandparents are more involved with children in stepfamilies than are white fathers and grandparents (Papernow 2013). Latino stepparents are more likely to take on a disciplinary role than are white fathers as a result of *respeto*, or respect for one's elders (Papernow 2013). Adding to this complexity is that racial, ethnic, and religious identities also overlap. For example, the majority of Hispanic stepfamilies consider themselves religious and identify as Catholic (Hoffman, Breck, and Beasley 2020). Stepfamilies are also more likely than first married stepfamilies to involve partners of different races and ethnicities

(Stewart and Limb 2020). This presents challenges but also opportunities. Sean Singleton, a black stepfather to white stepchildren, says, "I feel fortunate to be involved in the lives of my wife's sons as it gives me the ability to provide them with a positive black male role model, one who combats the images and stereotypes they may most often be exposed to in the media and pop culture (Singleton 2017). Immigration status is important to stepfamily dynamics as well (Coltrane, Gutierrez, and Parke 2008). Overall, however, stepfamily systems tend to operate similarly regardless of the race and ethnicity of the family (e.g., Jensen et al. 2017).

LGBTQ+ Stepfamilies The LGBTQ+ (lesbian, gay, bisexual, transgender, queer or questioning, and other non-conforming gender or sexual identity) community has historically been excluded from stepfamily research, mainly because these families were not married, were not counted in statistics, or did not report their partner's gender and/or their relationship status. As of 2015 same-sex couples can legally marry, so most government surveys today do not assume parents are of opposite genders. Nevertheless, gay and lesbian stepfamilies may be undercounted because many parents do not disclose their sexual identity, sometimes even to their children (Lynch 2000). Nevertheless, estimates suggest that about one-fifth (17 percent) of same-sex couple households contained children from a previous relationship or the current union (Lofquist et al. 2012). As stepfamily researcher Bart Stykes (2020) notes, "The majority of children raised by LGBTQ+ parents are from prior heterosexual and same-sex unions. This means that most childrearing in LGBTQ+ families occurs in the context of stepfamilies" (p. 27). Right now, there are more questions than answers about LGBTQ+ stepfamilies. According to Stykes, one difficulty is separating the confounding effects of non-biological ties from the gender composition of the parents. Failure to do so, he says, runs the risk of attributing "effects" of stepfamily dynamics to LGBTQ+ parenting and vice versa.

A particular problem for children with same-sex parents and stepparents is whether to tell their friends and which ones. Despite changing attitudes, children with same-sex parents are still subject to bullying and harassment. Loyalty binds can be magnified for children in LGBTQ+ stepfamilies. For example, one stepdaughter's biological mother was very hurt about "being left for a man," which affected her relationship with her new stepfather. She said, "it was sort of like I was letting my mom down if I even looked at Frank" (Papernow 2013, p. 131). Despite the legalization of same-sex marriage, LGBTQ+ stepfamilies still face legal barriers, such as in the case of adoption (Kazyak, Woodell, Scherrer, and Finken 2018). Table 15.1 compares how researchers traditionally defined stepfamilies with how they define them today.

TABLE 15.1 A New Model of Stepfamilies

The social and demographic trends described in this section have important implications for the way that stepfamilies are defined. This table compares the traditional definition of a stepfamily to a "revised" definition that incorporates these trends.

ASSUMPTION	TRADITIONAL	REVISED
Union type	Remarriage	First marriage, remarriage, cohabitation
Residence of children	Co-resident; static	Co-resident and nonresident; dynamic
Stage of family life cycle	Child rearing; children ages 0–18	Parenting across the life course (includes children ages 18+)
Race/ethnicity	White	White, African American, Hispanic, etc.
Social class	Middle class	All classes (lower, middle, upper)
Sexual orientation	Heterosexual ("straight")	Heterosexual or homosexual (gay/lesbian)

Source: Stewart 2007

Other Types of Stepfamilies There are other variations in the structure of stepfamilies. For example, studies of stepfamilies usually don't include people who *used to be* stepparents or *used to be* stepchildren but no longer are. This scenario is not infrequent given the instability of remarriage and cohabitation. Relationships with stepchildren generally end with the break-up of the marriage (Noël Miller 2013). However, roughly 300,000 U.S. children reside with just a stepparent and not with a biological parent (Kreider and Lofquist 2014). Studies also often don't consider the special case of stepchildren who have been legally adopted by their stepparents, which is one of the most common forms of adoption in the United States (Stewart 2010). Finally, a child may be a member of a stepfamily but not be a stepchild themselves if that child is a product of the new union. These are children whose family is technically "intact" (i.e., they have both biological parents in the home) but who have half-siblings from their parents' previous relationships. The remainder of this chapter explores what family life is like for stepfamily members.

We begin with a look at how stepfamilies are perceived in society.

Perceptions of Stepfamilies: Stereotypes and Stigmas

Stigmatization refers to the subjection of people to negative labels, stereotypes, and cultural myths that portray them as deviant and harmful simply because they have certain social characteristics (Coleman, Ganong, and Fine 2000). Stepfamilies are *stigmatized* in that they are perceived as being less functional and desirable than original two-parent families. A study of movie plot summaries showed that stepparents were portrayed negatively more than half (58 percent) of the time and were otherwise portrayed as neutral rather than positive (Claxton-Oldfield and Butler 1998). The media has also perpetuated the idea that stepparents are physically or sexually abusive, although data suggest otherwise (Claxton-Oldfield 2008). For example, the stepmothers spank children much less frequently than biological mothers (Smith 2008). Newer TV shows and films portray stepparents and stepfamilies, such as *Modern Family*, *Stepbrothers*, and *Daddy's Home*, in a more positive and realistic way and show the lighter side of stepfamily living.

Still, the stigmatization of stepfamilies is real and extends to stepchildren as well as stepparents. In one study, 211 university students were asked to examine an 8-year-old boy's report card. All the students saw the same report card, but some were told that the child lived with his biological parents whereas others were told that he lived with his mother and stepfather. Asked about their impressions of the boy, male (but not female) students rated the stepchild less positively than the biological child with respect to social and emotional behaviors (Claxton-Oldfield et al. 2002).

One self-help book, *The Blended Family: Achieving Peace and Harmony in the Christian Home* (Douglas and Douglas 2000), compares stepfamilies to other groups of "sinners" including drug addicts and adulterers (Coleman and Nickleberry 2009). Some religions, such as Catholicism, do not recognize a remarriage after divorce unless the first marriage is annulled (Hornik 2001). Cultural variation in the stepfamily concept can create difficulties for stepfamily members and programs designed to help stepfamilies. For instance, the facilitators of one such program had trouble recruiting Hispanics because *stepfamily* is rarely, if ever, used in the Spanish language (Papernow 2013; Reck et al. 2012). There was no word for stepfamily in Japanese until a new Japanese word—*suteppufamiri*—was created in 2001, adopted from the English word *stepfamily* (Nozawa 2020). African Americans are more likely to consider step- and half-siblings as simply "siblings" (Chalandra 2020). The word *stepchild* is often used pejoratively to refer to a person or thing that is treated shabbily. An Internet search of the phrase "poor stepchild of" by the authors yielded 451,000 hits. For example, in a town hall meeting jointly sponsored by the Canadian Psychiatric and Canadian Medical Associations, mental health was referred to as "the poor stepchild of medical care" (Rich 2011).

Some stepfamilies face multiple stigmas stemming from their race, ethnicity, or sexual identity. An interview with one African American stepfamily illustrates the "double stigma" that black stepfamilies experience:

Interviewer: Does living in a black stepfamily make your experience special in any way?

Donnamae (biological mother): It certainly does. Being black is difficult enough in this society. Being a stepfamily is not an easy experience either. Being both presents a serious challenge to your skills of dealing with unpleasant reactions from teachers, physicians, and practically everybody else. It, excuse my cynical expression, may color one's life in black. Sometimes it takes a great deal of patience and humor to deal with all the ignorance and attitudes that you have to take.

Malcolm (stepfather): Most of the occasions on which I feel the burden occur because of other people's reactions . . . mostly white people's reactions. Raising other people's kids is no big deal for me because this is not an unusual thing in my family. It is done all the time. When I was very young, my father was placed in Europe with the army and my mother joined him for a year, leaving me to live with my aunt and uncle. It was kind of fun growing up with my cousins. Later on, when I was an adolescent, my older sister moved in with us because she got into some marital and psychological trouble, so in a way I helped raise my niece and nephews. Currently, all of them still live with my parents. So when Lorenzo came into my life it was not an unfamiliar experience . . . and it is pretty common in my community too. (Berger 1998, p. 145)

Negative stereotypes can even be held by members of stepfamilies themselves (Planitz and Feeney 2009). However, stepfamily members who believed no or few myths about stepfamilies were more optimistic about the remarriage and experienced greater family, marital, and personal satisfaction (Kurdek and Fine 1991). Therapist Anne Bernstein has urged that we work toward "deconstructing the stories of failure, insufficiency, and neglect" and instead "collaboratively reconstruct stories that liberate steprelationships" from this legacy (1999, p. 415). Larry Ganong and Marilyn Coleman (1997), two of the United States' leading experts on stepfamilies for over 40 years, identify another issue. They say that in addition to being regarded negatively, stepfamilies are simply disregarded and ignored by society. They should know. The couple, each of whom had been previously

married, and who married each other in 1979, formed a stepfamily that included Marilyn's two sons. Back then, there were even fewer models of healthy stepfamilies and little information to draw upon. Their research is cited throughout this chapter.

CHOOSING PARTNERS THE NEXT TIME

For people who have been divorced or who have children from previous relationships, forming a new partnership is serious business. Parents have to consider their children's needs in relation to their own needs for companionship. Moreover, finding a partner who accepts children as part of the relationship can be challenging. Finally, if wedding planning weren't difficult enough, couples with children have to consider the expectations and desires of children and in-laws who may be ambivalent about the marriage. As discussed in the next section, parents navigate this unchartered territory.

Dating with Children

Courtship among people who have been previously married has not been a major topic of research. Nevertheless, counselors advise waiting until one has worked through grief and anger over the prior divorce before entering into another serious relationship (Marano 2000; Bonach 2007; Dupuis 2007). Nevertheless, most men and women remarry relatively quickly, usually between three and four years after divorce (Kreider 2006).

Meanwhile, dating with children may differ in many respects from traditional dating. Courtship may proceed much more rapidly, with individuals viewing themselves as mature adults who know what they are looking for. Others are more cautious, wary of repeating an unhappy marital experience. Dating websites such as Match.com and eHarmony are efficient tools with which single moms and dads can identify suitable partners. Members can simply select the characteristics they are looking for in a mate, including their desire for children (online dating is discussed in Chapter 5).

Dating may include outings with one or both partners' children, or couples may prefer to keep their dating relationships and home lives separate. The children might not even know that their parent is dating; for instance, they may be led to believe that their mother's or father's new dating partner is simply a "good friend" who likes to come over and do things with them, especially with new sexual relationships. Sociologist William Marsiglio's (2004) interviews with stepfathers reveal the

great lengths couples will go through to keep their sex life secret. He explains the strategy one couple used to keep the child from finding out the mother's boyfriend had stayed over: "Juan would wake up early and walk outside and knock on the door to wake up the child and signal to him that he had just arrived early that morning. This little trick apparently worked" (p. 36). Marsiglio claims that "A child's resistance to parents' dating is one of the most difficult scenarios faced by divorced, dating parents" (p. 34). Couples with children who are dating often struggle to determine the "right time" to introduce children to dating partners and when to introduce children to each other.

The fact that the majority of stepfamilies begin as cohabiting relationships is a complicating factor. As discussed in Chapter 1, we sometimes "slide" into a family situation rather than making a conscious decision (Stanley 2009). Couples with children from previous relationships can "drift" into cohabitation, with new partners spending increasing amounts of time in the home (Manning and Smock 2005). Others are more deliberate about forming new partnerships. Reid and Golub (2015) conducted interviews with fifteen low-income cohabiting black mothers regarding how they made the transition to living with their partner. The authors suggest that these women engage in a process of "vetting and letting." The women used various strategies to ensure their children's well-being ("vetting") before allowing their partner to move in ("letting"). For example, Dionne said,

> We was friends like three years prior. He was always around and I think one of the reasons I wanted to get with him was because I would always see him with his son and I liked the way he was with his child. I figured if he's good with his kid and he's trustworthy by how he cares for his son then he has to be a good dad. So I figured he was someone I could have around my daughter, because she is a girl and you have to watch who you have around your female kids. So I seen how he was. (Reid and Golub 2015, p. 1240–1241)

Stepfamily experts Karen Bonnel and Patricia Papernow (2018) in their book, *The Stepfamily Handbook*, say that when it comes to dating, it is important for parents to protect kids from the "emotional roller coaster" that comes with dating, as well as staying emotionally available, because many kids fear losing a parent to a new partner. Even though it's enticing, both emotionally and logistically, to introduce new partners and children, they say it's important to protect children from premature "We-ness." Talking with children about the possibility their parents dating, and telling them when they are, is important because children easily pick-up on their parents' dating behavior. Co-parents should be told as well, so they don't end up hearing this information from the kids. A first meeting between a new partner and children should be casual and short, should involve an activity of some kind, and children should always be told in advance that the new partner will be there.

What Kinds of People Become Stepparents?

People who remarry or become stepparents are not randomly selected. For most, they are either divorced themselves or have children from a previous relationship, or they are willing to form a partnership with someone who is divorced or who already has children. Although children from prior relationships can encourage a couple to marry, children can be a liability in the marriage market, especially for women, and more so for women whose children live in the home full time, which is the case for most (Schnor, Pasteels, and Van Bavel 2017; Vanassche, Corijn, Matthijs, and Swicegood 2015). Many single mothers don't have the time or the inclination to date, and potential mates may perceive her childrearing responsibilities as too burdensome. For example, McNamee, Amato, and King (2014) found that women whose children see their nonresident fathers more often are more likely to remarry, although other research shows father involvement has no effect on women's repartnering (Berger, Panico, and Solaz 2018).

Men with children (either co-resident or nonresident) of their own are significantly more likely than childless men to cohabit or marry a woman with children (Goldscheider and Sassler 2006; Stewart, Manning, and Smock 2003). According to the National Survey of Families and Households (NSFH), for men, marrying "someone with children" is less desirable than marrying "someone of a different race," "someone five years older," "someone of a different religion," or "someone who earns much more/much less than you" (Goldscheider and Kaufman 2006). Only marrying "someone unlikely to hold a steady job" ranked lower. Women were less adverse to children than men but still ranked the desirability of this characteristic as low. For both men and women, willingness to marry a partner with children was greater among people who were older, less educated, and white; had more egalitarian gender role attitudes; were from a nonintact family; and had been married or had children themselves. Several studies indicate that remarried spouses and stepparents have less desirable characteristics than first-married

Remarrying couples differ from first-marrying couples in their degree of homogamy. This is because choosing a remarriage partner differs from making a marital choice the first time inasmuch as there is a smaller pool of eligibles on any given attribute.

Mike Tauber/Getty Images

spouses and spouses with no prior children. They tend to be less integrated with parents and in-laws; are more willing to leave the marriage; are more likely to exhibit risky and immature behaviors; and have lower occupational status, education, and income (Killewald 2013; Schramm et al. 2012; Stewart 2007). Overall, previously married men and women are disadvantaged in the marriage market and must, according to demographers Qian and Lichter (2018), "cast a wider net" in their search for a mate. Divorced men who become stepfathers tend to have less education than do unmarried divorced fathers, and fathers who marry childless women (Schnor, Vanassche, and Van Bavel 2017).

Men have higher rates of remarriage and repartnering than do women, but this gender difference has become smaller (Payne 2018a). There are racial and ethnic differences in remarriage rates, with whites repartnering more quickly than Latina and Black women. Black-white intermarriages are less common in remarriages than first marriages, but white-Asian and white-Hispanic are more common, which the authors of the study attribute to a smaller pool of eligible partners available to the previously married (Choi and Teinda 2017; Qian and Lichter 2018).

Second Weddings

Americans tend to have rigid notions of what a wedding is supposed to look like. In her book *White*

Weddings, sociologist Chrys Ingraham (2008) describes the homogeneous way that media portray weddings—white, middle-class, heterosexual, and the *first* marriage of both partners. The modern wedding is not playing into this stereotype, however, and is in fact "celebrating" growing diversity in marriage pool. The wedding industry estimates that remarried couples are responsible for about 40 percent of all their revenues (Ingraham 2008). Second weddings are different from first weddings—they tend to have fewer guests, have few or no wedding attendants, and forgo traditions such as the tossing of the bouquet and garter (perhaps because the wedding guests are older and less likely to be single). Remarried couples have the freedom to design a wedding that is unique. One groom described his wedding: "After [the ceremony], everyone lined up to get tacos and cupcakes from food trucks then relaxed on the grass" (Wexler 2014). Couples may choose to get married away from home and are thought to be fueling the trend toward "destination weddings," which sometimes involve flying the wedding party and guests to an exotic location.

Remarriage wedding ceremonies are complicated, emotionally charged, and often awkward affairs that need to be handled delicately. Professor Leslie Baxter and colleagues asked thirty male and fifty female stepchildren from two Midwest colleges to describe their parents' remarriage ceremonies (Baxter et al. 2009). The wedding ceremonies the students described ranged from a downsized version of the traditional wedding to a courthouse civil ceremony to an informal event in a backyard or casual restaurant. Sometimes a couple left the area to remarry and then informed family and friends later.

Many of the students were critical of their stepparent's wedding ceremony. Some argued that "the relationship didn't deserve that" degree of celebration. Other students critiqued the ceremony as being too casual: "Well, if you want everyone to take [your remarriage] seriously, it needs to be a little more than um, a barbeque" (Baxter et al. 2009, pp. 476, 477). A male student said he felt that the ceremony had inadvertently insulted his family of origin:

> The only part that upset me was the pastor was talking about how life's events lead you up to this moment and how there's bumps in the road, and blah, blah, blah, but this is where you're supposed to be. And I got pissed, because I was like, was my mom the bump in the road? (p. 480)

The children of remarrying couples are likely to have mixed emotions about the event. Some couples design a family-centered ceremony, paying particular attention to their children's feelings. In this case, not only the bride and groom but also all members of the new stepfamily are celebrated. For example, one of Baxter's respondents described how all of the stepchildren as well as the remarrying couple "got little rings to show that we all got married" (Baxter et al. 2009, p. 475). Although the family-centered ceremony was the least common wedding type described by the students, it was the one most appreciated:

> Over and over again, participants told us that they had wanted far greater involvement with the remarriage event... [whether] it was being granted sufficient time to get to know the stepparent, being informed and consulted about the decision to marry, participating in the planning of the ritual, or creating a ceremony and/or artifacts that celebrated the family. (Baxter et al. 2009, pp. 481, 485)

A 20-year-old respondent offered the following advice: "If people are thinking about starting a stepfamily, they should take their time and, you know, keep everyone informed and pay attention to everyone's feelings. . . . Like [a mother should] talk to her daughters and see how we, um, feel, and take those feelings into consideration" (in Baxter et al. 2009, p. 481). Indeed, researchers advise that forming stepfamilies when a stepchild-to-be is in adolescence can be especially difficult (Bray and Easling 2005). On the other hand, remarriages formed when the children are older may be viewed quite positively. One adult stepchild from a British study says this about his mother's remarriage:

> Oh yeah, brilliant together. I mean mum never went abroad with my dad because dad wouldn't go anywhere and they didn't have the money anyway. It's been very good. It's been good for each other, you know. . . . If something had happened to one of them, I think the other one would just roll up because they go to their clubs four or five times a week. (Allan, Crow, and Hawker 2011, p. 136)

Because the re-wedding ceremony can influence how children feel about their future stepfamily, involving them may be more important than a couple realizes. Stepchildren's adjustment is a factor in a remarried couple's overall happiness.

HAPPINESS, SATISFACTION, AND STABILITY IN REMARRIAGE

Research consistently shows no difference in marital happiness, satisfaction, and other dimensions of marital quality between couples in first and subsequent marriages (DeLongis and Zwicker 2017; Lin, Brown, and Cupka 2018). Yet, despite similar levels of marital quality, remarried couples are more likely than first-married couples to divorce. There are several reasons for this. You might recall from previous chapters that cohabiting before marriage increases the likelihood of divorce (Rosenfeld and Roesler 2019). Remarried people are

more likely to have cohabited, either before their first marriage or before their second. This is partly due to *selectivity*. For one thing, people who divorce in the first place—and those who cohabit—are disproportionately from the lower-middle and lower classes, which generally have a higher tendency to divorce and redivorce. They also tend to have more liberal attitudes toward divorce.

Maybe you've heard people who are about to remarry say that they plan to work harder in their new marriage and not repeat the mistakes they made in their first. For some couples, this may be the case (Brimhall, Wampler, and Kimball 2008). Many remarried couples may go to great lengths to avoid conflict. Couples with less marital conflict have been shown to have higher marital quality, and in turn a lower risk of divorce, especially among remarried couples with stepchildren (Van Eeden-Moorefield and Pasley 2008). However, remarried couples who are reluctant to directly address problems that arise in their relationship may fare no better. "The experience of destructive conflict that often precedes the breakup of a first marriage can be highly stressful and . . . might prompt avoidance of communication about the difficulties that stepfamily couples have to negotiate" (Halford, Nicholson, and Sanders 2007, p. 480). As pointed out in Chapter 11, researchers and counselors advise directly addressing difficult issues. In one Canadian study (Saint-Jacques et al. 2011), researchers interviewed two groups of participants: one group comprised of stepfamilies who had been together for at least five years and a second group consisting of stepfamilies that had dissolved

within the first five years. The key difference between the two was not the nature, number, and intensity of their problems but the *way* in which they approached the problems. Whereas families that stayed together tried multiple solutions, families that did not either avoided talking about the problem or gave up entirely by ending the relationship.

A third reason that remarriages are more likely to end in divorce may be that if seemingly irresolvable problems do arise, remarrieds as a category are more accepting of divorce; they have already demonstrated their willingness to divorce. As one remarried husband said, "We're not going to tolerate the kind of crap we did the first time around. . . . I don't need it again. She doesn't either" (Brimhall, Wampler, and Kimball 2008, p. 378).

Fourth, although stepfamilies' increasing numbers and visibility have led to their growing social and cultural acceptance, remarried families continue to be stereotyped as "less than" or other than "normal" (Ganong and Coleman 2004). One indication of this situation involves re-weddings. As discussed in Chapter 7, weddings publicly announce a couple's commitment. Re-weddings are typically less extravagant than first weddings—indicating their culturally diminished importance. As a result of the relative devaluing of remarriage, remarrieds may receive less social support from friends or extended kin and be somewhat less integrated with parents and in-laws and thus not experience the encouragement or social pressures that can act as barriers to divorce (Coleman, Ganong, and Fine 2000).

Finally, the most important reason for higher divorce rates among remarried couples is stepchildren (DeLongis and Zeicker 2017; Aughinbaugh, Robles, and Sun 2013). This fact was first discovered over four decades ago. Sociologists Lynn White and Alan Booth interviewed a national sample of more than 2,000 married people under age 55 in 1980 and reinterviewed four-fifths of them in 1983. White and Booth found that the quality of the remarital relationship itself did not affect the odds of divorce, but the partners' overall satisfaction with family life did:

> [W]e interpret this as evidence that the stepfamily, rather than the marriage, is stressful. . . . These data suggest that . . . if it were not for the children these marriages would be stable. The partners manage to be relatively happy despite the presence of stepchildren, but they nevertheless are more apt to divorce because of child-related problems. (White and Booth 1985, p. 696)

When neither spouse enters a remarriage with children, the couple's union is

Although stepfamilies are, on average, less stable than traditional two-biological parent families, family cohesiveness can be built through shared activities. Positive family relationships are more important to children's outcomes than family structure itself.

usually much like a first marriage. But when at least one spouse has children from a previous marriage, family life often differs sharply from that of first marriages. Parents' relationships with their children predate the couple relationship and they often have developed particular habits and routines (e.g., favorite TV shows, styles of play, bedtime rituals) very new to stepparents. Here's one such scenario, based on Patricia Papernow's (2013, p. 92) work with stepfamilies:

Sandy Danforth and her ten nine-year-old daughter Sabina had always celebrated Christmas morning in their pajamas. On their first Christmas morning as a new stepfamily, Eric appeared for their holiday breakfast dressed in a suit and tie. Sabina recalls, "I started crying." Sandy smiled. "Actually, Sabina hollered, 'WHAT are you wearing? THAT'S NOT CHRISTMAS!' and ran upstairs sobbing hysterically." She went on to explain that Eric, seeing as how this was their first Christmas, had just wanted to dress in his "holiday best." The next day Sabina and Sandy went shopping and got Eric "the fanciest silk bathrobe we could find so he could still dress in his 'holiday best' on Christmas. The next year he came down for Christmas breakfast in his pajamas, wearing the bathrobe, plus his favorite silk necktie. And that's what Eric's worn for Christmas morning every year since."

Another issue is that stepfamilies are formed through loss, especially for children, whose lives may be completely upended and who lack control over their situation. Parents and new partners should openly acknowledge things are different, listen to their children's concern ("I miss our Saturday afternoons together") and reassure them that most things will remain the same ("How about we make sure you get some time with just Daddy on Saturdays. Deal?") (Bonnell and Papernow 2018; DeLongis and Zwicker 2017).

Couples with stepchildren experience more tension and conflict in their relationship than do first-marrieds, especially related to discipline. One of the biggest sources of conflict is when stepparents taking on a disciplinary role too soon. Bonnell and Papernow (2018) caution, "connection before correction!" and cite their colleague, Beverly Reifman who says, "relationships before rules" (p. 79). Directives like, "Clean up your room" and "No TV before homework" need to come from the biological parent. Similarly, Ganong and Coleman (2016) emphasize that developing a positive stepparent-stepchild relationship is a process, which they refer to as **affinity building** not unlike "courting" the child. They say to stepparents, "You have to be a big person. You've got all the costs, but none of the benefits. Stepparents need to be strong on delayed gratification because that's all they're going to get. Making an effort to build a friendship before assuming the parental role is the key to success." These efforts generally pay off in the form of reduced stepparent-stepchild

conflict, improved marital quality, marital confidence, and family cohesion (Ganong, Jensen, Sanner, Russel, and Coleman 2019).

DAY-TO-DAY LIVING IN STEPFAMILIES

Although the formal marriage or remarriage ceremony is typically considered the "start date" of a stepfamily, stepfamily formation is a process that unfolds over time. Stepfamily members do not become an "instant family," no matter how much they want to. Moreover, stepfamily dynamics will always be different from those of original two-parent families. In the following section, we discuss what it is like to live in a stepfamily and identify factors that affect the development of a stepfamily identity, stepfamily roles, relationships between family members, and changes in those relationships over time.

Challenges to Developing a Stepfamily Identity

A **cultural script** is a set of socially prescribed and understood guidelines for defining responsibilities and obligations and hence for relating to each other (Ganong and Coleman 2000). Society offers members of stepfamilies an underdeveloped script. As discussed previously, our traditional wedding "script" assumes the couple is marrying for the first time. Lacking a standard set of guidelines, couples planning a second wedding must make things up as they go. Noting the cultural ambiguity of stepfamily relationships, social scientist Andrew Cherlin (1978) called the remarried family an **incomplete institution**. Unlike first marriages, he said, remarriages lack social norms to guide behavior and therefore remarried couples do not have the tools to solve problems and "get along." Incomplete institutionalization is associated with greater stress, poorer relationship quality, lower levels of closeness, and overall lower well-being in stepfamilies than in traditional nuclear families. Moreover, "cohabiting stepfamilies are arguably even less institutionalized than married stepfamilies, which are formed through a tie that is legally binding" (Brown and Manning 2009, p. 88). The result is that in stepfamilies everyday living can be extremely challenging. In terms of evidence, he cites our language, laws, and customs (see "A Closer Look at Diversity: Do You Speak Stepfamily?").

Boundary Ambiguity in Stepfamilies In a stepfamily, "Whose picture goes on the mantle?" may be a difficult question to answer (Munroe 2009, p. 168). You may recall that family **boundary ambiguity** is a "state when family members are uncertain in their perception

A Closer Look at Diversity

Do You Speak Stepfamily?

Whereas children in traditional families call their parents "Mom" and "Dad," there is no standard way for stepchildren to refer to their stepparents. Communication researchers interviewed thirty-nine stepchildren at a large Midwestern university (Kellas, LeClaire-Underberg, and Normand 2008). The students described how they purposefully choose language to clarify their family form for others:

> Whenever I talk about [my stepfamily] with people it's always my stepdad, my stepmom, stepsister. . . . I always put those terms in there because I do have a biological, real sister and so I guess, I try to help people out because obviously my family's really confusing. (p. 251)

Note this respondent's reference to her "real" sister.

In addition to clarifying their family situation for others, the students deliberately used language that normalized stepfamily living for outsiders: "If I am outside the family and people ask me where I am going I say I am going to my mom and dad's house. So face-to-face, I call [my stepmother by her name], but with everybody else, it's just my mom" (p. 249).

The students in this study also described the way that language—interestingly, language associated with the nuclear-family model—symbolized and communicated stepfamily members' closeness and solidarity. One participant reported that as a young child she referred to her stepfather as "Daddy" to acknowledge that she felt close to him. Another student reported overhearing his younger stepbrother talking with his friends about how happy he was to have a new big brother: "[H]e called me his brother. . . . After I heard that it went from being a stepfamily . . . to being an actual family" (p. 249). Similarly, in a different study, a stepfather told an interviewer: "I have never, ever, thought of these two girls as my stepchildren. They're just my daughters, and I've always referred to them as such" (in Hans and Coleman 2009, p. 611).

On the other hand, some students reported strategically using language that communicated separateness: "At the very beginning I wouldn't even call her my stepmom. I would call her my dad's wife. . . . I didn't want that connection" (in Kellas, LeClaire-Underberg, and Normand 2008, p. 250). Other students pointed to consciously negotiated terms by which they referred to or addressed stepfamily members as they balanced relationships in an ambiguous family environment. Fairly common was the decision to call a stepfather by a different term than that reserved for the biological father—referring to a biological father as "Dad," for example, while calling a stepfather by his first name. One stepdaughter reported that when with her biological father, she "always has to be really careful" to refer to her stepfather as "Paul" although she usually calls him "Dad."

Another student confessed that he consistently addressed his stepfather as "Bill," although he wished he could have called him "Dad," but "it just never came"—even though "he really is the one that raised me" (Kellas, LeClaire-Underberg, and Normand 2008, p. 251). Whether the "stepparent" label is adopted is dependent upon the quality of the child's relationship with their biological parents. Whereas closeness with mothers increases their likelihood using "stepfather," closeness to fathers increases their likelihood of resisting that label in favor of "my mother's husband." The presence of stepsiblings in the home encouraged the adoption of the stepfather label (Thorson and King 2016). One father explains how his children have adopted special names for each of their eight grandparents including "Granny Pam," "Nana," Nona," and "Yaya" (Papernow 2013, p. 161). Finally, some members of stepfamilies may not call their step-relatives anything, engaging in a language pattern called *no-naming* or *zero forms of address* (Duvall 1954, p. 7; Schneider 1968, p. 84).

Critical Thinking

Have you ever considered how you refer to your family members? What makes a family member "not real" as opposed to "real"?

---

of who is in or out of the family or who is performing what roles and tasks within the family system" (Boss 1987, p. 709). Boundary ambiguity is common in stepfamilies. One stepfamily member, a wife and mother who complained about never knowing how much to fix for dinner on any given day, describes her family as an "accordion" that "shrinks and expands alternately" (Berger 1998).

Interviews with stepfamily members reveal that definitions of family often differ between parents and children and between siblings. Susan Stewart (2005a) analyzed 2,313 stepfamilies, which were defined as "married or cohabiting couples in which at least one partner has a biological or adopted child from a previous union living inside or outside the household." Stewart defined *boundary ambiguity* as "any discrepancy

in partners' reports of shared children (the biological or adopted children of both partners) and/or stepchildren (biological or adopted children from previous unions)" (p. 1009). Boundary ambiguity was present among 25 percent of couples with stepchildren and higher among couples with nonresident stepchildren than with resident stepchildren, especially when they were the biological children of the wife. Boundary ambiguity was also more common among cohabitors. This matters because couples with greater boundary ambiguity have significantly lower relationship quality and stability.

In another study, adolescents and their mothers were asked to list the members of their families. Brown and Manning (2009) compared boundary ambiguity in four family types: (1) families headed by two biological parents, (2) single-mother families, (3) married families in which one parent is a stepparent, and (4) cohabiting families in which one parent is a stepparent. As shown in Table 15.2, cohabiting stepparent families evidenced the greatest boundary ambiguity whereas families with two biological parents evidenced the least.

Swiss researchers Castrén and Widmer (2015) asked parents and children from stepfamilies to draw a picture of their families and then examined the drawings. Whereas mothers showed a higher level of "exclusiveness" in their drawings, children's pictures showed more "inclusiveness." For example, most children included their nonresident biological fathers in their pictures. Boundary ambiguity has been observed among step- and half-siblings as well (White 1998). Relationships with kin outside the immediate stepfamily are also complex and uncharted (Ganong and Coleman 2004). There are few mutually accepted ways of defining and relating to new extended and ex-kin relationships, but it appears that new relatives do not so much *replace* kin as *add* them (White and Riedmann 1992).

The Stepfamily System

All families, including stepfamilies, are part of a *system* of relationships. In other words, relationships between two stepfamily members are affected by relationships between all other stepfamily members.

Family Systems Theory A family systems approach (see Chapter 2) is especially useful for understanding stepfamily relationship dynamics. Family systems theory emphasizes *interdependence* in family relationships, the idea that the relationship between two family members is influenced by each one's relationship with all the other members of the family. This interplay is particularly important in stepfamilies (Raley and Sweeney 2020). In two studies of stepfather families by Valarie King and colleagues (King, Amato, and Lindstrom 2015; King and Lindstrom 2016), children's level of closeness to their stepfather was higher among those who were closer to their mothers and, somewhat surprisingly, their nonresident biological father. Yet closeness with the nonresident biological parents was associated with more stress. The authors speculate that the children might feel "torn between these two relationships, resulting in greater stress" (Jensen, Shafer, and Holmes 2017, p. 9). Amato, King, and Thorsen (2016) found children with close relationships with mothers tended to have close relationships with stepfathers; in that study, relationships with nonresident fathers were not dependent on that relationship. And Jensen et al. (2017) found that among emerging adults, close relationships with resident biological parents, resident stepparents, and non-resident parents each have independent effects on children.

Because stepfamily members are often uncertain about how they should behave toward one another, they tend to look to other family members for cues. For example, a new stepmother may look to her husband for guidance about how best to get the children to eat their vegetables. The strategy she employs will depend on her husband's relationships with the children and their relationships with each other. Perhaps they will decide that the husband will be the one to say something to them. As the children grow older and as the family becomes more comfortable in their relationships, she will need her husband's guidance less. Having the biological parent act as a "mediator" between the stepparent and stepchildren, at least initially, is a wise choice (Bonner and Papernow 2018).

TABLE 15.2 Boundary Ambiguity in Four Family Forms

No boundary ambiguity means that a mother's report of who is in the family coincided with that of her adolescent child.

FAMILY FORM	BOUNDARY AMBIGUITY (%)
Two-biological-parent family	0.6
Single-mother family	11.6
Married stepparent family	30.2
Cohabiting stepparent family	65.9

Source: Based on analysis of data from the National Longitudinal Study of Adolescent Health (*N* = 14,047). Adapted from Brown and Manning 2009, p. 92.

Triadic Communication Family dynamics can sometimes become set. In one study, researchers asked fifty university student stepchildren to describe the communication patterns in their stepfamilies (Baxter, Braithwaite, and Bryant 2006). As illustrated in Figure 15.3, the results showed four different relationship and communication patterns among a biological parent, stepparent, and child. In a *linked triad*, a child's interaction is connected with the stepparent through the child's biological parent: "A lot of stuff I

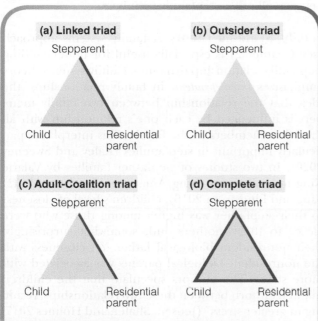

FIGURE 15.3 Perceived types of triadic communication structures in stepfamilies. "A darker line represents the presence of a direct, positive . . . line of / in a given stepfamily dyad, and a lighter line represents the absence of a direct, positive . . . line of communication" (Baxter, Braithwaite, and Bryant 2006, p. 388).

(a) Linked Triad. Communication between child and stepparent is linked though the residential biological or adoptive parent.

(b) Outsider Triad. Child's communication takes place primarily with the residential biological or adoptive parent; child feels little interdependence with the stepparent. In this triad, the step parent is an outsider.

(c) Adult-Coalition Triad. Stepparent and residential biological or adoptive parent are viewed as forming a coalition to the relative exclusion of the child.

(d) Complete Triad. Equal communication and interaction along all three sides of the triad, incorporating all three triadic components equally.

Source: Baxter, Braithwaite, and Bryant 2006, p. 388

communicate with my stepdad goes through my mom" (Baxter, et al. 2006, p. 389). In the *outsider triad*, the child and the biological parent maintain interaction, but the stepparent remains an outsider and pretty much irrelevant to the child's life: "I focus my talking towards my mom. And um like when we're [all] watching the TV, I just always turn to my mom and usually talk to her" (p. 391).

In the *adult-coalition triad*, the child views the biological parent and the stepparent as maintaining a couple relationship that ignores the child's concerns. As an example:

> My sister and I have some difficulties with my stepmom. . . . And [my dad will] say, "Well, honey, you have to understand this and this and this," and like makes excuses for her, and I kind of want to say to him, "Well, we're your *daughters*," you know. (p. 392)

In the *complete triad*, communication flows freely, involving all stepfamily members equally: "I'll call home and I'll talk to my stepfather and I'll talk to my mom. . . . He's like a father to me" (p. 393). Unanimously, the student respondents regarded the adult-coalition pattern as negative and saw the complete triad pattern as ideal. However, the researchers hastened to add that "the outsider triad can be functional for the stepfamily so long as it is compatible with expectations of the stepparent" (p. 395).

Other Challenges Facing Stepfamily Systems Disruptions associated with one or more stepchildren's comings and goings may be stressful (Kheshgi-Genovese and Genovese 1997). Stepchildren in joint custody arrangements may regularly move back and forth between two households with two sets of rules. A child's ties with the noncustodial parent can make the pre-divorced or separated family seem "more real" than the stepfamily. After returning from visits with the noncustodial parent, a stepchild may unintentionally undermine stepfamily definitions. As an example, a 5-year-old visited her biological mother's house where she mentioned a "brother" in her stepfamily. When the girl returned to her stepmother's home, she announced, "My mother says he's not my brother, he's my half-brother" (Bernstein 1997).

Furthermore, stepsiblings may not get along well. Especially in the case of multipartnered parenthood, the lives of stepchildren living in one household can vary greatly:

> [One] child may have a devoted nonresident father who sees her regularly, another child who has no contact with her father jealously watches her half-sister go away for weekends with her dad, and a third child—from the new partnership—has both of her parents in the household. . . . The inequalities among children in the same household can be stark. (Cherlin 2009a, p. 195)

In some cases, stepsibling rivalry is sparked by actual or perceived inequality of treatment by one parent. As one stepchild explained,

> [the] way my brothers and I saw it was that my dad treated us the same and he tried to treat her children the same, but we saw a difference in the way she treated her children and the way she treated us. . . . My brothers and I were being treated differently than her children were. (Baxter et al. 2006, p. 390)

Moreover, as noted by psychotherapist Susan Pacey (2005):

> Grandparents are powerful figures in the hierarchy of the stepfamily, and can help or hinder the couple in forming a new life together. For example, gifts or bequests made to the biological grandchildren only, when a long established stepfamily home includes step or half siblings, may prove divisive and detrimental to the stepfamily and the couple. (p. 368)

Another challenge to fashioning cohesive stepfamilies lies in children's lack of desire to see them work out. Children often harbor fantasies that their original parents will reunite (Bray 1999; Gamache 1997). This hope can last into young adulthood. As one college student described hearing of her parent's remarriage intentions,

> It didn't hit me until I hung up the phone and then I remember just crying and I couldn't understand why I was crying. You know? And I think it had just hit me that my parents are never going to get back together . . . . I had never thought that [their] getting divorced was the end of them. (Baxter et al. 2009, p. 480)

In the case of remarriage after widowhood, children may have idealized, almost sacred, memories of the parent who died and may not want another to take his or her place (Andersen 2002; Barash 2000).

Thus, although it is common to think of children as the "receptors" or even "victims" of family dynamics, children also shape their interactions with adult family members and adult family members' relationships with each other (Menning 2008). Whereas original two-parent families might be thought of as operating from the "top down" (from parents to children), stepfamilies are thought to operate from the "bottom up" (from children to parents). Rosenberg and Hajfal (1985) refer to these as **dripolator** and **percolator effects**. Several studies suggest that stepchildren have a greater impact on the marital relationship and overall family happiness than biological children (Bray, Berger, and Beothel 1994; Crosbie-Burnett 1984; Hetherington and Jodl 1994). One stepparent laments, "We didn't have a very long romance or honeymoon period. In fact, we never had a honeymoon period . . . when [stepchild] came home, the romance went out of our marriage" (Felker et al. 2002, p. 131).

Stepfamily Roles

Incomplete institutionalization of stepfamilies includes stepfamily roles themselves. As we have seen, there is no cultural script to show members of stepfamilies how to play their roles. Given this **role ambiguity**—that is, few clear guidelines regarding what responsibilities, behaviors, and emotions stepfamily members are expected to exhibit—it may not be surprising that relationship and communication patterns in stepfamilies are highly variable.

Lower role ambiguity has been associated with higher remarital satisfaction, especially for wives, and with greater parenting satisfaction, especially for stepfathers (Kurdek and Fine 1991; Munroe 2009). However, the role of the stepparent is "precarious; the relationship between a stepparent and stepchild only exists in law as long as the biological parent and stepparent are married" (Beer 1989, p. 11). Although some stepchildren certainly do maintain relations with a stepparent after a stepparental divorce, one must forge personalized ways to do so in the absence of commonly understood norms (Dickinson 2002).

One result of role ambiguity is that society seems to expect stepparents and children to love each other in much the same way as biologically related parents and children do. In reality, this is not often the case, and therapists point out that stepparents and stepchildren should not expect to feel the same as they would if they were biologically related (Barash 2000). Stepparents exhibit wide variation in their relationships with their stepchildren. At one extreme, there are stepparents who legally adopt their children, have their stepchildren legally take their last name, or pretend to be their child's biological parent (Stewart 2010; Filinson 1986). At the other end of the spectrum are stepparents who have little or no involvement in their stepchildren's lives or who are abusive. In an Australian study, children in stepfamilies were at greater risk of injuries than children in two-biological-parent families (Malvaso et al. 2015). Importantly, the stepparent was not necessarily the perpetrator. Rather, abuse was more common in families with greater residential instability, those in which mothers abused alcohol, and those that included boys—and stepfamilies have more of these traits.

Most stepchildren do not suffer abuse, but parental relationships with stepchildren *are* different. Among stepfathers living with minor stepchildren who were part of the National Survey of Families and Households, one-third felt that it is *definitely true* or *somewhat true* that a stepparent is "more like a friend than a parent," but half of the stepfathers felt that this statement was *somewhat false* or *definitely false* (Marsiglio 1992). A little more than half (55 percent) of stepfathers said that it was *somewhat true* or *definitely true* that "having stepchildren is just as

satisfying as having your own children," yet one-quarter (27 percent) said that it was *somewhat true* or *definitely true* that "it is harder to love stepchildren than it is to love your own children." A more recent study of stepfathers by the same author (Marsiglio 2004) reveals a similar ambivalence with respect to **paternal claiming**, the extent to which to stepfathers see their stepchildren as biological children.

In general, stepparents take a less active role in the lives of their children than do biological parents (Raley and Sweeney 2020). This is not necessarily bad. In one study, stepfathers and their spouses tended to agree that a low level of involvement of the stepfather was good (Kurdek and Fine 1991) and that too much involvement too soon can result in negative backlash from the stepchild (Hetherington and Kelly 2002). Other research indicates that the effect of stepparental time on stepchildren's well-being is small, especially for older children (Hofferth 2006). The quality of the stepparent-stepchild relationship is more important than quantity of time together in determining their degree of influence on their stepchildren (Carlson and Knoester 2011; White and Gilbreth 2001). Overall, stepparents find parenting more stressful than do biological parents (Raley and Sweeney 2020).

The Stepfather Role Many children have positive relationships with their stepfathers and men who may eventually become their stepfathers. At one end of the continuum are men who are deeply involved in the children's lives and marry the mothers. At the other end are mothers' short-term romantic liaisons who may have little to do with the children and may be around only for a few months. In between is a gray area of men in the household who are not taking on a parental role but who do spend some time with the children and may be present for a year or two (Cherlin 2009a, p. 101).

Speaking about his girlfriend's 7-month-old child, Jesse said this:

> We played with him for a day at the park and it was a blast! It was fun. . . . It was just a lot of fun going to the park with Shaun and everything. It was neat. . . . I was like—oh my goodness, this thing is—he's awesome. . . . I was still kind of uneasy because I wasn't around kids a lot . . . but it was a really fun day for both of us. . . . (Marsiglio 2004, p. 41)

Children's relationships with their stepfathers is dependent upon many factors, such as their gender and age, and how long the stepfather has been in the child's life. Some studies suggest girls develop closer relationships with their stepfathers than do boys (Jensen and Shafer 2013). There is more ambiguity in the stepfather role among stepfathers who are cohabiting as opposed to married (Manning, Smock, and Bergstrom-Lynch 2009). These men are less involved with their stepchildren, and their relationships are of shorter duration (Stewart 2007).

Men who decide to marry women with children come to their new responsibilities with varied emotions, typically quite different from those that motivate a man to assume responsibility for his biological children. "I was really turned on by her," said one stepfather of his second wife. "Then I met her kids." Along with feeling positive about what he is undertaking, a new husband may be anxious, fearful, or ambivalent.

Children and their stepfathers are more likely to feel positive about their relationship when the stepfather assumes a parental identity, when his parenting behavior meets his own and other family members' expectations, and when his parental demands are not contested by an involved, nonresidential biological father (Coleman, Ganong, and Fine 2000; MacDonald and DeMaris 2002; Marsiglio 2004). For teens, close ties to stepfathers are more likely to develop when the adolescent has close ties to his or her mother before the stepfather enters the family. Most stepchildren do not have to "choose" between their stepfather and biological father; they can have close relationships with both (King 2009). Moreover, children's relationships with their stepfathers are not static. Analysis of the National Longitudinal Study of Adolescent Health showed variation over time in how the children referred to their stepfathers. Only 75 percent used "stepfather" or "my mother's husband" consistently; 25 percent of children either delayed use of the term or stopped using it (Thorson and King 2016).

The Stepmother Role There has been far less research on mothers and stepmothers in stepfamilies than fathers and stepfathers (Sweeney 2010). Research indicates that the stepmother's role and relationships with stepchildren are more problematic than those of stepfathers (Stewart 2007). Stepmothers are given what many consider an impossible job, a situation described as the **stepmother trap**. On the one hand, mothers in our society are expected to be self-sacrificing and completely devoted to their children (Arendell 2000). On the other hand, stepmothers are often stigmatized—seen and portrayed as cruel, vain, selfish, competitive, and even abusive (Schrodt 2008; Whiting, Smith, and Barnett 2007). Stepmothers often find themselves in the position of a parent in the face of resistance from stepchildren. Mothers of stepchildren are more likely than mothers of biological children to feel that the division of childcare with their partner is unfair, and feel their partners are poorer parents (Guzzo, Hemez, Anderson, Manning, and Brown 2019).

Maureen McHugh (2007) is a stepmother who writes for the website Second Wives Café: Online Support for Second Wives and Stepmoms (secondwivescafe.com).

The following is an excerpt from her online article "The Evil Stepmother":

> My 9-year-old stepson Adam and I were coming home from Kung Fu. "Maureen," Adam said—he calls me "Maureen" because he was 7 when Bob and I got married and that was what he had called me before. "Maureen," Adam said, "are we going to have a Christmas tree?"
>
> "Yeah," I said, "of course." After thinking a moment, "Adam, why didn't you think we were going to have a Christmas tree?"
>
> "Because of the new house," he said, rather matter-of-fact. "I thought you might not let us." It is strange to find that you have become the kind of person who might ban Christmas trees.

In one study, stepmothers were asked about their expectations about the stepmother role. The researchers found that stepmothers expected to be included in stepfamily activities but certainly do not see themselves as replacing the stepchild's mother. The more time a stepmother spent with her stepchildren, the more she expected to be included in stepfamily functions and decisions, and the more she behaved as concerned parent rather than as a friend (Orchard and Solberg 2000). Biological mothers and stepmothers are often pitted against one another, leading many biological mothers to simply assume their relationship will be difficult. While there are certainly challenges, in an "open letter" to her child's stepmother, one biological mother said, "You weren't what I had in mind . . . You were supposed to be a mean old hag, remember? But you weren't, you were a sweet, young woman. I wanted to resent you but you made it impossible, and I quickly grew thankful for you" (Curry 2014).

Another important variable is whether or not the stepmother lives with her stepchildren. Most do not. Only 13 to 14 percent of stepmothers live with their stepchildren full-time (Guzzo 2015). Many nonresident stepmothers do not think of themselves as a "stepmother" and are resistant to using the term (Allan, Crow, and Hawker 2011). Orchard and Solberg's (2000) survey of stepmothers who were members of the Stepfamily Association of America found that stepmothers who had stepchildren who spent the majority of their time in their home had higher expectations for functional inclusion in the family (e.g., feeling welcomed into the stepfamily, sharing equally in discipline), parental love, household responsibility, and mother replacement (e.g., not competing for affection, not being a wicked stepmother) than stepmothers who spent less time with their stepchildren. Smith (1990) argues that "part-time stepmothering can sometimes be even more difficult and stressful than full-time; the role of the part-time stepmother is extremely ambiguous"

(p. 3). Indeed, in another study of full and part-time stepmothers, 30 percent thought that "spending more time together" would help improve their relationship with their stepchildren (Quick, McKenry, and Newman 1994).

Another part-time stepmother tells about her experience:

> In the early stages I [had feelings of hatred toward the children] and I felt ashamed of the hatred in myself. I did not want the children to be hurt by it. I used to go off until I could control the hatred better. I felt everyone hated me. I felt mean! But as time goes by I have learned that I don't need to invest so much. . . . They have two very good parents already. . . . I can have closeness and fun without the whole burden. But it is muddling and difficult to get the balance right. (Smith 1990, p. 20)

Nonresident stepmothers' relationships with their stepchildren may also depend on whether or not they have had a child with their spouse or partner. Stepchildren influence stepparents' intentions to have additional children, even if the stepchildren live elsewhere (Stewart 2002). Adding a new biological child may help nonresident stepmothers feel more secure in their role and allow them to better tolerate their stepchildren's visits (Ambert 1989). Yet some nonresidential stepmothers have reported that their partner's visiting children are bad influences on their own biological children (Henry and McCue 2009).

Stepmothers who live with their stepchildren may face somewhat different challenges. Social worker and stepmother Emily Bouchard tells her story:

> When I moved in with my husband and his two teenage daughters, he had a real "hands off" approach. . . . Sparks

Michael Newman/PhotoEdit

Conflicting expectations concerning a stepfather's—or stepmother's—role may make it stressful. When stepparents can ignore the myths and negative images of the role and maintain optimism about the remarriage, they are more likely to have high family, marital, and personal satisfaction.

began to fly as soon as I asserted what I needed to be different. . . . For example, when I noticed that my car had been "borrowed" (the odometer was different) without my knowledge or permission, I had to show up as a parent the way I needed to parent—setting limits, confronting the greater issues of lying and sneaking, and asserting the natural consequences for unacceptable behavior. This method was foreign to their family, and there were reactions all the way around! Thankfully, my husband supported me in front of his daughter, and then we discussed our differences privately and came to a mutual understanding about how to handle parenting together from then on. (Bouchard n.d.)

Stepfamily Relationships

Relationships between stepfamily members depend on many factors such as the age of the children at the time of stepfamily formation, whether the stepchildren are boys or girls, and whether the stepparent is a mother or a father.

Between Stepparents and Stepchildren Relationships between stepparents and stepchildren are complicated and diverse. Moreover, they are not static. As noted, stepfamily relationships evolve and stepparents generally become less involved with and close to stepchildren over time (Stewart 2005b). However, this decline may have more to do with natural developmental changes in children than the stepparent-stepchild relationship. Similar declines in involvement and closeness occur between biological parents and children with increasing conflict during adolescence (Cooksey and Fondell 1996; Marsiglio 1991; Rossi and Rossi 1990). How stepparent-stepchild relationships unfold depends on many factors. For example, stepchildren who lived with their stepfathers for a longer period of time as a child felt closer to them in adulthood (King and Lindstrom 2016). Closeness to mothers mattered as well. Children who became closer to their mothers over time were less likely to experience a decline in closeness to their stepfathers.

Stepparents have the potential to positively and negatively affect their stepchildren's lives. Kalmijn (2015b) compared how married biological fathers, divorced biological fathers, and stepfathers influenced their children's educational attainment and church attendance. Although stepfathers were less influential than divorced fathers regarding their education, they were more influential than divorced fathers with respect to church attendance. The author speculates that because the stepfathers lived with their stepchildren (the divorced fathers did not), they transmitted their values regarding religion on a more routine basis. These results suggest that stepparents be included in children's constellations of parental figures when studying child outcomes.

Less is known about LGBTQ+ stepfamilies, and most research was conducted prior to the legalization of gay marriage. Goldberg and Allen (2013) interviewed twenty young adults with lesbian mothers regarding how they felt about the dissolution of their parents' relationship and their formation of new partnerships. First was the problem that many people did not acknowledge the divorce because their parents weren't legally married. Meredith, age 19, explains:

> I remember telling people right when they had told me that they were separating that they were getting a divorce and being told by both teachers and my friends that they can't get a divorce because they weren't married. . . . There was another kid in my class [whose] parents were getting a divorce, and watching the ways that people saw that as opposed to how they saw my family, it really brought to light that they didn't see it as equal. My teacher actually announced [to the class] that mom and her partner were breaking up. . . . I said, "You mean my moms" and she was like, "No. She won't be your mom anymore. They're breaking up." (Goldberg and Allen 2013, p. 535)

In general, these stepchildren's relationships to their stepparents, biological parents, and nonbiological mothers were similar in nature to those in opposite-sex stepfamilies. When these kids' parents formed new partnerships, most of the twenty participants in the study referred to their mothers' new partners as a *stepparent* as opposed to some other term (e.g., "aunt"). Though several noted challenges such as ambiguity and confusion with respect to their new stepparents' role in their life and they experienced varying levels of closeness with them, overall most reported having "positive" relationships.

Between Biological Parents and Children Most stepfamily research has been devoted to the relationship between the stepparent and stepchild, but biological parent-child relationships in stepfamilies should not be overlooked. Most of what is known is limited to biological parent-child relationships immediately after divorce (Coleman, Ganong, and Fine 2000). As discussed in Chapter 14, some research suggests that the quality of mothers' parenting declines somewhat after divorce, a result of being the sole primary caretaker of children combined with busy work or school schedules and a loss of income. Fathers can help fill this gap. In a study of black single-mother families, the involvement of nonresident fathers was associated with better psychological well-being and lower parenting stress among mothers, which in turn led to fewer behavioral problems in the children (Jackson, Choi, and Preston 2015).

Yet, other studies find that divorced mothers in stepfamilies are similar to other mothers in relationship quality and time spent with children, and styles of discipline (Smith 2008). The author said "there was no evidence that mothers disengaged from parenting after

stepfamily formation" (p. 170). At the same time, mothers in stepfamilies had "an additional layer of parenting activities" in the form of defending, mediating, and facilitating the stepparents and stepchildren (Smith 2008, p. 171; Weaver and Coleman 2010). In a study of young adults in lesbian stepfamilies, young adults maintained stronger ties to their biological mothers than their nonbiological mothers (Goldberg and Allen 2013).

Most information on biological father-child relationships in stepfamilies focuses on children with nonresident fathers. There is not a lot of information specifically on resident biological father-child relationships in stepfamilies. It is thought that resident father-child relations undergo a more dramatic change with remarriage than resident mother-child relations because stepmothers take over many of the primary caretaking duties formerly performed by single fathers (Hetherington and Stanley-Hagan 1997). Resident biological fathers after divorce tend to have greater difficulty than mothers in the areas of communication and monitoring (especially with adolescent daughters), but they have fewer problems with control and are as warm and nurturing as mothers (Buchanan, Maccoby, and Dornbusch 1996; Hetherington and Stanley-Hagan 1999).

Between Full, Step-, and Half-Siblings Relationships between siblings, including stepsiblings, remain a neglected area of study (McHale, Updegraff, and Whiteman 2012; Sweeney 2010). Fifteen percent of children live with a stepsibling, half-sibling, or both (Manning, Brown, and Stykes 2014). Given that mothers tend to retain custody of biological children and fathers do not, many children in stepfamilies do not live with their half- or stepsiblings (Manning, Brown, and Stykes 2014). Stepsiblings who reside in separate households are more common than those who live together, although with parents increasingly opting to share custody, this could be changing. Step- and half-siblings are more prevalent in cohabitation and single-mother households than married ones (Manning, Brown, and Stykes 2014). A slightly higher percentage of African American and Hispanic children than white and Asian children reside with step- and half-siblings (Pew Research Center 2011b).

In general, biological siblings are closer and more engaged than stepsiblings, as are siblings (biological, half-, or step-) who live in the same household (Allan, Crow, and Hawker 2011; Sweeney 2010). Stepsibling relationships are less intense, competitive, and conflictual than biological and half-sibling relationships (Ganong and Coleman 1993; Hetherington and Stanley-Hagan 1999). However, there is a great deal of variability in stepsibling relationships. Some children make no distinction between biological and stepsiblings, but this appears to be related to whether the stepparent does so (Allan, Crow, and Hawker 2011).

Stepparents' Decisions about Having Children
When people remarry or form a new committed partnership, they have decisions to make about having children together. Does it make a difference whether one or both partners already have children? The answer to that question is yes.

A study of more than 2,000 couples drawn from the National Survey of Families and Households found that married and cohabiting couples in which one or both partners already had children were less likely to intend to have a child together compared to couples with no children (Stewart 2002). Where the children live also matters. Among couples who already have a child together, those whose prior children live in the household (as opposed to with their other parent) are less likely to intend to have a second child, as are couples who have more contact with their nonresident children (Hohmann-Marriott 2015). The author speculates that providing a sibling is a main motivation for having a second child and that stepchildren, especially those who are in the couple's life, can fulfill that role.

The biological children of both partners in a stepfamily are called *mutual*, *shared*, or *joint* children (Stewart 2007). These are the half-siblings of any stepchildren. One in ten children lives with a biological parent who had a child with a different partner (Manning, Brown, and Stykes 2014). Research shows that a principal reason for remarried couples to have a child together is the hope that the mutual child will "cement" the remarriage bond (Ganong and Coleman 2004). If one partner has not yet had a child, then having a shared child can provide them with the opportunity to parent a biological child (Hohmann-Marriott 2015).

Believing that they will now be ignored by the stepparent—or seeing the mutual child as having a privileged place in the family—stepchildren may feel threatened, jealous, or resentful (Cartwright 2008; Munroe 2009). Feelings of jealousy about a future mutual child can be initiated as early as a parent's wedding ceremony. For instance, a 19-year-old female whose father had been remarried for two years told an interviewer:

> I think at the wedding it would have been more ideal if like family members had, I don't know, I guess, paid more attention to me and my brother. 'Cause they were kind of like you're married and you're going to have kids and everything, but they kind of forgot that there's already two kids in this family. (in Baxter et al. 2009, p. 482)

Relationships with Grandparents Even if they do not maintain a relationship with one another, ex-spouses and partners do not always make a "clean break" from in-laws because children remain the grandchildren of former in-laws (Finch and Mason 1990). Step-grandparenting is increasingly common; estimates suggest that step-grandparenting accounts for 15 percent of total grandparenting time for those at age 65

(Yahirun, Park, Seltzer 2018). The extent of relationships with former in-laws after divorce depends on many things, including the quality of the relationship during the marriage, the amount of reciprocal support, how close the grandparents are to their grandchildren, whether the children's father remains involved with them, whether either partner remarries, and who has custody. After divorce, children have more contact with their maternal grandparents than paternal grandparents because children tend to reside with their mothers (Jappens and Van Bavel 2015). Contact with paternal grandparents depends on the level of conflict between parents, and how involved the nonresident father is in their children's lives post-divorce (Westphal, Poortman, and Van der Lippe 2015).

Many older adults find themselves "step-" grandparents to their children's stepchildren. Other become step-grandparents when their *stepchildren* have children. Grandparents can make a positive contribution to stepfamilies. They provide economic and social support to grandchildren, act as intermediaries to keep everyone informed and involved regarding the children, and provide a "neutral zone" for ex-spouses, childcare, and other assistance (Stewart 2007). For example, contact with, closeness with, and confiding in grandparents are associated with less depression in grandchildren, including step-grandchildren (Ruiz and Silverstein 2007).

Becoming a step-grandparent is a *process* not unlike the process of becoming a stepparent. According to Henry, Ceglian, and Matthews (1992), grandparents must (1) accept the loss of the fantasy of a "lifelong happy marriage" for their child, and traditional grandparenthood for themselves; (2) cope with ambiguous family relationships; and (3) accept new family members (p. 28). Allan, Crow, and Hawker's (2011) study of British stepfamilies included step-grandparents. They found that the stepgrandchild-grandparent relationship was dependent on the ages of the children, how much time they spend together, and the number of biologically related grandchildren in the family. Perhaps the most important factor is the quality of the relationship between the grandparent and their adult child (and with their spouse or partner). As a result, there is a great deal of variability in grandparent-grandchild relationships in stepfamilies Whereas some step-grandchildren and step-grandparents never come to view one another as "kin," others do not make a distinction between biological and step-grandchildren. Many step-grandchildren do not even realize that their step-grandparent is not a blood relative. Chapman, Coleman, and Ganong (2016) interviewed 25 young adults with step-grandparents. Kevin, a participant in their study, said "To be honest, I don't really know a difference between step- and biological grandparents. . . . [M]y stepgrandpa is my grandpa who I've always known.

Nobody in the family made the distinction to me ever that he wasn't my grandfather" (Chapman et al. 2016, p. 638). Similarly, many step-grandparents do not distinguish between step-grandchildren and grandchildren. Kendra stated, "If I would like go to church with them or something, [my step-grandfather] would literally drag me to every person that he could and say, 'This is my granddaughter, Kendra.'"

But there can be disappointment and hurt feelings when step-grandparents don't act how a grandparent "should." Lance said, "He was never the kind of guy that made the advance to try to be my grandpa. He didn't seem interested in building a relationship. So it always felt like there was tension between us. . . . [L]ike it would be weird to give him a hug or say "I love you" to him like I do my grandma. . . . He wasn't that kind of involved grandpa that you could be close to" and Bonnie stated, "My stepgrandma doesn't really do grandmotherly duties. . . . [S]he just doesn't provide that 'grandmother care.' . . . I just wish she would do grandmother things, like pass down her pearls or earrings or something like that." (p. 638)

Common issues that might face step-grandparents include whether to extend monetary gifts to stepgrandchildren and whether and to what extent to include stepgrandchildren in family holidays and vacations. Although having step-grandparents provides children with more "kin," transfers of money and time from step-grandparents to step-grandchildren are less than those of biological grandparents (Wiemers, Seltzer, Schoeni, Hotz, and Bianchi 2019).

Financial and Legal Issues

Two important yet understudied issues in stepfamilies have to do with how stepfamilies manage their finances and navigate the laws governing stepfamilies. These issues are intertwined because stepfamily finances are impacted by legal decisions about child support, provision of health insurance, and other income transferred between divorced parents living in separate households. Moreover, laws governing stepfamilies are complicated, are often ambiguous, and vary by state.

Financial Arrangements in Stepfamilies All families struggle with how to manage their money, and this is the primary source of conflict for couples (Bennett 2015). Couples who have stepchildren argue more about money than do couples with biological children only (DeLongis and Zwicker 2017). For example, some stepfamily couples believe that prior child support agreements should be modified to accommodate the needs of the current family (Hans 2009). In the 1970s, some states became concerned about child support and passed legislation designed to prevent the remarriage of people whose child support was not paid up. But the

Supreme Court ruled in *Zablocki v. Redhail* (1978) that marriage—including remarriage—was so fundamental a right that it could not be abridged in this way. Whether legally required to or not, most stepparents help to support the stepchildren, either through direct contributions to the child's personal expenses or through payments toward general household expenses such as food and shelter (Mason et al. 2002; Stewart 2007).

Many remarried husbands report feeling caught between what they see as the impossible financial demands of both their former family and their current one (Hans and Coleman 2009). Some second wives—more often those without children of their own—feel resentful about the portion of the husband's income that goes to his first partner to help support his children from that union (Hans and Coleman 2009). As an Australian nonresidential stepmother told an interviewer,

> I have always felt very cross that we have to give a large amount of our income [in child support] when I would really like that income to help . . . [my autistic son] . . . with therapy. . . . I am forced to work many hours to pay for [his] therapy. (Henry and McCue 2009, p. 196)

Other mothers may feel guilty about the burden of support that their own children place on their stepfather (Barash 2000). Mothers with children from a previous relationship may worry about receiving regular child support from an ex-partner. Research shows that financial support and visitation both decline when either parent has additional children with new partners (Manning, Stewart, and Smock 2003; Meyer and Cancian 2012).

Despite the fact that money problems are the primary issue that all couples argue about, there have been relatively few studies of stepfamilies' financial arrangements (Gold 2009). The best-known study of how stepfamilies manage their finances was published nearly 40 years ago by Barbara Fishman (1983). She found that stepfamilies tend to adopt one of two models of economic behavior: a common-pot or two-pot economic system. In the **common-pot system**, economic resources are pooled and distributed according to need regardless of biological relatedness. In the **two-pot system**, economic resources are divided and distributed along biological lines, and only secondarily distributed according to need. For example, common-pot families would put their money together in a joint account and share ownership of such assets as their home, whereas two-pot families would keep their money in separate savings accounts and might have a prenuptial agreement. Fishman suggests that a concern for the *common good* (family harmony, trust, and closeness) underlies the common-pot system, whereas rationality, economic independence, and personal autonomy underlie the two-pot system. She feels that whereas the common pot encourages household unity, separate pots discourage it.

Several studies indicate that the majority of stepfamilies are common-pot economies and show that, consistent with Fishman's findings (1983), joint banks accounts were associated with higher relationship quality. Contextual factors such as income, race and ethnicity, and gender play a role (Addo and Sassler 2010; Higginbotham, Tulane, and Skogrand 2012; Lowan and Dolan 1994; Mason et al. 2002). Most Hispanic couples adopt a shared system. Cohabiting stepfamilies, who tend to have fewer economic resources—might adopt a common pot simply out of convenience (Lin et al. 2018).

Whether a common-pot or two-pot system is adopted, having a clear financial management strategy is associated with greater financial security (Van Eeden-Moorefield et al. 2007). This is no easy feat, especially with economic resources moving in and out of the household in the form of child support. Couples must make decisions about allowances, college expenses, gifts, and what to do should adult stepchildren "boomerang" back into the family home. Should they be required to pay rent? And how much?

Legal Issues in Stepfamilies As discussed above, stepfamilies are considered an *incomplete institution*. You can see this most readily by looking at our laws and policies, which do little if anything to accommodate stepfamily relationships. Susan Stewart and Elcy Timothy (2020), provide a review of stepfamilies' current legal status:

Twenty-five years ago, Mason and Simon (1995, p. 447) wrote of the "ambiguous stepparent" and "a lack of coherent federal policy toward stepchildren" and stepfamilies seeking legal rights must navigate a "patchwork quilt" of federal laws, state laws, and court precedents (Pollet 2010). The situation is further complicated by a distinct *absence* of laws that would guide judges' and policymakers' decisions. The result is that stepparent and stepchildren's rights and responsibilities depend on largely on where they happen to reside, individual judges' definitions of "family," and their ability to understand the law, pay for court costs, and retain an attorney. Without explicit instructions on how to act, teachers, public officials, social workers, and others who interact with stepfamilies are forced to make quick case-by-case decisions that are likely to be influenced by personal opinions, leading to inconsistent treatment.

Laws and policies governing stepfamily relationships are based on three legal concepts. First, stepparents are considered **legal strangers** to one another under the law, which attaches no significance to the stepparent-stepchild relationship. This concept is based on the outcome of the 1988 worker's compensation case, *Mendoza v. B.H.L. Electronics*. The judge ruled that the stepparent's benefits, after getting hurt on the job, would not go to his stepchildren. The result of *Mendoza* is that stepparents have almost no legal say in the education, health, religion, and welfare of stepchildren, even if they raised

their stepchildren since birth. The majority of states do not require a stepparent to financially support a stepchild, even during their marriage. Adult stepchildren are typically not allowed to make medical or financial decisions for a stepparent who is ill or incapacitated. Employers are not required to extend health insurance to stepchildren. Should the parent and stepparents' relationship dissolve, stepchildren have no legal right to visitation or child support from their stepparent and stepparents have no legal right to see their stepchildren or gain custody. In hospitals, whether stepfamily members are allowed to be in areas designated "just family" can depend on the views of individual administrators, nurses, and doctors. Hundreds of thousands of children reside with a stepparent and no biological parent. They too are considered legal strangers under the law even if the stepparent is their sole caretaker.

Few legal provisions exist for the special circumstances of stepfamilies, especially if the remarried union should dissolve. Stepchildren whose stepparent and biological parent are not legally married have even fewer protections.

Should a stepparent pass away, assets would go to children who are biologically related as opposed to stepchildren, unless the stepparent has specified otherwise—which many stepparents have not (and this can still be contested). Stepparents are also faced with the dilemma of how to divide their assets. One stepparent explains:

> No, I haven't made a will but it wouldn't be equal. It is awfully complicated. . . . On the one hand, you sort of say well Nick's my son I should leave it all to my son, then blow the rest of [them]. But I don't feel that strongly enough to actually say it. I, I feel I owe it to Nick that he should get the lion's share or he should certainly get a bigger proportion, even if you just base it on the technicality that the others will inherit from their blood father. (Allan, Crow, and Hawker 2011, p. 45)

The second legal concept affecting stepfamilies is **the rule of two**, which means that children cannot have more than two legal parents. With few exceptions, the only way for stepparents to establish a legal relationship with their stepchild is to adopt them, which in most cases requires child's other biological parent to relinquish their parental rights. Stepchildren can change their last name to match their stepparent's but there is no legal significance attached to doing so. The "rule of two" permeates our educational system. School and extracurricular activity forms and on-line systems of communication typically allow space for two parties and teachers may be unprepared for meetings and correspondence with multiple parental figures.

The third concept guiding the legality of stepfamily relationships is the *de facto* ("in the place of") parent. A **de facto parent** under federal law is defined as "those stepparents legally married to a natural parent who primarily resides with their stepchildren or provide at least 50% of the child's support" (Mason and Simon 1995, p. 468-469). There have been cases in which the stepparent was granted custody in the case of the death of the custodial parent and several states have statutes that allow stepparent visitation if they are found to be *de facto* parents during the marriage. Yet, in the case of *Troxel v. Granville*, the Supreme Court struck down a Washington state law that allowed a third party to petition the courts for child visitation over parental objections.

How federal policies regard stepfamilies is highly inconsistent. FERPA, the Family Education Rights and Privacy Act, utilizes the *de facto* parent status. Under FERPA rules, a stepparent has rights to educational records but only if they are married to the child's natural parents and are present on a "day-to-day" basis. The Health Insurance Portability and Accountability Act (HIPAA) is the federal law that protects people's private health information. Under this law, a court order granted by a judge and/or a statement signed by a biological parent is required for stepparents to access a stepchild's information or make medical decisions for them.

Under U.S. federal income tax rules, stepfamily members can be claimed as dependents (if they meet certain criteria) and stepchildren can be claimed for the Earned Income Tax Credit, as long as the stepparent's household is the stepchild's primary place of residence.

Not everyone in a stepfamily shares a biological or legal relationship. Many stepfamilies struggle to fill out forms and paperwork designed with only "first families" in mind.

fizkes/Shutterstock.com

Temporary Assistance for Needy Families (TANF) provides cash assistance to poor families and stepchildren are included in benefit calculations. The Supplemental Nutrition Assistance Program (SNAP) provides food benefits to all household members who share food expenses and includes stepchildren. Supplemental Security Income (SSI) provides financial support to disabled adults and children whose parents are unable to support them financially. SSI, TANF, and SNAP assume stepchildren are supported by their stepparents and benefits are reduced in relation to stepparent's income. In contrast, Medicaid, a federal program that provides health care for low income individuals and families, does *not* assume stepchildren are supported by their stepparents and benefits are not reduced.

Stepchildren can receive social security income in the event of their stepparent's death, but only if the stepparent and natural parent were married and it can be proven that the stepparent provided at least 50% of their support before their death. The rules of federal government employee benefits programs (e.g., retirement, life insurance) vary with some imposing conditions on stepparent-stepchild relationships and some excluding stepchildren outright. With few exceptions, military benefits (e.g., housing, death benefits) are not extended to anyone other than spouses and biological or adopted children. On the other hand, the Family and Medical Leave Act (FMLA) defines "child" without qualification.

The federal student loan program provides grants and low interest loans for college. The Free Application for Federal Student Aid (FAFSA) assumes stepchildren are financially supported by their stepparent as long as the parents are married and the child lives with the stepparent at least half the time. The U.S. immigration system defines families broadly and stepfamily relations are considered "immediate relatives" for the purpose of family reunification if certain criteria are met (Gubernskaya and Drebey 2017).

Many states have laws regarding stepfamilies which can override federal ones. The number of states with laws specifically referring to "stepparents" and "stepchildren" is increasing and as of 2017, 12 states (Alaska, California, Delaware, Florida, Louisiana, Maine, New Jersey, New York, North Dakota, Oregon, Pennsylvania, Washington) had specific laws or court cases that allow children to have more than two parents (Kazyak, Woodell, Scherrer, and Finken 2018; Peltz 2017). However, most of the time stepparents' rights and responsibilities are decided in the courts. For example, *Spells v. Spells* in Pennsylvania was the first appellate court decision to address stepparent visitation rights and stated that "rejection of visitation privileges cannot be grounded in the mere status as a stepparent" (Gregory 1998, p. 364).

Compared to other Western countries, where stepparents and stepchildren are afforded many rights, the United States has been slow to recognize and incorporate stepfamily relationships into its laws, policies, and programs. Because family law is constantly evolving, members of stepfamilies are advised to check with an attorney regarding applicable laws in their state. Another good resource is the National Stepfamily Resource Center's "Frequently Asked Questions" on law and policy.

WELL-BEING IN STEPFAMILIES

One of the most researched areas concerning stepfamilies is the well-being of their members, especially children. As discussed in previous chapters, members of nontraditional families (not living in a married, two-parent household) generally do not fare as well on a range of economic, social, and emotional variables. As you will see in this section, the reasons for these differences are quite complicated.

The Well-Being of Parents and Stepparents

Like first marriage, remarriage has physical, emotional, and financial benefits for both men and women, with the benefits somewhat greater for men (Raley and Sweeney 2020). For women, the advantages are mostly financial. Single mothers who marry have significantly higher incomes than single mothers who cohabit, marry and later divorce, or remain unpartnered (Painter, Frech, and Williams 2015). The potential for increasing their children's standard of living can be an incentive for some single mothers to repartner. In one study, those who received child support were more likely to remain single (Cancian and Meyer 2014).

While men to do not, on average, benefit as much financially as do women, there are still many advantages to recoupling for men. As noted above, less educated, lower income men may benefit economically from partnering as opposed to living on their own. In general, single and divorced men have less healthy lifestyles than do married men. On average, they drink more alcohol, are more likely to smoke cigarettes, eat less balanced meals, and go to the doctor less often (Waite and Gallagher 2000). However, the extent to which remarriage is associated with healthier behaviors among men is unclear. In a Swedish study, remarriage had no effect on depression levels of divorced men (Hiyoshi, Fall, Netuveli, and Montgomary 2015).

Several studies indicate that parenting stepchildren is less satisfying than parenting biological children, especially for stepmothers (Levin 1997; Thoits 1992). Stepmothers have reported feeling isolated, resentful, and guilty, and they can suffer from low self-esteem (Smith 1990). There is also evidence that stepmothers have worse mental health than biological mothers

(Shapiro and Stewart 2011; Smith 2008). Stepchildren do not generally "substitute" for biological children. Adult stepchildren feel less obligation to help stepparents than biological parents. Whereas 85 percent said they would feel "very obligated" to help out a biological parent if they were faced with a serious problem, this number was only 56 percent for stepparents. The same study found that parents felt less obligated to help their adult stepchildren than their adult biological children, at 78 percent versus 62 percent, respectively (Pew Research Center 2011b).

Not unexpectedly, then, stepchildren provide less social support to their aging parents than do biological children, and this support to stepparents is more dependent on the quality of the relationship than support to biological parents (Ganong and Coleman 2006a, 2006b). However, stepparenting has many positive benefits for stepparents. Some benefits of stepmothering mentioned by stepmothers are the opportunity to engage in the maternal role, enjoying the challenge of living in a stepfamily, and personal growth (Quick, McKenry, and Newman 1994). Self-esteem is higher among stepmothers who have better relationships with their stepchildren (Quick, McKenry, and Newman 1994).

The Well-Being of Children

How does membership in a stepfamily affect children's well-being? Although the majority of children in stepfamilies are happy and successful in life, considerable research has found that, on average, stepchildren of all ages have somewhat worse outcomes than children from two- biological-parent homes. They have higher rates of smoking, alcohol and drug use, and juvenile delinquency; are less successful in school; are more likely to have experienced early sexual behavior and childbearing; may experience more family conflict; and score lower on measures of socioemotional health. However, as explained in Chapter 14, these effects are primarily the result of stepchildren having gone through multiple family transitions (their parents' divorce, cohabitation(s), remarriage) as opposed to simply living in a stepfamily per se (Raley and Sweeney 2020).

Such outcomes result whether or not the stepchild has been adopted by their stepparent (Stewart 2010). Mathew Bramlett of the National Center for Health Statistics compared the health of adopted children, stepchildren, and adopted stepchildren. He found that stepchildren are in better health based on a range of outcomes (e.g., dental health, asthma, learning disabilities) than adopted children, and that adopted stepchildren have health outcomes more similar to stepchildren than adopted children. The author states, "[This] is not surprising, as they were presumably stepchildren at some point, and what differentiates them from stepchildren

is that their stepparent adopted them" (Bramlett 2010, p. 262). Still, all three groups score lower on health outcomes than biological children.

As noted above, not all children living in stepfamilies are stepchildren. Some children in stepfamilies are the biological children of both parents and half-siblings to stepchildren. Research suggests that all children raised in stepfamilies, even if both of their biological parents are in the household, experience lower levels of well-being compared to children raised in continuously intact families (Thomson and McLanahan 2012). In one study, compared to children who resided with biological siblings, children who resided with half- and stepsiblings showed significantly more aggression (Fomby, Goode, and Mollborn 2016). Households that include half-siblings and stepsiblings are poorer and more likely to receive public assistance than other households (Brown, Manning, and Stykes 2015).

Overall, the effect of stepfamily living on children is not straightforward. Acquiring a stepparent may increase some problems but improve other areas of children's lives, and these effects may depend on the child's gender, age, race/ethnicity, and income (Fomby, Goode, and Mollborn 2016; Raley and Sweeney 2020). Some research suggests that the effect of stepfamilies on child well-being may be less negative among African Americans, because of its communal culture, more permeable family boundaries, and a history of parenting non-biological children traced back to slavery (Bryant 2020). Some African American children may be more accustomed to life transitions and stressful life conditions and therefore have greater capacity to adjust to changes, compared with nonHispanic white children. Analysis of data from a national sample of African American youth found that the presence of a father, whether biological or stepfather, served to increase the likelihood of positive outcomes (Adler-Baeder et al. 2010).

Further, children in stepfamilies grow up with fewer economic resources than children in original, two- parent families. Stepfamilies have lower earnings, have less savings, are less likely to be homeowners, and have less equity in their homes (Stewart 2001). The parents of stepchildren have lower aspirations for them in terms of going to college, are less involved in their schoolwork, and are less likely to help them go to college, establish a business, or buy a home (Astone and McLanahan 1991; Hetherington and Kelly 2002). The more limited financial resources of stepfathers relative to biological fathers explain a large portion of stepchildren's lower school achievement (Hofferth 2006). Less financial help goes to stepchildren than to biological children. Stepparents spend less on food, schooling, clothing, and miscellaneous items (gifts, hobbies, and pocket money) for children than do biological parents (Case, Lin, and McLanahan 2000). Data from the National Postsecondary Student Aid Study show that remarried parents' and

divorced parents' contribution to their children's college costs are similar, despite remarried parents' higher incomes (Turley and Desmond 2011).

Stepparents make fewer investments in their children's health than biological parents. Children with resident stepmothers are significantly less likely to complete routine doctor and dentist visits and have a consistent medical provider, less likely to wear seatbelts, and more likely to be living with a cigarette smoker than children living with a biological mother (Case and Paxson 2001). Good or bad, teenagers in remarried stepfamilies do significantly more housework than teens in original two-parent families (Gager, Cooney, and Call 1999).

Some researchers have found that remarriage lessens some negative effects for children of divorce—but only for those who experienced their parents' divorce at an early age and when the subsequent remarriage remains intact (Arendell 1997; Cherlin 2009a). Two of the main benefits of a parent's remarriage for children are improvements in economic status and in school achievement (Raley and Sweeney 2020). Additional research shows that younger children adjust better to a parent's remarriage than do older children, especially adolescents (Amato 2005; Carlson 2006). Siblings in the same family system may have different levels of exposure to marital conflict and therefore may have differential levels of well-being in the areas of self-blame, depression, and behavior problems (Richmond and Stocker 2003).

What about the well-being of stepchildren living in cohabiting stepfamilies? Chapters 6 and 7 discuss how the outcomes of children from cohabiting couple households are significantly worse compared to children with two married parents. These findings apply to children with both biological parents in the home and children living with one biological parent and one stepparent. In fact, children living with a mother and her cohabiting partner have outcomes more similar to those of children living with a single mother than children living with a stepparent (Thomson and McLanahan 2012). Nevertheless, children benefit economically from living with a cohabiting partner (Manning and Brown 2006).

The lower levels of well-being that stepchildren experience during childhood can follow them into adulthood. Stepchildren leave home earlier than biological children, and have a greater risk of nonmarital childbearing and cohabitation (Amato and Kane 2011). Zito and Coster (2016) found that adolescent girls whose mothers were dating had sex at an earlier age, even after controlling for dating mothers' more permissive attitudes toward teenage sex and pregnancy. Stepchildren who leave the parental home are more likely to give "friction at home" as the reason for their early departure than are other children (Kiernan 1992). This is referred to as *extrusion*, which is "defined as individuals' being 'pushed out' of their households earlier than normal for members of their cultural group, either because they are forced to leave or because remaining in their households is so stressful that they 'choose' to leave" (Crosbie-Burnett et al. 2005, p. 213).

Most stepchildren do very well and appreciate their stepfamily lives. Several studies have concluded that stepchildren's well-being and future outcomes largely depend on the quality of the relationships and communication among family members regardless of family structure (Crawford and Novak 2008; Doohan et al. 2009; Schoppe-Sullivan, Schermerhorn, and Cummings 2007). The extent to which parents or stepparents monitor their children's comings and goings is probably more important to positive child outcomes than family structure itself (Crawford and Novak 2008). Furthermore, recent research shows that a close, nonconflictual relationship with a stepfather enhances the overall well-being of adolescents (Booth, Scott, and King 2010; Yuan and Hamilton 2006). Among Hispanics, stepfather involvement was associated with fewer behavior and emotional problems in adolescents (Coltrane, Gutierrez, and Parke 2008). Positive relationships with siblings and half-siblings are correlated with higher levels of adjustment in adolescents in stepfamilies (Baham et al. 2008).

CREATING SUPPORTIVE STEPFAMILIES

Creating a supportive stepfamily is not automatic. Supportive stepfamily relationships must be forged. Patricia Papernow's (1993) book *Becoming a Stepfamily* uses a *developmental approach* to stepfamilies. Developmentalists focus on the complex processes that occur within family systems and how these aspects of family life unfold over time (O'Brien 2005). The **stepfamily cycle** is the process by which veritable strangers form "nourishing, reliable relationships," (Papernow 1993, p. 12). She argues that it is critically important for stepfamilies to have what she refers to as a *developmental map*. A developmental map helps family members recognize what is normal and predictable and what is a family crisis. She compares this to the experience of a child throwing a temper tantrum. Once parents understand that tantrums are to be expected of children of a certain age, then the family will be less likely to be stressed out and overreact to the incident. The stepfamily cycle therefore offers both guidance and reassurance to stepfamilies as they go about their daily lives, as well as a model of intervention points for professionals working with stepfamilies. The stages of the stepfamily cycle and the life cycle tasks associated with each stage are outlined in Table 15.3.

TABLE 15.3 Papernow's Stepfamily Cycle[a]

STAGE	DESCRIPTION	LIFE CYCLE TASKS
Early		
1. Fantasy	Parents have hopes of being an "instant family." Children harbor the fantasy that their parents will reunite.	Acknowledging fears and fantasies. For children, acknowledging loss of their "real family."
2. Immersion	Reality of the challenges of living in a stepfamily sets in. Increasing sense of unease among family members. Stepparents in particular are likely to have negative emotions.	Bearing disappointment of "instant family" fantasy not coming true. For stepparents, dealing with rejection and new routines. Keeping hopeful about the future.
3. Awareness	Stepparents become aware that they are "outsiders." Increasing pressure on the biological parent as the "insider."	Each stepfamily member gathering data about the family and his or her place in it. Naming feelings, lowering expectations.
Middle		
4. Mobilization	Differences are aired more openly. This state is marked by increased chaos and conflict.	Voicing unheard needs and perceptions. Increasing the focus on the couple.
5. Action	Power struggles between insiders and outsiders diminish.	Negotiating agreements about how the family will function. Drawing new boundaries around step-relationships.
Later		
6. Contact	The "honeymoon" period. The chaos of previous periods has stabilized. Stepparents and stepchildren forge a "real" relationship. The marital relationship improves.	Working on developing a "workable" stepparent role. Letting go of old hurts.
7. Resolution	Stepfamily norms are established. Step-relationships no longer require constant attention. Increased clarity, acceptance, and satisfaction. The stepparent is established as an "intimate outsider."	Continued acceptance of differences between family members. Staying on developmental track as new issues emerge (e.g., new babies, financial problems).

[a]Adapted from Papernow 1993. Reprinted from Stewart 2007.

Papernow cautions that the stepfamily cycle does not unfold in a neat and precise way. Moreover, it is possible for stepfamilies to get stuck and never reach maturation. For stepfamilies that do make it, the end result of the process is the **intimate outsider role** for the stepparent, which Papernow (1993) describes as, "intimate enough to be a confidante, and outside enough to provide support and mentoring in areas too threatening to share with biological parents: sex, career choices, drugs, relationships, remaining distress about the divorce" (pp. 16–17).

Many stepparent-stepchild relationships improve over time. One stepchild says the following about her stepfather: "He was [trying], and I was rebelling because I wasn't happy with him being there. . . . I wanted a reason not to like him, but eventually the harder he

tried, the wall broke down with me, and I saw what he was doing for my mom" (Ganong, Coleman, and Jamison 2011, p. 406). Ganong et al. (2011) interviewed young adults who had had one or more stepparents over the course of their lives. They found six patterns of step-relationship development: accepting as a parent, liking from the start, accepting with ambivalence, changing trajectory, rejecting, and coexisting. These patterns were dependent on the age of the stepchild when the stepfamily was formed, the child's gender, and the amount of time the stepparent and stepchild spent together. As we saw in Chapter 11, communication is vitally important to stepfamily functioning and well-being. In a study of Mexican and European families, adolescents whose stepfathers explained their actions or apologized rather than defending their actions had less

depression, exhibited fewer externalizing behaviors, and felt better about themselves and their relationship with their fathers (Cookston et al. 2015).

Stepfamily development does not proceed at a constant rate, with closeness sometimes surging in relation to *turning points*, which Dawn Braithwaite and colleagues (Braithwaite et al. 2018) define as "significant pivotal events or experiences at a particular moment or time in your life that were important in bringing your relationship with your stepparent to where it is today" (p. 5). As one 25-year-old stepdaughter explains,

I was in high school and I remember my stepdad was in the garage working on something. . . I had my first boyfriend breakup. I just needed to talk to one of my parents, and he was the first one I saw. So, I told him the whole story of what happened. And I remember crying, and was just giving me advice about, "It hurts right now, but there will be others. You will get through this." Just words of encouragement. (p. 7)

It helps to remember that the unrealistic "urge to blend the two biological families as quickly as possible" may lead to disappointment when one or more adult or child members "resist connecting" (Wark and Jobalia 1998, p. 70). You may have noticed that we have not used the once-familiar term *blended family* in this chapter. That's because family therapists and other experts have concluded that stepfamilies do not readily "blend" (Ganong and Coleman 2016). Playing with the language, stepmother and online columnist Dawn Miller refers to stepfamily living as "life in a blender" (Miller 2004).

It is important for members of stepfamilies (and the clinicians who work with them) to recognize that the "stepfamily architecture" presents special challenges (Papernow 2008). That is, creating supportive and cohesive stepfamilies involves recognizing and building on potential family strengths that are unique to stepfamilies. Relationships with new extended kin may be a potential source of new friendships. Beginning or renewing a sense of family history is another strength builder. Although holidays often divide the stepfamily because of visitation agreements with a noncustodial parent, it is possible to create new family holidays when the entire stepfamily is sure to be together. Well-designed empirically based relationship interventions for stepfamilies have been shown be effective at improving couple relationships, parenting, and child well-being (Lucier-Greer and Adler-Baeder 2012; Nicholson et al. 2008).

Technology is also transforming family dynamics and communication. Research is ongoing about the extent to which members of stepfamilies use technologies such as e-mail, FaceTime, texting, and social media apps. On the one hand, new technologies such as texting and e-mail allow parents greater control over their communications with ex-spouses and partners

Michael Newman/PhotoEdit

Many stepfamilies are especially complex, and include new biological children, the half-siblings of the stepchildren in the family. In the absence of a guiding cultural script, individuals in stepfamilies are free to fashion their own relationships.

(Ganong et al. 2012). On the other hand, technology may introduce new problems such as disagreements between parents and stepparents about limiting children's online surfing, texting, or gaming as well as their own use (Hertlein 2012). Technology also allows ex-spouses to communicate more frequently, and managing those interactions can create tension between partners. Given the complexity of stepfamily relationships, technology can help stepfamily members stay connected, especially those family members who reside in different households.

Increasing access to technology means that today's stepfamilies have access to more resources than in the past. For instance, the website of the Stepfamily Foundation is designed to give advice and report research findings concerning stepfamilies. The website I Do! Take Two offers not only helps design second weddings but provides advice on topics ranging from second wedding etiquette to religious, financial, and legal issues.

There are many books for stepfamilies. Living up to its title, Erin Munroe's *The Everything Guide to Stepparenting* (2009) addresses topics from dating a parent to the logistics of moving in together to questions about maintaining step-relationships after a second divorce. *The Stepfamily Handbook* mentioned previously provides tips and advice from "dating, to getting serious, to forming a 'blended family'." There are a number

of books or children. One directed to teens is *Stepliving for Teens: Getting along with Step-Parents, Parents, and Siblings* (Block and Bartell 2001). A book for preteens is *The Step-Tween Survival Guide: How To Deal with Life in a Stepfamily* (Cohn, Glasser, and Mark 2008). Sally Hewitt's *My Stepfamily* (2009) and *My SUPER Family: A Book for Blended Families* by Heather Orchard are written for younger children.

Stepfamily enrichment programs, support groups, and various other group counseling resources for step-families of various ethnicities are increasingly available and have been found to be helpful. One example is the *Active Parenting for Stepfamilies* program. Another is the 5-week Stepfamily Enrichment Program which has

been found to be helpful (Michaels 2010). We close this chapter with the paragraph that stepfamily scholar Susan Stewart uses to close her book *Brave New Stepfamilies* (2007):

One might conclude that Americans can maximize their well-being by getting married, staying married, reproducing their own biological offspring, and toughing it out. Yet an increasing number of Americans live increasing portions of their lives in increasingly diverse families that do not align with this idea. Perhaps Americans might do better by admitting the emerging normality of stepfamilies and building institutional supports to make their brave new stepfamilies strong. (p. 224)

Summary

- Remarriage rates are declining but most divorced men and women remarry.

- There is a great deal of diversity among stepfamilies as a result of of nonmarital childbearing and cohabitation, involvement of noncustodial parents, the aging of the population, racial and ethnic diversity, and societal support of same-sex couples.

- Remarriages are as happy as first marriages, but they tend to be slightly less stable.

- One reason for greater instability in remarriage is lack of a widely recognized cultural script for living in remarriages or stepfamilies.

- Couples with stepchildren often unconsciously try to approximate the nuclear-family model, but it does not work well for most stepfamilies.

- Our legal system and social policies make few accommodations for stepfamily relationships.

- Stepparents and children raised in stepfamilies, on average, have fewer financial resources and lower levels of socioemotional well-being and decisions about money are often fraught.

- Stepfamilies have many challenges but also have unique strengths.

Questions for Review and Reflection

1. Discuss some structural differences between stepfamilies and original two-parent families. Discuss different types of stepfamilies.

2. The remarried family has been called an incomplete institution. What does this mean? How does this affect remarried couples with stepchildren?

3. What evidence can you gather from observation, your own personal experience, or both to show that stepfamilies (a) may be more culturally acceptable today than in the past and (b) remain negatively stereotyped as not as functional or as normal as original two-parent families?

4. What are some challenges that stepparents face? What are some challenges faced particularly by stepfathers? Why might the role of stepmother be more difficult than that of stepfather? How might these challenges be confronted?

5. **Policy Question.** How could family laws and policies be changed to better accommodate stepfamily relationships?

Key Terms

affinity building   409
boundary ambiguity   409
common-pot system   419
cultural script   409
de facto parent   420
dripolator effect   413
incomplete institution   409

intimate outsider role   424
legal strangers   419
paternal claiming   414
percolator effect   413
role ambiguity   413
stepchildren   398
stepfamilies   398

stepfamily cycle   423
stepmother trap   414
stigmatization   404
the rule of two   420
two-pot system   419

16

AGING AND MULTIGENERATIONAL FAMILIES

Learning Objectives

1   Describe the changing age structure in the United States and other industrialized countries.

2   Discuss how the diversity of family forms among the elderly can be expected to affect caring for them in the future.

3   Describe economic circumstances and issues faced by the elderly.

4   Describe grandparent roles.

5   Explain how gender affects caregiving.

6   Discuss issues surrounding elder abuse and neglect.

7   List the costs and benefits of caring for an elderly family member.

Zeljkodan/Shutterstock.com

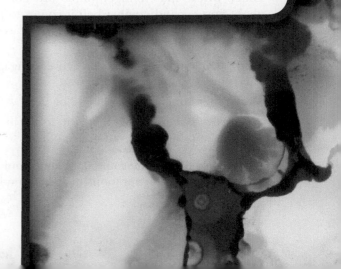

- "I feel very young," boasts an 82-year-old great grandmother who lifts weights weekly. She and her husband regularly take their great grandchildren to car shows

- In her seventies, Martha Stewart joins Match.com and shows off her new guys on television.

- After work Ramona visits her dad at his assisted-living facility, then rushes home to help her children with their homework.

- Jeanne intervenes to rescue her grandparents, whose caregiver son lives with them and is financially abusing and physically neglecting them.

Americans live longer than in decades past, a situation making for more **multigenerational families**—families that include several generations (Carr and Utz 2020). Compared with a few decades ago, many older Americans feel better and behave more youthfully today (Frizell 2015). Americans' happiness level, although highest for those in their early twenties and gradually dropping after that, begins to increase once more at about age 60 and does not drop again until after about age 75. At least into the twenty-first century we know that even into their nineties, a large majority of Americans have told pollsters they "experienced happiness, enjoyment, and smiling or laughter during a lot of the day" (Graham and Nikolova 2014; Newport and Pelham 2009).

This chapter examines multigenerational families, particularly issues focused on the contributions and needs of aging family members. We will explore older Americans' living arrangements. We'll discuss the grandparent role and then look at issues concerning caregiving to older family members. The ever-increasing diversity within today's older population—in racial and ethnicity and also in family form—provides a backdrop to these discussions. To begin, we note that many of the topics explored elsewhere in this text apply to aging and multigenerational families. For instance:

- Older families, like other families, comprise a diversity of family forms, including LGBTQ+ individuals, couples, and families.

- Older wives—like younger ones—are concerned about fairness and equity when it comes to power, decision making, housework, and other caregiving tasks.

- Aging family members may be actively engaged in parenting.

- More older families are stepfamilies.

- Supportive communication is important in older, younger, and multigenerational families.

To begin, we'll examine some facts about our aging population, the older generations in today's multigenerational families.

OUR AGING POPULATION

"The current growth of the population ages 65 and older is one of the most significant demographic trends in the history of the United States" (Mather, Jacobsen, and Pollard 2015, p. 2). In 1980, there were 25.5 million Americans age 65 or older; today more than 50 million Americans are 65 or older, and this number is expected to double over the next forty years (U.S. Administration on Aging 2018). Americans age 75 and older numbered close to 10 million in 1980; by 2010, there were more than 18.5 million. Of those age 85 and older, there were 2.2 million in 1980 compared to more than 6.5 million today. Projections are that, by the year 2050, there will be nearly 88.5 million Americans age 65 and older, with about 19 million of them age 85 and older (U.S. Administration on Aging 2018).

Not just the number of elderly has increased but also their proportion of the total U.S. population. This is especially true for those in the "older-old" (age

Changing the concept of aging itself, seniors are increasingly active into older ages. According to LeRoy Hanneman of Del Webb Retirement Communities, at least some "Boomers should be called Zoomers" (in "The Demographics of Aging" n.d.).

Dave & Les Jacobs/Alamy Stock Photo

75 through 84) and the "old-old" (85 and older) age groups. The proportion of Americans age 75 and older rose from 4.4 percent in 1980 to 6 percent in 2010, with projections up to 11 percent in 2050. The proportion of Americans age 85 and older is expected to rise from 1 percent in 1980 to 4.3 percent in 2050. Although the number of centenarians (those age 100 and older) is small at less than 1 percent, that percentage is growing (U.S. Administration on Aging 2018).

Aging Baby Boomers

Between 1946 and 1964, in the aftermath of World War II, more U.S. women married and had children than ever before. The high birthrate created what is commonly called the **baby boom**. Now Baby Boomers are retiring, and within the next decades they will generate an unprecedentedly large elderly population (see Figure 16.1). Meanwhile, the number of children under age 18 is about the same today as it has been for several decades (about 75 million).

As a result, children now make up a decreasing proportion—and older Americans a growing proportion—of the population. The proportion of the U.S. population under age 18 was 24 percent in 2012 compared to 36 percent in 1960 (U.S. Census Bureau 2003, Table 11; 2012a, Table 9). This is a global trend:

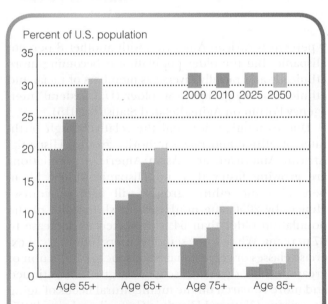

FIGURE 16.1 Older Americans as a percentage of the total U.S. population, 2000 and 2010, with projections for 2025 and 2050. Currently, Baby Boomers are in their fifties and sixties. As the baby boom cohort grows older, populations over ages 55, 65, 75, and 85 will increase.

Source: Calculated from U.S. Census Bureau 2002, Table 12; 2012a, Table 9.

There are not more people over age 65 than under age five (Nowakowski 2019). Along with the impact of the Baby Boomers' aging and the declining proportion of children in the population, longer life expectancy has contributed to the fact that our population is growing older as a whole (Carstensen 2015; Frizell 2015).

Longer Life Expectancy

Americans are now living long enough that demographers divide the aging population into three categories: the *young old* (age 65 through 74), the *older old* (age 75 through 85), and the *old old* (age 85 and older) (Carstensen 2015). Life expectancy at birth increased from 70.8 years in 1970 (67.1 years for men and 74.7 years for women) to 78.8 years in 2012 (76.4 years for men and 81.2 years for women) (Frizell 2015; Minino and Murphy 2012). Today life expectancy at birth is about 80 years (77.1 for men, and 81.9 for women) (U.S. Administration on Aging 2018).

Race, Ethnicity, and Life Expectancy Life expectancy differs by race and ethnicity. Life expectancy at birth for nonHispanic whites in 2010 was 78.8, compared with 74.7 for nonHispanic blacks (Minino and Murphy 2012; see also Jordan 2019). Much of the difference between blacks and whites has historically been associated with whites averaging higher incomes and lower poverty rates than blacks. Higher incomes, along with higher education levels, are associated with longer life expectancy, largely because people in higher socioeconomic groups are less likely to work or live in hazardous environments and have access to better health care (Conner 2000, p. 16). Not only life expectancy but also fertility and immigration patterns—with immigrants tending to be younger and having higher fertility rates—differ and result in greater diversity among the aging population (Jordan 2019).

The U.S. Hispanic population aged 65 and older is the largest other-than-nonHispanic white community in this age group. Hispanics have a higher average life expectancy—81.3 in 2012—than do nonHispanic whites. Hispanic females average the longest life expectancy of any racial, ethnic, or gender category—83.8 years in 2010 (Minino and Murphy 2012). Some experts attribute this situation to new immigrants' relatively healthy eating habits coupled with more walking and other physical activity. However, once they attain even a minimal level of economic success here, immigrants—and more so their American-born children—walk and exercise less while consuming unnecessarily high numbers of calories, situations that result in life-shortening obesity, diabetes, and high blood pressure (National Hispanic Council on Aging 2015; Tavernise 2013).

Gender and Life Expectancy On average, women live about five years longer than men. Because of this, the makeup of the elderly population differs by gender. In 2010, there were 22.9 million women age 65 and older compared to 17.3 million men. For Americans older than 84, there are 3.9 million women and about 1.9 million men (U.S. Census Bureau 2012a, Table 9). This gendered difference in life expectancy means that, among other things, elderly women are more likely than men to be widowed. Eighty percent of centenarians today are female (Meyer 2012). For reasons explained later in this chapter, elderly women are more likely than their male counterparts to be poor (U.S. Federal Interagency Forum on Aging-Related Statistics 2015).

Family Consequences of Longer Life Expectancy
Demographers point to at least two family-related consequences of our living longer. First, because more generations are alive at once, members of multigenerational families will increasingly have opportunities to maintain ties with grandparents, great-grandparents, and even great-great-grandparents (Bengston 2001).

A second consequence of longer life expectancy is that, on average, more Americans spend time near the end of their lives with chronic health problems or physical disabilities. We can think not just in terms of overall life expectancy but also in terms of **active life expectancy**—the period of life free of disability. After this, a period of being at least partly disabled may follow, partly because today's older Americans, while living longer, are not necessarily healthier and are more likely to be obese or to have diabetes or high blood pressure than previous generations of similar ages—situations that can lead to higher odds of stroke and resulting disability (Scommegna 2013). As Americans get older, more and more of us will be called on to provide care for a parent or other relative who is disabled (American Psychological Association 2017; Harris-Kojetin et al. 2013). We will return to issues surrounding giving care to aging family members later in this chapter. We turn now to the racial and ethnic composition of the older American population.

Racial and Ethnic Composition of the Older American Population

"While the elderly are often subsumed under the same umbrella as the 'over 65-generation,' it is important . . . to note that the aging population in the United States comes from a wide range of backgrounds and cultures" (Trask et al. 2009, p. 301). As a category, nonHispanic whites are older than people in other racial and ethnic categories (U.S. Census Bureau 2012a, Table 10). Figure 16.2 shows the nation's age distribution by race and ethnicity. About 78 percent of today's U.S. population over age 64 is nonHispanic white. Another

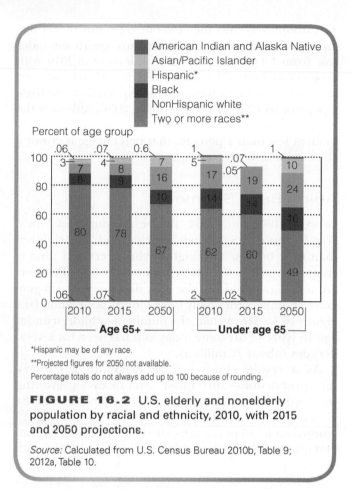

*Hispanic may be of any race.

**Projected figures for 2050 not available.

Percentage totals do not always add up to 100 because of rounding.

FIGURE 16.2 U.S. elderly and nonelderly population by racial and ethnicity, 2010, with 2015 and 2050 projections.

Source: Calculated from U.S. Census Bureau 2010b, Table 9; 2012a, Table 10.

9 percent is African American, with another 8 percent Hispanic. But the older population is becoming more ethnically and racially diverse as members of racial and ethnic minority groups grow older (U.S. Federal Interagency Forum on Aging-Related Statistics 2015).

Due to immigration and the relatively high birthrates of ethnic and racial minority groups, Hispanic, African American, and Asian American populations are growing faster than nonHispanic whites. Members of these ethnic groups will age, of course. Hence, by 2050, the nonHispanic white share of the population older than 64 is projected to decrease to 67 percent. While some senior centers "offer tai chi exercise classes or serve tamales for lunch, a reflection of greater ethnic diversity," scholars argue for continued and more research on the multicultural needs of aging Americans (National Hispanic Council on Aging 2015; Treas 1995, p. 8; Trask et al. 2009).

Older Americans and the Diversity of Family Forms

We have seen throughout this text that today's families are diverse in form. As the postmodern family (see Chapter 1) grows older, we can expect late-life family

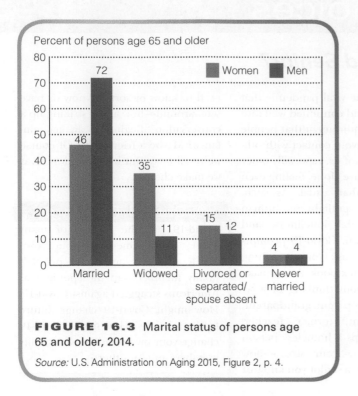

FIGURE 16.3 Marital status of persons age 65 and older, 2014.

Source: U.S. Administration on Aging 2015, Figure 2, p. 4.

forms to exhibit increasing diversity. You can see in Figure 16.3 that 72 percent of men and 46 percent of women age 65 and older were married in 2014. Fifteen percent of women and 12 percent of men in that age category were divorced or separated, with another 4 percent never married (U.S. Administration on Aging 2015, Figure 2).

The birthrate in the mid-1970s struck a record low while the divorce rate peaked. Americans who were in their twenties and thirties in the 1970s are now moving into later life. Increasingly, therefore, families will enter into older ages with fewer if any children and with histories of cohabitation, separation, divorce, and repartnering. In addition, now that same-sex families are increasingly visible, we can expect researchers and policy makers to pay more attention to aging same-sex families (Grant 2010). In fact, the Gay and Lesbian Association of Retired Persons formed in 1999 (www.gaylesbianretiring.org). The caregiving implications of the trend toward increased family diversity are addressed later in this chapter. Here we turn to a discussion of living arrangements among older Americans.

LIVING ARRANGEMENTS OF OLDER AMERICANS

About 2 million or 5 percent of Americans age 65 and older live in nursing homes (American Psychological Association 2017). The likelihood of living in an institutional setting such as a nursing home increases with

age: 1 percent of Americans between ages 65 and 74 reside in institutional settings. Among those between ages 75 and 84, the percentage is 3. Of those age 85 and older, 10 percent live in institutional settings (U.S. Administration on Aging 2018).

Among the vast majority of older adults not residing in institutional settings, some have moved to retirement communities, some of these in the "Sun Belt"—Florida and the Southwest—as well as to communities in countries south of the U.S. border where the cost of living may be less. Retirement communities have emerged specifically for lesbians and gays ("Birds of a Feather" 2015).

As the babyboom generation that legitimated cohabiting ages, it should be no surprise that the incidence of cohabiting among the elderly has increased. The number of adults over age fifty rose by 75 percent between 2007 and 2016 (Stepler 2017). While many older Americans cohabit, others live in cohousing arrangements, described in Chapter 6. Still others have established "living apart together" (LAT) relationships, also discussed in Chapter 6 (Carr and Utz 2020; Connidis, Borell, and Karlsson 2017).

Complementing the term *boomerang kids*, which refers to adult children who move back into their parents' homes, one journalist now writes of *boomerang seniors* (Kluger 2010). Since the onset of the recent economic downturn, a significant and growing number of older Americans—many of them healthy and active but needing to reduce expenses—have moved into their grown children's or grandchildren's homes (Lofquist et al. 2012, p. 15). Furthermore, more frail elders today reside in their children's homes when the family cannot afford assisted-living or other caregiving facilities. Recent studies have generally found that for the most part multigenerational households are happy (Donaldson 2012). In some instances, a grandparent's moving in benefits the grandchildren's school performance (Augustine and Raley 2013).

Nevertheless, historical trends in U.S. living arrangements show a long-term preference for separate households (Bures 2009). Currently, as shown in Chapter 1's Figure 1.2, about 28 percent of U.S. households are made up of people living alone. Many of them are older people. This situation represents a growing trend since about 1940. Among Americans age 65 and older, approximately one-third live alone. Because of the increased likelihood of being widowed as people age, about 40 percent of people over age 75 live by themselves. About half of women age 75 and older live alone (U.S. Administration on Aging 2015, p. 5).

The elderly who live alone "are usually within a close distance of relatives or only a phone call or e-mail away. Fewer than one out of twenty are socially isolated, and usually are so because they have lived that way most of their lives" (Moody 2006, p. 331; and see Bui and Miller

As We Make Choices

Want to Call or Visit an Isolated Senior?

Companionship is important to psychological well-being. Meanwhile, a recent survey finds that on average Americans over 65 spend half their waking hours alone (Livingston 2019). Nearly half of Americans over age 85 live alone (U.S. Administration on Aging 2018). It's true that tech adoption is up among older adults. About 40 percent of adults over age 65 now own smart phones while two-thirds use the Internet (Anderson and Perrin 2917). Nevertheless, digital connection, while helpful, cannot entirely replace face-to-face or voice-to-voice human contact and presence.

Actually, 22 percent of *millennials* say they have "no friends" (Resnick 2019).

But Covid-19, the viral pandemic that began in 2019 and continued well into 2020, with its requirement that people stay home and avoid contact with others, found many of us, especially older Americans who live alone, feeling even more isolated than usual. As such, Covid-19 put a spotlight on human isolation, potential loneliness, and psychological pain. Covid-19 prompts your authors to ask you something unusual for this textbook: As we make choices, how about thinking of a senior in your life—parent, grandparent, great-grandparent, former teacher, neighbor, perhaps a homeless person who walks along your street—how about thinking of a senior you know or used to know or sort of know or have seen around—how about striking up a conversation? Beginning an acquaintance? Maybe a friendship? Of course we're all busy. But it's an option....as we make choices.

Critical Thinking

Covid-19 put the welfare of many older family members—our own family members and other people's family members—in the spotlight as Americans struggled against Covid-19. How might Covid-19 change future family relations? How might Covid-19 change your own relations with family members and/or others?

2015). Nevertheless, a significant percent of elderly family members are isolated, at least at times. **As We Make Choices:** Want to Call or Visit an Isolated Senior? addresses this issue.

It remains to be seen whether preferences will change as our aging population grows more ethnically diverse with more elderly coming from a heritage of communal rather than individualistic values.

Racial and Ethnic Differences in Older Americans' Living Arrangements

Table 16.1 compares the living arrangements of non-Hispanic white, black, Asian American, and Hispanic adults age 65 and older. Due to both economic and cultural differences, the living arrangements of older Americans vary according to race and ethnicity. For instance, as shown in Figure 16.4, Asian Americans, blacks, Hispanics, American Indians, Alaska Natives, Native Hawaiians, and Pacific Islanders in the United States are far more likely than nonHispanic whites to live in multigenerational households (Lofquist 2012). Older family members may reside in the homes of their grown children, or they may have opened their doors to adult offspring who have moved in with them (Glick and Hook 2002).

One generalization we can make from the statistics in Table 16.1 and Figure 16.4 is that ethnic groups other than nonHispanic whites are significantly more likely to live in multigenerational households with people other than their spouse—grown children, siblings, or other relatives (Lofquist 2012). Partly as a result of economic necessity, coupled with social norms involving family members' obligations to one another, older Asian Americans and Hispanics are less likely than nonHispanic whites to live alone. This is true for Hispanics even though they are also less likely than nonHispanic whites to live with a spouse (U.S. Census Bureau 2012a, Table 58).

Older Hispanics and African Americans are less likely than nonHispanic whites to live with a spouse for two reasons. First, due to differences in patterns of marriage and divorce, African Americans are more likely than whites to enter older age without a spouse. Second, gender differences in life expectancy (with women living longer than men) are slightly higher among blacks and Hispanics (about seven years) than among nonHispanic whites (about five years) (U.S. Census Bureau 2010b, Table 102).

Gender Differences in Older Americans' Living Arrangements

Due mainly to differences in life expectancy, older heterosexual men are much more likely to be living with their spouse than are older heterosexual women (72 percent of men age 65 and older compared to about 45 percent of women). Older women (40 percent) are far more likely than men (13 percent) to be widowed.

TABLE 16.1 Living Arrangements of People 65 Years Old and Over, by Racial and ethnicity, 2010

	LIVING ARRANGEMENT	65–74 YEARS OLD (%)	75 AND OLDER (%)
Total Population	Alone	22	37
	With spouse	64	45
	With other people[b]	14	18
NonHispanic White	Alone	22	39
	With spouse	67	37
	With other people[b]	10	14
Black	Alone	32	39
	With spouse	42	27
	With other people[b]	26	33
Asian American	Alone	14	21
	With spouse	65	51
	With other people[b]	21	28
Hispanic Origin[a]	Alone	19	22
	With spouse	54	41
	With other people[b]	27	37

[a]People of Hispanic origin may be of any race.

[b]The category "with other people" includes relatives other than a spouse, as well as institutional settings, although the proportion of older Americans in institutional settings is relatively small.

Source: Calculated from U.S. Census Bureau 2012a, Table 58.

Of Americans between ages 65 and 74 who are living alone, two thirds are women. Of those 75 and older who are living alone, three quarters are women. Moreover, older women are significantly more likely than older men to live with people other than their spouse—a pattern that persists into old-old age (U.S. Administration on Aging 2015, Figure 3).

Generally, among older Americans without partners, living arrangements depend on a variety of factors, including the status of one's health, the availability of others to live with, social norms regarding obligations of other family members toward their elderly, the history of the family's relationship quality, personal preferences for privacy and independence, and economics (Bianchi and Casper 2000; Carr and Utz 2020). Older Americans with better health and higher incomes are more likely to live independently, a situation that suggests strong personal preferences—at least among today's mostly nonHispanic white elderly—for privacy and independence. However, those in financial need are more likely to live with relatives.

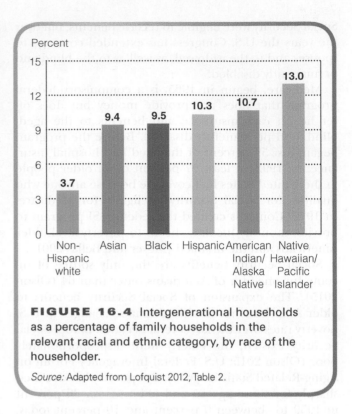

FIGURE 16.4 Intergenerational households as a percentage of family households in the relevant racial and ethnic category, by race of the householder.

Source: Adapted from Lofquist 2012, Table 2.

AGING IN TODAY'S ECONOMY

The corona virus pandemic, Covid-19, with resulting lost jobs and a devalued stock market, at least temporarily, no doubt changed retirement plans for many Americans, both of retirement age and younger. Even before the Covid-19 economic downturn, more and more older Americans were working—up to 20 percent of those 65 and older (Desilver 2016). At least through 2018 more were working into their eighties (Scommegna, Mather, and Kilduff 2020; Van Dam 2018).

Generally, today's older Americans live on a combination of employment income—investment income, Social Security benefits, private pensions from employers, personal savings, and social programs designed to meet the needs of the poor and disabled. About 40 percent of the income of Americans age 65 and older is from Social Security benefits and related federal programs such as Medicare, Medicaid, and Supplemental Security Income (SSI) (Larrimore et al. 2015).

During the Great Depression of the 1930s, millions of Americans lost their jobs and savings. In response, the federal government passed the 1935 Social Security Act, a new program designed to assist the elderly. The Social Security Act established the collection of taxes on income from one generation of workers to pay monthly pensions to an older generation who had retired from work. Initially, only those who contributed to

Social Security were eligible to receive benefits, but over the years the U.S. Congress has extended coverage to spouses and to the widowed, as well as to the blind and permanently disabled.

Medicare, begun in 1965, is a compulsory federal program that does not provide money but does offer health care insurance and benefits to the aged, blind, and permanently disabled. Before the program began, just 56 percent of the aged had hospital insurance. In 1992, at least 97 percent of all older people in the United States had coverage because anyone who qualifies for Social Security is eligible for Medicare. In 1972, Congress created the federal SSI program to provide monthly income checks to poverty-level older Americans and the disabled (Meyer and Bellas 2001).

Social Security benefits are the only source of income for one-fifth of Americans older than 64 (Olson 2015). The expansion of Social Security benefits to older Americans resulted in dramatic changes in U.S. poverty rates over the last several decades. Before Social Security existed, the elderly were disproportionately poor (Olson 2015; U.S. Federal Interagency Forum on Aging-Related Statistics 2015). But poverty has declined sharply for those age 65 and older—from 36 percent in 1959 to between 9 percent and 10 percent today. Due partly to older Americans' lobbying to protect Social Security benefits, the poverty rate for those older than 64 is now less than half that of children (U.S. Federal Interagency Forum on Child and Family Statistics 2012). Then, too, many of today's older Americans benefit from what was a generally stable or rising economy during most of their working years. However, having noted that today's older Americans have been generally better off than generations preceding them, we need to acknowledge that on average their income declines by as much as one-half on retirement, a situation that can lead to stress and relationship conflict (Semega, Kollar, Creamer, and Mohanty 2019).

Nearly 10 percent of adults age 65 and older are living in poverty—nearly 20 percent of blacks and Hispanics (U.S. Administration on Aging 2018). Between 6 percent and 10 percent of the homeless are age 65 and older, and this proportion may be modestly increasing (National Alliance to End Homelessness 2013; National Coalition for the Homeless 2009b). Along with the *near poor* (those with incomes at or below 125 percent of the poverty level, who make up another 9 percent of the elderly), these poverty-level older Americans are hardly enjoying the comfortable and leisurely lifestyle that we may imagine when we think of retirement (Larrimore et al. 2015).

Retirement?

From an historical standpoint, widespread retirement only became possible in the twentieth century when

Social Security and Medicare raised the incomes of older Americans, beginning in 1940, so that the proportion of elderly in the United States living in poverty today has declined and is less than that of children. Nevertheless, about 3.7 million, or 10 percent of older Americans—disproportionately the unmarried and women—are living in poverty. Another 2.4 million of the elderly are classified as "near poor," with incomes just 125 percent of the poverty level (U.S. Administration on Aging 2010).

the industrial economy was productive enough to support sizable numbers of nonworking adults. At the same time, the economy no longer needed so many workers in the labor force, and companies believed that older workers were not as quick or as productive as the young. Governments, corporations, labor unions, and older workers themselves saw retirement as a desirable policy, and it soon became the normal practice (Moody 2006).

Today, however, an unpredictable economy, the financial consequences of divorce, and a longer active life expectancy may render the policy of retiring around age 65 outdated (Gustke 2014; Larrimore et al. 2015). For some—although certainly not a majority—Americans' retirement results, not in the 24/7 vacation myth, but in poverty and bankruptcy (Bernard 2018).

Although most older people eventually retire, some do not—and many of those who don't are employed beyond age 70 (Desilver 2016). Then, too, many people retire gradually by steadily reducing their work hours or intermittently leaving, then returning to the labor force before retiring completely (Larrimore et al. 2015).

Not wanting to give up the psychological benefits associated with working—that is, feeling that one's life is meaningful and experiencing personal growth—is a reason that people give for not retiring. A less satisfying reason that is increasingly relevant in today's economy and particularly applicable to divorced older women

(and men, too, but fewer) is not being financially able to retire (Gustke 2014). Even before the Great Recession that began in 2009, the majority of aging Baby Boomers expected to work at least part-time after retirement age (Brougham and Walsh 2009). By 2009, 40 percent of those over 62 had delayed retirement due to the recession, according to the Pew Research Center. Among those between ages 50 and 61, 63 percent (54 percent of men and 72 percent of women) said that they might need to delay retirement because of the recession (Taylor 2009). Between 2009 and 2014, the number of fulltime workers over age 65 increased by more than one-third—from 3.5 million to 4.8 million. This trend toward more elderly employees partly results from the changing nature of work itself. Jobs requiring manual labor and physical strength are less numerous, while some careers are considered *age-appreciating*— that is, they require skills that actually improve with age (technical writing, for example, or human resources management) (Lam 2015; Van Dam 2018).

Gender Issues and Older Women's Finances

A study of Baby Boomers shows that among both women and men, those who married and stayed together are the best off financially, with never-marrieds faring worst (Addo and Lichter 2013; Lin and Brown 2013). Never-married, divorced, and separated women are worse off financially than widows, and many need to work for several years after traditional retirement age (Larrimore et al. 2015). In fact, the marriage gap between U.S. blacks and whites accounts for much of the wealth gap between the races in old age (Addo and Lichter 2013).

Nevertheless, on average, older men are considerably better off economically than older women—a warning particularly to younger women that they need to begin saving for retirement asap (Hannen 2018). In 2013, the median income of Americans age 65 and older was $29,327 for males and $16,301 for females (U.S. Administration on Aging 2015, p. 9). This dramatically unequal situation is partly due to the fact that men averaged higher earnings than did women throughout their employment years (U.S. Bureau of Labor Statistics 2012; and see Chapters 3 and 10). Consequently, older women today have smaller pensions, if any, from employers and lower Social Security benefits (Olson 2015). Furthermore, older women on average did not begin to save for retirement as early as did men (Even and Macpherson 2004; Herd 2009; Olson 2015).

Moreover, women are much more likely than men to rely on Social Security for at least 90 percent of their income. Maximum Social Security benefits ($3,680 monthly in 2018 if retiring at age 70) are available only to workers with lengthy and continuous labor force participation in higher-paying jobs. As addressed in Chapter 10, this situation works against older women, who were often raising children, were likely to have taken lower-paying jobs if they did work, or had dropped in and out of the labor force (Herd 2009; Olson 2015). "Thus, women are penalized for conforming to a role that they are strongly encouraged to assume—unpaid household worker—and their disadvantaged economic position is carried into old age" (Meyer and Bellas 2001, p. 193).

Unfortunately, the future does not look much better. Today, singles of both genders in their twenties and thirties—and particularly single mothers, many of whom cannot afford to put away money—have done little in the way of retirement-saving activities (Knoll, Tamborini, and Whitman 2012).

RELATIONSHIP SATISFACTION IN LATER LIFE

As shown in Figure 16.3, only about 4 percent of men or women 65 and older today have never married (U.S. Administration on Aging 2015, Figure 2). Some later-life marriages are remarriages, and the proportion of repartnered older Americans will increase as those who are now middle-aged grow older. Today, however, the majority of older married couples have been wed for quite some time—either in first or second unions. Retirement represents an important and usually temporarily stressful change for couples (Wickrama, Walker, and Lorenz 2013).

Role flexibility is important to successful adjustment (Bulanda 2011). Also, health is an important factor in morale in later life, and it has a substantial impact on marital quality as well (Connidis 2010; Radina 2013). The majority of older married couples place companionship and intimacy as central to their lives and describe their unions as happy (Iveniuk et al. 2014; Villar and Villamizar 2012). As a category, aging marrieds are happier and more satisfied with their lives than are their nonmarried counterparts (LaPierre 2009; Villar and Villamizar 2012).

This is not to say that *all* older couples are happy. A spouse's depression or otherwise poor mental or physical health can negatively affect couple happiness (Wickrama, O'Neal, and Lorenz 2013). Later-life couples who hold more egalitarian attitudes toward gender roles and who experience high levels of warm mutual interaction report significantly greater marital happiness (Schmitt, Kliegel, and Shapiro 2007).

Sexuality in Later Life

Frequency of sexual intercourse typically declines with age (Carr and Utz 2020). Nonetheless, many older Americans continue to be sexually active—even into old-old age and even in nursing homes (Forbes, Eaton,

Married older couples may be in first unions or in remarriages. Most older couples describe their marriages as happy. A retired husband may choose to spend more time doing homemaking tasks and give increased attention to being a companionate spouse. Role-sharing, feeling that work is fairly shared, and having supportive communication predict good adjustment for retiring couples.

and Krueger 2016; Schwartz 2012). A 2009 national survey of Americans found that among respondents age 70 and older, 11 percent of women and 22 percent of men reported having sexual intercourse at least once or twice a month (Fisher 2010, Tables 19 and 20). Whether it involved intercourse or other forms of sexual expression, 80 percent of men and 39 percent of women age 70 and older said that a sexual relationship was important to their quality of life (Fisher 2010, p. 9).

This is not to imply that older adults have no sexual problems (Forbes, Eaton, and Krueger 2016. Although older single women may be interested in sex, lack of a partner can be a problem. We have seen that, as they age, women are far more likely than men to be widowed. Moreover, as they grow older, women are adversely affected by the double standard of aging—that is, men aren't considered old or sexually ineligible as soon as women are (England and McClintock 2009). Beauty, "identified, as it is for women, with youthfulness, does not stand up well to age" (Sontag 1976, p. 352). For older single women, this situation can exacerbate more general feelings of loneliness (Narayan 2008).

Stress, dissatisfaction with one's partner, and health-related issues can inhibit sexual desire and activity for both sexes (Laumann, Das, and Waite 2008). According to psychiatrist Stephen Levine, "Over age 50, the quality of sex depends much more on the overall quality of a relationship than it does for young couples" (quoted in Jacoby 1999, p. 42; see also Carr and Utz 2020).

LATER-LIFE DIVORCE, WIDOWHOOD, AND REPARTNERING

The majority of couples who divorce do so before their retirement years. However, a growing proportion of couples divorce in later life, and Baby Boomers are divorcing at relatively high rates at older ages (Alfred 2019a, 2019c; Cohen 2019). As noted in Chapter 14, "silver" or later-life divorces are not necessarily easy on a couple's adult children. Family celebrations and holidays are disrupted. An adult child's graduation or wedding can be difficult when forced to accommodate recently divorced parents.

Adult children of divorcing parents may worry about having to become full-time caregivers to an aging parent in the absence of the parent's spouse. "Years after parents split, their children may wind up helping to sustain two households instead of one, and those households can be across town or across the country" (Span 2009a). Nevertheless, although some later-life marriages end in divorce, the vast majority do so with the death of a spouse.

Widowhood and Widowerhood

Adjustment to widowhood or widowerhood is an important common family transition in later life. A spouse's death brings the conjugal unit to an end—often a profoundly painful event. "The stress and emotional trauma of losing a spouse as a confidant might be greater now than in the past, and [for those who do not divorce] the average duration of marriage becomes longer with increasing life expectancy" (Liu 2009, p. 1,170).

Because women's life expectancy is longer and older men remarry more often than women do, widowhood is significantly more common than widowerhood. Just more than half (51 percent) of women between ages 75 and 84 are widowed compared with 17 percent of men. Among those age 85 and older, just under three-quarters of women (73 percent) are widowed compared with more than one-third (35 percent) of men (U.S. Federal Interagency Forum on Aging-Related Statistics 2015).

Typically, widowhood and widowerhood begin with **bereavement**, a period of mourning, followed by gradual adjustment to the loss and to the new, unmarried status. Bereavement manifests itself in physical, emotional, and intellectual symptoms. Recently widowed people perceive their health as declining and report feeling depressed. The bereaved experience various emotions—anger, guilt, sadness, anxiety, and preoccupation with

thoughts of the dead spouse—but these feelings tend to diminish over time (Connidis 2010; Jin and Chrisatakis 2009). Social support, adult children's help with housework and related tasks, and activities with friends, children, and siblings help (Cornwell and Waite 2009; Population Reference Bureau 2009).

There is evidence that being single in old age is more physically and emotionally detrimental for men than for women. Wives rather than husbands tend to be central "in the household production of health" (Jin and Chrisatakis 2009, p. 605). Put another way, females are more likely than males to concern themselves with the health of all household members, so a man's health is more likely to decline when a wife is absent than vice versa. Then, too, as sources of support, women more often have friends in addition to family members. Men are more often dependent solely on family. For some of the widowed, remarriage promises resumed intimacy and companionship.

Aging and Repartnering

About 14 percent of single elderly are in dating relationships (Brown and Shinohara 2013). Annually about one-half million Americans older than 65 remarry (Belkin 2010; Carr and Utz 2020). Repartnering elders, particularly those who are widowed, are likely to choose homogamous partners similar in social class, racial and ethnicity, and religious identity. A trend today is for more and more aging Baby Boomers—those who initiated the 1960s sexual revolution, after all—not necessarily to marry but to cohabit or create LAT relationships (described in Chapter 5) (Luxenberg 2014). Financial advisors suggest that cohabiting or LAT couples draw up formal agreements regarding who is legally allowed to make medical decisions as well as other aspects of the couple's living arrangements, such as specifying which partner is responsible for what expenses or who will inherit property when one partner dies (Luxenberg 2014).

However, middle-aged and older people may face considerable opposition to remarriage—from restrictive pension and Social Security regulations (Ebeling 2007) and sometimes from their adult children. In general, a widow or widower cannot receive Social Security survivors' benefits if remarrying before age 60. Remarrying after age 60 (or 50 if disabled) does not negatively affect receipt of survivor benefits (U.S. Social Security Administration 2015).

Chapter 15 explores adult children's attitudes about a parent's remarriage. Although some grown children may be supportive of a parent's remarriage, others may find it inappropriate (Gierveld and Merz 2013). Adult children may worry about the biological parent's potentially diminished interest in them or about their inheritance (Connidis 2010). Even when a will leaves everything to one's children, a second spouse may have the legal right to claim a share of the estate. A common option involves signing a prenuptial agreement according to which the spouse-to-be relinquishes any claim to the estate. Older couples with children from previous marriages are encouraged to see a financial adviser before marrying (Barnes 2009).

Especially for women, age reduces the likelihood of repartnering. For one thing, an uneven sex ratio (see Chapter 6) decreases the odds that older heterosexual women will repartner: Among Americans age 55 and older, there are approximately eighty-three men for every 100 women (U.S. Census Bureau 2012a, Table 9). The double standard of aging, described in Chapter 5, also works against older women (England and McClintock 2009). Then, too, women may be less interested than men in late-life remarriage (Levaro 2009).

Meanwhile, research shows that the myth that widowers are quick to replace a deceased wife is just that—an exaggerated stereotype. "Men with high levels of social support from friends are no more likely than women to report interest in repartnering" (Carr 2004, p. 1,065). During later life, morale and well-being frequently derive from relations with siblings, as well as from friends, neighbors, and other social contacts (Nelson 2011; Voorpostel and Blieszner 2008). Particularly for aging mothers, relationships with children and grandchildren are important.

MULTIGENERATIONAL TIES: OLDER PARENTS, ADULT CHILDREN, AND GRANDCHILDREN

More often than spousal relationships, those between parents and their biological children last a lifetime (Carr and Utz 2020). Many parents continue to aid their adult offspring virtually as long as possible. Often following traditional gender roles, aging mothers are likely to help with childcare, while older fathers may help adult offspring with home repairs, yardwork, or car repair (Kahn, McGill, and Bianchi 2011). In this section, we examine relations between older parents and their adult children and grandchildren.

Older Parents and Adult Children

Adults' relationships with their parents range from tight-knit, to sociable, to obligatory, to intimate but distant, to detached. Sociologists Merril Silverstein and Vern Bengston (2001) developed six indicators of relationship solidarity, or connection: geographic proximity, contact between members in a relationship, emotional closeness, similarity of opinions, providing

care, and receiving care. Based on survey evidence and using these six indicators, Silverstein and Bengston then developed a typology of these five kinds of parent-adult child relations.

Parent-adult child relations vary, depending on how family members combine—or, in the case of the detached relationship style, do not combine—the six indicators. For instance, in tight-knit relations, the parent and the adult child live near each other (geographic proximity), feel emotionally close, share similar opinions, and help each other (give and receive assistance). Sociable relations involve all these characteristics except that the parent and adult child do not exchange assistance. Table 16.2 defines all five relationship types.

There is no typical model for parent-adult child relationships (Arnett 2004; Silverstein and Bengston 2001). Parent-adult child relations might change over time, moving from one relationship type to another, depending on the parent's and the adult child's respective ages, the parent's changed marital status, and the presence or absence of grandchildren, among other factors. For instance, to be nearer to their aging parents, adult

children sometimes return to the area in which they grew up, or retired grandparents may decide to relocate to be near their grandchildren (Lee 2007). Both of these situations could move an intimate but distant relationship to a tight-knit one. Then, too, a parent-adult child relationship might change, depending only on emotional factors such as when an adult child chooses to forgive an aging parent for some past transgression, or vice versa.

Using national survey data from a sample of 971 adult children who had at least one surviving non-co-resident parent, Silverstein and Bengston (2001) made the following findings (among others):

- The majority of relations were neither tight-knit nor detached, but *variegated*—one of the three relationship styles in between (see Table 16.2). Variegated relations characterized 62 percent of adult children's interaction with their mothers and 53 percent with their fathers.

- Tight-knit relations are more likely to occur among lower socioeconomic groups and racial and ethnic minorities.

- NonHispanic whites were more likely than minority ethnic groups to have detached or obligatory relationships with their parents.

- The most common relationship between a mother and her adult child was tight-knit.

- The most common relationship between a father and his adult child was detached.

- Daughters were more likely than sons to have tight-knit relations with their mothers.

- Sons were more likely than daughters to have obligatory relations with their mothers.

- Adult children were more likely to have obligatory or detached relations with divorced or separated parents.

From these findings, we can conclude that daughters are more likely than sons to have close relationships with their parents, especially with their mothers. Even for mothers, a parent's divorce or separation often weakens the bond with adult children (Hans, Ganong, and Coleman 2009). However, the chance of having detached relations with one's adult children after divorce is nearly five times greater for fathers than for mothers. Partly, this is true because a divorced father is less likely than either a consistently married father or a divorced mother to live with his biological children and more likely to remarry (Pezzin, Pollak, and Schone 2008; Silverstein and Bengston 2001).

In some families, the reality of past abuse, a conflict-filled divorce, or simply fundamental differences in values or lifestyles makes it seem unlikely that parents and children will spend time together. Money matters

TABLE 16.2 Types of Intergenerational Relations

CLASS	DEFINITION
Tight-knit	Adult children are engaged with their parents based on geographic proximity, frequency of contact, emotional closeness, similarity of opinions, and providing and receiving assistance.
Sociable	Adult children are engaged with their parents based on geographic proximity, frequency of contact, emotional closeness, and similarity of opinions but not on providing or receiving assistance.
Obligatory	Adult children are engaged with their parents based on geographic proximity and frequency of contact but not on emotional closeness and similarity of opinions. Adult children are likely to provide or receive assistance or both.
Intimate but distant	Adult children are engaged with their parents based on emotional closeness and similarity of opinions but not on geographic proximity, frequency of contact, providing assistance, and receiving assistance.
Detached	Adult children are not engaged with their parents based on any of these six indicators of solidarity.

Source: Adapted from Silverstein and Bengston 2001, p. 55.

can also cause ambivalence and tension (Arnett 2004; Lendon, Silverstein, and Giarrusso 2014; and see Winter, Gitlin, and Dennis 2011). Middle-aged parents, aware that they should be saving for retirement, often find themselves helping to finance adult children's transportation, rent, medical care, student loans, and other expenses (Hymowitz 2015). Some aging parents sacrifice retirement comfort in order to save toward bequeathing an inheritance to their adult child(ren) (Lieber 2014).

Money-related tensions can be heightened in stepfamilies where "[a]dult children can feel resentful when they see a stepparent spending what they consider as their rightful inheritance" (Sherman 2006, p. F8). Especially if they have children, older persons are encouraged to thoroughly discuss and agree on money matters prior to a remarriage, living together, or LAT arrangement (Yip 2012). Overall, the majority of adult children's relationships with parents, although not necessarily tight-knit, continue to be meaningful and based on mutual feelings of relationship equity or fairness (Sechrist et al. 2014).

Grandparenthood

As noted in Chapter 1, technology and the desire for grandchildren can be strong enough that some people finance freezing the eggs of an adult daughter not yet ready to have children (Gootman 2012). Much more commonly, medical technology and grandparenthood

Financially independent adults' relationships with a parent can be of several types: tight-knit, sociable, obligatory, intimate but distant, or detached. Then, too, today's parents can find themselves in the "senior sandwich generation"—paying for an adult child's college tuition or other expenses, worrying about the financial burden of elder care for aging parents, while trying to save for their own retirement.

meet as grandparents-to-be study ultrasound images of their grandchild's developing fetus (Harpel and Hertzog 2010).

Thanks to longer life expectancy, which creates more opportunity for the role, grandparenting—and great-grandparenting—became increasingly important to families throughout the twentieth and into the twenty-first centuries. At the same time, the number of child-free families has grown, and so we can't assume that the "normal" family has grandchildren or grandparents (Margolis 2016). Nevertheless, many Americans are grandparents. Some are raising grandchildren, as discussed in Chapter 9. This section focuses on grandparents who are not primarily responsible for raising their grandchildren.

Grandparents play varied roles in their children's lives (Dunifon, Near, and Ziol—uest 2018). For many, the grandparent role is important enough to lessen their involvement in work while increasing their retirement-related concerns about their future health and finances (Wiese, Burk, and Jaeckel 2016). Younger grandparents are often employed and partnered while older grandparents are more likely to have physical disabilities. Hence, a grandchild's experiences with a younger grandparent is typically quite different from those with an older grandparent (Davey et al. 2009). Married grandparents have different experiences with grandchildren than do single grandparents. Once retired, however, both married and single grandfathers spend more time with grandchildren than they did while they were employed (Kahn, McGill, and Bianchi 2011). Geographical distance affects grandparent-grandchild relationships (Dunifon and Bajracharya 2012). You may be aware of grandparents who have relocated in order to live nearer to their grandchildren (Edleson 2015).

Grandparents tend to adopt a grandparenting style similar to one they experienced as a grandchild (Mueller and Elder 2003). Some grandparents (especially when they live far away) have remote relationships with their grandchildren, while other grandparent-grandchild relationships are more companionate, involving doing things with their grandchildren but exercising little authority in the grandparent-parent-grandchild relationship. Still other grandparents are highly involved (Cherlin and Furstenberg 1986; see also Dunifon and Bajracharya 2012).

Then, too, a grandparent may have different relationship styles with different grandchildren. Although grandparents do interact with their teenage grandchildren, they generally are most actively involved with (more available and emotionally responsive) preschoolers (Davey et al. 2009). After a typically disinterested adolescence, adult grandchildren often renew relationships with grandparents (Dunifon, Near, and Ziol-Guest 2018; Mansson, Myers, and Turner 2010).

Grandchildren give personal pleasure and a sense of immortality. Some elderly grandfathers see the role as an opportunity to be involved with babies and very young children, an activity that may have been discouraged when their own children were young (Cunningham-Burley 2001). Overall, the grandparent role is mediated by the parent. Not getting along with the parent dampens the grandparent's contact and, hence, the relationship with her or his grandchildren (Dunifon and Bajracharya 2012).

Grandparents may serve as valuable "family watchdogs" who are ready to provide assistance when needed (Troll 1985; Winerip 2009). For instance, the Interactive Autism Network (2010) conducted an online survey of individuals who had a grandchild with autism. This survey was hardly representative of all grandparents in this category, because to be aware of the survey one would have had to be interested enough in the topic to visit the website. However, nearly one-third of the grandparent respondents reported being the first in the family to notice anything out of the ordinary regarding their grandchild's development. Some grandparents (14 percent) had moved closer to their grandchild's family to help out. Many grandparents assisted with treatment-related costs, some dipping into retirement savings (Hamilton 2010; Interactive Autism Network 2010).

A close relationship with a grandparent can improve the grandchild's verbal skills as well as help facilitate a grandchild's adjustment after parental divorce (Arpino and Bordone 2014; Henderson et al. 2009). If an adult child of divorced parents becomes divorced, financial assistance and emotional support may be more readily available from a grandparent (Leopold and Schneider 2011; Timonen, Doyle, and O'Dwyer 2011). We might note, too, that **surrogate grandparents** (a form of fictive kin, not biologically related to the "grandchildren") sometimes attach to families (and vice versa) (Nelson 2013). Surrogate grandparents "fill some of the gaps in our mobile society" for both older people and children (O'Brien n.d.).

In low-income and ethnic-minority families, parents and children readily rely on grandparents for childcare and other help. Even among middle- and upper-middle-class families, it is not unusual for grandparents to help with childcare—or contribute to the cost of a grandchild's schooling, wedding, or first house. A *New York Times* article before the Great Recession featured several upper-middle-class grandparents who commuted by plane weekly to help with childcare (Lee 2007).

Race, Ethnicity, and Grandparenting Although there isn't yet much research on the subject—maybe you'll do some in the future—we know some ways in which racial and ethnicity affects grandparenting

(Karasik and Hamon 2007). For instance, a study in the mid-1980s found that 87 percent of black grandparents felt free to correct a grandchild's behavior compared to just 43 percent of white grandparents. As one black grandmother said of her 14-year-old grandson, "He can get around his mother, but he can't get around me so well" (quoted in Cherlin and Fursten berg 1986, p. 128).

As another example, Native American elders have traditionally served as cultural conservator grandparents—actively seeking contact and temporary co-residence with their grandchildren "for the expressed purpose of exposing them to the American Indian way of life" (Weibel-Orlando 2001, p. 143, quoted in Karasik and Hamon 2007, p. 145). Maintaining an ethnic-minority culture into future generations is typically a concern for ethnic-minority grandparents.

As a third example—and one that addresses transnational families—a study of 112 Asian-Indian grandchildren in the United States revealed that 40 percent had weekly phone conversations with grandparents in India; another one-third called one or more times a month. Seven percent maintained weekly e-mail contact. Forty-three percent of the grandchildren visited their grandparents in India every two years (Saxena and Sanders 2009).

Divorce, Remarriage, and Grandparenting How does a grown child's divorce affect a grandparent relationship? Evidence suggests that a grandparent may fret over whether to intervene on behalf of the grandchildren. As might be expected, effects of the divorce are different for the **custodial grandparent** (parent of the custodial parent) than for the **noncustodial grandparent** (parent of the noncustodial parent), with noncustodial grandparents significantly less likely to see their grandchildren as often as they had before the divorce (Connidis 2010). Because mothers more often get custody, the most common situation is for maternal grandparent relationships to be maintained or enhanced while paternal ones diminish (Henderson et al. 2009).

Due to pressure from noncustodial grandparents, states have passed laws giving grandparents the right to seek legalized visitation rights, but courts are reluctant to do so when parents object. Grandparents who go to court to seek visitation rights are successful between 30 percent and 40 percent of the time. When courts recognize visitation rights for grandparents in spite of parental objections, the reason typically involves the best interests of the child (Henderson 2005a, 2005b).

Remarriages and re-divorces create step-grandparents and ex-step-grandparents. Little research evidence exists on step-grandparents, but available data suggest they tend to distinguish their "real" grandchildren from

Grandparenting styles differ—shaped by the grandparent and grandchild's ages and personalities, as well as by the grandparent's health and employment status. Ties with a grandchild can be remote, companionate, or involved, and a grandparent may have different ties with different grandchildren. Maintaining one's culture into future generations may be of particular concern for an ethnic-minority, cultural conservator grandparent.

those of remarriages. Asked about his step-grandparents, one young man told this story:

> They were pretty good, but again, there was that line. And we knew. . . . I mean I remember a couple of Christmases ago . . . [my grandmother made] a family quilt and everyone was on it except me and my brother. . . . As we got older the fact that they weren't our grandparents became more predominant. (in Kemp 2007, p. 875)

Younger step-grandchildren and those who live with the grandparent's adult child are more likely to develop ties with the step-grandparent (Chapman, Coleman, and Ganong 2016). "As We Make Choices: Tips for Step-Grandparents" suggests ways to foster positive relationships with step-grandchildren. We turn now to an examination of caregiving to aging family members.

AGING FAMILIES AND CAREGIVING

When we associate caregiving with the elderly, we may tend to think only in terms of older generations as care recipients. However, about one-fifth of Americans age 75 and older are engaged in some form of caregiving, whether childcare or caring for other elders. In addition to assisting adult children and grandchildren financially and otherwise, older Americans give much to their communities. Many are volunteers in their churches, hospitals, schools, and various other settings. In this section, however, we focus on **elder care**— that is, care provided to the elderly.

Elder care involves emotional support, a variety of services, and sometimes financial assistance. A growing number of tax-funded, charity, and for-profit services provide elder care. Nevertheless, the persisting social expectation in the United States is that family members will either care for elderly relatives personally or organize and supervise the care provided by others. About 34.2 million adults age 18 and older (nearly 19 percent of all American adults) are engaged in **informal caregiving**—unpaid and personally provided care—to a family member or friend age 50 or older (National Alliance for Caregiving 2015). Adults responsible for aging family members often pay for some of the care recipient's expenses such as groceries, drugs, medical copayments, and transportation (Gross 2007).

Being concerned about elderly family members might involve nothing more than making a daily phone call to make sure they're okay, stopping by for a weekly visit, or accompanying them to appointments to translate or help explain. However, **gerontologists** (social scientists who study aging) specifically define **caregiving** as "assistance provided to persons who cannot, for whatever reason, perform the basic activities or instrumental activities of daily living for themselves" (Uhlenberg 1996, p. 682). Caregiving may be short term (taking care of someone who has recently had joint-replacement surgery, for example) or long term.

On average, long-term caregivers spend about twenty-four hours each week in this role, with another 11 percent spending more than forty hours per week. About 50 percent of caregivers are employed full-time and 11 percent part-time (National Alliance for Caregiving 2015).

As We Make Choices

Tips for Step-Grandparents

Not all step-grandparents assume the grandparent role. But for those who do, stepchildren can benefit from an older adult's genuine concern and support. The following advice to step-grandparents has been excerpted from the University of Florida Extension website.

You may [have] become an instant grandparent with step-grandchildren. You may have both grandchildren and step-grandchildren in the same family . . . . You probably have many thoughts and feelings about this role. You may think:

- I'm not old enough or ready to be a grandparent.
- This interferes with dreams about the birth of my first grandchild.
- Will my step-grandchild like me? Will I like my step-grandchild?
- Is it okay to feel differently toward my step-grandchildren than my real grandchildren?

- Will our family celebrations and traditions have to change? . . .

Remember that relationships are built over time. Your relationship and role as a step-grandparent will take time to develop. Communicate and spend time together in order to get to know each other.

Recognize the vital role of grandparents and step-grandparents in today's families. You can offer children in busy stepfamilies companionship, time, and a listening ear . . . children who are exposed to such contact are less fearful of old age and the elderly. They feel more connected to their families.

Create the grandparenting role that is comfortable to you and rewarding for your stepfamily. Step-grandparenting, like other stepfamily roles, is challenging and undefined. It is up to you to carve a role for yourself that fits your child's new family. Here

are some things to consider doing with step-grandchildren:

- Spend time one-on-one with them.
- Teach them a game or skill.
- Listen for their concerns, as well as their joys.
- Offer companionship for activities they enjoy.
- Share your history and family traditions.
- Show them acceptance.

Critical Thinking

How might a step-grandparent benefit from choosing to follow some of these suggestions? How might a step-grandchild benefit?

Source: Excerpted from Ferrer-Chancy 2009; and see Chapman, Coleman, and Ganong 2016.

The majority of the younger old need almost no help at all, but as individuals age, they may increasingly require assistance with tasks such as paying bills and later bathing or eating (National Alliance for Caregiving 2015). Severely ill, physically disabled older people and those with dementia often need a great deal of continuing care. An increasingly common alternative involves continuing care facilities that first offer small apartments for elderly able to care for themselves but with meals, housekeeping services, transportation to community events, and in-house social programs; then assisted-living services, such as help with bathing; then nursing home care. For families who can afford them, continuing care facilities ease some elder care decisions and needs. "Facts about Families: Community Resources for Elder Care" describes elder care options.

Many caregivers for the elderly are relatively young— in their thirties, with some in their twenties, partly because children born to older parents begin elder care at younger ages (Allen 2012; Bigda 2013). Then, too, grandchildren in their teens or twenties may be providing elder care (Brown 2020). As explored in Chapter 13's "A Closer Look at Family Diversity: Young Caregivers,"

some family caregivers are children, many of whom essentially put their life— including their childhood—on hold (Wilson 2012). However, the mean age for caregivers is around 50. Approximately one-third of caregivers are age 65 or older—and may themselves be in poor health or beginning to suffer from age-related disabilities (Johnson and Wiener 2006; National Alliance for Caregiving 2015).

Which family members provide elder care and how much they provide depend on the family's understanding of who is primarily responsible for giving the care, as well as the care receiver's preference (Connidis and Kemp 2008). The first choice for a caregiver is an available spouse, followed by adult children, siblings, grandchildren, nieces and nephews, friends and neighbors, and, finally, a formal service provider. This system of elderly care receivers' preference for caregivers is termed the **hierarchical compensatory model of caregiving** (Cantor 1979; Horowitz 1985; and see Nelson 2011).

Older LGBTQ+ individuals may be more inclined than straights to rely on friends, often fellows in their LGBTQ+ communities, who are more likely found in

Community Resources for Elder Care

More older Americans and their care providers are turning to professional elder care service providers for help. The table in this box defines several of these services and facilities. The following are three considerations regarding making decisions about these options:

1. Don't wait. Exploring options before they're needed helps all family members know what to expect and begin to prepare for the future (Greenwald 1999, p. 53).

2. Seek expert advice. Geriatric social workers can help assess an elderly person's needs and develop action plans.

3. Shop around. To compare cost and quality, families should visit as many facilities as possible on different days of the week and hours of the day. Ask for references (Shapiro 2001).

Critical Thinking

What kinds of community elder care services can you think of that are not included in this table? How might they be useful to caregivers and to the elderly as well?

THE OPTIONS	WHAT IS IT?
Home care	Wide range of services, including shopping and transportation, health aides who give baths, nurses who provide medical care, and physical therapy brought to the home (estimated average cost $35,000–$40,000 per year)
Adult day care	A place to get meals and spend the day, usually run by not-for-profit agencies
Congregate housing	A private home within a residential compound, providing shared activities and services; sometimes considered one type of assisted living
Assisted living	Numerous kinds of housing with services for people who do not have severe medical problems but who need help with personal care such as bathing, dressing, grooming, or meals (estimated average cost $48,000–$60,000 per year)
Continuing-care retirement community	A complex of residences that includes independent living, assisted living, and nursing home care so seniors can stay in the same general location as their housing needs change over time, beginning when they are still healthy and active
Nursing home (skilled nursing facilities)	Residential facilities with twenty-four-hour medical care available for those who need continual attention (estimated average cost $100,000-110,000 per year)

Source: "Eldercare Locator" at www.eldercare.gov, and "Glossary of Senior Housing Terms" (2010) at senioroutlook.com/glossary.asp.; "How Much Does Assisted Living Cost?" at www.whereyoulivematters.org; U.S. Administration on Aging 2015; "Aging," U.S. Department of Health and Human Services, http://www.hhs.gov.

urban than rural areas (Lee 2013). A small study of seventeen gay men and twenty lesbians over age 65 in Los Angeles concluded that, when relying on neighbors or acquaintances for social support, aging lesbians and gays differ in their approaches (Rosenfeld 1999). Those who formed a gay or lesbian identity before the 1970s, when homosexuality was highly stigmatized, tended to hide their sexual identity. Others—usually Baby Boomers who formed a lesbian or gay identity during the 1970s or later (in a "gay liberation atmosphere")—celebrate their identity, often becoming activists for LGBTQ+ rights regarding elderly housing, health care, and other services.

Whether gay or straight, older Americans provide a considerable amount of elder care to each other (Kelley 2018). Caregiving and receiving are expected components of marriage for most of today's older couples, who have developed a relationship of mutual exchange over many years (Machir 2003; Roper and Yorgason 2009). Among marrieds, spouses of either gender provide as much as 40 percent of elder care (Johnson and Weiner 2006). Caregivers in longer, emotionally close marriages with little ongoing conflict evidenced better overall well-being (investment) (Roper and Yorgason 2009). Caring for a disabled spouse in later years can involve mentally and emotionally reconstructing a new sense of marital closeness based on a shared history (Boylstein and Hayes 2012).

For the uncoupled and childfree, siblings are important in mutual caregiving (Eriksen and Gerstel 2002). After spouses, however, adult children are most likely to be providing elder care.

Issues for Thought

Filial Responsibility Laws

Should filial responsibility be a law? Consider the following argument by a legal scholar.

A filial responsibility statute is simply a law that "create[s] a statutory duty for adult children to financially support their parents who are unable to provide for themselves." Typically, such support includes an obligation to pay for "food, clothing, shelter, and medical attention." The rationale behind such laws arises from the reciprocal duty that parents have to care for their children; because parents extended voluntary care to their minor children, it is the filial responsibility of children to return that support to their parents . . . .

In 1601, filial responsibility was put into statutory form with the [British] enactment of the Elizabethan Poor Relief Act, from which most modern statutes are derived . . . . Carrying on the historical tradition of supporting indigent parents, [in] the United States . . . prior to the 1960s, federal legislation recognized this obligation as well. However, . . . the establishment of Medicare in the 1960s led to the repeal of the federal statute. Today, there are twenty-two states with filial responsibility statutes, and few—if any—of the states currently enforce these laws . . . . [C]hildren are only required to support their parents so long as "they are of sufficient ability". . . . Therefore, children should be excused from their filial responsibility if they have no economic means of supporting their parents. . . .

Children can also avoid filial responsibility if they can demonstrate the parent abandoned them. . . .

[According to one court decision] "The selection of the adult children is rational on the ground that the parents, who are now in need, supported and cared for their children during their minority and that such children should in return now support their parents to the extent to which they are capable . . . . [Filial responsibility] statutes would be beneficial to our society and provide desperately needed relief for our strained public treasury."

Critical Thinking

Other experts argue that (1) there is already considerable voluntary assistance from children to parents; (2) filial responsibility legislation can undermine parent-child relationships, creating resentment on the payee's part and guilt on the recipient's part; (3) government programs such as Social Security are preferable. Making Social Security contributions, children are paying for parents, but without the tension created by legislated direct payments (Callahan 1985). In your opinion, what would be some benefits of requiring filial responsibility to families and to society? What might be some drawbacks?

Source: Excerpted from Lundberg 2009, 534–582.

Adult Children as Elder Care Providers

Grown children caring for elderly parents generally provide most of the care themselves, although they also seek assistance from formal service providers (Kelley 2018). Motivated by **filial responsibility** (a child's obligation to parents), respect, and affection, adult children care for their folks "because they're my parents" (Anderson, Fields, and Dobb 2013). Highly religious adult children report higher levels of filial responsibility (Gans, Silverstein, and Lowenstein 2009).

As pointed out in "Issues for Thought: Filial Responsibility Laws," principles of reciprocity are at work regarding filial responsibility. As one caregiver explained to a researcher, "My mom's been there for me, helping me out, so I'm always going to be there for her" (Radina 2007, p. 158). And from another caregiver, "For me, it is a joy and a form of self-reward to care for my elders, see to it there is a roof over their heads, food in the [fridge]. All the things they did for me as a young one" (in Gitner 2012). One caregiver explained that her mother deserved care now because of all that her mother had previously done for an elderly aunt (Gitner 2012; see also Allen 2012).

Generally, aging parents expect to receive help from offspring in proportion to the aid that the parents had once given them, and offspring who have received more financial help from their parents are more likely than their siblings to be caring for the elderly parent (Bucx, Van Wel, and Knijn 2012; Lin and Wu 2014). An interesting note is that adult stepchildren receive less financial and other help than do biological or adopted children—a situation that may help to explain the lower level of felt filial responsibility among stepchildren (Berry 2008; Sherman, Webster, and Antonucci 2013). University of Michigan researchers using secondary analysis with three large data sets (see Chapter 2) found that elderly parents with stepchildren only—that is, with no biological children—were significantly more

likely to become disabled or reside in institutional settings and to die somewhat earlier (Pezzin, Pollak, and Schone 2013).

On average, parent-adult child ties with stepchildren are not as close as those with biological or adopted children (Ward, Spitze, and Deane 2009), so it may be no surprise that adult children feel less obligation to aging stepparents than they do to biological parents, especially when the stepparent was acquired later in the adult child's life (Carr and Utz 2020; Sherman, Webster, and Antonucci 2013). Feeling obligated to help one's remarried biological parent is often the principal reason for helping the stepparent (Ganong, Coleman, and Rothrauff 2009).

Shared Elder Care among Siblings There is some evidence that the oldest sibling in a family feels filial responsibility more strongly and thus provides more care to an aging parent than do younger sibs (Fontaine, Gramain, and Wittwer 2009). Today, it's often the case that siblings have geographically moved away from each other and from their aging parents; geographical distance typically excuses a sibling from hands-on, day-to-day caregiving. Even when siblings live near an elderly parent, however, the burden of elder care does not always fall on each one equally . A study based on forty focus groups (see Chapter 2) asked elder care givers with siblings to describe how they felt about the caregiving situation:

> Siblings who described an imbalance in caregiving responsibilities reported feeling considerable distress. . . . One participant confessed that she was straddling a "real thin line between just taking her [barely participating sister's] head off some day, because I'm so mad at the inequity of it." (Ingersoll-Dayton et al. 2003, p. 205)

However, the majority in this study defined the situation as more or less fair (see also Connidis and Kemp 2008; Kuperminc, Jurkovic, and Casey 2009). They took into account a sibling's geographical distance, employment responsibilities, and other obligations. Some participants (both men and women) called on gendered expectations to help justify women's inequitable elder care responsibilities. "I guess it's my gender," said a woman whose brother did little to help (Ingersoll-Dayton et al. 2003, p. 207). Feeling that the aging parent favored one sibling over another—and perhaps continues to favor one sibling over another—can add to conflict and tension among siblings caring for an aging parent or grandparent (Jensen et al. 2013).

Gender Differences in Providing Elder Care

Men are involved in elder care, especially when retired (Kahn, McGill, and Bianchi 2011; Kelley 2018). About one-third of unpaid caregivers are male. Moreover, the proportion of sons involved in elder care is expected to increase due to the growing number of only-child sons, smaller sibling groups from which to draw care providers, and changing gender roles that have begun to make elder caregiving more expected for adult males (National Alliance for Caregiving 2015). A recent study shows that even though daughters are most often caregivers to aging mothers, other factors also affect caregiving relationships, particularly regarding caring for fathers. Factors affecting decisions about which sibling takes on the major caregiving role involve geographical distance from the recipient, work obligations, and recipient preference (Leopold, Raab, and Engelhardt 2014).

For now, however, women account for about two-thirds of all unpaid caregivers (Glauber and Day 2018; National Alliance for Caregiving 2015, p. 14). This statistic partly reflects the fact that women live longer, so more caregivers are wives. Feelings of obligation to care for a disabled spouse are fairly equal between husbands and wives (Roper and Yorgason 2009). However, gender makes a significant difference in adult children's caregiving obligations (Barnett 2013; Hill 2011).

Norms in many Asian American families designate the oldest son as the responsible caregiver to aging parents (Kamo and Zhou 1994; Lin and Liu 1993). Sons of other ethnicities often provide elder care in the absence of available daughters (Lee, Spitze, and Logan 2003). Except in these cases, the adult child involved in a parent's care is considerably more likely to be a daughter or even a daughter-in-law than a son. This situation—one that raises issues of gender equity similar to those of parenting and other unpaid family labor, discussed in Chapters 9 and 10—is partly due to ongoing employment differences between women and men (Sarkisian and Gerstel 2004).

Then, too, in accordance with the findings discussed earlier—that relations with daughters are more often tight-knit than those with sons—a parent may prefer a daughter's help or assume that even an employed daughter will help while a son is more often perceived as too busy. A son-in-law explained his wife's caring for her mother this way: "Mom just, I think, calls on her more, so that's the way. . . . It's not that the brothers wouldn't help at all, but it's just . . . she gets called on more" (Ingersoll-Dayton et al. 2003, p. 207).

Then, too, men and women tend to provide care differently—in ways that reflect socially gendered expectations (Raschick and Ingersoll-Dayton 2004). Sons, grandsons, and other male caregivers (although not husbands) tend to perform occasional tasks such as cleaning gutters or mowing the lawn, whereas daughters more often provide continually needed services such as housekeeping, cooking, and doing laundry (Hequembourg and Brallier 2005). Sons more often

serve as financial managers, as well as organizers, negotiators, supervisors, and intermediaries between the care receiver and formal service providers. A son is more likely than a daughter to enlist help from his spouse, the elder care receiver's daughter-in-law (Raschick and Ingersoll-Dayton 2004). "Providing intimate, hands-on care is culturally defined as feminine, [and the] dirty parts of care work are mainly women's work" (Isaksen 2002, pp. 806, 809).

The Sandwich Generation

Many who are caring for aging parents have children at home. About thirty years ago, journalists and social scientists took note of an emergent **sandwich generation**: individuals who are sandwiched between the simultaneous responsibilities of caring for their dependent children and aging parents. More than one in 10 U.S. parents is also caring for a parent (Livingston 2018c). But "[t]he sandwich generation is not really a single generation" (Council on Contemporary Families 2011). The majority (55 percent) of people with elder and childcare responsibilities are between ages 28 and 42. However, many are older (38 percent between ages 43 and 61) and 7 percent of so-called sandwichers are younger than 28 (National Alliance for Caregiving 2015).

Gerontologist Neal Cutler, who studies the effect of aging on finances, coined the term *senior sandwich generation*. These individuals between middle- and near-retirement age are "facing the ultimate financial trifecta: college for their kids (either current tuition bills or paying back borrowed money), retirement for themselves, and at-home or nursing-home care for one or more parents. . . . [all] at the same time" (Chatzky 2006). About "one-in-seven middle-aged adults (15 percent) is providing financial support to both an aging parent and a child" (Parker and Patten 2013).

Today's sandwiched caregivers often house financially strapped young adults who have returned home. Sometimes young adults who return home serve as caregivers to their aging parents. Even when the parent is healthy and active, young adults who live with them can be quite helpful—doing the lawn, for instance, cleaning gutters, or carrying groceries in from the car. Nevertheless, providing economically for multigenerations can mean mounting financial burdens for the sandwich generation (Parker and Patten 2013).

The sandwich generation experiences all the hectic task juggling discussed in Chapters 9 and 10 (Martin 2019). For some in this category, work demands take the majority of their attention. Others emphasize childcare responsibilities, whereas still others place greater focus on care for their aging parents (Cullen et al. 2009). Regardless of what corner of this triangle predominates—work, children, or aging parents—stress

The majority of the young old need almost no help at all, but as they grow older they are likely to need more help, sometimes requiring a great deal of continuing care. Often employed, members of the sandwich generation—adults sandwiched between the simultaneous responsibilities of caring for their children and aging parents—feel the strains associated with juggling work, childcare, and elder care.

builds as family members maneuver chaotic everyday life ("The Sandwich Generation: Stress and Other Issues" 2019; Williams and Boushey 2010).

Elder Care—Joy, Ambivalence, Reluctance, and Conflict

The media typically present elder care as tenderly nursing "loved ones," and this is often the case—but not always (Kelley 2018). Sometimes individuals resent being called on to care for aging parents (or other family members) with whom they never felt close (Span 2013). Many caregiver relationships are best characterized as ambivalent (Connidis 2011; Seidel, Majeske, and Marshall 2019). Unfortunately, ambivalent or consistently problematic caregiving relationships result in diminished well-being for both caregiver and receiver (Rook et al. 2012). One study of spouse and adult childcare givers found that a better-quality caregiver-receiver relationship prior to the receiver's onset of dementia resulted in caregivers' being less willing to place the receiver in a nursing home (Winter, Gitlin, and Dennis 2011).

It may help to realize that caregiver and cared-for may have very different perceptions of the situation, as discussed regarding handling crises in Chapter 13 (Seidel, Majeste, and Marshall 2019). It may help, too, to remember that older people can be stubborn, demanding, and angry about receiving help. Needing assistance may threaten their sense of autonomy and

self-esteem (Bottke 2010; Halpern 2009). In the words of one psychotherapist, "For men in particular, their ego and sense of self can be hit very hard when they can no longer do what they used to do. It's safe to take it out on someone you love, not because you don't love her or respect her, but because it's safe" (in Brody 2012). Then, too, caregivers may expect a certain amount of deference, or courteous submission to their opinions and decisions, from an aging parent in need—and when this does not occur, "intergenerational relations become strained" (Pyke 1999, p. 661).

Some conflicts between caregiver and recipient are about finances—when the elderly parent seems to be letting go of money unwisely, for example. Whether an aging parent can drive safely or should move into a residential care facility are other potential matters of contention. Discussing a parent's need for lifestyle changes following a critical health event such as a heart attack can also be problematic (Frum 2012; Goldsmith, Bute, and Lindholm 2012). More routine issues such as the urgency of an errand that needs to be run can also cause conflict. Using positive communication skills (see Chapter 11) to set boundaries is important to a caregiver's well-being (Bottke 2010). Susan Halpern's *Finding the Words: Candid Conversations with Loved Ones* (2009) is a good resource.

Disagreements may also emerge among caregivers (Mills and Wilmoth 2002). For instance, conflicts may arise when siblings are "forced to make urgent, complex decisions for loved ones in intensive care units" (Siegel 2004) Adult siblings may have differing feelings about a parent (Flora 2007). Sometimes, old sibling rivalries or perceptions of a parent's favoritism emerge (Suitor et al. 2009). All else being equal, families that have developed a shared understanding of the caregiving situation make more effective caregiving teams (Halpern 2009). With regard to the decision to withdraw treatment from a dying parent, the older family member's having previously written an advanced directive on this issue is advised and generally most helpful.

Caregiver Stress Caregiving may enhance one's sense of purpose and overall life satisfaction due not only to enhanced intimacy but also to the hope that helping others will result in assistance with one's own needs when the time arrives (Shapiro 2006b). Nevertheless, caregiving can be overwhelming and physically, financially, and emotionally costly (Glauber and Day 2018; Kelley 2018).

A recent national poll found that among adults with a living parent 65 or older, nearly 32 percent contributed to a parent's financial support during the preceding year; the majority were contributing to ongoing expenses (Carrns 2020; Parker and Patten 2013). Caregivers not only spend their own money but also (especially women) may take extended time from work or pass up promotions (Geewax 2012b; Searcey 2014). Some family caregivers take early retirement or geographically relocate to provide elder care (Anderson, Fields, and Dobb 2013). Caregivers (usually women) whose jobs are not accommodating—do not allow for flexibility or phone calls from work, for example—are more likely than others to quit working altogether (Lahaie, Earle, and Heymann 2012; Porter 2019). As discussed in Chapter 3, feminist scholars point to the intersectionality of race, class, and gender when analyzing women's lives. As a case in point here, while professional women may relinquish promotions, less affluent women often experience work situations hostile enough to their caregiving needs that they quit work altogether (Lahaie, Earle, and Heymann 2012; Searcey 2014).

A caregiver's making choices such as these limits his or her own old-age resources (Carrns 2020; Hill 2011). Furthermore, providing elder care can be socially isolating, bring on depression, and further strain one's physical health (He, Weingartner, and Sayer 2018; Searcey 2014). Older caregivers report losing contact with friends and other family members (National Alliance for Caregiving 2015). One recent study found that, at least for women caregivers, part-time work was associated with less depression (Glauber and Day 2018).

Younger caregivers experience limitations on dating and other relationships. As one 26-year-old explained to a research team, "I would like to go camping with my husband once in a while, but I can't just get up and go away, because of taking care of my grandparents" (in Dellmann-Jenkins, Blankemeyer, and Pinkard 2000, p. 181).

Caregiver stress and depression among Americans result partly from the fact that in our individualistic culture adult offspring are expected to establish lives apart from their parents and to achieve success as individuals rather than (or as well as) working to benefit the family system (He, Weingartner, and Sayer 2018). Furthermore, because the chronically ill or disabled of all ages are decreasingly cared for in hospitals, today's informal caregivers are asked to perform complicated care regimens that have traditionally been handled only by health care professionals (Byrne, Orange, and Ward-Griffin 2011; Guberman et al. 2005). This situation places added demands on a family caregiver's time, energy, and emotional stamina (Richards 2009). (See "Issues for Thought: Caring for Patients at Home—A Family Stressor," in Chapter 13.)

On days when a chronically ill care recipient appears to be doing better, caregivers report being in a better mood (Roper and Yorgason 2009). Also, when the caregiver-recipient relationship feels more reciprocal—with the caregiver feeling that he or she is getting something in return—the caregiver is less often depressed (LeBlanc and Wright 2000). Receiving adequate training and learning specific caregiver skills can lessen caregiver

stress, as can various forms of social support—phone calls and other contacts with friends as well as finding information and connecting with other caregivers either in face-to-face support groups or on the Internet (Roper and Yorgason 2009; Wilkins, Bruce, and Sirey 2009). Among married caregivers to their elderly parents, it helps if spouses agree on basic beliefs about the desire and obligation to help aging parents (Polenick, Zart, Birditt, Bangerter, Seidel, and Fingerman 2017).

Racial and Ethnic Diversity and Family Elder Care

Adult children of all races and ethnicities feel responsible for their aging parents and to siblings with whom they likely share elder care responsibilities. However, racial and ethnic differences do exist regarding elder care. For instance, Asian Americans have traditionally tended to emphasize the centrality of filial obligations over conjugal relationships (Burr and Mutchler 1999). In addition, blacks are more likely than nonHispanic whites to expect their adult children to personally care for them in old age (Lee, Peek, and Coward 1998). As we saw earlier in this chapter, ethnic groups other than nonHispanic whites have been more likely to establish multigenerational households.

Although more than two-thirds (67 percent) of family caregivers today are nonHispanic whites, this situation will change as members of other racial and ethnic groups grow older. Research is beginning to accumulate on how race and ethnicity influence caregivers' emotional and mental health (Skarupski et al. 2009). One study compared eighty-nine Latinas with ninety-six nonHispanic white female caregivers of older relatives with Alzheimer's disease and found that, regardless of ethnicity, lack of financial resources negatively impacted the caregivers' emotional health. However, stress from lack of resources was mitigated or lessened by Latinas' relatively strong familistic values (Montoro-Rodriguez and Gallagher-Thompson 2009). A study of 307 caregivers compared blacks with nonHispanic whites and found that blacks coped better emotionally when exposed to caregiver stress (Skarupski et al. 2009; see also Wilkins, Bruce, and Sirey 2009). Prior research with African American wife caregivers found that receiving support from their churches lessened their stress and helped their marital relationship (Chadiha, Rafferty, and Pickard 2003).

Racial and ethnic minorities have not been as likely as nonHispanic whites to use community-based services such as senior centers. This discrepancy occurs because of language barriers, because individuals don't know that they qualify for government-funded services, or because they are reluctant to include paid service providers as members of their caregiving team (Levine 2008). **Fictive kin** (family-like relationships that are not based on blood or marriage but on close friendship ties) are often resources for elder care exchanges among the single, childfree, and LGBTQ+Q elderly—and among African Americans (sometimes called "going for sisters"), Hispanics (compadrazgos), Italians (compare), and other ethnic groups as well (Ebaugh and Curry 2000, Nelson 2013).

However, it is incorrect to assume that ethnic minority families "take care of their own" without needing to rely on community services. Among many immigrant ethnic groups, **acculturation** (the process whereby immigrant groups adopt the beliefs, values, and norms of their new culture) affects norms of filial obligation. Younger generations are more likely than their elders to become acculturated—a situation that creates the need for family members to renegotiate their respective roles (see Chapter 2) with the potential for intergenerational conflict (Rudolph, Cornelius-White, and Quintana 2005; Wong, Yoo, and Stewart 2006). A study of older Puerto Ricans found that filial obligation has declined in the younger generations (Zsembik and Bonilla 2000). And research on Chinese immigrant families in California found that sons often outsourced elder care:

> "I told her that I hire you to help me achieve my filial duty," Paul Wang, a 60-year-old Taiwanese immigrant owning a software company in Silicon Valley, California, described . . . his conversation with the in-home care worker he employed for his mother suffering from Alzheimer's disease. (Lan 2002, p. 812)

Acculturation has also meant that aging immigrants increasingly live in housing designed for the elderly rather than with their grown children (Kershaw 2003, p. A10). We turn now to a discussion of elder abuse and neglect.

ELDER ABUSE AND NEGLECT

Parallel to child abuse and neglect, discussed in Chapter 12, **elder abuse** involves overt acts of aggression, whereas **elder neglect** involves failure to give adequate care. Elder abuse includes physical assault, verbal abuse and other forms of emotional humiliation, purposeful social isolation (for example, forbidding use of the telephone), and financial exploitation ("Elder Abuse and Neglect" 2019).

The profile of the abused or neglected elderly person is a female, 70 years old or older, who has physical, mental, or emotional impairments and is dependent on the abuser or caregiver for companionship and help with daily activities (Leisey, Kupstas, and Cooper 2009; Newman 2018). Unfortunately, accurate statistics on the prevalence of elder abuse and neglect do not exist, partly because much elder abuse and neglect goes unreported

(U.S. Federal Interagency Forum on Aging-Related Statistics 2013). About 38,000 cases of elderly financial abuse are reported every year, and experts believe the actual number of occurrences is much higher (Bendix 2009; Miles 2008; see also Hull 2008). Depending on how broadly a researcher defines elder maltreatment, various studies have concluded that as many as 5 million—between 1 percent and about 10 percent—of individuals age 60 and older are abused or neglected annually by professional and family caregivers (Hildreth, Burke, and Glass 2009; Ramnarace 2010).

Data suggest that nonfamily members—typically paid caregivers either in the aged person's home or in an institutional setting—are responsible for more than half of all elder verbal, physical, and financial abuse (Laumann, Leitsch, and Waite 2008). Accordingly, education programs for family members and others who work with the elderly—social workers, physicians, nurses, dentists, attorneys—have been designed to facilitate detection and prevention of elder maltreatment in institutional settings (Newman 2018; Wagenaar 2009).

The majority of elderly Americans maintain their own homes. However, many of the frail elderly depend on or live with family members. Although care for an aging parent—most often by daughters—is often given with fondness and love, it can also bring stress, conflicting emotions, and great demands on time, energy, health, and finances.

Elder Maltreatment by Family Members

In 1996, the federal government sponsored the National Elder Abuse Incidence Study, a national study that combined reports from Adult Protective Services with interviews with people in the community who had contact with the elderly. Estimates from this study—still our best source for statistics—were that about one-half million Americans older than 60 and living in family households were abused or neglected (Ramnarace 2010). Neglect was the more common form of elder maltreatment (National Center on Elder Abuse 2013). Another national study found that among elderly who suffered maltreatment by a family member, 9 percent experienced verbal abuse, and 3.5 percent had suffered financial abuse. Less than 1 percent experienced physical abuse.

Common to cases of physical elder abuse by family members are shared living arrangements, the abuser's poor emotional health (often including alcohol or drug problems), and a pathological relationship between victim and abuser (Anetzberger, Korbin, and Austin 1994). Abusive acts may be "carried out by abusers to compensate for their perceived lack or loss of power" (Pillemer 1986, p. 244). "In many instances, both the

victim and the perpetrator [are] caught in a web of interdependency and disability, which [make] it difficult for them to seek or accept outside help or to consider separation" (Wolf 1986, p. 221).

However, the largest proportion of abuse is from a spouse or a romantic partner (Laumann, Leitsch, and Waite 2008). In fact, "there is reason to believe that a certain proportion of elder abuse is actually spouse abuse grown old" (Phillips 1986, p. 212; see also Leisey, Kupstas, and Cooper 2009; Shannon 2009). In some cases, marital violence among the elderly involves abuse of the caregiving spouse by a partner who is ill with Alzheimer's disease (Pillemer 1986). Adult children are also responsible for elder abuse, and an abuser is often financially dependent on the elderly victim (Robinson, Saisan, and Sega 2019).

Two Models to Explain Elder Abuse

Researching and combating elder abuse generally proceed from either of two models: the caregiver model or the domestic violence model. The **caregiver model of elder abuse and neglect** views abusive or neglectful caregivers as individuals who are simply overwhelmed by caregiving burdens, a situation that can cause the caregiver to lose control and verbally or physically abuse the receiver (Abbey 2009; Bainbridge et al. 2009). This model also points to ecological factors or environmental

factors—or absence of resources, combined with work demands, or one's own parenting demands—that make things harder (Norris et al. 2013).

Abusive or neglectful family caregivers are often stressed by socially structured conditions such as job conflicts. An interesting recent study attempted to explain intimate partner violence in later life by analyzing readers' comments posted after Internet news stories that reported it (Brossoie, Roberto, and Barrow 2012). Typical of one pattern found in the data was a spouse caregiver frustrated with available social services:

> I have begged and pleaded for help [from community services] and there simply is none. If [my husband] were to die I'm sure there would be charges of neglect against me even though I am not neglecting him. . . . His doctors know, the courts know, his family knows and no one cares as long as there is someone else to blame. (Brossoie, Roberto, and Barrow 2012, p. 798)

Professional care providers employed by community agencies are often trained to recognize potentially abusive family situations and then work to reduce dangerous levels of caregiver stress that may trigger abuse (Brandl 2007; Thobaben 2008).

A second model, the **domestic violence model of elder abuse and neglect**, views elder abuse as a form of unlawful family violence and focuses on negative personal characteristics of abusers and on possible criminal justice responses such as arrest and prosecution (Hagan 2010; Wallace 2008). The criminal justice response is especially likely in the case of financial abuse (Gross 2006b).

THE CHANGING AMERICAN FAMILY AND ELDER CARE IN THE FUTURE

In the future, families may be more and more creative in fashioning caregiving arrangements for elderly members. Multigenerational and communal households are examples, and architects are beginning to design homes that better accommodate a variety of family and friendship living options such as *granny flats*—additions or renovations to standard dwellings specifically for in-laws, including "Next Door Garage Apartments" or little units on a family property called *backyard living* (Gerace 2012).

Nevertheless, longer life expectancy and more chronically ill Americans in old-old age raise concerns about providing elder care now and in the future (He, Weingartner, and Sayer 2018; Levine 2008). For one thing, a considerable number of elderly live alone now and will age alone. "One hundred years ago, 70 percent of American widows and widowers moved in with their families. Today nearly the same proportion of widows and widowers live alone" (Klinenberg, Torres, and Portacolone 2012, p. 1).

The American family is changing in form, as we have seen throughout this text. Many of these changes could result in a diminishing caregiver "kin supply" (Bengston 2001, p. 5; He, Weingartner, and Sayer 2018). Also as noted throughout this text, aging Baby Boomers have had higher cohabitation rates, smaller families, and high divorce rates. Consequently, fewer spouses or grown children—the two most important caregiving resources—will be available for caregiving (Jacobsen, Mather, and Dupuis 2012).

A high sense of filial obligation has been positively related to caregiving (Anderson, Fields, and Dobb 2013; Kuperminc, Jurkovic, and Casey 2009). But parental divorce reduces the younger generation's sense of filial obligation, particularly to divorced fathers and (expectedly) from former daughters-in-law (Ganong, Coleman, and Rothrauff 2009; Kalmijn 2013). Research suggests that ties between parents and adult children are generally less close when the parents are divorced (Connidis 2009; Kalmijn 2013). Children raised in divorced families may have received fewer resources or less emotional support from a parent and hence feel less filial obligation.

Moreover, the extent to which one's cohabiting partner participates in elder care is a matter for future research, but since the majority of cohabiting relationships are short-lived, we might expect that elder care is or will not be part of the cohabitation equation.

We are gradually learning more about the ramifications of remarriage for elder care. In one qualitative study of late-life remarried caregivers,

> [w]ives revealed how little support or assistance they received from adult stepchildren for [the children's] ailing father. Often these caregivers endured a kind of amplified stress, isolation, and conflict in their caregiving role, which they attributed to their remarried, stepmother status. By the same token, adult children and stepchildren are not always granted access to the critical decision making or caretaking. (Sherman 2006, p. F8; see also Sherman and Boss 2007)

Furthermore, females' increased participation in the labor force decreases the time that women, the principal providers of elder care, have available to engage in elder care. Then, too, geographical mobility negatively affects hands-on elder care. Moreover, in an era of more transnational families (see Chapter 1), aging parents may be in the country of origin while their adult offspring struggle to help to care for their aging parents from abroad (Bodrug-Lungu and Kostina-Ritchey 2013; Senyurekli and Detzner 2008; Zhou 2012).

In addition to demographic changes, some have suggested that American society may be becoming increasingly individualistic, a situation that—in this view—threatens to weaken the family.

Today's older Americans are likely to have sisters and brothers who may be able to help them. However, as

American families are changing, and the economy does too. Today more and more older adults remain in the labor force many years after what we used to think of as retirement age. Typically they do so for the income. Also, often living alone, elderly individuals may continue to work for the comradery that a job can provide.

Jetta Productions/Getty Images

younger parents choose to have fewer or only children, more of the elderly in the more future will have fewer or no siblings. Then, too, more siblings in a family can ease each one's burden of care for an aging parent because they can share necessary tasks (Wolf, Freedman, and Soldo 1997; see also Fontaine, Gramain, and Wittwer 2009). However, increasingly, fewer siblings will be available to share in elder care. Some childfree and single elderly are likely to fashion fictive-kin families of mutually caring friends (Spencer and Pahl 2006). Others may prove to be "less likely than are parents to have robust network types capable of maintaining independent living" (Wenger et al. 2007, p. 1419; and see Bures, Koropeckyj-Cox, and Loree 2009).

Same-Sex Families and Elder Care

Same-sex couples and families face elder care challenges. "But along with getting older, they also have to face the prejudices of being gay or lesbian. . . . Nursing homes and private retirement centers . . . [often] make assumptions that their residents are heterosexual and structure activities on the basis of these assumptions" (Powell 2004, p. 60). Partly so as not to be separated from their partners by well-meaning relatives who may put them in separate nursing homes, gay men and lesbian couples (who can afford it) have begun to create assisted-living communities of their own ("Birds of a Feather" 2010; Grant 2010; Mitchell 2007).

Although many policy analysts express concern about families' ability to effectively provide elder care in the future, some also point to the **latent kin matrix**, which is defined as "a web of continually shifting linkages that provide the potential for activating and intensifying close kin relationships" (Riley 1983, p. 441; see also Spencer and Pahl 2006). An important feature of the latent matrix is that, although they may remain dormant for long periods, family relations with adult siblings and extended kin emerge as a resource when the need arises (Silverstein and Bengston 2001).

Moreover, while family forms become increasingly diverse, so do the many ways that members of the postmodern family deal with elder care. For instance, in research involving in-depth interviews with forty-five older respondents, a 67-year-old separated wife, whose husband had become diabetic and asthmatic and had heart trouble, reported that she still loved him very much: "If something happened to him and this gal [his new romantic partner] didn't take care of him, I would go and take care of him myself" (quoted in Allen et al. 1999, p. 154). It is not entirely unusual for a divorced mate to provide assistance to an aging ex-spouse who needs it:

> Hospice workers, academics, and doctors say they are seeing more such cases. . . . Often a person feels deep ties to a former husband or wife or feels a responsibility born of common experience and child rearing. . . . "They are acting more like a brother or sister, or cousin, or extended family member." (in Richtel 2005, p. ST1-2)

TOWARD BETTER CAREGIVING

Family sociologist and demographer Andrew Cherlin (1996) distinguishes between the "public" and the "private" faces of families. The **private face of family** "provides individuals with intimacy, emotional support, and love" (p. 19) and is especially important to the well-being of aging family members (Lowe and McBride-Henry 2012). The **public face of family** produces public goods and services by educating children and caring for the ill and elderly. "While serving each other, members of the 'public' family also serve the larger community" (Conner 2000, p. 36; Gross 2004).

The Private Face of Family Caregiving

We've seen throughout this text that providing family members with intimacy, emotional support, and love involves the positive communication techniques described in Chapter 11. Making use of these techniques, families can discuss many topics regarding an aging member—from who will go for groceries to money matters to advanced care planning and end-of-life care to the possibility of organ donation when the family member dies (Boerner, Carr, and Moorman 2013).

With regard to multigenerational living, counselors advise that before moving in together, family members of all ages (even young children) are encouraged to talk about how they expect life to change: what they want, what they are excited about, and what they're nervous about regarding the changing situation. For instance, how much will grandparents help with childcare? How will meals be handled? Families are encouraged to schedule fairly regular meetings to discuss issues before they become problems (M. Alvarez 2009; Gitner 2012).

Discussing things may be particularly important in families where cultural expectations differ widely between generations. Immigrants arriving from more familistic cultures put a strong emphasis on filial obligations. However, children of immigrants become more individualistic as they assimilate into American society (Lee and Smith 2012; Ruiz and Ransford 2012). For instance, a study of Korean Americans found that older caregivers were much more likely to want to personally care for a spouse with dementia at home, while (next-generation) adult children more often favored hiring professionals (Lee and Smith 2012).

There are many good resources that can help in these discussions. Several Internet examples: American Association of Retired Persons, Center for Retirement Research at Boston College, National Family Caregivers Alliance, National Alliance for Caregiving, National Elder Law Network, National Institute on Aging, Rand Center for the Study of Aging, Social Security Administration, and the U.S. Administration on Aging. For those who still like paper, the National Institute on Aging, a division of the U.S. Department of Health and Human Services, will send various pamphlets free of charge on topics such as aging in one's home, choosing a nursing home, caregiving from a geographical distance, and elder abuse, among others.

The Public Face of Family Caregiving

More than at any time since the 1935 onset of Social Security, contemporary families serve a **multigenerational safety net function** whereby older family members care

Many older Americans remain active even into very old age. Nonetheless, longer life span does mean that more and more Americans will spend the last months or years of their lives with chronic health problems or disabilities. The aging of the population raises concerns about how a changing family structure will be able to care for tomorrow's elderly and to what extent community and government resources can or will be engaged to help.

for younger ones (even young adults), and vice versa. Ironically, however,

> the Great Economic Recession has increased the need for the family safety net, while at the same time, family members have become less able to act as a safety net because parents and children are often exposed to the same disadvantages, such as unemployment. The public safety net is also fraying, thereby increasing the need for kin support and simultaneously reducing the resources available to share among kin. (Seltzer 2011, F2)

The public face looks to government leaders, who ironically tend to emphasize the "paramount importance" of "personal responsibility and accountability for planning for one's longevity" (White House Conference on Aging 2005, p. 18). However, the financial and emotional costs of providing good care for ill and disabled family members are often too high for family caregivers to manage without help (Parker and Patten 2013; Piercy 2010; Span 2009b). Many point to the increasing need for public services to assist family caregivers—for example, with transportation, personal care services, and adult day care (Rosen 2007).

However,

> The media constantly reinforce the conventional wisdom that the care crisis is an individual problem. Books, magazines, and newspapers offer American women an endless stream of advice about how to maintain their "balancing act," how to be better organized and more efficient or how to meditate, exercise, and pamper themselves to

relieve their mounting stress. Missing is the very pragmatic proposal that American society needs new policies that will restructure the workplace and reorganize family life. (Rosen 2007, p. 13)

Social scientists Francesca Cancian and Stacey Oliker (2000) have proposed the following strategies for moving our society toward better elder care coupled with greater gender equity in providing elder care:

- Provide government funds that support more care outside the family, such as government-funded day care centers for the elderly and respite (time off) services for caregivers.

- Increase social recognition of caregiving—both paid and unpaid—as productive and valuable work.

- Make caregiving more economically rewarding or at least less economically costly to the caregiver (p. 130).

How these policy changes might be accomplished may be difficult to imagine, but this fact negates the utility neither of the vision nor of the political debate that needs to emerge (Gross 2007). Elder care (as well as childcare) is indeed a responsibility not only of individual families but also of an entire society.

Summary

- The number of elderly, as well as their proportion of the total U.S. population, is growing.

- Along with the impact of the Baby Boomers' aging and the declining proportion of children in the population, longer life expectancy has contributed to the aging of our population.

- Due mainly to differences in life expectancy, older men are much more likely to be living with their spouse than are older women.

- Among older Americans without partners, living arrangements depend on one's health, the availability of others with whom to reside, social norms regarding obligations of other family members toward their elderly, personal preferences for privacy and independence, and economics.

- Growth in Social Security benefits has resulted in dramatic reductions in U.S. poverty rates for the elderly over the last several decades, although 9 percent of older adults are living in poverty.

- Due to differences in work patterns, wage differentials, and Social Security regulations, older men are considerably better off financially than are older women.

- Most older married couples place intimacy as central to their lives, describe their unions as happy, and continue to be interested in sex into old age.

- Even when it is not an abrupt event, retirement represents a great change for individuals and couples.

- Adjustment to widowhood or widowerhood is an important family transition that married couples often must face in later life. Bereavement manifests itself in physical, emotional, and intellectual symptoms.

- Daughters are more likely than sons to have close relationships with their parents, especially with their mothers. However, a parent's divorce, separation, or repartnering often weakens the bond with adult children.

- Partly due to longer life expectancy, grandparenting became increasingly important to families throughout the twentieth century. In the twenty-first century, we see an increasing number of great-grandparents.

- As members of our families age, elder care is becoming an important feature in family life, with women providing the bulk of it.

- After spouses, adult children (usually daughters) are a preferred choice of an older family member as elder care providers.

- Elder care in families typically follows a caregiving trajectory as the care receiver ages, and it involves not only benefits but also stresses for caregivers.

- Elder maltreatment—that is, abuse and neglect—by family caregivers exists in a small percentage of aging families and is addressed in public policy by either the caregiver model or the domestic violence model.

- Changes in the American family lead some policy analysts to be concerned that families will have greater difficulty providing elder care in the future. However, others point out that family relationships, though latent for long periods, can be activated when needed.

- Better elder care in the future will necessitate involving more men as caregivers and developing public policy that adequately supports expanded community services to assist families in providing elder care.

Questions for Review and Reflection

1. Discuss ways that society's age structure today affects American families.

2. Describe the living arrangements of older Americans today, and give some reasons for these arrangements.

3. Discuss some ways that the growing diversity of family forms among the older population can be expected to affect caregiving.

4. Apply the exchange, interactionist, structure-functionalist, or ecological perspective (see Chapter 2) to the process of providing elder care.

5. **Policy Question.** Describe two suggestions for policy changes that would make family elder care less difficult.

Key Terms

acculturation 450
active life expectancy 432
baby boom 431
bereavement 438
caregiver model of elder abuse and neglect 451
caregiving 443
custodial grandparent 442
domestic violence model of elder abuse and neglect 452

elder abuse 450
elder care 443
elder neglect 450
fictive kin 450
filial responsibility 446
gerontologists 443
hierarchical compensatory model of caregiving 444
informal caregiving 443
latent kin matrix 453

multigenerational family 430
multigenerational safety net function 454
noncustodial grandparent 442
private face of family 453
public face of family 453
sandwich generation 448
surrogate grandparent 442

GLOSSARY

ABC-X model A model of family crisis in which A (the stressor event) interacts with B (the family's resources for meeting a crisis) and with C (the definition the family formulates of the event) to produce X (the crisis).

abortion The surgically or pharmaceutically caused expulsion of a fertilized embryo or fetus from the uterus.

abstinence The standard that maintains that nonmarital intercourse is wrong or inadvisable for both women and men regardless of the circumstances. Many religions espouse abstinence as a moral imperative, while some individuals are abstinent as a temporary or permanent personal choice.

abstinence-only-until-marriage (AOUM) programs Sex education programs first established in 1996 and reauthorized in 2010 that must adhere to strict federal guidelines regarding the information they provide. The overall goal of these programs is to teach the positive social, psychological and overall physical health outcomes found by abstaining from sexual activity.

accomplishment of natural growth parenting model Educational model in which children's abilities are allowed to develop naturally; this includes working-class children spending more time watching television and playing video games than children of highly educated parents.

acculturation Process whereby immigrant groups adopt the beliefs, values, and norms of their new culture.

acquaintance rape Forced or unwanted sexual contact between people who know each other, sometimes—although not necessarily—taking place on a date. *See also* **date rape**.

active life expectancy The period of life free of disability in activities of daily living, after which may follow a period of being at least somewhat disabled.

adulting Being skillful in the art of living.

affectional orientation Newer term used in conjunction with sexual identity and sexual orientation; includes emotional and physical attractions beyond sexual attraction.

affinity building Relationship building involving high levels of intimacy, personal disclosure, and vulnerability around shared interests and other commonalities.

affirmative consent Situation in which a partner gets clear and unambiguous oral consent from a potential partner before having sex.

agentic character traits Traits such as confidence, assertiveness, and ambition that enable a person to accomplish difficult tasks or goals; also known as *instrumental character traits*.

AIDS *See* **HIV/AIDS**.

arranged marriage Unions in which parents choose their children's marriage partners.

asexual, asexuality A person who is asexual does not experience sexual desire. This is different from abstinence or celibacy, which is a choice to not engage in sexual activity despite feelings of sexual desire. Asexuality may be considered a sexual orientation/identity.

assisted reproductive technology (ART) Advanced reproductive technology, such as artificial insemination, in vitro fertilization, or embryo transplantation, that enables infertile couples or individuals, including gay and lesbian couples, to have children.

assortative mating Social psychological filtering process in which individuals gradually narrow down their pool of eligible individuals for long-term committed relationships, removing those who would not make the best spouse or partner.

attachment disorder An emotional disorder in which a person defensively shuts off the willingness or ability to make emotional attachments to anyone.

attachment theory A psychological theory that holds that, during infancy and childhood, a young person develops a general style of attaching to others; once an individual's attachment style is established, she or he unconsciously applies that style to later, adult relationships. The three basic styles are *secure, insecure/anxious*, and *avoidant*.

authoritarian parenting style All decision making is in parents' hands, and the emphasis is on compliance with rules and directives. Parents are more punitive than supportive, and use of physical punishment is likely.

authoritative parenting style (positive parenting) Parents accept the child's personality and talents and are emotionally supportive. At the same time, they consciously set and enforce rules and limits, whose rationale is usually explained to the child. Parents provide guidance and direction and state expectations for the child's behavior. Parents are in charge, but the child is given responsibility and must take the initiative in completing schoolwork and other tasks and in solving child-level problems.

baby boom The unusually large cohort of U.S. children born after the end of World War II, between 1946 and 1964.

battered woman syndrome When low self-esteem interacts with fear, depression, confusion, anxiety, feelings of self-blame, and loss of a sense of personal control to keep a physically or emotionally abused women from finding a way out of her situation.

belligerence A negative communication/relationship behavior that challenges a partner's power and authority.

bereavement A period of mourning after the death of a loved one.

bifurcated consciousness Divided perception in which someone is aware of and often troubled by two conflicting messages.

binational An immigrant family in which some members are citizens or legal residents of the country they migrate to, while others are undocumented—that is, they are not legal residents.

binuclear family One family in two household units. A term created to describe a postdivorce family in which both parents remain involved and children have a home in both households.

biosocial perspective Theoretical perspective based on concepts linking psychosocial factors to anatomy, physiology, genetics, or hormones as shaped by evolution.

bisexual An individual who is sexually attracted to both males and females.

boomerangers Adults who leave home and then return to live with their parents.

borderwork Interaction rituals that are based on and reaffirm boundaries and differences between girls and boys.

boundary ambiguity When applied to a family, a situation in which it is unclear who is in and who is out of the family.

bystander education Program that teaches people safe ways to intervene if they think a sexual assault is about to occur.

caesarean section Surgical procedure in which incisions are made in a woman's abdomen and uterus to deliver her baby.

caregiver model of elder abuse and neglect A view of elder abuse or neglect that highlights stress on the caregiver as important to the understanding of abusive behavior.

caregiving "Assistance provided to persons who cannot, for whatever reason, perform the basic activities or instrumental activities of daily living for themselves" (Cherlin 1996, p. 762).

causal feedback loop When the workplace impacts within-family interactions and decision making—and within-family interactions and decision-making affect the workplace.

child abuse Overt acts of aggression against a child, such as beating or inflicting physical injury or excessive verbal derogation. Sexual abuse is a form of physical child abuse. *See also* **emotional neglect.**

childcare The care and education of children by people other than their parents. Child care may include before- and after-school care for older children and overnight care when employed parents must travel, as well as day care for preschool children.

child maltreatment Defined by the 1974 Child Abuse Prevention and Treatment Act as the "physical or mental injury, sexual abuse, or negligent treatment of a child under the age of 18 by a person who is responsible for the child's welfare under circumstances that indicate that the child's health or welfare is harmed or threatened."

child neglect Failure to provide adequate physical or emotional care for a child. *See also* **emotional neglect.**

child support Money paid by the noncustodial parent to the custodial parent to financially support children of a former marital, cohabiting, or sexual relationship.

child-to-parent violence A form of family violence involving a child's (especially an adolescent's) physical and emotional abuse of a parent.

cisgender Describes an individual whose gender identity aligns with the sex he or she was assigned at birth; opposite of transgender.

civil union Legislation that allows any two single adults—including same-sex partners or blood relatives, such as siblings or a parent and adult child—to have access to virtually all marriage rights and benefits on the state level, but none on the federal level. Designed to give same-sex couples many of the legal benefits of marriage while denying them the right to legally marry.

closeness Emotional bonding of couples and family members with one another; also known as *togetherness.*

coercive control A single act or pattern of acts that are used with the intention of hurting, punishing, or scaring one's partner in order to control them. These acts can involve threatening, intimidating, and/or humiliating the victim.

coercive power One of the six power bases or sources of power. This power is based on the dominant person's ability and willingness to punish the partner either with psychological–emotional or physical abuse or with more subtle methods of withholding affection.

cohabitation Living together in an intimate, sexual relationship without traditional, legal marriage. Sometimes referred to as *living together* or *marriage without marriage,* cohabitation can be a courtship process or an alternative to legal marriage, depending on how partners view it.

cohousing An intentional living environment classified by multiple private homes grouped together around shared space featuring a common house.

collaborative divorce A legal process where both spouses have an attorney and sign a contract agreeing to work together to resolve pertinent issues without needing to go to court.

collectivist society A society in which people identify with and conform to the expectations of their relatives or clan, who look after their interests in return for their loyalty. The group has priority over the individual. A synonym is *communal society.*

commitment Willingness to work through problems and conflicts as opposed to calling it quits when problems arise.

common-pot system When economic resources for a family are pooled and distributed according to need regardless of biological relatedness.

communal society *See* **collectivist society**.

communes Groups of adults and perhaps children who live together, sharing aspects of their lives. Some communes are group marriages, in which members share sex; others are communal families, with several monogamous couples, who share everything except sexual relations and their children.

community-based resources Characteristics, competencies, and means of people, groups, and institutions outside the family that the family may call upon, access, and use to meet their demands.

companionate marriage The single-earner, breadwinner–homemaker marriage that flourished in the 1950s. Although husbands and wives in the companionate marriage usually adhered to a sharp division of labor, they were supposed to be each other's companion—friends, lovers—in a realization of trends beginning in the 1920s.

comprehensive sex education (CSE) A gender-focused, age-appropriate approach to teaching sex education over the course of several years. Programs include exploration of values regarding sex and reproductive health, and discussions about family life and relationships in addition to providing information about human and sexual development, reproductive health, contraception, childbirth, and sexually transmitted infections.

concerted cultivation The parenting model or style according to which parents often praise and converse with their children, engage them in extracurricular activities, take them on outings, and so on, with the goal of cultivating their child's talents and abilities.

conflict perspective Theoretical perspective that emphasizes social conflict in a society and within families. Power and dominance are important themes.

Conflict Tactics Scale A scale developed by sociologist Murray Straus to assess how couples handle conflict. Includes detailed items on various forms of physical violence.

Confucian training doctrine Concept used to describe Asian and Asian American parenting philosophy that emphasizes blending parental love, concern, involvement, and physical closeness with strict and firm control.

conjugal power The ability to exercise one's will or autonomy in a marital relationship.

consensual marriages Heterosexual, conjugal unions that have not gone through a legal marriage ceremony.

consummate love A complete love, in terms of Sternberg's triangular theory of love, in which the components of passion, intimacy, and commitment come together.

contempt A partner's feeling that his or her spouse is inferior or undesirable; one of the **Four Horsemen of the Apocalyps**e.

co-parenting, co-parents Shared decision making and parental supervision in such areas as discipline and schoolwork or shared holidays and recreation. Can refer to parents working together in a marriage or other ongoing relationship or after divorce or separation.

courtesy stigma Refers to a situation in which not only the initially stigmatized individual but also her or his intimates are stigmatized by association.

covenant marriage A type of legal marriage in which the bride and groom agree to be bound by a marriage contract that will not let them get divorced as easily as is allowed under no-fault divorce laws.

criminal justice response The punitive approach for perpetrators of child maltreatment; advocates believe that one or both parents should be held legally responsible for child abuse.

criticism Disapproving judgments or evaluations of one's partner; one of the **Four Horsemen of the Apocalypse**.

cross-national marriages Marriages in which spouses are from different countries.

crude divorce rate The number of divorces per 1,000 population. *See also* **refined divorce rate**.

cultural script Set of socially prescribed and understood guidelines for relating to others or for defining role responsibilities and obligations.

custodial grandparent When a grandparent has custody of a grandchild.

cycle of violence Three consecutive phases in which there is the violent episode itself (phase 1), a calm period (phase 2), and tension buildup during which a victim feels increasingly disappointed and intimidated while the abuser's behavior is unpredictable and threatening (phase 3).

data-collection techniques Ways that data are gathered when doing research; these include interviews and questionnaires, naturalistic observation, focus groups, experiments and laboratory observation, and case studies, among others.

date rape Forced or unwanted sexual contact between people who are on a date. *See also* **acquaintance rape**.

dating scripts Highly gendered expectations that govern behavior in the getting-to-know-you stage of dating relationships, with men and women having far different expectations about what happens during and after a date.

deciding versus sliding Two ways of dealing with choices.

de facto parent A person found by the court to have fulfilled the role of the parent on a daily basis by taking care of a child's physical and emotional needs and well being for a significant amount of time.

defensiveness Attitude characterized by preparing to defend oneself against an anticipated attack; one of the **Four Horsemen of the Apocalypse**.

deinstitutionalization of marriage A situation in which time-honored family definitions are changing and family-related social norms are weakening so that they "count for far less" than in the past.

disengaged couple or family Couple or family characterized by their loss of desire or willingness to put any energy, time, and emotion into the relationship or family system to keep it or make it better.

displacement A passive-aggressive behavior in which a person expresses anger with another by being angry at or damaging people or things the other cherishes. *See also* **passive-aggression**.

divorce divide The gap in divorce rates between college-educated and less-educated men and women. The divorce rate has declined substantially among those who are college educated but not among the less educated.

divorce fallout Ruptures of relationships and changes in social networks that come about as a result of divorce.

divorce mediation A nonadversarial means of dispute resolution by which the couple, with the assistance of a mediator or mediators (frequently a lawyer–therapist team), negotiate the terms of their settlement of custody, support, property, and visitation issues.

domestic partnership Arrangement in which an unmarried couple registers their partnership with a civil authority and then enjoys some (although not necessarily all) rights, benefits, and entitlements traditionally reserved for marrieds.

domestic violence model of elder abuse and neglect A model that conceptualizes elder abuse as a form of family violence.

double message *See* mixed message.

double standard The standard according to which nonmarital sex or multiple partners are more acceptable for males than for females.

dramaturgy A theoretical subcategory within the interactionist-constructionist perspective, sees individuals as enacting culturally constituted scripts and socially prescribed roles in front of others (everyday-life audiences).

dripolator effect The "bottom up" operation of a stepfamily (from children to parents); *see* **percolator effect**.

economic hardship perspective One of the theoretical perspectives concerning the negative outcomes among children of divorced parents. From this perspective, it is the economic hardship brought about by marital dissolution that is primarily responsible for problems faced by children.

egalitarian norm The marital norm (cultural rule) that husband and wife should have equal power in a marriage.

egalitarian relationship Equal relationship in which resources and power are relatively gender-free.

elder abuse Overt acts of aggression toward the elderly, in which the victim may be physically assaulted, emotionally humiliated, purposefully isolated, or materially exploited.

elder care Care provided to older generations.

elder neglect Acts of omission in the care and treatment of the elderly.

emerging adulthood Fairly new life cycle stage typically defined as young adults ages 18 to 29 who spend more time in higher education or exploring options regarding work, career, and family making than in the past.

emotional abuse Verbal threats and routine comments that damage a partner's self esteem.

emotional intelligence (1) Awareness of what we're feeling so that we can express our feelings more authentically; (2) ability and willingness to repair our moods, not unnecessarily nursing our hurt feelings; (3) healthy balance between controlling rash impulses and being candid and spontaneous; and (4) sensitivity to the feelings and needs of others.

emotional neglect When a parent or guardian is overly harsh and critical with a child, failing to provide guidance or being uninterested in the child's needs.

endogamy Marrying within one's own social group. *See also* **exogamy**.

enmeshed couple or family Couple or family characterized by a lack of clear personal boundaries.

equity A standard for distribution of power or resources of partners according to the contribution each person has made to the unit. Another way of characterizing an equitable result is that it is "fair."

estrangement No longer feeling a sense of belonging to or being welcomed by a particular social group (i.e., one's family).

ethnicity A group's identity based on a sense of a common culture and language.

exchange model Historical tendency for women to trade their ability to bear and raise children, perform domestic duties, be sexually accessible, and be physically attractive in exchange for a man's protection, status, and economic support.

exchange theory Theoretical perspective that sees relationships as determined by the exchange of resources and the reward–cost balance of that exchange. This theory predicts that people tend to marry others whose social class, education, physical attractiveness, and even self-esteem are similar to their own.

exogamy Marrying a partner from outside one's own social group. *See also* **endogamy**.

expectations of permanence One component of the marriage premise, according to which individuals enter marriage expecting that mutual affection and commitment will be lasting.

expectations of sexual exclusivity The cultural ideal according to which spouses promise to have sexual relations with only each other.

experience hypothesis The idea that the independent variable in a hypothesis is responsible for changes to a dependent

variable. With regard to marriage, the experience hypothesis holds that something about the experience of being married itself causes certain results for spouses. *See also* the antonym, **selection hypothesis**.

experiment One tool of scientific investigation in which behaviors are carefully monitored or measured under controlled conditions. Participants are randomly assigned to treatment or control groups.

expert power One of the six power bases or sources of power. This power stems from the dominant person's superior judgment, knowledge, or ability.

expressive character traits Relationship-oriented characteristics of warmth, sensitivity, the ability to express tender feelings, and placing concern about others' welfare above self-interest.

expressive sexuality The view of human sexuality in which sexuality is basic to the humanness of both women and men, all individuals are free to express their sexual selves, and there is no one-sided sense of ownership.

extended family Family including relatives besides parents and children, such as aunts or uncles. *See also* **nuclear family**.

familistic values Communal or collective values that focus on the family group as a whole and on maintaining family identity and cohesiveness.

family Any sexually expressive or parent-child or other kin relationship in which people live together with a commitment in an intimate interpersonal relationship. Family members see their identity as importantly attached to the group, which has an identity of its own. Families today take several forms: single-parent, remarried, dual-career, communal, homosexual, traditional, and so forth. *See also* **extended family, nuclear family**.

family cohesion That intangible emotional quality that holds groups together and gives members a sense of common identity. *See also* **closeness**.

family crisis A situation (resulting from a stressor) in which the family's usual behavior patterns are ineffective and new ones are called for.

family-change perspective Characterization by some family scholars and policy makers that the family can continue to play a strong role in society by adapting to such recent changes as increased age at first marriage, divorce, cohabitation, and nonmarital births and the declines in fertility.

family-decline perspective View that late-twentieth-century developments had put the family into decline through such changes as increases in the age at first marriage, divorce, cohabitation, and nonmarital births and the decline in fertility.

family ecology perspective Theoretical perspective that explores how a family influences and is influenced by the environments that surround it. A family is interdependent first with its neighborhood, then with its social–cultural environment, and ultimately with the human-built and physical–biological environments. All parts of the model are interrelated and influence one another.

family fluidity Frequency and rate of changes in family-related experiences and outcomes; measure of children's adjustment to family changes.

family-friendly workplace policies Workplace policies that are supportive of employee efforts to combine family and work commitments.

family identity Ideas and feelings about the uniqueness and value of one's family unit.

family impact lens Way of looking at how a policy in question impacts families.

family instability perspective The thesis that a negative impact of divorce on children is primarily caused by the number of changes in family structure, not by any particular family form. A stable single-parent family may be less harmful to children than a divorce followed by a single-parent family followed by cohabitation, then remarriage, and perhaps a redivorce.

family leave A leave of absence from work granted to family members to care for new infants, newly adopted children, ill children, or aging parents, or to meet similar family needs or emergencies.

family life course development framework Theoretical perspective that follows families through fairly typical stages in the life course, such as through marriage, childbirth, stages of raising children, adult children's leaving home, retirement, and possible widowhood.

family of orientation The family in which an individual grows up. Also called *family of origin*.

family of procreation The family that is formed when an individual marries and has children.

family policy All the actions, procedures, regulations, attitudes, and goals of government that affect families.

family preservation A program of support for families in which children have been abused. The support is intended to enable the child to remain in the home safely rather than being placed in foster care.

family stress State of tension that arises when demands tax a family's resources.

family structure The form a family takes such as nuclear family, extended family, single-parent family, stepfamily, and the like.

family systems theory An umbrella term for a wide range of specific theories. This theoretical perspective examines the family as a whole. It looks to the patterns of behavior and relationships within the family in which each member is affected by the behavior of others. Systems tend toward equilibrium and will react to change in one part by seeking equilibrium either by restoring the old system or by creating a new one.

family transitions Expected or predictable changes in the course of family life that often precipitate family stress and can result in a family crisis.

female-demand/male-withdraw interaction pattern A cycle of negative verbal expression by one partner, followed by the other partner's withdrawal in the face of the other's demands.

femininities Culturally defined ways of being a woman. The plural conveys the idea that there are varied models of appropriate behavior.

feminist theory Feminist theories are conflict theories. The primary focus of the feminist perspective is male dominance in families and society as oppressive to women. The mission of this perspective is to end this oppression of women (or related pattern of subordination based on social class, race/ethnicity, age, or sexual orientation) by developing knowledge and action that confront this disparity. *See also* **conflict perspective**.

fertility Births to a woman or category of women (actual births, not reproductive capacity).

fictive kin Family-like relationships that are not based on blood or marriage but on close friendship ties.

filial responsibility A child's obligation to a parent.

flexible scheduling A type of employment scheduling that includes scheduling options such as **job sharing** and **flextime**.

flextime A policy that permits an employee some flexibility to adjust working hours to suit family needs or personal preference.

formal kinship care Out-of-home placement with biological relatives of children who are in the custody of the state.

foster care Care provided to children by other than their parents as a result of state intervention.

Four Horsemen of the Apocalypse Contempt, criticism, defensiveness, and stonewalling—marital communication behaviors delineated by John Gottman that often indicate a couple's future divorce.

free-choice culture Culture or society in which individuals choose their own marriage partners, a choice usually based at least somewhat on partners' love for one another.

friends with benefits Sexual activity between friends or acquaintances with no expectation of romance or emotional attachment; typically practiced by unattached people who want to have a sexual outlet without "complications."

gay A person whose sexual attraction is to people of the same sex. Used especially for males, but may include both sexes. This term is usually used rather than *homosexual*.

gender Attitudes and behavior associated with and expected of the two sexes. The term *sex* denotes biology, while *gender* refers to social role.

gender differentiation Cultural expectations about how men and women should behave. Humans are to some extent differentiated, or thought of as separate and different, according to gender.

gender expectations Set of societal norms that prescribe what is and is not acceptable behavior for individuals, according to their gender. Gender expectations, taken together, constitute a gender role.

gender identity the degree to which an individual sees her- or himself as feminine or masculine, some other gender, or no gender.

gender-modified egalitarian model Family model in which absolute equality is diminished by the symbolic importance of maintaining fairly traditional, comfortable, and familiar gender roles.

gender schema theory Theoretical framework of knowledge and beliefs about gender socialization differences or similarities between males and females. Gender schema shape socialization into gender roles.

gender structure Way in which gender roles are influenced by a society's sociocultural environment.

gender variance Expansive concept that includes people dressing and behaving more like the "other" gender.

geographic availability Traditionally known in the marriage and family literature as propinquity or proximity and referring to the fact that people tend to meet potential mates who are present in their regional environment.

gerontologists Social scientists who study aging and the elderly.

good providers A specialized masculine role that emerged in this country around the 1830s and that emphasized the husband as the only or primary economic provider for his family. The good provider role had disappeared as an expected masculine role by the 1970s. *See also* **provider role**.

grandfamilies Families headed by grandparents.

habituation The decreased interest in sex over time that results from the increased accessibility of a sexual partner and the predictability of sexual behavior with that partner.

habituation hypothesis Hypothesis that the decline in sexual frequency over a marriage results from habituation.

heterogamy Marriage between partners who differ in race, age, education, religious background, or social class. Compare with **homogamy**.

heteronormative bias—the idea that heterosexuality is the only normal, acceptable, or "real" marriage option.

heteronormativity The idea that gender is binary—a person is either male or female—and that heterosexuality is the only normal, acceptable, or "real" option for all individuals.

heterosexism The taken-for-granted system of beliefs, values, and customs that places superior value on heterosexual behavior (as opposed to homosexual) and denies or stigmatizes nonheterosexual relations. This tendency also sees the heterosexual family as standard.

heterosexual Term used for people who prefer sexual partners of the opposite sex.

hierarchical compensatory model of caregiving The idea that elderly people prefer their caregivers in ranked order as follows: an available spouse, adult children, siblings, grandchildren, nieces and nephews, friends, neighbors, and, finally, a formal service provider.

hierarchical parenting Concept used to describe a Hispanic parenting philosophy that blends warm emotional support for children with demand for significant respect for parents and other authority figures, including older extended-family members.

HIV/AIDS HIV is *human immunodeficiency virus*, the virus that causes AIDS, or *acquired immune deficiency syndrome*. AIDS is a sexually transmitted disease involving breakdown of the immune system defense against viruses, bacteria, fungi, and other diseases.

homogamy Marriage between partners of similar race, age, education, religious background, and social class. *See also* **heterogamy**.

homophobia Fear, dread, aversion to, and often hatred of homosexuals.

homosexual An individual who is sexually attracted to people of the same sex. Preferred terms are *gay* or *gay man* for men and *lesbian* for women. *See also* **gay** and **lesbian**.

hooking up A sexual encounter between men and women with the understanding that there is no obligation to see each other again or to endow the sexual activity with emotional meaning. Usually there is a group or network context for hooking up; that is, the individuals meet at a social event or have common acquaintances. On some college campuses and elsewhere, hooking up has replaced dating, which is courtship-oriented socializing and sexual activity.

hormonal processes Chemical processes within the body regulated by such hormones as testosterone (a "male" hormone) and estrogen (a "female" hormone). Hormonal processes are thought to shape behavior, as well as physical development and reproductive functions, although experts disagree as to their impact on behavior.

hormones Chemical substances secreted into the bloodstream by the endocrine glands.

household As a Census Bureau category, a household is any group of people residing together.

hyperparenting A parenting style; once dubbed "helicopter parents," hyperparents hover over and meddle excessively in their children's lives; also called *intensive parenting*.

incest Sexual relations between close and genetically related individuals.

income-to-needs ratio An assessment of income as to the degree it meets the needs of the individual, family, or household.

incomplete institution Cherlin's description of a remarried family due to cultural ambiguity.

independence effect Occurs when an increase in a married woman's income leads to marital dissolution because she is better able to afford to live separately.

individualism The cultural milieu that emerged in Europe with industrialization and that values personal self-actualization and happiness along with individual freedom.

individualistic society Society in which the main concern is with one's own interests (which may or may not include those of one's immediate family).

individualistic (self-fulfillment) values Values that encourage personal growth, autonomy, and independence over commitment to family or other communal needs.

individualized marriage Concept associated with the argument that contemporary marriage in the United States and other fully industrialized Western societies is no longer institutionalized. Four interrelated characteristics distinguish individualized marriage: (1) it is optional; (2) spouses' roles are flexible—negotiable and renegotiable; (3) its expected rewards involve love, communication, and emotional intimacy; and (4) it exists in conjunction with a vast diversity of family forms.

informal adoption Children are taken into a home and considered to be children of the parents, although the "adoption" is not legally formalized.

informal caregiving Unpaid caregiving, provided personally by a family member.

informational power One of the six power bases or sources of power. This power is based on the persuasive content of what the dominant person tells another individual.

institutional marriage Marriage as a social institution based on dutiful adherence to the time-honored marriage premise, particularly the norm of permanence. Also referred to as *institutionalized marriage*.

instrumental character traits *See* agentic character traits.

intensive parenting *See* **hyperparenting**.

interaction–constructionist perspective Theoretical perspective that focuses on internal family dynamics; the ongoing action among and response to one another of family members.

interactionist perspective on human sexuality A perspective, derived from symbolic interaction theory, which holds that sexual activities and relationships are shaped by the sexual scripts available in a culture.

interethnic unions Marriages between spouses who are not defined as of different races but do belong to different ethnic groups.

interfaith marriages Marriages in which partners are of different religions.

intermittent cohabitation Relationships in which parenting couples move in together, then out, then back in.

interparental conflict perspective One of the theoretical perspectives concerning the negative outcomes among children of divorced parents. From the interparental conflict perspective, the conflict between parents before, during, and after the divorce is responsible for the lowered well-being of the children of divorce.

interpersonal exchange model of sexual satisfaction A view of sexual relations, derived from exchange theory, that sees sexual satisfaction as shaped by the costs, rewards, and expectations of a relationship and the alternatives to it.

interracial unions Marriages of a partner of one (socially defined) race to someone of a different race.

intersectionality Often associated with gender theory, the idea that various social categories such as race, sex/gender, social class, sexual orientation, and immigrant status, among others, are interconnected and cannot really be examined separately from one another. In critical feminist theory, intersectionality emphasizes ways

in which these social categories interact to create systems either of privilege or oppression.

intersex A person whose genitalia, secondary sex characteristics, hormones, or other physiological features are not unambiguously male or female.

intimate outsider role Stepparent role when stepfamilies reach maturation; end result is a stepparent who can be a confidante, trusted adviser, and mentor for a child on subjects and areas too threatening to share with biological parents such as sex and drug use.

intimate partner violence (IPV) Violence against current or former spouses, cohabitants, or sexual or relationship partners.

intimate terrorism Abuse that is almost entirely male and that is oriented to controlling the partner through fear and intimidation.

involuntary infertility Situation of a couple or individual who would like to have a baby but cannot. Involuntary infertility is medically diagnosed when a woman has tried for twelve months to become pregnant without success.

involved fathers Men who work fewer hours than do childless men in order to spend more time with their children.

job sharing Two people sharing one job.

joint custody A situation in which both divorced parents continue to take equal responsibility for important decisions regarding their child's general upbringing. Joint custody can include joint physical custody, which involves the children living at least part of the time with each parent.

kin Parents and other relatives such as in-laws, grandparents, aunts and uncles, and cousins. *See also* **extended family**.

kin keeping Maintaining contact with family members and remembering anniversaries and birthdays, sending cards, shopping for gifts, and organizing family activities; more frequently done by women than by men.

labor force A social invention that arose with the industrialization of the nineteenth century, when people characteristically became wage earners, hiring out their labor to someone else.

latent kin matrix "A web of continually shifting linkages that provide the potential for activating and intensifying close kin relationships" (Riley 1983, p. 441).

legal custody Court determination of who has the right to make decisions with respect to a child's upbringing.

legal strangers A third party who is not privy to a particular legal action or established agreement.

legitimate power One of the six power bases or sources of power. Legitimate power stems from the more dominant individual's ability to claim authority, or the right to request compliance.

lesbian A woman who is sexually attracted to other women. This term is usually used rather than *homosexual*.

Levinger's model of divorce decisions This model, derived from exchange theory, presents a decision to divorce as involving a calculus of the *barriers* to divorce (e.g., concerns about children and finances; religious prohibitions), the *rewards* of the marriage, and *alternatives* to the marriage (e.g., can the divorced person anticipate a new relationship, career development, or a single life that will be more rewarding and less stressful than the marriage?).

LGBTQ+ An acronym for *lesbian, gay, bisexual, transgendered, queer, and additional gender identities*; a term commonly used when discussing sexual minorities.

life chances The opportunities that exist for a social group or an individual to pursue education and economic advancement, to secure medical care and preserve health, to marry and have children, to have material goods and housing of desired quality, and so forth.

life stress perspective One of the theoretical perspectives concerning the negative outcomes among children of divorced parents. From the life stress perspective, divorce involves the same stress for children as for adults. Divorce is not one single event but a process of stressful events—moving, changing schools, and so on.

limerence A typically involuntary state of being infatuated or obsessed with someone, characterized by a strong desire for one's feelings to be returned. This is generally not primarily for a sexual relationship.

living apart together (LAT) Emerging lifestyle choice in which a couple is committed to a long-term relationship but each partner maintains a separate dwelling.

manipulating Seeking to control the feelings, attitudes, and behavior of one's partner or partners in underhanded ways rather than by assertively stating one's case.

marital rape A husband's forcing a wife to submit to sexual contact that she does not want or that she finds offensive.

marital sanctification A religious belief system that encourages spouses to see their marriages as ordained by God and hence having divine significance.

marriage gap Disparity in marriage rates between the poor and those who are not poor.

marriage market The sociological concept that potential mates take stock of their

personal and social characteristics and then comparison shop or bargain for the best buy (mate) they can get.

marriage premise By getting married, partners accept the responsibility to keep each other primary in their lives and to work hard to ensure that their relationship continues.

martyring Doing all one can for others while ignoring one's own legitimate needs. Martyrs often punish the person to whom they are martyring by letting the person know "just how much I put up with."

masculine dominance View that masculine males exercise authority over females and people they perceive as not masculine enough; usually involves wielding greater power in a heterosexual relationship as well as greater control and influence over society's institutions and benefits.

masculinities Idea that there are varied ways to demonstrate masculinity; three major culturally defined obligations for men involve group leadership, protecting group territory and weaker or dependent others, and providing resources, typically by means of occupational success.

mate selection The process of selecting a partner to form a committed relationship with. Personal preferences and social expectations play a role in this process.

matching Tendency of individuals to select partners with characteristics similar to their own.

menarche A girl's first menstrual cycle; age at which occurs dropped during the twentieth century for all races and ethnicities.

microaggressions Commonplace and subtle verbal, behavioral, or environmental indignities, whether intentional or unintentional, that communicate hostile, derogatory, or negative feelings toward a marginalized group.

minority/minority group Group that is distinguishable and in some way disadvantaged within a society regardless of its size.

misogynistic Behavior that exhibits hatred, dislike, mistrust, mistreatment, or general disregard for women.

mixed message Two simultaneous messages that contradict each other; also called a *double message*. For example, society gives us mixed messages regarding family values and individualistic values and about premarital sex. People, too, can send mixed messages, as when a partner says, "Of course I always like to talk with you" while turning up the TV.

money allocation system Arrangements couples make for handling their income, wealth, and expenditures. They may involve pooling partners' resources or keeping them separate. Who controls pooled resources is another dimension of an allocation system.

monogamy Electing to be in a relationship with, sexually involved with, and/or married to only one person at a time.

motherhood penalty Negative lifetime impact on earnings for women who raise children.

multigenerational family Family that includes several generations.

multigenerational safety net function Different generations in a family providing each other with a cushion, or safety net, against potential or actual economic or health adversities.

multipartnered fertility (MPF) Having children with more than one biological partner.

mutual dependency Stage of a relationship in which two people desire to spend more time together and thereby develop interdependence.

nadir (of family disorganization) Low point of family disorganization when a family is going through a family crisis.

National Family Violence Surveys Early and continuing research of Murray Straus, Richard Gelles, and their colleagues in shaping scientific research on family violence among the general population.

naturalistic observation A technique of scientific investigation in which a researcher lives with a family or social group or spends extensive time with them, carefully recording their activities, conversations, gestures, and other aspects of everyday life.

need fulfillment Stage of relationship development in which two people find they satisfy a majority of each other's emotional needs with the result that rapport increases and leads to deeper self-revelation, more mutually dependent habits, and still greater need satisfaction.

negative affect Showing emotion(s) defined as negative, such as anger, sadness, whining, disgust, tension, fear, and/or belligerence.

neotraditional family Family that values traditional gender roles and organizes its life in these terms as far as practicable. Formal male dominance is softened by an egalitarian spirit.

noncustodial grandparent A parent of a divorced, noncustodial parent.

nuclear family A family group comprising only a wife, husband, and their children. *See also* **extended family**.

nuclear-family model monopoly The cultural assumption that the first-marriage family is the "real" model of family living, with all other family forms viewed as deficient.

occupational segregation The distribution of men and women into substantially different occupations. Women are overrepresented in clerical and service work, for example, whereas men dominate the higher professions and the upper levels of management.

opportunity costs The economic opportunities for wage earning and investments that parents forgo when raising children.

pansexual/pansexuality Identity of someone who has the potential to be sexually attracted to various gender expressions, including those outside the gender-conforming binary.

parallel parents Postdivorce parental relationship in which former partners parent alongside each other with minimal contact, communication, or conflict; associated with worse child outcomes.

parental adjustment perspective One of the theoretical perspectives concerning the negative outcomes among children of divorced parents. From the parental adjustment perspective, the parent's child-raising skills are impaired as a result of the divorce, with probable negative consequences for the children.

parental loss perspective One of the theoretical perspectives concerning the negative outcomes among children of divorced parents. From the parental loss perspective, divorce involves the absence of a parent from the household, which deprives children of the optimal environment for their emotional, practical, and social support.

parentification When children of divorce or those from a single parent home are forced to take on adult responsibilities before they are developmentally mature enough to handle them; linked to worse child outcomes.

passive-aggression Expressing anger at some person or situation indirectly, through nagging, nitpicking, or sarcasm, for example, rather than directly and openly. *See also* **displacement, sabotage**.

paternal claiming Extent to which stepfathers see their stepchildren as their own biological children.

patriarchal norm The marital norm (cultural rule) that the man should be dominant in a marital relationship.

patriarchal sexuality The view of human sexuality in which men own everything in the society, including women and women's sexuality. Males' sexual needs are emphasized while females' needs are minimized.

patriarchy A social system in which males are dominant.

percolator effect The "bottom up" operations of a stepfamily (from children to parents); *see* **dripolator effect**.

period of family disorganization That period in a family crisis, after the stressor

event has occurred, during which family morale and organization slump and habitual roles and routines become nebulous.

permissive parenting style One of three parenting styles in this schema, permissive parenting gives children little parental guidance.

physical custody Legal determination of where a child will live and with which parent.

polyamory A marriage system in which one or both spouses retain the option to sexually love others in addition to their spouses.

polygamy A marriage system in which a person takes more than one spouse.

pool of eligibles A group of individuals who, by virtue of background or social status, are most likely to be considered eligible to make culturally compatible marriage partners.

pornification Broad social phenomenon in which sending, receiving, and viewing sexually explicit material is seen as normal activity.

pornography Sexually explicit images and descriptions as found in books, magazines, film, and cyberspace.

positive affect The expression, either verbal or nonverbal, of one's feelings of affection toward another.

postmodern family Term used to describe the situation in which (1) families today exhibit multiple forms, and (2) new or altered family forms continue to emerge or develop.

postmodern theory Theoretical perspective that largely analyzes social interaction (discourse or narrative) in order to demonstrate that a phenomenon is socially constructed.

power The ability to exercise one's will. *Personal power*, or *autonomy*, is power exercised over oneself. *Social power* is the ability to exercise one's will over others.

power politics Power struggles between spouses in which each seeks to gain a power advantage over the other; the opposite of a no-power relationship.

private face of family The aspect of the family that provides individuals with intimacy, emotional support, and love.

private safety net Social support from family and friends rather than from public sources.

private sphere World inside someone's home.

pronatalist bias A cultural attitude that takes having children for granted.

provider role A term for the family role involving wage work to support the family.

May be carried out by one spouse or partner only or by both.

psychological parent The parent, usually but not necessarily the mother, who assumes principal responsibility for raising the child.

public face of family The aspect of the family that produces public goods and services.

public sphere World outside someone's home.

queer kinship Social network of choice characterized by uniting a group of people through connection and cooperation based on similar gender identity.

race A group or category thought of as representing a distinct biological heritage. In reality, there is only one human race. "Racial" categories are social constructs; the so-called races do not differ significantly in terms of basic biological makeup. But "racial" designations nevertheless have social and economic effects and cultural meanings.

race socialization The socialization process that involves developing a child's pride in his or her cultural heritage while warning and preparing him or her about the possibilities of encountering discrimination.

rape myths Beliefs about rape that function to blame the victim and exonerate the rapist.

rapport Feelings of mutual trust and respect; often established by similarity of values, interests, and background.

rapport talk In Deborah Tannen's terms, this is conversation engaged in by women aimed primarily at gaining or reinforcing rapport or intimacy. *See also* **report talk**.

referent power One of the six power bases or sources of power. In a marriage or relationship, this form of power is based on one partner's emotional identification with the other and his or her willingness to agree to the other's decisions or preferences.

refined divorce rate Number of divorces per 1,000 married women over age fifteen. *See also* **crude divorce rate**.

reframing In the family stress and coping literature, redefining stressful events to make them more manageable.

relationship-focused coping Process in relationships in which partners help or hinder one another in dealing with stresses and strain.

relationship ideologies Expectations for closeness and/or distance as well as ideas about how partners should play their roles.

replacement level fertility The average number of births per woman (a total fertility rate of 2.1) necessary to replace the population.

report talk In Deborah Tannen's terms, this is conversation engaged in by men aimed primarily at conveying information. *See also* **rapport talk**.

reproductive coercion Behavior related to reproductive health that is used to maintain power and control in a relationship.

resilient The ability to recover from challenging situations.

resilient families Families that emphasize mutual acceptance, respect, and shared values; members rely on one another for emotional support.

resource hypothesis Hypothesis (originated by Robert Blood and Donald Wolfe) that the relative power between wives and husbands results from their relative resources as individuals.

resources in cultural context The effect of resources on marital power depends on the cultural context. In a traditional society, norms of patriarchal authority may override personal resources. In a fully egalitarian society, a norm of intimate partner and marital equality may override personal resources. It is in a transitional society that the resource hypothesis is most likely to shape marital power relations.

reward power One of the six power bases or sources of power. With regard to marriage or partner relationships, this power is based on an individual's ability to give material or nonmaterial gifts and favors to the partner.

role ambiguity The situation in which there are few clear guidelines regarding what responsibilities, behaviors, and emotions family members are expected to exhibit.

role conflict When individuals with roles in two institutions experience conflict when they attempt to meet demands of one institution that conflict with different demands from another institution.

role-making Improvising a course of action as a way of enacting a role. In role-making, we may use our acts to alter the traditional expectations and obligations associated with a role. This concept emphasizes the variability in the ways different individuals enact a particular role.

sabotage A passive-aggressive action in which a person tries to spoil or undermine some activity another has planned. Sabotage is not always consciously planned. *See also* **passive-aggression**.

sandwich generation Middle-aged (or older) individuals, usually women, who are sandwiched between the simultaneous responsibilities of caring for their dependent children (sometimes young adults) and aging parents.

science "A logical system that bases knowledge on . . . systematic observation,

empirical evidence, facts we verify with our senses" (Macionis 2006, p. 15).

second shift Sociologist Arlie Hochschild's term for the domestic work that employed women must perform after coming home from a day on the job.

selection effect When individuals "select" themselves into a category being investigated.

selection hypothesis The idea that many of the changes found in a dependent variable, which might be assumed to be associated with the independent variable, are really due to sample selection. For instance, the selection hypothesis posits that many of the benefits associated with marriage—for example, higher income and wealth, along with better health—are not necessarily due to the fact of being married but, rather, to the personal characteristics of those who choose—or are selected into—marriage. Similarly, the selection hypothesis posits that many of the characteristics associated with cohabitation result not from the practice of cohabiting itself but from the personal characteristics of those who choose to cohabit. *See also* the antonym, **experience hypothesis**.

selection perspective Posits that individuals have preexisting traits called *selection factors* that sort them into groups with higher divorce risks as well as negative outcomes after divorce.

self-care An approach to child care for working parents in which the child is at home or out without an adult caretaker. Parents may be in touch by phone.

self-concept The basic feelings people have about themselves, their characteristics and abilities, and their worth; how people think of or view themselves.

self-identification theory Theory of gender socialization developed by psychologist Lawrence Kohlberg that begins with a child's categorization of self as male or female. The child goes on to identify sex-appropriate behaviors in the family, media, and elsewhere and to adopt those behaviors.

self-revelation Gradually sharing intimate information about oneself.

serial monogamy Form of sexual exclusivity in which someone has only one sexual partner at a time.

sex Refers to biological characteristics—that is, male or female anatomy or physiology. The term *gender* refers to the social roles, attitudes, and behavior associated with males or females.

sex ratio The number of men per 100 women in a society. If the sex ratio is above 100, there are more men than women; if it is below 100, there are more women than men.

sex reassignment/gender affirming surgery Involves one or more surgical procedures that change a transgender individual's genitalia or other biological sex characteristics to correspond with the person's gender identity; also called *gender reassignment surgery,* and more often now *gender confirmation surgery.*

sexism Prejudice or discrimination based on one's sex.

sexting Using cell phones to send sexually explicit images or messages to others.

sexual abuse A form of child abuse that involves forced, tricked, or coerced sexual behavior—exposure, unwanted kissing, fondling of sexual organs, intercourse, rape, and incest—between a minor and an older person.

sexual assault Defined by U.S. legal code as any type of sexual contact or behavior that occurs without the explicit consent of the recipient and includes fondling, groping, digital penetration (with fingers), forced oral and anal sex (sodomy), forced sexual intercourse (rape), and attempted rape.

sexual coercion Use of verbal or emotional pressure (threatening violence, tricking, lying, or using guilt), one's position of power (being a boss, teacher, coach, or other adult), or other means to manipulate a victim into sexual activity. Common behavior among perpetrators of acquaintance rape.

sexual double standard The recognition that men and women are judged differently by people and in society for engaging in the same sexually-based behaviors.

sexual identity Whether one is attracted to one's own gender or a different gender; often described as *sexual orientation* or *sexual orientation identity*.

sexual scripts *Scripts* are culturally written patterns or "plots" for human behavior. Sexual scripts offer reasons for having sex and designate who should take the sexual initiative, how long an encounter should last, what positions are acceptable, and so forth.

shelters Physically protected and safe spaces that provide a woman (and often her children) with temporary housing, food, and clothing to alleviate the problems of economic dependency and physical safety.

shift work As defined by the Bureau of Labor Statistics, any work schedule in which more than half of an employee's hours are before 8 a.m. or after 4 p.m.

sibling violence Family violence that takes place between siblings (brothers and sisters).

silver divorce Phenomenon seen among older couples following long-term marriages;

particularly true for baby boomers born between 1946 and 1964; also known as *gray divorce.*

single mothers by choice Women who intentionally become mothers, although they are not married or with a partner. They are typically older, with economic and educational resources that enable them to be self-supporting.

situational couple violence Mutual violence between partners that often occurs in conjunction with a specific argument. It involves fewer instances, is not likely to escalate, and tends to be less severe in terms of injuries.

social class Position in the social hierarchy, such as *upper class, middle class, working class,* or *lower class.* Can be viewed in terms of such indicators as education, occupation, and income or analyzed in terms of status, respect, and lifestyle.

social fathers Males who are not biological fathers but perform the roles of a father, such as a stepfather.

social institution A system of patterned and predictable ways of thinking and behaving—beliefs, values, attitudes, and norms—concerning important aspects of people's lives in society. Examples of major social institutions are the family, religion, government, the economy, and education.

social learning theory According to this theory of of gender socialization, children learn gender roles as they are taught or modeled by parents, schools, and the media.

socialization The process by which society influences members to internalize attitudes, beliefs, values, and expectations.

social media A group of Internet applications that allow the creation and exchange of *user-generated* content; examples include Facebook, Twitter, blogs, Internet forums, photo sharing, social gaming, video-sharing services, and virtual worlds.

socioeconomic status (SES) One's position in society, measured by educational achievement, occupation, and/or income.

sociological imagination Placing an individual's or family's private troubles within a society-wide context.

spillover How pleasures or stresses associated with work affect interaction within the family, and vice versa.

starter marriage First marriage that ends in divorce within the first few years, typically before a couple has children.

status exchange hypothesis Regarding interracial/interethnic marriage, the argument that an individual might trade his or her socially defined superior racial/ethnic status for the economically or educationally superior status of a partner in a less-privileged racial/ethnic group.

stay-at-home dads Men who stay at home to care for their houses and families while their wives work.

stepchildren Children from a previous marriage or relationship.

stepfamilies Families that include stepchildren.

stepfamily cycle Process by which new family members move from being strangers to forming nourishing and reliable relationships.

stepmother trap The conflict between two views: Society sentimentalizes the stepmother's role and expects her to be unnaturally loving toward her stepchildren but at the same time views her as a wicked witch.

Sternberg's triangular theory of love Robert Sternberg's theory that consummate love involves three components: intimacy, passion, and commitment.

stigmatization When people are subjected to negative labels, stereotyping, and cultural myths that portray them as deviant and harmful simply because they have certain social characteristics.

stonewalling Refusing to listen to a partner's complaints; one of the **Four Horsemen of the Apocalypse**.

stress State of tension that results from the need to respond to change.

stress model of parental effectiveness The idea that stress experienced by parents causes parental frustration, anger, and depression, increasing the likelihood of household conflict and leading to poorer parenting practices.

stressors Precipitating events that cause a crisis; they are often situations for which the family has had little or no preparation. *See also* **ABC-X model**.

stressor overload A situation in which an unrelenting series of small crises adds up to a major crisis; also known as *pileup*.

stress-related growth Personal growth and maturity attained in the context of a stressful life experience such as divorce.

structural antinatalism The structural, or societal, conditions in which bearing and raising children is discouraged either overtly or—as may be the case in the United States—covertly through inadequate support for parenting.

structural constraints Economic and social forces that limit options and, hence, personal choices.

structure–functional perspective Theoretical perspective that looks to the functions that institutions perform for society and the structural form of the institution.

surrogate grandparent A form of fictive kin, an older individual not biologically related to the "grandchildren" who, attached to a nonbiologically related family, plays a role ordinarily designated for biological grandparents.

swinging A marriage agreement in which couples exchange partners to engage in purely recreational sex.

theoretical perspective A way of viewing reality, or a lens through which analysts organize and interpret what they observe. Researchers on the family identify those aspects of families that are of interest to them, based on their own theoretical perspective.

total fertility rate (TFR) For a given year, the number of births that women would have over their reproductive lifetimes if all women at each age had babies at the rate for each age group that year; can be calculated for social or age categories as well as for nations as a whole.

transfamilies Families in which one or more family members is or are transgendered.

transgender Describes an individual who has a gender identity different from conventional ideas about what is female or male or different from the gender assigned at birth.

transition to parenthood The circumstances involved in assuming the parent role.

transnational families Families of immigrants or immigrant stock that maintain close ties with their sending country. Identity and behavior connect an immigrant family to the new country and the old, and their social networks cross national boundaries.

transsexuals Individuals who switch physical sex through surgery, hormone therapy, electrolysis (hair removal), and other treatments.

two-career relationship Relationship now seen as an available and workable option; for two-career couples with children, however, family life can be hectic as partners juggle schedules, chores, and child care.

two-earner partnerships Two-earner and provider–housewife couples in which both partners work; arrangements are ever-changing and flexible, varying with the arrival and ages of children and with partners' job options and preferences.

two-pot system When economic resources for a family economic resources are divided and distributed along biological lines and only secondarily distributed according to need.

unilateral divorce A divorce can be obtained under the no-fault system by one partner even if the other partner objects. The term *unilateral divorce* emphasizes this feature of current divorce law.

unpaid family work The necessary tasks of attending to both the emotional needs of all family members and the practical needs of dependent members, such as children or elderly parents, and maintaining the family domicile.

value of children perspective Motivation for parenthood because of the rewards, including symbolic rewards, that children bring to parents.

Violence Against Women Act Federal law enacted in 1996 that has resulted in greater perpetrator accountability for rape and stalking, as well as increasing rates of prosecution, conviction, and sentencing of offenders; 2010 expansion of the act includes same-gender couples, and 2013 expansion includes additional categories such as transgender individuals.

voluntarily childfree The deliberate choice not to become a parent; often used now instead of the more negative-sounding term *childless*.

voluntary childlessness *See* voluntarily childfree.

vulnerable families Families that have a low sense of common purpose, feel in little control over what happens to them, and tend to cope with problems by showing diminished respect and/or understanding for each other.

wage gap The persistent difference in earnings between men and women.

War on Poverty Series of federal programs and initiatives put forth by President Lyndon Johnson in the 1960s; included the Job Corps or the Neighborhood Youth Corps, Head Start, and Adult Basic Education. Although most measures have ended, Head Start and the Job Corps continue to exist.

wheel of love An idea developed by Ira Reiss in which love is seen as developing through a four-stage, circular process, including rapport, self-revelation, mutual dependence, and personality need fulfillment.

REFERENCES

AAMFT. 2020. "About Marriage and Family Therapists." Retrieved January 15, 2020 (https://www.aamft.org/About_AAMFT /About_Marriage_and_Family_Therapists .aspx?hkey=1c77b71c-0331 -417b-b59b-34358d32b909).

Abbey, L. 2009. "Elder Abuse and Neglect: When Home Is Not Safe." *Clinics in Geriatric Medicine* 25(1):47–61.

Aber, J. Lawrence. 2007. *Child Development and Social Policy.* Washington, DC: American Psychological Association.

Abma, Joyce C., and Gladys M. Martinez. 2006. "Childlessness among Older Women in the United States: Trends and Profiles." *Journal of Marriage and Family* 68(4):1045–1056.

Abrahamson, Iona, Rafat Hussain, Adeel Khan, and Margot J. Schofield. 2011. "What Helps Couples Build Their Relationship After Infidelity." *Journal of Family Issues* 33:1494.

Abram, Carolyn. 2006. "Welcome to Facebook, Everyone."Retrieved February 28, 2016 (https://www.facebook.com /notes/facebook/welcome-to -facebook-everyone/2210227130/).

"The Abstinence-Only Delusion" (editorial). 2007. *The New York Times,* April 28.

Acevedo, Gabriel A., Christopher G. Ellison, and Murat Yilmaz. 2015. "Religion and Child-Rearing Values in Turkey." *Journal of Family Issues* 36(12):1595–1623.

Adam, Barry D. 2007. "Relationship Innovation in Male Couples." pp. 122–140 in *The Sexual Self: The Construction of Sexual Scripts,* edited by Michael S. Kimmel. Nashville: Vanderbilt University Press.

Adamczyk, Katarzyna, and Chris Segrin. 2015. "Direct and Indirect Effects of Young Adults' Relationship Status on Life Satisfaction through Loneliness and Perceived Social Support." *Psychologica Belgica* 55.

Adamo, Shelley A. 2013. "Attrition of Women in the Biological Sciences: Workload, Motherhood, and Other Explanations Revisited." *BioScience* 63(1):43–48.

Adams, Lynn. 2010. *Parenting on the Autism Spectrum: A Survival Guide.* San Diego: Plural Publishers.

Adamsons, Kari, and Kay Pasley. 2006. "Coparenting Following Divorce and Relationship Dissolution." pp. 241–261 in *Handbook of Divorce and Relationship Dissolution,* edited by Mark A. Fine and John H. Harvey. Mahwah, NJ: Erlbaum.

Addo, Fenaba R. 2014. "Debt, Cohabitation, and Marriage in Young Adulthood." *Demography* 51:1677–1701.

Addo, Fenaba R., and Daniel T. Lichter. 2013. "Marriage, Marital History, and Black-White Wealth Differentials among Older Women." *Journal of Marriage and Family* 75(2):342–362.

Addo, Fenaba R., and S. Sassler. 2010. "Financial Arrangements and Relationship Quality in Low Income Couples." *Family Relations* 59(4):408–423.

Adelman, Rebecca A. 2009. "Sold(i)ering Masculinity: Photographing the Coalition's Male Soldiers." *Men and Masculinities* 11(3):259–285.

Adler-Baeder, Francesca, Christiana Russell, Jennifer Kerpelman, Joe Pittman, Scott Ketring, Thomas Smith, Mallory Lucier-Green, Angela Bradford, and Kate Stringer. 2010.

Admission Reports/Arrivals by Region. 2019. Retrieved November 16, 2019, (https://www .wrapsnet.org/archives).

African American Youth." *Journal of Adolescent Health* 46:396–398.

Afifi, Tamara D. 2008. "Communication in Stepfamilies: Stressors and Resilience." pp. 299–317 in *The International Handbook of Stepfamilies,* edited by J. Pryor. Hoboken, NJ: John Wiley & Sons.

Afifi, Tracie, Natalie Mota, Patricia Dasiewicz, Harriet MacMillan, and Jitender Sareen. 2012. "Physical Punishment and Mental Disorders: Results from a Nationally Representative U.S. Sample." *Pediatrics* 130(2):184–192.

African Wedding Guide. n.d. Retrieved October 15, 2006 (www.africanweddingguide.com).

After School. 2018. "45% of Teens Say They're Stressed 'All the Time'." February 21. Retrieved February 6, 2020 (https://www.globenewswire.com).

Agee, Mark D., Scott E. Atkinson, and Thomas D. Crocker. 2008. "Multiple-Output Child Health Production Functions: The Impact of Time-Varying and Time-Invariant Inputs." *Southern Economic Journal* 75(2):410–428.

Agerbo, Esben, Presben Bo Mortensen, and Trine Munk-Olsen. 2012. "Childlessness, Parental Mortality and Psychiatric Illness: A Natural Experiment Based on In vitro Fertility Treatment and Adoption." *Journal of Epidemiology and Community Health, December* 5:201–245.

Ahn, Annie, Bryan Kim, and Park Yong. 2008. "Asian Cultural Values Gap, Cognitive Flexibility, Coping Strategies, and Parent-Child Conflicts among Korean Americans." *Cultural Diversity and Ethnic Minority Psychology* 14(4):353–363.

Ahrold, Tierney K., Melissa Farmer, Paul D. Trapnell, and Cindy M. Meston (2011). "The Relationship among Sexual Attitudes, Sexual Fantasy, and Religiosity." *Archives of Sexual Behavior* 40:619–630.

Ahrold, Tierney K., and Cindy M. Meston. 2011. "Ethnic Differences is Sexual Attitudes of U.S. College Students: Gender, Acculturation, and Religiosity Factors." *Archives of Sexual Behavior* 39:190–202.

Ahrons, Constance. 1994. *The Good Divorce: Raising Your Family Together When Your Marriage Comes Apart.* New York: HarperCollins.

Ahtone, Tristan. 2011. "Native American Intermarriage Puts Benefits at Risk." Retrieved May 31, 2013 (http://www .mixcloud.com/NPR/native-american -intermarriage-puts-benefits-at-risk/).

Aichner, Thomas, and Frank H. Jacob. 2015. "Measuring the Degree of Corporate Social Media Use." *International Journal of Market Research* 57:257–275.

AIDS.gov. 2015. "Overview." Retrieved December 11, 2015 (https://www.aids .gov /federal-resources/national-hiv-aids -strategy /overview/).

Ainsworth, Mary D. S., M. C. Blehar, E. Waters, and S. Wall. 1978. *Patterns of Attachment: A Psychological Study of the Strange Situation.* Hillsdale, NJ: Erlbaum.

Aizer, Anna. 2004. "Home Alone: Supervision After School and Child Behavior." *Journal of Public Economics* 88:1835–1848.

Alarie, Milaine, and Jason T. Carmichael. 2015. "The "Cougar" Phenomenon: An Examination of the Factors That Influence Age-Hypogamous Sexual Relationships among Middle-Aged Women." *Journal of Marriage and Family* 77:1250–1265.

Alan Guttmacher Institute. 2006. "Facts on Induced Abortion in the United States." May. New York: Guttmacher Institute.

Alan Guttmacher Institute. 2019a. "Sex and HIV Education." Retrieved September 21 2019 (https://www.guttmacher.org /state-policy/explore/sex-and-hiv-education).

Alan Guttmacher Institute. 2019b. "Unintended Pregnancy in the United States." Retrieved November 2, 2019 (https://www.guttmacher. org/fact-sheet /unintended-pregnancy-united-states).

Alan Guttmacher Institute. 2019. "United States Teens." Retrieved October 28, 2019 (https://www.guttmacher.org /united-states/teens).

Albrecht, Chris, and Jay D. Teachman. 2003. "Childhood Living Arrangements and the Risk of Premarital Intercourse." *Journal of Family Issues* 24(7):867–894.

Aldous, Joan. 1978. *Family Careers: Developmental Change in Families.* New York: Wiley.

Aleccia, Jonel. 2014. "Teens More Stressed-Out Than Adults, Survey Shows." February 11. NBC News. Retrieved December 4, 2015 (http://www .nbcnews.com).

Alexander, Brian. 2010. "Lovesick: Hooking Up Over a Shared Disease." *MSNBC,* February 12. www.msnbc.msn .com.

Alexander, Michelle. 2014. "Telling My Son About Ferguson." *The New York Times,* November 26. Retrieved December 2, 2014 (http://www. nytimes.com).

Ali, Lorraine, and Raina Kelley. 2008. "The Curious Lives of Surrogates." *Newsweek, April* 7: 45–51.

Ali, Lorraine, and Julie Scelfo. 2002. "Choosing Virginity." *Newsweek,* December 9, pp. 61–71.

Allan, Graham, Graham Crow, and Sheila Hawker. 2011. *Stepfamilies.* New York: Palgrave Macmillan.

Allen, Elizabeth S., and David C. Atkins. 2012. "The Association of Divorce and Extramarital Sex in a Representative U.S. Sample." *Journal of Family Issues* 33(11):1477–1493.

Allen, Jane E. 2012. "Early Burdens: Eldercare Falls on Young Shoulders." ABC News, May 4. Retrieved May 7, 2013 (http://www.abcnews.go.com).

Allen, Katherine R. 1997. "Lesbian and Gay Families." pp. 196–218 in *Contemporary Parenting: Challenges and Issues,* edited by Terry Arendell. Thousand Oaks, CA: Sage.

Allen, Katherine R. 2016. "Feminist Theory in Family Studies: History, Reflection, and Critique." *Journal of Family Theory & Review* 8(2):207–224.

Allen, Kathleen R., R. Blieszner, and K. A.Roberto. 2011. "Perspectives on Extended Family and Fictive Kin in the Later Years Strategies and Meanings of Kin Reinterpretation." *Journal of Family Issues* 32(9):1156–1177.

Allen, Katherine R., Rosemary Bleiszner, Karen A. Roberto, Elizabeth B. Farnsworth, and Karen L. Wilcox. 1999. "Older Adults and Their Children: Family Patterns of Structural Diversity." *Family Relations* 48(2):151–157.

Allen, Katherine R., Christine E. Kaestle, and Abbie E. Goldberg. 2011. "More than Just a Punctuation Mark: How Boys and Young Men Learn about Menstruation." *Journal of Family Issues* 32(2):129–156.

Allen, Katherine R., and Erin S. Lavender-Stott. 2015. "Family Contexts of Informal Sex Education: Young Men's Perceptions of First Sexual Images," *Family Relations* 64:393–406.

Allen, Katherine, and Ana Jaramillo-Sierra. 2015. "Feminist Theory and Research on Family Relationships: Pluralism and Complexity." *Sex Roles* 73(3/4):93–99.

Allen, Kimberly, and Nichole L. Huff. 2014. "Family Coaching: An Emerging Family Science Field." *Family Relations* 63:569–582.

Allendorf, Keera, and Roshan K. Pandian. 2016. "The Decline of Arranged Marriage? Marital Change and Continuity in India. "*Population and Development Review* 42:435.

Allred, Colette. 2019a. *Age Variation in the Divorce Rate, 1990 & 2017.* Family Profiles, FP-19-13. Bowling Green, OH: National Center for Family & Marriage Research. Retrieved February 5, 2020 (https://www.bgsu.edu/content/dam/BGSU/college-of-arts-and-sciences/NCFMR/documents/FP/fp-19-13-age-var-div.pdf).

Allred, Collette A. (2018). "Attitudes on Women's Roles in the Home." Family Profiles, FP-18-10 Bowling Green, OH: National Center for Family & Marriage Research.

Allred, Collette A. (2019). "High School Seniors' Expectations to Marry, 2017." Family Profiles, FP-19-11. Bowling Green, OH: National Center for Family & Marriage Research.

Allred, Colette. 2019b. *Divorce Rate in the U.S.: Geographic Variation, 2018.* Family Profiles, FP—19-23. Bowling Green, OH: National Center for Family & Marriage Research. Retrieved February 5, 2020 (https://www.bgsu.edu/content/dam/BGSU/college-of-arts-and-sciences/NCFMR/documents/FP/fp-19-23-divorce-rate-geo-var-2018.pdf).

Allred, Colette. 2019c. *Gray Divorce Rate in the U.S.: Geographic Variation, 2017.* Family Profiles, FP—19-20. Bowling Green, OH: National Center for Family & Marriage Research. Retrieved February 5, 2020. (https://www.bgsu.edu/ncfmr/resources/data/family-profiles/allred-gray-divorce-rate-geo-var-2017-fp-19-20.html).

Allred, C. A. (2018). *Marriage: More Than a Century of Change, 1900–2016.* Family Profiles, FP-18-17. Bowling Green, OH: National Center for Family & Marriage Research. Retrieved March 2, 2020 (https://doi.org/10.25035/ncfmr/fp-18-17).

Altintas, Evrim, and Oriel Sullivan. 2016. "Fifty Years of Change Updated: Cross-national Gender Convergence in Housework." *Demographic Research* 35:455–469.

Alman, Isadora. 2018. "Who Has the Power in Your Relationship?" October 20. Retrieved February 5, 2020 (https://www.psychologytoday.com/us/blog/sex-sociability/201810/who-has-the-power-in-your-relationship

Alternatives to Marriage Project 2012 (http://www.unmarried.org/).

Altman, Barbara M., and Debra L. Blackwell. 2014. "Disability in U.S. Households, 2000–2010." *Family Relations* 63(1):20–38.

Altman, Irwin, and Dalmas A. Taylor. 1973. *Social Penetration: The Development of Interpersonal Relations.* New York: Holt, Rinehart & Winston.

Altman, Lawrence K. 2004. "Study Finds That Teenage Virginity Pledges Are Rarely Kept." *The New York Times,* March 10.

Alvarez, Jaime F. 2010. "Columbian Family Farmers' Adaptations to New Conditions in the World Coffee Market." *Latin American Perspectives* 37(2):93–110.

Alvarez, Lizette. 2009. "G.I. Jane Stealthily Breaks the Combat Barrier." *The New York Times,* August 16.

Alvarez, Michelle. 2009. "Exclusive AARP Bulletin Poll Reveals New Trends in Multigenerational Housing." March 3. AARP Press Center. Retrieved November 21, 2009 (www.aarp.org).

Amato, Paul R. 1993. "Children's Adjustment to Divorce: Theories, Hypotheses, and Empirical Support." *Journal of Marriage and Family* 55(1): 23–28.

Amato, Paul R., and Alan Booth. 1997. *A Generation at Risk: Growing Up in an Era of Family Upheaval.* Cambridge, MA: Harvard University Press.

Amato, Paul R., Alan Booth, David R. Johnson, and S. J. Rogers. 2007. *Alone Together: How Marriage in America is Changing.* Cambridge, MA: Harvard University Press.

Amato, Paul R., and Jacob E. Cheadle. 2008. "Parental Divorce, Marital Conflict and Children's Behavior Problems: A Comparison of Adopted and Biological Children." *Social Forces* 86(3):1139–1161.

Amato, Paul R., and Bryndl Hohmann-Marriott. 2007. "A Comparison of High- and Low-Distress Marriages That End in Divorce." *Journal of Marriage and Family* 69:621–638.

Amato, Paul R., David R. Johnson, Alan Booth, and Stacy J. Rogers. 2003. "Continuity and

Change in Marital Quality between 1980 and 2000." *Journal of Marriage and Family* 65(1):1–22.

Amato, Paul R., and Jennifer B. Kane. 2011. "Life Course Pathways and the Psychosocial Adjustment of Young Adult Women." *Journal of Marriage and Family* 73(1):279–295.

Amato, Paul R., Jennifer B. Kane, and Spencer James. 2011. "Reconsidering the 'Good Divorce.'" *Family Relations* 60:511–524.

Amato, Paul R., Valarie King, and Maggie L. Thorsen. 2016. "Parent–Child Relationships in Stepfather Families and Adolescent Adjustment: A Latent Class Analysis." *Journal of Marriage and Family* 78:482–497.

Amato, Paul R., Nancy S. Landale, Tara C. Havasevich-Brooks, and Alan Booth. 2008. "Precursors of Young Women's Family Formation Pathways." *Journal of Marriage and Family* 70:1271–1286.

Amato, Paul R., Catherine E. Meyers, and Robert E. Emery. 2009. "Changes in Nonresident Father-Child Contact from 1976 to 2002." *Family Relations* 58(1):41–53.

Ambert, A. M. 1989. *Ex-Spouses and New Spouses: A Study of Relationships.* Greenwich, CT: Jai Press.

American Academy of Pediatrics. 2013. "American Academy of Pediatrics Support Same Gender Civil Marriage." Press Release, March 21. Retrieved March 21, 2013 (http://www.aap.org/en-us/about -the-aap/aap-press-room/Pages/American -Academy-of-Pediatrics -Supports-Same -Gender-Civil-Marriage.aspx).

American Association for the Advancement of Science. 2019. "Texas A&M to Lead NSF Study of Sexual Harassment in STEM Fields." Retrieved October 4, 2019 (https://www.eurekalert.org/pub_releases/2019-02/tau-tat020719.php).

American Association of University Women. 2019. "Barriers and Bias: The Status of Women in Leadership." Retrieved October 4, 2019 (https://www.aauw.org/files/2017/03/barriersbias-one-pager-nsa.pdf)

American College of Obstetricians and Gynecologists. 2012. "Committee Opinion: Intimate Partner Violence." February. Number 518.

American College of Pediatricians. 2007. "Corporal Punishment: A Scientific Review of Its Use in Discipline." Retrieved from http://www.acpeds.org.

———. 2013. "Guidelines for Parental Use of Disciplinary Spanking." Retrieved from http://www.BestforChildren.org.

American College of Pediatricians. 2018. "Press Release: Spanking: A Valid Option for Parents." November 7. Retrieved December 9, 2019. (www.acpeds.org).

American Immigration Council. 2018. Retrieved December 10, 2019. (https://www.americanimmigrationcouncil.org/research/us-citizen-children-impacted-immigration-enforcement

American Moslem Society. 2010. www.masjiddearborn.org.

American Principles Project. 2015. "Statement Calling for Constitutional Resistance to *Obergefell v. Hodges.*" October 8. Retrieved November 3, 2015 from http://www.nytimes.com.

American Psychological Association 2005. "The Impact of Abortion on Women: What Does

the Psychological Research Say?" APA Briefing Paper, January 31. Washington, DC: American Psychological Association. Retrieved November 10, 2006 (www.apa .org).

———. 2007. "Answers to Your Questions about Sexual Orientation and Homosexuality." Washington, DC: American Psychological Association. Retrieved June 12, 2007 (www.apa.org).

———. 2013. *Stress in America*, February 22. Washington, DC: American Psychological Association.

———. 2015. "Sexual Orientation and Homosexuality." Retrieved December 3, 2015 (http://www.apa.org/helpcenter /sexual-orientation.aspx).

American Psychological Association. 2017. "Older Adults' Health and Age-Related Changes: Reality versus Myth." 2017. Retried January 15, 2019 (https://www .apa.org/pi/aging).

American Psychological Association Task Force on Appropriate Therapeutic Responses to Sexual Orientation. 2009. *Appropriate Therapeutic Responses to Sexual Orientation*. Retrieved on June 24, 2013 (http://www.apa.org/pi/lgbt /resources /therapeutic-response.pdf).

American Society for Reproductive Medicine. 2014. *2012 Assisted Reproductive Technology Fertility Clinic Success Rates Report*. Atlanta: U.S. Department of Health and Human Services.

Amirchaghmaghi, Elham, Faridch Malekzadeh, Mohammad Chehrazi, Zahra Ezabadi, and Shokufeh Sabeti. 2020. "A Comparison of Postpartum Depression in Mothers Conceived by Assisted Reproductive Technology and Those Naturally Conceived." *Int J Fertil Steril* 13, no. 4.

Amiraian, Dana E., and Jeffery Sobal. 2009. "Dating and Eating. Beliefs about Foods among University Students." *Appetite* 53:226–232.

Andersen, Julie Donner. 2002. "His Kids: Becoming a W.O.W. Stepmother." SelfGrowth.com (www.selfgrowth.com /articles/Andersen3.html).

Anderson, Jenny. 2018. "Lessons from the Overworked Mom Who Wants to Pay Someone to Teach Her Family How to Do Chores." *Quartz*. September 21. Retrieved January 30, 2020 (https://qz.com/1396842/lessons-from-the -overworked-mom-who-wants-to-pay -someone-to-teach-her-family-how-to-do -chores/).

Anderson, Lydia R. 2016 " FP-16-13 High School Seniors' Attitudes on Cohabitation as a Testing Ground for Marriage." Retrieved November 19, 2019

Anderson, Lydia R. 2016. *High School Seniors' Expectations to Marry*. Family Profiles, FP-16-14. Bowling Green, OH: National Center for Family & Marriage Research.

(https://www.bgsu.edu/ncfmr/resources /data/family-profiles/anderson-hs-seniors -attitudes-cohab-test-marriage-fp-16-13 .html).

Anderson, Lydia R. 2016. *High School Seniors' Expectations to Marry*. Family Profiles, FP-16-14. Bowling Green, OH: National Center for Family & Marriage Research.

Anderson, Monica, and Andrew Perrin. 2018. "Nearly one-in-five teens can't always finish their homework because of the digital divide." *Facttank: News in the Numbers*. October 26. Pew Research Organization. Retrieved January 30, 2019 (www.pewresearch.org).

Anderson, Monica, and Andrew Perrin. 2017. "Tech Adoption Climbs among Older Adults." May 17. Pew Research Organization. Retrieved February 17, 2019 (www.pewresearch.org).

Anderson, Elijah. 1999. *The Code of the Street*. New York: Norton.

Anderson, Janna, and Lee Rainie. 2018. "Stories from Experts about the Impact of Digital Life." Pew Research Center. Retrieved January 25, 2020 (file:///C:/Users/stewa/Downloads /PI_2018.07.03_Stories-About-Digital-Life _FINAL-with-table%20(1).pdf).

Anderson, Keith A., Noelle L. Fields, and Lynn A. Dobb. 2013. "Caregiving and Early Life Trauma: Exploring the Experiences of Family Caregivers to Aging Holocaust Survivors." *Family Relations* 62(2):366–377.

Anderson, Kristin J. 2015. *Feminism Is Now: Fighting Modern Misogyny and the Myth of the Post-Feminist Era*. New York: Oxford University Press.

Anderson, Kristin L. 2010. "Conflict, Power, and Violence in Families." *Journal of Marriage and Family* 72(3):726–742.

———. 2013. "Why Do We Fail to Ask 'Why' about Gender and Intimate Partner Violence?" *Journal of Marriage and Family* 75(2):314–318.

Anderson, Michael A., Paulette Marie Gillig, Marilyn Sitaker, Kathy McCloskey, Katherine Malloy, and Nancy Grigsby. 2003. "'Why Doesn't She Just Leave?' A Descriptive Study of Victim Reported Impediments to Her Safety." *Journal of Family Violence* 18:151–155.

Anderson, Sarah E., and Aviva Must. 2005. "Interpreting the Continued Decline in the Average Age at Menarche: Results from Two Nationally Representative Surveys of U.S. Girls Studied 10 Years Apart." *Journal of Pediatrics* 147(6):753–760.

Anderson, Stephen A., and Ronald M. Sabatelli. 2007. *Family Interaction: A Multigenerational Developmental Perspective*, 4th edition. Boston and New York: Pearson.

Andersson, G. (2002). "Children's Experience of Family Disruption and Family Formation: Evidence from 16 FFS Countries." *Demographic Research* 7:343–364.

Andersson, Gunnar, and Dimiter Philipov. 2002. Life-Table Representations of Family Dynamics in Sweden, Hungary, and 14 Other FFS Countries: A Project Description of Demographic Behavior. *Demographic Research* 7:67–144.

Andersson, Gunnar, Elizabeth Thomson, and Aija Duntava. 2017. "Life-Table Representations of Family Dynamics in the 21st Century." *Demographic Research* 37:1081–1230.

Anetzberger, Georgia, Jill Korbin, and Craig Austin. 1994. "Alcoholism and Elder Abuse." *Journal of Interpersonal Violence* 9(2):184–193.

Angelo, Megan. 2010. "Fortune 500 Women CEOs." *Fortune*, April 23.

Angier, Natalie. 2013. "The Changing American Family." *The New York Times*, November 25. Retrieved May 6, 2014 (http://www.nytimes.com).

Animal Legal & Historical Center. 2019. West's Alaska Statutes Annotated. Title 25. Marital and Domestic Relations. Chapter 24. Divorce and Dissolution of Marriage. Article 1. Divorce and Annulment. § 25.24.160. Judgment. Retrieved February 12, 2020 (https://www.animallaw.info /statute/ak-divorce-%C2%A7-2524160-judgment).

Annie E. Casey Foundation 2019. 2019 *Kids Count Data Book*. Retrieved December 20, 2019 (https://www.aecf.org/m/resourcedoc/aecf -2019kidscountdatabook-2019.pdf).

Appleby, Julie. 2012. "Many Businesses Offer Health Benefits to Same-Sex Couples Ahead of Laws." *The Rundown: A Blog of News and Insight*. PBS News, May 14. Retrieved September 3, 2012 (http://www .pbs.org/newshour).

Arditti, Joyce A. 2003. "Incarceration Is a Major Source of Family Stress." *Family Focus* (June):F15–F17. Minneapolis: National Council on Family Relations.

Arditti, Joyce A., and Timothy Z. Keith. 1993. "Visitation Frequency, Child Support Payment, and the Father–Child Relationship Postdivorce." *Journal of Marriage and Family* 55(3):699–712.

Arditti, Joyce A., Jennifer Lambert-Shute, and Karen Joest. 2003. "Saturday Morning at the Jail: Implications of Incarceration for Families and Children." *Family Relations* 52(3):195–204.

Arbel, Reout, Aubrey J. Rodriguez, and Gayla Margolin. 2015. "Cortisol Reactions During Family Conflict Discussions: Influences of Wives' and Husbands' Exposure to Family-of-Origin Aggression." *Psychology of Violence* 10:1037–1049.

Arendell, Terry. 1997. "Divorce and Remarriage." pp. 154–195 in *Contemporary Parenting: Challenges and Issues*, edited by Terry Arendell. Thousand Oaks, CA: Sage.

———. 2000. "Conceiving and Investigating Motherhood: The Decade's Scholarship." *Journal of Marriage and Family* 62(4):1192–1207.

Arias, Elizabeth and Jaiquan Xu. 2019. "United States Life Tables, 2017." *National Vital Statistics Reports* 68(7). June 24. Hyattsville MD: National Center for Health Statistics.

Ariès, Phillipe. 1962. *Centuries of Childhood: A Social History of Family Life*. New York: Knopf.

Arif, Nadia, and Iram Fatima. 2015. "Marital Satisfaction in Different Types of Marriage." *Pakistan Journal of Social and Clinical Psychology* 13:36.

Armstrong, Larry. 2003. "Your Mouse Knows Where Your Car Is." *Business Week* 16.

Armstrong, Elizabeth A., Paula England, and Alison C. K. Fogarty. 2012. "Accounting for Women's Orgasm and Sexual Enjoyment in College Hookups and Relationships." *American Sociological Review* 72(3):435–462.

Arocho, Rachel and Claire M. Kamp Dush. 2016. "Anticipating the 'Ball and Chain'? Reciprocal Associations between Marital Expectations and Delinquency." *Journal of Marriage and Family* 78 (5):1371–1781.

Arnett, Jeffrey Jensen. 2000. "Emerging Adulthood: A Theory of Development from the Late Teens through the Twenties." *American Psychologist* 55(5):469–480. Retrieved March 28, 2003 (PsycARTICLES 0003-006X).

———. 2004. *Emerging Adulthood: The Winding Road from the Late Teens through the Twenties*. London: Oxford University Press.

———. 2014. *Emerging Adulthood: The Winding Road from the Late Teens through the Twenties.* Oxford, England: Oxford University Press.

Arnott, Teresa, and Julie Matthaei. 2007. "Race, Class, and Gender and Women's Works." pp. 283–292 in *Race, Class, and Gender*, 6th ed., edited by Margaret Andersen and Patricia Hill Collins. Belmont, CA: Wadsworth.

Aron, Arthur, Christine C. Norman, Elaine N. Aron, and Gary Lewandowski. 2002. "Shared Participation in self-expanding Activities: Positive Effects on Experienced Marital Quality." pp. 177–194 in *Understanding Marriage: Developments in the Study of Couple Interaction*, edited by Patricia Noller and Judith A. Feeney. Cambridge University Press, 2002.

Aronson, Pamela. 2003. "Feminists or 'Postfeminists'? Young Women's Attitudes toward Feminism and Gender Relations." *Gender and Society* 17:903–922.

Arpino, Bruno, and Valeria Bordone. 2014. "Does Grandparenting Pay Off? The Effect of Child Care on Grandparents' Cognitive Functioning." *Journal of Marriage and Family* 76(2):337–351.

Artis, Julie E. 2004. "Judging the Best Interests of the Child: Judges' Accounts of the Tender Years Doctrine." *Law and Society Review* 38(4):769–806.

Artis, Julie E., and Eliza K. Pavalko. 2003. "Explaining the Decline in Women's Household Labor: Individual Change and Cohort Differences." *Journal of Marriage and Family* 65:746–761.

Artz, Benjamin, Amanda H. Goodall, and Andrew J. Oswald. 2018. "Do Women Ask?" *Industrial Relations: A Journal of Economy and Society* 57:611–636.

Asay, Sylvia, John DeFrain, Marcee Metzger, and Bob Moyer. 2014. "Intimate Partner Violence Worldwide: A Strengths-Based Approach." *Family Focus.* Issue FF61: F1-F4.

Asencio, Marysol. 2009. "Migrant Puerto Rican Lesbians Negotiating Gender, Spituality, and Ethnonationality." *NWSA Journal* 21(3):2–23.

Asexual Visibility and Education Network (AVEN). "Overview." 2015. Retrieved December 3, 2015 (http://www.asexuality.org/home/?q=overview.html).

Astone, N. M., and S. McLanahan. 1991. "Family Structure, Parental Practices, and High School Completion." *American Sociological Review* 56(3):309–320.

Ashwin, Sarah, and Olga Isupova. 2014. "'Behind Every Great Man . . . ': The Male Marriage Wage Premium Examined Qualitatively." *Journal of Marriage and Family* 76(1):37–55.

Associated Press. 2019. "Hundreds of Migrant Children Held in Unsafe Conditions at Texas Facility, Lawyers Warn." June 21. *The Guardian.* Retrieved November 5, 2019 (https://www.theguardian.com/us-news/2019/jun/21/child-migrant-facilities-shocking-conditions-revealed).

Athenstaedt, Ursula, Gerold Mikula, and Cornelia Bredt. 2009. "Gender Role Self-Concept and Leisure Activities of Adolescents." *Sex Roles* 60(5/6):399–409.

Atkin, B. (2008). "Legal Structures and Re-formed Families: The New Zealand Example." pp. 522–544 in *The International Handbook of Stepfamilies: Policy and Practice in Legal, Research, and Clinical Environments*, edited by Jan Pryor. Hoboken, NJ: Wiley.

Atkins David C., and Furrow, James. 2009. "Infidelity Is on the Rise: But for Whom and Why?" Paper presented at the annual meeting of the Association for Behavioral and Cognitive Therapies, Orlando, FL.

Attar-Schwartz, Shalhevet, and Esme Fuller-Thomson. 2017. "Adolescents' Closeness to Paternal Grandmothers in the Face of Parents' Divorce." *Children and Youth Services Review* 77:118–126.

Aughinbaugh, Alison, Omar Robles, and Hugette Sun. 2013. "Marriage and Divorce: Patterns by Gender, Race, and Educational Attainment." *Monthly Labor Review* 136:1.

Augustine, Jennifer March, Kate C. Prickett, and Rachel Tolbert Kimbro. 2017. "Health-Related Parenting among U.S. Families and Young Children's Physical Health." *Journal of Marriage and Family* 79 (3): 816–832.

Augustine, Jennifer March, and R. Kelly Raley. 2012. "Multigenerational Households and the School Readiness of Children Born to Unmarried Mothers." *Journal of Family Issues* 34(4):431–459.

Aune, Kristin. 2015. "Feminist Spirituality As Lived Religion: How UK Feminists Forge Religio-spiritual Lives." *Gender and Society* 29(1):122–145.

Austin, Jennifer L., and Mariana K. Falconier. 2012. "Spirituality and Common Dyadic Coping: Protective Factors from Psychological Aggression in Latino Immigrant Couples." *Journal of Family Issues* 34(3):323–346.

Avellar, Sarah, and Pamela J. Smock. 2003. "Has the Price of Motherhood Declined Over Time? A Cross-Cohort Comparison of the Motherhood Wage Penalty." *Journal of Marriage and Family* 65:597–607.

Avishai, Orit, Afshan Jafar, and Rachel Rinaldo. 2015. "A Gender Lens on Religion." *Gender and Society* 29(1):5–25.

Azad, Sonia. 2018. "Same-Sex Couple Carries Same 'Miracle' Baby in What May Be Fertility World First." *USA Today Network.* October 29. Retrieved August 16, 2019 (www.usatoday.com).

Babbie, Earl. 2007. *The Practice of Social Research.* 11th ed. Belmont, CA: Wadsworth/Cengage.

———. 2014. *The Practice of Social Research*, 14th ed. Belmont, CA: Wadsworth/Cengage.

Babbitt, Charles E., and Harold J. Burbach. 1990. "A Comparison of Self-Orientation among College Students across the 1960s, 1970s and 1980s." *Youth and Society* 21(4):472–482.

Baca Zinn, Maxine, Pierette Hondagneu-Sotelo, and Michael A. Messner. 2004. "Gender through the Prism of Difference." pp. 166–174 in *Race, Class, and Gender*, 5th ed., edited by Margaret L. Andersen and Patricia Hill Collins. Belmont, CA: Wadsworth.

———. 2007. "Sex and Gender through the Prism of Difference." pp. 147–155 in *Race, Class, and Gender: An Anthology*, 6th ed, edited by Margaret L. Andersen and Patricia Hill Collins. Belmont, CA: Wadsworth.

———. 2010. *Gender Trough the Prism of Difference*, 4th ed. New York: Oxford University Press.

Baca Zinn, Maxine, and Barbara Wells. 2007. "Diversity within Latino Families: New Lessons for Family Social Science." pp. 422–447 in *Family in Transition,* 14th ed., edited by Arlene S. Skolnick and Jerome H. Skolnick. Boston: Allyn & Bacon.

Bachrach, Christine, Patricia F. Adams, Soledad Sambrano, and Kathryn A. London. 1990. "Adoption in the 1980s." *Advance Data*, No. 181. Hyattsville, MD: U.S. National Center for Health Statistics, January 5.

Backstrom, Laura, Elizabeth A. Armstrong, and Jennifer Puentes. 2012. "Women's Negotiation of Cunnilingus in College Hookups and Relationships." *Journal of Sex Research* 49(1):1–12.

Badgett, M. V. Lee, and Jody L. Herman. 2011. *Patterns of Relationship Recognition by Same-Sex Couples in the United States.* Retrieved April 14, 2013 (http://williamsinstitute.law.ucla.edu/wp-content/uploads/Badgett-Herman-Marriage-Dissolution-Nov-2011.pdf).

Baer, Judith C., and Mark F. Schmitz. 2007. "Ethnic Differences in Trajectories of Family Cohesion for Mexican American and Non-Hispanic White Adolescents." *Journal of Youth and Adolescence* 36:583–592.

Bagley, Nate. 2019. "8 Important Questions to Ask during Your Relationship." *Blog:Growth Marriage.* Retrieved March 1, 2020 (https://podcasts.apple.com/us/podcast/growth-marriage).

Baham, Melinda E., Amy A. Weimer, Sanford L. Braver, and William V. Fabricius. 2008. "Sibling Relationships in Blended Families." pp. 175–207 in *The International Handbook of Stepfamilies*, edited by J. Pryor. Hoboken, NJ: John Wiley & Sons.

Bailey, Jo Daugherty, and Dawn McCarty. 2009. "Assessing Empowerment in Divorce Mediation." *Negotiation Journal* 25(3):327–336.

Bailey, Sandra J., Bethany L. Letiecq, and Mara Vannatta. 2011. "So Much for the Empty Nest: The Transitions of Grandparents Rearing Grandchildren." *Family Focus* FF49 (Summer): F17–F18.

Bailey, Sarah Pulliam. 2017. "Poll Shows Dramatic Generational Divide in White Evangelical Attitudes on Gay Marriage." Retrieved October 17, 2019 (https://www.washingtonpost.com/news/acts-of-faith/wp/2017/06/27/there-is-now-a-dramatic-generational-divide-over-white-evangelical-attitudes-on-gay-marriage/).

Bainbridge, Daryl, Paul Krueger, Lynne Lohfeld, and Kevin Brazil. 2009. "Stress Processes in Caring for an End-of-Life Family Member: Application of a Theoretical Model." *Aging and Mental Health* 13(4):537–545.

Baio, J., L. Wiggins, D.L. Christensen et al. 2018. "Prevalence of Autism Spectrum Disorder among Children Aged 8 Years." Retrieved March 5, 2020 (www.cdc.gov/mmwr).

Baitar, Rachid, Ann Buysse, Ruben Brondeel, Jan De Mol, and Peter Rober. 2012. "Post-Divorce Well-Being in Flanders: Facilitative Professionals and Quality of Arrangements Matter." *Journal of Family Studies* 18(1):62–75.

Bakalar, Nicholas. 2007. "Optional Caesareans Carry Higher Risks, Study Finds." *The New York Times*, March 27. Retrieved June 26, 2013 (http://www.nytimes.com)

———. 2010. "Premature Birth Rate Drops for 2nd Year." *The New York Times*, May 25.

Baker, Regina S. 2015. "The Changing Association among Marriage, Work, and Child Poverty in the United States, 1974–2010." *Journal of Marriage and Family* 77(5):1166–1178

Balbim, Guilherme M., Isabela G. Marques, Claudia Cortez, Melissa Magallanes, Judith Rocha, and David X. Marquez. 2019. "Coping Strategies Utilized by Middle-Aged and Older Latino Caregivers of Loved Ones with Alzheimer's Disease and Related Dementia." *Journal of Cross-Cultural Gerontology* 34(4):355–371.

Baldwin, Aleta, Debby Herbenick, Vanessa R. Schick, Brenda Light, Brian Dodge, Crystal A. Jackson, and J. Dennis Fortenberry. 2019. "Sexual Satisfaction in Monogamous, Nonmonogamous, and Unpartnered Sexual Minority Women in the U.S." *Journal of Bisexuality* 19:103-119.

Baldwin, Katherine. 2019. "I Feel Grief and Relief that I've Never Had Children. Other Women Must Share This." Retrieved December 19, 2019 (https://www.theguardian.com/commentisfree/2019/apr/29/grief-relief-children-women-choice-motherhood-ambivalence).

Balistreri, Kelly Stamper, Kara Joyner, and Grace Kao. 2017. "Trading Youth for Citizenship? The Spousal Age Gap in Cross-Border Marriages." *Population and Development Review* 43:443–466.

Ball, Derek, and Peter Kivisto. 2006. "Couples Facing Divorce." pp. 145–161 in *Couples, Kids, and Family Life*, edited by Jaber F. Gubrium and James A. Holstein. New York: Oxford University Press.

Ball, Victoria, and Kristin A. Moore. 2008. "What Works For Adolescent Reproductive Health: Lessons from Experimental Evaluations of Programs and Interventions." Fact Sheet No. 2008-20. Bethesda, MD: Child Trends.

Balsam, Kimberly F., Theodore P. Beauchaine, Esther D. Rothblum, and Sondra E. Solomon. 2008. "Three-Year Follow-Up of Same-Sex Couples Who Had Civil Unions in Vermont, Same-Sex Couples Not in Civil Unions, and Heterosexual Married Couples." *Developmental Psychology* 44(1):102–116.

Balsam, Kimberly F., Yamile Molina, Blair Beadnell, Jane Simoni, and Karina Walters. 2011. "Measuring Multiple Minority Stress: The LGBT People of Color Microaggressions Scale." *Cultural Diversity and Ethnic Minority Psychology* 17(2):163–174.

Baltzly, Vaughn Bryan. 2012. "Same-Sex Marriage, Polygamy, and Disestablishment." *Social Theory and Practice* 38(2):333–354.

Bandura, Albert, and Richard H. Walters. 1963. *Social Learning and Personality Development*. New York: Holt, Rinehart & Winston.

Banks, Ralph Richard. 2011. *Is Marriage for White People? How the African American Marriage Decline Affects Everyone*. New York: Penguin.

Banyard, Victoria L., Valerie J. Edwards, and Kathleen Kendall-Tackett, eds. 2009. *Trauma and Physical Health: Understanding the Effects of Extreme Stress and of Psychological Harm*. New York: Routledge.

Barash, Susan Shapiro. 2000. *Second Wives: The Pitfalls and Rewards of Marrying Widowers and Divorced Men*. Far Hills, NJ: New Horizon.

Barber, Bonnie L., and David H. Demo. 2006. "The Kids Are Alright (at Least Most of Them): Links Between Divorce and Dissolution and Child Well-Being." pp. 289–311 in *Handbook of Divorce and Relationship Dissolution*, edited by Mark A. Fine and John H. Harvey. Mahwah, NJ: Erlbaum.

Barber, Jennifer S., Jennifer Eckerman Yarger, and Heather H. Gatny. 2015. "Black-White Differences in Attitudes Related to Pregnancy among Young Women." *Demography* 52.1–30.

Barbosa, Peter, Hector Torres, Marc Anthony Silva, and Nosh Khan. 2010. "Agapé Christian Reconciliation Conversations: Exploring the Intersections of Culture, Religiousness, and Homosexual Identity in Latino and European Americans." *Journal of Homosexuality* 57(1):98–116.

Barker Meg-John. 2018. "Forward." pp. xvi in *Researching Sex and Sexualities*, edited by Charlotte Morris, Paul Boyce, Andrea Cornwall, Hannah Frith, Laura Harvey, and Yingying Huang. London, UK: Zed Books.

Barlinska, Julia, Anna Szuster, and Mikolaj Winiewski. 2012. "Cyberbullying among Adolescent Bystanders: Role of the Communication Medium, Form of Violence, and Empathy." *Journal of Community and Applied Social Psychology* 23:37–51.

Barnes, Medora. 2013. "Having a First Versus a Second Child: Comparing Women's Maternity Leave Choices and Concerns." *Journal of Family Issues* 34(1):85–112.

Barnes, Norine R. 2006. "Children Need Guidance. . . . Developmentally Appropriate Interaction and Discipline." Retrieved December 30, 2009 (http://www.parentingweb.com).

Barnes, Richard E. 2009. *Estate Planning for Blended Families: Providing for Your Spouse and Children in a Second Marriage*. Berkeley, CA: Nolo Press.

Barnett, Amanda E. 2013. "Pathways of Adult Children Providing Care to Older Parents." *Journal of Marriage and Family* 75(1):178–190.

Barnett, Melissa A. 2008. "Mother and Grandmother Parenting in Low-Income Three-Generation Rural Households." *Journal of Marriage and Family* 70(5):1241–1257.

Barnett, Rosalind C., Karen C. Gareis, Laura Sabattini, and Nancy M. Carter. 2009. "Parental Concerns about After-School Time: Antecedents and Correlates among Dual-Earner Parents." *Journal of Family Issues* 31(5):606–625.

Barnett, Rosalind C., and Caryl Rivers. 2014. "Advice for Ambitious Women Is Needed, Not Elitist." *We-News*, April 30. Retrieved February 3, 2015 (http://womensnews.org).

Barone, Emily. 2019. "Many American Men Have a Skewed View of Gender Inequality, TIME Poll Finds." Retrieved September 29, 2019 (https://time.com/5667397/gender-equality-opinions/).

Barr, Ashley B., Ronald L. Simons, and Eric A. Stewart 2013. "The Code of the Street and Romantic Relationships." *Personal Relationships* 20:84–106.

Barr, Ashley Brooke, and Ronald L. Simons. 2018. "Marital Beliefs Among African American Emerging Adults: The Roles of Community Context, Family Background, and Relationship Experiences." *Journal of Family Issues* 39:352–382.

Barrionuevo, Alexei. 2011. "Upwardly Mobile Nannies Move into the Brazilian Middle Class." *The New York Times*, May 20:A7.

Barron, David J. 2010. "Whether the Criminal Provisions of the Violence Against Women Act Apply to Otherwise Covered Conduct when the Offender and Victim are the Same Sex." *Memorandum Opinion for the Acting Deputy Attorney General*, April 27. Washington, DC: United States Department of Justice Office of Legal Counsel.

Barrow, Karen. 2010. "Difference Is the Norm on These Dating Sites." Retrieved July 14, 2013 (http://www.nytimes.com/2010/12/28/health/28dating.html).

Barry, Ellen. 2018. "In Sweden's Preschools, Boys Learn to Dance and Girls Learn to Yell." Retrieved September 30, 2019 (https://www.nytimes.com/2018/03/24/world/europe/sweden-gender-neutral-preschools.html).

Barth, Richard P. and Marianne Berry. 1988. *Adoption and Disruption: Rates, Risks, and Responses*. New York: Aldine.

Bartkowski, John P., and Jen'nan Ghazal Read. 2003. "Veiled Submission: Gender, Power, and Identity among Evangelical and Muslim Women in the United States." *Qualitative Sociology* 26(1):71–92.

Bartoli, Angela M., and M. Diane Clark. 2006. "The Dating Game: Similarities and Differences in Dating Scripts among College Students." *Sexuality and Culture* 10:54–80.

Barton, Allen W., and Robert C. Bishop. 2014. "Paradigms, Processes, and Values in Family Research." *Journal of Family Theory and Review* 6(3):241–256.

Barton, Allen W., Tera R. Hurt, Ted G. Futris, Kameron F. Sheats, Stacey E. McElroy, and Antoinette M. Landor. 2017. "Being Committed: Conceptualizations of Romantic Relationship Commitment among Low-Income African American Adolescents." *Journal of Black Psychology* 43:111–134.

Bass, Brenda L., Adam B. Butler, Joseph G. Grzywacz, and Kirsten D. Linney. 2009. "Do Job Demands Undermine Parenting? A Daily Analysis of Spillover and Crossover Effects." *Family Relations* 58(April):201–215.

Bastaits, Kim, Koen Ponnet, and Dimitri Mortelmans. 2012. "Parenting of Divorced Fathers and the Association with Children's Self-Esteem." *Journal of Youth and Adolescence* 41:1643–1656.

Batson, Christie D., Zhenchao Qian, and Daniel T. Lichter. 2006. "Interracial and Intraracial Patterns of Mate Selection among America's Diverse Black Populations." *Journal of Marriage and Family* 68(3):658–672.

"Battered Woman Syndrome." 2017. Healthline. July 5. Retrieved February 15, 2020 (www.healthline.com).

Baucom, Donald H., Kristina C. Gordon, Douglas K. Snyder, David C. Atkins, and Andrew Christensen. 2006. "Treating Affair Couples: Clinical Considerations and Initial Findings." *Journal of Cognitive Psychotherapy* 20:375–392.

Baucom, Donald H., Douglas K. Snyder, and Kristina Coop Gordon. 2009. *Helping Couples Get Past the Affair: A Clinician's Guide*. New York: The Guilford Press.

Bauer, Robin. 2014. *Queer BDSM Intimacies: Critical Consent and Pushing Boundaries.* New York: Palgrave Macmillan.

Baum, Angela C., Sedahlia Jasper Crase, and Kirsten Lee Crase. 2001. "Influences on the Decision to Become or Not Become a Foster Parent." *Families in Society* 82(2):202–221.

Baumrind, Diana. 1978. "Parental Disciplinary Patterns and Social Competence in Children." *Youth and Society* 9:239–276.

Baumrind, Diana, Robert E. Larzelere, and Philip A. Cowan. 2002. "Ordinary Physical Punishment: Is it Harmful? Comment on Gershoff." *Psychological Bulletin* 128(4):580–589.

Bauserman, Robert. 2002. "Child Adjustment in Joint-Custody versus Sole Custody Arrangements: A Meta-Analytic Review." *Journal of Family Psychology* 16:91–102.

———. 2012. "A Meta-Analysis of Parental Satisfaction, Adjustment, and Conflict in Joint Custody and Sole Custody Following Divorce." *Journal of Divorce and Remarriage* 53(6):464–488.

Baxter, Christine C. 1989. "Investigating Stigma as Stress in Social Interactions of Parents." *Journal of Intellectual Disability Research* 33(6):455–466.

Baxter, Janeen, Belinda Hewitt, and Michele Haynes. 2008. "Life Course Transitions and Housework: Marriage, Parenthood, and Time on Housework." *Journal of Marriage and Family* 70(2):259–269.

Baxter, Leslie A., Dawn O. Braithwaite, and Leah E. Bryant. 2006. "Types of Communication Triads Perceived by Young-Adult Stepchildren in Established Stepfamilies." *Communication Studies* 57(4):381–400.

Baxter, Leslie A., Dawn O. Braithwaite, Jody K. Kellas, Cassandra LeClaire-Underberg, Emily Lamb Normand, Tracy Routsong, and Matthew Thatcher. 2009. "Empty Ritual: Young-Adult Stepchildren's Perceptions of the Remarriage Ceremony." *Journal of Social and Personal Relationships* 26(4):467–487.

Bay-Cheng, Laina Y. and Nicole M. Fava. 2011. "Young Women's Experiences and Perceptions of Cunnilingus During Adolescence." *Journal of Sex Research* 48(6):531–542.

BBC News. 2018. "India Court Legalises Gay Sex in Landmark Ruling." Retrieved October 16, 2019 (https://www.bbc.com/news/world-asia-india-45429664).

Beaman, Lori G. 2001. "Molly Mormons, Mormon Feminists and Moderates: Religious Diversity and the Latter Day Saints Church." *Sociology of Religion* 62:65–86.

Bearman, Peter. 2008. "Exploring Genetics and Social Structure." *American Journal of Sociology* 114(Supplement):v–x.

Bearman, Peter S., and Hannah Brückner. 2001. "Promising the Future: Virginity Pledges and First Intercourse." *American Journal of Sociology* 106:859–912.

Beaubien, Jason. 2018. "Stay-At-Home Dads Still Struggle with Diapers, Drool, Stigma, and Isolation." *All Things Considered.* National Public Radio. June 17.

Bech-Sørensen, Jens, and Thomas V. Pollet. 2016. "Sex Differences in Mate Preferences: A Replication Study, 20 Years Later. " *Evolutionary Psychological Science* 2:171–176.

Beck, Audrey N., Carey E. Cooper, Sara McLanahan, Jeanne Brooks-Gunn. 2010. "Partnership Transitions and Maternal Parenting." *Journal of Marriage and Family* 72(2):219–233.

Beck, Kelli Dougal, Cassidy Debenham, and Dustin Jones. 2015. "The Influence of Religion on the Partner Selection Strategies of Emerging Adults." *Journal of Family Issues* 36:212–231.

Becker, Arránz O. 2012. "Effects of Similarities of Life Goals, Values, and Personality on Relationship Satisfaction and Stability: Findings from a Two-Wave Panel Study." *Personal Relationships.* Retrieved May 26, 2013 (http://onlinelibrary.wiley.com/doi/10.1111/j.1475-6811.2012.01417.x /abstract).

Becker, Gail. 2014. "Forget Conscious Uncoupling; Try Conscious Unscheduling." *The Huffington Post*, April 7. Retrieved May 8, 2015 (http://www.huffingtonpost.com).

Becker, Gay. 2000. *The Elusive Embryo: How Women and Men Approach New Reproductive Technologies.* Berkeley and Los Angeles: University of California Press.

Beer, William R. 1989. *Strangers in the House: The World of Stepsiblings and Half Siblings.* New Brunswick, NJ: Transaction Books.

Begley, Ann, and Susan Piggott. 2012. "Exploring Moral Distress in Potential Sibling Stem Cell Donors." *Nursing Ethics* 20(2):178–188.

Behnke, Andrew O., Shelley M. MacDermid, Scott L. Coltrane, Ross D. Parke, Sharon Duffy, and Keith F. Widaman. 2008. "Family Cohesion in the Lives of Mexican American and European American Parents." *Journal of Marriage and Family* 70(4):1045–1059.

Beidler, Jeannie Jennings. 2012. "We Are Family: When Elder Abuse, Neglect, and Financial Exploitation Hit Home." *Generations: Journal of the American Society on Aging* 36(3):21–25.

Beins, Bernard. 2008. *Research Methods: A Tool for Life.* Boston: Allyn & Bacon.

Beitin, Ben, Katherine Allen, and Maureen Bekheet. 2010. "A Critical Analysis of Western Perspectives on Families of Arab Descent." *Journal of Family Issues* 31(2):211–233.

Bélanger, Danièle, and Andrea Flynn. 2018. "Gender and Migration: Evidence from Transnational Marriage Migration. " Pp. 183–201 in *International Handbook on Gender and Demographic Processes*, edited by Nancy E. Riley, and Jan Brunson. Dordrecht: Springer.

Belkin, Lisa. 2010. "The Marrying Kind." *The New York Times*, March 22. Retrieved April 23, 2010 (www.nytimes.com).

Belknap, Joanne. 2012. "Types of Intimate Partner Homicides Committed by Women: Self-Defense, Proxy/Retaliation, and Sexual Proprietariness." *Homicide Studies* 16(4):359–374.

Bell, David C. 2012. "Next Steps in Attachment Theory." *Journal of Family Theory and Review* 4(4):275–281.

Bell, Maya. 2003. "More Gays and Lesbians than Ever Are Becoming Parents." Knight Ridder /Tribune News Service, October 1.

Bellafante, Ginia. 2004. "Two Fathers, with One Happy to Stay at Home." *The New York Times*, January 12.

Bellah, Robert N., Richard Madsen, William M. Sullivan, Ann Swidler, and Steven M. Tipton.

1985. *Habits of the Heart: Individualism and Commitment in American Life.* Berkeley and Los Angeles: University of California Press.

Bellamy, Jennifer L., Matthew Thullen, and Sydney Hans. 2015. "Effect of Low-Income Unmarried Fathers' Presence at Birth on Involvement." *Journal of Marriage and Family* 77(3):647–661.

Bellido, Héctor, José Alberto Molina, Anne Solaz, and Elena Stancanelli. 2016. "Do Children of the First Marriage Deter Divorce?" *Economic Modelling* 55:15–31.

Bellis, Rich. 2017. "Here's Everywhere in The U.S. You Can Still Get Fired For Being Gay or Trans." Retrieved September 29, 2019 (https://www.fastcompany.com/40456937/heres-everywhere-in-the-u-s-you-can-still-get-fired-for-being-gay-or-trans)

Belluck, Pam. 2009. "In Turnabout, Children Take Caregiver Role." *The New York Times*, February 23. Retrieved March 16, 2010 (www.nytimes.com).

———. 2011. "Scientific Advances on Contraceptive for Men." *The New York Times*, July 23. Retrieved September 13, 2011 (http://www.nytimes.com).

———. 2014. "'Thinking of Ways to Harm Her': New Findings on Timing and Range of maternal Mental Illness." *The New York Times*, June 15. Retrieved June 23, 2014 (http://www.nytimes.com).

Bellware, Kim. 2015. "As Same-Sex Couples Line Up to Wed, Others Celebrate the Right to Divorce." Retrieved March 31, 2016 (http://www.huffingtonpost.com /2015/07/01 /gay-divorce-new-orleans _n_7707968.html).

Belsky, Jay. 2002. "Quantity Counts: Amount of Child Care and Children's Socioemotional Development." *Developmental and Behavioral Pediatrics* 23:167–170.

Bem, Sandra Lipsitz. 1981. "Gender Schema Theory: A Cognitive Account of Sex Typing." *Psychological Review* 88:354–364.

Ben-Asher, Noa. 2017. "In the Shadow of a Myth: Bargaining for Same-Sex Divorce." Ohio St. LJ 78: 1345.

Bengston, Vern L. 2001. "Beyond the Nuclear Family: The Increasing Importance of Multigenerational Bonds." *Journal of Marriage and Family* 63(1):1–16.

Bengston, Vern L., Timothy J. Biblarz, and Robert E. L. Roberts. 2007. "How Families Still Matter: A Longitudinal Study of Youth in Two Generations." pp. 315–324 in Family in Transition, 14th ed., edited by Arlene S. Skolnick and Jerome H. Skolnick. Boston: Allyn & Bacon.

Benkel, I., H. Wijk, and U. Molander. 2009. "Family and Friends Provide Most Social Support for the Bereaved." *Palliative Medicine* 23(2):141–149.

Benner, Aprile D., and Su Yeong Kim. 2009. "Intergenerational Experiences of Discrimination in Chinese American Families: Influences of Socialization and Stress." *Journal of Marriage and Family* 71(4):862–877.

———. 2010. "Understanding Chinese American Adolescents' Developmental Outcomes: Insights from the Family Stress Model." *Journal of Research on Adolescence* 20(1):1–12.

Bennett, Fran. 2013. "Researching Within-Household Distribution: Overview, Developments, Debates, and Methodological Challenges." *Journal of Marriage and Family* 75(3):582–597.

———. 2015. "Opening Up the Black Box: Researching the Distribution of Resources Within the Household." *National Council on Family Relations Report*, 60:1.

Bennett, Linda A., Steven J. Wolin, and David Reiss. 1988. "Deliberate *Family Process*: A Strategy for Protecting Children of Alcoholics." *British Journal of Addiction* 83:821–829.

Beras, Erika. 2019. "Condom Sales Lag as Millennials Have Less Sex." Retrieved October 24, 2019 (https://www.marketplace.org/2019/05/03/condom-sales-lag-millennials-having-less-sex/)

Berenbaum, Sheri A., Judith E. Owen Blakemore, and Adriene M. Belz. 2011. "A Role for Biology in Gender-Related Behavior." *Sex Roles* 64:804–825.

Berg, E. C. 2003. "The Effects of Perceived Closeness to Custodial Parents, Stepparents, and Nonresident Parents on Adolescent Self-Esteem." *Journal of Divorce and Remarriage* 40:69–86.

Berge, Jerica M., and Kristen E. Holm. 2007. "Boundary Ambiguity in Parents with Chronically Ill Children: Integrating Theory and Research." *Family Relations* 56(2):123–134.

Bergen, Karla Mason. 2017. "Bridging the Distance of Commuter Families." *Family Focus*, FF71, p. F4–F5.

Bergen, Raquel Kennedy. 2006. "Marital Rape: New Research and Directions." The National Online Resource Center on Violence Against Women. Retrieved June 9, 2010 (new.vawnet.org).

Berger, John. 2015. "Unequal Gender Ratios at Colleges are Driving Hookup Culture." Retrieved October 29, 2019 (http://money.com/money/4072951/college-gender-ratios-dating-hook-up-culture/).

Berger, Lawrence M., Lidia Panico, and Anne Solaz. 2018. "Maternal Repartnering: Does Father Involvement Matter? Evidence from United Kingdom." *European Journal of Population* 34:1–31.

Berger, Lawrence M., and Marcia J. Carlson. 2020. "Family Policy and Complex Contemporary Families: A Decade in Review and Implications for the Next Decade of Research and Policy Practice." *Journal of Marriage and Family* 82:478–507.

Berger, Lawrence M., Theresa Heintze, Wendy Naidich, and Marcia Meyers. 2008. "Subsidized Housing and Household Hardship among Low-Income Single-Mother Households." *Journal of Marriage and Family* 70(4):934–949.

Berger, Lawrence M., and Sara S. McLanahan. 2015. "Income, Relationship Quality, and Parenting: Associations with Child Development in Two-Parent Families." *Journal of Marriage and Family* 77(4):996–1015.

Berger, Peter L., and Hansfried Kellner. 1970. "Marriage and the Construction of Reality." pp. 49–72 in *Recent Sociology No. 2*, edited by Hans Peter Dreitzel. New York: Macmillan.

Berger, Peter L., and Thomas Luckman. 1966. *The Social Construction of Reality: A Treatise in the Sociology of Knowledge*. Garden City, NY: Anchor Books.

Berger, Roni. 1998. *Stepfamilies: A Multi-Dimensional Perspective*. New York: Haworth.

———. 2000. "Gay Stepfamilies: A Triple-Stigmatized Group." *Families in Society* 81(5):504–516.

Bergmann, Barbara R. 2011. "Sex Segregation in the Blue-Collar Occupations: Women's Choices or Unremedied Discrimination?" *Gender and Society* 25(1):88–93.

Berkowitz, Dana, and William Marsiglio. 2007. "Gay Men: Negotiating Procreative, Father, and Family Identities." *Journal of Marriage and Family* 69(2):366–381.

Berlin, Gordon. 2007. "Rewarding the Work of Individuals: A Counterintuitive Approach to Reducing Poverty and Strengthening Families." *The Future of Children* 17(2):17–42.

Berlin, Lisa, Jean Ispa, Mark Fine, Patrick Malone, Jeanne Brooks-Gunn, Christy Brady-Smith, Catherine Ayoub, and Yu Bai. 2009. "Correlates and Consequences of Spanking and Verbal Punishment for Low-Income White, African American, and Mexican American Toddlers." *Child Development* 80(5):1403–1420.

Bermudez, J. Maria, Elizabeth Sharp, and Narumi Taniguchi. 2015. "Tapping into the Complexity: Ambivalent Sexism, Dating, and Familial Beliefs among Hispanics." *Journal of Family Issues* 36(10):1274–1295.

Bermúdez, J. Maria, Elizabeth A. Sharp, and Narumi Taniguchi. 2015. "Tapping into the Complexity: Ambivalent Sexism, Dating, and Familial Beliefs among Young Hispanics." *Journal of Family Issues* 36:1274–1295.

Bernard, Tara Siegel. 2012. "Marriage Maintenance When Money Is Tight." *The New York Times*, March 30. Retrieved August 28, 2012 (http://www.nytimes.com).

———. 2013. "Fired for Being Gay? Protections are Piecemeal." Retrieved December 8, 2015 (http://www.nytimes.com/2013/06/01/your-money/protections-for-gays-in-workplace-are-piecemeal.html?_r=0).

———. 2014. "Insurance Coverage for Fertility Treatments Varies Widely." *The New York Times*, July 25. Retrieved January 14, 2015 (http://www.nytimes.com).

Bernard, Tara S. 2018a. "When She Earns More: As Roles Shift, Old Ideas on Who Pays the Bills Persist." *The New York Times* July 6. Retrieved December 15, 2019 (https://www.nytimes.com).

Bernard, Tara S. 2018b. "Too Little Too Late: Bankruptcy Booms among Older Americans." August 5. Retrieved December 15, 2019 (https://www.nytimes.com).

Bernardi, Fabrizio, and Diederik Boertien. 2016. "Understanding Heterogeneity in the Effects of Parental Separation on Educational Attainment in Britain: Do Children from Lower Educational Backgrounds Have Less to Lose?" *European Sociological Review* 32:807–819. Berrington, Ann, Francesco C. Billari, Olivier Thevenon, and Daniela Vono de Vilhena. 2017. *Becoming an Adult in Europe*. Population Europe, Policy Brief No. 13. Berlin, Germany: Max Planck Society for the Advancement of Sciences.

Bernasek, Anna. 2014. "The Typical Household, Now Worth a Third Less." *The New York Times*, July 26. Retrieved January 14, 2015 (http://www.nytimes.com).

Bernstein, Anne C. 1997. "Stepfamilies from Siblings' Perspectives." pp. 153–175 in *Stepfamilies: History, Research, and Policy*, edited by Irene Levin and Marvin B. Sussman. New York: Haworth.

———. 1999. "Reconstructing the Brothers Grimm: New Tales for Stepfamily Life." *Family Process* 38(4):415–430.

———. 2007. "Revisioning, Restructuring, and Reconciliation: Clinical Practice With Complex Postdivorce Families." *Family Process* 46(1):67–78.

Bernstein, Jeffrey, and Susan Magee. 2004. *Why Can't You Read My Mind? Overcoming the 9 Toxic Thought Patterns That Get in the Way of a Loving Relationship*. New York: Marlowe.

Berridge, K., and T. Robinson. 1998. "What Is the Role of Dopamine in Reward: Hedonic Impact, Reward Learning, or Incentive Salience?" *Brain Research Review* 28(3):309–369.

Berry, Brent. 2008. "Financial Transfers from Living Parents to Young Adult Children: Who Is Helped and Why?" *American Journal of Economics and Sociology* 67(2):207–239.

Bertram, Rosalyn M., and Jennifer L. Dartt. 2009. "Post Traumatic Stress Disorder: A Diagnosis for Youth from Violent, Impoverished Communities." *Journal of Child and Family Studies*, 294–302.

Between Two Homes. 2020. "Between Two Homes. LLC." Retrieved February 3, 2020 (http://www.childreninthemiddle.com/).

Beutel, Ann M., Stephanie W. Burge, and B. Ann Borden. 2018. "Femininity and Choice of College Major." *Gender Issues* 35:113–136.

Bhalla, Vibha. 2008. "Couch Potatoes and Super-Women: Gender, Migration, and the Emerging Discourse on Housework among Asian Indian Immigrants." *Journal of American Ethnic History* 27(4):71–99.

Bialik, Kristen. 2018. "Middle Children Have Become Rarer, but a Growing Share of Americans Now Say Three or More Kids Are 'Ideal'." *Facttank: News in the Numbers*. August 9. Pew Research Organization. Retrieved January 30, 2019 (www.pewresearch.org).

Bianchi, Suzanne M. 2000. "Maternal Employment and Time with Children: Dramatic Change or Surprising Continuity?" *Demography* 37:401–614.

Bianchi, Suzanne M., and Lynne M. Casper. 2000. "American Families." *Population Bulletin* 55(4). Washington, DC: Population Reference Bureau.

Bianchi, Suzanne M., John P. Robinson, and Melissa A. Milkie. 2006. *Changing Rhythms of American Family Life*. New York: Russell Sage.

Biblarz, Timothy J., and Evren Savci. 2010. "Lesbian, Gay, Bisexual, and Transgender Families." *Journal of Marriage and Family* 72(3):480–497.

Bick, Debra. 2010. "Media Portrayal of Birth and the Consequences of Misinformation." *Midwifery* 26:147–148.

Bickman, Leonard, and Debra J. Rog. 2009. *The SAGE Handbook of Applied Social Research Methods*. 2nd ed. Los Angeles: Sage.

Bierman, Alex, Elena M. Fazio, and Melissa A. Milkie. 2006. "A Multifaceted Approach to

the Mental Health Advantage of the Married." *Journal of Family Issues* 27(4):554–582.

Bierman, Noah, and David G. Savage. 2019. "Trump Administration Abruptly Gives Up Fight Over Citizenship Question on Census." *Los Angeles Times.* July 2. Retrieved July 7, 2019 (https://www.latimes.com).

Bigda, Carolyn. 2013. "Young Adult Caregivers." Retrieved March 2013 (http://www.seniorlivingexperts.com).

Bigler, Rebecca, Amy R. Hayes, and Veronica Hamilton. 2013. "The Role of Schools in the Early Socialization of Gender Differences." pp. 1–5 in R. E. Tremblay, M.

Björkqvist, Kaj. 2018. "Gender Differences in Aggression." *Current Opinion in Psychology* 19:39–42.

Bilefsky, Dan. 2012. "Hard Times in Spain Force Feuding Couples to Delay Divorce." Retrieved April 12, 2013 (http://www.nytimes.com/2012/12/18/world/europe /hard-times -in-spain-force-feuding-couples -to-delay-divorce .html?_r=0&pagewanted =print).

Billingsley, Andrew. 1968. *Black Families in White America.* Englewood Cliffs, NJ: Prentice Hall.

Billingsley, Sunnee, and Tommy Ferrarini. 2014. "Family Policy and Fertility Intentions in Twenty-One European Countries." *Journal of Marriage and Family* 76(2):428–445.

Bindley, K. 2011, "The Mommy Wars Continue: Relations Between Nannies, Working Moms, and Stay-at-Home Moms." *Huffington Post Parents*, October 13. Retrieved February 20, 2012 (http://www .huffingtonpost.com).

Bingenheimer, Jeffrey B., Elizabeth A. Asante, and Clement Ahiadeke. 2015. "Reliability, Validity, and Associations with Sexual Behavior among Ghanaian Teenagers of Scales Measuring Four Dimensions of Relationships with Parents and Other Adults." *Journal of Family Issues* 36(5):647–668.

Bird, Gloria W., Rick Peterson, and Stephanie Hotta Miller. 2002. "Factors Associated with Distress among Support-Seeking Adoptive Parents." *Family Relations* 51(3):215–220.

"Birds of a Feather—More than a Place to Live, a Way to Live." 2010. Retrieved May 17, 2010 (flock2it.com).

———. 2015. Retrieved December 26, 2015 (flock2it.com).

Birnbaum, Gurit E., Mario Mikulincer, and Michal Austerlitz. 2013 "A Fiery Conflict: Attachment Orientations and the Effects of Relational Conflict on Sexual Motivation." *Personal Relationships* 20(2):294–310.

Birnbaum, Rachel, and Michael Saini. 2015. "A Qualitative Synthesis of Children's Experiences of Shared Care Post-Divorce." *International Journal of Children's Rights* 23:109–132.

Birnie-Porter, Carolyn, and John E. Lydon. 2013. "A Prototype Approach to Understanding Sexual Intimacy Through Its Relationship to Intimacy." *Personal Relationships* 20(2):236–258.

Biro, Frank M., Ashley Pajak, Mary S. Wolff, Susan M. Pinney, Gayle C. Windham, Maida P. Galvez, Louise C. Greenspan, Larry H. Kushi, and Susan L. Teitelbaum. 2018. "Age of Menarche in a Longitudinal U.S. Cohort." *Journal of Pediatric and Adolescent Gynecology* 31:339–345.

Biro, Frank M., Maida P. Galvez, Louise C. Greenspan, Paul A. Succop, Nita Vangeepuram, Susan M. Pinney, Susan Teitelbaum, Lawrence H. Kushi, and Mary S. Wolff. 2010. "Pubertal Assessment Method and Baseline Characteristics in a Mixed Longitudinal Study of Girls." *Pediatrics* 126(3):e583–e590.

Black, Michael Ian. 2018. "The Boys are Not All Right." Retrieved September 29, 2019 (https://www.nytimes.com/2018/02/21/opinion/boys -violence-shootings-guns.html).

Blackman, Lorraine, Obie Clayton, Norval Glenn, Linda Malone-Colon, and Alex Roberts. 2006. *The Consequences of Marriage for African Americans: A Comprehensive Literature Review.* New York: Institute for American Values.

Blakemore, Judith, and Craig Hill. 2008. "The Child Gender Socialization Scale: A Measure to Compare Traditional and Feminist Parents." *Sex Roles* 58(3/4):192–207.

Blackstone, Amy, Heather McLaughlin, and Christopher Uggen. 2018. "Workplace Sexual Harassment." *Stanford Center on Poverty and Inequality, State of the Union 2018.* Retrieved October 24, 2019 (https://inequality.stanford.edu/publications/pathway/state-union-2018).

Blake, Lucy. 2017. "Parents and Children Who Are Estranged in Adulthood: A Review and Discussion of the Literature." *Journal of Family Theory and Review* 9 (4): 521–536.

Blanchard, Victoria L., Alan J. Hawkins, Scott A. Baldwin, and Elizabeth B. Fawcett. 2009. "Investigating the Effects of Marriage and Relationship Education on Couples' Communication Skills: A Meta-Analytic Study." *Journal of Family Psychology* 23(2):203–214.

Blanck, J. L. 2007. "Are helicopter Parents Landing in Graduate School?" *Journal of Career Planning and Employment* 68(2):35–39.

Blankenhorn, David, William Galston, Jonathan Rauch, and Barbara Dafoe Whitehead. 2015. "Can Gay Wedlock Break Political Gridlock?" *Washington Monthly*, April/May. Retrieved October 21, 2015 (http://www .washingtonmonthly.com).

Blaustein, Jonathan. 2015. "Bangladesh's Third Gender." *The New York Times*, March 18. Retrieved October 14, 2015 (http://www .nytimes.com).

Bledsoe, Carolinem and Papa Sow. 2011. "Back to Africa: Second Chances for the Children of West African Immigrants." *Journal of Marriage and Family* 73(4): 747–762.

Block, Joel D., and Susan S. Bartell 2001. *Stepliving for Teens: Getting Along with Step-Parents, Parents, and Siblings.* New York: Penguin Young Readers Group.

Blood, Robert O. Jr., and Donald M. Wolfe. 1960. *Husbands and Wives: The Dynamics of Married Living.* New York: The Free Press.

Blossfeld, Hans-Peter, and Kathleen Kiernan. 2019. *The New Role of Women: Family Formation in Modern Societies.* New York: Routledge.

Blow, Adrian J., and Kelley Hartnett. 2005. "Infidelity in Committed Relationships II: A Substantive Review." *Journal of Marital and Family Therapy* 31(2):217–233.

Blumberg, Antonia. 2015. "What You Need to Know about the 'Quiverfull' Movement." Retrieved February 5, 2016 (http://www .huffingtonpost.com/2015 /05/26/quiverfull -movement-facts_n _7444604.html).

Blumberg, Rae Lesser, and Marion Tolbert Coleman. 1989. "A Theoretical Look at the Gender Balance of Power in the American Couple." *Journal of Family Issues* 10:225–250.

Blumenstock, Shari, and Lauren Papp. 2017. "Sexual Distress and Marital Quality of Newlyweds." *Family Relations* 66 (5):794–808.

Blumstein, Philip, and Pepper Schwartz. 1983. *American Couples: Money, Work, Sex.* New York: Morrow.

Bluth, Karen, Patricia Roberson, Rhett Billen, and JuliSams. 2013. "A Stress Model for Couples Parenting Children with Autism Spectrum Disorders and the Introduction of a Mindfulness Intervention." *Journal of Family Theory and Review* 5(3):194–213.

Bobbitt-Zeher, Donna. 2011. "Connecting Gender Stereotypes, Institutional Policies, and Gender Composition of Workplace." *Gender and Society* 25(6):764–786.

Bock, Jane D. 2000. "'Doing the Right Thing?' Single Mothers by Choice and the Struggle for Legitimacy." *Gender and Society* 14:62–86.

Bodenmann, Guy, Thomas Ledermann, and Thomas N. Bradbury. 2007. "Stress, Sex, and Satisfaction in Marriage." *Personal Relationships* 14(4):551–569.

Bodrug-Lungu, Valentina, and Erin Kostina-Ritchey. 2013. "Impact of Parents' Migration on Moldavian Youths' Perception of Family." *Family Focus* FF56: F13–F15.

Boerner, Kathrin, Deborah Carr, and Sara Moorman. 2013. "Family Relationships and Advance Care Planning: Do Supportive and Critical Relations Encourage or Hinder Planning?" *Journals of Gerontology, Series B: Psychological and Social Sciences* 68(2):246–256.

Boertien, Diederik, and Juho Härkönen. 2018. "Why Does Women's Education Stabilize Marriages? The Role of Marital Attraction and Barriers to Divorce." *Demographic Research* 38:1241–1276.

Bogenschneider, Karen, Olivia M. Little, Theodora Ooms, Sara Benning, Karen Cadigan, and Thomas Corbett. 2012. "The Family Impact Lens: A Family-Focused, Evidence-Informed Approach to Policy and Practice." *Family Relations* 61(3): 514–531.

Boghani, Priyanka. 2015. "Frontline. When Transgender Kids Transition: Medical Risks Are Both Known and Unknown." Public Broadcasting Corporation. Aired June 30. Retrieved October 10, 2015 (http://www.pbs.org).

Bogle, Kathleen A. 2004. "From Dating to Hooking Up: The Emergence of a New Sexual Script." Unpublished PhD dissertation, Department of Sociology, University of Delaware. Newark, DE.

———. 2008. *Hooking Up: Sex, Dating, and Relationships on Campus.* New York: New York University Press.

Boivin, and R. Peters (Eds.) Encyclopedia on Early Childhood Development—online. Montreal, Quebec: Centre of Excellence for

Early Childhood Development and Strategic Knowledge Cluster on Early Child Development. Retrieved October 10, 2015 (http://www .child -encyclopedia.com).

Bolick, Kate. 2011a. "Let's Hear It for Aunthood." *The New York Times*, September 16. Retrieved September 23, 2011 (http://www .nytimes.com).

Bolick, Kate. 2015. *Spinster: Making a Life of One's Own*. New York: Broadway Books.

Bolick, Kate. 2011b. "What, Me Marry? All the Single Ladies." *The Atlantic*, November: 116–136.

Bonach, Kathryn. 2007. "Forgiveness Intervention Model: Application to Coparenting Post-Divorce." *Journal of Divorce and Remarriage* 48(1/2):105–123.

"Boomerang Kids Help." n.d. Retrieved November 4, 2015 (http://www .boomerangkidshelp.com).

Boon, Susan D., Vicki L. Deveau, and Alishia M. Allbhal. 2009. "Payback: The Parameters of Revenge in Romantic Relationships." *Journal of Social and Personal Relationships* 26(6–7):747–768.

Boon, Susan D., Sarah J. Watkins, and Rowan A. Sciban. 2014. "Pluralistic Ignorance and Misperception of Social Norms Concerning Cheating in Dating Relationships." *Personal Relationships* 21(3):482–496.

Boonstra, Heather D. 2015. "Meeting the Sexual and Reproductive Health Needs of Adolescents in School-Based Health Centers." *Guttmacher Policy Review* 18(1):21–26.

Boonstra, Heather D., Rachel Benson Gold, Cory L. Richards, and Lawrence B. Finer. 2006. *Abortion in Women's Lives*. New York: Guttmacher Institute.

Booth, Alan, Karen Carver, and Douglas A. Granger. 2000. "Biosocial Perspectives on the Family." *Journal of Marriage and Family* 62(4):1018–1034.

Booth, Alan, Ann C. Crouter, and Mari Clements, eds. 2001. *Couples in Conflict*. Mahwah, NJ: Erlbaum.

Booth, Alan, Douglas Granger, Allan Mazur, and Katie Kivlighan. 2006. "Testosterone and Social Behavior." *Social Forces* 85(1):167–191.

Booth, Alan, David R. Johnson, and Douglas A. Granger. 2005. "Testosterone, Marital Quality, and Role Overload." *Journal of Marriage and Family* 67(2):483–498.

Booth, Alan, Elisa Rustenbach, and Susan McHale. 2008. "Early Family Transitions and Depressive Symptom Changes from Adolescence to Early Adulthood." *Journal of Marriage and Family* 70(1):3–14.

Booth, Alan, Mindy E. Scott, and Valarie King. 2010. "Father Residence and Adolescent Problem Behavior: Are Youth Always Better Off in Two-Parent Families?" *Journal of Family Issues* 31(5):585–605.

Booth, Cathy. 1977. "Wife-Beating Crosses Economic Boundaries." *Rocky Mountain News*, June 17.

Bordere, Tashel. 2017. "Disenfranchisement and Ambiguity in the Face of Loss: The Suffocated Grief of Sexual Assault Survivors." *Family Relations* 66 (1): 29–45.

Boss, Pauline. 1980. "Normative Family Stress: Family Boundary Changes across the Lifespan." *Family Relations* 29:445–452.

Boss, Pauline. 2015. "On the Usefulness of Theory: Applying Family Therapy and Family Science to the Relational Developmental Systems Metamodel." *Journal of Family Theory and Review* 7(2):105–108.

Boss, Pauline. 2016. "The Context and Process of Theory Development: The Story of Ambiguous Loss." Boustani, Maya Mroué, Stacy L. Frazier, Chelsey Hartley, Michael Meinzer, and Erin Hedemann. 2015. "Perceived Benefits and Proposed Solutions for Teen Pregnancy: Qualitative Interviews with Youth Care Workers." *American Journal of Orthopsychiatry* 85:80.

Bottke, Allison. 2010. *Setting Boundaries with Your Aging Parents*. Eugene, OR: Harvest House Publishers.

Bouchard, Emily. n.d. "Navigating Parenting Differences." SelfGrowth.com. Retrieved September 21, 2004. (www .selfgrowth.com /articles/Bouchard2 .html).

Bouchard, Genevieve, and Danielle Doucet. 2011. "Parental Divorce and Couples' Adjustment During the Transition to Parenthood: The Role of Parent-Adult Child Relationships." *Journal of Family Issues* 32(4):507–527.

Bourdieu, Pierre. 1977 [1972]. *Outline of a Theory of Practice*. New York: Cambridge.

Bouton, Katherine. 1987. "Fertility and Family." *Ms.*, April, p. 92.

Bowen, Alison. 2017. "1 in 3 U.S. Women Have C-Sections. How Chicago Doctors Are Working to Change That." Retrieved December 24, 2019 (https://www.chicagotribune.com/lifestyles /health/ct-cesarean-sections-births-health-0515 -20170515-story.html).

Bowen, Gary L., Roderick A. Rose, Joelle D. Powers, and Elizabeth J. Glennie. 2008. "The Joint Effects of Neighborhoods, Schools, Peers, and Families on Changes in the School Success of Middle School Children." *Family Relations* 57(4):504–516.

Bowlby, John. 1988. *A Secure Base*. London: Routledge.

Bowleg, Lisa, Jennifer Huang, Kelly Brooks, Amy Black, and Gary Burkholder. 2003. "Triple Jeopardy and Beyond: Multiple Minority Stress and Resilience among Black Lesbians." *Journal of Lesbian Studies* 7(4):87–108.

Bowleg, Lisa, Michelle Teti, Jenné S. Massie, Aditi Patel, David J. Malebranche, and Jeanne M. Tschann. 2011. "'What Does It Take to Be a Man? What Is a Real Man?': Ideologies of Masculinity and HIV Sexual Risk among Black Heterosexual Men." *Culture, Health, and Sexuality* 13(5):545–559.

Bowles, Nellie. 2018. "The Digital Gap Between Rich and Poor Kids Is Not What We Expected." Retrieved January 25, 2020 (https://newsela.com /read/social-media-teens-mental-health-benefits /id/36844/).

Boxer, Christie F., Mary C. Noonan, and Christine B. Whelan. 2015. "Measuring Mate Preferences: A Replication and Extension. " *Journal of Family Issues* 36:163–187.

Boylstein, Craig, and Jeanne Hayes. 2012. "Reconstructing Marital Closeness While Caring for a Spouse with Alzheimer's." *Journal of Family Issues* 33(5):584–612.

Brackc, Pict, Wendy Christiaens, and Naomi Wauterickx. 2008. "The Pivotal Role of Women in Informal Care." *Journal of Family Issues* 29(10):1348–1378.

Bradbury, Thomas N., Frank D. Fincham, and Steven R. H. Beach. 2000. "Research on the Nature and Determinants of Marital Satisfaction: A Decade in Review." *Journal of Marriage and Family* 62(4):964–980.

Bradbury, Thomas N., and Benjamin R. Karney. 2004. "Understanding and Altering the Longitudinal Course of Marriage." *Journal of Marriage and Family* 66(6):862–879.

Braithwaite, Dawn O., and Leslie A. Baxter. 2006. *Engaging Theories in Family Communication: Multiple Perspectives*. Thousand Oaks, California: Sage.

Braithwaite, Scott R., Sean C. Aaron, Krista K. Dowdle, Kersti Spjut, and Frank D. Fincham. 2015. "Does Pornography Consumption Increase Participation in Friends with Benefits Relationships?" *Sexuality and Culture* 19:1–20.

Brandt, Justin S., Mayra A. Cruz Ithier, Todd Rosen, and Elena Ashkinadze. 2019. "Advanced Paternal Age, Infertility, and Reproductive Risks: A Review of the Literature." *Prenatal Diagnosis* 39:81–87.

Bramlett, Matthew D. 2010. "When Stepparents Adopt: Demographic, Health and Health Care Characteristics of Adopted Children, Stepchildren, and Adopted Stepchildren." *Adoption Quarterly* 13:248–267.

Bramlett, Matthew D., and William D. Mosher. 2002. *Cohabitation, Marriage, Divorce, and Remarriage in the United States*. National Center for Health Statistics. *Vital Health Statistics* 23(22).

Branden, Nathaniel. 1988. "A Vision of Romantic Love." pp. 218–231 in *The Psychology of Love*, edited by Robert J. Sternberg and Michael L. Barnes. New Haven, CT: Yale University Press.

Brandl, Bonnie. 2007. *Elder Abuse Detection and Intervention: A Collaborative Approach*. New York: Springer.

Bratter, Jenifer L. and Karl Eschbach. 2006. "'What about the Couple?' Interracial Marriage and Psychological Distress." *Social Science Research* 35(4):1025–1047.

Bratter, Jenifer L. M., and Rosalind B. King. 2008. "'But Will It Last?' Marital Instability among Interracial and Same-Race Couples." *Family Relations* 57(April):160–171.

Braver, Sanford, Jennesa R. Shapiro, and Matthew R. Goodman. 2006. "Consequences of Divorce for Parents." pp. 313–337 in *Handbook of Divorce and Relationship Dissolution*, edited by Mark A. Fine and John H. Harvey. Mahwah, NJ: Erlbaum.

Braver, Sanford L., Pamela J. Fitzpatrick, and R. Curtis Bay. 1991. "Noncustodial Parent's Report of Child Support Payments." *Family Relations* 40(2):180–185.

Bray, James H. 1999. "From Marriage to Remarriage and Beyond." pp. 253–271 in *Coping with Divorce, Single Parenting and Remarriage*, edited by E. Mavis Hetherington. Mahwah, NJ: Erlbaum.

Bray, J. H., S. H. Berger, and C. L. Boethel. 1994. "Role Integration and Marital Adjustment in Stepfather Families." pp. 69–86 in *Stepparenting: Issues in Theory, Research, and Practice*, edited by K.Pasley and M.Ihinger-Tallman. Westport, CT: Greenwood Press.

Bray, J. H., and I. Easling. 2005. pp. 267–294 in *Family Psychology: The Art of the Science*, edited by W. M.Pinsof and J. L. Lebow. New York: Oxford Press.

Brazelton, T. Berry, and Stanley Greenspan. 2000. "Our Window to the Future." *Newsweek* Special Issue, Fall/Winter, pp. 34–36.

Breitenbecher, Kimberly Hanson. 2006. "The Relationships among Self-blame, Psychological Distress, and Sexual Victimization." *Journal of Interpersonal Violence* 21(5):597–611.

Brenan, Megan. 2019. "Birth Control Still Tops List of Morally Acceptable Issues." Retrieved October 17, 2019 (https://news.gallup.com/poll/257858/birth-control-tops-list-morally-acceptable-issues.aspx).

Brennan, Bridget. 2003. "No Time. No Sex. No Money." *First Years and Forever: A Monthly Online Newsletter for Marriages in the Early Years*. Chicago: Archdiocese of Chicago, Family Ministries. Retrieved September 8, 2006 (www.familyministries.org).

Brewer, Graham Lee. 2019. "Cherokee Nation Names First Delegate to Congress." Retrieved October 2, 2019 (https://www.npr.org/2019/09/03/756048206/cherokee-nation-names-first-delegate-to-congress).

Brewis, Alexandra, and Mary Meyer. 2005. "Marital Coitus across the Life Course." *Journal of Biosocial Science* 37:499–518.

Brewster, Karin L., and Irene Padavic. 2002. "No More Kin Care? Change in Black Mothers' Reliance on Relatives for Child Care, 1977–94." *Gender and Society* 16:546–563.

Bricker, Jesse, Brian Bucks, Arthur Kennickell, Traci Mach, and Kevin Moore. 2012. "The Financial Crisis from the Family's Perspective." *Journal of Consumer Affairs* 46(3):537–555.

Bridges, Ana J., Chyng F. Sun, Matthew B. Ezzell, and Jennifer Johnson. 2016. "Sexual Scripts and the Sexual Behavior of Men and Women Who Use Pornography." *Sexualization, Media, & Society* 2: 2374623816668275.

Bridges, Tristan, and Jesse M. Philbin. 2019. "Gender Convergence Over 'Cheap Sex.'" Retrieved February 14, 2020

Brimhall, Andrew, Karen Wampler, and Thomas Kimball. 2008. "Learning from the Past, Altering the Future: A Tentative Theory of the Effect of Past Relationships on Couples Who Remarry." *Family Process* 47(3):373–387.

Brink, Susan. 2008. "Modern Puberty." *Los Angeles Times*, January 21.

Britt-Lutter, Sonya, Cassandra Dorius, and Derek Lawson. 2018. "The Financial Implications of Cohabitation Among Young Adults." *Journal of Financial Planning* 31:38-45.

Brock, Rebecca L., and Erika Lawrence. 2009. "Too Much of a Good Thing: Underprovision Versus Overprovision of Partner Support." *Journal of Family Psychology* 23(2):181–192.

Broderick, Carlfred B. 1979. *Marriage and the Family*. Englewood Cliffs, NJ: Prentice Hall.

Brody, Gene, Yi Fu Chen, Steven Kogan, Velma McBride Murray, Patricia Logan, and Zupei Luo. 2008. "Linking Perceived Discrimination to Longitudinal Changes in African American Mothers' Parenting Practices." *Journal of Marriage and Family* 70(2):319–31.

Brody, Jane E. 2004. "Abstinence-Only: Does It Work?" *The New York Times*, June 3.

———. 2008. "When Families Take Care of Their Own." *The New York Times*, November 11. Retrieved March 16, 2010 (www.nytimes.com).

———. 2012. "Caregiving as a Roller-Coaster Ride from Hell." *The New York Times*, April 10. Retrieved May 21, 2013 (http://well.blogs.nytimes.com/2012/04/09/caregiving-as-a-roller-coaster-ride-from-hell/?_r=0).

Broman, Clifford L., Li Xin, and Mark Reckase. 2008. "Family Structure and Mediators of Adolescent Drug Use." *Journal of Family Issues* 29(12):1625–1649.

Bronfenbrenner, Urie. 1979. *The Ecology of Human Development: Experiments by Nature and Design*. Cambridge, MA: Harvard University Press.

Bronte-Tinkew, Jacinta, Jennifer Carrano, Allison Horowitz, and Akemi Kinukawa. 2008. "Involvement among Resident Fathers and Links to Infant Cognitive Outcomes." *Journal of Family Issues* 29(9):1211–1244.

Bronte-Tinkew, Jacinta, and Allison Horowitz. 2010. "Factors Associated with Unmarried, Nonresident Fathers' Perceptions of Their Coparenting." *Journal of Family Issues* 31(1):31–65.

Bronte-Tinkew, Jacinta, Allison Horowitz, and Mindy E. Scott. 2009. "Fathering with Multiple Partners: Links to Children's Well-Being in Early Childhood." *Journal of Marriage and Family* 71(3):608–631.

Brooke, Jill. 2004. "Close Encounters with a Home Barely Known." *The New York Times*, July 22.

———. 2006. "Home Alone Together." *The New York Times*, May 4. Retrieved May 7, 2006 (www.nytimes.com).

Brooks, Kim. 2019. "We Have Ruined Childhood." *The New York Times*. August 17. Retrieved August 27, 2019 (https://www.nytimes.com).

Brooks, Robert, and Sam Goldstein. 2001. *Raising Resilient Children: Fostering Strength, Hope, and Optimism in Your Child*. New York: Contemporary Books.

Brossoie, Nancy, Karen Roberto, and Katie Barrow. 2012. "Making Sense of Intimate Partner Violence in Later Life: Comments from Online News Readers." *The Gerontologist* 52(6):792–801.

Brotherson, Sean. 2003. "Time, Sex, and Money: Challenges in Early Marriage." *The Meridian*. Retrieved September 8, 2006 (www.meridianmagazine.com).

Brougham, Ruby R. and David A. Walsh. 2009. "Early and Late Retirement Exits." *International Journal of Aging and Human Development* 69(4):267–286.

Browder, Laura. 2010. "Impact of War." *Washington Post*, May 24. Retrieved February 21, 2013 (http://www.voices.washingtonpost.com).

Brown, Alyssa, and Jeffrey M. Jones. 2012. "Separation, Divorce Linked to Sharply Lower Wellbeing: Married Americans Have Highest Wellbeing." Gallup Poll, April 20. Retrieved December 4, 2012 (http://www.gallup.com).

Brown, Andrew, Cerith S. Waters, and Katherine H. Shelton. 2017. "A Systematic Review of the School Performance and Behavioural and Emotional Adjustments of Children Adopted from Care. " *Adoption & Fostering* 41:346–368.

Brown, Carrie. 2008. "Gender-Role Implications on Same-Sex Intimate Partner Abuse." *Journal of Family Violence* 23:475–462.

Brown, Chris, Heather B. Transgrud, and Rachel M. Linnemeyer. 2009. "Battered Women's Process of Leaving." *Journal of Career Assessment* 17(4): 439–456.

Brown, Edna, Terri L. Orbuch, and Artie Maharaj. 2010. "Social Networks and Marital Stability among Black American and White American Couples." pp. 318–334 in *Support Processes in Intimate Relationships*, edited by Kieran T. Sullivan and Joanne Davila. New York: Oxford University Press.

Brown, Hana E. 2013. "Race, Legality, and the Social Policy Consequences of Anti-Immigration Mobilization." *American Sociological Review* 78(2):290–314.

Brown, Patricia Leigh. 2004. "For Children of Gays, Marriage Brings Joy." *The New York Times*, March 19.

Brown, Jenna. 2020. "The Young Caregiver." Retrieved March 1, 2020 (https://gentwenty.com/hidden-demographic-young-caregiver).

Brown, Susan L. 2004. "Family Structure and Child Well-Being: The Significance of Parental Cohabitation." *Journal of Marriage and Family* 66(2):351–367.

———. 2013. "A 'Gray Divorce' Boom." *Los Angeles Times*. Retrieved April 15, 2013 (http://www.latimes.com/news/opinion/commentary/la-oe-brown-gray-divorce-20130331,0,7982785.story).

Brown, Susan L., Jennifer Roebuck Bulanda, and Gary R. Lee. 2012. "Transitions Into and Out of Cohabitation in Later Life." *Journal of Marriage and Family* 74(4) (August): 774–793.

Brown, Susan L., and I-Fen Lin. 2012. "The Gray Divorce Revolution: Rising Divorce among Middle-Aged and Older Adults." *Journal of Gerontology Series B: Psychological Sciences and Social Sciences* 67(6):731–741.

———. 2013. "Age Variation in the Remarriage Rate, 1990-2011" (FP-13-17). Retrieved April 12, 2016 (https://www.bgsu.edu/content/dam/BGSU/college-of-arts-and-sciences/NCFMR/documents/FP/FP-13-17.pdf).

Brown, Susan L., I-Fen Lin, and Krista K. Payne. 2014a. "Gray Divorce: A Growing Risk Regardless of Class or Education." Council on Contemporary Families. Retrieved March 18, 2016 (https://contemporaryfamilies.org/growing-risk-brief-report/).

Brown, Susan L., I-Fen Lin, and Krista K. Payne. 2014b. "Age Variation in the Divorce Rate, 1990–2012" (FP-14-16). Bowling Green, OH: National Center for Family and Marriage Research.

Brown, Susan L., and Wendy D. Manning. 2009. "Family Boundary Ambiguity and the Measurement of Family Structure: The Significance of Cohabitation." *Demography* 46(1):85–101.

Brown, Susan L., Wendy D. Manning, and Krista K. Payne. 2015. "Relationship Quality among Cohabiting Versus Married Couples." *Journal of Family Issues*: 0192513X15622236.

Brown, Susan L., Wendy D. Manning, and J. Bart Stykes. 2015. "Family Structure and Child Well-Being: Integrating Family Complexity." *Journal of Marriage and Family* 77(1):177–190.

Brown, Susan L., and Sayaka K. Shinohara. 2013. "Dating Relationships in Older Adulthood: A National Portrait." *Journal of Marriage and Family* 75:1194–1202.

Brown, Tiffany, Miriam Linver, Melanie Evans, and Donna DeGennaro. 2009. "African American Parents' Racial and Ethnic Socialization and Adolescent Academic Grades: Teasing Out the Role of Gender." *Journal of Youth and Adolescence* 38(2):214–227.

Brownridge, Douglas, Tamara Tallieu, Kimberly Tyler, Agnes Tiwari, Ko Ling Chan, and Susy Santos. 2011. "Pregnancy and Intimate Partner Violence: Risk Factors, Severity, and Health Effects." *Violence Against Women* 17(7):858–881.

Bruch, Elizabeth, Fred Feinberg, and Kee Yeun Lee. 2016. "Extracting Multistage Screening Rules from Online Dating Activity Data." *Proceedings of the National Academy of Sciences* 113:10530–10535.

Brush, Lisa D. 2008. "Book Review." *Gender and Society* 22(1):126–142.

Bryan, Willie V. 2010. *Sociopolitical Aspects of Disabilities: The Social Perspectives and Political History of Disabilities and Rehabilitation in the United States*. Springfield, IL: Charles C. Thomas.

Bryant, Chalandra. 2020. "African American Stepfamilies," pp. 45-76 in *Multicultural Stepfamilies*, edited by Susan D. Stewart and Gordon Limb. San Diego, CA: Cognella.

Bubolz, Margaret M., and M. Suzanne Sontag. 1993. "Human Ecology Theory." pp. 419–448 in *Sourcebook of Family Theories and Methods: A Contextual Approach*, edited by Pauline G. Boss, William J. Doherty, Ralph LaRossa, Walter R. Schumm, and Suzanne K. Steinmetz. New York: Plenum.

Buchanan, Christy M., Eleanor E. Maccoby, and Sanford M. Dornbusch. 1996. *Adolescents After Divorce*. Cambridge, MA: Harvard University Press.

Bucx, Freek, Fritz Van Well, and Trudie Knijn. 2012. "Life Course Status and Exchanges of Support between Young Adults and Parents." *Journal of Marriage and Family* 74(1):101–115.

Buehler, Cheryl, Ambika Krishnakumar, Gaye Stone, Christine Anthony, Sharon Pemberton, Jean Gerard, and Brian K. Barber. 1998. "Interpersonal Conflict Styles and Youth Problem Behaviors." *Journal of Marriage and Family* 60(1):119–132.

Buehler, Cheryl, and Marion O'Brien. 2011. "Mothers' Part-Time Employment: Associations with Mother and Family Well-Being." *Journal of Family Psychology* 25(6):895–906.

Bufkin, Sarah. 2012. "Domestic Workers Bill Killed in California by Jerry Brown Veto." *The Huffington Post Politics*, October 1. Retrieved February 20, 2013 (http://www.huffingtonpost.com).

Bui, Quoctrung, and Claire Cain Miller. 2018. "The Age that Women Have Babies: How a Gap Divides America." *The New York Times*. August 4. Retrieved December 9, 2019 (www.nytimes.com).

Bui, Quoctrung, and Claire Cain Miller. 2015. "The Typical American Lives Only 18 Miles from Mom." *The New York Times, December* 23. Retrieved December 28, 2015 (http://www.nytimes.com).

Bukhari, Zahid Hussain. 2004. *Muslims' Place in the American Public Square: Hope, Fears, and Aspirations*. Walnut Creek, CA: AltaMira.

Buiano, Madeline. 2018 "Deportations of Parents of U.S. Kids Are Likely to Rise with a Wider Trump Net." Center for Public Integrity. Retrieved November 16, 2019 (https://publicintegrity.org /business/immigration/ice-data-tens-of -thousands-of-deported-parents-have-u-s-citizen -kids/).

Bulanda, Jennifer Roebuck. 2011. "Doing Family, Doing Gender, Doing Religion: Structured Ambivalence and the Religion-Family Connection." *Journal of Family Theory and Review* 3(3)(September):179–197.

Bulanda, Jennifer R., and S. L. Brown. 2007. "Race-Ethnic Differences in Marital Quality and Divorce." *Social Science Research* 36:945–967.

Bulcroft, Kris, Richard Bulcroft, Linda Smeins, and Helen Cranage. 1997. "The Social Construction of the North American Honeymoon, 1800–1995." *Journal of Family History* 22(4):462–491.

Bumpass, Larry L., and K. Raley. 2007. "Measuring Separation and Divorce." pp. 125–144 in *Handbook of Measurement Issues in Family Research*, edited by S. L. Hofferth and L. M. Casper. Mahwah, NJ: Erlbaum.

Bumpass, Larry L., James A. Sweet, and Andrew Cherlin. 1991. "The Role of Cohabitation in Declining Rates of Marriage." *Journal of Marriage and Family* 53(4):913–927.

Burchinal, Margaret, Nathan Vandergrift, Robert Pianta, and Andrew Mashburn. 2010. "Threshold Analysis of Association Between Child Care Quality and Child Outcomes for Low-Income Children in Pre-Kindergarten Programs." *Early Childhood Research Quarterly* 25(2):166–176.

Bures, Regina M. 2009. "Living Arrangements Over the Life Course." *Journal of Family Issues* 30(5):579–585.

Bures, Regina M., Tanya Koropeckyj-Cox, and Michael Loree. 2009. "Childlessness, Parenthood, and Depressive Symptoms among Middle-Aged and Older Adults." *Journal of Family Issues* 30(5):670–687.

Burgess, Ernest, and Harvey Locke. 1953 [1945]. *The Family: From Institution to Companionship*. New York: American.

Burgess, Wes. 2008. *The Bipolar Handbook for Children, Teens, and Families: Real-Life Questions with Up-to-Date Answers*. New York: Avery.

Burgos, Evan, Gayatri Kaul, and Alpha Newberry. 2015. "India's Third Gender." NBC News, June 22.

Burke, Minyvonne. 2019. "Alabama Girl Wore Tux for Senior Portrait. Her School Yearbook Left her Picture Out." Retrieved October 5, 2019 (https://www.nbcnews.com/news/us-news /alabama-girl-wore-tux-senior-portrait-her -school-yearbook-left-n1043611).

Burnette, Josh, and Pete Hardesty. *Adulting 101:# Wisdom4Life*. BroadStreet Publishing Group LLC, 2018.

Burns, George W. 2010. *Happiness, Healing, Enhancement: Your Casebook Collection for Applying Positive Psychology in Therapy*. Hoboken, NJ: Wiley.

Burpee, Leslie C., and Ellen J. Langer. 2005. "Mindfulness and Marital Satisfaction." *Journal of Adult Development* 12(1):1281–1287.

Burr, Jeffrey A., and Jan E. Mutchler. 1999. "Race and Ethnic Variation in Norms of Filial Responsibility among Older Persons." *Journal of Marriage and Family* 61(3):674–687.

Burr, Wesley R., Shirley Klein, and Marilyn McCubbin. 1995. "Reexamining Family Stress: New Theory and Research." *Journal of Marriage and Family* 57(3):835–846.

Burrell, Ginger L., and Mark W. Roosa. 2009. "Mothers' Economic Hardship and Behavior Problems in Their Early Adolescents." *Journal of Family Issues* 30(4):511–531.

Burton, Linda M., Andrew Cherlin, Donna-Marie Winn, Angela Estacion, and Clara Holder-Taylor. 2009. "The Role of Trust in Low-Income Mothers' Intimate Unions." *Journal of Marriage and Family* 71(December):1107–1124.

Burton, Linda M., and M. Belinda Tucker. 2009. "Romantic Unions in an Era of Uncertainty: A Post-Moynihan Perspective on African American Women and Marriage." *The Annals of the American Academy of Political and Social Science* 621(1):132–148.

Busby, Dean M., and Thomas B. Holman. 2009. "Perceived Match or Mismatch on the Gottman Conflict Styles: Associations with Relationship Outcome Variables." *Family Process* 48(4):531–545.

Busby, Dean M., Brian J. Willoughby, and Jason S. Carroll. 2013. "Sowing Wild Oats: Valuable Experience or a Field Full of Weeds?" *Personal Relationships* 20:706–718.

Buss, D. M., Todd K. Shackelford, Lee A. Kirkpatrick, and Randy J. Larsen. 2001. "A Half Century of Mate Preferences: The Cultural Evolution of Values." *Journal of Marriage and Family* 63(2):491–503.

Butler, J. (1988). Performance Acts and Gender Constitution: An Essay in Phenomenology and Feminist Theory. *Theatre Journal*, 40(4), 519–531.

Butler, Katy. 2006. "Beyond Rivalry: A Hidden World of Sibling Violence." New York Times, February 28.

Butler, Martha, and Cynthia Kirkby. 2013. "Same-Sex Marriage, Divorce and Families: Selected Recent Developments." Parliamentary Information and Research Service.

Butler, Susan Bulkeley. 2015. "New Year, New Chance to Tackle the Old Problem of Workplace Inequality." The SBB Institute for the Development of Women Leaders, January 20. Retrieved February 3, 2015 (http://www .sbbinstitute.org).

Buttigieg, Pete. 2018. "The Key to Happiness Might Be as Simple as a Library or a Park." *The New York Times*. September 14. Retrieved December 28, 2019 (www .nytimes.com).

Byers, E. Sandra. 2005. "Relationship Satisfaction and Sexual Satisfaction: A Longitudinal Study of Individuals in Long-Term Relationships." *Journal of Sex Research* 42(2):113–118.

Byers, E. Sandra, and Jacqueline N. Cohen. 2017. "Validation of the Interpersonal Exchange Model of Sexual Satisfaction with Women in a Same-Sex Relationship." *Psychology of Women Quarterly* 41:32–45.

Byers, E. Sandra, and Heather A. Sears. 2012. "Mothers Who Do and Do Not Intend to Discuss Sexual Health with Their Young Adolescents." *Family Relations* 61(5):851–863.

Byrd, Stephanie Ellen. 2009. "The Social Construction of Marital Commitment." *Journal of Marriage and Family* 71(2):318–336.

Byrne, Kerry, Joseph Orange, and Catherine Ward-Griffin. 2011. "Care Transition Experiences of Spousal Caregivers: From a Geriatric Rehabilitation Unit to Home." *Qualitative Health Research* 21(10):1371–1387.

Byrnes, Hilary, and Brenda Miller. 2012. "The Relationship Between Neighborhood Characteristics and Effective Parenting Behaviors: The Role of Social Support." *Journal of Family Issues* 33(12):1658–1687.

Bzostek, Sharon H. 2008. "Social Fathers and Child Well-Being." *Journal of Marriage and Family* 70(4):950–961.

Bzostek, Sharon, and Christine Percheski. 2016. "Children Living with Uninsured Family Members: Differences by Family Structure." 2016. *Journal of Marriage and Family* 78 (5): 1208–1223.

Bzostek, Sharon, M. J. Carlson, and Sara McLanahan. 2007. "Repartnering after a Nonmarital Birth: Does Mother Know Best?" Working Paper No. 2006-27-FF, Center for Research on Child Wellbeing. Princeton, NJ: Princeton University.

Bzostek, Sharon H., Sara S. McLanahan, and Marcia J. Carlson. 2012. "Mothers' Repartnering after a Nonmarital Birth." *Social Forces* 90(3):817–841.

Cabanes, Jason V., and Kristel A. Acedera. 2012. "Of Mobile Phones and Mother-Fathers: Calls, Text Messages, and Conjugal Power Relations in Mothers-away Filipino Families." *New Media and Society* 14(6):916–930.

Cabaniss, Dale. 2019. "Paid Family Leave for Federal Employees." December 27. U.S. Office of Personnel Management. Retrieved January 30, 2020 (https://chcoc.gov/content/paid-parental-leave-federal-employees).

Cabello, Rosario, Miguel A. Sorrel, Irene Fernández-Pinto, Natalio Extremera, and Pablo Fernández-Berrocal. 2016. "Age and Gender Differences in Ability Emotional Intelligence in Adults: A Cross-sectional Study." *Developmental Psychology* 52:1486.

Cabrera, Natasha, Hiram Fitzgerald, Robert Bradley, and Lori Roggman. 2014. "The Ecology of Father-Child Relationships: An Expanded Model." *Journal of Family Theory and Review* 6(4):336–354.

Calabrese, Sarah K., Joshua G. Rosenberger, Vanessa R. Schick, and David S. Novak. 2015. "Pleasure, Affection, and Love among Black Men Who Have Sex with Men (MSM) versus MSM of Other Races: Countering Dehumanizing Stereotypes via Cross-Race Comparisons of Reported Sexual Experience at Last Sexual Event." *Archives of Sexual Behavior* (44):1–14.

Caldwell, John. 1982. *Theory of Fertility Decline.* London: Academic Press.

Caldwell, Nicole. 2016. "Why Small-Town Dating Is so Much Better than City Dating." Retrieved November 21, 2019 (https://www.thrillist.com/sex-dating/nation/rural-dating-in-america-what-its-really-like)

California Coalition Against Violence. n.d. "The Cycle of Violence: Why Do Women Stay in Violent Relationships?" Retrieved March 14, 2013(http://www .coalitionagainstviolence.ca).

California Secretary of State's Business Programs Division. 2011. *Terminating a California Registered Domestic Partnership.* Sacramento, CA: State Government Offices, Notary Public and Special Filings Section. Retrieved December 5, 2012 (http://www.sos.ca.gov/dpregistry/forms/sf-dp2.pdf).

Call, Vaughn, Susan Sprecher, and Pepper Schwartz. 1995. "The Incidence and Frequency of Marital Sex in a National Sample." *Journal of Marriage and Family* 57(3):639–652.

Callahan, Daniel. 1985. "What Do Children Owe Elderly Parents? Toward a Policy that Promotes, not Corrupts, Family Bonds." *Hastings Center Report* (April):32–37.

Calms, Jackie. 2013. "Obama Signs Expanded Anti-Violence Law." *The New York Times*, March 7. Retrieved March 14, 2013 (http://thecaucus.blogs.nytimes .com).

Calvette, Esther, Izaskun Orue, Lorena Bertino, Zahira Gonzalez, Yadira Montes, Patricia Padilla, and Roberto Pereira. 2014. "Child-to-Parent Violence in Adolescents." *Journal of Family Violence* 29(3):343–352.

Calzo, Jerel P., and L. Monique Ward. 2009. "Contributions of Parents, Peers, and Media to Attitudes Toward Homosexuality: Investigating Sex and Ethnic Differences." *Journal of Homosexuality* 56(8):1101–1116.

Cameron, Nicole, 2018. "The Effects of Exposure to Consequences of Bystander Intervention on Intentions to Intervene in Intimate Partner Violence Situations." Ph.D. dissertation, Washington State University.

Canary, D. J., and K. Dindia. 1998. *Sex Differences and Similarities in Communication.* Mahwah, NJ: Erlbaum.

Cancian, Francesca M. 1985. "Gender Politics: Love and Power in the Private and Public Spheres." pp. 253–264 in *Gender and the Life Course,* edited by Alice S. Rossi. New York: Aldine.

———. 1987. *Love in America: Gender and Self-Development.* New York: Cambridge University Press.

Cancian, Francesca M., and Stacey J. Oliker. 2000. *Caring and Gender.* Walnut Creek, CA: AltaMira.

Cancian, M., and D. Meyer. 2014. "Testing the Economic Independence Hypothesis: The Effect of an Exogenous Increase in Child Support on Subsequent Marriage and Cohabitation." *Demography* 51:857–880.

Cancian, Maria, Daniel R. Meyer, Patricia R. Brown, and Steven T. Cook. 2014. "Who Gets Custody Now? Dramatic Changes in Children's Living Arrangements after Divorce." *Demography* 51:1381–1396.

Cancian, Maria, Daniel R. Meyer, and Robert Wood. 2019. "Final Impact Findings from the Child Support Noncustodial Parent Employment Demonstration (CSPED). No. 93c8996bb09444a292e2598d2d315aa3. Mathematica Policy Research, 2019.

Cancian, Maria, Daniel R. Meyer, and Eunhee Han. 2012. "Full-Time Father or 'Deadbeat Dad'? Does the Growth in Father Custody Explain the Declining Share of Single Parents with a Child Support Order?" Institute for Research on Poverty, University of Wisconsin. Retrieved April 26, 2013 (http://paa2012 .princeton .edu/papers/120291).

Cantor, M. H. 1979. "Neighbors and Friends: An Overlooked Resource in the Informal Support System." *Research on Aging* 1:434–463.

Cao, H., N. Zhou, M. Fine, Y. Liang, J. Li, and W. R. Mills-Koonce. 2017. "Sexual Minority Stress and Same-Sex Relationship Well-Being: A Meta Analysis." *Journal of Marriage and Family* 79 (5): 1258 –1277.

Capizzano, Jeffrey, Gina Adams, and Jason Ost. 2006. *Caring for Children of Color: The Child Care Patterns of White, Black and Hispanic Children.* Washington, DC: Urban Institute.

Cardiff, Ashley. 2013. *Don't Be That Guy.* Retrieved May 22, 2013 (http://www .thegloss.com/2012/12/04/sex-and -dating/canada-date -rape-campaign - 991/#ixzz2U32E1Qk4).

Cardinal, Sherry. 2019. Stress Proofing Your Life. Retrieved February 25, 2020 (https://www.criticalincidentstress.com/stress_proofing_your_life).

Cardoso, Jodi Berger, Jennifer Scott, Monica Faulkner, and Liza Barnes Lane. 2018. "Parenting in the Context of Deportation Risk." *Journal of Marriage and Family* 80 (2):301–316.

Caregiver Support Blog. 2008 (October 13). Retrieved March 13, 2010 (caregiversupport. wordpress.com).

Carey, Benedict. 2012. "Debate on a Study Examining Gay Parents." *The New York Times*, June 11. Retrieved August 28, 2012 (http://www .nytimes.com).

Carey v. Population Services International. 1977. 431 U.S. 678, 52 L. Ed. 2d 675, 97 S. Ct. 2010.

Carlson, Bonnie E., Katherine Maciol, and Joanne Schneider. 2006. "Sibling Incest: Reports from Forty-One Survivors." *Journal of Child Sexual Abuse* 15(4):19–34.

Carlson, Daniel L. 2015. "The Roles of Paid and Unpaid Labor in the Production of Gender Inequality in Families." *Family Focus.* Issue FF63: F7-F8.

Carlson, Daniel L., Sarah Hanson, and Andrea Fitzroy. 2015. "The Division of Childcare, Sexual Intimacy, and Relationship Quality in Couples." *Sociology Faculty Publications Paper 4.* Retrieved January 7, 2015 (http://scholarworks.gsu .edu/sociology_facpub/4).

Carlson, Daniel L., and Chris Knoester. 2011. "Family Structure and the Intergenerational Transmission of Gender Ideology." *Journal of Family Issues* 32(6):709–734.

Carlson, Daniel L. and Jamie L. Lynch. 2017. "Purchases, Penalties, and Power: The Relationship between Earnings and Housework." *Journal of Marriage and Family* 79 (1): 199–224.

Carlson, Marcia, Alicia VanOrman, and Kimberly Turner. 2017. "Fathers' Investments of Money and Time Across Residential Contexts." *Journal of Marriage and Family* 79 (1): 10–23.

Carlson, M. J., VanOrman, A. G., & Turner, K. J. (2017). Fathers' Investments of Money and Time across Residential Contexts." *Journal of Marriage and Family* 79:10–23.

Carlson, Elwood. 2009. "20th-Century U.S. Generations." *Population Bulletin* 64(1). Retrieved June 10, 2009 (www.prb.com).

Carlson, Marcia J. 2006. "Family Structure, Father Involvement, and Adolescent Behavior Outcomes." *Journal of Marriage and Family* 68(1):137–154.

———. 2011. "Adults Who Have Biological Children with More than One Partner: Patterns and Implications for U.S. Families." *Population Reference Bureau Policy Seminar*, April 15.

Carlson, Marcia J., and Frank F. Furstenberg Jr. 2006. "The Prevalence and Correlates of Multipartnered Fertility among Urban U.S. Parents." *Journal of Marriage and Family* 68(3):718–732.

Carlson, Marcia J., Natasha V. Pilkauskas, Sara S. McLanahan, and Jeanne Brooks-Gunn. 2011. "Couples as Partners and Parents Over Children's Early Years." *Journal of Marriage and Family* 73(2):317–334.

Carlson, Marcia J., Alicia G. VanOrman, and Natasha V. Pilkauskas. 2013. "Examining the Antecedents of US Nonmarital Fatherhood." *Demography* 50:1421–1447.

Carlson, Ryan G., Andrew P. Daire, Matthew D. Munyon, and Mark E. Young. 2012. "A Comparison of Cohabiting and Noncohabiting Couples Who Participated in Premarital Counseling Using the PREPARE Model." *The Family Journal* 20(2):123–130.

Carmalt, Julie H., John Cawley, Kara Joyner, and Jeffery Sobal. 2008. "Body Weight and Matching With a Physically Attractive Romantic Partner." 2008. *Journal of Marriage and the Family* 70:1287–1296.

Carney, Michelle, Fred Buttell, and Don Dutton. 2007. "Women Who Perpetrate Intimate Partner Violence: A Review of the Literature with Recommendations for Treatment." *Aggression and Violent Behavior* 12(1):108–115.

Carr, Anne, and Mary Stewart Van Leeuwen, eds. 1996. *Religion, Feminism, and the Family*. Louisville, KY: Westminster John Knox Press.

Carr, Deborah. 2004. "The Desire to Date and Remarry among Older Widows and Widowers." *Journal of Marriage and Family* 66(4):1051–1068.

Carr, Deborah, and Rebecca Utz. 2020. "Families in Later Life: A Decade in Review." *Journal of Marriage and Family* 82 (1): 346–363.

Carré, Justin M., and Cheryl M. McCormick. 2008. "Aggressive Behavior and Change in Salivary Testosterone Concentrations Predict Willingness to Engage in a Competitive Task." *Hormones and Behavior* 54(3):403–409.

Carrigan, William D., and Clive Webb. 2009. "Repression and Resistance: The Lynching of Persons of Mexican Origin in the United States, 1848–1928." pp. 69–86 in *How the United States Racializes Latinos: White Hegemony and Its Consequences*, edited by Jose A. Cobas,

Jorge Duany, and Joe R. Feagin. Boulder, CO: Paradigm Publishers.

Carroll, Helen. 2015. "Think Grown-Up Children Can't Be Hurt by Divorce? These Tales Prove You're WRONG." Retrieved March 19, 2016 (http://www.dailymail.co.uk /femail/article-3110137/Think -grown -children-t-hurt-parents-getting -divorced -haunting-stories-prove-WRONG .html).

Carroll, Jason S., Sarah Badger, and Chongming Yang. 2006. "The Ability to Negotiate or the Ability to Love? Evaluating the Developmental Domains of Marital Competence." *Journal of Family Issues* 27(7):1001–1032.

Carroll, Jason S., Chad D. Olson, and Nicolle Buckmiller. 2007. "Family Boundary Ambiguity: A 30-Year Review of Theory, Research, and Measurement." *Family Relations* 56(2):210–230.

Carroll, Joseph. 2007a. "Public: 'Family Values' Important to Presidential Vote." December 26. Retrieved May 20, 2010 (www.gallup.com/poll).

———. 2007b. "Stress More Common among Younger Americans, Parents, Workers." Gallup News Service, January 24. Retrieved February 19, 2009 (www.gallup .com/poll).

———. 2008. "Time Pressures, Stress Common for Americans." *Gallup Poll*, January 2. www. gallup.com.

Carrns, Ann. 2020. "Many Adults Are Helping Their Parents Financially Despite Strain." *The New York Times*. January 31. Retrieved February 10, 2020 (https://www.nytimes .com/2020/01/31/your-money/finances-adult -children-parents.html).

Carroll, Megan. 2018. "Gay Fathers on the Margins: Race, Class, Marital Status, and Pathway to Parenthood." *Family Relations* 67 (1): 104–117.

Carson, R., Dunstan, E., Dunstan, J., & Roopani, D. (2018). *Children and Young People in Separated Families: Family Law System Experiences and Needs*. Melbourne: Australian Institute of Family Studies.

Carstensen, Laura L. 2015. "The New Age of Much Older Age." *Time*, February 23:69–97.

Carter, Allison, Jessie V. Ford, Maya Luetke, Tsung-chieh Jane Fu, Ashley Townes, Devon J. Hensel, Brian Dodge, and Debby Herbenick. 2019. "'Fulfilling His Needs, Not Mine': Reasons for Not Talking About Painful Sex and Associations with Lack of Pleasure in a Nationally Representative Sample of Women in the United States." *The Journal of Sexual Medicine* 16:1953–1965.

Carter, Betty, and Monica McGoldrick. 1988. *The Changing Family Life Cycle: A Framework for Family Therapy*, 2nd ed. New York: Gardner.

Carter, Gerard A. 2009. "Book Review: Unmarried Couples with Children." *Journal of Marriage and Family* 71(2):432–434.

Carwile, J.L., Willett, W.C., Spiegelman, D., Hertzmark, E., Rich-Edwards, J., Frazier, A.L. and Michels, K.B. 2015. "Sugar-sweetened Beverage Consumption and Age at Menarche in a Prospective Study of U.S. Girls." *Human Reproduction* 30:675–683.

Cartwright, Claire. 2008. "Resident Parent-Child Relationships in Stepfamilies." pp. 208–230 in *The International Handbook of Stepfamilies*, edited by J. Pryor. Hoboken, NJ: John Wiley & Sons.

Case, A., and C. Paxson. 2001. "Mothers and Others: Who Invests in Children's Health." *Journal of Health Economics* 20:301–328.

Case, Anne, I-Fen Lin, and Sara McLanahan. 2000. "How Hungry Is the Selfish Gene?" *The Economic Journal* 110(October):781–804.

Case *Van Kuck v. Germany*. 2003, June 12. Retrieved October 8, 2015 (http://www .menschenrechte.ac.at/orig/03_3/Kuck .pdf).

Casey, Erin C., Rebecca J. Shlafer, and Ann S. Masten. 2015. "Parental Incarceration as a risk Factor for Children in Homeless Families." *Family Relations* 64(4):490–504.

Casper, Lynne M., and Suzanne M. Bianchi. 2002. *Continuity and Change in the American Family*. Thousand Oaks, CA: Sage.

Cassano, Michael C., and Janice L. Zeman. 2010. "Parental Socialization of Sadness Regulation in Middle Childhood: The Role of Socialization and Gender." *Developmental Psychology* 46(5):1214–1226.

Casselman, Ben, Patricia Cohen, and Doris Burke. 2018. "The Great Recession Knocked Them Down. Only Some Got Up Again." *The New York Times*. September 12. Retrieved September 15, 2019 (www .nytimes.com).

Castillo, Linda G., Rachel L. Navarro, Jo Ellyn O. Y. Walker, Seth J. Schwartz, Byron L. Zamboanga, Susan Krauss Whitbourne, Robert S. Weisskirch et al. 2015. "Gender Matters: The Influence of Acculturation and Acculturative Stress on Latino College Student Depressive Symptomatology." *Journal of Latina/o Psychology* 3:40.

Castrén, Anna-Maija, and Eric D. Widmer. 2015. "Insiders and Outsiders in Stepfamilies: Adults' and Children's Views on Family Boundaries." *Current Sociology* 63:35–56.

Catalano, Shannan. 2012. *Intimate Partner Violence, 1993–2010*. NCJ 239203 Washington, DC: U.S. Bureau of Justice Statistics, November. Retrieved March 13, 2013 (www.ojp.usdoj.gov /bjs).

———. 2013. *Intimate Partner Violence, 1993– 2011*. NCJ 243300 Washington, DC: U.S. Bureau of Justice Statistics, November. Retrieved March November 20, 2015 (http://www.ojp.usdoj.gov /bjs).

Catalpa, Jory. M., and Jasmine M. Routon. 2018. "Queer Kinship: Family Networks Among Sexual and Gender Minoriites." Family Focus FFS75:F12-F13.

Catron, Mandy Len. 2015. "To Fall in Love with Anyone, Do This." *The New York Times*, January 9. Retrieved January 17, 2016 (http://www .nytimes.com/2015/01/11/fashion/modern -love-to-fall-in-love-with -anyone-do-this.html).

Caughy, Margaret. 2011. "Profiles of Racial Socialization among African American Parents: Correlates, Context, and Outcome." *Journal of Child and Family Studies* 20(4):491–502.

Cavanagh, Shannon E. 2008. "Family Structure History and Adolescent Adjustment." *Journal of Family Issues* 29(7):944–980.

———. 2011. "Early Pubertal Timing and the Union Formation. Behaviors of Young Women" *Social Forces* 89(4):1217–1238.

Cave, Damien. 2012. "American Children, Now Struggling to Adjust to Life in Mexico." *The New York Times*, June 18. Retrieved June 20, 2012 (http://www.nytimes.com).

CBS News. 2016. "One Sperm Donor's Extended Family." Retrieved February 14, 2016 (http://www.cbsnews.com/news/one-sperm-donors-extended-family/).

Ceballo, Rosario, Jennifer E. Lansford, Antonia Abbey, and Abigail J. Stewart. 2004. "Gaining a Child: Comparing the Experiences of Biological Parents, Adoptive Parents, and Stepparents." *Family Relations* 53(1):38–48.

Celock, John. 2013. "Kansas Sperm Donor for Lesbian Couple Faces Child Support Suit from State." *The Huffington Post, January* 2. Retrieved January 15, 2013 (http://www.huffingtonpost.com).

Center for the Advancement of Women. 2003. "Progress and Perils: New Agenda for Women." (www.advancewomen.org).

Center for American Women and Politics. 2019. Women in the U.S. Congress 2019. Retrieved October 2, 2019 (https://www.cawp.rutgers.edu/women-us-congress-2019).

Center for the Improvement of Child Caring. n.d. "Systematic Training for Effective Parenting Programs." Retrieved February 12, 2007 (www.ciccparenting.org).

Center for LGBTQIZ+ Student Success, Iowa State University. 2019. "Terms and Definitions." Retrieved September 24, 2019 (https://center.dso.iastate.edu/sites/default/files/Terms%20and%20Definitions%20for%20website%20(Oct%202017).pdf).

Center for Women and Business at Bentley University. 2017. Multi-generational Impacts on the Workplace. Bentley University.

Centers for Disease Control. 2019a. "Assisted Reproductive Technology (ART)." Retrieved December 23, 2019 (https://www.cdc.gov/art/artdata/index.html).

Centers for Disease Control. 2019b. "Contraception." Retrieved December 22, 2019 (https://www.cdc.gov/reproductivehealth/contraception/index.htm#Contraceptive-Effectiveness).

Centers for Disease Control and Prevention. 2018. "Estimated HIV Incidence and Prevalence in the United States, 2010–2015." *HIV Surveillance Supplemental Report 2018*. Retrieved November 2, 2019 (http://www.cdc.gov/hiv/library/reports/hiv-surveillance.html).

Centers for Disease Control. 2016. "19 Critical Sexual Education Topics." Retrieved October 21, 2019 (https://www.cdc.gov/healthyyouth/data/profiles/pdf/19_criteria_landscape.pdf).

Centers for Disease Control. 2018. *HIV Surveillance Report, 2016, vol. 28*. Retrieved November 2, 2019 (http://www.cdc.gov/hiv/library/reports/hiv-surveillance.html).

Centers for Disease Control. 2019a. "Pre-Exposure Prophylaxis (PrEP)." Retrieved November 2, 2019 (https://www.cdc.gov/hiv/risk/prep/index.html).

Centers for Disease Control. 2019b. "Reproductive Health: Teen Pregnancy." Retrieved October 29, 2019 (https://www.cdc.gov/teenpregnancy/about/index.htm).

Center for Evidence & Innovation 2019. *Pulse*. Retrieved December 20, 2019 (https://www.healthyteennetwork.org/projects/pulse/).

Cha, Ariana Eunjung. 2018a. "Discounts, Guarantees, and the Search for 'Good' Genes: The Booming Fertility Business. " Retrieved December 25, 2019 (https://www.washingtonpost.com/national/health-science/donor-eggs-sperm-banks-and-the-quest-for-good-genes/2017/10/21/64b9bdd0-aaa6-11e7-b3aa-c0e2e1d41e38_story.html).

Cha, Ariana Eunjung. 2018b. "From Sex Selection to Surrogates, American IVF Clinics Provide Services Outlawed Elsewhere. " https://www.washingtonpost.com/national/health-science/from-sex-selection-to-surrogates-american-ivf-clinics-provide-services-outlawed-elsewhere/2018/12/29/0b596668-03c0-11e9-9122-82e98f91ee6f_story.html).

Cha, Ariana Eunjung. 2018c. "How Religion is Coming to Terms with Modern Fertility Methods. " Retrieved December 25, 2019 (https://www.washingtonpost.com/graphics/2018/national/how-religion-is-coming-to-terms-with-modern-fertility-methods/).

Cha, Ariana Eunjung. 2018a. "44 Siblings and Counting." *The Washington Post*. September 12. Retrieved August 16, 2019 (www.washingtonpost.com).

Cha, Ariana Eunjung. 2019d. "Who Gets the Embryos? " Retrieved December 25, 2019 (https://www.washingtonpost.com/national/health-science/who-gets-the-embryos-whoever-wants-to-make-them-into-babies-new-law-says/2018/07/17/8476b840-7e0d-11e8-bb6b-c1cb691f1402_story.html).

Chabot, Jennifer M., and Barbara D. Ames. 2004. "'It Wasn't "Let's Get Pregnant and Go Do It"': Decision Making in Lesbian Couples Planning Motherhood Via Donor Insemination." *Family Relations* 53(4):348–356.

Chadiha, Letha A., Jane Rafferty, and Joseph Pickard. 2003. "The Influence of Caregiving Stressors, Social Support, and Caregiving Appraisal on Marital Functioning among African American Wife Caregivers." *Journal of Marital and Family Therapy* 29(4):479–490.

Chafetz, Janet Saltzman. 1989. "Marital Intimacy and Conflict: The Irony of Spousal Equality." pp. 149–156 in *Women: A Feminist Perspective*, 4th ed., edited by Jo Freeman. Mountain View, CA: Mayfield.

Chalabi, Mona. 2015. "I Love You, You Love Me, We've Probably Never Said That to More Than Three." Retrieved January 12, 2016 (http://fivethirtyeight.com/datalab/i-love-you-you-love-me-weve-probably-never-said-that-to-more-than-three/).

Chambers, Anthony L., and Aliza Kravitz. 2011. "Understanding the Disproportionately Low Marriage Rate among African Americans: An Amalgam of Sociological and Psychological Constraints." *Family Relations* 60(5): 648–660.

Champion, Jennifer E., Sarah S. Jaser, Dristen L. Reeslund, Lauren Simmons, Jennifer E. Potts, Angela R. Shears, and Bruce E. Compas. 2009. "Caretaking Behaviors by Adolescent Children of Mothers with and without a History of Depression." *Journal of Family Psychology* 23(2):156–166.

Chandra, A., Copen, C. E., and Mosher, W. D. 2013. "Sexual Behavior, Sexual Attraction, and Sexual Identity in the United States: Data from the 2006–2010 National Survey of Family Growth." pp. 45–66 in the *International Handbook on the Demography of Sexuality*, edited by Amanda K. Baumle. Netherlands: Springer.

Chandra, Anjani, Casey E. Copen, and Elizabeth Hervey Stephen. 2013. *Infertility and Impaired Fecundity in the United States, 1982–2010: Data from the National Survey of Family Growth*. National Health Statistics Reports, no. 67. Hyattsville, MD: National Center for Health Statistics.

Chandra, Anjani, Gladys M. Martinez, William D. Mosher, Joyce C. Abma, and Jo Jones. 2005. "Fertility, Family Planning, and Reproductive Health of U.S. Women: Data from the 2002 National Survey of Family Growth." *Vital and Health Statistics* 23(25), December. Hyattsville, MD: U.S. National Center for Health Statistics.

Chandra, Anjani, William D. Mosher, Casey Copen, and Catlainn Sionean. 2011. *Sexual Behavior, Sexual Attraction, and Sexual Identity in the United States: Data from the 2006–2008 National Survey of Family Growth*. National Health Statistics Reports No. 36. Hyattsville, MD: National Center for Health Statistics. 2011.

Chandra, Anjani, and Elizabeth Hervey Stephen. 2010. "Infertility Service Use among U.S. Women: 1995 and 2002." *Fertility and Sterility* 93(3):725–736.

Chandra, Sho, and Sarah Foster. 2018. "Why So Many Americans Still Work Multiple Jobs in Strong Market." *Los Angeles Times*. October 4. Retrieved January 3, 2019 (www.latimes.com).

Chang, Juju, and Dan Lieberman. 2012. "Swingers: Inside the Secret World of Provocative Parties and Couples Who 'Swap.'" *ABC Nightline*, May 21. Retrieved December 3, 2012 (http://www.abcnews.go.com).

Chao, R. K. 1994. "Beyond Parental Control and Authoritarian Parenting Style: Understanding Chinese Parenting through the Cultural Notion of Training." *Child Development* 65(4):1111–1119.

Chaplin, Tara M., Pamela M. Cole, and Carolyn Zahn-Waxler. 2005. "Parental Socialization of Emotion Expression: Gender Differences and Relations to Child Adjustment." *Emotion* 5(1):80–88.

Chapman, Ashton, Marilyn Coleman, and Lawrence Ganong. 2016. "'Like my Grandparent, but Not': A Qualitative Investigation of Skip-Generation Stepgrandchild-Stepgrandparent Relationships." *Journal of Marriage and Family* 78 (3): 634–643.

Chapman, Gary. 2014. *The Five Love Languages*. Chicago: Northfield Publishing.

Chappell, Crystal Lee Hyun Joo. 1996. "Korean-American Adoptees Organize for Support." *Minneapolis Star Tribune*, December 29, p. E7.

Charles, Laurie L., Dina Thomas, and Matthew L. Thornton. 2005. "Overcoming Bias toward Same-Sex Couples." *Journal of Marital and Family Therapy* 31(3):239–249.

Chatters, Linda M., Robert Joseph Taylor, Amanda Toler Woodward, and Emily J. Nicklett. 2015. "Social Support from Church and Family Members and Depressive Symptoms among Older African Americans." *American Journal of Geriatric Psychiatry* 23:559–567.

Chatzky, Jeann. 2006. "Just When You Thought It Was Safe to Retire. . . ." CNN Money.com, September 21. Retrieved October 29, 2006 (money.cnn.com).

Chaves, Mark, Shawna Anderson, and Jason Byassee. 2009. "American Congregations at the Beginning of the 21st Century." *National Congregations Study*.

Cheadle, Jacob E., Paul R. Amato, and Valerie King. 2010. "Patterns of Nonresident Father Contact." *Demography* 47(1):205–225.

Chemaly, Soraya. 2018. *Rage Becomes Her: The Power of Women's Anger*. Simon and Schuster.

Chen, Y., 2015. "Does a Nonresident Parent have the Right to Make Decisions for his Nonmarital Children? Trends in Legal Custody among Paternity Cases." *Children and Youth Services Review* 51:55–65.

Chen, Yingyu, and Sarah E. Ullman. 2010. "Women's Reporting of Sexual and Physical Assaults to Police in the National Violence Against Women Survey." *Violence Against Women* 16(3):262–279.

Cheng, Simon, Laura Hamilton, Stacy Missari, and Josef (Kuo-Hsun) Ma. 2014. "Sexual Subjectivity among Adolescent Girls: Social Disadvantage and Young Adult Outcomes." *Social Forces* 93(2):515–544.

Cherlin, Andrew J. 2020. "Degrees of Change: An Assessment of the Deinstitutionalization of Marriage Thesis." *Journal of Marriage and Family* 82:62–80.

Cherlin, Andrew J. 2017. "Introduction to the Special Collection on Separation, Divorce, Repartnering, and Remarriage around the World." *Demographic Research* 37:1275–1296.

Cherlin, Andrew, Erin Cumberworth, S. Philip Morgan, and Christopher Wimer. 2013. "The Effects of the Great Recession on Family Structure and Fertility." *Annals of the American Academy of Political and Social Science* 650:214–231.

Cherlin, Andrew J., and Frank F. Furstenberg Jr. 1986. *The New American Grandparent: A Place in the Family, a Life Apart*. New York: BasicBooks.

Chesler, Phyllis. 2005 [1972]. *Women and Madness*. New York: Palgrave/Macmillan.

Chesley, Noelle. 2011. "Stay-at-Home Fathers and Breadwinning Mothers." *Gender and Society* 25(5):642–664.

Chesser, Casandra. 2017. "It's Not Okay For You To Pass Judgement On How Many Kids I Have." Retrieved December 11, 2019 (https://thefederalist.com/2017/01/20/its-not-okay-for-you-to-pass-judgment-on-how-many-kids-i-have/).

Cheung, Felix, and Richard E. Lucas. 2015. "When Does Money Matter Most? Examining the Association between Income and Life Satisfaction over the Life Course." *Psychology and Aging* 30:120.

"Child Care Costs on the Upswing, Census Bureau Reports." 2013. Report Number CB13-62. U.S. Census Bureau. Retrieved November 17, 2015 from http://www .census.gov.

Child Care Aware. 2019. *The U.S. and the High Cost of Child Care: Examination of a Broken System, 2019 Report*. Retrieved December 10, 2019 (www.usa.childcareaware.org).

Child Trends, 2018. "The Rate of High School-aged Youth Considering and Committing Suicide Continues to Rise, Particularly among Female Students." Retrieved November 12, 2018 (www.childtrends.org).

Child Trends. 2019. "Dating. " Retrieved November 21, 2019 (https://www.childtrends.org/indicators/dating).

Child Trends. 2018a. *Teen Abortions*. Retrieved December 20, 2019 (https://www.childtrends.org/indicators/teen-abortions).

Child Trends. 2018b. *Teen Pregnancy*. Retrieved December 20, 2019 (https://www.childtrends.org/indicators/teen-pregnancy).

Child Trends. 2019. "Children in Poverty." Retrieved February 8, 2020 (https://www.childtrends.org/indicators/children-in-poverty).

Child Welfare Information Gateway. 2019. *Foster Care Statistics 2017*. Retrieved December 9, 2019 (www.childwelfare.gov).

Children in the Middle. 2013. "Rules for Co-Parenting." Retrieved April 13, 2013. (http://www.childreninthemiddle.com/rulescoparent.htm).

Children's Bureau. 2014. "Trends in Foster Care and Adoption." Retrieved February 14, 2016 (http://www.acf.hhs.gov/sites/default/files/cb/trends_fostercare_adoption2014.pdf).

Children's Defense Fund. 2017. *The State of America's Children 2017*. Retrieved December 9, 2019 (www.childrensdefensefund.org).

Childs, Erica Chito. 2008. "Listening to the Interracial Canary: Contemporary Views on Interracial Relationships among Blacks and Whites." *Fordham Law Review* 76(6):2772–2786.

Chin, Helen B., Theresa Ann Sipe, Randy Elder, Shawna L. Mercer, Sajal K. Chattopadhyay, Verughese Jacob, Holly R. Wethington et al. 2012. "The Effectiveness of Group-based Comprehensive Risk-reduction and Abstinence Education Interventions to Prevent or Reduce the Risk of Adolescent Pregnancy, Human Immunodeficiency Virus, and Sexually Transmitted Infections: Two Systematic Reviews for the Guide to Community Preventive Services." *American Journal of Preventive Medicine* 42:272–294.

Chiu, Allyson. 2019. "Their Only Son Died Unexpectedly, but a Court Order to Retrieve His Sperm Means His Legacy Could Live On." *The Washington Post*. March 5. Retrieved August 14, 2019 (www.washingtonpost.com).

Cho, Rosa M. 2011. "Understanding the Mechanism behind Maternal Imprisonment and Adolescent School Dropout." *Family Relations* (603):272–289.

Choi, Heejeong, and Nadine F. Marks. 2013. "Marital Quality, Socioeconomic Status, and Physical Health." *Journal of Marriage and Family* 75(4):903–919.

Choi, Kate H., and Marta Tienda. 2017. "Marriage-Market Constraints and Mate-Selection Behavior: Racial, Ethnic, and Gender Differences in Intermarriage." *Journal of Marriage and Family* 79:301–317.

Choice, Pamela, and Leanne K. Lamke. 1997. "A Conceptual Approach to Understanding Abused Women's Stay/Leave Decisions." Journal of Family Issues 18:290–314.

Chokshi, Niraj. 2019. "Your Kids Think You're Addicted to Your Phone." Retrieved January 25, 2020 (https://www.nytimes.com/2019/05/29/technology/cell-phone-usage.html).

Chotpitayasunondh, Varoth, and Karen M. Douglas. 2016. "How "Phubbing" Becomes the Norm: The Antecedents and Consequences of Snubbing via Smartphone." *Computers in Human Behavior* 63:0 18.

Chotpitayasunondh, Varoth, and Karen M. Douglas. 2018. "The Effects of "Phubbing" on Social Interaction." *Journal of Applied Social Psychology* 48(6):304–316.

Chrisler, Alison J. 2017. "Understanding Parent Reactions to Coming Out as Lesbian, Gay, or Bisexual: A Theoretical Framework. *Journal of Family Theory and Review* 9 (2): 165 –181.

Christopher, F. Scott, and Susan Sprecher. 2000. "Sexuality in Marriage, Dating, and Other Relationships." *Journal of Marriage and Family* 62:999–1017.

Christopherson, Brian. 2006. "Some Buck Trends, Marry Before Finishing College." *Lincoln Journal Star*, October 3. Retrieved October 4, 2006 (www.journalstar.com).

Christie-Mizell, C. Andre, Erin M. Pryor, and Elizabeth Grossman. 2008. "Child Depressive Symptoms, Spanking, and Emotional Support: Differences Between African American and European American Youth." *Family Relations* 57(June):335–350.

Chuck, Elizabeth. 2018. "From Sperm Donor to 'Dad': When Strangers with Shared DNA Become a Family." *U.S. News*. Retrieved August 18, 2019 (www.usnews.com).

Ciabattari, Teresa. 2004. "Cohabitation and Housework: The Effects of Marital Intentions." *Journal of Marriage and Family* 66(1):118–125.

Ciaramigoli, Arthur P., and Katherine Ketcham. 2000. *The Power of Empathy: A Practical Guide to Creating Intimacy, Self-Understanding, and Lasting Love in Your Life*. New York: Dutton.

Cieraad, Irene. 2006. At Home: *An Anthropology of Domestic Space*. Syracuse, NY: Syracuse University Press.

Claire Masurel 2002. *Two Homes*. Summerville, MA: Candlewick Press.

Clark, Jane Bennett. 2010. "Finance Basics for Partners." *Kiplinger's Personal Finance* 64(8):50–52.

Clark, Michele C., and Pamela M. Diamond. 2010. "Depression in Family Caregivers of Elders: A Theoretical Model of Caregiver Burden, Sociotropy, and Autonomy." *Research in Nursing and Health* 33:20–34.

Clark, Vicki, Catherine Huddleston-Casas, Susan Churchill, Denise O'Neil Green, and Amanda Garrett. 2008. "Mixed Methods Approaches in Family Science Research." *Journal of Family Issues* 29(11):1543–1566.

Clark University. 2012a. "New Clark Poll: 18 to 29-year-olds are Traditional about Roles in Sex, Marriage, and Raising Children." Retrieved January 26, 2016 (http://news.clarku.edu/news/2012/08/07/new-clark-poll-18-to-29-year-olds-are-traditional-about-roles-in-sex-marriage-and-raising-children/).

———. 2012b. "New Clark Poll of Emerging Adults belies 'Freeloader' Stereotype." Retrieved

January 26, 2016 (http://news.clarku.edu/news/2012/09/11/new-clark-poll-of-emerging-adults-belies-freeloader-stereotype/).

———. 2013a. "What is the Key to Being an Adult? Clark Releases New Poll Findings." Retrieved January 26, 2016 (http://news.clarku.edu/news/2013/06/17/what-is-the-key-to-being-an-adult-clark-releases-new-poll-findings/).

———. 2013b. "Parents of Emerging Adults." Retrieved January 26, 2016 (http://www.clarku.edu/clark-poll-emerging-adults/).

Clarke, L. 2005. "Remarriage in Later Life: Older Women's Negotiation of Power, Resources, and Domestic Labor." *Journal of Women and Aging* 17(4):21–41.

Claxton-Oldfield, S., and Butler, B. (1998). "Portrayal of Stepparents in Movie Plot Summaries." *Psychological Reports* 82:879–882.

Claxton-Oldfield, Stephen. 2008. "Stereotypes of Stepfamilies and Stepfamily Members." pp. 30–52 in *The International Handbook of Stepfamilies*, edited by J. Pryor. Hoboken, NJ: John Wiley & Sons.

Claxton-Oldfield, Stephen, Carla Goodyear, Tina Parsons, and Jane Claxton-Oldfield. 2002. "Some Possible Implications of Negative Stepfather Stereotypes." *Journal of Divorce and Remarriage* Spring-Summer:77–89.

Clayton, Any, and Maureen Perry-Jenkins. 2008. "No Fun Anymore: Leisure and Marital Quality Across the Transition to Parenthood." *Journal of Marriage and Family* 70(1):28–43.

Clayton, Obie, and Joan Moore. 2003. "The Effects of Crime and Imprisonment on Family Formation." pp. 84–102 in *Black Fathers in Contemporary American Society: Strengths, Weaknesses, and Strategies for Change*, edited by Obie Clayton, Ronald B. Mincy, and David Blankenhorn. New York: Russell Sage

Cleek, Elizabeth, Matt Wofsy, Nancy Boyd-Franklin, Brian Mundy, and Ramika Lowell. 2012. "The Family Empowerment Program." *Family Process* 51(2):207–217.

Clemetson, Lynette. 2006. "Adopted in China: Seeking Identity in America." *The New York Times*, March 23.

———. 2007. "Working on Overhaul, Russia Halts Adoption Applications." *The New York Times*, April 12.

Clemetson, Lynette, and Ron Nixon. 2006. "Overcoming Adoption's Racial Barriers." *The New York Times*, August 17.

Cloud, Henry, and John Sims Townsend. 2005. *Rescue Your Love Life: Changing Those Dumb Attitudes and Behaviors That Will Sink Your Marriage*. Nashville, TN: Integrity Publishers.

Cloud, John. 2007. "Busy Is O.K." *Time*, January 29, p. 51.

Clunis, D. Merilee, and G. Dorsey Green. 2005. *Lesbian Couples: A Guide to Creating Healthy Relationships*. Emeryville, CA: Seal Press.

CNN.com. 2010. "Facebook a 'Tool' for Cheating Spouses, Some Say." Retrieved December 15, 2015 (http://www.cnn.com/2010/TECH/social.media/07/14/facebook.cheating/).

Coalition for Marriage, Family, and Couples Education. 2009. "Smart Marriages." Retrieved October 11, 2009 (www.smartmarriages.com).

Coan, James A., and John M. Gottman. 2007. "Sampling, Experimental Control, and Generalizability in the Study of Marital Process Models." *Journal of Marriage and Family* 69(1):73–80.

Coates, Jennifer. 2015. *Women, Men and Language: A Sociolinguistic Account of Gender Differences in Language*. Routledge.

Coates, Karen, and Valeria Fernandez. 2019. "The Young Hands That Feed Us." *Pacific Standard*. July 9. Retrieved August 27, 2019 (https://psmag.com).

Coates, Ta-Nehisi. 2015. "The Black Family in the Age of Incarceration." *The Atlantic*, October. Retrieved October 22, 2015 (http://www.theatlantic.com).

Cochran, Joshua, Sonja Siennick, and Daniel Mears. 2018. "Social Exclusion and Parental Incarceration Impacts on Adolescents' Networks and School Engagement." *Journal of Marriage and Family* 80 (2): 478–498.

Cockerham, William C. 2007. "A Note on the Fate of Postmodern Theory and Its Failure to Meet the Basic Requirements for Success in Medical Sociology." *Social Theory and Health* 5(4):285–296.

Coffman, Ginger, and Carol Markstrom-Adams. 1995. "A Model for Parent Education among Incarcerated Adults." Presented at the annual meeting of the National Council on Family Relations, November 15–19, Portland, OR.

Cogan, Rosemary, and Bud C. Ballinger III. 2006. "Alcohol Problems and the Differentiation of Partner, Stranger, and General Violence." *Journal of Interpersonal Violence* 21(7):924–935.

Cohen, Alex, and A. Martinez. 2012. "How Parents Unwittingly Fall into 'The Gender Trap' When Raising Children." *Take Two*, October 25. Retrieved November 8, 2012 (http://www.scpr.org).

Cohen, Leonard. 2008. *Runaway Youth and Multisystemic Therapy (MST): A Program Model*. West Hartford, CT: University of Hartford.

Cohen, Neil A., Thanh Tran, and Siyon Rhee. 2007. *Multicultural Approaches in Caring for Children, Youth, and Their Families*. Boston: Pearson, Allyn, and Bacon Publishers.

Cohen, Matthew, Kimberly Pentel, Sara Boeding, and Donald Baucom. 2019. "Postpartum Role Satisfaction in Couples." *Journal of Family Issues* 40 (9): 1181–1200.

Cohen, Patricia. 2007. "As Ethics Panels Expand Grip, No Research Field Is Off Limits." *The New York Times*, February 28.

Cohen, Philip N. 2019. "The Coming Divorce Decline." *Socius* 5 (doi: 2378023119873497).

Cohen, Philip N. 2014. "Recession and Divorce in the United States, 2008–2011." *Population Research and Policy Review* 33:615–628.

Cohen, Robin A., and Barbara Bloom. 2005. "Trends in Health Insurance and Access to Medical Care for Children under Age 19 Years: United States, 1998–2003." *Advance Data from Vital and Health Statistics*, No. 355. Hyattsville, MD: U.S. National Center for Health Statistics.

Cohen, S., et al. 2007. "Sexual Impairment in Psychiatric Inpatients: Focus on Depression." *Pharmacopsychiatry* 40(2):58–63.

Cohen, Sheldon, and Denise Janicki-Deverts. 2012. "Who's Stressed? Distributions of Psychological Stress in the United States in Probability Samples from 1983, 2006, and 2009." *Journal of Applied Social Psychology* 42:1320–1334.

Cohn, D'Vera, Gretchen Livingston, and Wendy Wang. 2014. "After Decades of Decline, A Rise in Stay-at-Home Mothers. " Retrieved December 19, 2019 (https://www.pewsocialtrends.org/2014/04/08/after-decades-of-decline-a-rise-in-stay-at-home-mothers/).

Cohn, D'Vera, and Jeffrey S. Passel. 2018. "A Record 64 Million Americans Live in Multigenerational Households." Retrieved November 30, 2019 (https://www.pewresearch.org/fact-tank/2018/04/05/a-record-64-million-americans-live-in-multigenerational-households/).

Cohn, D'Vera. 2014a. "Census Says it Will Count Same-Sex Marriages but with Caveats." Retrieved March 5, 2016 (http://www.pewresearch.org/fact-tank/2014/05/29/census-says-it-will-count-same-sex-marriages-but-with-caveats/).

———. 2014b. "Census Struggles to Reach an Accurate Number on Gay Marriages." Pew Research Center, May 13. Retrieved May 20, 2014 (http://www.pewresearch.org).

Cohn, D'Vera, Jeffrey S. Passel, Wendy Wang, and Gretchen Livingston. 2011. "Barely Half of U.S. Adults Are Married: A Record Low." Retrieved January 26, 2016 (http://www.pewsocialtrends.org/2011/12/14/barely-half-of-u-s-adults-are-married-a-record-low/).

Cohn, Lisa, Debbie Glasser, and Steve Mark. 2008. *The Step-Tween Survival Guide: How to Deal with Life in a Stepfamily*. Minneapolis: Free Spirit Publishers.

"Cohousing in Today's Real Estate Market." 2006. *Cohousing Magazine*. The Cohousing Association of the United States. Retrieved October 3, 2006 (www.cohousing.org).

Coker, A.L., Fisher, B.S., Bush, H.M., Swan, S.C., Williams, C.M., Clear, E.R. and DeGue, S., 2015. "Evaluation of the Green Dot Bystander Intervention to Reduce Interpersonal Violence Among College Students Across Three Campuses." *Violence Against Women* 21:1507–1527.

Colapinto, John. 2000. *As Nature Made Him: The Boy Who Was Raised as a Girl*. New York: HarperCollins.

Colby, Sandra L., and Jennifer M. Ortman. 2015. *Projections of the Size and Composition of the U.S. Population: 2014 to 2060*. Current Population Reports, P25-1143. Washington, DC: U.S. Census Bureau.

Cole, Charles Lee. 2011. "The Search for Emotional Connection in Marriage and Marriage-Like Relationships." *Family Focus* FF48 (spring): F6–F7.

Cole, Thomas. 1983. "The 'Enlightened' View of Aging." *Hastings Center Report* 13:34–40.

Coleman, Marilyn, Lawrence H. Ganong, and Susan M. Cable. 1997. "Beliefs about Women's Intergenerational Family Obligations to Provide Support Before and After Divorce and Remarriage." *Journal of Marriage and Family* 59(1):165–176.

Coleman, Marilyn, Lawrence H. Ganong, and Mark Fine. 2000. "Reinvestigating Remarriage: Another Decade of Progress." *Journal of Marriage and Family* 62:1288–1307.

Coleman, Marilyn, and Lynette Nickleberry. 2009. "An Evaluation of the Remarriage and Stepfamily Self-Help Literature." *Family Relations* 58(December):549–561.

Coleman, Patrick A. 2017. " 'Work-Life Balance' and 'Having It All' Are Not Women's Issues." *Fatherly* August 4. Retrieved January 15, 2020 (https://www.fatherly.com/news/work-life-balance-gender-myth).

Coles, M. E., L. M. Cook, and T. R. Blake. 2007. "Assessing Obsessive Compulsive Symptoms and Cognitions on the Internet: Evidence for the Comparability of Paper and Internet Administration." *Behaviour Research and Therapy* 45:2232–2240.

Coles, Roberta L. 2009. "Just Doing What They Gotta Do: Single Black Custodial Fathers Coping with the Stressors and Reaping the Rewards of Parenting." *Journal of Family Issues* 30(10):1311–1338.

Coley, Rebekah, L. Ribar, and Elizabeth Votruba-Drzal. 2011. "Do Children's Behavior Problems Limit Poor Women's Labor Market Success?" *Journal of Marriage and Family* 73(1):1–13.

Colaianni, Graziana, Li Sun, Mone Zaidi, and Alberta Zallone. 2015. "The "Love Hormone" Oxytocin Regulates the Loss and Gain of the Fat–Bone Relationship." *Frontiers in Endocrinology* 6.

Collaborative Group on Hormonal Factors in Breast Cancer. 2004. "Breast Cancer and Abortion: Collaborative Reanalysis of Data from 53 Epidemiological Studies, Including 83,000 Women with Breast Cancer from 16 Countries." *Lancet* 363(9414):1007–1016.

Collins, Patricia Hill. 2000. "Gender, Black Feminism, and Black Political Economy." *The Annals of the American Academy of Political and Social Science* 568:41–53.

Collins, Randall, and Scott Coltrane. 1995. *Sociology of Marriage and the Family: Gender, Love, and Property*, 4th ed. Chicago: Nelson-Hall.

Collins, Tara J., and Omri Gillath. 2012. "Attachment, Breakup Strategies, and Associated Outcomes: The Effects of Security Enhancement on the Selection of Breakup Strategies." *Journal of Research in Personality* 46:210–222.

Collymore, Yvette. 2002. "Risk of Homicide Is High for U.S. Infants." *Population Today*, May/June, p. 10–10.

Coltrane, Scott. 1998. "Gender, Power, and Emotional Expression: Social and Historical Contexts for a Process Model of Men in Marriages and Families." pp. 193–211 in *Men in Families: When Do They Get Involved? What Difference Does It Make?* edited by Alan Booth and Ann C. Crouter. Mahwah, NJ: Erlbaum.

———. 2000. "Research on Household Labor: Modeling and Measuring the Social Embeddedness of Routine Family Work." *Journal of Marriage and Family* 62:1208–1233.

Coltrane, Scott, and Randall Collins. 2001. *Marriage and the Family: Gender, Love, and Property.* Belmont, CA: Wadsworth /Cengage.

Coltrane, Scott, Erika Gutierrez, and Ross D. Parke. 2008. "Stepfathers in Cultural Context: Mexican American Families in the United States." pp. 100–121 in *The International Handbook of Stepfamilies*, edited by J. Pryor. Hoboken, NJ: John Wiley & Sons.

Colvin, Jan, Lillian Chenoweth, Mary Bold, and Cheryl Harding. 2004. "Caregivers of Older Adults: Advantages and Disadvantages of Internet-Based Social Support." *Family Relations* 53(1):49–57.

Comerford, Lynn. 2006. "The Child Custody Mediation Policy Debate." *Family Focus* (March): F7–F12. National Council on Family Relations.

Comfort, Megan. 2008. *Doing Time Together: Love and Family in the Shadow of Prison.* Chicago: University of Chicago Press.

Compton, Julie. 2018. " 'Boy or Girl?' Parents Raising 'Theybies' Let Kids Decide." NBC News. July 19. Retrieved August 29, 2019 (www.nbcnews.com).

Compton, Julie. 2019. "The 'Orgasm Gap': Why it Exists and What Women Can Do About it." Retrieved November 4, 2019 (https://www.nbcnews.com/better/lifestyle/orgasm-gap-why-it-exists-what-women-can-do-about-ncna983311).

Compton, July. 2018. " 'You Can't Undo Surgery': More Parents of Intersex Babies are Rejecting Operations." Retrieved September 29, 2019 (https://www.nbcnews.com/feature/nbc-out/you-can-t-undo-surgery-more-parents-intersex-babies-are-n923271).

Concise Columbia Encyclopedia. 1994. New York: Columbia University Press.

"Conclusions Are Reported on Teaching of Abstinence." 2007. *The New York Times*, April 15.

Condon, Stephanie. 2010. "Reid: Unemployment Leads to Domestic Violence." CBS News, February 23 (www .cbsnews.com).

Conference for American Indian Women of Proud Nations. 2012. Retrieved October 31, 2012 (http://www.aiwpn.org/).

Conger, Rand D., Katherine J. Conger, and Monica J. Martin. 2010. "Socioeconomic Status, Family Processes, and Individual Development." *Journal of Marriage and Family* 72(3):685–704.

Connect Safely. 2018. "Tips for Dealing with Teen Sexting." Retrieved October 30, 2019 (https://www.connectsafely.org/tips-for-dealing-with-teen-sexting/).

Connidis, Ingrid Arnet. 2007. "Negotiating Inequality among Adult Siblings: Two Case Studies." *Journal of Marriage and Family* 69(2):482–499.

———. 2009. *Family Ties and Aging.* Thousand Oaks, CA: Pine Forge Press.

———. 2010. *Family Ties and Aging*, 2nd ed. Los Angeles: Pine Forge Press.

———. 2011. "Reflections on Intergenerational Relations." *Family Focus* FF50 (Fall)): F3–F5.

Connidis, Ingrid Arnet, and Candace L. Kemp. 2008. "Negotiating Actual and Anticipated Parental Support: Multiple Sibling Voices in Three-Generation Families." *Journal of Aging Studies* 22:229–238.

Connidis, Ingrid, Klas Borell, and Sofie Karlsson. 2017. "Ambivalence and Living Apart Together in Later Life: A Critical Research Proposal." *Journal of Marriage and Family* 79 (5): 1404–1418.

Connley, Courtney. 2019. "The Number of Women Running Fortune 500 Companies Is at a Record High." CNBC May 16, 2019. Retrieved January 15, 2020 (www.cnbc.com).

Connor, James. 2007. *The Sociology of Loyalty.* New York: Springer-Verlag.

Conrad, Kate, Travis Dixon, and Yuanyuan Zhang. 2009. "Controversial Rap Themes, Gender Portrayals, and Skin Tone Distortion: A Content Analysis of Rap Music Videos." *Journal of Broadcasting and Electronic Media* 53(1):134–156.

Conway, Jane R., Nyala Noë, Gert Stulp, and Thomas V. Pollet. 2015. "Finding your Soulmate: Homosexual and Heterosexual Age Preferences in Online Dating." *Personal Relationships* 22:666–678.

Conway, Lynn. 2012. "Transsexual Women's Successes: Links and Photos." Retrieved June 25, 2013 (http://ai.eecs .umich.edu/people/conway/TSsuccesses /TSsuccesses.html).

Conway, Tiffany, and Rutledge Q. Hutson. 2007. "Is Kinship Care Good for Kids?" Center for Law and Social Policy, March 2. Retrieved September 14, 2009 (www.clasp .org).

Cook, Judith, Lynne Mock, Jessica Jonikas, Jane Burke-Miller, Tina Carter, Amanda Taylor, Carol Petersen, Dennis Grey, and David Gruenfelder. 2009. "Prevalence of Psychiatric and Substance Use Disorders among Single Mothers Nearing Lifetime Welfare Eligibility Limits." *Archives of General Psychiatry* 66(3):249–260.

Cook, P. W. 2009. *Abused Men: The Hidden Side of Domestic Violence.* Westport, CT: Praeger.

Cooke, Lynn Prince, and Janeen Baxter. 2010. " 'Families in International Context: Comparing Institutional Effects Across Western Societies." *Journal of Marriage and Family* 72(3) (June):516–536.

Cooke, Lynn Prince, and Sylvia Fuller. 2018. "Class Differences in Establishment Pathways to Fatherhood Wage Premiums." *Journal of Marriage and Family* 80 (3): 737–751.

Cooke, Lynn Prince, and Jennifer L. Hook. 2018. "Productivity or Gender? The Impact of Domestic Tasks across the Wage Distribution." *Journal of Marriage and Family* 80 (3): 721–736.

Cooksey, E. C., and M. M. Fondell. 1996. "Spending Time with His Kids: Effects of Family Structure on Fathers' and Children's Lives." *Journal of Marriage and the Family* 58:693–707.

Cooksey, Elizabeth. 2016. "Introduction to Special National Longitudinal Surveys Issue." *Journal of Marriage and Family* 78 (5):1250–1251.

Cookston, Jeffrey T., Andres Olide, Ross D. Parke, William V. Fabricius, Delia S. Saenz, and Sanford L. Braver. 2015. "He Said What? Guided Cognitive Reframing About the Co-Resident Father/Stepfather–Adolescent Relationship." *Journal of Research on Adolescence* 25:263–278.

Cooney, Rosemary, Lloyd H. Rogler, Rose Marie Hurrel, and Vilma Ortiz. 1982. "Decision Making in Intergenerational Puerto Rican Families." *Journal of Marriage and Family* 44:621–631.

Coontz, Stephanie. 2014. "Overview: Family Trends You Might Not Have Expected." Retrieved February 8, 2020 (https://thesocietypages.org/ccf/2014/10/08/overview-family-trends/).

Coontz, Stephanie. 2015. "Revolution in Intimate Life and Relationships." *Journal of Family Theory & Review* 7:5–12.

Cooper, Carey E., Robert Crosnoe, Marie-Anne Suizzo, and Keenan A. Pituch. 2009. "Poverty, Race, and Parental Involvement During the Transition to Elementary School." *Journal of Family Issues* (October).

Cooper, Carey E., Sara McLanahan, Sarah Meadows, and Jeanne Brooks-Gunn. 2009. "Family Structure Transitions and Maternal Parenting Stress." *Journal of Marriage and Family* 71(3):558–574.

Cooper, Marianne, and Allison J. Pugh. 2020. "Families across the Income Spectrum: A Decade in Review." *Journal of Marriage and Family* 82:272–299.

Cooper, Shauna M., Clara Smalls-Glover, Isha Metzger, and Charity Griffin. 2015. "African American Fathers' Racial Socialization Patterns: Associations with Racial Identity Beliefs and Discrimination Experiences." *Family Relations* 64(2):278–290.

Coopersmith, Jared. 2009. *Characteristics of Public, Private, and Bureau of Indian Education Elementary and Secondary School Teachers in the United States: Results from the 2007–08 Schools and Staffing Survey.* NCES 2009-324. Washington, DC: National Center for Education Statistics, Institute of Education Sciences, U.S. Department of Education.

Copeland, Valire Carr, and Kimberly Snyder. 2011. "Barriers to Mental Health Treatment Services for Low-Income African American Women Whose Children Receive Behavioral Health Services: An Ethnographic Investigation." *Social Work in Public Health* 26:78–95.

Copen, Casey E., Kimberly Daniels, Jonathan Vespa, and William Mosher. 2012. *First Marriages in the United States: Data from the 2006–2010 Survey of Family Growth.* National Health Statistics Reports no. 49. Hyattsville, MD: National Center for Health Statistics, March 22.

Copp, Jennifer E., Peggy C. Giordano, Monica A. Longmore, and Wendy D. Manning. 2017. "Living with Parents and Emerging Adults' Depressive Symptoms." *Journal of Family Issues* 38:2254–2276.

Cordes, Henry J. 2009. "'He-Cession' Reshuffles Roles." *Omaha World-Herald*, November 29. www.omaha.com. Retrieved May 2, 2009.

Cordova, David, Amanda Ciofu, and Richard Cervantes. 2014. "Exploring Culturally Based Intrafamilial Stressors among Latino Adolescents." *Family Relations* 63(5):693–706.

Corliss, Richard, and Sonja Steptoe. 2004. "The Marriage Savers." *Time*, January 19.

Cornwell, Benjamin, and Edward O. Laumann. 2011. "Network Position and Sexual Dysfunction: Implications of Partner Betweenness for Men." *American Journal of Sociology* 117(1):172–208.

Cornwell, Erin York, and Linda J. Waite. 2009. "Social Disconnectedness, Perceived Isolation, and Health among Older Adults." *Journal of Health and Social Behavior* 50(March):31–48.

Corra, Mamadi, Shannon K. Carter, J. Scott Carter, and David Knox. 2009. "Trends in Marital Happiness by Gender and Race, 1973 to 2006." *Journal of Family Issues* 30(10):1379–1404.

Corrigan, Maureen. 2013. " 'Lean In': Not Much of a Manifesto, But Still a Win for Women."

National Public Radio, March 12. Retrieved March 14, 2013 (http://www.npr.org).

Cott, Nancy F. 2000. *Public Vows: A History of Marriage and the Nation.* Cambridge, MA: Harvard University Press.

Cotter, David, Joan Hermsen, Reeve Vanneman, Sharon Sassler, Christine Schwartz, and Youngloo Cha. 2014. Council on Contemporary Families Gender Rebound Symposium, University of Miami School of Education and Human Development.

Cotton, Sheila R., Russell Burton, and Beth Rushing. 2003. "The Mediating Effects of Attachment to Social Structure and Psychosocial Resources on the Relationship between Marital Quality and Psychological Distress." *Journal of Family Issues* 24(4):547–577.

Cottrell, Barbara, and Peter Monk. 2004. "Adolescent-to-Parent Abuse: A Qualitative Overview of Common Themes." *Journal of Family Issues* 25(8):1072–1095.

Couch, Kenneth A., Christopher R. Tamborini, and Gayle L. Reznik. 2015. "The Long-Term Health Implications of Marital Disruption: Divorce, Work Limits, and Social Security Disability Benefits among Men." *Demography* 52:1487–1512.

Coughlin, Patrick, and Jay Wade. 2012. "Masculinity Ideology, Income Disparity, and Romantic Relationship Quality among Men with Higher Earning Female Partners." *Sex Roles* 67: 311–322.

Council on Contemporary Families. 2011. "Fact Sheet: Sandwich Generation Month." Retrieved July 14, 2011 (http://www .sandwichgenerationmonth.com).

Council on Foreign Relations. "Demographics of the U.S. Military." Retrieved October 7, 2019 (https://www.cfr.org/article/ demographics-us-military).

Covert, Juanita J., and Travis L. Dixon. 2008. "A Changing View: Representation and Effects of the Portrayal of Women of Color in Mainstream Women's Magazines." *Communication Research* 35(2):232–256.

Cowan, Gloria. 2000. "Beliefs about the Causes of Four Types of Rape." *Sex Roles* 42(9/10):807–823.

Cowan, Philip, and Carolyn Cowan. 2009. "News You Can Use: Are Babies Bad for Marriage?" Press Release, January 9. Chicago: Council on Contemporary Families (www. contemporaryfamilies.org).

Cowan, Philip A., Carolyn Pape Cowan, Marsha K. Pruett, Kyle Pruett, and Peter Gillette. 2014. "Evaluating a Couples Group to Enhance Father Involvement in Low-Income Families Using a Benchmark Comparison." *Family Relations* 63(3):356–370.

Cowan, Philip A., Carolyn Pape Cowan, Marsha Kline Pruett, Kyle Pruett, and Jessie J. Wong. 2009. "Promoting Fathers' Engagement with Children: Preventive Interventions for Low-Income Families." *Journal of Marriage and Family* 71(4):663–679.

Cowan, Ruth Schwartz. 1983. *More Work for Mother: The Ironies of Household Technology from the Open Hearth to the Microwave.* New York: Basic Books.

Cowdery, Randi S., Norma Scarborough, Carmen Knudson-Martin, Gita Seshadri,

Monique E. Lewis, and Anne Rankin Mahoney. 2009. "Gendered Power in Cultural Contexts: Part II. Middle Class African American Heterosexual Couples with Young Children." *Family Process* 48(1):25–39.

Cox, Erin. 2008. *Intimate Partner Violence among Pregnant and Parenting Women: Local Health Department Strategies for Assessment, Intervention, and Prevention.* Washington, DC: National Association of County and City Health Officials.

Coy, Peter, Michelle Conlin, and Moira Herbst. 2010. "The Disposable Worker." *Bloomberg Businessweek*, January 18:33–39.

Cox, Deborah, Sally Stabb, and Karin Bruckner. 2016. *Women's Anger: Clinical and Developmental Perspectives.* Routledge.

Cox, Ronald. 2027. "Promoting Resilience: the !Unidos Se Puede! Program: An Example of Translational Research for Latino Families." *Family Relations* 66 (4):712–728.

Coyne, Sarah M., Laura Stockdale, Dean Busby, Bethany Iverson, and David M. Grant. 2011. "'I luv u:)': A Descriptive Study of the Media Use of Individuals in Romantic Relationships." *Family Relations* 60:150–162.

Craig, Lyn. 2015. "How Mothers and Fathers Allocate Time to Children: Trends, Resources, and Policy Context." *Family Focus.* Spring:F9–F10.

Craig, Lyn, and Judith E. Brown. 2017. "Feeling Rushed: Gendered Time Quality, Work Hours, Nonstandard Work Schedules, and Spousal Crossover." *Journal of Marriage and Family* 79 (1): 225–242.

Crary, David. 2007a. "U.S. Divorce Rate Lowest Since 1979." Associated Press, May 10. Retrieved May 10, 2007 (www.breitbart .com).

———. 2007b. "More Couples Seeking Kinder, Gentler Divorces." *MSNBC*, December 18. Retrieved June 21, 2010 (www.msnbc.com).

———. 2008. "Housework Gets You Laid." *The Huffington Post*, March 6. Retrieved May 20, 2010 (www.huffingtonpost.com).

Crawford, D., D. Feng, and J. Fischer. 2003. "The Influence of Love, Equity, and Alternatives on Commitment in Romantic Relationships." *Family and Consumer Sciences Research Journal* 31(3):253–271.

Crawford, Mary, and Danielle Popp. 2003. "Sexual Double Standards: A Review and Methodological Critique of Two Decades of Research." *Journal of Sex Research* 40(1):13–26.

Crawford, Lizabeth A., and Katherine B. Novak. 2008. "Parent-Child Relations and Peer Associations as Mediators of the Family Structure-Substance Use Relationship." *Journal of Family Issues* 29(2):155–184.

Crawford, Natalie M., and Anne Z. Steiner. 2015. "Age-Related Infertility." *Obstetrics and Gynecology Clinics* 42:15–25.

Crawford, Susan P. 2011. "The New Digital Divide." *The New York Times*, December 3. Retrieved September 15, 2012 (http://www .nytimes.com).

Crespo, Carla. 2012. "Families as Contexts for Attachment: Reflections on Theory, Research, and the Role of Family Rituals." *Journal of Family Theory and Review* 4(4):290–298.

Crile, Susan. 2011. "Obama Eliminates Abstinence-Only Funding in Budget." Retrieved January 25, 2013 (http://www.huffingtonpost.com/2009/05/07/obama-climinates-abstinen_n_199205}.html?view=print&comm_ref=false).

Criss, Michael, Carolyn Henry, Amanda Harrist, and Robert Larzelere. 2015. "Introduction to Special Issue on Family and Individual Resilience." *Family Relations* 64(1):1–4.

Crittenden, Ann. 2001. *The Price of Motherhood: Why the Most Important Job in the World Is Still the Least Valued*. New York: Metropolitan.

Crockett, Emily. 2015. "Your TV is Lying to You about Who Has Abortions." Retrieved February 14, 2016 (http://www.vox.com/2015/12/29/10683610/tv-lying-abortion-demographics).

Crook, Tylon, Chippewa M. Thomas, and Debra C. Cobia. 2009. "Masculinity and Sexuality: Impact on Intimate Relationships of African American Men." *Family Journal* 17(4):360–366.

Crooks, Robert, and Karla Baur. 2005. *Our Sexuality*. 9th ed. Belmont, CA: Wadsworth.

Crosbie-Burnett, M. 1984. "The Centrality of the Step Relationship: A Challenge to Family Theory and Practice." *Family Relations* 459–463.

Crosbie-Burnett, Margaret, Edith A. Lewis, Summer Sullivan, Jessica Podolsky, Rosane Mantilla de Souza, and Victoria Mitrani. 2005. "Advancing Theory through Research: The Case of Extrusion in Stepfamilies." pp. 213–230 in *Sourcebook of Family Theory and Research*, edited by Vern L. Bengston, Alan C. Acock, Katherine R. Allen, Peggye Dilworth-Anderson, and David M. Klein. Thousand Oaks, CA: Sage.

Crosby, John F. 1991. *Illusion and Disillusion: The Self in Love and Marriage*, 4th ed. Belmont, CA: Wadsworth.

Crosnoe, Robert and Arya Ansari. 2016. "Family Socioeconomic Status, Immigration, and Children's Transitions into School." *Family Relations* 65 (1):73–84.

Crosnoe, Robert, and Shannon E. Cavanagh. 2010. "Families with Children and Adolescents: A Review, Critique, and Future Agenda." *Journal of Marriage and Family* 72(3) (June):594–611.

Cross, Christina J. 2019. "Racial/Ethnic Differences in the Association Between Family Structure and Children's Education." *Journal of Marriage and Family* 81 (https://doi.org/10.1111/jomf.12625).

Cross-Barnet, Caitlin, Andrew Cherlin, and Linda Burton. 2011. "Bound By Children: Intermittent Cohabitation and Living Together Apart." *Family Relations* 60(December): 633–647.

Crosse, Marcia. 2008. "Abstinence Education: Assessing the Accuracy and Effectiveness of Federally Funded Programs." *Testimony Before the Committee on Oversight and Government Reform, House of Representatives* April 23. Washington, DC: United States Government Accountability Office.

Crouch, Elizabeth, and Lori Dickes. 2016. "Economic Repercussions of Marital Infidelity." *International Journal of Sociology and Social Policy* 36:53–65.

Crouter, Ann C., and Alan Booth, eds. 2003. *Children's Influence on Family Dynamics: The Neglected Side of Family Relationships*. Mahwah, NJ: Erlbaum.

Crowder, Kyle D., and Stewart E. Tolnay. 2000. "A New Marriage Squeeze for Black Women: The Role of Racial Intermarriage by Black Men." *Journal of Marriage and Family* 62(3):792–807.

Crowley, Jocelyn E., and Stephanie Curenton. 2011. "Organizational Social Support and Parenting challenges among Mothers of Color: The Case of Mocha Moms." *Family Relations* 60(1): 1–14.

Crowley, Martha, Daniel T. Lichter, and Zhenchao Qian. 2006. "Beyond Gateway Cities: Economic Restructuring and Poverty among Mexican Immigrant Families and Children." *Family Relations* 55(3):345–360.

Cruz, J. 2012. *Remarriage Rate in the U.S., 2010* (FP-12-14). Bowling Green, OH: National Center for Family and Marriage Research. Retrieved September 7, 2012 (http://ncfmr.bgsu.edu.pdf/family_profiles/file114853.pdf).

Cui, Ming, and M. Brent Donnellan. 2009. "Trajectories of Conflict over Raising Adolescent Children and Marital Satisfaction." *Journal of Marriage and Family* 71(3):478–494.

Cui, Ming, Frederick O. Lorenz, Rand D. Conger, Janet N. Melby, and Chalandra M. Bryant. 2005. "Observer, Self-, and Partner Reports of Hostile Behaviors in Romantic Relationships." *Journal of Marriage and Family* 67(5):1169–1181.

Cui, Ming, Koji Ueno, Mellissa Gordon, and Frank D. Fincham. 2013. "The Continuation of Intimate Partner Violence from Adolescence to Young Adulthood." *Journal of Marriage and Family* 75:300–313.

Cullen, Jennifer C., Leslie B. Hammer, Margaret B. Neal, and Robert R. Sinclair. 2009. "Development of a Typology of Dual-Earner Couples Caring for Children and Aging Parents." *Journal of Family Issues* 30(4):458–483.

Cullen, Lisa T. 2007. "Till Work Do Us Part." *Time Magazine*, September 27. Retrieved May 27, 2010 (www.time.com).

Cummings, Elijah E. 2019. *Child Separations by the Trump Administration*. Staff Report: U.S. House of Representatives Committee on Oversight and Reform. July. Retrieved September 9, 2019 (http://oversight.house.gov).

Cunningham-Burley, Sarah. 2001. "The Experience of Grandfatherhood." pp. 92–96 in *Families in Later Life: Connections and Transitions*, edited by Alexis J. Walker, Margaret Manoogian-O'Dell, Lori A. McGraw, and Diana L. G. White. Thousand Oaks, CA: Pine Forge Press.

Curington, Celeste, Ken-Hou Lin, and Jennifer Lundquist. 2015. "Dating Partners Don't Always Prefer 'Their Own Kind': Some Multiracial Daters Get Bonus Points in the Dating Game," Brief Report, Council on Contemporary Families, July 1. Retrieved January 12, 2016 (https://contemporaryfamilies.org/multiracial-dating-brief-report/).

Curran, Winifred. 2015. "'That's My Sister!' Taking a Twin's Gender Fluidity in Stride." *The New York Times*, January 18. Retrieved February 16, 2015 (http://www.nytimes.com).

Currie, Janet, and Cathy Spatz Widom. 2010. "Long-Term Consequences of Child Abuse and Neglect on Adult Economic Well-Being." *Child Maltreatment* 15(2):111–120.

Curry, Candace. 2014. "To My Daughter's Stepmom: I Never Wanted You Here, but Thank You." Retrieved April 16, 2016 (http://www.today.com/parents/letter-my-daughters-stepmom-i-never-wanted-you-here-1D80341783).

Curtin, Sally C., Stephanie J. Ventura, and Gladys M. Martinez. 2014. "Recent Declines in Nonmarital Childbearing in the United States." NCHS Data Brief no. 162. Retrieved February 1, 2016 (http://www.cdc.gov/nchs/data/databriefs/db162.pdf).

Curtis, David S., Norman B. Epstein, and Brandan Wheeler. 2015. "Relationship Satisfaction Mediates the Link Between Partner Aggression and Relationship Dissolution The Importance of Considering Severity." *Journal of Interpersonal Violence* : 0886260515588524.

Cypress, Brigitte, and Keville Frederickson. 2017. "Family Presence in the Intensive Care Unit and Emergency Department: A Metasynthesis." *Journal of Family Theory & Review* 9 (2): 201–218.

Cyr, Francine, Gessica Di Stefano, and Bertrand Desjardins. 2013. "Family Life, Parental Separation, and Child Custody in Canada: A Focus on Québec." *Family Court Review* 51:522–541.

Cyr, Mireille, Pierre McDuff, and John Wright. 2006. "Prevalence and Predictors of Dating Violence among Adolescent Female Victims of Child Sexual Abuse." *Journal of Interpersonal Violence* 21(8):1000–1017.

Dahl, Svenn-Åge, Hans, Tore Hansen, and Bo Vignes. 2015. "His, Her, or Their Divorce? Marital Dissolution and Sickness Absence in Norway." *Journal of Marriage and Family* 77:461–479.

Dahms, Alan M. 1976. "Intimacy Hierarchy." pp. 85–104 in *Process in Relationship: Marriage and Family*, 2nd ed., edited by Edward A. Powers and Mary W. Lees. New York: West.

Dailard, Cynthia. 2003. "Understanding 'Abstinence': Implications for Individuals, Programs and Policies." *The Guttmacher Report* 6(5), December (www.agi-usa.org).

Dailey, Rene M. 2009. "Confirmation from Family Members: Parent and Sibling Contributions to Adolescent Psychosocial Adjustment." *Western Journal of Communication* 73(3):273–299.

Dailey, Rene M., Nicholas Brody, and Jessica Knapp. 2015."Friend Support of Dating Relationships: Comparing Relationship Type, Friend and Partner Perspectives." *Personal Relationships* 22:368–385.

Dalla, Rochelle, Susan Jacobs-Hagen, Betsy Jareske, and Julie Sukup. 2009. "Examining the Lives of Navajo Native American Teenage Mothers in Context: A 12- to 15-Year Follow-Up." *Family Relations* 58(April):148–161.

Daly, Martin, and Margo I. Wilson. 2000. "The Evolutionary Psychology of Marriage and Divorce." pp. 91–110 in L. Waite, C. Bachrach, M. Hindin, E. Thomson, and A. Thornton, eds., *The Ties That Bind: Perspectives on Marriage and Cohabitation*. Hawthorne, NY: Aldine de Gruyter.

Danesi, Marcel. 2016. *The Semiotics of Emoji: The Rise of Visual Language in the Age of the Internet*. Bloomsbury Publishin.

Daniel, Brigid, Julie Taylor, and Jane Scott. 2010. "Recognition of Neglect and Early Response: Overview of a Systematic Review of the Literature." *Child and Family Social Work* 15(2):248–257.

Dao, James. 2005. "Grandparents Given Rights by Ohio Court." *The New York Times*, October 11.

Darghouth, Sarah, Leslie Brody, and Margarita Alegría. 2015. "Does Marriage Matter? Marital Status, Family Processes, and Psychological Distress among Latino Men and Women." *Hispanic Journal of Behavioral Sciences*: 0739986315606947.

Datta, J., M. J. Palmer, C. Tanton, L. J. Gibson, K. G. Jones, W. Macdowall, A. Glasier et al. 2016. "Prevalence of Infertility and Help Seeking Among 15,000 Women and Men. " *Human Reproduction* 31:2108 –2118.

Davey, Adam, Jvoti Savla, Megan Janke, and Shayne Anderson. 2009. "Grandparent-Grandchild Relationships: From Families in Context to Families as Contexts." *Aging and Human Development* 69(4):311–325.

Davey, Monica. 2006. "As Tribal Leaders, Women Still Fight Old Views." *The New York Times*, February 4.

David, Deborah, and Robert Brannon. 1979. *The Forty-Nine Percent Majority: The Male Sex Role.* Addison-Wesley.

David, Prabu, and Laura Stafford. 2015. "A Relational Approach to Religion and Spirituality in Marriage: The Role of Couples' Religious Communication in Marital Satisfaction." *Journal of Family Issues* 36(2):232–249.

Davidson, Adam. 2014. "It's Official: The Boomerang Kids Won't Leave." *The New York Times*, June 20. Retrieved June 23, 2014 (http://www.nytimes.com).

Davies, Lorraine, Marilyn Ford-Gilboe, Joanne Hammerton. 2009. "Gender Inequality and Patterns of Abuse Post Leaving." *Journal of Family Violence* 24(1):27–39.

Davis, Kelly D., W. Benjamin Goodman, Amy E. Pirretti, and David M. Almeida. 2008. "Nonstandard Work Schedules, Perceived Family Well-Being, and Daily Stressors." *Journal of Marriage and Family* 70(4):991–1003.

Davis, Melanie. 2015. "Tips for Talking about Masturbation." Retrieved December 3, 2015 (http://www.advocatesforyouth .org /parents/2027-tips-kids-mast).

Davis, Shannon, Theodore Greenstein, and Jennifer Gertelsen Marks. 2007. "Effects of Union Type on Division of Household Labor." *Journal of Family Issues* 28(5):1260–1271.

Davis-Sowers, Regina. 2012. "'It Just Kind of Like Falls in Your Hands': Factors that Influence Black Aunts' Decisions to Parent Their Nieces and Nephews." *Journal of Black Studies* 43(3):231–250.

Dawkins, Richard. 2006. *The Selfish Gene: The 30th Anniversary Edition.* New York: Oxford University Press.

Dawn, Laura. 2006. *It Takes a Nation: How Strangers Became Family in the Wake of Hurricane Katrina.* San Rafael, CA: Earth Aware Editions.

Dawson, Christopher. 2018. "When Students Were Bullied Because of Dirty Clothes." CNN.

December 4. Retrieved August 29, 2019 (www. cnn.com).

Day, Jennifer Cheeseman. 2019. "Among the Educated, Women Earn 74 Cents for Every Dollar Men Make." *America Counts: Stories behind the Numbers.* May 29. U.S. Census Bureau. Retrieved November 4, 2019 (https://census.gov).

Day, Randal D., and Alan Acock. 2013. "Marital Well-Being and Religiousness as Mediated by Relational Virtue and Equality." *Journal of Marriage and Family* 75(1):164–177.

"Daycare Centers." 2015. Retrieved November 5, 2015 from http://www .babycenter.com.

DeBel, V., M. Kalmijn and M. A. J. Duijn. 2019. "Balance in Family Triads: How Intergenerational Relationships Affect the Adult Sibling Relationship." *Journal of Family Issues* (July). Retrieved September 25, 2019.

Decker, Melissa, Heather L. Littleton, and Katie M. Edwards. 2018. "An Updated Review of the Literature on LGBTQ+ Intimate Partner Violence.   " *Current Sexual Health Reports* 10:265–272.

Declercq, Eugene. 2012. "Trends in Midwife-Attended Births in the United States, 1989–2009." *Journal of Midwifery and Women's Health* 57:321–326.

Declercq, Eugene, Fay Menacker, and Marian MacDorman. 2004. "Rise in 'No Indicated Risk' Primary Caesareans in the United States, 1991–2001: Cross-Sectional Analysis." *British Medical Journal* (doi:10.1136/bmj.38279.705336OB). Online First BMJ.com.

Declercq, Eugene. 2015. "Midwife-Attended Births in the United States, 1990–2012: Results from Revised Birth Certificate Data. " *Journal of Midwifery & Women's Health* 60:10–15.

DeFrain, John. 2002. *Creating a Strong Family: American Family Strengths Inventory.* Nebraska Cooperative Extension NF01-498 (ianrpubs.unl. edu/family/nf498.htm).

De Henau, Jerome, and Susan Himmelweit. 2013. "Unpacking Within-Household Gender Differences in Partners' Subjective Benefits from Household Income." *Journal of Marriage and Family* 75(3):611–624.

De Jong, David C., and Harry T. Reis. 2015. "Sexual Similarity, Complementarity, Accuracy, and Overperception in Same-Sex Couples." *Personal Relationships* 22:647–665.

De La Lama, Luisa Batthyany, Luis De La Lama, and Ariana Wittgenstein. 2012. "The Soul Mates Model: A Seven-Stage Model for Couples Long-Term Relationship Development and Flourishing." *The Family Journal.* doi (10.1177/1066480712449797).

DeLamater, John D., and Morgan Sill. 2005. "Sexual Desire in Later Life." *Journal of Sex Research* 42(2):138–149.

DeLeire, Thomas, and Ariel Kalil. 2005. "How Do Cohabiting Couples with Children Spend Their Money?" *Journal of Marriage and Family* 67(2):286–295.

della Cava, Marco R. 2009. "Women Step Up As Men Lose Jobs." *USA Today*, March 19, 1D.

Dell'Antonia, K. J. 2011. "Spacing Children Farther Apart Benefits Older Siblings." *The New York Times, Motherlode : Adventures in Parenting,*

November 21. Retrieved December 14, 2012 (http://parenting.blogs.nytimes.com).

Dellmann-Jenkins, Mary, Maureen Blankemeyer, and Odessa Pinkard. 2000. "Young Adult Children and Grandchildren in Primary Caregiver Roles to Older Relatives and Their Service Needs." *Family Relations* 49(2):177–186.

DeLongis, Anita, and Amy Zwicker. 2017. "Marital Satisfaction and Divorce in Couples in Stepfamilies." *Current Opinion in Psychology* 13:158–161.

DeMaris, Alfred. 2009. "Distal and Proximal Influences on the Risk of Extramarital Sex: A Prospective Study of Longer Duration Marriages." *Journal of Sex Research* 46(6):597–607.

DeMaris, Alfred, Annette Mahoney, and Kenneth Pargament. 2011. "Doing the Scut Work of Infant Care: Does Religiousness Encourage Father Involvement?" *Journal of Marriage and Family* 73(2): 354–368.

Demby, Kimberly P., Shelley A. Riggs, and Patricia L. Kaminski. 2017. "Attachment and Family Processes in Children's Psychological Adjustment in Middle Childhood." *Family process* 56(1):234–249.

D'Emilio, John and Estelle B. Freedman. 1988. *Intimate Matters: A History of Sexuality in America.* New York: HarperCollins.

Demo, David H., William S. Aquilino, and Mark A. Fine. 2005. "Family Composition and Family Transitions." pp. 119–134 in *Sourcebook of Family Theory and Research*, edited by Vern L. Bengston, Alan C. Acock, Katherine R. Allen, Peggye Dilworth-Anderson, and David M. Klein. Thousand Oaks, CA: Sage.

Demo, David H., and Mark A. Fine. 2010. *Beyond the Average Divorce.* Thousand Oaks, CA: Sage.

"The Demographics of Aging." n.d. Transgenerational Design Matters. Retrieved May 5, 2010 (www.transgenerational.org /aging /demographics.htm).

Demos, J. (1966). *A Little Commonwealth: Family Life in Plymouth Colony.* New York: Oxford University Press.

DeNavas-Walt, Carmen, and Bernadette D. Proctor. 2015. *Income, Poverty, and Health Insurance Coverage in the United States : 2014.* U.S. Census Bureau, Current Population Reports, P60-252. Washington, DC: U.S. Government Printing Office.

DeNavas-Walt, Carmen, Bernadette D. Proctor, and Jessica C. Smith. 2011. *Income, Poverty, and Health Insurance Coverage in the United States : 2010.* U.S. Census Bureau, Current Population Reports, P60–243. Washington, DC: U.S. Government Printing Office, September.

———. 2012. *Income, Poverty, and Health Insurance Coverage in the United States: 2011.* U.S. Census Bureau, Current Population Reports, P60–243, September. Washington, DC: U.S. Government Printing Office.

Denes, Amanda, and Tamara D. Afifi. 2014. "Pillow Talk and Cognitive Decision-Making Processes: Exploring the Influence of Orgasm and Alcohol on Communication after Sexual Activity." *Communication Monographs* 81(3):333–358.

Denes, Amanda, Tamara D. Afifi, and Douglas A. Granger. 2017. "Physiology and pillow

talk: Relations between testosterone and communication post sex." *Journal of Social and Personal Relationships* 34(3):281–308.

Denes, Amanda. 2018. "Toward a post-sex disclosures model: Exploring the associations among orgasm, self-disclosure, and relationship satisfaction." *Communication Research* 45(3):297–318.

eBizMBA. 2020. "Top 15 Most Popular Social Networking Sites & Apps." Retrieved January 15, 2020 (http://www.ebizmba.com/articles/social-networking-websites).

Denizet-Lewis, Benoit. 2003."Double Lives on the Down Low." *The New York Times Magazine*, August 3.

———. 2004. "Friends, Friends with Benefits, and the Benefits of the Local Mall." *The New York Times*, May 30.

———. 2014. "The Scientific Quest to Prove that Bisexuality Exists." Retrieved December 15, 2015 (http://www.nytimes.com/2014/03/23/magazine/the-scientific-quest-to-prove-bisexuality-exists.html?_r=0).

Denny, Justin T. 2014. "Families, Resources, and Suicide: Combined Effects on Mortality." *Journal of Marriage and Family* 76(1):218–231.

DePaulo, Bella. 2006. *Singled Out: How Singles Are Stereotyped, Stigmatized, and Ignored, and Still Live Happily Ever After.* New York: St. Martin's Press.

———. 2012. "A New American Experiment." *The New York Times*, February 13. Retrieved November 6, 2012 (http://www.nytimes.com).

DePillis, Lydia. 2015. "The Next Labor Fight Is Over When You Work, Not How Much You Make." *Washington Post Workblog*, May 8. Retrieved November 14, 2015 (http://www.washingtonpost.com).

Deprez, Esmé. 2015. "The Vanishing U.S. Abortion Clinic." Retrieved Februrary 14, 2016 (http://www.bloombergview.com/quicktake/abortion-and-the-decline-of-clinics).

DeRose, Laura M., Mariya P. Shiyko, Holly Foster, and Jeanne Brooks-Gunn. 2011. "Associations Between Menarcheal Timing and Behavioral Developmental Trajectories for Girls from Age 6 to Age 15." *Journal of Youth and Adolescence* 40(10):1329–1342.

DeSilver, Drew. 2014. "Rising Cost of Child Care May Help Explain Recent increase in Stay-At-Home Moms." Retrieved February 8, 2016 (http://www.pewresearch.org/fact-tank/2014/04/08/rising-cost-of-child-care-may-help-explain-increase-in-stay-at-home-moms/).

Del Valle, Gaby. 2018. "'For Conservatives, by Conservatives': The Rise of Right-Wing Dating Apps." Retrieved November 21, 2019 (https://www.vox.com/the-goods/2018/12/26/18150322/righter-donald-daters-patrio-conservative-dating-apps).

DeMaris, Alfred, and Annette Mahoney. 2017. "Equity Dynamics in the Perceived Fairness of Infant Care." *Journal of Marriage and Family* 79 (1): 261–276.

Deng, Yongchun, Hua Xu, and XiaoHua Zeng. 2018. "Induced Abortion and Breast Cancer: An Updated Meta-Analysis. " *Medicine* 97.

Desilver, Drew. 2016. "More Older Americans Are Working, and Working More, Than They Used To." *Facttank: News in the Numbers.* June 20. Pew Research Organization. Retrieved January 30, 2019 (www.pewresearch.org)

De Vaus, David, Matthew Gray, Lixia Qu, and David Stanton. 2017. "The Economic Consequences of Divorce in Six OECD Countries." *Australian Journal of Social Issues* 52:180–199.

Deveny, Kathleen. 2008. "Why Only-Children Rule." *Newsweek*, June 2. www.newsweek.com.

DeVisser, Richard, and Dee McDonald. 2007. "Swings and Roundabouts: Management of Jealousy in Heterosexual 'Swinging' Couples." *British Journal of Social Psychology* 46(2):459–476.

Devitt, Kerry, and Debi Roker. 2009. "The Role of Mobile Phones in Family Communication." *Children and Society* 23:189–202.

Devor, Camron S., Susan D. Stewart, and Cassandra Dorius. 2018. "Parental Divorce, Social Capital, and Postbaccalaurate Educational Attainment among Young Adults." *Journal of Family Issues* 39:2806–2835.

Dew, Jeffrey. 2008. "Debt Change and Marital Satisfaction Change in Recently Married Couples." *Family Relations* 57(1):60–71.

———. 2011. "Financial Issues and Relationship Outcomes among Cohabiting Individuals." *Family Relations* 60(2):178–190.

———. 2015. "The Many Interfaces of Money and Marriage." *Family Focus.* Spring:F11–F12.

Dew, Jeffrey P., Bonnie L. Anderson, Linda Skogrand, and Cassandra Chaney. 2017. "Financial Issues in Strong African American Marriages: A Strengths-Based Qualitative Approach." *Family Relations* 66 (2):287–301.

Dew, Jeffrey, Sonya Britt, and Sandra Huston. 2012. "Examining the Relationship Between Financial Issues and Divorce." *Family Relations* 61:615–628.

Dew, Jeffrey, and Jeremy Yorgason. 2010. "Economic Pressure and Marital Conflict in Retirement-Aged Couples." *Journal of Family Issues* 31(2):164–188.

Dewan, Shaila, and Robert Gebeloff. 2012. "More Men Enter Fields Dominated by Women." *The New York Times*, May 21: A1, A3.

Dewey, Caitlin. 2014. "'Back-Up Husbands,' 'Emotional Affairs,' and the Rise of Digital Infidelity." Retrieved December 15, 2015 (https://www.washingtonpost.com/news/the-intersect/wp/2014/10/03/back-up-husbands-emotional-affairs-and-the-rise-of-digital-infidelity/).

Dewilde, Caroline, and Wilfred Uunk. 2008. "Remarriage As a Way to Overcome the Financial Consequences of Divorce— A Test of the Economic Need Hypothesis for European Women." *European Sociological Review* 24(3):393–407.

Diamond-Smith, Nadia G., Minakshi Dahal, Mahesh Puri, and Sheri D. Weiser. 2019. "Semi-Arranged Marriages and Dowry Ambivalence: Tensions in the Changing Landscape of Marriage Formation in South Asia. " *Culture, Health & Sexuality*: 1–16.

Diaz, Christina J., and Jeremy E. Fiel. 2016. "The Effect (s) of Teen Pregnancy: Reconciling

Theory, Methods, and Findings." *Demography* 53:1–32.

Dickinson, Amy. 2002. "An Extra-Special Relation." *Time*, November 18, pp.A1+.

Diefendorf, Sarah. 2015. "After the Wedding Night Sexual Abstinence and Masculinities over the Life Course." *Gender and Society*, September 9. doi: 0891243915591597

Didonato, Theresa. 2015. "4 Truths About Power in Relationships (Including Yours)." *Psychology Today*. February 29. Retrieved February 8, 2020 (www.psychologytoday.com).

Dierckx, Myrte, Dimitri Mortelmans, Joz Motmans and Guy T'Sjoen. 2017. "Resilience in Families in Transition: What Happens When a Parent Is Transgender?" *Family Relations* 66 (3):399–411.

Dimock, Michael. 2019. "Defining Generations: Where Millennials End and Generation Z Begins." *Facttank: News in the Numbers.* Pew Research Organization. Retrieved January 30, 2019 (www.pewresearch.org).

Dion, Karen K. 1995. "Delayed Parenthood and Women's Expectations about the Transition to Parenthood." *International Journal of Behavioral Development* 18(2):315–333.

Dion, Karen K., and Kenneth L. Dion. 1991. "Psychological Individualism and Romantic Love." *Journal of Social Behavior and Personality* 6:17–33.

Dion, M. Robin, Heather Zaveri, and Pamela Holcomb. 2015. "Responsible Fatherhood Programs in the Parents and Children Together (PACT) Evaluation." *Family Court Review* 53:292–303.

DiStefano, Joseph. 2001. "Jumping the Broom." Retrieved October 2, 2006 (www.randomhouse.com).

Ditzen, Beate, Marcel Schacr, Barbara Gabriel, Guy Bodenmann, Ulrike Ehlert, and Markus Heinrichs. 2009. "Intranasal Oxytocin Increases Positive Communication and Reduces Cortisol Levels During Couple Conflict." *Biological Psychiatry* 65(9):728–732.

Dixon, Lee J., K. C. Gordon, N. Frousakis, and J. A. Schumm. 2012. "A Study of Expectations and the Marital Quality of Participants of a Marital Enrichment Seminar." *Family Relations* 61(1):75–89.

Dixon, Nicholas. 2007. "Romantic Love, Appraisal, and Commitment." *The Philosophical Forum* 38(4):373–386.

Doan, Long, Lisa R. Miller, and Annalise Loehr. 2015. "The Power of Love: The Role of Emotional Attributions and Standards in Heterosexuals' Attitudes toward Lesbian and Gay Couples." *Social Forces*, August 6. doi: 10.1093/sf/sov047.

Dockterman, Eliana. 2019. "'A Doll for Everyone': Meet Mattel's Gender-Neutral Doll." Retrieved September 29, 2019 (https://time.com/5684822/mattel-gender-neutral-doll/?fbclid=IwAR0qQ7LwiySrXGF0e6YumnOI6Jf_Z3rdzyjKXqjbDzBoRpASxW1xhUvVS40).

Dodge, Brian, and Michael Reece, Debby Herbenick, Vanessa Schick, Stephanie A. Sanders, and J. Dennis Fortenberry. 2010. "Sexual Health among U.S. Black and Hispanic Men and Women: A Nationally Representative

Study." *Journal of Sexual Medicine* 7(suppl 5):330–345.

Dodson, Jualynne E. 2007. "Conceptualization and Research of African American Family Life in the United States: Some Thoughts." pp. 51–68 in *Black Families*, 4th ed., edited by Harriette Pipes McAdoo. Thousand Oaks, CA: Sage.

Doherty, William J. 1992. "Private Lives, Public Values." *Psychology Today* 25(3):32–39.

Doherty, William, Jenet Jacob, and Beth Cutting. 2009. "Community Engaged Parent Education: Strengthening Civic Engagement among Parents and Parent Educators." *Family Relations* 58(3):303–315.

Dolan, Elizabeth M., Bonnie Braun, and Jessica C. Murphy. 2003. "A Dollar Short: Financial Challenges of Working-poor Rural Families." *Family Focus* (June):F13–F15. National Council on Family Relations.

Dolan, Frances Elizabeth. 2008. *Marriage and Violence: The Early Modern Legacy*. Philadelphia: University of Pennsylvania Press.

Dolbin-MacNab, Megan L. 2006. "Just Like Raising Your Own? Grandmothers' Perceptions of Parenting a Second Time Around." *Family Relations* 55(5):564–575.

Dolbin-MacNab, Megan I., and Margaret K. Keiley. 2009. "Navigating Interdependence: How Adolescents Raised Solely by Grandparents Experience Their Family Relationships." *Family Relations* 58(April):162–175.

"Domestic Partnership Benefits and Obligations Act." 2015. Human Rights Campaign. Retrieved September 25, 2015 (http://www.hrc.org).

Dominguez, Silvia, and Amy Lubitow. 2008. "Transnational Ties, Poverty, and Identity: Latin American Immigrant Women in Public Housing." *Family Relations* 57(October): 419–430.

Don, Brian P., and Kristin D. Mickelson. 2014. "Relationship Satisfaction Trajectories Across the Transition to Parenthood among Low-Risk Parents." *Journal of Marriage and Family* 76(3):677–692.

Donaldson, Doug. 2012. "The New American Super-Family." *Saturday Evening Post*, July–August. Retrieved April 10 (http://www.saturdayeveningpost.com).

D'Onofrio, Brian M., and Benjamin B. Lahey. 2010. "Biosocial Influences on the Family: A Decade Review." *Journal of Marriage and Family* 72(3)(June):762–782.

Donovan, Jack. 2012. "The Way of Men: Masculinity Explained." March 26. Retrieved October 14, 2015 (http://www .jack-donovan .com).

Doohan, Eve-Anne M., Sybil Carrere, Chelsea Siler, and Cheryl Beardslee. 2009. "The Link Between the Marital Bond and Future Triadic Family Interactions." *Journal of Marriage and Family* 71(4):892–904.

Dore, Margaret K. 2004. "The 'Friendly Parent' Concept: A Flawed Factor for Child Custody." *Loyola Journal of Public Interest Law* 6:41–56.

Dorius, Cassandra R., Alan Booth, Jacob Hibel, Douglas Granger, and David Johnson. 2011. "Parents' Testosterone and Children's Perception of Parent-Child Relationship Quality." *Hormones and Behavior* 60:512–519.

Dorius, Cassandra J. 2018. "Revisiting the Changing Face of Teenage Parenthood in the United States: Evidence from the NLSY79 and NLSY97. " *Child & Youth Care Forum* 47:343–350.

Doss, Brian D., Galena K. Rhoads, Scott M. Stanley, and Howard J. Markman. 2009. "The Effect of the Transition to Parenthood on Relationship Quality: An 8-Year Prospective Study." *Journal of Personality and Social Psychology* 96(3):601–619.

Doss, Brian D., and Galena K. Rhoades. 2017. "The Transition to Parenthood: Impact on Couples' Romantic Relationships. " *Current Opinion in Psychology* 13:25–28.

Doss, Brian D., Larisa N. Cicila, Annie C. Hsueh, Kristen R. Morrison, and Kathryn Carhart. 2014. "A Randomized Controlled Trial of Brief Coparenting and Relationship Interventions During the Transition to Parenthood. " *Journal of Family Psychology* 28:483.

Doucet, Andrea, and Robyn Lee. 2014. "Fathering, Feminism(s), Gender, and Sexualities: Connections, Tensions, and New Pathways." *Journal of Family Theory and Review* 6(4):355–373.

Doucleff, Michaeleen. 2018. "Rate of C-Sections is Rising At An 'Alarming Rate.'" Retrieved December 24, 2019 (https://www.npr.org /sections/goatsandsoda/2018/10/12/656198429 /rate-of-c-sections-is-rising-at-an-alarming-rate).

Doughty, Susan E., Susan M. McHale, and Mark E. Feinberg. 2015. "Sibling Experiences as Predictors of Romantic Relationship Qualities in Adolescence." *Journal of Family Issues* 36:589–608.

Douglas, Edward, and Sharon Douglas. 2000. *The Blended Family: Achieving Peace and Harmony in the Christian Home*. Franklin, TN: Providence House.

Dowd, Maureen. 2012. "Don't Tread on Us." *The New York Times, March* 13. Retrieved October 26, 2012 (http://www .nytimes.com).

Downey, Douglas B., and Dennis J. Condron. 2004. "Playing Well with Others in Kindergarten: The Benefit of Siblings at Home." *Journal of Marriage and Family* 66(2):333–350.

Downing-Matibag, Teresa, and Brandi Geisinger. 2009. "Hooking Up and Sexual Risk Taking among College Students: A Health Belief Model Perspective." *Qualitative Health Research* 19(9):1196–1209.

Downs, Barbara. 2003. *Fertility of American Women: June 2002*. Current Population Reports P20–548, October. Washington, DC: U.S. Census Bureau.

Doyle, Alison. 2015. "Federal and State Minimum Wage Rates for 2015, 2016, and 2017." Retrieved September 29, 2015 (http:// jobsearch.about.com).

Drash, Wayne. 2018. "What Parents Should Know about the 'Huge Epidemic' of Vaping." CNN. November 17. Retrieved August 3, 2019 (www.cnn.com).

Dreger, Alice D., and April Herndon. 2009. "Progress and Politics in the Intersex Rights Movement: Feminist Theory in Action." *GLQ: A Journal of Lesbian and Gay Studies* 15(2):199–224.

Driscoll, Anne K. 2018. "Asian-American Mothers: Demographic Characteristics by Maternal Place of Birth and Asian Subgroup, 2016." *National Vital Statistics Reports* 67(2). April 18. Hyattsville MD: National Center for Health Statistics.

Driscoll, Anne K., Stephen R. Russell, and Lisa J. Crockett. 2008. "Parenting Styles and Youth Well-Being across Immigrant Generations." *Journal of Family Issues* 29(2):185–209.

Driver, Janice L., and John M. Gottman. 2004. "Daily Marital Interactions and Positive Affect During Marital Conflict among Newlywed Couples." *Family Process* 43(3):301–314.

Drouin, Michelle, Manda Coupe, and Jeff R. Temple. 2017. "Is Sexting Good for Your Relationship? It Depends...." *Computers in Human Behavior* 75:749–756.

Drummet, Amy R., Marilyn Coleman, and Susan Cable. 2003. "Military Families Under Stress: Implications for Family Life Education." *Family Relations* 52(3):279–287.

Duba, Jill, A. Hughey, Tracy Lara, and M. Burke. 2012. "Areas of Marital Dissatisfaction among Long-Term Couples." *Adultspan Journal* 11(1): 39–54.

Duba, Jill D., and Richard E. Watts. 2009. *Journal of Clinical Psychology* 65(2):210–223.

Dubbs, Shelli L., Abraham P. Buunk, and Jessica Li. 2012. "Parental Monitoring, Sensitivity Toward Parents, and a Child's Mate preferences." *Personal Relationships* 19:712–722.

Dubin, Minna. 2019. "The Rage Mothers Don't Talk About." Retrieved September 27, 2019 (https://parenting.nytimes.com /parent-life/mother-rage).

Ducharme, Jamie. 2018. "More Than 90% of Generation Z Is Stressed Out, and Gun Violence Is Partly to Blame." *Time*. October 30. Retrieved November 15, 2018 (www.time.com).

Ducharme, Jamie. 2019. "Why Do So Many Couples Look Alike? Here's the Psychology Behind this Weird Phenomenon. " Retrieved November 21, 2019

Duenwald, Mary. 2005. "For Them, Just Saying No Is Easy." *The New York Times*, June 9.

Duffin, Erin. 2019. *Birth Rate in the United States in 2017, by Household Income*. Retrieved December 17, 2019 (https://www.statista.com /statistics/241530/birth-rate-by-family -income-in-the-us/).

Duffin, Erin. 2019. Language Spoken (at Home) Other Than English in the United States by Number of Speakers in 2018. Retrieved October 16, 2019 (www.statistica.com).

Dugan, Andrew. 2015a. "Men, Women Differ on Morals of Sex, Relationships." Retrieved December 3, 2015 (http://www .gallup.com /poll/183719/men-women -differ-morals-sex -relationships.aspx?g _source=marriage& g_medium=search&g _campaign=tiles).

———. 2015b. "Once Taboo, Some Behaviors Are Now More Acceptable in the U.S." Gallup Poll Topics, Social Issues. Retrieved February 8, 2016 from http://www.gallup.com.

Dugan, Laura, Daniel S. Nagin, and Richard Rosenfeld. 2004. "Do Domestic Violence Services Save Lives?" *NIJ Journal* 250:20–25. Washington, DC: National Institute of Justice and U.S. Department of Justice.

Dugger, Celia W. 1998. "In India, an Arranged Marriage of Two Worlds." *The New York Times*, July 20, pp. A1, A10.

Duncan, Simon, and Miranda Phillips. 2011. "People Who Live Apart Together (LATs): New Family Form or Just a Stage?" *International Review of Sociology* 21 (3):513–532.

Duncan, Stephen F., Jeffry H. Larson, and Shelece McAllister. 2014. "Characteristics of Individuals Associated with Involvement in Different Types of Marriage Preparation Interventions." *Family Relations* 63:680–692.

Dunifon, Rachel, and Ashish Bajracharya. 2012. "The Role of Grandparents in the Lives of Youth." *Journal of Family Issues* 33(9):1168–1194.

Dunifon, Rachel, Christopher Near, and Kathleen Ziol-Guest. 2018. "'Backup Parents, Playmates, Friends: Grandparents' Time with Grandchildren." *Journal of Marriage and Family* 78 (3): 752–767.

Dunifon, Rachel, and Lori Kowaleski-Jones. 2007. "The Influence of Grandparents in Single-Mother Families." *Journal of Marriage and Family* 69(2):456–481.

Dunnewind, Stephanie. 2003. "Book Helps Impart Coping Skills, Self-Esteem to Multiracial Children." Knight Ridder /Tribune News Service, August 5.

Dupre, M. E., and S. O. Meadows. 2007. "Disaggregating the Effects of Marital Trajectories on Health." *Journal of Family Issues* 28:623–652.

DuPree, Devin, Jason Whiting, and Steven Harris. 2016. "A Person-Oriented Analysis of Couple and Relationship Education." *Family Relations* 65 (5): 635–646.

Dupuis, Sara. 2007. "Examining Remarriage: A Look at Issues Affecting Remarried Couples and the Implications Towards Therapeutic Techniques." *Journal of Divorce and Remarriage* 48(1–2):91–104.

Duquaine-Watson, Jillian M. 2007. "'Pretty Darned Cold': Single Mother Students and the Community College Climate in Post-Welfare Reform America." *Equity and Excellence in Education* 40:229–240.

Duran-Aydintug, C. 1993. "Relationships with Former In-Laws: Normative Guidelines and Actual Behavior." *Journal of Divorce and Remarriage* 19:69–81.

Durham, Ricky. 2010. Prescription 4 Love. (http://www.prescription4love.com).

Durodoye, Beth A., and Angela D. Coker. 2008. "Crossing Culture in Marriage: Implications for Counseling African American/African Couples." *International Journal of Counseling* 30:25–37.

Dush, Claire M. Kamp. 2011. "Relationship-Specific Investments, Family Chaos, and Cohabitation dissolution Following a Nonmarital Birth." *Family Relations* 60(5):586–601.

Dush, Claire M. Kamp, Catherine L. Cohan, and Paul R. Amato. 2003. "The Relationship between Cohabitation and Marital Quality and Stability: Change Across Cohorts?" *Journal of Marriage and Family* 65(3):539–549.

Dush, Claire M. Kamp, Miles G. Taylor, and Rhiannon A. Kroeger. 2008. "Marital Happiness and Psychological Well-Being across the Life Cycle." *Family Relations* 57(April):211–226.

Dutton, Donald G., and Tanya L. Nicholls. 2005. "The Gender Paradigm in Domestic Violence Research and Theory: The Conflict of Theory and Data." *Aggression and Violent Behavior* 10:680–714.

Duvall, E. M. 1954. *In-Laws: Pro and Con—An Original Study of Inter-Personal Relations.* New York: Association Press.

Dyer, W. Justin, Joseph H. Pleck, and Brent McBride. 2012. "Imprisoned Fathers and Their Family Relationships: A 40-Year Review from a Multi-Theory View." *Journal of Family Theory and Review* 4(1):20–47.

Dykstra, Pearl A., and Gunhild O. Hagestad. 2007. "Roads Less Taken: Developing a Nuanced View of Older Adults Without Children." *Journal of Family Issues* 28(10):1275–1310.

Eaklor, Vicki L. 2008. *Queer America: A GLBT History of the 20th Century.* Westport, CT: Greenwood Press.

Early, Theresa J., Thomas K. Gregoire, and Thomas P. McDonald. 2002. "Child Functioning and Caregiver Well-Being in Families of Children with Emotional Disorders." *Journal of Family Issues* 23(3):374–391.

East, Patricia L., Nina C. Chien, and Jennifer S. Barber. 2012. "Adolescents' Pregnancy Intentions, Wantedness, and Regret: Cross-Legged Relations with Mental Health and Harsh Parenting." *Journal of Marriage and Family* 74(1):167–185.

Easterlin, Richard. 1987. *Birth and Fortune: The Impact of Numbers on Personal Welfare,* 2nd rev. ed. Chicago: University of Chicago Press.

Eaton, Danice K., Laura Kann, Steve Kinchen, Shari Shanklin, James Ross, Joseph Hawkins, William A. Harris, Richard Lowry, Tim McManus, David Chyen, Connie Lim, Nancy D. Brener, and Howell Wechsler. 2008. "Youth Risk Behavior Surveillance—United States, 2007." Surveillance Summaries. *Morbidity and Mortality Weekly Report* 57, No. SS-04, June 6.

Ebaugh, Helen Rose, and Mary Curry. 2000. "Fictive Kin as Social Capital in New Immigrant Communities." *Sociological Perspectives* 43(2):189–209.

Ebeling, Ashlea. 2007. "The Second Match." *Forbes,* November 12. Retrieved April 15, 2010 (www.forbes.com).

Eccles, Jacquelynne. 2011. "Gendered Educational and Occupational Choices: Applying the Eccles et al. Model of Achievement-Related Choices." *International Journal of Behavioral Development* 35(3):195–201.

Eckholm, Eric. 2008. "Special Session Called on Nebraska Safe-Haven Law." *The New York Times,* October 30. Retrieved December 12, 2008 (www.nytimes.com).

———. 2012. "'Ex-Gay' Men Fight Back Against View That Homosexuality Can't Be Changed." Retrieved June 25, 2012 (http://www.nytimes.com/2012/11/01/us/ex-gay-men-fight-view-that-homosexuality-cant-be-changed.html?_r=0).

Economist, The. 2016. "Bare Branches, Redundant Males." Retrieved January 11, 2016 (http://www.economist.com/news/asia/21648715-distorted-sex-ratios-birth-generation-ago-are-changing-marriage-and-damaging-societies-asias).

Eddy, J. Mark, and Julie Poehlmann-Tynan (Eds). 2019. *Handbook on Children with Incarcerated Parents.* Springer, Switzerland: Springer Publishers.

Edin, Kathryn, and Rebecca Joyce Kissane. 2010. "Poverty and the American Family: A Decade in Review." *Journal of Marriage and Family* 72(3) (June):460–479.

Edin, K., & Shaefer, H. L. (2015). *$2.00 a Day: Living on Almost Nothing in America.* Boston, MA: Houghton Mifflin Harcourt.

Edleson, Harriet. 2015. "Grandparents Who Move to Be Closer to their Grandchildren." *The New York Times, June* 26. Retrieved December 30, 2015 (http://www.nytimes.com).

Edwards, Margie L. K. 2004. "We're Decent People: Constructing and Managing Family Identity in Rural Working-Class Communities." *Journal of Marriage and Family* 66(2):515–529.

Edwards, Tom 2009. "As Baby Boomers Age, Fewer Families Have Children Under 18 at Home." *U.S. Census Bureau News,* February 25. Retrieved June 20, 2009 (www.census.gov /Press-Release/www /releases/).

Egan, Timothy. 2014. "Sex and the Saints." *The New York Times,* November 29. Retrieved December 14, 2014 (http://www .nytimes.com).

Egelko, Bob. 2008. "Churches on Both Sides of Marriage Law Debate." *San Francisco Chronicle,* February 18: A1, A10.

Eggebeen, David. J. 2012. "What Can We Learn from Studies of Children Raised by Gay or Lesbian Parents?" *Social Science Research* 41: 775–778.

Eggebeen, David, and Jeffrey Dew. 2009. "The Role of Religion in Adolescence for Family Formation in Young Adulthood." *Journal of Marriage and Family* 71(1):108–121.

Ehrenfeld, Temma. 2002. "Infertility: A Guy Thing." *Newsweek,* March 25, pp. 60–61.

Ehrenreich, Barbara. 2001. *Nickel and Dimed: On (Not) Getting By in America.* New York: Henry Holt.

Ehrenreich, Barbara. 2018. *Natural Causes: An Epidemic of Wellness, the Certainty of Dying, and Killing Ourselves to Live Longer.* Hachette UK.

Eicher-Catt, Deborah. 2004. "Noncustodial Mothers and Mental Health: When Absence Makes the Heart Break." *Family Focus* (March):F7–F8. Minneapolis: National Council on Family Relations.

Eickmeyer, Kasey J. 2014. "Divorce Rate in the U.S." Retrieved March 14, 2016 (https://www .bgsu.edu/content/dam /BGSU/college-of -arts-and-sciences /NCFMR/documents/FP /eickmeyer -divorce-rate-us-geo-2014-fp-15-18 .pdf).

———. 2015. "Generation X and Millennials' Attitudes toward Marriage and Divorce" (FP-15-12). National Center for Marriage and Family Research.

———. 2016. "Divorce Rate in the U.S." NCFMR Family Profiles, FP-15-18. Retrieved January 26, 2016 (https://www .bgsu.edu/content/dam /BGSU/college -of-arts-and-sciences/NCFMR /documents /FP/eickmeyer-divorce-rate-us -geo-2014 -fp-15-18.pdf).

Eickmeyer, Kasey J. 2017. "American Children's Family Structure: Single Parent Families." Family Profiles, FP-17-17. Bowling Green, OH: National

Center for Family & Marriage Research. Retrieved February 9, 2020 (https://www.bgsu.edu/ncfmr/resources/data/family-profiles/eickmeyer-single-parent-families-fp-17-17.html).

Einhorn, Erin. 2019. "Kicking Kids Out of Preschool Is Damaging, Experts Say. So Why Is It Still Happening?" NBC News. August 3. Retrieved October 14. (http://www.nbcnews.com). Elliott, Sinikka, and Sarah Bowen 2018. "Defending Motherhood: Morality, Responsibility, and Double Binds in Feeding Children." *Journal of Marriage and Family* 80 (2): 499–520.

Elkind, David. 1988. *The Hurried Child: Growing Up Too Fast Too Soon.* Reading, MA: Addison-Wesley.

———. 2007a. *The Hurried Child: Growing Up Too Fast Too Soon,* 25th anniversary ed. Cambridge, MA: Da Capo Lifelong.

———. 2007b. *The Power of Play: How Spontaneous, Imaginative Activities Lead to Happier, Healthier Children.* Cambridge, MA: Da Capo Lifelong.

———. 2010. "Playtime Is Over." *The New York Times, March* 26. Retrieved July 15, 2013 (http://nytimes.com).

Eller, Jack David. 2015. *Culture and Diversity in the United States: So Many Ways to be American.* New York: Routledge.

Ellin, Abby. 2009. "The Recession. Isn't It Romantic?" *The New York Times,* February 12. Retrieved June 8, 2009 (www.nytimes.com).

Elliott, Diana B., and Tavia Simmons. 2011. *Marital Events of Americans: 2009.* American Community Survey Reports, ACS-13. Washington, DC: U.S. Census Bureau.

Elliott, Sinikka, and Debra Umberson. 2008. "The Performance of Desire: Gender and Sexual Negotiation in Long-Term Marriages." *Journal of Marriage and Family* 70(2):391–406.

Ellis, D. 2006. "Male Abuse of a Married or Cohabiting Female Partner: The Application of Sociological Theory to Research Findings." *Violence and Victims* 4:235–255.

Ellison, Christopher G., and Matt Bradshaw. 2009. "Religious Beliefs, Sociopolitical Ideology, and Attitudes Toward Corporal Punishment." *Journal of Family Issues* 30(3):330–340.

Ellison, Christopher G., Andrea K. Henderson, Norval D. Glenn, and Kristine E. Harkrider. 2011. "Sanctification, Stress, and Marital Quality." *Family Relations* 60(4): 404–420.

Ellison, Christopher G., Marc A. Musick, and George W. Holden. 2011. "Does Conservative Protestantism Moderate the Association Between Corporal Punishment and Child Outcomes?" *Journal of Marriage and Family* 73(5) (October): 946–961.

Ellison, Christopher G., Jenny A. Trinitapoli, Kristin L. Anderson, and Byron R. Johnson. 2007. "Race/Ethnicity, Religious Involvement, and Domestic Violence." *Violence Against Women* 13(11):1094–1112.

Ellison, Christopher G., Nicholas H. Wolfinger, and Aida I. Ramos-Wada. 2012. "Attitudes Toward Marriage, Divorce, Cohabitation, and Casual Sex among Working-Age Latinos: Does Religion Matter?" *Journal of Family Issues* 34(3):295–322.

Ellison, Peter T., and Peter B. Gray, eds. 2009. *Endocrinology of Social Relationships.* Cambridge, MA: Harvard University Press.

Else-Quest, Nicole M., and Janet Shibley Hyde. 2016. "Intersectionality in Quantitative Psychological Research: II. Methods and Techniques." *Psychology of Women Quarterly* 40:319–336.

El-Sheikh, Mona, and Elizabeth Flanagan. 2001. "Parental Problem Drinking and Children's Adjustment: Family Conflict and Parental Depression as Mediators and Moderators of Risk." *Journal of Abnormal Child Psychology* 29(5):417–435.

Emery, Lydia F., Amy Muise, Elizabeth Alpert, and Benjamin Le. 2015. "Do We Look Happy? Perceptions of Romantic Relationship Quality on Facebook." *Personal Relationships* 22:1–7.

Eckstein, Joseph Battocletti, and Carley Gardener. 2010. "First Date Sexual Expectations: The Effects of Who Asked, Who Paid, Date Location, and Gender." *Communication Studies* 61(3):339–355.

Elsen-Rooney. 2020. "Two Boys with the Same Disability Tried to Get Help." *USA Today.* February 10. Retrieved February 27, 2020 (https://www.usatoday.com).

Emlet, Charles A. 2016. "Social, Economic, and Health Disparities Among LGBT Older Adults." *Generations* 40:16–22.

Encyclopedia Britannica. 2019. "Feminism: The Fourth Wave." Retrieved October 5, 2019 (https://www.britannica.com/explore/100women/issues/feminism-the-fourth-wave/).

Engel, Marjorie. 2000. "The Financial (In)Security of Women in Remarriages." *Research Findings.* Stepfamily Association of America (www.saafamilies.org).

Engels, Friedrich. 1942 [1884]. *The Origin of the Family, Private Property, and the State.* New York: International.

England, Diane. 2009. *The Post Traumatic Stress Disorder Relationship: How to Support Your Partner and Keep Your Relationship Healthy.* Avon, MA: Adams Media.

England, Paula, and Kathryn Edin (eds.). 2007. *Unmarried Couples with Children.* New York: Russell Sage Foundation.

England, Paula, and Elizabeth A. McClintock. 2009. "The Gendered Double Standard of Aging in U.S. Marriage Markets." *Population and Development Review* 35(4):797–816.

Englander, Elizabeth, T. Milosevic, and E. Staksrud. 2019. "Sexting: Healthy or Harmful? Comparative Analysis of Teens in Colorado, Massachusetts, Norway, and Serbia." Symposia presented at World Anti-Bullying Forum, June 5, Dublin, Ireland.

Englander, Elizabeth. 2019. "What Do We Know About Sexting, and When Did We Know It?" *Journal of Adolescent Health* 65:577–578.

Ennis, Dawn. 2019. "Amazon Stopped Selling Books Promoting Conversion Therapy, and Some People Are Mad." *Forbes.* July 21. Retrieved August 13, 2019 (https://www.forbes.com).

Enochs, Kevin. 2017. "In the U.S., 'Interpolitical' Marriage Increasingly Frowned Upon." Retrieved November 21, 2019 (https://www.voanews.com/usa/us-interpolitical-marriage-increasingly-frowned-upon).

Enos, Sandra. 2001. *Mothering from the Inside: Parenting in a Women's Prison.* Albany: SUNY Press.

Enriquez, Laura E. 2015. "Multigenerational Punishment: Shared Experiences of Undocumented Immigration Status Within Mixed-Status Families." *Journal of Marriage and Family* 77(4):939–953.

Epstein, Ann S. 2007. *The Intentional Teacher: Choosing the Best Strategies for Young Children's Learning.* Washington, DC: National Association for the Education of Young Children.

Epstein, Marina. 2011. "Exploring Parent-Adolescent Communication about Gender: Results from Adolescent and Emerging Adult Samples." *Sex Roles* 65(1/2):108–118.

Epstein, Marina, Jerel P. Calzo, Andrew P. Smiler, and L. Monique Ward. 2009. "Anything from Making Out to Having Sex: Men's Negotiations of Hooking Up and Friends with Benefits Scripts." *Journal of Sex Research* 46(5)414–442.

Equality Can't Wait. 2019. "Equality Can't Wait." Retrieved October 5, 2019 (https://equalitycantwait.evoke.org/).

Eriksen, Shelley, and Naomi Gerstel. 2002. "A Labor of Love or Labor Itself: Care Work among Brothers and Sisters." *Journal of Family Issues* 23(7):836–856.

Erickson, Nancy S. 1991. "Battered Mothers of Battered Children: Using Our Knowledge of Battered Women to Defend Them against Charges of Failure to Act." *Current Perspectives in Psychological, Legal, and Ethical Issues,* Vol. 1A, *Children and Families: Abuse and Endangerment,* pp. 197–218.

Essex, Elizabeth L., and Junkuk Hong. 2005. "Older Caregiving Parents: Division of Household Labor, Marital Satisfaction, and Caregiver Burden." *Family Relations* 54(3):448–460.

Etcheverry, Paul E., and Benjamin Le. 2005. "Thinking about Commitment: Accessibility of Commitment and Prediction of Relationship Persistence, Accommodation, and Willingness to Sacrifice." *Personal Relationships* 23(1):103–123.

Evans, Gary W., and Theodore D. Wachs. 2010. *Chaos and Its Influence on Children's Development: An Ecological Perspective.* Washington, DC: American Psychological Association.

Evans, Stephen. 2013. "Germany Allows 'Indeterminate' Gender at Birth." BBC News, November 1. Retrieved October 16, 2015 (http://www.bbc.com).

Even, William E., and David A. Macpherson. 2004. "When Will the Gender Gap in Retirement Income Narrow?" *Southern Economic Journal* 71(1):182–201.

Even-Zohar, Ahuva, and Shlomo Sharlin. 2009. "Grandchildhood: Adult Grandchildren's Perception of Their Role towards Their Grandparents from an Intergenerational Perspective." *Journal of Comparative Family Studies* 40(2):167–185.

Ezzell, Matthew B. 2012. "'I'm in Control': Compensatory Manhood in a Therapeutic Community." *Gender and Society* 26(2):190–215.

Faber, Adele, and Elaine Mazlish. 2006. *How to Talk So Kids Will Listen and Listen So Kids Will Talk*. New York: Collins.

"Factsheet: The Violence Against Women Act." n.d. Retrieved March 14, 2013 (http://www.whitehouse.gov/sites /default/files/docs /vawa_factsheet.pdf).

Fagan, Jay. 2009. "Relationship Quality and Changes in Depressive Symptoms among Urban, Married African Americans, Hispanics, and Whites." *Family Relations* 58 (July):259–274.

———. 2011. "Effect on Preschoolers' Literacy When Never-Married Mothers Get Married." *Journal of Marriage and Family* 73(5):1001–1014.

———. 2014. "A Review of How Researchers Have Used Theory to Address Research Questions About Fathers in Three Large Data Sets." *Journal of Family Theory and Review* 6(4):374–389.

Fagan, Jay, and Marina Barnett. 2003. "The Relationship between Maternal Gatekeeping, Paternal Competence, Mothers' Attitudes about the Father Role, and Father Involvement." *Journal of Family Issues* 24(8):1020–1043.

Fagan, Jay, Randal Day, Michael Lamb, and Natasha Cabrera. 2014. "Should Researchers Conceptualize Differently the Dimensions of Parenting for Fathers and Mothers?" *Journal of Family Theory and Review* 6(4):390–405.

Fagan, Jay, and Y. Lee. 2011. "Do Coparenting and Social Support Have a Greater Effect on Adolescent Fathers than Adult Fathers?" *Family Relations* 60(3):247–258.

Fagan, Jay, and Rebecca Kaufman. 2015. "Co-Parenting Relationships among Low-Income, Unmarried Parents: Perspectives of Fathers in Fatherhood Programs." *Family Court Review* 53:304–316.

Fahrlander, Rebecca S. 2015. "Go Ahead, Get Grandma Out of Your Feed," Retrieved February 25, 2016 (https://www.washingtonpost.com/opinions /unfriending -your-family/2015/05/15 /f3b3463e-dca7 -11e4-be40-566e2653afe 5_story.html).

Fahs, Breanne, and Adrielle Munger. 2015. "Friends with Benefits? Gendered Performances in Women's Casual Sexual Relationships." *Personal Relationships* 22:188–203.

Fairchild, Emily. 2006. "'I'm Excited to Be Married, But . . .': Romance and Realism in Marriage." pp. 1–19 in *Couples, Kids, and Family Life*, edited by Jaber F. Gubrium and James A. Holstein. New York: Oxford University Press.

Fairchild, Kimberly. 2014. "Feminism is Now: Fighting Modern Misogyny and the Myth of the Post-Feminist Era." *Sex Roles* 73:453–455.

———. 2015. "Feminism Is Now: Fighting Modern Misogyny and the Myth of the Post-Feminist Era." *Sex Roles* 73(9–10):451–467.

Falconer, Mary Kay, Mary E. Haskett, Linda McDaniels, Thelma Dirkes, and Edward C. Siegel. 2008. "Evaluation of Support Groups for Child Abuse Prevention: Outcomes of Four State Evaluations." *Social Work with Groups* 31(2):165–182.

Falconier, Mariana K., and Norman B. Epstein. 2011. "Couples Experiencing Financial Strain: What We Know and What We Can Do." *Family Relations* 60(3): 303–317.

Fallahzadeh, Hossein, Hasan Zareei Mahmood Abadi, Mahdieh Momayyezi, Hakimeh Malaki Moghadam, and Naeimeh Keyghobadi. 2019. "The Comparison of Depression and Anxiety Between Fertile and Infertile Couples: A Meta-Analysis Study. " *International Journal of Reproductive BioMedicine* 17:153–162.

Fallik, Dawn. 2018. "What to Do About Lonely Older Men? Put Them to Work." Retrieved January 24, 2020 (https://www.washingtonpost.com/national/health -science/what-to-do-about-lonely-older-men -put-them-to-work/2018/06/22/0c07efc8 -53ab-11e8-a551-5b648abe29ef_story.html).

Families with Children from China. 1999. "Mongolian Spots." Retrieved April 26, 2006 (www.fwcc.org).

Family Equality Council. 2017. "LGBTQ Family Fact Sheet." Retrieved December 1, 2019 (https://www2.census.gov/cac/nac /meetings/2017-11/LGBTQ-families -factsheet.pdf?).

Family Success Consortium. 2016. "Home Page." Retrieved February 25, 2016 (http://www.gotofsc.com/).

Family Success Consortium. 2020. "Psychological Care for the Whole Family." Retrieved January 15, 2020 (https://gotofsc.com/).

Fanshel, David. 1972. *Far from the Reservation: The Transracial Adoption of American Indian Children*. Metuchen, NJ: Scarecrow.

Farr, Diane. 2011. "Bringing Home the Wrong Race." Retrieved May 31, 2013 (http://www.nytimes.com/2011 /06/05/fashion/modern-love-breaking -our-parents-rules-for-love.html?pagewanted=all).

Farr, Rachel, Yelena Ravvina, and Harold Grotevant. 2018. "Birth Family Contact Experiences among Lesbian, Gay, and Heterosexual Adoptive Parents with School-Age Children." *Family Relations* 67 (1): 180 146.

Farrell, Alison, Jeffry Simpson, and Alexander Rothman. 2015. "The Relationship Power Inventory." *Journal of the International Association of Relationship Research, Personal Relationships* 22 (3): 387–413.

Farrell, Anne F., Gary L. Bowen, and Danielle C. Swick. 2014. "Network Supports and Resiliency among U.S. Military Spouses with Children with Special Health Care Needs." *Family Relations* 63(1):55–70.

Farrell, Betty, Alicia VandeVusse, and Abigail Ocobock. 2012. "Family Change and the State of Family Sociology." *Current Sociology* 60(3):283–301.

Farrell, Warren. 1974. *The Liberated Man: Beyond Masculinity; Freeing Men and Their Relationships with Women*. University of Berkeley: Berkeley Books.

Farver, JoAnn M.,Yiyuan Xu, Bakhtawar R. Bhadha, Sonia Narang, and Eli Lieber. 2007. "Ethnic Identity, Acculturation, Parenting Beliefs, and Adolescent Adjustment: A Comparison of Asian Indian and European American Families." *Merrill-Palmer Quarterly* 53(2):184–215.

Faulkner, R. A., M. Davey, and A. Davey. 2005. "Gender-Related Predictors of Change in Marital Satisfaction and Marital Conflict." *American Journal of Family Therapy* 33:61–83.

Fausto-Sterling, Anne. 2000. "The Five Sexes Revisited." *Sciences*, July/August, pp.19–23.

Federal Bureau of Investigation (FBI). 2012. "Sexting: Risky Actions and Overreactions." Retrieved September 1, 2012 (http://www.fbi.gov/stats-services /publications/law-enforcement -bulletin /july-2010/sexting).

Federal Interagency Forum on Aging Statistics. 2017. *Older Americans 2016: Key Indicators of Well-Being*. Washington, DC: U.S. Government Printing Office.

Federal Interagency Forum on Child and Family Statistics. 2019. *America's Children: Key National Indicators of Well Being, 2019*. Washington, DC: U.S. Government Printing Office.

Federal Reserve Bank of New York. 2013. "Student Loan Debt by Age Group." Retrieved September 28, 2015 (http://www.newyorkfed.org).

Fehr, Beverley, Cheryl Harasymchuk, and Susan Sprecher. 2014. "Compassionate Love in Romantic Relationships A Review and Some New Findings." *Journal of Social and Personal Relationships*: 0265407514533768.

Feigelman, Susan, Howard Dubowitz, Wendy Lane, LesliePrescott, Walter Meyer, Kathleen Tracy, and Jeongeun Kim. 2009. "Screening for Harsh Punishment in a Pediatric Primary Care Clinic." *Journal of Child Abuse and Neglect* 33(5):269–277.

Feigelman, W. 2000. "Adjustments of Transracially and Inracially Adopted Young Adults." *Child and Adolescent Social Work Journal* 17:165–183.

Fein, Esther B. 1997. "Failing to Discuss Dying Adds to Pain of Patient and Family." *The New York Times*, March 5, pp. A1–A14.

Feinberg, Jessica. 2016. "Consideration of Genetic Connections in Child Custody Disputes Between Same-Sex Parents: Fair or Foul. *Mo. L. REv.* 81:331.

Feinberg, Mark E., Marni L. Kan, and E. Mavis Hetherington. 2007. "The Longitudinal Influence of Coparenting Conflict on Parental Negativity and Adolescent Maladjustment." *Journal of Marriage and Family* 69(3):687–702.

Feldhaus, Michael, and Valarie Heintz-Martin. 2015. "Long-Term Effects of Parental Separation: Impacts of Parental Separation During Childhood on the Timing and the Risk of Cohabitation, Marriage, and Divorce in Adulthood." *Advances in Life Course Research* 26:22–31.

Felker, J. A., D. K. Fromme, G. L. Arnaut, and B. M. Stoll. 2002. "A Qualitative Analysis of Stepfamilies: The Stepparent." *Journal of Divorce and Remarriage* 38(1–2):125–142.

Felkey, Amanda J., and Kristina M. Lybecker. 2018. "Do Restrictions Beget Responsibility? The Case of U.S. Abortion Legislation. " *The American Economist* 63:59–70.

Ferdman, Roberto A. 2014. "Americans Aren't Getting Married, and Researchers Think Porn May Be the Problem." Retrieved December 17, 2015

(https://www.washingtonpost.com/news /wonk/wp/2014/12/21/americans-arent -getting-married-and-researchers-think -porn-is-part-of-the-problem/).

Fergusson, David M., Joseph M. Boden, and L. John Horwood. 2007. "Abortion among Young Women and Subsequent Life Outcomes." *Perspectives on Sexual and Reproductive Health* 39(1):6–12.

Fernandez, Manny. 2018. "Texas Fetal Burial Law Struck Down in Another Blow to Abortion Restrictions. " Retrieved December 23, 2019

Ferrari, J. R., and R. A. Emmons. 1994. "Procrastination as Revenge: Do People Report Using Delays as a Strategy for Vengeance?" *Personality and Individual Differences* 17(4):539–542.

Ferree, Myra Marx. 2010. "Filling the Glass: Gender Perspectives on Families." *Journal of Marriage and Family* 72(3):420–439.

Ferrer-Chancy, Millie. 2009. "Stepping Stones for Stepfamilies: For Step-Grandparents." University of Florida Extension. Retrieved May 5, 2010 (www .edis .ifas.ufl.edu).

Ferro, Christine, Jill Cermele, and Ann Saltzman. 2008. "Current Perceptions of Marital Rape: Some Good and Not-So-Good News." *Journal of Interpersonal Violence* 23(6):764–779.

Festinger, Trudy B. 2005. "Adoption and Disruption." pp. 452–468 in *Child Welfare for the 21st Century: A Handbook of Practices, Policies, and Programs*, edited by G.Mallon and P. Hess. New York: Columbia University Press.

Fetsch, Robert J., Raymond K. Yang, and Matthew J. Pettit. 2008. "The RETHINK Parenting and Anger Management Program." *Family Relations* 57(5):543–552.

Few, April L., and Karen H. Rosen. 2005. "Victims of Chronic Dating Violence: How Women's Vulnerabilities Link to Their Decisions to Stay." *Family Relations* 54(2):265–279.

Few-Demo, April L. 2014. "Intersectionality as the 'New' Critical Approach in Feminist Family Studies: Evolving Racial/Ethnic Feminisms and Critical Race Theories." *Journal of Family Theory and Review* 6(2):169–183.

Few-Demo, April L., and Katherine R. Allen. 2020. "Gender, Feminist, and Intersectional Perspectives on Families: A Decade in Review." *Journal of Marriage and Family* 82 (1): 326–345.

Fiebert, Martin S. 2012. "References Examining Assaults by Women on Their Spouses or Male Partners: An Annotated Bibliography." Retrieved March 14, 2013 (http://www.csulb.edu).

Fields, Jason. 2004. *America's Families and Living Arrangements: 2003*. Current Population Reports P20–553, November. Washington, DC: U.S. Census Bureau.

Fields, Julianna. 2010. *Families Living with Mental and Physical Challenges*. Broomall, PA: Mason Crest Publishers.

File, Thom, and Camille Ryan. 2014. "Computer and Internet Use in the United States: 2013." *American Community Survey Reports*, ACS-28. Washington, DC: U.S. Census Bureau.

Filinson, R. 1986. "Relationship in Stepfamilies: An Examination of Alliances." *Journal of Comparative Family Studies* 17:43–61.

Finch, J., and J. Mason. 1990. "Divorce, Remarriage, and Family Obligations." *Sociological Review* 38:219–246.

Fincham, Frank D., and Ross W. May. 2017. "Infidelity in Romantic Relationships." *Current Opinion in Psychology* 13:70–74.

Fincham, Frank D., Ming Cui, Mellissa Gordon, and Koji Uemo. 2013. "What Comes Before Why: Specifying the Phenomenon of Intimate Partner Violence." *Journal of Marriage and Family* 75(2):319–324.

Fincham, Frank D., Julie Hall, and Steven R. H. Beach. 2006. "Forgiveness in Marriage: Current Status and Future Directions." *Family Relations* 55(4):415–427.

Fincham, Frank D., Scott M. Stanley, and Steven R. H. Beach. 2007. "Transformative Processes in Marriage: An Analysis of Emerging Trends." *Journal of Marriage and Family* 69(2):275–292.

Finer, Lawrence B., Lori Frohwirth, Lindsay A. Dauhiphinee, Sushella Singh, and Ann M. Moore. 2005. "Reasons U.S. Women Have Abortions: Quantitative and Qualitative Perspectives." *Perspectives on Sexual and Reproductive Health* 37(3):110–118.

Fingerman, Karen L., and Frank F. Furstenberg. 2012. "You Can Go Home Again." *The New York Times*, May 30. Retrieved August 28, 2012 (http://www .nytimes.com).

Fingerman, Karen L., Kim Kyungmin, Eden M. Davis, Frank Furstenberg, Kira Birditt, and Steven H. Zerit. 2015. "'I'll Give You the World': Socioeconomic Differences in Parental Support of Adult Children." *Journal of Marriage and Family* 77(4):844–865.

Fingerman, Karen, Laura Miller, Kira Birditt, and Steven Zarit. 2009. "Giving to the Good and the Needy: Parental Support of Grown Children." *Journal of Marriage and Family* 71(5):1220–1233.

Finkel, David. 2013. *Thank You for Your Service*. New York: Picador Press.

Finkelhor, David, Richard Ormrod, Heather Turner, and Sherry Hamby. 2005. "The Victimization of Children and Youth: A Comprehensive National Survey." *Child Maltreatment* 10:5–25.

Firmin, Michael, and Annie Phillips. 2009. "A Qualitative Study of Families and Children Possessing Diagnoses of ADHD." *Journal of Family Issues* 30(9):1155–1174.

Fisher, William A., Raymond C. Rosen, Ian Eardley, Michael Sand, and Irwin Goldstein 2005. "Sexual Experience of Female Partners of Men with Erectile Dysfunction: The Female Experience of Men's Attitudes to Life Events and Sexuality (FEMALES) Study." *Journal of Sexual Medicine* 2(5):675–684.

Fisher, Linda L. 2010. *Sex, Romance, and Relationships: AARP Survey of Midlife and Older Adults*. Washington, DC: American Association of Retired Persons. Retrieved May 10, 2010 (www.aarp.org).

Fisher, Terri D., Clive M. Davis, William L. Yarber, and Sandra L. Davis. 2010. *Handbook of Sexuality-Related Measures*. Thousand Oaks, CA: Sage.

Fishman, B. 1983. "The Economic Behavior of Stepfamilies." *Family Relations* 32:359–366.

Fishman, Margie. 2013. "Divorce Parties Mark Milestones in Uncoupling." *USA Today*. Retrieved April 22, 2013 (http://www.usatoday.com/story/news /nation/2013/01/31/divorce-party /1881623/).

Fitzpatrick, Jacki, Elizabeth Sharp, and Alan Reifman. 2009. "Midlife Singles' Willingness to Date Partners with Heterogeneous Characteristics." *Family Relations* 58(1):121–133.

Fitzsimmons, Bemma G. 2014. "A Scourge is Spreading. M.T.A.'s Cure? Due, Close Your Legs." Retrieved October 1, 2019 (https://www. nytimes.com/2014/12/21 /nyregion/MTA-targets-manspreading -on-new-york-city-subways.html).

Fivush, Robyn, Kelly Marin, Kelly McWilliams, and Jennifer Bohanek. 2009. "Family Reminiscing Style: Parent Gender and Emotional Focus in Relation to Child Well-Being." *Journal of Cognition and Development* 10(3):210–235.

Flaherty, Colleen. 2016. "More Faculty Diversity, Not on Tenure Track." Retrieved October 4, 2019 (https://www .insidehighered.com/news/2016/08/22 / study-finds-gains-faculty-diversity-not-tenure-track).

Flaherty, Colleen. 2019. "Faculty Salaries Stay Flat." Retrieved October 4, 2019 (https://www. insidehighered.com/news/2019/04/10/aaup-study-finds-small -gains-faculty-salaries-offset-inflation).

Fleeson, W., and E. Noftle. (2008). "The End of the Person-Situation Debate: An Emerging Synthesis in the Answer to the Consistency Question." *Social and Personality Psychology Compass* 2:1667–1684.

Fleming, C. J. Eubanks, and James V. Cordorva. 2012. "Predicting Relationship Help Seeking Prior to a Marriage Checkup." *Family Relations* 61:90–100.

Fletcher, Anne C., and Bethany L. Blair. 2014. "Maternal Authority Regarding Early Adolescents' Social Technology Use." *Journal of Family Issues* 35:54–74.

Fletcher, Garth. 2002. *The New Science of Intimate Relationships*. Malden, MA: Blackwell.

Flood, Sarah H., and Katie R. Genadek. 2016. "Time for Each Other: Work and Family Constraints among Couples." *Journal of Marriage and Family* 78 (1): 142–164.

Flora, Carlin. 2007. "Can Grown-Up Siblings Learn to Get Along?" *Psychology Today*, March /April:48–49.

Flores, Andrew R., Jody L. Herman, Gary J. Gates, and Taylor N. T. Brown. 2016. How Many Adults Identify as Transgender in the United States? Los Angeles, CA: The Williams Institute.

Florian, Sandra M. 2918. "Motherhood and Employment among Whites, Hispanics, and Blacks: A Life Course Approach." *Journal of Marriage and Family* 80 (1): 134–149.

Fogle, Asher. 2015. "15 Sneaky Signs Marriage May End in Divorce." Retrieved February 8, 2020 (https://www.goodhousekeeping .com/life/relationships/a33733 /surprising-predictors-of-divorce/).

Fomby, Paula, and Andrew J. Cherlin. 2007. "Family Instability and Child Well-Being." *American Sociological Review* 72(2):181–204.

Fomby, Paula, and Angela Estacion. 2011. "Cohabitation and Children's Externalizing

Behaviors in Low-Income Latino Families." *Journal of Marriage and Family* 73(1):46–66.

Fomby, Paula, Joshua A. Goode, and Stefanie Mollborn. 2016. "Family Complexity, Siblings, and Children's Aggressive Behavior at School Entry." *Demography* 53:1–26.

Fomby, Paula, and Cynthia Osborne. 2017. "Family Instability, Multipartner Fertility, and Behavior in Middle Childhood." *Journal of Marriage and Family* 79 (1): 75–93.

Fontaine, Romeo, Agnes Gramain, and Jerome Wittwer. 2009. "Providing Care for an Elderly Parent: Interactions among Siblings." *Health Economics* 18:1011–1029.

Foran, Heather M., and K. Daniel O'Leary. 2008. "Problem Drinking, Jealousy, and Anger Control: Variables Predicting Physical Aggression Against a Partner." *Journal of Family Violence* 23:141–148.

Forbes, Miri, Nicholas Eaton, and Robert Krueger. 2016. "Sex Quality of Life Doesn't Have to Decline as We Age." *CNN*. December 23. Retrieved January 15, 2017 (https://www.cnn.com).

Ford, Melissa. 2009. *Navigating the Land of If: Understanding Infertility and Exploring Your Options*. Berkeley, CA: Seal Press.

Fortuny, Karina, Randy Capps, Margaret Simms, and Ajay Chaudry. 2009. *Children of Immigrants: National and State Characteristics*. Washington, DC: Urban Institute.

Fosco, Gregory M., and John H. Grych. 2012. "Capturing the Family Context of Emotion Regulation: A Family Systems Model Comparison Approach." *Journal of Family Issues* 34(4):557–578.

Foster, Diana Green, Julia R. Steinberg, Sarah CM Roberts, John Neuhaus, and M. Antonia Biggs. 2015. "A Comparison of Depression and Anxiety Symptom Trajectories between Women Who Had an Abortion and Women Denied One." *Psychological Medicine* 45:2073–2082.

Foster, E. Michael, Damon Jones, and Saul D. Hoffman. 1998. "The Economic Impact of Nonmarital Childbearing: How Are Older, Single Mothers Faring?" *Journal of Marriage and Family* 60(1):163–174.

Foster, Holly, and Jeanne Brooks-Gunn. 2009. "Toward a Stress Process Model of Children's Exposure to Physical Family and Community Violence." *Clinical Child and Family Psychology Review* 12:71–94.

Fountain, Kim, Maryse Mitchell-Brody, Stephanie A. Jones, and Kaitlin Nichols. 2009. *Lesbian, Gay, Bisexual, Transgender and Queer Domestic Violence in the United States in 2008*. New York: The National Coalition of Anti-Violence Programs. Available at www.avp.org/documents/2008NCAVPLGBTQDVReportFINAL.pdf.

Fowler, Sarah. 2019. "Where Are My Parents? School on Standby to Help Children in Aftermath of ICE Raids." *Mississippi Clarion Ledger*. August 8.

Fox, Ashley M., Georgia Himmelstein, Hina Khalid, and Elizabeth A. Howell. 2019. "Funding for abstinence-only education and adolescent pregnancy prevention: does state ideology affect outcomes?" *American Journal of Public Health* 109: 497–504.

Fox, Bonnie. 2009. *When Couples Become Parents: The Creation of Gender in the Transition to Parenthood*. Buffalo, NY: University of Toronto Press.

Fox, Greer Litton, Michael L. Benson, Alfred A. DeMaris, and Judy Van Wyk. 2002. "Economic Distress and Intimate Violence: Testing Family Stress and Resources Theory." *Journal of Marriage and Family* 64:793–807.

Fox, Greer L., Vey M. Nordquist, Rhett M. Billen, and Emily Furst Savoca. 2015. "Father Involvement and Early Intervention: Effects of Empowerment and Father Role Identity." *Family Relations* 64(4):461–475.

Fracher, Jeffrey, and Michael S. Kimmel. 1992. "Hard Issues and Soft Spots: Counseling Men about Sexuality." pp. 438–450 in *Men's Lives*, 2nd ed., edited by Michael S. Kimmel and Michael A. Messner. New York: Macmillan.

Fraga, Juli. 2017. "'When I Was Your Age' and Other Pitfalls of Talking to Teens about Stress." Retrieved January 25, 2020 (https://www.npr.org/sections/health-shots/2017/04/16/523592625/-when-i-was-your-age-and-other-conversational-pitfalls-of-talking-to-teens).

Fraga, Lynette M. 2019. *The U.S. and the High Price of Child Care: An Examination of a Broken System*. 2019 Report. Retrieved January 15, 2020 (www.usa.childcareaware.org).

Frank, Kristyn, and Feng Hou. 2015. "Source-Country Gender Roles and the Division of Labor within Immigrant Families." *Journal of Marriage and Family* 77(2):557–574.

Frank, Robert H. 2011. "Gauging the Pain of the Middle Class." *The New York Times*, April 2. Retrieved September 13, 2011 (http://www.nytimes.com).

Franke-Ruta, Garance. 2013. "Obama's Minimum-Wage Gamble." *The Atlantic*. Retrieved February 20, 2013 (http://www.theatlantic.com).

Franklin, Cortney A., Travis W. Franklin, Matt R. Nobles, and Glen A. Kercher. 2012. "Assessing the Effect of Routine Activity Theory and Self-Control on Property, Personal, and Sexual Assault Victimization." *Criminal Justice and Behavior* 39(10):1296–1315.

Frech, Adrianne, and Rachel Tolbert Kimbro. 2011. "Maternal Mental Health, Neighborhood Characteristics, and Time Investments in Children." *Journal of Marriage and Family* 73(June):605–620.

Freedman, Gili, Darcey N. Powell, Benjamin Le, and Kipling D. Williams. 2019. "Ghosting and Destiny: Implicit Theories of Relationships Predict Beliefs about Ghosting." *Journal of Social and Personal Relationships* 36(3):905–924.

Freeland, Chrystia. 2013. "Sexist Mores of Super-Rich Hurt Us All." Retrieved September 29, 2019 (https://www.cnn.com/2018/07/21/opinions/abortion-fertility-miscarriage-stories-matter-saujani-opinion/index.html).

Freeman, Amanda L., and Lisa Dodson. 2014. "Social Network Development among Low-Income Single Mothers." *Family Relations* 63(5):589–601.

Freeman, Amanda L. 2017. "Moving 'Up and Out' Together: Exploring the Mother-Child Bond in Low-Income, Single-Mother-Headed Families." Friedman, Zack. 2019. "Student Loan Debt Statistics in 2019: A $1.5 Trillion Crisis." Retrieved November 25, 2019 (https://www.forbes.com/sites/zackfriedman/2019/02/25/student-loan-debt-statistics-2019/#20ed25f2133f).

Freeman, David W. 2011. "Same-Sex Affairs: Men More Forgiving than Women." CBS News. Retrieved November 20, 2012 (http://www.cbsnews.com/8301-504763_162-20030042-10391704.html).

Freeman, Marsha B. 2008. "Love Means Always Having to Say You're Sorry: Applying the Realities of Therapeutic Jurisprudence to Family Law." *UCLA Women's Law Journal* 17:215–241.

Freeman, Melissa, and Sandra Mathison. 2009. *Researching Children's Experiences*. New York: Guilford Press.

French, J. R. P., and Bertram Raven. 1959. "The Basis of Power." In *Studies in Social Power*, edited by D.Cartwright. Ann Arbor: University of Michigan Press.

Fricke, Hans, Jeffrey Grogger, and Andreas Steinmayr. 2018. "Exposure to Academic Fields and College Major Choice." *Economics of Education Review* 64:199–213.

Fried, Carla. 2016. "Don't Let Divorce Derail Your Retirement." Retrieved February 8, 2020 (https://finance.yahoo.com/news/don-t-let-divorce-derail-110004031.html).

Friedan, Betty. 1963. *The Feminine Mystique*. New York: Dell.

Friedman, Richard A. 2015. "How Changeable Is Gender?" *The New York Times*, August 22. Retrieved September 19, 2015 (http://www.nytimes.com).

Friedman, Zack. 2019. "Student Loan Debt Statistics in 2019: A $1.5 Trillion Crisis." Retrieved November 25, 2019 (https://www.forbes.com/sites/zackfriedman/2019/02/25/student-loan-debt-statistics-2019/#20ed25f2133f).

Fridel, Emma, and Edward A. Fox. 2019. "Gender Differences in Patterns and Trends in U.S. Homicide, 1976–2017." *Violence and Gender* 6 (1).

Frisco, Michelle L., Chandra Muller, and Kenneth Frank. 2007. "Parents' Union Dissolution and Adolescents' School Performance: Comparing Methodological Approaches." *Journal of Marriage and Family* 69(3):721–741.

Frisco, Michelle L., and Kristi Williams. 2003. "Perceived Housework Equity, Marital Happiness, and Divorce in Dual Earner Families." *Journal of Family Issues* 24:51–73.

Frith, Lucy, Eric Blyth, Marilyn Paul, and Roni Berger. 2011. "Conditional Embryo Relinquishment: Choosing to Relinquish Embryos for Family-Building though a Christian Embryo 'Adoption' Programme." *Human Reproduction* 26(12):3327–3338.

Frith, H., Kitzinger, C. 2001. Reformulating Sexual Script Theory Developing a Discursive Psychology of Sexual Negotiation. *Theory & Psychology* 11:209–232.

Fritsch, Jane. 2001. "A Rise in Single Dads." *The New York Times*, May 20.

Frizell, Sam 2015. "Americans Can Now Expect to Live Longer Than Ever." *Time*, September 5. Retrieved September 29, 2015 (http://time .com).

Fromm, Erich. 1956. *The Art of Loving*. New York: Harper & Row.

Frontline. 2012. "The Allure of Adult Content Users: American Porn." Public Broadcasting System. Retrieved July 24, 2012 (http://www .pbs.org/wgbh/pages /frontline/shows/porn /business /haveallure.html).

Frosch, Dan. 2013. "Dispute on Transgender Rights Unfolds at a Colorado School." Retrieved June 25, 2013 (http://www .nytimes .com/2013/03/18/us/in-colorado -a-legal -dispute-over-transgender-rights.html

?ref=danfrosch&gwh=3886D9DE6212CE5856D9 9C7BF57F03BF).

Froyen, Laura, Lori Skibbe, Ryan Bowles, Adrian Blow, and Hope Gerde. 2013. "Marital Satisfaction, Family Emotional Expressiveness, Home Learning Environments, and Children's Emergent Literacy." *Journal of Marriage and Family* 75(1):42–55.

Frum, David. 2012. "Get Off the Road." *Newsweek*, July 2 & 9:46–49.

Fruhauf, Christine A., Shannon E. Jarrott, and Katherine R. Allen. 2006. "Grandchildren's Perceptions of Caring for Grandparents." *Journal of Family Issues* 27(7):887–911.

Fry, Richard. 2018. "Gen X Rebounds as the Only Generation to Recover the Wealth Lost after the Housing Crash." *Facttank, News in the Numbers*. July 23. Pew Research Center. Retrieved September 14, 2018 (www.pewresearch.org).

Fry, Richard. 2017. "It's Becoming More Common for Young Adults to Live at Home—and for Longer Stretches." Retrieved November 26, 2019 (https://www.pewresearch.org /fact-tank/2017/05/05/its-becoming -more-common-for-young-adults-to-live -at-home-and-for-longer-stretches/).

Fry, Richard. 2019. "U.S. Women Near Milestone in the College-Educated Labor Force." Retrieved October 2, 2019 (https://www.pewresearch.org /fact-tank/2019/06/20/u-s-women -near-milestone-in-the-college-educated -labor-force/).

Fry, Richard, 2013. "Young Adults after the Recession: Fewer Homes, Fewer Cars, Less Debt." *Pew Research Social and Demographic Trends*, February 21. Retrieved March 20, 2013 (http:// www.pewsocialtrends.org).

———. 2014. "Young Adults, Student Debt, and Economic Well-Being." *Pew Research Center: Social and Demographic Trends*. Retrieved September 28, 2015 (http://www.pewsoecialtrends.org).

Fry, Richard, and D'Vera Cohn. 2010. *Women, Men and the New Economics of Marriage*. Washington, DC: Pew Research Center.

———. 2011. "Living Together: The Economics of Cohabitation." Pew Research Center Publications, June 27. Retrieved September 24, 2011 (http://pewresearch.org).

Fryer, Roland G. 2007. "Guess who's Been Coming to Dinner: Trends in Interracial Marriage over the Twentieth Century." *Journal of Economic Perspectives* 21(2):71–90.

Fu, Tsung-chieh, Debby Herbenick, Brian Dodge, Christopher Owens, Stephanie A. Sanders, Michael Reece, and J. Dennis Fortenberry. 2019. "Relationships Among Sexual Identity, Sexual Attraction, and Sexual Behavior: Results from a Nationally Representative Probability Sample of Adults in the United States." *Archives of Sexual Behavior* 48:1483–1493.

Fu, Vincent Kang, and Nicholas H. Wolfinger. 2011. "Broken Boundaries or Broken Marriages? Racial Intermarriage and Divorce in the United States." *Social Science Quarterly* 92:1096-1117.

Fu, Xuanning, and Tim B. Heaton. 2000. "Status Exchange in Intermarriage among Hawaiians, Japanese, Filipinos and Caucasians in Hawaii: 1983 –1994." *Journal of Comparative Family Studies* 31(1):45–64.

———. 2008. "Racial and Educational Homogamy: 1980 to 2000." *Sociological Perspectives* 51(4):735–758.

Fuhrmans, Vanessa. 2018. "As More New Dads Get Paternity Leave, Companies Push Them to Take It." *Wall Street Journal*. July 11. Retrieved August 29, 2019 (www.wsj.com).

Fujiura, Glenn T. 2010. "Aging Families and the Demographics of Family Financial Support of Adults with Disabilities." *Journal of Disability Policy Studies* 20(4):241–250.

Fulda, Barbara E., and Philipp M. Lersch. 2018. "Planning Until Death Do Us Part: Partnership Status and Financial Planning Horizon." *Journal of Marriage and Family* 80 (2): 409–425.

Furdyna, Holly E., M. Belinda Tucker, and Angela D. James. 2008. "Relative Spousal Earnings and Marital Happiness among African American and White Women." *Journal of Marriage and Family* 70(2):332–344.

Furman, Wyndol, and Laura Shaffer. 2011. "Romantic Partners, Friends, Friends with Benefits, and Casual Acquaintances as Sexual Partners." *Journal of Sex Research* 48(6):554–564.

Furstenberg, Frank F. Jr. 2000. "The Sociology of Adolescence and Youth in the 1990s: A Critical Commentary." *Journal of Marriage and Family* 62(4):896–910.

———. 2005. "Banking on Families: How Families Generate and Distribute Social Capital." *Journal of Marriage and Family* 67(4):809–821.

———. 2008. "The Changing Landscape of Early Adulthood in the U.S." *Family Focus*, March:F2-F3, F18.

Furstenberg, Frank F. 2020. "Kinship Reconsidered: Research on a Neglected Topic." *Journal of Marriage and Family* 82:364–382.

Furstenberg, Frank F. Jr., J. Brooks-Gunn, and S. Philip Morgan. 1987. *Adolescent Mothers in Later Life*. New York: Cambridge University Press.

Furstenberg, Frank F. Jr., Sheela Kennedy, Vonnie C. McLoyd, Rubén G. Rumbaut, and Richard A. Settersten Jr. 2004. "Growing Up Is Harder to Do." *Contexts* 3(3):33–41.

Furstenberg, Frank F. Jr., and Kathleen E. Kiernan. 2001. "Delayed Parental Divorce: How Much Do Children Benefit?" *Journal of Marriage and Family* 63(2):446–457.

Furstenberg, Frank F. Jr, Christine Winquist Nord, James L., Peterson, and Nicholas Zill. 1983. "The Life Course of Children of Divorce." *American Sociological Review* 48(5):656–668.

Furukawa, Ryoko, and Martha Driessnack. 2013. "Video-Mediated Communication to Support Distant Family Connectedness." *Clinical Nursing Research* 22(1):82–94.

Future of Sex Education Initiative. 2012. "National Sexuality Education Standards: Core Content and Skills, K-12." *Journal of School Health*. Retrieved November 30, 2015 (http://www .futureofsexeducation).

Gable, Shelly L., Harry T. Reis, Emily A. Impett, and Evan R. Asher. 2004. "Interpersonal Relations and Group Processes—What Do You Do When Things Go Right? The Intrapersonal and Interpersonal Benefits of Sharing Positive Events." *Journal of Personality and Social Psychology* 87(2):228–245.

Gaertner, Bridget M., Tracy I. Spinrad, Nancy Eisenberg, and Karissa A. Greving. 2007. "Parental Childrearing Attitudes as Correlates of Father Involvement During Infancy." *Journal of Marriage and Family* 69(4):962–976.

Gage, Kris. 2017. "Who Has the Power in Your Relationship" July 7. Retrieved February 5, 2020 (https://krisgage.com/2017/07/07 /who-has-the-power-in-your-relationship).

Gager, C. T., T. M. Cooney, and K. T. Call. 1999. "The Effects of Family Characteristics and Time Use on Teenagers' Household Labor." *Journal of Marriage and the Family* 61:982–994.

Gager, Constance T., and Laura Sanchez. 2003. "Two as One? Couples' Perceptions of Time Spent Together, Marital Quality, and the Risk of Divorce." *Journal of Family Issues* 24(1):21–50.

Gager, Constance T., and Scott T. Yabiku. 2010. "Who Has the Time? The Relationship between Household Labor Time and Sexual Frequency." *Journal of Family Issues* 31(2):135–163.

Gagnon, John H., and William Simon. 2005. *Sexual Conduct: The Social Sources of Human Sexuality (Social Problems and Social Issues)*. 2nd ed. New Brunswick, NJ: Transaction Books.

Gaille, Brandon. 2017. "23 Important Male victims of Domestic Violence Statistics." Retrieved February 15, 2020 (https:// brandongaille.com/22-important-male -victims-of-domestic-violence-statistics).

Gale, Jason. 2015. "How Thailand Became a Global Gender-Change Destination." Retrieved September 25, 2019 (https://www.bloomberg .com/news/features/2015-10-26/how -thailand-became-a-global-gender -change-destination).

Galinsky, Ellen, Kerstin Aumann, and James T. Bond. 2009. "Times are Changing: Gender and Generation at Work and at Home." *2008 National Study of the Changing Workforce*. New York: Families and Work Institute.

Gallagher, Charles A. 2006. "Interracial Dating and Marriage: Fact, Fantasy, and the Problem of Survey Data." pp. 141–153 in *African Americans and Whites: Changing Relationships on College*

Campuses, edited by Robert M. MooreIII. New York: University Press of America.

Gallup Poll. 2012. "Marriage." Retrieved September 10, 2012 (http://www.gallup .com).

———. 2015. "Marriage." Retrieved December 1, 2015 (http://www.gallup .com/poll/117328 /marriage.aspx).

———. 2016. "Abortion." Retrieved February 7, 2016 (http://www.gallup.com /poll/1576 /abortion.aspx?version=print).

Gallup. 2019. "Moral Issues." Retrieved February 4, 2020 (https://news.gallup.com/poll/1681/ moral-issues.aspx).

Gallup Poll. 2019. "In Depth: Topics from A to Z, Abortion. " Retrieved December 14, 2019 (https://news.gallup.com /poll/1576/abortion.aspx).

Galvin, Kathleen M., and Dawn O. Braithwaite. 2014. "Theory and Research from the Communication Field: Discourses that Constitute and Reflect Families." *Journal of Family Theory and Review* 6:97–111.

Gamache, Susan J. 1997. "Confronting Nuclear Family Bias in Stepfamily Research." *Marriage and Family Review* 26(1–2):41–50.

Games-Evans, Tina. 2009. "Finding Your Personal Identity as a Mom." Babyzone, June 1. www.babyzone.com.

Ganna, A., Verweij, K.J., Nivard, M.G., Maier, R., Wedow, R., Busch, A.S., Abdellaoui, A., Guo, S., Sathirapongsasuti, J.F., Lichtenstein, P. and Lundström, S., 2019. "Large-scale GWAS Reveals Insights into the Genetic Architecture of Same-Sex Sexual Behavior." *Science* 365, no. 3456, eaat7693 (Aug. 30, 2019).

Ganong, Lawrence, and Marilyn Coleman. 2018. "Studying Stepfamilies: Four Eras of Family Scholarship." *Family Process* 57:7–24.

Ganong, Lawrence, and Marilyn Coleman. 2017. "The Cultural Context of Stepfamilies." pp. 21–36 in *Stepfamily Relationships,* edited by L. Ganong and M. Coleman. Springer, Boston, MA.

Ganong, Lawrence, Todd Jensen, Caroline Sanner, Luke Russell, and Marilyn Coleman. 2019. "Stepfathers' Affinity-Seeking with Stepchildren, Stepfather-Stepchild Relationship Quality, Marital Quality, and Stepfamily Cohesion Among Stepfathers and Mothers." *Journal of Family Psychology* 33:521.

Ganong, Lawrence H., Marilyn Coleman, and Tyler Jamison. 2011. "Patterns of Stepchild-Stepparent Relationship Development." *Journal of Marriage and Family* 73:396–413.

Ganong, Lawrence H., Marilyn Coleman, and Tanja Rothrauff. 2009. "Patterns of Assistance between Adult Children and Their Older Parents: Resources, Responsibilities, and Remarriage." *Journal of Social and Personal Relationships* 26(2–3):161–178.

Gans, Daphne, Merril Silverstein, and Anela Lowenstein. 2009. "Do Religious Children Care More and Provide More Care for Older Parents? A Study of Filial Norms and Behaviors across Five Nations." *Journal of Comparative Family Studies* 40(2):187–201.

Gansen, Heidi M. 2017. "Reproducing (and Disrupting) Heteronormativity: Gendered Sexual Socialization in Preschool Classrooms. " *Sociology of Education* 90:255–272.

Garasky, Steven, Craig Gundersen, Susan D. Stewart, and Brenda J. Lohman. 2010. "Toward a Fuller Understanding of Nonresident Father Involvement: A Joint Examination of Child Support, In-Kind Support, and Visitation." *Population Research and Policy Review* 29:363–393.

Gao, George. 2015. "American Ideal Family Size Is Smaller than It Used to Be." Retrieved February 11, 2016 (http://www .pewresearch .org/fact-tank/2015/05/08 /ideal-size-of-the-american-family/).

Garcy, Pamela D. 2013. "Are You a Relationship Martyr? " Retrieved November 16, 2019 (https://www.psychologytoday.com /us/blog/fearless-you/201307 /are-you-relationship-martyr).

Garcia, Justin R., Susan M. Seibold-Simpson, Sean G. Massey, and Ann M. Merriwether. 2015. "Casual Sex: Integrating Social, Behavioral, and Sexual Health Research." pp. 203–222 In *Handbook of the Sociology of Sexualities,* edited by John DeLamater and Rebecca F. Plante. New York: Springer.

Gardner, Jonathan, and Andrew Oswald. 2006. "Do Divorcing Couples Become Happier by Breaking Up?" *Journal of the Royal Statistical Society,* Series A 169(2):319–336.

Gardner, Margo, Christopher Browning, and Jeanne Brooks-Gunn. 2012. "Can Organized Youth Activities Protect Against Internalizing Problems among Adolescents Living in Violent Homes?" *Journal of Research on Adolescence* 22(4):662–677.

Garey, Anita I., and Karen V. Hansen(eds.) 2011. *At the Heart of Work and Family.* New Brunswick, NJ: Rutgers University Press.

Garfield, Robert. 2010. "Male Emotional Intimacy: How Therapeutic Men's Groups Can Enhance Couples Therapy." *Family Process* 49(1):109–122.

Garson, David G. n.d. "Economic Opportunity Act of 1964." Retrieved October 9, 2006 (wps. prenhall.com).

Garton, Stephen. 2004. *Histories of Sexuality: Antiquity to Sexual Revolution.* London, UK: Equinox Publishing.

Gartrell, Nanette, Henny Bos, Heidi Peyser, Amalia Deck, and Carla Rodas. 2011. "Family Characteristics, Custody Arrangements, and Adolescent Psychological Well-Being After Lesbian Mothers Break Up." *Family Relations* 60(5):572–585.

Gassman-Pines, Anna. 2011. "Low-Income Mothers' Nighttime and Weekend Work: Daily Associations with Child Behavior, Mother-Child Interactions, and Mood." *Family Relations* 60(1):15–29.

———. 2013. "Daily Spillover of Low-Income Mothers' Perceived Workload to Mood and Mother-Child Interactions." *Journal of Marriage and Family* 75(5):1304–1318.

Gates, Gary J. 2009. *Same-Sex Spouses and Unmarried Partners in the American Community Survey, 2008.* Los Angeles: The Williams Institute. Retrieved November 20, 2009 (www .law.ucla.edu/williamsinstitute).

———. 2011. "Family Formation and Raising Children among Same-Sex Couples." *Family Focus* FF51 (Winter): F2–F4.

———. 2013. "LGBT Adult Immigrants in the United States." Retrieved December 3, 2015 (http://escholarship.org/uc/item /2cj0k29c).

Gates, Gary J., and Frank Newport. 2012. "Special Report: 3.4% of U.S. Adults Identify as LGBT." Retrieved June 25, 2013 (http://www. odec.umd.edu/CD/LGBT /Special%20Report%203.4%25%20of%20 U.S.%20Adults%20Identify%20as%20LGBT. pdf).

Gates, Gary J., and Taylor N. T. Brown. 2015. Marriage and Same-sex Couples after Obergefell." Retrieved November 25, 2019 (https://williamsinstitute.law.ucla.edu/wp-content/uploads/Marriage-and -Same-sex-Couples-after-Obergefell -November-2015.pdf).

Gauchat, Gordon, Maura Kelly, and Michael Wallace. 2012. "Occupational Gender Segregation, Globalization, and Gender Earnings Inequality in U.S. Metropolitan Areas." *Gender and Society* 26(5):718–747.

Gaughan, Monica. 2002. "The Substitution Hypothesis: The Impact of Premarital Liaisons and Human Capital on Marital Timing." *Journal of Marriage and Family* 64(2):407–419.

Gault-Sherman, Martha. 2012. "What Will the Neighbors Think? The Effect of Moral Communities on Cohabitation." *Review of Religious Research* 54(1):45–67.

Gaunt, Ruth. 2013. "Breadwinning Moms, Caregiving Dads: Double Standard in Social Judgments of Gender Norm Violators." *Journal of Family Issues* 34(1):3–24.

Gayles, Jochebed G., J. Douglas Coatsworth, Hilda M. Pantin, and Jose Szapocznik. 2009. "Parenting and Neighborhood Predictors of Youth Problem Behaviors within Hispanic Families: The Moderating Role of Family Structure." *Hispanic Journal of Behavioral Sciences* 31(3):277–296.

Gearing, Maeve. 2015. "Father Engagement in Home Visiting Programs: Promising Strategies, Benefits, and Challenges." In 2015 Fall Conference: The Golden Age of Evidence-Based Policy." Association for Public Policy Analysis and Management.

Geewax, Marilyn. 2012a. "Paying for College: More Tough Decisions." National Public Radio, May 15. Retrieved May 15, 2012 (http://www. npr.org).

———. 2012b. "Preparing for a Future That Includes Aging Parents." National Public Radio, April 24. Retrieved May 15, 2012 (http://www. npr.org).

Geiger, A. W. 2018. "Sharing Chores a Key to Good Marriage, Say Majority of Married Adults." *Facttank: News in the Numbers.* November 30. Retrieved July 11, 2019 (www.pewresearch.org).

Geiger, A.W., and Gretchen Livingston. 2019. "8 Facts about Love and Marriage in America." *Facttank: News in the Numbers.* February 13. Retrieved July 11, 2019 (www.pewresearch.org).

Geiger, A.W., Gretchen Livingston, and Kristen Bialik. 2019. "6 Facts about U.S. Moms." *Facttank: News in the Numbers.* May 8. Retrieved July 11, 2019 (www.pewresearch.org).

Geist, Claudia, and Leah Ruppanner. 2018. "Mission Impossible? New Housework Theories

for Changing Families." *Journal of Family Theory & Review* 10 (1): 242–262.

Gelatt, Vicky A., Francesca Adler-Baeder, and John R. Seeley. 2010. "An Interactive Web-Based Program for Stepfamilies: Development and Evaluation of Efficacy." *Family Relations* 59:572–586.

Geller, Amanda. 2013. "Paternal Incarceration and Father-Child Contact in Fragile Families." *Journal of Marriage and Family* 75(5):1288–1303.

Geller, Amanda, and Allyson Walker Franklin. 2014. "Paternal Incarceration and the Housing Security of Urban Mothers." *Journal of Marriage and Family* 76(2):411–427.

Gelles, Richard J. 1974. *The Violent Home: A Study of Physical Aggression Between Husbands and Wives.* Beverly Hills, CA: Sage.

———. 2005. "Protecting Children Is More Important than Preserving Families." pp. 329–340 in *Current Controversies on Family Violence*, 2nd ed., edited by Donileen R. Loseke, Richard J. Gelles, and Mary M. Cavanaugh. Thousand Oaks, CA: Sage.

Gelles, Richard J., and Mary M. Cavanaugh. 2005. "Violence, Abuse, and Neglect in Families and Intimate Relationships." pp. 129–154 in *Families and Change: Coping with Stressful Events and Transitions*, 3rd ed., edited by Patrick C. McKenryand Sharon J. Price. Thousand Oaks, CA: Sage.

Gelles, Richard J., and Murray A. Straus. 1988. *Intimate Violence: The Definitive Study of the Causes and Consequences of Abuse in the American Family.* New York: Simon and Schuster.

Gemelli, Marcella. 2008. "Understanding the Complexity of Attitudes of Low-Income Single Mothers Toward Work and Family in the Age of Welfare Reform." *Gender Issues* 25:101–113.

Gerace, Alyssa. 2012. "How Will Senior Housing Development Adapt to Multigenerational Trends?" National Public Radio. Senior Living News Wire. Retrieved May 15, 2012 (http://www .seniorlivingnewswire.com).

Gerber, Lynne. 2015. "Grit, Guts, and Vanilla Beans Godly Masculinity in the Ex-Gay Movement." *Gender and Society* 29:26–50.

Gerbrandt, Roxanne. 2007. *Exposing the Unmentionable Class Barriers in Graduate Education.* Unpublished doctoral dissertation, University of Oregon.

Gershoff, Elizabeth T. 2018. "Corporal Punishment Associated with Dating Violence. " *Journal of Pediatrics* 198:322–325.

Gershoff, Elizabeth, and Andrew Grogan-Kaylor. 2016. "Race as a Moderator of Associations between Spanking and Child Outcomes." *Family Relations* 65 (3): 490–501.

Gerson, Kathleen. 2010. *The Unfinished Revolution: How a New Generation is Reshaping Family, Work, and Gender in America.* New York: Oxford University Press.

Gerstein, Julie. 2015. "Boys in Puerto Rico Are Now Allowed to Wear Skirts to School." Buzzfeed, October 13. Retrieved October 13, 2015 (http://www.buzzfeed.com).

Gewertz, Catherine. 2009. "Report Probes Educational Challenges Facing Latinas." *Education Week* 29(2):12.

Gibson-Davis, Christina M., Kathryn Edin, and Sara McLanahan. 2005. "High Hopes but Even Higher Expectations: The Retreat from Marriage Among Low-Income Couples." *Journal of Marriage and Family* 67:1301–1312.

Gibson, Jennifer E. 2012. "Interviews and Focus Groups with Children: Methods That Match Children's Developing Competencies." *Journal of Family Theory and Review* 4(2) (June):148–159.

Gibson, Megan. 2012. "Parents Who Hid Child's Gender for Five Years Now Face Backlash." Newsfeed. Retrieved April 14, 2015 from http:// newsfeed.time.com.

Gibson-Davis, Christina M. 2008. "Family Structure Effects on Maternal and Paternal Parenting in Low-Income Families." *Journal of Marriage and Family* 70(2):452–465.

———. 2009. "Money, Marriage, and Children: Testing the Financial Expectations and Family Formation Theory." *Journal of Marriage and Family* 71(1):146–160.

Gibson-Davis, Christina, and Heather Rackin. 2014. "Marriage or Carriage? Trends in Union Context and Birth Type by Education." *Journal of Marriage and Family* 76(3):506–519.

Giddens, Anthony. 2007. "The Global Revolution in Family and Personal Life." pp. 26–31 in *Family in Transition*, edited by Arlene S. Skolnickand Jerome H. Skolnick. Boston: Allyn & Bacon.

Giele, Janet Z. 2007. "Decline of the Family: Conservative, Liberal, and Feminist Views." pp. 76–91 in *Family in Transition*, edited by Arlene S. Skolnickand Jerome H. Skolnick. Boston: Allyn & Bacon.

Gierveld, Jenny de Jong, and Eva-Maria Merz. 2013. "Parents' Partnership Decision Making after Divorce or Widowhood: The Role of (Step)Children." *Journal of Marriage and Family* 75(5):1098–1113.

Gilbert, Gizelle L., M. Diane Clark, and Melissa L. Anderson. 2012. "Do Deaf Individuals' Dating Scripts Follow the Traditional Dating Script?" *Sexuality and Culture* 16:90–99.

Gilbert, Neil. 2008. *A Mother's Work: How Feminism, the Market, and Policy Shape Family Life.* New Haven, CT: Yale University Press.

Gilbert, Sophie. 2015. "Why Women Aren't Having Children." *The Atlantic.* Retrieved February 13, 2016 (http://www .theatlantic .com/entertainment/archive /2015/04/why -women-arent-having -children/390765/).

Gilbert, Susan. 1997. "Two Spanking Studies Indicate Parents Should Be Cautious." *The New York Times*, August 20.

Gilkey, So'Nia L., JoAnne Carey, and Shari L. Wade. 2009. "Families in Crisis: Considerations for the Use of Web-Based Treatment Models in Family Therapy." *Families in Society: Journal of Contemporary Human Services* 90(1):37–46.

Gil-Llario, M. Dolores, Cristina Giménez, Rafael Ballester-Arnal, Georgina Cárdenas-López, and Ximena Durán-Baca. 2017. "Gender, Sexuality, and Relationships in Young Hispanic People." *Journal of Sex & Marital Therapy* 43:456–462.

Gillespie, Dair. 1971. "Who Has the Power? The Marital Struggle." *Journal of Marriage and Family* 33:445–458.

Gilligan, Megan, J. Jill Suitor, Scott Feld, and Karl Pillemer. 2015. "Do Positive Feelings Hurt? Disaggregating Positive and Negative Components of Intergenerational Ambivalence." *Journal of Marriage and Family* 77(1):261–276.

Gilligan, Megan, J. Jill Suitor, and Karl Pillemer. 2015. "Estrangement between Mothers and Adult Children: The Role of Norms and Values." *Journal of Marriage and Family* 77: 908–920.

Gillis, Aurelie, Barbara Gabriel, Sarah Galdiolo, and Isabelle Roskam. 2019. "Partner Support as a Protection against Distress During the Transition to Parenthood." *Journal of Family Issues* 40 (9): 1107–1125.

Gillmore, Mary Rogers, Jungeun Lee, Diane M. Morrison, and Taryn Lindhorst. 2008. "Marriage Following Adolescent Parenthood: Relationship to Adult Well-Being." *Journal of Marriage and Family* 70(5):1136–1144.

Gilrane-McGarry, Ursula, and Tom O'Grady. 2012. "Forgotten Grievers: An Exploration of the Grief Experiences of Bereaved Grandparents (Part 2)." *International Journal of Palliative Nursing* 18(4):179–187.

Ginott, Haim G., Alice Ginott, and Wallace Goddard. 2003. *Between Parent and Child: The Bestselling Classic That Revolutionized Parent-Child Communication.* New York: Three Rivers Press.

Giordano, Peggy C., Monica A. Longmore, and Wendy D. Manning. 2006. "Gender and the Meanings of Adolescent Romantic Relationships: A Focus on Boys." *American Sociological Review* 71(2):260–287.

Gitner, Jess. 2012. "'All about Family': Listeners' Stories on Living in Multigenerational Households." National Public Radio, May 15. Retrieved May 15, 2012 (http://www.npr.org).

Glanz, Jen. 2019. "How to Cope When a Friend Breaks up with You." Retrieved December 4, 2019 (https://www.nbcnews .com/better/lifestyle/how-cope-when -friend-breaks-you-ncna988516).

Glass, Jennifer, and Philip Levchak. 2014. "Red States, Blue States, and Divorce: Understanding the Impact of Conservative Protestantism on Regional Variation in Divorce Rates." *American Journal of Sociology* 119:1002–1046.

Glass, Shirley. 1998. "Shattered Vows." *Psychology Today* 31(4):34–52.

———. 2003. *Not "Just Friends": Rebuilding Trust and Recovering Your Sanity After Infidelity.* New York: Atria Books.

Glass, Valerie Q., and April L. Few-Demo. 2013. "Complexities of Informal Social Support Arrangements for Black Lesbian Couples." *Family Relations* 62(5):714–726.

Glassner, Barry. 1999. *The Culture of Fear: Why Parents Are Afraid of the Wrong Things.* New York: Basic Books.

Glauber, Rebecca, and Kristi L. Gozjolko. 2011. "Do Traditional Fathers Always Work More? Gender, Ideology, Race, and Parenthood." *Journal of Marriage and Family* 73(5):1133–1148.

Glauber, Rebecca, and Melissa D. Day. 2018. "Gender, Spousal Caregiving, and Depression: Does Paid Work Matter?" *Journal of Marriage and Family* 80 (2): 537–554.

Glenn, Norval. 1998. "The Course of Marital Success and Failure in Five American 10-Year

Marriage Cohorts." *Journal of Marriage and Family* 60(3):569–576.

Glenza, Jessica. 2019. "Ohio Bill Orders Doctors to 'Reimplant Ectopic Pregnancy' or Face 'Abortion Murder' Charges." Retrieved December 23, 2019 (https://www.theguardian.com/us-news/2019/nov/29/ohio-extreme-abortion-bill-reimplant-ectopic-pregnancy).

Glick, Jennifer E., and Jennifer Van Hook. 2002. "Parents' Coresidence with Adult Children: Can Immigration Explain Racial and Ethnic Variation?" *Journal of Marriage and Family* 64(1):240–253.

Glick, Peter, and Susan T. Fiske. 2018. "The Ambivalent Sexism Inventory: Differentiating Hostile and Benevolent Sexism." In *Social Cognition*, pp. 116–160. Routledge.

Goble, Priscilla, Carol L. Martin, Laura D. Hanish, and Richard A. Fabes. 2012. "Children's Gender-Typed Activity Choices Across Preschool Social Contexts." *Sex Roles* 67(7– 8):435–451.

Goeke-Morey, Marcie, E. Mark Cummings, and Lauren M. Papp. 2007. "Children and Marital Conflict Resolution: Implications for Emotional Security and Adjustment." *Journal of Family Psychology* 21(4):744–753.

Godwin, Emilie, Herman Lukow II, and Stephanie Lichiello. 2015. "Promoting Resilience Following Traumatic Brain Injury: Application of an Interdisciplinary Evidence-Based Model for Intervention." *Family Relations* 64(3):347–362.

Goisis, Alice, Hanna Remes, Pekka Martikainen, Reija Klemetti, and Mikko Myrskylä. 2019. "Medically Assisted Reproduction and Birth Outcomes: A Within-Family Analysis Using Finnish Population Registers." *The Lancet* 393:1225–1232.

Golash-Boza, Tanya. 2012. *Due Process Denied: Detentions and Deportations in the United States.* New York: Routledge, Taylor and Francis.

Golbeck, Jennifer. 2014. "Why You Might Want to have Difficult Conversations On-Line." Retrieved February 24, 2016 (https://www.psychologytoday.com/blog/your-online-secrets/201409/why-you-might-want-have-difficult-conversations-online).

Gold, Joshua M. 2009. "Negotiating the Financial Concerns of Stepfamilies: Directions for Family Counselors." *The Family Journal: Counseling and Therapy for Couples and Families* 17(2):185–188.

———. 2012. "Typologies of Cohabitation: Implications for Clinical Practice and Research." *The Family Journal: Counseling and Therapy for Couples and Families* 20(3):315–321.

Gold, Steven J. 1993. "Migration and Family Adjustment: Continuity and Change among Vietnamese in the United States." pp. 300–314 in *Family Ethnicity: Strength in Diversity,* edited by Harriette Pipes McAdoo. Newbury Park, CA: Sage.

Gold, Steven J., and Mehdi Bozorgmehr. 2007. "Middle East and North Africa." pp. 518–533 in *The New Americans: A Guide to Immigration Since 1965,* edited by Mary C. Waters, Reed Ueda, and Helen B. Marrow. Cambridge, MA: Harvard University Press.

Goldberg, Abbie E. 2007. "Talking about Family." *Journal of Family Issues* 28(1):100–131.

———. 2010. *Lesbian and Gay Parents and Their Children: Research on the Family Life Cycle.* Washington DC: American Psychological Association.

Goldberg, Abbie E., and Katherine R. Allen. 2013. "Same-Sex Relationship Dissolution and LGB Stepfamily Formation: Perspectives of Young Adults with LGB Parents." *Family Relations* 62:529–544.

Goldberg, Abbie E., Jordan B. Downing, and April M. Moyer. 2012. "Why Parenthood, and Why Now? Gay Men's Motivations for Pursuing Parenthood." *Family Relations* 61(1):157–174.

Goldberg, Abbie E., Deborah A. Kasby, and JuliAnna Z. Smith. 2012. "Gender-Typed Play Behavior in Early Childhood: Adopted Children with Lesbian, Gay, and Heterosexual Parents." *Sex Roles* 67:503–515.

Goldberg, Abbie E., Lori A. Kinkler, Hannah B. Richardson, and Jordan B. Downing. 2011. "Lesbian, Gay, and Heterosexual Couples in Open Adoption Arrangements: A Qualitative Study." *Journal of Marriage and Family* 73(2): 502–518.

Goldberg, Abbie E., and Katherine A. Kuvalanka. 2012. "Marriage (In)equality: The Perspectives of Adolescents and Emerging Adults with Lesbian, Gay, and Bisexual Parents." *Journal of Marriage and Family* 74(1):34–52.

Goldberg, Abbie E., and Aline Sayer. 2006. "Lesbian Couples' Relationship Quality across the Transition to Parenthood." *Journal of Marriage and Family* 68(1):87–100.

Goldberg, Abbie E., and JuliAnna Z. Smith. 2013. "Work Conditions and Mental Health in Lesbian and Gay Dual-Earner Parents." *Family Relations* 62(5):727–740.

Goldberg, Abbie, Katherine Allen, Kaitlin Black, Reihonna Frost, and Melissa Manley. 2018. "'There Is No Perfect School': The Complexity of School Decision-Making among Lesbian and Gay Adoptive Parents." *Journal of Marriage and Family* 80 (3): 684–703.

Goldberg, Susan. 2019. "Humanity in Motion." *National Geographic.* August, p. 4 and entire special issue.

Goldberg, J. S., and Carlson, M. J. 2015. "Patterns and Predictors of Coparenting after Unmarried Parents Part." *Journal of Family Psychology* 29:416.

Goldberg, Julia S. 2015. "Coparenting and Nonresident Fathers' Monetary Contributions to their Children." *Journal of Marriage and Family* 77:612–627.

Goldfarb, Samantha S., Will L. Tarver, Julie L. Locher, Julie Preskitt, and Bisakha Sen. 2015. "A Systematic Review of the Association between Family Meals and Adolescent Risk Outcomes." *Journal of Adolescence* 44:134–149.

Goldscheider, Francis, Sandra Hofferth, Carrie Spearin, and Sally Curtin. 2009. "Fatherhood Across Two Generations: Factors Affecting Early Family Roles." *Journal of Family Issues* 30(5):586–604.

Goldscheider, Frances, and Gayle Kaufman. (2006). "Willingness to Stepparent: Attitudes toward Partners Who Already Have Children." *Journal of Family Issues* 27(10):1415–1436.

Goldscheider, Frances, Gayle Kaufman, and Sharon Sassler. 2009. "Navigating the 'New' Marriage Market." *Journal of Family Issues* 30(6):719–737.

Goldscheider, Frances, and Sharon Sassler. 2006. "Creating Stepfamilies: Integrating Children into the Study of Union Formation." *Journal of Marriage and Family* 68(2):275–291.

Goldsmith, Daena J., Jennifer J. Bute, and Kristin A. Lindholm. 2012. "Patient and Partner Strategies for Talking about Lifestyle Change Following a Cardiac Event." *Journal of Applied Communication Research* 40(1):65–86.

Goldstein, Arnold P., Harold Keller, and Diane Erne. 1985. *Changing the Abusive Parent.* Champaign, IL: Research Press.

Goleman, Daniel. 1985. "Patterns of Love Charted in Studies." *The New York Times,* September 10.

———. 1992. "Family Rituals May Promote Better Emotional Adjustment." *The New York Times,* March 11.

Golombok, Susan, Sophie Zadeh, Susan Imrie, Venessa Smith, and Tabitha Freeman. 2016. "Single Mothers by Choice: Mother–Child Relationships and Children's Psychological Adjustment." *Journal of Family Psychology* 30:409.

Gomes, Peter J. 2004. "For Massachusetts, a Chance and a Choice." *Boston Globe,* February 8. Retrieved October 6, 2006 (www.boston.com/news/globe).

Gonzalez, Cindy. 2006a. "Latino Leader Urges End to Concept of 'Minorities.'" *Omaha World-Herald,* November 17.

———. 2006b. "Short Supply of Visas Adds to Illegal Immigration." *Omaha World-Herald,* May 16.

Gonzalez, Cindy, and Michael O'Connor. 2002. "Dialogue Key in Blending Cultures." *Omaha World-Herald,* May 4.

Gonzalez, Henry, and Melissa A. Barnett. 2014. "Romantic Partner and Biological Father Support: Associations with Maternal Distress in Low-Income Mexican-Origin Families." *Family Relations* 63(3):371–383.

Gonzalez, Jose-Michael. 2016. "Conceptualizing the Transformative Sibling Process toward a Framework for Understanding Alternative Coparenting Relationships in Diverse Family Systems." *Family Focus* (FF67, Spring):F13–F14. Minneapolis: National Council on Family Relations.

Good, Maria, and Teena Willoughby. 2006. "The Role of Spirituality Versus Religiosity in Adolescent Psychosocial Adjustment." *Journal of Youth and Adolescence* 35:41–55.

Goode, Erica. 2012. "Fewer Children Are Found Exposed to Violent Crime." *The New York Times,* September 19.

Goode, William J. 1971. "Force and Violence in the Family." *Journal of Marriage and Family* 33:624–636.

———. 1982. "Why Men Resist." pp. 131–150 in *Rethinking the Family: Some Feminist Questions,* edited by Barrie Thorne and Marilyn Yalom. New York: Longman.

———. 2007 [1982]. "Theoretical Importance of the Family." pp. 14 –25 in *Family in Transition,* 14th ed., edited by Arlene S. Skolnick and Jerome H. Skolnick. Boston: Allyn & Bacon.

Goode-Cross, and David T. Tager. 2011. "Negotiating Multiple Identities: How African-American Gay and Bisexual Men Persist at a Predominantly White Institution." *Journal of Homosexuality* 58(90):1235–1254.

Goodluck, Charlotte, and Angela A. A. Willeto. 2009. "Seeing the Protective Rainbow: How Families Survive and Thrive in the American Indian and Alaska Native Community." The Annie E. Casey Foundation.

Goodman, W. Benjamin, Ann C. Crouter, Stephanie Lanza, Martha Cox, Lynne Vernon-Feagans, and the Family Life Project Key Investigators. 2011. "Parental Work Stress and Latent Profiles of Father-Infant Parenting Quality." *Journal of Marriage and Family* 73(3): 588–604.

Goodsell, Todd L., Spencer L. James, Jeremy B. Yorgason, and Vaughn R.A. Call. 2015. "Intergenerational Assistance to Adult Children: Gender and Number of Sisters and Brothers." *Journal of Family Issues* 36(8):979–1000.

Goodstein, Laurie. 2015. "Mormons Sharpen Stand Against Same-Sex Marriage." *The New York Times*, November 6. Retrieved November 8, 2015 (http://www.nytimes.com).

Goodwin, Paula, Brittany McGill, and Anjani Chandra. 2009. *Who Marries and When? Age at First Marriage in the United States, 2002.* NCHS data brief, no. 19. Hyattsville, MD: National Center for Health Statistics.

Google Books. 2020. "*Two Homes*: Book Description." Retrieved February 12, 2020 (https://books.google.com/books/about/Two_Homes.html?id=cv2mBAAAQBAJ&source=kp_book_description).

Goosby, Bridget J. 2007. "Poverty Duration, Maternal Psychological Resources, and Adolescent Socioemotional Outcomes." *Journal of Family Issues* 28(8):1113–1134.

Gootman, Elissa. 2012. "So Eager for Grandchildren, They're Paying the Egg-Freezing Clinic." *The New York Times*, May 13. Retrieved August 28, 2012 (http://www.nytimes.com).

Gordon, K. C., D. H. Baucom, and D. K. Snyder, D. K. (2004). "An Integrative Intervention for Promoting Recovery from Extramarital Affairs." *Journal of Marital and Family Therapy*, 30(2):213–231.

Gordon, Larry. 2016. "Young Men Outnumbered in College Readiness Efforts; More Male Recruits Sought." Retrieved October 4, 2019 (https://edsource.org/2016/young-men-outnumbered-in-college-readiness-efforts-more-male-recruits-sought/572835).

Gordon, Mellissa, and Ming Cui. 2014. "School-Related Parental Involvement and Adolescent Academic Achievement: The Role of Community Poverty." *Family Relations* 63(5):616–626.

Gordon, Rachel A., Hillary L. Rowe, and Karina Garcia. 2015. "Promoting Family Resilience Through Evidence-based Policy Making: Reconsidering the Link Between Adult-Infant Bedsharing and Infant Mortality." *Family Relations* 64(1):134–152.

Gordon, Rachel A., Margaret L. Usdansky, Xue Wang, and Anna Gluzman. 2011. "Child Care and Mothers' Mental Health: Is High-Quality Care Associated with Fewer Depressive Symptoms?" *Family Relations* 60(4):446–460.

Gordon, Thomas. 2000. *Parent Effectiveness Training: The Proven Program for Raising Responsible Children.* New York: Three Rivers Press.

Gorman, Bridget K., Justin T. Denney, Hilary Dowdy, and Rose Anne Medeiros. 2015. "A New Piece of the Puzzle: Sexual Orientation, Gender, and Physical Health Status." *Demography* 52: 1357–1382.

Gormley, Barbara, and Frederick G. Lopez. 2010. "Psychological Abuse Perpetration in College Dating Relationships: Contributions of Gender, Stress, and Adult Attachment Orientations." *Journal of Interpersonal Violence* 25(2):204–218.

Gornick, Janet C., Harriet B. Presser and Caroline Batzdorf. 2009. "Outside the 9-to-5." *The American Prospect*, June 9. Retrieved May 26, 2010 (www.prospect.org).

Gosselin, Julie, Elisa Romano, Tessa Bell, Lyzon Babchishin, Isabelle Hudon-ven der Buhs, Annie Gagné, and Natasha Gosselin. 2014. "Canadian Portrait of Changes in Family Structure and Preschool Children's Behavioral Outcomes." *International Journal of Behavioral Development* 38:518–528.

Gott, Natalie. 2010. "Clergy Women Make Connections." *Faith and Leadership*, February 16.

Gottesdiener, Laura. 2012. "Feminism's Next Big Step." *Alternet News and Politics*, August 23. Retrieved November 8, 2012 (http://www.salon.com).

Gottlieb, Laurie N., Ariella Lang, and Rhonda Amsel. 1996. "The Long-term Effects of Grief on Marital Intimacy Following an Infant's Death." *Omega* 33(1):1–9.

Gottlieb, Lori. 2006. "How Do I Love Thee?" *The Atlantic*, March, pp. 58–70.

Gotta, Gabrielle, Robert-Jay Green, Esther Rothblum, Sondra Solomon, Kimberly Balsam, and Pepper Schwartz. 2011. "Heterosexual, Lesbian, and Gay Male Relationships: A Comparison of Couples in 1975 and 2000." *Family Process* 50(3):353–376.

Gottlieb, Aaron, Natasha Pilkauskas, and Irwin Garfinkel. 2014. "Private Financial Transfers, Family Income, and the Great Recession." *Journal of Marriage and Family* 76(5):1011–1024.

Gottman, John, and Julie Gottman. 2017. "The Natural Principles of Love." *Journal of Family Theory & Review* 9 (1): 7–26.

Gottman, John M. 1979. *Marital Interaction: Experimental Investigations.* New York: Academic.

———. 1994. *Why Marriages Succeed or Fail.* New York: Simon and Schuster.

———. 1996. *What Predicts Divorce? The Measures.* Hillsdale, NJ: Erlbaum.

———. 2011. "John Gottman on Trust and Betrayal." Retrieved January 29, 2013 (http://greatergood.berkeley.edu).

Gottman, John M., James Coan, Sybil Carrere, and Catherine Swanson. 1998. "Predicting Marital Happiness and Stability from Newlywed Interactions." *Journal of Marriage and Family* 60(1):5–22.

Gottman, John M., and Joan DeClaire. 2001. *The Relationship Cure: A Five-step Guide for Building Better Connections with Family, Friends, and Lovers.* New York: Crown.

Gottman, John M., and L. J. Krotkoff. 1989. "Marital Interaction and Satisfaction: A Longitudinal View." *Journal of Consulting and Clinical Psychology* 57:47–52.

Gottman, John M., and Robert W. Levenson. 2000. "The Timing of Divorce: Predicting When a Couple Will Divorce Over a 14-Year Period." *Journal of Marriage and Family* 62(3):737–745.

Gottman, John M., Robert W. Levenson, James Gross, Barbara Frederickson, Leah Rosenthal, Anna Ruef, and Dan Yoshimoto. 2003. "Correlates of Gay and Lesbian Couples' Relationship Satisfaction and Relationship Dissolution." *Journal of Homosexuality* 45(1):23–45.

Gottman, John M., and Clifford I. Notarius. 2000. "Decade Review: Observing Marital Interaction." *Journal of Marriage and Family* 62(4):927–947.

———. 2003. "Marital Research in the 20th Century and a Research Agenda for the 21st Century." *Trends in Marriage, Family, and Society* 25(2):283–297.

Gottman, John M., and Nan Silver. 1999. *The Seven Principles for Making Marriage Work.* New York: Crown.

———. 2015. *The Seven Principles for Making Marriage Work.* New York: Random House.

Gottman Institute. 2015. *Research Agendas and Publications.* Retrieved October 3, 2015 (http://www.Gottman.com).

Gough, Brendan, Nicky Weyman, Julie Alderson, Gary Butler, and Mandy Stoner. 2008. "'They Did Not Have a Word': The Parental Quest to Locate a 'True Sex' For Their Intersex Children." *Psychology and Health* 23(4):493–507.

Gouin, J., C. Da Estrela, K. Desmarais, and E.T. Barker. 2016. "The Impact of Formal and Informal Support on Health in the Context of Caregiving Stress." *Family Relations* 65 (1): 191–206.

Gove, Walter R., Carolyn Briggs Style, and Michael Hughes. 1990. "The Effect of Marriage on the Well-Being of Adults." *Journal of Family Issues* 11(1):4–35.

Gowan, Annie. 2009. "Immigrants' Children Look Closer for Love: More Young Adults Are Seeking Partners of the Same Ethnicity." *Washington Post*, March 8:A01.

Grace, Gerald. 2002. *Catholic Schools: Mission, Markets and Morality.* London: Routledge Farmer.

Grady, Bill. 2009. "A Marriage Blending Family and Race." pp. 90–93 in *Social Issues First Hand: Blended Families*, edited by Stefan Kiesbye. New York: Greenhaven Press/Cengage Learning.

Graf, Nikki. 2019. "Key Findings on Marriage and Cohabitation in the U.S." Retrieved December 1, 2019 (https://www.pewresearch.org/fact-tank/2019/11/06/key-findings-on-marriage-and-cohabitation-in-the-u-s/).

Graefe, Deborah R., and Daniel T. Lichter. 2007. "When Unwed Mothers Marry." *Journal of Family Issues* 28(5):595–622.

Graham, Carol, and Milena Nikolova. 2014. "Why Aging and Working Makes Us Happy." *Brookings*, March 28. Retrieved December 28, 2015 (http://www .brookings.edu).

Graham, Elspeth, and Lucy P. Jordan. 2011. "Migrant Parents and the Psychological Well-Being of Left-Behind Children in Southeast Asia." *Journal of Marriage and Family* 73(4): 763–787.

Graham, Jamie, L., Elizabeth Keneski, and Timothy J. Loving. 2014. "Mental and Physical Health Correlates of Nonmarital Relationship Dissolution." *National Council on Family Relations Report* 59.4:F8-F9. Minneapolis: National Council on Family Relations.

Graham, Kathy T. 2008. "Same-Sex Couples: Their Rights as Parents, and Their Children's Rights as Children." *Santa Clara Law Review* 48:999–1037.

Graham, Susanna. 2018. "Being a 'Good' Parent: Single Women Reflecting Upon 'Selfishness' and 'Risk' When Pursuing Motherhood Through Sperm Donation. " *Anthropology & Medicine* 25:249–264.

Grall, Timothy. 2018. "Custodial Mothers and Fathers and Their Child Support: 2015." Retrieved February 10, 2020 (https://www.youngwilliams.com/sites/default/files/pdf-resource/custodialmothersandfathersandtheir childsupport2015.pdf).

———. 2016. "Custodial Mothers and Fathers and Their Child Support." Retrieved March 19, 2016 (http://www .census.gov/content/dam /Census/library /publications/2016/demo /P60-255.pdf).

Grange, Christina M., Sarah Jane Brubaker, and Maya A. Corneille. 2011. "Direct and Indirect Messages African American Women Receive from Their Family Networks about Intimate Relationships and Sex: The Intersecting Influence of Race, Gender, and Class." *Journal of Family Issues* 32(5):605–628.

Granka, Patrick R. 2014. *Intersectionality*. Boulder, CO: Westview Press.

Grant, Jaime M. 2010. *Outing Age 2010*. Washington, DC: National Gay and Lesbian Task Force Policy Institute. Retrieved May 15, 2010 (www.thetaskforce.org.).

Greeley, Andrew. 1991. *Faithful Attraction: Discovering Intimacy, Love, and Fidelity in American Marriage*. New York: Doherty.

Green Adam I., Jenna Valleriani, and Barry Adam. 2016. "Marital Monogamy as Ideal and Practice: The Detraditionalization Thesis in Contemporary Marriages." *Journal of Marriage and Family* 78 (2): 416–430.

Green Jonathan, and Michael Addis. 2012. "Individual Differences in Masculine Gender Socialization as Predictive of Men's Psychophysioogical Responses to Negative Affect." *International Journal of Men's Health* 11(1):63–82.

Green, R. J. 2009. "From Outlaws to In-laws: Gay and Lesbian Couples in Contemporary Society." pp. 197–213, 488–489, 527– 530 in *Families as They Really Are*, edited by B. Risman. New York: Norton.

Greenberg, Jerrold S., Clint E. Bruess, and Debra W. Haffner. 2002. *Exploring the Dimensions of Human Sexuality*. Sudbury, MA: Jones and Bartlett.

Greenblatt, Cathy Stein. 1983. "The Salience of Sexuality in the Early Years of Marriage." *Journal of Marriage and Family* 45:289–299.

Greenhouse, Steven. 2012. "Equal Opportunity Panel Updates Hiring Policy." *The New York Times*, April 25. Retrieved August 28, 2012 (http://www.nytimes.com).

Greenstein, Theodore N. 2009. "National Context, Family Satisfaction, and Fairness in the Division of Household Labor." *Journal of Marriage and Family* 71(4):1039–1051.

Greenwald, John. 1999. "Elder Care: Making the Right Choice." *Time*, August 30, pp. 52–56.

Gregory, John DeWitt. 1998. "Blood Ties: A Rationale for Child Visitation by Legal Strangers." *Wash. & Lee L. Rev.* 55:351.

Gregory, Sean. 2017. "Kid Sports." *Time*. September 4. Pp. 40–51.

Griffin, Riley. 2018. "Almost Half of U.S. Births Happen Outside Marriage, Signaling Cultural Shift." Bloomberg News. October 17. Retrieved November 15, 2019 (www.bloomberg.com).

Grogan-Kaylor, Andrew, and Melanie D. Otis. 2007. "The Predictors of Parental Use of Corporal Punishment." *Family Relations* 56(1):80–91.

Gromoske, Andrea, and Kathryn Maguire-Jack. 2012. "Transactional and Cascading Relations Between Early Spanking and Children's Social-Emotional Development." *Journal of Marriage and Family* 74(5):1054–1068.

Gross, Jane. 2002. "U.S. Fund for Tower Victims Will Aid Some Gay Partners." *The New York Times*, May 22.

———. 2004. "Alzheimer's in the Living Room: How One Family Rallies to Cope." *The New York Times*, September 16.

———. 2006a. "Seeking Doctor's Advice in Adoptions from Afar." *The New York Times*, January 3.

———. 2006b. "Forensic Skills Seek to Uncover Hidden Patterns of Elder Abuse." *The New York Times*, September 27.

———. 2007. "A Taste of Family Life in U.S., but Adoption Is in Limbo." *The New York Times*, January 13.

Grossman, Arnold H., and Anthony R. D'Augelli. 2006. "Transgender Youth: Invisible and Vulnerable." *Journal of Homosexuality* 51(1):111–128.

Grov, C., Hirshfield, S., Remien, R. H., Humberstone, M., and Chiasson, M. A. 2013. "Exploring the Venue's Role in Risky Sexual Behavior among Gay and Bisexual Men: An Event-Level Analysis from a National Online Survey in the U.S." *Archives of Sexual Behavior* 42: 291–302.

Grzywacz, Joseph G., Stephanie S. Daniel, Jenna Tucker, Jill Walls, and Esther Leerkes. 2011. "Nonstandard Work Schedules and Developmentally Generative Parenting Practices." *Family Relations* 60(1):45–59.

Grzywacz, Joseph G, and Wendy Middlemiss. 2017. "Looking Backward, Around, and Forward: Family Science Has Always Been Translational Science." *Family Relations* 66 (4):547–549.

Grzywacz, Joseph G, and Amy M. Smith. 2016. "Work-Family Conflict and Health among Working Parents." *Family Relations* 65 (1): 176–190.

Guberman, Nancy, Eric Gagnon, Denyse Cote, Claude Gilbert, Nicole Thivierge, and Marielle Tremblay. 2005. "How the Trivialization of the Demands of High-Tech Care in the Home Is Turning Family Members into Para-Medical Personnel." *Journal of Family Issues* 20(2).247–272.

Gubernskaya, Zoya, and Joanna Dreby. 2017. "US Immigration Policy and the Case for Family Unity." *Journal on Migration and Human Security* 5::417–430.

Gubrium, Jaber F., and James A. Holstein. 1990. *What Is Family?* Mountain View, CA: Mayfield Press.

Gudmunson, Clinton G., Sharon M. Danes, James D. Werbel, and Johnben Teik-Cheok Loy. 2009. "Spousal Support and Work—Family Balance in Launching a Family Business." *Journal of Family Issues* 30(8):1098–1121.

Gueler, Aysel, André Moser, Alexandra Calmy, Huldrych F. Günthard, Enos Bernasconi, Hansjakob Furrer, Christoph A. Fux et al. 2017. "Life Expectancy in HIV-Positive Persons in Switzerland: Matched Comparison with General Population." *AIDS* 31:427.

Guilamo-Ramos, Vincent, James Jaccard, Patricia Dittus, and Alida M. Bouris. 2006. "Parental Experience, Trustworthiness, and Accessibility: Parent-Adolescent Communication and Adolescent Risk Behavior." *Journal of Marriage and Family* 68(5):1229–1246.

Gullickson, Aaron, and Florencia Torche. 2014. "Patterns of Racial and Educational Assortative Mating in Brazil." *Demography* 51:835–856.

Gültekin, Laura. 2012. "Family Homelessness, Housing Insecurity, and Health: Understanding and Acting on What We Know." *Family Focus* FF54(Fall)·F4–F6

Gurrentz, Benjamin T. 2017. "Family Formation and Close Social Ties Within Religious Congregations." *Journal of Marriage and Family* 79 (4):1125–1143.

Gurrentz, Benjamin T. 2018. "Living with an Unmarried Partner Now Common for Young Adults." Retrieved December 1, 2019. (https:// www.census.gov/library/stories/2018/11 /cohabitaiton-is-up-marriage-is-down-for-young -adults.html).

Gurrentz, Benjamin T. 2019. "Cohabiting Partners Older, More Racially Diverse, More Educated, Higher Earners." September 23. U.S. Census Bureau. Retrieved November 14, 2019 (www.census.gov).

Gustke, Constance. 2014. "Retirement Plans thrown into Disarray by a Divorce." *The New York Times*, June 28: B5.

Guttmacher Institute. 2019a. *Induced Abortion in the United States*. Retrieved December 22, 2019 (https://www.guttmacher.org/fact-sheet /induced-abortion-united-states).

Guttmacher Institute. 2019b. *State Facts About Abortion: North Dakota. Retrieved December 27, 2019 (Guttmacher Institute 2019c. Teen Pregnancy. Retrieved December 20, 2019 (https://www .guttmacher.org/united-states/teens /teen-pregnancy).

Guttmacher Institute. 2019c. *State Policies on Abortion.* Retrieved December 23, 2019 (https://www.guttmacher.org/united-states/abortion/state-policies-abortion).

Guttmacher Institute 2019d. *Teen Pregnancy.* Retrieved December 20, 2019 (https://www.guttmacher.org/united-states/teens/teen-pregnancy).

Guttmacher Institute. 2019e. *Unintended Pregnancy in the United States.* Retrieved December 14, 2019 (https://www.guttmacher.org/fact-sheet/unintended-pregnancy-united-states).

Guynn, Jessica. 2013. "Yahoo CEO Marissa Mayer Causes Uproar with Telecommuting Ban." *Los Angeles Times*, February 26. Retrieved February 26, 2013 (http://www.latimes.com).

Guzzo, Karen Benjamin. 2018. *Childbearing Desires, Intentions, and Attitudes Among Women 40–44.* Retrieved December 19, 2019 (https://www.bgsu.edu/ncfmr/resources/data/family-profiles/guzzo-childbearing-desires-intent-attitudes-women-40-44-fp-18-09.html).

Guzzo, Karen B. 2016. *Stepfamilies in the U.S.* Family Profiles, FP-16-09. Bowling Green, OH: National Center for Family & Marriage Research. Retrieved February 19, 2020 (https://www.bgsu.edu/content/dam/BGSU/college-of-arts-and-sciences/NCFMR/documents/FP/guzzo-stepfamilies-women-fp-16-09.pdf).

Guzzo, Karen Benjamin. 2017. "Shifts in Higher-Order Unions and Stepfamilies Among Currently Cohabiting and Married Women of Childbearing Age." *Journal of Family Issues* 38:1775–1799.

Guzzo, Karen Benjamin, Paul Hemez, Lydia Anderson, Wendy D. Manning, and Susan L. Brown. 2019. "Is Variation in Biological and Residential Ties to Children Linked to Mothers' Parental Stress and Perceptions of Coparenting"" *Journal of Family Issues* 40:488–517.

Guzzo, Karen Benjamin, Sarah R. Hayford, Vanessa Wanner Lang, Hsueh-Sheng Wu, Jennifer Barber, and Yasamin Kusunoki. 2019. "Dimensions of Reproductive Attitudes and Knowledge Related to Unintended Childbearing Among U.S. Adolescents and Young Adults." *Demography* 56:201–228.

Guzzo, Karen Benjamin, Vanessa Wanner Lang, and Sarah R. Hayford. 2019. "Teen Girls' Reproductive Attitudes and the Timing and Sequencing of Sexual Behaviors." *Journal of Adolescent Health* 65:507–513.

Ha, Thao, and Douglas A. Granger. 2016. "Family Relations, Stress, and Vulnerability: Biobehavioral Implications for Prevention and Practice." *Family Relations* 65 (1): 9–23.

Hackstaff, Karla B. 2007. "Divorce Culture: A Quest for Relational Equality in Marriage." pp. 188–222 in *Family in Transition*, 4th ed., edited by Arlene S. Skolnickand Jerome H. Skolnick. Boston: Pearson.

Hadfield, Kristin, Margaret Amos, Michael Ungar, Julie Gosselin, and Lawrence Ganong. 2018. "Do Changes to Family Structure Affect Child and Family Outcomes? A Systematic Review of the Instability Hypothesis." *Journal of Family Theory & Review* 10 (1): 87–110.

Hafner, Katie. 2016. "Researchers Confront an Epidemic of Loneliness." Retrieved January 24, 2020 (https://www.nytimes.com/2016/09/06/health/lonliness-aging-health-effects.html).

Hagan, Frank E. 2010. *Crime Types and Criminals.* Thousand Oaks, CA: Sage.

Hagerman, Margaret. 2017. "White Racial Socialization: Progressive Fathers on Raising 'Antiracist' Children." *Journal of Marriage and Family* 79 (1): 60–74.

Hagestad, G. 1986. "The Family: Women and Grandparents as Kin Keepers." pp. 141–160 in *Our Aging Society*, edited by A. Piferand L. Bronte. New York: Norton.

———. 1996. "On-Time, Off-Time, Out of Time? Reflections on Continuity and Discontinuity from an Illness Process." pp. 204–222 in *Adulthood and Aging*, edited by V. L. Bengston. New York: Springer.

Hagewen, Kellie J., and S. Philip Morgan. 2005. "Intended and Ideal Family Size in the United States, 1970-2002." *Population and Development Review* 31(3):507–527.

"The Hague Convention on Intercountry Adoption." 2010. *Child Welfare Information Gateway*, U.S. Department of Health and Human Services (www.childwelfare.gov).

Haines, Victor Y. III, Steve Harvey, Pierre Durand, and Alain Marchand. 2013. "Core Self-Evaluations, Work-Family Conflict, and Burnout." *Journal of Marriage and Family* 75(3):778–793.

Halford, Kim, Jan Nicholson, and Matthew Sanders. 2007. "Couple Communication in Stepfamilies." *Family Process* 46(4):471–483.

Hall, Edie Jo, and E. Mark Cummings. 1997. "The Effects of Marital and Parent-Child Conflicts on Other Family Members: Grandmothers and Grown Children." *Family Relations* 46(2):135–143.

Hall, Heather, and J. Carolyn Graff. 2012. "Maladaptive Behaviors of Children with Autism: Parent Support, Stress, and Coping." *Issues in Comprehensive Pediatric Nursing* 35(3 –4):194–214.

Hall, Jamie Bryan. 2015. "The Research on Same-Sex Parenting: 'No Differences' No More." Issue brief no 4393. The Heritage Foundation. Retrieved January 13, 2016 (http://heritage.org).

Hall, Scott S., and Shelley M. MacDermid. 2009. "A Typology of Dual Earner Marriages Based on Work and Family Arrangements." *Journal of Family and Economic Issues* 30(3):215–225.

Hall, Sharon K. 2008. *Raising Kids in the 21st Century.* West Sussex, UK: Wiley-Blackwell.

Halpern, Susan P. 2009. *Finding the Words: Candid Conversations with Loved Ones.* Berkeley, CA: North Atlantic Books.

Halpern-Felsher, Bonnie L., Jodi L. Cornell, Rhonda Y. Kropp, and Jeanne M. Tschann. 2005. "Oral versus Vaginal Sex among Adolescents: Perceptions, Attitudes, and Behavior." *Pediatrics* 115(4):845–851.

Halpern-Meekin, Sarah, Wendy D. Manning, Peggy C. Giordano, and Monica A. Longmore. 2013. "Relationship Churning, Physical Violence, and Verbal Abuse in Young Adult Relationships." *Journal of Marriage and Family* 75:2–12.

Halpert, Julie. 2018. "Late to Launch: The Post-Collegiate Struggle." Retrieved November 30, 2019 (https://www.nytimes.com/2018/12/04/well/family/late-to-launch-the-post-collegiate-struggle.html).

Halpin, John, and Ruy Teixeira. 2009. "Battle of the Sexes Gives Way to Negotiations." In *The Shriver Report: A Woman's Nation Changes Everything*, edited by Heather Boushey and Ann O'Leary. Washington, DC: Maria Shriver and the Center for American Progress.

Halsall, Paul. 2001. *Internet Medieval Sourcebook.* Retrieved October 6, 2006 (www.fordham.edu).

Halton, Mary. 2019. "We Need to Talk About the Orgasm Gap—and How to Fix it." Retrieved November 4, 2019 (https://ideas.ted.com/we-need-to-talk-about-the-orgasm-gap-and-how-to-fix-it/).

Hamamci, Zeynep. 2005. "Dysfunction Relationship Beliefs in Marital Conflict." *Journal of Relational-Emotive and Cognitive-Behavior Therapy* 23(3):245–261.

Hamel, John C., and Tanya L. Nicholls. 2006. *Family Interventions in Domestic Violence: A Handbook of Gender-Inclusive Theory and Treatment.* New York: Springer.

Hamilton, Brady E., Joyce A. Martin, and Stephanie J. Ventura. 2009. *Births: Preliminary data for 2007.* National Vital Statistics Reports 57(12), March 18.

Hamilton, Brady E., and Stephanie J. Ventura. 2012. "Birth Rates for U.S. Teenagers Reach Historic Lows for All Age and Ethnic Groups." *NCHS Data Brief Number 89.* Hyattsville, MD: National Center for Health Statistics.

Hammel, Paul. 2012. "Court: No Inheritance for Child Conceived by Artificial Insemination." *World-Herald Bureau*, November 16. Retrieved March 31, 2013 (http://omaha.com).

Hammen, Constance, Patricia A. Brennan, and Josephine H. Shih. 2004. "Family Discord and Stress Predictors of Depression and Other Disorders in Adolescent Children of Depressed and Nondepressed Women." *Journal of the American Academy of Child and Adolescent Psychiatry* 43(8):994–1003.

Hamon, Raeann R., and Bron B. Ingoldsby. 2003. *Mate Selection across Cultures.* Thousand Oaks, CA: Sage.

Hamoudi, Amar, and Jenna Nobles. 2014. "Do Daughters Really Cause Divorce? Stress, Pregnancy, and Family Composition." *Demography* 51:1423–1449.

Hamplova, Dana, and Celine LeBourdais. 2009. "One Pot or Two Pot Strategies? Income Pooling in Married and Unmarried Households in Comparative Perspective." *Journal of Comparative Family Studies* 40(3): 355–385.

Hampton, Keith, Lee Rainie, W. Lu, I. Shin, and K. Purcell. 2015. "Social Media and the Cost of Caring." Pew Research Center. January 15. Retrieved August 10, 2019 (www.pewresearch.org).

Han, Chong-suk. 2008. "A Qualitative Exploration of the Relationship Between Racism and Unsafe Sex among Asian Pacific Islander Gay Men." *Archives of Sexual Behavior* 37(5):827–837.

Han, Wen-Jui, and Liana E. Fox. 2011. "Parental Work Schedules and Children's Cognitive Trajectories." *Journal of Marriage and Family* 73(5):962–980.

Hancock, A. N. 2012. "'It's a Macho Thing, Innit?' Exploring the Effects of Masculinity on Career Choice and Development." *Gender, Work, and Organization* 19(4):392–415.

Hanna, Sharon L., Rose Suggett, and Doug Radtke. 2008. *Person to Person: Positive Relationships Don't Just Happen*, 5th ed. Upper Saddle River, NJ: Pearson/Prentice Hall.

Hannon, Kerry. 2018. "Young Women Have to Play Catch-Up in Retirement Savings." *The New York Times*. December 4. Retrieved January 10, 2019 (https://www.nytimes.com).

Hans, Jason D. 2002. "Stepparenting after Divorce: Stepparents' Legal Position Regarding Custody, Access, and Support." *Family Relations* 51(4):301–307.

———. 2009. "Beliefs about Child Support Modification Following Remarriage and Subsequent Childbirth." *Family Relations* 58(February):65–78.

Hans, Jason D., and Marilyn Coleman. 2009. "The Experiences of Remarried Stepfathers Who Pay Child Support." *Personal Relationships* 16:597–618.

Hans, Jason D., Lawrence H. Ganong, and Marilyn Coleman. 2009. "Financial Responsibilities Toward Older Parents and Stepparents Following Divorce and Remarriage." *Journal of Family Economic Issues* 30:55–66.

Hans, Jason D., Martie Gillen, and Katrina Akande. 2010. "Sex Redefined: The Reclassification of Oral-Genital Contact." *Perspectives on Sexual and Reproductive Health* 42(2):74–78.

Hansen, Donald A., and Reuben Hill. 1964. "Families Under Stress." pp. 782–819 in *The Handbook of Marriage and the Family*, edited by Harold Christensen. Chicago: Rand McNally.

Hansen, Matthew. 2012. "The Talk: A North Omaha Rite of Passage." *The Omaha World Herald*, March 25.

Hansen, Matthew, and Karyn Spencer. 2009. "Safe Haven Meant Kids Finally Got Right Help." *Omaha World Herald*, February 1. Retrieved February 3, 2009 (www.omaha .com).

Hansen, Thorn, Torbjorn Moum, and Adam Shapiro. 2007. "Relational and Individual Well-Being among Cohabitors and Married Individuals in Midlife." *Journal of Family Issues* 28(7):910–933.

Hankivsky, Olena, & Cormier, Renee. 2011. "Intersectionality and Public Policy: Some Lessons from Existing Models." *Political Research Quarterly* 64:217–229.

Hanson, Hilary. 2016. "Finally, Unwed Couples Can Legally Live Together in Florida." Retrieved December 1, 2019 (https://www.huffpost.com /entry/unwed-unmarried-couples-florida-law_n _57068015e4b0a506064e5bca).

Hanson, Hilary. 2018. "Neighbor Calls Police, Child Services about 8-Year-Old Girl Walking Dog." August 25. Retrieved November 23, 2019 (https://www.huffpost.com).

Hardesty, Jennifer L., and Grace H. Chung. 2006. "Intimate Partner Violence, Parental Divorce, and Child Custody: Directions for Intervention and Future Research." *Family Relations* 55:200–216.

Hardesty, Jennifer L., Kimberly A. Crossman, Megan L. Haselschwerdt, Marcela Raffaelli, Brian G. Ogolsky, and Michael P. Johnson. 2015. "Toward a Standard Approach to Operationalizing Coercive Types." *Journal of Marriage and Family* 77(4):833–843.

Hardesty, Jennifer, and Brian Ogolsky. 2020. "A Sociological Perspective on Intimate Partner Violence Research: A Decade In Review." *Journal of Marriage and Family* 82 (1): 454–477.

Harding, Jessica, Pamela Morris, and Diane Hughes. 2015. "The Relationship Between Maternal Education and Children's Academic Outcomes: A Theoretical Framework." *Journal of Marriage and Family* 77(1):60–76.

Harding, Rosie. 2007. "Sir Mark Potter and the Protection of the Traditional Family: Why Same Sex Marriage Is (Still) a Feminist Issue." *Feminist Legal Studies* 15(2):223–234.

Hardy, Rich. 2018. "The Right to Disconnect: The New Laws Banning After-hours Work E-mails." *New Atlas*. August 4. Retrieved February 2, 2020 (www .newatlas.com).

Harknett, Kristen. 2006. "The Relationship between Private Safety Nets and Economic Outcomes among Single Mothers." *Journal of Marriage and Family* 69(1):172–191.

Harknett. Kristen S., and Caroline Sten Hartnett. 2011. "Who Lacks Support and Why? An Examination of Mothers' Personal Safety Nets." *Journal of Marriage and Family* 73(4):861–875.

Harknett, Kristen, and Jean Knab. 2007. "More Kin, Less Support: Multipartnered Fertility and Perceived Support among Mothers." *Journal of Marriage and Family* 69(1):237–253.

Harmanci, Reyhan. 2006. "The Neighbordaters." *San Francisco Chronicle Magazine*, February 12, pp. 13–14.

Harmon, Amy. 2007a. "Prenatal Test Puts Down Syndrome in Hard Focus." *The New York Times*, May 9.

———. 2007b. "Sperm Donor Father Ends His Anonymity." *The New York Times*, February 14.

Harpel, Tammy S., and Jodie Hertzog. 2010. "'I Thought My Heart Would Burst': The Role of Ultrasound Technology on Expectant Grandmotherhood." *Journal of Family Issues* 31(2):257–274.

Harper, Nevin J. 2009. "Family Crisis and the Enrollment of Children in Wilderness Treatment." *Journal of Experimental Education* 31(3):447–450.

Harper, Scott E., and Alan M. Martin. 2014. "Transnational Migratory Labor and Filipino Fathers." *Journal of Family Issues* 34(2):272–292.

Harrington, Suzanne. 2015. "Are We All Fake Feminists Now?" *Irish Republic*, June 11, 30–31.

Harris, Deborah, and Domenico Parisi. 2008. "Looking for 'Mr. Right': The Viability of Marriage Initiatives for African American Women in Rural Settings." *Sociological Spectrum* 28(4):338–356.

Harris, Emily. 2015. "Israeli Dads Welcome Surrogate-Born Baby in Nepal on Earthquake Day." National Public Radio, April 29.

Harris, Marian S., and Ada Skyles. 2008. "Kinship Care for African American Children: Disproportionate and Disadvantageous." *Journal of Family Issues* 29(8):1013–1030.

Harris, Micki, and Nicole Marie Dilts. 2015. "Social Media and Its Changes on Student's Formal Writing." *CRIUS* 3.

Harris, Paul. 2009. "Revealed: The Shocking Rise of 'New Slavery' in US Midwest." *The Observer* November 22, p. 37.

Harris-Kojetin, L., M. Sengupt, E. Park-Lee, and R. Valverde. 2013. *Long-Term Care Services in the United States: 2013 Overview*. Atlanta: Centers for Disease Control and Prevention, National Center for Health Statistics.

Harris, Richard. 2018. "Record High Number of STD Infections in U.S., As Prevention Funding Declines." Retrieved October 24, 2019 (https://www.npr.org/sections/health -shots/2018/08/28/642664883/record -high-number-of-std-infections-in-u-s-as -prevention-funding-declines).

Harrison, David, and Soo Oh. 2019. "Women Working Longer Hours, Sleeping Less, as They Juggle Commitments." *Wall Street Journal*. June 19 Retrieved October 3, 2019 (www.wjs.com).

Harrist, Amanda W., and Ricardo C. Ainslie. 1998. "Marital Discord and Child Behavior Problems." *Journal of Family Issues* 19(2):140–163.

Hart, Joshua, Jacqueline A. Hung, Peter Glick, and Rachel E. Dinero. 2012. "He Loves Her, He Loves Her Not: Attachment Style As a Personality Antecedent to Men's Ambivalent Sexism." *Personality and Social Psychology Bulletin* 38(11):1495–1505.

Harter, James, and Raksha Arora. 2008. "Social Time Crucial to Daily Emotional Well-Being in U.S." June 5. Retrieved November 24, 2009 (www.gallup.com/poll).

Hartig, Hannah, and Carroll Doherty. 2018. "More in U.S. See Drug Addiction, College Affordability, and Sexism As 'Very Big' National Problems." *Facttank News in the Numbers: Pew Research Center Analysis*. Pew Research Center. October 22. Retrieved August 10, 2019 (www .pewresearch.org).

Hartman, Karen. 2011. "Bound in a Gay Union by a State Denying It." *The New York Times*, July 15. Retrieved September 13, 2011 (http://www .nytimes.com).

Hartman, Margaret. 2011. "Cohabitation Is Illegal in Florida, and Conservatives Want to Keep It That Way." August 31. Retrieved November 24, 2012 (http://jezebel.com /5836431/cohabitation-is-illegal-in-florida -and-conservatives-want-to-keep-it-that-way).

Hartman, Susan. 2015. "Home From Afghanistan, and Learning to Be a Couple Again." *The New York Times*. May 15. Retrieved May 21, 2015 (www.nyt.com.)

Hartmann, Heidi, Ashley English, and Jeffrey Hayes. 2010. "Women and Men's Employment and Unemployment in the Great Recession." *Institute for Women's Research Briefing Paper C373*. Washington, DC: Institute for Women's Research.

Hartnett, Carolyn Sten, and Emilio Parrado. 2012. "Hispanic Familism Reconsidered: Ethnic Differences in the Perceived Value of Children and Fertility Decisions." *Sociological Quarterly* 53(4):636–653.

Hartog, Henrik. 2000. *Man and Wife in America: A History*. Cambridge, MA: Harvard University Press.

Hasan, Tabinda, and Mahmood Fauzi. 2012. "If 'Women Are from Venus and Men Are from Mars,' Does an Answer Lie with Neuroanatomy?" *International Journal of Collaborative Research on Internal Medicine and Public Health* 4(5):566–577.

Haselschwerdt, Megan. 2012. "Who Cares about the Rich Folk? An Argument for More Research on Affluent Families and Communities." *Family Focus* FF54(Fall):F14–F16.

Hassrick, Elizabeth McGhee, and Barbara Schneider. 2009. "Parent Surveillance in Schools: A Question of Social Class." *American Journal of Education* 115(February):195–211.

Hatch, Alison. 2015. "Saying 'I Don't' to Matrimony: An Investigation of Why Long-Term Heterosexual Cohabitors Choose Not to Marry." *Journal of Family Issues:* 0192513X15576200.

Hatfield, Elaine. 2013. Passionate Love Scale. Retrieved May 26, 2013, from http://www .elaine-hatfield.com /Passionate%20Love%20 Scale.pdf.

Hatfield, Elaine, and Sprecher, Susan. 2010. "The Passionate Love Scale." pp 466–468 in *Handbook of Sexuality-Related Measures: A Compendium*, 3rd ed., edited by T. D. Fisher, C. M. Davis, W. L. Yaberand S. L. Davis (Eds.). Thousand Oaks, CA: Taylor & Francis.

Haughney, Kathleen. 2011. "Unmarried? Living Together? You're Breaking the Law in Florida." *SunSentinel*, August 31. Retrieved November 24, 2012 (http://articles.sun-sentinel.com).

Havermans, Nele, Sofie Vanassche, and Koen Matthijs. 2017. "Children's Post-Divorce Living Arrangements and School Engagement: Financial Resources, Parent–Child Relationship, Selectivity and Stress." *Journal of Child and Family Studies* 26:3425–3438.

Hawkins, Alan J. 2014. "Continuing the Important Debate on Government-Supported Healthy Marriages and Relationships Initiatives: A Brief Response to Johnson's (2014) Comment." *Family Relations* 63(2):305–308.

Hawkins, Alan J., Paul R. Amato, and Andrea Kinghorn. 2013. "Are Government-Supported Healthy Marriage Initiatives Affecting Family Demographics? A State-Level Analysis." *Family Relations* 62(3):501–513.

Hawkins, Daniel N., and Alan Booth. 2005. "Unhappily Ever After: Effects of Long-Term, Low-Quality Marriages on Well-Being." *Social Forces* 84(1):451–471.

Hawkins, Stacy, Gabriel Schlomer, Leslie Bosch, Deborah Casper, Christine Wiggs, Noel Card, and Lynne Borden. 2012. "A Review of the Impact of U.S. Military Deployments during Conflicts in Afghanistan and Iraq on Children's Functioning." *Family Science* 3(2):99–108.

Hayden, Dolores. 1981. *The Grand Domestic Revolution: A History of Feminist Designs for American Homes, Neighborhoods, and Cities*. Cambridge: MIT Press.

Hayford, Sarah R., Karen Benjamin Guzzo, and Pamela J. Smock. 2014. "The Decoupling of Marriage and Parenthood? Trends in the Timing of Marital First Births, 1945–2002." *Journal of Marriage and Family* 76(June):520–538.

Hayghe, Howard. 1982. "Dual Earner Families: Their Economic and Demographic Characteristics." pp. 27–40 in *Two Paychecks*, edited by Joan Aldous. Newbury Park, CA: Sage.

Healy, Jack. 2015. "Mormons Say Duty to Law on Same-Sex Marriage Trumps Faith." *The New York Times, October* 22. Retrieved November 3, 2015 (http://www.nytimes.com).

Heard, Holly E. 2007. "The Family Structure Trajectory and Adolescent School Performance: Differential Effects by Race and Ethnicity." *Journal of Family Issues* 28(3):319–354.

Hearn, Jeff. 2013. "The Sociological Significance of Domestic Violence: Tensions, Paradoxes, and Implications." *Current Sociology* 61(2):152–170.

Heaton, Tim B. 2002. "Factors Contributing to Increasing Marital Stability in the United States." *Journal of Family Issues* 23(3):392–409.

Heene, Els, Ann Buysse, and Paulette Van Oost. 2007. "An Interpersonal Perspective on Depression: The Role of Marital Adjustment, Conflict Communication, Attributions, and Attachment within a Clinical Sample." *Family Process* 46(4):499–514.

Heffner, Christopher L. 2003. "Counseling the Gay and Lesbian Client." *AllPsych Journal*, August 12. Retrieved March 3, 2010 (allpsych .com/journal).

Hegewisch, Ariane, and Hannah Liepmann. 2010. "The Gender Wage Gap by Occupation." *Fact Sheet IWPR no. C350a*. Washington, DC: Institute for Women's Policy Research.

Heggeness, Misty. 2018. "Spouses Report Earnings Differently When Wives Earn More." Retrieved October 5, 2019 (https://www.census .gov/library/stories/2018/07/wives-earning -more-than-husbands.html).

Heilmann, Ann. 2011. "Gender and Essentialism: Feminist Debates in the Twenty-First Century." *Critical Quarterly* 53(4):78–89.

Heisler, Jennifer M. 2014. "They Need to Sow Their Wild Oats Mothers' Recalled Memorable Messages to Their Emerging Adult Children Regarding Sexuality and Dating." *Emerging Adulthood* 2: 280–293.

Helms, Heather M., Andrew J. Supple, and Christine M. Proulx. 2011. "Mexican-Origin Couples in the Early Years of Parenthood: Marital Well-Being in Ecological Context." *Journal of Family Theory and Review* 3(June):67–95.

Helm, Herbert W. Jr., Stephanie D. Gondra, and Duane C. McBride. 2015. "Hook-Up Culture among College Students: A Comparison of Attitudes toward Hooking-Up Based on Ethnicity and Gender." *North American Journal of Psychology* 17:221.

Helmstetter, Shad. 2013. *The Power of Neuroplasticity*. Gulf Breeze, FL: Park Avenue Press. Cardinal, Sherry. 2019. Stress Proofing Your Life. Retrieved February 25, 2020 (https://www.criticalincidentstress.com /stress_proofing_your_life).

Hemesath, Crystal Wilhite. 2020. *Falling Out of Romanic Love*. New York: Routledge.

Hench, David. 2004. "Is Anger Management a Remedy for Batterers?" *Portland Press Herald* (Maine), October 10.

Henderson, Craig E., Bert Hayslip Jr., Leah M. Sanders, and Linda Louden. 2009. "Grandmother-Grandchild Relationship Quality Predicts Psychological Adjustment among Youth from Divorced Families." *Journal of Family Issues* 30(9):1245–1264.

Henderson, Tammy L., and Patricia B. Moran. 2001. "Grandparent Visitation Rights." *Journal of Family Issues* 22(5):619–638.

Henderson, Tammy, Ann Shigeto, James Ponzetti, Jr., Anne Edwards, Jessica Stanley and Chandra Story. 2017. "A Cultural-Variant Approach to Community-Based Participatory Research." *Family Relations* 66 (4): 629–643.

Hendrick, Clyde, Susan S. Hendrick, and Amy Dicke. 1998. "The Love Attitudes Scale: Short Form." *Journal of Social and Personal Relationships* 15(2):147–159.

Hendy, Helen, Mary Burns, Hakan Can, and Cory Scherer. 2012. "Adult Violence with the Mother and Sibling as Predictors of Partner Violence." *Journal of Interpersonal Violence* 27(11):2276–2297.

Hengstebeck, Natalie, Heather Helms, and Yuliana Rodriguez. 2015. "Spouses' Gender Role Attitudes, Wives' Employment Status, and Mexican-Origin Husbands' Marital Satisfaction." *Journal of Family Issues* 36(1):111–132.

Henig, Robin Marantz. 2014. "Your Adult Siblings May Be the Secret to a Long, Happy Life." National Public Radio, November 27. Retrieved January 14, 2015 (http://www /npr.org).

Henly, Julia R., Sandra K. Danziger, and Shira Offer. 2005. "The Contribution of Social Support to the Material Well-Being of Low-Income Families." *Journal of Marriage and Family* 67(1):122–140.

Henry, Carolyn, Amanda Morris, and Amanda Harrist. 2015. "Family Resilience: Moving into the third Wave." *Family Relations* 64(1):22–43.

Henry, C. S., C. P. Ceglian, and D. W. Matthews. 1992. "The Role Behaviors, Role Meanings, and Grandmothering Styles of Grandmothers and Stepgrandmothers: Perceptions of the Middle Generation." *Journal of Divorce and Remarriage* 17:1–22.

Henry, Pamela J., and James McCue. 2009. "The Experience of Nonresidential Stepmothers." *Journal of Divorce and Remarriage* 50:185–205.

Henshaw, Stanley K., and Kathryn Kost. 2008. *Trends in the Characteristics of Women Obtaining Abortions, 1974 to 2004*. New York: Guttmacher Institute.

Hequembourg, Amy L. 2007. "Becoming Lesbian Mothers." *Journal of Homosexuality* 53(3):153–180.

Herbenick, Debby, Margo Mullinax, and Kristen Mark. 2014. "Sexual Desire Discrepancy as a Feature, Not a Bug, of Long-Term Relationships: Women's Self-Reported Strategies for Modulating Sexual Desire." *Journal of Sexual Medicine* 11: 2196-206.

Herbenick, Debby, Elizabeth Bartelt, Tsung-Chieh Fu, Bryant Paul, Ronna Gradus, Jill Bauer, and Rashida Jones. 2019. "Feeling Scared During Sex: Findings from a U.S. Probability Sample of Women and Men Ages 14 to 60." *Journal of Sex & Marital Therapy* 45:424–439.

Herbenick, Debby, Jessamyn Bowling, Tsung-Chieh Jane Fu, Brian Dodge, Lucia Guerra-Reyes, and Stephanie Sanders. 2017. "Sexual Diversity in the United States: Results from a Nationally Representative Probability Sample of Adult Women and Men." *PloS one* 12:e0181198 (July 20, 2017).

Herbenick, Debby, Tsung-Chieh Fu, Christopher Owens, Elizabeth Bartelt, Brian Dodge, Michael Reece, and J. Dennis Fortenberry. 2019. "Kissing, Cuddling, and Massage at Most Recent Sexual Event: Findings From a U.S. Nationally Representative Probability Sample." *Journal of Sex & Marital Therapy* 45:159-172.

Herbenick, Debby, Tsung-Chieh Fu, Jennifer Arter, Stephanie A. Sanders, and Brian Dodge. 2018. "Women's Experiences with Genital Touching, Sexual Pleasure, and Orgasm: Results from a U.S. Probability Sample of Women Ages 18 to 94." *Journal of Sex & Marital Therapy* 44:201–212.

Herbenick, Debby, Michael Reece, Vanessa Schick, Stephanie A. Sanders, Brian Dodge, and J. Dennis Fortenberry. 2010a. "Sexual Behavior in the United States: Results from a National Probability Sample of Men and Women Ages 14–94." *Journal of Sexual Medicine* 7(Suppl. 5):255–265.

Herbenick, Debby, Michael Reece, Vanessa Schick, Stephanie A. Sanders, Brian Dodge, and J. Dennis Fortenberry. 2010b. "An Event-Level Analysis of the Sexual Characteristics and Composition among Adults Ages 18–59: Results from a National Probability Sample in the United States." *Journal of Sexual Medicine* 7 (Suppl. 5):346–361.

Herbenick, Debby, Vanessa Schick, Michael Reece, Stephanie A. Sanders, Nicole Smith, Brian Dodge, and J. Dennis Fortenberry. 2013. "Characteristics of Condom and Lubricant Use among a Nationally Representative Probability Sample of Adults Ages 18–59 in the United States." *Journal of Sexual Medicine* 10: 171–183.

Herd, Pamela. 2009. "Women, Public Pensions, and Poverty: What Can the United States Learn from Other Countries?" *Journal of Women, Politics, and Policy* 30(2–3):301–334.

Herman-Giddens, Marcia E. 2007. " Comment on: 'The decline in the age of menarche in the United States: should we be concerned?'" *Journal of Adolescent Health* 40(3):201–203.

Herman-Giddens, Marcia E., Steffes, Jennifer., Harris, Donna., Slora, Eric., Hussey, Michael et al. 2012. "Secondary Sexual Characteristics in Boys: Data from the Pediatric Research in Office Settings Network." *Pediatrics* 130:e1058–e1068.

Hermez, Paul, and Wendy D. Manning. 2017. "Over Twenty-Five Years of Change in Cohabitation Experience in the U.S., 1987–2013." Retrieved December 1, 2019 (https://www.bgsu.edu/content/dam/BGSU/college-of-arts-and-sciences/NCFMR/documents/FP/FP-15-01-twenty-five-yrs-cohab-us.pdf).

Hermez, Paul, and Wendy D. Manning. 2017. "Thirty Years of Change in Women's Premarital Cohabitation Experience. " Retrieved November 23, 2019 (https://magic.piktochart.com/output/19755947-hemez-manning-30-yrs-change-women-premarital-cohab-fp-17-05).

Hernandez, Raymond. 2001. "Children's Sexual Exploitation Underestimated, Study Finds." *The New York Times*, September 10.

Heron, Melonie. 2019. "Deaths: Leading Causes for 2017." *National Vital Statistics Reports*, vol 68 no 6. Hyattsville, MD: National Center for Health Statistics.

Herring, Jeff. 2005. "Affairs Don't Have to Be Physical." Knight/Ridder Newspapers, November 6.

Herszenhorn, David M. 2013. "In Russia, Ban on U.S. Adoptions Creates Rancor and Confusion." *The New York Times*, January 15. Retrieved January 16, 2013 (http://www.nytimes.com).

Herszenhorn, David M., and Andrew E. Kramer. 2012. "Russian Adoption Ban Brings Uncertainty and Outrage." *The New York Times*, December 28. Retrieved January 8, 2012 (http://www.nytimes.com).

Hertlein, Katherine M. 2012. "Digital Dwelling: Technology in Couple and Family Relationships." *Family Relations* 61(3) (July):374–387.

Hertlein, K. M., and F. P. Piercy. 2006. "Internet Infidelity: A Critical Review of the Literature." *Family Journal: Counseling and Therapy for Couples and Families* 14:366–371.

Hertz, Rosanna. 1997. "A Typology of Approaches to Child Care." *Journal of Family Issues* 18(4):355–385.

Hertz, Rosanna, Ana María Rivas, and María Isabel Rubio Jociles. 2016. "Single Mothers by Choice in Spain and the United States. " *Encyclopedia of Family Studies*:1–5.

Hess, Cynthia, and Alona Del Rosario. 2018. *Dreams Deferred: A Survey of Intimate Partner Violence on Survivors' Education, Careers, and Economic Security*. Institute for Women's Policy Research. Retrieved February 13, 2020 (https://iwpr.org).

Hess, Cynthia, Tanima Ahmed McPhil, and Jeff. Hayes. 2020. "Providing Household and Care Work in the United States: Uncovering Inequality." Institute for Women's Policy Research. Retrieved February 4, 2020 (www.iwpr.org).

Hesse-Biber, Sharlene Nagy. 2007. *Handbook of Feminist Research: Theory and Praxis*. Thousand Oaks, CA: Sage.

Hetherington, E. Mavis. 1987. "Family Relations Six Years after Divorce." pp. 185–205 in *Remarriage and Stepparenting*, edited by K. Pasley and M. Ihinger-Tallman. New York: Guilford Press.

Hetherington, E. Mavis, and K. M. Jodl. 1994. "Stepfamilies as Settings for Child Development." pp. 55–80 in *Stepfamilies: Who Benefits? Who Does Not?* edited by Alan Booth and J. Dunn. Hillsdale, NJ: Erlbaum.

Hetherington, E. Mavis, and John Kelly. 2002. *For Better or for Worse: Divorce Reconsidered*. New York: Norton.

Hetherington, E. Mavis, and M. M. Stanley-Hagan. 1999. "Stepfamilies." pp. 137–159 in *Parenting and Child Development in "Nontraditional" Families*, edited by M. E. Lamb. Mahwah, NJ: Lawrence Erlbaum Associates.

Hetter, Katia. 2011. "What's Fueling Bible Belt Divorces?" CNN. Retrieved April 17, 2013 (http://www.cnn.com/2011/LIVING/08/25/divorce.bible.belt/index.html).

Hewitt, Belinda, and Gavin Turrell. 2011. "Short-Term Functional Health and Well-Being after Marital Separation: Does Initiator Status Make a Difference?" *American Journal of Epidemiology* 173(11):1308–1318.

Heuveline, Patrick, Jeffrey M. Timberlake, and Frank F. Furstenberg, F. F. (2003). Shifting Childrearing to Single Mothers: Results from 17 Western Countries. *Population and Development Review* 29.47–71.

Hewlett, Sylvia Ann. 2002. *Creating a Life: Professional Women and the Quest for Children*. New York: Hyperion.

Heyman, Richard, Ashley Hunt-Martorano, Jill Malik, and Amy M. Smith. 2009. "Desired Change in Couples: Gender Differences and Effects on Communication." *Journal of Family Psychology* 23(4):474–484.

Heyman, Richard A., and Amy M. Smith Slep. 2002. "Do Child Abuse and Interparental Violence Lead to Adulthood Family Violence?" *Journal of Marriage and Family* 64:864–870.

Hicken, Melanie. 2014. "Average Cost of Raising a Child Hits $245,000." Retrieved February 8, 2016 (http://money.cnn.com/2014/08/18/pf/child-cost/).

Higginbotham, Brian J., Julie J. Miller, and Sylvia Niehuis. 2009. "Remarriage Preparation: Usage, Perceived Helpfulness, and Dyadic Adjustment." *Family Relations* 58(July):316–329.

Higginbotham, Brian J., Sarah Tulane, and Linda Skogrand. 2012. "Stepfamily Education and Changes in Financial Practices." *Journal of Family Issues* 33(1):1398–1420.

Hildingsson, Ingegerdm and Jan Thomas. 2014. "Parental Stress in Mothers and Fathers One Year After Birth." *Journal of Reproductive and Infant Psychology* 32(2014):41–56.

Hildreth, Carolyn J., Alison Burke, and Richard Glass. 2009. "Elder Abuse." *Journal of the American Medical Association* 302(5):588.

Hill, Amelia. 2015. "Princess Awesome: The Fight Against 'Pinkification.'" *The Guardian*, February 20. Retrieved March 14, 2015 (http://www.theguardian.com).

Hill, Catey. 2015. "Child Care Costs More Than College in These 24 States." Marketwatch, October 13. Retrieved November 17, 2015 (http://www.marketwatch.com).

Hill, Catherine. 2015. *The Simple Truth about the Gender Pay Gap (Fall 2015)*. American Association of University Women. Retrieved October 8, 2015 (http://www.aauw.org).

Hill, Rachelle, Eric Tranby, Erin Kelly, and Phyllis Moen. 2013. "Reliving the Time Squeeze? Effects of a White-Collar Workplace change on Parents." *Journal of Marriage and Family* 75(4):1014–1029.

Hill, Reuben. 1958. "Generic Features of Families Under Stress." *Social Casework* 49:139–150.

Hill, Robert B. 2003 [1972]. *The Strengths of Black Families*, 2nd ed. with "Epilogue: Thirty Years Later." Lanham, MD: University Press of America.

Hill, Shirley A. 2004. *Black Intimacies: A Gender Perspective on Families and Relationships*. Lanham, MD: Rowman and Littlefield.

Hill, Twyla. 2011. "Spousal Caregiving in Later Life: Predictors and Consequences." *Family Focus* FF48:F15–F16.

Hilliard, L. J., and L. S. Liben. 2010. "Differing Levels of Gender Salience in Preschool Classrooms: Effects of Children's Gender Attitudes and Intergroup Bias." *Child Development* 81(6)(Nov–Dec):1787–1798.

Hilmantel, Robin 2013. "What Do You Think a Normal Woman Looks Like Down There?" Retrieved December 15, 2015 (http://www.womenshealthmag.com /sex-and-love /normal-vagina).

Himanek, Celeste Eckman. 2011. "Transitioning to Parenthood: The Experience of Infertility." *Family Focus* FF49:F9–F11.

Himes, Christine L. 2001. "Social Demography of Contemporary Families and Aging." pp. 47–50 in *Families in Later Life: Connections and Transitions*, edited by Alexis J. Walker, Margaret Manoogian-O'Dell, Lori A. McGraw, and Diana L. G. White. Thousand Oaks, CA: Pine Forge Press.

Hinders, Dana. 2012. "Average Cost of Surrogacy." *Love to Know: Advice Women Can Trust*. Retrieved January 14, 2013 (http://www.pregnancy.lovetoknow.com).

Hines, Denise A., and Emily M. Douglas. 2009. "Women's Use of Intimate Partner Violence against Men: Prevalence, Implications, and Consequences." *Journal of Aggression, Maltreatment and Trauma* 18(6):572–586.

Hintz, Elizabeth A., and Clinton L. Brown. 2019. "Childfree by Choice: Stigma in Medical Consultations for Voluntary Sterilization." *Women's Reproductive Health* 6:62–75.

Hirsch, Jennifer S. 2008. "Catholics Using Contraceptives: Religion, Family Planning, and Interpretive Agency in Rural Mexico." *Studies in Family Planning* 39(2):93–104.

Hirsch, Jennifer S., Miguel Muñoz-Laboy, Christina M. Nyhus, Kathryn M. Yount, and José A. Bauermeister. 2009. "'They Miss More than Anything Their Normal Life Back Home': Masculinity and Extramarital Sex among Mexican Migrants in Atlanta." *Perspectives on Sexual and Reproductive Health* 41(1):23–32.

Hirji, Aliya. 2012. "Islam's Role in Family (and Life) Education." *Family Focus*. Issue FF53. Summer: F11–F12.

HIV.gov. 2019. "What is 'Ending the HIV Epidemic: A Plan for America'?" Retrieved November 2, 2019 (https://www.hiv.gov/federal-response /ending-the-hiv-epidemic/overview).

Hiyoshi, Ayako, Katja Fall, Gopalakrishnan Netuveli, and Scott Montgomery. 2015. "Remarriage after Divorce and Depression Risk." *Social Science & Medicine* 141:109–114.

Hoch, Kacie. "Building an Affirmative Counseling Practice for the Transgender and Gender Nonconforming Communities." (2019). Retrieved January 15, 2020 (https://scholarworks.boisestate.edu/gss_2019/139/).

Hochschild, Arlie. 1989. *The Second Shift: Working Parents and the Revolution at Home*. New York: Viking/Penguin.

Hofferth, S. L. 2006. "Residential Father Family Type and Child Well-Being: Investment versus Selection." *Demography* 43:53–77.

Hoffman, Jan. 2011. "Boys Will Be Boys? Not in These Families." *The New York Times*, June 10. Retrieved September 13, 2011 (http://www.nytimes.com).

Hoffman, Kristin, and John N. Edwards. 2004. "An Integrated Theoretical Model of Sibling Violence and Abuse." *Journal of Family Violence* 19(3):185–200.

Hoffman, Lois W., and Jean B. Manis. 1979. "The Value of Children in the United States: A New Approach to the Study of Fertility." *Journal of Marriage and Family* 41:583–596.

Hoffman, Steven, Bethany Breck, and Lauren Beasley. 2020. "Hispanic Stepfamilies. " In *Multicultural Stepfamilies*, edited by Susan D. Stewart and Gordon Limb. San Diego, CA: Cognella, pp. 77–110.

Hogan, Dennis P. and Nan M. Astone. 1986. "The Transition to Adulthood." *Annual Review of Sociology* 12:109–130.

Hoge, Charles W. 2010. *Once a Warrior, Always a Warrior: Navigating the Transition from Combat to Home—Including Combat Stress, PTSD, and MTBI*. Guilford, CT: GPP Life.

Hohmann-Marriott, Bryndl. 2009. "Father Involvement Ideals and the Union Transitions of Unmarried Parents." *Journal of Family Issues* 30(7):898–920.

Hohmann-Marriott, Bryndl, and Paul Amato. 2008. "Relationship Quality in Interethnic Marriages and Cohabitations." *Social Forces* 87(2):825–855.

Holland, Rochelle. 2009. "Perceptions of Mate Selection for Marriage among African American, College-Educated, Single Mothers." *Journal of Counseling and Development* 87(Spring):170–178.

Hollander, D. 2015. "Growing Evidence Links Sexting to Teenagers' Sexual Activity; Link to Risky Behavior Less Clear." *Perspectives on Sexual & Reproductive Health* 47:51–52.

Holley, Sarah R., Claudia M. Haase, and Robert W. Levenson. 2013. "Age-Related Changes in Demand-Withdraw Communication Behaviors." *Journal of Marriage and Family* 75:822–836.

Hollingsworth, William-Glenn, Megan Dolbin-MacNab, and Lydia Marek. 2016. "Boundary Ambiguity and Ambivalence in Military Family Reintegration." *Family Relations* 65(4): 603-615.

Holmberg, Diane, and Karen L. Blair. 2009. "Sexual Desire, Communication, Satisfaction, and Preferences of Men and women in Same-Sex versus Mixed-Sex Relationships." *Journal of Sex Research* 46:57–66.

Holmberg, Diane, Karen Blair, and Maggie Phillips. 2010. "Women's Sexual Satisfaction as a Predictor of Well-Being in Same-Sex Versus Mixed-Sex Relationships." *Journal of Sex Research* 47(1):1–11.

Holmes, Erin Kramer, Takayuki Sasaki, and Nancy L. Hazen. 2013. "Smooth Versus Rocky Transitions to Parenthood: Family systems in Developmental Context." *Family Relations* 62(5):824–837.

Holmes, Erin K., Alan J. Hawkins, Braquel M. Egginton, Nathan Robbins, and Kevin Shafer.

2018. *Summary Report: Do Responsible Fatherhood Programs Work? Fatherhood Research and & Practice Network*. Retrieved February 12, 2020 (https://www.frpn.org/sites/default/files /FRPN_MetaAnalysis_Summary_121418_v3.pdf).

Holmes, Leonard. 2018. "Managing the Balance of Power in Relationships." Verywell Mind. July 13. Retrieved February 4, 2020 (https://www.verywellmind.com).

Holmes, Steven A. 1990. "Day Care Bill Marks a Turn toward Help for the Poor." *The New York Times*, January 25.

Holohan, Meghan. 2019. "'Adulting' Class at Kentucky High School Teaches Crucial Life Skills." Retrieved November 26, 2019 (https://www.today.com/parents /adulting-class-kentucky-high-school -teaches-crucial-life-skills-t151240).

Holstein, James A. and Jaber F. Gubrium. 2008. *Handbook of Constructionist Research*. New York: Guilford Press.

Holt, Amanda. 2013. *Adolescent-to-Parent Abuse*. New York: Policy Press.

Holtzman, Mellisa. 2011. "Nonmarital Unions, Family Definitions, and Custody Decision Making." *Family Relations* 60:617–632.

Holtzworth-Munroe, Amy, Amy G. Applegate, and Brian D'Onofrio. 2009. "For the Sake of the Children: Collaborations between Law and Social Science to Advance the Field of Family Dispute Resolution: Family Dispute Resolution: Charting a Course for the Future." *Family Court Review* (Special Issue) 47(3):493–505.

Homan, Timothy, and Frank Bass. 2012. "Number of U.S. Same-Sex Households Increases by 80 Percent Since 2000." *Bloomberg Business Report*, September 27. Retrieved November 3, 2015 (http://www .Bloomberg.com).

Hoppe, Trevor. 2013. "Controlling Sex in the Name of "Public Health": Social Control and Michigan HIV Law." *Social Problems* 60:27–49.

Hopper, Joseph. 1993. "The Rhetoric of Motives in Divorce." *Journal of Marriage and Family* 55:801–813.

Hopper, Joseph. 2001. "The Symbolic Origins of Conflict in Divorce." *Journal of Marriage and Family* 63:430–445.

Hornfeck, Fabienne, Ina Bovenschen, Sabine Heene, Janin Zimmermann, Annabel Zwönitzer, and Heinz Kindler. 2019. "Emotional and Behavior Problems in Adopted Children—the Role of Early Adversities and Adoptive Parents' Regulation and Behavior. " *Child Abuse & Neglect* 98:104221.

Hornik, Donna. 2001. "Can the Church Get in Step with Stepfamilies?" *U.S. Catholic* 66(7):30–41.

Horowitz, A. 1985. "Sons and Daughters as Caregivers to Older Parents: Differences in Role Performance and Consequences." *The Gerontologist* 25:612–617.

Horwitz, Briana N., and Jenae M. Neiderhiser. 2011. "Gene-Environment Interplay, Family Relationships, and Child Adjustment." *Journal of Marriage and Family* 73(4)(August): 804–816.

Horowitz, Juliana. 2019a. "Race in America 2019." April 9. Pew Research Center. Retrieved November 5, 2019 (www.pewsocialtrends.org).

Horowitz, Juliana. 2019b. "Despite Challenges at Home and Work, Most Working Moms and Dads Say Being Employed Is What's Best for Them." *Facttank: News in the Numbers.* September 12. Pew Research Organization. Retrieved January 30, 2019 (www.pewresearch.org).

Horowitz, Juliana, and Nikki Graf. 2019. "Most U.S. Teens See Anxiety and Depression as a Major Problem among their Peers." Pew Research Center. February 20. Retrieved August 10, 2019 (www.pewresearch.org).

Horowitz, Juliana Menasce, Ruth Igielnik, and Kim Parker. 2018. "Women and Leadership 2018." Retrieved October 2, 2019 (https://www.pewsocialtrends.org/2018/09/20/women-and-leadership-2018/).

House, Anthony. 2002. *A Problematic Solution: Responses to the Marriage Reform Act of 1753.* Retrieved October 5, 2006 (users.ox.ac.uk).

"Household Income Quintiles." 2019. Tax Policy Center Statistics. August 5. Retrieved October 16, 2019 (www.taxpolicycenter.org).

Houts, Carrie R., and Sharon G. Horne. 2008. "The Role of Relationship Attributions in Relationship Satisfaction among Cohabiting Gay Men." *The Family Journal: Counseling and Therapy for Couples and Families* 16(3):240–248.

Houts, Leslie A. 2005. "But Was It Wanted? Young Women's First Voluntary Sexual Intercourse." *Journal of Family Issues* 26(8):1082–1102.

Howard, Hilary. 2012. "A Confederacy of Bachelors." *The New York Times, August* 3. Retrieved November 6, 2012 (http://www.nytimes.com).

Howard, Lakin. 2018. "9 Ways to Take Your Power Back in a Relationship, According to an Expert." May 2.Retrieved February 5, 2020 (https://www.bustle.com/p/9-ways-to-take-your-power-back-in-a-relationship-according-to-expert-8943124).

Hoy, Aaron, and Andrew S. London. 2017. "Same-Sex Sexuality and the Duration of First Marriages." *Population Review* 56:136–163.

Hsia, Annie. 2002. "Considering Grandparents' Rights and Parents' Wishes: 'Special Circumstances' Litigation Alternatives." *The Legal Intelligencer* 227(83):7.

Huang, Chien-Chung, Ronald B. Mincy, and Irwin Garfinkel. 2005. "Child Support Obligations and Low-Income Fathers." *Journal of Marriage and Family* 67(5):1213–1225.

Huang, Hanyun, and Louis Leung. 2009. "Instant Messaging Addiction among Teenagers in China: Shyness, Alienation, and Academic Performance Decrement." *CyberPsychology and Behavior* 12:675–679.

Huber, Joan. 1980. "Will U.S. Fertility Decline toward Zero?" *Sociological Quarterly* 21:481–492.

Huber, Lindsay Pérez, and Daniel G. Solorzano. 2013. "Visualizing Everyday Racism Critical Race Theory, Visual Microaggressions, and the Historical Image of Mexican Banditry." *Qualitative Inquiry* 21:223–238.

Hubin, Donald C. 2014. "Fractured Fatherhood: An Analytic Philosophy Perspective on Moral and Legal Paternity." *Journal of Family Theory and Review* 6:76–90.

Huckaby, Jody. M. 2009. "Statement of Parents, Families and Friends of Lesbians and Gays (PFLAG) National to the United States Senate Committee on Health, Education, Labor and Pensions Regarding the Employment Non-Discrimination Act: Ensuring Opportunity for All Americans." Retrieved April 14, 2016 (http://community.pflag.org/page.aspx?pid=194).

Hudziak, Jim, and Masha Y. Ivanova. 2016. "The Vermont Family-Based Approach: Family-Based Health Promotion, Illness Prevention, and Intervention." *Child and Adolescent Psychiatric Clinics* 25(2):167–178.

Huebner, Angela J., Jay A. Mancini, Ryan M. Wilcox, Saralyn R. Grass, and Gabriel A. Grass. 2007. "Parental Deployment and Youth in Military Families: Exploring Uncertainty and Ambiguous Loss." *Family Relations* 56(2):112–122.

Hughes, Mary E., and Linda J. Waite. 2009. "Marital Biography and Health at Mid-Life." *Journal of Health and Social Behavior* 50:344–358.

Hughes, Michael, and Walter R. Gove. 1989. "Explaining the Negative Relationship between Social Integration and Mental Health: The Case of Living Alone." Presented at the annual meeting of the American Sociological Association, August, San Francisco, CA.

Hughes, Patrick C., and Fran C. Dickson. 2005. "Communication, Marital Satisfaction, and Religious Orientation in Interfaith Marriages." *Journal of Family Communication* 5(1):25–41.

Hughes, Rachel, Katie Heiden-Rootes, Ashley Weingard, and Danielle Bono. 2016. "Suicide Risk and Protective Factors in the Family for LGBTQ Adults and Adolescents." *Family Focus* (FF67, Spring):F13–F14. Minneapolis: National Council on Family Relations.

Hughes, Robert Jr., Jill Bowers, Elissa T. Mitchell, Sarah Curtiss, and Aaron Ebata. 2012. "Developing Online Family Life Prevention and Education Programs." *Family Relations* 61(5):711–727.

Hull, K. G. 2008. "Broken Trust: Pursuing Remedies for Victims of Elder Financial Abuse by Agents Under Power-of-Attorney Agreements." *Clearinghouse Review* 42(5/6):223–231.

Humble, Aine M. 2009. "The Second Time 'Round: Gender Construction in Remarried Couples' Wedding Planning." *Journal of Divorce and Remarriage* 50:260–281.

Hunt, Gail, Carol Levine, and Linda Naiditch. 2005. *Young Caregivers in the U.S: Report of Findings September 2005.* National Alliance for Caregiving.

Hunt, Lucy L., Paul W. Eastwick, and Eli J. Finkel. 2015. "Leveling the Playing Field: Acquaintance Length Predicts Reduced Assortative Mating on Attractiveness." *Psychological Science* 26:1046–1053.

Huston, Ted L., and Heidi Melz. 2004. "The Case for (Promoting) Marriage: The Devil Is in the Details." *Journal of Marriage and Family* 66(4):943–958.

Hutchinson, Katherine M. and Julie A. Cederbaum. 2011. "Talking to Daddy's Little Girl about Sex: Daughters' Reports of Sexual Communication and Support from Fathers." *Journal of Family Issues* 32(4):550–572.

Hyde, Abbey, Jonathan Drennan, Michelle Butler, Etaoine Howlett, Marie Carney, and Maria Lohan. 2013. "Parents' Constructions of Communication with their Children about Safer Sex." *Journal of Clinical Nursing* 22:3438-3446.

Hyde, Janet Shibley. 2005. "The Gender Similarities Hypothesis." *American Psychologist* 60(6):581–592.

Hyman, Mark. 2010. "Sports Training Has Begun for Babies and Toddlers." *The New York Times,* November 30. Retrieved January 9, 2011 (http://www.nytimes.com).

Hymowitz, Carol. 2015. "Time to Be Selfish: Parents Are Risking their Retirement to Subsidize Their Kids." *Bloomberg Business.* Retrieved December 27, 2015 (http://www.bloomberg.com).

Hymowitz, Kay. 2006. *Marriage and Caste in America: Separate and Unequal Families in a Post-Marital Age.* Chicago: Ivan R. Dee.

Iacuone, David. 2005. "'Real Men Are Tough Guys': Hegemonic Masculinity and Safety in the Construction Industry." *Journal of Men's Studies* 13(2):247–267.

Iceland, John, and Kyle Anne Nelson. 2008. "Hispanic Segregation in Metropolitan America: Exploring the Multiple Forms of Spatial Assimilation." *American Sociological Review* 73(5):741–765.

Idstad, Mariann, Fartein Ask Torvik, Ingrid Borren, Kamilla Rognmo, Espen Røysamb, and Kristian Tambs. 2015. "Mental Distress Predicts Divorce Over 16 Years: The HUNT Study." *BMC Public Health* 15:1.

Ihinger-Tallman, Marilyn, and Kay Pasley. 1997. "Stepfamilies in 1984 and Today—A Scholarly Perspective." *Marriage and Family Review* 26(1–2):19–41.

Iman, Ben C., Patricia Cohen, and Doris Burke. 2018. "The Great Recession Knocked Them Down. Only Some Got Up Again." *The New York Times.* September 12. Retrieved August 29, 2019 (www.nytimes.com).

"Improvements in Teen Sexual Risk Behavior Flatline." 2006. Press Release. Washington, DC: Advocates for Youth.

Ingber, Hanna. 2015. "Muslim Parents on How They Talk to Their Children about Hatred and Extremism." *The New York Times.* December 15. Retrieved May 30, 2018 (www.nytimes.com).

Ingersoll-Dayton, Berit, Margaret B. Neal, Jung-Hwa Ha, and Leslie B. Hammer. 2003. "Redressing Inequity in Parent Care among Siblings." *Journal of Marriage and Family* 65(1):201–212.

Ingoldsby, Bron B. 2006a. "Family Origin and Universality." pp. 67–78 in *Families in Global and Multicultural Perspective*, 2nd ed., edited by Bron B. Ingoldsby and Suzanna D. Smith. New York: Guilford.

Ingoldsby, Bron B., and Suzanna D. Smith, eds. 2005. *Families in Global and Multicultural Perspective*, 2nd ed. Thousand Oaks, CA: Sage.

Ingraham, Chrys. 2008. *White Weddings: Romancing Heterosexuality in Popular Culture*, 2nd ed. New York: Routledge.

Ingram, Stephanie, Jay L. Ringle, Kristen Hallstrom, David E. Schill, Virginia M. Gohr, and

Ronald W. Thompson. 2008. "Coping with Crisis across the Lifespan: The Role of a Telephone Hotline." *Journal of Child and Family Studies* 17:663–74.

Internet Watch Foundation. 2018. *Trends in Online Child Sexual Exploitation.* May. Retrieved February 15, 2020 (www.iwf.org).

Isaksen, Lise W. 2002. "Toward a Sociology of (Gendered) Disgust." *Journal of Family Issues* 23(7):791–811.

Ishii-Kuntz, Masako. 2000. "Diversity within Asian American Families." pp. 274–292 in *Handbook of Family Diversity,* edited by David H. Demo, Katherine R. Allen, and

Mark Fine. New York: Oxford University Press.

Ishii-Kuntz, Masako, Jessica N. Gomei, Barbara J. Tinsley, and Rose D. Parks. 2009. "Economic Hardship and Adaptation among Asian American Families." *Journal of Family Issues* 31(3):407–420.

Italie, Leanne. 2008. "Parents of Teens Watch Nebraska Safe Law." Associated Press, November 11. Retrieved December 12, 2008 (ap.google .com).

Iveniuk, James, Linda J. Waite, Edward Laumann, Martha K. McClintock, and Andres D. Tiedt. 2014. "Marital Conflict in Older Couples: Positivity, Personality, and Health." *Journal of Marriage and Family* 76(1):130–144.

Jacewicz, Natalie. 2017. "Social Media Can Be Both a Bummer and Boon for the Brain." Retrieved January 25, 2020 (https://newsela.com/read /social-media-teens-mental-health-benefits /id/36844/).

Jackson, Aubrey. 2015. "State Contexts and the Criminalization of Marital Rape Across the United States." *Social Science Research* 51:290–306.

Jackson, Aubrey L. 2016. "The Combined Effect of Women's Neighborhood Resources and Collective Efficacy on IPV." *Journal of Marriage and Family* 78 (4):890–907.

Jackson, Aurora P., Jeong-Kyun Choi, and Peter Bentler. 2009. "Parenting Efficacy and the Early School Adjustment of Poor and Near-Poor Black Children." *Journal of Family Issues* 30(10):1339–1355.

Jackson, Aurora P., Jeong-Kyun Choi, and Kathleen S. J. Preston. 2015. "Nonresident Fathers' Involvement with Young Black Children: A Replication and Extension of a Mediational Model." *Social Work Research* 39:245.

Jackson, Jeffrey, Richard Miller, Megan Oka, and Ryan Henry. 2014. "Gender Differences in Marital Satisfaction: A Meta Analysis." *Journal of Marriage and Family* 76(1):105–129.

Jackson, Pamela Braboy. 2004. "Role Sequencing: Does Order Matter for Mental Health?" *Journal of Health and Social Behavior* 45:132–154.

Jackson, Pamela Braboy, Sibyl Kleiner, Claudia Geist, and Kara Gebulko. 2011. "Conventions of Courtship: Gender and Race Differences in the Significance of Dating Rituals." *Journal of Family Issues* 32: 629–652.

Jackson, Robert Max. 2006. "Opposing Forces: How, Why, and When Will Gender Inequality Disappear?" pp. 215–244 in *The Declining Significance of Gender?* edited by Francine

D. Blau, Mary C. Brinton, and David B. Grusky. New York: Russell Sage.

Jackson, Robert M. 2007. "Destined for Equality." pp. 109–116 in *Family in Transition,* 14th ed., edited by Arlene S. Skolnick and Jerome H. Skolnick. Boston, MA and New York: Pearson.

Jackson-Newsom, Julia, Christy M. Buchanan, and Richard M. McDonald. 2008. "Parenting and Perceived Warmth in European American and African American Adolescents." *Journal of Marriage and Family* 70(1):62–75.

Jacobs, Andres. 2006. "Extreme Makeover, Commune Edition." *The New York Times,* June 11. Retrieved June 11, 2006 (www .nytimes.com).

Jacobs, Jerry A., and Kathleen Gerson. 2004. *The Time Divide: Work, Family, and Gender Inequality.* Cambridge, MA: Harvard University Press.

Jacobsen, Linda A., Mary Kent, Marlene Lee, and Mark Mather. 2011. "America's Aging Population." *Population Bulletin* 66(1): February. Washington, DC: Population Reference Bureau.

Jacobsen, Linda A., and Mark Mather. 2011. "A Post-Recession Update on U.S. Economic and Social Trends." *Population Bulletin Update, December.* Washington, DC: Population Reference Bureau. Retrieved August 29, 2012 (http://www .prb.org).

Jacobsen, Linda A., Mark Mather, and Genevieve Dupuis. 2012. "Household Change in the United States." *Population Bulletin* 67 (1): February. Washington, DC: Population Reference Bureau.

Jacoby, Susan. 1999. "Great Sex: What's Age Got to Do with It?" *Modern Maturity* (September-October), pp. 41–47.

Jahromi, Laudan B. 2018. "Mother-Grandmother and Mother-father Coparenting across Time among Mexican-Origin Adolescent Mothers and their Families." *Journal of Marriage and Family* 80 (2):349-366.

James, Anthony. 2016. "Parenting and Protecting: Advocating Microprotections through Loving and Supporting Black Parent-Child Relationships." *Family Focus* (FF67, Spring):F2–F3. Minneapolis: National Council on Family Relations.

James, E. L. 2011. *Fifty Shades of Grey.* London: Vintage Books.

James, Susan Donaldson. 2012. "Transgender Parents: Child's Journey When Mom Becomes Dad." ABC News, March 27. Retrieved October 13, 2015 (http://www.abcnews.com).

Jamison, Tyler B., and Lawrence Ganong. 2011. "'We're Not Living Together': Stayover Relationships among College-Educated Emerging Adults." 2011. *Journal of Social and Personal Relationships* 28(4):536–557.

Jang, Soo Jung, Allison Zippay, and Rhokeun Park. 2012. "Family Roles as Moderators of the Relationship Between Schedule Flexibility and Stress." *Journal of Marriage and Family* 74(4):897–912.

Janning, Michelle. 2015. "An Unexpected Box of Love Research." *Contexts* 14:76–76.

Janning, Michelle, and Helen Scalise. 2015. "Gender and Generation in the Home Curation of Family Photography." *Journal of Family Issues* 36(12):1702–1725.

Janofsky, Michael. 2001. "Conviction of a Polygamist Raises Fears among Others." *The New York Times,* May 24.

Jansen, Mieke, Dimitri Mortelmans, and Laurent Snoeckx. 2009. "Repartnering and (Re)employment: Strategies to Cope with the Economic Consequences of Partnership Dissolution." *Journal of Marriage and Family* 71:1271–1293.

Janz, Philip, Christopher A. Pepping, and W. Kim Halford. 2015. "Individual Differences in Dispositional Mindfulness and Initial Romantic Attraction: A Speed Dating Experiment." *Personality and Individual Differences* 82:14–19.

Jappens, Maaike, and Jan Van Bavel. 2015. "Parental Divorce, Residence Arrangements, and Contact between Grandchildren and Grandparents." *Journal of Marriage and Family* 77:424–440.

Jarrett, Laura. 2017. "Sessions Says Civil Rights Law Doesn't Protect Transgender Workers." Retrieved September 29, 2019 (https://www .cnn.com/2017/10/05/politics/jeff-sessions -transgender-title-vii/index.html).

Jaschik, Scott. 2009. "Probe of Extra Help for Men." *Inside Higher Ed,* November 2 (www .insidehighered.com).

Jatlaoui, et al. 2019. *Abortion Surveillance—United States, 2016.* Retrieved December 23, 2019 (https://www.cdc.gov /mmwr/volumes/68/ss/ss6811a1 .htm?s_cid=ss6811a1_w).

Jayakody, R., and A. Kalil. 2002. "Social Fathering in Low-income, African American Families with Preschool Children." *Journal of Marriage and Family* 64(2):504–516.

Jenkins, Nate. 2008. "Nebraska Parents Rush to Leave Kids Before Law Changes." Associated Press, November 13. Retrieved December 12, 2008 (news.yahoo.com).

Jenkins-Guarnieri, Michael A., Stephen L. Wright, and Lynette M. Hudiburgh. 2012. "The Relationships among Attachment Style, Personality Traits, Interpersonal Competency, and Facebook Use." *Journal of Applied Developmental Psychology,* 33(6):294–301.

Jensen, Alexander, Shawn Whiteman, Karen Fingerman, and Kira Birditt. 2013. "'Life Still Isn't Fair': Parental Differential Treatment of Young Adult Siblings." *Journal of Marriage and Family* 75(2):438–452.

Jensen, Todd M., Kevin Shafer, and Erin K. Holmes. 2015. "Transitioning to Stepfamily Life: The Influence of Closeness with Biological Parents and Stepparents on Children's Stress." *Child and Family Social Work.* doi: 10.111 /cfs.12237.

Jeong, Jae Y., and Sharon G. Horne. 2009. "Relationship Characteristics of Women in Interracial Same-Sex Relationships." *Journal of Homosexuality* 56(4):443–456.

Jepsen, Lisa K., and Christopher A. Jepsen. 2002. "An Empirical Analysis of the Matching Patterns of Same-Sex and Opposite-Sex Couples?" *Demography* 39(3):435–453.

Jhally, Sut. 2007. *Dreamworlds 3: Desire, Sex and Power in Music Video.* Northampton, MA: Media Education Foundation.

Jin, Lei, and Nicholasa Chrisatakis. 2009. "Investigating the Mechanism of Marital

Mortality Reduction: The Transition to Widowhood and Quality of Health Care." *Demography* 46(3):605–625.

Jenkins, Cameron. 2018. "The American Academy of Pediatrics on Spanking Children: Don't Do It Ever." National Public Radio. November 11. Retrieved December 10, 2019 (www.npr.org.).

Jensen, Todd M., and Kevin Shafer. 2013. "Stepfamily Functioning and Closeness: Children's Views on Second Marriages and Stepfather Relationships." *Social Work* 58:127–136.

Jensen, Todd M., and Matthew O. Howard. 2015. "Perceived Stepparent–Child Relationship Quality: A Systematic Review of Stepchildren's Perspectives." *Marriage & Family Review* 51:99–153.

Jensen, Todd M., Kevin Shafer, and Erin K. Holmes. 2017. "Transitioning to Stepfamily Life: The Influence of Closeness with Biological Parents and Stepparents on Children's Stress." *Child & Family Social Work* 22:275–286.

Jiang, Jingjng. 2018. "How Teens and Parents Navigate Screen Time and Device Distractions." Pew Research Center. August 22. Retrieved August 29, 2019 (www.pewinternet.org).

Jo, Moon H. 2002. "Coping with Gender Role Strains in Korean American Families." pp. 78–83 in *Contemporary Ethnic Families in the United States*, edited by Nijole V. Benekraitis. Upper Saddle River, NJ: Prentice Hall.

John, Robert. 1998. "Native American Families." pp. 382–421 in *Ethnic Families in America: Patterns and Variations*, edited by Charles H. Mindel, Robert W. Haberstein, and Roosevelt Wright Jr. Upper Saddle River, NJ: Prentice Hall.

Johnson, Dirk. 1991. "Polygamists Emerge from Secrecy, Seeking Not Just Peace but Respect." *The New York Times*, April 9.

Johnson, Jason B. 2000. "Something Akin to Family: Struggling Parents, Kids, Move in with Their Mentors." *San Francisco Chronicle*, November 10.

Johnson, Matthew D., Joanne Davila, Ronald D. Rogge, Kieran T. Sullivan, Catherine L. Cohan, Erika Lawrence, Benjamin R. Karney, and Thomas N. Bradbury. 2005. "Problem-Solving Skills and Affective Expressions as Predictors of Change in Marital Satisfaction." *Journal of Consulting and Clinical Psychology* 73(1):15–27.

Johnson, Matthew D. 2014. "Government-Supported Healthy Marriage Initiatives Are Not Associated with Changes in Family Demographics: A Comment on Hawkins, Amato, and Kinghorn (2013)." *Family Relations* 63(2):300–304.

Johnson, Matthew D., Nancy L. Galambos, and Jared R. Anderson. 2015. "Skip the Dishes? Not So Fast! Sex and Housework Revisited." *Journal of Family Psychology*, October 12.

Johnson, Matthew, Rebecca Horne, and Franz Neyer. 2018. "The Development of Willingness to Sacrifice and Unmitigated Communion in Intimate Partnerships." *Journal of Marriage and Family* 80 (3):637 –652.

Johnson, Michael P. 2008. *A Typology of Domestic Violence: Intimate Terrorism, Violent Resistance, and Situational Couple Violence.* Lebanon, NH: Northeastern University Press.

Johnson, Michael P., and Kathleen J. Ferraro. 2000. "Research on Domestic Violence in the 1990s: Making Distinctions." *Journal of Marriage and Family* 62(4):948–963.

Johnson, Michael P., and Janel M. Leone. 2005. "The Differential Effects of Intimate Terrorism and Situational Couple Violence: Findings from the National Violence Against Women Survey." *Journal of Family Issues* 20(3).922–949.

Johnson, Monica Kirkpatrick. 2013. "Parental Financial Assistance and Young Adults' Relationships with Parents and Well-Being." *Journal of Marriage and Family* 75(3):713–733.

Johnson, Phyllis J. 1998. "Performance of Household Tasks by Vietnamese and Laotian Refugees." *Journal of Family Issues* 10(3):245–273.

Johnson, Phyllis J., and Kathrin Stoll. 2008. "Remittance Patterns of Southern Sudanese Refugee Men: Enacting the Global Breadwinner Role." *Family Relations* 57(4):431–443.

Johnson, Richard W., and Joshua M. Wiener.2006. "A Profile of Frail Older Americans and Their Caregivers." Urban Institute, March 1. Retrieved May 18, 2007 (www.urban.org/publications).

Johnson, Wendi L., Wendy D. Manning, Peggy C. Giordano, and Monica A. Longmore. 2015. "Relationship Context and Intimate Partner Violence from Adolescence to Young Adulthood." *Journal of Adolescent Health* 57:631–636.

Jones, A. J., and M. Galinsky. 2003. "Restructuring the Stepfamily: Old Myths, New Stories." *Social Work* 48(2):228–237.

Jones, A. R., R. M. Hyland, K. N. Parkinson, and A. J. Adamson. 2009. "Developing a Focus Group Approach for Exploring Parents' Perspectives on Childhood Overweight." *Nutrition Bulletin* 34(2):214–214.

Jones, Antwan. 2010. "Stability of Men's Interracial First Unions: A Test of Educational Differentials and Cohabitation History." *Journal of Family and Economic Issues* 31:241 256.

Jones, Del. 2006. "One of USA's Exports: Love, American Style." *USA Today*, February 14.

Jones, Jeffrey M. 2008. "Most Americans Not Willing to Forgive Unfaithful Spouse." Washington, DC: Gallup Poll News Service. Retrieved December 2, 2008 (www.gallup .com /poll).

Jones, Jeffrey, M. 2017. "In U.S., 10.2% of LGBT Adults Now Married to Same-Sex Spouse." Retrieved October 17, 2019 (https://news. gallup.com/poll/212702 /lgbt-adults-married-sex-spouse.aspx).

Jones, Jo, and Paul Placek. 2017. *Adoption by the Numbers, A Comprehensive Report of U.S. Adoption Statistics*. National Council on Adoption.

Jones, Nicholas A., and J. Bullock. 2012. *The Two or More Races Population 2010*. U.S. Census Brief No. C2010BR-13, September. Washington DC: U.S. Census Bureau.

Jones, Rachel K., Jacqueline E. Darroch, and Stanley K. Henshaw. 2002. "Patterns in the Socioeconomic Characteristics of Women Obtaining Abortions in 2000–2001." *Perspectives on Sexual and Reproductive Health* 34:226–235.

Jones, Rachel L., and Magan L. Kavanaugh. 2011. "Changes in Abortion Rates between 2000 and 2008 and Lifetime Incidence of Abortion." *Obstetrics and Gynecology* 117:1358–1366.

Jones, Rachel K., Elizabeth Witwer, and Jenna Jerman. 2019. *Abortion Incidence and Service Availability in the United States, 2017*. Retrieved December 27, 2019 (https://www.guttmacher. org/report/abortion -incidence-service-availability-us-2017).

Jordan, Jewel. 2019. "Population Estimates Show GIN Across Race Groups Differs." June 20. U.S. census bureau Public Information Office. Retrieved September 4, 2019 (www.census.gov).

Jorgensen, Bryce, and Damon Rappleyea. 2015. "Combating Family Financial Strain." *Family Focus* Spring:F13–F14.

Jose, Anita, K. Daniel O'Leary, and Anne Moyer. 2010. "Does Premarital Cohabitation Predict Subsequent Marital Stability and Marital Quality? A Meta-Analysis." *Journal of Marriage and Family* 72:105–116.

Joseph, Elizabeth. 1991. "My Husband's Nine Wives." *The New York Times*, May 23.

Joshi, Pamela, and Karen Bogen. 2007. "Nonstandard Schedules and Young Children's Behavioral Outcomes among Working Low-Income Families." *Journal of Marriage and Family* 60(1):139–156.

Joshi, Pamela, James M. Quane, and Andrew J. Cherlin. 2009. "Contemporary Work and Family Issues Affecting Marriage and Cohabitation among Low-Income Single Mothers." *Family Relations* 58(5):647–661.

Joyce, Amy. 2006. "Kid-Friendly Policies Don't Help Singles." *Washington Post*, September 16.

Joyner, Kara, and Grace Kao. 2005. "Interracial Relationships and the Transition to Adulthood." *American Sociological Review* 70(4):563–581.

Joyner, Kara, Wendy Manning, and Ryan Bogle. 2013. *The Stability and Qualities of Same-Sex and Different-Sex Couples in Young Adulthood.* The Center for Family and Demographic Research Working Paper Series, Bowling Green State University. Retrieved April 14, 2013 (http:// www.bgsu .edu/organizations/cfdr/file127059 .pdf).

Juby, Heather, Jean-Michel Billette, Benoît Laplante, and Céline Le Bourdais. 2007. "Nonresident Fathers and Children: Parents' New Unions and Frequency of Contact." *Journal of Family Issues* 28(9):1220–1245.

Juffer, Femmie, and Marinus H. van Uzendoorn. 2005. "Behavior Problems and Mental Health Referrals of International Adoptees: A Meta-Analysis." *Journal of the American Medical Association* 293(20):2501–2515.

Julian, Tiffany. 2012. *Work-Life Earnings by Field of Degree and Occupation for People with a Bachelor's Degree: 2011.* American Community Survey Briefs, October. Retrieved October 15, 2015 (http://www .census.gov).

Junn, Ellen Nan, and Chris J. Boyatzis. 2005. *Child Growth and Development 05/06*, 12th ed. Guilford, CT: McGraw-Hill/Dushkin.

Kader, Samuel. 1999. *Openly Gay, Openly Christian: How the Bible Really Is Gay Friendly.* Leyland Publications.

Kagan, Marilyn, and Neil Einbund. 2008. *Defenders of the Heart: Managing the Habits and*

Attitudes That Block You from a Richer, More Satisfying Life. Carlsbad, CA: Hay House.

Kahn, Joan R., Frances Goldscheider, and Javier García-Manglano. 2013. "Growing Parental Economic Power in Parent–Adult Child Households: Coresidence and Financial Dependency in the United States, 1960–2010." *Demography* 50:1449–1475.

Kahn, Joan R., Brittany S. McGill, and Suzanne M. Bianchi. 2011. "Help to Family and Friends: Are There Gender Differences at Older Ages?" *Journal of Marriage and Family* 73(1):77–92.

Kalata, Dominique, and Autumn M. Bermea. 2019. "Centering the Experiences and Needs of Individuals in Consensually Nonmonogamous Relationships in Family Science." *Family Focus* FF80:F12-F13.

Kale, Sirin. 2019. "A Second Chance at First Love: Meet The Couples Who Marry, Divorce, Then Remarry." Retrieved February 8, 2020 (https://www.theguardian.com /lifeandstyle/2019/aug/13/second-chance -first-love-meet-couples-marry-divorce-remarry).

Kalil, Ariel, Rachel Dunifon, Danielle Crosby, and Jessica Houston Su. 2014. "Work Hours, Schedules, and Insufficient Sleep among Mothers and Their Young Children." *Journal of Marriage and Family* 76(5):891–904.

Kallivayalil, Diya. 2004. "Gender and Cultural Socialization in Indian Immigrant Families in the United States." *Feminism and Psychology* 14(4):535–559.

Kalmijn, Matthijs. 2010. "Educational Inequality, Homogamy, and Status Exchange in Black-White Intermarriage: A Comment on Rosenfeld." *American Journal of Sociology* 115:1252–1263.

Kalmijn, Matthijs, and Frank Van Tubergen. 2010. "A Comparative Perspective on Intermarriage: Explaining Differences among National-Origin Groups in the United States." *Demography* 47(2):459–479.

Kamo, Yoshinori, and Min Zhou. 1994. "Living Arrangements of Elderly Chinese and Japanese in the United States." *Journal of Marriage and Family* 56(3):544–558.

Kamp Dush, Claire M. 2011. "Relationship-Specific Investments, Family Chaos, and Cohabitation Dissolution Following a Nonmarital Birth." *Family Relations* 60:586–601.

Kan, Marni L., Susan M. McHale, and Ann C. Crouter. 2008. "Interparental Incongruence in Differential Treatment of Adolescent Siblings: Links with Marital Quality." *Journal of Marriage and Family* 70(May):466–479.

Kane, Emily W. 2012a. *The Gender Trap: Parents and the Pitfalls of Raising Boys and Girls.* New York: New York University Press.

Kane, Jennifer B. 2016. "Marriage Advantages in Prenatal Health: Evidence of Marriage Selection or Marriage Protection?" *Journal of Marriage and Family* 78(1):212–229.

Kane, Jennifer B., S. Philip Morgan, Kathleen Mullan Harris, and David K. Guilkey. 2013. "The Educational Consequences of Teen Childbearing." *Demography* 50:2129–2150.

Kane, Jennifer B., Timothy J. Nelson, and Kathryn Edin. 2015. "How Much In-Kind Support Do Low-Income Nonresident Fathers Provide? A Mixed-Method Analysis." *Journal of Marriage and Family* 77:591–611.

Kann, Laura, Steve Kinchen, Shari L. Shanklin, Katherine H. Flint, Joseph Kawkins, William A. Harris, Richard Lowry, et al. 2014. Youth Risk Behavior Surveillance—United States, 2013. *MMWR Surveillance Summary* 63:1–168.

Kann, Laura, Tim McManus, William A. Harris, Shari L. Shanklin, Katherine H. Flint, Barbara Queen, Richard Lowry et al. 2018. "Youth Risk Behavior Surveillance—United States, 2017." *MMWR Surveillance Summaries* 67:1.

Kao, Grace, Kelly Stamper Balistreri, and Kara Joyner. 2018. " Asian American Men in Romantic Dating Markets." *Contexts* 17:48–53.

Kantrowitz, Barbara. 2000. "Busy around the Clock." *Newsweek,* July 17, pp. 49–50.

Kantrowitz, Barbara, and Karen Springen. 2005. "A Peaceful Adolescence." *Newsweek,* April 25, pp. 58–61.

Kantrowitz, Barbara, and Peg Tyre. 2006. "The Fine Art of Letting Go." *Newsweek,* May 22, pp. 49–61.

Kaplan, Andreas M., and Haenlein Michael (2010). "Users of the World, Unite! The Challenges and Opportunities of Social Media." *Business Horizons* 53:61.

Kappeler, Evelyn M., and Amy Feldman Farb. 2014. "Historical Context for the Creation of the Office of Adolescent Health and the Teen Pregnancy Prevention Program." *Journal of Adolescent Health* 54:S3–S9.

Kapur, Devesh, and John McHale. 2009. "International Migration and the World Income Distribution."*Journal of International Development* 21:1102–1110.

Karney, Benjamin. 2015. "Stress is Bad for Couples, Right?" *Family Focus* (FF66, Winter):F2. Minneapolis: National Council on Family Relations.

Karney, Benjamin R., David S. Loughran, and Michael S. Pollard. 2012. "Comparing Marital Status and Divorce Status in Civilian and Military Populations." *Journal of Family Issues* 33(12):1572–1594.

Karney, Benjamin R., and Thomas N. Bradbury. 2020. "Research on Marital Satisfaction and Stability in the 2010s: Challenging Conventional Wisdom." *Journal of Marriage and Family* 82: 100–116.

Karraker, Amelia, and John DeLamater. 2013. "Past-Year Sexual Inactivity among Older Married Persons and Their Partners." *Journal of Marriage and Family* 75:142–163.

Karraker, Amelia, and Kenzie Latham. 2015. "In Sickness and in Health? Physical Illness as a Risk Factor for Marital Dissolution in Later Life." *Journal of Health and Social Behavior* 56:420–435.

Karasik, Rona J., and Raeann R. Hamon. 2007. "Cultural Diversity and Aging Families." pp. 136–153 in *Cultural Diversity and Families,* edited by Bahira Sherif Traskand Raeann R. Hamon. Thousand Oaks, CA: Sage.

Karimzadeh, Mohammad Ali, and Sedigheh Ghandi. 2008. "Early Marriage: a Policy for Infertility Prevention." *Journal of Family and Reproductive Health* 2(2):61–64.

Karre, Jennifer K., and Nina S. Mounts. 2012. "Nonresidents' Fathers' Parenting Style and the Adjustment of Late Adolescent Boys." *Journal of Family Issues* 33(12):1642–1657.

Kastbom, Åsa A., Gunilla Sydsjö, Marie Bladh, Gisela Priebe, and Carl-Göran Svedin. 2015. "Sexual Debut Before the Age of 14 Leads to Poorer Psychosocial Health and Risky Behaviour in Later Life." *Acta Paediatrica* 104(1):91–100.

Kate, Brooke Whitfield, and Jennifer Manlove. 2019. "Differences in Reproductive Knowledge, Attitudes, and Intentions Among Adolescents by Pre-sexual Experience." Retrieved October 28, 2019 (https://www.childtrends.org/wp-content /uploads/2019/10/HTNBrief_Pre-sex -behaviors_ChildTrends_September2019-1 .pdfWelte).

Katz, Jennifer, Vanessa Tirone, and Erika van der Kloet. 2012. "Moving In and Hooking Up: Women's and Men's Casual Sexual Experiences During the First Two Months of College." *Electronic Journal of Human Sexuality* 15. Retrieved October 19, 2012 (http://www.ejhs.org /volume15 /Hookingup.html).

Katz, Jonathan Ned. 2007. *The Invention of Heterosexuality.* Chicago: University of Chicago Press.

Kaufman, Gayle, and Eva Bernhardt. 2012. "His and Her Job: What Matters Most for Fertility Plans and Actual Childbearing?" *Family Relations* 61(4):686–697.

Kaufman, Gayle, and Peter Uhlenberg. 2000. "The Influence of Parenthood on the Work Effort of Married Men and Women." Social Forces 78:931–949.

Kaufman, Leslie. 2006. "Facing Hardest Choice in Child Safety, New York Tilts to Preserving Families." *The New York Times,* February 4.

Kaukinen, Catherine. 2004. "Status Compatibility,Physical Violence, and Emotional Abuse in Intimate Relationships." *Journal of Marriage and Family* 66:452–471.

Kaukinen, Catherine E., Silke Meyer, and Caroline Akers. 2013. "Status Compatibility and Help-Seeking Behaviors among Female Intimate Partner Violence Victims." *Journal of Interpersonal Violence* 28(3):577–601.

Kaye, Sarah. 2005. "Substance Abuse Treatment and Child Welfare: Systematic Change Is Needed." *Family Focus on . . . Substance Abuse across the Life Span:* FF25: F15–F16. Minneapolis: National Council of Family Relations.

Kaye, Kelleen, Katherine Suellentrop, and Corinna Sloup. 2009. *The Fog Zone: How Misperceptions, Magical Thinking, and Ambivalence Put Young Adults at Risk for Unplanned Pregnancy.* Washington, DC: National Campaign to Prevent Teen and Unplanned Pregnancy.

Kazyak, Emily, Brandi Woodell, Kristin Scherrer, and Emma Finken. 2018. "Law and Family Formation Among LGBQ-Parent Families." *Family Court Review* 56:364–373.

Kearney, Christopher A. 2010. *Casebook in Child Behavior Disorders,* 4th ed. Belmont, CA: Wadsworth/Cengage Learning.

Keaten, James, and Lynne Kelly. 2008. "Emotional Intelligence as a Mediator of Family Communication Patterns and Reticence." *Communication Reports* 21(2):104–116.

Kee, Caroline. 2018. "Fertility Rates Are Dropping and the Age of New Moms Is Rising across the US." *BuzzFeed News.* October 17. Retrieved February 12, 2019 (www. buzzfeednews.com).

Keely, Charles B. 1979. "The Development of U.S. Immigration Policy Since 1965." *Journal of International Affairs* 33 (2): 265–289.

Kellas, Jody Koenig, Cassandra LeClair-Underberg, and Emily Lamb Normand. 2008. "Stepfamily Address Terms: 'Sometimes They Mean Something and Sometimes They Don't.'" *Journal of Family Communication* 8:238–263.

Kelley, Karen Dunn. 2018. *Subjective Well-Being of Eldercare Providers: 2012–2013.* Current Population Survey Reports. February. U.S. Department of Commerce. Report P23-215.

Kelly, Joan B. 2007. "Children's Living Arrangements Following Separation and Divorce: Insights from Empirical and Clinical Research." *Family Process* 46(1):35–52.

Kelly, Joan B., and Michael E. Lamb. 2003. "Developmental Issues in Relocation Cases Involving Young Children: When, Whether, and How?" *Journal of Family Psychology* 17:193–205.

Kelly, Katherine Patterson, and Lawrence Ganong. 2011a. "Moving to Place: Childhood Cancer Treatment Decision Making in Single-Parent and Repartnered Family Structures." *Qualitative Health Research* 21(3):349–364.

———. 2011b. "Shifting Family Boundaries' After the Diagnosis of Childhood Cancer in Stepfamilies." *Journal of Family Nursing* 17(1):105–132.

Kelly, Maura. 2009. "Women's Voluntary Childlessness: A Radical Rejection of Motherhood?" *WSQ: Women's Studies Quarterly* 37(3/4):157–172.

Kemp, Candace L. 2007. "Grandparent-Grandchild Ties: Reflections on Continuity and Change across Three Generations." *Journal of Family Issues* 28(7):855–881.

Kempe, C. Henry, Frederic N. Silverman, Brandt F. Steele, William Droegemuller, and Henry K. Silver. 1962. "The Battered Child Syndrome." *Journal of the American Medical Association* 181:17–24.

Kennedy, Kimberley D., and Harriett D. Romo. 2013. "'All Colors and Hues': An Autobiography of a Multiethnic Family's Strategies for Bilingualism and Multiculturalism." *Family Relations* 62(1):109–124.

Kenney, Catherine. 2006. "The Power of the Purse: Allocative Systems and Inequality in Couple Households." *Gender and Society* 20(3):354–381.

Kenney, Catherine, and Sara S. McLanahan. 2006. "Why Are Cohabiting Relationships More Violent than Marriages?" *Demography* 43(1):127–140.

Kennedy, Denise E., and Laurie Kramer. 2008. "Improving Emotion Regulation and Sibling Relationship Quality: The More Fun with Sisters and Brothers Program." *Family Relations* 57(December):567–578.

Kennedy, G. E., and C. E. Kennedy. 1993. "Grandparents: A Special Resource for Children in Stepfamilies." *Journal of Divorce and Remarriage*, 19:45–68.

Kennedy, Sheela, and Steven Ruggles. 2014. "Breaking Up Is Hard to Count: The Rise of Divorce in the United States, 1980–2010." *Demography* 51:587–598.

Kephart, William. 1971. "Oneida: An Early American Commune." pp. 481–492 in *Family in Transition: Rethinking Marriage, Sexuality, Child Rearing, and Family Organization,* edited by Arlene S. Skolnick and Jerome H. Skolnick. Boston: Little, Brown.

Kern, Louis J. 1981. *An Ordered Love: Sex Roles and Sexuality in Victorian Utopias—The Shakers, the Mormons, and the Oneida Community.* Chapel Hill. University of North Carolina Press.

Kerpelman, Jennifer. 2014. "Healthy Adolescent Relationships." *National Council on Family Relations Report* 59.4:F1–F2. Minneapolis: National Council on Family Relations.

Kershaw, Sarah. 2003. "Many Immigrants Decide to Embrace Homes for Elderly." *The New York Times,* October 20, pp. A1–A10.

Khaleeli, Homa. 2014. "Hijra: India's Third Gender Claims Its Place in Law." *The Guardian,* April 16. Retrieved October 8, 2015 (http://www/theguardian.com).

Keskivaara, Anna-liisa Japvenpaa, and Timo E. Strandberg.2006. "Continuity of Temperament from Infancy to Middle Childhood." *Infant Behavior and Development* 29(4):494–508.

Keshner, Andrew. 2019. "Child-care Costs in America Have Soared to Nearly $10K Per Year." Retrieved December 14, 2019 (https://www.marketwatch.com/story/child-care-costs-just-hit-a-new-high 2018 10 22).

Kettrey, Heather Hensman, and Robert A. Marx. 2019. "The Effects of Bystander Programs on the Prevention of Sexual Assault Across the College Years: A Systematic Review and Meta-analysis." *Journal of Youth and Adolescence* 48:212–227.

Khazan, Olga. 2014. "Why Most Brazilian Women get C-Sections." *The Atlantic.* Retrieved February 14, 2016 (http://www.theatlantic.com/health/archive/2014/04/why-most-brazilian-women-get-c-sections/360589/)

Kheshgi-Genovese, Zareena, and Thomas A. Genovese. 1997. "Developing the Spousal Relationship within Stepfamilies." *Families in Society* 78(3):255–264.

Khrais, Reema. 2012. "Phone Home: Tech Draws Parents, College Kids Closer." Capitol Public Radio, September 25.

Kibria, Nazli. 2000. "Race, Ethnic Options,and Ethnic Binds: Identity Negotiations of Second-Generation Chinese and Korean Americans." *Sociological Perspectives* 43(1):77–95.

———. 2007. "Vietnamese Americans and the Rise of Women's Power." pp. 220–227 in *Race, Class, and Gender: An Anthology,* 6th ed., edited by Margaret L. Andersen and Patricia Hill Collins. Belmont, CA: Wadsworth.

Kids Count. 2019. "Children in Foster Care in the United States." Retrieved December 9, 2019 (https://datacenter.kidscount.org).

Kiecolt, K. Jill, Rosemary Blieszner, and Jyoti Savla. 2011. "Long-Term Influences of Intergenerational Ambivalence on Midlife Parents' Psychological Well-Being." *Journal of Marriage and Family* 73(2):369–382.

Kiernan, K. 1992. "The Impact of Family Disruption in Childhood on Transitions Made in Young Adult Life." *Population Studies* 46:213–234.

Kiernan, Kathleen. 2002. "Cohabitation in Western Europe: Trends, Issues, and Implications." pp. 3–31 in *Just Living Together: Implication of Cohabitation on Families, Children, and Social Policy,* edited by Alan Boothand A. C.Crouter. Mahwah, NJ:Erlbaum.

Kiersz, Andy. 2014. "Seven Stats We Just Learned About Same-Sex Couples." *Business Insider,* September 20. Retrieved October 7, 2015 (http://www.busines-sinsider.com).

Kilen, Mike. 2011. "'Kardashian Marriages' Rare but Aren't Unheard of in Iowa." *Des Moines Register.* Retrieved April 19, 2013 (http://www.public.iastate.edu/~nscentral/mr/11/1111/kardashian .html).

Killewald, Alexandra. 2011. "Opting Out and Buying Out: Wives' Earnings and Housework Time." *Journal of Marriage and Family* 73:459–471.

———. 2013. "A Reconsideration of the Fatherhood Premium: Marriage, Coresidence, Biology, and Fathers' Wages." *American Sociological Review* 78:96–116.

Killewald, Alexandra. 2016. "Money, Work, and Marital Stability: Assessing Change in the Gendered Determinants of Divorce." *American Sociological Review* 81:696–719.

Killian, Timothy, and Lawrence H. Ganong. 2002. "Ideology, Context, and Obligations to Assist Older Persons." *Journal of Marriage and Family* 64(4):1080–1088.

Killoren, Sarah E., Shawna M. Thayer, and Kimberly A. Updegraff. 2008. "Conflict Resolution Between Mexican Origin Adolescent Siblings." *Journal of Marriage and Family* 70(December):1200–1212.

Killoren, Sarah, Kimberly Updegraff, F. Scott Christopher, and Adriana J. Umana-Taylor. 2011. "Mothers, Fathers, Peers, and Mexican-Origin Adolescents' Sexual Intentions." *Journal of Marriage and Family* 73(1):209–220.

Kilmer, Beau, Nancy Nicosia, Paul Heaton, and Greg Midgette. 2013. "Efficacy of Frequent Monitoring." *American Journal of Public Health* 103(1):E37–E43.

Kim, Hyoun K., Deborah M. Capaldi, and Lynn Crosby. 2007. "Generalizability of Gottman and Colleagues' Affective Process Models of Couples' Relationship Outcomes." *Journal of Marriage and Family* 69(1):55–72.

Kim, Hyoun K., and Patrick C. McKenry. 2002. "The Relationship between Marriage and Psychological Well-Being." *Journal of Family Issues* 23(8):885–911.

Kim, Hyun Sik. 2011. "Consequences of Divorce for Child Development." *American Sociological Review* 76(3):487–511.

Kim, Ji-Yeon, Susan M. McHale, Ann C. Crouter, and D. Wayne Osgood. 2007. "Longitudinal Linkages Between Sibling Relationships and Adjustment from Middle Childhood through Adolescence." *Developmental Psychology* 43(4):960–973.

Kim, Jeounghee. 2012. "Educational Differences in Marital Dissolution: Comparison of White and African American Women." *Family Relations* 61:811–824.

Kim, Joongbaeck, and Hyeyoung Woo. 2011. "The Complex Relationship Between Parental Divorce and Sense of Control." *Journal of Family Issues* 32(8):1050–1072.

Kimbro, Rachel Tolbert. 2008. "Together Forever? Romantic Relationship Characteristics and Prenatal Health Behaviors." *Journal of Marriage and Family* 70(3):756–757.

Kimbro, Rachel Tolbert, and Ariela Schachter. 2011. "Neighborhood Poverty and Maternal Fears of Children's Outdoor Play." *Family Relations* 60(4)(October):461–475.

Kimerling, Rachel, Jennifer Alvarez, Joanne Pavao, Katelyn P. Mack, Mark W. Smith, and Nikki Baumrind. 2009. "Unemployment among Women: Examining the Relationship of Physical and Psychological Intimate Partner Violence and Posttraumatic Stress Disorder." *Journal of Interpersonal Violence* 24(3):450–463.

Kimmel, Michael S. 1995. "Misogynists, Masculinist Mentors, and Male Supporters: Men's Responses to Feminism." pp. 561–572 in *Women: A Feminist Perspective*, 5th ed., edited by Jo Freeman. Mountain View, CA: Mayfield.

Kimmel, Michael S., and Michael A. Messner. 1998. *Men's Lives.*4th ed. Boston: Allyn & Bacon.

King, Valarie. 2007. "When Children Have Two Mothers: Relationships with Nonresident Mothers, Stepmothers, and Fathers." *Journal of Marriage and Family* 69(5):1178–1193.

———. 2009. "Stepfamily Formation: Implications for Adolescent Ties to Mothers, Nonresident Fathers, and Stepfathers." *Journal of Marriage and Family* 71(November):954–968.

King, Valarie, Paul R. Amato, and Rachel Lindstrom. 2015. "Stepfather–Adolescent Relationship Quality During the First Year of Transitioning to a Stepfamily." *Journal of Marriage and Family* 77:1179–1189.

King, Valarie, and Rachel Lindstrom. 2016. "Continuity and Change in Stepfather–Stepchild Closeness Between Adolescence and Early Adulthood." *Journal of Marriage and Family* 78:730–743.

King, Valarie, Katherine Stamps Mitchell, and Daniel N. Hawkins. 2010. "Adolescents with Two Nonresident Biological Parents: Living Arrangements, Parental Involvement, and Well-Being." *Journal of Family Issues* 31(1):3–30.

King, Valarie, and Juliana M. Sobolewski. 2006. "Nonresident Fathers' Contributions to Adolescent Well-Being." *Journal of Marriage and Family* 68(3):537–557.

Kingsbury, Kathleen Burns. 2014. "The Economic Power of Women." *Financial Planning*, February 18. New York: SourceMedia.

Kingston, Anne. 2009. "No Kids, No Grief." *Maclean's* 122(29/30):38–41.

Kinkler, Lori A., and Abbie E. Goldberg. 2011. "Working with What We've Got: Perceptions of Barriers and Supports among Small-Metropolitan-Area Same-Sex Adopting Couples." *Family Relations* 60(4): 387–403.

Kinsey, Alfred, Wardell B. Pomeroy, and Clyde E. Martin. 1948. *Sexual Behavior in the Human Male.* Philadelphia, PA: Saunders.

———. 1953. *Sexual Behavior in the Human Female.* Philadelphia, PA: Saunders.

Kirby, Douglas, B. A. Laris, and Lori Rolleri. 2006. *Sex and HIV Education Programs for Youth: Their Impact and Important Characteristics*, May 13. Research Triangle Park, NC: Family Health International.

Kirmeyer, Sharon E., and Brady E. Hamilton. 2011. "Childbearing Differences among Three Generations of U.S. Women." *NCHS Data Brief, No. 68, August.* U.S. Department of health and Human Services, Centers for Disease Control and Prevention, National Center for health Statistics.

Kiselica, Mark S., and Matt Englar-Carlson. 2010. "Identifying, Affirming, and Building upon Male Strengths: The Positive/Psychology Positive/Masculinity Model of Psychotherapy with Boys and Men." *Psychotherapy: Theory, Research, Practice, and Training* 47(3):276–287.

Kiselica, Mark S., and Mandy Morrill-Richards. 2007. "Sibling Maltreatment: The Forgotten Abuse." *Journal of Counseling and Development* 85(2):148–61.

Kisler, Tiffani S., and F. Scott Christopher. 2008. "Sexual Exchanges and Relationship Satisfaction: Testing the Role of Sexual Satisfaction as a Mediator and Gender as a Moderator." *Journal of Social and Personal Relationships* 25(4):587–602.

Kislev, Elyakim. 2019. *Happy Singlehood: The Rising Acceptance and Celebration of Solo Living.* Oakland, CA: University of California Press.

Kjobli, John, and Kristine Amlund Hagen. 2009. "A Mediation Model of Interparental Collaboration, Parenting Practices, and Child Externalizing Behavior in a Clinical Sample." *Family Relations* 58(July):275–288.

Klaassen, Marleen J. E., and Jochen Peter. 2014. "Gender (In)equality in Internet Pornography: A Content Analysis of Popular Pornographic Internet Videos." *Journal of Sex Research* 10:1080/00224499.2014.976781.

Klein, Rebecca. 2018. "Teachers Are Serving as First Responders to the Opioid Crisis." November 3. *Huffpost News. Huffington Post.* Retrieved December 10, 2019 (www.huffpost.com).

Kleinfield, N. R. 2003. "Around Tree, Smiles Even for Wives No. 2 and 3." *The New York Times*, December 24.

———. 2011. "Baby Makes Four, and Complications." *The New York Times*, June 19. Retrieved September 13, 2011 (http://www.nytimes.com).

Klein, Rebecca. 2014. "This Is Why 12 Percent of High School Graduates Don't Go to College." *The Huffington Post*, September 29. Retrieved October 10, 2015 (http://www.huffingtonpost.com).

Kleinplatz, P. J., A. D. Me'nard, N. Paradis, M. Campbell, T. Dalgleish, A. Segovia, and K. Davis. 2009. "From Closet to Reality: Optimal Sexuality among the Elderly." *The Irish Psychiatrist* 10:15–18.

Kliff, Sarah. 2010. "Outing Abortion, from Town Halls to Twitter." *Newsweek*, March 3. www.newsweek.com.

Klimstra, Theo A., Koen Luyckx, Susan Branje, Eveline Teppers, Luc Goossens, and Wim H. J. Meeus. 2013. "Personality Traits, Interpersonal Identity, and Relationship Stability: Longitudinal Linkages in Late Adolescence and Young Adulthood." *Journal of Youth and Adolescence* 42:1661–1673.

Klinenberg, Eric. 2012. *Going Solo: The Extraordinary Rise and Surprising Appeal of Living Alone.* New York: Penguin.

Klinenberg, Eric. 2016. "Social Isolation, Loneliness, and Living Alone: Identifying the Risks for Public Health." *American Journal of Public Health* 1106:786–787.

Klinenberg, Eric, Stacy Torres, and Elena Portacolone. 2012. "Aging Alone in America." A Briefing Paper Prepared for the Council on Contemporary Families for Older Americans Month May 2012. Retrieved May 21, 2013 (http://www.contemporaryfamilies.org/aging/aging.html).

Klohnen, E. C., and S. J. Bera. 1998. "Behavioral and Experiential Patterns of Avoidantly and Securely Attached Women Across Adulthood: A 30-Year Longitudinal Perspective." *Journal of Personality and Social Psychology* 74:211–223.

Klohnen, Eva C., and Gerald A. Mendelsohn. 1998. "Partner Selection for Personality Characteristics: A Couple-Centered Approach." *Personality and Social Psychology Bulletin* 24(3):268–277.

Kluger, Jeffrey. 2004. "The Power of Love." *Time*, January 19.

Kluwer, Esther S. 2011. "Psychological Perspectives on Gender Deviance Neutralization." *Journal of Family Theory and Review* 3(1): 14–17.

Knapp, Jennie, Briana Muller, and Alicia Quros. 2009. *Women, Men, and the Changing Roles of Gender in Immigration.* Notre Dame, IN: Institute for Latin Studies. Retrieved October 9, 2015 (https://latinostudies.nd.edu).

Kneip, Thorsten, Gerrit Bauer, and Steffen Reinhold. 2014. "Direct and Indirect Effects of Unilateral Divorce Law on Marital Stability." *Demography* 51:2103–2126.

Knight, George, Cady Berkel, Adriana Umana-Taylor, Nancy Gonzales, Idean Ettekal, Maryanne Jaconis, and Brenna Boyd. 2011. "The Familial Socialization of Culturally Related Values in Mexican American Families." *Journal of marriage and Family* 73(5): 913–925.

Knobloch, Leanne, and Lynne Knobloch-Fedders. 2010. "The Role of Relational Uncertainty in Depressive Symptoms and Relationship Quality: An Actor-Partner Interdependence Model." *Journal of Social and Personal Relationships* 27(1):137–159.

Knobloch, Leanne K., Denise H. Solomon, and Jennifer A. Theiss. 2013. "Experiences of U.S. Military Couples During the Post-Deployment Transitions: Applying the Relational Turbulence Model." *Journal of Social and Personal Relationships* 29(4): 423–450.

Knoester, Chris, and Alan Booth. 2000. "Barriers to Divorce: When Are They Effective? When Are They Not?" *Journal of Family Issues* 21:78–99.

Knoll, Melissa, Christopher Tamborini, and Kevin Whitman. 2012. "I Do . . . Want to Save: Marriage and Retirement Savings in Young Households." *Journal of Marriage and Family* 74(1):86–100.

Knop, B., and Karin Brewster, K. 2016. "Family Flexibility in Response to Recession: Fathers' Involvement in Child Care Tasks." *Journal of Marriage and Family*, 78(2): 283–292.

Knox, David, and Marty E. Zusman. 2009. "Sexuality in Black and White: Data from 783 Undergraduates." *Electronic Journal of Human Sexuality* 12, June 26 (www.ejhs.org).

Knowlton, Brian. 2010. "Muslim Women Gain Higher Profile in U.S." *The New York Times,* December 27. Retrieved December 31, 2010 (http://www.nytimes.com).

Knudson-Martin, Carmen. 2012. "Attachment in Adult Relationships: A Feminist Perspective." *Journal of Family Theory and Review* 4(4): 299–305.

Knudson-Martin, Carmen, and Martha J. Laughlin. 2005. "Gender and Sexual Orientation in Family Therapy: Toward a Postgender Approach." *Family Relations* 54(1):101–115.

Knudson-Martin, Carmen, and Anne Rankin Mahoney. 2009. *Couples, Gender, and Power: Creating Change in Intimate Relationships.* New York:Springer.

Kobliner, Beth. 2010. "Guiding a Child to Financial Independence." *The New York Times,* November 4. Retrieved January 9, 2011 (http://www.nytimes.com).

Koch, Wendy. 2009. "Savings Plan Benefits Teens Leaving Foster Care." *USA Today,* June 15. Retrieved February 10, 2010 (www.usatoday.com).

Koerner, Claudia. 2019. "Kids Describe in Their Own Words the Dire Conditions Inside a Border Detention Center. June 27. Retrieved October 8, 2019 (https://www.theguardian.com/us-news/2019/jun/21/child-migrant-facilities-shocking-conditions-revealedCenter).

Kontula, Osmo, and Anneli Miettinen. 2016. "Determinants of Female Sexual Orgasms." *Socioaffective Neuroscience & Psychology* 6:31624.

Koesten, Joy, Paul Schrodt, and Debra Ford. 2009. "Cognitive Flexibility as a Mediator of Family Communication Environments and Young Adults' Well-Being." *Health Communication* 24:82–94.

Kogan, Steven M., Leslie G. Simons, Yi fu Chen, Stephanie Burwell, and Gene H. Brody. 2013. "Protective Parenting, Relationship Power Equity, and Condom Use among Rural African American Emerging Adult Women." *Family Relations* 62:341–353.

Kohl, Gillian Ferris. 2012. "Keeping Kids Connected with Their Jailed Parents." National Public Radio. Retrieved July 18, 2012 (http://www.npr.org).

Kohlberg, Lawrence. 1966. "A Cognitive–Developmental Analysis of Children's Sex-Role Concepts and Attitudes." pp. 82–173 in *The Development of Sex Differences,* edited by Eleanor E. Maccoby. Palo Alto, CA: Stanford University Press.

Kohut, Taylor, Jodie L. Baer, and Brendan Watts. 2015. "Is Pornography Really about 'Making Hate to Women'? Pornography Users Hold More Gender Egalitarian Attitudes Than Nonusers in a Representative American Sample." *Journal of Sex Research* August 26:1–11.

Kolata, Gina. 2009. "Picture Emerging on Genetic Risks of IVF." *The New York Times,* February 17, D1.

Konigsberg, Ruth Davis. 2011. "Chore Wars." *Time,* August 8. Retrieved February 4, 2012 (http://www.time.com).

Kontula, Osmo, and Elina Haavio-Mannila. 2009. "The Impact of Aging on Human Sexual Activity and Sexual Desire." *Journal of Sex Research* 46(1):46–56.

Kools, S. 2008. "From Heritage to Postmodern Grounded Theorizing: Forty Years of Grounded Theory." *Studies in Symbolic Interaction* 32:73–86.

Kornrich, Sabino, and Allison Roberts. 2018. "Household Income, Women's Earnings, and Spending on Household Services, 1980–2010." 2018. *Journal of Marriage and Family* 80 (1): 150–165.

Kornicha, Sabino, Julie Brines, and Katrina Leupp. 2013. "Egalitarianism, Housework, and Sexual Frequency in Marriage." *American Sociological Review* 78(1):26–50.

Koro-Ljungberg, Mirka, and Regina Bussing. 2009. "The Management of Courtesy Stigma in the Lives of Families with Teenagers with ADHD." *Journal of Family Issues* 30(9):1175–1200.

Koropeckyj-Cox, Tanya, and Gretchen Pendell. 2007. "The Gender Gap in Attitudes about Childlessness in the United States." *Journal of Marriage and Family* 69(4):899–915.

Kosanovich, Karen. 2018. "Workers in Alternative Employment Arrangements." U.S. Bureau of Labor Statistics. November. Retrieved February 10, 2019 (www.bls.gov).

Koslow, Sally. 2012. *Slouching Toward Adulthood: Observations from the Not-So-Empty Nest.* New York: Penguin Books.

Kossinets, Gueorgi, and Duncan J. Watts. 2009. "Origins of Homophily in an Evolving Social Network." *American Journal of Sociology* 115(2):405–450.

Kost, Kathryn, and Laura Lindberg. 2015. "Pregnancy Intentions, Maternal Behaviors, and Infant Health: Investigating Relationships with New Measures and Propensity Score Analysis." *Demography* 52:83–111.

Kotila, Letitia E., Sarah J. Schoppe-Sullivan, and Claire M. Kamp Dush. 2013. "Time Parenting Activities in Duel-Earner Families at the Transition to Parenthood." *Family Relations* 62(5):795–807.

Kouros, Chrystyna, Christine Merrilees, and E. Mark Cummings. 2008. "Marital Conflict and Children's Emotional Security in the Context of Parental Depression." *Journal of Marriage and Family* 70(3):684–697.

Kramer, Elena. 2011. "Where the Women Are—And Aren't." *Harvard Magazine, January*–February. Retrieved February 11, 2013 (http://www.harvardmagazine.com).

Kowarski, Ilana. 2018. "How Gender Influences College Admissions." Retrieved October 4, 2019 (https://www.usnews.com/education/best-colleges/articles/2018-11-02/how-gender-influences-college-admissions).

Kramer, Sarah. 2011. "Three Generations Under One Roof." *The New York Times,* September 23. Retrieved September 27, 2011 (http://www.nytimes.com).

Krampe, Edythe M. 2009. "When Is the Father Really There? A Conceptual Reformulation of Father Presence." *Journal of Family Issues* 30(7):875–897.

Krampe, Edythe M., and Rae R. Newton. 2012. "Reflecting on the Father: Childhood Family Structure and Women's Paternal Relationships." *Journal of Family Issues* 33(6):773–800.

Kreager, Derek A., Shannon E. Cavanagh, John Yen, and Mo Yu. 2014. "'Where Have All the Good Men Gone?' Gendered Interactions in Online Dating." *Journal of Marriage and Family* 76.907–110.

Kreider, Rose M. 2003. *Adopted Children and Stepchildren: 2000.* Census 2000 Special Reports CENSR-6, August. Washington, DC: U.S. Census Bureau.

Kreider, Rose M., and Renee Ellis. 2011a. *Living Arrangements of Children,* June.Washington, DC: U.S. Census Bureau, Department of Commerce.

Kreider, Rose M., and Jason M. Fields. 2005. *Living Arrangements of Children: 2001.*Current Population Reports P70-104.Washington, DC: U.S. Census Bureau.

Kreider, Rose, and Daphne A. Lofquist. 2014. *Adopted Children and Stepchildren, P20-572.* Washington, DC: United States Census Bureau.

Kreider, Tim. 2012. "The 'Busy' Trap." *The New York Times,* June 30. Retrieved May 8, 2015 (http://opinionator.blogs.nytimes.com).

Kringelbach, Morten L., and Kent C. Berridge. 2015. "Motivation and Pleasure in the Brain." *The Psychology of Desire.* 129.

Krogstad, Jens Manuel, and Jeffrey S. Passel. 2015. *Five Facts about Illegal Immigration in the U.S.* Pew Research Center, July 24.

Kruger, Daniel J., and Susan M. Hughes. 2011. "Tendencies to Fall Asleep First after Sex Are Associated with Greater Partner Desires for Bonding and Affection." *Journal of Social, Evolutionary, and Cultural Psychology* 5(4):239.

Kruk, E. 2015. "What Exactly is the 'Best Interest of the Child?'" *Psychology Today.* Retrieved March 30, 2020 (https://www.psychologytoday.com/us/blog/co-parenting-after-divorce/201502/what-exactly-is-the-best-interest-the-child).

Kruk, Edward. 2015. "The Lived Experiences of Non-Custodial Parents in Canada: A Comparison of Mothers and Fathers." *International Journal for Family Research and Policy* 1.

Krumrei, Elizabeth J., Annette Mahoney, and Kenneth I. Pargament. 2011. "Demonization of Divorce: Prevalence Rates and Links to Postdivorce Adjustment." *Family Relations* 60:90–103.

Kulkin, Heidi S., June Williams, Heath F. Borne, Dana de la Bretonne, Judy Laurendine. 2008. "A Review of Research on Violence in Same-Gender Couples." *Journal of Homosexuality* 53(4):71–87.

Kulu, Hill. 2014. "Marriage Duration and Divorce: The Seven-Year Itch or a Lifelong Itch?" *Demography* 51:881–893.

Kumpfer, Karol, and Connie Tait. 2000. *Family Skills Training for Parents and Children.* Juvenile Justice Bulletin, April.

Kuperberg, A. 2014. "Age at Coresidence, Premarital Cohabitation, and Marriage Dissolution: 1985–2009." *Journal of Marriage and Family:* 352–369.

Kuperberg, Arielle. 2012. "Reassessing Differences in Work and Income in

Cohabitation and Marriage." *Journal of Marriage and Family* 74(4): 688–707.

Kuperberg, Arielle, and Alicia M. Walker. 2018. "Heterosexual College Students who Hookup with Same-Sex Partners." *Archives of Sexual Behavior* 47:1387–1403.

Kuperminc, Gabriel P., Gregory J. Jurkovic, and Sean Casey. 2009. "Relation of Filial Responsibility to the Personal and Social Adjustment of Latino Adolescents from Immigrant Families." *Journal of Family Psychology* 23(1):14–22.

Kurdek, Lawrence A. 1994. "Areas of Conflict for Gay, Lesbian, and Heterosexual Couples: What Couples Argue about Influences Relationship Satisfaction." *Journal of Marriage and Family* 56(4):923–934.

Kurdek, Lawrence A., and Mark A. Fine. 1991. "Cognitive Correlates of Satisfaction for Mothers and Stepfathers in Stepfather Families." *Journal of Marriage and Family* 53(3):565–572.

Khurshid, Ayesha. 2015. "Islamic Traditions and Modernity: Gender, Class, and Islam in a Transnational Women's Education Project. *Gender and Society* 29(1):98–121.

Kurtz, Stanley. 2006. "Big Love, from the Set." *National Review*, March 13.

Kurz, Demie. 1993. "Physical Assaults by Husbands: A Major Social Problem." pp. 88–103 in *Current Controversies on Family Violence*, edited by Richard J. Gelles and Donileen R. Loseke. Newbury Park, CA: Sage.

Kushlev, Kostadin, Jason D.E. Proulx, and Elizabeth W. Dunn. 2017. "Digitally Connected, Socially Disconnected: The Effects of Relying on Technology Rather Than Other People." *Computers in Human Behavior* 76:68–74.

Kushlev, Kostadin, John F. Hunter, Jason Proulx, Sarah D. Pressman, and Elizabeth Dunn. 2019. "Smartphones Reduce Smiles between Strangers." *Computers in Human Behavior* 91:12–16.

Kushlev, Kostadin, Ryan Dwyer, and Elizabeth W. Dunn. 2019. "The Social Price of Constant Connectivity: Smartphones Impose Subtle Costs on Well-Being." *Current Directions in Psychological Science* 28(4)347–352.

Kutner, Lawrence. 1988. "Parent and Child: Working at Home; or, The Midday Career Change." *The New York Times*, December 8.

Kuvalanka, Katherine A., Samuel H. Allen, Cat Munroe, Abbie E. Goldberg, and Judith L. Weiner. 2018. "The Experiences of Sexual Minority Mothers with Trans* Children." *Family Relations* 67 (1): 70–87.

Kwiatkowski, Joelle, Nancy A. Burrell, Lindsay Timmerman, and Mike Allen. 2014. "Meta-Analysis of Sex Differences in Process and Outcome." *Managing Interpersonal Conflict: Advances Through Meta-Analysis* 140.

LaBorde, Monique. 2019. "'I Didn't Think It Was Possible': North Carolina City Rings in Its First LGBTQ Pride." National Public Radio. Retrieved October 28, 2019 (www .npr.org).

Lacey, Rachel Saul, Alan Reifman, Jean Pearson Scott, Steven M. Harris, and Jacki Fitzpatrick. 2004. "Sexual-Moral Attitudes, Love Styles, and Mate Selection." *Journal of Sex Research* 41(2):121–129.

Lacy, Karyn R. 2007. *Blue-Chip Black: Race, Class, and Status in the New Black Middle Class*. Berkeley: University of California Press.

Laczko, Frank. 2010. *Migration, Environment, and Climate Change: Assessing the Evidence*. Geneva: International Organization for Migration.

Lahaie, Claudia, Alison Earle, and Jody Heymann. 2012. "An Uneven Burden: Social Disparities in Adult Caregiving Responsibilities, Working Conditions, and Caregiver Outcomes." *Research on Aging* 35(3):243–274.

Lam, Bourree. 2015. "Nine to Five, After 65." *The Atlantic*, April 22. Retrieved May 8, 2015 (http://www.theatlantic.com).

Lam, Brian Trung. 2005. "Self-Esteem among Vietnamese American Adolescents: The Role of Self-Construal, Family Cohesion, and Social Support." *Journal of Ethnic and Cultural Diversity in Social Work* 14(3/4):21–34.

Lam, Ching Man, and Wai Man Kwong. 2012. "The 'Paradox of Empowerment' in Parent Education: A Reflexive Examination of Parents' Pedagogical Expectations." *Family Relations* 61(1):65–74.

Lam, Chun, Susan McHale, and Ann Crouter. 2012. "The Division of Household Labor: longitudinal Changes and Within-Couple Variation." *Journal of Marriage and Family* 74(5):944–952.

Lam, Chun, Susan McHale, and Kimberly Updergraff. 2012. "Gender Dynamics in Mexican American Families: Connecting Mothers', Fathers', and Youths' Experiences." *Sex Roles* 67(1/2):17–28.

Lamanna, Mary Ann. 1977. "The Value of Children to Natural and Adoptive Parents." PhD dissertation, Department of Sociology, University of Notre Dame, Notre Dame, IN.

Lambert, Nathaniel M., and David C. Dollahite. 2006. "How Religiosity Helps Couples Prevent, Resolve, and Overcome Marital Conflict." *Family Relations* 55(4):439–449.

Lambert, Nathaniel M., Seth Mulder, and Frank Fincham. 2014. "Thin Slices of Infidelity: Determining Whether Observers Can Pick out Cheaters from a Video Clip Interaction and What Tips Them Off." *Personal Relationships* 21:612–619.

Lamidi, Esther. 2015. "Trends in Cohabitation" (FP15-21). Bowling Green, OH: National Center for Family and Marriage Research. Retrieved April 12, 2016 (https://www.bgsu.edu/content /dam/BGSU/college-of-arts-and-sciences / NCFMR/documents/FP/FP-13-12.pdf).

———. 2016. "Trends in Cohabitation: The Never Married and Previously Married, 1995–2014." NCFMR Family Profiles, FP-15-21. Retrieved January 25, 2016 (https://www.bgsu. edu/ncfmr /resources/data/family-profiles/lamidi -cohab-trends-never-previously-married -fp-15-21.html).

Lamidi, Esther, Susan L. Brown, and Wendy D. Manning. 2015a. "Assortative Mating: Age Heterogramy in U.S. Marriages, 1965-2014" (FP-15-14). National Center for Marriage and Family Research. Retrieved January 12, 2016 (https:// www.bgsu.edu/ncfmr /resources /data/family-profiles /lamidi-brown -manning-assortative -mating-age -heterogamy-fp-15-14.html).

Lamidi, Esther, Susan L. Brown, and Wendy D. Manning. 2015b. "Assortative Mating: Educational Heterogramy in U.S. Marriages, 1965-2014" (FP-15-15). National Center for Marriage and Family Research. Retrieved January 12, 2016 (https://www.bgsu.edu/ncfmr /resources /data/family-profiles/lamidi -brown -manning-assortative-mating-edu -homogamy-fp-15-15.html).

Lamidi, Esther, Susan L. Brown, and Wendy D. Manning. 2015c. "Assortative Mating: Racial Heterogramy in U.S. Marriages, 1965-2014" (FP-15-16). National Center for Marriage and Family Research. Retrieved January 12, 2016 (https:// www .bgsu.edu/content /dam/BGSU/college -of-arts-and-sciences/NCFMR/documents /FP /lamidi-brown-manning-assortative -mating -racial-homogamy-fp-15-16.pdf).

Lamidi, Esther, and Wendy D. Manning. 2016. *Marriage and Cohabitation Experiences Among Young Adults*. Family Profiles, FP-16-17. Bowling Green, OH: National Center for Family & Marriage Research. Retrieved November 26, 2019 (https://www.bgsu.edu/content/dam/BGSU /college-of-arts-and-sciences/NCFMR /documents/FP/lamidi-manning-marriage -cohabitation-young-adults-fp-16-17.pdf).

Lan, Pei-Chia. 2002. "Subcontracting Filial Piety: Elder Care in Ethnic Chinese Immigrant Families in California." *Journal of Family Issues* 23(7):812–835.

Landale, Nancy S., Robert Schoen, and Kimberly Daniels. 2009. "Early Family Formation among White, Black, and Mexican American Women." *Journal of Family Issues* 31(4):445–474.

Landau, Iddo. 2008. "Problems with Feminist Standpoint Theory in Science Education." *Science and Education* 17(10):1081–1088.

Landers-Potts, Melissa, K.A.S. Wickrama, Leslie Simons, Carolyn Cutrona, Frederick Gibbons, Ronald Simons, and Rand Conger. 2015. "An Extension and Moderational Analysis of the Family Stress Model Focusing on African American Adolescents." *Family Relations* 64(2):233–248.

Lane, Marian, Laurel Hourani, Robert Bray, and Jason Williams. 2012. "Prevalence of Perceived Stress and Mental Health Indicators among Reserve-Component and Active-Duty Military Personnel." *American Journal of Public Health* 102(6):1213–1220.

Lane, Wendy G., David M. Rubin, Ragin Monteith, and Cindy Christian. 2002. "Racial Differences in the Evaluation of Pediatric Fractures for Physical Abuse." *Journal of the American Medical Association* 288(13) (www.jama .org).

Lane-Steele, Laura. 2011. "Studs and Protest-Hypermasculinity: The Tomboyism with Black Lesbian Female Masculinity." *Journal of Lesbian Studies* 15(4):480–492.

Langeslag, Sandara J. E., Peter Muris, and Ingmar H. A. Franken. 2012. *Journal of Sex Research*. Retrieved May 25, 2013 (http://www .tandfonline.com/doi/abs/10.1080 /00224499.2012.714011?journalCode=hjsr 20#preview).

Langeslag, Sandra J. E., Frederik M van der Veen, and Durk Fekkes. 2012. "Blood Levels of Serotonin Are Differentially Affected by Romantic Love in Men and Women." *Federation of European Psychophysiology Societies* 26(2):92–98.

Langhinrichsen-Rohling, J., and D. Capaldi. 2012. "Clearly We've Only Just Begun: Developing Effective Prevention Programs for Intimate Partner Violence." *Prevention Science* 13(4):410–414.

Lannutti, Pamela J. 2007. "The Influence of Same-Sex Marriage on the Understanding of Same-Sex Relationships." *Journal of Homosexuality* 53(3):135–151.

Lara, Cristina. 2013. "A Letter to College Women: On (Not) Finding Your Husband." Retrieved May 30, 2013, from http://www.huffingtonpost.com /cristina-lara/a-letter-to-college-women_b_3034014 .html.

Lareau, Annette. 2003a. "The Long-Lost Cousins of the Middle Class." *The New York Times,* December 20.

———. 2003b. *Unequal Childhoods: Class, Race, and Family Life.* Berkeley: University of California Press.

———. 2006. "Unequal Childhoods: Class, Race, and Family Life." pp. 537–548 in *The Inequality Reader: Contemporary and Foundational Readings in Class, Race, and Gender,* edited by David B. Grusky and Szonja Szelenyi. Boulder, CO: Westview.

———. 2011. *Unequal Childhoods: Class, Race, and Family Life,* 2nd ed. Berkeley: University of California Press.

———. 2012. "Using the Terms Hypothesis and Variable for Qualitative Work: A Critical Reflection," *Journal of Marriage and Family* 74(4):671–677.

LaRossa, Ralph. 2009. "Single-Parent Family Discourse in Popular Magazines and Social Science Journals." *Journal of Marriage and Family* 71(2):235–239.

Larrimore, Jeff, Mario Arthur-Bentil, Sam Dodini, and Logan Thomas. 2015. *Report on the Economic Well-Being of U.S. Households in 2014.* Washington, DC: Board of Governors of the Federal Reserve System. Retrieved December 15, 2015 (http://www.federalreserve.gov).

Larrivée, Marie-Claude, Louise Hamelin Brabant, and Genevieve Lessard. 2012. "Knowledge Translation in the Field of Violence against Women and Children." *Children and Youth Services Review* 34:2381–2391.

Larson, Jeffry H., and Rachel Hickman. 2004. "Are College Marriage Textbooks Teaching Students the Premarital Predictors of Marital Quality?" *Family Relations* 53(4):385–392.

Larzelere, Robert E., and Ronald Cox Jr. 2013. "Making Valid Causal Inferences About Corrective Actions by Parents from Longitudinal Data." *Journal of Family Theory and Review* 5(4):282–299.

Lasch, Christopher. 1977. *Haven in a Heartless World: The Family Besieged.* New York: Basic Books.

———. 1980. *The Culture of Narcissism.* New York: Warner Books.

Lasen, Amparo, and Elena Casado. 2012. "Mobile Telephony and the Remediation of Couple Intimacy." *Feminist Media Studies* 12(4):550–559.

Lasky, Marjorie Penn (Ed.). 2018. *You're Doing What? Older Women's Tales of Achievement & Adventure.* Berkeley, CA: Regent Press.

Latham, Melanie. 2008. "The Shape of Things to Come: Feminism, Regulation and Cosmetic Surgery." *Medical Law Review* 16:437–457.

Lau, Anna S., David T. Takeuchi, and Margarita Alegria. 2006. "Parent-to-Child Aggression among Asian American Parents: Culture, Context, and Vulnerability." *Journal of Marriage and Family* 68(5):1261–1275.

Lauby, Mary R., and Sue Else. 2008. "Recession Can Be Deadly for Domestic Abuse Victims." *Boston Globe,* December 25. Retrieved June 8, 2009 (bostonglobe .com).

Lauer, Sean R., and Carrie Yodanis. 2011. "Individualized Marriage and the Integration of Resources." *Journal of Marriage and Family* 73(June):669–683.

Laughlin, Lynda. 2011. "Maternity Leave and Employment Patterns of First-Time Mothers: 1961–2008," October. *Household Economic Studies.* Washington, DC: U.S. Census Bureau and U.S. Department of Commerce.

Laughlin, Lynda, and Kristin Smith. 2015. "Father-Provided Child Care among Married Couples in a Recessionary Context." SEHSD Working Paper 2015-16. Washington, DC: U.S. Census Bureau.

Lauletta, Tyler. 2018. "A 9-year-old Warriors Fan got Stephen Curry and Under Armour to Make the NBA Star's Shoes Available for Girls." Retrieved September 23, 2019 (https://www.businessinsider.com/ stephen-curry-letter-girls-shoes-riley-2018-11).

Laumann, Edward, Aniruddha Das, and Linda Waite. 2008. "Sexual Dysfunction among Older Adults: Prevalence and Risk Factors from a Nationally Representative U.S. Probability Sample of Men and Women 57–85 Years of Age." *Journal of Sexual Medicine* 5(10):2300–2311.

Laumann, Edward, John H. Gagnon, Robert T. Michael, and Stuart Michaels. 1994. *The Social Organization of Sexuality: Sexual Practices in the United States.* Chicago: University of Chicago Press.

Laumann, Edward, Sara Leitsch, and Linda Waite. 2008. "Elder Mistreatment in the United States: Prevalence Estimates from a Nationally Representative Study." *Journals of Gerontology Series B: Social Sciences* 63(4):S248–S254.

Laumann-Billings, L., and R. E. Emery. 2000. "Distress among Young Adults from Divorced Families." *Journal of Family Psychology* 14(4):671.

Lavery, Diana. 2012. "More Mothers of Young Children in Labor Force." Population Reference Bureau. Retrieved November 13, 2012 (http:// www.prb.org).

Lavin, Judy. 2003. "Smoothing the Step-Parenting Transition." SelfGrowth.com. Retrieved April 26, 2007 (www.selfgrowth .com).

Lavy, Shiri, Mario Mikulincer, Phillip R. Shaver, and Omri Gillath. 2009. "Intrusiveness in Romantic Relationships: A Cross-cultural Perspective on Imbalances Between Proximity and Autonomy." *Journal of Social and Personal Relationships* 26(6–7):989–1008.

Lawrence, K., and E. S. Byers. 1995. "Sexual Satisfaction in Long-Term Heterosexual Relationships: The Interpersonal Exchange Model of Sexual Satisfaction." *Personal Relationships* 2:267–285.

Lawson, Willow. 2004a. "Encouraging Signs: How Your Partner Responds to Your Good News Speaks Volumes." *Psychology Today* (January /February):22.

———. 2004b. "The Glee Club: Positive Psychologists Want to Teach You to Be Happier." *Psychology Today* (January /February):34–40.

Le, Thao N. 2005. "Narcissism and Immature Love As Mediators of Vertical Individualism and Ludic Love Style." *Journal of Social and Personal Relationships* 22(4):543–560.

Leamaster, Reid J, and Rachel L. Einwohner. 2018. "I'm Not Your Stereotypical Mormon Girl ': Mormon Women's Gendered Resistance." *Review of Religious Research* 60:161–181.

Leamaster, Reid J., and Andres Bautista. 2018. "Understanding Compliance in Patriarchal Religions: Mormon Women and the Latter Day Saints Church as a Case Study." *Religions* 9:143.

Leavitt, Keith, Christopher M. Barnes, Trevor Watkins, and David T. Wagner. 2019. "From the Bedroom to the Office: Workplace Spillover Effects of Sexual Activity at Home." *Journal of Management* 45:1173–1192.

LeBlanc, Allen J., David M. Frost, and Kayla Bowen. 2018. "Legal Marriage, Unequal Recognition, and Mental Health among Same-Sex Couples." *Journal of Marriage and Family* 80 (2):397–408.

LeBlanc, Allen J., and Richard G. Wright. 2000. "Reciprocity and Depression in AIDS Caregiving." *Sociological Perspectives* 43(4):631–649.

Lebow, Jay L., Anthony L. Chambers, Andrew Christensen, and Susan M. Johnson. 2012. "Research on the Treatment of Couple Distress." *Journal of Marital and Family Therapy* 38:145–168.

Ledbetter, Andrew M. 2009. "Family Communication Patterns and Relational Maintenance Behavior: Direct and Mediated Associations with Friendship Closeness." *Human Communication Research* 36:130–147.

Ledermann, Thomas, Guy Bodenmann, Myriam Rudaz, and Thomas N. Bradbury. 2010. "Stress, Communication, and Marital Quality in Couples." *Family Relations* 59(2):195–206.

Ledwell, Maggie, and Valarie King. 2015. "Bullying and Internalizing Problems: Gender Differences and the Buffering Role of Parental Communication." *Journal of Family Issues* 36(5):543–566.

Lee, Arlene F., Philip M. Genty, and Mimi Laver. 2005. *The Impact of the Adoption and Safe Families Act on Children of Incarcerated Parents.* Washington, DC: Child Welfare League of America.

Lee, Cameron, and Judith Iverson-Gilbert. 2003. "Demand, Support, and Perception in Family-related Stress among Protestant Clergy." *Family Relations* 52(3):249–257.

Lee, Carol E. 2006. "Sibling Seeks Same to Share Apartment." *The New York Times,* January 29.

Lee, Chu-Yuan, Jared Anderson, Jason Horowitz, and Gerald August. 2009. "Family Income and Parenting: The Role of Parental Depression and Social Support." *Family Relations* 58(October):417–430.

Lee, Elizabeth A. Ewing, and Wendy Troop-Gordon. 2011. "Peer Socialization of Masculinity and Femininity: Differential Effects of Overt and Relational Forms of Peer Victimization." *British Journal of Developmental Psychology* 29(2):197–213.

Lee, Eunju, Glenna Spitze, and John R. Logan. 2003. "Social Support to Parents-in-Law: The Interplay of Gender and Kin Hierarchies." *Journal of Marriage and Family* 65(2):396–403.

Lee, Eunjin, and Linda Roberts. 2018. "Between Individual and Family Coping: A Decade of Theory and Research on Couples Coping with Health-Related Stress." *Journal of Family Theory & Review* 10 (1):141–164.

Lee, Gary R., Chuck W. Peek, and Raymond T. Coward. 1998. "Race Differences in Filial Responsibility Expectations among Older Parents." *Journal of Marriage and Family* 60(2):404–412.

Lee, Jacrim, Mary Jo Katras, and Jean W. Bauer. 2009. "Children's Birthday Celebrations from the Experiences of Low-Income Rural Mothers." *Journal of Family Issues* 30(4):532–553.

Lee, Jennifer. 2007. "The Incredible Flying Granny Nanny." *The New York Times*, May 10. Retrieved May 11, 2007 (www.nytimes.com).

Lee, Jin-Kyung, and Sarah J. Schoppe-Sullivan. 2017. "Resident Fathers' Positive Engagement Family Poverty, and Change in Child Behavior Problems." *Family Relations* 66(3): 4484–496.

Lee, John Alan. 1973. *The Colours of Love.* Toronto: New Press.

———. 1981. "Forbidden Colors of Love: Patterns of Gay Love." pp. 128–139 in *Single Life: Unmarried Adults in Social Context,* edited by Peter J. Stein. New York: St. Martin's.

Lee, Kristen Schultz, and Hiroshi Ono. 2012. "Marriage, Cohabitation, and Happiness: A Cross-National Analysis of 27 Countries." *Journal of Marriage and Family* 74(5):953–972.

Lee, Michael. 2013. "Comparing Supports for LGBT Aging in Rural versus Urban Areas." *Journal of Gerontological Social Work* 56(2):112–124.

Lee, Mo-Yee. 2002. "A Model of Children's Postdivorce Behavioral Adjustment in Maternal- and Dual-Residence Arrangements." *Journal of Family Issues* 23(5):672–697.

Lee, Rosalyn, Mikel Walters, Jeffrey Hall, and Kathleen Basile. 2012. "Behavioral and Attitudinal Factors Differentiating Male Intimate Partner Violence Perpetrators with and without a History of childhood Family Violence." *Journal of Family Violence* 28:85–94.

Lee, Wai-Yung, Man-Lun Ng, Ben Cheung, and Joyce Wayung. 2010. "Capturing Children's Response to Parental Conflict and Making Use of It." *Family Process* 49(1):43–58.

Lee, Youjung, and Laura Smith. 2012. "Qualitative Research on Korean American Dementia Caregivers' Perception of Caregiving: Heterogeneity between Spouse Caregivers and Child Caregivers." *Journal of Human Behavior in the Social Environment* 22(2):115–129.

Lees, Janet, and Jan Horwath. 2009. "'Religious Parents . . . Just Want the Best for Their Kids': Young People's Perspectives on the Influence of Religious Beliefs on Parenting." *Children and Society* 23(3):162–175.

LeFebvre, Leah E., Mike Allen, Ryan D. Rasner, Shelby Garstad, Aleksander Wilms, and Callie Parrish. 2019. "Ghosting in Emerging Adults' Romantic Relationships: The Digital Dissolution Disappearance Strategy." *Imagination, Cognition and Personality* 39(2):125–150.

Lefkowitz, Eva S., Cindy L. Shearer, Meghan M. Gillen, and Graciela Espinosa-Hernandez. 2014. "How Gendered Attitudes Relate to Women's and Men's Sexual Behaviors and Beliefs." *Sexuality and Culture* 18:833–846.

Lehmann, Vicky, Marrit A. Tuinman, Johan Braeken, A. J. Vingerhoets, Robbert Sanderman, and Mariët Hagedoorn. 2015. "Satisfaction with Relationship Status: Development of a New Scale and the Role in Predicting Well-Being." *Journal of Happiness Studies* 16:169–184.

Lehmann-Haupt, Rachel. 2009. "Why I Froze My Eggs." *Newsweek*, May 18, pp. 50–52.

Leidy, Melinda, Ross D. Parke, Mina Cladis, Scott Coltrane, and Sharon Duffy. 2009. "Positive Marital Quality, Acculturative Stress, and Child Outcomes among Mexican Americans." *Journal of Marriage and Family* 71(4):833–47.

Leigh, Suzanne. 2004. "Fertility Patients Deserve to Know the Odds—and Risks." *USA Today*, July 7.

Leisey, Monica, Paul Kupstas, and Aly Cooper. 2009. "Domestic Violence in the Second Half of Life." *Journal of Elder Abuse and Neglect* 21(2):141–155.

Leisenring, Amy. 2008. "Controversies Surrounding Mandatory Arrest Policies and the Police Response to Intimate Partner Violence." *Sociology Compass* 2(2):451–466.

Leite, Randall. 2007. "An Exploration of Aspects of Boundary Ambiguity among Young, Unmarried Fathers during the Prenatal Period." *Family Relations* 56(2):162–174.

Lendon, Jessica P., Merril Silverstein, and Roseann Giarrusso. 2014. "Ambivalence in Older Parent-Adult Child Relationships: Mixed Feelings, Mixed Measures." *Journal of Marriage and Family* 76(2):272–284.

Lenhart, Amanda. 2012. "Teens, Smartphones, and Texting." Retrieved March 3, 2016 (http://www.pewinternet .org/2012/03/19/teens -smartphones -texting/).

Lenhart, Amanda, Monica Anderson, and Aaron Smith. 2015. "Teens, Technology, and Romantic Relationships." Retrieved January 16, 2016 (http://www.pewinternet .org/2015/10/01 /teens-technology-and -romantic-relationships/).

Lento, Jennifer. 2006. "Relational and Physical Victimization by Peers and Romantic Partners in College Students." *Journal of Social and Personal Relationships* 23(3):331–348.

Leonard, Kenneth E., Philip H. Smith, and Gregory G. Homish. 2014. "Concordant and Discordant Alcohol, Tobacco, and Marijuana Use as Predictors of Marital Dissolution." *Psychology of Addictive Behaviors* 28:780.

Leonhardt, David. 2014. "A Link Between Fidgety Boys and a Sputtering Economy." *The New York Times*, April 29. Retrieved April 29, 2014 (http://www.nytimes.com).

Leopold, Thomas, Marcel Raab, and Henriette Engelhardt. 2014. "The Transition to Parent Care: Costs, Commitments, and Caregiver Selection among Children." *Journal of Marriage and Family* 76(2):300–318.

Leopold, Thomas, and Thorsten Schneider. 2011. "Family Events and the Timing of Intergenerational Transfers." *Social Forces* 90(2):595–616.

Leopold, Thomas. 2018. "Gender Differences in the Consequences of Divorce: A Study of Multiple Outcomes." *Demography* 55:769–797.

Lepore, Jill. 2011. "Birthright: What's Next for Planned Parenthood?" *The New Yorker*, November 4:44– 52.

Lepper, John M. 2009. *When Crisis Comes Home.* Macon, GA: Smyth & Helwys.

Leprince, Chloé, Fabienne D'Arripe-Longueville, and Julie Doron. 2018. "Coping in Teams: Exploring Athletes' Communal Coping Strategies to Deal with Shared Stressors." *Frontiers in Psychology* 9:1908.

Lerner, Harriet. 2001. *The Dance of Connection: How to Talk to Someone When You're Mad, Hurt, Scared, Frustrated, Insulted, Betrayed, or Desperate.* New York: HarperCollins.

Lersch, Philipp M. 2016. "Family Migration and Subsequent Employment: The Effect of Gender Ideology." *Journal of Marriage and Family* 78 (1): 230–245.

Leslie, Barri, and Mandy Morgan. 2011. "Soulmates, Compatibility, and Intimacy: Allied Discursive Resources in the Struggle for Relationship Satisfaction in the New Millennium. " *New Ideas in Psychology* 29:10–23.

Lessane, Patricia Williams. 2007. "Women of Color Facing Feminism—Creating Our Space at Liberation's Table: A Report on the Chicago Foundation for Women's "F" Series." *Journal of Pan African Studies* 1(7):3–10.

Letarte, Marie-Josée, Sylvie Normandeau, and Julie Allard. 2010. "Effectiveness of a Parent Training Program 'Incredible Years' in a Child Protection Service." *Child Abuse and Neglect* 34(4):253–261.

Letiecq, Bethany L., Sandra J. Bailey, and Marcia A. Kurtz. 2008. "Depression among Rural Native American and European American Grandparents Rearing Their Grandchildren." *Journal of Family Issues* 29(3):334–356.

Letiecq, Bethany L., Sandra J. Bailey, and Fonda Porterfield. 2008. " 'We Have No Rights, We Get No Help': The Legal and Policy Dilemmas Facing Grandparent Caregivers." *Journal of Family Issues* 29(8):995–1012.

Levaro, Liz Bayler. 2009. "Living Together or Living Apart Together: New Choices for Old Lovers." *Family Focus* (Summer): F9–F10. Minneapolis: National Council on Family Relations.

Levchenko, Polina, and Catherine Solheim. 2013. "International Marriages between Eastern European-Born Women and US-Born Men." *Family Relations* 62:30–41.

Levin, Irene. 1997. "Stepfamily as Project." pp. 123–133 in *Stepfamilies: History, Research, and Policy,* edited by Irene Levinand Marvin B. Sussman. New York: Haworth.

Levin, Sam. 2015. "Racial Profiling Via Nextdoor.com." *Eastbay Express,* July 13: 14-18.

Levine, Carol. 2008. "Family Caregiving." pp. 63– 68 in *From Birth to Death and Bench to Clinic. The Hastings Center Bioethics Briefing Book for Journalists, Policymakers, and Campaigns,* edited by Mary Crowley. Garrison, NY: The Hastings Center. Retrieved May 17, 2010 (www .thehastingscenter .org).

Levine, Ethan Czuy, Debby Herbenick, Omar Martinez, Tsung-Chieh Fu, and Brian Dodge. 2018. "Open Relationships, Nonconsensual Nonmonogamy, and Monogamy Among U.S. Adults: Findings from the 2012 National Survey of Sexual Health and Behavior." *Archives of Sexual Behavior* 47:1439–1450.

Levine, Kathryn. 2009. "Against All Odds: Resilience in Single Mothers of Children with Disabilities." *Social Work in Health Care* 48(4):402–419.

Levine, Robert, Suguru Sato, Tsukasa Hashimoto, and Jyoti Verma. 1995. "Love and Marriage in Eleven Cultures." *Journal of Cross-Cultural Psychology* 26(5):554–571.

Levinger, George. 1976. "A Social Psychological Perspective on Marital Discord." *Journal of Social Issues* 32:21–47.

Levy, Donald P. 2005. "Hegemonic Complicity, Friendship, and Comradeship: Validation and Causal Processes among White, Middle-class, Middle-aged Men." *Journal of Men's Studies* 13(2):199–225.

Levy, Pema. 2011. "How Kansas Banned Abortion." *Prospect,* July 1. Retrieved September 13, 2011 (http://www .prospect.com).

Lewin, Ellen. 2009. *Gay Fatherhood: Narratives of Family and Citizenship in America.* Chicago: University of Chicago Press.

Lewin, Tamar. 2001. "Study Says Little Has Changed." *The New York Times,* September 10.

———. 2005. "A Marriage of Unequals: When Richer Weds Poorer, Money Isn't the Only Difference." *The New York Times,* May 19.

———. 2011. "Study Undercuts View of College as a Place of Same-Sex Experimentation." *The New York Times,* March 17. Retrieved October 10, 2012 (http://www.nytimes.com/2011/03/18 /education/18sex.html?_r=0).

———. 2014a. "A Surrogacy Agency that Delivered Heartache." *The New York Times,* July 27. Retrieved January 14, 2015 (http://www. nytimes.com).

———. 2014b. "Surrogates and Couple Face a Maze of Laws, State by State." Retrieved February 14, 2016 (http://www .nytimes. com/2014/09/18/us/surrogates -and-couples-face-a-maze-of-laws-state-by -state.html?_r=0).

Lewis, Hilary. 2008. "Rush Limbaugh Gets $400 Million to Rant Trough 2016." *Business Insider.* Retrieved October 16, 2015 (http://www .businessinsider.com).

Lewis, Jamie M., and Rose M. Kreider. 2015. "Remarriage in the United States," *American Community Survey Reports,* ACS-30. Washington, DC: U.S. Census Bureau.

Lewis, Thomas, Fari Amini, and Richard Lannon. 2000. *A General Theory of Love.* New York: Random House.

LGBTSS, Iowa State University. 2015. "Sexual Orientation." Retrieved December 3, 2015 (http://www.dso.iastate.edu /lgbtss/library/sexual-orientation).

Li, Xuan, and Jason Meier. 2017. "Father Love and Mother Love: Contributions of Parental Acceptance to Children's Psychological Adjustment." *Journal of Family Theory and Review* 9 (1). 159–190.

Lichter, Daniel T., J. Brian Brown, Zhenchao Qian, and Julie H. Carmalt. 2007. "Marital Assimilation among Hispanics: Evidence of Declining Cultural and Economic Incorporation?" *Social Science Quarterly* 88(3):745–765.

Lichter, Daniel T., and Julie H. Carmalt. 2009. "Cohabitation and the Rise in Out-of-Wedlock Childbearing." *Family Focus* (Summer):F11–F13. Minneapolis: National Council on Family Relations.

Lichter, Daniel T. and Zhenchao Qian. 2008. "Serial Cohabitation and the Marital Life Course." *Journal of Marriage and Family* 70(4):861–878.

Lichter, Daniel T., Zhenchao Qian, and Leanna M. Mellott. 2006. "Marriage or Dissolution? Union Transitions among Poor Cohabiting Women." *Demography* 43(2):223–241.

Lichter, D. T., Qian, Z., & Tumin, D. 2015. "Whom do Immigrants Marry? Emerging Patterns of Intermarriage and Integration in the United States. *Annals of the American Academy of Political and Social Science* 662:57–78.

Lieber, Ron. 2014. "Parents, the Children Will Be Fine. Spend Their Inheritance Now." *The New York Times,* September 19. Retrieved February 27, 2015 (http://www .nytimes.com).

———, 2015, "Bringing Paternity Leave into the Mainstream." *The New York Times,* August 7. Retrieved September 19, 2015 (http://www .nytimes.com).

Liechty, Toni, Sarah M. Coyne, Kevin M. Collier, and Aubrey D. Sharp. 2018. "It's Just Not Very Realistic": Perceptions of Media Among Pregnant and Postpartum Women. " *Health Communication* 33:85–859.

Liefbroer, Aart C., and Edith Dourleijn. 2006. "Unmarried Cohabitation and Union Stability: Testing the Role of Diffusion Using Data from 16 European Countries." *Demography* 43(2):203–221.

Limbaugh, Rush. 1992. *The Way Things Ought to Be.* New York: Pocket Books.

Lin, Chien, and William T. Liu. 1993. "Relationships among Chinese Immigrant Families." pp. 271–286 in *Family Ethnicity: Strength in Diversity,* edited by Harriette Pipes McAdoo. Newbury Park, CA: Sage.

Lin, I-Fen, and Susan L. Brown. 2013. "Unmarried Boomers Confront Old Age: A National Portrait." Working Paper Series 2012–2013. Bowling Green, OH: Bowling Green State University Center for Family and Demographic Research.

Lin, I-Fen, Susan L. Brown, and Anna M. Hammersmith. 2015. "Marital Biography, Social Security, and Poverty" (WP-15-01). Bowling Green, OH: National Center for Family and Marriage Research.

Lin, I-Fen, and Hsueh-Sheng Wu. 2014. "Intergenerational Exchange and Expected Support among the Young Old." *Journal of Marriage and Family* 76(2):261–271.

Lin, I-Fen, Susan L. Brown, and Anna M. Hammersmith. 2017. "Marital Biography, Social Security Receipt, and Poverty." *Research on Aging* 39:86–110.

Lin, I-Fen, Susan L. Brown, and Cassandra Jean Cupka. 2018. "A National Portrait of Stepfamilies in Later Life." *The Journals of Gerontology: Series B* 73:1043-1054.

Lin, K. H., and J. Lundquist. (2013). "Mate Selection in Cyberspace: The Intersection of Race, Gender, and Education." *American Journal of Sociology,* 119:183–215.

Lincoln, Karen, Robert Taylor, and James Jackson. 2008. "Romantic Relationships among Unmarried African Americans and Caribbean Blacks: Findings from the National Survey of American Life." *Family Relations* 57(April):254–266.

Lind, Ranveig, Geir Lorem, Per Nortvedt, and Olav Hevroy. 2012. "Intensive Care Nurses' Involvement in the End-of-Life Process: Perspectives of Relatives." *Nursing Ethics* 19(5):666–676.

Lindau, Stacy Tessler, and Natalia Gavrilova. 2010. "Sex, Health, and Years of Sexually Active Life Gained Due to Good Health: Evidence from Two U.S. Population Based Cross-Sectional Surveys of Ageing." *British Medical Journal* 340(7746):580.

Lindau, Stacy Tessler, L. Philip Schumm, Edward O. Laumann, Wendy Levinson, Colm A. O'Muircheartaigh, and Linda J. Waite. 2007. "A Study of Sexuality and Health among Older Adults in the United States." *New England Journal of Medicine* 357(8):762–674.

Lindberg, Laura Duberstein, Rachel Jones, and John S. Santelli. 2008. "Non-Coital Sexual Activities among Adolescents," *Journal of Adolescent Health* 43(3):231–238.

Lindberg, Laura D., Katheryn Kost, and Isaac Maddow-Zimet. 2017. "The Role of Men's Childbearing Intentions in Father Involvement." *Journal of Marriage and Family* 79 (1): 44–59.

Lindell, Anna K., Nicole Campione-Barr, and Sarah E. Killoren. 2015. "Technology-Mediated Communication with Siblings during the Transition to College: Associations with Relationship Positivity and Self-Disclosure." *Family Relations* 64:563–578.

Lindner Gunnoe, Marjorie, E. Mavis Hetherington, and David Reiss. 2006. *Journal of Family Psychology* 20(4):589–596.

Lindo, Endia, Karen Kliemann, Bertina Combs, and Jessica Frank. 2016. "Managing Stress Levels of Parents of Children with Developmental Disabilities: A Meta-Analytic Review of Interventions." *Family Relations* 65 (1): 207–224.

Lindsey, Elizabeth W. 1998. "The Impact of Homelessness and Shelter Life on Family Relationships." *Family Relations* 47(3):243–252.

Lindsey, Eric W., Yvonne M. Caldera, and Laura Tankersley. 2009. "Marital Conflict and the Quality of Young Children's Peer Play Behavior: The Mediating and moderating Role of Parent-Child Emotional Reciprocity and Attachment Security." *Journal of Family Psychology* 23(2):130–145.

Lindsey, Eric W., Jessico Campbell Chambers, James Frabutt, and Carol Mackinnon-Lewis. 2009. "Marital Conflict and Adolescents' Peer Aggression: The Mediating and Moderating Role of Mother-Child Emotional Reciprocity." *Family Relations* 58(December):593–606.

Lino, Mark. 2017. *The Cost of Raising a Child.* Retrieved December 17, 2019 (https://www .usda.gov/media/blog/2017/01/13 /cost-raising-child).

Liptak, Adam. 2019. "Supreme Court Reviews Transgender Ban for Military Service." Retrieved September 25, 2019 (https://www.nytimes. com/2019/01/22/us/politics/transgender-ban -military-supreme-court.html).

Little, Betsi, and Cheryl Terrance. 2010. "Perceptions of Domestic Violence in Lesbian Relationships: Stereotypes and Gender Role Expectations." *Journal of Homosexuality* 57(3):429–440.

Littleton, Heather, Holly Tabernik, Erika J. Canales, and Tamika Backstrom. 2009. "Risky Situations or Harmless Fun? A Qualitative Examination of College Women's Bad Hook-Up and Rape Scripts." *Sex Roles* 60:793–804.

Liu, Chien. 2000. "A Theory of Marital Sexual Life." *Journal of Marriage and Family* 62(2): 363–374.

Liu, Hui. 2009. "Till Death Do Us Part: Marital Status and U.S. Mortality Trends, 1986–2000." *Journal of Marriage and Family* 71(December):1158–1173.

Liu, Hui, and Corinne Reczek. 2012. "Cohabitation and U.S. Adult Mortality: An Examination by Gender and Race." *Journal of Marriage and Family* 74(4):794–811.

Liu, Siwei, and Kathryn Hynes. 2012. "Are Difficulties Balancing Work and Family Associated with Subsequent Fertility?" *Family Relations* 61(1): 16–30.

Liu, Hui, and Lindsey Wilkinson. 2017. "Marital Status and Perceived Discrimination among Transgender People." *Journal of Marriage and Family* 79 (5): 1295–1313.

Livingston, Gretchen. 2014. "Growing Number of Dads Home with the Kids." June 5. *Social and Demographic Trends* (Pew Research Center). Retrieved November 5, 2015 (http://www .pewsocialtrends.org).

———. 2015. "Childlessness Fall, Family Size Grows among Highly Educated Women." Retrieved February 8, 2016 (http://www .pewsocialtrends.org/2015/05/07 /childlessness-falls -family-size-grows-among -highly-educated -women/).

Livingston, Gretchen, and D'Vera Cohn. 2010. "The New Demography of American Motherhood." Retrieved February 8, 2016 (http://www.pewsocialtrends .org/2010/05/06/the-new-demography-of -american-motherhood/).

Livingston, Gretchen, and Kim Parker. 2011. "A Tale of Two Fathers." Pew Research Center. Retrieved September 13, 2011 (http://www .pewresearch.org).

Livingston, Gretchen, Kim Parker, and Molly Rohal. 2014. "Four-in-Ten Couples Are Saying 'I Do,' Again." Retrieved April 12, 2016 (http:// www.pewsocialtrends.org/files/2014/11/2014 -11-14 _remarriage-final.pdf).

Livingston, Gretchen. 2017. "Among U.S. Cohabitors, 18% Have a Partner of a Different Race." Retrieved December 1, 2019 (https:// www.pewresearch.org/fact-tank/2017/06/08 /among-u-s-cohabiters-18-have-a-partner-of-a -different-race-or-ethnicity/).

Livingston, Gretchen. 2017. "The Rise of Multiracial and Multiethnic Babies in the U.S." *Facttank: News in the Numbers.* June 6. Pew Research Organization. Retrieved January 30, 2019 (www.pewresearch.org).

Livingston, Gretchen. 2018a. "Most Dads Say They Spend Too Little Time with Their Children." *Facttank: News in the Numbers.* January 8. Pew Research Organization. Retrieved January 30, 2019 (www.pewresearch.org).

Livingston, Gretchen. 2018b. "Stay-at-Home Moms and Dads Account for about One-in-Five U.S. Parents." *Facttank: News in the Numbers.* September 24. Pew Research Organization. Retrieved January 30, 2019 (www.pewresearch .org).

Livingston, Gretchen. 2018c. "More Than One in Ten U.S. Parents Are Also Caring for an Adult." *Facttank: News in the Numbers.* November 28. Pew Research Organization. Retrieved January 30, 2019 (www.pewresearch.org).

Livingston, Gretchen. 2018d. "Adult Caregiving Often Seen as Very Meaningful by Those Who Do It." *Facttank: News in the Numbers.* December19. Pew Research Organization . Retrieved August 15, 2019 (www.pewresearch .org).

Livingston, Gretchen. 2019. "On Average, Older Adults Spend Over Half Their Waking Hours Alone." *Facttank: News in the Numbers.* July 3. Pew Research Organization. Retrieved August 15, 2019 (www.pewresearch.org).

Livingston, Gretchen. 2018a. "They're Waiting Longer, but U.S. Women Today More Likely to Have Children than a Decade Ago. " Retrieved December 17, 2019 (file:///C:/Users/stewa /Downloads/Pew-Motherhood-report -FINAL.pdf).

Livingston, Gretchen. 2018b. "U.S. Women Are Postponing Motherhood, but Not as Much as Those in Most Other Developed Nations. " Retrieved December 14, 2019 (https://www .pewresearch.org/fact-tank/2018/06/28 /u-s-women-are-postponing-motherhood-but -not-as-much-as-those-in-most-other-developed -nations/).

Livingston, Gretchen, and Ana Brown. 2017. "Intermarriage in the U.S. 50 Years After Loving v. Virginia. " Retrieved November 18, 2019 (https://www.pewsocialtrends.org/2017/05/18 /intermarriage-in-the-u-s-50-years-after -loving-v-virginia/).

Livingston, Gretchen, and Juliana Menasce Horowitz. 2018. "Most Parents—and Many Non-Parents—Don't Expect to Have Kids in the Future. " Retrieved December 19, 2019 (https://www.pewresearch.org /fact-tank/2018/12/12/most-parents -and-many-non-parents-dont-expect-to -have-kids-in-the-future/).

Livingston, Gretchen, and Kim Parker. 2019. "8 Facts About American Dads." *Facttank: News in the Numbers.* June 19. Pew Research Organization. Retrieved August 15, 2019 (www .pewresearch.org).

Livingston, Gretchen, Kim Parker, and Molly Rohal. 2015. "Childlessness Falls, Family Size Grows Among Highly Educated Women. " Retrieved December 19, 2019 (http://assets .pewresearch.org/wp-content/uploads /sites/3/2015/05/2015-05-07_children -ever-born_FINAL.pdf).

Lleras, Christy. 2008. "Employment, Work Conditions, and the Home Environment in Single-Mother Families." *Journal of Family Issues* 29(10):1268–1297.

Lloyd, Kim M. 2006. "Latinas' Transition to First Marriage: An Examination of Four Theoretical Perspectives." *Journal of Marriage and Family* 68(4):993–1014.

Lloyd, Sally A., April L. Few, and Katherine R. Allen. 2007. "Feminist Theory, Methods, and Praxis in Family Studies: An Introduction to the Special Issue." *Journal of Family Issues* 28(4):447–451.

Lloyd-Thomas, Matthew, and Amy Wang. 2012. "Women Underrepresented in Faculty, Report Finds." *Yale Daily News,* September 11. Retrieved October 27, 2012 (http://www.yaledailynews. com).

LoBiondo-Wood, Geri, Laurel Williams, and Charles McGhee. 2004. "Liver Transplantation in Children: Maternal and Family Stress, Coping, and Adaptation." *Journal of the Society of Pediatric Nurses* 9(2):59–67.

Lodge, Amy C., and Debra Umberson. 2012. "All Shook Up: Sexuality of Mid- to Later Life Married Couples." *Journal of Marriage and Family* 74:428–443.

Lofquist, Daphne. 2011. "Same-Sex Couple Households: American Community Survey Briefs." U.S. Census Bureau ACSBR 10-03, September.

———. 2012. "Same-Sex Couples' Consistency in Reports of Marital Status." Paper presented at the annual meeting of the Population Association of America, San Francisco: May 3–5.

Loewus, Liana. 2017. "The Nation's Teaching Force is Mostly White and Female," Retrieved October 2, 2019 (https://www.edweek.org/ew /articles/2017/08/15/the-nations -teaching-force-is-still-mostly.html).

Lofquist, Daphne, Terry Lugaila, Martin O'Connell, and Sarah Feliz. 2012. *Households and Families: 2010.* U.S. Census Briefs Publication Number C2010BR–14, April. Washington, DC: U.S. Census Bureau and U.S. Department of Commerce.

Loftin, Colin, David McDowall, and Matthew Fetzer. 2008. "The Accuracy of Supplementary Homicide Report Data for Large U.S. Cities." Paper presented at the Annual Meeting of the American Society of Criminology, November 12. St. Louis Adam's Mark, St. Louis, Missouri.

Loftus, Jeni. 2001. "America's Liberalization in Attitudes toward Homosexuality, 1973 to 1998." *American Sociological Review* 66:762–782.

Logan, T. K., Robert Walker, and William Hoyt. 2012. "The Economic Costs of Partner Violence and Cost-Benefit of Civil Protective Orders." *Journal of Interpersonal Violence* 27(6):1137–1154.

Lois, Jennifer. 2010. "Gender and Emotion Management in the Stages of Edgework."

pp. 333–344 in *The Kaleidoscope of Gender: Prisms, Patterns, and Possibilities*, 3rd ed., edited by Joan Z. Spadea and Catherine G. Valentine. Newbury Park, CA: Pine Forge Press.

Lombardi, Caitlin McPherran, and Rebekah Levine Coley. 2013. "Low-Income Mothers' Employment Experiences: Prospective Links with Young Children's Development." *Family Relations* 62(3):514–520.

London, Andrew S., Elizabeth Allen, and Janet M. Wilmoth. 2013. "Veteran Status, Extramarital Sex, and Divorce: Findings from the 1992 National Health and Social Life Survey." *Journal of Family Issues* 34(11):1452–1473.

Long, Heather. 2019. "This Doesn't Look Like the Best Economy Ever: 40% of Americans Say They Still Struggle to Pay Bills." *The Washington Post.* July 4. Retrieved August 13, 2019 (www.washingtonpost .com).

Longmore, Monica A., Abbey L. Eng, Peggy C. Giordano, and Wendy D. Manning. 2009. "Parenting and Adolescents' Sexual Initiation." *Journal of Marriage and Family* 71(4):969–982.

Loper, A., C. Clarke, and D. Dallaire. 2019. "Parenting Programs for Incarcerated Fathers and Mothers: Current Research and New Directions." September 14. Pp. 183–203 in J. Eddy and J. Poehlmann-Tynan (Eds.) *Handbook on Children with Incarcerated Parents. Springer International Publishing.*

Lopez, Mark Hugo, and Ana Gonzalez-Barrera. 2014. "Women's College Enrollment Gains Leave Men Behind." Pew Research Center (March 6). Retrieved October 8, 2015 (http:// www.pewresearch .org).

Lopez, Mark Hugo, Ana Gonzalez-Barrera, and Seth Motel. 2011. "As Deportations Rise to Record Levels, Most Latinos Oppose Obama's Policy." December 28. Pew Research Center /Pew Hispanic Center. Retrieved February 16, 2012 (http://www.pewhispanic.org).

Lopez, Omar S. 2015. "Averting Another Lost Decade: Moving Hispanic Families from Outlier to Mainstream Family Research." *Journal of Family Issues* 36(1):133–159.

Loquist, Daphne. 2011. *Same-Sex Couple Households: American Community Survey Briefs.* Washington, DC: U.S. Census Bureau.

Lorenzo-Blanco, Elma, Cristina Bares, and Jorge Delva. 2013. "Parenting, Family Processes, Relationships, and Parental Support in Multiracial and Multiethnic Families: An Exploratory Study of Youth Perceptions." *Family Relations* 62(1):125–139.

Loseke, Donileen R., and Demie Kurz. 2005. "Men's Violence toward Women Is the Serious Social Problem." pp. 79–95 in *Current Controversies on Family Violence,* edited by Donileen R. Loseke, Richard J. Gelles, and Mary M. Cavanaugh. Thousand Oaks, CA: Sage.

Love, Patricia. 2001. *The Truth about Love.* New York: Simon and Schuster.

Love, Patricia, and Steven Stosny. 2007. *How to Improve Your Marriage Without Talking about It: Finding Love Beyond Words.* New York: Broadway Books.

Lovell, Vicky, Elizabeth O'Neill, and Skylar Olsen. 2007. *Maternity Leave in the United States.* Washington, DC: Institute for Women's Policy Research.

Lovett, Ian. 2012. "Measure Opens Door to Three Parents, or Four." *The New York Times, July* 13. Retrieved August 28, 2012 (http://www.nytimes .com).

Lovett, Ian. 2013. "Changing Sex, Changing Teams." Retrieved September 29, 2019 (https:// www.nytimes.com/2013/05/07/us/transgender -high-school-students-gain-admission-to -sports-teams.html).

"Loving Your Partner as a Package Deal." 2000. *Newsweek,* March 20, p. 78.

Lowan, J. M., and E. M. Dolan. 1994. "Remarried Families' Economic Behavior: Fishman's Model Revisited." *Journal of Divorce and Remarriage* 22:103–119.

Lowe, Chelsea, and Bruce M. Cohen. 2010. *Living with Someone Who's Living with Bipolar Disorder: A Practical Guide for Family, Friends, and Coworkers.* San Francisco: Jossey-Bass.

Lowe, Kendra, Katharine Adams, Blaine Browne, and Kerry Hinkle. 2012. "Impact of Military Deployment on Family Relationships." *Journal of Family Studies* 18(1):17–27.

Lowe, Pauline, and Karen McBride-Henry. 2012. "What Factors Impact upon the Quality of Life of Elderly Women with Chronic Illnesses." *Contemporary Nurse* 41(1):18–27.

Lowrey, Annie. 2014. "Can Marriage Cure Poverty?" *The New York Times Magazine,* February 4. Retrieved April 17, 2014 (http://www .nytimesmag).

Lubrano, Alfred. 2003. *Blue-Collar Roots, White-Collar Dreams.* New York: Wiley.

Lucas, Demitria L. 2013. "Princeton Mom to Female Students: Find Husband in College." Retrieved May 30, 2013 (http://www. clutchmagonline.com/2013/04 /princeton-mom-to-female-students -find-husband-in-college/).

Lucas, Kristen. 2007. "Anticipatory Socialization in Blue-Collar Families: The Social Mobility-Reproduction Dialectic." Paper presented at the annual meeting of the International Communication Association, San Francisco CA, May 23. Retrieved January 19, 2010 (www .allacademic.com).

Luce, Ann, Marilyn Cash, Vanora Hundley, Helen Cheyne, Edwin Van Teijlingen, and Catherine Angell. 2016. "'Is it Realistic' The Portrayal of Pregnancy and Childbirth in the Media." *BMC Pregnancy and Childbirth* 16:40.

Lucier-Greer, Mallory, and Francesca Adler-Baeder. 2012. "Does Couple and Relationship Education Work for Individuals in Stepfamilies? A Meta-Analytic Study." *Family Relations* 61(5):756–769.

Lucier-Greer, Mallory, Amy Arnold, Jay Mancini, James Ford, and Chalandra Bryant. 2015. "Influences of Cumulative risk and Protective Factors on the Adjustment of Adolescents in Military Families." *Family Relations* 64(3):363–377.

Ludden, Jennifer. 2011. "Ask for a Raise? Most Women Hesitate." February 14. Retrieved February 14, 2011 (http://www .npr.org).

Luerssen, Anna, Gugan Jote Jhita, and Ozlem Ayduk. 2017. "Putting Yourself on the Line: Self-Esteem and Expressing Affection in Romantic Relationships." *Personality and Social Psychology Bulletin* 43(7):940–956.

Lugo Steidel, Angel G., and Josefina M. Contreras. 2003. "A New Familism Scale for Use with Latino Populations." *Hispanic Journal of Behavioral Sciences* 25(3):312–330.

Lumpkin, James R. 2008. "Grandparents in a Parental or Near-Parental Role: Sources of Stress and Coping Mechanisms." *Journal of Family Issues* 29(3):357–372.

Lundberg, Michael. 2009. "Our Parents' Keepers: The Current Status of American Filial Responsibility Laws." *Utah Law Review* 11(2):534–582.

Lundquist, Jennifer Hickes. 2004. "When Race Makes No Difference: Marriage and the Military." *Social Forces* 83(2):731–757.

Lundquist, Jennifer Hickes, Michelle J. Budig, and Anna Curtis. 2009. "Race and Childlessness in America, 1988–2002." *Journal of Marriage and Family* 71(3):741–755.

Lundquist, Jennifer, and Zhun Xu. 2014. "Reinstitutionalizing Families: Life Course Policy and Marriage in the Military." *Journal of Marriage and Family* 76(5):1063–1081.

Luo, Shanhong. 2014. "Effects of Texting on Satisfaction in Romantic Relationships: The Role of Attachment." *Computers in Human Behavior* 33:145–152.

Luscombe, Belinda. 2011. "Latchkey Parents." *Time,* Retrieved May 13, 2013 (http:// www.time.com/time/magazine /article/0,9171,2093312,00.html).

———. 2013. "Confidence Woman: Facebook's Sheryl Sandberg Is on a Mission to Change the Balance of Power." *Time,* March 18:36–42.

Lustig, Daniel C. 1999. "Family Caregiving of Adults with Mental Retardation: Key Issues for Rehabilitation Counselors." *Journal of Rehabilitation* 65(2):26–15.

Luscombe, Madison. 2019. "I am a Person of Faith Who Had an Abortion." Retrieved December 23, 2019 (https://www.scarymommy.com /religious-but-had-abortion/).

Lusinski, Natalia. 2018. "11 Women Share Why They Don't Want to Get Married." Retrieved November 30, 2019 (https://www.bustle .com/p/11-women-share-why-they-dont -want-to-get-married-9230430).

Luthra, Rohini, and Christine A. Gidycz. 2006. "Dating Violence among College Men and Women: Evolution of a Theoretical Model." *Journal of Interpersonal Violence* 21(6):717–731.

Luxenberg, Stanley. 2014. "Welcoming Love at an Older Age, but Not Necessarily Marriage." *The New York Times, April* 25. Retrieved April 30, 2014 (http://www .nytimes.com).

Lyons, Heidi, Peggy C. Giordano, Wendy D. Manning, and Monica A. Longmore. 2011. "Identity, Peer Relationships, and Adolescent Girls' Sexual Behavior: An Exploration of the Contemporary Double Standard." *Journal of Sex Research* 48(5):437–449.

Lyons, Heidi A., Wendy D. Manning, Monica A. Longmore, and Peggy C. Giordano. 2014. "Young Adult Casual Sexual Behavior Life-Course-Specific Motivations and Consequences." *Sociological Perspectives* 57:79–101.

Lyons, Linda. 2004. "How Many Teens and Cool with Cohabitation?" Retrieved

January 16, 2016 (http://www.gallup.com/poll/11272/How-Many-Teens-Cool-Cohabitation.aspx?g_source=cohabitation&g_medium=search&g_campaign=tiles).

M.E. Herman-Giddens. 2006. "Recent Data on Pubertal Milestones in United States Children: the Secular Trend Toward Earlier Development." *International Journal of Andrology.* 29:241–246

Maas, Megan K., Brandon T. McDaniel, Mark E. Feinberg, and Damon E. Jones. 2015. "Division of Labor and Multiple Domains of Sexual Satisfaction among First-Time Parents." *Journal of Family Issues:* 0192513X15604343.

Maccoby, E. E., and J. A. Martin. 1983. "Socialization in the Context of the Family: Parent-Child Interaction." pp. 1–101 in *Handbook of Child Psychology,* edited by P. Mussen. New York: Wiley.

Maccoby, Eleanor E. 1998. *The Two Sexes: Growing Up Apart; Coming Together.* Cambridge, MA: Belknap/Harvard University Press.

———. 2002. "Gender and Group Process: A Developmental Perspective." *Current Directions in Psychological Science* 11(2):54–58.

Maccoby, Eleanor E., and Carol Nagy Jacklin. 1974. *The Psychology of Sex Differences.* Stanford, CA: Stanford University Press.

Maccoby, Eleanor E., and Robert Mnookin. 1992. *Dividing the Child: Social and Legal Dilemmas of Custody.* Cambridge, MA: Harvard University Press.

Macdonald, Cameron L. 2011. *Shadow Mothers: Nannies, Au Pairs, and the Micropolitics of Mothering.* Berkeley CA: University of California Press.

MacDonald, William L., and Alfred DeMaris. 2002. "Stepfather-Stepchild Relationship Quality: The Stepfather's Demand for Conformity and the Biological Father's Involvement." *Journal of Family Issues* 23(1):121–137.

MacDorman, M. F., and S. Kirmeyer. 2009. "The Challenge of Fetal Mortality." *NCHS Data Brief,* April 16:1–8.

MacDorman, Marian F., Eugene Declercq, Howard Cabral, and Christine Morton. 2016. "Is the United States Maternal Mortality Rate Increasing? Disentangling Trends from Measurement Issues. " *Obstetrics and Gynecology* 128:447.

MacFarquhar, Neil. 2006. "It's Muslim Boy Meets Girl, but Don't Call It Dating." *The New York Times,* September 19. Retrieved September 19, 2006 (www.nytimes.com).

MacFarquhar, Larissa. 2019. "The Radical Transformation of a Battered Women's Shelter." *The New Yorker.* August 12. Retrieved November 5, 2019 (https://www.newyorker.com/magazine/2019/08/19/the-radical-transformations-of-a-battered-womens-shelter).

Machir, John. 2003. "The Impact of Spousal Caregiving on the Quality of Marital Relationships in Later Life." *Family Focus* (September):F11–F13. Minneapolis: National Council on Family Relations.

Macionis, John J. 2006. *Society: The Basics.* 6th ed. Upper Saddle River, NJ: Prentice Hall.

MacInnes, Maryhelen D. 2008. "One's Enough for Now: Children, Disability, and the Subsequent Childbearing of Mothers." *Journal of Marriage and Family* 70(3):758–771.

Mack, Jessica. 2011. "One Feminist Asks, 'Is Polygamy Inherently Bad for Women?" *Ms.Blog,* January 5. Retrieved October 23, 2015 (http://www.msmagazine.com).

Mackey, Richard A., Matthew A. Diemer, and Bernard A. O'Brien. 2000. "Psychological Intimacy in the Lasting Relationships of Heterosexual and Same-gender Couples." *Sex Roles* (August):201–215.

Mackler, Jennifer S., Rachael T. Kelleher, Lilly Shanahan, Susan D. Calkins, Susan Keane, and Marion O'Brien. 2015. "Parenting Stress, Parental Reactions, and Externalizing Behavior From Ages 4 to 10." *Journal of Marriage and Family* 77(2):388–406.

MacMillan, Amanda. 2017. "Birth Control for Men? Researchers Will Test a Hormonal Gel in 2018. " Retrieved December 26, 2019 (https://time.com/5077942/male-contraceptive-hormonal-gel/).

MacNeil, Sheila, and Sandra E. Byers. 2009. "Role of Sexual Self-Disclosure in the Sexual Satisfaction of Long-Term Heterosexual Couples." *Journal of Sex Research* 46(1):3–14.

MacPhee, David, Erika Lunkenheimer, and Nathaniel Riggs. 2015. "Resilience as Regulation of Developmental and Family Processes." *Family Relations* 64(1):153–175.

Madathil, Jayamala, and James M. Benshoff. 2008. "Importance of Marital Characteristics and Marital Satisfaction: A Comparison of Asian Indians in Arranged Marriages and Americans in Marriages of Choice." *The Family Journal: Counseling and Therapy for Couples and Families* 16(3):222–230.

Magnuson, Katherine, and Lawrence Berger. 2009. "Family Structure States and Transitions: Associations with Children's Well-Being During Middle Childhood." *Journal of Marriage and Family* 71(3):575–591.

Mahay, Jenna, and Alisa C. Lewin. 2007. "Age and the Desire to Marry." *Journal of Family Issues* 28(5):706–723.

Mahoney, Annette. 2005. "Religion and Conflict in Marital and Parent-Child Relationships." *Journal of Social Issues* 61(4):689–717.

Maier, Thomas. 1998. "Everybody's Grandfather." *U.S. News & World Report,* March 30, p. 59.

Maillard, Kevin Noble. 2008. "The Multiracial Epiphany of *Loving.*" *Fordham Law Review* 76(6):2709–2732.

Mainemer, Henry, Lorraine C. Gilman, and Elinor W. Ames. 1998. "Parenting Stress in Families Adopting Children from Romanian Orphanages." *Journal of Family Issues* 19(2):164–180.

Major, Brenda, Mark Appelbaum, Linda Beckman, Mary Ann Dutton, Nancy Felipe Russo, and Carolyn West. 2009. "Abortion and Mental Health." *American Psychologist* 64(9):863–890.

Makino, Momoe. 2019. "Marriage, Dowry, and Women's Status in Rural Punjab, Pakistan." *Journal of Population Economics* 32:769–797.

Malakh-Pines, Ayala. 2005. *Falling in Love: Why We Choose the Lovers We Choose.* New York: Routledge.

Malebranche, D. 2007. *Black Bisexual Men and HIV: Time to Think Deeper.* Paper presented at the Center for Sexual Health Promotion Sexual Health Seminar Series, Bloomington, Indiana.

Maleck, Sarah, and Lauren M. Papp. 2015. "Childhood Risky Family Environments and Romantic Relationship Functioning among Young Adult Dating Couples." *Journal of Family Issues* 36:567–588.

Malia, Sarah E. C. 2005. "Balancing Family Members' Interests Regarding Stepparent Rights and Obligations: A Social Policy Challenge." *Family Relations* 54(2):298–319.

Malia, Sarah E. 2008. "How Relevant Are U.S. Family and Probate Laws to Stepfamilies?" pp. 545–572 in *The International Handbook of Stepfamilies: Policy and Practice in Legal, Research, and Clinical Environments,* edited by Jan Pryor. John Wiley & Sons.

Malik, Rasheed. 2019. "Working Families Are Spending Big Money on Childcare." Center for American Progress. June 20. Retrieved January 30, 2020 (www.americanprogress.org).

Malinen, Kaisa, Asko Tolvanen, and Anna Ronka. 2012. "Accentuating the Positive, Eliminating the Negative? Relationship Maintenance as a Predictor of Two-Dimensional Relationship Quality." *Family Relations* 61(5):784–797.

Mallette, Jacquelyn K., Ted G. Futris, Geoffrey L. Brown, and Assaf Oshri. 2015. "The Influence of Father Involvement and Interparental Relationship Quality on Adolescent Mothers' Maternal Identity." *Family Relations* 64(4):476–489.

Malvaso, Catia, Paul Delfabbro, Michael Proeve, and Gavin Nobes. 2015. "Predictors of Child Injury in Biological and Stepfamilies." *Journal of Child and Adolescent Trauma* 8: 149–159.

Manago, Adriana M., Christia Spears Brown, and Campbell Leaper. 2009. "Feminist Identity among Latina Adolescents." *Journal of Adolescent Research* 24(6):750–776.

Mancillas, Adrean. 2006. "Challenging the Stereotypes about Only Children: A Review of the Literature and Implications for Practice. " *Journal of Counseling & Development* 84: 268–275.

Mandara, Jelani, Jamie S. Johnston, Carolyn B. Murray, and Fatima Varner. 2008. "Marriage, Money, and African American Mothers' Self-Esteem." *Journal of Marriage and Family* 70(5):1188–1199.

Mandara, Jelani, Carolyn Murray, James Telesford, Fatima Varner, and Scott Richman. 2012. "Observed Gender Differences in African American Mother-Child Relationships and Child Behavior." *Family Relations* 61(1):129–141.

Mandara, Jelani, Sheba Y. Rogers, and Richard Zinbarg. 2011. "The Effects of Family Structure on African American Adolescents' Marijuana Use." *Journal of Marriage and Family* 73(June):557–569.

Mangla, Ismat Sarah. 2013. "The Talk: How to Tackle Your Spouse's Overspending." *Money* 42(1). Retrieved January 29, 2013 (http://web .ebscohost.com).

Manisses Communications Group. 2000. "When It Comes to Handling Your Hard-to-Handle Child, Are You an Authoritative, Authoritarian or Permissive Parent?" *The Brown University Child and Adolescent Behavior Letter* 16(3):81–82.

Manlove, Jennifer, Elizabeth Cook, Brooke Whitfield, Makedah Johnson, Genevieve Martínez-García, and Milagros Garrido. 2019. "Short-Term Impacts of Pulse: An App-Based Teen Pregnancy Prevention Program for Black and Latinx Women. " *Journal of Adolescent Health.*

Mann, Brian. 2008. "Military Moms Face Tough Choices." National Public Radio (NPR), May 27. Retrieved May 27, 2008 (http://www.npr.org).

Manning, Jennifer E., and Ida A. Brudnick. 2015. *Women in Congress, 1917–2015.* Congressional Research Services Report. Retrieved October 15, 2015 (http://www .crs.gov).

Manning, Margaret M., Laurel Wainwright, and Jillian Bennett. 2011. "The Double ABCX Model of Adaptation in Racially Diverse Families with a School-Age Child with Autism." *Journal of Autism Development Disorder* 41:320–331.

Manning, Wendy D. 2001. "Childbearing in Cohabiting Unions: Racial and Ethnic Differences." *Family Planning Perspectives* 33(5):217–234.

Manning, Wendy D., and Jessica A. Cohen. 2015. "Teenage Cohabitation, Marriage, and Childbearing. " *Population Research and Policy Review* 34:161–177.

Manning, Wendy D., and Kara Joyner. 2019. "Demographic Approaches to Same-Sex Relationship Dissolution and Divorce." pp. 35–48 in *LGBTQ Divorce and Relationship Dissolution: Psychological and Legal Perspectives and Implications for Practice,* edited by Abbie E. Goldberg and Adam P. Romero. Oxford: Oxford University Press.

Manning, Wendy D., and Susan Brown. 2006. "Children's Economic Well-Being in Married and Cohabiting Parent Families." *Journal of Marriage and Family* 68(2):345–362.

Manning, Wendy D., Susan L. Brown, and J. Bart Stykes. 2014. "Family Complexity among Children in the United States." *The Annals of the American Academy of Political and Social Science* 654:48–65.

Manning, Wendy D., and Jessica A. Cohen. 2012. "Premarital Cohabitation and Marital Dissolution: An Examination of Recent Marriages." *Journal of Marriage and Family* 74(2):377–387.

Manning, Wendy D., Jessica A. Cohen, and Pamela J. Smock. 2011. "The Role of Romantic Partners, Family, and Peer Networks in Dating Couples' Views about Cohabitation." *Journal of Adolescent Research* 26(1): 115–149.

Manning, Wendy D., Peggy C. Giordano, and Monica A. Longmore. 2006. "Hooking Up the Relationship Contexts of 'Nonrelationship' Sex." *Journal of Adolescent Research* 21(5):459–483.

Manning, Wendy D., and Kathleen A. Lamb. 2003. "Adolescent Well-Being in Cohabiting, Married, and Single-Parent Families." *Journal of Marriage and Family* 65(4):876–893.

Manning, Wendy D., and Nancy S. Landale. 1996. "Racial and Ethnic Differences in the Role of Cohabitation in Premarital Childbearing." *Journal of Marriage and Family* 58(1):63–77.

Manning, Wendy D., and Pamela J. Smock. 1999. "New Families and Nonresident Father-Child Visitation." *Social Forces* 78:87–116.

Manning, Wendy, Pamela J. Smock, and C. Bergstrom-Lynch. 2009. "Cohabitation and Parenthood: Lessons from Focus Groups and In-Depth Interviews." In *Marriage and Family: Complexity and Perspectives,* edited by E. Peters and C. Kamp-Dush. New York: Columbia University Press.

Manning, Wendy D., Susan D. Stewart, and Pamela J. Smock. 2003. "The Complexity of Fathers' Parenting Responsibilities and Involvement with Nonresident Children." *Journal of Family Issues* 24(5):645–667.

Manning, Wendy D., and Bart Stykes. 2016. "Twenty Five Years of Change in Cohabitation in the U.S., 1987–2013." NCFMR Family Profiles, FP-15-01. Retrieved January 25, 2016 (https:// www .bgsu.edu/content/dam/BGSU/college -of-arts-and-sciences/NCFMR/documents /FP /FP-15-01-twenty-five-yrs-cohab-us.pdf).

Manning, Wendy D., Deanna Trella, Heidi Lyons, and Nola Cora DuToit. 2010. "Marriageable Women: A Focus on Participants in a Community Healthy Marriage Program." *Family Relations* 59(1):87–102.

Mansson, Daniel H., Scott A. Myers, and Lynn H. Turner. 2010. "Relational Maintenance Behaviors in the Grandchild-Grandparent Relationship." *Communication Research Reports* 27(1):68–79.

Marano, Hara Estroff. 2000. "Divorced? (Remarriage in America)." *Psychology Today* 33(2):56–60.

Marano, Hare Estroff. 2018. "Love and Power." *Psychology Today.* January 1. Retrieved February 10, 2020 (http://www.psychologytoday.com).

Marcotte, Amanda. 2014. "Republicans are Quietly Trying to Kill No-Fault Divorce." Retrieved April 1, 2016 (http://www. democraticunderground.com/10024821344).

Margalit, Liraz. 2014. *Psychology Today,* November 28. Retrieved February 3, 2015 (https://www.psychologytoday.com/blog /behind-online-behavior/201411/is-all -you-need-know-about-potential-partner).

Margelisch, Katja, Klaus A. Schneewind, Jeanine Violette, and Pasqualina Perrig-Chiello. 2017. "Marital Stability, Satisfaction and Well-Being in Old Age: Variability and Continuity in Long-Term Continuously Married Older Persons." *Aging & Mental Health* 21:389–398.

Margolis, Rachel. 2016. "The Changing Demography of Grandparenthood." *Journal of Marriage and Family* 78 (3): 610–622.

Margolis, R., and Myrskylä, M. (2015). "Parental Well-Being Surrounding First Birth as a Determinant of Further Parity Progression." *Demography* 52:1147–1166.

Marikar, Sheila. 2009. "Cher is Supporting Chasity's Sex Change, Though She Doesn't Understand It." ABC News, June 18. Retrieved January 12, 2010 (www.abcnews .go.com).

Mark, Kristen P., Justin R. Garcia, and Helen E. Fisher. 2015. "Perceived Emotional and Sexual Satisfaction Across Sexual Relationship Contexts: Gender and Sexual Orientation Differences and Similarities." *Canadian Journal of Human Sexuality* 24:120–130.

Mark, Kristen, Debby Herbenick, Dennis Fortenberry, Stephanie A. Sanders, and Michael Reece, M. 2011. "The Object of Sexual Desire: Examining the "What" in "What do you Desire?" *Journal of Sexual Medicine* 11:2709–2719.

Mark, Kristen P., and Sarah H. Murray. 2012. "Gender Differences in Desire Discrepancy as a Predictor of Sexual and Relationship Satisfaction in a College Sample of Heterosexual Romantic Relationships." *Journal of Sex and Marital Therapy* 38(2):198–215.

Mark, Kristen P., and Kristen N. Jozkowski. 2013. "The Mediating Role of Sexual and Nonsexual Communication between Relationship and Sexual Satisfaction in a Sample of College-Age Heterosexual Couples." *Journal of Sex and Marital Therapy* 39:410–427.

Markham, Melinda Stafford, and Marilyn Coleman. 2012. "The Good, The Bad, and The Ugly: Divorced Mothers' Experiences with Coparenting." *Family Relations* 61:586–600.

Markham, Melinda Stafford. 2017. "Communication Technology Use Following Divorce or Separation." *Family Focus,* FF71, p. F10–F11.

Markman, Howard J., Galena K. Rhoades, Scott M. Stanley, Erica P. Ragan , and Sarah W. Whitton. 2010. "The Premarital Communication Roots of Marital Distress and Divorce: The First Five Years of Marriage." *Journal of Family Psychology* 24(3):289–298.

Markman, Howard, Scott Stanley, and Susan L. Blumberg. 2001. *Fighting for Your Marriage: Positive Steps for Preventing Divorce and Preserving a Lasting Love.* San Francisco: Jossey-Bass.

Markon, Jerry. 2010. "The Baby He's Never Met; Va. Father Fights for Child His Girlfriend Sent to Utah for Adoption." *Washington Post,* April 14, p. A1.

Marks, Loren. 2012. "Same-Sex Parenting and Children's Outcomes: A Closer Examination of the American Psychological Association's Brief on Lesbian and Gay Parenting." *Social Science Research* 41:735–751.

Marks, Loren, Katrina Hopkins, Cassandra Chaney, Pamela Monroe, Olena Nesteruk, and Diane Sasser. 2008. "'Together We Are Strong': A Qualitative Study of Happy, Enduring, African American Marriages." *Family Relations* 57(2):172–185.

Marlow, Lenard, and S. Richard Sauber. 2013. *The Handbook of Divorce Mediation.* New York: Springer Science and Business Media.

Marquardt, Elizabeth. n.d. *The Revolution in Parenthood: The Emerging Global Clash between Adult Rights and Children's Needs.* Commission on Parenthood's Future. Retrieved October 4, 2006 (www .americanvalues.org).

Marriage Encounter. 2020. "Believe in a Better Marriage." Retrieved January 15, 2020 (http:// www.agme.org).

Marshall, Barbara L. 2009. "Science, Medicine and Virility Surveillance: 'Sexy Seniors' in the Pharmaceutical Imagination." *Sociology of Health and Illness* 32(2):211–224.

Marshall, Catherine A. 2010. *Surviving Cancer as a Family and Helping Co-Survivors Thrive.* Santa Barbara, CA: Praeger.

Marshall, Nancy, and Allison Tracy. 2009. "After the Baby: Work-Family Conflict and Working Mothers' Psychological Health." *Family Relations* 58 (October):380–391.

Marshall, Tara C., Kathrine Bejanyan, Gaia Castro, and Ruth A. Lee. 2013. "Attachment Styles as Predictors of Facebook-Related Jealousy and Surveillance in Romantic Relationships." *Personal Relationships* 20:1–22.

Marsiglio, William. 1991. "Paternal Engagement in Activities with Minor Children." *Journal of Marriage and the Family* 53:973–986.

———. 1992. "Stepfathers with Minor Children Living at Home." *Journal of Family Issues* 13:195–214.

———. 2004. *Stepdads: Stories of Love, Hope, and Repair.* Boulder, CO: Rowman & Littlefield.

———. 2012. "Interpreting a Fatherhood Legacy." *Family Focus* FF54 (Fall): F2–F3.

Martin, Brittny A., Ming Cui, Koji Ueno, and Frank D. Fincham. 2013. "Intimate Partner Violence in Interracial and Monoracial Couples." *Family Relations* 62 (February):202–211.

Martin JA, Hamilton BE, Osterman MJK, Driscoll AK, Drake P. 2018. *Births: Final data for 2017.* National Vital Statistics Reports, Vol 67 No 8. Hyattsville, MD: National Center for Health Statistics.

Martin, Joyce A., Brady E. Hamilton, and Michelle J. K. Osterman. 2019. *Births in the United States, 2018.* Retrieved December 24, 2019 (https://www.cdc.gov/nchs/products/databriefs/db346.htm).

Martin, Joyce A., Brady E. Hamiton, Michelle J. K. Osterman, Sally C. Curtin, and T. J. Matthews. 2015. "Births: Final Data for 2013." *National Vital Statistics Reports* 64(1). Retrieved February 5, 2016 (http://www.cdc.gov/nchs/data/nvsr/nvsr64/nvsr64_01.pdf).

Martin, Joyce A., Brady E. Hamilton, Paul D. Sutton, Stephanie J. Ventura, Fay Menacker, Sharon Kirmeyer, and T. J. Mathews. 2009. "Births: Final Data for 2006." *National Vital Statistics Report* 57(7). Hyattsville, MD: National Center for Health Statistics, January 7.

Martin, Joyce A., Michelle J. K. Osterman, and Paul D. Sutton. 2010. "Are Preterm Births on the Decline in the United States? Recent Data from the National Vital Statistics System." *National Vital Statistics Report* 57(7), January 7.

Martin, Joyce, Brady E. Hamilton, Michelle J. K. Osterman, Anne K. Driscoll, and Patrick Drake. 2018. *Births: Final Data for 2017.* National Vital Statistics Report, Vol. 67, No. 8. Retrieved December 11, 2019 (https://www.cdc.gov/nchs/data/nvsr/nvsr67/nvsr67_08-508.pdf).

Martin, Joyce, Brady Hamilton, Michelle Osterman, Anne Driscoll, and Patrick Drake. 2018. *Births: Final Data for 2017.* November 7. National Vital Statistics Reports 67(8). Hyattsville, MD: National Center for Health Statistics.

Martin, Joyce, Brady Hamilton, Michelle Osterman, and Anne Driscoll. 2019. *Births: Final Data for 2018.* November 27. National Vital Statistics Reports 68 (13). Hyattsville, MD: National Center for Health Statistics.

Martin, Julie. 2019. "This Is What No One Told Me about Suddenly Joining the Sandwich Generation." *Huffpost Personal.* July 13. Retrieved September 5, 2019 (www.huffpost.com).

Martin, Nina, and Renee Montagne. 2017. "Lost Mothers: Maternal Mortality in the U.S. " Retrieved December 24, 2019 (https://www.npr.org/2017/05/12/527806002/focus-on-infants-during-childbirth-leaves-u-s-moms-in-danger).

Martin, Todd F. 2018. "Family Development Theory 30 Years Later." *Journal of Family Theory & Review* 10(1):49–69.

Martin, Molly A., and Adam M. Lippert. 2012. "Feeding Her Children, But Risking Her Health: The Intersection of Fender, Household Food Insecurity, and Obesity." *Social Science and Medicine* 74:1754–1764.

Martin, Philip, and Elizabeth Midgley. 2006. "Immigration: Shaping and Reshaping America." *Population Bulletin* 61(4). Washington, DC: Population Reference Bureau.

Martin, Sandra L., April Harris-Britt, Yun Li, Kathryn E. Moracco, Lawrence L. Kupper, and Jacquelyn C. Campbell. 2004. "Change in Intimate Partner Violence during Pregnancy." *Journal of Family Violence* 19:243–247.

Martinez, Gladys M., Anjani Chandra, Joyce C. Abma, Jo Jones, and William D. Mosher. 2006. "Fertility, Contraception, and Fatherhood: Data on Men and Women from Cycle 6 (2002) of the National Survey of Family Growth." *Vital and Health Statistics* 23(26), May. Retrieved January 27, 2007 (www.nchs.gov).

Martinez, G., C. E. Copen, and J. C. Abma. 2011. *Teenagers in the United States: Sexual Activity, Contraceptive Use, and Childbearing, 2006–2010 National Survey of Family Growth.* National Center for Health Statistics, Vital Health Stat 23(31).

Martinez, Gladys, Kimberly Daniels, and Anjani Chandra. 2012. "Fertility of Men and women Aged 15–44 Years in the United States: National Survey of Family Growth, 2006–2010." *National Health Statistics Reports* 51 (April 12). Washington, DC: U.S. Department of Health and Human Services and Centers for Disease Control and Prevention.

Martyn, Kristy, Carol Loveland-Cherry, Antonia Villarruel, Esther Gallegos Cabriales, Yan Zhou, David Ronis, and Brenda Eakin. 2009. "Mexican Adolescents' Alcohol Use, Family Intimacy, and Parent-Adolescent Communication." *Journal of Family Nursing* 15(2):152–170.

Masanori, Ishimori, Ikuo Daibo, and Yuji Kanemasa. 2004. "Love Styles and Romantic Love Experiences in Japan." *Social Behavior and Personality* 32(3):265–281.

Masarik, April, Monica Martin, Emilio Ferrer, Frederick Lorenz, Katherine Conger, and Rand Conger. 2016. "Couple Resilience to Economic Pressure over Time and across Generations." *Journal of Marriage and Family* 78 (2): 326–345.

Masci, David, Anna Brown, and Jocelyn Kiley. 2019. "5 Facts about Same-Sex Marriage." *Facttank: News in the Numbers.* September 14. Pew Research Organization. Retrieved January 30, 2019 (www.pewresearch.org).

Masci, David and Drew DeSilver. 2019. " A Global Snapshot of Same-Sex Marriage." Retrieved October 17, 2019 (https://www.pewresearch.org/fact-tank/2019/06/21/global-snapshot-same-sex-marriage/).

Mason, Ashley, E., Rita W. Law, Amanda E. B. Bryan, Robert M. Portley, and David A. Sbarra. 2012. "Facing a Breakup: Electromyographic Responses Moderate Self-Concept Recovery Following Romantic Separation." *Personal Relationships* 19:551–568.

Mason, Mary Ann, Mark A. Fine, and Sarah Carnochan. 2001. "Family Law in the New Millennium: For Whose Families?" *Journal of Family Issues* 22(7):859–881.

Mason, Mary Ann, Sydney Harrison-Jay, Gloria Messick Svare, and Nicholas H. Wolfinger. 2002. "Stepparents: De Facto Parents or Legal Strangers?" *Journal of Family Issues* 23(4):507–522.

Mason, Mary Ann, and David W. Simon. 1995. "The Ambiguous Stepparent: Federal Legislation in Search of a Model." *FAM. lq* 29:445.

Masten, Ann S., and Amy R. Monn. 2015. "Conceptualizing Family and Individual Resilience." *Family Relations* 64(1):5–21.

Masten, Ann S. 2018. "Resilience Theory and Research on Children and Families: Past, Present, and Promise." *Journal of Family Theory & Review* 10 (1): 12–31.

Masters, N. Tatiana, Erin Casey, Elizabeth A. Wells, and Diane M. Morrison. 2013. "Sexual Scripts among Young Heterosexually Active Men and Women: Continuity and Change." *Journal of Sex Research* 50(5):409–420.

Masters, William H., and Virginia E. Johnson. 1966. *Human Sexual Response.* Boston: Little, Brown.

———. 1976. *The Pleasure Bond: A New Look at Sexuality and Commitment.* New York: Bantam.

Masters, William H., Virginia E. Johnson, and Robert C. Kolodny. 1994. *Heterosexuality.* New York: HarperCollins.

Mather, Mark. 2009. "Children in Immigrant Families Chart New Path." Washington, DC: Population Reference Bureau.

———. 2011. "More Young Adults in U.S. Postponing Marriage, Living at Home, Disconnected from Work and School." *Population Reference Bureau,* December. Retrieved August 29, 2012 (http://www.prb.org).

———. 2012. "Fact Sheet: The Decline of U.S. Fertility." *Population Reference Bureau, July.* Retrieved December 7, 2012 (http://www.prb.org).

Mather, Mark, Linda Jacobsen, and Kelvin Pollard. 2015. "Aging in the United States." *Population Bulletin* 70 (2). Population Reference Bureau. Retrieved March 10, 2016 (https://www.prb.org/).

Mathews, T. J., and Brady E. Hamilton. 2019. *Total Fertility Rates by State and Race and Hispanic Origin: United States, 2017.* National Vital Statistics Report, Vol. 68, No. 1. Retrieved December 11, 2019 (https://stacks.cdc.gov/view/cdc/61878).

Matiasko, Jennifer, Leslie Grunden, and Jody Ernst. 2007. "Structural and Dynamic Process Family Risk Factors: Consequences for Holistic Adolescent Functioning." *Journal of Marriage and Family* 69(3):654–674.

Matsunaga, Masaki, and Tadasu Todd Imahori. 2009. "Profiling Family Communication Standards." *Communication Research* 36(1):3–31.

Matta, William J. 2006. *Relationship Sabotage: Unconscious Factors That Destroy Couples, Marriages, and Family.* Westport, CT: Praeger.

Matthews, Sarah H. 2012. "Enhancing the Qualitative-Research Culture in Famil Studies." *Journal of Marriage and Family* 74(4) (August):666–670.

Mattingly, Marybeth J., and Suzanne M. Blanchi. 2003. "Gender Differences in the Quantity and Quality of Free Time: The US Experience." *Social Forces* 81:999–1030.

Matzek, A. E., C. G. Gudmunson, and S. M. Danes. 2010. "Spousal Capital as a Resource for Couples Starting a Business." *Family Relations* 59:58–71.

May, Rollo. 1975. "A Preface to Love." pp. 114–119 in *The Practice of Love*, edited by Ashley Montagu. Englewood Cliffs, NJ: Prentice Hall.

May, Ross W., Shanmukh V. Kamble, and Frank D. Fincham. 2015. "Forgivingness, Forgivability, and Relationship-Specific Effects in Responses to Transgressions in Indian Families." *Family Relations* 64:332–346.

Mayes, Tessa. 2001. "Career Women Rent Wombs to Beat Hassles of Pregnancy." *The Sunday Times—London*, August 7. Retrieved January 14, 2013 (http://www.sunday-times.co.uk).

May, Ashley. 2017. "What Your Religion Has to Say about How You Become a Parent." *USA Today Network.* April 22. Retrieved August 16, 2019 (www.usatoday.com).

Mayo Clinic Staff. 2017. "Domestic Violence against Women: Recognize Patterns, Seek Help." Retrieved February 16, 2020 (https://www.mayoclinic.org).

Mayo Clinic Staff. 2020. "Domestic Violence." Retrieved February 17, 2020 (https://www.mayoclinic.org).

Mayo Clinic Staff. 2017. "Family Planning: Get the Facts about Pregnancy Spacing." Retrieved December 20, 2019 (https://www.mayoclinic.org/healthy-lifestyle/getting-pregnant/in-depth/family-planning/art-20044072).

Mays, Vickie M., and Susan D. Cochran. 1999. "The Black Woman's Relationship Project: A National Survey of Black Lesbians." pp. 59–66 in *The Black Family: Essays and Studies*, 6th ed., edited by Robert Staples. Belmont, CA: Wadsworth.

Mazelis, Joan Maya, and Laryssa Mykyta. 2011. "Relationship Status and Activated Kin Support: The Role of Need and Norms." *Journal of Marriage and Family* 73:430–445.

McAdoo, Harriette Pipes. 2007. *Black Families,* 4th edition. Thousand Oaks, CA: Sage.

McBride-Chang, Catherine, and Lei Chang. 1998. "Adolescent–Parent Relations in Hong Kong: Parenting Styles, Emotional Autonomy, and School Achievement." *Journal of Genetic Psychology* 159(4):421–435.

McCabe, Janice, Emily Fairchild, Liz Grauerholz, Bernice A. Pescosolido, and Daniel Tope. 2011. "Gender in Twentieth-Century Children's Books." *Gender and Society* 25(2):197–226.

McCabe, Marita P., and Denise Goldhammer. 2012. "Demographic and Psychological Factors Related to Sexual Desire among Heterosexual Women in a Relationship." *Journal of Sex Research* 49(1):78–87.

McCann, Brandy Renee. 2012. "The Persistence of Gendered Kin Work in Maintaining Family Ties: A Review Essay." *Journal of Family Theory and Review* 1(3) (September):219–251.

McCarthy, Justin. 2014. Americans' Views on Origins of Homosexuality Remain Split. Retrieved December 3, 2015 (http://www.gallup.com/poll/170753/americans-views-origins-homosexuality-remain-split.aspx?g_source=do you think gay or lesbian relations between cons&g_medium=search&g_campaign=tiles).

———. 2015a. "U.S. Support for Gay Marriage Stable after High Court Ruling." Gallup, July 17. Retrieved September 19, 2015 (http://www.gallup.com.poll).

———. 2015b. "Record High of 60 Percent of Americans Support Same-Sex Marriage." Retrieved March 3, 2016 (http://www.gallup.com/poll/183272/record-high-americans-support-sex-marriage.aspx).

McCarthy, Justin. 2019. "Americans Still Greatly Overestimate U.S. Gay Population." Retrieved October 21, 2019 (https://news.gallup.com/poll/259571/americans-greatly-overestimate-gay-population.aspx).

McCarthy, Justin. 2019. "U.S. Support for Gay Marriage Stable, at 63%" Retrieved October 17, 2019 (https://news.gallup.com/poll/257705/support-gay-marriage-stable.aspx).

McCarthy, Justin. 2019. "Gallup First Polled on Gay issues in '77. What Has Changed?" Retrieved December 1, 2019 (https://news.gallup.com/poll/258065/gallup-first-polled-gay-issues-changed.aspx).

McClain, Lauren Rinelli. 2011. "Better Parents, More Stable Partners: Union Transitions among Cohabiting Parents." *Journal of Marriage and Family* 73(5):889–901.

McClendon, David. 2016. "Religion, Marriage Markets, and Assortative Mating in the United States." *Journal of Marriage and Family* 78(5):1399–1421.

McClendon, David and Aleksandra Sandstrom. 2016. "Child Marriage is Rare in U.S., Though This Varies by State." Retrieved November 16, 2019 (https://www.pewresearch.org/fact-tank/2016/11/01/child-marriage-is-rare-in-the-u-s-though-this-varies-by-state/).

McLanahan, Sara, and Isabel Sawhill. 2015. "Marriage and Child Wellbeing Revisited: Introducing the Issue." *The Future of Children*, 3–9.

McClintock, Elizabeth Aura. 2011. "Handsome Wants as Handsome Does: Physical Attractiveness and Gender Differences in Revealed Sexual Preferences." *Biodemograhy and Social Biology* 57:221–257.

McClintock, Elizabeth Aura. 2017. "Occupational Sex Composition and Gendered Housework Performance: Compensation or Conventionality." *Journal of Marriage and Family* 79 (2): 475–510.

———. 2014a. "Beauty and Status: The Illusion of Exchange in Partner Selection?" *American Sociological Review* 79: 575-604.

———. 2014b. "Why Break-ups are Actually Tougher on Men." *Psychology Today*, December 19. Retrieved January 17, 2015 (https://www.psychologytoday.com/blog/it-s-man-s-and-woman-s-world/201412/why-breakups-are-actually-tougher-men).

McClure, Robin. 2010. "Afterschool Child Care—Number of Kids Home Alone After School Has Risen." Retrieved February 13, 2013 (http://childcare.about.com).

McCoy, Kelsey, and James Oelschager. 2013. *Sexual Coercion Awareness and Prevention.* Retrieved May 27, 2013 (http://www.fit.edu/caps/documents/SexualCoercion_000.pdf).

McCormick, Meghan, JoAnn Hsueh, Christine Merrilees, Patricia Chou, and E. Mark Cummings. 2017. "Moods, Stressors, and Severity of Marital Conflict: A Daily Diary Study of Low-Income Families." *Family Relations* 66 (3): 425–440.

McCubbin, Hamilton I., and Marilyn A. McCubbin. 1991. "Family Stress Theory and Assessment: The Resiliency Model of Family Stress, Adjustment and Adaptation." pp. 3–32 in *Family Assessment Inventories for Research and Practice*, 2nd ed., edited by Hamilton I. McCubbin and Anne I. Thompson. Madison: University of Wisconsin, School of Family Resources and Consumer Services.

———. 1994. "Families Coping with Illness: The Resiliency Model of Family Stress, Adjustment, and Adaptation." Chapter 2 in *Families, Health, and Illness.* St. Louis: Mosby.

McCubbin, Hamilton I., and Joan M. Patterson. 1983. "Family Stress and Adaptation to Crisis: A Double ABCX Model of Family Behavior." pp. 87–106 in *Family Studies Review Yearbook*, Vol. 1, edited by David H. Olson and Brent C. Miller. Newbury Park, CA: Sage.

McCubbin, Hamilton I., Anne I. Thompson, and Marilyn A. McCubbin, eds. 1996. *Family Assessment: Resiliency, Coping and Adaptation: Inventories for Research and Practice.* Madison: University of Wisconsin Press.

McCubbin, Hamilton I., Elizabeth Thompson, Anne Thompson, Jo A. Futrell, and Suniya Luthar. 2001. "The Dynamics of Resilient Families." *Contemporary Psychology* 48(2):154–156.

McCubbin, Laurie D., Hamilton I. McCubbin, Wei Zhang, Lisa Kehl, and Ida Strom. 2013. "Relational Well-Being: An Indigenous Perspective and Measure." *Family Relations* 62(2):354–365.

McCubbin, Marilyn A. 1995. "The Typology Model of Adjustment and Adaptation: A Family Stress Model." *Guidance and Counseling* 10(4):31–39.

McCubbin, Marilyn A., and Hamilton I. McCubbin. 1989. "Theoretical Orientations to Family Stress and Coping." pp. 3–43 in *Treating Families Under Stress*, edited by Charles Figley. New York: Brunner/Mazel.

McCue, Margi Laird. 2008. *Domestic Violence: A Reference Handbook*, 2nd ed. Santa Barbara, CA: ABC-CLIO, Inc.

McDaniel, Brandon T., and Sarah M. Coyne. 2016. "'Technoference': The Interference of Technology in Couple Relationships and Implications for Women's Personal and Relational Well-Being." *Psychology of Popular Media Culture* 5:85.

McDaniel, Brandon T., and Michelle Drouin. 2015. "Sexting among Married Couples: Who Is Doing It, and Are They More Satisfied?" *Cyberpsychology, Behavior, and Social Networking* 18:628–634.

McDermott, Brett, and Vanessa E. Cobham. 2012. "Family Functioning in the Aftermath of a Natural Disaster." *BMC Psychiatry* 12:55.

McDonough, Tracy A. 2010. "A Policy Capturing Investigation of Battered Women's Decisions to Stay in Violent Relationships." *Violence and Victims* 25(2):165–184.

McDowell, Margaret A, Debra J. Brody, and Jeffery P. Hughes. 2007. "Has Age at Menarche Changed? Results from the National Health and Nutrition Examination Survey (NHANES) 1999–2004." *Journal of Adolescent Health* 40(3):227–233.

McFarland, Daniel A., Dan Jurafsky, and Craig Rawlings. 2013. "Making the Connection: Social Bonding in Courtship Situations." *American Journal of Sociology* 118:1596–1649.

McGene, Juliana, and Valarie King. 2012. "Implications of New Marriages and Children for Coparenting in Nonresident Father Families." *Journal of Family Issues* 33(12):1619–1641.

McGough, Lucy S. 2005. "Protecting Children in Divorce: Lessons from Caroline Norton." *Maine Law Review* 57:13–37.

McGowin, Emily Hunter. 2018. *Quivering Families: The Quiverfull Movement and Evangelical Theology of the Family.* Minneapolis, MN: Fortress Press.

McGraw, Lori A., Anisa M. Zvonkovic, and Alexis J. Walker. 2000. "Studying Postmodern Families: A Feminist Analysis of Ethical Tensions in Work and Family Research." *Journal of Marriage and the Family* 62(February):68–77.

McHale, James, Maureen R. Waller, and Jessica Pearson. 2012. "Coparenting Interventions for Fragile Families: What Do We Know and Where Do We Need to Go Next?" *Family Process* 51(3):284–306.

McHale, Susan, Kimberly Updegraff, and Shawn Whiteman. 2012. "Sibling Relationships and Influences in Childhood and Adolescence." *Journal of Marriage and Family* 74(5):913–930.

McHugh, Maureen F. 2007. "The Evil Stepmother." Second Wives Café: Online Support for Second Wives and Stepmothers. Retrieved May 2, 2007 (secondwivescafe.com).

McLaren, Leah. 2017. "I Can't Forget the Horror of My Son's Birth. " Retrieved December 24, 2019 (https://www.theguardian.com /lifeandstyle/2017/may/07/i-cant-forget -the-horror-of-my-sons-birth -post-traumatic-stress-disorder-childbirth).

McIntosh, Peggy. 1988. "White Privilege and Male Privilege: A Personal Account of Coming to See Correspondences Through Work in Women's Studies." Working paper no. 189. Wellesley, MA: Wellesley College Center for Research on Women.

McIntyre, Matthew H., and Carolyn Pope Edwards. 2009. "The Early Development of Gender Differences." *Annual Review of Anthropology* 38(1):83–97.

McKay, Alexander, E. Sandra Byers, Susan D. Voyer, Terry P. Humphreys, and Chris Markham. 2014. "Ontario Parents' Opinions and Attitudes towards Sexual Health Education in the Schools." *Canadian Journal of Human Sexuality* 23(3):159–166.

McKinley, Jesse, and John Schwartz. 2010. "Court Rejects Same-Sex Marriage Ban in California." *The New York Times, August* 4. Retrieved August 10, 2010 (www.nytimes .com).

McLanahan, Sarah (ed.). 2010. "Transition to Adulthood." *The Future of Children* 20(1). Princeton, NJ: Princeton University and the Brookings Institution.

McLanahan, Sara, and Marcia J. Carlson. 2002. "Welfare Reform, Fertility, and Father Involvement." *The Future of Children* 12(1):147–165.

McLanahan, Sara, Irwin Garfinkel, Nana E. Reichman, and Julien O. Teitler. 2001. "Unwed Parents or Fragile Families? Implications for Welfare and Child Support Policy." pp. 202–228 in *Out of Wedlock: Causes and Consequences of Nonmarital Fertility*, edited by Lawrence L. Wuand Barbara Wolfe. New York: Russell Sage.

McLanahan, S. S., and Adams, J. (1987). "Parenthood and Psychological Well-Being." *Annual Review of Immunology* 5:237–257.

McLaren, Rachel M., Denise H. Solomon, and Jennifer S. Priem. 2012. "The Effect of Relationship Characteristics and Relational Communication on Experiences of Hurt from Romantic Partners." *Journal of Communication* 62:950–971.

McLaughlin, Katy. 2018. "Wealthy Parents Help Child Athletes Go Pro in Their Own Backyards." *The Wall Street Journal.* August 16. Retrieved August 20, 2019 (https://www.waj.com).

McLoyd, Vonnie C., Ana Mari Cauce, David Takeuchi, and Leon Wilson. 2000. "Marital Processes and Parental Socialization in Families of Color: A Decade Review of Research." *Journal of Marriage and Family* 62:1070–1093.

McMahon, Thomas, Justin Winkel, Nancy Suchman, and Bruce Rounsaville. 2007. "Drug-Abusing Fathers: Patterns of Pair Bonding, Reproduction, and Paternal Involvement." *Journal of Substance Abuse Treatment* 33:295–302.

McMahon-Howard, Jennifer, Jody Clay-Warner, and Linda Renzulli. 2009. "Criminalizing Spousal Rape: The Diffusion of Legal Reforms." *Sociological Perspectives* 52(4):505–531.

McManus, Mike. 2006. "Inside/Out Dads: Helping Prisoners Reenter Society." Retrieved November 17, 2006 (smartmarriages.com).

McManus, Patricia A., and Thomas DiPrete. 2001. "Losers and Winners: The Financial Consequences of Separation and Divorce for Men." *American Sociological Review* 66:246–268.

McNamara, Melissa. 2009. "Elder Care Benefits." CBS Evening News, February 11. Retrieved February 11, 2013 (http://www .cbsnews.com).

McNamara, Robert Hartmann. 2008. *Homelessness in America.* Westport, CT: Praeger.

McNamee, Catherine B., Paul Amato, and Valarie King. 2014. "Nonresident Father Involvement With Children and Divorced Women's Likelihood of Remarriage." *Journal of Marriage and Family* 76:862–874.

McQuillan, Julia, Arthur L. Greil, and Karina M. Shreffler. 2011. "Pregnancy Intentions among Women Who Do Not Try: Focusing on Women who are Okay Either Way." *Maternal and Child Health Journal* 15:178–187.

McQuillan, Julia, Arthur Greil, Karina Shreffler, Patricia Wonch-Hill, Kari Gentzler, and John Hathcoat. 2012. "Does the Reason Matter? Variations in Childlessness Concerns among U.S. Women." *Journal of Marriage and Family* 74(5):1166–1181.

McWey, Lenore M., Amy M. Claridge, Armeda Stevenson Wojciak, and Cassandra G. Lettenberger -Klein. 2015. "Parent-Adolescent Relationship Quality as an Intervening Variable on Adolescent Outcomes among Families at Risk: Dyadic Analyses." *Family Relations* 64(2):249–262.

McWey, Lenore M., Tammy L. Henderson, and Jenny Burroughs Alexander. 2008. "Parental Rights and the Foster Care System." *Journal of Family Issues* 29(8):1031–1050.

McWilliams, Summer, and Anne E. Barrett. 2014. "Online Dating in Middle and Later Life Gendered Expectations and Experiences." *Journal of Family Issues* 35:411–436.

Mead, George Herbert. 1934. *Mind, Self, and Society.* Chicago: University of Chicago Press.

Meadows, Sarah O., Sara S. McLanahan, and Jeanne Brooks-Gunn. 2007. "Parental Depression and Anxiety and Early Childhood Behavior Problems Across Family Types." *Journal of Marriage and Family* 69(5):1162–1177.

Meadows, Sarah O., Sara S. McLanahan, and Jean T. Knab. 2009. "Economic Trajectories in Non-Traditional Families with Children." *The Rand Corporation*, August 21, pp. 1–41.

Medeiros, Rose A., and Murray A. Straus. 2006. "Risk Factors for Physical Violence Between Dating Partners: Implications for Gender-Inclusive Prevention and Treatment of Family Violence." pp. 59–87 in *Family Interventions in Domestic Violence A Handbook of Gender-Inclusive Theory and Treatment*, edited by John C.Hamel and Tanya L. Nicholls. New York: Springer.

Meerwijk, Esther L., and Jae M. Sevelius. 2017. "Transgender Population Size in the United States: A Meta-Regression of Population-Based Probability Samples." *American Journal of Public Health* 107:e1–e8.

Mehta, Pranjal H., Amanda C. Jones, and Robert A. Josephs. 2008. "The Social Endocrinology of Dominance: Basal Testosterone Predicts Cortisol Changes and Behavior Following Victory and Defeat." *Journal of Personality and Social Psychology* 94(6):1078–1093.

Meier, Ann, and Gina Allen. 2009. "Romantic Relationships from Adolescence to Young Adulthood." *The Sociological Quarterly* 50(2):308–335.

Meier, Ann, Kathleen E. Hull, and Timothy A. Ortyl. 2009. "Young Adult Relationship Values at the Intersection of Gender and Sexuality." *Journal of Marriage and Family* 71(3):510–525.

Meier, Ann, and Kelly Musick. 2014. "Variation in Associations between Family Dinners and Adolescent Well-Being." *Journal of Marriage and Family* 76(1):13–23.

Meier, S. Colton, Carla Sharp, Jared Michonski, Julia C. Babcock, and Kara Fitzgerald. 2013. "Romantic Relationships of Female-to-Male Trans Men: A Descriptive Study." *International Journal of Transgenderism* 11.75–05.

Mekouar, Dora. 2019. "What Americans Really Earn After Taxes. " Retrieved December 14, 2019 (https://www .voanews.com/usa/all-about-america /what-americans-really-earn-after-taxes).

Melby, Todd. 2010. "Sexuality for the Young and Old." *Contemporary Sexuality* 44(1):1, 4– 6.

Meltzer, Andrea L., and James K. McNulty. 2014. "'Tell Me I'm Sexy . . . and Otherwise Valuable': Body Valuation and Relationship Satisfaction." *Personal Relationships* 21:68–87.

Mendes, Elizabeth, Lydia Saad, and Kyley McGeeney. 2012. "Stay-at-Home Moms Report More Depression, Sadness, Anger." Gallup, May 18. Retrieved July 15, 2013 (http://www.gallup .com).

Mendoza, Angela Nancy, and Christine A. Fruhauf. 2015. "Grandparents Raising Grandchildren: Risk and Resilience Through Conflict." *Family Focus* Issue FF65 (Fall):P.F11.

Menning, Chadwick L. 2008. "'I've Kept It That Way on Purpose': Adolescents' Management of Negative Parental Relationship Traits after Divorce and Separation." *Journal of Contemporary Ethnography* 37:586–618.

Merton, Robert K. 1973 [1942]. "The Normative Structure of Science." Chapter 3 in *The Sociology of Science: Theoretical and Empirical Investigations,* edited by Robert K. Merton. Chicago: University of Chicago Press.

———. 1968 [1949]. *Social Theory and Social Structure.* New York: The Free Press.

Merz, Eva-Marie, and Aart C. Liefbroer. 2012. "The Attitude Toward Voluntary Childlessness in Europe: Cultural and Institutional Explanations." *Journal of Marriage and Family* 74(3):587–600.

Messner, Michael A. 1997. *The Politics of Masculinity: Men in Movements.* Thousand Oaks, CA: Sage.

Meyer, Cathy. 2011. "Should Premarital Counseling Be Required by Law?" March 20. Retrieved December 7, 2012 (http:// divorcesupport.about.com/b /2011/03/20 /should-premarital -counseling-be-required-by -law.htm).

———. 2013. "The Difference Between Domestic Abuse and Normal Marital Conflict." Retrieved March 14, 2013 (http:// divorcesupport.about.com).

Meyer, D. R., and M. Cancian. 2012. "'I'm Not Supporting His Kids': Nonresident Fathers' Contributions Given Mothers' New Fertility." *Journal of Marriage and Family,* 74(1):132–151.

Meyer, Daniel R., Maria Cancian, and Steven T. Cook. 2017. "The Growth in Shared Custody in the United States: Patterns and Implications." *Family Court Review* 55:500–512.

Meyer, Harris. 2009. "Getting Back in the Game." *The Oregonian,* November 4.

Meyer, Jennifer. 2007. "Making Good Decisions." *Omaha World Herald,* January 22.

Meyer, Julie. 2012. *Centenarians: 2010.* Special Reports 2010, No. C2010SR-03. Washington, DC: U.S. Census Bureau.

Meyer, Madonna H., and Marcia L. Bellas. 2001. "U.S. Old-age Policy and the Family." pp. 191–201 in *Families in Later Life: Connections and Transitions,* edited by Alexis J. Walker, Margaret Manoogian-O'Dell, Lori A. McGraw, and Diana L. G. White. Thousand Oaks, CA: Pine Forge.

Meyers, Dvora. 2011. "Virtual Visitation Rights." *The New York Times,* March 18. Retrieved from http://www.nytimes.com /2011/03/20 /fashion/20Facebook .html?pagewanted=all.

Michaels, Marcia L. 2006. "Stepfamily Enrichment Program: A Preventive Intervention for Remarried Couples." *Journal for Specialists in Group Work* 31(2):135–152.

Middlemiss, Wendy. 2016. "Building a Foundation for Resiliency from the Inside Out." *Family Relations* 65 (1):7–8.

Middlemiss, Wendy, and William McGuigan. 2005. "Ethnicity and Adolescent Mothers' Benefit from Participation in Home-Visitation Services." *Family Relations* 54(2):212–224.

Mikulincer, Mario, and Phillip R. Shaver. 2012. "Adult Attachment Orientations and Relationship Processes." *Journal of Family Theory and Review* 4(4):239–274.

Milardo, Robert M. 2005. "Generative Uncle and Nephew Relationships." *Journal of Marriage and Family* 67(5):1226–1236.

Miles, Donna. 2010. "Federal Employees' Same-Sex Domestic Partners Garn New Benefits." American Forces Press Service, June 3. Retrieved September 3, 2012 (http://www.af.mil/news).

Miles, Leonora. 2008. "The Hidden Toll— Financial Abuse Is one of the Most Common Forms of Elder Abuse." *Adults Learning* 19(9):28–30.

Milevsky, Avidan. 2019. "Parental Factors, Psychological Well-Being, and Sibling Dynamics: A Mediational Model in Emerging Adulthood." *Marriage & Family Review* 55:476–492.

Miller, Amanda J., Sharon Sassler, and Dela Kusi-Appouh. 2011. "The Specter of Divorce: Views from Working- and Middle-Class Cohabitors." *Family Relations* 60(5): 602–616.

Miller, Claire Cain. 2014a. "Paternity Leave: The Rewards and the Remaining Stigma." *The New York Times,* November 7. Retrieved October 10, 2015 (http://www .nytimes.com).

———. 2014b. "Can Family Leave Policies Be Too Generous? It Seems So." *The New York Times* August 9. Retrieved April 22, 2015 (http://www .nytimes.com).

———. 2015a. "More Than Their Mothers, Young Women Plan Career Pauses." *The New York Times,* July 22. Retrieved October 4, 2015 (http://www.nytimes.com).

———. 2015b. "Stressed, Tired, Rushed: A Portrait of the Modern Family." *The New York Times,* November 4. Retrieved November 8, 2015 (http://www.nytimes.com).

———. 2015c. "Mounting Evidence of Advantages for Children of Working Mothers." *The New York Times,* May 15. Retrieved May 19, 2015 (http://www .nytimes.com).

———. 2015d. "When Family-Friendly Policies Backfire." *The New York Times,* May 26, 2015. Retrieved May 28, 2015 (http:// www.nytimes.com).

Miller, Andrew H., and James Eli Adams. 1996. *Sexualities in Victorian Britain.* Bloomington, IN: Indiana University Press.

Miller, Claire Cain. 2018. "Americans are Having Fewer Babies: They Told Us Why." Retrieved December 14, 2019 (https://www.nytimes .com/2018/07/05/upshot/americans-are -having-fewer-babies-they-told-us-why.html).

Miller, Claire C. 2019. "Women Did Everything Right. Then Work Got 'Greedy.'" *The New York Times.* April 26. Retrieved December 9, 2019 (www.nytimes.com).

Miller, Claire C. 2018a. "The Relentlessness of Modern Parenting." *The New York Times.* December 25.   Retrieved December 9, 2019 (www.nytimes.com).

Miller, Claire C. 2018b "When Wives Earn More than Husbands, Neither Partner Likes to Admit It." *The New York Times.* July 17. Retrieved December 9, 2019 (www.nytimes.com).

Miller, Claire C. 2018c. "The Costs of Motherhood Are Rising, and Catching Women off Guard." *The New York Times.* August 17. Retrieved December 9, 2019 (www.nytimes .com).

Miller, Claire C. 2018d. "Americans Value Equality at Work More than Equality at Home." *The New York Times.* December 3. Retrieved December 9, 2019 (www.nytimes.com).

Miller, Claire C. 2018e. "A 'Generationally Perpetuated' Pattern: Daughters Do More Chores." *The New York Times.* April 8. Retrieved December 9, 2019 (www.nytimes.com).

Miller, Claire C. 2015. "Millennial Men Aren't the Dads They Thought They'd Be." *The New York Times.* July 30. Retrieved December 9, 2019 (www.nytimes.com).

Miller, Claire Cain. 2019. "Sweden Finds a Simple Way to Improve New Mothers' Health. It Involves Fathers. " Retrieved December 27, 2019 (https://www.nytimes.com/2019/06/04 /upshot/sweden-finds-a-simple-way-to-improve -new-mothers-health-it-involves-fathers.html).

Miller, Courtney Waite, and Michael E. Roloff. 2005. "Gender and Willingness to Confront Hurtful Messages from Romantic Partners." *Communication Quarterly* 53(3):323–338.

Miller, Daniel P., and Jina Chang. 2015. "Parental Work Schedules and Child Overweight or Obesity: Does Family Structure Matter?" *Journal of Marriage and Family* 77(5):1266–1281.

Miller, Dawn. 2004. "From the Author. The Stepfamily Life: A Column from Life in the Blender" (www.thestepfamilylife.com).

———. n.d. "Don't Go Nuclear—Negotiate." SelfGrowth.com. Retrieved September 21, 2004 (www.selfgrowth.com /articles/Miller).

Miller, Elizabeth. 2000. "Religion and Families over the Life Course." pp. 173–186 in *Families Across Time: A Life Course Perspective,* edited by Sharon J. Price, Patrick C. McKenry, and Megan J. Murphy. Los Angeles: Roxbury.

Miller, Laurie C. 2005a. *Handbook of International Adoptive Medicine.* New York: Oxford University Press.

———. 2005b. "International Adoption, Behavior, and Mental Health." *Journal of the American Medical Association* 293(20):2533–2535.

Miller, Richard, Cody Hollist, Joseph Olsen, and David Law. 2013. "Marital Quality and Health Over 20 Years: A Growth Curve Analysis." *Journal of Marriage and Family* 75(3):667–680.

Mills, C. Wright. 2000 [1959]. *The Sociological Imagination.* 40th anniversary ed. New York: Oxford University Press.

Mills, Terry L., and Janet M. Wilmoth. 2002. "Intergenerational Differences and Similarities in Life-Sustaining Treatment Attitudes and Decision Factors." *Family Relations* 51(1):46–54.

Milne-Tyte, Ashley. 2014. "Why Women Don't Ask for More Money." National Public Radio, April 8. Retrieved April 27, 2015 (http://npr .org).

Milner, Joel S., Cynthia J. Thomsen, Julie L. Crouch, Mandy M. Rabenhorst, Patricia M. Martens, Christopher W. Dyslin, Jennifer M. Guimond, Valerie A. Stander, and

Ming, Dan. 2019. "Female Runners with High Testosterone Must Take Hormone Suppressants to Compete, Sports Court Rules." Retrieved September 29, 2019 (https://www.vice.com /en_us/article/wjvda4/female-runners-with -high-testosterone-must-take-hormone-blockers -to-compete-sports-court-rules).

Mind Tools. 2020. "How Emotionally Intelligent Are You?" Retrieved January 25, 2020 (https:// www.mindtools.com/pages/article/ei-quiz .htm).

Min, Pyong Gap. 2002. "Korean American Families." pp. 193–211 in *Minority Families in the United States: A Multicultural Perspective,* 3rd ed., edited by Ronald L. Taylor. Upper Saddle River, NJ: Prentice Hall.

Minino, Arialdi M., and Sherry L. Murphy. 2012. "Death in the United States, 2010." NCHS *Data Brief* 99 (July). Centers for Disease Control and Prevention. Retrieved May 20, 2013 (http:// www.cdc.gov/nchs).

Mintz, Laurie. 2015. "The Orgasm Gap: Simple Truths & Sexual Solutions." Retrieved November 4, 2019 (https://www.psychologytoday.com/us /blog/stress-and-sex/201510 /the-orgasm-gap-simple-truth-sexual-solutions).

Mirzaie, Rose. 2016. "Divorce Mediation for Women: An Examination of Feminist Critiques." *Dispute Resolution Journal* 71:161.

Mishel, Lawrence, Jared Bernstein, and Heidi Shierholz. 2009. *The State of Working America 2008/2009.* New York: Cornell University Press.

Mistry, Rashmita S., Edward D. Lowe, Aprile Benner, and Nina Chien. 2008. "Expanding the Family Economic Stress Model: Insights from a Mixed-Methods Approach." *Journal of Marriage and Family* 70(1):196–209.

Mitchell, Colter, Sara McLanahan, Daniel Notterman, John Hobcraft, Jeanne Brooks-Gunn, and Irwin Garfinkel. 2015. "Family Structure Instability, Genetic Sensitivity, and Child Well-Being." *American Journal of Sociology* 120:1195–1225.

Mitchell, Katherine Stamps, Alan Booth, and Valarie King. 2009. "Adolescents with Nonresident Fathers: Are Daughters More Disadvantaged than Sons?" *Journal of Marriage and Family* 71:650–662.

Mitchell, Melanthia. 2007. "Cohousing for Lesbians Planned in Bremerton." Associated Press, July 8. Retrieved May 17, 2010 (www. seattlepi.com).

Moen, Phyllis. 2016. *Encore Adulthood: Boomers on the Edge of Risk, Renewal, & Purpose.* New York, NY: Oxford University Press.

Mohney, Gillian. 2015. "Rhode Island Finds Increase in STDs After Rise of Social Media Dating." Retrieved December 9, 2015 (http:// abcnews.go.com/Health /rhode -island-finds-increase-stds-rise -social-media /story?id=31339541).

Mollborn, Stefanie. 2009. "Norms about Nonmarital Pregnancy and Willingness to Provide Resources to Unwed Parents." *Journal of Marriage and Family* 71(1):122–134.

Monestero, Nancy. 1990. Personal communication.

"Monogamy: Is It for Us?" 1998. *The Advocate,* June 23, p. 29.

Monk, J. Kale, Brian G. Ogolsky, and Angela M. Whittaker. 2019. "Power in Families. Pp. 143–172 in Christopher Agnew and Jennifer Harman. 2019. *Power in Close Relationships.* Cambridge, UK: Cambridge University Press.

Monroe, Barbara, and Frances Kraus. 2010. *Brief Interventions with Bereaved Children,* 2nd ed. New York: Oxford University Press.

Monroe, Rachel. 2016. "What It's Like to Play Tinder in Rural America. " Retrieved November 20, 2019 (https://splinternews .com/what-its-like-to-play-tinder-in-rural -america-1793855722).

Monserud, Maria A. 2008. "Intergenerational Relationships and Affectual Solidarity Between Grandparents and Young Adults." *Journal of Marriage and Family* 71(1):182–195.

Monte, Lindsay M. 2017. "Multiple Partner Fertility Research Brief." Retrieved March 1, 2020 (https://www.census.gov/library /publications/2017/demo/p70br-146.html).

Monte, Lindsay M. 2019. "Multiple-Partner Fertility in the United States: A Demographic Portrait. " *Demography* 56:103–127.

Monte, Lindsay M. 2017. *Multiple Partner Fertility.* Retrieved December 22, 2019 (https://www. census.gov/content/dam /Census/library/publications/2017 /demo/p70br-146.pdf).

Montgomery, Marilyn J., and Gwendolyn T. Sorell. 1997. "Differences in Love Attitudes across Family Life Stages." *Family Relations* 46:55–61.

Montoro-Rodriguez, J., and D. Gallagher-Thompson. 2009. "The Role of Resources and Appraisals in Predicting Burden among Latina and Non-Hispanic White Female Caregivers." *Aging and Mental Health* 13(5):648–658.

Moody, Harry R. 2006. *Aging: Concepts and Controversies,* 5th ed. Thousand Oaks, CA: Pine Forge Press.

Moore, Abigail Sullivan. 2010. "Failure to Communicate." *The New York Times,* July 22. Retrieved September 8, 2010 (http://www .nytimes.com).

Moore, Elena. 2012. "Paternal Banking and Maternal Gatekeeping in Postdivorce Families." *Journal of Family Issues* 33(6):745–772.

Moore, Ginger A., and Jenae M. Neiderhiser. 2014. "Behavioral Genetic Approaches and Family Theory." *Journal of Family Theory and Review* 6(1):18–30.

Moore, Kathleen A., Marita P. McCabe, and Roger B. Brink. 2001. "Are Married Couples Happier in Their Relationships than Cohabiting Couples? Intimacy and Relationship Factors." *Sexual and Relationship Therapy* 16(1):35–46.

Moore, Kristin Anderson, Zakia Redd, Mary Burkhauser, Kassim Mbwana, and Ashleigh Collins. 2009. "Research Brief: Children in Poverty: Trends, Consequences, and Policy Options." April. Washington, DC: Child Trends. Retrieved December 12, 2009 (www.childtrends .org).

Moore, Mignon R. 2008. "Gendered Power Relations among Women: A Study of Household Decision Making in Black, Lesbian Stepfamilies." *American Sociological Review* 73(2):335–356.

Moorman, Elizabeth, and Eva Pomerantz. 2008. "The Role of Mothers' Control in Children's Mastery Orientation: A Time Frame Analysis." *Journal of Family Psychology* 22(5):734–741.

Morey, Jennifer N., Amy L. Gentzier, Brian Creasy, Ann M. Oberhauser, and David Westerman. 2013. "Young Adults' Use of Communication Technology within their Romantic Relationships and Associations with Attachment Style." *Computers in Human Behavior* 29(4):1771–1778.

Morgan, Charlie V. 2015. "Crossing Borders, Crossing Boundaries: How Asian Immigrant Backgrounds Shape Gender Attitudes About Interethnic Partnering." *Journal of Family Issues* 36(10):1324–1350.

Morgan, Rachel, and Barbara Oudekerk. 2019. "Criminal Victimization, 2018." U.S. Department of Justice, Bureau of Justice Statistics. September. Retrieved February 9, 2020 (www.djs.gov).

Morgan, Elizabeth M. 2011. "Associations Between Young Adults' Use of Sexually Explicit Materials and Their Sexual Preferences, Behaviors, and Satisfaction." *Journal of Sex Research* 48(6):520–530.

Morgan, S. Philip, and R. B. King. 2001. "Why Have Children in the 21st Century? Biological Predispositions, Social Coercion, Rational Choice." *European Journal of Population* 17:3–20.

Morgan, S. Philip, and Heather Rackin. 2010. "The Correspondence Between Fertility Intentions and Behavior in the United States." *Population and Development Review* 36(1):91–118.

Moriarty, Kate. 2014. "What Is Conscious Uncoupling?" Retrieved March 19, 2016 (http://www.womenshealthmag.com /sex-and-love/conscious-uncoupling).

Morin, Erica A. 2012. "No Vacation for Mother: Traditional Gender Roles in Outdoor Travel

Literature, 1940–1965." *Women's Studies* 41(4) (June):436–456.

Morin, Rich. 2011. "The Public Renders a Split Verdict on Changes in Family Structure." *Social and Demographic Trends.* Pew Research Center (February 16). Retrieved February 23, 2011 (http://pewsocialtrends.org http://pewsocialtrends.org/).

Morrow, Lance. 1992. "Family Values." *Time,* August 31, pp. 22–27.

Moses, Tally. 2010. "Exploring Parents' Self-Blame in Relation to Adolescents' Mental Disorders." *Family Relations* 59(2):103–120.

Mosher, William D., Anjani Chandra, and Jo Jones. 2005. "Sexual Behavior and Selected Health Measures: Men and Women 15–44 Years of Age, United States, 2002." *Advance Data from Vital and Health Statistics,* No. 362, September 15. Hyattsville, MD: U.S. National Center for Health Statistics.

Moss, Aaron J., Alison Blodorn, Amanda R. Van Camp, and Laurie T. O'Brien. 2019. " Gender Equality, Value Violations, and Prejudice toward Muslims." *Group Processes & Intergroup Relations* 22:288 –301.

Mnookin, Robert H., and D. Kelly Weisberg, D. K. (2014). *Child, Family, and State: Problems and Materials on Children and the Law.* NY: Wolters Kluwer Law & Business.

Mroz, Jacqueline. 2011. "One Sperm Donor, 150 Offspring." *The New York Times,* September 5. Retrieved September 12, 2011 (http://www.nytimes.com).

Muehlenhard, Charlene L., and Sheena K. Shippee. 2010. "Men's and Women's Reports of Pretending Orgasm." *Journal of Sex Research* 47(6):552–567.

Mueller, Margaret M., and Glen H. Elder Jr. 2003. "Family Contingencies across the Generations: Grandparent-Grandchild Relationships in Holistic Perspective." *Journal of Marriage and Family* 65(2):404–417.

Muise, Amy, Cheryl Harasymchuk, Lisa C. Day, Chantal Bacev-Giles, Judith Gere, and Emily A. Impett. 2019. "Broadening your Horizons: Self-expanding Activities Promote Desire and Satisfaction in Established Romantic Relationships." *Journal of Personality and Social Psychology* 116:237.

Muise, Amy, Elaine Giang, and Emily A. Impett. 2014. "Post Sex Affectionate Exchanges Promote Sexual and Relationship Satisfaction." *Archives of Sexual Behavior* 43(7):1391–1402.

Mullins, Larry, Elizabeth Molzon, Kristina Suorsa, Alayna Tackett, AhnaPai, and John Chaney. 2015. "Models of Resilience: Developing Psychosocial Interventions for Parents of Children with Chronic Health Conditions." *Family Relations* 64(1):176–189.

Mulqueen, Maggie. 2019. "Texting Really Is Ruining Personal Relationships." Retrieved January 24, 2020 (https://www.nbcnews.com/think/opinion/texting-really-ruining-personal-relationships-ncna1097461).

Mulvaney, Matthew K., and Carolyn J. Mebert. 2007. "Parental Corporal Punishment Predicts Behavior Problems in early Childhood." *Journal of Family Psychology* 21(3):389–397.

Mumola, Christopher J. 2000. *Incarcerated Parents and Their Children.* Washington, DC: U.S. Bureau of Justice Statistics.

Mundy, Liza. 2012. "Women, Money, and Power." *Time,* March 26: 28–31.

Munroe, Erin A. 2009. *The Everything Guide to Stepparenting.* Avon, MA: Aadams Media.

Munsch, Christin L. 2015. "Her Support, His Support: Money, Masculinity, and Marital Infidelity." *American Sociological Review* 80(3):469–495.

Murdock, George P. 1949. *Social Structure.* New York: The Free Press.

Murphey, David. 2012. "The More We Eat Together." Retrieved February 21, 2016 (http://www.childtrends.org/wp-content/uploads/2012/01/Child_Trends-2012_01_01_FS_SharedMeals.pdf).

Murphy Sherry L., Jianquan Xu, Kenneth D. Kochanek, and Elizabeth Arias. 2018. "Mortality in the United States, 2017." NCHS Data Brief, no 328. Hyattsville, MD: National Center for Health Statistics.

Murphy, Tim, and Loriann Hoff Oberlin. 2006. *Overcoming Passive-Aggression: How to Stop Hidden Anger from Spoiling Your Relationships, Career, and Happiness.* New York: Marlowe & Company.

Murray, Christine E., and A. Keith Mobley. 2009. "Empirical Research about Same-Sex Intimate Partner Violence: A Methodological Review." *Journal of Homosexuality* 56(3):361–386.

Murry, Velma M., P. Adama Brown, Gene H. Brody, Carolyn E. Cutrona, and Ronald L. Simons. 2001. "Racial Discrimination as a Moderator of the Links among Stress, Maternal Psychological Functioning, and Family Relationships." *Journal of Marriage and Family* 63(4):915–926.

Musick, Kelly. 2002. "Planned and Unplanned Childbearing among Unmarried Women." *Journal of Marriage and Family* 64(4):915–929.

Musick, Kelly, and Larry Bumpass. 2012. "Reexamining the Case for Marriage: Union Formation and Changes in Well-Being." *Journal of Marriage and Family* 74(1):1–18.

Mustillo, Sarah, John Wilson, and Scott M. Lynch. 2004. "Legacy Volunteering: A Test of Two Theories of Intergenerational Transmission." *Journal of Marriage and Family* 66(2):530–541.

Mutiso, Steve Kyende, Alfred Murage, and Abraham Mwaniki Mukaindo. 2018. "Prevalence of Positive Depression Screen among Post Miscarriage Women—a Cross Sectional Study." *BMC Psychiatry* 18:32.

Myers, Jane E., Jayamala Madathil, andLynne R. Tingle. 2005. "Marriage Satisfaction and Wellness in India and the United States: A Preliminary Comparison of Arranged Marriages and Marriages of Choice." *Journal of Counseling and Development* 83(2):183–190.

Myers, Scott A. 2011. "'I Have to Love Her, Even if Sometimes I May Not Like Her': The Reasons Why Adults Maintain Their Sibling Relationships." *North American Journal of Psychology* 13(1):51–62.

Mykyta, Laryssa, and Suzanne Macartney. 2012. "Sharing a Household: Household Composition and Economic Well-Being: 2007–2010." *Consumer Population Report* P60–242, June.

Washington, DC: U.S. Census Bureau and U.S. Department of Commerce.

Nadir, Aneesah. 2009. "Preparing Muslims for Marriage." Retrieved October 11, 2009 (www.soundvision.com).

Nagasawa, Miho, Shota Okabe, Kazutaka Mogi, and Takefumi Kikusui. 2012. "Oxytocin and Mutual Communication in Mother-Infant Bonding." *Towards a Neuroscience of Social Interaction* 98.

Nakonezny, Paul A., and Wayne H. Denton. 2008. "Marital Relationships: A Social Exchange Theory Perspective." *American Journal of Family Therapy* 36:402–412.

Nanji, Azim A. 1993. "The Muslim Family in North America." pp. 229–242 in *Family Ethnicity: Strength in Diversity,* edited by Harriette Pipes McAdoo. Newbury Park, CA: Sage.

Napolitano, Laura. 2015. "'I'm Not Going to Leave Her High and Dry': Young Adult Support to Parents during the Transition to Adulthood." *Sociological Quarterly* 56:329–354.

NARAL. 2015. "Americans Support Responsible Sex Education." Retrieved December 6, 2015 (http://www.prochoiceamerica.org/media/fact-sheets/americans-support-responsible.pdf).

Narayan, Chetna. 2008. "Is There a Double Standard of Aging?" *Educational Gerontology* 34(9):782–787.

National Alliance for Caregiving. 2015. *Caregiving in the United States: A Focused Look at Those Caring for Someone Age 50 and Older, Executive Summary.* Retrieved December 30, 2015 (http://www.aarp.org).

National Alliance to End Homelessness. 2013. "Demographics of Homeless Series: The Rising Elderly Population." April 1. Retrieved May 13, 2013 (http://www.endhomelessness.org).

National Alliance to End Sexual Violence. 2012. "2012 Rape Crisis Center Survey." Retrieved March 18, 2013 (http://www.endsexualviolence.org).

National Association for Sick Child Day Care. 2010. Birmingham, AL: National Association for Sick Child Day Care (www.nascd.com).

National Association of Child Care Resource and Referral Agencies (NACCRRA). 2015. "Is This the Right Place for My Child? A Research-based Checklist." Retrieved November 5, 2015 (http://www.naccrra.org).

National Association of Child Care Resource and Referral Agencies (NACCRRA) 2019. "Finding a Daycare Center." Retrieved December 10, 2019 (http://naccrra.org).

National Center for Education Statistics. 2019a. "Fast Facts: Degrees Conferred by Race and Sex." Retrieved October 2, 2019 (https://nces.ed.gov/fastfacts/display.asp?id=72).

National Center for Education Statistics. 2019b. "Fast Facts: Enrollment." Retrieved October 2, 2019 (https://nces.ed.gov/fastfacts/display.asp?id=98).

National Center for Family and Marriage Research. (2017). *Fast Facts on American families.* Retrieved February 23. 2020 (https://www.bgsu.edu/content/dam/BGSU/college-of-arts-and-sciences/NCFMR/documents/Marketing/fast-facts-american-families.pdf).

National Center for Health Statistics. 2019. *Infertility*. Retrieved December 23, 2019 (https://www.cdc.gov/nchs/fastats/infertility.htm).

National Center for Health Statistics. 2017. "Marriage and Divorce, 2017 Provisional Data." Retrieved November 15, 2019 (www.cdc.gov/nchs).

National Center for Health Statistics. 2018. "Provisional Number of Divorces and Annulments and Rate: 2000-2018. Retrieved February 5, 2020 (https://www.cdc.gov/nchs/data/dvs/national-marriage-divorce-rates-00-18.pdf).

National Coalition for the Homeless. 2019. "Homelessness in America." Retrieved December 10, 2019 (www.nationalhomeless.org).

National Institute on Drug Abuse. 2019. "Overdose Death Rates." Retrieved October 2, 2019 (https://www.drugabuse.gov/related-topics/trends-statistics/overdose-death-rates).

National Right to Life. 2019. "Is Abortion Safe? " Retrieved December 23, 2019 (https://www.nrlc.org/abortion/medicalfacts/safety/).

National Campaign to Prevent Teen and Unplanned Pregnancy. 2008. *Sex and Tech: Results from a Survey of Teens and Young Adults*. Washington, DC: Author.

National Center on Elder Abuse. 2013. "Types of Elder Abuse in Domestic Settings." (http://www.ncea.aoa.gov/Resources/Publication/docs/fact1.pdf).

National Center for Education Statistics. 2015. *Digest of Education Statistics: 2013*. NCES, Washington, DC. Retrieved October 9, 2015 (http://www.nces.gov).

National Center for Health Statistics. 2014. "National Marriage and Divorce Rate Trends." Retrieved March 9, 2016 (http://www.cdc.gov/nchs/nvss/marriage_divorce_tables.htm).

———. 2015. "Deaths: Final Data for 2013." Retrieved October 8, 2015 (http://www.nchs.gov).

National Center for Marriage and Family Research (NCFMR). 2010. "Thirty Years of Change in Marriage and Cohabitation Attitudes, 1976–2008." Retrieved January 28, 2016 (https://www.bgsu.edu/content/dam/BGSU/college-of-arts-and-sciences/NCFMR/documents/FP/FP-10-03.pdf).

National Center for Missing and Exploited Children and U.S. Department of Justice. 2013. "Know the Rules . . . for Children Who Are Home Alone." Washington, DC: Office of Justice and Delinquency Prevention. Retrieved February 15, 2013 (http://www.missingkids.com).

National Child Traumatic Stress Network. 2014. "Military and Veteran Families and Children." March 20. Retrieved December 4, 2015 (http://www.nctsn.org).

National Clearinghouse on Marital and Date Rape. n.d. Retrieved from http://www.ncmdr.org/

National Coalition Against Domestic Violence. 2009. "Domestic Violence Facts." Retrieved March 18, 2013 (http://www.ncadv.org).

National Coalition for the Homeless. 2009a. "How Many People Experience Homelessness?" Washington, DC: Author. Retrieved January 20, 2010 (www.nationalhomeless.org).

———. 2009b. "Why are People Homeless?" Washington, DC: Author. Retrieved January 20, 2010 (www.nationalhomeless.org).

"National Compensation Survey: Employee Benefits in Private Industry in the United States, March 2007." 2007. Summary 07–05. Washington, DC: U.S. Bureau of Labor Statistics (www.bls.gov).

National Congress of American Indians (NCAI). 2015. *A Spotlight on Native Women and Girls*. Policy Research Center, May. Retrieved October 9, 2015 (http://www.ncai.org).

National Employment Law Project (NELP). 2012. "The Low-Wage Recovery and Growing Inequality." Data Brief, August. Retrieved September 10, 2012 (http://www.nelp.org).

———. 2013. "Raise the Minimum Wage: Rebuilding an Economy that Works for All of Us." Retrieved February 24, 2013 (http://raisetheminimumwage.org).

National Hispanic Council on Aging. 2015. *Status of Hispanic Older Adults: Recommendations from the Field*. Washington, DC. Retrieved December 28, 2015 (http://www.nhcoa.org).

National Institute of Justice. 2007. *Commercial Sexual Exploitation of Children: What Do We Know and What Do We Do about It?* Washington, DC: U.S. Department of Justice.

National Marriage Project. 2009. "Figures Supplement to *The State of Our Unions: The Social Health of Marriage in America, 2008*." Rutgers University. Retrieved October 12, 2009 (http://marriage.rutgers.edu).

National Marriage Project and the Institute for American Values. 2012. *The State of Our Unions: Marriage in America 2012*. Charlottesville, VA.

National Public Radio. 2011. "How A Divorce Helped One Couple Grow Closer." Retrieved April 17, 2013 (http://www.npr.org/2011/03/04/134236042/how-a-divorce-helped-one-couple-grow-closer).

National Resource Center on Children and Families of the Incarcerated. 2009a. "An Overview of Statistics" Retrieved March 13, 2010 (www.fcnetwork.org).

———. 2009b. "What Happens to Children?" Retrieved March 13, 2010 (www.fcnetwork.org).

National Right to Life. 2005. "Abortion's Physical Complications," July. Washington, DC: National Right to Life Educational Trust Fund.

———. 2006. "Abortion's Psycho-Social Consequences." Washington, DC: National Right to Life Educational Trust Fund, December 6. Retrieved March 29, 2007 (www.nrlc.org).

National Women's Law Center. 2010. *Congress Must Act to Close the Wage Gap for Women*. Washington, DC: National Women's Law Center.

———. 2015. "Fact Sheet: Recently Introduced and Enacted State and Local Fair Scheduling Legislation." May. Retrieved November 18, 2015 from http://www.nwlc.org.

Nazario, Sonia L. 1990. "Identity Crisis: When White Parents Adopt Black Babies, Race Often Divides." *Wall Street Journal*, September 20.

Neal, Margaret B., and Leslie B. Hammer. 2007. *Working Couples Caring for Children and Aging Parents: Effects on Work and Well-Being*. Mahwah, NJ: Erlbaum.

Neel, Aly. 2013. "Surrogate Mother Refused Abortion." *Washington Post*, March 6. Retrieved March 26, 2013 (http://www.washingtonpost.com).

Neely-Barnes, Susan L., and J. Carolyn Graff. 2011. "Are There Adverse Consequences to Being a Sibling of a Person with a Disability? A Propensity Score Analysis." *Family Relations* 60(3):331–341.

Neff, Kristen D., and S. Natasha Beretvas. 2013. "The Role of Self-Compassion in Romantic Relationships." *Self and Identity* 12:78–98.

Negrusa, Sebastian, Brighita Negrusa, and James Hosek. 2016. "Deployment and Divorce: An In-Depth Analysis by Relevant Demographic and Military Characteristics." pp. 35–54 in Shelley MacDermid Wadsworth and David S. Riggs (Eds.), *War and Family Life*. Gewerbestrasse, Switzerland: Springer International Publishing.

Negy, Charles, and Douglas K. Snyder. 2000. "Relationship Satisfaction of Mexican American and Non-Hispanic White American Interethnic Couples: Issues of Acculturation and Clinical Intervention." *Journal of Marital and Family Therapy* 26(3):293–305.

Neimark, Jill. 2003. "All You Need Is Love: Why It's Crucial to Your Health—and How to Get More in Your Life." *Natural Health* 33(8):109–113.

Nelson, Jennifer. 2012. "Feminism Gave Rise to Superwoman in Advertising." *WeNews online*. Retrieved November 8, 2012 (http://womensenews.org).

Nelson, Margaret K. 2006. "Families in Not-So-Free Fall: A Response to Comments." *Journal of Marriage and Family* 68(4):817–823.

———. 2008. "Watching Children." *Journal of Family Issues* 29(4):516–538.

———. 2011. "Between Family and Friendship: The Right to Care for Anna." *Journal of Family Theory and Review* 3(4):241–255.

———. 2013. "Fictive Kin, Families We Choose, and Voluntary Kin: What Does the Discourse Tell Us?" *Journal of Family Theory and Review* 5(4):259–281.

Nemko, Marty. n.d. "The New Double Standard." Retrieved October 28, 2012 (http://www.martynemko.com).

Neppl, Tricia K., Sinyoung Jeon, Tomas J. Schofield, and M. Brent Donnellan. 2015. "The Impact of Economic Pressure on Parent Positivity, Parenting, and Adolescent Positivity into Emerging Adulthood." *Family Relations* 64(1):80–92.

Neuman, Scott. 2015. "Couples Who Choose Not to Have Children Are 'Selfish,' Pope Says." Retrieved February 13, 2016 (http://www.npr.org/sections/thetwo-way/2015/02/12/385735269/couples-who-chose-not-to-have-children-are-selfish-pope-says).

Newman, Amie. 2009. "'Laboring Under An Illusion': Rewire Talks to Filmmaker Vicki Elson." Retrieved December 25, 2019 (https://rewire.news/article/2009/12/21/laboring-under-an-illusion-rh-reality-check-talks-filmmaker-vicki-elson/).

Newman, Leah. 2018. "How to Spot Elder Abuse." Philips Lifeline. June 1. Retrieved August 17, 2018 (https://www.lifeline .philips.com/resources/blog/2018/06 /caring-for-seniors-recognizing-signs-of -elder-abuse.html).

Newman, Judith. 2014. "But I Want to Do Your Homework." *The New York Times*, June 21. Retrieved November 0, 2015 (http://www. nytimes.com).

Newman, Katherine. 2012. *The Accordion Family: Boomerang Kids, Anxious Parents, and the Private Toll of Global Competition*. Boston: Beacon Press.

Newman, Katherine, and James B. Knapp. 2013. "The Accordion Family." *Family Focus* FF56:F2–F3.

Newman, Susan. 2011. *The Case for the Only Child: Your Essential Guide*. Deerfield Beach, FL: Health Communications, Inc.

Newport, Frank. 2008. "Wives Still Do Laundry, Men Do Yard Work." Gallup Poll, April 4. Retrieved December 2, 2009 (www .gallup.com/poll).

———. 2015a. "Americans Greatly Overestimate Percent Gay, Lesbian in U.S." Retrieved December 3, 2015 (http://www .gallup.com /poll/183383/americans -greatly -overestimate-percent-gay-lesbian.aspx).

———. 2015b. "Five Things We've Learned About Americans and Moral Values." Retrieved December 3, 2015 (http://www.gallup.com/ opinion/polling -matters/183518/five-things- learned -americans-moral-values.aspx?g_source =do you think gay or lesbian relations between cons&g_medium=search&g _campaign=tiles).

———. 2015c. "Americans Continue to Shift Left on Key Moral Issues." Retrieved January 26, 2016 (http://www.gallup .com /poll/183413/americans-continue -shift -left-key-moral-issues.aspx?g_source = moral issues&g_medium=search&g _campaign=tiles).

———. 2015d. "Gallup: Religion, Race and Same-Sex Marriage." Gallup Poll. Retrieved September 19, 2015 (http://www.gallup.com).

Newport, Frank, and Brett Pelham. 2009. "Americans Least Happy in Their 50s and Late 80s." Gallup Poll, October 5. Retrieved May 6, 2010 (www.gallup.com /poll).

Newport, Frank. 2018. "Slight Preference for Having Boy Children Persists in U.S. " Retrieved December 14, 2019 (https://news.gallup.com /poll/236513/slight-preference-having-boy -children-persists.aspx).

Newport, Frank. 2019. "In U.S., Estimate of LGBT Population Rises to 4.5%" Retrieved October 17, 2019 (https://news .gallup.com/poll/234863/estimate-lgbt -population-rises.aspx).

Ni, Preston. 2014. "How to Recognize and Handle a Master Manipulator. " Retrieved November 19, 2019 (https://www .psychologytoday.com/us/blog /communication-success/201407 /how-recognize-and-handle-manipulative -relationships).

Nibley, Lydia. 2011. *Two Spirits*. PBS Television. First broadcast June 14.

Nicholson v. Scopetta. 2004. N.Y. No. 113.

Nichols, Tom. 2017. "Americans Are Now Utterly Intolerant of Ever Being Told They're Wrong about Almost Anything." Retrieved January 25, 2020 (https://www.marketwatch.com /story/the-real-reason-americans-cant-agree -on-unemployment -or-just-about-anything-else-2017-03-29

Nicholson, Jan M., Matthew R. Sanders, W. Kim Halford, Maddy Phillips, and Sarah W. Whitton. 2008. "The Prevention and Treatment of Children's Adjustment Problems in Stepfamilies." pp. 485–521 in *The International Handbook of Stepfamilies*, edited by J. Pryor. Hoboken, NJ: John Wiley & Sons.

Niehuis, Sylvia, Kyung-Hee Les, A. Reifman, A. Swenson, and S. Hunsaker. 2011. "Idealization and Disillusionment in Intimate Relationships: A Review of Theory, Method, and Research." *Journal of Family Theory and Review* 3(4):273–302.

Nielsen, Linda. 2018a "Joint Versus Sole Physical Custody: Outcomes for Children Independent of Family Income or Parental Conflict." *Journal of Child Custody* 15:35–54.

Nielsen, Linda. 2018b. "Preface to the Special Issue: Shared Physical Custody: Recent Research, Advances, and Applications." *Journal of Child Custody* 15:237–246.

Nielsen, Rasmus. 2009. "Adaptionism—30 Years After Gould and Lewontin." *Evolution* 63(10):2487–2490.

Nierenberg, Gerard, and Henry H. Calero. 1973. *Meta-Talk: Guide to Hidden Meanings in Conversations*. New York: Trident Press.

"9 Signs of Stress in Children and How to Help Them." 2018. *Curejoy Editorial*. March 29. Retrieved February 2, 2020 (https://www .curejoy.com/content/signs-of -childhood-stress-and-tips-to-reduce-it/).

Nir, Sarah Maslin. 2013. "You May Kiss the Computer Screen." *The New York Times*, March 5. Retrieved March 7, 2013 (http://www.nytimes).

Nissim, R., S. Hales, C. Zimmermann, A. Deckert, B. Edwards and G. Rodin. 2017. "Supporting Family Caregivers of Advanced Cancer Patients: A Focus Group Study." *Family Relations* 66 (5):867–879.

Nixon, Elizabeth, Sheila Greene, and Diane M. Hogan. 2012. "Negotiating Relationships in Single-Mother Households: Perspectives of Children and Mothers." *Family Relations* 61(1):142–156.

———. 2015. "'It's What's Normal for Me': Children's Experiences of Growing Up in a Continuously Single-Parent Household." *Journal of Family Issues* 36(8):1043–1061.

Noah, Aggie J. and Nancy S. Landale. 2018. "Parenting Strain among Mexican-Origin Mothers: Differences by Parental Leal Status and Neighborhood." *Journal of Marriage and Family* 80 (2):317–333.

Nobles, Jenna. 2013. "Migration and Father Absence: Shifting Family Structure in Mexico." *Demography* 50:1303–1314.

Nock, Matthew K., Bethany D. Michel, and Valerie I. Photos. 2008. "Single-Case Research Designs." pp. 337–349 in Dean McKay (Ed.) *Handbook of Research Methods in Clinical and Abnormal Psychology*. Thousand Oaks, CA: Sage.

Nock, Steven L. 2001. "The Marriages of Equally Dependent Spouses." *Journal of Family Issues* 22:755–775.

———. 2005. "Marriage as a Public Issue." *The Future of Children* 15(2):13–32.

Nock, Steven M., Laura A. Sanchez, and James D. Wright. 2008. *Covenant Marriage: The Movement to Reclaim Tradition in America*. New Brunswick, NJ: Rutgers University Press.

Noël-Miller, Claire M. 2013. "Repartnering Following Divorce: Implications for Older Fathers' Relations with Their Adult Children." *Journal of Marriage and Family* 75:697–712.

Noguchi, Yuki. 2015. "Some Companies Fight Pay Gap by Eliminating Salary Negotiations." National Public Radio, April 20. Retrieved April 27, 2015 (http://www.npr.org).

Noland, Virginia J., Karen D. Liller, Robert J. McDermott, Martha Coulter, and Anne E. Seraphine. 2004. "Is Adolescent Sibling Violence a Precursor to Dating Violence?" *American Journal of Health Behavior* 28(Supp. 1):813–823.

Noller, Patricia, and Mary Anne Fitzpatrick. 1991. "Marital Communication in the Eighties." pp. 42–53 in *Contemporary Families: Looking Forward, Looking Back*, edited by Alan Booth. Minneapolis: National Council on Family Relations.

Nomaguchi, Kei M. 2008. "Gender, Family Structure, and Adolescents' Primary Confidants." *Journal of Marriage and Family* 70(5):1213–1227.

Nomaguchi, Kei M., and Susan L. Brown. 2011. "Parental Strains and Rewards among Mothers: The Role of Education." *Journal of Marriage and Family* 73(3):621–636.

Nomaguchi, Kei M., and Melissa A. Milkie. 2003. "Costs and Rewards of Children: The Effects of Becoming a Parent on Adults' Lives." *Journal of Marriage and Family* 65:356–374.

Nomaguchi, Kei, Wendi Johnson, Mallory Minter, and Lindsey Aldrich 2017. "Clarifying the Association between Mother-Father Relationship Aggression and Parenting." *Journal of Marriage and Family* 79 (1): 161–178.

Nomaguchi, Kei, and Melissa A. Milkie. 2020. "Parenthood and Well-Being: A Decade in Review." *Journal of Marriage and Family* 82 (1): 198–223.

Nomaguchi, Kei, Susan Brown, and Tanya M. Leyman. 2017. "Fathers' Participation in Parenting and Maternal Parenting Stress: Variation by Relationship Status." *Journal of Family Issues* 38:1132–1156.

Nordqvist, Petra. 2015. "'I've Redeemed Myself by Being a 1950s Housewife': Parent-Grandparent Relationships in the Context of Lesbian Childbirth." *Journal of Family Issues* 36(4):480–500.

Nordwall, Smita P., and Paul Leavitt. 2004. "Court Rules for Battered Women's Rights." *USA Today*, October 24.

Norona, Jerika C., Alexander Khaddouma, Deborah P. Welsh, and Hannah Samawi. 2015. "Adolescents' Understandings of Infidelity." *Personal Relationships* 22:431–448.

Norris, Jeanette, et al. 2013. "How do Alcohol and Relationship Type Affect Women's Risk

Judgment of Partners with Differing Risk Histories?" *Psychology of Women Quarterly* 37(2):209–223.

Notter, Megan L., Katherine A. MacTavish, and Devora Shamah. 2008. "Pathways Toward Resilience among Women in Rural Trailer Parks." *Family Relations* 57(5):613–624.

Novaco, Raymond W. "Anger." 2016. *In Stress: Concepts, Cognition, Emotion, and Behavior*, pp. 285–292. Academic Press.

Novak, J., J. Anderson, M. Johnson, A. Walker, A. Wilcox, V. Lewis, and D. Robbins. 2017. "Associations between Economic Pressure and Diabetes Efficacy in Couples with Type 2 Diabetes." *Family Relations* 66 (2):273–286.

Nowakowski, Kelsey. 2019. "There Are Now More People Over Age 65 Than Under Five—What That Means." *National Geographic*. July 11. Retrieved January 4, 2020 https://www.nationalgeographic.com/culture/2019/07/global-population/).

Nowinski, Joseph. 2014. "Reading between the Lines of Your Partner's Texting." *Psychology Today*, April 1. Retrieved January 17, 2016 (https://www.psychologytoday.com/blog/the-almost-effect/201404/reading-between-the-lines-your-partners-texting).

Nozawa, Shinji. 2008. "The Social Context of Emerging Stepfamilies in Japan: Stress and Support for Parents and Stepparents." In J. Pryor (Ed.), *The International Handbook of Stepfamilies* (pp. 79–99). Hoboken, NJ: John Wiley & Sons.

Nozawa, Shinji. 2020. "East Asian Stepfamilies." pp. 149–178 in *Multicultural Stepfamilies*, edited by Susan D. Stewart and Gordon Limb. San Diego, CA: Cognella.

Nugent Colleen N., Jill Daugherty. 2018. *A Demographic, Attitudinal, and Behavioral Profile of Cohabiting Adults in the United States, 2011–2015*. National Health Statistics Reports, No 111. Hyattsville, MD: National Center for Health Statistics.

"Numbers and Trends." 2015. U.S. Department of Health and Human Services, Administration for Children and Families. Retrieved January 24, 2016 from http://www.childwelfare.gov.

Nuñez, Alicia, Patricia González, Gregory A. Talavera, Lisa Sanchez-Johnsen, Scott C. Roesch, Sonia M. Davis, William Arguelles et al. 2016. "Machismo, Marianismo, and Negative Cognitive-emotional Factors: Findings from the Hispanic Community Health Study/Study of Latinos Sociocultural Ancillary Study." *Journal of Latina/o Psychology* 4:202.

Nunn, Brittany. 2012. "Statistics Suggest Bleak Futures for Children Who Grow Up in Foster Care." *Amarillo Globe*, June 24. Retrieved January 27, 2012 (http://amarillo.com).

Nwoye, Irene C. 2015. "New York's Bravest Is Trans FDNY Firefighter Brooke Guinan." Retrieved February 25, 2015 (http://blogs.villagevoice.com).

Nyman, Charlott, Lasse Reinikainen, and Janet Stocks. 2013. "Reflections on a Cross-National Qualitative Study of Within-Household Finances." *Journal of Marriage and Family* 75(3):640–650.

O'Brien, Brendan. 2012. "Wisconsin's Baldwin Becomes First Openly Gay Senator." *Chicago Tribune*. Retrieved November 9, 2012 (http://articles.chicagotribune.com/2012-11-07/news/sns-rt-us-usa-campaign-wisconsin-senatebre8a60by-20121106_1_governor-tommy-thompson-tammy-baldwin-expensive-senate-race).

O'Brien, Karen M. and Kathy P. Zamostny. 2003. "Understanding Adoptive Families: An Integrative Review of Empirical Research and Future Directions for Counseling Psychology." *The Counseling Psychologist* 31(6):679–710.

O'Brien, M. 2005. "Studying Individual and Family Development: Linking Theory and Research." *Journal of Marriage and the Family* 67:880–890.

O'Brien, Sharon. n.d. "All about Grandparents Day." *Senior Living*. Retrieved May 9, 2013 (http://www.seniorliving.org).

Ocobock, Abigail. 2018. "Status or Access? The Impact of Marriage on Lesbian, Gay, Bisexual, and Queer Community Change." *Journal of Marriage and Family* 80 (2): 367–382.

O'Connor, Joe. 2012. "Full Comment. Joe O'Connor: Trend of Couples Not Having Children Just Plain Selfish." *The National Post*, September 19. Retrieved December 13, 2012 (http://fullcomment.nationalpost.com).

Odom, Samuel L. 2009. *Handbook of Developmental Disabilities*. New York: Guilford Press.

OECD. 2019. *OECD Family Database*. Retrieved December 19, 2019 (http://www.oecd.org/els/family/database.htm).

Offer, Shira. 2013. "Family Time Activities and Adolescents' Emotional Well-Being." *Journal of Marriage and Family* 75(1): 26–41.

Ojeda, Norma. 2011. "Living Together Without Being Married: Perceptions of Female Adolescents in the Mexico-United States Border Region." *Journal of Comparative Family Studies* 42(4):439–454.

Okun, Barbara S., and Liat Raz-Yurovich. 2019. "Housework, Gender Role Attitudes, and Couples' Fertility Intentions: Reconsidering Men's Roles in Gender Theories of Family Change. " *Population and Development Review* 45:169–196.

"Older Moms' Birth Risks Called Greater." 1999. *Omaha World-Herald*, January 2.

Oldham, J. Thomas. 2008. "What If the Beckhams Move to L.A. and Divorce—Marital Property Rights of Mobile Spouses When They Divorce in the United States." *Family Law Quarterly* 42(2):263–293.

Olmer, Ruth M., and Kristina S. Brown. 2016. "Divorce Mediation in the United States." *Encyclopedia of Family Studies*: 1-4.

Olson, David H., and Dean M. Gorall. 2003. "Circumplex Model of Marital and Family Systems." pp. 514–547 in *Normal Family Processes*,3rd ed., edited by F.Walsh. New York: Guilford.

Olson, Elizabeth. 2015. "For Widows, Social Security System Can Provide Rude Shocks." *The New York Times*, September 11. Retrieved September 19, 2015 (http://www.nytimes.com).

Olson, Jonathan R., James P. Marshall, H. Wallace Goddard, and David G. Schramm. 2015. "Shared Religious Beliefs, Prayer, and Forgiveness as Predictors of Marital Satisfaction." *Family Relations* 64:519–533.

Olson, Kristina, Lily Durwood, Madeleine DeMeules, and Katie McLaughlin. 2016. "Mental Health of Transgender Children Who Are Supported in Their Identities." *Pediatrics* 137 (3).

Olson, M., C. S. Russell, M. Higgins-Kessler, and R. B. Miller. 2002. "Emotional Processes Following Disclosure of an Extramarital Fidelity." *Journal of Marital and Family Therapy* 28:423–434.

O'Malley, Katie. 2019. "Four People Tell Their Stories on Living with an Ex…" Retrieved November 20, 2019 (https://www.independent.co.uk/life-style/dating/living-with-your-ex-break-up-love-island-amy-hart-curtis-pritchard-a8998291.html).

O'Neal, Catherine W., and Jay A. Mancini. 2017. "Families at Arm's Length." *Family Focus*, FF71, p. F1–F2.

O'Neil, Adrienne, Victor Sojo, Bianca Fileborn, Anna J. Scovelle, and Allison Milner. 2018. "The #MeToo Movement: An Opportunity in Public Health?" *The Lancet* 391:2587–2589.

Ono, Hiromi. 2009. "Husbands' and Wives' Education and Divorce in the United States and Japan, 1946–2000." *Journal of Family History* 34(3):292–322.

Ontai, Lenna, Yoshie Sano, Holly Hatton, and Katherine Conger. 2008. "Low-Income Rural Mothers' Perceptions of Parent Confidence: The Role of Family Health Problems and Partner Status." *Family Relations* 57(July):324–334.

Oppenheim, Keith. 2007. "Soldier Fathers Child Two Years after Dying in Iraq." CNN.com, March 20. Retrieved March 20, 2007 (http://cnn.usnews.com).

Orbuch, T. L., A. Thornton, and J. Cancio. 2000. "The Impact of Marital Quality, Divorce, and Remarriage on the Relationships between Parents and Their Children." *Marriage and Family Review* 29:221–246.

Orchard, Ann L., and Kenneth B. Solberg. 2000. "Expectations of the Stepmother's Role." *Journal of Divorce and Remarriage* 31(1/2):107–124.

Orenstein, Peggy. 1994. *School Girls: Young Women, Self-Esteem, and the Confidence Gap*. New York: Doubleday.

———. 2011. *Cinderella Ate My Daughter: Dispatches from the Front Lines of the New Girlie-Girl Culture*. New York: HarperCollins.

———. 2012. "Peggy Orenstein on the Gender Trap." Retrieved November 8, 2012 (http://thebrowser.com).

Ornelas, India J., Krista M. Perreira, Linda Beeber, and Lauren Maxwell. 2009. "Challenges and Strategies to Maintaining Emotional Health: Qualitative Perspectives of Mexican Immigrant Mothers." *Journal of Family Issues* 30(11):1556–1575.

Oropesa, R. S. 1996. "Normative Beliefs about Marriage and Cohabitation: A Comparison of Non-Latino Whites, Mexican Americans, and Puerto Ricans." *Journal of Marriage and Family* 58:49–62.

Oropesa, R. S. and Nancy S. Landale. 2004. "The Future of Marriage and Hispanics." *Journal of Marriage and Family* 66(4):901–920.

Orr, Amy J. 2011. "Gendered Capital: Childhood Socialization and the 'Boy Crisis' in Education." *Sex Roles* 65:271–284.

Orthner, Dennis K., Hinckley Jones-Sanpei, and Sabrina Williamson. 2004. "The Resilience and Strengths of Low-Income Families." *Family Relations* 53(2):159–167.

Ory, Brett, and Tim Huijts. 2015. "Widowhood and Well-Being in Europe: The Role of National and Regional Context." *Journal of Marriage and Family* 77(3):730–746.

Osborne, Cynthia. 2012. "Further Comments on the Papers by Marks and Regnerus." *Social Science Research* 41:779–783.

Osborne, Cynthia, and Lawrence M. Berger. 2009. "Parental Substance Abuse and Child Well-Being." *Journal of Family Issues* 30(3):341–370.

Osborne, Cynthia, Lawrence Berger, and Katherine Magnuson. 2012. "Family Structure Transitions and Changes in Maternal Resources and Well-Being." *Demography* 49(1):23–47.

Oshri, Assaf, Mallory Lucier-Greer, Catherine Walker O'Neal, Amy Arnold, Jay Mancini, and James Ford. 2015. "Adverse Childhood Experiences, Family Functioning, and Resilience in Military Families: A pattern-Based Approach." *Family Relations* 64(1):44–63.

Oswald, Ramona F., and Vanja Lazarevic. 2011. " 'You Live Where?!' Lesbian Mothers' Attachment to Nonmetropolitan Communities." *Family Relations* 60(4):373–386.

Oswald, Ramona Faith, and Brian P. Masciadrelli. 2008. "Generative Ritual among Nonmetropolitan Lesbians and Gay Men: Promoting Social Inclusion." *Journal of Marriage and Family* 70:1060–1073.

Overall, Christine. 2013. *Why Have Children? The Ethical Debate.* Cambridge: MIT Press.

Overall, Nickola, Matthew Hammond, James McNulty, and Eli Finkel. 2016. "When Power Shapes Interpersonal Behavior: Low Relationship Power Predicts Men's Aggressive Responses to Low Situational Power." *Journal of Personal and Social Psychology* 111(2):195–217.

Oyamot, Clifton M. Jr., Paul T. Fuglestad, and Mark Snyder. 2010. "Balance of Power and Influence in Relationships: The Role of Self-monitoring." *Journal of Social and Personal Relationships* 27(1):23–46.

Ozawa, Martha N., and Yongwoo Lee. 2006. "The Net Worth of Female-Headed Households: A Comparison to Other Types of Households." *Family Relations* 55(1):132–134.

Pacey, Susan. 2005. "Step Change: The Interplay of Sexual and Parenting Problems When Couples Form Stepfamilies." *Social and Relationship Therapy* 20(3):359–369.

Padawer, Ruth. 2012. "What's So Bad about a Boy Who Wants to Wear a Dress?" *The New York Times, August* 8. Retrieved August 27, 2012 (http://www.nytimes).

Padilla-Walker, Laura, Sarah M. Coyne, and Ashley M. Fraser. 2012. "Getting a High-Speed Connection: Associations between Family Media Use and Family Connection." *Family Relations* 61(3):426–440.

Pahl, Jan M. 1989. *Marriage and Money.* Basingstoke, UK: Macmillan.

Paik, Anthony. 2011. "Adolescent Sexuality and Risk of Marital Dissolution." *Journal of Marriage and Family* 73:472–485.

Painter, Matthew, Adrianne Frech, and Kristi Williams. 2015. "Nonmarital Fertility, Union History, and Women's Wealth." *Demography* 52:153–182.

Palkovitz, Rob, Bahira Serif Trask, and Kari Adamsons. 2014. "Essential Differences in the Meaning and Processes of Mothering and Fathering." *Journal of Family Theory and Review* 6(4):406–420.

Palm, Glen. 2014. "Attachment Theory and Fathers: Moving from 'Being There' to 'Being With.'" *Journal of Family Theory and Review* 6(4):282–297.

Papadimitriou, Anastasios. 2016. "The Evolution of the Age at Menarche from Prehistorical to Modern Times." *Journal of Pediatric and Adolescent Gynecology* 29:527–530.

Papernow, Patricia. 1993. *Becoming a Stepfamily: Patterns of Development in Remarried Families.* San Francisco: Jossey-Bass.

———. 2008. "A Clinician's View of 'Stepfamily Architecture': Strategies for Meeting the Challenges." pp. 423–454 in *The International Handbook of Stepfamilies*, edited by J. Pryor. Hoboken, NJ: John Wiley & Sons.

———. 2015. *Becoming a Stepfamily: Patterns of Development in Remarried Families.* Boca Raton, FL: CRC Press.

Papernow, Patricia L. 2013. *Surviving and Thriving in Stepfamily Relationships: What Works and What Doesn't.* Abingdon, UK: Routledge.

Papp, Lauren, E. Mark Cummings, and Marcie C. Goeke-Morey. 2009. "For Richer, for Poorer: Money as a Topic of Marital Conflict in the Home." *Family Relations* 55(February):91–103.

Parade, Stephanie H., A. Nayena Blankson, Esther M. Leerkes, Susan C. Crockenberg, and Richard Faldowski. 2014. "Close Relationships Predict Curvilinear Trajectories of Maternal Depressive Symptoms over the Transition to Parenthood." *Family Relations* 63(2):206–218.

Parade, Stephanie, Esther Leerkes, and Heather Helms. 2013. "Remembered Parental Rejection and Postpartum Declines in Marital Satisfaction." *Family Relations* 62(2):298–311.

Parcell, Erin Sahlstein, Benjamin M. A. Baker, and Ronald L. Johnson. 2017. "Military Families and Deployment: Communicating at a Distance." *Family Focus*, FF71, p. F8–F9.

Paris, Ruth, and Nicole Dubus. 2005. "Staying Connected while Nurturing an Infant: A Challenge of New Motherhood." *Family Relations* 54(1):72–83.

Park, Kristin. 2002. "Stigma Management among the Voluntarily Childless." *Sociological Perspectives* 45:21–45.

———. 2005. "Choosing Childlessness: Weber's Typology of Action and Motives of the Voluntarily Childless." *Sociological Inquiry* 75(3):372–402.

Park, Madison. 2017. "Election Night Brings Historic Wins for Minority and LGBT Candidates." Retrieved September 29, 2019 (https://www.cnn.com/2017/11/08/us/election-firsts-lgbt-minorities/index.html).

Park, Yong, Leyna Vo, and Yuying Tsong. 2009. "Family Affection As a Protective Factor Against Effects of Perceived Asian Values Gap on the Parent-Child Relationship for Asian American Male and Female College Students." *Cultural Diversity and Ethnic Minority Psychology* 51(1):18–26.

Parker, Kim. 2012. "The Boomerang Generation: Feeling OK about Living with Mom and Dad." *Pew Social and Demographic Trends.* Pew Research Center: March 15. Retrieved August 28, 2012 (http://www.pewsocialtrends.org).

———. 2015. "Women More than Men Adjust Their Careers for Family Life." Pew Research Center, October 15. Retrieved October 8, 2015 (http://www.pewresearch.org).

Parker, Kim, and Eileen Patten. 2013. "The Sandwich Generation: Rising Financial Burdens for Middle-Aged Americans." *Pew-Research Social and Demographic Trends*, January 30. Retrieved March 20, 2013 (http://www.pewsocialtrends.org).

Parker, Kim, and Wendy Wang. 2013. "Modern Parenthood: Roles of Moms and Dads Converge as They Balance Work and Family." *Pew Research Social and Demographic Trends*, March 14. Washington, DC: Pew Research Center.

Parker, Kim. 2011. "A Portrait of Stepfamilies." Retrieved February 21, 2020 (https://www.pewsocialtrends.org/2011/01/13/a-portrait-of-stepfamilies/).

Parker, Kim. 2019. "The American Veteran Experience and the Post-9/11 Generation." Pew Research Center, Social and Demographic Trends. November 10.

Parker, Kim, Juliana Menasce Horowitz, and Molly Rohal. 2015. *Parenting in America.* Retrieved December 20, 2019 (https://www.pewresearch.org/wp-content/uploads/sites/3/2015/12/2015-12-17_parenting-in-america_FINAL.pdf).

Parker, Kim, and Renee Stepler. 2017. "Americans See Men as the Financial Providers, Even as Women's Contributions Grow." Retrieved November 16, 2019 (https://www.pewresearch.org/fact-tank/2017/09/20/americans-see-men-as-the-financial-providers-even-as-womens-contributions-grow/).

Parker, Kim and Renee Stepler. 2017. "Americans see Men as the Financial Providers, Even as Women's Contributions Grow." (https://www.pewresearch.org/fact-tank/2017/09/20/americans-see-men-as-the-financial-providers-even-as-womens-contributions-grow/).

Parker, Kim, and Renee Stepler. 2017. "As U.S. Marriage Rate Hovers at 50%, Education Gap in Marital Status Widens." *Facttank: News in the Numbers.* September 14. Pew Research Organization. Retrieved January 30, 2019 (www.pewresearch.org).

Parker-Pope, Tara. 2008. "Love, Sex and the Changing Landscape of Infidelity." *The New York Times, October* 27. Retrieved November 20, 2012 (http://www.nytimes.com).

Parkinson, Patrick. 2011. "Another Inconvenient Truth: Fragile Families and the Looming Financial Crisis for the Welfare State." *Family Law Quarterly* 45(3):329–352.

Parra-Cardona, Jose Ruben, Emily Meyer, Lawrence Schiamberg, and Lori Post. 2007. "Elder Abuse and Neglect in Latino Families: An Ecological and Culturally Relevant Theoretical Framework for Clinical Practice." *Family Process* 46(4):451–470.

Parsons, Talcott. 1943. "The Kinship System of the Contemporary United States." *American Anthropologist* 45:22–38.

Parsons, Talcott and Robert F. Bales. 1955. *Family, Socialization, and Interaction Process.* Glencoe, IL: The Free Press.

Partners Task Force for Gay and Lesbian Couples: An International Resource for Same-Sex Couples. 2015. Retrieved January 4, 2016 (http://buddybuddy .com).

Pascoal, Patrícia M., E. Sandra Byers, Maria-João Alvarez, Pablo Santos-Iglesias, Pedro J. Nobre, Cicero Roberto Pereira, and Ellen Laan. 2018. "A Dyadic Approach to Understanding the Link Between Sexual Functioning and Sexual Satisfaction in Heterosexual Couples." *The Journal of Sex Research* 55:1155–1166.

Paset, Pamela S., and Ronald D. Taylor. 1991. "Black and White Women's Attitudes toward Interracial Marriage." *Psychological Reports* 69:753–754.

Patten, Eileen and Kim Parker. 2011. "Women in U.S. Military: Growing Share, Distinctive Profile." *Pew Research Social and Demographic Trends,* December 22. Retrieved June 13, 2013 (http://www .pewsocialtrends.org).

Patterson, Charlotte. 2000. "of Lesbians and Gay Men." *Journal of Marriage and Family* 62(4):1052–1069.

Patterson, J. M. 2002a. "Integrating Family Resilience and Family Stress Theory." *Journal of Marriage and Family* 64(2):349–360.

———. 2002b. "Family Caregiving for Medically Fragile Children." *Family Focus* (December):F5–F7. Minneapolis: National Council on Family Relations.

Patterson, Jennifer, B. C. Gardner, B.K. Burr, D.S. Hubler, and K.M. Roberts. 2012. "Nonverbal Behavioral Indicators of Negative Affect in Couple Interaction." *Contemporary Family Therapy* 34:11–28.

Paul, Aditi. 2014. "Is Online Better than Offline for Meeting Partners? Depends: Are You Looking to Marry or to Date?" *Cyberpsychology, Behavior, and Social Networking* 17:664–667.

Paul, Annie Murphy. 2011. "Older Parents Find More Joy in Their Bundles." *The New York Times,* April 7. Retrieved September 13, 2011 (http://www.nytimes.com).

Paul, Pamela. 2002. *The Starter Marriage and the Future of Matrimony.* New York: Villard.

———. 2005. *Pornified: How Pornography Is Damaging Our Lives, Our Relationships, and Our Families.* New York: Henry Holt & Co.

———. 2009. *Parenting, Inc.: How the Billion-Dollar Baby Business Has Changed the Way We Raise Our Children.* New York: Holt Paperbacks.

———. 2010. "A Young Man's Lament: Love Hurts!" Retrieved Mary 25, 2013 (http://www .nytimes.com/2010/07/25 /fashion/25Studied.html).

———. 2012. "Divorce, Custody, and Co-Parenting in the Digital Age." *The New York Times.* Retrieved from http://weinrieblaw .com/divorce-custody-and-co-pareting-in -the-digital-age/.

Paulson, James F., Sarah E. Dauber, Jenn A. Leiferman. 2011. "Parental Depression, Relationship Quality, and Nonresident Father Involvement with Their Infants." *Journal of Family Issues* 32(4):528–549.

Payne, Krista K. 2014a. Demographic Profile of Same-Sex Couple Households with Minor Children, 2012 (FP-14-03). National Center for Family and Marriage Research. Retrieved April 25, 2016 (http://www.bgsu.edu/content/dam / BGSU/college-of-arts-and-sciences /NCFMR /documents/FP/FP-14-03 _DemoSSCoupleHH .pdf).

Payne, Krista K. 2018a. *Change in the U.S. Remarriage Rate, 2008 and 2016.* National Center for Family & Marriage Research, Bowling Green State University. Retrieved February 21, 2020 (https://www.bgsu.edu/ncfmr/resources/data /family-profiles/payne-change-remarriage -rate-fp-18-16.html).

Payne, Krista K. (2018b). *First Marriage Rate in the U.S., 2016.* National Center for Family & Marriage Research, Bowling Green State University. Retrieved February 21, 2020 (https://www.bgsu.edu/ncfmr /resources/data/family-profiles/payne -first-marriage-rate-fp-18-14.html).

Payne, Krista. K. 2014. *The Divorce Rate and the Great Recession.* Family Profiles, FP-14-19. Bowling Green, OH: National Center for Family & Marriage Research. Retrieved February 5, 2020 (http://www.bgsu. edu/content/dam / BGSU/college-of-arts-and-sciences/ NCFMR /documents/ FP/FP-14-19-divorcerate -recession.pdf).

Payne, Krista K. 2019. *Young Adults in the Parental Home, 2007 & 2018.* Family Profiles, FP-19-04. Bowling Green, OH: National Center for Family & Marriage Research. Retrieved November 29, 2019 (https://www.bgsu.edu/ncfmr/resources /data/family-profiles/payne-young-adults -parental-home-2007-2018-fp-19-04.html).

PBS News Hour. 2019. "How Would Trump's Stamp Plan Affect Low-Income Americans?" July 23. Retrieved August 19, 2019 (https://www .pbs.orf/newshour).

Pearce, Lisa D. 2002. "The Influence of Early Life Course Religious Exposure on Young Adults' Dispositions Toward Childbearing." *Journal for the Scientific Study of Religion* 41(2):325–340.

Pearce, Lisa D. and Shannon N. Davis. 2016. "How Early Life Religious Exposure Relates to the Timing of First Birth." *Journal of Marriage and Family* 78 (5):1422–1438.

Peck, Emily. 2015. "Proof That Working from Home Is Here To Stay: Even Yahoo Still Does It." *Huffington Post Business,* March 18. Retrieved November 15, 2015 (http://www. huffingtonpost.com).

Peck, M. Scott. 1978. *The Road Less Traveled: A New Psychology of Love, Traditional Values and Spiritual Growth.* New York: Simon & Schuster.

Peck, M. Scott, and Sharon Peck. 2006. *The Top 60 Love Skills You Were Never Taught.* Solana Beach, CA: Lifepath Publishers.

Pelham, Brett. 2008. "Relationships, Financial Security Linked to Well-Being." November 17.

Retrieved December 2, 2008 (www.gallup.com /poll).

Pelletiere, Nicole. 2019. "The Mom Stands in at LGBTQ+ Weddings When Families Refuse to Attend. Now, Her Story is Being Made into a Movie." Retrieved October 17, 2019 (https:// www.goodmorningamerica.com/family/story /mom-stands-lgbtq-weddings-families-refuse -attend-now-64021444).

Peltz, Jennifer. 2017. "Courts and 'Tri-Parenting': A State-by-State Look." Retrieved March 2, 2020 (https://www .boston.com/news/national -news/2017/06/18/courts-and-tri -parenting-a-state-by-state-look).

Pena, J. Vicente, Carmen R. Menendez, and Susana Torio. 2010. "Family and Socialization Processes: Parental Perception and Evaluation of Their Children's Household Labor." *Journal of Comparative Family Studies* 41(1):131–148.

Pendley, Elisabeth. 2006. *Marriage Works: Before You Say "I Do."* Bellevue, WA: Merril Press.

Penning, Margaret, and Zheng Wu. 2013. "Intermarriage and Social Support among Canadians in Middle and Later Life." *Journal of Marriage and Family* 75:1044–1064.

Pepin, Joanna R., and David A. Cotter. 2018. "Separating Spheres? Diverging Trends in Youth's Gender Attitudes about Work and Family." *Journal of Marriage and Family* 80 (1): 7–24.

Peplau, Letitia Anne., A. Fingerhut, and K. P. Beals. 2004. "Sexuality in the Relations of Lesbians and Gay Men." pp. 349–369 in *The Handbook of Sexuality in Close Relationships,* edited by J. H. Harvey, A. Wenzel, and S. Sprecher. Mahwah, NJ: Erlbaum.

Peralta, Robert L., and J. Michael Cruz. 2006. "Conferring Meaning onto Alcohol-Related Violence: An Analysis of Alcohol Use and Gender in a Sample of College Youth." *Journal of Men's Studies* 14(1):109–136.

Perel, Esther. 2017. *The State of Affairs: Rethinking Infidelity.* Hachette UK.

Perelli-Harris, Brienna, and Nora Sanchez Gassen. 2012. "How Similar Are Cohabitation and Marriage? Legal Approaches to Cohabitation across Western Europe." *Population and Development Review* 38(3):435–467.

Perelli-Harris, Brienna, M. Kreyenfeld, W. Sigle-Rushton, R. Keizer, T. Lappengard, A. Jasilioniene, C. Berghammer, and P. DiGiulio. 2012. "Changes in Union Status during the Transition to Parenthood in Eleven European Countries, 1970s to Early 2000s." *Population Studies* 66(2):167–182.

Pérez-Peña, Richard. 2013. "College Health Plans Respond as Transgender Students Gain Visibility." Retrieved June 25, 2013 (http:// mobile.nytimes .com/2013/02/13/education /12sexchange.html?f=21).

Perez-Pena, Richard, and Ian Lovett. 2014. "California Law on Sexual Consent Pleases Many but Leaves Some Doubters." *The New York Times,* September 29. Retrieved January 12, 2016 (http://www.nytimes .com/2014/09/30/us /california-law-on -sex-consent-pleases-many -but-leaves-some -doubters.html).

Perlstein, Linda. 2012. "Do-It-(All)-Yourself Parents." 2012. *Newsweek*, February 6:47–51.

Perper, Kate, Kristen Peterson, and Jennifer Manlove. 2010. "Diploma Attainment among Teen Mothers." *Child Trends Fact Sheet* 2010–01. Washington, DC: Child Trends.

Pernice-Duca, Francesca, David Biegel, Heather Hess, Chia-Chung, and Ching-Wen Chang. 2015. "Family Members' Perceptions of How They Benefit When Relatives Living with Serious Mental Illness Participate in Clubhouse Community Programs." *Family Relations* 64(3):446-459.

Perrow, Susan. 2009 (2003). "The Hurried Child Syndrome." *Kindred Natural Parenting Magazine.* Retrieved January 15, 2010 (www.kindredmedia .com).

Perry, Samuel. 2015. "A Match Made in Heaven? Religion-Based Marriage Decisions, Marital Quality, and the Moderating Effects of Spouses' Religious Commitment." *Social Indicators Research* 123(1):203–225.

Perry-Jenkins, Maureen, Julianna Smith, Abbie Goldberg, and Jade Logan. 2011. "Working-Class Jobs and New Parents' Mental Health." *Journal of Marriage and Family* 73(5):1117–1132.

Perry-Jenkins, Maureen, and Shelley MacDermid Wadsworth. 2017. "Work and Family Research and Theory: Review and Analysis from an Ecological Perspective." *Journal of Family Theory & Review* 9 (2): 219–237).

Peter, Tracey. 2009. "Exploring Taboos: Comparing Male- and Female-Perpetrated Child Sexual Abuse." *Journal of Interpersonal Violence* 24(7):1111–1128.

Peter-Hagene, Liana C., and Sarah E. Ullman. 2018. "Longitudinal Effects of Sexual Assault Victims' Drinking and Self-blame on Posttraumatic Stress Disorder." *Journal of Interpersonal Violence* 33:83–93.

Peters, Marie F. 2007. "Parenting of Young Children in Black Families." pp. 203–218 in *Black Families*, 4th ed., edited by Harriette Pipes McAdoo. Thousand Oaks, CA: Sage.

Petersen, Larry R. 1994. "Education, Homogamy, and Religious Commitment." *Journal for the Scientific Study of Religion* 33(2):122–128.

Peterson, Jennifer L., and Janet Shibley Hyde. 2011. "Gender Differences in Sexual Attitudes and Behaviors: A Review of Meta-Analytic Results and Large Datasets." *Journal of Sex Research* 42(1–2):149–165.

Pettit, Gregory S., Patrick S. Malone, Jennifer E. Lansford, Kenneth A. Dodge, and John E. Bates. 2010. "Domain Specificity in Relationship History, Social-Information Processing, and Violent Behavior in Early Adulthood." *Journal of Personality and Social Psychology* 98(2):190–200.

Pew Forum on Religion and Public Life. 2008. *U.S. Religious Landscape Survey: Religious Affiliation: Diverse and Dynamic.* Washington, DC: Pew Research Center.

Pew Research Center. 2013. "A Survey of LGBT Americans." Retrieved November 19, 2019 (file:///C:/Users/stewa/Downloads /SDT_LGBT-Americans_06-2013%20(2) .pdf).

Pew Research Center. 2016. "Changes in the American Workplace." October 6. Retrieved September 4, 2019 (www.pewforum.org).

Pew Research Center. 2019. "In U.S., Decline of Christianity Continues at Rapid Pace: An Update on America's Changing Religious Landscape." October 17. Retrieved October 19, 2019 (www .pewforum.org).

Pew Research Center. 2015. "Parenting in America." December 17. Retrieved October 19, 2019 (www.pewforum.org).

Pew Research Center. 2018a. "'The Changing Profile of Unmarried Parents." April 25. Retrieved October 19, 2019 (www.pewforum .org).

Pew Research Center. 2010. "The Decline of Marriage and Rise of New Families. " Retrieved November 16, 2019 (https://www .pewsocialtrends.org/2010/11/18 /the-decline-of-marriage-and-rise-of-new -families/).

Pew Research Center. 2019. "U.S. Public Continues to Favor Legal Abortion, Oppose Overturning Roe v. Wade. " Retrieved December 23, 2019 (https://www.people-press .org/2019/08/29/u-s-public-continues-to-favor -legal-abortion-oppose-overturning-roe-v-wade/).

Pew Research Center. 2018b. "Where Americans Find Meaning in Life." November 20. Retrieved October 19, 2019 (www.pewforum.org).

Pew Research Center. 2019. "Generation Z Looks a Lot Like Millennials on Key Social and Political Issues." Retrieved September 23, 2019 (https://www.pewsocialtrends. org/2019/01/17/generation-z-looks-a-lot -like-millennials-on-key-social-and-political -issues/).

Pew Research Center. 2019. "The Narrowing, but Persistent, Gender Gap in Pay." Retrieved September 24, 2019 (https://www.pewresearch .org/factsrank/2019/03/22/gender-pay-gap-facts/)

Pew Research Center. 2014. "Couples, the Internet, and Social Media." Retrieved January 26, 2020 (https://www .pewresearch.org/internet/2014/02/11 /couples-the-internet-and-social-media/).

Pflum, Mary. 2018. "A Year Ago, Alyssa Milano Started a Conversation About #MeToo. These Women Replied." Retrieved October 1, 2019 (https://www.nbcnews.com/news/us-news /year-ago-alyssa-milano-started-conversation -about-metoo-these-women-n920246

Phan, Mai B., Nadine Blumer, and Erin Demaiter. 2009. "Helping Hands: Neighborhood Diversity, Deprivation, and Reciprocity of Support in Non-kin Networks." *Journal of Sexual and Personal Relationships* 26(6–7):899–918.

Phelan, Amanda. 2009. "Practice Development—Preventing Neglect in Formal Care Settings: Elder Abuse and Neglect: The Nurse's Responsibility in Care of the Older Person." *International Journal of Older People Nursing* 4(2):115–119.

Phillips, Julie A., and Megan M. Sweeney. 2005. "Premarital Cohabitation and Marital Disruption among White, Black, and Mexican American Women." *Journal of Marriage and Family* 67(2):296–314.

Phillips, Linda R. 1986. "Theoretical Explanations of Elder Abuse: Competing Hypotheses and Unresolved Issues." pp. 197–217 in *Elder Abuse: Conflict in the Family*, edited by Karl A. Pillemer and Rosalie S. Wolf. Dover, MA: Auburn.

Phillips, Lisa A. 2019. "The Endless Breakup. " Retrieved November 21, 2019 (https://www .psychologytoday.com/us /articles/201905/the-endless-breakup)

Phillips, Roderick. 1997. "Stepfamilies from a Historical Perspective." pp. 5–18 in *Stepfamilies: History, Research, and Policy*, edited by Irene Levin and Marvin B. Sussman. New York: Haworth.

Pickert, Kate. 2012. "Are You Mom Enough?" *Time*, May 21:30–39.

———. 2013. "What Choice? Abortion-rights Activists Won an Epic Victory in *Roe v. Wade*. They've Been Losing Ever Since." *Time*, January 14:40–46.

Piercy, Kathleen W. 2010. *Working with Aging Families: Therapeutic Solutions for Caregivers, Spouses, and Adult Children.* New York: W.W. Norton.

Pietropinto, Anthony, and Jacqueline Simenauer. 1977. *Beyond the Male Myth: What Women Want to Know about Men's Sexuality; A National Survey.* New York: Times Books.

Pilkauskas, Natasha V. 2012. "Three-Generation Family Households: Differences by Family Structure at Birth." *Journal of Marriage and Family* 74(5):931–943.

Pilkauskas, Natasha V., Colin Campbell, and Christopher Wimer. 2017. "Giving unto Others: Private Financial Transfers and Hardship among Families with Children." *Journal of Marriage and Family* 79 (3): 705–722.

Pilarz, Alejandra Ros, and Heather D. Hill. 2017. "Child-Care Instability and Behavior Problems: Does Parenting Stress Mediate the Relationship?" *Journal of Marriage and Family* 79 (5): 1353–1368.

Pillemer, Karl A. 1986. "Risk Factors in Elder Abuse: Results from a Case-Control Study." pp. 239–264 in *Elder Abuse: Conflict in the Family*, edited by Karl A. Pillemer and Rosalie S. Wolf. Dover, MA: Auburn.

Piña-Watson, Brandy, Elma I. Lorenzo-Blanco, Marianela Dornhecker, Ashley J. Martinez, and Julie L. Nagoshi. 2016. "Moving Away From a Cultural Deficit to a Holistic Perspective: Traditional Gender Role Values, Academic Attitudes, and Educational Goals for Mexican Descent Adolescents." *Journal of Counseling Psychology* 63:307.

Pincheta, Rob. 2019. "Emma Watson's 'Self-Partnership' Shows We're in the Golden Age of Singlehood." Retrieved December 4, 2019 (https://www.cnn.com/2019/11/10/uk/emma -watson-singlehood-gbr-scli-intl/index.html).

Pines, Maya. 1981. "Only Isn't Lonely (or Spoiled or Selfish)." *Psychology Today* 15: 15–19.

Pinto, Consuela A. 2009. "Eliminating Barriers to Women's Advancement: Focus on the Performance Evaluation Process." Chicago: American Bar Association. Retrieved January 16, 2010 (www.abanet .org).

Pinto, Katy, and Scott Coltrane. 2009. "Divisions of Labor in Mexican Origin and Anglo

Families: Structure and Culture." *Sex Roles* 60(7/8):482–495.

Piper, Kaitlin N., Tyler J. Fuller, Amy A. Ayers, Danielle N. Lambert, Jessica M. Sales, and Gina M. Wingood. 2019. "A Qualitative Exploration of Religion, Gender Norms, and Sexual Decision-Making within African American Faith-Based Communities." *Sex Roles*:1–17.

Pitt, David. 2018. "Iowa Surrogacy Contracts Case Appealed to the U.S. Supreme Court." Retrieved December 25, 2019 (https://apnews.com/cf123a0287a44790a1c3b99cb9c26c39/Iowa-surrogacy-contracts-case-appealed-to-US-Supreme-Court).

Pittman, Joe F. 2012. "Attachment Orientations: A Boon to Family theory and Research." *Journal of Family Theory and Review* 4(4):306–310.

Pittman, Joe F., Margaret K. Keiley, Jennifer L. Kerpelman, and Brian E. Vaughn. 2011. "Attachment, Identity, and Intimacy: Parallels Between Bowlby's and Erikson's Paradigms." *Journal of Family Theory and Review* 3(1):32–46.

Pittman, Joe F., Jennifer L. Kerpelman, and Jennifer M. McFadyen. 2004. "Internal and External Adaptation in Army Families: Lessons from Operations Desert Shield and Desert Storm." *Family Relations* 53(3):249–260.

Pitts, Leonard Jr. 2008. "Overstressed Parents, Kids Need Help." *Miami Herald*, November 29. Retrieved December 8, 2008 (www.miamiherald.com).

Pitzer, Ronald L. 1997a. "Change, Crisis, and Loss in Our Lives." University of Minnesota Extension Service (www.extension.umn.edu).

———. 1997b. "Perception: A Key Variable in Family Stress Management." University of Minnesota Extension Service (www.extension.umn.edu).

Plan International USA. 2018. "The State of Gender Equality for U.S. Adolescents." Retrieved September 29, 2019 (https://www.planusa.org/docs/state-of-gender-equality-summary-2018.pdf).

Plan International. 2018. "The State of Gender Equality for U.S. Adolescents." Warwick, RI: Plan International.

Planitz, Judith M., and Judith A. Feeney. 2009. "Are Stepsiblings Bad, Stepmothers Wicked, and Stepfathers Evil? An Assessment of Australian Stepfamily Stereotypes." *Journal of Family Studies* 15(1):82–97.

Poehlmann, Julie. 2005. "Children's Family Environments and Intellectual Outcomes During Maternal Incarceration." *Journal of Marriage and Family* 67(5):1275–1285.

Pofeldt, Elaine. 2013. "Can Yahoo Really Be Doing This?" *Forbes*. Retrieved February 26, 2013 (http://www.forbes.com).

Polenick, Courtney, Steven Zart, Kira Birditt, Lauren Bangerter, Amber Seidel, and Karen Fingerman. 2017. "Intergenerational Support and Marital Satisfaction: Implications of Beliefs about Helping Aging Parents." *Journal of Marriage and Family* 79 (1): 131–146.

Polgreen, Erin. 2014. "Muslim Feminist Activism in Indonesia: Seeking (and Finding) Gender Equality in the Quran." Sociologists for Women in Society, November 24, 2014. Retrieved October 5, 2015 (https://www.socwomen.org/wp-content/uploads/pr8_MuslimFemActivism.pdf).

Pollack, Andrew. 2012a. "Before Birth, Dad's ID." *The New York Times, June* 19. Retrieved August 28, 2012 (http://www.nytimes.com).

———. 2012b. "DNA Blueprint for Fetus Built Using Tests of Parents." *The New York Times, June* 6. Retrieved August 28, 2012 (http://www.nytimes.com).

Pollack, William. 2006. "The 'War' For Boys: Hearing 'Real Boys' Voices, Healing Their Pain." *Professional Psychology: Research and Practice* 37(2):190–195.

Pollard, Amy. 2018. "Living Behind the Wheel." *Slate*. August 20. Retrieved September 15, 2019 (http://slate.com).

Pollet, Susan L., and Melissa Lombreglia. 2008. "A Nationwide Survey of Mandatory Parent Education." *Family Court Review* 46(2):375–394.

Pollet, Susan L. 2010. "Still a Patchwork Quilt: A Nationwide Survey of State Laws Regarding Stepparent Rights and Obligations." *Family Court Review* 48:528–540.

Pollitt, Amanda, Joel Muraco, Arnold Grossman, and Stephen Russell. 2017. "Disclosure Stress, Social Support, and Depressive Symptoms among Cisgender Bisexual Youth." *Journal of Marriage and Family* 79 (5): 1278–1294.

Pollmann-Schult, Matthias. 2014. "Parenthood and Life Satisfaction: Why Don't Children Make People Happy?" *Journal of Marriage and Family* 76:319–336.

Polyamory Society. n.d. "Children Education Branch." Retrieved October 4, 2006 (www.polyamorysociety.org).

Pong, Suet-ling, Jamie Johnston, Vivien Chen. 2010. "Authoritarian Parenting and Asian Adolescent School Performance: Insights from the US and Taiwan." *International Journal of Behavioral Development* 34(1):62–72.

Poniewozik, James. 2002. "The Cost of Starting Families." *Time*, April 15, pp. 56–58.

———. 2012. "Daddy Issues: What's So Funny about Men Taking Care of Babies?" *Time*, June 18:60.

Poortman, Anne-Rigt, and Judith A. Seltzer. 2007. "Parents' Expectations about Children after Divorce: Does Anticipating Difficulty Deter Divorce?" *Journal of Marriage and Family* 69(1):254–269.

Popenoe, David. 1993. "American Family Decline, 1960–1990: A Review and Appraisal." *Journal of Marriage and Family* 55(3):527–555.

———. 1994. "The Evolution of Marriage and the Problem of Stepfamilies: A Biosocial Perspective." pp. 3–28 in *Stepfamilies: Who Benefits? Who Does Not?* Edited by Alan Booth and Judy Dunn. Hillsdale, NJ: Erlbaum.

———. 2008. "Cohabitation, Marriage, and Child Wellbeing: A Cross-National Perspective." The National Marriage Project at Rutgers University. Retrieved October 12, 2009 (marriage.rutgers.edu).

Popenoe, David, and Barbara Dafoe Whitehead. 2005. *The State of Our Unions 2005: The Social Health of Marriage in America*. New Brunswick, NJ: Rutgers University, National Marriage Project. Retrieved August 23, 2006 (http://marriage.rutgers.edu).

Popkin, Michael H. and Elizabeth Einstein. 2006. *Active Parenting: A New Program to Help Create Successful Stepfamilies*. Kennesaw, GA: Active Parenting Publishers.

Porche, Michelle, and Diane Purvin. 2008. "'Never in Our Lifetime': Legal Marriage for Same-Sex Couples in Long-Term Relationships." *Family Relations* 37(April):144–159.

Porter, Eduardo. 2014. "Time to Try Compassion, Not Censure, for Families." *The New York Times*, March 4. Retrieved March 6, 2014 (http://www.nytimes.com).

Porter, Eduarcio. 2019. "Why Aren't More Women Working? They're Caring for Parents." *The New York Times* August 29. Retrieved October, 2019 (www.nytimes.com).

Portero, Ashley. 2012. "U.S. Has Second-Highest Rate of Child Poverty in Developed World, Only Romania is Worse." *International Business Times*, May 30. Retrieved December 28, 2012 (http://www.ibtimes.com).

Portes, Alejandro, and M. Zhou. 1993. "The New Second Generation: Segmented Assimilation and Its Variants." *Annals of the American Academy of Political and Social Sciences* 530:74–96.

"Post-Traumatic Stress Disorder in Children." 2019. Centers for Disease Control and Prevention. Retrieved February 3, 2020

Potârcă, Gina, Melinda Mills, and Wiebke Neberich. 2015. "Relationship Preferences among Gay and Lesbian Online Daters: Individual and Contextual Influences." *Journal of Marriage and Family* 77:523–541.

Poteat, V. Paul, Laura M. O'Dwyer, and Ethan H. Mereish. 2012. "Changes in How Students Use and Are Called Homophobic Epithets Over Time: Patterns Predicted by Gender, Bullying, and Victimization Status." *Journal of Educational Psychology* 104(2):393–406.

Potter-Efron, Ronald T., and Patricia S. Potter-Efron. 2008. *The Emotional Affair: How to Recognize Emotional Infidelity and What to Do about It*. Oakland, CA: New Harbinger Publications.

Potter, Sharyn, Rebecca Howard, Sharon Murphy, and Mary M. Moynihan. 2018. "Long-term Impacts of College Sexual Assaults on Women Survivors' Educational and Career Attainments." *Journal of American College Health* 66:496–507.

Powell, Darcey N., and Katherine Karraker. 2017. "Prospective Parents' Knowledge about Parenting and Their Anticipated Child-Rearing Decisions." *Family Relations* 66 (3): 453–467.

Powell, Jean W. 2004. *Older Lesbian Perspectives on Advance Care Planning*. University of Rhode Island (http://digitalcommons.uri.edu/dissertations/AAI3160037/).

Power, Paul W. 1979. "The Chronically Ill Husband and Father: His Role in the Family." *Family Coordinator* 28:616–621.

Prahlad, Anand. 2006. *The Greenwood Encyclopedia of African American Folklore*. Westport, CT: Greenwood Press.

Pratt, Laura A., Debra J. Brody, and Qiuping Gu. 2011. *Antidepressant Use in Persons Aged 12 and Over: United States, 2005-2008*. NCHS data brief, no 76. Hyattsville, MD: National Center for Health Statistics.

Prendergast, Sarah, and David MacPhee. 2018. "Family Resilience Amid Stigma and Discrimination: A Conceptual Model for Families Headed by Same-Sex Parents." *Family Relations* 67 (1): 26–40.

PREPARE/ENRICH. 2020. "Embrace the Journey." Retrieved January 15, 2020 (https://www.prepare-enrich.com/)

Presser, Harriet B. 2000. "Nonstandard Work Schedules and Marital Instability." *Journal of Marriage and Family* 62:93–110.

Preston, Julia. 2007. "As Pace of Deportation Rises, Illegal Families Are Digging In." *The New York Times,* May 1.

———. 2011. "Risks Seen for Children of Illegal Immigrants." *The New York Times,* June 15. Retrieved September 23, 2011 (http://www.nytimes.com).

———. 2015. "Toys R Us Brings Temporary Foreign Workers to U.S. to Move Jobs Overseas." *The New York Times, September* 29. Retrieved October 2, 2015 (http://www.nytimes.com).

Preston, Julia, and John H. Cushman Jr. 2012. "Obama to Permit Young Migrants to Remain in the U.S." *The New York Times, June* 15. Retrieved September 20, 2012 (http://www.nytimes.com).

Preston, Samuel H., and John McDonald. 1979. "The Incidence of Divorce Within Cohorts of American Marriages Contracted Since the Civil War." *Demography* 16:1–25.

Preves, Sharon E. 2010. "Intersex Variations of Sex Development: Implications for Families." *Family Focus* FF46:F3–F6.

Previti, Denise, and Paul R. Amato. 2003. "Why Stay Married? Rewards, Barriers, and Marital Stability." *Journal of Marriage and Family* 65:561–573.

Prickett, Pamela J. 2015. "Negotiating Gendered Religious Space: The Particularities of Patriarchy in an African American Mosque." *Gender and Society* 29(1):51–72.

Priem, Jennifer S., Rachel McLaren, and Denise Haunani Solomon. 2010. "Relational Messages, Perceptions of Hurt, and Biological Stress Reactions to a Disconfirming Interaction." *Communication Research* 37(1):48–72.

Priem, Jennifer S., Denise Haunani Solomon, and Keli Ryan Steuber. 2009. "Accuracy and Bias in Perceptions of Emotionally Supportive Communication in Marriage." *Personal Relationships* 16:531–551.

Proctor, Bernadette D., and Joseph Dalaker. 2003. *Poverty in the United States: 2002.* Current Population Reports P60–222. Washington, DC: U.S. Census Bureau.

Pryor, Jan. 2011. "Commentary on 'Reconsidering the "Good Divorce"' by Paul Amato et al. *Family Relations* 60:525–527.

"The Psychological Impact of Infertility and Its Treatment." 2009. *Harvard Mental Health Letter* 24(11):1–3.

Psychology Today. 2012. "Does Couple Therapy Work? Keys to Success?" Retrieved February 25, 2016 (https://www.psychologytoday.com/blog/headshrinkers-guide-the-galaxy/201203/does-couples-therapy-work-keys-success).

"PTSD and Relationships: A National Center for PTSD Fact Sheet." 2006. United States Department of Veterans Affairs National Center for PTSD. Retrieved August 16, 2006 (www.ncptsd.va.gov).

Pugh, Allison J. 2009. *Longing and Belonging: Parents, Children, and Consumer Culture.* Berkeley: University of California Press.

Purcell, Patrick. 2009. *Income and Poverty among Older Americans in 2008.* Congressional Research Service, October 2. Retrieved May 8, 2010 (www.crs.gov).

Purcell, Tom. 2007. "Baby Name Game Brings Shame." *Omaha World Herald,* August 6.

Purnine, Daniel M., and Michael P. Carey. 1998. "Age and Gender Differences in Sexual Behavior Preferences: A Follow-Up Report." *Journal of Sex and Marital Therapy* 24:93–102.

Pyke, Karen D. 1999. "The Micropolitics of Care in Relationships between Aging Parents and Adult Children: Individualism, Collectivism, and Power." *Journal of Marriage and Family* 61(3):661–672.

Pyke, Karen, and Michelle Adams. 2010. "What's Age Got to Do With It? A Case Study Analysis of Power and Gender in Husband-Older Marriages." *Journal of Family Issues* 31(6):748–777.

Pyke, Karen D., and Vern L. Bengston. 1996. "Caring More or Less: Individualistic and Collectivist Systems of Family Eldercare." *Journal of Marriage and Family* 58(2): 379–392.

Qian, Zhenchao, Jennifer E. Glick, and Christie D. Batson. 2012. "Crossing Boundaries: Nativity, Ethnicity, and Mate Selection." *Demography* 49:651–675.

Qian, Zhenchao, and Daniel T. Lichter. 2007. "Social Boundaries and Marital Assimilation: Interpreting Trends in Racial and Ethnic Intermarriage." *American Sociological Review* 72(1):68–94.

———. 2011. "Changing Patterns of Interracial Marriage in a Multiracial Society." *Journal of Marriage and Family* 73(5)(October):1065–1084.

Qian, Zhenchao, and Daniel T. Lichter. 2018. "Marriage Markets and Assortative Mating in First Marriages and Remarriages in the USA." Retrieved February 23, 2020 (http://www.niussp.org/article/marriage-markets-assortative-mating-first-marriages-remarriages-usa/).

Qian, Zhenchao, and Daniel T. Lichter. 2018. "Marriage Markets and Intermarriage: Exchange in First Marriages and Remarriages." *Demography* 55:849–875.

Quart, Alissa. 2012. "Why Women Hide Their Pregnancies." *The New York Times* October 6. Retrieved November 2, 2012 (http://www.nytimes.com).

Quealy, Kevin, and Claire Cain Miller. 2019. "Young Adulthood in America: Children Are Grown, but Parenting Doesn't Stop." *The New York Times.* March 13. Retrieved August 29, 2019 (https://www.nytimes.com) .

Quick, D. S., P. C. McKenry, and B. M. Newman. 1994. "Stepmothers and Their Adolescent Children: Adjustment to New Family Roles." pp. 105–125 in *Stepparenting: Issues in Theory, Research, and Practice,* edited by K. Pasley and M. Ihinger-Tallman. Westport, CT: Greenwood Press.

Quinn, Tom. 2018. "Romance and Financial Compatibility: Questions to Think About." Retrieved January 25, 2020 (https://blog.myfico.com/romance-and-financial-compatibility-questions/).

Rabin, Roni. 2007. "It Seems the Fertility Clock Ticks for Men, Too." *The New York Times,* February 27.

Rackin, Heather, and Christina M. Gibson-Davis. 2012. "The Role of Pre- and Postconception Relationships for First-Time Parents." *Journal of Marriage and Family* 74(3):526–539.

Radey, Melissa, and Karen A. Randolph. 2009. "Parenting Sources: How Do Parents Differ in Their Efforts to Learn about Parenting?" *Family Relations* December: 536–548.

Radina, M. Elise. 2007. "Mexican American Siblings Caring for Aging Parents: Processes of Caregiver Selection/Designation." *Journal of Comparative Family Studies* 38(1):143–168.

———. 2013. "Toward a Theory of Health-Related Family Quality of Life." *Journal of Family Theory and Review* 5(1):35–50.

Raffaelli, Marcela, Steve P. Tran, Angela R. Wiley, Maria Galarza-Heras, and Vanja Lazarevic. 2012. "Risk and Resilience in Rural Communities: The Experiences of Immigrant Latina Mothers." *Family Relations* 61(4)(October):559–570.

Rahman, Towhidur, Mahmuda Kulsum Moni, Md Samaun Khalid, Farhana Begum, TamannaTabassum Khan, and Nisa Chakma. 2018. "The Dowry as a Crime in Disguise of Gifts: Evidences from Sadar and Faridganj Upazilla of Chandpur District." *Bangladesh Journal of Public Administration (BJPA)* 26:57–77.

Raj, Anita, Nicole Johns, and Rupa Jose. 2019. "Racial/ethnic Disparities in Sexual Harassment in the United States, 2018." *Journal of Interpersonal Violence.* DOI:0886260519842171.

Raley, R. Kelly, Megan M. Sweeney, and Danielle Wondra. 2015. "The Growing Racial and Ethnic Divide in US Marriage Patterns." *The Future of Children* 89–109.

Raley, R. Kelly, and Megan M. Sweeney. 2020. "Divorce, Repartnering, and Stepfamilies: A Decade in Review." *Journal of Marriage and Family* 82:81–99.

Ramisch, Julie. 2012. "Marriage and Family Therapists Working with Couples Who Have Children with Autism." *Journal of Marriage and Family Therapy* 38(2):305–316.

Ramnarace, Cynthia. 2010. "Congress Passes Elder Justice Act." *AARP Bulletin Today,* March 25. Retrieved May 15, 2010 (bulletin.aarp.org).

Ramos Salazar, Leslie. 2015. "The Negative Reciprocity Process in Marital Relationships: A Literature Review." *Aggression and Violent Behavior* 24:113–119.

Ramos-Sánchez, Lucila, and Donald R. Atkinson. 2009. "The Relationships Between Mexican American Acculturation, Cultural Values, Gender, and Help-Seeking Intentions." *Journal of Counseling and Development* 87(1):62–71.

Randles, J., and K. Woodward. 2018. "Learning to Labor, Love, and Live: Shaping the Good Neoliberal Citizen in State Work and Marriage Programs." *Perspectives* 61: 39–56.

Rankin, Jane. 2005. *Parenting Experts: Their Advice, the Research and Getting It Right*. Westport, CT: Praeger.

Rao, Patricia A., and Deborah C. Beidel. 2009. "The Impact of Children with High-Functioning Autism on Parental Stress, Sibling Adjustment, and Family Functioning." *Behavior Modification* 33(4):437–451.

Rapp, Ingmar. 2018. "Partnership Formation in Young and Older Age. " *Journal of Family Issues* 39:3363–3390.

Rappleyea, Damon L., Alan C. Taylor, and Xiangming Fang. 2014. "Gender Differences and Communication Technology Use among Emerging Adults in the Initiation of Dating Relationships." *Marriage and Family Review* 50:269–284.

Rasberry, Catherine N., and Patricia Goodson. 2009. "Predictors of Secondary Abstinence in U.S. College Undergraduates." *Archives of Sexual Behavior* 38:74–86.

Raschick, Michael, and Berit Ingersoll-Dayton. 2004. "Costs and Rewards of Caregiving among Aging Spouses and Adult Children." *Family Relations* 53(3):317–325.

Rasheed, Janice M., Mikal N. Rasheed, and James A. Marley. 2010. *Readings in Family Therapy: From Theory to Practice*. Los Angeles: Sage.

Rasmussen Reports. 2009. "80% Say Parents Should Teach Their Children about Sex." *Rasmussen Reports*, January 14. Retrieved March, 2010 (www .rasmussenreports.com).

Rauer, Amy J., and Brenda L. Volling. 2007. "Differential Parenting and Sibling Jealousy: Developmental Correlates of Young Adults' Romantic Relationships." *Personal Relationships* 14:495–511.

Rauhala, Emily. 2018. "Comrade: Meet Cupid: China's Communist Party Plays Matchmaker to Millennials. " Retrieved November 11, 2019 (https://www.washingtonpost.com/world /asia_pacific/comrade-meet-cupid–young -communists-play-matchmaker-for-chinas -millennials/2018/01/11/3da12de4-f2c4 -11e7-90ed-77167c6861f2_story.html).

Rauscher, Emily A., and Mark A. Fine. 2012. "The Role of Privacy in Families Created Through Assisted Reproductive Technology." *Journal of Family Theory and Review* 4(3):220–234.

Ravitz, Jessica. 2018. "Two Dads, an Egg Donor and a Surrogate." CNN. June 29. Retrieved August 16, 2019 (www.cnn.com)

Ravitz, Paula, Robert Maunder, and Carolina McBride. 2008. "Attachment, Contemporary Interpersonal Theory, and IPT: An Integration of Theoretical, Clinical, and Empirical Perspectives." *Journal of Contemporary Psychotherapy* 38(1):11–21.

Ray, Barbara. 2012. "Parents Give Young Adult Children about $7500 Annually, New Report Finds." *The MacArthur Network on Transitions to Adulthood*. Retrieved May 15, 2012 (http://www .transitions2adulthood .com).

Ray, Julie. 2019. "Americans' Stress, Worry and Anger Intensified in 2018." *Gallup*. April 25. Retrieved February 12, 2020 (https://news. gallup.com).

Rayman, Noah. 2013. "Supreme Court Rules Indian Law Doesn't Apply to Controversial Adoption Case." *Time*,

June 25. Retrieved June 27, 2013 (http://nation .time.com).

Raymond, Joan. 2015. "Could Our Social Media Connections Actually Hurt Our Relationships?" Retrieved February 24, 2016 (http://www.today.com/series /wired /could-our-social-media-connection s-actually-hurt-our-relationships-t60121).

Read, Amanda. 2011. "International Women's Day and the Post-Feminist Era." *Washington Times Communities: Social Journalism from Independent Voices, March* 11. Retrieved October 15, 2012 (http://communities.washingtontimes.com.).

Reader, Steven K., Lindsay M. Stewart, and James H. Johnson. 2009. *Journal of Clinical Psychology in Medical Settings* 16:148–160.

Real, Terrence. 2002. *How Can I Get Through to You? Reconnecting Men and Women*. New York: Scribner's.

Reck, Katie, Brian Higginbotham, Linda Skogrand, and Patricia Davis. 2012. "Facilitating Stepfamily Education for Latinos." *Marriage and Family Review* 48:170–187.

Rector, Robert. 2012. *Marriage: America's Greatest Weapon Against Child Poverty, September* 5. Domestic Policy Studies Special Report No. 117. Washington, DC: Heritage Foundation. Retrieved December 10, 2012 (http://report. heritage.org/sr117).

Rector, Robert, and Kirk A. Johnson. 2005. "Adolescent Virginity Pledges and Risky Sexual Behaviors." Presented at the Eighth Annual National Welfare Research and Evaluation Conference of the Administration for Children and Families, June 15, Washington, DC. Retrieved March 28, 2007 (www.heritage.org).

Reczek, Corinne. 2020. "Sexual- and Gender-Minority Families: A 2010 to 2020 Decade in Review." *Journal of Marriage and Family* 82:300–325.

Reczek, Corinne, Sinikka Elliott, and Debra Umberson. 2009. "Commitment Without Marriage." *Journal of Family Issues* 30(6):738–756.

Reczek, Corinne, Tetyana Pudrovska, Deborah Carr, Debra Umberson, and Mieke Beth Thomeer. 2016. "Marital Histories and Heavy Alcohol Use among Older Adults." *Journal of Health and Social Behavior* 57:77.

Reczek, Corinne, and Debra Umberson. 2012. "Gender, Health Behavior, and Intimate Relationships." *Social Science and Medicine* 74(11): 1783–1790.

Redden, Molly. 2015. "Gender Pay Gap Closing Partially because of Men's Declining Wages, Report Says." *The Guardian*, November 18. Retrieved January 26, 2016 (http://www .theguardian.com /us-news/2015/nov/18 /gender-pay-gap -men-wages-economic-policy -institute -report).

Reddock, Ebony, Cleopatra Caldwell, and Toni Antonucci. 2015. "African American Paternal Grandmothers' Satisfaction with the Fathering Practices of Their Teenage Sons." *Journal of Family Issues* 36(7):831–851.

Reed, Acacia and Philippa Strum. 2008. *New Scholarship in Race and Ethnicity—The Long Shadow of the GI Bill: U.S. Social Policy and the Black-White Gap*. Washington, DC: Woodrow Wilson International Center for Scholars.

Reed, Joanna, Paula England, Krystale Littlejohn, Brooke Conroy Bass, and Monica L. Caudillo. 2014. "Consistent and Inconsistent Contraception among Young Women: Insights from Qualitative Interviews." *Family Relations* 63(2):244–258.

Regnerus, Mark. 2012. "Parental same-sex Relationship, Family Instability, and Subsequent Life Outcomes for Adult Children: Answering Critics of the New Family Structures Study with Additional Analyses." *Social Science Research* 41: 1367–1377.

Regnerus, Mark, and David Gordon. 2013. "Social, Emotional, and Relational Distinctions in Patterns of Recent Masturbation among Young Adults." Austin, TX: Austin Institute for the Study of Family and Culture.

Rehman, Uzma, Amy Holtzworth-Munroe, Katherine Herron, and Kahni Clements. 2009. "'My Way or No Way': Anarchic Power, Relationship Satisfaction, and Male Violence." *Personal Relationships* 16:475–488.

Reid, Megan, and Andrew Golub. 2015. "Vetting and Letting: Cohabiting Stepfamily Formation Processes in Low-Income Black Families." *Journal of Marriage and Family* 77:1234–1249.

Reilly, Katie. 2018. " 'I Work 3 Jobs and Donate Blood Plasma to Pay the Bills.' This Is What It's Like to Be a Teacher in America." *Time*. September 13. Retrieved October 10, 2019 (www.time.com).

Reinert, Duane F. 2005. "Spirituality, Self-Representations, and Attachment to Parents: A Longitudinal Study of Roman Catholic College Seminarians." *Counseling and Values* 49(3):226–238.

Reinhold, Steffen. 2010. "Reassessing the Link between Premarital Cohabitation and Marital Instability." *Demography* 47:719–733.

Reiss, David, Sandra Gonzalez, and Norman Kramer. 1986. "Family Process, Chronic Illness, and Death: On the Weakness of Strong Bonds." *Archives of General Psychiatry* 43:795–804.

Reiss, Ira L. 1980. *Family Systems in America*, 3rd ed. Belmont, CA: Wadsworth.

————. 1986. *Journey into Sexuality: An Exploratory Voyage*. Englewood Cliffs, NJ: Prentice Hall.

Remnick, Noah. 2015. "Activists Say Police Abuse of Transgender People Persists Despite Reforms." *The New York Times*, September 6. Retrieved September 19, 2015 (http://www .nytimes.com).

Renner, Lynette, and Stephen D. Whitney. 2012. "Risk Factors for Unidirectional and Bidirectional Intimate Partner Violence among Young Adults." *Child Abuse and Neglect* 36(1):40–52.

Renteln, Alison Dundes. 2004. *The Cultural Defense*. New York: Oxford University Press.

Repa, Barbara. 2014. *Your Rights in the Workplace*. Berkeley, CA: Nolo.

Repetti, Rena, and Theodore F. Robles. 2016. "Nontoxic Family Stress: Potential Benefits and Underlying Biology." *Family Relations* 65 (1): 163–175.

Resnick, Brian. 2019. "22 Percent of Millennials Say They Have 'No Friends.'" *Vox*. August 1. Retrieved August 27, 2019 (https://www.vox .com).

Ressler, Robert W., Chelsea Smith, Shannon Cavanagh, and Robert Crosnoe. 2017. "Mothers' Union Statuses and Their Involvement in Young Children's Schooling." *Journal of Marriage and Family* 79 (1): 94–109.

Retz, Wolfgang, and Rachel G. Klein. 2010. *Attention-Deficit Hyperactivity Disorder (ADHD) in Adults.* New York: Karger.

Reuter, Tyson R., and Sarah W. Whitton. 2018. "Adolescent Dating Violence Among Lesbian, Gay, Bisexual, Transgender, and Questioning Youth. " Pp. 215–231 in *Adolescent Dating Violence,* edited by David A. Wolf and Jeff R. Temple. Academic Press.

Reuters. 2019. "China's Parliament Rules Out Allowing Sam-Sex Marriage." Retrieved October 16, 2019 (https://www.reuters.com/article/us-china-lgbt-marriage/chinas-parliament-rules-out-allowing-same-sex-marriage-idUSKCN1VB09E).

Reyes, H., Vangie A. Foshee, Jane R. Conway, Nyala Noe, Gert Stulp, and Thomas V. Pollet. 2015. "Finding your Soulmate: Homosexual and Heterosexual Age Preferences in Online Dating." *Personal Relationships* 22:666–678.

Reynolds, Jamila, and Melinda Gonzales-Backen. 2017. "Ethnic-Racial Socialization and the Mental Health of African Americans: A Critical Review." *Journal of Family Theory and Review* 9 (2): 182–200.

Rhoades, Galena K., Scott M. Stanley, and Howard J. Markman. 2009. "Couples' Reasons for Cohabitation." *Journal of Family Issues* 30(2):233–258.

Ribar, David C. 2015. "Why Marriage Matters for Child Wellbeing." *The Future of Children* 11–27.

Rice, C. 2007. "Becoming 'the Fat Girl': Acquisition of an Unfit Identity." *Women's Studies International Forum* 30(2):158–174.

Rich, Motoko. 2014. "School Data Finds Pattern of Inequality Along Racial Lines." *The New York Times.* March 21. Retrieved December 10, 2019 (www.nytimes.com).

Rich, Paul. 2011. "Mental Health the Poor Stepchild of Medical Care." Retrieved September 24, 2012 (http://www.cma.ca/mental-health-poor-stepchild).

Richards, Marty. 2009. *Caresharing: A Reciprocal Approach to Caregiving and Care Receiving in the Complexities of Aging, Illness, or Disability.* Woodstock, VT: Skylight Paths Publishers.

Richards, Meghan A. and Kirsten A. Oinonen. 2011. "Age at Menarche Is Associated with Divergent Alcohol Use Patterns in Early Adolescence and Early Adulthood." *Journal of Adolescence* 34(5):1065–1076.

Richards, Sarah. 2014. "Should a Woman Be Allowed to Hire a Surrogate Because She Fears a Pregnancy Will Hurt Her Career?" *Elle,* April 17. Retrieved November 13, 2015 (http://www.elle.com).

Richards, Sarah Elizabeth. 2019. "What Happened to All Those Frozen Eggs? " (https://www.nytimes.com/2019/12/21/opinion/sunday/egg-freezing-numbers.html).

Richardson, Evin W., and Ted Futris. 2019. "Foster Caregivers' Marital and Coparenting Relationship Experience: A Dyadic Perspective."

Family Relations 68 (2). Retrieved April 12, 2019 (https://onlinelibrary.wiley.com).

Richardson, Joseph B. Jr. 2009. "Men Do Matter: Ethnographic Insights on the Socially Supportive Role of the African American Uncle in the Lives of Inner-City African American Male Youth." *Journal of Family Issues* 30(8):1041–1069.

Richmond, M. K., and G. M. Stocker. 2003. "Siblings' Differential Experiences of Marital Conflict and Differences in Psychological Adjustment." *Journal of Family Psychology* 17:339–350.

Richtel, Matt. 2015. "Push, Don't Crush, the Students." *The New York Times.* April 24. Retrieved April 26, 2015 (http://nytimes.com).

Ricon, Maria, and Brian Trung Lam. 2011. "The Perspectives of Latina Mothers on Latina Lesbian Families." *Journal of Human Behavior in the Social Environment* 21(4):334–349.

Richtel, Matt. 2005. "Past Divorce, Compassion at the End." *The New York Times,* May 19.

———. 2012. "Wasting Time Is New Divide in Digital Era." *The New York Times, May* 29. Retrieved September 15, 2012 (http://www.nytimes.com).

Riedmann, Agnes, and Lynn White. 1996. "Adult Sibling Relationships: Racial and Ethnic Comparisons." pp. 105–126 in *Sibling Relationships: Their Causes and Consequences,* edited by Gene H. Brody. Norwood, NJ: Ablex.

Riegle-Crumb, Catherine, and Melissa Humphries. 2012. "Exploring Bias in Math Teachers' Perceptions of Students' ability by Gender and Race/Ethnicity." *Gender and Society* 26(2):290–322.

Riemer, Abigail R., Sarah J. Gervais, Jeanine LM Skorinko, Sonya Maria Douglas, Heather Spencer, Katherine Nugai, Anastasia Karapanagou, and Andreas Miles-Novelo. 2019. "She Looks like She'd Be an Animal in Bed: Dehumanization of Drinking Women in Social Contexts." *Sex Roles* 80:617–629.

Riley, Glenda. 1991. *Divorce: An American Tradition.* New York: Oxford University Press.

Riley, Matilda W. 1983. "The Family in an Aging Society: A Matrix of Latent Relationships." *Journal of Family Issues* (4):439–454.

Riley, Moira, Laura V. Scaramella, and Lucy McGoron. 2014. "Disentangling the Associations Between Contextual Stress, Sensitive Parenting, and Children's Social Development." *Family Relations* 63(2):287–299.

Riley, Naomi Schaefer. 2013. "Interfaith Unions: A Mixed Blessing." Retrieved May 31, 2013 (http://www.nytimes.com/2013/04/06/opinion/interfaith -marriages-a-mixed-blessing.html).

Rimer, Sara. 1988. "Child Care at Home: Two Women, Complex Roles." *The New York Times,* December 26. Retrieved February 20, 2013 (http://www.nytimes.com).

Rinaldo, Rachel. 2014. "Pious and Critical: Muslim Women Activists in Indonesia." November 26. Retrieved February 4, 2015 (https://gendersociety.wordpress.com).

Rinker, Austin G. 2009. "Recognition and Perception of Elder Abuse by Prehospital and

Hospital-Based Care Providers." *Archives of Gerontology and Geriatrics* 48(1):110–116.

Risch, Gail S., Lisa A. Riley, and Michael G. Lawler. 2004. "Problematic Issues in the Early Years of Marriage: Content for Premarital Education." *Journal of Psychology and Theology* 31(1):253–269.

Riska, Elianne. 2011. "Gender and Medical Careers." *Maturitas* 68(3):264–267.

Risman, Barbara J. 2011. "Gender As Structure or Trump Card?" *Journal of Family Theory and Review* 3(1)(March):18–22.

Rittenour, Christine, and Jordon Soliz. 2009. "Communicative and Relational Dimensions of Shared Family Identity and Relational Intentions in Mother-in-Law /Daughter-in-Law Relationships." *Western Journal of Communication* 73(1):67–90.

Robbins C.L. et al., 2011. "Prevalence, Frequency and Associations of Masturbation with Partnered Sexual Behaviors among U.S. Adolescents," *Archives of Pediatrics & Adolescent Medicine* 165:1087–1093.

Roberts, Brandon, Deborah Povich, and Mark Mather. 2011–2012. *Overworked and Underpaid: Number of Low-Income Working Families Increases to 10.2 Million.* Policy Brief. Winter. Washington, DC: Working Poor Families Project. Retrieved September 19, 2012 (http://www.workingpoorfamilies.org).

Roberts, James C., Loreen Wolfer, and Marie Mele. 2008. "Why Victims of Intimate Partner Violence Withdraw Protection Orders." *Journal of Family Violence* 23(5):369–375.

Roberts, Linda J. 2000. "Fire and Ice in Marital Communication: Hostile and Distancing Behaviors as Predictors of Marital Distress." *Journal of Marriage and Family* 62(3):693–707.

———. 2005. "Alcohol and the Marital Relationship." *Family Focus on . . . Substance Abuse across the Life Span:* FF25: F12–F13. Minneapolis: National Council of Family Relations.

Roberts, Lisen C., and Priscilla White Blanton. 2001. "" I Always Knew Mom and Dad Loved Me Best": Experiences of Only Children. " *Journal of Individual Psychology* 57:125   –140.

Roberts, Sam. 2012. "How Prisoners Make Us Look Good." *The New York Times,* October 27. Retrieved November 2, 2012 (http://www.nytimes.com).

———. 2013. "As Child Care Costs Rise, Families Seek Alternatives." *The New York Times,* April 3. Retrieved June 4, 2013 (http://www.nytimes.com).

Robins, Simon. 2015. "Ambiguous Loss and Addressing Legacies of Disappearance in Conflict." *Family Focus* (FF65, Fall):F1-2. Minneapolis: National Council on Family Relations.

Robinson, Brandon Andrew. 2018. "Conditional Families and Lesbian, Gay, Bisexual, Transgender, and Queer Youth Homelessness." *Journal of Marriage and Family* 80 (2): 383–396.

Robinson, Lawrence, and Jeanne Segal. 2019. "Help for Men Who Are Being Abused." HelpGuide. Retrieved February 13, 2020 (https://www.helpguide.org/articles/abuse/help-for-men-who-are -being-abused.htm).

Robinson, Lawrence, Joanna Saisan, and Jeanne Segal. 2019. "Elder Abuse and Neglect." June. *HelpGuide.* Retrieved September 9, 2019 (https://www.helpguide.org/articles/abuse/elder-abuse-and-neglect.htm).

Robinson, Kelly J., and Jessica J. Cameron. 2011. "Self-Esteem Is a Shared Resource." *Journal of Research in Personality* 46:227–230.

Robinson, Lawrence, and Jeanne Segal. 2012. "Help for Abused Men." Retrieved March 14, 2013 (http://www.helpguide .org).

Robinson, Robert Burton. 2009. "Seven Steps to Adult Family Conflict Resolution." Retrieved February 21, 2010 (www .mindovermania.com).

Robinson, Russell K. 2008. "Structural Dimensions of Romantic Preferences." *Fordham Law Review* 76(6):2787–2819.

Robison, Jennifer. 2002. 2003. "Young Love, First Love, True Love?" The Gallup Poll, February 11. Retrieved August 28, 2006 (www.gallup.com/content).

Rockquemore, Kerry Ann, and Tracey A. Laszloffy. 2005. *Raising Biracial Children.* Lanham, MD: AltaMira.

Rodprasert, W., H. E. Virtanen, S. Sadov, A. Perheentupa, N. E. Skakkebæk, N. Jørgensen, and J. Toppari. 2019. "An Update on Semen Quality Among Young Finnish Men and Comparison with Danish Data." *Andrology* 7:15–23.

Roehr, Bob. 2015. "FDA Committee Recommends Approval for 'Female Viagra.'" *British Medical Journal* 350: h3097.

Rogers, Michelle L., and Dennis P. Hogan. 2003. "Family Life with Children with Disabilities: The Key Role of Rehabilitation." *Journal of Marriage and Family* 65(4):818–833.

Rogers, S., and L. K. White. 1998. "Satisfaction with Parenting: The Role of Marital Happiness, Family Structure, and Parents' Gender." *Journal of Marriage and the Family* 60:293–308.

Rollins, Alethea, and Andrea G. Hunter. 2013. "Racial Socialization of Biracial Youth: Maternal Messages and Approaches to Address Discrimination." *Family Relations* 62(1):140–153.

Romero, Mary. 2002. *Maid in the U.S.A. Tenth Anniversary Edition.* New York: Routledge.

Rook, Karen, Gloria Luong, Dara Sorkin, Jason Newsom, and Neal Krause. 2012. "Ambivalent versus Problematic Social Ties: Implications for Psychological Health, Functional Health, and Interpersonal Coping." *Psychology and Aging* 27(4):912–923.

Roper, Susanne Olsen, and Jeremy B. Yorgason. 2009. "Older Adults with Diabetes and Osteoarthritis and Their Spouses: Effects of Activity Limitations, Marital Happiness, and Social Contacts on Partners' Daily Mood." *Family Relations* 58(October):460–474.

Røsand, Gun-Mette B., Kari Slinning, Espen Røysamb, and Kristian Tambs. 2014. "Relationship Dissatisfaction and Other Risk Factors for Future Relationship Dissolution: A Population-Based Study of 18,523 Couples." *Social Psychiatry and Psychiatric Epidemiology* 49:109–119.

Roscoe, Lori, Elizabeth Corsentino, Shirley Watkins, Marcia McCall, and Juan Sanchez-Ramos. 2009. "Well-Being of Family Caregivers of Persons with Late-Stage Huntington's Disease: Lessons in Stress and Coping." *Health Communication* 24(3):239–248.

Rose, A. J., and K. D. Rudolph. 2006. "A Review of Sex Differences in Peer Relationship Processes: Potential Trade-Offs for the Emotional and Behavioral Development of Girls and Boys." *Psychological Bulletin* 132:98–131.

Rose, India, Mary Prince, Shannon Flynn, Sarah Kershner, and Doug Taylor. 2014. "Parental Support for Teenage Pregnancy Prevention Programmes in South Carolina Public Middle Schools." *Sex Education* 14(5):510–524.

Rose-Greenland, Fiona, and Pamela J. Smock. 2013. "Living Together Unmarried: What Do We Know About Cohabiting Families?" pp. 255–273 In *Handbook of Marriage and the Family,* edited by Gary W. Peterson and Kevin R. Bush. New York: Springer.

Rosen, Ruth. 2007. "The Care Crisis." *The Nation,* March 12, pp. 11–16.

Rosenbaum, Janet Elise. 2009. "Patient Teenagers? A Comparison of the Sexual Behavior of Virginity Pledgers and Matched Nonpledgers." *Pediatrics* 123(1):e110–120.

Rosenberg, E. B., and F. Hajfal. 1985. "Stepsibling Relationships in Remarried Families." *Social Casework: The Journal of Contemporary Social Work* 66:287–292.

Rosenberg, J. 2014. "Sexting Is Positively Linked to Sexual Experience Among Middle School Students." Hoboken, N.J.

Rosenblatt, Paul C., and Linda Hammer Burns. 1986. "Long-term Effects of Perinatal Loss." *Journal of Family Issues* 7:237–254.

Rosenbloom, Stephanie. 2006. "Here Come the Great-Grandparents." *The New York Times,* November 2. Retrieved November 3, 2006 (http://www.nytimes .com).

Rosenblum, Emma. 2014. "Later, Baby." *Bloomberg Business Week,* May 5–11, pp. 45–49.

Rosenfeld, Dana. 1999. "Identity Work among Lesbian and Gay Elderly." *Journal of Aging Studies* 13(2):121–144.

Rosenfeld, Michael J. 2008. "Increasing Percentage of Marriages in the U.S. that are Interracial." Stanford, CA: Stanford University, Department of Sociology. Retrieved November 1, 2009 (www .stanford.edu).

Rosenfeld, Michael J., and Kim Byung-Soo. 2005. "The Independence of Young Adults and the Rise of Interracial and Same-Sex Unions." *American Sociological Review* 70(4):541–562.

Rosenfeld, Michael J., and Reuben J. Thomas. 2012. Searching for a Mate: The Rise of the Internet as a Social Intermediary. *American Sociological Review* 77(4):523–547.

Rosenfeld, Michael J., and Katharina Roesler. 2019. "Cohabitation Experience and Cohabitation's Association with Marital Dissolution." *Journal of Marriage and Family* 81:42–58.

Rosenthal, Andrew. 2013. "Women in Combat." *The New York Times,* January 24. Retrieved April 5, 2016 (www.nytimes .com).

Ross, Catherine E. 1995. "Reconceptualizing Marital Status as a Continuum of Social Attachment." *Journal of Marriage and Family* 57(1):129–140.

Ross, Mary Ellen Trail, and Lu Ann Aday. 2006. "Stress and Coping in African American Grandparents Who Are Raising Their Grandchildren." *Journal of Family Issues* 27(7):912–932.

Ross, Michael W., Sven-Axel Månsson, and Kristian Daneback. 2012. "Prevalence, Severity, and Correlates of Problematic Sexual Internet Use in Swedish Men and Women." *Archives of Sexual Behavior* 41(2):459–466.

Rossi, Alice S. 1968. "Transition to Parenthood." *Journal of Marriage and Family* 30:26–39.

———. 1973. *The Feminist Papers.* New York: Bantam.

———. 1984. "Gender and Parenthood." *American Sociological Review* 49:1–19.

Rossi, A., and P. Rossi. 1990. *Of Human Bonding.* New York: Aldine de Gruyter.

Ross-Sheriff, Fariyal. 2008. "Aging and Gender, Feminist Theory, and Social Work Practice Concerns." *Journal of Women and Social Work* 23(4):309–311.

Roth, Ilona, and Chris Barson. 2010. *The Autism Spectrum in the 21st Century: Exploring Psychology, Biology, and Practice.* London: Jessica Kingsley Publishers.

Rothman, Barbara Katz. 1999. "Comment on Harrison: The Commodification of Motherhood." pp. 435 –438 in *American Families: A Multicultural Reader,* edited by Stephanie Coontz. New York: Routledge.

Rovers, Martin W. 2006. *Healing the Wounds That Hurt Relationships.* Peabody, MA: Hendrickson.

Rowe, Matthew. 2014. "Becoming and Belonging in Gay Men's Life Stories A Case Study of a Voluntaristic Model of Identity." *Sociological Perspectives* 57:434-449.

Rubin, Lillian B. 1976. *Worlds of Pain: Life in the Working-Class Family.* New York: Basic Books.

———. 2007. "The Approach-Avoidance Dance: Men, Women, and Intimacy." pp. 319–324 in *Men's Lives,* 7th ed., edited by Michael S. Kimmel and Michael A. Messner. Boston: Pearson.

Rubin, Roger H. 2001. "Alternative Lifestyles Revisited, or Whatever Happened to Swingers, Group Marriages, and Communes?" *Journal of Family Issues* 22(6):711–726.

Rudman, Laurie A., and Peter Glick. 2008. *The Social Psychology of Gender: How Power and Intimacy Shape Gender Relations.* New York: Guilford Press.

Rudolph, Bonnie, Cecily Cornelius-White, and Fernando Quintana. 2005. "Filial Responsibility among Mexican American College Students: A Pilot Investigation and Comparison." *Journal of Hispanic Higher Education* 4(1):64–78.

Rueter, Martha A., and Ascan F. Koerner. 2008. "The Effect of Family Communication Patterns on Adopted Adolescent Adjustment." *Journal of Marriage and Family* 70(3):715–727.

Ruggles, Steven. 2011. "Intergenerational Coresidence and Family Transitions in the United States, 1850–1880." *Journal of Marriage and Family* 73(1):136–148.

Ruiz, Maria Elena, and H. Edward Ransford. 2012. "Latino Elders Regraming *Familismo:* Implications for Health and Caregiving Support." *Journal of Cultural Diversity* 19(2):50–57.

Ruiz, Sarah A., and Merril Silverstein. 2007. "Relationships with Grandparents and the Emotional Well-Being of Late Adolescent and Young Adult Grandchildren." *Journal of Social Issues* 63(4):793–808.

Rumney, Avis. 2009. *Dying to Please: Anorexia, Treatment and Recovery*. Jefferson, NC: McFarland & Co.

Runner, Michael, Mieko Yoshihama, and Steve Novick. 2009. *Intimate Partner Violence in Immigrant and Refugee Communities: Challenges, Promising Practices and Recommendations*. Princeton, NJ: Family Violence Prevention Fund for the Robert Wood Johnson Foundation.

Ruppanner, Leah. 2013. "Conflict Between Work and Family: An Investigation of Four Family Measures." *Social Indicators Research* 110:327–347.

Rush, Michael. 2016. "Parental Leave in the U.S.: International Outlier Riding a New Wave among Cross-Currents." *Family Focus* (FF67, Spring):F16. Minneapolis: National Council on Family Relations.

Rushe, Dominic. 2018. "McDonald's Workers Walk Out in 10 U.S. Cities Over Sexual Harassment Epidemic." September 18. *The Guardian*. Retrieved January 20, 2019 (www.theguardian.com).

Russell, Stephen T., Thomas J. Clarke, and Justin Clary. 2009. "Are Teens 'Post-Gay'? Contemporary Adolescents' Sexual Identity Labels." *Journal of Youth and Adolescence* 38(7):884–890.

Russell, Stephen, Amanda Pollitt, Gu Li, and Arnold Grossman. 2018. "Chosen Name Use Is Linked to Reduced Depressive Symptoms, Suicidal Ideation, and Suicidal Behavior among Transgender Youth." *Journal of Adolescent Health* 63: 503–505.

Russell, Stephen T., and Jessica N. Fish. 2016. "Mental Health in Lesbian, Gay, Bisexual, and Transgender (LGBT) Youth." *Annual Review of Clinical Psychology* 12:465–487.

Rutgers University School of Criminal Justice and the New Jersey Institute for Social Justice. 2006. *Bringing Families In: Recommendations of the Incarceration, Reentry and the Family Roundtables*, December. Retrieved March 11, 2010 (www.reentry .net/library/item.126583-Bringing _Families_In_Recommendations_of_the _Incarceration_Reentry_and_the_F).

Rutter, Michael. 2002. "Nature, Nurture, and Development: From Evangelism through Science toward Policy and Practice." *Child Development* 73(1):1–21.

Rutter, Virginia, and Pepper Schwartz. 2012. *The Gender of Sexuality*, 2nd ed. Lanham, MD: Rowman & Littlefield.

Ruvalcaba, Yanet, and Asia A. Eaton. 2019. "Nonconsensual Pornography among U.S. Adults: A Sexual Scripts Framework on Victimization, Perpetration, and Health Correlates for Women and Men." *Psychology of Violence* 10:68–78.

Ryan, Kathryn M., Kim Weikel, and Gene Sprechini. 2008. "Gender Differences in Narcissism and Courtship Violence in Dating Couples." *Sex Roles* 58:802–813.

Ryan, Rebecca M., Ariel Kalil, and Lindsey Leininger. 2009. "Low-Income Mothers' Private Safety Nets and Children's Socioemotional Well-Being." *Journal of Marriage and Family* 71(2):278–297.

Ryan, Rebecca M., Ariel Kalil, and Kathleen M. Ziol-Guest. 2008. "Longitudinal Patterns of Nonresident Fathers' Involvement: The Role of Resources and Relations: Using Data from the Fragile Families and Child Wellbeing Study." *Journal of Marriage and Family* 70(November):962–977.

Ryan, Suzanne, Kerry Franzetta, Erin Schelar, and Jennifer Manlove. 2009. "Family Structure History" Links to Relationship Formation Behaviors in Young Adulthood." *Journal of Marriage and Family* 71(November):935–953.

Ryle, Robyn. 2015. *Questioning Gender: A Sociological Exploration*. Thousand Oaks, CA: Sage.

Saad, Lydia. 2003. "*Roe v. Wade* Has Positive Public Image." Gallup News Service, January 20 (www.gallup.com/poll).

———. 2004. "No Time for R & R." Gallup Poll Tuesday Briefing, May 11 (www.gallup .com).

———. 2006. "Blacks Committed to the Idea of Marriage." The Gallup Poll, July 14. Retrieved October 10, 2006 (poll.gallup.com).

———. 2012a. "In U.S., Half of Women Prefer a Job Outside the Home." Gallup Poll, September 7. Retrieved February 7, 2013 (http://www .gallup.com).

———. 2012b. "Stay-at-Home Moms in U.S. Lean Independent, Lower-Income." Gallup Poll, April 19. Retrieved August 29, 2012 (http:// www.gallup.com).

Saad, Lydia. 2018. "Americans, in Theory, Think Larger Families Are Ideal. " Retrieved December 14, 2019 (https://news.gallup.com /poll/236696/americans -theory-think-larger-families-ideal.aspx).

Saad, Lydia. 2017. "Eight in 10 Americans Afflicted by Stress." *Gallup*. December 20. Retrieved February 12 (https://news .gallup.com).

Saad, Lydia. 2015. "Fewer Young People Say I Do—to Any Relationship. " Retrieved November 19, 2019 (https://news.gallup .com/poll/183515/fewer-young-people -say-relationship.aspx).

Saad, Lydia. 2019. "Majority in U.S. Still Want Abortion Legal, with Limits." Retrieved October 21, 2019 (https:// news.gallup.com/poll/259061/majority -abortion-legal-limits.aspx).

Saadeh, Wasim, Christopher P. Rizzo, and David G. Roberts. 2002. "This Month's Debate: Spanking." *Clinical Pediatrics* (March):87–91.

Sabatelli, Ronald M., Hyanghee Lee and Karen Ripoll-Nunez. 2018. "Placing the Social Exchange Framework in an Ecological Context." *Journal of Family Theory & Review* 10(1):32–48.

Sabia, Joseph J., and Daniel I. Rees. 2009."The Effect of Sexual Abstinence on Females' Educational Attainment." *Demography* 46(4):695–715.

Sachs, Andrea, Dorian Solot, and Marshall Miller. 2003. "Happily Unmarried." *Time*, March 3.

Sabol, William J., and Heather C. West. 2009. *Prison and Jail Inmates at Midyear 2008*. Bureau of Justice Statistics Bulletin. Washington, DC:

U.S. Department of Justice, March 31. Retrieved March 13, 2010 (bjs.ojp.usdoj.gov).

Sachs-Ericsson, Natalie, Mathew D. Gayman, Kathleen Kendall-Tackett, Donald A. Lloyd, Amanda Medley, Nicole Collins, Elizabeth Corsentino, and Kathryn Sawyer. 2010. "The Long-Term Impact of Childhood Abuse on Internalizing Disorders among Older Adults: The Moderating Role of Self-Esteem. *Aging and Mental Health* 14(4):489–501.

Safilios-Rothschild, Constantina. 1970. "The Study of Family Power Structure: A Review 1960-1969." *Journal of Marriage and Family* 32:539–543.

Said, Edward W. 1978. *Orientalism*. New York: Vintage Books.

Saint-Jacques, Marie-Christine, Caroline Robitaille, Elisabeth Godbout, Claudine Parent, Dylvie Drapeau, and Marie-Helen Gagne. 2011. "The Processes Distinguishing Stable from Unstable Stepfamily Couples." *Family Relations* 60:545–561.

Saint Louis, Catherine. 2012. "Letting Children Share in Grief." *The New York Times*, September 20:D1, D7.

———. 2015. "Many Children Under Five Are Left to Their Mobile Devices, Survey Finds." *The New York Times*, November 2. Retrieved November 9, 2015 (http://www .nytimes.com).

Salisbury, Emily J., Kris Henning, and Robert Holdford. 2009. "Fathering by Partner-Abusive Men." *Child Maltreatment* 14(3):232–242.

Sallee, Margaret W. 2011. "Performing Masculinity: Considering Gender in Doctoral Student Socialization." *Journal of Higher Education* 82(2)(March/April): 187–216.

"Salsa Beats Out Ketchup as America's Preferred Condiment." 2013. *Daily News*, October 17. Retrieved March 25, 2016 (http://www .nydailynews.com).

Saluter, Arlene F. and Terry A. Lugaila. 1998. *Marital Status and Living Arrangements: March 1996*. Current Population Reports P20–496. Washington, DC: U.S. Bureau of the Census.

Salvatore, Jessica E., and Danielle M. Dick. 2015. "Gene-Environment Interplay: Where We Are, Where We Are Going." *Journal of Marriage and Family* 77(2):344–350.

Samek, Diana, Bibiana D. Koh, and Martha A. Rueter. 2013. "Overview of Behavioral Genetics Research for Family Researchers." *Journal of Family Theory and Review* 5(3):214–233.

Samuels, Gina Miranda. 2009. "'Being Raised by White People': Navigating Racial Difference among Adopted Multiracial Adults." *Journal of Marriage and Family* 71(1):80–94.

Samji, Hasina, Angela Cescon, Robert S. Hogg, Sharada P. Modur, Keri N. Althoff, Kate Buchacz, Ann N. Burchell et al. 2013. "Closing the Gap: Increases in Life Expectancy among Treated HIV-Positive Individuals in the United States and Canada." Retrieved January 7, 2015 (http:// journals.plos.org/plosone/article?id=10.1371 /journal.pone .0081355).

Sandberg, Jonathan G., James M. Harper, E. Jeffrey Hill, Richard B. Miller, Jeremy B. Yorgason, and Randal D. Day. 2013. "'What Happens at Home Does Not Necessarily Stay at Home": The Relationship of Observed Negative Couple Interaction with Physical Health, Mental

Health, and Work Satisfaction." *Journal of Marriage and Family* 75:808–821.

Sandberg, Jonathan, Jeremy Yorgason, Richard Miller, and E. Jeffrey Hill. 2012. "Family-to-Work Spillover in Singapore: Marital Distress, Physical and Mental Health, and Work Satisfaction." *Family Relations* 61(1):1–15.

Sandberg, Sheryl. 2013a. *Lean In: Women, Work, and the Will to Lead.* New York: Random House.

———. 2013b. "Why I Want Women to Lean In." *Time,* March 18:44–45.

———. 2013c "Welcome to the Lean In Community." *Lean In,* March 5. Retrieved March 14, 2013 from http://leanin.org.

Sandberg-Thoma, Sara E., Anastasia R. Snyder, and Bohyun Joy Jang. 2015. "Exiting and Returning to the Parental Home for Boomerang Kids." *Journal of Marriage and Family* 77(3):806–818.

Sandfort, Theo G. M., and Brian Dodge. 2008. ". . . And Then There was the Down Low: Introduction to Black and Latino Male Bisexualities." *Archives of Sexual Behavior* 37(5):675–682.

Sandler, Lauren. 2011. "The Mother Majority." *Slate,* October 17. Retrieved December 13, 2012 (http://www.slate.com).

———. 2013. "None is Enough." *Time,* August 12, 2013.

"A Sane Approach to an Emotional Issue: Adult Children Living at Home with Parents Agreement." 2015. Retrieved January 17, 2016 (http://www.asaneapproach.com).

"The Sandwich Generation: Stress and Other Issues." 2019. Griswold Home Care. June 20. Retrieved August 18, 2019 (https://www.GriswoldHomecare.com/blog).

Saner, Emine. 2018. "The Breakup Guru Who Invented Conscious Uncoupling: 'I Understand The Backlash.'" Retrieved February 9, 2020 (https://www.theguardian.com/lifeandstyle/2018/apr/22/the-breakup-guru-who-invented-conscious-uncoupling-i-understand-the-backlash).

Sanford, Keith. 2006. "Communication during Marital Conflict: When Couples Alter Their Appraisal, They Change Their Behavior." *Journal of Family Psychology* 20(2):256–266.

Saujani, Reshma. 2018. "Why We Should Ditch the 'Perfect Woman' Myth." Retrieved September 29, 2019 (https://www.cnn.com/2018/07/21/opinions/abortion-fertility-miscarriage-stories-matter-saujani-opinion/index.html).

Sandstrom, Gillian M., and Elizabeth W. Dunn. 2014. "Is efficiency overrated? Minimal social interactions lead to belonging and positive affect." *Social Psychological and Personality Science* 5(4):437–442.

Sandstrom, Gillian M., Vincent Wen-Sheng Tseng, Jean Costa, Fabian Okeke, Tanzeem Choudhury, and Elizabeth W. Dunn. 2016. "Talking Less during Social Interactions Predicts Enjoyment: A Mobile Sensing Pilot Study." *PloS one* 11(7).

Sano, Yoshie, Leslie N. Richards, and Anisa M. Zvonkovic. 2008. "Are Mothers Really 'Gatekeepers' of Children?: Rural Mothers' Perceptions of Nonresident Fathers' Involvement in Low-Income Families." *Journal of Family Issues* 29(12):1701–1723.

Santelli, John S., Laura Duberstein Lindberg, Lawrence B. Finer, and Susheela Singh. 2007. "Explaining Recent Declines in Adolescent Pregnancy in the United States: The Contribution of Abstinence and Improved Contraceptive Use." *American Journal of Public Health* 97(1):150–157.

Santelli, John S., and Stephanie A. Grilo. 2019. "Ideology or Evidence? Examining the Population-Level Impact of U.S. Government Funding to Prevent Adolescent Pregnancy." *American Journal of Public Health* 109:356–357.

Santelli, John, Stephanie A. Grilo, Laura D. Lindberg, Ilene Speizer, Amy Schalet, Jennifer Heitel, Leslie Kantor et al. 2017. "Abstinence-only-until-marriage Policies and Programs: An Updated Position Paper of the Society for Adolescent Health and Medicine." *The Journal of Adolescent Health: Official Publication of the Society for Adolescent Medicine* 61:400.

Sarkar, N. N. 2008. "The Impact of Intimate Partner Violence on Women's Reproductive Health and Pregnancy Outcome." *Journal of Obstetrics and Gynecology* 28(3):266–271.

Sarkis, Stephanie. 2011. "Overscheduled and Overextended: How To Stop." *The Huffington Post,* August 23. Retrieved May 8, 2015 (http://www.huffingtonpost.com).

Sarkisian, Natalia, Mariana Gerena, and Naomi Gerstel. 2006. "Extended Family Ties among Mexicans, Puerto Ricans, and Whites: Superintegration or Disintegration?" *Family Relations* 55(3):331–344.

Sarkisian, Natalia and Naomi Gerstel. 2004. "Explaining the Gender Gap in Help to Parents: The Importance of Employment." *Journal of Marriage and Family* 66(2):431–451.

Sassler, Sharon. 2004. "The Process of Entering into Cohabiting Unions." *Journal of Marriage and Family* 66(2):491–505.

———. 2010. "Partnering Across the Life Course: Sex, Relationships, and Mate Selection." *Journal of Marriage and Family* 72(3) (June):557–575.

Sassler, Sharon, and Anna Cunningham. 2008. "How Cohabitors View Childbearing." *Sociological Perspectives* 51(1):3–28.

Sassler, Sharon, and Frances Goldscheider. 2004. "Revisiting Jane Austen's Theory of Marriage Timing: Changes in Union Formation among Men in the Late 20th Century." *Journal of Family Issues* 25(2):139–166.

Sassler, Sharon, and Amanda J. Miller. 2014. "'We're Very Careful . . .': The Fertility Desires and Contraceptive Behaviors of Cohabiting Couples." *Family Relations* 63:538–553.

Sassler, Sharon, Amanda Miller, and Sarah Favinger. 2009. "Planned Parenthood?" *Journal of Family Issues* 30(2):206–232.

Sattler, David N. 2006. "Family Resources, Family Strains, and Stress Following the Northridge Earthquake." *Stress, Trauma, and Crisis: An International Journal* 9(3-4):187–202.

Saulny, Susan. 2006. "In Baby Boomlet, Preschool Derby Is the Fiercest Yet." *The New York Times,* March 3. Retrieved March 15, 2006 (www.nytimes.com).

———. 2011a. "Black? White? Asian? More Young Americans choose All of the Above." *The New York Times, January* 29. Retrieved September 12, 2011 (http://www.nytimes.com).

———. 2011b. "In Strangers' Glances at Family Tensions Linger." *The New York Times,* October 12. Retrieved October 17, 2011 (http://www.nytimes.com).

"Save the Date: Relationships Ward Off Disease and Stress." 2004. *Psychology Today* (January/February):32.

Sawhill, Isabel V. 2014a. "Beyond Marriage." *The New York Times, September* 13. Retrieved February 27, 2015 (http://www.nytimes.com).

Sawhill, Isabel V. 2014b. *Generation Unbound: Drifting into Sex and Parenthood without Marriage.* Washington, DC: Brookings Institution Press.

Sawhill, Isabel, and Joanna Venator. 2015. "Is There a Shortage of Marriageable Men?" *Center on Children and Families* Retrieved November 23, 2019 (https://pdfs.semanticscholar.org/a051/eb28b75a12253034006a1b7120dbb22e0cb0.pdf).

Sax, Leonard. 2002. "How Common is Intersex? A Response to Anne Fausto-Sterling." *Journal of Sex Research* 39:174–178.

Saxena, Divya, and Gregory F. Sanders. 2009. "Quality of Grandparent-Grandchild Relationship in Asian-Indian Immigrant Families." *International Journal of Aging and Human Development* 68(4):321–337.

Saxena, Mamta, and Kari Adamsons. 2013. "Siblings of Individuals with Disabilities: Reframing the Literature Through a Bioecological Lens." *Journal of Family Theory and Review* 5(4):300–316.

Sayer, Liana. 2006. "Economic Aspects of Divorce and Relationship Dissolution." pp. 385–406 in *Handbook of Divorce and Relationship Dissolution,* edited by Mark A. Fine and John H. Harvey. Mahwah, NJ: Erlbaum.

Sayer, Liana C., Paula England, Paul D. Allison, and Nicole Kangas. 2011. "She Left, He Left: How Employment and Satisfaction Affect Women's and Men's Decisions to Leave Marriages." *American Journal of Sociology* 116(6):1982–2018.

Scanzoni, John H. 1972. *Sexual Bargaining: Power Politics in the American Marriage.* Englewood Cliffs, NJ: Prentice Hall.

Scelfo, Julie. 2015. "A University Recognizes a Third Gender: Neutral." *The New York Times,* February 3. Retrieved February 4, 2015 (http://www.nytimes.com).

Schade, Lori Cluff, Jonathan Sandberg, Roy Bean, Dean Busby, and Sarah Coyne. 2013. "Using Technology to Connect in Romantic Relationships: Effects on Attachment, Relationship Satisfaction, and Stability in Emerging Adults." *Journal of Couple and Relationship Therapy* 12:314–338.

Schafer, Markus H., Kenneth F. Ferraro, and Sarah A. Mustillo. 2011. "Children of Misfortune: Early Adversity and Cumulative Inequality in Perceived Life Trajectories." *American Journal of Sociology* 116(4):1053–1091.

Schalet, Amy. 2011. *Not Under My Roof: Parents, Teens, and the Culture of Sex*. Chicago: University of Chicago Press.

————. 2012. "Caring, Romantic, American Boys." *The New York Times*, April 6. Retrieved October 6, 2012 (http://www.nytimes.com).

Schecter, E., A. Tracy, K. Page, and G. Luong. 2008. "Shall We Marry?" Legal Marriage as a Commitment Event in Same-Sex Relationships." *Journal of Homosexuality* 54: 400–422.

Schechtman, Morris R., and Arleah Schechtman. 2003. *Love in the Present Tense: How to Have a High Intimacy, Low Maintenance Marriage*. Boulder, CO: Bull Publishing.

Scheiber, Noam. 2015a. "Attitudes Shift on Paid Leave: Dads Sue, Too." *The New York Times*, September 16. Retrieved September 19, 2015 (http://www.nytimes.com).

————. 2015b. "Attitudes Shift on Paid Leave: Dads Sue, Too." *The New York Times*, September 15. Retrieved September 19, 2015 from http://www.nytimes.com.

————. 2015c. "The Perils of Ever-Changing Work Schedules Extend to Children's Well-Being." *The New York Times*, August 13. Retrieved November 14, 2015, from http://www.nytimes.com.

————. 2015d. "Starbucks Falls Short After Pledging Better Labor Practices." *The New York Times*, September 23. Retrieved November 14, 2015, from http://www.nytimes.com.

Scherrer, Kristin S., Emily Kazyak, and Rachel Schmitz. 2015. "Getting 'Bi' in the Family: Bisexual People's Disclosure Experiences." *Journal of Marriage and Family* 77:680–696.

Schick, Vanessa, et al. 2010. "Sexual Behaviors, Condom Use, and Sexual Health of Americans Over 50: Implications for Sexual Health Promotion for Older Adults." *Journal of Sexual Medicine* 7(suppl 5):315–329.

Schick, Vanessa R., Joshua G. Rosenberger, Debby Herbenick, Erika Collazo, Stephanie A. Sanders, and Michael Reece. 2016. "The Behavioral Definitions of 'Having Sex with a Man' and 'Having Sex with a Woman' Identified by Women who have Engaged in Sexual Activity with both Men and Women." *The Journal of Sex Research* 53:578–587.

Schindler, Holly S. 2010. "The Importance of Parenting and Financial Contributions in Promoting Fathers' Psychological Health." *Journal of Marriage and Family* 72(2):318–32.

Schlomer, Gabriel L., H. H. Cleveland, David J. Vandenbergh, Gregory M. Fosco, and Mark E. Feinberg. 2015. "Looking Forward in Candidate Gene Research: Concerns and Suggestions." *Journal of Marriage and Family* 77(2):351–354.

Schlomer, Gabriel L., Gregory M. Fosco, H. H. Cleveland, David J. Vandenbergh, and Mark E. Feinberg. 2015. "Interparental Relationship Sensitivity Leads to Adolescent Internalizing Problems: Different Genotypes, Different Pathways." *Journal of Marriage and Family* 77(2):329–343.

Schlomer, Gabriel, Stacy Ann Hawkins, Christine Wiggs, Leslie Bosch, Deborah Casper, Noel Card, and Lynne Borden. 2012. "Deployment and Family Functioning: A Literature Review of US Operations in Afghanistan and Iraq." *Family Science* 3(2):86–98.

Schmeeckle, Maria. 2007. "Gender Dynamics in Stepfamilies: Adult Stepchildren's Views." *Journal of Marriage and Family* 69(1):174–189.

Schmidt, Samantha. 2019. "Merriam-Webster Adds Non-Binary Pronoun to Dictionary." Retrieved September 23, 2019 (https://www.washingtonpost.com/dc-md-va/2019/09/17/merriam-webster-adds-non-binary-pronoun-they-dictionary/).

Schmitt, Marina, Matthias Kliegel, and Adam Shapiro. 2007. "Marital Interaction in Middle and Old Age." *International Journal of Aging and Human Development* 65(4):283–300.

Schneider, Barbara Lynn, Gregory Wallsworth, and Iliya Gutin. 2014. "Family Experiences of Competition and Adolescent Performance." *Journal of Marriage and Family* 76(3):665–676.

Schneider, D. M. 1968. *American Kinship: A Cultural Account*. Englewood Cliffs, NJ: Prentice-Hall.

Schneider, Daniel. 2011. "Market Earnings and Household Work: New Tests of Gender Performance Theory." *Journal of Marriage and Family* 73(4):845–860.

————. 2015. "The Great Recession, Fertility, and Uncertainty: Evidence From the United States." *Journal of Marriage and Family* 77:1144–1156.

Schneider, Daniel, Kristen Harknett, and Matthew Stimpson. 2018. "What Explains the Decline in First Marriage in the United States? Evidence from the Panel Study of Income Dynamics, 1969 to 2013." *Journal of Marriage and Family* 80:791–811.

Schneider, Jennifer P., Robert Weiss, and Charles Samenow. 2012. "Is it Really Cheating? Understanding the Emotional Reactions and Clinical Treatment of Spouses and Partners Affected by Cybersex Infidelity." *Sexual Addiction and Compulsivity* 19(1–2):123–139.

Schneiderman, Inna, Oma Zagoory-Sharon, James F. Leckman, and Ruth Felding. 2012. "Oxytocin During the Initial Stages of Romantic Attachment: Couples' Interactive Reciprocity." *Psychoneuroendocrinology* 37(8):1277–1285.

Schneiderman, Inna, Yaniv Kanat-Maymon, Richard P. Ebstein, and Ruth Feldman. 2014. "Cumulative Risk on the Oxytocin Receptor Gene (OXTR) Underpins Empathic Communication Difficulties at the First Stages of Romantic Love." *Social Cognitive and Affective Neuroscience* 9:1524–1529.

Schnittker, Jason, Jeremy Freese, and Brian Powell. 2003. "Who Are the Feminists and What Do They Believe? The Role of Generations." *American Sociological Review* 68:607–622.

Schnor, Christine, Inge Pasteels, and Jan Van Bavel. 2017. "Sole Physical Custody and Mother's Repartnering After Divorce." *Journal of Marriage and Family* 79:879–890.

Schnor, Christine, Sofie Vanassche, and Jan Van Bavel. 2017. "Stepfather or Biological Father? Education-Specific Pathways of Postdivorce Fatherhood." *Demographic Research* 37:1659–1694.

Schoen, Robert, Nan Marie Astone, Kendra Rothert, Nicola J. Standish, and Young J. Kim. 2002. "Women's Employment, Marital Happiness, and Divorce." *Social Forces* 81:643–662.

Schoen, Robert, and Yen-Hsin Alice Cheng. 2006. "Partner Choice and the Differential Retreat from Marriage." *Journal of Marriage and Family* 68(1):1–10.

Schoen, Robert, Nancy S. Landale, Kimberly Daniels, and Yen-Hsin Alice Cheng. 2009. "Social Background Differences in Early Family Behavior." *Journal of Marriage and Family* 71(2):984–995.

Schoen, Robert, Stacy J. Rogers, and Paul R. Amato. 2006. "Wives' Employment and Spouses' Marital Happiness: Assessing the Direction of Influence Using Longitudinal Couple Data." *Journal of Family Issues* 27(4):506–528.

Schoenberg, Nara. 2016. "For Some IVF Patients, a Choice: Do You Want a Boy or a Girl?" Retrieved February 14, 2016 (http://www.chicagotribune.com/lifestyles/health/sc-gender-selection-health-1021-20151014-story.html).

Schoonover, Katie, and Bree McEwan R. 2014. "Are You Really Just Friends? Predicting the Audience Challenge in Cross-Sex Friendships." *Personal Relationships* 21:387–403.

Schoppe-Sullivan, Sarah J., and Jay Fagan. 2020. "The Evolution of Fathering Research in the 21st Century: Persistent Challenges, New Directions." *Journal of Marriage and Family* 82:175–197.

Schoppe-Sullivan, Sarah J., Sarah C. Mangeisdorf, Geoffrey L. Brown, and Margaret Sokolowski-Szewczyk. 2007. "Goodness-of-Fit in Family Context: Infant Temperament, Marital Quality, and Early Coparenting Behavior." *Infant Behavior and Development* 30(1):82–97.

Schoppe-Sullivan, Sarah J., Alice C. Schermerhorn, and E. Mark Cummings. 2007. "Marital Conflict and Children's Adjustment: Evaluation of the Parenting Process Model." *Journal of Marriage and Family* 69(5):1118–1134.

Schottenbauer, Michelle A., Stephanie M. Špernak, and Ingrid Hellstrom. 2007. "Relationship between Family Religious Behaviors and Child Well-Being among Third-Grade Children." *Mental Health, Religion, and Culture* 10(2):191–198.

Schramm, David G., James P. Marshall, Victor W. Harris, and Thomas R. Lee. 2012. "Religiosity, Homomgamy, and Marital Adjustment: An Examination of Newlyweds in First Marriages and Remarriages." *Journal of Family Issues* 33(2):246–268.

Schrobsdorff, Susanna. 2016. "Anxiety, Depression and the American Adolescent: The Kids Are Not All Right." *Time*. November 7: 42–51.

Schrodt, Paul. 2008. "Sex Differences in Stepchildren's Reports of Stepfamily Functioning." *Communication Reports* 21(1):46–58.

————. 2009. "Family Strength and Satisfaction as Functions of Family Communication Environments." *Communication Quarterly* 57(2): 171–186.

Schrodt, Paul, Andrew M. Ledbetter, Kodiane A. Jernberg, Lara Larson, Nicole Brown, and Katie Glosnek. 2009. "Family Communication Patterns As Mediators of Communication Competence in the Parent-Child Relationship." *Journal of Social and Personal Relationships* 26(6–7):853–874.

Schrodt, Paul, Jordan Soliz, and Dawn O. Braithwaite. 2008. "A Social Relations Model of Everyday Talk and Relational Satisfaction in Stepfamilies." *Communication Monographs* 75(2):190–217.

Schuman, Michael. 2009. "Going Home: On the Road Again." *Time*, April 27, pp. 18–26.

Schumm, Walter R. 2012. "Methodological Decisions and the Evaluation of Possible Effects of Different Family Structures on Children: The New Family Structures Survey (NFSS)." *Social Science Research* 41:1357–1366.

Schwartz, Christine R. 2014. "Brief: It's Not Just Attitudes: Marriage Is Also Becoming More Egalitarian." Council on Contemporary Families Gender Rebound Symposium, July 30.

Schwartz, Christine R., and Hongyun Han. 2014. "The Reversal of the Gender Gap in Education and Trends in Marital Dissolution." *American Sociological Review* 79(4):605–629.

Schwartz Christine R., and Nikki L. Graf. 2009. "Assortative Mating Among Same-Sex and Different-Sex Couples in the United States, 1990–2000. *Demographic Research* 21:843–878

Schwartz, Christine R., and Pilar Gonalons-Pons. 2016. "Trends in Relative Earnings and Marital Dissolution: Are Wives Who Outearn Their Husbands Still More Likely to Divorce?" *RSF: The Russell Sage Foundation Journal of the Social Sciences* 2:218–236.

Schwartz, John. 2010. "In Same-Sex Ruling, and Eye on the Supreme Court." *The New York Times*, August 4. Retrieved August 10, 2010 (www.nytimes.com).

Schwartz, Seth, Byron Zamboanga, Russell Ravert, Su Yeong Kim, Robert Weisskirch, Michelle Williams, Melina Bersamin, and Gordon Finley. 2009. "Perceived Parental Relationships and Health-Risk Behaviors in College-Attending Emerging Adults." *Journal of Marriage and Family* 71(3):727–740.

Schwarz, Hunter. 2014. "For the First Time, There Are More Single American Adults than Married Ones, and Here's Where they Live." Retrieved January 26, 2016 (https://www .washingtonpost.com/blogs/govbeat /wp/2014/09/15/for -the-first-time-there-are -more-single -american-adults-than-married -ones-and -heres-where-they-live/).

Schweiger, Wendi K., and Marion O'Brien. 2005. "Special Needs Adoption: An Ecological Systems Approach." *Family Relations* 54(4):512–522.

Schweingruber, David, Sine Ahahita, and Nancy Berns. 2004. "Popping the Question When the Answer is Known: The Engagement Proposal as Performance." *Sociological Focus* 37(2):143–161.

Schweizer, Valerie. 2019. *FP-19-17 The Retreat from Remarriage, 1950–2017*. National Center for Marriage and Family Research, Bowling Green State University. Retrieved February 21, 2020 (https://scholarworks.bgsu.edu/cgi /viewcontent.cgi?article=1199&context=ncfmr _family_profiles).

Schweizer, Valerie J. 2019. *Marriage to Divorce Ratio in the U.S.: Geographic Variation, 2018*. Family Profiles, FP-19-24. National Center for Family & Marriage Research. Retrieved February 5, 2020 (https://www.bgsu.edu/ncfmr/resources/data /family-profiles/schweizer-marriage-divorce-ratio -geo-var-2018-fp-19-24.html).

Scommegna, Paola. 2013. "Aging U.S. Baby Boomers Face More Disability." *Population Reference Bureau*, March. Retrieved May 7, 2013 (http://www.prb .org).

Scott, Katreena, and Murray Straus. 2007. "Denial, Minimization, Partner Blaming, and Intimate Aggression in Dating Partners." *Journal of Interpersonal Violence* 22(7):851–871.

Scommegna, Paola, Mark Mather, and Lillian Kilduff. 2020. "Eight Demographic Trends Transforming America's Older Population." Population Reference Bureau. Retrieved February 13, 2020 (https://www.prb.org/eight -demographic-trends-transforming-americas -older-population/).

Scott, Christina L., Siri Wilder, and Justine Bennett. 2019. "Going it Alone: A Multigenerational Investigation of Women ᾽s Perceptions of Single Mothers by Choice Versus Circumstance. " In *Childbearing and the Changing Nature of Parenthood: The Contexts, Actors, and Experiences of Having Children*, edited by Rosalina Pisco Costa and Sampson Lee Blair. Bingley, UK: Emerald Publishing Limited.

Scott, Paula. 2018. "How America Lives: Creative Housing Options for Boomers, Veterans, Millennials, and More." Retrieved November 30, 2019 (https://parade.com/690685 /paulaspencer/how-america-lives-creative -housing-options-for-boomers-veterans -millennials-and-more/).

Seabrook, Rita C., L. Monique Ward, Lilia M. Cortina, Soraya Giaccardi, and Julia R. Lippman. 2017. "Girl Power or Powerless Girl? Television, Sexual Scripts, and Sexual Agency in Sexually Active Young Women." *Psychology of Women Quarterly* 41: 240–253.

Seager, Joni. 2009. "Murders of Women by Intimate Partners." *Environment and Planning* 41(10):2287.

Seal, David Wyatt, and Anke A. Ehrhardt. 2003. "Masculinity and Urban Men: Perceived Scripts for Courtship, Romantic, and Sexual Interactions with Women." *Culture, Health, and Sexuality* 5:295–319.

Seal, David Wyatt, Lucia F. O'Sullivan, and Anke A. Ehrhardt. 2007. "Miscommunications and Misinterpretations: Men's Scripts about Sexual Communication and Unwanted Sex in Interactions with Women." pp. 141–161 in *The Sexual Self: The Construction of Sexual Scripts*, edited by Michael Kimmel. Nashville: Vanderbilt University Press.

Searcey, Dionne. 2014. "For Women in Midlife, Career Gains Slip Away." *The New York Times*, June 23. Retrieved December 26, 2014 (http:// www.nytimes.com).

Seaton, Eleanor K., and Ronald D. Taylor. 2003. "Exploring Familial Processes in Urban, Low-income African American Families." *Journal of Family Issues* 24(5):627–644.

Seccombe, Karen. 2007. *Families in Poverty*. New York: Pearson Education.

Sechrist, Jori, J. Jill Suitor, Abigail R. Howard, and Karl Pillemer. 2014. "Perceptions of Equity, Balance of Support Exchange, and Mother-Adult Child Relations." *Journal of Marriage and Family* 76(2):285–299.

Sedlak, Andrea, Jane Mettenburg, Monica Basena, Ian Petta, Karla McPherson, Angela Greene, and Spencer Li. 2010. *Fourth National Incidence Study of child Abuse and Neglect (NIS-4):Report to Congress*. Washington, DC: U.S. Department of Health and Human Services and Administration for Children and Families.

Segal-Engelchin, Dorit, Pauline I. Erera, and Julie Cwikel. 2012. "Having It All? Unmarried Women Choosing Hetero-Gay Families." *Journal of Women and Social Work* 27(4):391–405.

Seidel, Amber J., Karl Majeske, and Mary Marshall. 2019. "Factors Associated with Support by Middle-Aged Children to Their Parents." *Family Relations*. November 28. Retrieved February 13, 2020 (https://onlinelibrary.wiley. com/action/showCitFormats?doi=10.1111%2F fare.12413).

Seideman, Ruth Young, Roma Williams, Paulette Burns, Sharon Jacobson, Francene Weatherby, and Martha Primeaux. 1994. "Cultural Sensitivity in Assessing Urban Native American Parenting." *Public Health Nursing* 11(2):98–103.

Seidman, Steven. 2003. *The Social Construction of Sexuality*. New York: W.W. Norton.

Self, Sharmistha, and Richard Grabowski. 2009. "Modernization, Inter-Caste Marriage, and Dowry." *Journal of Asian Economics* 20(1):69–77.

Seligson, Hannah. 2019. "The New 30-Something." Retrieved November 30, 2019 (https://www.nytimes.com/2019/03/02/style /financial-independence-30s.html).

Seltzer, Judith A. 2000. "Families Formed Outside of Marriage." *Journal of Marriage and Family* 62(4):1247–1268.

———. 2011. "Intergenerational Relationships: New Questions Call for New Data." *Family Focus* FF50:F2–F5.

Sellers, Anna. 2015. "Reddit CEO Ellen Pao Bans Salary Negotiations." PBS News Hour, April 7. Retrieved April 20, 2015 (http://www.pbs.org).

Seltzer, Judith A., Charles Q. Lau, and Suzanne M. Bianchi. 2012. "Doubling Up When Times Are Tough: A Study of Obligations to Share A Home in Response to Economic Hardship." *Social Science Research* 41:1307–1319.

Seltzer, Marsha, and Tamar Heller, eds. 1997. "Family Caregiving for Persons with Disabilities." *Family Relations* Special Issue 46(4).

Seminara, David. 2008. *Hello, I Love You, Won't You Tell Me Your Name: Inside the Green Card Marriage Phenomenon*. Washington, DC: Center for Immigration Studies.

Semega, Jessica, Melissa Kollar, John Creamer, and Abinash Mohanty. 2019. *Income and Poverty in the United States: 2018*. United States Census Bureau.

Senguttuvan, Umadevi, Shawn D. Whiteman, and Alexander C. Jensen. 2014. "Family Relationships and Adolescents' Health Attitudes and Weight: The Understudied Role of Sibling Relationships." *Family Relations* 63(3):384–396.

Senior, Jennifer. 2014. *All Joy and No Fun: The Paradox of Modern Parenthood*. New York: HarperCollins.

Senyurekli, Aysem R., and Daniel F. Detzner. 2008. "Intergenerational Relationships in a Transnational Context: The Case of Turkish Families." *Family Relations* 57(4):457–467.

Settersten, Richard, and Barbara Ray. 2010. *Not Quite Adults: Why 20-Somethings Are Choosing a Slower Path to Adulthood, and Why It's Good for*

Everyone. New York: Random House, Bantam Books.

"7 Helpful Programs for Children of Incarcerated Parents." 2016. Retrieved February 4, 2020 (https://web .connectnetwork.com/programs-for -children-of-incarcerated-parents).

Severson, Kim. 2011. "Moving to the City, but Clinging to Native Ways." *The New York Times,* April 7. Retrieved May 2, 2013 (http://www. nytimes.com).

Sexual Identity and Gender Identity Glossary. 2015. Retrieved December 3, 2015 (http:// feminism.eserver.org/sexual -gender-identity .txt).

Shafer, Emily Fitzgibbons. 2011. "Wives' Relative Wages, Husbands' Paid Work Hours, and Wives' Labor-Force Exit." *Journal of Marriage and Family* 73(1):250–263.

Shafer, Kevin, Todd M. Jensen, and Jeffry H. Larson. 2014. "An Actor-Partner Model of Relationship Effort and Marital Quality." *Family Relations* 63:654–666.

Shaffer, Anne, and Byron Egeland. 2011. "Intergenerational Transmission of Familial Boundary Dissolution: Observations and Psychosocial Outcomes in Adolescence." *Family Relations* 60:290–302.

Shanahan, Lilly, Susan M. McHale, Ann C. Crouter, and D. Wayne Osgood. 2008. "Linkages between Parents' Differential Treatment, Youth Depressive Symptoms, and Sibling Relationships." *Journal of Marriage and Family* 70(May):480–494.

Shandra, Carrie, and Anna Penner. 2017. "Benefactors and Beneficiaries? Disability and Care to Others." *Journal of Marriage and Family* 79 (4): 1160–1183.

Shanker, Deena, and Polly Mosendz. 2019. "Border Detainees Are Fed 'Appalling' Menu of Slimy Sandwiches and Unhealthy Ramen." *Bloomberg News.* June 28. Retrieved October 5, 2019 (www.bloomberg.com).

Shannon, Joyce B. 2009. *Learning Disabilities Sourcebook,* 3rd ed. Detroit: Omnigraphics.

Shapiro, Adam, and R. Corey Remle. 2010. "Generational Jeopardy? Parents' Marital Transitions and the Provision of Financial Transfers to Adult Children." *Journal of Gerontology: Social Sciences* 66R(1):99–108.

Shapiro, Danielle N., and Abigail J. Stewart. 2011. "Parenting Stress, Perceived Child Regard, and Depressive Symptoms among Stepmothers and Biological Mothers." *Family Relations* 60:533–544.

Shapiro, Joseph P. 2001. "Growing Old in a Good Home." *U.S. News & World Report,* May 21, pp. 57–61.

———. 2006a. "Caregiver Role Brings Purpose—and Risk—to Kids." National Public Radio, March 23. Retrieved March 23, 2006 (www.npr.org).

———. 2006b. "Family Ties Source of Strength for Elderly Caregivers." National Public Radio, March 23. Retrieved March 23, 2006 (www.npr. org).

———. 2009. "Donor-Conceived Kids Connect with Half Siblings." National Public Radio,

March 2. Retrieved February 3, 2009 (www.npr .org).

Sharkey, Colleen. 2019. "Fewer Unintended Pregnancies Contribute to All-Time Low U.S. Fertility Rate, New Research Says. " Retrieved December 14, 2019 (https://medicalxpress.com /news/2019-02-unintended-pregnancies -contribute-all-time-fertility.html).

Sharp, Elizabeth, Anisa Zvonkovic, Aine M. Humble, and M. Elise Radina. 2014. "Cultivating the Family Studies Terrain: A Synthesis of Qualitative Conceptual Articles." *Journal of Family Theory and Review* 6(2):139–168.

Sharp, Elizabeth A. and Shannon E. Weaver. 2015. "Feeling Like a Feminist Fraud: Theorizing Feminist Accountability in Feminist Family Studies Research Within a Neoliberal, Postfeminist Context." *Journal of Family Theory & Review* 7(3):299–320.

Shaughnessy, Krystelle, E. Sandra Beyers, and Lindsay Walsh. 2011. "Online Sexual Activity Experiences of Heterosexual Students: Gender Similarities and Differences." *Archives of Sexual Behavior* 40:419–427.

Shaw, Elyse, Ariane Hegewisch, and Cynthia Hess. 2018. "Sexual Harassment and Assault at Work: Understanding the Costs." *Institute for Women's Policy Research Publication, IWPR B 376.*

Shear, Michael D. 2014. "Obama Starts Initiative for Young Black Men, Noting His Own Experience." *The New York Times,* February 27. Retrieved November 3, 2015 (http://www. nytimes.com).

Sheff, Elisabeth. 2014. *The Polyamorists Next Door: Inside Multiple-Partner Relationships and Families.* New York: Rowman & Littlefield.

Shellenbarger, Sue. 2009. "Extreme Child-Care Maneuvers." *Wall Street Journal,* May 20, D1.

Sheridan, Mary, Marne Sherman, Tamarha Pierce, and Bruce Compas. 2010. "Social Support, Social Constraint, and Affect in Spouses of Women with Breast Cancer: The Role of Cognitive Processing." *Journal of Social and Personal Relationships* 27(1):5–22.

Sherif-Trask, Bahira. 2003. "Marriage from a Cross-Cultural Perspective." *National Council on Family Relations Report* 48(3):F13–F14. Minneapolis: National Council on Family Relations.

Sherkat, Darren E. 2017. "'That they be keepers of the home': The effect of conservative religion on early and late transitions into housewifery." In *Maintaining Our Differences,* pp. 29–47. Routledge.

Shen, Yun, Deepthi S. Varma, Yi Zheng, Jenny Boc, and Hui Hu. 2019. "Age at Menarche and Depression: Results from the NHANES 2005– 2016." *PeerJ* 7: e7150.

Sheridan, Susan, and Lorey Wheeler. 2017. "Building Strong Family-School Partnerships: Transitioning from Basic Findings to Possible Practices." *Family Relations* 66 (4): 670–683.

Sherman, Carey Wexler. 2006. "Remarriage and Stepfamily in Later Life." *Family Focus on . . . Families and the Future* FF32:F8–F9. Minneapolis: National Council on Family Relations.

Sherman, Carey Wexler, and Pauline Boss. 2007. "Spousal Dementia Caregiving in the Context of Late-Life Remarriage." *Dementia* 6(2):245–270.

Sherman, Carey Wexler, Noah J. Webster, and Toni C. Antonucci. 2013. "Dementia Caregiving in the Context of Late-Life Remarriage: Support Networks, Relationship Quality, and Well-Being." *Journal of Marriage and Family* 75(5):1149–1163.

Sherman, Michelle, Kyle Hawkey, and Lynne Borden. 2015. "The Experience of Reintegration for Military Families." *Family Focus* Fall:F21–F22.

Sherman, Michelle, Kyle Hawkey, and Lynne Borden. 2015. "The Experience of Reintegration for Military Families." *Family Focus* (FF67, Fall):F22. Minneapolis: National Council on Family Relations.

Shifren, Kim, ed. 2009. *How Caregiving Affects Development: Psychological Implications for Child, Adolescent, and Adult Caregivers.* Washington, DC: American Psychological Association.

Shinn, Lauren Keel, and Marion O'Brien. 2008. "Parent-Child Conversational Styles in Middle Childhood: Gender and Social Class Differences." *Sex Roles* 59:61–67.

"Shocking Student Debt Statistics." 2013. Retrieved September 28, 2015 (http://www .faastweb.com).

Shorter, Edward. 1975. *The Making of the Modern Family.* New York: Basic Books.

Shortt, Joann, Deborah Capaldi, Kim Hyoun, David Kerr, Lee Owen, and Alan Feingold. 2013. "Stability of Intimate Partner Violence by Men across 12 Years in Young Adulthood: Effects of Relationship Transitions." *Prevention Science* 13(4):360–369.

Shortt, Joann Wu, Stacey S. Tiberio, Deborah Capaldi, and Sabina Low. 2019. "Child Exposure to Intimate Partner Violence and Parent Aggression in Two Generations." National Criminal Justice Reference Service. Retrieved February 13, 2020 (www.ncjrs.gov).

Showden, Carisa R. 2009. "What's Political about the New Feminisms?" *Frontiers: A Journal of Women Studies* 30(2):166–198.

Shpancer, Noam. 2014. "Looking for a Soul Mate?' You Can Do Better." *Psychology Today,* September 1. Retrieved November 4, 2014 (https://www.psychologytoday.com /blog/insight-therapy/201409 /looking-soul-mate-you-can-do-better.

Shramm, David G., and Francesca Adler-Baeder. 2012. "Marital Quality for Men and Women in Stepfamilies: Examining the Role of Economic Pressure, Common Stressors, and Stepfamily- Specific Stressors." *Journal of Family Issues* 33(10):1373–1397.

Shreffler, Karina, Arthur L. Greil, and Julia McQuillan. 2011. "Pregnancy Loss and Distress among U.S. Women." *Family Relations* 60(3):342–355.

Shreffler, Karina, Julia McQuillan, Arthur L. Greil, and Katherine M. Johnson. 2011. "On the Fence: New Insights on Pregnancy Ambivalence." *Family Focus* FF49: F7–F9.

Shreffler, Karina M. 2017. "Contextual Understanding of Lower Fertility Among U.S. Women in Professional Occupations." *Journal of Family Issues* 38:204–224.

Shriner, Michael. 2009. "Marital Quality in Remarriage A Review of Methods and Results." *Journal of Divorce and Remarriage* 50:81–99.

Shulman, Julie L., Gabrielle Gotta, and Robert-Jay Green. 2012. "Will Marriage Matter? Effects of Marriage Anticipated by Same-Sex Couples." *Journal of Family Issues* 33(2):158–181.

Siddiqui, Faiz. 2018. "Americans' Commutes Keep Getting Longer." *The Washington Post.* September 17. Retrieved June 3, 2019 (www.washingtonpost.com).

Sidelinger, Robert J., and Melanie Booth-Butterfield. 2007. "Mate Value Discrepancy as Predictor of Forgiveness and Jealousy in Romantic Relationships." *Communication Quarterly* 55(2):207–223.

SIECUS. 2019. "Trump attempts to shift TPPP to Promote Abstinence-Only Ideology." Retrieved October 21, 2019 (https://siecus.org/wp-content/uploads/2019/02/TPPP-Timeline-Doc-Feb-2019.pdf).

Siegel, Mark D. 2004. "To the Editor: Making Decisions about How to Die." *The New York Times,* October 3.

Siegel, Michele, Judith Brisman, and Margot Weinshel. 2009. *Surviving An Eating Disorder: Strategies for Families and Friends,* 3rd ed. New York: Collins Living.

Signorella, Margaret L., and Irene Hanson Frieze. 2008. "Interrelations of Gender Schemas in Children and Adolescents: Attitudes, Preferences, and Self-Perceptions." *Social Behavior and Personality* 36(7):941–954.

"Signs and Symptoms of Stress in Children." nd. Our Family Wizard. Retrieved March 1, 2020 www.ourfamilywizard.com/blog/signs-and-symptoms-stress-children).

Silva, Tony J. 2018. "'Helpin'a Buddy Out': Perceptions of Identity and Behaviour Among Rural Straight Men that Have sex with Each Other." *Sexualities* 21:68–89.

Silva, Tony. 2017. "Bud-Sex: Constructing Normative Masculinity Among Rural Straight Men that Have Sex with Men." *Gender & Society* 31:51–73.

Silverstein, Merril, and Vern L. Bengston. 2001. "Intergenerational Solidarity and the Structure of Adult Child-Parent Relationships in American Families." pp. 53–61 in *Families in Later Life: Connections and Transitions,* edited by Alexis J. Walker, Margaret Manoogian-O'Dell, Lori A. McGraw, and Diana L. G. White. Thousand Oaks, CA: Pine Forge Press.

Simmel, Cassandra, Richard P. Barth, and Devon Brooks. 2007. "Adopted Foster Youths' Psychosocial Functioning: A Longitudinal Perspective." *Child and Family Social Work* 12(4):336–348.

Simmons-Duffin, Selena. 2019. "When Hospitals Sue for Unpaid Bills, it Can Be 'Ruinous' for Patients." National Public Radio. June 25. Retrieved June 26, 2019 (www.npr.org).

Simon, W., Gagnon, J. H. (1986). "Sexual scripts: Permanence and Change." *Archives of Sexual Behavior* 15:97–120.

Simon, Rita J. 1990. "Transracial Adoptions Can Bring Joy: Letters to the Editor." *Wall Street Journal,* October 17.

Simon, Rita J., and Howard Altstein. 2002. *Adoption, Race, and Identity: From Infancy to Young Adulthood.* New Brunswick, NJ: Transaction Books.

Simon, Robin E., and Anne E. Barrett. 2010. "Nonmarital Romantic Relationships and Mental Health in Early Adulthood: Does the Association Differ for Women and Men?" *Journal of Health and Social Behavior* 51(2):168–182.

Simons, Leslie Gordon, and R and D. Conger. 2009. "Linking Mother-Father Differences in Parenting to a Typology of Family Parenting Styles and Adolescent Outcomes." *Journal of Family Issues* 28(2):212–241.

Simons, Leslie Gordon, Ronald L. Simons, Gene Brody, and Carolyn Cutrona. 2006. "Parenting Practices and Child Adjustment in Different Types of Households: A Study of African American Families." *Journal of Family Issues* 27(6):803–825.

Simons, Leslie Gordon, K.A.S. Wickrama, T.K. Lee, Melissa Landers-Potts, Carolyn Cutrona, and Rand Conger. 2016. "Testing Family Stress and Family Investment Explanations for Conduct Problems among African American Adolescents." *Journal of Marriage and Family* 78(2):498–515.

Simpson, Emily K., and Christine A. Helfrich. 2005. "Lesbian Survivors of Intimate Partner Violence: Provider Perspectives on Barriers to Accessing Services." *Journal of Gay and Lesbian Social Services* 18(2):39–59.

Simpson, Ian. 2015. "Judge Upholds Arizona 'Show Your Papers' Immigration Law." Reuters, September 5. Retrieved October 3, 2015 (http://www.reuters .com).

Simpson, Victoria J., Gayle Brewer, and Colin A. Hendrie. 2014. "Evidence to Suggest that Women's Sexual Behavior is Influenced by Hip Width Rather than Waist-to-Hip Ratio." *Archives of Sexual Behavior* 43:1367–1371.

Sinclair, Stacey L., and Gerald Monk. 2004. "Couples—Moving Beyond the Blame Game: Toward a Discursive Approach to Negotiating Conflict within Couple Relationships." *Journal of Marital and Family Therapy* 30(3):335–349.

Singh, Maanvi. 2015. "Breaking Up Is Hard to Do, but Science Can Help." Retrieved January 17, 2015 (http://www.npr.org /sections/health-shots/2015/01/13/376804930/breaking-up-is-hard-to-do -but-science-can-help).

Singleton, Sean. 2017. "Blending Families of Different Races Brings Challenges, Blessings." Retrieved March 1, 2020 (https://citydadsgroup.com/blog/blending-families-mixed-race/).

Sirjamaki, John. 1948. "Cultural Configurations in the American Family." *American Journal of Sociology* 53(6):464–470.

Skarupski, Kimberly, Judy McCann, Julia Bienias, and Denis Evans. 2009. "Race Differences in Emotional Adaptation of Family Caregivers." *Aging and Mental Health* 13(5):715–724.

Skogrand, Linda, Abril Barrios-Bell, and Brian Higginbotham. 2009. "Stepfamily Education for Latino Families: Implications for Practice." *Journal of Couple and Relationship Therapy* 8(2):113–128.

Sky, Natasha. 2009. "Leaving Behind Traditional Views of Belonging." pp. 73–76 in *Social Issues First Hand: Blended Families,* edited by Stefan Kiesbye. New York: Greenhaven Press/Cengage Learning.

Slater, Dan. 2013. *Love in the Time of Algorithms.* New York: Penguin Books.

Slater, Lauren. 2006. "Love: The Chemical Reaction." *National Geographic,* February, pp. 34–49.

Slaughter, Anne-Marie. 2012. "Why Women Still Can't Have It All." *The Atlantic,* July. Retrieved August 29, 2012 (http://www.theatlantic.com).

Sloan, Kathleen. 2017. "Surrogacy Reaches the Supreme Court. " Retrieved December 25, 2019 (https://www.thepublicdiscourse.com/2017/09/20130/).

Smalley, Gary. 2000. *Secrets to Lasting Love: Uncovering the Keys to Life-long Intimacy.* New York: Simon and Schuster.

Smallwood, Beverly. 2013. "Love Myths: Debunking Five Common Misconceptions on Love and Romance." Retrieved May 26, 2013, from http://www.sideroad.com/Relationships/love-myths.html.

Smart Marriages. 2020. "Smart Marriages: The Coalition for Marriage, Family, and Couples Education." Retrieved January 15, 2019 (https://www.smartmarriages.com/index.html).

Smiler, Andrew P. 2008. "'I Wanted to Get to Know Her Better': Adolescent Boys' Dating Motives, Masculinity Ideology, and Sexual Behavior." *Journal of Adolescence* 31(1):17–32.

Smith, Brendan. 2011. "Are Internet Affairs Different?" American Psychological Association. Retrieved March 26, 2013 (http://www.apa.org/monitor/2011/03 /internet.aspx).

Smith, Christian. 2003. "Religious Participation and Network Closure among American Adolescents." *Journal for the Scientific Study of Religion* 42(2):259–267.

Smith, Donna. 1990. *Stepmothering.* New York: St. Martin's.

Smith, Dorothy. 1987. *The Everyday World As Problematic: A Feminist Sociology.* Northeastern University Press.

Smith, Emily Esfahani. 2013. "There's No Such Thing as Everlasting Love." *The Atlantic,* January 24. Retrieved April 24, 2015 (http://www .theatlantic.com/sexes /archive/2013/01 /theres-no-such-thing -as-everlasting-love-according-to-science /267199/).

Smith, Gregory C., Kelly E. Cichy, and Julian Montoro-Rodriguez. 2015. "Impact of Coping Resources on the Well-Being of Custodial Grandmothers and Grandchildren." *Family Relations* 64(3):378–392.

Smith, Jeremy Adam. 2009. *The Daddy Shift: How Stay-at-Home Dads, Breadwinning Moms, and Shared Parenting Are Transforming the American Family.* Boston: Beacon Press.

Smith, Kevin M., Patti A. Freeman, and Ramon B. Zabriskie. 2009. "An Examination of Family Communication within the Core and Balance model of Family Leisure Functioning." *Family Relations* 58(February):79–90.

Smith, Kristin. 2008. "Working Hard for the Money Trends in Women's Employment: 1970 to 2007." *A Carsey Institute Report on Rural America.* Durham: University of New Hampshire.

Smith, Melinda, and Jeanne Segal. 2012a. "Domestic Violence and Abuse." Retrieved March 14, 2013 (http://www.helpguide .org).

———. 2012b. "Help for Abused and Battered Women." Retrieved March 14, 2013 (http://www.helpguide.org).

Smith, Sandra M., Yvonne Amanor-Boadu, Marjorie S. Miller, Erin Menhusen, and April Few-Demo. 2011."Vulnerabilities, Stressors, and Adaptations in Situationally Violent Relationships." *Family Relations* 60(1):73–89.

Smith, Suzanna D. 2006. "Global Families." pp. 3–24 in *Families in Global and Multicultural Perspective*, 2nd ed., edited by Bron B. Ingoldsby and Suzanna D. Smith. Thousand Oaks, CA: Sage.

Smith, Tim. 2006. *The Danger of Raising Nice Kids: Preparing Our Children to Change Their World.* Downers Grove, IL: IVP Books.

Smith, Tom W. 1999. *The Emerging 21st Century American Family.* GSS Social Change Report No. 42. Chicago: University of Chicago,National Opinion Research Center.

Smith, Chelsea, Robert Crosnoe, and Shannon Cavanagh. 2017. "Family Instability and Children's Health." *Family Relations* 66 (4): 601–613.

Smith, Samuel. 2019. "Strong majority of Americans Say Premarital Sex, Divorce Are 'Morally Acceptable': Gallup." Retrieved October 28, 2019 (https://www.christianpost.com/news/strong-majority-of-americans-say -premarital-sex-divorce-are-morally-acceptable -gallup.html).

Smith, Sharon, Xinjian Zhang, Kathleen Basille, Melissa Merrick, Jing Wang, M. Kresnow, and Jieru Chen. 2018. *National Intimate Partner and Sexual Violence Survey: 2015 Data Brief— Updated Release.* Centers for Disease Control and Prevention; National Center for Injury Prevention and Control. November. Retrieved February 14, 2020 (https://www.cdc.gov /violenceprevention/datasources/nisvs /index.html).

Smock, Pamela J. 2000. "Cohabitation in the United States: An Appraisal of Research Themes, Findings, and Implications." *Annual Review of Sociology,* 26:1–20.

Smock, Pamela J., and Fiona Rose Greenland. 2010. "Diversity in Pathways to Parenthood: Patterns, Implications, and Emerging Research Directions." *Journal of Marriage and Family* 72(3) (June):576–593.

Smock, Pamela J. and Sanjiv Gupta. 2002. "Cohabitation in Contemporary North America." pp. 53–84 in *Just Living Together: Implications of Cohabitation on Families,* edited by Alan Booth and Ann C. Crouter. Mahwah, NJ: Erlbaum.

Smock, Pamela J., Penelope Huang, Wendy D. Manning, and Cara A. Bergstrom. 2016. "Heterosexual Cohabitation in the United States: Motives for Living Together among Young Men and Women." Center for Family and Demographic Research, Bowling Green State University. Retrieved February 1, 2016 (http://www.bgsu.edu/content/dam /BGSU/college-of-arts-and-sciences/center -for-family-and-demographic-research /documents/working-papers/2006/CFDR -Working-Paper-2006-10-Heterosexual -Cohabitation-in-the-United-States-Motives -for-Living-Together-Among-Young-Men -and-Women.pdf).

Smock, Pamela J., Wendy D. Manning, and Meredith Porter. 2005. "'Everything's There Except Money': How Money Shapes Decisions to Marry among Cohabitors." *Journal of Marriage and Family* 67:680–696.

Snapp, Shannon, Rene Lento, Ehri Ryu, and Karen S. Rosen. 2014. "Why Do They Hook Up? Attachment Style and Motives of College Students." *Personal Relationships* 21:468–481.

Snider, Carolyn, Daniel Webster, Chris S. O'Sullivan, and Jacquelyn Campbell. 2009. "Intimate Partner Violence: Development of a Brief Risk Assessment for the Emergency Department." *Academic Emergency Medicine* 16(11):1208–1216.

Snipp, C. Matthew, and Sin Yi Cheung. 2016. "Changes in Racial and Gender Inequality Since 1970." *The ANNALS of the American Academy of Political and Social Science* 663:80–98.

Snoeckx, Laurent, Britt Dehertogh, and Dimitri Mortelmans. 2008. "The Distribution of Household Tasks in First-Marriage Families and Stepfamilies in Europe." pp. 277–298 in *The International Handbook of Stepfamilies,* edited by J. Pryor. Hoboken, NJ: John Wiley & Sons.

Somer, Sarah J. Holdt, Rachel G. Sinkey, and Allison S. Bryant. 2017. "Epidemiology of Racial /Ethnic Disparities in Severe Maternal Morbidity and Mortality. " *Seminars in Perinatology* 41:258–265.

Sopelsa, Brooke, and Julie Moreau. 20019. "Trans Workers Not Protected by Civil Rights Law, Trump Administration Tells Supreme Court." August 16. Retrieved August 18, 2019 (https://www.nbcnews.com).

Sorokowski, Piotr, Agnieszka Sorokowska, Marina Butovskaya, Maciej Karwowski, Agata Groyecka, Bogdan Wojciszke, and Bogusław Pawłowski. 2017. "Love Influences Reproductive Success in Humans. " *Frontiers in Psychology* 8:1922.

Snyder, Douglas K., Donald H. Baucom, and Kristina C. Gordon. 2008. "An Integrative Approach to Treating Infidelity." *The Family Journal: Counseling and Therapy for Couples and Families* 16(4):300–304.

Snyder, Karrie Ann. 2007. "A Vocabulary of Motives: Understanding How Parents Define Quality Time." *Journal of Marriage and Family* 69(2):320–340.

Snyder, Karrie Ann, and Adam Isaiah Green. 2008. "Revisiting the Glass Escalator: The Case of Gender Segregation in a Female Dominated Occupation." *Social Problems* 55(2):271–299.

Sobolewski, Juliana, and Paul R. Amato. 2005. "Economic Hardship in the Family of Origin and Children's Psychological Well-Being in Adulthood." *Journal of Marriage and Family* 67(1):141–156.

Socha, Thomas J., and Glen H. Stamp. 2009. *Parents and Children Communicating with Society: Managing Relationships Outside of Home.* New York: Routledge.

Society for Human Sexuality. 2015. "The Swing Community: A Profile—and Comparisons with Other Sex-Positive Communities." Retrieved October 24, 2015 (http://sexuality.org).

Sohn, Heeju. 2015. "Health Insurance and Risk of Divorce: Does Having Your Own Insurance Matter?" *Journal of Marriage and Family* 77:982–995.

Soliz, Jordan, Allison R. Thorson, and Christine E. Rittenour. 2009. "Communicative Correlates of Satisfaction, Family Identity, and Group Salience in Multiracial/Ethnic Families." *Journal of Marriage and Family* 71(4):819–832.

Soli, Anna R., Susan M. McHale, and Mark E. Feinberg. 2009. "Risk and Protective Effects of Sibling Relationships among African American Adolescents." *Family Relations* 58(December):578–592.

Solomon, Andrew. 2011. "Meet My Real Modern Family." *Newsweek, February* 7:32–37.

Sommers, Christina Hoff. 2013. "Boys at the Back." *The New York Times,* February 2. Retrieved February 4, 2013 (http://opinionator.blogs. nytimes.com).

Song, Zhimin, Katharine E. McCann, John K. McNeill, Tony E. Larkin, Kim L. Huhman, and H. Elliott Albers. 2014. "Oxytocin Induces Social Communication by Activating Arginine-Vasopressin V1a Receptors and not Oxytocin Receptors." *Psychoneuroendocrinology* 50:14–19.

Sontag, Susan. 1976. "The Double Standard of Aging." pp. 350–366 in *Sexuality Today and Tomorrow,* edited by Sol Gordon and Roger W. Libby. North Scituate, MA: Duxbury.

Soons, Judith, and Aart Liefbroer. 2009. "The Long-Term Consequences of Relationship Formation for subjective Well-Being." *Journal of Marriage and Family* 71(December):1254–1270.

Sorbring, Emma. 2014. "Parents' Concerns About Their Teenage Children's Internet Use." *Journal of Family Issues* 35:75–96.

Sorensen, Elaine. 2010. "Rethinking Public Policy Toward Low-Income Fathers in the Child Support Program." *Journal of Policy Analysis and Management* 29(3):604–610.

Soukhanov, Anne H. 1996. "Watch." *The Atlantic,* August: 96ff.

South, Scott J., and Kyle D. Crowder. 2010. "Neighborhood Poverty and Nonmarital Fertility: Spatial and Temporal Dimensions." *Journal of Marriage and Family* 72(1):89–104.

South-Derose, J. Lynn. n.d. "How to Handle Unresolved Conflict in Your Family." Retrieved January 20, 2013 (http://www.ehow.com).

Spack, Norman P., Laura Edwards-Leeper, Henry A. Feldman, Scott Liebowitz, Francie Mandel, David Diamond, and Stanley R. Vance. 2012. "Children and Adolescents with Gender Identity Disorder Referred to a Pediatric Medical Center." *Pediatrics* 129(3):418–425.

Spade, Joan Z., and Catherine G. Valentine. 2010. *The Kaleidoscope of Gender: Prisms, Patterns, and Possibilities,* 3rd ed. Newbury Park, CA: Pine Forge Press.

Spagnola, Mary, and Barbara Fiese. 2010. "Preschoolers with Asthma: Narratives of Family Functioning Predict Behavior Problems." *Family Process* 49 (1):74–91.

Span, Paula. 2009a. *When the Time Comes: Families with Aging Parents Share Their Struggles and Solutions.* New York: Springboard Press.

———. 2009b. "Years Later, Divorce Complicates Caregiving." *The New York Times.* New Old Age Blog: Caring and Coping.

Retrieved May 5, 2010 (newoldage.blogs. nytimes.com).

———. 2013. "The Reluctant Caregiver." *The New York Times*, February 20. Retrieved May 13, 2013 (http://www.nytimes.com).

Spangler, A., Susan L. Brown, I-Fen Lin, Anna Hammersmith, and Matthew Wright. 2016. "Divorce Timing and Economic Well-Being" (FP-16-01). Bowling Green, OH: National Center for Family and Marriage Research.

Spar, Debora L. 2006. *The Baby Business: How Money, Science, and Politics Drive the Commerce of Conception*. Boston: Harvard Business School Press.

Speer, Rebecca B., and April R. Trees. 2007. "The Push and Pull of Stepfamily Life: the Contribution of Stepchildren's Autonomy and Connection-Seeking Behaviors to Role Development in Stepfamilies." *Communication Studies* 58(4):377–394.

Spencer, Liz, and R. E. Pahl. 2006. *Rethinking Friendship: Hidden Solidarities Today*. Princeton, NJ: Princeton University Press.

Spengler, Elliot S., Elliott N. DeVore, Paul M. Spengler, and Nicholas A. Lee. 2019. "What Does "Couple" Mean in Couple Therapy Outcome Research? A Systematic Review of the Implicit and Explicit, Inclusion and Exclusion of Gender and Sexual Minority Individuals and Identities." *Journal of Marital and Family Therapy*.

Spiegel, Alix. 2008a. "Parents Consider Treatment to Delay Son's Puberty." National Public Radio, May 8. Retrieved February 21, 2012 (http://www.npr.org).

———. 2008b. "Two Families Grapple with Sons' Gender Identity." National Public Radio, May 7. Retrieved February 21, 2012 (http://www.npr.org).

Spielmann, Stephanie S., Geoff MacDonald, and Jennifer L. Tackett. 2012. "Social Threat, Social Reward, and Regulation of Investment in Romantic Relationships." *Personal Relationships* 19(4):601–622.

Spiro, Melford. 1956. *Kibbutz: Venture in Utopia*. New York: Macmillan.

Spitze, Glenna, and Katherine Trent. 2006. "Gender Differences in Adult Sibling Relations in Two-Child Families." *Journal of Marriage and Family* 68(4):977–992.

Spratling, Cassandra. 2009. "A Grand Living Arrangement." *Detroit Free Press*, January 18. Retrieved February 3, 2009 (www.modbee.com).

Sprecher, Susan, Maria Schmeeckle, and Diane Felmlee. 2006. "The Principle of Least Interest: Inequality in Emotional Involvement in Romantic Relationships." *Journal of Family Issues* 27(9):1255–1280.

Sprecher, Susan, and Stanislav Treger. 2015. "The Benefits of Turn-Taking Reciprocal Self-Disclosure in Get-Acquainted Interactions." *Personal Relationships* 22:460–475.

Sprigg, Peter. 2015. "Protect Client and Therapist Freedom of Choice Regarding Sexual Orientation Change Efforts." Retrieved December 3, 2015 (http://www.frc.org /socetherapyban).

Srinivasan, Padma, and Gary R. Lee. 2004. "The Dowry System in Northern India: Women's

Attitudes and Social Change." *Journal of Marriage and Family* 66(5):1108–1117.

St. George, Donna. 2006. "Home but Still Haunted." *Washington Post*, August 20.

———. 2010. "More Wives Are the Higher-Income Spouse, Pew Report Says." *Washington Post*, January 9. Retrieved January 9, 2010 (www.washingtonpost.com).

Stacey, Judith. 1990. *Brave New Families: Stories of Domestic Upheaval in Late Twentieth Century America*. New York: Basic Books.

———. 1996. *In the Name of the Family: Rethinking Family Values in the Postmodern Age*. Boston: Beacon Press.

———. 2006. "Feminism and Sociology in 2005: What Are We Missing?" *Social Problems* 53(4):479–482.

———. 2011. *Unhitched: Love, Marriage, and Family Values from West Hollywood to Western China*. New York and London: New York University Press.

Stacey, Judith, and Tey Meadow. 2009. "New Slants on the Slippery Slope: The Politics of Polygamy and Gay Family Rights in South Africa and the United States." *Politics and Society* 37(2):167–202).

Stack, Carol B. 1974. *All Our Kin: Strategies for Survival*. New York: Harper & Row.

Stanczyk, Alexandria B., Julia R. Henly, and Susan J. Lamert. 2017. "Enough Time for Housework? Low-Wage Work and Desired Housework Time Adjustments." *Journal of Marriage and Family* 79 (1): 243–260.

Stanik, Christine E., and Chalandra M. Bryant. 2012. "Sexual Satisfaction, Perceived Availability of Alternative Partners, and Marital Quality in Newlywed African American Couples." *Journal of Sex Research* 49(4):400–407.

Stanik, Christine E., Susan M. McHale, and Ann C. Crouter. 2013. "Gender Dynamics Predict Changes in Marital Love among African American Couples." *Journal of Marriage and Family* 75(4):795–807.

Stanley, Angela. 2011. "Black, Female, and Single." Retrieved July 14, 2013 (http://www .nytimes.com/2011/12/11 /opinion /sunday/black-and-female-the -marriage -question.html?pagewanted =all&_r=0).

Stanley, Scott M. 2009. "'Sliding vs. Deciding': Understanding a Mystery." *Family Focus* (Summer):F1–F3. Minneapolis: National Council on Family Relations.

Stanley, Scott M., Galena K. Rhoades, P. R. Amato, H. J. Markman, and C. A. Johnson. 2010. "The Timing of Cohabitation and Engagement: Impact on First and Second Marriages." *Journal of Marriage and Family* 72(4):906–918.

Stanley, Scott M., Galena Kline Rhoades, and Howard J. Markman. 2006. "Sliding versus Deciding: Inertia and the Premarital Cohabitation Effect." *Family Relations* 55:499–509.

Stanley-Stevens, Leslie, and Karen C. Kaiser. 2011. "Decisions of First Time Expectant Mothers in Central Texas Compared to Women in Great Britain and Spain." *Journal of Comparative Family Studies* 46(1):113–130.

Stanton, Glenn T. 2004a. "Why Marriage Matters for Adults." June 21. *Focus on Social*

Issues: Marriage and Family. Focus on the Family. Retrieved September 30, 2006 (www.family.org /forum).

———. 2004b. "Why Marriage Matters for Children." June 21. *Focus on Social Issues: Marriage and Family*. Focus on the Family. Retrieved September 30, 2006 (www.family .org/forum).

Staples, Robert. 1994. *The Black Family: Essays and Studies*. 5th ed. Belmont, CA: Wadsworth.

———. 1999. *The Black Family: Essays and Studies*. 6th ed. Belmont, CA: Wadsworth.

Stark, Oded. 2009. "Do Religious Children Care More and Provide More Care for Older Parents? A Study of Filial Norms and Behaviors across Five Nations." *Journal of Comparative Family Studies* 40(2):629–631.

Stark, Patrick, Amber Noel, and Joel McFarland. 2015. *Trends in High School Dropout and Completion Rates in the United States: 1972–2012*. National Center for Education Statistics. Washington, DC: U.S. Department of Education.

Stearns, Peter N. 2003. *Anxious Parents: A History of Modern Childrearing in America*. New York: NYU Press.

Stebbins, Samuel. 2018. "Here are the Cities Where the Most People Live Alone. Retrieved January 24, 2020 (https:// www.usatoday.com/story/money /personalfinance/2018/11/02 /cities-where-the-most-people-live -alone/38255689/).

Stein, Abby. 2013. "Intimate Partner Violence: A Psychological Diagnosis of Social Pathology." *Journal of Psychohistory* 40(3):187–192.

Stein, Loren. 2012. "Sex and Seniors: The 70-Year Itch." *HealthDay*. Retrieved November 21, 2012 (http://consumer .healthday.com /encyclopedia/article .asp?AID=647575).

Steinbach, Anja. 2019. "Children's and Parents' Well-Being in Joint Physical Custody: A Literature Review." *Family Process* 58:353–369.

Steinberg, Julia Renee, and Nancy F. Russo. 2008. "Abortion and Anxiety: What's the Relationship?" *Social Science and Medicine* 67:238–252.

Steinberg, Robert J., and Michael L. Barnes (Eds.). 1988. *The Psychology of Love*. New Haven, CT: Yale University Press.

Steinmetz, Katy. 2014. "The Transgender Tipping Point: America's Next Civil Rights Frontier." *Time*, June 9, pp. 38–46.

Stempel, Jonathon. 2010. "Wal-Mart in $86 Million Settlement of Wage Lawsuit." *Reuters*, May 12. Chicago: Thompson Reuters.

Stephens, William N. 1963. *The Family in Cross-Cultural Perspective*. New York: Holt, Rinehart & Winston.

Stephens-Davidowitz, Seth. 2014. "Google, Tell me. Is My Son a Genius?" *The New York Times*, January 18. Retrieved August 31, 2014 (http:// www.nytimes.com).

Stepler, Renee. 2019. "Americans Say a Man Should be Able to Support a Family…" 2019. Pew Research Center. Retrieved February 28, 2020 (www.preresearch.org).

Stepler, Renee. 2017. "Number of U.S. Adults Cohabiting with a Partner Continues to

Rise, Especially Among those 50 and Older." Retrieved November 26, 2019 (https://www.pewresearch.org/fact-tank/2017/04/06/number-of-u-s-adults-cohabiting-with-a-partner-continues-to-rise-especially-among-those-50-and-older/).

Stepp, Laura Sessions. 2007. *Unhooked: How Young Women Pursue Sex, Delay Love and Lose At Both*. New York: Riverhead Books.

Sternberg, Robert J. 1988a. "Triangular Love." pp. 119–138 in *The Psychology of Love*, edited by Robert J. Sternberg and Michael L. Barnes. New Haven, CT: Yale University Press.

———. 1988b. *The Triangle of Love: Intimacy, Passion, Commitment*. New York: Basic Books.

Sternberg, Robert J., and Karin Sternberg. 2008. *The New Psychology of Love*. New Haven, CT: Yale University Press.

Stevenson, Betsey, and Justin Wolfers. 2004. *Bargaining in the Shadow of the Law: Divorce Laws and Family Distress*. National Bureau of Economic Research Working Paper 10175. Cambridge, MA: National Bureau of Economic Research, October 4.

———. 2007. *Marriage and Divorce: Changes and Their Driving Forces*. National Bureau of Economic Research Working Paper 12944. Cambridge, MA: National Bureau of Economic Research, March. Retrieved April 27, 2007 (www.nber.org).

Steverman, Ben. 2018. "Millennials Are Causing the U.S. Divorce Rate to Plummet." Retrieved February 5, 2020 (https://www.bloomberg.com/news/articles/2018-09-25/millennials-are-causing-the-u-s-divorce-rate-to-plummet).

Stewart, David W., and Prem N. Shamdasani. 2015. *Focus Groups: Theory and Practice*. Thousand Oaks, CA: Sage.

Stewart, Lisa M. 2013. "Family Care Responsibilities: Exploring the Impact of Type of Family Care on Work-Family and Family-Work Conflict." *Journal of Family Issues* 34(1):113–138.

Stewart, Mary White. 1984. "The Surprising Transformation of Incest: From Sin to Sickness." Presented at the annual meeting of the Midwest Sociological Society, April 18, Chicago, IL.

Stewart, Susan D. 1999a. "Disneyland Dads, Disneyland Moms?" *Journal of Family Issues* 20:539–556.

———. 1999b. "Nonresident Mothers' and Fathers' Social Contact with Children." *Journal of Marriage and the Family* 61:894–907.

———. 2001. Contemporary American Stepparenthood: Integrating Cohabiting and Nonresident Stepparents. *Population Research and Policy Review* 20:345–364.

———. 2002. "The Effect of Stepchildren on Childbearing Intentions and Births." *Demography* 39:181–197.

———. 2003. "Nonresident Parenting and Adolescent Adjustment." *Journal of Family Issues* 24(2):217–244.

———. 2005a. "Boundary Ambiguity in Stepfamilies." *Journal of Family Issues* 26(7):1002–1029.

———. 2005b. "How the Birth of a Child Affects Involvement with Stepchildren." *Journal of Marriage and Family* 67(2): 461–473.

———. 2007. *Brave New Stepfamilies: Diverse Paths toward Stepfamily Living*. Thousand Oaks, CA: Sage.

———. 2010. "Stepchildren Who Are Adopted by Their Stepparents: Prevalence, Characteristics, and Well-Being." *Family Relations* 59:558–571.

Stewart, Susan D., W. D. Manning, and P. J. Smock. 2003. "Union Formation among Men in the U.S.: Does Having Prior Children Matter?" *Journal of Marriage and the Family* 65:90–104.

Stewart, Susan D. 2010. *Family Dinners, Family Structure, and Adolescent Well-Being*. Population Association of America, Dallas, TX.

Stewart, Susan D. 2017. *Co-Sleeping: Parents, Children, and Musical Beds*. New York, NY: Roman and Littlefield.

Stewart, Susan D. 1998. "Economic and Personal Factors Affecting Women's Use of Nurse-Midwives in Michigan." *Family Planning Perspectives*: 231–235.

Stewart, Susan D. 2010. "Stepchildren Who are Adopted by Their Stepparents: Prevalence, Characteristics, and Well-Being." *Family Relations* 59:558–571.

Stewart, Susan D., and Elcy E. Timothy. 2020. "Stepfamilies: U.S. Laws and Policies." *Marriage and Divorce in America: Issues, Trends, and Controversies*, edited by Jamie L. Hartenstein. Santa Barbara, CA: ABC-CLIO.

Stewart, Susan D., and Gordon Limb. 2020. *Multicultural Stepfamilies*. San Diego: Cognella.

Stinnett, Nick. 1985. *Secrets of Strong Families*. New York: Little, Brown.

———. 2008. *Fantastic Families: Six Proven Steps to Building a Strong Family*. West Monroe, LA: Howard Publishing.

Stinnett, Nick, Donnie Hilliard, and Nancy Stinnett. 2000. *Magnificent Marriage: Ten Beacons Show the Way to Marriage Happiness*. Montgomery, AL: Pillar Press.

Stoddard, Martha. 2006. "Grandparents Prevail in Visitation Challenge." *Omaha World-Herald*, June 6.

Stokes, Charles E., and Christopher G. Ellison. 2010. "Religion and Attitudes Toward Divorce Laws among U.S. Adults." *Journal of Family Issues* XX(X):1–26. doi:10.1177/0192513X10363887.

Stone, Juliet, Ann Berrington, and Jane Falkingham. 2014. "Gender, Turning Points, and Boomerangs: Returning Home in Young Adulthood in Great Britain." *Demography* 51:257–276.

Stone, Lawrence. 1980. *The Family, Sex, and Marriage in England, 1500–1800*. New York: Harper & Row.

Stone, Lyman. 2018. "American Women are Having Fewer Children Than They'd Like." Retrieved December 14 (https://www.nytimes.com/2018/02/13/upshot/american-fertility-is-falling-short-of-what-women-want.html).

Stop Street Harassment. 2018. *The Facts Behind the #metoo Movement*. Retrieved October 18, 2019 (http://www.stopstreetharassment.org/our-work/nationalstudy/2018-national-sexual-abuse-report/).

Stotland, Nada L. 2018. "Abortion in America: The War on Women is Not Hyperbole." *The Lancet Psychiatry* 5:862–864.

Stout, David. 2007. "Supreme Court Upholds Ban on Abortion Procedure." *The New York Times*, April 18.

Stout, Hilary. 2011. "Effort to Restore Children's Play Gains Momentum." *The New York Times*, January 5. Retrieved January 9, 2011 (http://www.nytimes.com).

Stratton, Peter. 2003. "Causal Attributions during Therapy: Responsibility and Blame." *Journal of Family Therapy* 25(2):136–160.

Straus, Martha. 2009. "Bungee Families—While Some Warn That the Conveyor Belt That Once Transported Adolescents into Adulthood Has Broken Down, Others Insist the Increasing Numbers of Adult Children Living at Home is Less about Dysfunction than the Changing Function of Family life Today." *Psychotherapy Networker* 33(5):30–38.

Straus, Murray A. 1993. "Physical Assaults by Wives: A Major Social Problem." pp. 67–87 in *Current Controversies on Family Violence*, edited by Richard J. Gelles and Donileen R. Loseke. Newbury Park, CA: Sage.

———. 1999. "The Benefits of Avoiding Corporal Punishment: New and More Definitive Evidence." Paper No. CP40-59/CP41B.P. University of New Hampshire: Family Research Laboratory.

———. 2007. "Do We Need a Law to Prohibit Spanking?" *Family Focus on . . . Adolescence*. FF34:F7. Minneapolis: National Council on Family Relations.

———. 2008. "From Ideology to Inclusion." National Family Violence Legislative Center Conference. Sacramento, California, February.

———. 2010. "Ending Spanking Can Make a Major Contribution to Preventing Physical Abuse." *Family Focus*. Winter: F35, F40.

Straus, Murray A., Sherry L. Hambey, Sue Boney-McCoy, and David B. Sugarman. 1996. "The Revised Conflict Tactics Scales (CTS2): Development and Preliminary Psychometric Data." *Journal of Family Issues* 17:283–316.

Straus, Murray A., and Ignacio Luis Ramirez. 2007. "Gender Symmetry in Prevalence, Severity, and Chronicity of Physical Aggression Against Dating Partners by University students in Mexico and USA." *Aggressive Behavior* 33:281–290.

Strauss, Anselm. and Barney Glaser. 1975. *Chronic Illness and the Quality of Life*. St. Louis: Mosby.

Streep, Peg. 2014. "What Porn Does to Intimacy." *Psychology Today*. Retrieved April 24, 2015 (https://www.psychologytoday.com/blog/tech-support/201407/what-porn-does-intimacy).

Streib, Jessi. 2015. "For Richer or Poorer: The Challenges of Marrying Outside your Class." *Washington Post*, March 26. Retrieved January 12, 2016 (https://www.washingtonpost.com/opinions/for-richer-or-poorer-the-challenges-of-marrying-outside-your-class/2015/03/26/cd7ccf72-ccac-11e4-8a46-b1dc9be5a8ff_story.html).

Streitfeld, Rachel. 2010. "Families Struggle after Cuts for Disabled." CNN.com. Retrieved January 22, 2010 (http://www.cnn.com).

Strohschein, Lisa. 2005. "Household Income Histories and Child Mental Health Trajectories." *Journal of Health and Social Behavior* 46(4):359–375.

Strokoff, Johanna, Jesse Owen, and Frank D. Fincham. 2015. "Diverse Reactions to Hooking Up among U.S. University Students." *Archives of Sexual Behavior* 4: 935–943.

Stroud, Angela. 2012. "Good Guys with Guns: Hegemonic Masculinity and Concealed Handguns." *Gender and Society* 26(2):216–238.

Strow, Claudia W., and Brian K. Strow. 2006. "A History of Divorce and Remarriage in the United States." *Humanomics* 22(4):239–251.

Strully, Kate. 2014. "Racially and Ethnically Diverse Schools and Adolescent Romantic Relationships." *American Journal of Sociology* 120:750–797.

Sturge-Apple, Melissa, Patrick Davies, Dante Cicchetti, and E. Mark Cummings. 2009. "The Role of Mothers' and Fathers' Adrenocortical Reactivity in Spillover Between Interparental Conflict and Parenting Practices." *Journal of Family Psychology* 23(2):215–225.

Stykes, Bart. 2015a. "Three Out of Four Women Supportive of Bearing and Rearing Children in Cohabiting Unions." Retrieved February 14, 2016 (http://archive.constantcontact. com /fs157/1102326693843/archive /1121521813372.html).

Stykes, Bart. 2015b. "Marital Stability Following Mother's 1st Marital Birth" (FP-15). Bowling Green, OH: National Center for Family and Marriage Research.

Stykes, Bart, and Karen Benjamin Guzzo. 2015. "Remarriage and Stepfamilies" (FP-15-10). Bowling Green, OH: National Center for Family and Marriage Research. Retrieved April 11, 2016 (https://www .bgsu.edu/ncfmr/resources /data/family -profiles/stykes-guzzo-remarriage -stepfamilies-fp-15-10.html).

Stykes, Bart, and Seth Williams. 2013. "Diverging Destinies: Children's Family Structure Variation by Maternal Education" (FP-13-16). Bowling Green, OH: National Center for Family and Marriage Research.

Stykes, J. Bart, Wendy D. Manning, and Susan L. Brown. 2013. "Nonresident Fathers and Formal Child Support: Evidence from the CPS, the NSFG, and the SIPP." *Demographic Research* 29:1299.

Stykes, B., & Guzzo, Karen B. (2015). *Remarriage and Stepfamilies*. National Center for Family & Marriage Research, Bowling Green State University. Retrieved February 21,2020 (https://www.bgsu.edu/ncfmr /resources/data/family-profiles/stykes -guzzo-remarriage-stepfamilies-fp-15-10.html).

Stykes, Bart. 2020. "What We Really Know About Stepfamilies." pp. 21–44 in *Multicultural Stepfamilies*, edited by Susan D. Stewart and Gordon Limb. San Diego, CA: Cognella.

Su, Jessica Houston, Rachel Dunifon, and Sharon Sassler. 2015. "Better for Baby? The Retreat from Mid-Pregnancy Marriage and Implications for Parenting and Child Well-Being." *Demography* 52:1167–1194.

Sue, Derald Wing. 2010. *Microaggressions in Everyday Life: Race, Gender, and Sexual Orientation.* John Wiley & Sons.

Suitor, J. Jill, Jori Sechrist, Mari Plikuhn, Seth T. Pardo, Megan Gilligan, and Karl Pillemer. 2009. "The Role of Perceived Maternal Favoritism in Sibling Relations in Midlife." *Journal of Marriage and Family* 71 (November):1026–1038.

Suizzo, Marie-Anne, Karen Moran Jackson, Erin Pahlke, Yesenia Marroquin, Lauren Blondeau, and Anthony Martinez. 2012. "Pathways to Achievement: How Low-Income Mexican-Origin Parents Promote Their Adolescents Through School." *Family Relations* 61(4) (October):533–547.

Sullivan, Harmony B., and Maureen C. McHugh. 2009. "The Critical Eye: Whose Fantasy is This?" *Sex Roles* 60:745–747.

Sullivan, Kieran T., Lauri A. Pasch, Matthew D. Johnson, Thomas N. Bradbury. 2010. "Social Support, Problem Solving, and the Longitudinal Course of Newlywed Marriage." *Journal of Personality and Social Psychology* 98(4):631–644.

Sullivan, Oriel. 2011a. "An End to Gender Display Through the Performance of Housework?" *Journal of Family Theory and Review* 3(1):1–13.

———. 2011b. "Gender Deviance Neutralization Through Housework: Where Does It Fit in the Bigger Picture?" *Journal of Family Theory and Review* 3(1): 27–31.

Sullivan, Oriel, and Scott Coltrane. 2008. "Men's Changing Contribution to Housework and Childcare." Prepared for the *11th Annual Conference of the Council on Contemporary Families*, April 25–26.

Sullivan, Oriel, Jonathan Gershuny, and John P. Robinson. 2018. "Stalled or Uneven Gender Revolution? A Long-Term Processual Framework for Understanding Why Change Is Slow." *Journal of Family Theory & Review* 10 (1): 263–279.

Sullivan, Paul. 2019. "5 Tips on Managing the 'Boomerang Generation.'" *The New York Times*. June 28. Retrieved August 29, 2019 (www. nytimes.com).

Sullivan, Paul. 2015. "Work-Life Balance Poses Challenges Regardless of Wealth." *The New York Times*, October 9. Retrieved October 13, 2015 (http://www.nytimes .com).

Sumra, Monika K., and Michael A. Schillaci. 2015. "Stress and the Multiple-Role Woman: Taking a Closer Look at the 'Superwoman.'" *PloS one* 10:e0120952.

Sun, Hua, and Lei Gao. 2019. "Lending Practices to Same-sex Borrowers." *Proceedings of the National Academy of Sciences* 116:9293–9302.

Sun, Yongmin, and Yuanzhang Li. 2011. "Effects of Family Structure Type and Stability on Children's Academic Performance Trajectories." *Journal of Marriage and Family* 73(June):541–556.

Suro, Roberto. 1992. "Generational Chasm Leads to Cultural Turmoil for Young Mexicans in U.S." *The New York Times*, January 20.

Surra, Catherine A., and Debra K. Hughes. 1997. "Commitment Processes in Accounts of the Development of Premarital Relationships." *Journal of Marriage and Family* 59:5–21.

Sue, Derald Wing, David Sue, Helen A. Neville, and Laura Smith. 2019. *Counseling the Culturally Diverse: Theory and Practice.* John Wiley & Sons.

Sussman, L. J. 2006. "A 'Delicate Balance': Interfaith Marriage, Rabbinic Officiation, and Reform Judaism in America, 1870–2005." *CCAR Journal: A Reform Jewish Quarterly* 53(2):38–67.

Sussman, Marvin B., Suzanne K. Steinmetz, and Gary W. Peterson. 1999. *Handbook of Marriage and the Family*, 2nd ed. New York: Plenum.

Suter, Elizabeth A., Karen L. Daas, and Karla Mason Bergen. 2008. "Negotiating Lesbian Family Identity via Symbols and Rituals." *Journal of Family Issues* 29(1):26–47.

Sutorius, Margo. 2015. "Communicating with Individuals with Disabilities: A Reconsideration of Competency and Awareness." *Family Focus.* Summer:F19–F21.

Sutphin, Suzanne. 2006. "The Division of Household Labor in Same Sex Couples." Paper presented at the annual meeting of the *American Sociological Association*, August 11.

Svevo-Cianci, K., and S. C. Velazquez. 2010. "Convention on the Rights of the Child Special Protection Measures: Overview of Implications and Value for Children in the United States." *Child Welfare* 89(5):139–157.

Swain, Carol M. 2002. *The New White Nationalism: Its Challenge to Integration.* Cambridge, UK: Cambridge University Press.

Swanbrow, Diane. 2000. "Home Alone? New Study Shows What Kids Do After School." *The University Record*, April 3.University of Michigan. Retrieved February 13, 2013 (http://www. ur.umich.edu).

Swarns, Rachel L. 2012a. "More Americans Rejecting Marriage in 50s and Beyond." *The New York Times*, March 1. Retrieved November 24, 2012 (http://www.nytimes .com).

———. 2012b. "Testing Autism and Air Travel." *The New York Times*, October 26. Retrieved May 8, 2013 (http://www .nytimes.com).

Swartz, Teresa. 2008. "Family Capital and the Invisible Transfer of Privilege: Intergenerational Support and Social Class in Early Adulthood." *New Directions for Child and Adolescent Development* 119:11–24.

Swartz, Teresa T., Minzee Kim, Mayumi Uno, Jeylan Mortimer, and Kristen B. O'Brien. 2011. "Safety Nets and Scaffolds: Parental Support in the Transition to Adulthood." *Journal of Marriage and Family* 73(2):414–429.

Sweeney, Megan M. 2010. "Remarriage and Stepfamilies: Strategic Sites for Family Scholarship in the 21st Century." *Journal of Marriage and Family*, 72:667–684.

Swenson, Sue, and Charlie Lakin. 2014. "A Wicked Problem: Can Governments Be Fair to Families Living with Disabilities?" *Family Relations* 63(1):185–191.

Swiss, Liam, and Céline Le Bourdais. 2009. "Father-Child Contact After Separation: The Influence of Living Arrangements." *Journal of Family Issues* 30(5):623–652.

Szalavitz, Maria. 2011. "Q&A: Why 50 Is Not the New 30, and How to Make the Best of It." *Time*, Retrieved November 21, 2012 (http://healthland.time.com/2011/07/07 /mind-reading-why-50-is-not-the-new-30 -and-how-to-make-the-best-of-it/).

Tach, Laura M., and Alicia Eads. 2015. "Trends in the Economic Consequences of Marital and Cohabitation Dissolution in the United States." *Demography* 52:401–432.

Tach, Laura M., and Alicia Eads. 2015. "Trends in the Economic Consequences of Marital and Cohabitation Dissolution in the United States." *Demography* 52:401–432.

Tach, Laura, and Sarah Halpern-Meekin. 2009. "How Does Premarital Cohabitation Affect Trajectories of Marital Quality?" *Journal of Marriage and Family* 71(2):298–317.

Tach, Laura, Ronald Mincy, and Kathryn Edin. 2010. "Parenting as a 'Package Deal': Relationships, Fertility, and Nonresident Father Involvement among Unmarried Parents." *Demography* 47(1).181–204.

Taft, Casey, Candice Monson, Jeremiah Schumm, Laura Watkins, Jillian Panuzio, and Patricia Resick. 2009. "Posttraumatic Stress Disorder Symptoms, Relationship Adjustment, and Relationship Aggression in a Sample of Female Flood Victims." *Journal of Family Violence* 24:389–396.

Tak, Young Ran, and Marilyn McCubbin. 2002. "Family Stress, Perceived Social Support, and Coping Following the Diagnosis of a Child's Congenital Heart Disease." *Journal of Advanced Nursing* 39(2):190–198.

Takagi, Dana Y. 2002. "Japanese American Families." pp. 164–180 in *Multicultural Families in the United States*, 3rd ed., edited by Ronald L. Taylor. Upper Saddle River, NJ: Prentice Hall.

Takahashi, Kayo, Kei Mizuno, Akihiro T. Sasaki, Yasuhiro Wada, Masaaki Tanaka, Akira Ishii, Kanako Tajima, et al. 2015. "Imaging the Passionate Stage of Romantic Love by Dopamine Dynamics." *Frontiers in Human Neuroscience* 9.

Talan, Kenneth H. 2009. *Help Your Child or Teen Get Back on Track: What Parents and Professionals Can Do for Childhood Emotional and Behavioral Problems*. London & Philadelphia: Jessica Kingsley Publisher.

Talbot, Margaret. 2001. "Open Sperm Donation." *The New York Times Magazine*, December 8, p. 88–88.

Tamborini, Christopher R., Kenneth A. Couch, and Gayle L. Reznik. 2015. "Long-Term Impact of Divorce on Women's Earnings across Multiple Divorce Windows: A life Course Perspective." *Advances in Life Course Research* 26:44–59.

Tanne, Janice Hopkins. 2009. "Virginity Pledge Ineffective Against Teen Sex Despite Government Funding, US Study Finds." *British Medical Journal* 338(7686):69–69.

Tannen, Deborah. 1990. *You Just Don't Understand*. New York: Morrow.

———. 2006. *You're Wearing That? Understanding Mothers and Daughters in Conversation*. New York: Random House.

———. 2013. *You Just Don't Understand*. New York: William Morrow.

Tanner, Lindsey. 2005. "Study: Children Adopted from Foreign Countries Adjust Surprisingly Well." Associated Press, May 24. Retrieved July 22, 2007 (www.associatedpress.org).

Tanton, Clare, Kyle G. Jones, Wendy Macdowall, Soazig Clifton, Kirsten R. Mitchell, Jessica Datta, et al. 2015. "Patterns and Trends in Sources of Information About Sex among Young People in Britain: Evidence from Three National Surveys of Sexual Attitudes and Lifestyles." *BMJ Open* 5:e007834.

Tashiro, Ty, Patricia Frazier, and Margit Berman. 2006. "Stress-Related Growth Following Divorce and Relationship Dissolution." pp. 361–384 in *Handbook of Divorce and Relationship Dissolution*, edited by Mark A. Fine and John H. Howard. Mahwah, NJ: Erlbaum.

Tavernise, Sabrina. 2012. "Day Care Centers Adapt to Round-the-Clock Demand." *The New York Times, January* 16.

———. 2013. "The Health Toll of Immigration." *The New York Times*, May 18. Retrieved May 20, 2013 (http://www.nytimes.com)

Tavernise, Sabrina, and Robert Gebeloff. 2011. "Once Rare in Rural America, Divorce is Changing the Face of Its Families." *The New York Times*. Retrieved April 17, 2013 (http://www.nytimes.com /2011/03/24/us/24divorce.html?pagewanted=all&_r=0).

Taylor, Alan. 2015. "Alaska's Climate Refugees." *The Atlantic*. July 7. Retrieved August 4, 2019 (https://amp.theatlantic.com).

Taylor, Laura C. 2018. "The Experience of Infertility Among African American Couples." *Journal of African American Studies* 22:357–372.

Taylor, Paul. 2009. *America's Changing Workforce: Recession Turns a Graying Office Grayer*, September 3. Pew Research Center, Social and Demographic Trends Project. Retrieved May 8, 2010 (pewresearch.org).

Taylor, Paul, Kim Parker, D'Vera Cohn, Jeffrey Passel, Gretchen Livingston, Wendy Wang, and Eileen Patten. 2011. "New Marriages Down 5% from 2009 to 2010: Barely Half of U.S. Adults Are Married—A Record Low." *Social and Demographic Trends*, December 14. Washington, DC: Pew Research Center.

Taylor, Robert J., Karen D. Lincoln, and Linda M. Chatters. 2005. "Supportive Relationships with Church Members among African Americans." *Family Relations* 54(4):501–511.

Taylor, Robert Joseph, Linda M. Chatters, Amanda Toler Woodward, and Edna Brown. 2013. "Racial and Ethnic Differences in Extended Family, Friendship, Fictive Kin, and Congregational Informal Support Networks." *Family Relations* 62:609–624.

Taylor, Ronald L. 2002a. "Black American Families." pp. 19–47 in *Minority Families in the United States: A Multicultural Perspective*, 3rd ed., edited by Ronald L. Taylor. Upper Saddle River, NJ: Prentice Hall.

———. 2002b. *Minority Families in the United States: A Multicultural Perspective*. 3rd ed. Upper Saddle River, NJ: Prentice Hall.

———. 2007. "Diversity within African American Families." pp. 398–421 in *Family in Transition*, 14th ed., edited by Arlene S. Skolnick and Jerome H. Skolnick. Boston: Allyn & Bacon.

Taylor, Verta, Katrina Kimport, Nella Van Dyke, and Ellen Ann Andersen. 2009. "Culture and Mobilization: Factional Repertoires, Same-Sex Weddings, and the Impact on Gay Activism." *American Sociological Review* 74(6):865–890.

Teachman, Jay D. 2002. "Stability across Cohorts in Divorce Risk Factors." *Demography* 39(2):331–351.

———. 2008a. "Complex Life Course Patterns and the Risk of Divorce in Second Marriages." *Journal of Marriage and Family* 70(2):294–305.

———. 2008b. "The Living Arrangements of Children and Their Educational Well-Being." *Journal of Family Issues* 29(6):734–761.

———. 2009. "Military Service, Race, and the Transition to Marriage and Cohabitation." *Journal of Family Issues* 30(10):1433–1454.

———. 2010. "Wives' Economic Resources and Risk of Divorce." *Journal of Family Issues* 31(4):1–19.

Teachman, Jay D., Lucky M. Tedrow, and Matthew Hall. 2006. "The Demographic Future of Divorce and Dissolution." pp. 59–82 in *Handbook of Divorce and Relationship Dissolution*, edited by Mark A. Fine and John H. Howard. Mahwah, NJ: Erlbaum.

Teitler, Julien, and Nancy Reichman. 2008. "Mental Illness as a Barrier to Marriage among Unmarried Mothers." *Journal of Marriage and Family* 70(3):772–782.

Temkin, Deborah, Jonathan Belford, Tyler McDaniel, Brandon Stratford, and Dominique Parris. 2017. "Improving Measurement of Sexual Orientation and Gender Identity Among Middle and High School Students." Retrieved October 28, 2019 (https://www.childtrends.org/wp-content/uploads/2017/06/2017-23LGBTSurveyMeasuresExecSum.pdf).

Tennov, Dorothy. 1999 [1979]. *Love and Limerence: The Experience of Being in Love*. 2nd ed. New York: Scarborough House.

Tepperman, Lorne, and Susannah J. Wilson, eds. 1993. *Next of Kin: An International Reader on Changing Families*. New York: Prentice Hall.

The American College of Obstetricians and Gynecologists 2017. "Sterilization of Women: Ethical Issues and Considerations. " Retrieved December 19, 2019 (https://www.acog.org/Clinical-Guidance-and-Publications/Committee-Opinions/Committee-on-Ethics/Sterilization-of-Women-Ethical-Issues-and-Considerations).

The Cut. 2017. "What We Learned About Sexual Desire from 10 Years of Pornhub User Data." Retrieved November 2, 2019 (https://www.thecut.com/2017/06/pornhub-data-sexual-habits.html

The Endocrine Society. 2019. "Male Birth Control Pill Passes Human Safety Tests." Retrieved December 26, 2019 (https://www.technologynetworks.com/drug-discovery/news/male-birth-control-pill-passes-human-safety-tests-317223).

The Mayo Clinic Staff. 2020. "Mental Health Providers: Tips on Finding One." Retrieved January 15, 2020 (https://www.mayoclinic.org/diseases-conditions/mental-illness/in-depth/mental-health-providers/art-20045530).

The World Bank (2019). *Fertility Rate, Total (Births per Woman)—United States*. Retrieved December 11, 2019 (https://data.worldbank.org/indicator/SP.DYN.TFRT.IN?locations=US).

Theran, Sally A. 2009. "Predictors of Level of Voice in Adolescent Girls: Ethnicity, Attachment, and Gender Role Socialization." *Journal of Youth and Adolescence* 38(8):1027–1037.

Therborn, Göran. 2004. *Between Sex and Power: Family in the World, 1900–2000*. London: Routledge.

Thobaben, Marshelle. 2008. "Elder Abuse Prevention." *Home Health Care Management and Practice* 20(2):194–196.

Thoits, P. A. 1992. "Identity Structures and Psychological Well-Being: Gender and Marital Status Comparisons." *Social Psychology Quarterly* 55:236–256.

Thomas, Adam, and Isabel Sawhill. 2005. "For Love *and* Money? The Impact of Family Structure on Family Income." *The Future of Children* 15(2):57–74.

Thomas, Alexander, Stella Chess, and Herbert G. Birch. 1968. *Temperament and Behavior Disorders in Children*. New York: New York University Press.

Thomas, Brett W., and Mark W. Roberts. 2009. "Sibling Conflict Resolution Skills: Assessment and Training." *Journal of Child and Family Studies* 18:447–453.

Thomas, J. 2009. "Virginity Pledgers Are Just as Likely as Matched Nonpledgers to Report Premarital Intercourse." *Perspectives on Sexual and Reproductive Health* 41(1):63.

Thomas, Jerry R., and Karen E. French. 1985. "Gender Differences Across Age in Motor Performance: A Meta-Analysis." *Psychological Bulletin* 98:260.

Thomas Katherine Woodward. 2015. *Conscious Uncoupling: The Five Steps to Living Happily Even After*. London: Yellow Kite.

Thomas, Susan Gregory. 2012. "The Gray Divorcés." *Wall Street Journal*. Retrieved on April 17, 2013 (http://online.wsj.com /article/SB100014240529702037537045772552 30471480276.htm).

Thomason, Deborah J. 2005. "Natural Disasters: Opportunity to Build and Reinforce Family Strengths." *Family Focus on . . . Family Strengths and Resilience* FF28:F11. Minneapolis: National Council on Family Relations.

Thompson, Kacie M. 2009. "Sibling Incest: A Model for Group Practice with Adult Female Victims of Brother-Sister Incest." *Journal of Family Violence* 24(7):531–537.

Thomsen, Jacqueline. 2019. "Supreme Court Rules against Trump on Census Citizenship Question." *The Hill*. July 27. Retrieved July 30, 2019 (https://thehill .com).

Thomson, Elizabeth, and Ugo Colella. 1992. "Cohabitation and Marital Stability: Quality or Commitment?" *Journal of Marriage and Family* 54(2):368–378.

Thomson, Elizabeth, and Sara S. McLanahan. 2012. "Reflections on 'Family Structure and Child Well-Being: Economic Resources vs. Parental Socialization.'" *Social Forces* 91(1):45–53.

Thomson, Elizabeth, Jane Mosley, Thomas L. Hanson, and Sara S. McLanahan. 2001. "Remarriage, Cohabitation, and Changes in Mothering Behavior." *Journal of Marriage and Family* 63(2):370–380.

Thorne, Marie E., Sheree Boulet, Joyce A. Martin, and Dmitry Kissin. 2014. "Births Resulting from Assisted Reproductive Technology: Comparing Birth Certificate and National ART Surveillance System Data, 2011." *National Vital Statistics Reports* 63(8).

Thornock, Carly M., Larry J. Nelson, Clyde C. Robinson, and Craig H. Hart. 2013. "The Direct and Indirect Effects of Home Clutter on Parenting." *Family Relations* 62(5):783–794.

Thornton, Arland. 2009. "Framework for Interpreting Long-Term Trends in Values and Beliefs Concerning Single-Parent Families." *Journal of Marriage and Family* 71(2):230–234.

Thornton, Arland, William G. Axinn, and Yu Xie. 2007. *Marriage and Cohabitation*. Chicago: University of Chicago Press.

Thornton, Arland, and Deborah Freedman. 1983. "The Changing American Family." Population Bulletin 38. Washington, DC: Population Reference Bureau.

Thorsen, Maggie L., and Valarie King. 2016. "My Mother's Husband: Factors Associated with How Adolescents Label Their Stepfathers." *Journal of Social and Personal Relationships* 33:835–851.

Tian, Felicia F., and S. Philip Morgan. 2015. "Gender Composition of Children and the Third Birth in the United States." *Journal of Marriage and Family* 77: 1157–1165.

Tichenor, Veronica Jaris. 2005. *Earning More and Getting Less: Why Successful Wives Can't Buy Equality*. New Brunswick, NJ: Rutgers University Press.

———. 2010. "Thinking about Gender and Power in Marriage." pp. 415–425 in *The Kaleidoscope of Gender: Prisms, Patterns, and Possibilities*, 3rd ed., edited by Joan Z. Spade and Catherine G. Valentine. Newbury Park, CA: Pine Forge Press.

Tiefer, Leonore. 2007. "Sexuopharmacology: A Fateful New Element in Sexual Scripts." pp. 239–248 in *The Sexual Self: The Construction of Sexual Scripts*, edited by Michael Kimmel. Nashville, TN: Vanderbilt University Press.

———. 2008. "Female Genital Cosmetic Surgery: Freakish or Inevitable? Analysis from Medical Marketing, Bioethics, and Feminist Theory." *Feminism and Psychology* 18(4):466–479.

Tierney, John. 2013. "Prison and the Poverty Trap." *The New York Times, February* 19. Retrieved November 7, 2015 (http://www.nytimes.com).

Tillman, Kathryn Harker. 2007. "Family Structure Pathways and Academic Disadvantage among Adolescents in Stepfamilies." *Sociological Inquiry* 77(3):383–424.

Timonen, Virpi, Martha Doyle, and Ciara O'Dwyer. 2011. "'He Really Leant on Me a Lot': Parents' Perspective on the Provision of Support to Divorced and Separated Adult Children in Ireland." *Journal of Family Issues* 32(2):1622–1646.

Tolin, David F., Randy O. Frost, Gail Steketee, and Kristin E. Fitch. 2008. "Family Burden of Compulsive Hoarding: Results of an Internet Survey." *Behaviour Research and Therapy* 46:334–344.

Tonelli, Bill. 2004. "Thriller Draws on Oppression of Italians in Wartime U.S." *The New York Times*, August 2.

Tondo, Luigi, M. Pinna, Giulia Serra, L. De Chiara, and Ross J. Baldessarini. 2017. "Age at Menarche Predicts Age at Onset of Major Affective and Anxiety Disorders." *European Psychiatry* 39:80–85.

Toner, Robin. 2007. "Women Feeling Freer to Suggest 'Vote for Mom.'" *The New York Times*, January 29.

Tong, Benson. 2004. *Asian American Children: An Historical Guide*. Westport, CT: Greenwood Press.

Tonnu, Tony. 2014. "My Mother's Child, First and Always." *Hyphen: Asian American Unabridged*, June 26. Retrieved October 13, 2015 (http:// hyphenmagazine.com).

Toomey, Russell B., Amy K. Syvertsen, and Maura Shramko. 2018. "Transgender Adolescent Suicide Behavior." *Pediatrics* 142:e20174218

Toot, Sandeep, Juanita Hoe, Ritchard Ledgerd, Karen Burnell, Mike Devine, and Martin Orrell. 2013. "Causes of Crises and Appropriate Interventions: The Views of People with Dementia, Carers, and Healthcare Professionals." *Aging and Mental Health* 17(3):328–335.

Totura, Christine M. Wienke, Carol MacKinnon-Lewis, Ellis L. Gesten, Ray Gadd, Katherine P. Divine, Sherri Dunham, and Dimitra Kamboukos. 2009. "Bullying and Victimization among Boys and Girls in Middle School: The Influence of Perceived Family and School Contexts." *Journal of Early Adolescence* 29(4):571–609.

Townsend, Marshall John, and Timothy H. Wasserman. 2011. "Sexual Hookups among College Students: Sex Differences in Emotional Reactions." *Archives of Sexual Behavior* 40:1173–1181.

Townsend, Nicholas W. 2002. *The Package Deal: Marriage, Work, and Fatherhood in Men's Lives*. Philadelphia, PA: Temple.

Tozzi, John. 2010. "Home-Based Businesses Increasing." *Bloomberg Businessweek*, January 25. Retrieved May 26, 2010 (www .businessweek. com).

Tracey, Marlon R., and Solomon W. Polachek. 2015. "Does Father Visitation Affect Child Health? The Role of Baby Looks and Maternal Earnings." Retrieved April 4, 2016 (http://www .sole-jole.org /16088.pdf).

Tracy, J. K., and J. Junginger. 2007. "Correlates of Lesbian Sexual Functioning." *Journal of Women's Health* 16:499–509.

Trail, Thomas E., and Benjamin R. Karney. 2012. "What (Not) Wrong with Low-Income Marriages." *Journal of Marriage and Family* 74(3): 413–427.

Tramonte, Lucia, Anne Gauthier, and J. Douglas Wilms. 2015. "Engagement and Guidance: The Effects of Maternal Parenting Practices on Children's Development." *Journal of Family Issues* 36(3):396–420.

Trandafir, Mircea. 2015. "Legal Recognition of Same-Sex Couples and Family Formation." *Demography* 52:113–151.

Trask, Bahira Sherif. 2013. "Locating Multiethnic Families in a Globalizing World." *Family Relations* 62 (February):17–29.

Trask, Bahira Sherif, Jocelyn D. Taliaferro, Margaret Wilder, and Raheemah Jabbar-Bey. 2005. "Strengthening Low-Income Families through Community-Based Family Support Initiatives." . . . *Family Strengths and Resilience* FF28. Minneapolis: National Council on Family Relations.

Trask, Bahira Sherif, Bethany Willis Hepp, Barbara Settles, and Lilianah Shabo. 2009. "Culturally Diverse Elders and Their Families: Examining the Need for Culturally Competent Services." *Journal of Comparative Family Studies* 40(2):293–303.

Travis, Carol. 2006. "Letters to the Editor: The New Science of Love." *The Atlantic,* May.

Treas, Judith. 1995. "Older Americans in the 1990s and Beyond." *Population Bulletin* 50(2). Washington, DC. Population Reference Bureau.

Treas, Judith, and Deirdre Giesen. 2000. "Sexual Infidelity among Married and Cohabiting Americans." *Journal of Marriage and Family* 62(1):48–60.

Treas, Judith, and Tsui-o Tai. 2012. "How Couples Manage the Household: Work and Power in Cross-National Perspective." *Journal of Family Issues* 33(8): 1088–1116.

Trent, Maria, Danielle Dooley, and Jacqueline Douge. 2019. "The Impact of Racism on Child and Adolescent Health." *Pediatrics* 144   (2). American Academy of Pediatrics. Retrieved February 18, 2020 (www.aappublications.org /news).

Trevillion, K., S. Oram, G. Feder, and L.M. Howard. 2012. "Experiences of Domestic Violence and Mental Disorders: A Systematic Review and Meta-Analysis." *Plos One* 7(12). Retrieved March 12, 2013 (http://www.plosone. org).

Trimble, Megan. 2019. "New Jersey Adds Neutral Gender to Birth Certificates." Retrieved September 25, 2019 (https://www. usnews.com/news /national-news/articles/2019-01-30/new -jersey-adds-neutral-gender-option-to-birth -certificates).

Trillingsgaard, Tea, Katherine Baucom, and Richard Heyman. 2014. "Predictors of Change in Relationship Satisfaction during the Transition to Parenthood." *Family Relations* 63(5):667–679.

Trillingsgaard, Tea, K. Baucom, R. Heyman, and A Elklit. 2012. "Relationship Interventions During the Transition to Parenthood: Issues of Timing and Efficacy." *Family Relations* 61(5): 770–783.

Trimberger, E. Kay. 2005. *The New Single Woman.* New York: Beacon Press.

Troilo, Jessica. 2011. "Stepfamilies and the Law: Legal Ambiguities and Suggestions for Reform." *Journal of Divorce and Remarriage* 52(8):610–621.

Troilo, Jessica, and Marilyn Coleman. 2008. "College Students' Perceptions of the Content of Father Stereotypes." *Journal of Marriage and Family* 70(1):218–227.

——. 2012. "Full-Time, Part-Time Full-Time, and Part-Time Fathers: Father Identities Following Divorce." *Family Relations* 61:601–614.

Troll, Lillian E. 1985. "The Contingencies of Grandparenting." pp. 135–150 in *Grandparenthood,* edited by Vern L. Bengston and Joan F. Robertson. Newbury Park, CA: Sage.

Tromholt, Morten. 2016. "The Facebook Experiment: Quitting Facebook Leads to Higher Levels of Well-Being." *Cyberpsychology, Behavior, and Social Networking* 19(11):661–666.

Troy, Adam B., Jamie Lewis-Smith, and Jean-Phillippe Laurenceau. 2006. "Interracial and Intraracial Romantic Relationships: The Search for Differences in Satisfaction, Conflict, and Attachment Style." *Journal of Social and Personal Relationships* 23(1):65–80.

Trudge, J. Ayse Payir, Elisa Mercon-Vargas, H. Cao, Yue Liang, J. Li and Lia O'Brien. 2016. "Still Misused after All These Years? A Reevaluation of the Uses of Bronfenbrenner's Bioecological Theory of Human Development." *Journal of Family Theory & Review* 8(4):427–445.

Truman, Jennifer L., and Lynn Langton. 2015. *Criminal Victimization, 2014.* U.S. Department of Justice. Report NCJ 248973.

Tshann, Jeane M., Lauri A. Pasch, Elena Flores, Barbara VanOss Marin, E. Marco Baisch, and Charles J. Wibbelsman. 2009. "Nonviolent Aspects of Interparental Conflict and Dating Violence among Adolescents." *Journal of Family Issues* 30(3):295–319.

Tsui, Anjali. 2017. "In Fight Over Child Marriage Laws, States Resist Calls for a Total Ban. " Retrieved November 16, 2019 (https://www .pbs.org/wgbh/frontline/article/in-fight-over -child-marriage-laws-states-resist-calls-for-a-total -ban/).

Tucker, Corinna J., Susan M. McHale, and Ann C. Crouter. 2003. "Conflict Resolution: Links with Adolescents' Family Relationships and Individual Well-Being." *Journal of Family Issues* 24(6):715–736.

Tugend, Alina. 2011. "Family Happiness and the Overbooked Child." *The New York Times,* August 12. Retrieved September 13, 2011 (http://www. nytimes.com).

Tuller, David. 2001. "Adoption Medicine Brings New Parents Answers and Advice." *The New York Times,* September 4.

Tumin, Dimitry, and Hui Zheng. 2018. "Do the Health Benefits of Marriage Depend on the Likelihood of Marriage? *Journal of Marriage and Family* 80 (3):622–636.

Tumin, Dmitry, and Zhenchao Qian. 2015. "Unemployment and the Transition from Separation to Divorce." *Journal of Family Issues:* 0192513X15600730.

Turley, Ruth N. Lopez, and Matthew Desmond. 2011. "Contributions to College Costs by Married, Divorced, and Remarried Parents." *Journal of Family Issues* 32(6):767–790.

Turnbull, A., and H. Turnbull. 1997. *Families, Professionals, and Exceptionality: A Special Partnership,* 3rd ed. Upper Saddle River, NJ: Merrill.

Turner, Lynn H., and Richard L. West. 2006. *The Family Communication Sourcebook.* Thousand Oaks, California: Sage.

Turner, Ralph H. 1976. "The Real Self: From Institution to Impulse." *American Journal of Sociology* 81:989–1016.

Turner, Ryan, Gordon Limb, and Susan Stewart. 2020. "American Indian Stepfamilies." In *Multicultural Stepfamilies,* edited by Susan D. Stewart and Gordon Limb, pp. 111–148. San Diego, CA: Cognella.

Turney, Kristin. 2011. "Chronic and Proximate Depression among Mothers: Implications for Child Well-Being." *Journal of Marriage and Family* 73(1):149–163.

——. 2015. "Hopelessly Devoted? Relationship Quality During and After Incarceration." *Journal of Marriage and Family* 77(2):480–495.

Turney, Kristin, and Jessica Hardie. 2018. "Health Limitations among Mothers and

Fathers: Implications for Parenting." *Journal of Marriage and Family* 80 (1): 219–238.

Turney, Kristin, and Sarah Halpern-Meekin. 2017. "Parenting in On/Off relationships: The Link between Relationship Churning and Father Involvement. " *Demography* 54:861–886.

Turney, Kristin, and Yader Lanuza. 2017. "Parental Incarceration and the Transition to Adulthood." *Journal of Marriage and Family* 79 (5): 1314–1330.

Twenge, Jean M. 1997. "Attitudes toward Women, 1970-1995: A Meta-Analysis." *Psychology of Women Quarterly* 21(1):35–51.

Twenge, Jean M., W. Keith Campbell, and Brittany Gentile. 2012. "Male and Female Pronoun Use in U.S. Books Reflects Women's Status, 1900–2008." *Sex Roles* 67(9–10):488–493.

Twenge, Jean M., Ryne A. Sherman, and Brooke E. Wells. 2015. "Changes in American Adults' Sexual Behavior and Attitudes, 1972–2012." *Archives of Sexual Behavior* 44:1–13.

Twenge, Jean. 2017. "Why Couples are Having So Much Less Sex." Retrieved October 24, 2019 (https://www.psychologytoday.com/us/blog /our-changing-culture/201703 /why-couples-are-having-so-much-less-sex).

Twenge, Jean M. 2017. *IGen: Why Today's Super-connected Kids are Growing up Less Rebellious, More Tolerant, Less Happy—and Completely Unprepared for Adulthood—and What That Means for the Rest of Us.* NY: Simon and Schuster.

Twenge, Jean M. 2017. *iGen.* Simon and Schuster, 2017.

Tyre, Peg. 2004. "A New Generation Gap." *Newsweek,* January 19, pp. 68–71.

Udry, J. Richard. 1994. "The Nature of Gender." *Demography* 31(4):561–573.

——. 2000. "The Biological Limits of Gender Construction." *American Sociological Review* 65:443–457.

Uecker, Jeremy E., and Christopher G. Ellison. 2012. "Parental Divorce, Parental Religious Characteristics, and Religious Outcomes in Adulthood." *Journal for the Scientific Study of Religion* 51(4):777–794.

"UF Study: Sibling Violence Leads to Battering in College Dating." 2004. *UF News.* Gainesville: University of Florida.

Uhlenberg, Peter. 1996. "Mortality Decline in the Twentieth Century and Supply of Kin over the Life Course." *The Gerontologist* 36:681–685.

Umberson, Debra, Kristin Anderson, Jennifer Glick, and Adam Shapiro. 1998. "Domestic Violence, Personal Control, and Gender." *Journal of Marriage and Family* 60(2):442–452.

Umberson, Debra, Meichu D. Chen, James S. House, Kristine Hopkins, and Ellen Slaten. 1996. "The Effect of Social Relationships on Psychological Well-Being: Are Men and Women Really So Different?" *American Sociological Review* 61:837–857.

Umberson, Debra, Tetyana Pudrovska, and Corinne Reczek. 2010. "Parenthood, Childlessness, and Well-Being: A Life Course Perspective." *Journal of Marriage and Family* 72(3) (June): 612–629.

Umberson, Debra, Mieke Beth Thomeer, Rhiannon A. Kroeger, Amy C. Lodge, and Minle

Xu. 2015. "Challenges and Opportunities for Research on Same-Sex Relationships." *Journal of Marriage and Family* 77(1):96–111.

Umberson, Debra, Mieke Beth Thomeer, and Amy C. Lodge. 2015. "Intimacy and Emotion Work in Lesbian, Gay, and Heterosexual Relationships." *Journal of Marriage and Family* 77: 542-556.

UNICEF (United Nations Children's Fund). 2012. *Innocenti Report Card 10*, Florence, Italy: UNICEF Innocenti Research Centre.

Unitarian Universalist Association. n.d. *Premarital Counseling Guide for Same Gender Couples*. Boston: Unitarian Universalist Association Office of Bisexual, Gay, Lesbian, and Transgender Concerns. Retrieved March 3, 2010 (www.uua.org /obgltc).

UN News. 2016. "New UN Initiative Aims to Protect Millions of Girls from Child Marriage." Retrieved November 16,2019 (https://news .un.org/en/story/2016/03/523802-new-un -initiative-aims-protect-millions-girls-child -marriage#.Vt8pqfkrK71).

Updegraff, Kimberly A., Norma J. Perez-Brena, Megan E. Baril, Susan M. McHale, and Adriana J. Umana-Taylor. 2012. "Mexican-Origin Mothers' and Fathers' Involvement in Adolescents' Peer Relationships: A Pattern-Analytic Approach." *Journal of Marriage and Family* 74(5):1069–1083.

Uruk, Ayse, Thomas Sayger, and Pamela Cogdal. 2007. "Examining the Influence of Family Cohesion and Adaptability on Trauma Symptoms and Psychological Well-Being." *Journal of College Student Psychotherapy* 22(2):51–63.

U.S. Administration on Aging. 2018. *2018 Profile of Older Americans*. Washington, DC: U.S. Department of Health and Human Services. Retrieved May 15, 2019 (www.aoa.gov).

U.S. Bureau of Labor Statistics. 2018. "Highlights of Women's Earnings in 2017." Retrieved October 4, 2019 (https:// www.bls.gov/opub/reports/womens -earnings/2017/pdf/home.pdf)

U.S. Bureau of Labor Statistics. 2014b. *Women in the Labor Force: A Databook*. Report 1052. Washington, DC: U.S. Bureau of Labor Statistics (www.bls.gov).

U.S. Bureau of Labor Statistics. 2019. *Women in the Labor Force: A Databook*. Report 1084. Washington, DC: U.S. Bureau of Labor Statistics (www.bls.gov).

U.S. Census Bureau. 2017. American Community Survey 1-Year Estimates.

U.S. Census Bureau. 2018a. "Characteristics of Same-Sex Couple Households: 2005 to Present." Retrieved December 1, 2019 (https://www .census.gov/data/tables/time-series/demo /same-sex-couples/ssc-house-characteristics .html).

U.S. Census Bureau. 2019b. "Current Population Survey, 2019: Annual Social and Economic Supplement." Washington, DC: U.S. Census Bureau. Retrieved December 10, 2019 (www.census.gov).

U.S. Census Bureau. 2018a. "America's Families and Living Arrangements: 2018." Retrieved October 18, 2019 (www.census.gov).

U.S. Census Bureau. 2018b. "Current Population Survey, 2018: Annual Social and Economic Supplement." Washington, DC: U.S. Census Bureau. Retrieved June 29, 2019 (www .census.gov).

U.S. Census Bureau. 2019a. "Estimates of Same Sex Couples." Retrieved December 1, 2019 (https://www.census.gov/library /visualizations/2019/comm/same-sex -couple-households.html).

U.S. Census Bureau. 2018. "Historical Marital Status Tables. Table MS-2. Estimated Age at First Marriage, by Sex: 1890 to the Present. " Retrieved November 15, 2019 (https://www.census.gov /data/tables/time-series/demo/families /marital.html).

U.S. Census Bureau. 2019b. "Historical Marital Status Tables." Retrieved November 25, 2019 (https://www.census.gov/data/tables/time -series/demo/families/marital.html)

U.S. Census Bureau. 2018c. "Historical Poverty Tables." Retrieved November 16, 2019 (https://www.census.gov/data /tables/time-series/demo/income -poverty/historical-poverty-people.html).

U.S. Census Bureau. 2018d. "Population Estimates July 1, 2018." Retrieved October 15, 2019 (www.census.gov /quickfacts).

U.S. Census Bureau. 2018b. "Table C3. Living Arrangements of Children Under 18 Years and Marital Status of Parents, by Age, Sex, Race, and Hispanic Origin and Selected Characteristics of the Child for All Children: 2018." Retrieved November 30, 2019 (https://www.census.gov /data/tables/2018/demo/families/cps-2018 .html).

U.S. Census Bureau. 2016. "Table C8. Poverty Status, Food Stamp Receipt, and Public Assistance for Children Under 18 Years by Selected Characteristics: 2016." Retrieved February 8, 2020 (https://www .census.gov/data/tables/2016/demo /families/cps-2016.html).

U.S. Census Bureau. 2018c. "Table FG3. Married Couple Family Groups, by Presence of Own Children Under 18, and Age, Earnings, Education, and Race and Hispanic Origin of Both Spouses: 2018." Retrieved November 30, 2019 (https://www.census.gov/data /tables/2018/demo/families/cps-2018.html).

U.S. Census Bureau. 2018d. "Table H1. Households by Type and Tenure of Householder for Selected Characteristics: 2018." Retrieved November 30, 2019 (https://www.census.gov /data/tables/2018/demo/families/cps-2018 .html).

U.S. Census Bureau. 2018e. "Table UC3. Opposite Sex Unmarried Couples by Presence of Biological Children Under 18, and Age, Earnings, Education, and Race and Hispanic Origin of Both Partners: 2018." Retrieved November 30, 2019 (https://www.census.gov /data/tables/2018/demo/families/cps-2018 .html).

U.S. Census Bureau. 2018f. "U.S. Census Bureau Releases 2018 Families and Living Arrangements Tables." Retrieved November 25, 2019 (https:// www.census.gov/newsroom/press-releases/2018 /families.html).

U.S. Census Bureau. 2019a. "Historical Living Arrangements of Children." Retrieved February 8, 2020 (https://www.census.gov/data/tables /time-series/demo/families/children.html).

U.S. Census Bureau. 2019b. "Living Arrangements of Children Under 18 Years and Marital Status of Parents, by Age, Sex, Race, and Hispanic Origin and Selected Characteristics of the Child for All Children: 2019." Retrieved February 8, 2020 (https://www.census.gov/data /tables/2019/demo/families/cps-2019 .html).

U.S. Census Bureas. 2019c. "Table A1. Marital Status of People 15 Years and Over, by Age, Sex, and Personal Earnings: 2019." Retrieved February 9, 2020 (https://www .census.gov/data/tables/2019/demo /families/cps-2019.html).

U.S. Citizenship and Immigration Services 2020. "Adoption Information: Russia. " Retrieved January 1, 2020 (U.S. Citizenship and Immigration Services 2020).

U.S. Department of Education. 2019. "National Federal Student Loan Cohort Default Rate Continues to Decline." Retrieved November 25, 2019 (https://www.ed.gov/news/press-releases /national-federal-student-loan-cohort-default -rate-continues-decline).

U.S. Department of Health and Human Services. 2020. *Child Maltreatment 2018*. Retrieved February 14, 2020. (https://www.acf.hhs.gov/cb/research -data-technology/statistics-research /child-maltreatment).

U.S. Department of Health and Human Services, Administration for Children and Families, Administration on Children, Youth and Families, Children's Bureau. 2020. *Child Maltreatment 2018*. Retrieved February 7, 2020 (https://acf.hhs.gov).

U.S. Department of State, Office of the Historian. n.d. "The Immigration Act of 1924 (the Johnson-Reed Act)." Retrieved November 16, 2019 (https://history.state .gov/milestones).

U.S. Equal Employment Opportunity Commission. 2019. "Sexual Harassment." Retrieved October 18, 2019 (https://www .eeoc.gov/laws/types/sexual_harassment .cfm).

"U.S. Population by Race 2018." 2019. Livepopulation. Retrieved November 10, 2019 (www.livepopulationof.com).

U.S. Citizenship and Immigration Services. 2011. "Battered Spouse, Children, and Parents." Retrieved March 14, 2013 (http://www.uscis .gov).

U.S. Department of Health and Human Services. 2010. *Child Maltreatment 2008*. Washington, DC: Administration for Children and Families, Administration on Children, Youth and Families, Children's Bureau (www .acf.hhs.gov /programs/cb /stats_research/index .htm#can).

———. 2012a. The AFCARS Report for 2011. No. 19. Washington, DC: U.S. Government Printing Office.

———. 2012b. Child Maltreatment 2011. Washington, DC: U.S. Government Printing Office. Retrieved April 28, 2013 (http://www .acf.hhs.gov/programs/cb /research -data-technology/statistics -research /child-maltreatment).

———. 2013. *Preventing Child Maltreatment and Promoting Well-Being: A Network for Action—2013*

Resource Guide. Washington, DC: U.S. Government Printing Office. Retrieved April 28, 2013 (https://www.childwelfare.gov/pubs /guide2013/guide .pdf).

————. 2015. *Child Maltreatment 2013.* Washington, DC: U.S. Government Printing Office. Retrieved November 20, 2015 (http:// www.acf.hhs.gov).

U.S. Department of Health, Education, and Welfare. 1975. *Child Abuse and Neglect. Vol. I, An Overview of the Problem.* Publication (OHD) 75 –30073. Washington, DC: U.S. Government Printing Office.

U.S. Department of Justice. 2013. "Sexual Assault." Washington, DC: U.S. Government Printing Office. Retrieved May 27, 2013 (http:// www.ovw.usdoj.gov/sexassault.htm).

U.S. Federal Interagency Forum on Aging-Related Statistics. 2013. *Older Americans 2012: Key Indicators of Well-Being.* Washington, DC: U.S. Government Printing Office.

————. 2015. *Older Americans 2014: Key Indicators of Well-Being.* Washington, DC: U.S. Government Printing Office.

U.S. Federal Interagency Forum on Child and Family Statistics. 2012. *America's Children in Brief: Key National Indicators of Well-Being, 2012.* Washington, DC: U.S. Government Printing Office. Retrieved September 3, 2012 (http:// childstats.gov /americaschildren).

————. 2014. *America's Young Adults: Special Issue, 2014.* Washington, DC: U.S. Government Printing Office.

————. 2015. *America's Children: Key National Indicators of Well-Being, 2015.* Washington, DC: U.S. Government Printing Office. Retrieved September 25, 2015 (http://www.childstats .gov).

U.S. Food and Drug Administration. 2006. "Mifeprex (mifepristone) Information." Washington, DC: U.S. Government Printing Office, April 10. Retrieved March 29, 2007 (www.fda.gov).

"U.S. Recession Causing Increase in Child Abuse Reports." 2009. Red Orbit News, April 16. Retrieved June 8, 2009 (www .redorbit.com).

U.S. Social Security Administration. 2015. "Fact Sheet: 2016 Social Security Changes." Washington, DC: U.S. Government Printing Office. Retrieved December 28, 2015 (www .socialsecurity.gov/).

Upadhyay, Ushma D., M. Antonia Biggs, and Diana Greene Foster. 2015. "The Effect of Abortion on Having and Achieving Aspirational One-Year Plans. " *BMC Women's Health* 15:102.

Usdansky, Margaret L. 2009a. "Ambivalent Acceptance of Single-Parent Families: A Response to Comments." *Journal of Marriage and Family* 71(2):240–246.

————. 2009b. "A Weak Embrace: Popular and Scholarly Depictions of Single-Parent Families, 1900–1998." *Journal of Marriage and Family* 71(2):209–225.

————. 2011. "The Gender-Equality Paradox: Class and Incongruity Between Work-Family Attitudes and Behaviors." *Journal of Family Theory and Review* 3(3)(September):163–178.

Vaaler, Margaret L., Christopher G. Ellison, and Daniel A. Powers. 2009. "Religious Influences on the Risk of Marital Dissolution." *Journal of Marriage and Family* 71(4):917–934.

Valencia, Lorena. 2015. "*Being* a Mother, Practicing Motherhood, Mothering Someone: The Impact of Psy-Knowledge and Processes of Subjectification." *Journal of Family Issues* 36(9):1233–1252.

Valenti, Jessica. 2015. "Women Deserve Orgasm Equality." Retrieved November 4. 2019 (https:// www.theguardian.com/commentisfree /2015/jun/05/women-deserve-orgasm -equality

Valiente, Carlos, Richard A. Fabes, Nancy Eisenberg, and Tracy L. Spinrad. 2004. "The Relations of Parental Expressivity and Support to Children's Coping with Daily Stress." *Journal of Family Psychology* 18(1):97–107.

Valkila, Joni, Pertti Haaparanta, and Niina Niemi. 2010. "Empowering Coffee Traders? The Coffee Value Chain from Nicaraguan Fair Trade Farmers to Finnish Consumers." *Journal of Business Ethnics* 97(2):257–270.

Valle, Giuseppina, and Kathryn Harker Tillman. 2014. "Childhood Family Structure and Romantic Relationships during the Transition to Adulthood." *Journal of Family Issues* 35:97–124.

Vanassche, Sofie, Martine Corijn, Koen Matthijs, and Gray Swicegood. 2015. "Repartnering and Childbearing after Divorce: Differences According to Parental Status and Custodial Arrangements." *Population Research & Policy Review* 34:761–784.

Van Dam, Andrew. 2018. "A Record Number of Folks Age 85 and Older Are Working." *The Washington Post.* July 5. Retrieved August 24, 2018 (www.washingtonpost.com).

Vandeleur, C. L., N. Jeanpretre, M. Perrez, and D. Schoebi. 2009. "Cohesion, Satisfaction with Family Bonds, and Emotional Well-Being in Families with Adolescents." *Journal of Marriage and Family* 71(5):1205–1219.

Vandell, Deborah Lowe, Jay Belsky, Margaret Burchinal, Laurence Steinberg, and Nathan Vandergrift. 2010. "Do Effects of Early Child Care Extend to Age 15 Years? Results from the NICHD Study of Early Child Care and Youth Development." *Child Development* 81(3):737–756.

Vanderbilt-Adriance, Ella, Daniel Shaw, Lauretta Brennan, Thomas Dishion, Frances Gardner, and Melvin Wilson. 2015. "Child, Family, and Community Protective Factors in the Development of Children's Early Conduct Problems." *Family Relations* 64(1):64–79.

van der Heijden, Franciëlla, Anne-Rigt Poortman, and Tanja Van der Lippe. 2016. "Children's Postdivorce Residence Arrangements and Parental Experienced Time Pressure." *Journal of Marriage and Family* 78:468–481.

Vander Borght, Mélodie, and Christine Wyns. 2018. "Fertility and Infertility: Definition and Epidemiology. " *Clinical Biochemistry* 62:2–10.

van der Heijden, Franciëlla, Anne-Rigt Poortman, and Tanja van der Lippe. 2016. "Children's Postdivorce Residence Arrangements and Parental Experienced Time Pressure." *Journal of Marriage and Family* 78:468–481.

Van Hook, Jennifer, and Jennifer Glick. 2020. "Spanning Borders, Cultures, and Generations: A Decade of Research on Immigrant Families." *Journal of Marriage and Family* 82 (1): 224–243.

VanderLaan, Doug P., and Paul L. Vasey. 2009. "Patterns of Sexual Coercion in Heterosexual and Non-Heterosexual Men and Women." *Archives of Sexual Behavior* 38(6):987–999.

Van der Lippe, Tanja. 2010. "Women's Employment and Housework." In *Dividing the Domestic: Men, Women, and Household Work in Cross-National Perspective,* edited by Judith Treasand Sonja Drobnič. Stanford, CA: Stanford University Press.

Vandewater, Elizabeth A., and Jennifer E. Lansford. 2005. "A Family Process Model of Problem Behaviors in Adolescents." *Journal of Marriage and Family* 67(1):100–109.

Vandivere, Sharon, Karin Malm, and Laura Radel. 2009. *Adoption USA: A Chartbook Based on the 2007 National Survey of Adoptive Parents.* Washington, DC: U.S. Department of Health and Human Services, Office of the Assistant Secretary for Planning and Evaluation.

VanDorn, Richard A., Gary L. Bowen, and Judith R. Blau. 2006. "The Impact of Community Diversity and Consolidated Inequality on Dropping Out of High School." *Family Relations* 55(1):105–118.

Van Eeden-Moorefield, Brad, Christopher R. Martell, Mark Williams, and Marilyn Preston. 2011. "Same-Sex Relationships and Dissolution: the Connection Between Heteronormativity and Homonormativity." *Family Relations* 60(5):562–571.

Van Eeden-Moorefield, Brad, and Kay Pasley. 2008. "A Longitudinal Examination of Marital Processes Leading to Instability in Remarriages and Stepfamilies." pp. 231–249 in *The International Handbook of Stepfamilies,* edited by J. Pryor. Hoboken, NJ: John Wiley & Sons.

Van Eeden-Morrefield, Brad, Kay Pasley, Elizabeth Dolan, and Margorie Engel. 2007. "Financial Management and Security among Remarried Women." *Journal of Divorce and Remarriage* 47(3–4):21–42.

Van Gelderen, Loes, Nanette N. Gartrell, Henny M.W. Box, and Jo M.A. Hermanns. 2013. "Stigmatization and Promotive Factors in relation to Psychological Health and Life Satisfaction of Adolescents in Planned Lesbian Families." *Journal of Family Issues* (6):809–827.

VanNatta, Michelle. 2005. "Constructing the Battered Woman." *Feminist Studies* 31(2):416–429.

Van Ouytsel, Joris, Michel, Walrave, and Koen Ponnet. 2019. "An Exploratory Study of Sexting Behaviors Among Heterosexual and Sexual Minority Early Adolescents." *Journal of Adolescent Health* 65:621–626.

"Vasectomy." 2012. Planned Parenthood. Retrieved January 7, 2012 (http://www .plannedparenthood.org).

Vaughan, Ellen, Richard Feinn, Stanley Bernard, Maria Brereton, and Joy Kaufman. 2012. "Relationships Between Child Emotional and Behavioral Symptoms and Caregiver Strain and Parenting Stress." *Journal of Family Issues* 34(4):534–556.

Veaux, Franklin, and Eve Rickert. 2015. *More Than Two: A Practical Guide to Ethical Polyamory.* Portland, OR: Thorntree Press (www.indiegogo .com).

Vennum, Amber, Nathan Hardy, D. Scott Sibley, and Frank D. Fincham. 2015. "Dedication and Sliding in Emerging Adult Cyclical and Non-Cyclical Romantic Relationships." *Family Relations* 64:407–419.

Ventura, Stephanie J. 2009. "Changing Patterns of Nonmarital Childbearing in the United States." *NCHS Data Brief* 18. Hyattsville, MD: U.S. National Center for Health Statistics.

Ventura, Stephanie J., Brady E. Hamilton, and T. J. Mathews. 2014. "National and State Patterns of Teen Births in the United States, 1940–2013." *National Vital Statistics Reports* 63. Hyattsville, MD: National Center for Health Statistics.

Verhofstadt, Lesley L., William Ickes, and Ann Buysse. 2010. "I Know What You Need Right Now:" Empathic Accuracy and Support Provision in Marriage." pp. 71–88 in *Support Processes in Intimate Relationships*, edited by Kieran T. Sullivan and Joanne Davila. New York: Oxford University Press.

Vera, Amir. 2019. "Pennsylvania School District Tells Parents to Pay Their Lunch Debt, or Their Kids Will Go Into Foster Care." CNN. July 21. Retrieved August 27, 2019 (www.cnn.com).

Vesely, Colleen K., Bethany L. Letiecq and Rachael D. Goodman. 2017. "Immigrant Family Resilience in Context: Using a Community-Based Approach to Build a New Conceptual Model." *Journal of Family Theory & Review* 9(1):93–110.

Vespa, Jonathan, Jamie M. Lewis, and Rose M. Kreider. 2013. *America's Families and Living Arrangements: 2012.* Current Population Reports, P20-570. Washington, DC: U.S. Census Bureau.

Vestal, Christine. 2009. "Gay Marriage Legal in Six States." Stateline.org, June 4. Retrieved November 20, 2009 (www .stateline.org).

Viglianco-VanPelt, Michelle, and Kyla Boyse. 2015. "Masturbation." Retrieved December 3, 2015 (http://www.med .umich.edu/yourchild/topics/masturb .htm).

Villar, F., and D. J. Villamizar. 2012. "Hopes and Concerns in Couple Relationships Across Adulthood and Their Association with Relationship Satisfaction." *Aging and Human Development* 75(2):115–139.

Vinciguerra, Thomas. 2007. "He's Not My Grandpa, He's My Dad." *The New York Times*, April 12.

Vinjamuri, Mohan. 2015. "Reminders of Heteronormativity: Gay Adoptive Fathers Navigating Uninvited Social Interactions." *Family Relations* 64(2):263–277.

Vogan, Vanessa, Johanna Lake, Jonathan Weiss, Suzanne Robinson, Ami Tint, and Yona Lunsky. 2014. "Factors Associated with Caregiver Burden among Parents of Individuals with ASD." *Family Relations* 63(4):554–567.

Vogel, Erin A., Jason P. Rose, Lindsay R. Roberts, and Katheryn Eckles. 2014. "Social Comparison, Social Media, and Self-Esteem." *Psychology of Popular Media Culture* 3:206.

Vogler, Carolyn. 2005. "Cohabiting Couples: Rethinking Money in the Household at the Beginning of the Twenty-First Century." *Sociological Review* 53(1):1–29.

Vogler, Carolyn, Clare Lyonette, and Richard D. Wiggins. 2008. "Money, Power and Spending Decisions in Intimate Relationships." *Sociological Review* 56(1):117–143.

Volck, William, Zachary A. Ventress, Debby Herbenick, Paula J. Adams Hillard, and Jill S. Huppert. 2013. "Gynecologic Knowledge Is Low in College Men and Women." *Journal of Pediatric and Adolescent Gynecology* 26:161–166.

Von Oldershausen, Sasha. 2012. "Iran's Sex-Change-Operations Provided Nearly Free-of-Cost." Huffington Post Worldpost. Retrieved October 15, 2015 (http://www .huffpost.com).

Von Rosenvinge, Kristina. n.d. "Seven Tips to Improve Couple Communication." Retrieved January 26, 2010 (ezinearticles .com).

Von Schlippe, Arist, and Frank Hermann. 2013. "The Theory of Social Systems as a Framework for Understanding Family Businesses." *Family Relations* 62(3):384–398.

Voorpostel, M., and Rosemary Blieszner. 2008. "Intergenerational Support and Solidarity between Adult Siblings." *Journal of Marriage and Family* 70(1):157–167.

Vrangalova, Zahana, and Rachel E. Bukberg. 2015. "Are Sexually Permissive Individuals More Victimized and Socially Isolated?" *Personal Relationships* 22:230–242.

Wade, Lisa. 2015a. "A Quarter of College Students Think that Love Brainwashes Women." *Sociological Images*. Retrieved April 3 2015. (http://socimages.tumblr .com/post/117387968233/a-quarter-of -college-students-think-that-love).

———. 2015b. "What Do Women (Seeking Men) Want?" Retrieved April 24, 2015 (http://thesocietypages.org /socimages/2015/02/27 /what-do-women -want/).

Vera, Amir. 2019. "Pennsylvania School District Tells Parents to Pay Their Lunch Debt, or Their Kids Will Go Into Foster Care." CNN. July 21. Retrieved August 27, 2019 (www.cnn.com).

Vesely, Colleen K., Bethany L. Letiecq and Rachael D. Goodman. 2017. "Immigrant Family Resilience in Context: Using a Community-Based Approach to Build a New Conceptual Model." *Journal of Family Theory & Review* 9(1):93–110.

Wade, T. Joel, Ryan Kelley, and Dominique Church. 2011. "Are There Sex Differences in Reaction to Different Types of Sexual Infidelity?" *Psychology* 3(2):161–164.

Wadsworth, Shelley, and M. MacDermid. 2010. "Family Risk and Resilience in the Context of War and Terrorism." *Journal of Marriage and Family* 72(3):537–556.

Wagenaar, Deborah B. 2009. "Elder Abuse Education in Residency Programs: How Well Are We Doing?" *Academic Medicine: Journal of the Association of American Medical Colleges* 84(5):611–619.

Wagner-Raphael, Lynne I., David Wyatt Seal, and Anke A. Ehrhardt. 2001. "Close Emotional Relationships with Women versus Men." *Journal of Men's Studies* 9(2):243–256.

Wahl, David W. 2020. *Speaking Through the Silence: Narratives, Interaction, and the Construction of Sexual Selves.* Iowa State University, 2020.

Waite, Evelyn B, Lilly Shanahan, Susan D. Calkins, Susan P. Keane, and Marion O'Brien. 2011. "Life Events, Sibling Warmth, and Youths' Adjustment." *Journal of Marriage and Family* 73(5) (October):902–912.

Waite, Linda J. 1995. "Does Marriage Matter?" *Demography* 32(4):483–507.

Waite, Linda J., and Maggie Gallagher. 2000. *The Case for Marriage: Why Married People Are Happier, Healthier, and Better Off Financially.* New York: Doubleday.

Waite, Linda J., Ye Luo, and Alisa C. Lewin. 2009. "Marital Happiness and Marital Stability: Consequences for Psychological Well-Being." *Social Science Research* 38(1):201–212.

Wakeland, Shannon. n.d. "How to Manage Conflict Between Your Siblings." Retrieved January 29, 2013 (http://www.ehow.com).

Waldfogel, Jane. 2006. "What Do Children Need?" *Public Policy Research* 13(1):26–34.

Walker, Alexis J. 1985. "Reconceptualizing Family Stress." *Journal of Marriage and Family* 47(4):827–837.

Walker, Lenore E. 2009. *The Battered Woman Syndrome*, 3rd ed. New York: Springer.

Wallace, Harvey. 2008. *Family Violence: Legal, Medical, and Social Perspectives.* Boston: Pearson /Allyn & Bacon.

Wallace, Stephen G. 2007. "Hooking Up, Losing Out?" *Healthy Teens Camping Magazine* (March /April):26–30.

Wallach, Ian. 2010. "Mourning Has Broken." *Oprah Magazine*, October: 190–194.

Waller, Maureen R., and Sara S. McLanahan. 2005. "'His' and 'Her' Marriage Expectations: Determinants and Consequences." *Journal of Marriage and Family* 67:53–67.

Waller, Maureen R., and H. Elizabeth Peters. 2008. "The Risk of Divorce As a Barrier to Marriage among Parents of Young Children." *Social Science Research* 37(4):1188–1199.

Waller, Willard. 1951. *The Family: A Dynamic Interpretation*, revised by Reuben Hill. New York: Dryden.

Wallerstein, Judith S., and Sandra Blakeslee. 1989. *Second Chances: Men, Women, and Children a Decade After Divorce.* New York: Ticknor and Fields.

———. 1995. *The Good Marriage: How and Why Love Lasts.* Boston: Houghton Mifflin.

Wallerstein, Judith S., and Joan Kelly. 1980. *Surviving the Break-Up: How Children Actually Cope with Divorce.* New York: Basic Books.

Wallerstein, Judith S., and Julia M. Lewis. 2007. "Disparate Parenting and Step-Parenting with Siblings in the Post-Divorce Family: Report from a 10-Year Longitudinal Study." *Journal of Family Studies* 13(2):224–235.

———. 2008. "Divorced Fathers and Their Adult Offspring: Report from a Twenty-Five-Year Longitudinal Study." *Family Law Quarterly* 42(4). Available through Academic Search Elite.

Walter, Carolyn Ambler. 1986. *The Timing of Motherhood.* Lexington, MA: Heath.

Walters, Mikel L., Jieru Chen, and Matthew J. Breidig. 2013. "The National Intimate Partner and Sexual Violence Survey." Atlanta: U.S. Centers for Disease Control and Prevention (http://www .cdc.gov/violenceprevention/pdf /nisvs _sofindings.pdf).

Walvoord, Emily C. 2010. "The Timing of Puberty: Is It Changing? Does It Matter?" *Journal of Adolescent Health* 47(5):433–439.

Wang, Feihong, Martha J. Cox, Roger Mills-Koonce, and Patricia Snyder. 2015. "Parental Behaviors and Beliefs, Child Temperament, and Attachment Disorganization." *Family Relations* 64(2):191–204.

Wang, Wendy. 2012. "The Rise of Intermarriage." Retrieved January 9, 2016 (http://www.pewsocialtrends.org/2012 /02/16/the-rise-of-intermarriage/).

———. 2015. "Interracial Marriage: Who is 'Marrying Out'?" Retrieved January 9, 2016 (http://www.pewresearch.org/fact -tank/2015/06/12/interracial-marriage -who-is-marrying-out/).

Wang, Wendy, and Rich Morin. 2009. "Home for the Holidays . . . and Every Other Day: Recession Brings Many Young Adults Back to the Nest." Washington, DC: Pew Research Center. Retrieved December 21, 2009 (pewresearch. org).

Wang, Wendy, and Kim Parker. 2014. "Record Share of Americans Have Never Married." Retrieved January 26, 2016 (http://www. pewsocialtrends.org/2014 /09/24/record-share- of-americans-have -never-married/).

Wang, Wendy, Kim Parker, and Paul Taylor. 2013. "Breadwinner Moms." *Pew Social and Demographic Trends*. Washington, DC: Pew Research Center.

Wang, Wendy, and Paul Taylor. 2011. "For Millennials, Parenthood Trumps Marriage." Retrieved January 26, 2016 (http://www .pewsocialtrends.org/2011 /03/09/for -millennials-parenthood -trumps-marriage/).

Wang, Wendy. 2018. "The State of our Unions: Marriage Up Among Older Americans, Down Among the Younger." Retrieved November 26, 2019 (https://ifstudies.org/blog/the-state-of -our-unions-marriage-up-among-older -americans-down-among-the-younger).

Wang, Wendy, and Kim Parker. 2014. *Public Views on Marriage*. September 24. Pew Research Organization. Retrieved January 30, 2019 (www.pewresearch.org).

Ward, Brian W., James M. Dahlhamer, Adena M. Galinsky, and Sarah S. Joestl. 2014. "Sexual Orientation and Health among U.S. Adults: National Health Interview Survey, 2013." *National Health Statistics Reports* 77. Hyattsville, MD: National Center for Health Statistics.

Ward, Jane. 2015. *Not Gay: Sex Between Straight White Men*. NYU Press

Ward, Margaret. 1997. "Family Paradigms and Older-child Adoption: A Proposal for Matching Parents' Strengths to Children's Needs." *Family Relations* 46(3):257–262.

Ward, Russell A., Glenna Spitze, and Glenn Deane. 2009. "The More the Merrier? Multiple Parent-Adult Child Relations." *Journal of Marriage and Family* 71(1):161–173.

Wark, Linda, and Shilpa Jobalia. 1998. "What Would It Take to Build a Bridge? An Intervention for Stepfamilies." *Journal of Family Psychotherapy* 9(3):69–77.

Warne, Garry L. and Vijayalakshmi Bhatia. 2006. "Intersex, East and West." pp. 183–205 in *Ethics*

and Intersex, edited by Sharon E. Sytsma. New York: Springer.

Warner, Judith. 2006. *Perfect Madness: Motherhood in the Age of Anxiety*. New York: Riverhead Books.

Warren, Robert. 2015. *U.S. Born Children of Undocumented Residents: Numbers and Characteristics in 2013*. Center for Migration Studies. Retrieved November 17, 2019 (https://cmsny.org/publications /warren-usbornchildren/).

Wasikowska, Mia. 2011. "A Minimum Wage Increase." *The New York Times*, March 27. Retrieved February 20, 2013 (http://nytimes. com).

Waskul, Dennis D., Phillip Vannini, and Desiree Wiesen. 2007. "Women and their Clitoris: Personal Discovery, Signification, and Use." *Symbolic Interaction* 30:151–174.

Wasserman, Jason Adam. 2009. "But Where Do We Go from Here: A Reply to Tomso on the State and Direction of Postmodern Theory." *Social Theory and Health* 7(1):78–80.

Watson, Bruce. 2010. "A Hidden Crime: Domestic Violence Against Men Is a Growing Problem." Retrieved March 14, 2013 (http:// www.dailyfinance.com).

Watson, Rita. 2011. "Low Infidelity, Shock Statistics, and the Forgiveness Factor." *Psychology Today*. Retrieved November 17, 2012 (http:// www.psychologytoday .com/blog/love-and- gratitude/201109 /low-infidelity-shock-statistics- and-the -forgiveness-factor).

Watson, Stephanie. 2012. "Why It's Ok to Have Just One Child." *WebMD*, February 1. Retrieved January 8, 2013 (http://www .webmd.com /parenting/features/just -one-child).

Weaver, Shannon E., and Marilyn Coleman. 2010. "Caught in the Middle: Mothers in Stepfamilies." *Journal of Social and Personal Relationships* 27(3):305–326.

Weaver, Shannon E., Marilyn Coleman, and Lawrence H. Ganong. 2003. "The Sibling Relationship in Young Adulthood." *Journal of Family Issues* 24(2):245–263.

Webb, Amy Pieper, Christopher G. Ellison, Michael J. McFarland, Jerry W. Lee, Kelly Morton, and James Walters. 2010. "Divorce, Religious Coping, and Depressive Symptoms in a Conservative Protestant Religious Group." *Family Relations* 59:544–557.

Webster, Murray, and Lisa Rashotte. 2009. "Fixed Roles and Situated Actions." *Sex Roles* 61(5/6):325–337.

Weeks, John R. 2002. *Population: An Introduction to Concepts and Issues*, 8th ed. Belmont, CA: Wadsworth.

Weeks, Linton. 2010. "A Temporary Solution for a New American Worker." National Public Radio. Retrieved February 15, 2011 (http:// www.npr.org).

Weger, H. 2005. "Disconfirming Communication and Self-Verification in Marriage: Associations among the Demand/Withdraw Interaction Pattern, Feeling Understood, and Marital Satisfaction." *Journal of Social and Personal Relationships* 22(1):19–31.

Wehmeyer, Michael L. 2014. "Self-Determination: A Family Affair." *Family Relations* 63 (February):178–184.

Weibel-Orlando, J. 2001. "Grandparenting Styles: Native American Perspectives." pp. 139–145 in *Families in Later Life: Connections and Transitions*, edited by Alexis J. Walker, Margaret Manoogian-O'Dell, Lori A. McGraw, and Diana L. G. White. Thousand Oaks, CA: Pine Forge Press.

Weidman, Aaron, Joey T Cheng, Chandra Chisholm, and Jessica L. Tracy. 2015. "Is She the One? Personality Judgments from Online Personal Advertisements." *Personal Relationships* 22:591–603.

Weigel, Daniel J. 2008. "The Concept of Family: An Analysis of Laypeople's Views of Family." *Journal of Family Issues* 29(11):1426–1447.

Weil, Elizabeth. 2012. "Puberty Before Age 10: A New 'Normal'?" *The New York Times Magazine*. Retrieved March 25, 2013 (http://www.nytimes .com/2012/04/01 /magazine /puberty-before-age-10-a-new -normal. html?pagewanted=all&_r=0).

Weill, Kelly. 2017. "New Law Lets Dads Veto Abortions. " Retrieved December 23, 2019 (https://www.thedailybeast.com /new-law-lets-dads-veto-abortions).

Weiner Jonah. 2010. "Married? A Bit Bored? See a Shootout." *The New York Times*. Retrieved November 21, 2012 (http://www.nytimes. com/2010/04/04 /movies/04date.html?_r=0).

Weininger, Elliot B., and Annette Lareau. 2009. "Paradoxical Pathways: An Ethnographic Extension of Kohn's Findings on Class and Childrearing." *Journal of Marriage and Family* 71(3):680–695.

Weissbourd, Richard, Trisha Ross Anderson, Alison Cashin, and Joe McIntyre. 2019. *The Talk*. Retrieved Novmber 1, 2019 (https://static1.squarespace.com /static/5b7c56e255b02c683659fe43 /t/5bd51a0324a69425bd079b59/1540692500558 /mcc_the_talk_final.pdf

Weissman, M. M., J. C. Markowitz, and G. L. Klerman. 2007. *Clinician's Quick Guide to Interpersonal Psychotherapy*. New York: Oxford University Press.

Weitzman, Lenore J. 1985. *The Divorce Revolution: The Unexpected Social and Economic Consequences for Women and Children in America*. New York: The Free Press.

Wejnert, Cyprian. 2008. "Strategies for Measuring and Promoting Mothers' Social Support Networks." *Marriage and Family Review* 44(2/3):380–388.

Weiser, Dana A. 2017. "Confronting Myths about Sexual Assault: A Feminist Analysis of the False Report Literature." *Family Relations* 66 (1):46–60.

Weissbrodt, David, and Laura Danielson. 2005. *Immigration Law and Procedure*. West Publishing Company.

Welborn, Vickie. 2006. "Black Students Ordered to Give Up Seats to Whites." *Shreveport Times*, August 24. Retrieved October 2, 2006 (www. shreveporttimes.com).

Welch, Liz. 2011. "The Exact Words that Could Help a Friend in an Abusive Relationship." Huffpost Healthy Living, May 9. Retrieved May 8, 2013 (http://www .huffingtonpost.com).

Wells, Mary S., Mark A. Widmer, and J. Kelly McCoy. 2004. "Grubs and Grasshoppers: Challenge-based Recreation and the Collective

Efficacy of Families with At-Risk Youth." *Family Relations* 53(3):326–333.

Wells, Robert V. 1985. *Uncle Sam's Family: Issues in and Perspectives on American Demographic History.* Albany: State University of New York Press.

Wen, Ming. 2008. "Family Structure and Children's Health and Behavior." *Journal of Family Issues* 29(11):1492–1519.

Wenger, G. Clare, Pearl Dykstra, Tuula Melkas, and Kees C.P.M. Knipscheer. 2007. "Social Embeddedness and Late-Life Parenthood." *Journal of Family Issues* 28(11):1419–1456.

Wentland, Jocelyn J., and Elke D. Reissing. 2011. "Taking Casual Sex Not Too Casually: Exploring Definitions of Casual Sex Relationships." *Canadian Journal of Human Sexuality* 20(3):75–91.

Wergin, Clemens. 2015. "The Case for Free-Range Parenting." *The New York Times*, March 15. Retrieved March 20, 2015 (http://www.nytimes.com).

West, Abby. 2012. "'Sesame Street' and Abby Cadabby Talk about Divorce." Retrieved February 12, 2020 (https://ew.com /article/2012/12/11/sesame-street-divorce/).

West, Candace, and Don H. Zimmerman. 1987 "Doing Gender." *Gender & Society* 1:125–151.

Western, Bruce, and Sara McLanahan. 2000. "Fathers Behind Bars: The Impact of Incarceration on Family Formation." pp. 309–324 in *Families, Crime, and Criminal Justice*, edited by Greer Litton Fox and Michael L. Benson. New York: Elsevier Science.

Westphal, Sarah Katharina, Anne Rigt Poortman, and Tanja Van der Lippe. 2015. "What About the Grandparents? Children's Postdivorce Residence Arrangements and Contact with Grandparents." Journal of Marriage and Family 77:424–440.

Wetzstein, Cheryl. 2015. "James Langevin Pushes Legislation to Curb 'Rehoming' of Adopted Children." Retrieved February 14, 2016 (http://www.washingtontimes.com /news/2015/aug/4/rep-james -langevin-pushes-legislation-to-curb-reho /print/).

Wexler, Laura. 2014. "The Freedom of 'I Do,' Take Two." Retrieved April 12, 2016 (http:// www.nytimes.com/2015/05/17 /fashion /weddings/the-freedom-of-i-do -take-two .html?_r=0).

Wexler, Richard. 2005. "Family Preservation Is the Safest Way to Protect Most Children." pp. 311–327 in *Current Controversies on Family Violence*, 2nd ed., edited by Donileen R. Loseke, Richard J. Gelles, and Mary M. Cavanaugh. Thousand Oaks, CA: Sage.

Wharff, Elizabeth, Katherine Ginnis, and Abigail Ross. 2012. "Family-Based Crisis Intervention with Suicidal Adolescents in the Emergency Room." *Social Work* 57(2):133–143.

Whealin, Julia, and Ilona Pivar. 2006. "Coping when a Family Member Has Been Called to War: A National Center for PTSD Fact Sheet." U.S. Department of Veterans Affairs National Center for PTSD. Retrieved August 16, 2006 (www .ncptsd.va.gov).

Whisman, Mark A., Kristina Coop Gordon, and Yael Chatav. 2007. "Predicting Sexual Infidelity in a Population-Based Sample of Married Individuals." *Journal of Family Psychology* 21(2): 320–324.

Whisman, Mark A., and Douglas K. Snyder. 2007. "Sexual Infidelity in a National Survey of American Women: Difference in Prevalence and Correlates as a Function of Method of Assessment." *Journal of Family Psychology* 21(2):147–154.

Whitaker, Daniel J., Tadesse Haileyesus, Monica Swahn, and Linda S. Saltzman. 2007. "Differences in Frequency of Violence and Reported Injury between Relationships With Reciprocal and Nonreciprocal Intimate Partner Violence." *American Journal of Public Health* 97(5):941–947.

White House Conference on Aging. 2005. *Report to the President and the Congress: The Booming Dynamics of Aging—from Awareness to Action.* Retrieved May 16, 2007 (www .whcoa.gov).

White, James M., and David M. Klein. 2008. *Family Theories: An Introduction.* 3rd ed. Thousand Oaks, CA: Sage.

White, Lynn K. 1998. "Who's Counting? Quasi-Facts and Stepfamilies in Reports of Number of Siblings." *Journal of Marriage and Family* 60(August):725–733.

———. 1999. "Contagion in Family Affection: Mothers, Fathers, and Young Adult Children." *Journal of Marriage and Family* 61(2):284–294.

White, Lynn K., and Alan Booth. 1985. "The Quality and Stability of Remarriages: The Role of Stepchildren." *American Sociological Review* 50:689–698.

White, Lynn K., and Joan G. Gilbreth. 2001. "When Children Have Two Fathers: Effects of Relationships with Stepfathers and Noncustodial Fathers on Adolescent Outcomes." *Journal of Marriage and Family* 63:155–167.

White, Lynn K., and Agnes Riedmann. 1992. "When the Brady Bunch Grows Up: Step/Half-and Full-Sibling Relationships in Adulthood." *Journal of Marriage and Family* 54(1):197–208.

White, Martha C. 2012. "The Booming Business of Divorce Parties." Retrieved April 22, 2013 (http://business.time .com/2012/10/15 /the-booming-business -of-divorce-parties/).

White, Rebecca M. B., Mark W. Roosa, Scott R. Weaver, and Rajni L. Nair. 2009. "Cultural and Contextual Influences on Parenting in Mexican American Families." *Journal of Marriage and Family* 71(1):61–79.

Whitehead, Barbara Dafoe, and David Popenoe. 2001. "Who Wants to Marry a Soul Mate?" In *The State of Our Unions 2001: The Social Health of Marriage in America.* Piscataway, NJ: RutgersUniversity, National Marriage Project.

———. 2003. "Did a Family Turnaround Begin in the 1990s?" In *The State of Our Unions 2003: The Social Health of Marriage in America.* Piscataway, NJ: Rutgers University, National Marriage Project. Retrieved September 20, 2006 (marriage.rutgers.edu).

———. 2008. "Life Without Children: The Social Retreat from Children and How It Is Changing America." Piscataway, NJ: Rutgers University, National Marriage Project. Retrieved April 16, 2010 (marriage.rutgers.edu).

Whitehurst, Lindsay. 2012. "Polygamy." Retrieved June 26, 2013 (http://www .lindsaywhitehurst.com/polygamy.html).

Whiting, Jason, Donna Smith, and Tammy Barnett. 2007. "Overcoming the Cinderella Myth: A Mixed Methods Study of Successful Stepmothers." *Journal of Divorce and Remarriage* 47(1/2):95–109.

Whiting, Jason B., and Lisa V. Merchant. 2014. "Intimate Partner Violence in the United States: The Role of Distortion and Desistance." *Family Focus* Issue FF61:F9–F11.

Whitley, Bernard E., Christopher E. Childs, and Jena B. Collins. 2011. "Differences in Black and White American College Students' Attitudes Toward Lesbians and Gay Men." *Sex Roles: A Journal of Research* 64(5–6):299–310.

Whitton, Sarah W., Jan M. Nicholson, and Howard J. Markman. 2008. "Research on Interventions for Stepfamily Couples: The State of the Field." pp. 455–484 in *The International Handbook of Stepfamilies*, edited by J. Pryor. Hoboken, NJ: John Wiley & Sons.

Whitty, M. T. 2005. "The Realness of Cybercheating: Men's and Women's Representations of Unfaithful Internet Relationships." *Social Science Computer Review* 23:5–67.

"Why Interracial Marriages Are Increasing." 1996. *Jet*, June 3, pp. 12–15.

Whyte, Martin King. 1990. *Dating, Mating, and Marriage.* New York: Aldine.

Wickersham, Joan. 2008. *The Suicide Index: Putting My Father's Death in Order.* Orlando, FL: Harcourt.

Wickrama, K. A. S., Catherine Walker O'Neal, and Fred Lorenz. 2013. "Marital Functioning from Middle to Later Years: A Life Course-Stress Process Framework." *Journal of Family Theory and Review* 5(1):15–34.

Widarsson, Margareta, Gabriella Engström, Andreas Rosenblad, Birgitta Kerstis, Birgitta Edlund, and Pranee Lundberg. 2013. "Parental Stress in Early Parenthood among Mothers and Fathers in Sweden." *Scandinavian Journal of Caring Sciences* 27:839–847.

Widmer, Eric D., Francesco Giudici, Jean-Marie LeGoff, and Alexandre Pollien. 2009. "From Support to Control: A Configurational Perspective on Conjugal Quality." *Journal of Marriage and Family* 71(3):437–448.

Wiemers, Emily E., Judith A. Seltzer, Robert F. Schoeni, V. Joseph Hotz, and Suzanne M. Bianchi. 2019. "Stepfamily Structure and Transfers between Generations in U.S. Families." *Demography* 56:229–260.

Wienke, Chris, and Gretchen J. Hill. 2009. "Does the 'Marriage Benefit' Extend to Partners in Gay and Lesbian Relationships?" *Journal of Family Issues* 30(2):259–289.

Wiersma, Jacquelyn D., H. Harrington Cleveland, Veronica Herrera, and Judith L. Fischer. 2010. "Intimate Partner Violence in Young Adult Dating, Cohabiting, and Married Drinking Partnerships." *Journal of Marriage and Family* 72(2):360–374.

Wiese, Bettina, Christian Burk, and Dalit Jaeckel. 2016. "Transition to Grandparenthood and Job-Related Attitudes." *Journal of Marriage and Family* 78 (3): 830–847.

Wight, Vanessa, Suzanne Bianchi, and Bijou Hunt. 2012. "Explaining Racial/Ethnic Variation in Partnered Women's and Men's Housework:

Does One Size Fit All?" *Journal of Family Issues* 34(3):394–427.

Wight, Vanessa R., and Sara B. Raley. 2009. "When Home Becomes Work: Work and Family Time among Workers at Home." *Social Indicators Research* 93(1):197–202.

Wightman, Patrick, Robert Schoeni, and Keith Robinson. 2012. "Familial Financial Assistance to Young Adults." Paper Presented at the Annual Meeting of the Population Association of America, May 3.

Wilcox, W. Bradford. 2004. *Soft Patriarchs, New Men: How Christianity Shapes Fathers and Husbands.* Chicago: University of Chicago Press.

———. 2009. "The Evolution of Divorce." *National Affairs* 1(Fall):81–94.

———. 2011. "A Shaky Foundation for Families." *The New York Times,* August 30. Retrieved December 11, 2012 (http://www .nytimes.com).

Wilcox, W. Bradford, and Elizabeth Marquardt. 2009. *The State of Our Unions 2009: Marriage in America: Money and Marriage.* Virginia: National Marriage Project and the Institute for American Values.

Wilcox, W. Bradford, Elizabeth Marquardt, David Popenoe, and Barbara Dafoe Whitehead. 2011. *The State of Our Unions—Marriage in America 2011: How Parenthood Makes Life Meaningful and How Marriage Makes Parenthood Bearable.* University of Virginia, Institute for American Values: The National Marriage Project.

Wilcox, W. Bradford, Jared Anderson, William Doherty, David Eggebeen, Christopher Ellison, William Galston, Neil Gilbert, John Gottman, Ron Haskins, Robert Lerman, Linda Malone-Colon, Loren Marks, Rob Palkovitz, David Popenoe, Mark Regnerus, Scott Stanley, Linda Waite, and Judith Wallerstein. 2011. *Why Marriage Matters, Third Edition: Thirty Conclusions from the Social Sciences.* Virginia: National Marriage Project and the Institute for American Values.

Wilcox, W. Bradford and Steven L. Nock. 2006. "What's Love Got to Do with It? Equality, Equity, Commitment and Women's Marital Quality." *Social Forces* 84(3):1321–1345.

Wildeman, Christopher. 2009. "Parental Imprisonment, the Prison Boom, and the Concentration of Childhood Disadvantage." *Demography* 46(2):265–280.

———. 2012. "Despair by Association? The Mental Health of Mothers with Children by Recently Incarcerated Fathers." *American Sociological Review* 27(2):216–243.

Wildeman, Christopher, and Christine Percheski. 2009. "Associations of Childhood Religious Attendance, Family Structure, and Nonmarital Fertility Across Cohorts." *Journal of Marriage and Family* 71(5):1294–1308.

Wildman, Sarah. 2011. "A Showpiece of Communal Living in Berlin." *The New York Times,* November 9. Retrieved December 5, 2012 (http://nytimes.com).

Wildsmith, Elizabeth, and R. Kelley Raley. 2006. "Race-Ethnic Differences in Nonmarital Fertility: A Focus on Mexican American Women." *Journal of Marriage and Family* 68(2):491–508.

Wiley, Angela R., Henriette B. Warren, and Dale S. Montanelli. 2002. "Shelter in a Time of Storm:

Parenting in Poor Rural African American Communities." *Family Relations* 51(3):265–273.

Wilk, Kenneth Aarskaug, and Eva Bernhardt. 2017. "Cohabiting and Married Individuals' Relations with Their Partner's Parents." *Journal of Marriage and Family.* 79 (4): 111–1124.

Wilke, Joy, and Lydia Saad. 2013. "Older Americans' Moral Attitudes Changing." Retrieved January 26, 2016 (http://www .gallup.com/poll/162881/older-americans -moral-attitudes-changing.aspx).

Wilkins, Amy C. 2012. "Stigma and Status: Interracial Intimacy and Intersectional Identities among Black College Men." *Gender and Society* 26(2):165–189.

Wilkins, Victoria M., Martha L. Bruce, and Jo Anne Sirey. 2009. "Caregiving Tasks and Training Interest of Family Caregivers of Medically Ill Homebound Older Adults." *Journal of Aging and Health* 21(3):528–542.

Wilkinson, Deanna, Amanda Magora, Marie Garcia, and Atika Khurana. 2009. "Fathering at the Margins of Society: Reflections from Young, Minority, Crime-Involved Fathers." *Journal of Family Issues* 30(7):945–967.

Wilkinson, Doris. 2000. "Rethinking the Concept of 'Minority': A Task for Social Scientists and Practitioners." *Journal of Sociology and Social Welfare* 27:115–132.

Wilkinson, Will. 2011. "Why You Don't Believe that Kids Don't Make You Happier." *Forbes,* March 4. Retrieved December 13, 2012 (http:// www.forbes.com).

Willetts, Marion C. 2003. "An Exploratory Investigation of Heterosexual Licensed Domestic Partners." *Journal of Marriage and Family* 65(4):939–952.

———. 2006. "Union Quality Comparisons between Long-Term Heterosexual Cohabitation and Legal Marriage." *Journal of Family Issues* 27(1):110–127.

Williams, Alex. 2010. "The New Math on Campus." *The New York Times.* Retrieved December 14, 2012 (http://www.nytimes .com/2010/02/07/fashion/07campus .html?pagewanted=all&_r=0).

Williams, Amanda, and Rebecca A. Sheehan. 2015. "Emerging Adulthood in Time and Space: The Geographic Context of Homelessness." *Journal of Family Theory and Review* 7(2):126–143.

Williams, Amber, Meeta Banerjee, Fantasy Lozada-Smith, Danny Lambrouths III, and Stephanie Rowley. 2017. "Black Mothers' Perceptions of the Role of Race in Children's Education." *Journal of Marriage and Family* 79 (4): 932–946.

Williams, Joan C., and Heather Boushey. 2010. *The Three Faces of Work-Family Conflict: The Poor, the Professionals, and the Missing Middle.* Washington, DC: Center for American Progress. Retrieved May 8, 2010 (www.americanprogress.org).

Williams, K. 2003. "Has the Future of Marriage Arrived? A Contemporary Examination of Gender, Marriage, and Psychological Well-Being." *Journal of Health and Social Behavior* 44:470–487.

Williams, Lee M., and Michael G. Lawler. 2003. "Marital Satisfaction and Religious Heterogamy." *Journal of Family Issues* 24(8):1070–1092.

Williamson, Hannah C., Teresa P. Nguyen, Thomas N. Bradbury, and Benjamin R. Karney. 2015. "Are Problems that Contribute to Divorce Present at the Start of Marriage, or Do They Emerge Over Time?" *Journal of Social and Personal Relationships* 0265407515617705.

Willoughby, Brian J., and Jason S. Carroll. 2012. "Correlates of Attitudes Toward Cohabitation: Looking at the Associations with Demographics, Relational Attitudes, and Dating Behavior." *Journal of Family Issues* 33(11):1450–1476.

Willoughby, Brian J., Jason S. Carroll, and Dean M. Busby. 2011. "The Different Effects of 'Living Together': Determining and Comparing Types of Cohabiting Couples." *Journal of Social and Personal Relationships* 29(3):397–419.

Willoughby, Brian J., Scott S. Hall, and Heather P. Luczak. 2015. "Marital Paradigms: A Conceptual Framework for Marital attitudes, Values, and Beliefs." *Journal of Family Issues* 36(2):188–211.

Wilson, April C., and Ted L. Huston. 2013. "Shared Reality and Grounded Feelings During Courtship: Do They Matter for Marital Success." *Journal of Marriage and Family* 75:681–696.

Wilson, Brenda. 2009. "Sex Without Intimacy: No Dating, No Relationship." National Public Radio, May 18. Retrieved June 10, 2009 (www. npr.org).

Wilson, Jacque. 2012. "Young Caregivers Put Life on Hold." CNN (www.cnn.com), October 3.

Wilson, James Q. 2001. "Against Homosexual Marriage." pp. 123–127 in *Debating Points: Marriage and Family Issues,* edited by Henry L. Tischler. Upper Saddle River, NJ: Prentice Hall.

Wimmer, Andreas, and Kevin Lewis. 2010. "Beyond and Below Racial Homophily: ERG Models of Friendship Network Documented on Facebook." *American Journal of Sociology* 116(2):583–642.

Winch, Robert F. 1958. *Mate Selection: A Study of Complementary Needs.* New York: Harper & Row.

Wind, Rebecca. 2015a. "Teen Pregnancy Rates Declined in Many Countries between the Mid-1990s and 2011." Retrieved December 6, 2015. (https://www .guttmacher.org/media/nr /print/2015 /01/23/).

Wind, Rebecca. 2015b. "U.S. Teen Pregnancy, Birth, and Abortion Rates Reach Historic Lows." Retrieved December 6, 2015 (http://www .guttmacher.org /media/nr/2014/05/05/).

Wineberg, Howard. 1996. "The Resolutions of Separation: Are Marital Reconciliations Attempted?" *Population Research and Policy Review* 15:297–310.

Winerip, Michael. 2009. "Anything He Can Do, She Can Do." *The New York Times,* November 15.

Winestone, Jennifer. 2015. "Mandatory Mediation: A Comparative Review of How Legislatures in California and Ontario are Mandating the Peacemaking Process in Their Adversarial Systems." Retrieved April 4, 2016 (http://www.mediate.com/articles /WinestoneJ4.cfm).

Wingert, Pat, and Barbara Kantrowitz. 2010. "The Rise of the 'Silver Divorce.'" *Newsweek,* June 7. Retrieved June 19, 2010 (www.newsweek .com).

Winkelman, Sloane Burke, Karen Vail Smith, Jason Brinkley, and David Knox. 2014. "Sexting on the College Campus." *Electronic Journal of Human Sexuality* 17:1.

Winter, Laraine, Laura Gitlin, and Marie Dennis. 2011. "Desire to Institutionalize a Relative with Dementia: Quality of Premorbid Relationship and Caregiver Gender." *Family Relations* 60(2):221–230.

Wirecutter Staff. 2019. "How to Work from Home with Children." *The New York Times.* March 5. Retrieved August 14, 2019 (https://www.nytimes.com).

Wodtke, Geoffrey T. 2013. "Duration and Timing of Exposure to Neighborhood Poverty and the Risk of Adolescent Parenthood." *Demography* 50:1765–1788.

Wolak, Janis, David Finkelhor, and Kimberly J. Mitchell. 2012. "How Often are Teens Arrested for Sexting? Data from a National Sample of Police Cases." *Pediatrics* 129:4–12.

Woldt, Veronica. 2010. "Elder Care Benefits: Retention and Recruitment Tools." *Corporate Wellness Magazine,* May 7. Retrieved May 30, 2010 (www .corporatewellnessmagazine.com).

Wolf, D. A., V. Freedman, and B. J. Soldo. 1997. "The Division of Family Labor: Care for Elderly Parents." *Journal of Gerontology* 52B:102–109.

Wolf, Katthe. 2018. "What Does Family Mean to You?" *We Are Family.* Retrieved September 9, 2019 (www.bestrongfamilies.org).

Wolf, Marsha E., Uyen Ly, Margaret A. Hobart, and Mary A. Kernic. 2003. "Barriers to Seeking Police Help for Intimate Partner Violence." *Journal of Family Violence* 18:121–129.

Wolf, Naomi. 1991. *The Beauty Myth: How Images of Beauty Are Used Against Women.* New York: W. Morrow.

———. 2012. "Naomi Wolf on Third Wave Feminism." *Big Think.* Retrieved November 8, 2012 (http://bigthink.com/ideas).

Wolf, Rosalie S. 1986. "Major Findings from Three Model Projects on Elderly Abuse." pp. 218–238 in *Elder Abuse: Conflict in the Family,* edited by Karl A. Pillemerand Rosalie S. Wolf. Dover, MA: Auburn.

Wolfers, Justin. 2006. "Did Unilateral Divorce Raise Divorce Rates? A Reconciliation and New Results." *American Economic Review* 96(5):1802–1820.

———. 2015. "Fewer Women Run Big Companies Than Men Named John." *The New York Times,* March 2. Retrieved March 3, 2015 (http://www.nytimes.com).

Wolfers, Justin, David Leonhardt, and Kevin Quealy. 2015. "1.5 Million Missing Black Men." *The New York Times, April* 20. Retrieved April 21, 2015 (http://www .nytimes.com).

Wolff, Jennifer M., Kathleen M. Rospenda, and Anthony S. Colaneri. 2017. "Sexual Harassment, Psychological Distress, and Problematic Drinking Behavior Among College Students: An Examination of Reciprocal Causal Relations." *The Journal of Sex Research* 54:362–373.

Wolfinger, Nicholas H., and Raymond E. Wolfinger. 2008. "Family Structure and Voter Turnout." *Social Forces* 86(4):1513–1528.

"Women in S&P 500 Companies." 2020. *Catalyst* January 15, 2020. Retrieved January 29, 2010 (https:www.catalyst.org).

Women's Law Project. 2012. "Census Bureau: Mom Is Designated 'Parent,' Dad Is 'Childcare Arrangement.'" Retrieved February 18, 2013 (http://womenslawproject.wordpress.com).

Womersley, Kate. 2017. "Why Giving Birth is Safer in Britain Than in the U.S. Retrieved December 24, 2019 (https://www.propublica.org/article/why-giving-birth-is-safer-in-britain-than-in-the-u-s).

Wong, Jen, Marsha Mailick, Jan Greenberg, JinkukHong, and Christopher Coe. 2014. "Daily Work Stress and Awakening Cortisol in Mothers of Individuals with Autism Spectrum Disorders or Fragile X Syndrome." *Family Relations* 63(1):135–147.

Wong, Kristin. 2017. "A Beginner's Guide to Finding the Right Therapist." Retrieved January 15, 2020 (https://www.thecut.com/2017/12/a-beginners-guide-to-finding-the-right-therapist.html).

Wong, Mooly Mei-Ching, Joyce Lai-Chong Ma, and Lily Lili Xia. 2019. "A Qualitative Study of Parents' and Children's Views on Mediation." *Journal of Divorce & Remarriage* 60:418–435.

Wong, Sabrina, Grace Yoo, and Anita Stewart. 2006. "The Changing Meaning of Family Support among Older Chinese and Korean Immigrants." *Journal of Gerontology: Social Sciences* 61B(1):S4–S9.

Wood, Robert G., Quinn Moore, Andrew Clarkwest, and Alexandra Killewald. 2014. "The Long-Term Effects of Building Strong Families: A Program for Unmarried Parents." *Journal of Marriage and Family* 76(2):446–463.

Wood, Wendy, and Alice H. Eagly. 2002. "A Cross-cultural Analysis of Behavior of Women and Men: Implications for the Origins of Sex Differences." *Psychology Bulletin* 128:699–727.

Wooding, G. Scott. 2008. *Stepparenting and the Blended Family: Recognizing the Problems and Overcoming the Obstacles.* Markham, Ontario: Fitzhenry and Whiteside.

Woodman, Ashley C. 2014. "Trajectories of Stress among Parents of Children with Disabilities: A Dyadic Analysis." *Family Relations* 63(1):39–54.

Woodward, Kenneth L. 2001. "A Mormon Moment." *Newsweek,* September 10, pp. 44–51.

Workman, Simon, and Steven Jessen-Howard. 2018. "Understanding the True Cost of Child Care for Infants and Toddlers." Retrieved December 14, 2019 (https://www .americanprogress.org/issues/early -childhood/reports/2018/11/15/460970 /understanding-true-cost-child-care-infants -toddlers/).

World Economic Forum. 2018. "The Global Gender Gap Report 2018." Retrieved October 5, 2019 (http://www3.weforum .org/docs/WEF_GGGR_2018.pdf).

World Health Organization (WHO). 2015. *WHO Statement on Caesarean Section Rates.* Retrieved February 8, 2016 (http://apps .who.int/iris /bitstream/10665/161442 /1 /WHO_RHR_15.02_eng.pdf).

"World's 1st 'Test-Tube' Baby Gives Birth." 2007. CNN.com, January 15. Retrieved January 15, 2007 (http://cnn.health.com).

Worley, Karen. 2016. "Steps in the Right Direction: Stepfamily Researchers Share Personal Advice on Combining Households." Retrieved March 1, 2020 (https://www. columbiatribune.com /article/20160401/Lifestyle/304019979).

Wright, H. Norman. 2006. *How to Speak Your Spouse's Language: Ten Easy Steps to Great Communication from One of America's Foremost Counselors.* New York: Center Street.

Wright, David M., Michael Rosato, and Dermot O'Reilly. 2017. "Influence of Heterogamy by Religion on Risk of Marital Dissolution: A Cohort Study of 20,000 Couples." *European Journal of Population* 33:87–107.

Wright, Lindsay E., and Alicia Deaver. 2017. "The Impact of Sleep on Marital Satisfaction During the Transition to Parenthood." *Family Focus.* Issue FF72: F10.

Wright, Matthew R., and Susan L. Brown. 2017. "Psychological Well-being among Older Adults: The Role of Partnership Status." *Journal of Marriage and Family* 79 (3): 833–849.

Wright, Paul J., Debby Herbenick, and Bryant Paul. 2019. "Adolescent Condom Use, Parent-adolescent Sexual Health Communication, and Pornography: Findings from a U.S. Probability Sample." *Health Communication.* Retrieved February 14, 2020 (https://www.tandfonline. com/doi/abs/10.1080/10410236.2019.1652392 ?journalCode=hhth20).

Wright, John Paul, Rebecca Schnupp, Kevin M. Beaver, Matt Delisi, and Michael Vaughn. 2012. "Genes, Maternal Negativity, and Self-Control: Evidence of a Gene x Environment Interaction." *Youth, Violence, and Juvenile Justice* 10(3):245–260.

Wright, Paul J. 2015. "A Longitudinal Analysis of U.S. Adults' Pornography Exposure." *Journal of Media Psychology* 24:67–76.

Wright, Robert. 1994. *The Moral Animal.* New York: Pantheon.

Wright, Susan. 2005. "Autism: Willing the World to Listen." *Newsweek,* February 28. Retrieved March 13, 2009.

Wu, Ed Y., Ben Reeb, Monica Martin, Frederick Gibbons, Ronald Simons, and Rand Conger. 2014. "Paternal Hostility and Maternal Hostility in European American and African American Families." *Journal of Marriage and Family* 76(3):638–651.

Wu, Huijing. 2017. *Age Variation in the Divorce Rate, 1990 & 2015.* Family Profiles, FP-17-20. Bowling Green, OH: National Center for Family & Marriage Research. Retrieved February 5, 2020 (https://www.bgsu.edu/content/dam /BGSU/college-of-arts-and-sciences/NCFMR /documents/FP/wu-age-variation-divorce -rate-1990-2015-fp-17-20.pdf).

Wu, H. 2018. "Grandchildren Living in a Grandparent-Headed Household." *Family Profiles,* FP-18-01. Bowling Green, Ohio: National Center for Family and Marriage Research.

Wu, Zheng, and Feng Hou. 2008. "Family Structure and Children's Psychosocial Outcomes." *Journal of Family Issues* 29(12):1600–1624.

Wu, Zheng, Christoph M. Schimmele, and Feng Hou. 2015. "Family Structure, Academic Characteristics, and Postsecondary Education." *Family Relations* 64(2):205–220.

Wurzel, Barbara J. n.d. "Extension Fact Sheet: Growing Up with Yours, Mine, and Ours in Stepfamilies." Ohio State University Family and Consumer Sciences. Retrieved April 26, 2007 (ohioline.osu.edu).

Xu, Xianohe, Clarke D. Hudspeth, and John P. Bartkowski. 2006. "The Role of Cohabitation in Remarriage." *Journal of Marriage and Family* 68(2):261–274.

Yabiku, Scott T., and Constance T. Gager. 2009. "Sexual Frequency and the Stability of Marital and Cohabiting Unions." *Journal of Marriage and Family* 71(November):983–1000.

Yahirun, Jenjira J., Sung S. Park, and Judith A. Seltzer. 2018. "Step-Grandparenthood in the United States." *The Journals of Gerontology: Series B* 73:1055–1065.

Yahya, Siham, and Simon Boag. 2014. "'My Family Would Crucify Me!' The Perceived Influence of Social Pressure on Cross-Cultural and Interfaith Dating and Marriage." *Sexuality and Culture* 18:759–772.

Yakushko, Oksana, and Oliva M. Espin. 2010. "The Experience of Immigrant and Refugee Women: Psychological Issues." pp. 535–558 in *Handbook of Diversity in Feminist Psychology*, edited by Hope Landrineand Nancy Felipe Russo. New York: Springer.

Yancey, G. 2007. "Experiencing Racism: Differences in the Experiences of Whites Married to Blacks and Non-Black Racial Minorities." *Journal of Comparative Family Studies* 38:197–213.

Yarrow, Andrew L. 2015. "Falling Marriage Rates Reveal Economic Fault Lines." *The New York Times, February* 6. Retrieved March 3, 2016 (http://www.nytimes.com/2015/02/08/fashion/weddings/falling-marriage-rates-reveal-economic-fault-lines.html).

Yarrow, Andrew L. 2018. "I Spoke to Hundreds of American Men Who Still Can't Find Work." *Vox.* September 17. Retrieved February 12, 2019 (www.vox.com).

Yavorsky, Jill E., Claire M. Kamp Dush, and Sarah J. Schoppe-Sullivan. 2015. "The Production of Inequality: The Gender Division of Labor Across the Transition to Parenthood." *Journal of Marriage and Family* 77(3):662–679.

Yavorsky, Jill E., Claire M. Kamp Dush, and Sarah J. Schoppe-Sullivan. 2015. "The Production of Inequality: The Gender Division of Labor across the Transition to Parenthood. " *Journal of Marriage and Family* 77:662–679.

Ybarra, Michele L., and Kimberly J. Mitchell. 2015. "A National Study of Lesbian, Gay, Bisexual (LGB), and Non-LGB Youth Sexual Behavior Online and In-Person." *Archives of Sexual Behavior* April 18:1–16.

Yellowbird, Michael, and C. Matthew Snipp. 2002. "American Indian Families." pp. 227–249 in *Multicultural Families in the United States, 3rd* ed., edited by Ronald L. Taylor. Upper Saddle River, NJ: Prentice Hall.

Yeung, King-To, and John Levi Martin. 2003. "The Looking Glass Self: An Empirical Test and Elaboration." *Social Forces* 81(3):843–879.

Yin, Sandra. 2007. "New Restrictions Could Limit U.S. Adoptions from Top Two Countries of Origin: China and Guatemala," March. Washington, DC: Population Reference Bureau. Retrieved March 26, 2007 (www.prb.org).

Yip, Pamela. 2012. "Older Sweethearts Should Talk Money." Retrieved May 12, 2013 (http://www.dallasnews.com).

Yoder, Jamie R., Daniel Brisson, and Amy Lopez. 2016. *Family Relations* 65 (3): 462–476.

Yodanis, Carrie, and Sean Lauer. 2007. "Managing Money in Marriage: Multilevel and Cross-National Effects of the Breadwinner Role." *Journal of Marriage and Family* 20(December):1307–1325.

Yodanis, Carrie, and Sean Lauer. 2014. "Is Marriage Individualized? What Couples Actually Do." *Journal of Theory and Review* 6 (June):184–197.

Yodanis, Carrie, Sean Lauer, and Risako Ota. 2012. "Interethnic Romantic Relationships: Enacting Affiliative Ethnic Identities." *Journal of Marriage and Family* 74:1021–1037.

Yorburg, Betty. 2002. *Family Realities: A Global View.* Upper Saddle River, NJ: Prentice Hall.

Yoshikawa, Hirokazu and Carola Suarez-Orozco. 2012. "Deporting Parents Hurts Kids." *The New York Times*, April 20. Retrieved August 28, 2012 (http://www .nytimes.com).

Yoshikawa, Hirokazum and Jenya Kholoptseva. 2013. *Unauthorized Immigrant Parents and Their Children's Development: A Summary of the Evidence*, March. Migration Policy Institute. Retrieved March 20, 2013 (http://mpi.org).

"Young Caregivers." 2009 (April 29). Retrieved March 13, 2010 (www .girlshealth.gov).

Young, Marisa, and Scott Schieman. 2018. "Scaling Back and Finding Flexibility: Gender Differences in Parents' Strategies to Manage Work-Family Conflict." *Journal of Marriage and Family* 80 (1): 99–118.

Young, Molly. n.d. "Native Women Move to the Front of Tribal Leadership." *Native Daughters.* Retrieved October 31, 2012 (http://cojmc.unl.edu/nativedaughters /leaders/native-women -move-to-the-front -of-tribal-leadership).

Young, Sarah. 2017. "Going to Bed Angry Could Be Ruining Your Sleep." Retrieved January 25, 2020 (https://www.independent.co.uk/life-style/going-bed-angry-ruin-sleep-iowa-state-university-rage-nights-quality-mood-a7965636.html).

Young, Sarah. 2017. "Going to Bed Angry Could Be Ruining Your Sleep." Retrieved January 25, 2020 (https://www.independent.co.uk/life-style/going-bed-angry-ruin-sleep-iowa-state-university-rage-nights-quality-mood-a7965636.html).

Young, Stacy L. 2010. "Positive Perceptions of Hurtful Communication: The Packaging Matters." *Communication Research Reports* 27(1):49–57.

Yu, Wei-Hsin. 2015. "Placing Families in Context: Challenges for Cross-National Family Research." *Journal of Marriage and Family* 77(1):23–39.

Yuan, Anastasia S. Vogt, and Hayley A. Hamilton. 2006. "Stepfather Involvement and Adolescent Well-Being." *Journal of Family Issues* 27(9):1191–1213.

Zablocki v. Redhail. 1978. 434 U.S. 374, 54 L. Ed. 2d 618, 98 S. Ct. 673.

Zacchilli, Tammy L., Clyde Hendrick, and Susan S. Hendrick. 2010. "The Romantic Partner Conflict Scale: A New Scale to Measure Relationship Conflict." *Journal of Social and Personal Relationships* 26(8):1073–1096.

Zakaria, Rafia. "Love and dowry." (2018).Retrieved November 19, 2019 (http://202.166.170.213:8080/xmlui/bitstream/handle/123456789/6421/Love%20and%20dowry%20by%20Rafia%20Zakaria.docx?sequence=1&isAllowed=y).

Zaman, Ahmed. 2008. "Gender Sensitive Teaching: A Reflective Approach for Early Childhood Education Teacher Training." *Education* 129(1):110–118.

Zang, Xiaowel. 2008. "Gender and Ethnic Variation in Arranged Marriages in a Chinese City." *Journal of Family Issues* 29(5):615–638.

Zehner Ourada, Verna, and Alexis Walker. 2014. "A Comparison of Physical Health Outcomes for Caregiving Parents and Caregiving Adult Children." *Family Relations* 63(1):163–177.

Zeitz, Joshua M. 2003. "The Big Lie about the Little Pill." *The New York Times,* December 27.

Zelman, Jessica, and Mark Ferro. 2018. "The Parental Stress Scale." *Family Relations* 67 (2): 240–252.

Zill, Nicholas. 2017. "The Changing Face of Adoption. " Retrieved December 25, 2019 (https://ifstudies.org/blog/the-changing-face-of-adoption-in-the-united-states).

Zemp, Martina, Christine E. Merrilees, and Guy Bodenmann. 2014. "How Much Positivity Is Needed to Buffer the Impact of Parental Negativity on Children?" *Family Relations* 63(5):602–615.

Zerubavel, Eaiatar. 2013. *Ancestors and Relatives: Genealogy, Identity, and Community.* New York: Oxford University Press.

Zerzan, John. 2014. "Patriarchy, Civilization, and the Origins of Gender." Retrieved October 8, 2015 (http://theanarchistlibrary.org).

Zernike, Kate. 2009. "And Baby Makes How Many?" *The New York Times*, February 9. Retrieved December 5, 2009 (www.nytimes.com).

———. 2011. "Fast-Tracking to Kindergarten?" *The New York Times*, May 15:1,10.

———. 2012. "Court's Split Decision Provides Little Clarity on Surrogacy." *The New York Times*, October 24. Retrieved November 2, 2012 (http://www.nytimes.com).

Zhang, Huiping, and Sandra Kit Man Tsang. 2013. "Relative Income and Marital Happiness among Urban Chinese Women: The Moderating Role of Personal Commitment." *Journal of Happiness Studies* 14: 1575–1584.

Zhang, Jing, Suzanna Smith, Marilyn Swisher, Danling Fu, and Kate Fogarty. 2011. "Gender Role Disruption and Marital Satisfaction among Wives of Chinese International Students in the United States." *Journal of Comparative Family Studies* 42(4):523–542.

Zhang, Shuangyue. 2009. "Sender-Recipient Perspectives of Honest but Hurtful Evaluative Messages in Romantic Relationships." *Communication Reports* 22(2):89–101.

Zhang, Shuangyue, and Susan L. Kline. 2009. "Can I Make My Own Decision? A Cross-Cultural Study of Perceived Social Network Influence in Mate Selection." *Journal of Cross-Cultural Psychology* 40(1):3–23.

Zhang, Yuanting, and Jennifer Van Hook. 2009. "Marital Dissolution among Interracial Couples." *Journal of Marriage and Family* 71(February):95–107.

Zhao, Yilu. 2002. "Immersed in Two Worlds, New and Old." *The New York Times*, July 22.

Zhou, Yanqui Rachel. 2012. "Space, Time, and Self: Rethinking Aging in the Contexts of Immigration and Transnationalism." *Journal of Aging Studies* 26:232–242.

Zielinski, David S. 2009. "Child Maltreatment and Adult Socioeconomic Well-Being." *Child Abuse and Neglect* 33(10):666–678.

Zimmerman, Gregory M., and Steven F. Messner. 2013. "Individual, Family Background, and Contextual Explanations of Racial and Ethnic Disparities in Youths' Exposure to Violence." *American Journal of Public Health* 103(3):435–442.

Zimmerman, Gregory M., and Greg Pogarsky. 2011. "The Consequences of Parental Underestimation and Overestimation of Youth Exposure to Violence." *Journal of Marriage and Family* 73(1):194–208.

Zion-Waldoks, Tanya. 2015. "Politics of Devoted Resistance: Agency, Feminism, and Religion among Orthodox Agunah Activists in Israel." *Gender and Society* 29(1):73–97.

Zito, Rena Cornell, and Stacy De Coster. 2016. "Family Structure, Maternal Dating, and Sexual Debut: Extending the Conceptualization of Instability." *Journal of Youth and Adolescence*. doi: 10:1007/s10964-016-0457-7.

Zobl, Sarah R., and Pamela J. Smock 2015. *Multiple Partner Fertility among White Married Couples in the U.S., 1955.* FP-15-02. Bowling Green, OH: National Center for Family and Marriage Family Research.

Zsembik, Barbara A., and Zobeida Bonilla. 2000. "Eldercare and the Changing Family in Puerto Rico." *Journal of Family Issues* 21(5):652–674.

Zuang, Yuanting. 2004. "Why Foreign Adoption?" Presented at the annual meeting of the American Sociological Association, August 15, San Francisco, CA.

Zweig, Janine M., Jennifer Yahner, Meredith Dank, and Pamela Lachman. 2014. "Can Johnson's Typology of Adult Partner Violence Apply to Teen Dating Violence?" *Journal of Marriage and Family* 76:808–825.

NAME INDEX

SUBJECT INDEX

Note: Page numbers with a *b* indicate box material, with an *f* indicate a figure, and with a *t* indicate a table.